Statistics Now™

You determine what you need to learn ...NOW!

Log on today using 1pass access!

This student website greatly improves your chances at success. Dynamic, interactive, and instructive, **StatisticsNow**™ allows you to create a personalized learning plan for each chapter of **Elementary Statistics**—and gain a better understanding of statistical concepts.

What does **StatisticsNow's** integrated learning system do?

This dynamic program allows you to build a learning plan for yourself...based on what you know and what you need to know to get a better grade in your course. Icons within the text suggest opportune times to visit **StatisticsNow**.

Just log on to **StatisticsNow** by using the 1pass access code packaged in front of this card. Once you log on, you will immediately notice the system's simple, browser-based format—as easy to use as surfing the web. Just a click of the mouse gives you the freedom to enter and explore the system at any point.

You can build a complete, personalized learning plan for yourself by taking advantage of all three of **StatisticsNow's** powerful components:

- The **Pre-Test**, authored by Deborah Rumsey of The Ohio State University, helps determine how well you've learned the chapter material

- The **Personalized Learning Plan**, based on your answers to the **Pre-Test**, lets you review chapter content according to your specific needs, use applets to visually learn concepts, view video examples, and think critically about statistics

- The **Post-Test**, also authored by Deborah Rumsey, helps you assess your mastery of core chapter concepts

Save time.
Improve understanding.
Get a better grade.

One password is all you need to access the study tools designed to help you get a better grade.

1pass™ gives you access to:

 • **StatisticsNow™**—the robust, personalized online learning companion, complete with videos, interactive applets, and a host of learning tools

 • **ThomsonNOW™**—a powerful online homework system

 • **InfoTrac® College Edition**—the online university library

• Michael Larsen's **Internet Companion**—a guide to using the Internet more efficiently

 • **vMentor™**—live, online tutoring

 • **Companion Website**—with tutorial quizzes and datasets

• Datasets, applets, and technology lab manuals found on the **Student's Suite CD-ROM**

The **1pass** code is all you need—simply register for 1pass by following the directions on the card.

DUXBURY

Elementary Statistics

TENTH EDITION

Robert Johnson
Patricia Kuby

Monroe Community College

THOMSON

Australia • Brazil • Canada • Mexico • Singapore • Spain   **BROOKS/COLE**
United Kingdom • United States

THOMSON

BROOKS/COLE

Elementary Statistics, Tenth Edition

Robert Johnson and Patricia Kuby

Senior Acquisitions Editor: Carolyn Crockett

Development Editor: Danielle Derbenti

Senior Assistant Editor: Ann Day

Technology Project Manager: Fiona Chong

Marketing Manager: Joseph Rogove

Marketing Assistant: Brian Smith

Marketing Communications Manager: Darlene Amidon-Brent

Project Manager, Editorial Production: Shelley Ryan

Creative Director: Rob Hugel

Art Director: Lee Friedman

Print Buyer: Doreen Suruki

Permissions Editor: Joohee Lee

Production Service: Graphic World Inc.

Text Designer: Lisa Devenish

Photo Researcher: Terri Wright Design

Copy Editor: Graphic World Inc.

Illustrator: Graphic World Inc.

Cover Designer: Lee Friedman

Cover Image: © Getty Images

Cover Printer: Courier-Kendallville

Compositor: Graphic World Inc.

Printer: RR Donnelley/Willard

Library of Congress Control Number: 2005938661

Student Edition: ISBN 0-495-01763-9

Annotated Instructor's Edition: ISBN 0-495-10534-1

Thomson Higher Education
10 Davis Drive
Belmont, CA 94002-3098
USA

For more information about our products, contact us at:
Thomson Learning Academic Resource Center
1-800-423-0563
For permission to use material from this text or product, submit a request online at **http://www.thomsonrights.com.** Any additional questions about permissions can be submitted by e-mail to **thomsonrights@thomson.com.**

Brief Contents

Detailed Contents

PART 4   More Inferential Statistics

Preface

Our Approach

Over the years, *Elementary Statistics* has developed into a readable introductory textbook that promotes learning, understanding, and motivation by presenting statistics in a real-world context for students without sacrificing mathematical rigor. In addition, *Elementary Statistics* has responded to the gradual acceptance by almost every discipline that statistics is a most valuable tool for them. As a result, the applications, examples, projects, and exercises contain data from a wide variety of areas of interest, including the physical and social sciences, public opinion and political science, business, economics, and medicine.

Now, more than 30 years after *Elementary Statistics* was first published, at least one statistics course is recommended for students in all disciplines, because statistics today is seen as reaching into multiple areas of daily life. Despite this change in perception, our approach has not changed. We continue to strive for readability and a common-sense tone that will appeal to students who are increasingly more interested in application than in theory.

Coverage in the New Edition

***NEW* Chapter 1—Statistics: This chapter has been rewritten** to place a greater emphasis on interpretation of statistical information when learning key statistical terms and procedures.

***Chapter 3—Descriptive Analysis and Presentation of Bivariate Data: Descriptive regression and correlation are introduced early** in the presentation for those who prefer this approach. Continuing with relationships between two variables makes for a logical progression of material and satisfies students' natural curiosity about two variables after studying descriptive statistics of one variable. In addition, this early introduction allows instructors to go through nearly all of the thought process for a hypothesis test without any of the technical names and procedures. Later, in Chapter 8, when it comes time to introduce the hypothesis test procedure, by reusing the correlation decision as an introductory example, students will feel comfortable with the "new" testing process.

***NEW* Chapter 4—Probability: This chapter has been completely revised** with increased focus on analysis as opposed to formulas to increase student interest and comprehension of this sometimes challenging topic.

***p*-Value and classical approaches to hypothesis testing** are introduced separately, but are thereafter presented side-by-side to offer pedagogical flexibility and to emphasize their comparability.

Tour of the New Edition

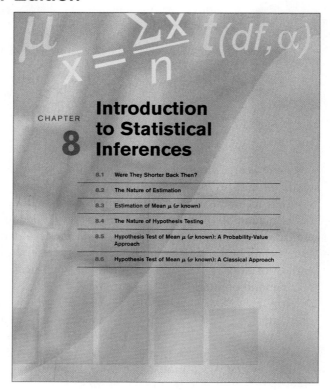

Chapter Outlines appear at the beginning of each chapter to give an overview of what is to be presented.

***NEW* and Updated Chapter Opening Sections** serve as an "example introduction," demonstrating statistics in action with respect to the specific chapter's material. Each example illustrates a familiar situation using statistics in a relevant, approachable manner for the student.

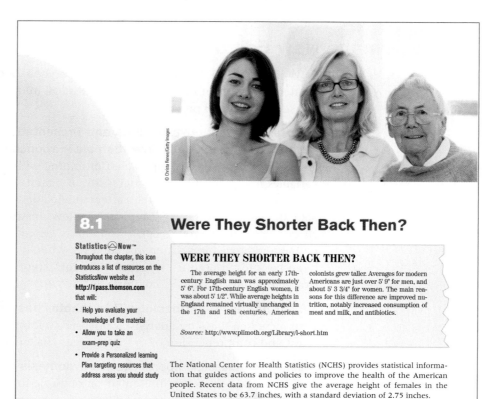

8.1   Were They Shorter Back Then?

Statistics⊘Now™
Throughout the chapter, this icon introduces a list of resources on the StatisticsNow website at **http://1pass.thomson.com** that will:

- Help you evaluate your knowledge of the material
- Allow you to take an exam-prep quiz
- Provide a Personalized learning Plan targeting resources that address areas you should study

WERE THEY SHORTER BACK THEN?

The average height for an early 17th-century English man was approximately 5′ 6″. For 17th-century English women, it was about 5′ 1/2″. While average heights in England remained virtually unchanged in the 17th and 18th centuries, American colonists grew taller. Averages for modern Americans are just over 5′ 9″ for men, and about 5′ 3 3/4″ for women. The main reasons for this difference are improved nutrition, notably increased consumption of meat and milk, and antibiotics.

Source: http://www.plimoth.org/Library/l-short.htm

The National Center for Health Statistics (NCHS) provides statistical information that guides actions and policies to improve the health of the American people. Recent data from NCHS give the average height of females in the United States to be 63.7 inches, with a standard deviation of 2.75 inches.

NEW and Updated Chapter Projects, at the end of each chapter, bring the Chapter Opening Sections full circle by incorporating the chapter's material. A complete mini-study is also provided for individual student or small group investigations.

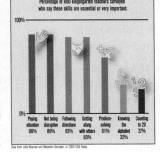

Chapter Project

Were They Shorter Back Then?

Data from the National Center for Health Statistics indicate that the average height of a female in the United States is 63.7 inches with a standard deviation of 2.75 inches. Use the data on heights of females in the health profession from Section 8.1, "Were They Shorter Back Then?" (p. 395), to answer the following questions.   [EX08-001]

65.0	66.0	64.0	67.0	59.0	69.0	66.0	69.0	64.0	61.5
63.0	62.0	63.0	64.0	72.0	66.0	65.0	64.0	67.0	68.0
70.0	63.0	63.0	68.0	58.0	60.0	63.5	66.0	64.0	62.0
64.5	69.0	63.5	69.0	62.0	58.0	66.0	68.0	59.0	56.0
64.0	66.0	65.0	69.0	67.0	66.5	67.5	62.0	70.0	62.0

Putting Chapter 8 to Work

8.199 a. Are the assumptions of the confidence interval and hypothesis test methods of this chapter satisfied? Explain.

   b. Using the sample data and a 95% level of confidence, estimate the mean height of females in the health profession. Use the given population standard deviation of 2.75 inches.

   c. Test the claim that the mean height of females in the health profession is different from 63.7 inches, the mean height for all females in the United States. Use a 0.05 level of significance.

   d. On the same histogram used in part b of Exercise 8.1 on page 396:
       (i)   Draw a vertical line at the hypothesized population mean value, 63.7.
       (ii) Draw a horizontal line segment showing the 95% confidence interval found in part b.

   e. Does the mean $\mu = 63.7$ fall in the interval? Explain what this means.

   f. Describe the relationship between the two lines drawn on your graph for part c of Exercise 8.2 on page 396 and the two lines drawn for part d of this exercise.

   g. On the basis of the results obtained earlier, does it appear that the females in this study, on average, are the same height as all females in the United States as reported by the NCHS? Explain.

Updated Plentiful **Examples,** throughout the text, present the step-by-step solution process for key statistical concepts and methods.

318   CHAPTER 6   *Normal Probability Distributions*

EXAMPLE 6.2   **Finding Area in the Right Tail of a Normal Curve**

Find the area under the normal curve to the right of $z = 1.52$: $P(z > 1.52)$.

SOLUTION The area to the right of the mean (all the shading in the figure) is exactly 0.5000. The problem asks for the shaded area that is not included in the 0.4357. Therefore, we subtract 0.4357 from 0.5000:

$$P(z > 1.52) = 0.5000 - 0.4357 = \mathbf{0.0643}$$

Note: As we have done here, always draw and label a sketch. It is most helpful.
   Make it a habit to write z with two decimal places and areas and probabilities with four decimal places, as done in Table 3.

Updated Relevant **Applied Examples** incorporate statistical concepts to demonstrate how statistics work in the real world.

APPLIED
EXAMPLE 1.1   **Telling Us about Our Early Behavior**

Remember going to kindergarten? Maybe, maybe not! If you do remember, your first concern was most likely whether you would have a good time and make some friends. What would your teacher's concerns have been?

   Consider the information included in the graphic "Even in kindergarten, social skills trump." It describes the skills that kindergarten teachers consider essential or very important. Eight hundred kindergarten teachers (only a fraction of all of them) were surveyed, producing the skills and percentages reported. Leading the list are "Paying attention" and "Not being disruptive." Of the 800 surveyed teachers, 86% considered these skills essential or very important. Looking at all the per-

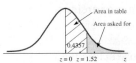

NEW Did You Know?
short history stories
and fun facts provide
an informative and
entertaining look at
related concepts or
methods being presented
in the corresponding
section.

The standard normal variable z is our test statistic for this hypothesis test.

Critical region: The set of values for the test statistic that will cause us to reject the null hypothesis. The set of values that are not in the critical region is called the **noncritical region** (sometimes called the _acceptance region_).

Recall that we are working under the assumption that the null hypothesis is true. Thus, we are assuming that the mean shearing strength of all rivets in the sampled population is 925. If this is the case, then when we select a random sample of 50 rivets, we can expect this sample mean, $\bar{x}$, to be part of a normal distribution that is centered at 925 and to have a standard error of $\sigma/\sqrt{n} = 18/\sqrt{50}$, or approximately 2.55. Approximately 95% of the sample mean values will be greater than 920.8 (a value 1.65 standard errors below the mean: $925 - (1.65)(2.55) = 920.8$). Thus, if H_o is true and $\mu = 925$, then we expect $\bar{x}$ to be greater than 920.8 approximately 95% of the time and less than 920.8 only 5% of the time.

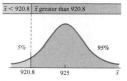

If, however, the value of $\bar{x}$ that we obtain from our sample is less than 920.8—say, 919.5—we will have to make a choice. It could be that either: (A) such an $\bar{x}$ value (919.5) is a member of the sampling distribution with mean 925 although it has a very low probability of occurrence (less than 0.05), or (B) $\bar{x} = 919.5$ is a member of a sampling distribution whose mean is less than 925, which would make it a value that is more likely to occur.

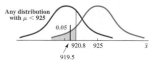

DID YOU KNOW

Disputes in Approach

Statistics is not just mathematics. There are different ways to approach statistical inferences and different ways to interpret what the data are telling. The more significant the differences, the more likely there are to be heated disagreements between those of opposing viewpoints. Just such a dispute erupted in 1935 at a Royal Statistical Society discussion when R. A. Fisher challenged Jerzy Neyman with regard to his being fully acquainted with the topic being discussed. The dispute centered on Pearson and Neyman's use of confidence intervals and approach to hypothesis testing versus Fisher's intervals and concept of p-values in significance testing. The feud lasted until Fisher's death in 1962.

NEW and Updated
With nearly 550 new
exercises and almost
100 updated exercises,
the new edition of
Elementary Statistics
provides instructors with
up-to-date, relevant
homework sets geared
toward students'
interests.

NEW and Updated
More than 300 classic
Exercises are also
available on the
Student's Suite CD-ROM,
as well as the solutions to
odd-numbered exercises.

With more than 2100
exercises, students will
have ample opportunity
for practice and instructors will have a greater
choice of exercises to
use in their course.

SECTION 8.3 EXERCISES

Statistics⬤Now™
Skillbuilder Applet Exercises must be worked using an accompanying applet found on your Student's Suite CD-ROM or at the StatisticsNow website at **http://1pass.thomson.com**.
Datasets can be found on your Student's Suite CD-ROM or at the StatisticsNow website at **http://1pass.thomson.com**.

8.19 Discuss the conditions that must exist before we can estimate the population mean using the interval techniques of formula (8.1).

8.20 Determine the value of the confidence coefficient $z_{(\alpha/2)}$ for each situation described:
a. $1 - \alpha = 0.90$ b. $1 - \alpha = 0.95$

8.21 Determine the value of the confidence coefficient $z_{(\alpha/2)}$ for each situation described:
a. 98% confidence   b. 99% confidence

8.23 Given the information, the sampled population is normally distributed, $n = 16$, $\bar{x} = 28.7$, and $\sigma = 6$:
a. Find the 0.95 confidence interval for μ.
b. Are the assumptions satisfied? Explain.

8.24 Given the information, the sampled population is normally distributed, $n = 55$, $\bar{x} = 78.2$, and $\sigma = 12$:
a. Find the 0.98 confidence interval for μ.
b. Are the assumptions satisfied? Explain.

8.25 Given the information, $n = 86$, $\bar{x} = 128.5$, and $\sigma = 16.4$:
a. Find the 0.90 confidence interval for μ.
b. Are the assumptions satisfied? Explain.

8.26 Given the information, $n = 22$, $\bar{x} = 72.3$, and

Chapter Exercises

Statistics⬤Now™
Go to the StatisticsNow website **http://1pass.thomson.com** to
• Assess your understanding of this chapter
• Check your readiness for an exam by taking the Pre-Test quiz and exploring the resources in the Personalized Learning Plan
Datasets can be found on your Student's Suite CD-ROM or at the StatisticsNow website at **http://1pass.thomson.com**.

6.101 According to Chebyshev's theorem, at least how much area is there under the standard normal distribution between $z = -2$ and $z = +2$? What is the actual area under the standard normal distribution between $z = -2$ and $z = +2$?

6.102 The middle 60% of a normally distributed population lies between what two standard scores?

6.103 Find the standard score (z) such that the area above the mean and below z under the normal curve is:
a. 0.3962   b. 0.4846   c. 0.3712

6.104 Find the standard score (z) such that the area below the mean and above z under the normal curve is:
a. 0.3212   b. 0.4788   c. 0.2700

***Updated* Skillbuilder Applet Exercises,** found in Section and Chapter Exercises, help students "see" statistical concepts and allow hands-on exploration of statistical concepts and calculations. To explore the accompanying applets, students can find them on the Student's Suite CD-ROM or in StatisticsNow.

SECTION 6.4 EXERCISES

Statistics ⌂ Now™

Skillbuilder Applet Exercises must be worked using an accompanying applet found on your Student's Suite CD-ROM or at the StatisticsNow website at **http://1pass.thomson.com.**

Datasets can be found on your Student's Suite CD-ROM or at the StatisticsNow website at **http://1pass.thomson.com.**

6.39 Skillbuilder Applet Exercise demonstrates that probability is equal to the area under a curve. Given that college students sleep an average of 7 hours per night with a standard deviation equal to 1.7 hours, use the scroll bar in the applet to find the following:

a. P(a student sleeps between 5 and 9 hours)

b. P(a student sleeps less than 4 hours)

6.41 Given $x = 58$, $\mu = 43$, and $\sigma = 5.2$, find z.

6.42 Given $x = 237$, $\mu = 220$, and $\sigma = 12.3$, find z.

6.43 Given that x is a normally distributed random variable with a mean of 60 and a standard deviation of 10, find the following probabilities:

a. $P(x > 60)$ b. $P(60 < x < 72)$ c. $P(57 < x < 83)$

d. $P(65 < x < 82)$ e. $P(38 < x < 78)$ f. $P(x < 38)$

6.44 Given that x is a normally distributed random variable with a mean of 28 and a standard deviation of 7, find the following probabilities:

a. $P(x < 28)$ b. $P(28 < x < 38)$ c. $P(24 < x < 40)$

d. $P(30 < x < 45)$ e. $P(19 < x < 35)$ f. $P(x < 48)$

6.45 Using the information given in Example 6.10 (p. 324):

New and Updated Expanded **Chapter Review,** tailored to reviewers' feedback and students' needs, functions as a chapter study guide at the end of each chapter. Each Chapter Review includes:

- **In Retrospect,** a summary of concepts learned in each chapter that points out the relationships between previously covered material.

> **In Retrospect**
>
> We have learned about the standard normal probability distribution, the most important family of continuous random variables. We have learned to apply it to all other normal probability distributions and how to use it to estimate probabilities of binomial distributions. We have seen a wide variety of variables that have this normal distribution or are reasonably well approximated by it.
>
> In the next chapter we will examine sampling distributions and learn how to use the standard normal probability to solve additional applications.

- **Vocabulary and Key Concept List,** which give students an idea of how much of the material they truly understand.

> **Vocabulary and Key Concepts**
>
> area representation for probability (p. 316)
> bell-shaped curve (p. 315)
> binomial distribution (p. 343)
> binomial probability (p. 343)
> continuity correction factor
>
> continuous random variable (pp. 315, 344)
> discrete random variable (pp. 315, 344)
> normal approximation of binomial (p. 343)
>
> percentage (p. 316)
> probability (p. 316)
> proportion (p. 316)
> random variable (p. 315)
> standard normal distribution (pp. 316, 323, 338)

- **Learning Outcomes,** a summary list outlining the key concepts that should have been learned through the course of the chapter accompanied by corresponding review exercises and section references to ensure comprehension of the chapter material.

> **Learning Outcomes**
>
> ✓ Understand the difference between a discrete and continuous random variable. — p. 315
>
> ✓ Understand the relationship between the empirical rule and the normal curve. — pp. 313–314, Ex. 6.1
>
> ✓ Understand that a normal curve is a bell-shaped curve, with total area under the curve equal to 1. — pp. 315–316, EXP 6.1, Ex. 6.40
>
> ✓ Understand that the normal curve is symmetrical about the mean, — pp. 315–317, EXP 6.2

- **Chapter Exercises,** which offer practice on all the concepts found in the chapter while also tying in comprehensive material learned from previous chapters.

Chapter Exercises

Statistics Now™
Go to the StatisticsNow website **http://1pass.thomson.com** to
- Assess your understanding of this chapter
- Check your readiness for an exam by taking the Pre-Test quiz and exploring the resources in the Personalized Learning Plan

Datasets can be found on your Student's Suite CD-ROM or at the StatisticsNow website at **http://1pass.thomson.com**.

6.101 According to Chebyshev's theorem, at least how much area is there under the standard normal distribution between $z = -2$ and $z = +2$? What is the actual area under the standard normal distribution between $z = -2$ and $z = +2$?

6.102 The middle 60% of a normally distributed population lies between what two standard scores?

6.103 Find the standard score (z) such that the area above the mean and below z under the normal curve is:

a. 0.3962   b. 0.4846   c. 0.3712

6.104 Find the standard score (z) such that the area below the mean and above z under the normal curve is:

a. 0.3212   b. 0.4788   c. 0.2700

- **Chapter Project,** which offers students the opportunity to revisit the Chapter Opening Sections to answer the questions proposed at the beginning of the chapter, using the knowledge gained from studying the chapter.

Chapter Project

Intelligence Scores

All normal probability distributions have the same shape and distribution relative to the mean and standard deviation. In this chapter we learned how to use the standard normal probability distribution to answer questions about all normal distributions. Let's return to distribution of IQ scores discussed in the Section 6.1, "Intelligence Scores" (p. 313), and try out some of our new knowledge.

k. What percentage of the SAT scores are below 450?

l. What percentage of the SAT scores are above 575?

m. What SAT score is at the 95th percentile? Explain what this means.

- **Chapter Practice Test,** which provides a formal self-evaluation of the mastery of the material before being tested in class. Correct responses are in the back of the textbook.

c. What percentage of the adult population has "superior" intelligence?

d. What is the probability of randomly selecting one person from this population who is classified below "average"?

e. What IQ score is at the 95th percentile? Explain what this means.

Chapter Practice Test

PART I: Knowing the Definitions

Answer "True" if the statement is always true. If the statement is not always true, replace the words shown in bold with words that make the statement always true.

6.1 The normal probability distribution is symmetric about **zero.**

6.2 The total area under the curve of any normal distribution is **1.0.**

6.3 The theoretical probability that a particular value of a **continuous** random variable will occur is exactly zero.

6.4 The unit of measure for the standard score is the **same as the unit of measure of the**

6.10 The most common distribution of a continuous random variable is the **binomial** probability.

PART II: Applying the Concepts

6.11 Find the following probabilities for z, the standard normal score:

a. $P(0 < z < 2.42)$ b. $P(z < 1.38)$

c. $P(z < -1.27)$ d. $P(-1.35 < z < 2.72)$

6.12 Find the value of each z-score:

a. $P(z > ?) = 0.2643$ b. $P(z < ?) = 0.17$

c. $z(0.04)$

6.13 Use the symbolic notation $z(\alpha)$ to give the symbolic name for each z-score shown in the figure at the bottom of the page.

6.14 The lifetimes of flashlight batteries are normally distributed about a mean of 35.6 hr with a standard deviation of 5.4 hr. Kevin selected one of these batteries at random and tested it. What is the probability that this one battery will last less than 40.0 hr?

6.15 The lengths of time, x, spent commuting daily, one-way, to college by students are believed to have a mean of 22 min with a stan-

NEW and Updated **MINITAB, Excel, and TI-83/84 instructions** are introduced in the text alongside appropriate material. This approach allows instructors to choose which statistical technology, if any, they would like to incorporate into their course.

NEW and Updated **With more than 400 data sets,** ranging from small to large, students have the opportunity to practice using their statistical calculator or computer.

Technology manuals offer additional instruction in these various statistical technologies. **Found on the Student's Suite CD-ROM, as well as in StatisticsNow,** our offerings include:

- **MINITAB Manual** by Diane L. Benner and Linda M. Myers, Harrisburg Area Community College.

- **Excel Manual** by Diane L. Benner and Linda M. Myers, Harrisburg Area Community College.

- **TI-83/84 Manual** by Kevin Fox, Shasta College.

Note: These technology manuals are available in both print and electronic formats. Instructors, contact your sales representative to find out how these manuals can be custom published for your course.

---

166   CHAPTER 3   *Descriptive Analysis and Presentation of Bivariate Data*

TECHNOLOGY INSTRUCTIONS: CORRELATION COEFFICIENT

MINITAB (Release 14)   Input the *x*-variable data into C1 and the corresponding *y*-variable data into C2; then continue with:

Choose: **Stat > Basic Statistics > Correlation...**
Enter: **Variables: C1 C2 > OK**

Excel   Input the *x*-variable data into column A and the corresponding *y*-variable data into column B, activate a cell for the answer; then continue with:

Choose: **Insert function, f_x > Statistical > CORREL > OK**
Enter: **Array 1: x data range > OK**
              Array 2: y data range > OK

TI-83/84 Plus   Input the *x*-variable data into L1 and the corresponding *y*-variable data into L2; then continue with:

Choose: **2nd > CATALOG > DiagnosticOn* > ENTER > ENTER**
Choose: **STAT > CALC > 8:LinReg(a + bx)**
Enter: **L1, L2**

*DiagnosticOn must be selected for *r* and r^2 to show. Once set, omit this step.

FIGURE 3.9
The Data Window

Understanding the Linear Correlation Coefficient

The following method will create (1) a visual meaning for correlation, (2) a visual meaning for what the linear coefficient is measuring, and (3) an estimate for *r*. The method is quick and generally yields a reasonable estimate when the "window of data" is approximately square.

Note: This estimation technique does not replace the calculation of *r*. It is very sensitive to the "spread" of the diagram. However, if the "window of data" is approximately square, this approximation will be useful as a mental estimate or check.

FIGURE 3.10
Focusing on Pattern

Procedure

1. Construct a scatter diagram of your data, being sure to scale the axes so that the resulting graph has an approximately square "window of data," as demonstrated in Figure 3.9 by the light green frame. The window may

---

SECTION 3.2 EXERCISES

Statistics⊘Now™

Datasets can be found on your Student's Suite CD-ROM or at the StatisticsNow website at **http://1pass.thomson.com**.

3.3 [EX03-003] In a national survey of 500 business and 500 leisure travelers, each was asked where they would most like "more space."

	On Airplane	Hotel Room	All Other
Business	355	95	50
Leisure	250	165	85

a. Express the table as percentages of the total.

b. Express the table as percentages of the row totals. Why might one prefer the table to be expressed this way?

c. Express the table as percentages of the column totals. Why might one prefer the table to be expressed this way?

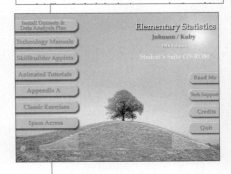

Working with Your Own Data sections, appearing at the end of each of the four major parts of the book, are designed to encourage further exploration, independent student learning, and critical thinking. These can be used as individual class projects or in small groups.

Working with Your Own Data

Putting Probability to Work

The sampling distribution of sample means and the central limit theorem are very important to the development of the rest of this course. The proof, which requires the use of calculus, is not included in this textbook. However, the truth of the SDSM and the CLT can be demonstrated both theoretically and by experimentation. The following activities will help to verify both statements.

A THE POPULATION

Consider the theoretical population that contains the three numbers 0, 3, and 6 in equal proportions.

1. a. Construct the theoretical probability distribution for the drawing of a single number, with replacement, from this population.

   b. Draw a histogram of this probability distribution.

   c. Calculate the mean, μ, and the standard deviation, σ, for this population.

B THE SAMPLING DISTRIBUTION, THEORETICALLY

Let's study the theoretical sampling distribution formed by the means of all possible samples of size 3 that can be drawn from the given population.

2. Construct a list showing all the possible samples of size 3 that could be drawn from this population. (There are 27 possibilities.)

3. Find the mean for each of the 27 possible samples listed in answer to question 2.

4. Construct the probability distribution (the theoretical sampling distribution of sample means) for these 27 sample means.

5. Construct a histogram for this sampling distribution of sample means.

6. Calculate the mean $\mu_{\bar{x}}$ and the standard error of the mean $\sigma_{\bar{x}}$ using the probability distribution found in question 4.

7. Show that the results found in questions 1c, 5, and 6 support the three claims made by the sampling distribution of sample means and the central limit theorem. Cite specific values to support your conclusions.

C THE SAMPLING DISTRIBUTION, EMPIRICALLY

Let's now see whether the sampling distribution of sample means and the central limit theorem can be verified empirically; that is, does it hold when the sampling distribution is formed by the sample means that result from several random samples?

8. Draw a random sample of size 3 from the given population. List your sample of three numbers and calculate the mean for this sample.

   You may use a computer to generate your samples. You may take three identical "tags" numbered 0, 3, and 6, put them in a "hat," and draw your sample using replacement between each drawing. Or you may use dice; let 0 be represented by 1 and 2; 3, by 3 and 4; and 6, by 5 and 6. You may also use random numbers to simulate the drawing of your samples. Or you may draw your sample from the list of random samples at the end of this section. Describe the method you decide to use. (Ask your instructor for guidance.)

9. Repeat question 8 forty-nine more times so that you have a total of 50 sample means that have resulted from samples of size 3.

---

10. Construct a frequency distribution of the 50 sample means found in questions 8 and 9.

11. Construct a histogram of the frequency distribution of observed sample means.

12. Calculate the mean $\bar{x}$ and standard deviation $s_{\bar{x}}$ of the frequency distribution formed by the 50 sample means.

13. Compare the observed values of $\bar{x}$ and $s_{\bar{x}}$ with the values of $\mu_{\bar{x}}$ and $\sigma_{\bar{x}}$ Do they agree? Does the empirical distribution of $\bar{x}$ look like the theoretical one?

Here are 100 random samples of size 3 that were generated by computer:

6 3 0	0 3 0	6 6 0	3 3 6	6 6 3	6 3 3
0 0 3	3 0 6	3 3 0	3 6 6	0 3 0	6 6 3
6 6 6	0 3 0	6 3 6	0 6 3	6 0 3	6 3 3
6 0 0	3 0 6	6 3 3	3 3 0	3 3 0	3 3 3
3 3 3	3 0 0	6 6 6	3 3 6	0 0 6	0 6 3
6 6 6	0 0 6	3 3 0	0 6 6	0 0 3	6 6 3
0 0 6	0 0 6	6 6 6	6 3 6	6 6 0	3 0 0
3 6 6	6 3 0	3 6 3	3 0 0	3 3 6	0 6 0
3 0 0	0 3 6	6 3 3	6 0 6	3 3 6	6 0 3
0 3 6	3 6 3	6 6 3	6 6 0	3 3 3	3 0 0
6 3 0	6 6 0	0 3 0	6 6 0	3 6 6	0 3 6
6 3 3	0 3 0	6 6 0	6 6 3	6 6 0	3 0 3
3 6 3	3 6 0	0 0 6	0 3 3	3 6 6	0 3 6
0 6 0	6 0 0	0 6 0	0 6 6	0 3 3	0 3 6
3 3 6	3 3 3	3 3 6	6 3 6	3 3 3	3 6 6
6 3 3	3 0 0	3 0 6	6 0 3	3 6 6	6 0 3
0 3 3	6 3 0	0 3 6	0 3 6		

Learning Resources

NEW StatisticsNow (0-495-10523-6) is a personalized learning companion that helps students gauge their unique study needs and makes the most of their study time by building focused, chapter-by-chapter, Personalized Learning Plans that reinforce key concepts.

- **Pre-Tests,** developed by Deborah Rumsey of The Ohio State University, give you an initial assessment of your knowledge.

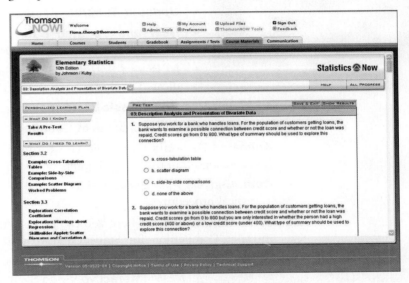

- **Personalized Learning Plans,** based on students' answers to the pre-test questions, outline key elements for review.
- **Post-Tests,** also developed by Deborah Rumsey of The Ohio State University, assess students' mastery of core chapter concepts; results can be e-mailed to the instructor!

Note: StatisticsNow also serves as a one-stop portal for nearly all your *Elementary Statistics* resources, which are also found on the Student's Suite CD-ROM, as well as the Interactive Video Skillbuilder CD-ROM. Throughout the text, StatisticsNow icons have been thoughtfully placed to direct students to the resources they need when they need them. Access StatisticsNow via http://1pass.thomson.com.

Learning Resources *(continued)*

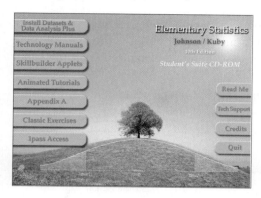

Student's Suite CD-ROM

(0-495-10533-3) This valuable learning resource includes:

- *Updated* **MINITAB Manual** by Diane L. Benner and Linda M. Myers, Harrisburg Community College.

- *Updated* **Excel Manual** by Diane L. Benner and Linda M. Myers, Harrisburg Area Community College.

- *Updated* **TI-83/84 Manual** by Kevin Fox, Shasta College.

- *NEW and Updated* **Data sets** formatted for MINITAB®, Microsoft® Excel®, SPSS®, JMP®, SAS®, TI-83/84, and ASCII.

- *NEW and Updated* **Skillbuilder Applets** to accompany indicated exercises from the text.

- *NEW and Updated* **Classic Exercises** for each chapter of the text.

- **Appendix A: Basic Principles of Counting.**

- **Animated Tutorials** that guide you through essential computational concepts presented in the early chapters of your text.

- **Data Analysis Plus Excel add-in** created to complement Excel's menu of Statistical procedures.

- *NEW* **Link to 1pass,** where you can access StatisticsNow, vMentor, Larsen's Internet Companion for Statistics, InfoTrac® College Edition, and much more.

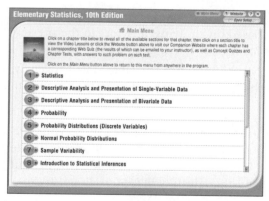

NEW **Interactive Video Skillbuilder CD-ROM** contains hours of helpful, interactive video instruction. Students can watch as an instructor walks them through key examples from the text, step by step—giving a foundation in the skills that they need to know. Each example found on the CD-ROM is identified by icons located in the margin of the text. Think of it as portable office hours!

vMentor™ allows students to talk (using their own computer microphones) to tutors who will skillfully guide them through a problem using an interactive whiteboard for illustration. Up to 40 hours of live tutoring a week is available with every new book and can be accessed through http://1pass.thomson.com.

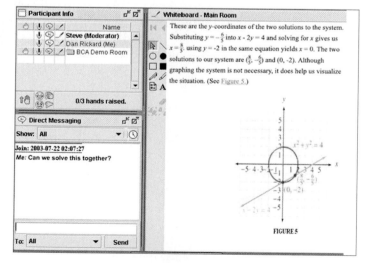

The **Book Companion Website** offers book- and course-specific resources, such as tutorial quizzes for each chapter and data sets for exercises. Students can access the website through http://1pass.thomson.com.

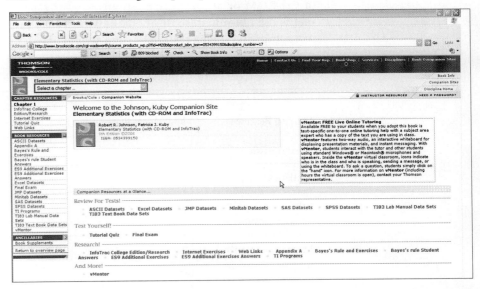

Updated **Student Solutions Manual** (0-495-10531-7), written by Patricia Kuby, includes fully worked out solutions for all odd-numbered exercises and also provides hints, tips, and additional interpretation for specific exercises.

Internet Companion for Statistics, written by Michael Larsen, from Iowa State University, offers practical information on how to use the Internet to increase students' understanding of statistics. Organized by key topics covered in the introductory course, the text offers a brief review of a topic, listings of appropriate websites, and study questions designed to build students' analytical skills. This can be accessed through http://1pass.thomson.com.

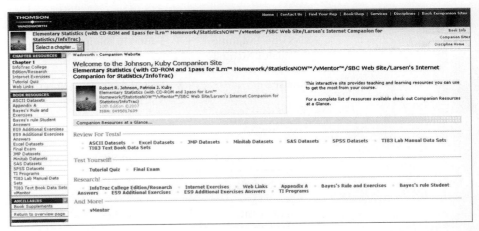

Student Access Information

To summarize, many of the items mentioned thus far are available either on a CD-ROM or at http://1pass.thomson.com via your 1pass access code. If students did not purchase this book new or if they purchased a Basic Select text—containing no media resources—but would like to use any of the resources mentioned, visit us online at http://1pass.thomson.com to purchase any of these study tools. The following chart summarizes this access information.

Learning Resources	http://1pass.thomson.com	Student's Suite CD-ROM	Interactive Video Skillbuilder CD-ROM
iLrn Statistics	X		
StatisticsNow	X		
InfoTrac® College Edition	X		
Michael Larsen's Internet Companion for Statistics	X		
vMentor™	X		
Link to Book Companion Website	X		X
Appendix A: Basic Principles of Counting		X	
MINITAB Manual	X (Found in StatisticsNow)	X	
Excel Manual	X (Found in StatisticsNow)	X	
TI-83/84 Graphing Calculator Manual	X (Found in StatisticsNow)	X	
Data sets	X (Found in StatisticsNow)	X	
Data Analysis Plus Excel Add-in		X	
Skillbuilder Applets	X (Found in StatisticsNow)	X	
Classic Exercises		X	
Animated Tutorials	X (Found in StatisticsNow)	X	
Link to 1pass		X	
Skillbuilder Videos	X (Found in StatisticsNow)		X

Instructor Resources

The Instructor's Suite CD-ROM contains everything found on the Student's
Suite CD-ROM, in addition to:

- Classic Exercises with complete solutions.

- Complete Solutions Manual.

- Test Bank in Microsoft® Word®.

- Multimedia Manager containing all of the figures from the book in PowerPoint® and as .jpg files.

A printed Test Bank contains more than 3900 true/false, multiple choice, short
answer, and applied and computational questions by section, authored by
Mohammed A. El-Saidi of Ferris State University.

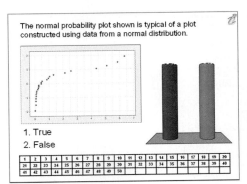

NEW **JoinIn™ on Turning Point®** offers instructors text-specific
JoinIn content for electronic response systems, written by Bryan
James and Joanna Pruden of Pennsylvania College of Technology. Instructors can transform their classroom and assess students' progress with instant in-class quizzes and polls. Turning
Point software lets you pose book-specific questions and display
students' answers seamlessly within Microsoft PowerPoint lecture slides, in conjunction with a choice of "clicker" hardware.
Enhance how your students interact with you, your lecture,
and each other.

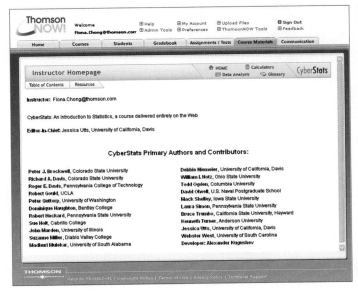

CyberStats offers complete online content
for your introductory statistics course. It promotes learning through interaction on the
web and can be used as the sole text for a
course or in conjunction with a traditional
text. Students internalize the behavior of statistical concepts by interacting with hundreds
of applets (simulations and calculations) and
receiving immediate feedback on practice
items. Effective for both distance and on-campus courses, CyberStats provides a learning opportunity that cannot be delivered in
print. Instructors interested in using this for
their course should contact their local sales
representative. CyberStats is available for
purchase by students through http://1pass
.thomson.com.

Updated **iLrn Statistics** is your system for homework, integrated testing, and course management on the web! Using iLrn, instructors can easily set up online courses; assign tests, quizzes, and homework; and monitor students' progress, enabling them to mentor students on the right points at the right time. Student responses are automatically graded and entered into the iLrn grade book, making it easy for you to assign and collect homework or offer testing over the web. Accessed through http://1pass.thomson.com, iLrn Statistics is comprised of two parts:

- **iLrn Testing, containing algorithmically generated test items,** can be used for testing, homework, or quizzing. You choose! Contact your sales representative to find out how get access to this valuable resource.

- **iLrn Homework, which contains the exercises from the book,** facilitates classroom management and assesses students on homework, quizzes, or exams, in the process of doing real data analysis on the web.

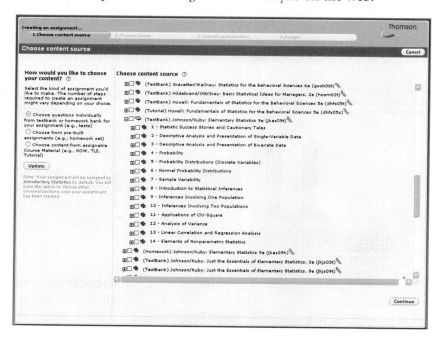

Acknowledgments

It is a pleasure to acknowledge the aid and encouragement we have received throughout the development of this text from students and colleagues at Monroe Community College. In addition, special thanks to all the reviewers who read and offered suggestions about this and previous editions:

Nancy Adcox, *Mt. San Antonio College*

Paul Alper, *College of St. Thomas*

William D. Bandes, *San Diego Mesa College*

Matrese Benkofske, *Missouri Western State College*

Tim Biehler, *Fingerlakes Community College*

Barbara Jean Blass, *Oakland Community College*

Austin Bonis, *Rochester Institute of Technology*

Nancy C. Bowers, *Pennsylvania College of Technology*

Shane Brewer, *College of Eastern Utah, San Juan Campus*

Robert Buck, *Slippery Rock University*

Louis F. Bush, *San Diego City College*

Ronnie Catipon, *Franklin University*

Rodney E. Chase, *Oakland Community College*

Pinyuen Chen, *Syracuse University*

Wayne Clark, *Parkland College*

David M. Crystal, *Rochester Institute of Technology*

Joyce Curry and Frank C. Denny, *Chabot College*

Larry Dorn, *Fresno Community College*

Shirley Dowdy, *West Virginia University*

Thomas English, *Pennsylvania State University, Erie*

Kenneth Fairbanks, *Murray State University*

Dr. William P. Fox, *Francis Marion University*

Joan Garfield, *University of Minnesota General College*

Monica Geist, *Front Range Community College*

David Gurney, *Southeastern Louisiana University*

Edwin Hackleman

Carol Hall, *New Mexico State University*

Silas Halperin, *Syracuse University*

Noal Harbertson, *California State University, Fresno*

Hank Harmeling, *North Shore Community College*

Bryan A. Haworth, *California State College at Bakersfield*

Harold Hayford, *Pennsylvania State University, Altoona*

Jim Helms, *Waycross College*

Marty Hodges, *Colorado Technical University*

John C. Holahan, *Xerox Corporation*

James E. Holstein, *University of Missouri*

Soon B. Hong, *Grand Valley State University*

Robert Hoyt, *Southwestern Montana University*

Peter Intarapanach, *Southern Connecticut State University*

T. Henry Jablonski, Jr., *East Tennessee State University*

Brian Jean, *Bakersfield University*

Jann-Huei Jinn, *Grand Valley State University*

Sherry Johnson

Meyer M. Kaplan, *The William Patterson College of New Jersey*

Michael Karelius, *American River College*

Anand S. Katiyar, *McNeese State University*

Jane Keller, *Metropolitan Community College*

Gayle S. Kent, *Florida Southern College*

Andrew Kim, *Westfield State College*

Amy Kimchuk, *University of the Sciences in Philadelphia*

Raymond Knodel, *Bemidji State University*

Larry Lesser, *University of Northern Colorado*

Natalie Lochner, *Rollins College*

Robert O. Maier, *El Camino College*

Linda McCarley, *Bevill State Community College*

Mark Anthony McComb, *Mississippi College*

Carolyn Meitler, *Concordia University Wisconsin*

John Meyer, *Muhlenberg College*

Jeffrey Mock, *Diablo Valley College*

David Naccarato, *University of New Haven*

Harold Nemer, *Riverside Community College*

John Noonan, *Mount Vernon Nazarene University*

Dennis O'Brien, *University of Wisconsin, LaCrosse*

Chandler Pike, *University of Georgia*

Daniel Powers, *University of Texas, Austin*

Janet M. Rich, *Miami-Dade Junior College*

Larry J. Ringer, *Texas A & M University*

John T. Ritschdorff, *Marist College*

John Rogers, *California Polytechnic Institute at San Luis Obispo*

Neil Rogness, *Grand Valley State University*

Thomas Rotolo, *University of Arizona*

Barbara F. Ryan and Thomas A. Ryan, *Pennsylvania State University*

Robert J. Salhany, *Rhode Island College*

Melody Smith, *Dyersburg State Community College*

Dr. Sherman Sowby, *California State University, Fresno*

Roger Spalding, *Monroe County Community College*

Timothy Stebbins, *Kalamazoo Valley Community College*

Howard Stratton, *State University of New York at Albany*

Larry Stephens, *University of Nebraska-Omaha*

Paul Stephenson, *Grand Valley State University*

Richard Stockbridge, *University of Wisconsin, Milwaukee*

Thomas Sturm, *College of St. Thomas*

Edward A. Sylvestre, *Eastman Kodak Co.*

Gwen Terwilliger

William K. Tomhave, *Concordia College, Moorhead, MN*

Bruce Trumbo, *California State University, Hayward*

Richard Uschold, *Canisius College*

John C. Van Druff, *Fort Steilacoom Community College*

Philip A. Van Veidhuizen, *University of Alaska*

John Vincenzi, *Saddleback College*

Kenneth D. Wantling, *Montgomery College*

Joan Weiss, *Fairfield University*

Mary Wheeler, *Monroe Community College*

Barbara Whitney, *Big Bend Community College*

Sharon Whitton, *Hofstra University*

Don Williams, *Austin College*

Rebecca Wong, *West Valley College*

Pablo Zafra, *Kean University*

Yvonne Zubovic, *Indiana University Purdue University, Fort Wayne*

Robert Johnson

Patricia Kuby

CHAPTER

1 Statistics

<table>
<tr><td>**1.1**</td><td># Americans, Here's Looking at You</td></tr>
</table>

The U.S. Census Bureau annually publishes the *Statistical Abstract of the United States,* a 1000+-page book that provides us with a statistical insight into many of the most obscure and unusual facets of our lives. This is only one of thousands of sources for all kinds of things you have always wanted to know about and never thought to ask about. Are you interested in how many hours we work and play? How much we spend on snack foods? How the price of Red Delicious apples has gone up? All this and more—much more—can be found in the *Statistical Abstract* (http://www.census.gov/statab/www).

The statistical tidbits that follow come from a variety of sources and represent only a tiny sampling of what can be learned about Americans statistically. Take a look.

COMMUNICATION METHOD PREFERRED BY WORKERS

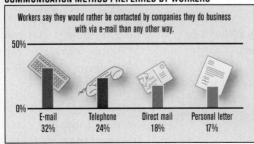

Workers say they would rather be contacted by companies they do business with via e-mail than any other way.

| E-mail | Telephone | Direct mail | Personal letter |
| 32% | 24% | 18% | 17% |

Data from Anne R. Carey and Ron Coddington. © 2004 USA Today.

SHOULD THE PENNY BE ELIMINATED?

Nearly 6 in 10 Americans want the penny to remain in circulation.

Yes 23%
No 59%
Not sure 18%

Data from Shannon Reilly and Chad Palmer. © 2004 USA Today.

WOULD YOU LIKE TO SEE YOUR 100TH BIRTHDAY?

Yes 63%
No 32%
Don't know 5%

Data from USA Today, 10/13/2003.

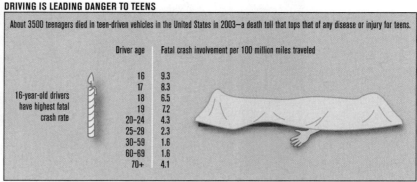

DRIVING IS LEADING DANGER TO TEENS

About 3500 teenagers died in teen-driven vehicles in the United States in 2003—a death toll that tops that of any disease or injury for teens.

Driver age	Fatal crash involvement per 100 million miles traveled
16	9.3
17	8.3
18	6.5
19	7.2
20–24	4.3
25–29	2.3
30–59	1.6
60–69	1.6
70+	4.1

16-year-old drivers have highest fatal crash rate

Data from USA Today. © 2003.

The preceding examples and thousands of other measures are used to describe life in the United States.

Consider the graphic "Would you like to see your 100th birthday?" If you had been asked, "Do you want to live to be 100?" how would you have answered? Do you believe the diagram accurately depicts your answer? Does it make you stop and wonder how the information was obtained and where it came from? Do you believe the "printed" material? As you work through Chapter 1, you will begin to learn how to read and analyze statistical measures to arrive at appropriate conclusions. Then you will be able to further investigate "Americans, Here's Looking at You" in the Chapter Project section with Exercises 1.88 and 1.89 (p. 35).

SECTION 1.1 EXERCISES

Statistics⌂Now™

Datasets can be found on your Student's Suite CD-ROM or at the StatisticsNow website at **http://1pass.thomson.com**.

1.1 a. Each of the statistical graphics presented in this section seem to suggest that information is about what population? Is that the case? Justify your answer.

b. Describe the information that was collected and used to determine the statistics reported in "Communication method preferred by workers."

c. "63%—Yes" was one specific statistic reported in the graphic "Would you like to see your 100th birthday?" Describe what that statistic tells you.

d. Consider the graphic "Should the penny be eliminated?" If you had been asked, how would you have responded? Do you believe your answer is represented accurately in the diagram? What does the percentage associated with your answer really mean? Explain.

e. How do you interpret the 7.2 that is listed for driver age of 19 in the graphic "Driving is leading danger to teens"?

1.2 a. Write a 50-word paragraph describing what the word *statistics* means to you right now.

b. Write a 50-word paragraph describing what the word *random* means to you right now.

c. Write a 50-word paragraph describing what the word *sample* means to you right now.

1.3 [EX01-003] Do you work hard for your money? Java professionals think they do, reporting long working hours at their jobs. Java developers from around the world were surveyed about the number of hours they work weekly. Listed here are the average number of hours worked weekly in various regions of the United States and the world.

Region	Hours Worked	Region	Hours Worked
U.S.	48	California	50
Northeast	47	Pacific NW	47
Mid-Atlantic	49	Canada	43
South	47	Europe	48
Midwest	47	Asia	47
Central Mt	51	South America and Africa	49

Source: Jupitermedia Corporation

a. How many hours do you work per week (or anticipate working after you graduate)?

b. What happened to the 40-hour workweek? Does it appear to exist for the Java professional?

c. Does the information in this chart make a career of being a Java professional seem attractive?

1.4 "What You Make Depends on Where You Work." When grouped by the type of organization they work for, the risk takers (self-employed) rise to the top again. For Java developers, the self-employed make the most money, followed by those at publicly held companies; both groups earn almost twice as much as those who work for educational institutions.

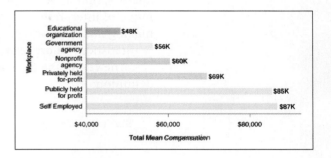

Source: Jupitermedia Corporation

a. Examine the graph and describe carefully the "picture" the graph has painted.

b. Does this graph make a career of being a self-employed Java professional seem attractive?

c. Can you conclude anything about the availability of jobs in these six groupings of the workplace?

d. Can you conclude anything about the number of hours a Java professional works each week to earn these incomes?

1.2   What Is Statistics?

As we embark on our journey into the study of statistics, we must begin with the definition of *statistics* and expand on the details involved.

Statistics has become the universal language of the sciences. As potential users of statistics, we need to master both the "science" and the "art" of using statistical methodology correctly. Careful use of statistical methods will enable us to obtain accurate information from data. These methods include (1) carefully defining the situation, (2) gathering data, (3) accurately summarizing the data, and (4) deriving and communicating meaningful conclusions.

Statistics involves information, numbers and visual graphics to summarize this information, and their interpretation. The word *statistics* has different meanings to people of varied backgrounds and interests. To some people it is a field of "hocus-pocus" in which a person attempts to overwhelm others with incorrect information and conclusions. To others it is a way of collecting and displaying information. And to still another group it is a way of "making decisions in the face of uncertainty." In the proper perspective, each of these points of view is correct.

The field of statistics can be roughly subdivided into two areas: descriptive statistics and inferential statistics. *Descriptive statistics* is what most people think of when they hear the word *statistics*. It includes the collection, presentation, and description of sample data. The term *inferential statistics* refers to the technique of interpreting the values resulting from the descriptive techniques and making decisions and drawing conclusions about the population.

Statistics is more than just numbers: it is data, what is done to data, what is learned from the data, and the resulting conclusions. Let's use the following definition:

> **Statistics:** The science of collecting, describing, and interpreting data.

Before going any further, let's look at a few illustrations of how and when statistics can be applied.

APPLIED EXAMPLE 1.1

Telling Us about Our Early Behavior

Remember going to kindergarten? Maybe, maybe not! If you do remember, your first concern was most likely whether you would have a good time and make some friends. What would your teacher's concerns have been?

Consider the information included in the graphic "Even in kindergarten, social skills trump." It describes the skills that kindergarten teachers consider essential or very important. Eight hundred kindergarten teachers (only a fraction of all of them) were surveyed, producing the skills and percentages reported. Leading the list are

EVEN IN KINDERGARTEN, SOCIAL SKILLS TRUMP

Percentage of 800 kingergarten teachers surveyed who say these skills are essential or very important:

| Paying attention 86% | Not being disruptive 86% | Following directions 83% | Getting along with others 83% | Problem-solving 61% | Knowing the alphabet 32% | Counting to 20 27% |

Data from Julia Neyman and Alejandro Gonzalez, © 2004 USA Today.

"Paying attention" and "Not being disruptive." Of the 800 surveyed teachers, 86% considered these skills essential or very important. Looking at all the percentages, it is noted that they add up to more than 100%. Apparently, the teachers surveyed were allowed to give more than one skill as an answer.

APPLIED EXAMPLE 1.2

Describing Our Softer Side

The spa industry is booming. The International SPA Association reports statistics demonstrating that pampering people can certainly produce a profit. Income from spa/salons has increased by 409% over the 1997–2003 years. In fact, the spa industry has become the fourth largest leisure industry in the United States. It surpasses amusement/theme parks and movie theaters.

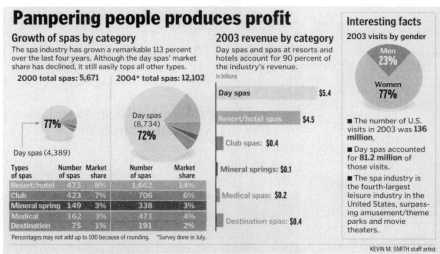

Pampering people produces profit

Growth of spas by category
The spa industry has grown a remarkable 113 percent over the last four years. Although the day spas' market share has declined, it still easily tops all other types.

2000 total spas: 5,671

77%

Day spas (4,389)

2004* total spas: 12,102

Day spas (8,734)
72%

Types of spas	Number of spas	Market share	Number of spas	Market share
Resort/hotel	473	8%	1,662	14%
Club	423	7%	706	6%
Mineral spring	149	3%	338	3%
Medical	162	3%	471	4%
Destination	75	1%	191	2%

Percentages may not add up to 100 because of rounding.   *Survey done in July.

2003 revenue by category
Day spas and spas at resorts and hotels account for 90 percent of the industry's revenue.

In billions

Day spas	$5.4
Resort/hotel spas	$4.5
Club spas:	$0.4
Mineral springs:	$0.1
Medical spas:	$0.2
Destination spas:	$0.4

Interesting facts
2003 visits by gender

Men **23%**

Women **77%**

- The number of U.S. visits in 2003 was **136 million.**
- Day spas accounted for **81.2 million** of those visits.
- The spa industry is the fourth-largest leisure industry in the United States, surpassing amusement/theme parks and movie theaters.

KEVIN M. SMITH staff artist

From Rochester *Democrat and Chronicle,* 12/5/2004. Reprinted with permission.

Much information is given in these graphs about the spa industry. Consider what information would have to be collected to formulate the charts and graphs—not just the number of spas but the type or category of spa as well as the gender of those visiting a spa. But where did these figures come from? Always note the source to published statistics. In this case the source is the International SPA Association. The association is recognized worldwide as a professional organization and the voice of the spa industry.

APPLIED EXAMPLE 1.3

Telling Us What Companies Think

Newspapers publish graphs and charts telling us how various organizations or peoples think as a whole. Do you ever wonder how much of what we think is directly influenced by the information we read in these articles?

The following graphic reports that 65% of companies do not worry that an increasingly obese workforce will have an impact on revenue or productivity. Where did this information come from? Note the source, Duffey Communications. How did they collect the information? They conducted a survey of 450 business and political figures. A margin of error is given at ±5 percentage points. (Remember to check the small print, usually at the bottom of a statistical graph or chart.) Based on this information, between 60% and 70% of companies do not worry that an increasingly obese workforce will have an impact on revenue or productivity.

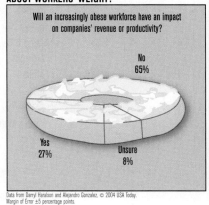

ARE COMPANIES CONCERNED ABOUT WORKERS' WEIGHT?

Will an increasingly obese workforce have an impact on companies' revenue or productivity?

No
65%

Yes
27%

Unsure
8%

Data from Darryl Haralson and Alejandro Gonzalez. © 2004 USA Today.
Margin of Error ±5 percentage points.

This seems rather amazing given the amount of information that appears in the news about obesity and its effect on health, as well as the amount of money and attention spent on diets and losing weight.

Statistics Is Tricky Business

"One ounce of statistics technique requires one pound of common sense for proper application."

Consider the **International Shark Attack File** (ISAF). The ISAF, administered by the American Elasmobranch Society and the Florida Museum of Natural History, is a compilation of all known shark attacks. It is shown in the graph and chart that follow.

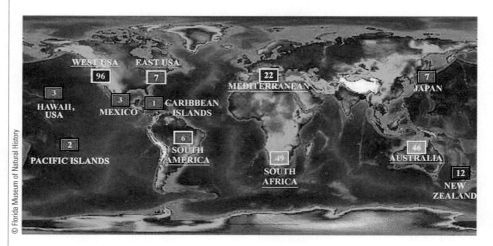

Source: http://www.sharkattackphotos.com/Shark_Attack_Map.htm

Territory	Total Attacks	Fatal Attacks	Last Fatality	Territory	Total Attacks	Fatal Attacks	Last Fatality
United States (w/out Hawaii)	761	39	2004	Antilles and Bahamas	59	19	1972
Australia	294	134	2004	Middle America	58	31	1997
Africa	264	69	2004	New Zealand	45	9	1968
Asia	116	55	2000	Europe	38	18	1984
Pacific/Oceania Islands (w/out Hawaii)	114	47	2003	Bermuda	4	0	
				General	20	6	1965
				World	1969	464	2004
Hawaii	100	15	2004				
South America	96	22	2004				

Source: http://gerber.iwarp.com/Attack/GAttack/World.htm

Common sense? Using common sense and reviewing the preceding graph, one would certainly stay away from the United States if they enjoyed the ocean. Nearly two-fifths of the world's shark attacks occur in the United States. U.S. waters must be full of sharks, and the sharks must be mad! Common sense—remember? Is the graph a bit misleading? What else could be influencing the statistics shown here? First, one must consider how much of a country's or continent's border comes in contact with an ocean. Second, who is tracking these attacks? In this case, it is stated at the top of the chart—The Florida Museum of Natural History, a museum in the United States. Apparently, the United States is trying to keep track of unprovoked shark attacks. What else is different about the United States compared with the other areas? Is the ocean a recreational area in the other places? What is the economy of these other areas, and/or who is keeping track of their shark attacks?

Remember to consider the source when reading a statistical report. Be sure you are looking at the complete picture.

The uses of statistics are unlimited. It is much harder to name a field in which statistics is not used than it is to name one in which statistics plays an integral part. The following are a few examples of how and where statistics are used:

- In education, descriptive statistics are frequently used to describe test results.
- In science, the data resulting from experiments must be collected and analyzed.
- In government, many kinds of statistical data are collected all the time. In fact, the U.S. government is probably the world's greatest collector of statistical data.

A very important part of the statistical process is that of studying the statistical results and formulating appropriate conclusions. These conclusions must then be communicated accurately—nothing is gained from research unless the findings are shared with others. Statistics are being reported everywhere: newspapers, magazines, radio, and television. We read and hear about all kinds of new research results, especially in the health-related fields.

To further continue our study of statistics, we need to "talk the talk." Statistics has its own jargon, terms beyond *descriptive statistics* and *inferential statistics,* that need to be defined and illustrated. The concept of a population is the most fundamental idea in statistics.

> **Population:** A collection, or set, of individuals, objects, or events whose properties are to be analyzed.

The population is the complete collection of individuals or objects that are of interest to the sample collector. The population of concern must be carefully defined and is considered fully defined only when its membership list of elements is specified. The set of "all students who have ever attended a U.S. college" is an example of a well-defined population.

Typically, we think of a population as a collection of people. However, in statistics the population could be a collection of animals, manufactured objects, whatever. For example, the set of all redwood trees in California could be a population.

There are two kinds of populations: finite and infinite. When the membership of a population can be (or could be) physically listed, the population is said

to be **finite.** When the membership is unlimited, the population is **infinite.** The books in your college library form a finite population; the OPAC (Online Public Access Catalog, the computerized card catalog) lists the exact membership. All the registered voters in the United States form a very large finite population; if necessary, a composite of all voter lists from all voting precincts across the United States could be compiled. On the other hand, the population of all people who might use aspirin and the population of all 40-watt light bulbs to be produced by Sylvania are infinite. Large populations are difficult to study; therefore, it is customary to select a *sample* and study the data in the sample.

Sample: A subset of a population.

A sample consists of the individuals, objects, or measurements selected from the population by the sample collector.

Variable (or response variable): A characteristic of interest about each individual element of a population or sample.

A student's age at entrance into college, the color of the student's hair, the student's height, and the student's weight are four variables.

Data value: The value of the variable associated with one element of a population or sample. This value may be a number, a word, or a symbol.

For example, Bill Jones entered college at age "23," his hair is "brown," he is "71 inches" tall, and he weighs "183 pounds." These four data values are the values for the four variables as applied to Bill Jones.

Data: The set of values collected from the variable from each of the elements that belong to the sample.

The set of 25 heights collected from 25 students is an example of a set of data.

Experiment: A planned activity whose results yield a set of data.

An experiment includes the activities for both selecting the elements and obtaining the data values.

Parameter: A numerical value summarizing all the data of an entire population.

The "average" age at time of admission for all students who have ever attended our college and the "proportion" of students who were older than 21 years of age when they entered college are examples of two population parameters. A parameter is a value that describes the entire population. Often a

Greek letter is used to symbolize the name of a parameter. These symbols will be assigned as we study specific parameters.

For every parameter there is a *corresponding sample statistic.* The statistic describes the sample the same way the parameter describes the population.

FYI Parameters describe the population; notice that both words begin with the letter **p**. A **s**tatistic describes the **s**ample; notice that both words begin with the letter **s**.

> **Statistic:** A numerical value summarizing the sample data.

The "average" height, found by using the set of 25 heights, is an example of a sample statistic. A statistic is a value that describes a sample. Most sample statistics are found with the aid of formulas and are typically assigned symbolic names that are letters of the English alphabet (for example, $\bar{x}$, s, and r).

EXAMPLE 1.5

Applying the Basic Terms

Statistics ⬡ Now™

Watch a video example at
http://1pass.thomson.com
or on your CD.

A statistics student is interested in finding out something about the average dollar value of cars owned by the faculty members of our college. Each of the eight terms just described can be identified in this situation.

1. The *population* is the collection of all cars owned by all faculty members at our college.
2. A *sample* is any subset of that population. For example, the cars owned by members of the mathematics department is a sample.
3. The *variable* is the "dollar value" of each individual car.
4. One *data value* is the dollar value of a particular car. Mr. Jones's car, for example, is valued at $9400.
5. The *data* are the set of values that correspond to the sample obtained (9400; 8700; 15,950; . . .).
6. The *experiment* consists of the methods used to select the cars that form the sample and to determine the value of each car in the sample. It could be carried out by questioning each member of the mathematics department, or in other ways.
7. The *parameter* about which we are seeking information is the "average" value of all cars in the population.
8. The *statistic* that will be found is the "average" value of the cars in the sample.

FYI Parameters are fixed in value, whereas statistics vary in value.

Note: If a second sample were to be taken, it would result in a different set of people being selected—say, the English department—and therefore a different value would be anticipated for the statistic "average value." The average value for "all faculty-owned cars" would not change, however.

There are basically two kinds of variables: (1) variables that result in *qualitative* information and (2) variables that result in *quantitative* information.

> **Qualitative, or attribute, or categorical, variable:** A variable that describes or categorizes an element of a population.

> **Quantitative, or numerical, variable:** A variable that quantifies an element of a population.

A sample of four hair-salon customers was surveyed for their "hair color," "hometown," and "level of satisfaction" with the results of their salon treatment. All three variables are examples of qualitative (attribute) variables, because they describe some characteristic of the person and all people with the same attribute belong to the same category. The data collected were {blonde, brown, black, brown}, {Brighton, Columbus, Albany, Jacksonville}, and {very satisfied, satisfied, somewhat satisfied}.

The "total cost" of textbooks purchased by each student for this semester's classes is an example of a quantitative (numerical) variable. A sample resulted in the following data: $238.87, $94.57, $139.24. [To find the "average cost," simply add the three numbers and divide by 3: (238.87 + 94.57 + 139.24)/3 = $157.56.]

Note: Arithmetic operations, such as addition and averaging, are meaningful for data that result from a quantitative variable.

Each of these types of variables (qualitative and quantitative) can be further subdivided as illustrated in the following diagram.

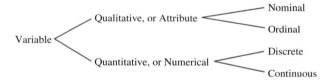

Qualitative variables may be characterized as nominal or ordinal.

> **Nominal variable:** A qualitative variable that characterizes (or describes, or names) an element of a population. Not only are arithmetic operations not meaningful for data that result from a nominal variable, but an order cannot be assigned to the categories.

In the survey of four hair-salon customers, two of the variables, "hair color" and "hometown," are examples of nominal variables because both name some characteristic of the person and it would be meaningless to find the sample average by adding and dividing by 4. For example, (blonde + brown + black + brown)/4 is undefined. Furthermore, color of hair and hometown do not have an order to their categories.

> **Ordinal variable:** A qualitative variable that incorporates an ordered position, or ranking.

In the survey of four hair-salon customers, the variable "level of satisfaction" is an example of an ordinal variable because it does incorporate an ordered ranking: "Very satisfied" ranks ahead of "satisfied," which ranks ahead of "somewhat satisfied." Another illustration of an ordinal variable is the ranking of five landscape pictures according to someone's preference: first choice, second choice, and so on.

Quantitative or numerical variables can also be subdivided into two classifications: *discrete* variables and *continuous* variables.

Discrete variable: A quantitative variable that can assume a countable number of values. Intuitively, the discrete variable can assume any values corresponding to isolated points along a line interval. That is, there is a gap between any two values.

Continuous variable: A quantitative variable that can assume an uncountable number of values. Intuitively, the continuous variable can assume any value along a line interval, including every possible value between any two values.

In many cases, the two types of variables can be distinguished by deciding whether the variables are related to a count or a measurement. The variable "number of courses for which you are currently registered" is an example of a discrete variable; the values of the variable may be found by counting the courses. (When we count, fractional values cannot occur; thus, there are gaps between the values that can occur.) The variable "weight of books and supplies you are carrying as you attend class today" is an example of a continuous random variable; the values of the variable may be found by measuring the weight. (When we measure, any fractional value can occur; thus, every value along the number line is possible.)

When trying to determine whether a variable is discrete or continuous, remember to look at the variable and think about the values that might occur. Do not look at only data values that have been recorded; they can be very misleading.

Consider the variable "judge's score" at a figure-skating competition. If we look at some scores that have previously occurred, 9.9, 9.5, 8.8, 10.0, and we see the presence of decimals, we might think that all fractions are possible and conclude that the variable is continuous. This is not true, however. A score of 9.134 is impossible; thus, there are gaps between the possible values and the variable is discrete.

Note: Don't let the appearance of the data fool you in regard to their type. Qualitative variables are not always easy to recognize; sometimes they appear as numbers. The sample of hair colors could be coded: 1 = black, 2 = blonde, 3 = brown. The sample data would then appear as {2, 3, 1, 3}, but they are still nominal data. Calculating the "average hair color" [(2 + 3 + 1 + 3)/4 = 9/4 = 2.25] is still meaningless. The hometowns could be identified using ZIP codes. The average of the ZIP codes doesn't make sense either; therefore, ZIP code numbers are nominal too.

Let's look at another example. Suppose that after surveying a parking lot, I summarized the sample data by reporting 5 red, 8 blue, 6 green, and 2 yellow cars. You must look at each individual source to determine the kind of information being collected. One specific car was red; "red" is the data value from that one car, and red is an attribute. Thus, this collection (5 red, 8 blue, and so on) is a summary of nominal data.

Another example of information that is deceiving is an identification number. Flight #249 and Room #168 both appear to be numerical data. The numeral 249 does not describe any property of the flight—late or on time, quality of snack served, number of passengers, or anything else about the flight. The flight number only identifies a specific flight. Driver's license numbers, Social Security numbers, and bank account numbers are all identification numbers used in the nominal sense, not in the quantitative sense.

Remember to inspect the individual variable and one individual data value, and you should have little trouble distinguishing among the various types of variables.

APPLIED EXAMPLE 1.6

Census Data

Census information often makes news whether it is the local census or the national census. The results of the census have a variety of uses—from helping to determine legislative seats and appropriation of taxes to visitor information (as shown here). We are all part of the population census and have all seen reports similar to this one.

Who We Are: Lee County, Florida
The News Press—Visitors 2001

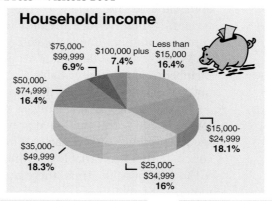

Household income

Less than $15,000
16.4%

$100,000 plus
7.4%

$75,000-$99,999
6.9%

$50,000-$74,999
16.4%

$35,000-$49,999
18.3%

$25,000-$34,999
16%

$15,000-$24,999
18.1%

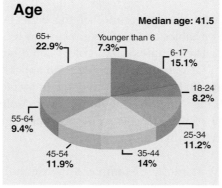

Age

Median age: 41.5

65+ **22.9%**

Younger than 6 **7.3%**

6-17 **15.1%**

18-24 **8.2%**

25-34 **11.2%**

35-44 **14%**

45-54 **11.9%**

55-64 **9.4%**

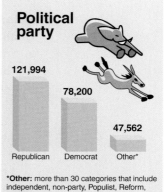

Political party

121,994

78,200

47,562

Republican   Democrat   Other*

Other: more than 30 categories that include independent, non-party, Populist, Reform, Statehood, Whig and U.S. Taxpayer

Source: The News-Press

Let's see how we can apply our new terminology to the preceding report. Considering the overall title to the reported graphs, the general population of interest would be the residents of Lee County, Florida. To be more specific to each graph, the population would be all households of Lee County for the "Household income" graph, all residents of Lee County for the "Age" graph, and all registered adults for the "Political party" graph. The variables needed to complete these graphs are income, age, and political party affiliation. Income and age are continuous variables, whereas party affiliation is a nominal variable. The circle graph data were most likely collected by categories and then the percentages calculated. Counts were used to create the Political party bar graph.

Statistics⚠Now™

FYI Watch a supplement video example at **http://1pass.thomson.com** or on your CD for more on basic terms.

SECTION 1.2 EXERCISES

Statistics⚠Now™

Skillbuilder Applet Exercises must be worked using an accompanying applet found on your Student's Suite CD-ROM or at the StatisticsNow website at **http://1pass.thomson.com.**

1.5 *Statistics* is defined on page 4 as "the science of collecting, describing, and interpreting data." Using your own words, write a sentence describing each of the three statistical activities. Retain your work for Exercise 1.87.

1.6 Determine which of the following statements is descriptive in nature and which is inferential. Refer to "Even in kindergarten, social skills trump" in Applied Example 1.1 (pp. 4–5).

a. Of all U.S. kindergarten teachers, 32% say that "Knowing the alphabet" is an essential skill.

b. Of the 800 U.S. kindergarten teachers polled, 32% say that "Knowing the alphabet" is an essential skill.

1.7 Determine which of the following statements is descriptive in nature and which is inferential. Refer to "Pampering people produces profit" in Applied Example 1.2 (p. 5).

a. Of the surveyed spas in 2004, 72% were categorized as day spas.

b. Of all visits to spas in 2003, 23% were visits by men.

1.8 Refer to the graphic "America's students by grade level."

The latest Census report on schools found about 70 million students (27.8% of the population) from nursery school through college.

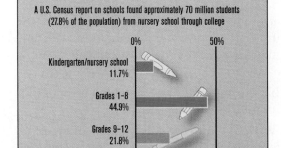

AMERICA'S STUDENTS BY GRADE LEVEL

A U.S. Census report on schools found approximately 70 million students (27.8% of the population) from nursery school through college

Kindergarten/nursery school 11.7%

Grades 1–8 44.9%

Grades 9–12 21.8%

College 15.2%

Data from USA Today, 5/9/2000.

a. What is the population?

b. What information was obtained from each person?

c. Using the information given, estimate the number of students in college.

d. Using the information given, estimate the size of the entire U.S. population.

1.9 International Communications Research (ICR) conducted the 2004 National Spring Cleaning Survey for The Soap and Detergent Association. ICR questioned 1000 American male and female heads of household regarding their house cleaning attitudes. The survey has a margin of error of plus or minus 5%.

a. What is the population?

b. How many people were polled?

c   What information was obtained from each person?

THOSE HARD TO CLEAN PLACES

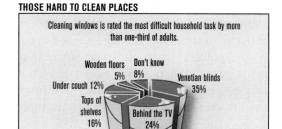

Cleaning windows is rated the most difficult household task by more than one-third of adults.

Wooden floors 5%
Don't know 8%
Under couch 12%
Venetian blinds 35%
Tops of shelves 16%
Behind the TV 24%

Data from Anne R. Carey and Gia Kereselidze, USA TODAY; Source: Swiffer

d. Using the information given, estimate the number of surveyed adults who think cleaning under the couch is the most difficult cleaning job.

e. What do you think the "margin of error of plus or minus 5%" means?

f. How would you use the "margin of error" in estimating the percentage of all adults who think that Venetian blinds are the hardest to clean?

1.10 Refer to the graphic "Driver cell phone distractions."

DRIVER CELL PHONE DISTRACTIONS

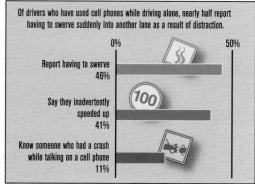

Of drivers who have used cell phones while driving alone, nearly half report having to swerve suddenly into another lane as a result of distraction.

0%   50%

Report having to swerve 46%

Say they inadvertently speeded up 41%

Know someone who had a crash while talking on a cell phone 11%

Data from Lori Joseph and Sam Ward. © 2001 USA Today.

a. What group of people was polled?

b. How many people were polled?

c. What information was obtained from each person?

d. Explain the meaning of "41% say they inadvertently speeded up."

e. How many people answered "Say they inadvertently speeded up"?

1.11

HOW WILL YOU SPEND YOUR TAX REFUND?

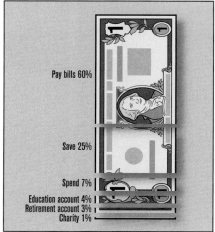

Pay bills 60%

Save 25%

Spend 7%

Education account 4%
Retirement account 3%
Charity 1%

Data from Darryl Haraison and Jerry Mosemak, USA TODAY; Source: turbotax.com

a. What is the population?

b. Describe the sample used for this report.

c. Identify the variables used to collect this information.

d. What is the majority of people going to do with their tax refund? How does this majority show on the graph?

1.12 During a radio broadcast on August 16, 1998, David Essel reported the following three statistics: (1) the U.S. divorce rate is 55%; and when married adults were asked whether they would remarry their spouse, (2) 75% of the women said yes and (3) 65% of the men said yes.

a. What is the "stay married" rate?

b. There seems to be a contradiction in this information. How is it possible for all three of these statements to be correct? Explain.

1.13 A working knowledge of statistics is very helpful when you want to understand the statistics reported in the news. The news media and our government often make statements like "Crime rate jumps 50% in your city."

a. Does an increase in the crime rate from 4% to 6% represent an increase of 50%? Explain.

b. Why would anybody report an increase from 4% to 6% as a "50% rate jump"?

1.14 Find a recent newspaper article that illustrates an "apples are bad" type of report.

1.15 Of the adult U.S. population, 36% has an allergy. A sample of 1200 randomly selected adults resulted in 33.2% reporting an allergy.

a. Describe the population.

b. What is the sample?

c. Describe the variable.

d. Identify the statistic and give its value.

e. Identify the parameter and give its value.

1.16 In your own words, explain why the parameter is fixed and the statistic varies.

1.17 Is a football jersey number a quantitative or a categorical variable? Support your answer with a detailed explanation.

1.18 a. Name two attribute variables about its customers that a newly opened department store might find informative to study.

b. Name two numerical variables about its customers that a newly opened department store might find informative to study.

1.19 a. Name two nominal variables about its customers that a newly opened department store might find informative to study.

b. Name two ordinal variables about its customers that a newly opened department store might find informative to study.

1.20 **Skillbuilder Applet Exercise** simulates taking a sample of size 10 from a population of 100 college students. Take one sample and note the outcome.

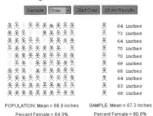

POPULATION: Mean = 66.9 inches   SAMPLE: Mean = 67.3 inches
Percent Female = 64.0%   Percent Female = 80.0%

a. Name the attribute variable involved in this experiment. Is it nominal or ordinal?

b. Name the numerical variable involved in this experiment. Is it discrete or continuous?

1.21 a. Explain why the variable "score" for the home team at a basketball game is discrete.

b. Explain why the variable "number of minutes to commute to work" is continuous.

1.22 The severity of side effects experienced by patients while being treated with a particular medicine is under study. The severity is measured on the scale: none, mild, moderate, severe, very severe.

a. Name the variable of interest.

b. Identify the type of variable.

1.23 The faculty members at Boise State University were surveyed on the question "How satisfied were you with the Summer 2002 Schedule?" Their responses were to be categorized as "very satisfied," "somewhat satisfied," "neither satisfied nor dissatisfied," "somewhat dissatisfied," or "very dissatisfied."

a. Name the variable of interest.

b. Identify the type of variable.

1.24 Students are being surveyed about the weight of books and supplies they are carrying as they attend class.

a. Identify the variable of interest.

b. Identify the type of variable.

c. List a few values that might occur in a sample.

1.25 A drug manufacturer is interested in the proportion of persons who have hypertension (elevated blood pressure) whose condition can be controlled by a new drug the company has developed. A study involving 5000 individuals with hypertension is conducted, and it is found that 80% of the individuals are able to control their hypertension with the drug. Assuming that the 5000 individuals are representative of the group who have hypertension, answer the following questions:

a. What is the population?

b. What is the sample?

c. Identify the parameter of interest.

d. Identify the statistic and give its value.

e. Do we know the value of the parameter?

1.26 The admissions office wants to estimate the cost of textbooks for students at our college. Let the variable x be the total cost of all textbooks purchased by a student this semester. The plan is to randomly identify 100 students and obtain their total textbook costs. The average cost for the 100 students will be used to estimate the average cost for all students.

a. Describe the parameter the admissions office wishes to estimate.

b. Describe the population.

c. Describe the variable involved.

d. Describe the sample.

e. Describe the statistic and how you would use the 100 data values collected to calculate the statistic.

1.27 A quality-control technician selects assembled parts from an assembly line and records the following information concerning each part:
A: defective or nondefective
B: the employee number of the individual who assembled the part
C: the weight of the part

a. What is the population?

b. Is the population finite or infinite?

c. What is the sample?

d. Classify the three variables as either attribute or numerical.

1.28 Select 10 students currently enrolled at your college and collect data for these three variables:
X: number of courses enrolled in
Y: total cost of textbooks and supplies for courses
Z: method of payment used for textbooks and supplies

a. What is the population?

b. Is the population finite or infinite?

c. What is the sample?

d. Classify the three variables as nominal, ordinal, discrete, or continuous.

1.29 A study was conducted by Aventis Pharmaceuticals Inc. to measure the adverse side effects of Allegra, a drug used for the treatment of seasonal allergies. A sample of 679 allergy sufferers in the United States was given 60 mg of the drug twice a day. The patients were to report whether they experienced relief from their allergies as well as any adverse side effects (viral infection, nausea, drowsiness, etc.).

Source: Good Housekeeping, February 2005, p. 120.

a. What is the population being studied?

b. What is the sample?

c. What are the characteristics of interest about each element in the population?

d. Are the data being collected qualitative or quantitative?

1.30 Skillbuilder Applet Exercise simulates taking a sample of size 10 from a population of 100 college students. Take a sample of size 10.

a. What is the population?

b. Is the population finite or infinite?

c. Name two parameters and give their values?

d. What is the sample?

e. Name the two corresponding statistics and give their values?

f. Take another sample of size 10. Which of the preceding items remain fixed and which changed?

1.31 Identify each of the following as an example of (1) attribute (qualitative) or (2) numerical (quantitative) variables:

a. the breaking strength of a given type of string

b. the hair color of children auditioning for the musical *Annie*

c. the number of stop signs in towns of less than 500 people

d. whether or not a faucet is defective

e. the number of questions answered correctly on a standardized test

f. the length of time required to answer a telephone call at a certain real estate office

1.32 Identify each of the following as examples of (1) nominal, (2) ordinal, (3) discrete, or (4) continuous variables:

a. A poll of registered voters as to which candidate they support

b. The length of time required for a wound to heal when a new medicine is being used

c. The number of televisions within a household

d. The distance first-year college women can kick a football

e. The number of pages per job coming off a computer printer

f. The kind of tree used as a Christmas tree

1.33 Suppose a 12-year-old asked you to explain the difference between a sample and a population.

a. What information should your answer include?

b. What reasons would you give for why one would take a sample instead of surveying every member of the population?

1.34 Suppose a 12-year-old asked you to explain the difference between a statistic and a parameter.

a. What information should your answer include?

b. What reasons would you give for why one would report the value of a statistic instead of a parameter?

1.3 Measurability and Variability

Within a set of measurement data, we always expect variation. If little or no variation is found, we would guess that the measuring device is not calibrated with a small enough unit. For example, we take a carton of a favorite candy bar and weigh each bar individually. We observe that each of the 24 candy bars weighs $7/8$ ounce, to the nearest $1/8$ ounce. Does this mean that the bars are all identical in weight? Not really! Suppose we were to weigh them on an analytical balance that weighs to the nearest ten-thousandth of an ounce. Now the 24 weights will most likely show **variability.**

It does not matter what the response variable is; there will most likely be variability in the data if the tool of measurement is precise enough. One of the primary objectives of statistical analysis is measuring variability. For example, in the study of quality control, measuring variability is absolutely essential. Controlling (or reducing) the variability in a manufacturing process is a field all its own—namely, statistical process control.

SECTION 1.3 EXERCISES

Statistics Now™

Skillbuilder Applet Exercises must be worked using an accompanying applet found on your Student's Suite CD-ROM or at the StatisticsNow website at **http://1pass.thomson.com.**

1.35 Suppose we measure the weights (in pounds) of the individuals in each of the following groups:
Group 1: cheerleaders for National Football League teams
Group 2: players for National Football League teams

For which group would you expect the data to have more variability? Explain why.

1.36 Suppose you were trying to decide which of two machines to purchase. Furthermore, suppose the length to which the machines cut a particular product part was important. If both machines produced parts that had the same length on the average, what other consideration regarding the lengths would be important? Why?

1.37 Consumer activist groups for years have encouraged retailers to use unit pricing of products. They argue that food prices, for example, should always be labeled in $/ounce, $/pound, $/gram, $/liter, and so on, in addition to $/package, $/can, $/box, $/bottle. Explain why.

1.38 A coin-operated coffee vending machine dispenses, on the average, 6 oz of coffee per cup. Can this statement be true of a vending machine that occasionally dispenses only enough to fill the cup half full (say, 4 oz)? Explain.

1.39 Teachers use examinations to measure students' knowledge about their subject. Explain how "a lack of variability in the students' scores might indicate that the exam was not a very effective measuring device." Thoughts to consider: What would it mean if all students attained a score of 100% on an exam? What would it mean if all attained a 0%? What would it mean if the grades ranged from 40% to 95%?

1.40 Skillbuilder Applet Exercise simulates sampling from a population of college students.

a. Take 10 samples of size 4 and keep track of the sample averages of hours per week that students study. Find the range of these averages by subtracting the lowest average from the highest average.

b. Take 10 samples of size 10 and keep track of the sample averages of hours per week that students study. Find the range of these averages by subtracting the lowest average from the highest average.

c. Which sample size demonstrated more variability?

d. If the population average is about 15 hours per week, which sample size demonstrated this most accurately? Why?

1.4 Data Collection

Because it is generally impossible to study an entire population (every individual in a country, all college students, every medical patient, etc.), researchers typically rely on *sampling* to acquire the information, or *data,* needed. It is important to obtain "good data" because the inferences ultimately made will be based on the statistics obtained from these data. These inferences are only as good as the data.

Although it is relatively easy to define "good data" as data that accurately represent the population from which they were taken, it is not easy to guarantee that a particular sampling method will produce "good data." We need to use sampling (data collection) methods that will produce data that are representative of the population and not *biased.*

Biased sampling method: A sampling method that produces data that systematically differ from the sampled population. An *unbiased sampling method* is one that is not biased.

Two commonly used sampling methods that often result in biased samples are the *convenience* and *volunteer samples.*

A convenience sample, sometimes called a *grab* sample, occurs when items are chosen arbitrarily and in an unstructured manner from a population, whereas a volunteer sample consists of results collected from those elements of the population that chose to contribute the needed information on their own initiative.

Did you ever buy a basket of fruit at the market based on the "good appearance" of the fruit on top, only to later discover the rest of the fruit was not as fresh? It was too inconvenient to inspect the bottom fruit, so you trusted a convenience sample. Has your teacher used your class as a sample from which to gather data? As a group, the class was quite convenient, but is it truly representative of the school's population? (Consider the differences among day, evening, and/or weekend students; type of course; etc.)

Have you ever mailed back your responses to a magazine survey? Under what conditions did (would) you take the time to complete such a questionnaire? Most people's immediate attitude is to ignore the survey. Those with strong feelings will make the effort to respond; therefore, representative samples should not be expected when volunteer samples are collected.

The Data-Collection Process

The collection of data for statistical analysis is an involved process and includes the following steps:

1. Define the objectives of the survey or study.
   Examples: compare the effectiveness of a new drug to the effectiveness of the standard drug; estimate the average household income in the United States.

2. Define the variable and the population of interest.
   Examples: length of recovery time for patients suffering from a particular disease; total income for households in the United States.

3. Define the data collection and data measuring schemes.
   This includes sampling frame, sampling procedures, sample size, and the data measuring device (questionnaire, telephone, and so on).

4. Collect your sample. Select the subjects to be sampled and collect the data.

5. Review of the sampling process upon completion of collection.
   Often an analyst is stuck with data already collected, possibly even data collected for other purposes, which makes it impossible to determine whether the data are "good." Using approved techniques to collect your own data is much preferred. Although this text is concerned chiefly with various data analysis techniques, you should be aware of the concerns of data collection.

The following example describes the population and the variable of interest for a specific investigation:

Two methods commonly used to collect data are *experiments* and *observational studies.* In an experiment, the investigator controls or modifies the environment and observes the effect on the variable under study. We often read about laboratory results obtained by using white rats to test different doses of a new medication and its effect on blood pressure. The experimental treatments were designed specifically to obtain the data needed to study the effect on the variable. In an observational study, the investigator does not modify the environment and does not control the process being observed. The data are obtained by sampling some of the population of interest. Surveys are observational studies of people.

APPLIED
EXAMPLE 1.7

Experiment or Observational Study?

SURGICAL INFECTION IS A MATTER OF TIME

Many surgical patients fail to get timely doses of the right medications, raising the risk of infection, researchers write in the Archives of Surgery. Of 30 million operations performed each year in the USA, about 2% are complicated by an on-site infection, the report says. The study of 34,000 surgical patients at nearly 3,000 hospitals in 2001 found that only 56% got prophylactic medications within an hour of surgery, when they can be effective.

Source: USA Today, February 22, 2005

This study is an example of an observational study. The researchers did not modify or try to control the environment. They observed what was happening and wrote up their findings.

If every element in the population can be listed, or enumerated, and observed, then a census is compiled. However, censuses are seldom used because they are often difficult and time-consuming to compile, and therefore very expensive. Imagine the task of compiling a census of every person who is a potential client at a brokerage firm. In situations similar to this, a *sample survey* is usually conducted.

When selecting a sample for a survey, it is necessary to construct a *sampling frame.*

> **Sampling frame:** A list, or set, of the elements belonging to the population from which the sample will be drawn.

Ideally, the sampling frame should be identical to the population with every element of the population included once and only once. In this case, a census would become the sampling frame. In other situations, a census may not be so easy to obtain, because a complete list is not available. Lists of registered voters or the telephone directory are sometimes used as sampling frames of the general public. Depending on the nature of the information being sought, the list of registered voters or the telephone directory may or may not serve as an unbiased sampling frame. Because only the elements in the frame have a chance to be selected as part of the sample, it is important that the sampling frame be representative of the population.

Part Better Than Whole

In the 1930s, Prasanta Chandra Mahalanobis placed a high priority on producing an appropriate representative sample. He wanted to determine the characteristics of large populations when it was almost impossible to obtain all measurements of a statistical population. Judgment samples seemed to be a good choice, but they have major faults–if enough is known about the population to collect a good judgment sample, there probably is no need for a sample; if the sample is wrong, there is no way to know how wrong it is. The answer to his quest was a random sample.

Once a representative sampling frame has been established, we proceed with selecting the sample elements from the sampling frame. This selection process is called the sample design. There are many different types of sample designs; however, they all fit into two categories: *judgment samples* and *probability samples.*

> **Judgment samples:** Samples that are selected on the basis of being judged "typical."

When a judgment sample is collected, the person selecting the sample chooses items that he or she thinks are representative of the population. The validity of the results from a judgment sample reflects the soundness of the collector's judgment. This is not an acceptable statisitical procedure.

> **Probability samples:** Samples in which the elements to be selected are drawn on the basis of probability. Each element in a population has a certain probability of being selected as part of the sample.

The inferences that will be studied later in this textbook are based on the assumption that our sample data are obtained using a probability sample. There are many ways to design probability samples. We will look at two of them, single-stage methods and multistage methods, and learn about a few of the many specific designs that are possible.

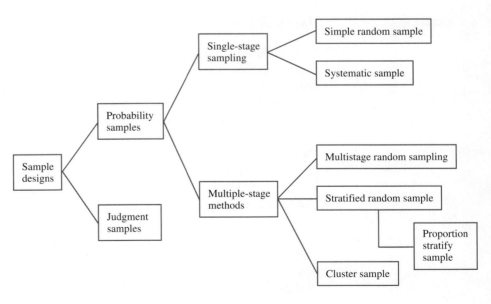

Single-Stage Methods

> **Single-stage sampling:** A sample design in which the elements of the sampling frame are treated equally and there is no subdividing or partitioning of the frame.

One of the most common single-stage probability sampling methods used to collect data is the *simple random sample.*

> **Simple random sample:** A sample selected in such a way that every element in the population or sampling frame has an equal probability of being chosen. Equivalently, all samples of size *n* have an equal chance of being selected.

Note: Random samples are obtained either by sampling with replacement from a finite population or by sampling without replacement from an infinite population.

Inherent in the concept of randomness is the idea that the next result (or occurrence) is not predictable. When a random sample is drawn, every effort must be made to ensure that each element has an equal probability of being selected and that the next result does not become predictable. The proper procedure for selecting a simple random sample requires the use of random numbers. Mistakes are commonly made because the term *random* (equal chance) is confused with haphazard (without pattern).

To select a simple random sample, first assign an identifying number to each element in the sampling frame. This is usually done sequentially using the same number of digits for each element. Then using random numbers with the same number of digits, select as many numbers as are needed for the sample size desired. Each numbered element in the sampling frame that corresponds to a selected random number is chosen for the sample.

EXAMPLE 1.8	**Using Random Numbers**

The admissions office at our college wishes to estimate the current "average" cost of textbooks per semester, per student. The population of interest is the "currently enrolled student body," and the variable is the "total amount spent for textbooks" by each student this semester. Because a random sample is desired, Mr. Clar, who works in the admissions office, has obtained a computer list of this semester's full-time enrollment. There were 4265 student names on the list. He numbered the students 0001, 0002, 0003, and so on, up to 4265; then, using four-digit random numbers, he identified a sample: 1288, 2177, 1952, 2463, 1644, 1004, and so on, were selected. (See the *Student Solutions Manual* for a discussion of the use of random numbers.)

A simple random sample is our first step toward an unbiased sample. Random samples are required for most of the statistical procedures presented in this book. Without a random design, the conclusions we draw from the statistical procedures may not be reliable.

APPLIED EXAMPLE 1.9	**Process for Collecting Data**

Consider the graphic "Even in kindergarten, social skills trump" on page 4 and the five steps of the data-collection process.

1. *Define the objectives of the survey or experiment.* Determine the opinion of U.S. kindergarten teachers regarding what they consider important skills for a kindergartner.

2. *Define the variable and the population of interest.* The variable is the opinion or response to a question concerning kindergarten skills. The population of interest is all U.S. kindergarten teachers.

3. *Define the data-collection and data-measuring schemes.* Based on the graphic itself, it can be seen that the source for the percentages presented was Mason-Dixon Polling. Upon further investigation, Mason-Dixon Polling & Research, Inc., of Washington, D.C., conducted the survey titled "The Fight Crime: Invest in Kids National Kindergarten Teacher Survey." It was a telephone survey of 800 randomly selected U.S. kindergarten teachers conducted from July 9 to July 14, 2004. The sampling frame was compiled from a list of state licensing records.

4. *Collect the sample.* The information collected from each teacher surveyed was the teachers' opinions on various skills they consider essential or very important for their students.

5. *Review of the sampling process upon completion of collection.* Since the sampling process was a telephone survey, what was the proportion of non-respondent? If the proportion was high, the validity of the percentages received would be questionable. Were the records available from all of the states so that each teacher had an equal chance of being selected?

In concept, the simple random sample is the simplest of the probability sampling techniques, but it is seldom used in practice because it often is an inefficient technique. One of the easiest to use methods for approximating a simple random sample is the *systematic sampling method.*

> **Systematic sample:** A sample in which every *k*th item of the sampling frame is selected, starting from a first element, which is randomly selected from the first *k* elements.

To select an x percent (%) systematic sample, we will need to randomly select 1 element from every $\frac{100}{x}$ elements. After the first element is randomly located within the first $\frac{100}{x}$ elements, we proceed to select every $\frac{100}{x}$th item thereafter until we have the desired number of data values for our sample.

For example, if we desire a 3% systematic sample, we would locate the first item by randomly selecting an integer between 1 and 33 ($\frac{100}{x} = \frac{100}{3} = 33.33$, which when rounded becomes 33). Suppose 23 was randomly selected. This means that our first data value is obtained from the subject in the 23rd position in the sampling frame. The second data value will come from subject in the 56th (23 + 33 = 56) position; the third, from the 89th (56 + 33); and so on, until our sample is complete.

The systematic technique is easy to describe and execute; however, it has some inherent dangers when the sampling frame is repetitive or cyclical in nature. For example, a systematic sample of every *k*th house along a long street might result in a sample disproportional with regard to houses on corner lots. The resulting information would likely be biased if the purpose for sampling is to learn about support for a proposed sidewalk tax. In these situations the results may not approximate a simple random sample.

Multistage Methods

When sampling very large populations, sometimes it is necessary to use a *multistage sampling* design to approximate random sampling.

> **Multistage random sampling:** A sample design in which the elements of the sampling frame are subdivided and the sample is chosen in more than one stage.

Multistage sampling designs often start by dividing a very large population into subpopulations on the basis of some characteristic. These subpopulations are called *strata*. These smaller, easier-to-work-with strata can then be sampled separately. One such sample design is the *stratified random sampling method*.

> **Stratified random sample:** A sample obtained by stratifying the population, or sampling frame, and then selecting a number of items from each of the strata by means of a simple random sampling technique.

A stratified random sample results when the population, or sampling frame, is subdivided into various strata, usually some already occurring natural subdivision, and then a subsample is drawn from each of these strata. These subsamples may be drawn from the various strata by using random or systematic methods. The subsamples are summarized separately first and then combined to draw conclusions about the entire population.

When a population with several strata is sampled, we often require that the number of items collected from each stratum be proportional to the size of the strata; this method is called a *proportional stratified sampling*.

> **Proportional stratified sample:** A sample obtained by stratifying the population, or sampling frame, and then selecting a number of items in proportion to the size of the strata from each strata by means of a simple random sampling technique.

A convenient way to express the idea of proportional sampling is to establish a quota. For example, the quota, "1 for every 150" directs you to select 1 data value for each 150 elements in each strata. That way, the size of the strata determines the size of the subsample from that strata. The subsamples are summarized separately and then combined to draw conclusions about the entire population.

Another sampling method that starts by stratifying the population, or sampling frame, is a *cluster sample*.

> **Cluster sample:** A sample obtained by stratifying the population, or sampling frame, and then selecting some or all of the items from some, but not all, of the strata.

The cluster sample is a multistage design. It uses either random or systematic methods to select the strata (clusters) to be sampled (first stage) and then uses either random or systematic methods to select elements from each identified cluster (second stage). The cluster sampling method also allows the possibility of selecting all of the elements from each identified cluster. Either way, the subsamples are summarized separately and the information then combined.

To illustrate a possible multistage random sampling process, consider that a sample is needed from a large country. In the first stage, the country is divided into smaller regions, such as states, and a random sample of these states is selected. In the second stage, a random sample of smaller areas within the selected states (counties) is then chosen. In the third stage, a random sample of even smaller areas (townships) is taken within each county. Finally in the fourth stage, if these townships are sufficiently small for the purposes of the study, the researcher might continue by collecting simple random samples from each of the identified townships. This would mean the entire sample was made up of several "local" subsamples identified as a result of the several stages.

Sample design is not a simple matter; many colleges and universities offer separate courses in sample surveying and experimental design. The topic of survey sampling is a complete textbook in itself. It is intended that the preceding information will provide you with an overview of sampling and put its role in perspective.

SECTION 1.4 EXERCISES

1.41 *USA Today* regularly asks its readers the following question: "Have a complaint about an airline baggage, refunds, advertising, customer service? Write: . . . ." What kind of sampling method is this? Are the results likely to be biased? Explain.

1.42 *USA Today* conducted a survey asking readers, "What is the most hilarious thing that has ever happened to you en route to or during a business trip?"

a. What kind of sampling method is this?

b. Are the results likely to be biased? Explain.

1.43 In a survey about families, Ann Landers asked parents if they would have kids again—70%

responded "No." An independent random survey, asking the same question, yielded a 90% "Yes" response. Give at least one explanation why the resulting percent from the Landers' survey is so much different than the random sample's percent?

1.44 Consider this question taken from CNN Quick Vote on the Internet on February 16, 2005: What should be done with the "Star Trek: Enterprise" show? The response was as follows: 45%, Let it boldly go on; 55%, Beam it out of here for good.

a. What kind of survey was used?

b. Do you think these results could be biased? Why?

1.45 We all know that exercise is good for us. But can exercise prevent or delay the symptoms of Parkinson's disease? A recent study by the Harvard School of Public Health studied 48,000 men and 77,000 women who were relatively healthy and middle-aged or older. During the course of the study, the disease developed in 387 people. The study found that men who had participated in some vigorous activity at least twice a week in high school, college, and up to age 40 had a 60% reduced risk of Parkinson's disease developing. The study found no such reduction for women. What type of sampling does this represent?

Source: Exercise may prevent Parkinson's, *USA Today,* February 22, 2005, p. 7D

1.46 A wholesale food distributor in a large metropolitan area would like to test the demand for a new food product. He distributes food through five large supermarket chains. The food distributor selects a sample of stores located in areas where he believes the shoppers are receptive to trying new products. What type of sampling does this represent?

1.47 Consider a simple population consisting of only the numbers 1, 2, and 3 (an unlimited number of each). Nine different samples of size 2 could be drawn from this population: (1, 1), (1, 2), (1, 3), (2, 1), (2, 2), (2, 3), (3, 1), (3, 2), (3, 3).

a. If the population consists of the numbers 1, 2, 3, and 4, list all the samples of size 2 that could possibly be selected.

b. If the population consists of the numbers 1, 2, and 3, list all the samples of size 3 that could possibly be selected.

1.48 a.   What is a sampling frame?

b. What did Mr. Clar use for a sampling frame in Example 1.8 (p. 22)?

c. Where did the number 1288 come from, and how was it used?

1.49 An article titled "Surface Sampling in Gravel Streams" (*Journal of Hydraulic Engineering,* April 1993) discusses grid sampling and areal sampling.

Grid sampling involves the removal by hand of stones found at specific points. These points are established on the gravel surface by using either a wire mesh or predetermined distances on a survey tape. The material collected by grid sampling is usually analyzed as a frequency distribution. An areal sample is collected by removing all the particles found in a predetermined area of a channel bed. The material recovered is most often analyzed as a frequency distribution by weight. Would you categorize these sample designs as judgment samples or probability samples?

1.50 A random sample may be very difficult to obtain. Why?

1.51 Why is the random sample so important in statistics?

1.52 Sheila Jones works for an established marketing research company in Cincinnati, Ohio. Her supervisor just handed her a list of 500 four-digit random numbers extracted from a statistical table of random digits. He told Sheila to conduct a survey by calling 500 Cincinnati residents on the telephone, provided the last four digits of their phone numbers matched one of the numbers on the list. If Sheila follows her supervisor's instructions, is he assured of obtaining a random sample of respondents? Explain.

1.53 Describe in detail how you would select a 4% systematic sample of the adults in a nearby large city in order to complete a survey about a political issue.

1.54 a.   What body of the federal government illustrates a stratified sampling of the people? (A random selection process is not used.)

b. What body of the federal government illustrates a proportional sampling of the people? (A random selection process is not used.)

1.55 Suppose you have been hired by a group of all-sports radio stations to determine the age distribution of their listeners. Describe in detail how you would select a random sample of 2500 from the 35 listening areas involved.

1.56 Explain why the polls that are so frequently quoted during early returns on Election Day TV coverage are an example of cluster sampling.

1.57 The telephone book might not be a representative sampling frame. Explain why.

1.58 The election board's voter registration list is not a census of the adult population. Explain why.

1.5 Comparison of Probability and Statistics

Probability and **statistics** are two separate but related fields of mathematics. It has been said that "probability is the vehicle of statistics." That is, if it were not for the laws of probability, the theory of statistics would not be possible.

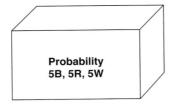
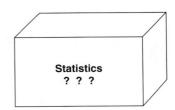

Let's illustrate the relationship and the difference between these two branches of mathematics by looking at two boxes. On one hand, we know the probability box contains five blue, five red, and five white poker chips. Probability tries to answer questions such as "If one chip is randomly drawn from this box, what is the chance that it will be blue?" On the other hand, in the statistics box we don't know what the combination of chips is. We draw a sample and, based on the findings in the sample, make conjectures about what we believe to be in the box. Note the difference: Probability asks you about the chance that something specific, such as drawing a blue chip, will happen when you know the possibilities (that is, you know the population). Statistics, in contrast, asks you to draw a sample, describe the sample (descriptive statistics), and then make inferences about the population based on the information found in the sample (inferential statistics).

SECTION 1.5 EXERCISES

1.59 Which of the following illustrates probability? statistics?

a. Determining how likely it is that a "6" will result when a die is rolled

b. Studying the weights of 35 babies to estimate weight gain in the first month after birth

1.60 Which of the following illustrates probability? statistics?

a. Collecting the number of credit hours from 100 students to estimate the average number of credit hours per student at a particular community college

b. Determining how likely it is to win the New York Lottery

1.61 Classify each of the following as a probability or a statistics problem:

a. Determining whether a new drug shortens the recovery time from a certain illness

b. Determining the chance that heads will result when a coin is tossed

c. Determining the amount of waiting time required to check out at a certain grocery store

d. Determining the chance that you will be dealt a "blackjack"

1.62 Classify each of the following as a probability or a statistics problem:

a. Determining how long it takes to handle a typical telephone inquiry at a real estate office

b. Determining the length of life for the 100-watt light bulbs a company produces

c. Determining the chance that a blue ball will be drawn from a bowl that contains 15 balls, of which 5 are blue

d. Determining the shearing strength of the rivets that your company just purchased for building airplanes

e. Determining the chance of getting "doubles" when you roll a pair of dice

1.6 | Statistics and Technology

In recent years, electronic technology has had a tremendous impact on almost every aspect of life. The field of statistics is no exception. As you will see, the field of statistics uses many techniques that are repetitive in nature: calculations of numerical statistics, procedures for constructing graphic displays of data, and procedures that are followed to formulate statistical inferences. Computers and calculators are very good at performing these sometimes long and tedious operations. If your computer has one of the standard statistical packages online or if you have a statistical calculator, then it will make the analysis easy to perform.

Throughout this textbook, as statistical procedures are studied, you will find the information you need to have a computer complete the same procedures using MINITAB (Release 14) and Excel software. Calculator procedures will also be demonstrated for the TI-83/84 Plus calculator.

An explanation of the most common typographical conventions that will be used in this textbook follows. As additional explanations or selections are needed, they will be given.

TECHNOLOGY INSTRUCTIONS: BASIC CONVENTIONS

MINITAB (Release 14)

FYI For information about obtaining MINITAB, check the Internet at http://www.minitab.com.

Choose: tells you to make a menu selection by a mouse "point and click" entry.
For example: **Choose: Stat > Quality Tools > Pareto Chart** instructs you to, in sequence, "point and click on" **Stat** on the menu bar, "followed by" **Quality Tools** on the pull-down, and then "followed by" **Pareto Chart** on the second pull-down.

Select: indicates that you should click on the small box or circle to the left of a specified item.

Enter: instructs you to type or select information needed for a specific item.

Excel

FYI Excel is part of Microsoft Office and can be found on many personal computers.

Choose: tells you to make a menu or tab selection by a mouse "point and click" entry.

For example: **Choose: Chart Wizard > XY(Scatter) > 1st graph picture > Next** instructs you to, in sequence, "point and click on" the **Chart Wizard** icon, followed by **XY(Scatter)** under Chart type, followed by **1st graph picture** on the Chart subtype, and then followed by **Next** on the dialog window.

Select: indicates that you should click on the small box or circle to the left of a specified item. It is often followed by a "point and click on" **Next** or **Finish** on the dialog window.

Enter: instructs you to type or select information needed for a specific item.

TI-83/84 Plus

FYI For information about obtaining TI-83/84 Plus, check the Internet at http://www.ti.com/calc.

Choose: tells you which keys to press or menu selections to make.

For example: **Choose: Zoom > 9:ZoomStat > Trace > > >** instructs you to press the **Zoom** key, followed by selecting **9:ZoomStat** from the menu, followed by pressing the **Trace** key; **> > >** indicates to press arrow keys repeatedly to move along a graph to obtain important points.

Enter: instructs you to type or select information needed for a specific item.

Screen
Capture: gives pictures of what your calculator screen should look like with chosen specifications highlighted.

Additional details about the use of MINITAB and Excel are available by using the Help system in the MINITAB and Excel software. Additional details for the TI-83/84 are contained in its corresponding *TI-83/84 Plus Graphing Calculator Guidebook.* Specific details on the use of computers and calculators available to you need to be obtained from your instructor or from your local computer lab person.

Your local computer center can provide you with a list of what is available to you. Some of the more readily available packaged programs are MINITAB, JMP-IN, and SPSS (Statistical Package for the Social Sciences).

Note: *There is a great temptation to use the computer or calculator to analyze any and all sets of data and then treat the results as though the statistics are correct. Remember the adage: "Garbage-in, garbage-out!" Responsible use of statistical methodology is very important. The burden is on the user to ensure that the appropriate methods are correctly applied and that accurate conclusions are drawn and communicated to others.*

SECTION 1.6 EXERCISES

1.63 How have computers increased the usefulness of statistics to professionals such as researchers, government workers who analyze data, statistical consultants, and others?

1.64 How might computers help you in statistics?

1.65 Did you ever hear someone say, "It must be right, that's what my calculator told me!" Explain why the calculator may or may not have given the correct answer.

1.66 What is meant by the saying "Garbage-in, garbage-out!" and how have computers increased the probability that studies may be victimized by the adage?

CHAPTER REVIEW

In Retrospect

You should now have a general feeling for what statistics is about, an image that will grow and change as you work your way through this book. You know what a sample and a population are and the distinction between qualitative (attribute) and quantitative (numerical) variables. You even know the difference between statistics and probability (although we will not study probability in detail until Chapter 4). You should also have an appreciation for and a partial understanding of how important random samples are in statistics.

Throughout the chapter you have seen numerous articles that represent various aspects of statistics. The statistical graphics picture a variety of information about ourselves as we describe ourselves and other aspects of the world around us. Statistics can even be entertaining. The examples are endless. Look around and find some examples of statistics in your daily life (see Exercises 1.85 and 1.86, p. 35).

Vocabulary and Key Concepts

attribute variable (p. 9)
biased sampling method (p. 18)
categorical variable (p. 9)
census (pp. 12, 20)
cluster sample (p. 25)
continuous variable (p. 11)
convenience sample (p. 19)
data (p. 8)
data collection (pp. 18, 22)
descriptive statistics (p. 4)
discrete variable (p. 11)
experiment (pp. 8, 19)
finite population (p. 7)
haphazard (p. 22)
inferential statistics (p. 4)
infinite population (p. 7)

judgment sample (p. 21)
multistage sampling, (p. 24)
nominal variable (p. 10)
numerical data (p. 10)
observational study (p. 19)
ordinal variable (p. 10)
parameter (p. 8)
population (p. 7)
probability (p. 27)
probability sample (p. 21)
proportional sample (p. 24)
qualitative variable (p. 9)
quantitative variable (p. 10)
random sample (p. 21)
representative sampling frame
  (p. 21)

sample (p. 8)
sample design (p. 21)
sampling frame (p. 20)
simple random sample (p. 22)
single-stage sampling, (p. 21)
statistic (p. 9)
statistics (pp. 3, 4, 27)
strata (p. 24)
stratified random sample (p. 24)
survey (p. 20)
systematic sample (p. 23)
unbiased sampling method
  (p. 18)
variability (p. 17)
variable (p. 8)
volunteer sample (p. 19)

Learning Outcomes

✓ Understand and be able to describe the difference between descriptive and inferential statistics.

pp. 3–4, Ex. 1.6, 1.7, 1.69

✓ Understand and be able to identify and interpret the relationships between sample and population, and statistic and parameter.

pp. 7–9, EXP 1.5

✓ Know and be able to identify and describe the different types of variables.

pp. 9–12, Ex. 1.31, 1.32

✓ Understand how convenience and volunteer samples result in biased samples.   pp. 18–19, Ex. 1.43

✓ Understand the differences among and be able to identify experiments, observational studies, and judgment samples.   pp. 19–21

✓ Understand and be able to describe the single-stage sampling methods of "simple random sample" and "systematic sampling."   pp. 21–24

✓ Understand and be able to describe the multistage sampling methods of "stratified sampling" and "cluster sampling."   pp. 24–25

✓ Understand and be able to explain the difference between probability and statistics.   p. 27, Ex. 1.61

✓ Understand that variability is inherent in everything and in the sampling process.   p. 17, Ex. 1.36

Chapter Exercises

Statistics⊘Now™
Go to the StatisticsNow website **http://1pass.thomson.com** to
- Assess your understanding of this chapter
- Check your readiness for an exam by taking the Pre-Test quiz and exploring the resources in the Personalized Learning Plan

1.67 We want to describe the so-called typical student at your college. Describe a variable that measures some characteristic of a student and results in:

a. Attribute data

b. Numerical data

1.68 A candidate for a political office claims that he will win the election. A poll is conducted, and 35 of 150 voters indicate that they will vote for the candidate, 100 voters indicate that they will vote for his opponent, and 15 voters are undecided.

a. What is the population parameter of interest?

b. What is the value of the sample statistic that might be used to estimate the population parameter?

c. Would you tend to believe the candidate based on the results of the poll?

1.69 A researcher studying consumer buying habits asks every 20th person entering Publix Su-

permarket how many times per week he or she goes grocery shopping. She then records the answer as T.

a. Is $T = 3$ an example of a sample, a variable, a statistic, a parameter, or a data value?

Suppose the researcher questions 427 shoppers during the survey.

b. Give an example of a question that can be answered using the tools of descriptive statistics.

c. Give an example of a question that can be answered using the tools of inferential statistics.

1.70 A researcher studying the attitudes of parents of preschool children interviews a random sample of 50 mothers, each having one preschool child. He asks each mother, "How many times did you compliment your child yesterday?" He records the answer as C.

a. Is $C = 4$ an example of a data value, a statistic, a parameter, a variable, or a sample?

b. Give an example of a question that can be answered using the tools of descriptive statistics.

c. Give an example of a question that can be answered using the tools of inferential statistics.

1.71 Harris Interactive conducted an online poll of U.S. adults during December 2004 for the *Wall Street Journal Online*'s *Health Industry Edition*.

These are some of the results of a Harris Interactive® poll of 2,013 U.S. adults conducted online between December 14 and 16, 2004 for the Wall Street Journal Online's Health Industry Edition.

Of all adults who have received a flu vaccine this year, 43 percent received it in a doctor's office. Other places adults have received a flu shot this year include vaccination clinics (18%), workplaces or schools (12%), and pharmacies (10%). Of note, nobody reported purchasing a flu vaccine via the Internet or in Canada.

Of the 83 percent of adults who have not had a flu shot this year, the majority (77%) say they are taking precautions to reduce their risk of catching the flu this season. The most common precaution reported being taken is to wash hands, wear gloves or use hand cleaning products more frequently (63%). Other precautions include taking vitamins or supplements (49%), trying to follow a healthy diet (42%), getting more rest (34%), and avoiding crowded places (24%).

Source: http://www.harrisinteractive.com/news/

a. What is the population?

b. Name at least four variables that must have been used.

c. Classify all the variables of the study as either attribute or numerical.

1.72 A *USA Today* Snapshot from June 4, 2002, described how executives feel about looking for a new job while still employed. According to the Snapshot, a survey of 150 executives from the nation's 1000 largest companies resulted in the following responses: 36% felt very comfortable, 33% felt somewhat comfortable, 26% felt somewhat uncomfortable, and 5% felt very uncomfortable. Would you classify the data collected and used to determine these percentages as qualitative (nominal or ordinal) or quantitative (discrete or continuous)?

1.73 Results from a study titled *Academic Atrophy: The Condition of the Liberal Arts in America's Public Schools* were released on March, 8, 2004. It was the first study on how the No Child Left Behind Act might be influencing instructional time in the social studies, that is, the arts, geography, history, and foreign languages. The study surveyed more than 1000 principals in four states and found that

47% of principals at high minority schools reported decreases in elementary social studies.

Source: http://music-for-all.org/CBESurvey.html

a. What is the population?

b. What is the sample?

c. Is this a judgment sample or a probability sample?

d. If this study is a probability sample, what type of sampling method do you think was used?

1.74 Based on a survey of more than 125,000 people, the National Center for Health Statistics reported that married people tend to be healthier than other groups. Among other things, the study looked at the number of people reporting they were in only fair or poor health. The study reported the following findings for all adults aged 18 and older: 11.9% reported they were in fair or poor health, including 10.5% of the married people, 19.6% of widows, 16.7% of divorced or separated, 12.5% of those never married, and 14% of those living with an unmarried partner.

Source: Finger Lakes Times, December 19, 2004

a. What is the population?

b. What is the sample?

c. Based on the size of the sample, what kind of sample do you suspect was taken?

1.75 The following graph shows the relationship between three variables: number of licensed drivers, number of registered vehicles, and the size of the resident population for the United States from 1961 to 2003.

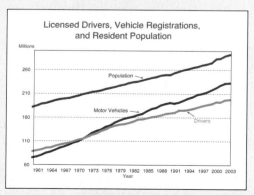

Source: United States Department of Transportation - Federal Highway Administration

Study the graph and answer the questions.

a. Does it seem reasonable that the Population line and the Drivers line run basically parallel to each other and that the Population line is above the Drivers line? Explain what it means for the lines to be parallel. What would it mean if they were not parallel?

b. The Drivers and Motor Vehicles lines cross. What does this mean? When did the lines intersect, and what does the point of intersection represent?

1.76 The 700 Club: Barry Bonds of the San Francisco Giants is on pace to become baseball's home-run king this season or next. Last year, he joined Hank Aaron and Babe Ruth as the only major-league players to have hit more than 700 home runs in their careers. Here is a look at how they amassed their totals.

a. Describe and compare the overall appearance of the three graphs. Include thoughts about such things as length of career, when the most home-runs per year were hit and their relationship to the aging process, and any others you think of.

b. Does it appear that one of them was more consistent with annual homerun production?

c. From the evidence presented here, who do you think should be called the "Homerun King"?

d. Was Barry Bonds' 73 homeruns in one season a fluke?

e. If you were a team owner and most interested in homerun production over the next several years, which one would you want on your team? Let's say you were signing him at age 21. At age 35.

1.77 In the autumn of 2003, the National Safe Kids Campaign conducted a study of helmet use among children ages 5 to 14 who participate in wheeled sports. Data were collected from various sites across the United States that were designated as places where children often engage in wheeled sports. Activity, apparent gender, and estimated age were recorded for each rider, along with information on helmet use. It was found that, overall, 41% of children were wearing a helmet while participating in a wheeled sport.

a. Was this study an experiment or an observational study?

b. Identify the parameter of interest.

c. Identify the statistic and give its value.

d. Classify the four variables as numerical or attribute.

1.78 *USA Today,* in a December 2004 article titled "There's no place like work for the holidays," presented the results of a study of 600 full-time U.S. workers done by Penn Schoen & Berland Associates. Results revealed that 33% of the respondents were taking no time off during the holidays. Of those surveyed, 28% reported taking 1 to 2 days off during the holidays.

a. What is the population?

b. What is the sample?

c. Is this a judgment sample or a probability sample?

Exercise 1.76

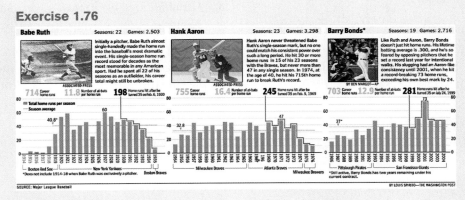

Source: The Washington Post

1.79 Two-thirds of older Americans take part in leisure-time physical activities, but poor nutrition remains a problem, especially when it comes to fruit and vegetables, according to a study by the Centers for Disease Control and Prevention and the Merck Institute of Aging & Health. Among the study's findings, one-third of Americans older than age 65 are not getting any leisure-time physical activity. Among all Americans, that figure is 24.5%. Two-thirds of older Americans do not eat five servings of fruits and vegetables a day. But they're doing better than Americans on the whole, of whom 73% do not meet the daily guidelines.

a. What is the population?

b. What are the characteristics of interest?

c. Classify all of the variables in the study as either attribute or numerical.

1.80 The 2001 National Aging Research Survey revealed that Americans have great expectations on living long, healthy, and independent lives. The X and Y generations, ages 18 to 36 years, are the most desirous of living to 100. Of this age group, 69% reported that they would like to live to be 100 years old. The sampling method used for collecting this information was random digit dialing (RDD). This RDD method gives every telephone-equipped household in the United States an equal chance of being contacted. RDD is a popular survey tool. Use the Internet to search for information and write a 100-word explanation of how it works.

1.81 Who takes the most drugs? The National Association of Chain Drug Stores provides some answers to that question.

a. What variable is used in preparation of this information?

b. Which gender takes the most drugs? By how much?

c. Which age group takes the most drugs? How much do they take?

d. Does geography matter? Which state has the highest usage? lowest?

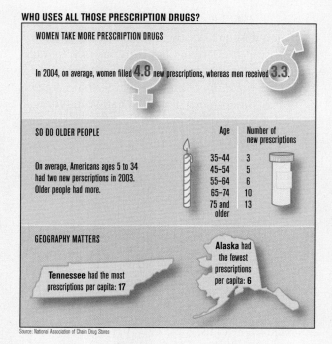

WHO USES ALL THOSE PRESCRIPTION DRUGS?

WOMEN TAKE MORE PRESCRIPTION DRUGS

In 2004, on average, women filled **4.8** new prescriptions, whereas men received **3.3**.

SO DO OLDER PEOPLE

On average, Americans ages 5 to 34 had two new perscriptions in 2003. Older people had more.

Age	Number of new prescriptions
35–44	3
45–54	5
55–64	6
65–74	10
75 and older	13

GEOGRAPHY MATTERS

Tennessee had the most prescriptions per capita: 17

Alaska had the fewest prescriptions per capita: 6

Source: National Association of Chain Drug Stores

1.82 "Drugs of choice," a chart in a March 2005 *Democrat & Chronicle* article titled "Pill unhooks heroin's claws, but few get it," presents the numbers of people abusing various drugs. The estimated number of people abusing marijuana in the United States during 2002 was 4.3 million. The source for this chart was the Substance Abuse and Mental Health Services Administration. What method of sampling do you think was used by the Administration to obtain this statistic? Explain.

1.83 Describe, in your own words, and give an example of each of the following terms. Your examples should not be ones given in class or in the textbook.

a. Variable   b. Data   c. Sample

d. Population   e. Statistic   f. Parameter

1.84 Describe, in your own words, and give an example of the following terms. Your examples should not be ones given in class or in the textbook.

a. Random sample   b. Probability sample

c. Judgment sample

1.85 Find an article or an advertisement in a newspaper or magazine that exemplifies the use of statistics.

a. Identify and describe one statistic reported in the article.

b. Identify and describe the variable related to the statistic in part a.

c. Identify and describe the sample related to the statistic in part a.

d. Identify and describe the population from which the sample in part c was taken.

1.86 a. Find an article in a newspaper or magazine that exemplifies the use of statistics in a way that might be considered "entertainment" or "recreational." Describe why you think this article fits one of these categories.

b. Find an article in a newspaper or magazine that exemplifies the use of statistics and presents an unusual finding as the result of a study. Describe why these results are (or are not) "newsworthy."

1.87 In Exercise 1.5, you were asked to write a sentence for each of the three statistical activities given in the definition of *statisitics*. Now that you have completed the chapter, review your work. Again, using your own words, change and/or enhance your work to complete a paragraph on the definition of *statistics*.

Chapter Project

Americans, Here's Looking at You

The chapter project takes us back to Section 1.1, "Americans, Here's Looking at You," as a way to assess what we have learned in this chapter. Study the statistical information presented by the graphs and charts, and ask yourself how the terms (population, sample, variable, statistic, type of variable) studied in this chapter apply to each and how you compare to the statistical story being told.

Putting Chapter 1 to Work

1.88 With respect to the four graphics in Section 1.1 on pages 1 and 2, complete the following:

a. What statistical population is of concern for all of these graphics?

b. Identify one specific graphic. What variables were used to collect the information needed to determine the statistics reported?

c. Name one statistic that is being reported in your graphic.

d. To obtain the data for your graphic, what methods do you think were used: convenience sample, volunteer sample, random sample, survey, observational study, experiment, or judgment sample?

e. Considering the method, how much faith do you have in the printed statistics? Describe possible biases.

Your Study

1.89 Select one of the "Americans, Here's Looking at You" graphics (p. 1); then, using the students at your school or college as the population of concern, collect sample data from 30 students and produce your own version of the graphic. Write a paragraph describing how your results compare to those reported in the selected graphic.

Chapter Practice Test

PART I: Knowing the Definitions

Answer "True" if the statement is always true. If the statement is not always true, replace the words printed in bold with words that make the statement always true.

1.1 **Inferential** statistics is the study and description of data that result from an experiment.

1.2 **Descriptive statistics** is the study of a sample that enables us to make projections or estimates about the population from which the sample is drawn.

1.3 A **population** is typically a very large collection of individuals or objects about which we desire information.

1.4 A statistic is the calculated measure of some characteristic of a **population.**

1.5 A parameter is the measure of some characteristic of a **sample.**

1.6 As a result of surveying 50 freshmen, it was found that 16 had participated in interscholastic sports, 23 had served as officers of classes and clubs, and 18 had been in school plays during their high school years. This is an example of **numerical data.**

1.7 The "number of rotten apples per shipping crate" is an example of a **qualitative** variable.

1.8 The "thickness of a sheet of sheet metal" used in a manufacturing process is an example of a **quantitative** variable.

1.9 A **representative** sample is a sample obtained in such a way that all individuals had an equal chance of being selected.

1.10 The basic objectives of **statistics** are obtaining a sample, inspecting this sample, and then making inferences about the unknown characteristics of the population from which the sample was drawn.

PART II: Applying the Concepts

The owners of Corner Convenience Store are concerned about the quality of service their customers receive. In order to study the service, they collected samples for each of several variables.

1.11 Classify each of the following variables as nominal, ordinal, discrete, or continuous:

  a. Method of payment for purchases (cash, credit card, check)

  b. Customer satisfaction (very satisfied, satisfied, not satisfied)

  c. Amount of sales tax on purchase

  d. Number of items purchased

  e. Customer's driver's license number

1.12 The mean checkout time for all customers at Corner Convenience Store is to be estimated by using the mean checkout time for 75 randomly selected customers. Match the items in column 2 with the statistical terms in column 1.

1	2
____ data value	(a) the 75 customers
____ data	(b) the mean time for all customers
____ experiment	
____ parameter	(c) 2 minutes, one customer's checkout time
____ population	
____ sample	
____ statistic	(d) the mean time for the 75 customers
____ variable	(e) all customers at Corner Convenience Store
	(f) the checkout time for one customer
	(g) the 75 times
	(h) the process used to select 75 customers and measure their times

PART III: Understanding the Concepts

Write a brief paragraph in response to each question.

1.13 The population and the sample are both sets of objects. Describe the relationship between them and give an example.

1.14 The variable and the data for a specific situation are closely related. Explain this relationship and give an example.

1.15 The data, the statistic, and the parameter are all values used to describe a statistical situation. How does one distinguish among these three terms? Give an example.

1.16 What conditions are required for a sample to be a random sample? Explain and include an example of a sample that is random and one that is not random.

CHAPTER 2

Descriptive Analysis and Presentation of Single-Variable Data

2.1 You and the Internet

Statistics⟨⟩Now™

Throughout the chapter, this icon introduces a list of resources on the StatisticsNow website at **http://1pass.thomson.com** that will:

- Help you evaluate your knowledge of the material
- Allow you to take an exam-prep quiz
- Provide a Personalized Learning Plan targeting resources that address areas you should study

Ever wonder what other people do when they are on the Internet? Well, you are not the only one. Stanford Institute for the Quantitative Study of Society (SIQSS) supported a study that looked at how people use the Internet. Four thousand respondents were asked to select which of 17 common Internet activities they did or did not do. E-mail was identified by 90% of the respondents as one of their many uses of the Internet. Other common uses included information gathering, entertainment, chat rooms, and business transactions.

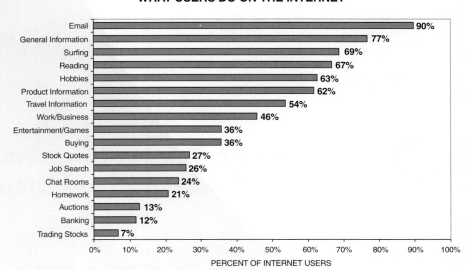

WHAT USERS DO ON THE INTERNET

Activity	Percent
Email	90%
General Information	77%
Surfing	69%
Reading	67%
Hobbies	63%
Product Information	62%
Travel Information	54%
Work/Business	46%
Entertainment/Games	36%
Buying	36%
Stock Quotes	27%
Job Search	26%
Chat Rooms	24%
Homework	21%
Auctions	13%
Banking	12%
Trading Stocks	7%

PERCENT OF INTERNET USERS

Source: Stanford Institute for the Quantitative Study of Society
[http://www.stanford.edu/group/siqss/Press_Release/Chart9gif].
Internet and Society–study by Mie and Erbring, funded by SIQSS, 2000 (see copy). Reprinted by permission.

The preceding graph summarizes all the information from the study of 4000 Internet users. Can you imagine if all of that information was written out in sentences? Graphical displays (pictures) can truly be worth a thousand words. Not only is the information in a clearer, more concise format, but the format also allows us to make some conclusions at the same time. We immediately know which activities most users engage in and which activities are not that popular.

If you had been asked, "What activity do you do the most on the Internet?" how would you have answered? Do you believe your answer is represented accurately in the diagram? As you work through Chapter 2, you will begin to learn how to organize and summarize data into graphical displays and numerical statistics in order to accurately and appropriately describe data. Then you will be able to further investigate "You and the Internet" in the Project section with Exercises 2.224 and 2.225 (p. 140).

SECTION 2.1 EXERCISES

Statistics⬡Now™

Datasets can be found on your Student's Suite CD-ROM or at the StatisticsNow website at **http://1pass.thomson.com**.

2.1 [EX02-001] Students in an online statistics course were asked how many different Internet activities they engaged in during a typical week. The following data show the number of activities:

6	7	3	6	9	10	8	9	9	6	4	9	4	9
4	2	3	5	13	12	4	6	4	9	5	6	9	
11	5	6	5	3	7	9	6	5	12	2	6	9	

a. If you were asked to present these data, how would you organize and summarize them?

b. This chapter will discuss a variety of methods for displaying and describing data. What type of information or conclusions would you like to know about these data if one of the pieces of data were from you?

2.2 a. How many different Internet activities did you engage in last week?

b. How do you think you compare to the 40 Internet users in the sample in Exercise 2.1?

c. How do you think you compare to all Internet users?

2.2   Graphs, Pareto Diagrams, and Stem-and-Leaf Displays

Once the sample data have been collected, we must "get acquainted" with them. One of the most helpful ways to become acquainted with the data is to use an initial exploratory data-analysis technique that will result in a pictorial representation of the data. The display will visually reveal patterns of behavior of the variable being studied. There are several graphic (pictorial) ways to describe data. The type of data and the idea to be presented determine which method is used.

Note: There is no single correct answer when constructing a graphic display. The analyst's judgment and the circumstances surrounding the problem play major roles in the development of the graphic.

Qualitative Data

Circle graphs and bar graphs: Graphs that are used to summarize qualitative, or attribute, or categorical data. Circle graphs (pie diagrams) show the amount of data that belong to each category as a proportional part of a circle. Bar graphs show the amount of data that belong to each category as a proportionally sized rectangular area.

EXAMPLE 2.1

Graphing Qualitative Data

Table 2.1 lists the number of cases of each type of operation performed at General Hospital last year.

TABLE 2.1

Operations Performed at General Hospital Last Year [TA02-01]

Type of Operation	Number of Cases
Thoracic	20
Bones and joints	45
Eye, ear, nose, and throat	58
General	98
Abdominal	115
Urologic	74
Proctologic	65
Neurosurgery	23
Total	498

The data in Table 2.1 are displayed on a circle graph in Figure 2.1, with each type of operation represented by a relative proportion of the circle, found by dividing the number of cases by the total sample size, namely, 498. The proportions are then reported as percentages (for example, 25% is ¼ of the circle). Figure 2.2 displays the same "type of operation" data but in the form of a

FIGURE 2.1 Circle Graph

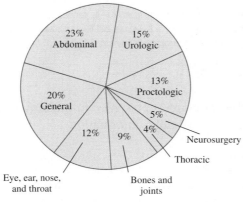

Operations Performed at General Hospital Last Year

FIGURE 2.2 Bar Graph

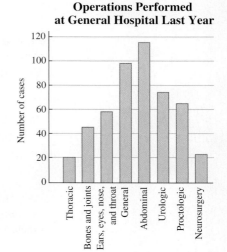

Operations Performed at General Hospital Last Year

FYI All graphic representations need to be completely self-explanatory. That includes a descriptive, meaningful title and proper identification of the quantities and variables involved.

bar graph. Bar graphs of attribute data should be drawn with a space between bars of equal width.

TECHNOLOGY INSTRUCTIONS: CIRCLE GRAPH

MINITAB (Release 14) Input the categories into C1 and the corresponding frequencies into C2; then continue with:

Choose:	**Graph > Pie Chart...**
Select:	**Chart values from a table**
Enter:	Categorical variable: **C1** Summary variables: **C2**
Select:	**Labels > Title/Footnotes** Enter: Title: **your title**
Select:	**Slice Labels** > Select desired labels > **OK** > **OK**

Excel Input the categories into column A and the corresponding frequencies into column B; then continue with:

Choose:	**Chart Wizard > Pie > 1st picture** (usually) > **Next**
Enter:	Data range: **(A1:B5 or select cells)**
Check:	Series in: **columns > Next**
Choose:	**Titles**
Enter:	Chart title: **Your title**
Choose:	**Data Labels**
Select:	**Category name and Percentage > Next > Finish**

To edit the pie chart:

Click On:	Anywhere clear on the chart
	—use handles to size
	Any cell in the category or frequency column
	and type in different name or amount > ENTER

TI-83/84 Plus Input the frequencies for the various categories into L1; then continue with:

Choose:	**PRGM > EXEC > CIRCLE***
Enter:	LIST: **L1** > **ENTER**
	DATA DISPLAYED?: **1:PERCENTAGES**
	OR
	2:DATA

*The TI-83/84 Plus program 'CIRCLE' and others can be downloaded from http://statistics.duxbury.com/jkes10e. The TI-83/84 Plus programs and data files are jkprogs.zip and jklists.zip. Save the files to your computer and uncompress them using a zip utility. Download the programs to your calculator using TI-Graph Link Software.

When the bar graph is presented in the form of a *Pareto diagram,* it presents additional and very helpful information.

Pareto diagram: A bar graph with the bars arranged from the most numerous category to the least numerous category. It includes a line graph displaying the cumulative percentages and counts for the bars.

EXAMPLE 2.2

Pareto Diagram of Hate Crimes

The FBI reported the number of hate crimes by category for 2003 (http://
www.fbigov/ucr/ucr.htm#hate). The Pareto diagram in Figure 2.3 shows the
8706 categorized hate crimes, their percentages, and cumulative percentages.

FIGURE 2.3 Pareto Diagram

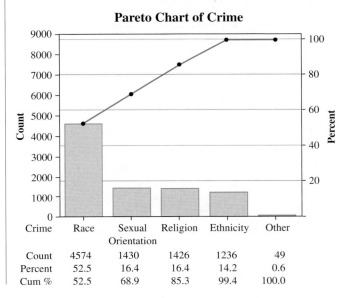

Pareto Chart of Crime

Crime	Race	Sexual Orientation	Religion	Ethnicity	Other
Count	4574	1430	1426	1236	49
Percent	52.5	16.4	16.4	14.2	0.6
Cum %	52.5	68.9	85.3	99.4	100.0

The Pareto diagram is popular in quality-control applications. A Pareto diagram of types of defects will show the ones that have the greatest effect on
the defective rate in order of effect. It is then easy to see which defects should
be targeted to most effectively lower the defective rate.

TECHNOLOGY INSTRUCTIONS: PARETO DIAGRAM

MINITAB (Release 14)   Input the categories into C1 and the corresponding frequencies into C2; then
continue with:

```
Choose:    Stat > Quality Tools > Pareto Chart
Select:    Chart defects table
Enter:     Labels in: C1   Frequencies in: C2
Select:    Options
Enter:     Title: your title > OK > OK
```

Excel   Input the categories into column A and the corresponding frequencies into
column B (column headings are optional); then continue with:
First, sorting the table:

```
Activate both columns of the distribution
Choose:    Data > Sort > Sort by: Column B (freq or rel freq col.)
Select:    Descending
           My list has: Header row or No Header row > OK
Choose:    Chart Wizard > Column > 1st picture (usually) > Next
```

```
Choose:      Data Range
Enter:       Data Range: (A1:B5 or select cells)
Select:      Series in:   Columns > Next
Choose:      Titles
Enter:       Chart title: your title
             Category (x) axis: title for x-axis
             Value (y) axis: title for y-axis > Next > Finish
```

To edit the Pareto diagram:

```
Click on:    Anywhere clear on the chart
             —use handles to size
             Any title name to change
             Any cell in the category column and type in a name > Enter
```

Excel does not include the line graph.

TI-83/84 Plus   Input the numbered categories into L1 and the corresponding frequencies into L2; then continue with:

```
Choose:      PRGM > EXEC > PARETO*
Enter:       LIST: L2 > ENTER
Ymax:        at least the sum of the frequencies > ENTER
Yscl:        increment for y-axis > ENTER
```

*Program 'PARETO' is one of many programs that are available for downloading from a website. See page 42 for specific instructions.

Quantitative Data

One major reason for constructing a graph of quantitative data is to display its *distribution*.

> **Distribution:** The pattern of variability displayed by the data of a variable. The distribution displays the frequency of each value of the variable.

One of the simplest graphs used to display a distribution is the *dotplot*.

> **Dotplot display:** Displays the data of a sample by representing each data value with a dot positioned along a scale. This scale can be either horizontal or vertical. The frequency of the values is represented along the other scale.

EXAMPLE 2.3   **Dotplot of Exam Grades**

Table 2.2 provides a sample of 19 exam grades randomly selected from a large class.

TABLE 2.2

Sample of 19 Exam Grades [TA02-02]

76	74	82	96	66	76	78	72	52	68
86	84	62	76	78	92	82	74	88	

Figure 2.4 is a dotplot of the 19 exam scores.

FIGURE 2.4
Dotplot

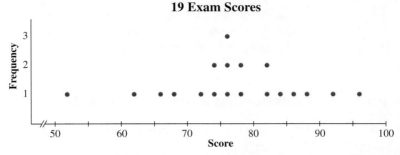

19 Exam Scores

Notice how the data in Figure 2.4 are "bunched" near the center and more "spread out" near the extremes.

The dotplot display is a convenient technique to use as you first begin to analyze the data. It results in a picture of the data that sorts the data into numerical order. (To *sort* data is to list the data in rank order according to numerical value.)

TECHNOLOGY INSTRUCTIONS: DOTPLOT

MINITAB (Release 14)   Input the data into C1; then continue with:

```
Choose:      Graph > Dotplot... > One Y. Simple > OK
Enter:       Graph Variables: C1 > OK
```

Excel   The dotplot display is not available, but the initial step of ranking the data can be done. Input the data into column A and activate the column of data; then continue with:

```
Choose:      Data > Sort
Enter:       Sort by: Column A
Select:      Ascending > My list has: Header row or No Header row
```

Use the sorted data to finish constructing the dotplot display.

TI-83/84 Plus   Input the data into L1; then continue with:

```
Choose:      PRGM > EXEC > DOTPLOT*
Enter:       LIST: L1 > ENTER
             Xmin: at most the lowest x value
             Xmax: at least the highest x value
             Xscl: 0 or increment
             Ymax: at least the highest frequency
```

*Program 'DOTPLOT' is one of many programs that are available for downloading from a website. See page 42 for specific instructions.

In recent years a technique known as the *stem-and-leaf display* has become very popular for summarizing numerical data. It is a combination of a graphic technique and a sorting technique. These displays are simple to create and use, and they are well suited to computer applications.

Stem-and-leaf display: Displays the data of a sample using the actual digits that make up the data values. Each numerical value is divided into two parts: The leading digit(s) becomes the stem, and the trailing digit(s) becomes the leaf. The stems are located along the main axis, and a leaf for each data value is located so as to display the distribution of the data.

EXAMPLE 2.4

Constructing a Stem-and-Leaf Display

FIGURE 2.5A
Unfinished Stem-and-Leaf Display

19 Exam Scores

5	2
6	6 8 2
7	6 4 6 8 2 6 8 4
8	2 6 4 2 8
9	6 2

FIGURE 2.5B
Final Stem-and-Leaf Display

19 Exam Scores

5	2
6	2 6 8
7	2 4 4 6 6 6 8 8
8	2 2 4 6 8
9	2 6

FIGURE 2.6 **Stem-and-Leaf Display**

19 Exam Scores

(50–54) 5	2
(55–59) 5	
(60–64) 6	2
(65–69) 6	6 8
(70–74) 7	2 4 4
(75–79) 7	6 6 6 8 8
(80–84) 8	2 2 4
(85–89) 8	6 8
(90–94) 9	2
(95–99) 9	6

Let's construct a stem-and-leaf display for the 19 exam scores given in Table 2.2 on page 45.

At a quick glance we see that there are scores in the 50s, 60s, 70s, 80s, and 90s. Let's use the first digit of each score as the stem and the second digit as the leaf. Typically, the display is constructed vertically. We draw a vertical line and place the stems, in order, to the left of the line.

$$
\begin{array}{c|}
5 \\
6 \\
7 \\
8 \\
9 \\
\end{array}
$$

Next we place each leaf on its stem. This is done by placing the trailing digit on the right side of the vertical line opposite its corresponding leading digit. Our first data value is 76; 7 is the stem and 6 is the leaf. Thus, we place a 6 opposite the stem 7:

$$7 \,|\, 6$$

The next data value is 74, so a leaf of 4 is placed on the 7 stem next to the 6.

$$7 \,|\, 6\ 4$$

The next data value is 82, so a leaf of 2 is placed on the 8 stem.

$$
\begin{array}{c|l}
7 & 6\ 4 \\
8 & 2 \\
\end{array}
$$

We continue until each of the other 16 leaves is placed on the display. Figure 2.5A shows the resulting stem-and-leaf display; Figure 2.5B shows the completed stem-and-leaf display after the leaves have been ordered.

From Figure 2.5B, we see that the grades are centered around the 70s. In this case, all scores with the same tens digit were placed on the same branch, but this may not always be desired. Suppose we reconstruct the display; this time instead of grouping 10 possible values on each stem, let's group the values so that only 5 possible values could fall on each stem, as shown in Figure 2.6. Do you notice a difference in the appearance of Figure 2.6? The general shape is approximately symmetrical about the high 70s. Our information is a little more refined, but basically we see the same distribution.

TECHNOLOGY INSTRUCTIONS: STEM-AND-LEAF DIAGRAM

MINITAB (Release 14) Input the data into C1; then continue with:

```
Choose:    Graph > Stem-and-Leaf . . .
Enter:     Graph variables: C1
           Increment:  stem width (optional) > OK
```

Excel Input the data into column A; then continue with:

```
Choose:    Tools > Data Analysis Plus* > Stem and Leaf Display > OK
Enter:     Input Range: (A2:A6 or select cells)
           Increment: Stem Increment
```

*Data Analysis Plus is a collection of statistical macros for Excel. They can be downloaded onto your computer from your Student's Suite CD-ROM.

TI-83/84 Plus Input the data into L1; then continue with:

```
Choose:    STAT > EDIT > 2:SortA(
Enter:     L1
```

Use sorted data to finish constructing the stem-and-leaf diagram by hand.

It is fairly typical of many variables to display a distribution that is concentrated (mounded) about a central value and then in some manner dispersed in one or both directions. Often a graphic display reveals something that the analyst may or may not have anticipated. Example 2.5 demonstrates what generally occurs when two populations are sampled together.

EXAMPLE 2.5 Overlapping Distributions

Statistics Now™

Watch a video example at **http://1pass.thomson.com** or on your CD.

A random sample of 50 college students was selected. Their weights were obtained from their medical records. The resulting data are listed in Table 2.3.

Notice that the weights range from 98 to 215 pounds. Let's group the weights on stems of 10 units using the hundreds and the tens digits as stems and the units digit as the leaf (see Figure 2.7). The leaves have been arranged in numerical order.

Close inspection of Figure 2.7 suggests that two overlapping distributions may be involved. That is exactly what we have: a distribution of female weights and a distribution of male weights. Figure 2.8 shows a "back-to-back" stem-and-leaf display of this set of data and makes it obvious that two distinct distributions are involved.

Figure 2.9, a "side-by-side" dotplot (same scale) of the same 50 weight data, shows the same distinction between the two subsets.

TABLE 2.3

Weights of 50 College Students [TA02-03]

Student	1	2	3	4	5	6	7	8	9	10
Male/Female	F	M	F	M	M	F	F	M	M	F
Weight	98	150	108	158	162	112	118	167	170	120
Student	11	12	13	14	15	16	17	18	19	20
Male/Female	M	M	M	F	F	M	F	M	M	F
Weight	177	186	191	128	135	195	137	205	190	120
Student	21	22	23	24	25	26	27	28	29	30
Male/Female	M	M	F	M	F	F	M	M	M	M
Weight	188	176	118	168	115	115	162	157	154	148
Student	31	32	33	34	35	36	37	38	39	40
Male/Female	F	M	M	F	M	F	M	F	M	M
Weight	101	143	145	108	155	110	154	116	161	165
Student	41	42	43	44	45	46	47	48	49	50
Male/Female	F	M	F	M	M	F	F	M	M	M
Weight	142	184	120	170	195	132	129	215	176	183

FIGURE 2.7 Stem-and-Leaf Display

**Weights of
50 College Students (lb)
Stem-and-Leaf of WEIGHT
$N = 50$ Leaf Unit = 1.0**

9	8
10	1 8 8
11	0 2 5 5 6 8 8
12	0 0 0 8 9
13	2 5 7
14	2 3 5 8
15	0 4 4 5 7 8
16	1 2 2 5 7 8
17	0 0 6 6 7
18	3 4 6 8
19	0 1 5 5
20	5
21	5

FIGURE 2.8 "Back-to-Back" Stem-and-Leaf Display

Weights of 50 College Students (lb)

Female		Male
8	09	
1 8 8	10	
0 2 5 5 6 8 8	11	
0 0 0 8 9	12	
2 5 7	13	
2	14	3 5 8
	15	0 4 4 5 7 8
	16	1 2 2 5 7 8
	17	0 0 6 6 7
	18	3 4 6 8
	19	0 1 5 5
	20	5
	21	5

Based on the information shown in Figures 2.8 and 2.9, and on what we know about people's weight, it seems reasonable to conclude that female college students weigh less than male college students. Situations involving more than one set of data are discussed further in Chapter 3.

FIGURE 2.9
Dotplots with
Common Scale

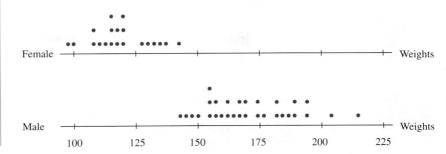

Weights of 50 College Students

TECHNOLOGY INSTRUCTIONS: MULTIPLE DOTPLOTS

MINITAB (Release 14) Input the data into C1 and the corresponding numerical categories into C2; then continue with:

```
Choose:   Graph > Dotplot...
Select:   One Y, With Groups > OK
Enter:    Graph variable: C1
          Categorical variables for grouping: C2 > OK
```

If the various categories are in separate columns, select Multiple Y's Simple and enter all of the columns under Graph variables.

Excel Multiple dotplots are not available, but the initial step of ranking the data can be done. Use the commands as shown with the dotplot display on page 45, then finish constructing the dotplots by hand.

TI-83/84 Input the data for the first dotplot into L1 and the data for the second dotplot into L3; then continue with:

```
Choose:   STAT > EDIT > 2:SortA(
Enter:    L1 > ENTER
          In L2, enter counting numbers for each category.
          Ex.      L1        L2
                   15         1
                   16         1
                   16         2
                   17         1
Choose:   STAT > EDIT > 2:SortA(
Enter:    L3 > ENTER
          In L4, enter counting numbers (a higher set*) for each category;
          *for example: use 10,10,11,10,10,11,12, ... (offsets the two dot-
          plots).
Choose:   2nd > FORMAT > AxesOff (Optional—
          must return to AxesOn)
Choose:   2nd > STAT PLOT > 1:PLOT1
```

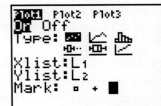

| Choose: | **2nd > STAT PLOT > 2:PLOT2** |

```
Plot1 Plot2 Plot3
On Off
Type: ▦ ⸝ ⊞
     ⊡  ⊞  ⊠
Xlist:L₃
Ylist:L₄
Mark: □ + ■
```

Choose:	**Window**
Enter:	**at most lowest value for both, at least highest value for both, 0 or increment, −2, at least highest counting number,1,1**
Choose:	**Graph > Trace > > > > (gives data values)**

SECTION 2.2 EXERCISES

Statistics △ Now™

Datasets can be found on your Student's Suite CD-ROM or at the StatisticsNow website at **http://1pass.thomson.com**.

2.3 As a statistical graph the circle graph has limitations. Examine the circle graph in Figure 2.1 and the bar graph in Figure 2.2.

a. What information do they both demonstrate?

b. What information is shown in the circle graph that cannot be shown in the bar graph?

c. "Generally speaking, the bar graph is a better choice for use than the circle graph." Justify this statement.

2.4 How Americans prefer to eat an apple was reported in *USA Today*, November 11, 2004: Bite into it—47%, Cut it into slices—39%, Peel—11%, Don't Know—3%.

a. Construct a circle graph showing how Americans prefer to eat an apple.

b. Construct a bar graph showing how Americans prefer to eat an apple.

c. In your opinion, does the circle graph in part a or the bar graph in part b result in a better representation of the information? Explain.

2.5 American Payroll Association got a big response to this question about company dress code: "The current dress code at my company is . . ."

Final results: a. A little too relaxed—27%
                b. A little too formal—15%
                c. Just right—58%

Most people mentioned the importance of "comfort" in their explanations. The vast majority of respondents were very happy with their company's dress code or policy.

a. Construct a circle graph depicting this information. Label completely.

b. Construct a bar graph depicting this same information. Label completely.

c. Compare the previous two graphs, describing what you see in each one now that the graphs have been drawn and completely labeled. Do you get the same impression about these people's feelings from both graphs? Does one emphasize anything the other one does not?

2.6 [EX02-006] The American Community Survey is limited to the household population and excludes the population living in institutions, college dormitories, and other group quarters. Montana's 2003 household make-up is listed here.

Household Population	
Householder	374,879
Spouse	197,379
Child	243,609
Other relatives	27,583
Nonrelatives	49,047
Total	892,497

Source: U.S. Census Bureau

a.   Construct a circle graph of this breakdown.

b.   Construct a bar graph of this breakdown.

c.   Compare the two graphs you constructed in parts a and b. Which one seems to be more informative? Explain why.

2.7 The number of points scored by the winning teams on November 2, 2004, the opening night of the 2004–2005 NBA season, are listed here.

Team	Detroit	Dallas	LA Lakers
Score	87	107	89

Source: http://www.nba.com/schedules/2004_2005_game_schedule/ November.html#scheds

a.   Draw a bar graph of these scores using a vertical scale ranging from 80 to 110.

b.   Draw a bar graph of the scores using a vertical scale ranging from 50 to 110.

c.   In which bar graph does it appear that the NBA scores vary more? Why?

d.   How could you create an accurate representation of the relative size and variation between these scores?

2.8 [EX02-008] A sample of student-owned vehicles produced by General Motors was identified and the make of each noted. The resulting sample follows (Ch = Chevrolet, P = Pontiac, B = Buick, O = Oldsmobile, Ca = Cadillac, G = GMC):

Ch	Ch	Ca	P	P	Ca	P	Ch	O	B
P	Ch	Ch	B	P	Ch	Ch	P	Ch	B
B	P	B	G	Ch	Ch	Ch	B	O	Ch
B	G	Ch	P	Ca	G	B	B	Ch	Ch
O	Ch	Ch	Ch	Ch	O	O	Ca	B	G

a.   Find the number of cars of each make in the sample.

b.   What percentage of these cars were Chevrolets? Pontiacs? Oldsmobiles? Buicks? Cadillacs? GMC?

c.   Draw a bar graph showing the percentages found in part b.

2.9 [EX02-009] The number of people, by age group, living in the 50 U.S. states and the District of Columbia in September 2004 was reported as follows.

Age Group	Number (millions)	Age Group	Number (millions)
0–17	73.45	35–49	66.62
18–24	28.86	50+	84.12
25–34	39.89		

Source: Sales & Marketing Management Survey of Buying Power, September 2004 for the 50 U.S. states and the District of Columbia

Draw a bar graph showing the number of people by age groups.

2.10 Cleaning countertops, disinfecting surfaces, and personal hygiene are among consumers' favorite uses for wipe products, according to The Soap and Detergent Association's (SDA) latest National Cleaning Survey. International Communications Research (ICR) completed the independent consumer research study in December 2004. The initial survey question was asked of 1021 American adults (509 men and 512 women).

Question asked: Have you ever used any type of cleaning, disinfectant, or antibacterial wipes?

Results: Yes—66%
No—34%
More women (72%) than men (60%) have ever used a wipe.

Posted at http://www.cleaning101.com/whatsnew/ 01-17-05.html

a.   Construct and fully label a bar graph showing the results from all adults surveyed.

b.   Construct and fully label a bar graph showing the results comparing the women and men separately.

c.   Discuss the graphs in parts a and b, being sure to comment on how accurately, or not, the graphs picture the information.

2.11 A shirt inspector at a clothing factory categorized the last 500 defects as follows: 67—missing button, 153—bad seam, 258—improperly sized, 22—fabric flaw. Construct a Pareto diagram for this information.

2.12 What NOT to get them on Valentine's Day!

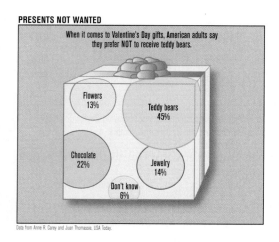

PRESENTS NOT WANTED

When it comes to Valentine's Day gifts, American adults say they prefer NOT to receive teddy bears.

Flowers 13%

Teddy bears 45%

Chocolate 22%

Jewelry 14%

Don't know 6%

Data from Anne R. Carey and Juan Thomassie, USA Today.

a. Draw a bar graph picturing the percentages of "Presents not wanted."

b. Draw a Pareto diagram picturing the "Presents not wanted."

c. If you want to be 80% sure you did not get your Valentine something unwanted, what should you avoid buying? How does the Pareto diagram show this?

d. If 300 adults are to be surveyed, what frequencies would you expect to occur for each unwanted item listed on the graphic?

2.13 [EX02-013] A study by Bruskin-Goldring for Whirlpool Corp. lists the major chores mothers say they would like to have the family help with. The most popular response was cleaning (53%), followed by laundry (18%), cooking (9%), dishes (8%), and other (12%).
Source: http://pqasb.pqarchiver.com/USAToday/

a. Construct a Pareto diagram displaying this information.

b. Because of the size of the "other" category, the Pareto diagram may not be the best graph to use. Explain why, and describe what additional information is needed to make the Pareto diagram more appropriate.

2.14 [EX02-014] The Office of Aviation Enforcement and Proceedings, U.S. Department of Transportation, posted this table listing the number of consumer complaints against top U.S. airlines by complaint category.

Complaint Category	Number	Complaint Category	Number
Advertising	68	Flight problems	2031
Baggage	1421	Oversales	454
Customer service	1715	Refunds	1106
Disability	477	Reservations/ticketing/boarding	1159
Fares	523	Other	322

Source: Office of Aviation Enforcement and Proceedings, U.S. Department of Transportation, Air Travel Consumer Report, http://www.infoplease.com/ipa/A0198353.html

a. Construct a Pareto diagram depicting this information.

b. What complaints would you recommend the airlines pay the most attention to correcting if they want to have the most effect on the overall number of complaints? Explain how the Pareto diagram from part a demonstrates the validity of your answer.

2.15 The final-inspection defect report for assembly line A12 is reported in a Pareto diagram.

a. What is the total defect count in the report?

b. Verify the 30.0% listed for "Scratch."

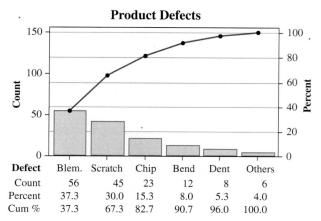

Product Defects

Defect	Blem.	Scratch	Chip	Bend	Dent	Others
Count	56	45	23	12	8	6
Percent	37.3	30.0	15.3	8.0	5.3	4.0
Cum %	37.3	67.3	82.7	90.7	96.0	100.0

c. Explain how the "cum % for bend" value of 90.7% was obtained and what it means.

d. Management has given the production line the goal of reducing their defects by 50%. What two defects would you suggest they give special attention to in working toward this goal? Explain.

2.16 [EX02-016] The world's cocoa production in 2004–2005 is predicted to decline following a record 3396 MT crop in 2003–2004. Most of the reduction is expected to occur in West Africa, with Asian and Latin American production remaining relatively unchanged. West Africa will continue to be the leader in world cocoa production in 2005.

World Cocoa Production (1000 tonnes)

Country	2003–2004	2004–2005 (forecast)
Brazil	163	164
Cameroons	150	150
Ecuador	100	93
Ghana	605	530
Indonesia	420	415
Ivory Coast	1500	1275
Malaysia	25	26
Nigeria	165	170
Other Africa	38	39
Other America	165	170
Other Asia	65	67

Source: World Cocoa Foundation, http://www.chocolateandcocoa.org/stats/supply/default.asp

a. By how much is the total production expected to decrease? What percent decrease is that?

b. Construct a Pareto diagram depicting the 2003–2004 production.

c. Construct a Pareto diagram depicting the 2004–2005 production forecast.

d. African countries are expected to supply what percentage of the world's cocoa for 2004–2005? The Americas? Asia?

2.17 [EX02-017] The number of points scored during each game by a high school basketball team last season was as follows: 56, 54, 61, 71, 46, 61, 55, 68, 60, 66, 54, 61, 52, 36, 64, 51. Construct a dotplot of these data.

2.18 [EX02-018] The table here lists the median house selling prices (in $1000s) for 20 U.S. cities as listed on Realtor.com.

120	120	120	119	117	117	116	116	115	115
114	114	112	112	112	111	109	107	105	105

Source: http://www.realtor.org/Research.nsf/pages/EconHousingData

a. Construct a dotplot of these data.

b. Describe the distribution displayed by the dotplot found in part a.

2.19 [EX02-019] HoopsHype.com regularly posts the latest on the NBA. Following are the heights (in inches) of the basketball players who were the first round picks by the professional teams on June 24, 2004:

82	82	74	79	75	79	80	83	78	79
83	85	71	81	81	78	80	78	79	72
89	81	80	74	76	79	78	75	84	

Source: http://www.hoopshype.com/draft.htm

a. Construct a dotplot of the heights of these players.

b. Use the dotplot to uncover the shortest and the tallest players.

c. What is the most common height, and how many players share that height?

d. What feature of the dotplot illustrates the most common height?

2.20 [EX02-020] As baseball players, Babe Ruth and Hank Aaron were well known for their ability to hit home runs. Mark McGwire and Sammy Sosa became well known for their ability to hit home runs during the "great home run chase" of 1998. Bobby Bonds gained his fame in 2001. Listed here are the number of home runs each player hit in each major-league season in which he played.

Exercise 2.20

Ruth	4	3	2	11	29	54	59	35	41	46	25	47	60	54	46	49	46	41	34	22	6	
Aaron	13	27	26	44	30	39	40	34	45	44	24	32	44	39	29	44	38	47	34	40	20	12	10
McGwire	3	49	32	33	39	22	42	9	9	39	52	58	70	65	32	29							
Sosa	4	15	10	8	33	25	36	40	36	66	63	50	64	49	40	35							
Bonds	16	25	24	19	33	25	34	46	37	33	42	40	37	34	49	73	46	45	45				

a. Construct a dotplot of the data for Ruth and Aaron, using the same axis.

b. Using the dotplots found in part a, make a case for each of the following statements with regard to past players: "Aaron is the homerun king!" "Ruth is the homerun king!"

c. Construct a dotplot of the data for McGwire, Sosa, and Bonds using the same axis.

d. Using the dotplots found in part c, make a case for the statements "McGwire is the homerun king!" "Bonds is the homerun king!" and "Sosa is not currently the homerun king!" with regard to the present players. In what way do the dotplots support each statement?

FYI If you use your computer or calculator, use the commands on page 49.

2.21 [EX02-021] Delco Products, a division of General Motors, produces commutators designed to be 18.810 mm in overall length. (A commutator is a device used in the electrical system of an automobile.) The following sample of 35 commutator lengths was taken while monitoring the manufacturing process:

18.802	18.810	18.780	18.757	18.824	18.827	18.825
18.809	18.794	18.787	18.844	18.824	18.829	18.817
18.785	18.747	18.802	18.826	18.810	18.802	18.780
18.830	18.874	18.836	18.758	18.813	18.844	18.861
18.824	18.835	18.794	18.853	18.823	18.863	18.808

Source: With permission of Delco Products Division, GMC

Use a computer to construct a dotplot of these data values.

2.22 A computer was used to construct the dotplot at the bottom of the page.

a. How many data values are shown?

b. List the values of the five smallest data.

c. What is the value of the largest data item?

d. What value occurred the greatest number of times? How many times did it occur?

2.23 [EX02-017] Construct a stem-and-leaf display of the number of points scored during each basketball game last season:

56	54	61	71	46	61	55	68
60	66	54	61	52	36	64	51

2.24 [EX02-024] Forbes.com posted the 5-year (2000 to 2004) total returns, in percents, for 17 banking industry companies.

Name	Return %	Name	Return %
Astoria Financial	23.9	Popular	15.3
Banknorth Group	18.6	State Street	5.0
Bank of America	13.2	Synovus Finl	8.5
BB&T	8.1	UnionBanCal	10.4
Compass Bancshares	16.7	Wachovia	10.5
Golden West Finl	29.1	Wells Fargo	8.8
M&T Bank	19.1	Westcorp	25.9
National City	12.6	Zions Bancorp	2.9
North Fork Bancorp	19.8		

Source: http://www.forbes.com/lists/results.jhtml

a. Construct a stem-and-leaf display of the data.

b. Based on the stem-and-leaf display, describe the distribution of percentages of profitability.

2.25 [EX02-025] The amounts shown here are the fees charged by Quik Delivery for the 40 small packages it delivered last Thursday afternoon:

4.03	3.56	3.10	6.04	5.62	3.16	2.93	3.82	4.30	3.86
4.57	3.59	4.57	6.16	2.88	5.03	5.46	3.87	6.81	4.91
3.62	3.62	3.80	3.70	4.15	2.07	3.77	5.77	7.86	4.63
4.81	2.86	5.02	5.24	4.02	5.44	4.65	3.89	4.00	2.99

a. Construct a stem-and-leaf display.

b. Based on the stem-and-leaf display, describe the distribution of the data.

Figure for Exercise 2.22

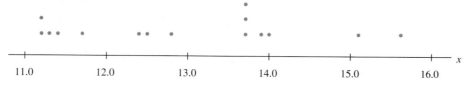

2.26 [EX02-026] One of the many things the U.S. Census Bureau reports to the public is the increase in population for various geographic areas within the country. The percentages of increase in population for the 24 fastest-growing counties in the United States from April 1, 2000, to July 1, 2003, are listed in the following table.

County, State	%
Collin County, TX	21.4

••• Remainder of data on Student's Suite CD-ROM

Source: http://www.census.gov/popest/counties/CO-EST2003-09.html

a. Construct a stem-and-leaf display.

b. Based on the stem-and-leaf display, describe the distribution of the data.

2.27 Given the following stem-and-leaf display:

```
Stem-and-Leaf of C1  N = 16
Leaf Unit = 0.010
  1       59    7
  4       60    148
 (5)      61    02669
  7       62    0247
  3       63    58
  1       64    3
```

a. What is the meaning of "Leaf Unit = 0.010"?

b. How many data are shown on this stem-and-leaf display?

c. List the first four data values.

d. What is the column of numbers down the left-hand side of the figure?

2.28 A term often used in solar energy research is *heating-degree-days*. This concept is related to the difference between an indoor temperature of 65°F and the average outside temperature for a given day. An average outside temperature of 5°F gives 60 heating-degree-days. The annual heating-degree-day normals for several Nebraska locations are shown on the accompanying stem-and-leaf display constructed using MINITAB.

a. What is the meaning of "Leaf Unit = 10"?

b. List the first four data values.

c. List all the data values that occurred more than once.

```
Stem-and-leaf of C1  N = 25
Leaf Unit = 10
   2      60    78
   7      61    03699
   9      62    69
  11      63    26
  (3)     64    233
  11      65    48
   9      66    8
   8      67    249
   5      68    18
   3      69    145
```

2.3

Frequency Distributions and Histograms

Lists of large sets of data do not present much of a picture. Sometimes we want to condense the data into a more manageable form. This can be accomplished with the aid of a *frequency distribution*.

> **Frequency distribution:** A listing, often expressed in chart form, that pairs values of a variable with their frequency.

To demonstrate the concept of a frequency distribution, let's use this set of data:

3	2	2	3	2	4	4	1	2	2
4	3	2	0	2	2	1	3	3	1

TABLE 2.4

Ungrouped Frequency Distribution

x	f
0	1
1	3
2	8
3	5
4	3

If we let x represent the variable, then we can use a frequency distribution to represent this set of data by listing the x values with their frequencies. For example, the value 1 occurs in the sample three times; therefore, the **frequency** for $x = 1$ is 3. The complete set of data is shown in the frequency distribution in Table 2.4.

The frequency, f, is the number of times the value x occurs in the sample. Table 2.4 is an *ungrouped frequency distribution*—"ungrouped" because each value of x in the distribution stands alone. When a large set of data has many different x values instead of a few repeated values, as in the previous example, we can group the values into a set of classes and construct a *grouped frequency distribution*. The stem-and-leaf display in Figure 2.5B (p. 46) shows, in picture form, a grouped frequency distribution. Each stem represents a class. The number of leaves on each stem is the same as the frequency for that same **class** (sometimes called a *bin*). The data represented in Figure 2.5B are listed as a grouped frequency distribution in Table 2.5.

TABLE 2.5

Grouped Frequency Distribution

		Class	Frequency
50 or more to less than 60	⟶	$50 \leq x < 60$	1
60 or more to less than 70	⟶	$60 \leq x < 70$	3
70 or more to less than 80	⟶	$70 \leq x < 80$	8
80 or more to less than 90	⟶	$80 \leq x < 90$	5
90 or more to less than 100	⟶	$90 \leq x < 100$	2
			19

The stem-and-leaf process can be used to construct a frequency distribution; however, the stem representation is not compatible with all **class widths.** For example, class widths of 3, 4, and 7 are awkward to use. Thus, sometimes it is advantageous to have a separate procedure for constructing a grouped frequency distribution.

EXAMPLE 2.6

Grouping Data to Form a Frequency Distribution

Statistics⊜Now™

Watch a video example at http://1pass.thomson.com or on your CD.

To illustrate this grouping (or classifying) procedure, let's use a sample of 50 final exam scores taken from last semester's elementary statistics class. Table 2.6 lists the 50 scores.

Procedure for Constructing a Grouped Frequency Distribution

1. Identify the high score ($H = 98$) and the low score ($L = 39$), and find the range:

$$\text{range} = H - L = 98 - 39 = 59$$

2. Select a number of classes ($m = 7$) and a class width ($c = 10$) so that the product ($mc = 70$) is a bit larger than the range (range = 59).

TABLE 2.6

Statistics Exam Scores [TA02-06]

60	47	82	95	88	72	67	66	68	98	90	77	86
58	64	95	74	72	88	74	77	39	90	63	68	97
70	64	70	70	58	78	89	44	55	85	82	83	
72	77	72	86	50	94	92	80	91	75	76	78	

3. Pick a starting point. This starting point should be a little smaller than the lowest score, L. Suppose we start at 35; counting from there by tens (the class width), we get 35, 45, 55, 65, . . . , 95, 105. These are called the **class boundaries**. The classes for the data in Table 2.6 are:

35 or more to less than 45	$\longrightarrow$	$35 \leq x < 45$
45 or more to less than 55	$\longrightarrow$	$45 \leq x < 55$
55 or more to less than 65	$\longrightarrow$	$55 \leq x < 65$
65 or more to less than 75	$\longrightarrow$	$65 \leq x < 75$
	⋮	$75 \leq x < 85$
		$85 \leq x < 95$
95 or more to and including 105	$\longrightarrow$	$95 \leq x \leq 105$

Notes:

1. At a glance you can check the number pattern to determine whether the arithmetic used to form the classes was correct (35, 45, 55, . . . , 105).
2. For the interval $35 \leq x < 45$, the 35 is the lower class boundary and 45 is the upper class boundary. Observations that fall on the lower class boundary stay in that interval; observations that fall on the upper class boundary go into the next higher interval, except for the last class.
3. The class width is the difference between the upper and lower class boundaries.
4. Many combinations of class widths, numbers of classes, and starting points are possible when classifying data. There is no one best choice. Try a few different combinations, and use good judgment to decide on the one to use.

Therefore, the following **basic guidelines** are used in constructing a grouped frequency distribution:

1. Each class should be of the same width.
2. Classes (sometimes called *bins*) should be set up so that they do not overlap and so that each data value belongs to exactly one class.
3. For the exercises given in this textbook, 5 to 12 classes are most desirable, because all samples contain fewer than 125 data values. (The square root of n is a reasonable guideline for the number of classes with samples of fewer than 125 data values.)
4. Use a system that takes advantage of a number pattern to guarantee accuracy.
5. When it is convenient, an even class width is often advantageous.

Once the classes are set up, we need to sort the data into those classes. The method used to sort will depend on the current format of the data: If the data are ranked, the frequencies can be counted; if the data are not ranked, we will **tally** the data to find the frequency numbers. When classifying data, it helps to use a standard chart (see Table 2.7).

TABLE 2.7

Standard Chart for Frequency Distribution

Class Number	Class Tallies	Boundaries	Frequency
1	||	$35 \le x < 45$	2
2	||	$45 \le x < 55$	2
3	||||| ||	$55 \le x < 65$	7
4	||||| ||||| |||	$65 \le x < 75$	13
5	||||| ||||| |	$75 \le x < 85$	11
6	||||| ||||| |	$85 \le x < 95$	11
7	||||	$95 \le x \le 105$	4
			50

Notes:
1. If the data have been ranked (list form, dotplot, or stem-and-leaf), tallying is unnecessary; just count the data that belong to each class.
2. If the data are not ranked, be careful as you tally.
3. The frequency, f, for each class is the number of pieces of data that belong in that class.
4. The sum of the frequencies should equal the number of pieces of data, n ($n = \Sigma f$). This summation serves as a good check.

Note: See the *Student Solutions Manual* for information about Σ **notation** (read **"summation notation"**).

TABLE 2.8

Frequency Distribution with Class Midpoints

Class Number	Class Boundaries	Frequency, f	Class Midpoints, x
1	$35 \le x < 45$	2	40
2	$45 \le x < 55$	2	50
3	$55 \le x < 65$	7	60
4	$65 \le x < 75$	13	70
5	$75 \le x < 85$	11	80
6	$85 \le x < 95$	11	90
7	$95 \le x \le 105$	4	100
		50	

Note: Now you can see why it is helpful to have an even class width. An odd class width would have resulted in a class midpoint with an extra digit. (For example, the class 45–54 is 9 wide and the class midpoint is 49.5.)

Each class needs a single numerical value to represent all the data values that fall into that class. The **class midpoint** (sometimes called the *class mark*) is the numerical value that is exactly in the middle of each class. It is found by adding the class boundaries and dividing by 2. Table 2.8 shows an additional column for the class midpoint, x. As a check of your arithmetic, successive class midpoints should be a class width apart, which is 10 in this illustration (40, 50, 60, . . . , 100 is a recognizable pattern).

APPLIED EXAMPLE 2.7

Cleaning House

The "Weekly hours devoted to housecleaning" graphic presents a circle graph version of a relative frequency distribution. Each sector of the circle represents the amount of time spent cleaning weekly by each person, and the "relative size" of the sector represents the percentage or relative frequency.

Now using statistical terminology, we can say that the *variable* "time spent cleaning" is represented in the graph by sectors of the circle. The *relative frequency* is represented by the size of the angle forming the sector. To form this information into a grouped "relative" frequency distribution, each interval of the variable will be expressed in the form $a \leq x < b$. For example, the 2- to 4-hour category would be expressed as $2 \leq x < 4$. (This way the lower boundary is part of the interval, but the upper boundary is part of the next larger interval.) The distribution chart for this circle graph would then appear as in the table shown to the left.

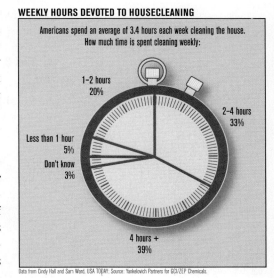

WEEKLY HOURS DEVOTED TO HOUSECLEANING

Americans spend an average of 3.4 hours each week cleaning the house. How much time is spent cleaning weekly:

1–2 hours 20%
2–4 hours 33%
Less than 1 hour 5%
Don't know 3%
4 hours + 39%

Data from Cindy Hall and Sam Ward, USA TODAY; Source: Yankelovich Partners for GCI/ZEP Chemicals.

Class Boundaries	Relative Frequency
$0 \leq x < 1$	0.05
$1 \leq x < 2$	0.20
$2 \leq x < 4$	0.33
$4 \leq x$	0.39
Don't know	0.03

When we classify data into classes, we lose some information. Only when we have all of the raw data do we know the exact values that were actually observed for each class. For example, we put a 47 and a 50 into class 2, with class boundaries of 45 and 55. Once they are placed in the class, their values are lost to us and we use the class midpoint, 50, as their representative value.

> **Histogram:** A bar graph that represents a frequency distribution of a quantitative variable. A histogram is made up of the following components:
> 1. A title, which identifies the population or sample of concern.
> 2. A vertical scale, which identifies the frequencies in the various classes.
> 3. A horizontal scale, which identifies the variable x. Values for the class boundaries or class midpoints may be labeled along the x-axis. Use whichever method of labeling the axis best presents the variable.

The frequency distribution from Table 2.8 appears in histogram form in Figure 2.10.

Sometimes the **relative frequency** of a value is important. The relative frequency is a proportional measure of the frequency for an occurrence. It is found by dividing the class frequency by the total number of observations. Relative frequency can be expressed as a common fraction, in decimal form, or as a percentage. For example, in Example 2.6 the frequency associated with the third class (55–65) is 7. The relative frequency for the third class is $\frac{7}{50}$, or 0.14, or 14%. Relative frequencies are often useful in a presentation because most people understand fractional parts when expressed as percents. Relative frequencies are particularly useful when comparing the frequency distributions of two different size sets of data. Figure 2.11 is a **relative frequency histogram** of the sample of the 50 final exam scores from Table 2.8.

FIGURE 2.10   Frequency Histogram

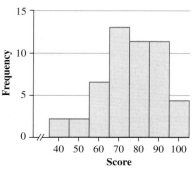

FIGURE 2.11   Relative Frequency Histogram

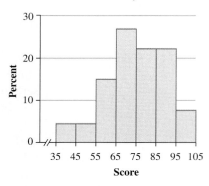

FYI Note that the frequency histogram and the relative frequency histogram have the same shape (assuming the same classes are used for both); only the vertical axis labels change.

FYI Be sure to identify both scales so that the histogram tells the complete story.

A stem-and-leaf display contains all the information needed to create a histogram. Figure 2.5B (p. 46) shows the stem-and-leaf display constructed in Example 2.4. In Figure 2.12A the stem-and-leaf display has been rotated 90° and labels have been added to show its relationship to a histogram. Figure 2.12B shows the same set of data as a completed histogram.

FIGURE 2.12A   Modified Stem-and-Leaf Display

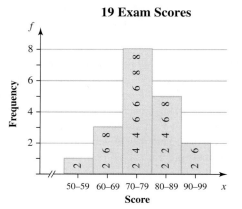

FIGURE 2.12B   Histogram

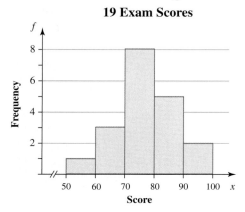

TECHNOLOGY INSTRUCTIONS: HISTOGRAM

MINITAB (Release 14)   Input the data into C1; then continue with:

```
Choose:      Graph > Histogram > Simple > OK
Enter:       Graph variables:C1
Choose:      Labels > Titles/Footnote
Enter:       Your title and/or footnote > OK
Choose:      Scale > Y-Scale Type
Select:      Y scale Type: Frequency or Percent or Density > OK > OK
To adjust histogram: Double click anywhere on bars of histogram.
Select:      Binning
Select:      Interval Type: Midpoint or Cutpoint
             Interval Definitions:
                 Automatic or. Number of intervals; Enter: N or, Midpt/cutpt
                 positions; Enter: A:B/C > OK
```

Notes:

1. Midpoints are the class midpoints, and cutpoints are the class boundaries.
2. Percent is relative frequency.
3. Automatic means MINITAB will make all the choices; N = number of intervals, that is, the number of classes you want used.
4. A = smallest class midpoint or boundary, B = largest class midpoint or boundary, C = class width you want to specify.

The following commands will draw the histogram of a frequency distribution. The end classes can be made full width by adding an extra class with frequency zero to each end of the frequency distribution. Input the class midpoints into C1 and the corresponding frequencies into C2.

```
Choose:      Graph > Scatterplot > With Connect Line > OK
Enter:       Y variables: C2   X variables: C1
Select:      Data View: Data Display: Symbols Connect > OK > OK
Double click on a connect line.
Select:      Options
             Connection Function: Step > OK
```

Excel   Input the data into column A and the upper class limits* into column B (optional) and (column headings are optional); then continue with:

```
Choose:      Tools > Data Analysis† > Histogram > OK
Enter:       Input Range: Data (A1:A6 or select cells)
             Bin Range: upper class limits (B1:B6 or select cells)
             [leave blank if Excel determines the intervals]
Select:      Labels (if column headings are used)
             Output Range
Enter:       area for freq. distr. & graph (C1 or select cell)
Select:      Chart Output
```

To remove gaps between bars:

```
Click on:    Any bar on graph
Click on:    Right mouse button
Choose:      Format Data Series > Options
Enter:       Gap Width: 0
```

To edit the histogram:

Click on: Anywhere clear on the chart
                —use handles to size
                Any title or axis name to change
                Any upper class limit§ or frequency in the frequency distribu-
                tion to change value > Enter

*If boundary = 50, then limit = 49.9 (depending on the number of decimal places in the data).
§If Data Analysis does not show on the Tools menu:
            Choose: Tools > Add-Ins
            Select: Analysis ToolPak
                       Analysis ToolPak-VBA
§Note that the upper class limits appear in the center of the bars. Replace with class midpoints. The "More" cell in the frequency distribution may also be deleted.

For tabled data, input the classes into column A (ex. 30–40) and the frequencies into column B; then continue with:

Choose: **Chart Wizard > Column > 1st picture** (usually) **> Next**
Enter: Data Range: **(A1:B4 or select cells)**
Select: Series in: **Columns > Next**
Choose: **Titles**
Enter: Chart title: **your title**
              Category (*x*) axis: **title for *x*-axis**
              Value (*y*) axis: **title for *y*-axis > Next > Finish**

Do as just described to remove gaps and adjust.

TI-83/84 Plus   Input the data into L1; then continue with:

Choose: **2nd > STAT PLOT > 1:Plot1**

Calculator selects classes:

Choose: **Zoom > 9:ZoomStat > Trace > > >**

Individual selects classes:

Choose: **Window**
Enter: **at most lowest value, at least highest value, class width, −1,**
             at least highest frequency, 1 (depends on frequency numbers), 1
Choose: **Graph > Trace** (use values to construct frequency distribution)

For tabled data, input the class midpoints into L1 and the frequencies into L2; then continue with:

Choose: **2nd > STAT PLOT > 1:Plot1**
Choose: **Window**
Enter: **smallest lower class boundary,**
             largest upper class boundary, class
             width, −ymax/4, highest frequency,
             0 (for no tick marks), 1
Choose: **Graph > Trace > > >**

To obtain a relative frequency histogram of tabled data instead:

Choose:	**STAT > EDIT > 1:EDIT...**
Highlight:	**L3**
Enter:	**L3 = L2/SUM(L2)** [SUM – 2nd LIST > MATH > 5:sum]
Choose:	**2nd > STAT PLOT > 1:Plot1**
Choose:	**Window**
Enter:	**smallest lower class boundary,**
	largest upper class boundary,
	class width, –ymax/4, highest rel.
	frequency, 0 (for no tick marks), 1
Choose:	**Graph > Trace > > >**

Histograms are valuable tools. For example, the histogram of a sample should have a distribution shape very similar to that of the population from which the sample was drawn. If the reader of a histogram is at all familiar with the variable involved, he or she will usually be able to interpret several important facts. Figure 2.13 presents histograms with specific shapes that suggest descriptive labels. Possible descriptive labels are listed under each histogram.

Briefly, the terms used to describe histograms are as follows:

Symmetrical: Both sides of this distribution are identical (halves are mirror images).

Normal: A symmetrical distribution is mounded up about the mean and becomes sparse at the extremes. (Additional properties are discussed later.)

Uniform (rectangular): Every value appears with equal frequency.

FIGURE 2.13 **Shapes of Histograms**

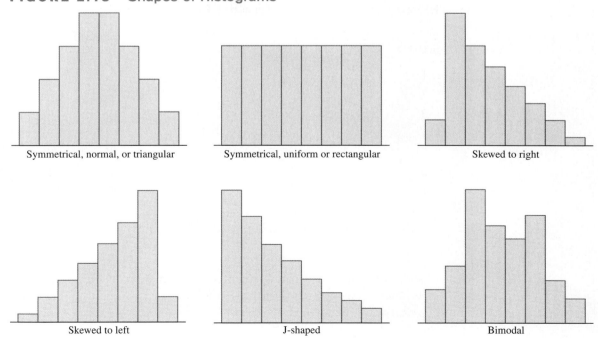

Skewed: One tail is stretched out longer than the other. The direction of skewness is on the side of the longer tail.

J-shaped: There is no tail on the side of the class with the highest frequency.

Bimodal: The two most populous classes are separated by one or more classes. This situation often implies that two populations are being sampled. (See Figure 2.7, p. 48.)

Notes:

1. The **mode** is the value of the data that occurs with the greatest frequency. (Mode will be discussed in Section 2.4, p. 76.)
2. The **modal class** is the class with the highest frequency.
3. A **bimodal distribution** has two high-frequency classes separated by classes with lower frequencies. It is not necessary for the two high frequencies to be the same.

Another way to express a frequency distribution is to use a *cumulative frequency distribution*.

> **Cumulative frequency distribution:** A frequency distribution that pairs cumulative frequencies with values of the variable.

The **cumulative frequency** for any given class is the sum of the frequency for that class and the frequencies of all classes of smaller values. Table 2.9 shows the cumulative frequency distribution from Table 2.8 (p. 58).

TABLE 2.9

Using Frequency Distribution to Form a Cumulative Frequency Distribution

Class Number	Class Boundaries	Frequency, f	Cumulative Frequency	
1	$35 \leq x < 45$	2	2	(2)
2	$45 \leq x < 55$	2	4	(2 + 2)
3	$55 \leq x < 65$	7	11	(7 + 4)
4	$65 \leq x < 75$	13	24	(13 + 11)
5	$75 \leq x < 85$	11	35	(11 + 24)
6	$85 \leq x < 95$	11	46	(11 + 35)
7	$95 \leq x \leq 105$	4	50	(4 + 46)
		50		

The same information can be presented by using a *cumulative relative frequency distribution* (see Table 2.10). This combines the cumulative frequency and the relative frequency ideas.

TABLE 2.10

Cumulative Relative Frequency Distribution

Class Number	Class Boundaries	Cumulative Relative Frequency	*Cumulative frequencies are for the interval 35 up to the upper boundary of that class.*
1	$35 \leq x < 45$	2/50, or 0.04	←——— *from 35 up to less than 45*
2	$45 \leq x < 55$	4/50, or 0.08	←——— *from 35 up to less than 55*
3	$55 \leq x < 65$	11/50, or 0.22	←——— *from 35 up to less than 65*
4	$65 \leq x < 75$	24/50, or 0.48	.
5	$75 \leq x < 85$	35/50, or 0.70	:
6	$85 \leq x < 95$	46/50, or 0.92	
7	$95 \leq x \leq 105$	50/50, or 1.00	←——— *from 35 up to and including 105*

Cumulative distributions can be displayed graphically.

Ogive (pronounced ō'jĪv): A line graph of a cumulative frequency or cumulative relative frequency distribution. An ogive has the following components:

1. A title, which identifies the population or sample.
2. A vertical scale, which identifies either the cumulative frequencies or the cumulative relative frequencies. (Figure 2.14 shows an ogive with cumulative relative frequencies.)
3. A horizontal scale, which identifies the upper class boundaries. (Until the upper boundary of a class has been reached, you cannot be sure you have accumulated all the data in that class. Therefore, the horizontal scale for an ogive is always based on the upper class boundaries.)

FIGURE 2.14
Ogive

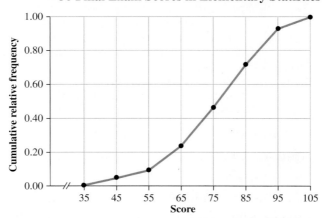

50 Final Exam Scores in Elementary Statistics

Note: Every ogive starts on the left with a relative frequency of zero at the lower class boundary of the first class and ends on the right with a cumulative relative frequency of 1.00 (or 100%) at the upper class boundary of the last class.

TECHNOLOGY INSTRUCTIONS: OGIVE

MINITAB (Release 14) Input the class boundaries into C1 and the cumulative percentages into C2 (enter 0 [zero] for the percentage paired with the lower boundary of the first class and pair each cumulative percentage with the class upper boundary). Use percentages; that is, use 25% in place of 0.25.

```
Choose:    Graph > Scatterplot > With Connect Line > OK
Enter:     Y variables: C2     X variables: C1
Select:    Data View: Data Display: Symbols Connect > OK
Select:    Labels > Titles/Footnotes
Enter:     your title or footnotes > OK > OK
```

Excel Input the data into column A and the upper class limits* into column B (include an additional class at the beginning).

```
Choose:    Tools > Data Analysis > Histogram > OK
Enter:     Input Range: data (A1:A6 or select cells)
           Bin Range: upper class limits (B1:B6 or select cells)
Select:    Labels (if column headings were used)
           Output Range
           Enter: area for freq. distr. & graph: (C1 or se-
           lect cell)
           Cumulative Percentage
           Chart Output
```

To close gaps and edit, see the histogram commands on pages 61–62.

For tabled data, input the upper class boundaries into column A and the cumulative relative frequencies into column B (include an additional class boundary at the beginning with a cumulative relative frequency equal to 0 [zero]); then continue with:

```
Choose:    Chart Wizard > Line > 4th picture (usually) > Next
Choose:    Series (have more control on input) > Remove (remove all
           columns except column B)
Enter:     Name: (B1 or select name cell — cum. rel. freq.)
           Values: (B2:B6 or select cells)
           Category (x) axis labels: (A2:A8 or select cells) > Next
Choose:    Titles
Enter:     Chart title: your title
           Category (x) axis: title for x-axis
           Value (y) axis: title for y-axis > Next > Finish
```

For editing, see the histogram commands on page 62.

*If the boundary = 50, then the limit = 49.9 (depending on the number of decimal places in the data).

TI-83/84 Plus Input the class boundaries into L1 and the frequencies into L2 (include an extra class boundary at the beginning with a frequency of zero); then continue with:

```
Choose:     STAT > EDIT > 1:EDIT...
Highlight:  L3
Enter:      L3 = 2nd > LIST > OPS > 6:cum sum(L2)
Highlight:  L4
Enter:      L4 = L3 / 2nd > LIST > Math > 5:sum (L2)
```

Choose: **2nd > STAT PLOT > 1:Plot**

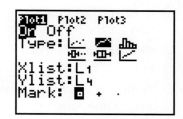

Choose: **Zoom > 9:ZoomStat > Trace > > >**

Adjust window if needed for better readability.

SECTION 2.3 EXERCISES

Statistics ⬡ Now™

Skillbuilder Applet Exercises must be worked using an accompanying applet found on your Student's Suite CD-ROM or at the StatisticsNow website at **http://1pass.thomson.com.**

Datasets can be found on your Student's Suite CD-ROM or at the StatisticsNow website at **http://1pass.thomson.com.**

2.29 a. Form an ungrouped frequency distribution of the following data:

1, 2, 1, 0, 4, 2, 1, 1, 0, 1, 2, 4

Referring to the preceding distribution:

b. Explain what $f = 5$ represents.

c. What is the sum of the frequency column?

d. What does this sum represent?

2.30 Bar graphs and histograms are not the same thing. Explain their similarities and differences.

2.31 [EX02-031] The players on the Rochester Raging Rhinos professional soccer team scored 38 goals during the 2002 season.

Player	1	2	3	4	5	6	7	8	9	10	11	12	13
Goals	2	8	1	2	2	6	2	1	5	2	3	2	2

Source: http://www.rhinossoccer.com/team.asp

a. If you want to show the number of goals scored by each player, would it be more appropriate to display this information on a bar graph or a histogram? Explain.

b. Construct the appropriate graph for part a.

c. If you wanted to show (emphasize) the distribution of scoring by the team, would it be

more appropriate to display this information on a bar graph or a histogram? Explain.

d. Construct the appropriate graph for part c.

2.32 [EX02-032] The California Department of Education gives an annual report on the Advanced Placement (AP) test results for each year. In the 2003–2004 school year, Modoc County had students with the following scores:

2	3	2	1	2	3	3	2	2	3	5	3	2	2
2	4	5	1	2	2	4	3	2	2	5	4	1	4
2	4	5	2	3	2	3	2	3	3	5	1	5	5

Source: http://data1.cde.ca.gov/dataquest/

a. Construct an ungrouped frequency distribution for the test scores.

b. Construct a frequency histogram of this distribution.

c. Prepare a relative frequency distribution for these same data.

d. If AP scores of at least 3 are often required for college transferability, what percentage of Modoc AP scores will receive college credit?

(Retain these solutions to use in Exercise 2.52 on p. 72.)

2.33 [EX02-033] The U.S. Women's Olympic Soccer team had a great year in 2004. One way to describe the players on that team is by their individual heights.

Height (inches)

68	67	65	66	65	67	64	69	69
65	64	71	66	67	68	66	65	71

Source: http://www.SoccerTimes.com

a. Construct an ungrouped frequency distribution for the heights.

b. Construct a frequency histogram of this distribution.

c. Prepare a relative frequency distribution for these same data.

d. What percentage of the team is at least 5 feet, 6 inches tall?

2.34 [EX02-034] The U.S. Census Bureau posted the following 2003 Report on America's Families and Living Arrangements for all races.

No. in Household	Percentage	No. in Household	Percentage
1	26.4%	5	6.3%
2	33.3%	6	2.3%
3	16.1%	7+	1.2%
4	14.3%		

Source: http://www.census.gov/population/www/socdemo/hh-fam/cps2003.html

a. Draw a relative frequency histogram for the number of people per household.

b. What shape distribution does the histogram suggest?

c. Based on the graph, what do you know about the households in the United States?

2.35 [EX02-035] The 2003 American Community Survey universe is limited to the household population and excludes the population living in institutions, college dormitories, and other group quarters. The accompanying table lists the number of rooms in each of the 8,658,290 housing units in Texas.

Rooms	Housing Units	Rooms	Housing Units
1 room	124,486	6 rooms	1,649,479
2 rooms	349,496	7 rooms	913,138
3 rooms	1,007,873	8 rooms	520,248
4 rooms	1,548,984	9+ rooms	485,506
5 rooms	2,059,080		

Source: U.S. Census Bureau, American Community Survey Office

a. Draw a relative frequency histogram for the number of rooms per household.

b. What shape distribution does the histogram suggest?

c. Based on the graph, what do you know about the number of rooms per household in Texas?

2.36 [EX02-036] Here are the ages of 50 dancers who responded to a call to audition for a musical comedy:

21	19	22	19	18	20	23	19	19	20
19	20	21	22	21	20	22	20	21	20
21	19	21	21	19	19	20	19	19	19
20	20	19	21	21	22	19	19	21	19
18	21	19	18	22	21	24	20	24	17

a. Prepare an ungrouped frequency distribution of these ages.

b. Prepare an ungrouped relative frequency distribution of the same data.

c. Prepare a relative frequency histogram of these data.

d. Prepare a cumulative relative frequency distribution of the same data.

e. Prepare an ogive of these data.

2.37 [EX02-037] The opening-round scores for the Ladies' Professional Golf Association tournament at Locust Hill Country Club were posted as follows:

69	73	72	74	77	80	75	74	72	83	68	73	75	78
76	74	73	68	71	72	75	79	74	75	74	74	68	79
75	76	75	77	74	74	75	75	72	73	73	72	72	71
71	70	82	77	76	73	72	72	72	75	75	74	74	74
76	76	74	73	74	73	72	72	74	71	72	73	72	72
74	74	67	69	71	70	72	74	76	75	75	74	73	74
74	78	77	81	73	73	74	68	71	74	78	70	68	71
72	72	75	74	76	77	74	74	73	73	70	68	69	71
77	78	68	72	73	78	77	79	79	77	75	75	74	73
73	72	71	68	70	71	78	78	76	74	75	72	72	72
75	74	76	77	78	78								

a. Form an ungrouped frequency distribution of these scores.

b. Draw a histogram of the first-round golf scores. Use the frequency distribution from part a.

2.38 Figuring *where* lightning strikes will occur is a near impossible task. *When* lightning strikes occur, however, has become more predictable based on research. For a small area in Colorado, data were collected and the results are displayed in the histogram that follows.

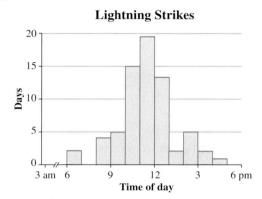

Lightning Strikes

Based on the histogram:

a. Data were collected for what variable?

b. What does each bar (interval) represent?

c. What conclusion can be reached about the "when" of lightning strikes in this small area of Colorado?

d. What characteristics of the graph support the conclusion?

2.39 [EX02-039] A survey of 100 resort club managers on their annual salaries resulted in the following frequency distribution:

Annual Salary ($1000s)	15–25	25–35	35–45	45–55	55–65
No. of Managers	12	37	26	19	6

a. The data value "35" belongs to which class?

b. Explain the meaning of "35–45."

c. Explain what "class width" is, give its value, and describe three ways that it can be determined.

d. Draw a frequency histogram of the annual salaries for resort club managers. Label class boundaries.

(Retain these solutions to use in Exercise 2.51 on p. 71.)

2.40 Skillbuilder Applet Exercise demonstrates the procedure of transforming a stem-and-leaf display into a histogram.

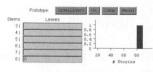

Type the leaves for the number of stories into the stem-and-leaf display. Click OK to see the corresponding histogram. Comment on the similarities and differences.

2.41 [EX02-041] The KSW computer science aptitude test was given to 50 students. The following frequency distribution resulted from their scores:

KSW Test Score	0–4	4–8	8–12	12–16	16–20	20–24	24–28
Frequency	4	8	8	20	6	3	1

a. What are the class boundaries for the class with the largest frequency?

b. Give all the class midpoints associated with this frequency distribution.

c. What is the class width?

d. Give the relative frequencies for the classes.

e. Draw a relative frequency histogram of the test scores.

2.42 [EX02-042] The USA Snapshot titled "Nuns an aging order" reports that the median age of the 94,022 Roman Catholic nuns in the United States is 65 years and the percentages of U.S. nuns by age group are as follows:

Under 50	51–70	Over 70	Refused to give age
16%	42%	37%	5%

This information is based on a survey of 1049 Roman Catholic nuns. Suppose the survey had resulted in the following frequency distribution (52 ages unknown):

Age	20–30	30–40	40–50	50–60	60–70	70–80	80–90
Frequency	34	58	76	187	254	241	147

a. Draw and completely label a frequency histogram.

b. Draw and completely label a relative frequency histogram of the same distribution.

c. Carefully examine the two histograms in parts a and b and explain why one of them might be easier to understand. (Retain these solutions to use in Exercise 2.166, p. 123.)

FYI Use the computer or calculator commands on pages 61–63 to construct a histogram of a frequency distribution.

2.43 [EX02-043] The speeds of 55 cars were measured by a radar device on a city street:

27	23	22	38	43	24	35	26	28	18	20
25	23	22	52	31	30	41	45	29	27	43
29	28	27	25	29	28	24	37	28	29	18
26	33	25	27	25	34	32	36	22	32	33
21	23	24	18	48	23	16	38	26	21	23

a. Classify these data into a grouped frequency distribution by using class boundaries 12–18, 18–24, . . . , 48–54.

b. Find the class width.

c. For the class 24–30, find the class midpoint, the lower class boundary, and the upper class boundary.

d. Construct a frequency histogram of these data.

FYI Use the computer or calculator commands on pages 61–63 to construct a histogram for a given set of data.

2.44 [EX02-044] The hemoglobin A_{1c} test, a blood test given to diabetics patients during their periodic checkups, indicates the level of control of blood sugar during the past 2 to 3 months. The following data values were obtained for 40 different diabetics patients at a university clinic:

6.5	5.0	5.6	7.6	4.8	8.0	7.5	7.9	8.0	9.2
6.4	6.0	5.6	6.0	5.7	9.2	8.1	8.0	6.5	6.6
5.0	8.0	6.5	6.1	6.4	6.6	7.2	5.9	4.0	5.7
7.9	6.0	5.6	6.0	6.2	7.7	6.7	7.7	8.2	9.0

a. Classify these A_{1c} values into a grouped frequency distribution using the classes 3.7–4.7, 4.7–5.7, and so on.

b. What are the class midpoints for these classes?

c. Construct a frequency histogram of these data.

2.45 [EX02-045] All of the third-graders at Roth Elementary School were given a physical-fitness strength test. The following data resulted:

12	22	6	9	2	9	5	9	3	5	16	1	22
18	6	12	21	23	9	10	24	21	17	11	18	19
17	5	14	16	19	19	18	3	4	21	16	20	15
14	17	4	5	22	12	15	18	20	8	10	13	20
6	9	2	17	15	9	4	15	14	19	3	24	

a. Construct a dotplot.

b. Prepare a grouped frequency distribution using classes 1–4, 4–7, and so on, and draw a histogram of the distribution. (Retain the solution for use in answering Exercise 2.75, p. 81.)

c. Prepare a grouped frequency distribution using classes 0–3, 3–6, 6–9, and so on, and draw a histogram of the distribution.

d. Prepare a grouped frequency distribution using class boundaries −2.5, 2.5, 7.5, 12.5, and so on, and draw a histogram of the distribution.

e. Prepare a grouped frequency distribution using classes of your choice, and draw a histogram of the distribution.

f. Describe the shape of the histogram found in parts b–e separately. Relate the distribution seen in the histogram to the distribution seen in the dotplot.

g. Discuss how the number of classes used and the choice of class boundaries used affect the appearance of the resulting histogram.

2.46 [EX02-046] People have marveled for years about the continuing eruptions of the geyser Old Faithful in Yellowstone National Park. The times of duration, in minutes, for a sample of 50 eruptions of Old Faithful are listed here.

4.00	3.75	2.25	1.67	4.25	3.92
4.53	1.85	4.63	2.00	1.80	4.00
4.33	3.77	3.67	3.68	1.88	1.97
4.00	4.50	4.43	3.87	3.43	4.13
4.13	2.33	4.08	4.35	2.03	4.57
4.62	4.25	1.82	4.65	4.50	4.10
4.28	4.25	1.68	3.43	4.63	2.50
4.58	4.00	4.60	4.05	4.70	3.20
4.60	4.73				

Source: http://www.stat.sc.edu/~west/javahtml/Histogram.html

a. Draw a dotplot displaying the eruption-length data.

b. Draw a histogram of the eruption-length data using class boundaries 1.6–2.0–2.4–·· ·–4.8.

c. Draw another histogram of the data using different class boundaries and widths.

d. Repeat part c.

e. Repeat parts a and b using the larger set of 107 eruptions found on the CD. [DS02-046]

f. Which graph, in your opinion, does the best job of displaying the distribution? Why?

g. Write a short paragraph describing the distribution.

2.47 [EX02-047] The Office of Coal, Nuclear, Electric and Alternate Fuels reported the following data as the costs (in cents) of the average revenue per kilowatt-hour for sectors in Arkansas:

6.61	7.61	6.99	7.48	5.10	7.56	6.65	5.93	7.92
5.52	7.47	6.79	8.27	7.50	7.44	6.36	5.20	5.48
7.69	8.74	5.75	6.94	7.70	6.67	4.59	5.96	7.26
5.38	8.88	7.49	6.89	7.25	6.89	6.41	5.86	8.04

a. Prepare a grouped frequency distribution for the average revenue per kilowatt-hour using class boundaries 4, 5, 6, 7, 8, 9.

b. Find the class width.

c. List the class midpoints.

d. Construct a relative frequency histogram of these data.

2.48 [EX02-048] Education has long been considered the ticket for upward mobility in the United States. In today's information age, a college education has become the minimum level of educational attainment to enter an increasingly competitive market for jobs with more than subsistence wages. A report from the SUNY Downstate Medical Center included a study of the suburban areas that surround U.S. cities. One variable that was reported was the percentage of suburban residents 25 years of age and older who attended at least some college:

49.3	75.2	64.7	66.1	51.8

••• Remainder of data on Student's Suite CD-ROM

Source: SUNY Downstate Medical Center, 2004

a. Prepare a grouped frequency distribution for the percent of suburban populations age 25 and older with any college attendance using class midpoints 25, 30, 35, . . . 75.

b. List the class boundaries.

c. Construct a relative frequency histogram of these data.

2.49 Can you think of variables whose distribution might yield the following different shapes? (See Figure 2.13 on p. 63 if necessary.)

a. A symmetrical, or normal, shape

b. A uniform shape

c. A skewed-to-the-right shape

d. A skewed-to-the-left shape

e. A bimodal shape

2.50 Skillbuilder Applet Exercise demonstrates the effect that the number of classes or bins has on the shape of a histogram.

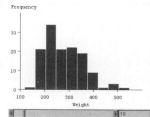

a. What shape distribution does using one class or bin produce?

b. What shape distribution does using two classes or bins produce?

c. What shape distribution does using 10 or 20 bins produce?

2.51 A survey of 100 resort club managers on their annual salaries resulted in the following frequency distribution. (See Exercise 2.39 on p. 69.)

Annual Salary ($1000s)	15–25	25–35	35–45	45–55	55–65
No. of Managers	12	37	26	19	6

a. Prepare a cumulative frequency distribution for the annual salaries.

b. Prepare a cumulative relative frequency distribution for the annual salaries.

c. Construct an ogive for the cumulative relative frequency distribution found in part b.

FYI Use the computer or calculator commands on page 66 to construct an ogive for a given set of data

2.52 [EX02-032] a.   Prepare a cumulative relative frequency distribution for the variable "AP score" in Exercise 2.32.

b.   Construct an ogive of the distribution.

2.53 [EX02-041] a.   Prepare a cumulative relative frequency distribution for the variable "KSW test score" in Exercise 2.41.

b.   Construct an ogive of the distribution.

2.54 Undergraduates who use loans to pay for college average $16,500 in debt when they graduate. The relative frequency distribution of their monthly debt after graduation is shown here:

Monthly debt, $	Less than 100	100–149	150–199	200–249	250–299	300 or more
Relative Frequency	0.17	0.17	0.17	0.19	0.1	0.2

Source: USA Today Snapshot, December 23, 2004

a.   Prepare a cumulative relative frequency distribution for the monthly debt.

b.   Construct an ogive for the cumulative relative frequency distribution found in part a.

2.55 [EX02-055] The Quality of Life in the Nation's 100 Largest Cities and Their Suburbs: New and Continuing Challenges for Improving Health and Well-Being, June 2004, reports on the percent of poor population living in a high-poverty neighborhood in 82 U.S. cities:

29.8	21.4	32.0	5.9	27.8

••• Remainder of data on Student's Suite CD-ROM

Source: SUNY Downstate Medical Center

a.   Prepare a grouped frequency distribution of the percent data using class midpoints of 0, 5, 10, . . . 45.

b.   Prepare a grouped relative frequency distribution of these data.

c.   Draw a relative frequency histogram of these data.

d.   Prepare a cumulative relative frequency distribution of the same data.

e.   Draw an ogive of these data.

2.56 The levels of various compounds resulted in the distribution graphs that follow. They all seem to be quite symmetrical about their centers, but they differ in their spreads.

a.   For which histogram, A, B, C, or D, would you anticipate the numerical measure of spread to be the largest? the smallest?

b.   Which two of the four histograms would you anticipate about the same difference between their smallest values and their largest values?

Figures for Exercise 2.56

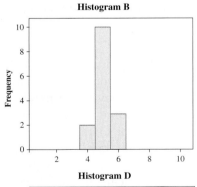

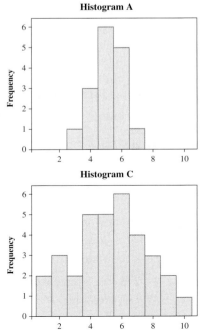

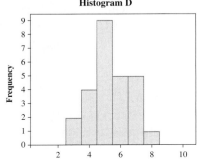

2.4 Measures of Central Tendency

Measures of central tendency are numerical values that locate, in some sense, the center of a set of data. The term *average* is often associated with all measures of central tendency.

Mean (arithmetic mean): The average with which you are probably most familiar. The sample mean is represented by $\bar{x}$ (read "x-bar" or "sample mean"). The mean is found by adding all the values of the variable x (this sum of x values is symbolized $\sum x$) and dividing the sum by the number of these values, n (the "sample size"). We express this in formula form as

$$\text{sample mean:} \quad \text{x-bar} = \frac{\text{sum of all } x}{\text{number of } x}$$

$$\bar{x} = \frac{\sum x}{n} \tag{2.1}$$

Notes:
1. See the *Student Solutions Manual* for information about $\sum$ notation ("summation notation").
2. The population mean, μ (lowercase mu, Greek alphabet), is the mean of all x values for the entire population.

EXAMPLE 2.8

Finding the Mean

FYI See Animated Tutorial Mean on CD for assistance with this calculation.

A set of data consists of the five values 6, 3, 8, 6, and 4. Find the mean.

SOLUTION Using formula (2.1), we find

$$\bar{x} = \frac{\sum x}{n} = \frac{6 + 3 + 8 + 6 + 4}{5} = \frac{27}{5} = 5.4$$

Therefore, the mean of this sample is **5.4**.

FYI The mean is the middle point by weight.

A physical representation of the mean can be constructed by thinking of a number line balanced on a fulcrum. A weight is placed on the number line at the number corresponding to each data value in the sample of Example 2.8. In Figure 2.15 there is one weight each on the 3, 8, and 4 and two weights on the 6, since there are two 6s in the sample. The mean is the value that balances the weights on the number line—in this case, 5.4.

FIGURE 2.15
Physical Representation of the Mean

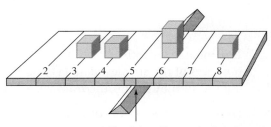

$\bar{x} = $ **5.4**   (the center of gravity, or balance point)

TECHNOLOGY INSTRUCTIONS: MEAN

MINITAB (Release 14) Input the data into C1; then continue with:

Choose: **Calc > Column Statistics**
Select: **Mean**
Enter: Input variable: **C1 > OK**

Excel Input the data into column A and activate a cell for the answer; then continue with:

Choose: **Insert Function, f_x > Statistical > AVERAGE > OK**
Enter: Number 1: **(A2:A6 or select cells)**
                 [Start at A1 if no header row (column title) is used.]

TI-83/84 Plus Input the data into L1; then continue with:

Choose: **2nd > LIST > Math > 3:mean(**
Enter: **L1**

> **Median:** The value of the data that occupies the middle position when the data are ranked in order according to size. The sample median is represented by $\tilde{x}$ (read "*x*-tilde" or "sample median").

Note: The population median, M (uppercase mu in the Greek alphabet), is the data value in the middle position of the entire ranked population.

Procedure for Finding the Median

STEP 1: **Rank the data.**

STEP 2: **Determine the depth of the median.** The **depth,** or position (number of positions from either end), of the median is determined by the formula

$$\textbf{depth of median:} \qquad depth\ of\ median = \frac{number + 1}{2}$$

$$d(\tilde{x}) = \frac{n + 1}{2} \qquad (2.2)$$

The median's depth (or position) is found by adding the position numbers of the smallest data (1) and the largest data (n) and dividing the sum by 2 (n is the number of pieces of data).

STEP 3: **Determine the value of the median.** Count the ranked data, locating the data in the $d(\tilde{x})$th position. The median will be the same regardless of which end of the ranked data (high or low) you count from. In fact, counting from both ends will serve as an excellent check.

The following two examples demonstrate this procedure as it applies to both odd-numbered and even-numbered sets of data.

EXAMPLE 2.9　　　**Median for Odd *n***

Find the median for the set of data {6, 3, 8, 5, 3}.

SOLUTION

FYI See Animated Tutorial Median on CD for assistance with this calculation.

STEP 1　The data, ranked in order of size, are 3, 3, 5, 6, and 8.

STEP 2　Depth of the median: $d(\tilde{x}) = \dfrac{n+1}{2} = \dfrac{5+1}{2} = 3$ (the "3rd" position).

FYI The value of $d(\tilde{x})$ is the depth of the median, NOT the value of the median, $\tilde{x}$.

STEP 3　The median is the third number from either end in the ranked data, or $\tilde{x} = $ **5.**

Notice that the median essentially separates the ranked set of data into two subsets of equal size (see Figure 2.16).

FIGURE 2.16
Median of {3, 3, 5, 6, 8}

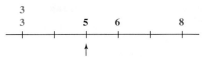

$\tilde{x} = 5$　(the middle value; **2 data values are smaller, 2 are larger**)

As in Example 2.9, when *n* is odd, the depth of the median, $d(\tilde{x})$, will always be an integer. When *n* is even, however, the depth of the median, $d(\tilde{x})$, will always be a half-number, as shown in Example 2.10.

EXAMPLE 2.10　　　**Median of Even *n***

Find the median of the sample 9, 6, 7, 9, 10, 8.

SOLUTION

STEP 1　The data, ranked in order of size, are 6, 7, 8, 9, 9, and 10.

STEP 2　Depth of the median: $d(\tilde{x}) = \dfrac{n+1}{2} = \dfrac{6+1}{2} = 3.5$ (the "3.5th" position).

FYI The median is the middle point by count.

STEP 3　The median is halfway between the third and fourth data values. To find the number halfway between any two values, add the two values together and divide the sum by 2. In this case, add the third value (8) and the fourth value (9) and then divide the sum (17) by

FYI See Animated Tutorial Median on CD for assistance with this calculation.

2. The median is $\tilde{x} = \dfrac{8+9}{3} = $ **8.5**, a number halfway between the "middle" two numbers (see Figure 2.17). Notice that the median again separates the ranked set of data into two subsets of equal size.

FIGURE 2.17
Median of {6, 7, 8, 9, 9, 10}

$\tilde{x} = 8.5$ (value in middle; **3 data values are smaller, 3 are larger**)

TECHNOLOGY INSTRUCTIONS: MEDIAN

MINITAB (Release 14) Input the data into C1; then continue with:

```
Choose:     Calc > Column Statistics
Select:     Median
Enter:      Input variable: C1 > OK
```

Excel Input the data into column A and activate a cell for the answer; then continue with:

```
Choose:     Insert Function, fx > Statistical > MEDIAN > OK
Enter:      Number 1: (A2:A6 or select cells)
```

TI-83/84 Plus Input the data into L1; then continue with:

```
Choose:     2nd > LIST > Math > 4:median(
Enter:      L1
```

> **Mode:** The mode is the value of x that occurs most frequently.

In the set of data from Example 2.9, {3, 3, 5, 6, 8}, the mode is 3 (see Figure 2.18).

FIGURE 2.18
Mode of {3, 3, 5, 6, 8}

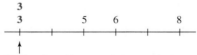

Mode = **3**   (the most frequent value)

In the sample 6, 7, 8, 9, 9, 10, the mode is 9. In this sample, only the 9 occurs more than once; in the data from Example 2.9, only the 3 occurs more than once. If two or more values in a sample are tied for the highest frequency (number of occurrences), we say there is **no mode**. For example, in the sample 3, 3, 4, 5, 5, 7, the 3 and the 5 appear an equal number of times. There is no one value that appears most often; thus, this sample has no mode.

> **Midrange:** The number exactly midway between a lowest-valued data, L, and a highest-valued data, H. It is found by averaging the low and the high values:
>
> $$\text{midrange} = \frac{low\ value + high\ value}{2}$$
>
> $$\text{midrange} = \frac{L + H}{2} \qquad (2.3)$$

For the set of data from Example 2.9, {3, 3, 5, 6, 8}, $L = 3$ and $H = 8$ (see Figure 2.19).

Therefore,

$$midrange = \frac{L + H}{2}$$

$$= \frac{3 + 8}{2} = 5.5$$

FIGURE 2.19
Midrange of {3, 3, 5, 6, 8}

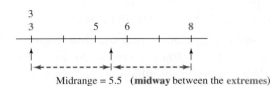

Midrange = 5.5   (**midway** between the **extremes**)

The four measures of central tendency represent four different methods of describing the middle. These four values may be the same, but more likely they will be different.

For the sample data from Example 2.10, the mean, $\bar{x}$, is 8.2; the median, $\tilde{x}$, is 8.5; the mode is 9; and the midrange is 8. Their relationship to one another and to the data is shown in Figure 2.20.

FIGURE 2.20
Measures of Central Tendency for {6, 7, 8, 9, 9, 10}

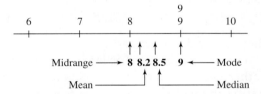

APPLIED EXAMPLE 2.11

"Average" Means Different Things

When it comes to convenience, few things can match that wonderful mathematical device called averaging. With an average, you can take a fistful of figures on any subject and compute one figure that will represent the whole fistful.

But there is one thing to remember. There are several kinds of measures ordinarily known as *averages,* and each gives a different picture of the figures it is called on to represent.

Take an example. Table 2.11 shows the annual incomes of 10 families.

TABLE 2.11

Annual Incomes of 10 Families [TA02-11]

$54,000	$39,000	$37,500	$36,750	$35,250	$31,500	$31,500	$31,500	$31,500	$25,500

What would this group's "typical" income be? Averaging would provide the answer, so let's compute the typical income by the simpler and most frequently used kinds of averaging.

- *The arithmetic mean.* This is the most common form of average, obtained by adding items in the data set, then dividing by the number of items;

for these data, the arithmetic mean is $35,400. The mean is representative of the data set in the sense that the sum of the amounts by which the higher figures exceed the mean is exactly the same as the sum of the amounts by which the lower figures fall short of the mean.

The higher incomes exceed the mean by a total of $25,650. The lower incomes fall short of the mean by a total of $25,650.

- *The median.* As you may have observed, six families earn less than the mean and four families earn more. You might wish to represent this varied group by the income of the family that is smack dab in the middle of the whole bunch. The median works out to $33,375.

- *The midrange.* Another number that might be used to represent the average is the midrange, computed by calculating the figure that lies halfway between the highest and lowest incomes: $39,750.

- *The mode.* So, three kinds of averages, and not one family actually has an income matching any of them. Say you want to represent the group by stating the income that occurs most frequently. That is called a mode. The modal income would be $31,500.

Four different averages are available, each valid, correct, and informative in its way. But how they differ!

arithmetic mean	*median*	*midrange*	*mode*
$35,400	$33,375	$39,750	$31,500

And they would differ still more if just one family in the group were millionaires—or one were jobless! The large value of $54,000 (extremely different from the other values) is skewing the data out toward larger data values. This skewing causes the mean and midrange to become much larger in value.

So there are three lessons. First, when you see or hear an average, find out which average it is. Then you will know what kind of picture you are being given. Second, think about the figures being averaged so that you can judge whether the average used is appropriate. Third, do not assume that a literal mathematical quantification is intended every time somebody says "average." It isn't. All of us often say "the average person" with no thought of implying a mean, median, or mode. All we intend to convey is the idea of other people who are in many ways a great deal like the rest of us.

Source: Reprinted by permission from *Changing Times,* March 1980. Copyright by The Kiplinger Washington Editors.

Now that we have learned how to calculate several sample statistics, the next question becomes, "How do I express my final answer?"

Round-off rule: When rounding off an answer, let's agree to keep one more decimal place in our answer than was present in the original information. To avoid round-off buildup, round off only the final answer, not the intermediate steps. That is, avoid using a rounded value to do further calculations. In our previous examples, the data were composed of whole numbers; therefore, those answers that have decimal values should be rounded to the nearest tenth. See the *Student Solutions Manual* for specific instructions on how to perform the rounding off.

SECTION 2.4 EXERCISES

2.57 Explain why it is possible to find the mean for the data of a quantitative variable but not for a qualitative variable.

2.58 The number of children, x, belonging to each of eight families registering for swimming was 1, 2, 1, 3, 2, 1, 5, 3. Find the mean, $\bar{x}$.

2.59 Skillbuilder Applet Exercise demonstrates the balancing effect of the mean. A plot is given with one data point at 10. Add more blocks by pointing and clicking on the desired location on the plot until a mean of 1 is achieved.

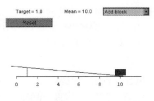

a. How many blocks were required to balance for a mean of 1?

b. At what value are these blocks located?

2.60 U.S. Interstate 64 runs between Portsmouth, VA, at I-264 on the eastern end to St. Louis, MO, at I-270 on the western end. While passing through six states and five more major cities, I-64 intersects nine more interstate highways. The number of miles in each state is as follows: Missouri—16 miles, Illinois—132 miles, Indiana—124 miles, Kentucky—191 miles, West Virginia—183 miles, Virginia—299 miles. (A diagram may be helpful.)
Source: http://www.ihoz.com/I90.html

a. Find the mean distance between major cities along I-64.

b. Find the mean distance between interchanges with interstate highways along I-64.

2.61 Interstate 29 intersects many other highways as it crosses four states in Mid-America, stretching from the southern end in Kansas City, MO, at I-35 to the northern end in Pembina, ND, at the Canadian border.

U.S. Interstate 29

State	Miles	Number of Intersections	State	Miles	Number of Intersections
Missouri	123	37	South Dakota	252	44
Iowa	161	32	North Dakota	217	40

Sources: Rand McNally and http://www.ihoz.com/ilist.html

Consider the variable "distance between intersections."

a. Find the mean distance between interchanges in Missouri.

b. Find the mean distance between interchanges in Iowa.

c. Find the mean distance between interchanges in North Dakota.

d. Find the mean distance between interchanges in South Dakota.

e. Find the mean distance between interchanges along U.S. I-29.

f. Find the mean of the four means found in answering parts a through d.

g. Compare the answers found to parts e and f. Did you expect them to be the same? Explain why they are different.

2.62 Find the median height of a basketball team: 73, 76, 72, 70, and 74 inches.

2.63 Find the median rate paid at Jim's Burgers if the workers' hourly rates are $4.25, $4.15, $4.90, $4.25, $4.60, $4.50, $4.60, $4.75.

2.64 Skillbuilder Applet Exercise demonstrates the effect one data value can have on the mean and on the median.

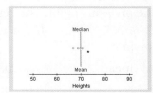

a. Move the red dot to the far right. What happens to the mean? What happens to the median?

b. Move the red dot to the far left. What happens to the mean? What happens to the median?

c. Which measure of central tendency, the mean or the median, gives a better sense of the center when a maverick (or outlier) is present in the data?

2.65 The number of cars per apartment owned by a sample of tenants in a large complex is 1, 2, 1, 2, 2, 2, 1, 2, 3, 2. What is the mode?

2.66 Each year, approximately 160 colleges compete in the National Concrete Canoe Competition. Each team must design a seaworthy canoe from a substance not known for its capacity to float. The canoes must weigh between 100 and 350 pounds. Find the midrange.
Source: Reader's Digest, March 2005

2.67 a. Find the mean, median, mode, and midrange for the sample data 9, 6, 7, 9, 10, 8.

    b. Verify and discuss the relationship between the answers in part a, as shown in Figure 2.20 on page 77.

2.68 Consider the sample 2, 4, 7, 8, 9. Find the following:

a. mean, $\bar{x}$ b. median, $\tilde{x}$

c. mode   d. midrange

2.69 Consider the sample 6, 8, 7, 5, 3, 7. Find the following:

a. mean, $\bar{x}$ b. median, $\tilde{x}$

c. mode   d. midrange

2.70 Fifteen randomly selected college students were asked to state the number of hours they slept the previous night. The resulting data values are 5, 6, 6, 8, 7, 7, 9, 5, 4, 8, 11, 6, 7, 8, 7. Find the following:

a. mean, $\bar{x}$ b. median, $\tilde{x}$

c. mode   d. midrange

2.71 [EX02-071] A random sample of 10 of the 2005 Nextel Cup NASCAR drivers produced the following ages: 33, 48, 41, 29, 40, 48, 44, 42, 49, 28.

a. Find the mean age for the 10 NASCAR drivers of the 2005 Nextel Cup.

b. Find the median age for the 10 NASCAR drivers of the 2005 Nextel Cup.

c. Find the midrange of age for the 10 NASCAR drivers of the 2005 Nextel Cup.

d. Find the mode, if one exists, for age for the 10 NASCAR drivers of the 2005 Nextel Cup.

2.72 [EX02-072] A constant objective in the manufacture of contact lenses is to improve those features that affect lens power and visual acuity. One such feature involves the tooling from which lenses are ultimately manufactured. The results of initial process development runs were examined for critical feature X. The resulting data are listed here:

0.026	0.027	0.024	0.023	0.034	0.035	0.035	0.033	0.034
0.033	0.032	0.038	0.041	0.041	0.021	0.022	0.027	0.032
0.023	0.023	0.024	0.017	0.023	0.019	0.027		

Source: Bausch & Lomb (variable not named and data coded at B&L's request)

a. Draw both a dotplot and a histogram of the critical feature X data.

b. Find the mean for critical feature X.

c. Find the median for critical feature X.

d. Find the midrange for critical feature X.

e. Find the mode, if one exists, for critical feature X.

f. What feature of the distribution, as shown by the graphs found in part a, seems unusual? Where do the answers found in parts b, c, and d fall relative to the distribution? Explain.

g. Identify at least one possible cause for this seemingly unusual situation.

2.73 [EX02-073] One measure of airline performance is overall flight on-time rates. For January 2005, the on-time arrival rates of domestic flights at the 31 largest U.S. airports were as follows.

| ATL | 69.09 | BWI | 74.01 | BOS | 62.14 |

••• Remainder of data on Student's Suite CD-ROM

Source: U.S. Department of Transportation, Bureau of Transportation Statistics

a. Find the mean on-time arrival rate for January 2005.

b. Find the median on-time arrival rate for January 2005.

c. Construct a stem-and-leaf display of the data.

d. Describe the relationship between the mean and the median and what properties of the data cause the mean to be lower than the median.

(Retain these solutions to use in Exercise 2.99 on p. 92.)

2.74 The "average" is a commonly reported statistic. This single bit of information can be very informative or very misleading, with the mean and median being the two most commonly reported.

a. The mean is a useful measure, but it can be misleading. Describe a circumstance when the mean is very useful as the average and a circumstance when the mean is very misleading as the average.

b. The median is a useful measure, but it can be misleading. Describe a circumstance when the median is very useful as the average and a circumstance when the median is very misleading as the average.

2.75 [EX02-075] All the third-graders at Roth Elementary School were given a physical fitness strength test. These data resulted:

12	22	6	9	2	9	5	9	3	5	16	1	22
18	6	12	21	23	9	10	24	21	17	11	18	19
17	5	14	16	19	19	18	3	4	21	16	20	15
14	17	4	5	22	12	15	18	20	8	10	13	20
6	9	2	17	15	9	4	15	14	19	3	24	

a. Construct a dotplot.

b. Find the mode.

c. Prepare a grouped frequency distribution using classes 1–4, 4–7, and so on, and draw a histogram of the distribution.

d. Describe the distribution; specifically, is the distribution bimodal (about what values)?

e. Compare your answers in parts a and c, and comment on the relationship between the mode and the modal values in these data.

f. Could the discrepancy found in the comparison in part e occur when using an ungrouped frequency distribution? Explain.

g. Explain why, in general, the mode of a set of data does not necessarily give us the same information as the modal values do.

2.76 [EX02-076] Consumers are frequently cautioned against eating too much food that is high in calories, fat, and sodium for numerous health and fitness reasons. *Nutrition Action HealthLetter* published a list of popular low-fat brands of hot dogs commonly labeled "fat-free," "reduced fat," "low-fat," "light," and so on, together with their calories, fat content, and sodium. All quantities are for one hot dog:

Hot Dog Brand	Calories	Fat (g)	Sodium (mg)
Ball Park Fat Free Beef Franks	50	0	460
Butterball Fat Free Franks	40	0	490

••• Remainder of data on Student's Suite CD-ROM

Source: Nutrition Action HealthLetter, "On the Links," July/August 1998, pp. 12–13

a. Find the mean, median, mode, and midrange of the calories, fat, and sodium contents of all the frankfurters listed. Use a table to summarize your results.

b. Construct a dotplot of the fat contents. Locate the mean, median, mode, and midrange on the plot.

c. In the summer of 2005, the winner of Nathan's Famous Fourth of July Hot Dog Eating Contest consumed 49 hot dogs in 12 minutes. If he had been served the median hot dog, how many calories, grams of fat, and milligrams of sodium did he consume in the single sitting? If the recommended daily allowance for sodium intake is 2400 mg, did he likely exceed it? Explain.

2.77 [EX02-077] The number of runs scored by major league baseball (MLB) teams is likely influenced by whether the game is played at home or at the opponents' ballpark. In an attempt to measure differences between playing at home or away, the average number of runs scored per game by each MLB team while playing at home and while playing away (at the opponents' fields) was recorded. The following table summarizes the data.

Team	Avg. Runs at Home	Avg. Runs Away
Angels	4.83	5.49
Red Sox	6.38	5.33
••• Remainder of data on Student's Suite CD-ROM		

Source: http://mlb.mlb.com

a. Find the mean, median, maximum, minimum, and midrange of the runs scored by the teams while playing at home.

b. Find the mean, median, maximum, minimum, and midrange of the runs scored by the teams while playing away.

c. Compare each of the measures you found in parts a and b. What can you conclude?

2.78 [EX02-078] Does everything increase, every year? Sometimes it seems like it! The annual percentage rate of increase in motor-fuel consumption for 2002 to 2003 is listed in the following table by U.S. state. Notice that the consumption did not increase in every state and not all states reported.

1.8	16.9	1.8	0.1	−2.8	−0.4	−2.6	−0.8	−6.3	2.7	1.1
0.1	−4.6	−0.4	0.2	0.9	−2.8	1.5	3.6	10.8	2.3	0.5
−1	−1	2.9	1.6	−1.2	−0.1	5.5	0.2	8.9	2.1	0.5
0.1	2.9	−0.5	3.7	1.1	−1.3	0	1.3	−2.4	0.9	0.1

Source: U.S. Department of Transportation, Federal Highway Administration

a. Explain the meaning of negative and positive values, large and small values, values near zero, and values not near zero.

b. Examine the data in the table. What do you anticipate the distribution of "percent change" will look like? What do you think the mean "percent change" will be? Justify your estimate, without any preliminary work or calculations.

c. If you expect very little or no change, what value will the mean have? Explain.

d. Construct a histogram of the percent of change.

e. Calculate the mean percent of changes in consumption from 2002 to 2003.

f. The Federal Highway Administration reported the percentage increase for the entire United States as 0.5586 of 1%. The value calculated for the mean in part e is not the same. Explain how this is possible.

2.79 [EX02-079] Students like to engage in the "Battle of the Sexes" when it comes to who's the better driver. But which gender outnumbers the other on the road? The numbers may surprise you. Listed here is the number of licensed male and female drivers in each of 18 randomly selected states.

State	Male	Female
KY	1,389,380	1,410,255
DE	286,144	298,992
••• Remainder of data on Student's Suite CD-ROM		

Source: U.S. Department of Transportation, Federal Highway Administration

a. Do the female drivers outnumber the male drivers? Study the table and see if the data seem to support your estimate. Explain your initial answer.

b. Define the variable "ratio M/F" as the number of licensed male drivers divided by the number of licensed female drivers in each state. Calculate "ratio M/F" for the states in the sample.

c. If a value of the "ratio M/F" is near 1.0, what does that mean? Greater than 1.0? Less than 1.0? Explain.

d. Construct a histogram.

e. Describe the distribution shown in the histogram found in part d.

f. Calculate the mean value of the "ratio M/F."

g. Explain the meaning of values in each of the tails of the histogram.

Optional: All 50 states plus DC are listed on the CD. [DS02-079]

h. List two states, not in the preceding table, that you expect to find near each tail of the distribution of M/F. Explain why you believe these states will have high or low ratios.

i. Answer questions of parts d and f using the data of all 51 items.

j. Compare the results found in part i to those found in parts d and f.

k. How accurate were your predications to part h? Explain.

2.80 You are responsible for planning the parking needed for a new 256-unit apartment complex, and you're told to base the needs on the statistic "average number of vehicles per household is 1.9."

a. Which average (mean, median, mode, midrange) will be helpful to you? Explain.

b. Explain why 1.9 cannot be the median, the mode, or the midrange for the variable "number of vehicles."

c. If the owner wants parking that will accommodate 90% of all the tenants who own vehicles, how many spaces must you plan for?

2.81 In what states do the residents pay the most taxes? the least? Perhaps it depends on the variable used to measure amount of taxes paid. In 2004 the Tax Policy Center reported the following statistics about the average annual 2002 taxes and percent of personal income paid per person by state.

	Taxes per Capita	Rank	% Personal Income	Rank
Hawaii	$2748	1	9.6	1
South Dakota	$1283	50	4.8	47
New Hampshire	$1478	45	4.4	50

Sources: Federation of Tax Administrators (2004) and U.S. Bureau of the Census and Bureau of Economic Analysis, http://taxpolicycenter.org/TaxFacts/TFDB/TFTemplate.cfm?Docid=309&Topic2id=90

a. Compare and contrast the variables "taxes per capita" and "percent of personal income." How do you account for the differences in ranks for South Dakota and New Hampshire?

b. Based on this information, using the highest and lowest per state amount of taxes paid per person, what was the "average" amount paid per person?

c. Based on this information, using the highest and lowest percent of income per state paid per person, what was the "average" percent paid per person?

d. Explain why your answers to parts b and c are the only average value you can determine from the given information. What is the name of this average?

2.82 Your instructor and your class have made a deal on the exam just taken and currently being graded. If the class attains a mean score of 74 or better, there will be no homework on the coming weekend. If the class mean is 72 or below, then not only will there be homework as usual but all of the class members will have to show up on Saturday and do 2 hours of general cleanup around the school grounds as a community service project. There are 15 students in your class. Your instructor has graded the first 14 exams, and their mean score is 73.5. Your exam is the only one left to grade.

a. What score must you get in order for the class to win the deal?

b. What score must you get in order that the class will not have to do the community service work?

2.83 Starting with the data values 70 and 100, add three data values to the sample so that the sample has the following: (Justify your answer in each case.)

a. Mean of 100   b. Median of 70

c. Mode of 87   d. Midrange of 70

e. Mean of 100 and a median of 70

f. Mean of 100 and a mode of 87

g. Mean of 100 and a midrange of 70

h. Mean of 100, a median of 70, and a mode of 87

2.84 Skillbuilder Applet Exercise matches means with corresponding histograms. After several practice rounds using "New Plots," explain your method of matching.

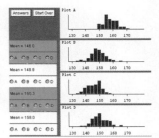

2.5   Measures of Dispersion

Having located the "middle" with the measures of central tendency, our search for information from data sets now turns to the measures of dispersion (spread). The **measures of dispersion** include the *range, variance,* and *standard deviation*. These numerical values describe the amount of spread, or variability, that is found among the data: Closely grouped data have relatively small values, and more widely spread-out data have larger values. The closest possible grouping occurs when the data have no dispersion (all data are the same value); in this situation, the measure of dispersion will be zero. There is no limit to how widely spread out the data can be; therefore, measures of dispersion can be very large. The simplest measure of dispersion is the range.

> **Range:** The difference in value between the highest-valued data, H, and the lowest-valued data, L:
>
> $$range = high\ value - low\ value$$
> $$range = H - L \tag{2.4}$$

The sample 3, 3, 5, 6, 8 has a range of $H - L = 8 - 3 = 5$. The range of 5 tells us that these data all fall within a 5-unit interval (see Figure 2.21).

FIGURE 2.21
Range of {3, 3, 5, 6, 8}

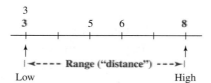

The other measures of dispersion to be studied in this chapter are measures of dispersion about the mean. To develop a measure of dispersion about the mean, let's first answer the question: How far is each x from the mean?

> **Deviation from the mean:** A deviation from the mean, $x - \bar{x}$, is the difference between the value of x and the mean, $\bar{x}$.

Each individual value x deviates from the mean by an amount equal to $(x - \bar{x})$. This deviation $(x - \bar{x})$ is zero when x is equal to the mean, $\bar{x}$. The deviation $(x - \bar{x})$ is positive when x is larger than $\bar{x}$ and negative when x is smaller than $\bar{x}$.

Consider the sample 6, 3, 8, 5, 3. Using formula (2.1), $\bar{x} = \dfrac{\sum x}{n}$, we find that the mean is 5. Each deviation, $(x - \bar{x})$, is then found by subtracting 5 from each x value:

Data, x	6	3	8	5	3
Deviation, $x - \bar{x}$	1	−2	3	0	−2

FIGURE 2.22

Deviations from the Mean

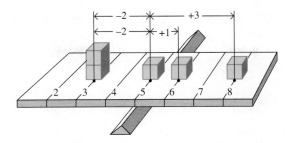

Figure 2.22 shows the four nonzero deviations from the mean.

To describe the "average" value of these deviations, we might use the mean deviation, the sum of the deviations divided by n, $\dfrac{\sum(x-\bar{x})}{n}$. However, because the sum of the deviations, $\sum(x-\bar{x})$, is exactly zero, the mean deviation will also be zero. In fact, it will always be zero, which means it is not a useful statistic. How does this happen, and why?

The sum of the deviations, $\sum(x-\bar{x})$, is always zero because the deviations of x values smaller than the mean (which are negative) cancel out those x values larger than the mean (which are positive). We can remove this neutralizing effect if we do something to make all the deviations positive. This can be accomplished in two ways. First, by using the absolute value of the deviation, $|x-\bar{x}|$, we can treat each deviation as its "size" or distance only. For our illustration we obtain the following *absolute deviations.*

Data, x	6	3	8	5	3
Absolute Value of Deviation, $\lvert x - \bar{x} \rvert$	1	2	3	0	2

> **Mean absolute deviation:** The mean of the absolute values of the deviations from the mean:
>
> $$\text{mean absolute deviation} = \frac{\text{sum of (absolute values of deviations)}}{\text{number}}$$
>
> $$\text{mean absolute deviation} = \frac{\sum |x - \bar{x}|}{x} \qquad (2.5)$$

For our example, the sum of the absolute deviations is 8 $(1 + 2 + 3 + 0 + 2)$ and

$$\text{mean absolute deviation} = \frac{\sum |x - \bar{x}|}{n} = \frac{8}{5} = 1.6$$

Although this particular measure of spread is not used very frequently, it is a measure of dispersion. It tells us the mean "distance" the data are from the mean.

A second way to eliminate the positive–negative neutralizing effect is to square each of the deviations; squared deviations will all be nonnegative (positive or zero) values. The squared deviations are used to find the *variance.*

Sample variance: The sample variance, s^2, is the mean of the squared deviations, calculated using $n - 1$ as the divisor:

sample variance: s squared $= \dfrac{sum\ of\ (deviations\ squared)}{number - 1}$

$$s^2 = \frac{\sum (x - \bar{x})^2}{n - 1} \qquad (2.6)$$

where n is the sample size—that is, the number of data in the sample.

FYI See page 90 for an explanation of these icons.

The variance of the sample 6, 3, 8, 5, 3 is calculated in Table 2.12 using formula (2.6).

TABLE 2.12

Calculating Variance Using Formula (2.6)

Step 1. Find $\sum x$	Step 2. Find $\bar{x}$	Step 3. Find each $x - \bar{x}$	Step 4. Find $\sum (x - \bar{x})^2$	Step 5. Find s^2
6	$\bar{x} = \dfrac{\sum x}{n}$	$6 - 5 = 1$	$(1)^2 = 1$	$s^2 = \dfrac{\sum (x - \bar{x})^2}{n - 1}$
3		$3 - 5 = -2$	$(-2)^2 = 4$	
8		$8 - 5 = 3$	$(3)^2 = 9$	
5	$\bar{x} = \dfrac{25}{5}$	$5 - 5 = 0$	$(0)^2 = 0$	$s^2 = \dfrac{18}{4}$
3		$3 - 5 = -2$	$(-2)^2 = 4$	
$\sum x = 25$	$\bar{x} = 5$	$\sum (x - \bar{x}) = 0$ (ck)	$\sum (x - \bar{x})^2 = 18$	$s^2 = 4.5$

Notes:

1. The sum of all the x values is used to find $\bar{x}$.

FYI See Animated Tutorial Variance on CD for assistance with this calculation.

2. The sum of the deviations, $\sum (x - \bar{x})$, is always zero, provided the exact value of $\bar{x}$ is used. Use this fact as a check in your calculations, as was done in Table 2.12 (denoted by (ck)).
3. If a rounded value of $\bar{x}$ is used, then $\sum (x - \bar{x})$ will not always be exactly zero. It will, however, be reasonably close to zero.
4. The sum of the squared deviations is found by squaring each deviation and then adding the squared values.

To graphically demonstrate what variances of data sets are telling us, consider a second set of data: {1, 3, 5, 6, 10}. Note that the data values are more dispersed than the data values in Table 2.12. Accordingly, its calculated variance is larger at $s^2 = 11.5$. An illustrative side-by-side graphical comparison of these two samples and their variances is shown in Figure 2.23.

FIGURE 2.23

Comparison of Data

Sample standard deviation: The standard deviation of a sample, s, is the positive square root of the variance:

sample standard deviation: $s = $ *square root of sample variance*

$$s = \sqrt{s^2} \qquad (2.7)$$

For the samples shown in Figure 2.23, the standard deviations are $\sqrt{4.5}$ or **2.1**, and $\sqrt{11.5}$ or **3.4**.

Note: The numerator for the sample variance, $\sum(x - \bar{x})^2$, is often called the *sum of squares for x* and symbolized by $SS(x)$. Thus, formula (2.6) can be expressed as

$$\text{sample variance:} \qquad s^2 = \frac{SS(x)}{n - 1} \qquad (2.8)$$

where $SS(x) = \sum(x - \bar{x})^2$.

The formulas for variance can be modified into other forms for easier use in various situations. For example, suppose we have the sample 6, 3, 8, 5, 2. The variance for this sample is computed in Table 2.13.

TABLE 2.13

Calculating Variance Using Formula (2.6)

Step 1. Find $\sum x$	Step 2. Find $\bar{x}$	Step 3. Find each $x - \bar{x}$	Step 4. Find $\sum(x - \bar{x})^2$	Step 5. Find s^2
6	$\bar{x} = \dfrac{\sum x}{n}$	$6 - 4.8 = 1.2$	$(1.2)^2 = 1.44$	$s^2 = \dfrac{\sum(x - \bar{x})^2}{n - 1}$
3		$3 - 4.8 = -1.8$	$(-1.8)^2 = 3.24$	
8		$8 - 4.8 = 3.2$	$(3.2)^2 = 10.24$	
5	$\bar{x} = \dfrac{24}{5}$	$5 - 4.8 = 0.2$	$(0.2)^2 = 0.04$	$s^2 = \dfrac{22.80}{4}$
2		$2 - 4.8 = -2.8$	$(-2.8)^2 = 7.84$	
$\sum x = 24$	$\bar{x} = 4.8$	$\sum(x - \bar{x}) = 0$ (ck)	$\sum(x - \bar{x})^2 = 22.80$	$s^2 = \mathbf{5.7}$

The arithmetic for this example has become more complicated because the mean contains nonzero digits to the right of the decimal point. However, the "sum of squares for x," the numerator of formula (2.6), can be rewritten so that $\bar{x}$ is not included:

Sum of Squares for x

$$SS(x) = \sum x^2 - \frac{(\sum x)^2}{n} \qquad (2.9)$$

FYI See page 90 for an explanation of icons.

Combining formulas (2.8) and (2.9) yields the "shortcut formula" for sample variance:

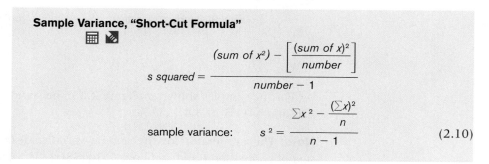

Sample Variance, "Short-Cut Formula"

$$s \text{ squared} = \frac{(\text{sum of } x^2) - \left[\frac{(\text{sum of } x)^2}{\text{number}}\right]}{\text{number} - 1}$$

sample variance: $\qquad s^2 = \dfrac{\sum x^2 - \dfrac{(\sum x)^2}{n}}{n - 1}$ (2.10)

FYI See Animated Tutorial Variance Shortcut on CD for assistance with this calculation.

Formulas (2.9) and (2.10) are called *shortcuts* because they bypass the calculation of $\bar{x}$. The computations for $SS(x)$, s^2, and s using formulas (2.9), (2.10), and (2.7) are performed as shown in Table 2.14.

TABLE 2.14

Calculating Standard Deviation Using the Shortcut Method

Step 1. Find $\sum x$	Step 2. Find $\sum x^2$	Step 3. Find $SS(x)$	Step 4. Find s^2	Step 5. Find s
6	$6^2 = 36$	$SS(x) = \sum x^2 - \dfrac{(\sum x)^2}{}$		$s = \sqrt{s^2}$
3	$3^2 = 9$		$s^2 = \dfrac{\sum x^2 - \dfrac{(\sum x)^2}{n}}{n}$	$s = \sqrt{5.7}$
8	$8^2 = 64$	$SS(x) = 138 - \dfrac{(24)^2}{5}$		$s = 2.4$
5	$5^2 = 25$			
2	$2^2 = 4$	$SS(x) = 138 - 115.2$	$s^2 = \dfrac{22.8}{4}$	
$\sum x = 24$	$\sum x^2 = 138$	$SS(x) = 22.8$	$s^2 = 5.7$	

The unit of measure for the standard deviation is the same as the unit of measure for the data. For example, if our data are in pounds, then the standard deviation, *s*, will also be in pounds. The unit of measure for variance might then be thought of as *units squared*. In our example of pounds, this would be *pounds squared*. As you can see, the unit has very little meaning.

TECHNOLOGY INSTRUCTIONS: STANDARD DEVIATION

MINITAB (Release 14)   Input the data into C1; then continue with:

```
Choose:     Calc > Column Statistics
Select:     Standard deviation
Enter:      Input variable: C1 > OK
```

Excel   Input the data into column A and activate a cell for the answer; then continue with:

```
Choose:     Insert Function, fₓ > Statistical > STDEV > OK
Enter:      Number 1: (A2:A6 or select cells)
```

TI-83/84 Plus   Input the data into L1; then continue with:

Choose: **2nd > LIST > Math > 7:StdDev(**
Enter: **L1**

TECHNOLOGY INSTRUCTIONS: ADDITIONAL STATISTICS

MINITAB (Release 14)   Input the data into C1, then continue with:

Choose: **Calc > Column Statistics**
             Then one at a time select the desired statistic
Select: **N total**   Number of data in column
             Sum   Sum of the data in column
             Minimum   Smallest value in column
             Maximum   Largest value in column
             Range   Range of values in column
             Sum of squares   Sum of squared x-values, $\sum x^2$
Enter: Input variable: **C1 > OK**

Excel   Input the data into column A and activate a cell for the answer; then continue with:

Choose: **Insert Function, f$_x$ > Statistical > COUNT**
                                              > MIN
                                              > MAX
             OR   **> All > SUM**
                                          > SUMSQ
Enter: Number 1: **(A2:A6 or select cells)**
For range, write a formula: **Max( ) − Min( )**

TI-83/84 Plus   Input the data into L1; then continue with:

Choose: **2nd > LIST > Math > 5:sum(**
                                   > 1:min(
                                   > 2:max(
Enter: **L1**

Standard deviation on your calculator: Most calculators have two formulas for finding the standard deviation and mindlessly calculate both, fully expecting the user to decide which one is correct for the given data. How do you decide?

The sample standard deviation is denoted by s and uses the "divide by $n − 1$" formula.

The population standard deviation is denoted by σ and uses the "divide by n" formula.

When you have sample data, always use the s or "divide by $n − 1$" formula. Having the population data is a situation that will probably never occur, other than in a textbook exercise. If you don't know whether you have sample data or population data, it is a "safe bet" that they are sample data—use the s or "divide by $n − 1$" formula!

Multiple formulas: Statisticians have multiple formulas for convenience—that is, convenience relative to the situation. The following statements will help you decide which formula to use:

1. When you are working on a computer and using statistical software, you will generally store all the data values first. The computer handles repeated operations easily and can "revisit" the stored data as often as necessary to complete a procedure. The computations for sample variance will be done using formula (2.6), following the process shown in Table 2.12.

2. When you are working on a calculator with built-in statistical functions, the calculator must perform all necessary operations on each data as the values are entered (most handheld nongraphing calculators do not have the ability to store data). Then after all data have been entered, the computations will be completed using the appropriate summations. The computations for sample variance will be done using formula (2.10), following the procedure shown in Table 2.14.

3. If you are doing the computations either by hand or with the aid of a calculator, but not using statistical functions, the most convenient formula to use will depend on how many data there are and how convenient the numerical values are to work with.

When a formula has multiple forms, look for one of these icons:

▣ is used to identify the formula most likely to be used by a computer.

▦ is used to identify the formula most likely to be used by a calculator.

▨ is used to identify the formula most likely to be convenient for hand calculations.

D is used to identify the "definition" formula.

SECTION 2.5 EXERCISES

Statistics ⬡ Now™

2.85 In 2004 the Tax Policy Center reported the following statistics about the average annual 2002 taxes and percent of personal income paid per person by state.

	Taxes per Capita	Rank	Percent of Personal Income	Rank
Hawaii	$2748	1	9.6	1
South Dakota	$1283	50	4.8	47
New Hampshire	$1478	45	4.4	50

Sources: Federation of Tax Administrators (2004) and U.S. Bureau of the Census and Bureau of Economic Analysis, http://taxpolicycenter.org/TaxFacts/TFDB/TFTemplate.cfm?Docid=309&Topic2id=90

a. Find the range for the amount of taxes paid per person.

b. Find the range for the percentage of personal income paid in taxes per person.

2.86 a. The data value $x = 45$ has a deviation value of 12. Explain the meaning of this.

b. The data value $x = 84$ has a deviation value of -20. Explain the meaning of this.

2.87 The summation $\Sigma(x - \bar{x})$ is always zero. Why? Think back to the definition of the mean (p. 73) and see if you can justify this statement.

2.88 All measures of variation are nonnegative in value for all sets of data.

a. What does it mean for a value to be "nonnegative"?

b. Describe the conditions necessary for a measure of variation to have the value zero.

c. Describe the conditions necessary for a measure of variation to have a positive value.

2.89 A sample contains the data {1, 3, 5, 6, 10}.

a. Use formula (2.6) to find the variance.

b. Use formula (2.10) to find the variance.

c. Compare the results from parts a and b.

2.90 Consider the sample 2, 4, 7, 8, 9. Find the following:

a. Range

b. Variance s^2, using formula (2.6)

c. Standard deviation, s

2.91 Consider the sample 6, 8, 7, 5, 3, 7. Find the following:

a. Range

b. Variance s^2, using formula (2.6)

c. Standard deviation, s

2.92 Given the sample 7, 6, 10, 7, 5, 9, 3, 7, 5, 13, find the following:

a. Variance s^2 using formula (2.6)

b. Variance s^2, using formula (2.10)

c. Standard deviation, s

2.93 Fifteen randomly selected college students were asked to state the number of hours they slept the previous night. The resulting data are 5, 6, 6, 8, 7, 7, 9, 5, 4, 8, 11, 6, 7, 8, 7. Find the following:

a. Variance s^2, using formula (2.6)

b. Variance s^2, using formula (2.10)

c. Standard deviation, s

2.94 [EX02-071] A random sample of 10 of the 2005 Nextel Cup NASCAR drivers produced the following ages: 33, 48, 41, 29, 40, 48, 44, 42, 49, 28.

a. Find the range.

b. Find the variance.

c. Find the standard deviation.

2.95 Adding (or subtracting) the same number from each value in a set of data does not affect the measures of variability for that set of data.

a. Find the variance of this set of annual heating-degree-day data: 6017, 6173, 6275, 6350, 6001, 6300.

b. Find the variance of this set of data (obtained by subtracting 6000 from each value in part a): 17, 173, 275, 350, 1, 300.

2.96 [EX02-096] One aspect of the beauty of scenic landscape is its variability. The elevations (feet above sea level) of 12 randomly selected towns in the Finger Lakes Regions of Upstate New York are recorded here.

559	815	767	668	651	895
1106	1375	861	1559	888	1106

Source: http://www.city-data.com

a. Find the mean.

b. Find the standard deviation.

2.97 [EX02-097] Recruits for a police academy were required to undergo a test that measures their exercise capacity. The exercise capacity (in minutes) was obtained for each of 20 recruits:

25	27	30	33	30	32	30	34	30	27
26	25	29	31	31	32	34	32	33	30

a. Draw a dotplot of the data.

b. Find the mean.

c. Find the range.

d. Find the variance.

e. Find the standard deviation.

f. Using the dotplot from part a, draw a line representing the range. Then draw a line starting at the mean with a length that represents the value of the standard deviation.

g. Describe how the distribution of data, the range, and the standard deviation are related.

2.98 [EX02-098] *Better Roads* magazine reported the percentage of interstate and state-owned bridges that were structurally deficient or functionally obsolete (%SD/FO) for each U.S. state in

2003. (Percentages are expressed in decimal form [e.g., 0.20 = 20%].)

State	SD/FO*	State	SD/FO*	State	SD/FO*
AK	0.20	AL	0.22	AR	0.20

••• Remainder of data on Student's Suite CD-ROM

Source: Better Roads, November 2003

*SD/FO = structurally deficient or functionally obsolete.

a. Construct a histogram.

b. Does the variable "%SD/FO" appear to have an approximately normal distribution?

c. Calculate the mean.

d. Find the median.

e. Find the range.

f. Find the standard deviation.

(Retain these solutions to use in Exercise 2.125 on p. 105.)

2.99 [EX02-099] One measure of airline performance is about overall flight on-time rates. For January 2005, the on-time arrival rates of domestic flights at the 31 largest U.S. airports were as follows.

ATL	69.1	BWI	74.0	BOS	62.1

••• Remainder of data on Student's Suite CD-ROM

Source: U.S. Department of Transportation, Bureau of Transportation Statistics

a. Find the range and the standard deviation for the on-time arrival rates.

b. Draw lines on the stem-and-leaf diagram drawn in answering Exercise 2.73 that represent the range and standard deviation. Remember: The standard deviation is a measure of the spread about the mean.

c. Describe the relationship among the distribution of the data, the range, and the standard deviation.

2.100 Consider these two sets of data:

Set 1	46	55	50	47	52
Set 2	30	55	65	47	53

Both sets have the same mean, 50. Compare these measures for both sets: $\sum(x - \bar{x})$, $\sum|x - \bar{x}|$, $SS(x)$, and range. Comment on the meaning of these comparisons.

2.101 Comment on the statement: "The mean loss for customers at First State Bank (which was not insured) was $150. The standard deviation of the losses was −$125."

2.102 Start with $x = 100$ and add four x values to make a sample of five data such that:

a. $s = 0$ b. $0 < s < 1$

c. $5 < s < 10$ d. $20 < s < 30$

2.103 Each of two samples has a standard deviation of 5. If the two sets of data are made into one set of 10 data, will the new sample have a standard deviation that is less than, about the same as, or greater than the original standard deviation of 5? Make up two sets of five data, each with a standard deviation of 5, to justify your answer. Include the calculations.

2.104 Skillbuilder Applet Exercise matches means and standard deviations with corresponding histograms. After several practice rounds using "Start Over," explain your method of matching.

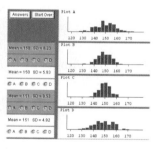

2.6 | Measures of Position

Measures of position are used to describe the position a specific data value possesses in relation to the rest of the data when in ranked order. *Quartiles* and *percentiles* are two of the most popular measures of position.

Quartiles: Values of the variable that divide the ranked data into quarters; each set of data has three quartiles. The *first quartile*, Q_1, is a number such that at most 25% of the data are smaller in value than Q_1 and at most 75% are larger. The *second quartile* is the median. The *third quartile*, Q_3, is a number such that at most 75% of the data are smaller in value than Q_3 and at most 25% are larger. (See Figure 2.24.)

FIGURE 2.24
Quartiles

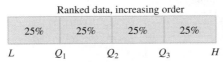

The procedure for determining the values of the quartiles is the same as that for percentiles and is shown in the following description of *percentiles*. Remember that your data must be ranked from low (L) to high (H).

Percentiles: Values of the variable that divide a set of ranked data into 100 equal subsets; each set of data has 99 percentiles (see Figure 2.25). The *k*th percentile, P_k, is a value such that at most k% of the data are smaller in value than P_k and at most $(100 - k)$% of the data are larger (see Figure 2.26).

FIGURE 2.25   Percentiles

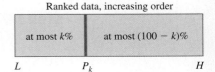

FIGURE 2.26   *k*th Percentile

Notes:
1. The first quartile and the 25th percentile are the same; that is, $Q_1 = P_{25}$. Also, $Q_3 = P_{75}$.
2. The median, the second quartile, and the 50th percentile are all the same: $\tilde{x} = Q_2 = P_{50}$. Therefore, when asked to find P_{50} or Q_2, use the procedure for finding the median.

The procedure for determining the value of any *k*th percentile (or quartile) involves four basic steps as outlined on the diagram in Figure 2.27. Example 2.12 demonstrates the procedure.

FIGURE 2.27
Finding P_k Procedure

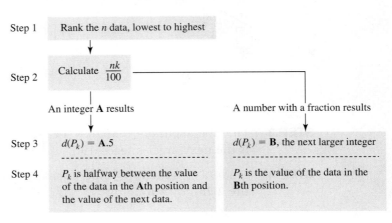

EXAMPLE 2.12 Finding Quartiles and Percentiles

Statistics Now™

Watch a video example at **http://1pass.thomson.com** or on your CD.

Using the sample of 50 elementary statistics final exam scores listed in Table 2.15, find the first quartile, Q_1; the 58th percentile, P_{58}; and the third quartile, Q_3.

TABLE 2.15

Raw Scores for Elementary Statistics Exam

60	47	82	95	88	72	67	66	68	98	90	77	86
58	64	95	74	72	88	74	77	39	90	63	68	97
70	64	70	70	58	78	89	44	55	85	82	83	
72	77	72	86	50	94	92	80	91	75	76	78	

SOLUTION

STEP 1 Rank the data: A ranked list may be formulated (see Table 2.16), or a graphic display showing the ranked data may be used. The dotplot and the stem-and-leaf display are handy for this purpose. The stem-and-leaf display is especially helpful, because it gives depth numbers counted from both extremes when it is computer generated (see Figure 2.28). Step 1 is the same for all three statistics.

Find Q_1:

STEP 2 Find $\frac{nk}{100}$: $\frac{nk}{100} = \frac{(50)(25)}{100} = \mathbf{12.5}$

($n = 50$ and $k = 25$, since $Q_1 = P_{25}$.)

TABLE 2.16

Ranked Data: Exam Scores

39	64	72	78	89
44	66	72	80	90
47	67	74	82	90
50	68	74	82	91
55	68	75	83	92
58	70	76	85	94
58	70	77	86	95
60	70	77	86	95
63	72	77	88	97
64	72	78	88	98

13th position from L

29th and 30th positions from L

13th position from H

FIGURE 2.28 Final Exam Scores

Stem-and-leaf of score $N = 50$

Leaf Unit = 1.0

1	3	9
2	4	4
3	4	7
4	5	0
7	5	5 8 8
11	6	0 3 4 4
15	6	6 7 8 8
24	7	0 0 0 2 2 2 2 4 4
(7)	7	5 6 7 7 7 8 8
19	8	0 2 2 3
15	8	5 6 6 8 8 9
9	9	0 0 1 2 4
4	9	5 5 7 8

STEP 3 Find the depth of Q_1: $d(Q_1) = \mathbf{13}$ (Since 12.5 contains a fraction, **B** is the next larger integer, 13.)

STEP 4 Find Q_1: Q_1 is the 13th value, counting from L (see Table 2.16 or Figure 2.28), $Q_1 = \mathbf{67}$

Find P_{58}:

FYI See Animated Tutorial Percentile on CD for assistance with these calculations.

STEP 2 Find $\dfrac{nk}{100}$: $\dfrac{nk}{100} = \dfrac{(50)(58)}{100} = \mathbf{29}$ ($n = 50$ and $k = 58$ for P_{58}.)

STEP 3 Find the depth of P_{58}: $d(P_{58}) = \mathbf{29.5}$ (Since **A** $= 29$, an integer, add 0.5 and use 29.5.)

STEP 4 Find P_{58}: P_{58} is the value halfway between the values of the 29th and the 30th pieces of data, counting from L (see Table 2.16 or Figure 2.28), so

$$P_{58} = \frac{77 + 78}{2} = \mathbf{77.5}$$

Therefore, it can be stated that "at most, 58% of the exam grades are smaller in value than 77.5." This is also equivalent to stating that "at most, 42% of the exam grades are larger in value than 77.5."

Optional technique: When k is greater than 50, subtract k from 100 and use $(100 - k)$ in place of k in Step 2. The depth is then counted from the highest-value data, H.

Find Q_3 using the optional technique:

STEP 2 Find $\dfrac{nk}{100}$: $\dfrac{nk}{100} = \dfrac{(50)(25)}{100} = \mathbf{12.5}$ ($n = 50$ and $k = 75$, since $Q_3 = P_{75}$, and $k > 50$; use $100 - k = 100 - 75 = 25$.)

STEP 3 Find the depth of Q_3 from H: $d(Q_3) = \mathbf{13}$

STEP 4 Find Q_3: Q_3 is the 13th value, counting from H (see Table 2.16 or Figure 2.28), $Q_3 = \mathbf{86}$

Therefore, it can be stated that "at most, 75% of the exam grades are smaller in value than 86." This is also equivalent to stating that "at most, 25% of the exam grades are larger in value than 86."

An additional measure of central tendency, the *midquartile*, can now be defined.

> **Midquartile:** The numerical value midway between the first quartile and the third quartile.
> $$\text{midquartile} = \frac{Q_1 + Q_3}{2} \tag{2.11}$$

EXAMPLE 2.13

Finding the Midquartile

Find the midquartile for the set of 50 exam scores given in Example 2.12.

SOLUTION

$Q_1 = 67$ and $Q_3 = 86$, as found in Example 2.12. Thus,

$$\text{midquartile} = \frac{Q_1 + Q_3}{2} = \frac{67 + 86}{2} = \mathbf{76.5}$$

The median, the midrange, and the midquartile are not necessarily the same value. Each is the middle value, but by different definitions of "middle." Figure 2.29 summarizes the relationship of these three statistics as applied to the 50 exam scores from Example 2.12.

FIGURE 2.29

Final Exam Scores

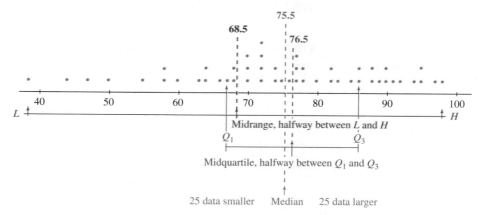

A *5-number summary* is very effective in describing a set of data. It is easy information to obtain and is very informative to the reader.

5-Number summary: The 5-number summary is composed of the following:

1. L, the smallest value in the data set
2. Q_1, the first quartile (also called P_{25}, the 25th percentile)
3. $\tilde{x}$, the median
4. Q_3, the third quartile (also called P_{75}, the 75th percentile)
5. H, the largest value in the data set

The 5-number summary for the set of 50 exam scores in Example 2.12 is

39	67	75.5	86	98
L	Q_1	$\tilde{x}$	Q_3	H

Notice that these five numerical values divide the set of data into four subsets, with one-quarter of the data in each subset. From the 5-number summary, we can observe how much the data are spread out in each of the quarters. We can now define an additional measure of dispersion.

Interquartile range: The difference between the first and third quartiles. It is the range of the middle 50% of the data.

The 5-number summary is even more informative when it is displayed on a diagram drawn to scale. A graphic display that accomplishes this is known as the *box-and-whiskers display.*

Box-and-whiskers display: A graphic representation of the 5-number summary. The five numerical values (smallest, first quartile, median, third quartile, and largest) are located on a scale, either vertical or horizontal. The box is used to depict the middle half of the data that lies between the two quartiles. The whiskers are line segments used to depict the other half of the data: One line segment represents the quarter of the data that is smaller in value than the first quartile, and a second line segment represents the quarter of the data that is larger in value than the third quartile.

Figure 2.30 is a box-and-whiskers display of the 50 exam scores.

FIGURE 2.30
Box-and-Whiskers
Display

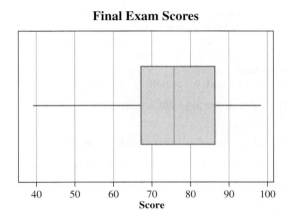

TECHNOLOGY INSTRUCTIONS: PERCENTILES

MINITAB (Release 14)   Input the data into C1; then continue with:

```
Choose:      Data > Sort ...
Enter:       Sort column(s): C1  By column: C1
Select:      Store sorted data in: Column(s) of current worksheet
Enter:       C2 > OK
```

A ranked list of data will be obtained in C2. Determine the depth position and locate the desired percentile.

Excel   Input the data into column A and activate a cell for the answer; then continue with:

```
Choose:      Insert Function, ƒ_x > Statistical > PERCENTILE > OK
Enter:       Array: (A2:A6 or select cells)
             k: K (desired percentile; ex. .95, .47)
```

TI-83/84 Plus   Input the data into L1; then continue with:

```
Choose:      STAT > EDIT > 2:SortA(
Enter:       L1
Enter:       percentile × sample size (ex. .25 × 100)
Based on product, determine the depth position; then continue with:
Enter:       L1(depth position) > Enter
```

TECHNOLOGY INSTRUCTIONS: 5-NUMBER SUMMARY

MINITAB (Release 14) Input the data into C1; then continue with:

Choose: **Stat > Basic Statistics > Display Descriptive Statistics ...**
Enter: **Variables: C1 > OK**

Excel Input the data into column A; then continue with:

Choose: **Tools > Data Analysis* > Descriptive Statistics > OK**
Enter: Input Range: **(A2:A6 or select cells)**
Select: **Labels in First Row** (if necessary)
            Output Range
              Enter: **(B1 or select cell)**
Select: **Summary Statistics > OK**
To make output readable:
Choose: **Format > Column > Autofit Selection**

*If Data Analysis does not show on the Tools menu; see page 62.

TI-83/84 Plus Input the data into L1; then continue with:

Choose: **STAT > CALC > 1:1-VAR STATS**
Enter: **L1**

TECHNOLOGY INSTRUCTIONS: BOX-AND-WHISKERS DIAGRAM

MINITAB (Release 14) Input the data into C1; then continue with:

Choose: **Graph > Boxplot ... > One Y, Simple > OK**
Enter: Graph variables: **C1**
Optional:
Select: **Labels > Titles/Footnoes**
Enter: **your title, footnotes > OK**
Select: **Scale > Axes and Ticks**
Select: **Transpose value and category scales > OK > OK**

For multiple boxplots, enter additional set of data into C2; then do as just described plus:

Choose: **Graph > Boxplot ... > Multiple Y's. Simple > OK**
Enter: Graph variables: **C1 C2 > OK**
Optional: See above.

Excel Input the data into column A; then continue with:

Choose: **Tools > Data Analysis Plus* > BoxPlot > OK**
Enter: **(A2:A6 or select cells)**

To edit the boxplot, review options shown with editing histograms on page 62.

*Data Analysis Plus is a collection of statistical macros for EXCEL. It can be downloaded onto your computer from your Student's Suite CD-ROM.

TI-83/84 Plus | Input the data into L1; then continue with:

Choose: **2nd > STAT PLOT >**
                1:Plot1...
Choose: **ZOOM > 9:ZoomStat > TRACE >>>**

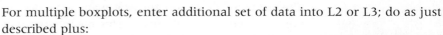

If class midpoints are in L1 and frequencies are in L2, do as just described except for:

Enter: Freq: **L2**

For multiple boxplots, enter additional set of data into L2 or L3; do as just described plus:

Choose: **2nd > STAT PLOT > 2:Plot2...**

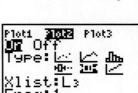

The position of a specific value can be measured in terms of the mean and standard deviation using the *standard score*, commonly called the *z-score*.

> **Standard score, or z-score:** The position a particular value of *x* has relative to the mean, measured in standard deviations. The *z*-score is found by the formula
>
> $$z = \frac{\text{value} - \text{mean}}{\text{st. dev.}} = \frac{x - \bar{x}}{s} \qquad (2.12)$$

EXAMPLE 2.14

Finding *z*-Scores

Find the standard scores for (a) 92 and (b) 72 with respect to a sample of exam grades that have a mean score of 75.9 and a standard deviation of 11.1.

SOLUTION

    a. $x = 92, \bar{x} = 75.9, s = 11.1$. Thus, $z = \dfrac{x - \bar{x}}{s} = \dfrac{92 - 75.9}{11.1} = \dfrac{16.1}{11.1} = \textbf{1.45.}$

    b. $x = 72, \bar{x} = 75.9, s = 11.1$. Thus, $z = \dfrac{x - \bar{x}}{s} = \dfrac{72 - 75.9}{11.1} = \dfrac{-3.9}{11.1} = \textbf{-0.35.}$

This means that the score 92 is approximately 1.5 standard deviations above the mean and the score 72 is approximately one-third of a standard deviation below the mean.

Notes:

1. Typically, the calculated value of *z* is rounded to the nearest hundredth.
2. *z*-scores typically range in value from approximately −3.00 to +3.00.

Because the *z*-score is a measure of relative position with respect to the mean, it can be used to help us compare two raw scores that come from separate populations. For example, suppose you want to compare a grade you received on a test with a friend's grade on a comparable exam in her course. You received a raw score of 45 points; she got 72 points. Is her grade better? We need more information before we can draw a conclusion. Suppose the mean on the exam you took was 38 and the mean on her exam was 65. Your grades are both 7 points above the mean, but we still can't draw a definite conclusion. The standard deviation on the exam you took was 7 points, and it was 14 points on your friend's exam. This means that your score is 1 standard deviation above the mean ($z = 1.0$), whereas your friend's grade is only 0.5 standard deviation above the mean ($z = 0.5$). Your score has the "better" relative position, so you conclude that your score is slightly better than your friend's score. (Again, this is speaking from a relative point of view.)

TECHNOLOGY INSTRUCTIONS: ADDITIONAL COMMANDS

MINITAB (Release 14)   Input the data into C1; then:

To sort the data into ascending order and store them in C2, continue with:

Choose:	**Data > Sort . . .**
Enter:	Sort column(s): **C1** By column: **C1**
Select:	Store sorted data in: **Column(s) of current worksheet**
Enter:	**C2 > OK**

To form an ungrouped frequency distribution of integer data, continue with:

Choose:	**Stat > Tables > Tally Individual Variables**
Enter:	Variables: **C1**
Select:	**Counts > OK**

To print data on the session window, continue with:

Choose:	**Data > Display Data**
Enter:	**Columns to display: C1** or **C1 C2** or **C1–C2 > OK**

Excel   Input the data into column A; then continue with the following to sort the data:

Choose:	**Data > Sort**
Enter:	Sort by: **(A2:A6 or select cells)**
Select:	**Ascending** or **Descending**
	Header row or **No header row**

TI-83/84 Plus   Input the data into L1; then continue with the following to sort the data:

Choose:	**2nd > STAT > OPS > 1:SortA(**
Enter:	**L1**

To form a frequency distribution of the data in L1, continue with:

Choose:	**PRGM > EXEC > FREQDIST***
Enter:	**L1 > ENTER**
	LW BOUND = **first lower class boundary**

UP BOUND = **last upper class boundary**
WIDTH = **class width** (use 1 for ungrouped distribution)

*The program 'FREQDIST' is one of many programs available for downloading from a website. See page 42 for specific instructions.

TECHNOLOGY INSTRUCTIONS: GENERATE RANDOM SAMPLES

MINITAB (Release 14)   The data will be put into C1:

Choose:	**Calc > Random Data > {Normal, Uniform, Integer, etc.}**
Enter:	Generate: **K** rows of data
	Store in column(s): **C1**
	Population parameters needed: (μ, σ, **L, H, A, or B**)
	> OK
	(Required parameters will vary depending on the distribution)

Excel

Choose:	**Tools > Data Analysis > Random Number Generation > OK**
Enter:	Number of Variables: **1**
	Number of Random Numbers: **(desired quantity)**
Select:	Distribution: **Normal, Integers, or others**
Enter:	Parameters: (μ, σ, **L, H, A, or B**)
	(Required parameters will vary depending on the distribution.)
Select:	**Output Range**
Enter:	**(A1 or select cell)**

TI-83/84 Plus

Choose:	**STAT > 1:EDIT**
Highlight:	**L1**
Choose:	**MATH > PRB > 6:randNorm(or 5:randInt(**
Enter:	μ, σ, **# of trials** or **L, H, # of trials**

TECHNOLOGY INSTRUCTIONS: SELECT RANDOM SAMPLES

MINITAB (Release 14)   The existing data to be selected from should be in C1; then continue with:

Choose:	**Calc > Random Data > Sample from Columns**
Enter:	Sample: **K** rows from column(s): **C1**
	Store samples in: **C2**
Select:	**Sample with replacement (optional) > OK**

Excel   The existing data to be selected from should be in column A; then continue with:

Choose:	**Tools > Data Analysis > Sampling > OK**
Enter:	Input range: **(A2:A10 or select cells)**
Select:	**Labels** (optional)
	Random
	Enter: Number of Samples: **K**
	Output range:
	Enter: **(B1 or select cell)**

APPLIED
EXAMPLE 2.15   The 85th Percentile Speed Limit

GOING WITH 85% OF THE FLOW

To the uninitiated, the "85th percentile rule" seems bizarre, unorthodox, and maybe even scary, but this speed limit benchmark has guided traffic engineers for decades and is even recognized as official policy in many government jurisdictions. The idea is that maximum speed limits should be set so that 85% of the vehicles on a particular stretch of road are at or below the limit. Under California policies, traffic engineers routinely measure how fast motorists drive and then often set the limit at the 85th percentile of traffic speed.

"The reasoning is that 85% of people drive reasonably and 15% do not," said David Roseman, city traffic engineer for Long Beach. "So we should be designing our speeds to accommodate reasonable drivers." Adds Tom Jones, principal traffic engineer for the city of Los Angeles, "The 85th percentile rule was established many years ago. It is a design criteria [sic], but it doesn't mean that it is necessarily OK."

Safety advocacy groups hate the 85th percentile rule, because they believe that speeding is a serious and growing highway hazard. Indeed, police are performing fewer routine traffic patrols and speeds are creeping up, according to studies published by safety groups. Barbara Harsha, executive director of the Governors Highway Safety Assn. in Washington, for example, is concerned that the 85th percentile rule can be used to legitimize unsafe speeding. When congestion is not limiting speeds, many sections of Southern California freeways have average speeds of more than 80 mph, well above the legal limit. Posted limits of 25 mph on residential streets are routinely ignored, according to neighborhood traffic studies.

"That just shows that legal speed limits are too low," says Chad Dornsife of the National Motorists Assn., a group that represents people generally unhappy with and often indignant about traffic laws and police enforcement. He says improperly set low speed limits actually increase accidents and cost lives, because they encourage unequal speeds, creating a hazard. He claims, for example, that when Montana imposed speed limits for the first time, fatal accidents doubled. A secondary problem Dornsife cites regarding artificially low speed limits is that yellow-light intervals are sometimes based on the posted limits, which leaves too little time for faster-moving cars to stop for a changing light before reaching an intersection. That, Dornsife claims, creates intersection collisions. "Every generation that has gone through this doesn't believe in the 85th percentile rule," he adds. "The law-enforcement community doesn't like the 85th percentile rule because they write fewer tickets. New traffic engineers aren't even taught the 85th percentile rule."

Source: Ralph Vartabedian, *Los Angeles Times* Staff Writer, March 9, 2005, http://www.latimes.com/classified/automotive/highway1/la-hy-wheels9mar09,1,6721856.story?ctrack=2&cset=true

DID YOU KNOW

The drivers in New York State drove a total of 135,046,000,000 miles on New York State roads in 2003. That's more than 5.4 million trips around the earth at the equator. That's a lot of driving!

SECTION 2.6 EXERCISES

Statistics⌓Now™

Datasets can be found on your Student's Suite CD-ROM or at the StatisticsNow website at **http://1pass.thomson.com**.

2.105 Refer to the table of exam scores in Table 2.16 on page 94 for the following.

a. Using the concept of depth, describe the position of 91 in the set of 50 exam scores in two different ways.

b. Find P_{20} and P_{35} for the exam scores in Table 2.16 on page 94.

c. Find P_{80} and P_{95} for the exam scores in Table 2.16.

2.106 [EX02-106] Following are the American College Test (ACT) scores attained by the 25 members of a local high school graduating class:

21	24	23	17	31	19	19	20	19	25	17	23	16
21	20	28	25	25	21	14	19	17	18	28	20	

a. Draw a dotplot of the ACT scores.

b. Using the concept of depth, describe the position of 24 in the set of 25 ACT scores in two different ways.

c. Find P_5, P_{10}, and P_{20} for the ACT scores.

d. Find P_{99}, P_{90}, and P_{80} for the ACT scores.

2.107 [EX02-107] The annual salaries (in $100) of the kindergarten and elementary school teachers employed at one of the elementary schools in the local school district are listed here:

574	434	455	413	391	471	458	269	501
326	367	433	367	495	376	371	295	317

a. Draw a dotplot of the salaries.

b. Using the concept of depth, describe the position of 295 in the set of 18 salaries in two different ways.

c. Find Q_1 for these salaries.

d. Find Q_3 for these salaries.

2.108 [EX02-108] Fifteen countries were randomly selected from the *World Factbook 2004* list of world countries, and the infant mortality per 1000 live births rate was recorded:

6.38	101.68	9.48	69.18	64.19	3.73	21.31	52.71
13.43	29.64	15.24	5.85	11.74	9.67	8.68	

Source: The World Factbook 2004

a. Find the first and third quartile for the infant mortality per 1000 rate.

b. Find the midquartile.

2.109 [EX02-109] The following data are the yields (in pounds) of hops:

3.9	3.4	5.1	2.7	4.4	7.0	5.6	2.6	4.8	5.6
7.0	4.8	5.0	6.8	4.8	3.7	5.8	3.6	4.0	5.6

a. Find the first and the third quartiles of the yields.

b. Find the midquartile.

c. Find and explain the percentiles P_{15}, P_{33}, and P_{90}.

2.110 [EX02-110] A research study of manual dexterity involved determining the time required to complete a task. The time required for each of 40 individuals with disabilities is shown here (data are ranked):

7.1	7.2	7.2	7.6	7.6	7.9	8.1	8.1	8.1	8.3	8.3	8.4	8.4	8.9
9.0	9.0	9.1	9.1	9.1	9.1	9.4	9.6	9.9	10.1	10.1	10.1	10.2	
10.3	10.5	10.7	11.0	11.1	11.2	11.2	11.2	12.0	13.6	14.7	14.9	15.5	

a. Find Q_1.　　　　b. Find Q_2.

c. Find Q_3.　　　　d. Find P_{95}.

e. Find the 5-number summary.

f. Draw the box-and-whisker display.

2.111 Draw a box-and-whiskers display for the set of data with the 5-number summary 42–62–72–82–97.

2.112 [EX02-112] The U.S. Geological Survey collected atmospheric deposition data in the Rocky Mountains. Part of the sampling process was to determine the concentration of ammonium ions (in percentages). Here are the results from the 52 samples:

2.9	4.1	2.7	3.5	1.4	5.6	13.3	3.9	4.0
2.9	7.0	4.2	4.9	4.6	3.5	3.7	3.3	5.7
3.2	4.2	4.4	6.5	3.1	5.2	2.6	2.4	5.2
4.8	4.8	3.9	3.7	2.8	4.8	2.7	4.2	2.9
2.8	3.4	4.0	4.6	3.0	2.3	4.4	3.1	5.5
4.1	4.5	4.6	4.7	3.6	2.6	4.0		

a. Find Q_1.　　b. Find Q_2.

c. Find Q_3.　　d. Find the midquartile.

e. Find P_{30}.　　f. Find the 5-number summary.

g. Draw the box-and-whiskers display.

2.113 [EX02-113] The NCAA men's basketball "Big Dance" kicks into full gear every March. If you look at the graduation rate of these athletes, however, you will find that many teams do not make the grade, according to a study released in

March 2005. Following are the graduation rates for 64 of the 2005 tournament teams.

Graduation Rates (%), 2005 Men's Teams, NCAA Division I Basketball Tournament

40	38	100	55	44	58	30	11	40	19	43	27	0
64	75	58	54	44	40	11	50	30	40	71	92	53
33	29	40	25	33	43	25	27	58	47	55	60	8
17	17	0	40	25	14	67	45	33	20	15	50	27
29	57	36	45	45	73	15	100	67	44	57	55	

Source: 2004 NCAA Graduation-Rates Report

a. Draw a dotplot of the graduation rate data.

b. Draw a stem-and-leaf display of these data.

c. Find the 5-number summary and draw a box-and-whiskers display.

d. Find P_5 and P_{95}.

e. Describe the distribution of graduation rates, being sure to include information learned in parts a through d.

f. Are there teams whose graduation rates appear to be quite different from the rest? How many? Which ones? Explain.

2.114 [EX02-114] The fatality rate on the nation's highways in 2003 was the lowest since record keeping began 29 years ago, but these numbers are still mind-boggling. The number of persons killed in motor vehicle traffic crashes, by state, including the District of Columbia, in 2003 is listed here:

1001	294	135	471	462	262	439	668	203	943
95	142	293	928	1283	293	1491	512	1193	600
1120	67	1453	894	657	368	1531	1577	3675	394
627	3169	834	207	871	127	105	104	309	848
4215	1603	441	649	1232	747	1277	968	69	165
632									

Source: Road & Travel Magazine, 2004

a. Draw a dotplot of the fatality data.

b. Draw a stem-and-leaf display of these data.

c. Find the 5-number summary and draw a box-and-whiskers display. Describe how the three large-valued data are handled.

d. Find P_{10} and P_{90}.

e. Describe the distribution of number of fatalities per state, being sure to include information learned in parts a through d.

f. Why might it be unfair to draw conclusions about the relative safety level of highways in the 50 states and District of Columbia based on these data?

2.115 [EX02-115] Are airline flight arrivals ever on time? The general public thinks they are always late—but are they? The U.S. Bureau of Transportation Statistics keeps records and periodically reports the findings. Listed here are the percentages of on-time arrivals at the 31 major U.S. airports for the period January 1, 2004, to October 31, 2004.

ATL	73.55	BOS	78.38	BWI	80.91

••• Remainder of data on Student's Suite CD-ROM

Source: U.S. Department of Transportation, Bureau of Transportation Statistics

a. Draw a dotplot of on-time performance data.

b. Draw a stem-and-leaf display of these data.

c. Find the 5-number summary and draw a box-and-whiskers display.

d. Find P_{10} and P_{20}.

e. Describe the distribution of on-time percentage, being sure to include information learned in parts a through d.

f. Why would you be more likely to talk about the top 80% or 90% of the performance percentages than the middle 80% or 90%?

g. Are there airports whose on-time percentages appear to be quite different from the rest? How many? Which ones? Explain.

2.116 [EX02-116] Major league baseball stadiums vary in age, style, number of seats, and many other ways, but to the baseball players, the size of the field is of the utmost importance. Suppose we agree to measure the size of the field by using the distance from home plate to the centerfield fence. Following is the distance (in feet) to the centerfield fence in the 30 major league stadiums.

422	405	400	400	400	402	404	435	399	410
400	400	400	400	408	401	395	410	410	401
420	408	405	410	402	415	400	404	405	400

Source: http://mlb.mlb.com

a. Construct a histogram.

b. The interquartile range is described by the bounds of the middle 50% of the data, Q_1 and Q_3. Find the interquartile range.

c. Are there any fields that appear to be considerably smaller or larger than the others?

d. Is there a great deal of difference in the size of these 30 fields as measured by the distance to centerfield? Justify your answer with statistical evidence.

2.117 What property does the distribution need for the median, the midrange, and the midquartile to all be the same value?

2.118 [EX02-118] Henry Cavendish, an English chemist and physicist (1731–1810), approached many of his experiments using quantitative measurements. He was the first to accurately measure the density of the Earth. Following are 29 measurements (ranked for your convenience) of the density of the Earth done by Cavendish in 1798 using a torsion balance. Density is presented as a multiple of the density of water. (Measurements are in g/cm³.)

4.88	5.07	5.10	5.26	5.27	5.29	5.29	5.30	5.34	5.34
5.36	5.39	5.42	5.44	5.46	5.47	5.50	5.53	5.55	5.57
5.58	5.61	5.62	5.63	5.65	5.68	5.75	5.79	5.85	

Source: The data and descriptive information are based on material from "Do robust estimators work with real data?" by Stephen M. Stigler, *Annals of Statistics* 5 (1977), 1055–1098.

a. Describe the data set by calculating the mean, median, and standard deviation.

b. Construct a histogram and explain how it demonstrates the values of the descriptive statistics in part a.

c. Find the 5-number summary.

d. Construct a box-and-whiskers display and explain how it demonstrates the values of the descriptive statistics in part c.

e. Based on the two graphs, what "shape" is this distribution of measurements?

f. Assuming that Earth density measurements have an approximately normal distribution, approximately 95% of the data should fall within 2 standard deviations of the mean. Is this true?

2.119 Find the z-score for test scores of 92 and 63 on a test that has a mean of 72 and a standard deviation of 12.

2.120 A sample has a mean of 50 and a standard deviation of 4.0. Find the z-score for each value of x:

a. $x = 54$ b. $x = 50$

c. $x = 59$ d. $x = 45$

2.121 An exam produced grades with a mean score of 74.2 and a standard deviation of 11.5. Find the z-score for each test score x:

a. $x = 54$ b. $x = 68$

c. $x = 79$ d. $x = 93$

2.122 A nationally administered test has a mean of 500 and a standard deviation of 100. If your standard score on this test was 1.8, what was your test score?

2.123 A sample has a mean of 120 and a standard deviation of 20.0. Find the value of x that corresponds to each of these standard scores:

a. $z = 0.0$ b. $z = 1.2$

c. $z = -1.4$ d. $z = 2.05$

2.124

a. What does it mean to say that $x = 152$ has a standard score of $+1.5$?

b. What does it mean to say that a particular value of x has a z-score of -2.1?

c. In general, the standard score is a measure of what?

2.125 [EX02-098] Consider the percentage of interstate and state-owned bridges that were structurally deficient or functionally obsolete (SD/FO) listed in Exercise 2.98 on page 92.

a. Omit the names of the states and rank the SD/FO values in ascending order, reading horizontally in each row.

b. Construct a 5-number summary table and the corresponding box-and-whiskers display.

c. Find the midquartile percentage and the interquartile range.

d. What are the z-scores for California, Hawaii, Nebraska, Oklahoma, and Rhode Island?

2.126 The ACT Assessment is designed to assess high school students' general educational development and their ability to complete college-level work. The following table lists the mean and standard deviation of scores attained by the 1,171,460 high school students of the 2004 graduating class who took the ACT exams.

2004	English	Mathematics	Reading	Science Reasoning	Composite
Mean	20.4	20.7	21.3	20.9	20.9
Std. dev.	5.9	5.0	6.0	4.6	4.8

Source: American College Testing

Convert the following ACT scores to z-scores for both English and math. Compare placement between the two tests.

a. $x = 30$ b. $x = 23$ c. $x = 12$

d. Explain why the relative positions in English and math changed for the ACT scores of 30 and 12.

e. If Jessica had a 26 on one of the ACT exams, on which one of the exams would she have the best possible relative score? Explain why.

2.127 Which x value has the higher position relative to the set of data from which it comes?

> A: $x = 85$, where mean $= 72$ and standard deviation $= 8$
>
> B: $x = 93$, where mean $= 87$ and standard deviation $= 5$

2.128 Which x value has the lower position relative to the set of data from which it comes?

> A: $x = 28.1$, where $\bar{x} = 25.7$ and $s = 1.8$
>
> B: $x = 39.2$, where $\bar{x} = 34.1$ and $s = 4.3$

2.7 Interpreting and Understanding Standard Deviation

Standard deviation is a measure of variation (dispersion) in the data. It has been defined as a value calculated with the use of formulas. Even so, you may be wondering what it really is and how it relates to the data. It is a kind of yardstick by which we can compare the variability of one set of data with that of another. This particular "measure" can be understood further by examining two statements that tell us how the standard deviation relates to the data: the *empirical rule* and *Chebyshev's theorem*.

The Empirical Rule and Testing for Normality

Empirical rule: If a variable is normally distributed, then (1) within 1 standard deviation of the mean, there will be approximately 68% of the data; (2) within 2 standard deviations of the mean, there will be approximately 95% of the data; and (3) within 3 standard deviations of the mean, there will be approximately 99.7% of the data. (This rule applies specifically to a normal [bell-shaped] distribution, but it is frequently applied as an interpretive guide to any mounded distribution.)

Figure 2.31 shows the intervals of 1, 2, and 3 standard deviations about the mean of an approximately normal distribution. Usually these proportions do

FIGURE 2.31
Empirical Rule

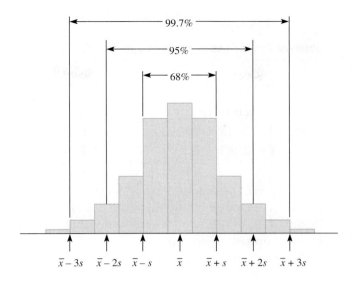

not occur exactly in a sample, but your observed values will be close when a large sample is drawn from a normally distributed population.

If a distribution is approximately normal, it will be nearly symmetrical and the mean will divide the distribution in half (the mean and the median are the same in a symmetrical distribution). This allows us to refine the empirical rule, as shown in Figure 2.32.

FIGURE 2.32
Refinement of
Empirical Rule

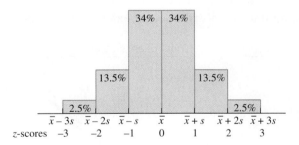

The empirical rule can be used to determine whether a set of data is approximately normally distributed. Let's demonstrate this application by working with the distribution of final exam scores that we have been using throughout this chapter. The mean, $\bar{x}$, was found to be 75.6, and the standard deviation, s, was 14.9. The interval from 1 standard deviation below the mean, $\bar{x} - s$, to 1 standard deviation above the mean, $\bar{x} + s$, is $75.6 - 14.9 = 60.7$ to $75.6 + 14.9 = 90.5$. This interval (60.7 to 90.5) includes 61, 62, 63, . . . , 89, 90. Upon inspection of the ranked data (Table 2.16, p. 94), we see that 35 of the 50 data, or 70%, lie within 1 standard deviation of the mean. Furthermore, $\bar{x} - 2s = 75.6 - (2)(14.9) = 75.6 - 29.8 = 45.8$ to $\bar{x} + 2s = 75.6 + 29.8 = 105.4$ gives the interval from 45.8 to 105.4. Of the 50 data, 48, or 96%, lie within 2 standard deviations of the mean. All 50 data, or 100%, are included within 3 standard deviations of the mean (from 30.9 to 120.3). This information can be placed in a table for comparison with the values given by the empirical rule (see Table 2.17).

TABLE 2.17

Observed Percentages versus the Empirical Rule

Interval	Empirical Rule Percentage	Percentage Found
$\bar{x} - s$ to $\bar{x} + s$	≈ 68	70
$\bar{x} - 2s$ to $\bar{x} + 2s$	≈ 95	96
$\bar{x} - 3s$ to $\bar{x} + 3s$	≈ 99.7	100

The percentages found are reasonably close to those predicted by the empirical rule. By combining this evidence with the shape of the histogram (see Figure 2.10, p. 60), we can safely say that the final exam data are approximately normally distributed.

There is another way to test for normality—by drawing a probability plot (an ogive drawn on probability paper*) using a computer or graphing calculator. For our illustration, a probability plot of the statistics final exam scores is shown on Figure 2.33. The test for normality, at this point in our study of statistics, is simply to compare the graph of the data (the ogive) with the straight line drawn from the lower left corner to the upper right corner of the graph. If the ogive lies close to this straight line, the distribution is said to be approximately normal. The vertical scale used to construct the probability plot is adjusted so that the ogive for an exactly normal distribution will trace the straight line. The ogive of the exam scores follows the straight line quite closely, suggesting that the distribution of exam scores is approximately normal.

FIGURE 2.33
Probability Plot of
Statistics Exam
Scores

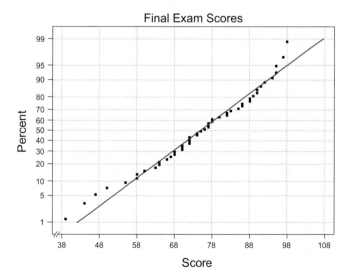

If using a computer, you will obtain an additional piece of information in determining normality. This piece of information comes in the form of a *p*-

---

*On probability paper the vertical scale is not uniform; it has been adjusted to account for the mounded shape of a normal distribution and its cumulative percentages.

value, and if its value is greater than 0.05, you can assume the sample was drawn from an approximately normal distribution (if p-value ≤ 0.05, not normal). (p-value will be more fully defined in Chapter 8, Section 8.5.)

TECHNOLOGY INSTRUCTIONS: TESTING FOR NORMALITY

MINITAB (Release 14)   Input the data into C1; then continue with:

```
Choose:     Stat > Basic Statistics > Normality Test
Enter:      Variable: C1
            Title: your title > OK
```

Excel   Excel uses a test for normality, not the probability plot.
Input the data into column A; then continue with:

```
Choose:     Tools > Data Analysis Plus > Chi-Squared Test of Normality >
            OK
Enter:      Input Range: data (A1:A6 or select cells)
Select:     Labels (if column headings were used) > OK
```

Expected values for a normal distribution are given versus the given distribution. If the p-value is greater than 0.05, then the given distribution is approximately normal.

TI-83/84 Plus   Input the data into L1; then continue with:

```
Choose:     Window
Enter:      at most the smallest data value, at least the largest data
            value, x scale, -5, 5, 1,1
Choose:     2nd > STAT PLOT > 1:Plot
```

Chebyshev's Theorem

In the event that the data do not display an approximately normal distribution, Chebyshev's theorem gives us information about how much of the data will fall within intervals centered at the mean for all distributions.

> **Chebyshev's theorem:** The proportion of any distribution that lies within k standard deviations of the mean is at least $1 - \dfrac{1}{k^2}$, where k is any positive number greater than 1. This theorem applies to all distributions of data.

This theorem says that within 2 standard deviations of the mean ($k = 2$), you will always find at least 75% (that is, 75% or more) of the data:

$$1 - \frac{1}{k^2} = 1 - \frac{1}{2^2} = 1 - \frac{1}{4} = \frac{3}{4} = 0.75, \text{ at least 75\%}$$

Figure 2.34 shows a mounded distribution that illustrates at least 75%.

If we consider the interval enclosed by 3 standard deviations on either side of the mean ($k = 3$), the theorem says that we will always find at least 89% (that is, 89% or more) of the data:

$$1 - \frac{1}{k^2} = 1 - \frac{1}{3^2} = 1 - \frac{1}{9} = \frac{8}{9} = 0.89, \text{ at least 89\%}$$

Figure 2.35 shows a mounded distribution that illustrates at least 89%.

FIGURE 2.34 Chebyshev's Theorem with $k = 2$ **FIGURE 2.35** Chebyshev's Theorem with $k = 3$

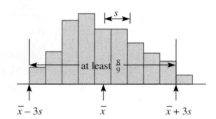

Let's revisit the results of the physical-fitness strength test given to the third-graders in Exercise 2.45 on page 70. Their test results are listed here in rank order and shown on the histogram. [EX02-045]

1	2	2	3	3	3	4	4	4	5	5	5	5	6	6	6
8	9	9	9	9	9	9	10	10	11	12	12	12	13	14	14
14	15	15	15	15	16	16	16	17	17	17	17	18	18	18	18
19	19	19	19	20	20	20	21	21	21	22	22	22	23	24	24

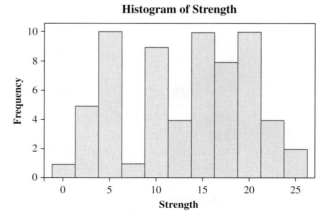

Histogram of Strength

Some questions of interest are: Does this distribution satisfy the empirical rule? Does Chebyshev's theorem hold true? Is this distribution approximately normal?

To answer the first two questions, we need to find the percent of data in each of the three intervals about the mean. The mean is 13.0, and the standard deviation is 6.6.

mean $\pm k$ (Std. Dev.)	Interval	Percentage Found	Empirical	Chebyshev
$13.0 \pm 1(6.6)$	6.4 to 19.6	$39/64 = 60.9\%$	68%	–
$13.0 \pm 2(6.6)$	-0.2 to 26.2	$64/64 = 100\%$	95%	At least 75%
$13.0 \pm 3(6.6)$	-6.8 to 32.8	$64/64 = 100\%$	99.70%	At least 89%

It is left to you to verify the values of the mean, standard deviation, the intervals, and the percentages.

The three percentages found (60.9, 100, and 100) do not approximate the 68, 95, and 99.7 percentages stated in the empirical rule. The two percentages found (100 and 100) do agree with Chebyshev's theorem in that they are greater than 75% and 89%. Remember, Chebyshev's theorem holds for all distributions.

The normality test, introduced on pages 108 and 109, yields a p-value of 0.009, and along with the distribution seen on the histogram and the three percentages found, it is reasonable to conclude that these test results are not normally distributed.

SECTION 2.7 EXERCISES

Statistics⬡Now™

Datasets can be found on your Student's Suite CD-ROM or at the StatisticsNow website at **http://1pass.thomson.com**.

2.129 Instructions for an essay assignment include the statement "The length is to be within 25 words of 200." What values of x, number of words, satisfy these instructions?

2.130 The empirical rule indicates that we can expect to find what proportion of the sample included between the following:

a. $\bar{x} - s$ and $\bar{x} + s$ b. $\bar{x} - 2s$ and $\bar{x} + 2s$

c. $\bar{x} - 3s$ and $\bar{x} + 3s$

2.131 Why is it that the z-score for a value that belongs to a normal distribution usually lies between -3 and $+3$?

2.132 The mean lifetime of a certain tire is 30,000 miles and the standard deviation is 2500 miles.

a. If we assume the mileages are normally distributed, approximately what percentage of all such tires will last between 22,500 and 37,500 miles?

b. If we assume nothing about the shape of the distribution, approximately what percentage of all such tires will last between 22,500 and 37,500 miles?

2.133 The average clean-up time for a crew of a medium-size firm is 84.0 hours and the standard deviation is 6.8 hours. Assume the empirical rule is appropriate.

a. What proportion of the time will it take the clean-up crew 97.6 hours or more to clean the plant?

b. Within what interval will the total clean-up time fall 95% of the time?

2.134 a. What proportion of a normal distribution is greater than the mean?

b. What proportion is within 1 standard deviation of the mean?

c.   What proportion is greater than a value that is 1 standard deviation below the mean?

2.135 Using the empirical rule, determine the approximate percentage of a normal distribution that is expected to fall within the interval described.

a.   Less than the mean

b.   Greater than 1 standard deviation above the mean

c.   Less than 1 standard deviation above the mean

d.   Between 1 standard deviation below the mean and 2 standard deviations above the mean

2.136 According to the empirical rule, almost all the data should lie between $(\bar{x} - 3s)$ and $(\bar{x} + 3s)$. The range accounts for all the data.

a.   What relationship should hold (approximately) between the standard deviation and the range?

b.   How can you use the results of part a to estimate the standard deviation in situations when the range is known?

2.137 Chebyshev's theorem guarantees that what proportion of a distribution will be included between the following:

a. $\bar{x} - 2s$ and $\bar{x} + 2s$ b. $\bar{x} - 3s$ and $\bar{x} + 3s$

2.138 According to Chebyshev's theorem, what proportion of a distribution will be within $k = 4$ standard deviations of the mean?

2.139 Chebyshev's theorem can be stated in an equivalent form to that given on page 109. For example, to say "at least 75% of the data fall within 2 standard deviations of the mean" is equivalent to stating "at most, 25% will be more than 2 standard deviations away from the mean."

a.   At most, what percentage of a distribution will be 3 or more standard deviations from the mean?

b.   At most, what percentage of a distribution will be 4 or more standard deviations from the mean?

2.140 The scores achieved by students in America make the news often, and all kinds of conclusions are drawn based on these scores. The ACT Assessment is designed to assess high school students' general educational development and their ability to complete college-level work. One of the categories tested is science reasoning. The mean ACT score for all high school graduates in 2004 in science reasoning was 20.9, with a standard deviation of 4.6.

a.   According to Chebyshev's theorem, at least what percent of high school graduates' ACT scores in science reasoning is between 11.7 and 30.1?

b.   If we know that ACT scores are normally distributed, what percent of ACT science reasoning scores are between 11.7 and 30.1?

2.141 **[EX02-141]** On the first day of class last semester, 50 students were asked for the one-way distance from home to college (to the nearest mile). The resulting data follow:

6	5	3	24	15	15	6	2	1	3
5	10	9	21	8	10	9	14	16	16
10	21	20	15	9	4	12	27	10	10
3	9	17	6	11	10	12	5	7	11
5	8	22	20	13	1	8	13	4	18

a.   Construct a grouped frequency distribution of the data by using 1–4 as the first class.

b.   Calculate the mean and the standard deviation.

c.   Determine the values of $\bar{x} \pm 2s$, and determine the percentage of data within 2 standard deviations of the mean.

2.142 **[EX02-142]** One of the many things the U.S. Census Bureau reports to the public is the increase in population for various geographic areas within the country. The percent of increase in population for the 100 fastest-growing counties in the United States from April 1, 2000, to July 1, 2003, are listed in the following table.

15.4	12.1	13.0	14.8	21.5

••• Remainder of data on Student's Suite CD-ROM

Source: http://www.census.gov/popest/counties/CO-EST2003-09.html

a. Calculate the mean and standard deviation.

b. Sort the data into a ranked list.

c. Determine the values of $\bar{x} \pm s$, $\bar{x} \pm 2s$, and $\bar{x} \pm 3s$, and determine the percentage of data within 1, 2, and 3 standard deviations of the mean.

d. Do the percentages found in part c agree with the empirical rule? What does that mean?

e. Do the percentages found in part c agree with Chebyshev's theorem? What does that mean?

f. Construct a histogram and one other graph of your choice. Do the graphs show a distribution that agrees with your answers in parts d and e? Explain.

2.143 [EX02-143] Each year, NCAA college football fans like to learn about the up-and-coming freshman class of players. Following are the heights (in inches) of the nation's top 100 high school football players for 2005, as rated by recruiting analyst Tom Lemming of ESPN.com:

75	70	71	75	76	76	70	72	70	75	75	68	73	75	74
74	73	75	77	75	72	77	73	73	72	71	78	79	80	74
73	79	78	77	73	74	74	72	73	75	68	72	72	73	72
70	76	73	74	76	74	76	74	78	75	77	77	78	74	73
76	70	76	77	77	70	73	75	76	73	75	76	78	75	71
78	75	76	77	78	75	76	74	74	79	73	74	76	71	74
71	76	76	74	76	76	76	72	76	73

Source: ESPN.com

a. Construct a histogram and one other graph of your choice that displays the distribution of heights.

b. Calculate the mean and standard deviation.

c. Sort the data into a ranked list.

d. Determine the values of $\bar{x} \pm s$, $\bar{x} \pm 2s$, and $\bar{x} \pm 3s$, and determine the percentage of data within 1, 2, and 3 standard deviations of the mean.

e. Do the percentages found in part d agree with the empirical rule? What does this imply? Explain.

f. Do the percentages found in part d agree with Chebyshev's theorem? What does that mean?

g. Do the graphs show a distribution that agrees with your answers in part e? Explain.

2.144 [EX02-144] Each year, NCAA college football fans like to learn about the size of the players in the current year's recruit class. Following are the weights (in pounds) of the nation's top 100 high school football players for 2005, as rated by recruiting analyst Tom Lemming of ESPN.com. Just because these data are taken from the same 100 football players as in Exercise 2.143 does not mean the distributions will be the same. In fact, they are quite different, as you are about to see.

207	220	218	215	215

••• Remainder of data on Student's Suite CD-ROM

Source: ESPN.com

a. Construct a histogram and one other graph of your choice that displays the distribution of weights.

b. Calculate the mean and standard deviation.

c. Sort the data into a ranked list.

d. Determine the values of $\bar{x} \pm s$, $\bar{x} \pm 2s$, and $\bar{x} \pm 3s$, and determine the percentage of data within 1, 2, and 3 standard deviations of the mean.

e. Do the percentages found in part d agree with the empirical rule? What does this imply? Explain.

f. Do the percentages found in part d agree with Chebyshev's theorem? What does that mean?

g. Do the graphs show a distribution that agrees with your answers in part e? Explain.

2.145 The empirical rule states that the 1, 2, and 3 standard deviation intervals about the mean will contain 68%, 95%, and 99.7%, respectively.

a. Use the computer or calculator commands on page 101 to randomly generate a sample of 100 data from a normal distribution with mean 50 and standard deviation 10. Construct a histogram using class boundaries that are multiples of the standard deviation 10; that is, use boundaries from 10 to 90 in intervals of 10 (see the commands on pp. 61–62). Calculate the mean and the standard deviation using the commands found on pages 74 and 88; then inspect the histogram to determine the percentage of the data that fell within each of the 1, 2, and 3 standard deviation intervals. How closely do the three percentages compare to the percentages claimed in the empirical rule?

b. Repeat part a. Did you get results similar to those in part a? Explain.

c. Consider repeating part a several more times. Are the results similar each time? If so, in what way?

d. What do you conclude about the truth of the empirical rule?

2.146 Chebyshev's theorem states that "at least $1 - \dfrac{1}{k^2}$" of the data of a distribution will lie within k standard deviations of the mean.

a. Use the computer commands on page 101 to randomly generate a sample of 100 data from a uniform (nonnormal) distribution that has a low value of 1 and a high value of 10. Con-struct a histogram using class boundaries of 0 to 11 in increments of 1 (see the commands on pp. 61–62). Calculate the mean and the standard deviation using the commands found on pages 74 and 88; then inspect the histogram to determine the percentage of the data that fell within each of the 1, 2, 3, and 4 standard deviation intervals. How closely do these percentages compare to the percentages claimed in Chebyshev's theorem and in the empirical rule?

b. Repeat part a. Did you get results similar to those in part a? Explain.

c. Consider repeating part a several more times. Are the results similar each time? If so, in what way are they similar?

d. What do you conclude about the truth of Chebyshev's theorem and the empirical rule?

2.8   The Art of Statistical Deception

"There are three kinds of lies—lies, damned lies, and statistics." These remarkable words spoken by Benjamin Disraeli (19th-century British prime minister) represent the cynical view of statistics held by many people. Most people are on the consumer end of statistics and therefore have to "swallow" them.

Good Arithmetic, Bad Statistics

Let's explore an outright statistical lie. Suppose a small business employs eight people who earn between $300 and $350 per week. The owner of the business pays himself $1250 per week. He reports to the general public that the average wage paid to the employees of his firm is $430 per week. That may be an example of good arithmetic, but it is bad statistics. It is a misrepresentation of the situation because only one employee, the owner, receives more than the mean salary. The public will think that most of the employees earn about $430 per week.

Graphic Deception

Graphic representations can be tricky and misleading. The frequency scale (which is usually the vertical axis) should start at zero in order to present a total picture. Usually, graphs that do not start at zero are used to save space. Nevertheless, this can be deceptive. Graphs in which the frequency scale starts at zero tend to emphasize the size of the numbers involved, whereas graphs that are chopped off may tend to emphasize the variation in the numbers without regard to the actual size of the numbers. The labeling of the horizontal scale can be misleading also. You need to inspect graphic presentations very carefully before you draw any conclusions from the "story being told."

The following two Applied Examples will demonstrate some of these misrepresentations.

Superimposed Misrepresentation

APPLIED
EXAMPLE 2.16

Claiming What the Reader Expects/Anticipated Bad News

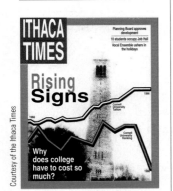

Courtesy of the Ithaca Times

Source: http://www.math.yorku.ca/SCS/Gallery/context.html

This "clever" graphic overlay, from the *Ithaca Times* (December 7, 2000), has to be the worst graph ever to make a front page. The cover story, "Why does college have to cost so much?" pictures two graphs superimposed on a Cornell University campus scene. The two broken lines represent "Cornell's Tuition" and "Cornell's Ranking," with the tuition steadily increasing and the ranking staggering and falling. A very clear image is created: Students get less, pay more!

Now view the two graphs separately. Notice: (1) The graphs cover two different time periods. (2) The vertical scales differ. (3) The "best" misrepresentation comes from the impression that a "drop in rank" represents a lower quality of education. Wouldn't a rank of 6 be better than a rank of 15?

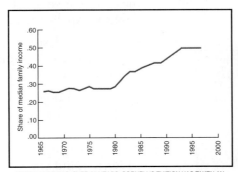

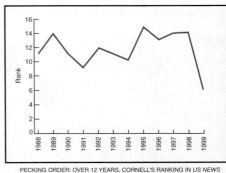

BY THE NUMBERS: OVER 35 YEARS, CORNELL'S TUITION HAS TAKEN AN INCREASINGLY LARGER SHARE OF ITS MEDIAN STUDENT FAMILY INCOME

PECKING ORDER: OVER 12 YEARS, CORNELL'S RANKING IN *US NEWS & WORLD REPORT* HAS RISEN AND FALLEN ERRATICALLY.

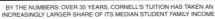

Source: http://www.math.yorku.ca/SCS/Gallery/context.html

What it all comes down to is that statistics, like all languages, can be and is abused. In the hands of the careless, the unknowledgeable, or the unscrupulous, statistical information can be as false as "damned lies."

Truncated Scale

APPLIED
EXAMPLE 2.17

Simple Is Not Always Best

This graphic is neat and very readable, but does it represent the information being shown? Truncating scales on graphs often leads to misleading visual impressions. For example, in "Service contractor complaints," it appears that "Take too long" is twice as likely to be the complaint as "Messy." Look for other visual misrepresentations.

SERVICE CONTRACTOR COMPLAINTS

Show up late
33%

Have to come back
30%

Take too long
27%

Messy
18%

Data from USA Today, 9/5/2001.

SECTION 2.8 EXERCISES

2.147 Is it possible for eight employees to earn between $300 and $350, a ninth to earn $1250 per week, and the mean to be $430? Verify your answer.

2.148 The graphic "Valentine's Day spending plan" shows a relative frequency distribution. This graph qualifies as a "tricky graph."

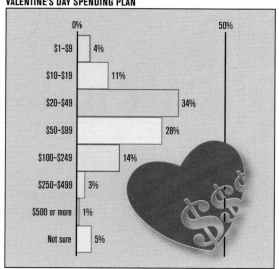

VALENTINE'S DAY SPENDING PLAN

0%   50%

$1–$9   4%
$10–$19   11%
$20–$49   34%
$50–$99   28%
$100–$249   14%
$250–$499   3%
$500 or more 1%
Not sure   5%

Data from Darryl Haralson and Julie Snider. © 2005 USA Today. Margin of Error ±3 percentage points.

a. Is the graph a bar graph or a histogram? Explain.

b. How does this graph violate the guidelines for drawing histograms?

2.149 "What's wrong with this picture?" That's the question one should be asking when viewing the graphs in Applied Example 2.16 on page 115.

a. Find and describe at least four features about the *Ithaca Times* front-page graph that are incorrectly used.

b. Find and describe at least two features about the "Pecking Order" graph that are misrepresenting.

2.150 a. Find and describe at least four incorrect impressions created by truncating the horizontal axis on the Applied Example 2.17's "Service contractor complaints" graphic on page 116.

b. Redraw the bar graph starting the horizontal scale at "zero."

c. Comment on the effect that your graph has on the impression presented.

2.151 The best value for price was most often reported by consumers as one of the draws when deciding where to do their holiday shopping. When asked, "What draws holiday shoppers to stores?" at Christmas time in 2004, they responded as follows.

What	Percent
Value	76
Convenient location	68
Quality	62
Selection	60

Source: USA Today and NPD Group

Prepare two bar graphs to depict the percentage data. Scale the vertical axis on the first graph from 50 to 80. Scale the second graph from 0 to 80. What is your conclusion concerning how the percentages of the four responses stack up based on the two bar graphs, and what would you recommend, if anything, to improve the presentations?

2.152 Find an article or an advertisement containing a graph that in some way misrepresents the information of statistics. Describe how this graph misrepresents the facts.

<table><tr><td>**2.9**</td><td></td></tr></table>

Mean and Standard Deviation of Frequency Distribution (optional)

When the sample data are in the form of a frequency distribution, we need to make a slight adaptation to formulas (2.1) and (2.10) in order to find the mean, the variance, and the standard deviation.

EXAMPLE 2.18

Calculations Using a Frequency Distribution

TABLE 2.18

Ungrouped Frequency Distribution

x	f
1	5
2	9
3	8
4	6
$\sum f = 28$	

Find the mean, the variance, and the standard deviation for the sample data represented by the frequency distribution in Table 2.18.

Note: This frequency distribution represents a sample of 28 values: five 1s, nine 2s, eight 3s, and six 4s.

In order to calculate the sample mean $\bar{x}$ and the sample variance s^2 using formulas (2.1) and (2.10), we need the sum of the 28 x values, $\sum x$, and the sum of the 28 x-squared values, $\sum x^2$.

The summations, $\sum x$ and $\sum x^2$, could be found as follows:

$$\sum x = \underbrace{1 + 1 + \cdots + 1}_{\text{5 of them}} + \underbrace{2 + 2 + \cdots + 2}_{\text{9 of them}} + \underbrace{3 + 3 + \cdots + 3}_{\text{8 of them}} + \underbrace{4 + 4 + \cdots + 4}_{\text{6 of them}}$$

$$= (5)(1) + (9)(2) + (8)(3) + (6)(4)$$

$$= 5 + 18 + 24 + 24 = \mathbf{71}$$

$$\sum x^2 = \underbrace{1^2 + \cdots + 1^2}_{\text{5 of them}} + \underbrace{2^2 + \cdots + 2^2}_{\text{9 of them}} + \underbrace{3^2 + \cdots + 3^2}_{\text{8 of them}} + \underbrace{4^2 + \cdots + 4^2}_{\text{6 of them}}$$

$$= (5)(1) + (9)(4) + (8)(9) + (6)(16)$$

$$= 5 + 36 + 72 + 96 = \mathbf{209}$$

However, we will use the frequency distribution to determine these summations by expanding it to become an *extensions table*. The extensions xf and x^2f are formed by multiplying across the columns row by row and then adding to find three column totals. The objective of the extensions table is to obtain these three column totals (see Table 2.19).

TABLE 2.19

Ungrouped Frequency Distribution: Extensions xf and x^2f

x	f	xf	x^2f
1	5	5	5
2	9	18	36
3	8	24	72
4	6	24	96

$\sum f = 28 \quad \sum xf = 71 \quad \sum x^2f = 209$

number of data   *sum of x, using frequencies*   *sum of x^2, using frequencies*

Notes:

1. The extensions in the xf column are the subtotals of the like x values.
2. The extensions in the x^2f column are the subtotals of the like x-squared values.
3. The three column totals, $\sum f$, $\sum xf$, and $\sum x^2f$, are the values previously known as n, $\sum x$, and $\sum x^2$, respectively. That is, $\sum f = n$, the number of pieces of data; $\sum xf = \sum x$, the sum of the data; and $\sum x^2f = \sum x^2$, the sum of the squared data.
4. Think of the f in the summation expressions $\sum xf$ and $\sum x^2f$ as an indication that the sums were obtained with the use of a frequency distribution.
5. The sum of the x column is NOT a meaningful number. The x column lists each possible value of x once, which does not account for the repeated values.

To find the **mean** of a frequency distribution, we modify formula (2.1) on page 73 to indicate the use of the frequency distribution:

Mean of Frequency Distribution

$$x\text{-bar} = \frac{sum\ of\ all\ x,\ using\ frequencies}{number\ using\ frequencies}$$

$$\bar{x} = \frac{\sum xf}{\sum f} \tag{2.13}$$

FYI See Animated Tutorial Frequency Distribution on CD for assistance with these calculations.

The mean value of x for the frequency distribution in Table 2.19 is found by using formula (2.13):

$$\text{mean:} \quad \bar{x} = \frac{\sum xf}{\sum f} = \frac{71}{28} = 2.536 = \mathbf{2.5}$$

To find the **variance** of the frequency distribution, we modify formula (2.10) on page 88 to indicate the use of the frequency distribution:

Variance of Frequency Distribution

$$s \text{ squared} = \frac{(\text{sum of } x^2, \text{ using frequencies}) - \left[\dfrac{(\text{sum of } x, \text{ using frequencies})^2}{\text{number, using frequencies}} \right]}{\text{number, using frequencies} - 1}$$

$$s^2 = \frac{\sum x^2 f - \dfrac{(\sum xf)^2}{\sum f}}{\sum f - 1} \tag{2.14}$$

The variance of x for the frequency distribution in Table 2.19 is found by using formula (2.14):

$$\text{variance:} \quad s^2 = \frac{\sum x^2 f - \dfrac{(\sum xf)^2}{\sum f}}{\sum f - 1} = \frac{209 - \dfrac{(71)^2}{28}}{28 - 1} = \frac{28.964}{27} = 1.073 = \mathbf{1.1}$$

The **standard deviation** of x for the frequency distribution in Table 2.19 is found by using formula (2.7), the positive square root of variance.

$$\text{standard deviation:} \quad s = \sqrt{s^2} = \sqrt{1.073} = 1.036 = \mathbf{1.0}$$

EXAMPLE 2.19

Calculations Using Grouped Frequencies

Find the mean, variance, and standard deviation of the sample of 50 exam scores using the grouped frequency distribution in Table 2.8 (p. 58).

SOLUTION We will use an extensions table to find the three summations in the same manner we did in Example 2.18. The class midpoints will be used as the representative values for the classes.

The mean value of x for the frequency distribution in Table 2.20 (p. 120) is found by using formula (2.13):

$$\text{mean:} \quad \bar{x} = \frac{\sum xf}{\sum f} = \frac{3780}{50} = \mathbf{75.6}$$

FYI See Animated Tutorial Frequency Distribution on CD for assistance with these calculations.

The variance of x for the frequency distribution in Table 2.20 (p. 120) is found by using formula (2.14):

$$\text{variance:} \quad s^2 = \frac{\sum x^2 f - \dfrac{(\sum xf)^2}{\sum f}}{\sum f - 1} = \frac{296{,}600 - \dfrac{3780^2}{50}}{50 - 1} = \frac{10{,}832}{49}$$

$$= 221.0612 = \mathbf{221.1}$$

The standard deviation of x for the frequency distribution in Table 2.20 is found by using formula (2.7):

$$\text{standard deviation:} \quad s = \sqrt{s^2} = \sqrt{221.0612} = 14.868 = \mathbf{14.9}$$

TABLE 2.20

Frequency Distribution of 50 Exam Scores

Class Number	Class Midpoints, x	f	xf	x^2f
1	40	2	80	3,200
2	50	2	100	5,000
3	60	7	420	25,200
4	70	13	910	63,700
5	80	11	880	70,400
6	90	11	990	89,100
7	100	4	400	40,000

$$\sum f = 50 \quad \sum xf = 3780 \quad \sum x^2f = 296{,}600$$

TECHNOLOGY INSTRUCTIONS: FREQUENCY DISTRIBUTION STATISTICS

MINITAB (Release 14)   Input the class midpoints or data values into C1 and the corresponding frequencies into C2; then continue with the following commands to obtain the extensions table:

```
Choose:      Calc > Calculator ...
Enter:       Store result in variable: C3
             Expression: C1*C2 > OK
```
Repeat the preceding commands, replacing the variable with **C4** and the expression with **C1*C3.**
```
Choose:      Calc > Column Statistics
Select:      Sum
Enter:       Input variable: C2
             Store result in: K1 > OK
```
Repeat preceding 'sum' commands, replacing variable with **C3** and result with **K2.**
Repeat preceding 'sum' commands, replacing variable with **C4** and result with **K3.**
```
Choose:      Data > Display data
Enter:       Columns to display: C1-C4 K1-K3 > OK
```

To find the mean, variance, and standard deviation, respectively, continue with:

```
Choose:      Calc > Calculator
Enter:       Store result in variable: K4
             Expression: K2/K1 > OK
```
Repeat preceding 'mean' commands, replacing variable with **K5** and expression with **(K3-(K2**2/K1))/(K1-1).**
Repeat preceding 'mean' commands, replacing variable with **K6** and expression with **SQRT(K5)** (select square root from functions).
```
Choose:      Data > Display data
Enter:       Columns to display: K4-K6 > OK
```

Excel   Input the class midpoints or data values into column A and the corresponding frequencies into column B; activate C1 or C2 (depending on whether column headings are used); then continue with the following commands to obtain the extensions table:

```
Enter:     = A2*B2 (if column headings are used)
Drag:      Bottom right corner of C2 down to give other products
Activate D2 and repeat preceding commands, replacing the formula with =
A2*C2.
Activate the data in columns B, C, and D.
Choose:    AutoSum (sums will appear at the bottom of the columns)
```

To find the mean, activate **E2**; then continue with:

```
Enter:     = (column C total/column B total) (ex. = C9/B9)
```

To find the variance, activate **E3** and repeat preceding 'mean' commands, replacing the formula with $=$ **(D9-(C9^2/B9))/(B9-1)**.

To find the standard deviation, activate **E4** and repeat preceding 'mean' commands, replacing the formula with $=$ **SQRT(E3)**.

TI-83/84 Plus   Input the class midpoints or data values into L1 and the frequencies into L2; then continue with:

```
Highlight:  L3
Enter:      L3 = L1*L2
Highlight:  L4
Enter:      L4 = L1*L3
Highlight:  L5(1) (first position in L5 column)
Enter:      L5(1) =sum(L2)      [∑ f ]
            [sum = 2nd LIST > MATH > 5:sum()
            L5(2)  = sum(L3)    [∑ xf]
            L5(3)  = sum(L4)    [∑ x²f]
            L5(4)  = L5(2)/L5(1) [to find mean]
            L5(5)  = (L5(3) − ((L5(2))²/L5(1)))/(L5(1)−1)
            [to find variance]
            L5(6)  = 2nd √‾ (L5(5))
            [to find standard deviation]
```

If the extensions table is not needed, just use:

```
Choose:    STAT > CALC > 1:1-VAR STATS
Enter:     L1, L2
```

SECTION 2.9 EXERCISES (OPTIONAL)

Statistics⬡Now™

Datasets can be found on your Student's Suite CD-ROM or at the StatisticsNow website at **http://1pass.thomson.com**.

2.153 A survey asked respondents to list the "number of telephones" per household, x; results are shown here as a frequency distribution.

x	0	1	2	3	4
f	1	3	8	5	3

a. Complete the extensions table.

b. Find the three summations, $\sum f$, $\sum xf$, $\sum x^2f$, for the frequency distribution.

c. Describe what each of the following represents: $x = 4$, $f = 8$, $\sum f$, $\sum xf$.

d. Explain why (i) the "sum of the x-column" has no relationship to the "sum of the data," and (ii) the "$\sum xf$" represents the "sum of the data" represented by the frequency distribution.

2.154 a. Find the mean of the data shown in the frequency distribution in Exercise 2.153.

b. Find the variance for the data shown in the frequency distribution in Exercise 2.153.

c. Find the standard deviation for the data shown in the frequency distribution in Exercise 2.153.

2.155 Pediatric dentists say a child's first dental exam should occur between ages 6 months and 1 year. The ages at first dental exam for a sample of children are shown in the distribution:

Age at First Dental Exam, x	1	2	3	4	5
Number of Children, f	9	11	23	16	21

a. Find the mean age of first dental exam for these children.

b. Find the median age.

c. Find the standard deviation.

2.156 A survey asked a group of medical doctors how many children they had fathered. The results are summarized by this ungrouped frequency distribution:

Number of Children	0	1	2	3	4	6
Number of Doctors	15	12	26	14	4	2

Calculate the sample mean, variance, and standard deviation for the number of children the doctors had fathered.

2.157 The weight gains (in grams) for chicks fed on a high-protein diet were as follows:

Weight Gain	12.5	12.7	13.0	13.1	13.2	13.8
Frequency	2	6	22	29	12	4

a. Find the mean.

b. Find the variance.

c. Find the standard deviation.

2.158 Find the mean, variance, and standard deviation of the data shown in the following frequency distribution.

Class	2-6	6-10	10-14	14-18	18-22
f	2	10	12	9	7

2.159 Find the mean, variance, and standard deviation for this grouped frequency distribution:

Class Boundaries	3-6	6-9	9-12	12-15	15-18
f	2	10	12	9	7

2.160 The following distribution of commuting distances was obtained for a sample of Mutual of Nebraska employees:

Distance (miles)	Frequency	Distance (miles)	Frequency
1.0–3.0	2	9.0–11.0	35
3.0–5.0	6	11.0–13.0	15
5.0–7.0	12	13.0–15.0	5
7.0–9.0	50		

Find the mean and the standard deviation for the commuting distances.

2.161 A quality-control technician selected twenty-five 1-pound boxes from a production process and found the following distribution of weights (in ounces):

Weight	Frequency	Weight	Frequency
15.95–15.98	2	16.04–16.07	3
15.98–16.01	4	16.07–16.10	1
16.01–16.04	15		

Find the mean and the standard deviation for this weight distribution.

2.162 It has been found that 35.2 million Americans 16 years and older fish our waters. A sample of freshwater fishermen produced the following age distribution:

Age of Fishermen, x	15–25	25–35	35–45	45–55	55–65	65–75
# Fishermen, f	13	20	28	20	10	9

Find the mean and the standard deviation for this distribution.

2.163 Private industry reported that more than 31,000 workers were absent from work in 2005 because of carpal tunnel syndrome (a nerve disorder causing arm, wrist, and hand pain). The length of time (in days) workers are absent as a result of this problem varies greatly.

Days of Absence, x	0–10	10–20	20–30	30–40	40–50
Number of Workers, f	37	24	38	32	27

Find the mean and the standard deviation for this distribution.

2.164 [EX02-164] The California Department of Education issues a yearly report on Scholastic Aptitude Test (SAT) scores for students in the various school districts. The following frequency table shows verbal test results for 2003–2004 for Merced County school districts.

District	Number Tested	Verbal Average
Delhi Unified	34	434
Dos Palos Oro Loma Jt. Unified	48	431
Gustine Unified	37	482
Hilmar Unified	43	488
Le Grand Union High	28	369
Los Banos Unified	109	479
Merced Co. Office of Education	0	0
Merced Union High	534	450

Source: http://data1.cde.ca.gov/dataquest/
SAT-I1.asp?cChoice=SAT1&cYear=2003-04&TheCount

a. What do the entries 34 and 434 for Delhi Unified mean?

b. What is the total for all student scores at Delhi Unified?

c. How many student test results are shown in this table?

d. What is the total for all student scores shown in the table?

e. Find the mean SAT verbal test result.

2.165 [EX02-165] A random sample of people of all ages was taken from the U.S. population, and the resulting 75 ages (in years) are listed in the following table:

22	7	72	32	18	4	9	48	49	18	18	58	47	39	48
1	27	61	48	25	34	75	29	53	37	25	42	49	29	31
10	27	8	78	63	50	39	32	5	39	8	15	8	50	39
46	38	4	9	43	3	65	25	67	19	9	34	8	36	48
56	73	24	20	34	38	45	40	11	40	37	17	63	9	91

a. Construct a grouped frequency distribution for the ages using class midpoints of 0, 10, 20, . . . 90. Show the class midpoints and the associated frequency counts in your table.

b. Construct a histogram.

c. Does the variable "age" appear to have an approximately normal distribution?

d. Calculate the mean age.

e. Find the median age.

f. Find the range of ages.

g. Find the standard deviation of ages.

h. Compare the values found in parts d through g against corresponding statistics calculated using the given ungrouped data. Use the percent error in each case, and present all results in a table to make your case.

2.166 [EX02-042] A USA Snapshot titled "Nuns an aging order" reports that the median age of 94,022 Roman Catholic nuns in the United States is 65 years and the percentages of U.S. nuns by age group are as follows:

Under 50	51–70	Over 70	Refused to give age
16%	42%	37%	5%

This information is based on a survey of 1049 Roman Catholic nuns. Suppose the survey had resulted in the following frequency distribution (52 ages unknown):

Age	20–30	30–40	40–50	50–60	60–70	70–80	80–90
Frequency	34	58	76	187	254	241	147

(See the histogram drawn in Exercise 2.42, p. 69.)

a. Find the mean, median, mode, and midrange for this distribution of ages.

b. Find the variance and standard deviation.

2.167 The number of sports reports a sports fan watches on television in a typical week was described in a USA Snapshot titled "Fans find sports in print and TV" (December 21, 2004).

Reports	0	1-2	3-4	5-6	7	8 or more
Percent	35.0%	24.8%	15.4%	11.4%	8.6%	4.8%

This information is based on an ESPN Sports Poll. Suppose the survey resulted in the following frequency distribution.

Reports	0	1-2	3-4	5-6	7	8 or more
Frequency	44	31	19	14	11	6

a. How many were surveyed?

b. Draw a histogram of these data.

c. Find the mean number of reports viewed per week. (Use 8.5 for "8 or more" midpoint.)

d. Find the median number of reports viewed per week.

e. Find the mode number of reports viewed per week.

2.168 [EX02-168] The USA Snapshot "Paydown after graduation" reports that undergraduates who take loans average $16,500 of debt at graduation. By percentage, the amount of their monthly debt is shown here.

Debt	Under $100	$100-$149	$150-$199	$200-$249	$250-$299	$300 or more
Percent	17%	17%	17%	19%	10%	20%

Suppose another survey had resulted in the frequency distribution shown here.

Debt	Under $100	$100-$149	$150-$199	$200-$249	$250-$299	$300 or more
Frequency	125	158	127	175	100	165

a. How many were surveyed?

b. Draw a histogram of these data.

c. Find the mean of the frequency distribution.

d. Find the median of the frequency distribution.

e. Find the mode of the frequency distribution.

2.169 [EX02-169] A random sample of 250 people living in the state of New York resulted in the following distribution of ages.

Age Group	Frequency	Age Group	Frequency
Younger than 5 years	18	45-54 years	48
5-14 years	35	55-64 years	21
15-24 years	20	65-74 years	17
25-34 years	35	75-84 years	16
35-44 years	38	85 years and older	2

a. What about this distribution is different than the distributions described in the text?

b. How can the procedures of this section be adapted to accommodate the classes at the extremes of this distribution?

c. Draw a histogram of the distribution of ages.

d. Find the mean age for the people included in this sample.

e. Find the standard deviation.

2.170 [EX02-170] A Champions Tour professional golfer is not expected to play in all the available tournaments in the course of a season. The number of tournaments played by each of the 2004 tour's top 50 money leaders is shown on page 126.

a. Construct a grouped frequency distribution showing the number of tournaments played using group intervals 9–11, 11–13, . . . 29–31; the class midpoints; and the associated frequency counts.

b. Find the mean, variance, and standard deviation of the number of tournaments played both with and without using the grouped distribution.

c. Compare the two sets of answers you obtained in part b. What percent is the error in each case?

Player	Events	Player	Events	Player	Events	Player	Events	Player	Events
Craig Stadler	21	Bruce Lietzke	20	D. A. Weibring	25	Bruce Summerhays	28	Dana Quigley	30
Mark James	20	John Jacobs	28	David Eger	28			Jay Sigel	28
Lonnie Nielsen	26	Bruce Fleisher	28	Jim Ahern	27	Dave Barr	28	Pete Oakley	12
Hale Irwin	23	Bob Gilder	28	Jim Thorpe	26	Doug Tewell	27	Morris Hatalsky	27
Jerry Pate	27	Gary McCord	14	Graham Marsh	30	Bobby Wadkins	26	Walter Hall	26
Don Pooley	21	Larry Nelson	25	Dave Stockton	21	Joe Inman	26	Hugh Baiocchi	26
Tom Kite	27	Fuzzy Zoeller	21	Allen Doyle	27	Tom Jenkins	27	Peter Jacobsen	9
Jose Maria Canizares	26	Gary Koch	18	Ed Fiori	28	Vicente Fernandez	26	John Harris	25
		Mark McNulty	20	Rodger Davis	20			Keith Fergus	18
John Bland	26	Andy Bean	28	Wayne Levi	27	Mike McCullough	28	Tom Purtzer	19
Gil Morgan	26	Tom Watson	12					Des Smyth	27

Source: PGA Tour, Inc.

CHAPTER REVIEW

In Retrospect

You have been introduced to some of the more common techniques of descriptive statistics. There are far too many specific types of statistics used in nearly every specialized field of study for us to review here. We have outlined the uses of only the most universal statistics. Specifically, you have seen several basic graphic techniques (circle and bar graphs, Pareto diagrams, dotplots, stem-and-leaf displays, histograms, and box-and-whiskers diagrams) that are used to present sample data in picture form. You have also been introduced to some of the more common measures of central tendency (mean, median, mode, midrange, and midquartile), measures of dispersion (range, variance, and standard deviation), and measures of position (quartiles, percentiles, and z-scores).

You should now be aware that an average can be any one of five different statistics, and you should understand the distinctions among the different types of averages. The article "'Average' Means Different Things" in Applied Example 2.11 (pp. 77–78) discusses four of the averages studied in this chapter. You might reread it now and find that it has more meaning and is of more interest. It will be time well spent!

You should also have a feeling for, and an understanding of, the concept of a standard deviation. You were introduced to the empirical rule and Chebyshev's theorem for this purpose.

The exercises in this chapter (as in others) are extremely important; they will reinforce the concepts studied before you go on to learn how to use these ideas in later chapters. A good understanding of the descriptive techniques presented in this chapter is fundamental to your success in the later chapters.

Vocabulary and Key Concepts

bar graph (p. 41)
bell-shaped distribution (p. 106)
bimodal frequency distribution (p. 63)
box-and-whiskers plot (p. 96)
Chebyshev's theorem (p. 109)
circle graph (p. 141)
class (p. 56)
class boundary (p. 57)
class midpoint (class mark) (p. 58)
class width (p. 56)

depth (pp. 74, 94)
deviation from the mean
  (p. 84)
distribution (p. 44)
dotplot (p. 44)
empirical rule (p. 106)
5-number summary (p. 96)
frequency (p. 44)
frequency distribution (p. 55)
frequency histogram (p. 60)
grouped frequency distribution
  (p. 56)
histogram (p. 59)
interquartile range (p. 96)
mean (pp. 73, 118)
measure of central tendency
  (p. 73)
measure of dispersion (p. 84)

measure of position (p. 92)
median (p. 74)
midquartile (p. 75)
midrange (p. 76)
modal class (p. 64)
mode (p. 76)
normal distribution
  (pp. 106, 108)
ogive (p. 65)
Pareto diagram (p. 42)
percentile (p. 93)
qualitative data (p. 41)
quantitative data (p. 44)
quartile (p. 93)
range (p. 84)
rectangular distribution (p. 63)
relative frequency (p. 59)

relative frequency distribution
  (p. 59)
relative frequency histogram
  (p. 69)
skewed distribution (p. 63)
standard deviation
  (pp. 87, 119)
standard score (p. 99)
stem-and-leaf display (p. 46)
summation (p. 58)
tally (p. 58)
ungrouped frequency distribution (p. 56)
variance (pp. 86, 119)
x-bar ($\bar{x}$) (p. 73)
z-score (p. 99)

Learning Outcomes

✓ Create and interpret graphical displays, including pie charts, bar graphs, Pareto diagrams, dotplots, and stem-and-leaf diagrams.	EXP 2.4, Ex. 2.5, 2.13, 2.15, 2.19, 2.25, 2.27
✓ Understand and be able to describe the difference between grouped and ungrouped frequency distributions, frequency and relative frequency, relative frequency and cumulative relative frequency.	pp. 55–56, 60, 64–65
✓ Identify and describe the parts of a frequency distribution: class boundaries, class width, and class midpoint.	EXP 2.6, Ex. 2.41, 2.43
✓ Create and interpret frequency histograms, relative frequency histograms, and ogives.	pp. 60, 64–65, Ex. 2.33, 2.36, 2.38
✓ Identify the shapes of distributions.	pp. 63–65
✓ Compute, describe, and compare the four measures of central tendency: mean, median, mode, and midrange.	EXP 2.11, Ex. 2.67
✓ Understand the effect of outliers on each of the four measures of central tendency.	Ex. 2.179, 2.180, 2.216
✓ Compute, describe, compare, and interpret the two measures of dispersion: range and standard deviation (variance).	pp. 84–87, Ex. 2.91, 2.97
✓ Compute, describe, and interpret the measures of position: quartiles, percentiles, and z-scores.	EXP 2.12, 2.14, Ex. 2.109, 2.119, 2.200
✓ Create and interpret boxplots.	Ex. 2.114
✓ Understand the empirical rule and Chebyshev's theorem and be able to assess a set of data's compliance to these rules.	Ex. 2.130, 2.137, 2.143
✓ Know when and when not to use certain statistics—graphic and numeric.	pp. 114–115, Ex. 2.148, 2.149
✓ Compute the mean and standard deviation for ungrouped and grouped frequency distributions. (Optional.)	EXP 2.18, 2.19, Ex. 2.155, 2.159

Chapter Exercises

Statistics⬛Now™

Go to the StatisticsNow website **http://1pass.thomson.com** to

- Assess your understanding of this chapter
- Check your readiness for an exam by taking the Pre-Test quiz and exploring the resources in the Personalized Learning Plan

Datasets can be found on your Student's Suite CD-ROM or at the StatisticsNow website at **http://1pass.thomson.com**.

2.171 "Who believes in the 5-second rule?" Most people say food dropped on the floor is not safe to eat.

WHO BELIEVES IN THE 5-SECOND RULE?

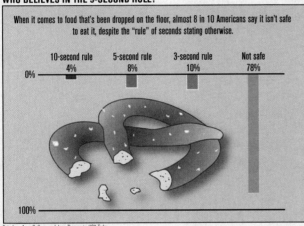

When it comes to food that's been dropped on the floor, almost 8 in 10 Americans say it isn't safe to eat it, despite the "rule" of seconds stating otherwise.

| 10-second rule 4% | 5-second rule 8% | 3-second rule 10% | Not safe 78% |

Data from Anne R. Carey and Juan Thomassie, USA Today.

a. Draw a circle graph depicting the percentages of adults for each response.

b. If 300 adults are to be surveyed, what frequencies would you expect to occur for each response on the "Do you eat food dropped on the floor" graph?

2.172 The supplies needed for a baby during his or her first year can be expensive—averaging $5000, as shown in this divided bar graph.

a. Construct a circle graph showing this same information.

b. Construct a bar graph showing this same information.

c. Compare the appearance of the divided bar graph given with the circle graph drawn in part a and the bar graph drawn in part b. Which

BUDGETING FOR BABY

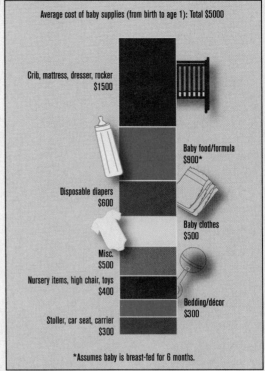

Average cost of baby supplies (from birth to age 1): Total $5000

Crib, mattress, dresser, rocker $1500

Baby food/formula $900*

Disposable diapers $600

Baby clothes $500

Misc. $500

Nursery items, high chair, toys $400

Bedding/décor $300

Stoller, car seat, carrier $300

*Assumes baby is breast-fed for 6 months.

Data from Julie Snider. © 2005 USA Today.

one best represents the relationship between various costs of baby supplies?

2.173 There are many types of statistical graphs one can choose from when picturing a set of data. The "divided bar graph" shown here is an alternative to the circle graph.

HOW WILL YOU SPEND YOUR TAX REFUND?

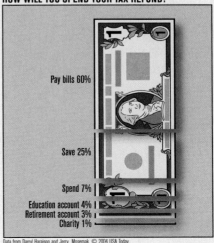

Pay bills 60%

Save 25%

Spend 7%

Education account 4%
Retirement account 3%
Charity 1%

Data from Darryl Haralson and Jerry Mosemak, © 2004 USA Today.

a. Construct a circle graph showing this same information.

b. Compare the appearance of the divided bar graph given and the circle graph drawn in part a. Which one is easier to read? Which one gives a more accurate representation of the information being presented?

2.174 One of the ways that students pay for college is to borrow money. These loans eventually must be paid back, and the accompanying divided bar graph shows the monthly debt that many face after graduation.

PAYING OFF THAT COLLEGE DEBT

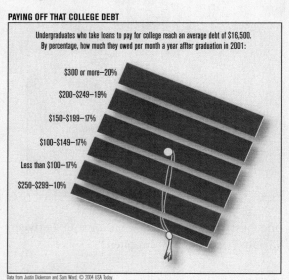

Undergraduates who take loans to pay for college reach an average debt of $16,500. By percentage, how much they owed per month a year affter graduation in 2001:

$300 or more—20%
$200-$249—19%
$150-$199—17%
$100-$149—17%
Less than $100—17%
$250-$299—10%

Data from Justin Dickerson and Sam Ward, © 2004 USA Today.

a. Construct a circle graph showing this same information.

b. Construct a bar graph showing this same information.

c. Compare the appearance of the divided bar graph given with the circle graph drawn in part a and the bar graph drawn in part b. Which one best represents the relationship between various monthly debt amounts?

2.175 [EX02-175] The 10 leading causes of death in the United States during 2002 were listed by the National Center for Statistics and Analysis in a January 2005 report. There were a total of 2,443,387 deaths recorded.

Cause of Death	Number (10,000s)
Alzheimer's disease	5.9
Chronic respiratory disease	12.5
Diabetes	7.3
Heart disease	69.7
Influenza/pneumonia	6.6
Malignant neoplasms	55.7
Motor vehicle traffic crashes	4.4
Nephritis/nephrosis	4.1
Septicemia	3.4
Stroke	16.3

Source: NHTSA's National Center for Statistics and Analysis

a. Construct a Pareto diagram of this information.

b. Write a paragraph describing what the Pareto diagram dramatically shows to its reader.

2.176 [EX02-176] The U.S. Census Bureau posted the following 2003 age distribution for the people of New York State. The 2003 American Community Survey is limited to the household population and excludes the population living in institutions, college dormitories, and other group quarters.

Sex and Age Distribution

Male	8,953,019	Younger than 5 years	1,205,816
Female	9,647,508	5–14 years	2,537,813
		15–24 years	2,353,665
		25–34 years	2,587,995
		35–44 years	2,991,609
		45–54 years	2,682,845
		55–64 years	1,897,521
		65–74 years	1,218,850
		75–84 years	857,177
		85 years and older	267,236

Source: U.S. Census Bureau

a. Construct a relative frequency distribution of both the gender and the age data.

b. Construct a bar graph of the gender data.

c. Construct a histogram of the age data.

d. Explain why the graph drawn in part b is not a histogram and the graph drawn in part c is a histogram.

2.177 Identify each of the following as examples of (1) attribute (qualitative) or (2) numerical (quantitative) variables.

a. Scores registered by people taking their written state automobile driver's license examination

b. Whether or not a motorcycle operator possesses a valid motorcycle operator's license

c. The number of television sets installed in a house

d. The brand of bar soap being used in a bathroom

e. The value of a cents-off coupon used with the purchase of a box of cereal

2.178 Identify each of the following as examples of (1) attribute (qualitative) or (2) numerical (quantitative) variables.

a. The amount of weight lost in the past month by a person following a strict diet

b. Batting averages of major league baseball players

c. Decisions by the jury in felony trials

d. Sunscreen usage before going in the sun (always, often, sometimes, seldom, never)

e. Reason a manager failed to act against an employee's poor performance

2.179 Consider samples A and B. Notice that the two samples are the same except that the 8 in A has been replaced by a 9 in B.

A:	2	4	5	5	7	8
B:	2	4	5	5	7	9

What effect does changing the 8 to a 9 have on each of the following statistics?

a. Mean b. Median c. Mode d. Midrange

e. Vange   f. Variance   g. Std. dev.

2.180 Consider samples C and D. Notice that the two samples are the same except for two values.

C:	20	60	60	70	90
D:	20	30	70	90	90

What effect does changing the two 60s to 30 and 90 have on each of the following statistics?

a. Mean b. Median c. Mode d. Midrange

e. Range   f. Variance   g. Std. dev.

2.181 The addition of a new accelerator is claimed to decrease the drying time of latex paint by more than 4%. Several test samples were conducted with the following percentage decreases in drying time:

5.2	6.4	3.8	6.3	4.1	2.8	3.2	4.7

a. Find the sample mean.

b. Find the sample standard deviation.

c. Do you think these percentages average 4 or more? Explain.

(Retain these solutions to use in Exercise 9.28, p. 490.)

2.182 [EX02-182] Gasoline pumped from a supplier's pipeline is supposed to have an octane rating of 87.5. On 13 consecutive days, a sample of octane ratings was taken and analyzed, with the following results:

88.6	86.4	87.2	88.4	87.2	87.6	86.8
86.1	87.4	87.3	86.4	86.6	87.1	

a. Find the sample mean.

b. Find the sample standard deviation.

c. Do you think these readings average 87.5? Explain.

(Retain these solutions to use in Exercise 9.56, p. 494.)

2.183 [EX02-183] These data are the ages of 118 known offenders who committed an auto theft last year in Garden City, Michigan:

11	14	15	15	16	16	17	18	19	21	25	36
12	14	15	15	16	16	17	18	19	21	25	39
13	14	15	15	16	17	17	18	20	22	26	43
13	14	15	15	16	17	17	18	20	22	26	46
13	14	15	16	16	17	17	18	20	22	27	50
13	14	15	16	16	17	17	19	20	23	27	54
13	14	15	16	16	17	18	19	20	23	29	59
13	15	15	16	16	17	18	19	20	23	30	67
14	15	15	16	16	17	18	19	21	24	31	
14	15	15	16	16	17	18	19	21	24	34	

a. Find the mean.   b. Find the median.

c. Find the mode.   d. Find Q_1 and Q_3.

e. Find P_{10} and P_{95}.

2.184 [EX02-184] A survey of 32 workers at building 815 of Eastman Kodak Company was taken last May. Each worker was asked: "How many hours of television did you watch yesterday?" The results were as follows:

0	0	½	1	2	0	3	2½	0	0	1
1½	5	2½	0	2	2½	1	0	2	0	
2½	4	06	2½	0	½	1	1½	0	2	

a. Construct a stem-and-leaf display.

b. Find the mean.   c. Find the median.

d. Find the mode.   e. Find the midrange.

f. Which measure of central tendency would best represent the average viewer if you were trying to portray the typical television viewer? Explain.

g. Which measure of central tendency would best describe the amount of television watched? Explain.

h. Find the range.   i. Find the variance.

j. Find the standard deviation.

2.185 [EX02-185] The stopping distance on a wet surface was determined for 25 cars, each traveling at 30 miles per hour. The data (in feet) are shown on the following stem-and-leaf display:

```
 6 | 3 7 6 3 9
 7 | 4 2 0 1 1 2 0 5
 8 | 5 4 5 5 6
 9 | 4 1 0 0 5
10 | 5 4
```

Find the mean and the standard deviation of these stopping distances.

2.186 [EX02-186] Forbes.com posted the 2004 EPS (earnings per share) in dollars for 17 banking industry companies.

Name	EPS ($)	Name	EPS ($)
Astoria Financial	2.92	Popular	1.71
Banknorth Group	2.20	State Street	3.13
Bank of America	3.67	Synovus Finl	1.36
BB & T	2.61	UnionBanCal	4.70
Compass Bancshares	2.86	Wachovia	3.68
Golden West Finl	3.97	Wells Fargo	4.00
M & T Bank	5.74	Westcorp	3.71
National City	3.75	Zions Bancorp	4.36
North Fork Bancorp	1.83		

Source: http://www.forbes.com/lists/results.jhtml

a. Find the mean EPS for the banks.

b. Find the median EPS for the banks.

c. Find the midrange of the EPS for the banks.

d. Write a discussion comparing the results from parts a, b, and c.

e. Find the standard deviation of the EPS for the banks.

f. Find the percentage of the data that are within 1 standard deviation of the mean.

g. Find the percentage of the data that are within 2 standard deviations of the mean.

h. Based on the preceding results, discuss whether you think the data are normally distributed, and why.

2.187 [EX02-187] The Office of Aviation Enforcement & Proceedings, U.S. Department of Transportation, reported the number of mishandled baggage reports filed per 1000 airline passengers during October 2004. The industry average was 4.02.

Airline	Reports	Passengers	Reports/1000
AirTran	2084	1,148,779	1.81
JetBlue	2295	1,057,510	2.17
••• Remainder of data on Student's Suite CD-ROM			

Source: Office of Aviation Enforcement & Proceedings, U.S. Department of Transportation

a. Define the terms *population* and *variable* with regard to this information.

b. Are the numbers reported (1.81, 2.17, . . . , 12.21) data or statistics? Explain.

c. Is the average, 4.02, a data value, a statistic, or a parameter value? Explain why.

d. Is the "industry average" the mean of the airline rates of reports per 1000? If not, explain in detail how the 19 airline values are related to the industry average.

2.188 [EX02-188] One of the first scientists to study the density of nitrogen was Lord Raleigh. He noticed that the density of nitrogen produced from the air seemed to be greater than the density of nitrogen produced from chemical compounds. Do his conclusions seem to be justified even though he has so little data?

Lord Raleigh's measurements, which first appeared in *Proceedings, Royal Society* (London, 55, 1894, pp. 340–344) are listed here. The data are the mass of nitrogen filling a certain flask under specified pressure and temperature.

Atmospheric		Chemical	
2.31017	2.31010	2.30143	2.29940
2.30986	2.31028	2.29890	2.29849
2.31010	2.31163	2.29816	2.29889
2.31001	2.30956	2.30182	2.30074
2.31024		2.29869	2.30054

Source: http://exploringdata.cqu.edu.au/datasets/nitrogen.xls

a. Construct side-by-side dotplots of the two sets of data, using a common scale.

b. Calculate mean, median, standard deviation, and first and third quartiles for each set of data.

c. Construct side-by-side boxplots of the two sets of data, using a common scale.

d. Discuss how these two sets of data compare. Do these two very small sets of data show convincing evidence of a difference?

FYI The differences between these sets of data helped lead to the discovery of argon.

2.189 [EX02-189] The top 2004 Nationwide Tour money leaders, together with their total earnings, are listed here:

Player	Money ($)	Player	Money ($)
Jimmy Walker	371,346	D. A. Points	332,815
••• Remainder of data on Student's Suite CD-ROM			

Source: PGA Tour, Inc.

a. Calculate the mean and standard deviation of the earnings of the Nike Tour golf players.

b. Find the values of $\bar{x} - s$ and $\bar{x} + s$.

c. How many of the 50 pieces of data have values between $\bar{x} - s$ and $\bar{x} + s$? What percentage of the sample is this?

d. Find the values of $\bar{x} - 2s$ and $\bar{x} + 2s$.

e. How many of the 50 pieces of data have values between $\bar{x} - 2s$ and $\bar{x} + 2s$? What percentage of the sample is this?

f. Find the values of $\bar{x} - 3s$ and $\bar{x} + 3s$.

g. What percentage of the sample has values between $\bar{x} - 3s$ and $\bar{x} + 3s$?

h. Compare the answers found in parts e and g to the results predicted by Chebyshev's Theorem.

i. Compare the answers found in parts c, e, and g to the results predicted by the empirical rule. Does the result suggest an approximately normal distribution?

j. Verify your answer to part i using one of the sets of technology instructions.

k. Does your answer to part j make sense? Explain.

2.190 Ask one of your instructors for a list of exam grades (15 to 25 grades) from a class.

a. Find five measures of central tendency.

b. Find the three measures of dispersion.

c. Construct a stem-and-leaf display. Does this diagram suggest that the grades are normally distributed?

d. Find the following measures of location: (i) Q_1 and Q_3, (ii) P_{15} and P_{60}, and (iii) the standard score z for the highest grade.

2.191 [EX02-191] The lengths (in millimeters) of 100 brown trout in pond 2-B at Happy Acres Fish Hatchery on June 15 of last year were as follows:

15.0	15.3	14.4	10.4

••• Remainder of data on Student's Suite CD-ROM

a. Find the mean.　b. Find the median.

c. Find the mode.　d. Find the midrange.

e. Find the range.　f. Find Q_1 and Q_3.

g. Find the midquartile.　h. Find P_{35} and P_{64}.

i. Construct a grouped frequency distribution that uses 10.0–10.5 as the first class.

j. Construct a histogram of the frequency distribution.

k. Construct a cumulative relative frequency distribution.

l. Construct an ogive of the cumulative relative frequency distribution.

m. Find the mean of the frequency distribution. (Optional.)

n. Find the standard deviation of the frequency distribution. (Optional.)

2.192 [EX02-192] The national highway system is made up of interstate and noninterstate highways. The Federal Highway Administration reported the number of miles of each type in each state. Listed is a random sample of 20.

Miles of Interstate and Noninterstate Highways by State

State	Interstate	Noninterstate	State	Interstate	Noninterstate
NE	235	590	TN	1,073	2,171
FL	1,471	2,897	NJ	1,000	1,935
MA	367	924	LA	904	1,701
HI	55	291	TX	3,233	10,157
MT	1,192	2,683	OH	1,574	2,812
MN	912	3,060	IN	782	2,434
GA	1,245	3,385	NM	1,674	3,476
OK	930	2,431	NC	482	2,496
NV	1,019	2,743	AR	1,167	1,566
RI	71	198	DE	13	70

Source: Federal Highway Administration, U.S. Department of Transportation

Define "ratio I/N" to be number of interstate miles divided by number of noninterstate miles.

a. Inspect the data. What do you estimate the "average" ratio I/N to be?

b. Calculate the "ratio I/N" for each of the 20 states listed.

c. Draw a histogram of the "ratio I/N."

d. Calculate the mean "ratio I/N" for the 20 states listed.

e. Use the 20-state total number of interstate and noninterstate miles to calculate the "ratio I/N" for the combined 20 states.

f. Explain why the answers to parts d and e are not the same.

g. Calculate the standard deviation for the "ratio I/N" for the 20 states listed.

h. The number of miles of interstate and noninterstate highway for all 50 states and Washington, DC, is listed on your CD. Answer the questions in parts b through g using all 51 values. (Optional.) [DS02-192]

2.193 [EX02-193] The National Environmental Satellite, Data, and Information Service, U.S. Department of Commerce, posted the area (sq. mi.) and the 2000 population for the 48 contiguous U.S. states.

State	Area (sq. mi.)	Population
AL	51,610	4,447,100
AZ	113,909	5,130,632

••• Remainder of data on Student's Suite CD-ROM

Source: U.S. Department of Commerce, http://www5.ncdc.noaa.gov/climatenormals/hcs/HCS_42.pdf

When studying how many people live in a country as vast and varied as the United States, perhaps a more interesting variable to study than the population of each state might be the population density of each state since the 48 contiguous states vary so much in area. Define "density" of a state to be the state's population divided by its area.

a. Name three states you believe will be among those with the highest density. Justify your choice.

b. Name three states you believe will be among those with the lowest density. Justify your choice.

c. Describe what you believe the distribution of density will look like. Include ideas of shape of distribution (normal, skewing, etc.).

d. Using the 48 state totals, calculate the overall density for the 48 contiguous U.S. states. Using each state's population and area, calculate the individual densities for the 48 contiguous U.S. states.

e. Calculate the measures of central tendency.

f. Construct a histogram.

g. Rank the density values. Identify the five states with the highest and the five with the lowest density.

h. Compare the distribution of density information (answers to parts e to g) to your expectation (answers to parts a to c). How did you do?

2.194 [EX02-194] The volume of Christmas trees sold annually in the United States has declined in recent decades according to a USDA National Agricultural Statistics Service report. All 50 states report contributions to the total U.S. sale of about 25 million Christmas trees annually. Furthermore, each state reports its crop by county. The top 20 producing counties in the United States come from seven states. The number of trees sold by the top 20 counties in 2002 are listed in the following table. This survey is done every 5 years.

Number of Christmas Trees Sold by County (10,000s)

42.8	25.4	84.8	36.5	16.7
21.3	87.6	65.4	140.0	15.2
41.2	17.2	15.0	103.0	22.1
25.6	20.3	259.0	64.2	19.1

Source: USDA National Agricultural Statistics Service

a. Calculate the mean, median, and midrange for the number of Christmas trees sold annually by the 20 top producing counties.

b. Calculate the standard deviation.

c. What do the answers to parts a and b tell you about the distribution for the number of trees? Explain.

d. Notice that the standard deviation is a larger number than the mean. What does that mean in this situation?

e. Draw a dotplot of the data.

f. Locate the values of the answers to parts a and b to the dotplot drawn for part e.

g. Answer parts c and d again, using the information learned from the dotplot.

2.195 [EX02-195] Following are the percentage of high school graduates per state from the class of 2003 who took the ACT exam.

0.73	0.07	0.16	0.10	0.52	0.62	0.69	0.70	0.12	0.76
0.32	0.05	0.60	0.69	0.73	0.15	0.12	0.74	0.16	0.73
0.27	0.30	1.00	0.67	0.34	0.15	0.08	0.33	0.63	0.80
0.73	0.41	0.21	0.88	0.08	0.80	0.06	0.67	0.69	0.07
0.15	0.22	0.66	0.69	0.06	0.64	0.34	0.11	0.62	0.12

Source: ACT Inc., The College Board

a. Examine the data in the table. What shape distribution do you anticipate these data will produce? Explain why you chose the shape distribution you did.

b. Construct the histogram of these percentages.

c. Describe, in detail, the distribution of your histogram in part b.

d. Compare your description in part c to your expectations in part a. How close were you? What did you not think about in part a that is apparent now that you see the actual distribution?

e. Find the mean percentage.

f. Where does the mean fall on the distribution? Locate the mean on the histogram constructed for part b. Is the mean percentage representative of these data? Explain.

g. Find the standard deviation.

h. On the histogram constructed for part b, locate the values 1 standard deviation above and below the mean. How much of the distribution is between these values?

i. Why is the standard deviation so large? Explain in detail.

2.196 [EX02-196] The dollar amounts listed here are the average hourly earnings of production or nonsupervisory workers on private nonfarm payrolls by major industry. Both five years of information and eleven years of information are listed on the CD located on the inside the back cover. Investigate this information, looking for any pattern that might exist. Find both numerical and graphic statistics by months and by years. Describe all patterns found:

Year	Jan.	Feb.	Mar.
2001	14.48	14.54	14.58
2002	15.05	15.11	15.15

••• Remainder of data on Student's Suite CD-ROM

Source: http://www.bls.gov/

a. Use the 5 years listed on the CD. [EX02-196]

b. Use the 11 years listed on the CD. [DS02-196]

2.197 [EX02-197] Who ate the M&M's? The following table gives the color counts and net weight (in grams) for a sample of 30 bags of M&M's. The advertised net weight is 47.9 grams per bag.

Case	Red	Gr.	Blue	Or.	Yel.	Br.	Weight
1	15	9	3	3	9	19	49.79
2	9	17	19	3	3	8	48.98

••• Remainder of data on Student's Suite CD-ROM

Source: http://www.math.uah.edu/stat/
Christine Nickel and Jason York, ST 687 project, Fall 1998

There is something about one case in this data set that is suspiciously inconsistent with the rest of the data. Find the inconsistency.

a.   Construct two different graphs for the weights.

b.   Calculate several numerical statistics for the weight data.

c.   Did you find any potential inconsistencies in parts a and b? Explain.

d.   Find the number of M&M's in each bag.

e.   Construct two different graphs for the number of M&M's per bag.

f.   Calculate several numerical statistics for the number of M&M's per bag.

g.   What inconsistency did you find in parts e and f? Explain.

h.   Give a possible explanation as to why the inconsistency does not show up in the weight data but does show up in the number data.

2.198 For a normal (or bell-shaped) distribution, find the percentile rank that corresponds to:

a. $z = 2$ b. $z = -1$

c.   Sketch the normal curve, showing the relationship between the z-score and the percentiles for parts a and b.

2.199 For a normal (or bell-shaped) distribution, find the z-score that corresponds to the kth percentile:

a. $k = 20$ b. $k = 95$

c.   Sketch the normal curve, showing the relationship between the z-score and the percentiles for parts a and b.

2.200 Bill and Rob are good friends, although they attend different high schools in their city. The city school system uses a battery of fitness tests to test all high school students. After completing the fitness tests, Bill and Rob are comparing their scores to see who did better in each event. They need help.

	Sit-ups	Pull-ups	Shuttle Run	50-Yard Dash	Softball Throw
Bill	$z = -1$	$z = -1.3$	$z = 0.0$	$z = 1.0$	$z = 0.5$
Rob	61	17	9.6	6.0	179 ft
Mean	70	8	9.8	6.6	173 ft
Std. Dev.	12	6	0.6	0.3	16 ft

Bill received his test results in z-scores, whereas Rob was given raw scores. Since both boys understand raw scores, convert Bill's z-scores to raw scores in order to make an accurate comparison.

2.201 Twins Jean and Joan Wong are in fifth grade (different sections), and the class has been given a series of ability tests. If the scores for these ability tests are approximately normally distributed, which girl has the higher relative score on each of the skills listed? Explain your answers.

Skill	Jean: z-Score	Joan: Percentile
Fitness	2.0	99
Posture	1.0	69
Agility	1.0	88
Flexibility	−1.0	35
Strength	0.0	50

2.202 The scores achieved by students in America make the news often, and all kinds of conclusions are drawn based on these scores. The ACT Assessment is designed to assess high school students' general educational development and their ability to complete college-level work. The following table shows the mean and standard deviation for the scores of all high school graduates in 2001 and in 2004 on the four ACT tests and their composite.

	English	Mathematics	Reading	Science Reasoning	Composite
2001					
Mean	20.5	20.7	21.3	21.0	21.0
Std. Dev.	5.6	5.0	6.0	4.6	4.7
2004					
Mean	20.4	20.7	21.3	20.9	20.9
Std. Dev.	5.9	5.0	6.0	4.6	4.8

Source: American College Testing

Based on the information in the table:

a. Discuss how the five distributions are similar and different from each other with regard to central value and spread.

b. Discuss any shift in the scores between 2001 and 2004. Include in your answer specifics about how each test distribution has, or has not, changed according to central value and spread.

2.203 [EX02-203] Manufacturing specifications are often based on the results of samples taken from satisfactory pilot runs. The following data resulted from just such a situation, in which eight pilot batches were completed and sampled. The resulting particle sizes are in angstroms (where $1 \text{ Å} = 10^{-8}$ cm):

3923	3807	3786	3710	4010	4230	4226	4133

a. Find the sample mean.

b. Find the sample standard deviation.

c. Assuming that particle size has an approximately normal distribution, determine the manufacturing specification that bounds 95% of the particle sizes (that is, find the 95% interval, $\bar{x} \pm 2s$).

2.204 [EX02-204] Delco Products, a division of General Motors, produces a bracket that is used as part of a power doorlock assembly. The length of this bracket is constantly being monitored. A sample of 30 power door brackets had the following lengths (in millimeters):

11.86	11.88	11.88	11.91	11.88	11.88	11.88	11.88	11.88	11.86
11.88	11.88	11.88	11.88	11.86	11.83	11.86	11.86	11.88	11.88
11.88	11.83	11.86	11.86	11.86	11.88	11.88	11.86	11.88	11.83

Source: With permission of Delco Products Division, GMC

a. Without doing any calculations, what would you estimate for the sample mean?

b. Construct an ungrouped frequency distribution.

c. Draw a histogram of this frequency distribution.

d. Use the frequency distribution and calculate the sample mean and standard deviation.

e. Determine the limits of the $\bar{x} \pm 3s$ interval and mark this interval on the histogram.

f. The product specification limits are 11.7–12.3. Does the sample indicate that production is within these requirements? Justify your answer.

2.205 [EX02-205] Americans love soups, and soups remain one of the most popular foods for lunch and as an appetizer before dinner. Manufacturers provide the calorie and sodium content on the label. The data for 40 popular multiserving (8 oz.) cans and mixes, most of which were low-fat varieties, appear in the following table.

Soup Brand	Calories	Sodium (mg)
Arrowhead Mills Red Lentil	100	230
Baxters Italian Bean & Pasta	80	430
••• Remainder of data on Student's Suite CD-ROM		

a. Compute the mean and standard deviation of both calorie and sodium content of the soups listed in the table.

b. Use your answers to part a to test Chebyshev's theorem that at least 75% of the soups' calorie and sodium content will fall within ± 2 standard deviations from the mean. Is this the case?

c. Find the limits for ± 1 standard deviation from the mean for the soups' sodium content. Does sodium content of soups appear to follow the empirical rule? Explain.

2.206 [EX02-206] The manager of Jerry's Barber Shop recently asked his last 50 customers to punch a time card when they first arrived at the shop and to punch out right after they paid for their haircut. He then used the data on the cards to measure how long it took Jerry and his barbers to cut hair and used that information to schedule their appointment intervals. The following times (in minutes) were tabulated:

50	21	36	35	35	27	38	51	28	35
32	32	27	25	24	38	43	46	29	45
40	27	36	38	35	31	28	38	33	46
35	31	38	48	23	35	43	31	32	38
43	32	18	43	52	52	49	53	46	19

a. Construct a stem-and-leaf plot of these data.

b. Compute the mean, median, mode, range, midrange, variance, and standard deviation of the haircut service times.

c. Construct a 5-number summary table.

d. According to Chebyshev's theorem, at least 75% of the haircut service times will fall between what two values? Is this true? Explain why or why not.

e. How far apart would you recommend that Jerry schedule his appointments to keep his shop operating at a comfortable pace?

2.207 [EX02-207] Each year, stock car drivers compete for NASCAR. Points are earned on the basis of finishes in sanctioned races scheduled on the circuit. At the end of the 2004 season, the standings posted at NASCAR.com; the top 32 drivers are shown in the following table.

Driver	Points	Driver	Points
Kurt Busch	6506	Jimmie Johnson	6498

••• Remainder of data on Student's Suite CD-ROM

Source: NASCAR

a. Draw a dotplot.

b. Calculate the mean and standard deviation of the points accumulated by the NASCAR drivers.

c. Construct a 5-number summary table and draw a box-and-whiskers display.

d. According to Chebyshev's theorem, at least 75% of the points will fall between what two amounts? Is this the case?

e. According to the empirical rule, approximately 68% of the points will fall between what two amounts? Is this the case?

f. Compare your answers to parts d and e to the results predicted by the empirical rule. Does your comparison suggest that the distribution

of NASCAR points approximates the normal distribution? Explain.

g. (Optional.) The 2004 season had a total of 88 drivers earning points. The complete list of 88 drivers is included in the data file for this exercise. Using all 88 pieces of data, answer the questions in parts a through e. [DS02-207]

2.208 The following dotplot shows the number of attempted passes thrown by the quarterbacks for 22 of the NFL teams that played on one particular Sunday afternoon.

a. Describe the distribution, including how points *A* and *B* relate to the others.

b. If you remove point *A*, and maybe point *B*, would you say the remaining data have an approximately normal distribution? Explain.

c. Based on the information about distributions that Chebyshev's theorem and the empirical rule give us, how typical an event do you think point *A* represents? Explain.

2.209 Starting with the data values of 70 and 85, add three data values to your sample so that the sample has the following: (Justify your answer in each case.)

a. A standard deviation of 5

b. A standard deviation of 10

c. A standard deviation of 15

d. Compare your three samples and the variety of values needed to obtain each of the required standard deviations.

2.210 Make up a set of 18 data (think of them as exam scores) so that the sample meets each of these sets of criteria:

a. Mean is 75, and standard deviation is 10.

b. Mean is 75, maximum is 98, minimum is 40, and standard deviation is 10.

Figure for Exercise 2.208

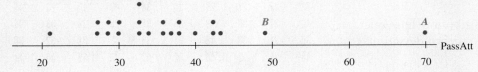

c. Mean is 75, maximum is 98, minimum is 40, and standard deviation is 15.

d. How are the data in the sample for part b different from those in part c?

2.211 Construct two different graphs of the points (62, 2), (74, 14), (80, 20), and (94, 34).

a. On the first graph, along the horizontal axis, lay off equal intervals and label them 62, 74, 80, and 94; lay off equal intervals along the vertical axis and label them 0, 10, 20, 30, and 40. Plot the points and connect them with line segments.

b. On the second graph, along the horizontal axis, lay off equally spaced intervals and label them 60, 65, 70, 75, 80, 85, 90, and 95; mark off the vertical axis in equal intervals and label them 0, 10, 20, 30, and 40. Plot the points and connect them with line segments.

c. Compare the effect that scale has on the appearance of the graphs in parts a and b. Explain the impression presented by each graph.

2.212 [EX02-212] When the Internet study "You and the Internet" (Section 2.1, p. 39) was performed, it appeared that the variable x, the number of Internet activities in a week, had an approximately normal distribution. That distribution is approximated by this relative frequency distribution:

Internet Activities/ Week, x	Relative Frequency	Internet Activities/ Week, x	Relative Frequency
1	0.01	8	0.14
2	0.03	9	0.11
3	0.05	10	0.08
4	0.09	11	0.05
5	0.10	12	0.04
6	0.14	13	0.03
7	0.13		

a. Select a random sample of size 40 from this relative frequency representation of the population of all Internet users.

b. Construct a histogram of the sample obtained in part a. Do not group the data. (See instructions that follow.)

MINITAB (Release 14)

Input the x values into C1 and the corresponding relative frequencies into C2; then continue with:

```
Choose:      Calc > Random Data > Discrete
Enter:       Generate: 40 rows of data
             Store in column(s): C3
             Values (of x) in: C1
             Probabilities in: C2 > OK
```

Excel

Input the x values into column A and the corresponding relative frequencies into column B; then continue with:

```
Choose:      Tools > Data Analysis > Random Number
             Generation > OK
Enter:       Number of Variables: 1
             Number of Random Numbers: 40
             Distribution: Discrete
             Value & Prob. Input Range:
             (A2:B5 select data cells not labels)
Select:      Output Range
Enter:       (C1 or select cell)
```

c. Find the mean, median, and standard deviation of the sample obtained in part a.

d. Repeat parts a–c three more times, being sure to keep the answers for each set of data together.

e. Describe the similarities and differences between the distributions shown on the four histograms.

f. Make a chart displaying the numerical statistics for each of the four samples and describe the variability from sample to sample of each statistic.

g. The four samples were all drawn randomly from the same distribution. Write a statement describing the overall variability between these four random samples.

2.213 Use a computer to generate a random sample of 500 values of a normally distributed variable x with a mean of 100 and a standard deviation of 20. Construct a histogram of the 500 values.

a. Use the computer commands on page 101 to randomly generate a sample of 500 data from a normal distribution with mean 100 and standard deviation 20. Construct a histogram using class boundaries that are multiples of the standard deviation 20; that is, use boundaries from 20 to 180 in intervals of 20 (see commands on pp. 61–62).

Let's consider the 500 *x* values found in part a as a population.

b. Use the computer commands on pages 101–102 to randomly select a sample of 30 values from the population found in part a. Construct a histogram of the sample with the same class intervals used in part a.

c. Repeat part b three times.

d. Calculate several values (mean, median, maximum, minimum, standard deviation, etc.) that describe the population and each of the four samples. (See p. 89 for commands.)

e. Do you think a sample of 30 data adequately represents a population? (Compare each of the four samples found in parts b and c to the population.)

2.214 Repeat Exercise 2.213 using a different sample size. You might try a few different sample sizes: $n = 10$, $n = 15$, $n = 20$, $n = 40$, $n = 50$, $n = 75$. What effect does increasing the sample size have on the effectiveness of the sample in depicting the population? Explain.

2.215 Repeat Exercise 2.213 using populations with different shaped distributions.

a. Use a uniform or rectangular distribution. (Replace the subcommands used in Exercise 2.213; in place of NORMAL use: UNIFORM with a low of 50 and a high of 150, and use class boundaries of 50 to 150 in increments of 10.)

b. Use a skewed distribution. (Replace the subcommands used in Exercise 2.213; in place of NORMAL use: POISSON 50 and use class boundaries of 20 to 90 in increments of 5.)

c. Use a J-shaped distribution. (Replace the subcommands used in Exercise 2.213; in place of NORMAL use: EXPONENTIAL 50 and use class boundaries of 0 to 250 in increments of 10.)

d. Does the shape of the distribution of the population have an effect on how well a sample of size 30 represents the population? Explain.

e. What effect do you think changing the sample size has on the effectiveness of the sample to depict the population? Try a few different sample sizes. Do the results agree with your expectations? Explain.

2.216 [EX02-216] *Outliers!* How often do they occur? What do we do with them? Complete part a to see how often outliers can occur. Then complete part b to decide what to do with outliers.

a. Use the technology of your choice to take samples of various sizes (10, 30, 100, 300 would be good choices) from a normal distribution (mean of 100 and standard deviation of 20 will work nicely) and see how many outliers a randomly generated sample contains. You will probably be surprised. Generate 10 samples of each size for a more representative result. Describe your results—in particular comment on the frequency of outliers in your samples.

MINITAB

Choose:	**Calc > Random Data > Normal**
Enter:	Generate **10** rows of data
	(Use $n = 10$, 30, 100, 300)
	Store in column(s): **C1–C10**
	Mean: **100**
	Stand. Dev.: **20**
Choose:	**Graph > Boxplot > Multiple Y's Simple >**
	OK
Enter:	Graph variables: **C1–C10**
Choose:	**Data View**
Select:	**Interquartile range box**
	Outlier symbols

In practice, we want to do something about the data points that are discovered to be outliers. First, the outlier should be inspected: if there is some obvious reason why it is incorrect, it should be corrected. (For example, a woman's height of 59 inches may well be entered incorrectly as 95 inches, which would be nearly 8 feet tall and is a very unlikely height. If the data value can be corrected, fix it! Otherwise, you must weigh the choice between discarding good data (even if they are different) and keeping erroneous data. At this level, it is probably best to make a note about the outlier and continue with using the solution. To help understand the effect of removing an outlier value, let's look at this set of data, randomly generated from a normal distribution $N(100, 20)$.

b. Construct a boxplot and identify any outliers.

74.2	84.5	88.5	110.8	97.6	100.2	116.4	78.3	154.8	144.7
110.6	93.7	113.3	96.1	86.7	97.3	102.8	91.8	58.5	120.1
102.8	82.5	107.6	91.1	95.7	98	98.4	81.9	58.5	118.1

c. Remove the outlier and construct a new box-plot.

d. Describe your findings and comment on why it might be best and less confusing while studying introductory statistics not to discard outliers.

2.217 [EX02-217] The distribution of credit hours, per student, taken this semester at a certain college was as follows:

Credit Hours	Frequency	Credit Hours	Frequency
3	75	15	400
6	150	16	1050
8	30	17	750
9	50	18	515
12	70	19	120
14	300	20	60

a. Draw a histogram of the data.

b. Find the five measures of central tendency.

c. Find Q_1 and Q_3.

d. Find P_{15} and P_{12}.

e. Find the three measures of dispersion (range, s^2, and s).

2.218 [EX02-218] An article in *Therapeutic Recreation Journal* reports a distribution for the variable "number of persistent disagreements." Sixty-six patients and their therapeutic recreation specialist each answered a checklist of problems with yes or no. Disagreement occurs when the specialist and the patient did not respond identically to an item on the checklist. It becomes a persistent disagreement if the item remains in disagreement after a second interview.

x	0	1	2	3	4	5	6	7	8	9	10	11
f	2	2	4	10	7	9	8	11	7	3	1	2

Source: Data reprinted with permission of the National Recreation and Park Association, Alexandria, VA, from Pauline Petryshen and Diane Essex-Sorlie, "Persistent Disagreement Between Therapeutic Recreation Specialists and Patients in Psychiatric Hospitals," *Therapeutic Recreation Journal*, Vol. XXIV, Third Quarter, 1990.

a. Draw a dotplot of these sample data.

b. Find the median number of persistent disagreements.

c. Find the mean number of persistent disagreements.

d. Find the standard deviation of the number of persistent disagreements.

e. Draw a vertical line on the dotplot at the mean.

f. Draw a horizontal line segment on the dotplot whose length represents the standard deviation (start at the mean).

2.219 [EX02-219] *USA Today* (October 25, 1994) reported in the USA Snapshot "Mystery of the remote" that 44% of the families surveyed never misplaced the family television remote control, 38% misplaced it one to five times weekly, and 17% misplaced it more than five times weekly. One percent of the families surveyed didn't know. Suppose you took a survey that resulted in the following data. Let x be the number of times per week that the family's television remote control gets misplaced.

x	0	1	2	3	4	5	6	7	8	9
f	220	92	38	21	24	30	34	20	16	5

a. Construct a histogram.

b. Find the mean, median, mode, and midrange.

c. Find the variance and standard deviation.

d. Find Q_1, Q_3, and P_{90}.

e. Find the midquartile.

f. Find the 5-number summary and draw a box-and-whiskers display.

2.220 [EX02-220] The following table shows the age distribution of heads of families:

Age of Head of Family	Number	Age of Head of Family	Number
20–25	23	50–55	48
25–30	38	55–60	39
30–35	51	60–65	31
35–40	55	65–70	26
40–45	53	70–75	20
45–50	50	75–80	16

a. Find the mean age of the heads of families.

b. Find the standard deviation.

2.221 [EX02-221] The lifetimes of 220 incandescent 60-watt lamps were obtained and yielded the frequency distribution shown in this table:

Class Limits	f	Class Limits	f
500–600	3	1000–1100	57
600–700	7	1100–1200	23
700–800	14	1200–1300	13
800–900	28	1300–1400	7
900–1000	64	1400–1500	4

a. Construct a histogram of these data using a vertical scale for the relative frequencies.

b. Find the mean lifetime.

c. Find the standard deviation of the lifetimes.

2.222 Do your monthly car payments prevent you from spending money on other things? More than 56% say, "yes." The distribution of the amount spent on monthly car payments is as follows: 32% spend less than $300, 43% spend $300 to $499, 17% spend $500 to $699, and 8% spend $700 or more. Suppose that this information was obtained from a sample of 1000 people with car payments. Use values of $150, $400, $600, and $800 as class midpoints, and estimate the sample mean and the standard deviation for the variable x, amount spent.

2.223 [EX02-223] The earnings per share for 40 firms in the radio and transmitting equipment industry follow:

4.62	0.10	1.29	7.25	6.04	3.20	9.56	4.90	4.22	3.71
0.25	1.34	2.11	5.39	0.84	−0.19	3.72	2.27	2.08	1.12
1.07	2.50	2.14	3.46	1.91	7.05	5.10	1.80	0.91	0.50
5.56	1.62	1.36	1.93	2.05	2.75	3.58	0.44	3.15	1.93

a. Prepare a frequency distribution and a frequency histogram for these data.

b. Which class of your frequency distribution contains the median?

Chapter Project

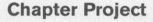

You and the Internet

Let's return to Section 2.1, "You and the Internet" (p. 39), as a way to assess what we have learned in this chapter. Based on the percents stated in the "What Users Do on the Internet" graph, respondents were able to pick more than one Internet activity. If you were asked, how many of the listed activities would you select as something you do? Suppose a sample of students was asked about their Internet activities. Would their answers differ from yours? Would they differ from the 4000 respondents? "Putting Chapter 2 to Work" will help us answer these questions.

Putting Chapter 2 to Work

2.224 [EX02-001] Students in a statistics course offered over the Internet were asked how many different Internet activities they engaged in during a typical week. The following data show the number of activities:

6	7	3	6	9	10	8	9	9	6	4	9	4	9
4	2	3	5	13	12	4	6	4	9	5	6	9
11	5	6	5	3	7	9	6	5	12	2	6	9

a. List all types of charts and graphs shown in Chapter 2 that would be appropriate for use with the set of 40 data listed.

b. What types of graphs would not be appropriate? Explain why.

c. Display the data using each of the charts and graphs listed in part a.

d. Which graph do you think best represents the data? Explain why.

e. Find the five measures of central tendency for these data (mean, median, mode, midrange, and midquartile).

f. Find the three measures of dispersion for the data (range, variance, and standard deviation).

g. Find the value of several measures of position: P_5, P_{10}, Q_1, Q_3, P_{90}, and P_{98}.

h. How many different Internet activities do you engage in during a typical week? Using the mean and standard deviation calculated in parts e and f, determine your z-score. What is this telling you about yourself with respect to statistics students' Internet usage?

i. Use one graph from part c plus at least one measure of central tendency and one measure of dispersion, and write a description of statistics students' Internet usage, number of Internet activities per week.

j. According to the empirical rule, if the distribution is normal, approximately 68% of the number of different Internet activities engaged in by statistics students will fall between what two values? Is this true? Why or why not?

k. According to Chebyshev's theorem, approximately 75% of the number of different Internet activities engaged in by statistics students will fall within what two values? Is this true? Why or why not?

l. The sample information pictured in Section 2.1's "What Users do on the Internet" graph on page 39 is different than, but related to, the sample information you have been working with in parts a through k. Describe the data collected for the graph in Section 2.1 and explain how they differ from the data listed here.

Your Study

2.225 a. Design your own study of Internet usage. Define a specific population that you will sample, describe your sampling plan, collect your data, and answer parts c through l in "Putting Chapter 2 to Work," Exercise 2.224.

b. Discuss the differences and similarities between the Internet usage described by the sample of 40 statistics students (given in Exercise 2.224) and your sample.

Chapter Practice Test

PART I: Knowing the Definitions

Answer "True" if the statement is always true. If the statement is not always true, replace the words in bold with the words that make the statement always true.

2.1 The **mean** of a sample always divides the data into two halves (half larger and half smaller in value than itself).

2.2 A measure of **central tendency** is a quantitative value that describes how widely the data are dispersed about a central value.

2.3 The sum of the squares of the deviations from the mean, $\sum (x - \bar{x})^2$, will **sometimes** be negative.

2.4 For any distribution, the sum of the deviations from the mean equals **zero**.

2.5 The standard deviation for the set of values 2, 2, 2, 2, and 2 is **2**.

2.6 On a test, John scored at the 50th percentile and Jorge scored at the 25th percentile; therefore, John's test score was **twice** Jorge's test score.

2.7 The frequency of a class is the number of pieces of data whose values fall within the **boundaries** of that class.

2.8 **Frequency distributions** are used in statistics to present large quantities of repeating values in a concise form.

2.9 The unit of measure for the standard score is always **standard deviations**.

2.10 For a bell-shaped distribution, the range will be approximately equal to **6 standard deviations**.

PART II: Applying the Concepts

2.11 The results of a consumer study completed at Corner Convenience Store are reported in the accompanying histogram. Answer each question.

a. What is the class width?

b. What is the class midpoint for the class 31–61?

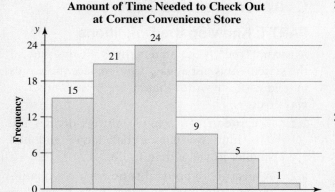

Amount of Time Needed to Check Out at Corner Convenience Store

c. What is the upper boundary for the class 61–91?

d. What is the frequency of the class 1–31?

e. What is the frequency of the class that contains the largest observed value of x?

f. What is the lower boundary of the class with the largest frequency?

g. How many pieces of data are shown in this histogram?

h. What is the value of the mode?

i. What is the value of the midrange?

j. Estimate the value of the 90th percentile, P_{90}.

2.12 A sample of the purchases of several Corner Convenience Store customers resulted in the following sample data (x = number of items purchased per customer):

x	1	2	3	4	5
f	6	10	9	8	7

a. What does the 2 represent?

b. What does the 9 represent?

c. How many customers were used to form this sample?

d. How many items were purchased by the customers in this sample?

e. What is the largest number of items purchased by one customer?

Find each of the following (show formulas and work):

f. Mode g. Median h. Midrange

i. Mean j. Variance k. Standard deviation

2.13 Given the set of data 4, 8, 9, 8, 6, 5, 7, 5, 8, find each of the following sample statistics:

a. Mean   b. Median   c. Mode

d. Midrange   e. First quartile   f. P_{40}

g. Variance   h. Standard deviation

i. Range

2.14 a. Find the standard score for the value $x = 452$ relative to its sample, where the sample mean is 500 and the standard deviation is 32.

b. Find the value of x that corresponds to the standard score of 1.2, where the mean is 135 and the standard deviation is 15.

PART III: Understanding the Concepts

Answer all questions.

2.15 The Corner Convenience Store kept track of the number of paying customers it had during the noon hour each day for 100 days. The resulting statistics are rounded to the nearest integer:

mean = 95	midrange = 93
median = 97	range = 56
mode = 98	standard
first quartile = 85	deviation = 12
third quartile = 107	

a. The Corner Convenience Store served what number of paying customers during the noon hour more often than any other number? Explain how you determined your answer.

b. On how many days were there between 85 and 107 paying customers during the noon hour? Explain how you determined your answer.

c. What was the greatest number of paying customers during any one noon hour? Explain how you determined your answer.

d. For how many of the 100 days was the number of paying customers within 3 standard deviations of the mean ($\bar{x} \pm 3s$)? Explain how you determined your answer.

2.16 Mr. VanCott started his own machine shop several years ago. His business has grown and become very successful in recent years.

Currently he employs 14 people, including himself, and pays the following annual salaries:

Owner, President	$80,000	Worker	$25,000
Business Manager	50,000	Worker	25,000
Production Manager	40,000	Worker	25,000
Shop Foreman	35,000	Worker	20,000
Worker	30,000	Worker	20,000
Worker	30,000	Worker	20,000
Worker	28,000	Worker	20,000

a. Calculate the four "averages": mean, median, mode, and midrange.

b. Draw a dotplot of the salaries and locate each of the four averages on it.

c. Suppose you were the feature writer assigned to write this week's feature story on Mr. VanCott's machine shop, one of a series on local small businesses that are prospering. You plan to interview Mr. VanCott, his business manager, the shop foreman, and one of his newer workers. Which statistical average do you think each will give when asked, "What is the average annual salary paid to the employees here at VanCott's?" Explain why each person interviewed has a different perspective and why this viewpoint may cause each to cite a different statistical average.

d. What is there about the distribution of these salaries that causes the four "average values" to be so different?

2.17 Create a set of data containing three or more values in the following cases:

a. Where the mean is 12 and the standard deviation is 0

b. Where the mean is 20 and the range is 10

c. Where the mean, median, and mode are all equal

d. Where the mean, median, and mode are all different

e. Where the mean, median, and mode are all different and the median is the largest and the mode is the smallest

f. Where the mean, median, and mode are all different and the mean is the largest and the median is the smallest

2.18 A set of test papers was machine scored. Later it was discovered that 2 points should be added to each score. Student A said, "The mean score should also be increased by 2 points." Student B added, "The standard deviation should also be increased by 2 points." Who is right? Justify your answer.

2.19 Student A stated, "Both the standard deviation and the variance preserve the same unit of measurement as the data." Student B disagreed, arguing, "The unit of measurement for variance is a meaningless unit of measurement." Who is right? Justify your answer.

CHAPTER

3

Descriptive Analysis and Presentation of Bivariate Data

AP/Wide World Photos

The Kid Is All Grown Up

MINNEAPOLIS—The Kid is all grown up, and he has an NBA MVP award to prove it. Kevin Garnett got 120 of 123 first-place votes to beat two-time winner Tim Duncan for the honor Monday, three days after his Minnesota Timberwolves won a playoff series for the first time. Garnett's teammates attended a packed news conference at the Timberwolves' arena, and he praised them repeatedly.

Playing everywhere from center to point guard, the 7-footer averaged 24.2 points, a league-leading 13.9 rebounds and 5.0 assists this season—and his playoff stats are even better. Garnett joined Larry Bird as the only players to average 20 points, 10 rebounds and five assists for five consecutive years.

Nicknamed "The Kid," Garnett made the All-Star team in his second season, and his success helped fuel the wave of preps-to-pros players.

GARNETT EARNS 120 OF 123 FIRST-PLACE VOTES
Associated Press
Monday, May 4, 2004

Minnesota Timberwolves, 2003–2004 Regular Season [EX03-001]

Player	Personal Fouls per Game	Points per Game	Player	Personal Fouls per Game	Points per Game
Garnett	2.5	24.2	Madsen	2.4	3.6
Cassell	3	19.8	Martin	1.4	3.4
Sprewell	1.2	16.8	McLeod	1.2	2.7
Szczerbiak	1.5	10.2	Goldwire	1.0	2.6
Hudson	1.1	7.5	Miller	1.9	2.5
Hoiberg	1.7	6.7	Johnson	2.4	1.9
Olowokandi	3.2	6.5	Lewis	0.7	1.1
Trent	1.9	5.6	Ebi	0.4	0.8
Hassell	2.5	5			

Source: http://sports.espn.go.com/nba/teams

Do you play basketball or at least follow the sport? Does it seem like those that make more baskets also have the most personal fouls? Think of those players who do not score any points, do they still make personal fouls? Now if you do not play basketball or even follow it, you still know about relationships. Just think about yourself. Does it seem that as you grew taller, your shoe size was also increasing? Is there a relationship between a person's height and his or her shoe size? Does it seem that those students who study more get better grades? Is there a relationship between hours studied and grades? Does it seem that those students who travel farther one way to school also need more time to travel to school? As you work through Chapter 3, you will learn how to display two variable data in such a form that exhibits their relationship. From there you will be able to determine the strength of the relationship, called *correlation,* and the equation of a line used for predicting, called *regression analysis.* Once you complete the three main topics just outlined, you will be able to further investigate "The Kid Is All Grown Up" in the Chapter Project on page 199.

SECTION 3.1 EXERCISES

Statistics⟁Now™

Datasets can be found on your Student's Suite CD-ROM or at the StatisticsNow website at **http://1pass.thomson.com**.

3.1 [EX03-001] Refer to the 2003–2004 Timberwolves data on page 145 to answer the following questions:

a. Is there a relationship (pattern) between the two variables, points scored per game and number of personal fouls committed per game? Explain why or why not.

b. Do you think it is reasonable (or possible) to predict the number of points scored based on the number of personal fouls committed per game for a Minnesota Timberwolves' player? Explain why or why not.

3.2 a. Is there relationship between a person's height and shoe size as he or she grows from an infant to age 16? As one variable gets larger, does the other also get larger? Explain your answers.

b. Is there a relationship between height and shoe size for people who are older than 16 years of age? Do taller people wear larger shoes? Explain your answers.

3.2   Bivariate Data

In Chapter 2, we learned how to graphically display and numerically describe sample data for one variable. We will now expand these techniques to cover sample data that involve two paired variables.

Bivariate data: The values of two different variables that are obtained from the same population element.

Each of the two variables may be either *qualitative* or *quantitative*. As a result, three combinations of variable types can form bivariate data:

1. Both variables are qualitative (attribute).
2. One variable is qualitative (attribute), and the other is quantitative (numerical).
3. Both variables are quantitative (both numerical).

In this section we present tabular and graphic methods for displaying each of these combinations of bivariate data.

Two Qualitative Variables

When bivariate data result from two qualitative (attribute or categorical) variables, the data are often arranged on a **cross-tabulation** or **contingency table.** Let's look at an example.

EXAMPLE 3.1

Statistics ⊖ Now™

Watch a video example at **http://1pass.thomson.com** or on your CD.

FYI $m = n$(rows) $n = n$(cols) for an $m \times n$ contingency table.

Constructing Cross-Tabulation Tables

Thirty students from our college were randomly identified and classified according to two variables: gender (M/F) and major (liberal arts, business administration, technology), as shown in Table 3.1. These 30 bivariate data can be summarized on a 2×3 cross-tabulation table, where the two rows represent the two genders, male and female, and the three columns represent the three major categories of liberal arts (LA), business administration (BA), and technology (T). The entry in each cell is found by determining how many students fit into each category. Adams is male (M) and liberal arts (LA) and is classified in the cell in the first row, first column. See the red tally mark in Table 3.2. The other 29 students are classified (tallied, shown in black) in a similar fashion.

The resulting 2×3 cross-tabulation (contingency) table, Table 3.3, shows the frequency for each cross-category of the two variables along

TABLE 3.1

Genders and Majors of 30 College Students [TA03-01]

Name	Gender	Major	Name	Gender	Major	Name	Gender	Major
Adams	M	LA	Feeney	M	T	McGowan	M	BA
Argento	F	BA	Flanigan	M	LA	Mowers	F	BA
Baker	M	LA	Hodge	F	LA	Ornt	M	T
Bennett	F	LA	Holmes	M	T	Palmer	F	LA
Brand	M	T	Jopson	F	T	Pullen	M	T
Brock	M	BA	Kee	M	BA	Rattan	M	BA
Chun	F	LA	Kleeberg	M	LA	Sherman	F	LA
Crain	M	T	Light	M	BA	Small	F	T
Cross	F	BA	Linton	F	LA	Tate	M	BA
Ellis	F	BA	Lopez	M	T	Yamamoto	M	LA

TABLE 3.2

Cross-Tabulation of Gender and Major (tallied)

Gender	Major LA		Major BA		Major T	
M	‖‖‖	(5)	‖‖‖‖	(6)	‖‖‖‖ ‖	(7)
F	‖‖‖‖ ‖	(6)	‖‖‖‖	(4)	‖	(2)

TABLE 3.3

Cross-Tabulation of Gender and Major (frequencies)

Gender	Major LA	BA	T	Row Total
M	5	6	7	18
F	6	4	2	12
Col. Total	11	10	9	30

with the row and column totals, called *marginal totals* (or *marginals*). The total of the marginal totals is the *grand total* and is equal to *n*, the *sample size*.

Contingency tables often show percentages (relative frequencies). These percentages can be based on the entire sample or on the subsample (row or column) classifications.

Percentages Based on the Grand Total (Entire Sample)

The frequencies in the contingency table shown in Table 3.3 can easily be converted to percentages of the grand total by dividing each frequency by the grand total and multiplying the result by 100. For example, 6 becomes 20% $\left[\left(\dfrac{6}{30}\right) \times 100 = 20\right]$. See Table 3.4.

From the table of percentages of the grand total, we can easily see that 60% of the sample were male, 40% were female, 30% were technology majors, and so on. These same statistics (numerical values describing sample results) can be shown in a bar graph (see Figure 3.1).

TABLE 3.4

Cross-Tabulation of Gender and Major (relative frequencies; % of grand total)

Gender	Major LA	BA	T	Row Total
M	17%	20%	23%	60%
F	20%	13%	7%	40%
Col. Total	37%	33%	30%	100%

FIGURE 3.1 Bar Graph

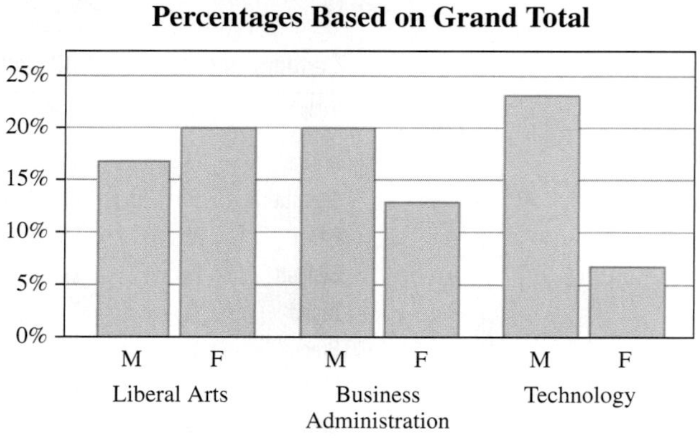

Percentages Based on Grand Total

Table 3.4 and Figure 3.1 show the distribution of male liberal arts students, female liberal arts students, male business administration students, and so on, relative to the entire sample.

Percentages Based on Row Totals

The frequencies in the same contingency table, Table 3.3, can be expressed as percentages of the row totals (or gender) by dividing each row entry by that row's total and multiplying the results by 100. Table 3.5 is based on row totals.

From Table 3.5 we see that 28% of the male students were majoring in liberal arts, whereas 50% of the female students were majoring in liberal arts. These same statistics are shown in the bar graph in Figure 3.2.

TABLE 3.5

Cross-Tabulation of Gender and Major (% of row totals)

| Gender | Major | | | |
	LA	BA	T	Row Total
M	28%	33%	39%	100%
F	50%	33%	17%	100%
Col. Total	37%	33%	30%	100%

FIGURE 3.2 Bar Graph

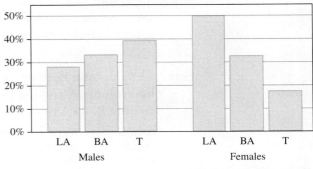

Percentages Based on Gender

Table 3.5 and Figure 3.2 show the distribution of the three majors for male and female students separately.

Percentages Based on Column Totals

The frequencies in the contingency table, Table 3.3, can be expressed as percentages of the column totals (or major) by dividing each column entry by that column's total and multiplying the result by 100. Table 3.6 is based on column totals.

From Table 3.6 we see that 45% of the liberal arts students were male, whereas 55% of the liberal arts students were female. These same statistics are shown in the bar graph in Figure 3.3.

TABLE 3.6

Cross-Tabulation of Gender and Major (% of column totals)

| Gender | Major | | | |
	LA	BA	T	Row Total
M	45%	60%	78%	60%
F	55%	40%	22%	40%
Col. Total	100%	100%	100%	100%

FIGURE 3.3 Bar Graph

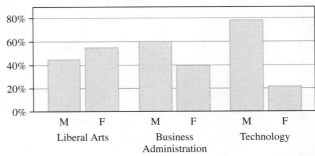

Percentages Based on Major

Table 3.6 and Figure 3.3 show the distribution of male and female students for each major separately.

TECHNOLOGY INSTRUCTIONS: CROSS-TABULATION TABLES

MINITAB (Release 14)   Input the row-variable single categorical values into C1 and the corresponding single column-variable categorical values into C2; then continue with:

```
Choose:   Stat > Tables > Cross Tabulation and Chi-Square
Enter:    Categorical variables: For rows: C1 For columns: C2
Select:   Counts
          Row Percents
          Column Percents
          Total Percents > OK
```

Suggestion: The four subcommands that are available for 'Display' can be used together; however, the resulting table will be much easier to read if one subcommand at a time is used.

Excel   Using column headings or titles, input the row-variable categorical values into column A and the corresponding column-variable categorical values into column B; then continue with:

```
Choose:   Data > Pivot Table and PivotChart Report ...
Select:   Microsoft Excel list or database > Next
Enter:    Range: (A1:B5 or select cells) > Next
Select:   Existing Worksheet
Enter:    (C1 or select cell) > Finish
Drag:     Headings to row or column (depends on preference)
          One heading into data area*
```

*For other summations, double click "Count of" in data area box; then continue with:
Choose: **Summarize by: Count > Options**
          Show data as: % of row or % of column or % of total > OK

TI-83/84 Plus   The categorical data must be numerically coded first; use 1, 2, 3, . . . for the various row variables and 1, 2, 3, . . . for the various column variables. Input the numeric row-variable values into L1 and the corresponding numeric column-variable values into L2; then continue with:

```
Choose:   PRGM > EXEC > CROSSTAB*
Enter:    ROWS: L1 >ENTER
          COLS: L2 >ENTER
```

The cross-tabulation table showing frequencies is stored in matrix [A], the cross-tabulation table showing row percentages is in matrix [B], column percentages in matrix [C], and percentages based on the grand total in matrix [D]. All matrices contain marginal totals. To view the matrices, continue with:

```
Choose:   MATRX > NAMES
Enter:    1:[A] or 2:[B] or 3:[C] or 4:[D] > ENTER
```

*Program 'CROSSTAB' is one of many programs that are available for downloading from the Duxbury website. See page 42 for specific instructions.

One Qualitative and One Quantitative Variable

When bivariate data result from one qualitative and one quantitative variable, the quantitative values are viewed as separate samples, each set identified by levels of the qualitative variable. Each sample is described using the techniques from Chapter 2, and the results are displayed side by side for easy comparison.

EXAMPLE 3.2

Statistics ⬡ Now™

Watch a video example at
http://1pass.thomson.com
or on your CD.

Constructing Side-by-Side Comparisons

The distance required to stop a 3000-pound automobile on wet pavement was measured to compare the stopping capabilities of three tire tread designs (see Table 3.7). Tires of each design were tested repeatedly on the same automobile on a controlled wet pavement.

TABLE 3.7

Stopping Distances (in feet) for Three Tread Designs [TA03-07]

Design A ($n = 6$)			Design B ($n = 6$)			Design C ($n = 6$)		
37	36	38	33	35	38	40	39	40
34	40	32	34	42	34	41	41	43

The design of the tread is a qualitative variable with three levels of response, and the stopping distance is a quantitative variable. The distribution of the stopping distances for tread design A is to be compared with the distribution of stopping distances for each of the other tread designs. This comparison may be made with both numerical and graphic techniques. Some of the available options are shown in Figure 3.4, Table 3.8, and Table 3.9.

FIGURE 3.4 Dotplot and Box-and-Whiskers Display Using a Common Scale

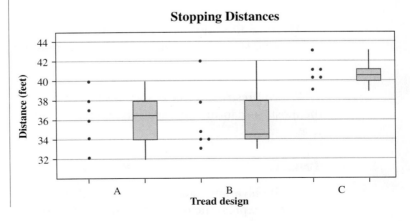

TABLE 3.8

5-Number Summary for Each Design

	Design A	Design B	Design C
High	40	42	43
Q_3	38	38	41
Median	36.5	34.5	40.5
Q_1	34	34	40
Low	32	33	39

TABLE 3.9

Mean and Standard Deviation for Each Design

	Design A	Design B	Design C
Mean	36.2	36.0	40.7
Standard deviation	2.9	3.4	1.4

TECHNOLOGY INSTRUCTIONS: SIDE-BY-SIDE BOXPLOTS AND DOTPLOTS

MINITAB (Release 14) Input the numerical values into C1 and the corresponding categories into C2; then continue with:

```
Choose:    Graph > Boxplot... > One Y, With Groups > OK
Enter:     Graph variables: C1 Categorical variables: C2 > OK
```

MINITAB commands to construct side-by-side dotplots for data in this form are located on page 49.

If the data for the various categories are in separate columns, use the MINITAB commands for multiple boxplots on page 98. If side-by-side dotplots are needed for data in this form, continue with:

```
Choose:    Graph > Dotplots
Select:    Multiple Y's, Simple > OK
Enter:     Graph variables: C1 C2 > OK
```

Excel Excel commands to construct a single boxplot are on page 98.

TI-83/84 Plus TI-83/84 commands to construct multiple boxplots are on page 99.
TI-83/84 commands to construct multiple dotplots are on page 49.

Much of the information presented here can also be demonstrated using many other statistical techniques, such as stem-and-leaf displays or histograms.

We will restrict our discussion in this chapter to descriptive techniques for the most basic form of correlation and regression analysis—the bivariate linear case.

Two Quantitative Variables

When the bivariate data are the result of two quantitative variables, it is customary to express the data mathematically as **ordered pairs** (x, y), where x is the **input variable** (sometimes called the **independent variable**) and y is the **output variable** (sometimes called the **dependent variable**). The data are said to be *ordered* because one value, *x*, is always written first. They are called *paired* because for each *x* value, there is a corresponding *y* value from the same source. For example, if *x* is height and *y* is weight, then a height and a corresponding weight are recorded for each person. The input variable *x* is measured or controlled in order to predict the output variable *y*. Suppose some research doctors are testing a new drug by prescribing different dosages and observing the lengths of the recovery times of their patients. The researcher can control the amount of drug prescribed, so the amount of drug is referred to as *x*. In the case of height and weight, either variable could be treated as input and the other as output, depending on the question being asked. However, different results will be obtained from the regression analysis, depending on the choice made.

In problems that deal with two quantitative variables, we present the sample data pictorially on a *scatter diagram*.

Scatter diagram: A plot of all the ordered pairs of bivariate data on a coordinate axis system. The input variable, *x*, is plotted on the horizontal axis, and the output variable, *y*, is plotted on the vertical axis.

Note: When you construct a scatter diagram, it is convenient to construct scales so that the range of the *y* values along the vertical axis is equal to or slightly shorter than the range of the *x* values along the horizontal axis. This creates a "window of data" that is approximately square.

EXAMPLE 3.3

Constructing a Scatter Diagram

Statistics Now™

Watch a video example at http://1pass.thomson.com or on your CD.

In Mr. Chamberlain's physical fitness course, several fitness scores were taken. The following sample is the numbers of push-ups and sit-ups done by 10 randomly selected students:

(27, 30) (22, 26) (15, 25) (35, 42) (30, 38)
(52, 40) (35, 32) (55, 54) (40, 50) (40, 43)

Table 3.10 shows these sample data, and Figure 3.5 shows a scatter diagram of the data.

TABLE 3.10

Data for Push-ups and Sit-ups [TA03-10]

Student	1	2	3	4	5	6	7	8	9	10
Push-ups, *x*	27	22	15	35	30	52	35	55	40	40
Sit-ups, *y*	30	26	25	42	38	40	32	54	50	43

The scatter diagram from Mr. Chamberlain's physical fitness course shows a definite pattern. Note that as the number of push-ups increased so did the number of sit-ups.

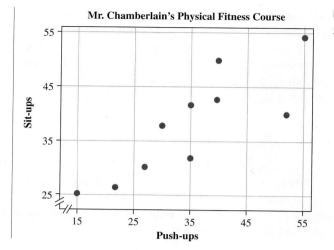

FIGURE 3.5
Scatter Diagram

APPLIED EXAMPLE 3.4

Northwest Ohio Schools and How They Rate

It has long been known that a student's ability to pass the state's fourth-grade proficiency tests is closely related to the income level in the student's home.

The accompanying chart shows how individual elementary schools performed in the March 2000 proficiency tests in fourth-grade math and reading—and whether the schools performed better or worse than could be predicted based on the poverty level of the students attending the school.

The percentage of children receiving free or reduced-price lunches was used as the measure of poverty.

Poverty predicts scores

Each of 2,025 elementary schools in Ohio analyzed by The Blade is represented on this chart as a single dot. The dots were located on the chart based on each school's poverty level compared with each school's overall passage rate on the state's fourth-grade reading proficiency test.

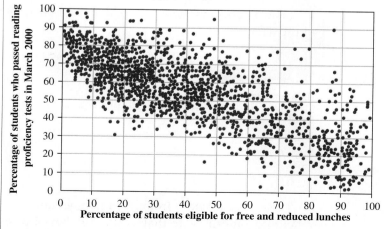

Source: Reprinted with permission of *The (Toledo) Blade,* August 5, 2001

Based on the preceding paragraph and the graph, the two variables that are examined in this example are "school's poverty level" and "passage rate." The scatter diagram clearly shows that a pattern exists. Note that the dots are all clustered together, sloping in a downward manner. Therefore, as the poverty level increased (read *x*-axis from left to right), the passage rate decreased.

TECHNOLOGY INSTRUCTIONS: SCATTER DIAGRAM

MINITAB (Release 14)   Input the *x*-variable values into C1 and the corresponding *y*-variable values into C2; then continue with:

Choose:	**Graph > ScatterPlot...> Simple > OK**
Enter:	Y variables: **C2** X variables: **C1**
Select:	**Labels > Titles/Footnotes**
Enter:	Title: **your title > OK > OK**

Excel   Input the *x*-variable values into column A and the corresponding *y*-variable values into column B; then continue with:

Choose:	**Chart Wizard > XY(Scatter) > 1st picture** (usually) **> Next**
Enter:	Data Range: **(A1:B12 or select cells(if necessary)) > Next**
Choose:	Titles
Enter:	Chart title: **your title;** Value(*x*) axis: **title for x axis;** Value(*y*) axis: **title for y axis* >** **Finish**

*To remove gridlines:

Choose:	**Gridlines**
Unselect:	**Value(Y) axis: Major Gridlines > Finish**

To edit the scatter diagram, follow the basic editing commands shown for a histogram on page 62.
To change the scale, double click on the axis; then continue with:

Choose:	**Scale**
Unselect:	**any Auto boxes**
Enter:	**new values > OK**

TI-83/84 Plus   Input the *x*-variable values into L1 and the corresponding *y*-variable values into L2; then continue with:

Choose:	**2nd > STATPLOT > 1:Plot1**
Choose:	**ZOOM > 9:ZoomStat**
	> TRACE > > >
	or
	WINDOW
	Enter: **at most lowest x value,**
	at least highest x value,
	x-scale, − y-scale, at least
	highest y value, y-scale,1
	TRACE > > >

SECTION 3.2 EXERCISES

Statistics⌃Now™

Datasets can be found on your Student's Suite CD-ROM or at the StatisticsNow website at **http://1pass.thomson.com**.

3.3 [EX03-003] In a national survey of 500 business and 500 leisure travelers, each was asked where they would most like "more space."

	On Airplane	Hotel Room	All Other
Business	355	95	50
Leisure	250	165	85

a. Express the table as percentages of the total.

b. Express the table as percentages of the row totals. Why might one prefer the table to be expressed this way?

c. Express the table as percentages of the column totals. Why might one prefer the table to be expressed this way?

3.4 The "Outlook for business travelers" graphic shows two circle graphs, each with four sections. This same information could be represented in the form of a 2 × 4 contingency table of two qualitative variables.

OUTLOOK FOR BUSINESS TRAVELERS

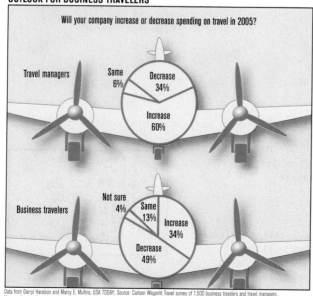

Data from Darryl Haralson and Marcy E. Mullins, USA TODAY; Source: Carlson Wagonlit Travel survey of 1,500 business travelers and travel managers. Margin of error ±3 percentage points.

a. Identify the population and name the two variables.

b. Construct the contingency table using entries of percentages based on row totals.

3.5 "The perfect age" graphic shows the results from a 9 × 2 contingency table for one qualitative and one quantitative variable.

"THE PERFECT AGE"

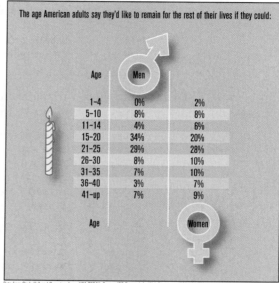

Data from Cindy Hall and Genevieve Lynn, USA TODAY; Source: IRC Research for Walt Disney. © 1998 USA TODAY, reprinted by permission.

a. Identify the population and name the qualitative and quantitative variables.

b. Construct a bar graph showing the two distributions side by side.

c. Does there seem to be a big difference between the genders on this subject?

3.6 [EX03-006] Under the National Highway System Designation Act of 1995, states are allowed to set their own highway speed limits. Most of the states raised the limits. The 2005 maximum speed limits on interstate highways for cars and trucks by each state are given in the following table (in miles per hour).

State	Cars	Trucks	State	Cars	Trucks	State	Cars	Trucks
AL	70	70	LA	70	70	OH	65	55
AK	65	65	ME	65	65	OK	75	75
AZ	75	75	MD	65	65	OR	65	65
AR	70	65	MA	65	65	PA	65	65
CA	70	55	MI	70	55	RI	65	65
CO	75	75	MN	70	70	SC	70	70
CT	65	65	MS	70	70	SD	75	75
DE	65	65	MO	70	70	TN	70	70
FL	70	70	MT	75	65	TX	75	75
GA	70	70	NE	75	75	UT	75	75
HI	55	55	NV	75	75	VT	65	65
ID	75	70	NH	65	65	VA	65	65
IL	65	55	NJ	65	65	WA	70	60
IN	65	60	NM	75	75	WV	70	70
IA	65	65	NY	65	65	WI	65	65
KS	70	70	NC	70	70	WY	75	75
KY	65	65	ND	75	75			

Source: The National Motorists Association, http://www.motorists.com/issues/speed/StateSpeeds.html

a. Build a cross-tabulation table of the two variables, vehicle type and maximum speed limit on an interstate highway. Express the results in frequencies, showing marginal totals.

b. Express the contingency table you derived in part a in percentages based on the grand total.

c. Draw a bar graph showing the results from part b.

d. Express the contingency table you derived in part a in percentages based on the marginal total for speed limit.

e. Draw a bar graph showing the results from part d.

FYI If you are using a computer or a calculator, try the cross-tabulation table commands on page 150.

3.7 [EX03-007] A statewide survey was conducted to investigate the relationship between viewers' preferences for ABC, CBS, NBC, PBS, or FOX for news information and their political party affiliation. The results are shown in tabular form:

Television Station

Political Affiliation	ABC	CBS	NBC	PBS	FOX
Democrat	203	218	257	156	226
Republican	421	350	428	197	174
Other	156	312	105	57	90

a. How many viewers were surveyed?

b. Why are these bivariate data? Name the two variables. What type of variable is each one?

c. How many viewers preferred to watch CBS?

d. What percentage of the survey was Republican?

e. What percentage of the Democrats preferred ABC?

f. What percentage of the viewers was Republican and preferred PBS?

3.8 [EX03-008] Consider the accompanying contingency table, which presents the results of an advertising survey about the use of credit by Martan Oil Company customers.

Preferred Method of Payment	Number of Purchases at Gasoline Station Last Year					
	0–4	5–9	10–14	15–19	≥20	Sum
Cash	150	100	25	0	0	275
Oil-Company Card	50	35	115	80	70	350
National or Bank Credit Card	50	60	65	45	5	225
Sum	250	195	205	125	75	850

a. How many customers were surveyed?

b. Why are these bivariate data? What type of variable is each one?

c. How many customers preferred to use an oil-company credit card?

d. How many customers made 20 or more purchases last year?

e. How many customers preferred to use an oil-company credit card and made between five and nine purchases last year?

f. What does the 80 in the fourth cell in the second row mean?

3.9 [EX03-009] The January 2005 unemployment rate for eastern and western U.S. states was as follows:

Eastern	4.7	4.1	4.8	3.5	3.9	5.0	5.1	4.4
Western	4.1	5.8	4.9	4.3	4.3	6.4	5.5	3.4

Source: U.S. Bureau of Labor Statistics

Display these rates as two dotplots using the same scale; compare means and medians.

3.10 [EX03-010] What effect does the minimum amount have on the interest rate being offered on 3-month certificates of deposit (CDs)? The following are advertised rates of return, y, for a minimum deposit of $500, $1000, $2000, $2500, $5000, or $10,000, x. (Note that x is in $100 and y is annual percent of return.)

Min Deposit	Rate	Min Deposit	Rate	Min Deposit	Rate
10	2.81	10	2.52	20	2.38
10	2.70	25	2.49	100	2.37
50	2.68	50	2.49	25	2.35
10	2.71	5	2.48	10	2.32
50	2.66	5	2.42	10	2.30
20	2.65	10	2.37	5	2.27
5	2.62	10	2.38	100	2.27
25	2.60	20	2.38		

Source: Bankrate.com, March 10, 2005

a. Prepare a dotplot of the rates associated with each of the six different minimum deposit requirements, using a common scale.

b. Prepare a 5-number summary and a boxplot of the six sets of data. Use the same scale as for the boxplots.

c. Describe any differences you see between the six sets of data.

FYI If you are using a computer or calculator for Exercise 3.10, try the commands on page 152.

3.11 [EX03-011] Can a woman's height be predicted from her mother's height? The heights of some mother–daughter pairs are listed; x is the mother's height and y is the daughter's height.

x	63	63	67	65	61	63	61	64	62	63	
y	63	65	65	65	64	64	63	62	63	64	
x	64	63	64	64	63	67	61	65	64	65	66
y	64	64	65	65	62	66	62	63	66	66	65

a. Draw two dotplots using the same scale and showing the two sets of data side by side.

b. What can you conclude from seeing the two sets of heights as separate sets in part a? Explain.

c. Draw a scatter diagram of these data as ordered pairs.

d. What can you conclude from seeing the data presented as ordered pairs? Explain.

3.12 [EX03-012] The following table lists the heights (in meters), weights (in kilograms), and ages of the players on the two teams that played in the 2002 World Cup finals: Brazil and Germany.

Player	Brazil			Germany		
	Height	Weight	Age	Height	Weight	Age
1	1.93	86	28	1.88	88	33
2	1.95	85	29	1.90	87	33

••• Remainder of data on Student's Suite CD-ROM

Source: http://worldcup.espnsoccernet.com/index

a. Compare each of the three variables—height, weight, and age—using either a dotplot or a histogram (use the same scale).

b. Based on what you see in the graphs in part a, can you detect a substantial difference between the two teams in regard to these three variables? Explain.

c. Explain why the data, as used in part a, are not bivariate data.

3.13 Consider the two variables of a person's height and weight. Which variable, height or weight, would you use as the input variable when studying their relationship? Explain why.

3.14 Draw a coordinate axis and plot the points (0, 6), (3, 5), (3, 2), (5, 0) to form a scatter diagram. Describe the pattern that the data show in this display.

3.15 Does studying for an exam pay off?

a. Draw a scatter diagram of the number of hours studied, x, compared with the exam grade received, y.

x	2	5	1	4	2
y	80	80	70	90	60

b. Explain what you can conclude based on the pattern of data shown on the scatter diagram drawn in part a. (Retain these solutions to use in Exercise 3.58, p. 185.)

3.16 Refer to "Northwest Ohio Schools and How They Rate" (Applied Example 3.4 on p. 154) to answer the following questions:

a. What are the two variables used?

b. Does the scatter diagram suggest a relationship between the two variables? Explain.

c. What conclusion, if any, can you draw from the appearance of the scatter diagram?

3.17 Growth charts are commonly used by a child's pediatrician to monitor a child's growth. Consider the growth chart that follows.

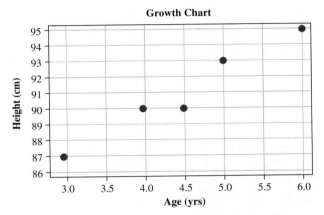

Growth Chart

a. What are the two variables shown in the graph?

b. What information does the ordered pair (3, 87) represent?

c. Describe how the pediatrician might use this chart and what types of conclusions might be based on the information displayed by it.

3.18 a. Draw a scatter diagram showing height, x, and weight, y, for the Brazilian World Cup soccer team using the data in Exercise 3.12.

b. Draw a scatter diagram showing height, x, and weight, y, for the German World Cup soccer team using the data in Exercise 3.12.

c. Explain why the data, as used in parts a and b, are bivariate data.

FYI If you are using a computer or calculator, try the commands on page 155.

3.19 [EX03-019] The accompanying data show the number of hours, x, studied for an exam and the grade received, y (y is measured in tens; that is, $y = 8$ means that the grade, rounded to the nearest 10 points, is 80). Draw the scatter diagram. (Retain this solution to use in Exercise 3.38, p. 170.)

x	2	3	3	4	4	5	5	6	6	6	7	7	7	8	8
y	5	5	7	5	7	7	8	6	9	8	7	9	10	8	9

3.20 [EX03-020] An experimental psychologist asserts that the older a child is, the fewer irrelevant answers he or she will give during a controlled experiment. To investigate this claim, the following data were collected. Draw a scatter diagram. (Retain this solution to use in Exercise 3.39, p 170.)

Age, x	2	4	5	6	6	7	9	9	10	12
Irr Answers, y	12	13	9	7	12	8	6	9	7	5

3.21 [EX03-021] The following table lists the percents of students who receive free or reduced-price lunches compared with the percents who passed the reading portion of a state exam. The results are for Sandusky County, Ohio, and were reported in *The Blade*, a Toledo newspaper, on August 5, 2001. Sandusky County has a combination of 13 rural and urban schools.

School	% Free/ Reduced-Price Lunches	% Passing Reading	School	% Free/ Reduced-Price Lunches	% Passing Reading
1	29	66	8	47	58
2	29	59	9	29	88
3	23	62	10	17	68
4	60	53	11	22	60
5	57	53	12	38	47
6	50	57	13	15	62
7	49	54			

Construct a scatter diagram of these data. (Retain this solution to use in Exercise 3.34, p. 170.)

3.22 [EX03-022] A sample of 15 upper-class students who commute to classes was selected at registration. The students were asked to estimate the distance, *x* (nearest mile), and the time, *y* (nearest 5 min.), required to commute each day to class (see accompanying table). Construct a scatter diagram depicting these data.

Distance, x	Time, y	Distance, x	Time, y
18	20	2	5
8	15	15	25
20	25	16	30
5	20	9	20
5	15	21	30
11	25	5	10
9	20	15	20
10	25		

3.23 [EX03-023] Baseball stadiums vary in age, style, size, and many other ways. Fans might think of the size of the stadium in terms of the number of seats, whereas players might measure the size of the stadium by the distance (in feet) from home plate to the center field fence.

Seats	CF	Seats	CF	Seats	CF
40,000	422	49,166	400	43,000	400
45,050	400	45,200	410	42,000	435
33,871	420	44,321	400	56,500	410
43,368	405	57,545	408	40,800	404
40,625	400	48,500	402	38,127	399
48,678	408	49,625	402	42,531	410
43,662	400	43,500	401	56,133	405
48,876	400	50,381	415	55,777	410
47,000	405	42,059	404	50,062	401
50,516	400	56,000	395	38,902	400

CF = distance from home plate to center field fence

Source: http://mlb.mlb.com

a. Is there a relationship between these two measurements for the "size" of the 30 major league baseball stadiums?

b. What do you think you will find? Bigger fields have more seats? Smaller fields have more seats? No relationship between field size and the number of seats? A strong relationship between field size and the number of seats? Explain.

c. Construct a scatter diagram.

d. Describe what the scatter diagram tells you, including a reaction to your answer in part b.

3.24 [EX03-024] Most adult Americans drive. But do you have any idea how many licensed drivers there are in each U.S. state? The following table lists the number of male and female drivers licensed in each of 15 randomly selected U.S. states.

Licensed Drivers per State (×100,000)

Male	Female	Male	Female	Male	Female
2.77	2.78	59.5	54.07	9.92	9.96
37.1	39.46	1.94	1.85	30.13	30.02
19.5	20.16	7.76	7.12	9.95	10.03
13.19	13.41	15.45	15.76	20.56	21.49
4.41	3.94	6.41	6.31	4.87	4.81

Source: Federal Highway Administration, U.S. Department of Transportation

a. Do you expect to find a linear (straight-line) relationship between number of male and number of female licensed drivers per state? How strong do you anticipate this relationship to be? Describe.

b. Construct a scatter diagram using *x* for the number of male drivers and *y* for the number of female drivers.

c. Compare the scatter diagram to your expectations in part a. How did you do? Explain.

d. Are there data points that look like they are separate from the pattern created by the rest of the ordered pairs? If they were removed from the data set, would the results change? What caused these points to be separate from the others but yet still part of the extended pattern? Explain.

e. (Optional) The data set listed on the CD contains this information for all 50 states and the District of Columbia. Construct a scatter diagram of all 51 ordered pairs. Compare the pattern of the sample of 15 to the pattern shown by all 51. Describe in detail. [DS03-024]

f. (Optional) Did the sample provide enough information to understand the relationship between the two variables in this situation? Explain.

3.25 [EX03-025] Are people stronger today than they used to be? Can they run faster? Let's compare the performances of the Olympic gold medal winners for the last century as a way to decide. The distances (in inches) for the gold medal performances in the long jump, high jump, and discus throw are given in the following table. The event year is coded, with 1900 = 0.

Year	Long Jump	High Jump	Discus Throw
−4	249.75	71.25	1147.5
0	282.875	74.8	1418.9
••• Remainder of data on Student's Suite CD-ROM			

Source: http://www.ex.ac.uk/cimt/data/olympics/olymindx.htm

a. Plot the data for each event on a separate scatter diagram using year, x.

b. Describe the shape of the distribution. For each scatter diagram, does the relationship between year and performance appear to follow a straight line?

c. How do the three scatter diagrams answer the question: Are people stronger today? Explain.

d. On each of the three scatter diagrams, draw the straight line that seems to best trace the pattern of points from 1896 to 2004. Use this line as an aid to predict the Olympic gold-medal-winning performance for each event at the Beijing 2008 games.

e. Investigate the relationship between high and long jumps with the aid of a scatter diagram. Describe your findings.

3.26 [EX03-026] The following table lists the height (in inches), weight (in pounds), and date of birth of the members of the 2004 Rochester Raging Rhinos professional soccer team.

Player	Height	Weight	DOB
1	68	160	12/7/1978
2	71	170	2/2/1970
••• Remainder of data on Student's Suite CD-ROM			

Source: http://www.RhinosSoccer.com

a. Is it true that taller players weigh more? What do you expect a scatter diagram will show for a soccer team?

b. Construct the scatter diagram of height, x, versus weight, y.

c. Does the scatter diagram in part b support your thoughts in part a? Explain why or why not.

d. Is it true that as players age, they tend to weigh more? What do you expect a scatter diagram will show for a soccer team?

e. Construct the scatter diagram of age, x, versus weight, y. (*Note:* You will need to convert date of birth to age in years. Solutions given are based on ages on January 1, 2006.)

f. Does the scatter diagram in part e support your thoughts in part d? Explain why or why not.

g. If you were to find the players' ages on January 1, 2008, what effect would this have on the scatter diagram constructed in part e? Explain.

h. (Optional) Construct the scatter diagram of age, x, versus weight, y, based on ages on January 1, 2008.

3.27 [EX03-027] Ronald Fisher, an English statistician (1890–1962), collected measurements for a sample of 150 irises. Of concern were five variables: species, petal width (PW), petal length (PL), sepal width (SW), and sepal length (SL) (all in mm). Sepals are the outermost leaves that encase the flower before it has opened. The goal of Fisher's experiment was to produce a simple function that could be used to classify flowers correctly. A random sample of his complete data set is given in the accompanying table.

Type	PW	PL	SW	SL	Type	PW	PL	SW	SL
0	2	15	35	52	1	24	51	28	58
2	18	48	32	59	1	19	50	25	63
1	19	51	27	58	0	1	15	31	49
0	3	13	35	50	1	23	59	32	68
0	3	15	38	51	2	13	44	23	63
2	12	44	26	55	2	15	42	30	59
1	20	64	38	79	1	25	57	33	67
2	15	49	31	69	1	21	57	33	67
2	15	45	29	60	0	2	15	37	54
2	12	39	27	58	1	18	49	27	63
1	22	56	28	64	1	17	45	25	49
1	13	52	30	67	1	24	56	34	63
0	2	14	29	44	0	2	14	36	50
2	16	51	27	60	2	10	50	22	60
0	5	17	33	51	0	2	12	32	50

a. Construct a scatter diagram of petal length, *x*, and petal width, *y*. Use different symbols to represent the three species.*

b. Construct a scatter diagram of sepal length, *x*, and sepal width, *y*. Use different symbols to represent the three species.

c. Explain what the scatter diagrams in parts a and b portray.

Let's see how well a random sample represents the data from which it was selected.

d. Complete parts a and b using all 150 of Fisher's data on your CD. [DS03-027]

e. Aside from the fact that the scatter diagrams in parts a and b have fewer data, comment on the similarities and differences between the distributions shown for 150 data and for the 30 randomly selected data.

---

*In addition to using the commands on page 155, use:
For MINITAB: Data display: For each: Select:
                                     Group
                  Group variable: Select: Type
For TI-83-84:   Enter different groups into separate x,
                  y columns. Use a separate Stat Plot and
                  "Mark" for each group.

3.28 [EX03-028] Total solar eclipses actually take place nearly as often as total lunar eclipses, but they are visible over a much narrower path. Both the path width and the duration vary substantially from one eclipse to the next. The following table shows the duration (in seconds) and path width (in miles) of 44 total solar eclipses measured in the past and those projected through the year 2010:

Date	Duration(s)	Width (mi)
1950	73	83
1952	189	85

••• Remainder of data on Student's Suite CD-ROM

Source: The World Almanac and Book of Facts 1998, p. 296

a. Draw a scatter diagram showing duration, *y*, and path width, *x*, for the total solar eclipses.

b. How would you describe this diagram?

3.3   Linear Correlation

The primary purpose of **linear correlation analysis** is to measure the strength of a linear relationship between two variables. Let's examine some scatter diagrams that demonstrate different relationships between input, or independent variables, *x*, and output, or dependent variables, *y*. If as *x* increases there is no definite shift in the values of *y*, we say there is **no correlation,** or no relationship between *x* and *y*. If as *x* increases there is a shift in the values of *y*, then there is a correlation. The correlation is **positive** when *y* tends to increase and **negative** when *y* tends to decrease. If the ordered pairs (*x*, *y*) tend to follow a straight-line path, there is a linear correlation. The preciseness of the shift in *y* as *x* increases determines the strength of the **linear correlation.** The scatter diagrams in Figure 3.6 demonstrate these ideas.

Perfect linear correlation occurs when all the points fall exactly along a straight line, as shown in Figure 3.7. The correlation can be either positive or negative, depending on whether *y* increases or decreases as *x* increases. If the data form a straight horizontal or vertical line, there is no correlation, because one variable has no effect on the other, as shown in Figure 3.7.

FIGURE 3.6   Scatter Diagrams and Correlation

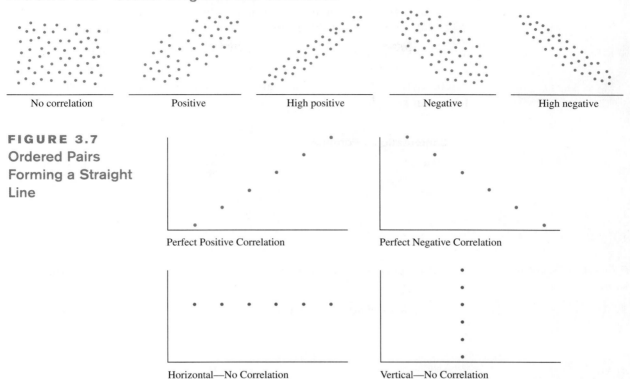

| No correlation | Positive | High positive | Negative | High negative |

FIGURE 3.7
Ordered Pairs
Forming a Straight
Line

Perfect Positive Correlation   Perfect Negative Correlation

Horizontal—No Correlation   Vertical—No Correlation

FIGURE 3.8
No Linear Correlation

Scatter diagrams do not always appear in one of the forms shown in Figures 3.6 and 3.7. Sometimes they suggest relationships other than linear, as in Figure 3.8. There appears to be a definite pattern; however, the two variables are not related linearly, and therefore there is no linear correlation.

The **coefficient of linear correlation, *r*,** is the numerical measure of the strength of the linear relationship between two variables. The coefficient reflects the consistency of the effect that a change in one variable has on the other. The value of the linear correlation coefficient helps us answer the question: Is there a linear correlation between the two variables under consideration? The linear correlation coefficient, r, always has a value between -1 and $+1$. A value of $+1$ signifies a perfect positive correlation, and a value of -1 shows a perfect negative correlation. If as x increases there is a general increase in the value of y, then r will be positive in value. For example, a positive value of r would be expected for the age and height of children because as children grow older, they grow taller. Also, consider the age, x, and resale value, y, of an automobile. As the car ages, its resale value decreases. Since as x increases, y decreases, the relationship results in a negative value for r.

The value of r is defined by **Pearson's product moment formula:**

Definition Formula

$$r = \frac{\sum (x - \bar{x})(y - \bar{y})}{(n - 1)s_x s_y}$$

(3.1)

Notes:
1. s_x and s_y are the standard deviations of the x and y variables.
2. The development of this formula is discussed in Chapter 13.

To calculate r, we will use an alternative formula, formula (3.2), that is equivalent to formula (3.1). As preliminary calculations, we will separately calculate three sums of squares and then substitute them into formula (3.2) to obtain r.

Computational Formula

$$\text{linear correlation coefficient} = \frac{\text{sum of squares for } xy}{\sqrt{(\text{sum of squares for } x)(\text{sum of squares for } y)}}$$

$$r = \frac{SS(xy)}{\sqrt{SS(x)SS(y)}} \tag{3.2}$$

FYI $SS(x)$ is the numerator of the variance.

Recall the $SS(x)$ calculation from formula (2.9) for sample variance (p. 87):

$$\text{sum of squares for } x = \text{sum of } x^2 - \frac{(\text{sum of } x)^2}{n}$$

$$SS(x) = \sum x^2 - \frac{(\sum x)^2}{n} \tag{2.9}$$

We can also calculate:

$$\text{sum of squares for } y = \text{sum of } y^2 - \frac{(\text{sum of } y)^2}{n}$$

$$SS(y) = \sum y^2 - \frac{(\sum y)^2}{n} \tag{3.3}$$

$$\text{sum of squares for } xy = \text{sum of } xy - \frac{(\text{sum of } x)(\text{sum of } y)}{n}$$

$$SS(xy) = \sum xy - \frac{\sum x \sum y}{n} \tag{3.4}$$

EXAMPLE 3.5

Calculating the Linear Correlation Coefficient, *r*

FYI See Animated Tutorial Preliminary Calculations on CD for assistance with these calculations.

Find the linear correlation coefficient for the push-up/sit-up data in Example 3.3 (p. 153).

SOLUTION First, we construct an extensions table (Table 3.11) listing all the pairs of values (x, y) to aid us in finding x^2, xy, and y^2 for each pair and the five column totals.

TABLE 3.11

Extensions Table for Finding Five Summations [TA03-10]

Student	Push-ups, x	x^2	Sit-ups, y	y^2	xy
1	27	729	30	900	810
2	22	484	26	676	572
3	15	225	25	625	375
4	35	1,225	42	1,764	1,470
5	30	900	38	1,444	1,140
6	52	2,704	40	1,600	2,080
7	35	1,225	32	1,024	1,120
8	55	3,025	54	2,916	2,970
9	40	1,600	50	2,500	2,000
10	40	1,600	43	1,849	1,720
	$\sum x = 351$	$\sum x^2 = 13{,}717$	$\sum y = 380$	$\sum y^2 = 15{,}298$	$\sum xy = 14{,}257$
	sum of x	*sum of x^2*	*sum of y*	*sum of y^2*	*sum of xy*

Second, to complete the preliminary calculations, we substitute the five summations (the five column totals) from the extensions table into formulas (2.9), (3.3), and (3.4), and calculate the three sums of squares:

$$SS(x) = \sum x^2 - \frac{(\sum x)^2}{n} = 13{,}717 - \frac{(351)^2}{10} = 1396.9$$

FYI The $\sum$ and SS values will be needed for regression in Section 3.4. Be sure to save them!

$$SS(y) = \sum y^2 - \frac{(\sum y)^2}{n} = 15{,}298 - \frac{(380)^2}{10} = 858.0$$

FYI See Animated Tutorial Correlation on CD for assistance with this calculation.

$$SS(xy) = \sum xy - \frac{\sum x \sum y}{n} = 14{,}257 - \frac{(351)(380)}{10} = 919.0$$

Third, we substitute the three sums of squares into formula (3.2) to find the value of the correlation coefficient:

$$r = \frac{SS(xy)}{\sqrt{SS(x)SS(y)}} = \frac{919.0}{\sqrt{(1396.9)(858.0)}} = 0.8394 = \mathbf{0.84}$$

Note: Typically, r is rounded to the nearest hundredth.

The value of the linear correlation coefficient helps us answer the question: Is there a linear correlation between the two variables under consideration? When the calculated value of r is close to zero, we conclude that there is little or no linear correlation. As the calculated value of r changes from 0.0 toward either $+1.0$ or -1.0, it indicates an increasingly stronger linear correlation between the two variables. From a graphic viewpoint, when we calculate r, we are measuring how well a straight line describes the scatter diagram of ordered pairs. As the value of r changes from 0.0 toward $+1.0$ or -1.0, the data points create a pattern that moves closer to a straight line.

FYI See this in action with Exercise 3.29 on page 169.

TECHNOLOGY INSTRUCTIONS: CORRELATION COEFFICIENT

MINITAB (Release 14) Input the *x*-variable data into C1 and the corresponding *y*-variable data into C2; then continue with:

```
Choose:     Stat > Basic Statistics > Correlation...
Enter:      Variables: C1 C2 > OK
```

Excel Input the *x*-variable data into column A and the corresponding *y*-variable data into column B, activate a cell for the answer; then continue with:

```
Choose:     Insert function, fx > Statistical > CORREL > OK
Enter:      Array 1: x data range
            Array 2: y data range > OK
```

TI-83/84 Plus Input the *x*-variable data into L1 and the corresponding *y*-variable data into L2; then continue with:

```
Choose:     2nd > CATALOG > DiagnosticOn* > ENTER > ENTER
Choose:     STAT > CALC > 8:LinReg(a + bx)
Enter:      L1, L2
```

*DiagnosticOn must be selected for *r* and *r²* to show. Once set, omit this step.

FIGURE 3.9
The Data Window

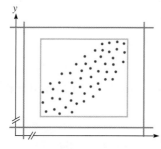

Understanding the Linear Correlation Coefficient

The following method will create (1) a visual meaning for correlation, (2) a visual meaning for what the linear coefficient is measuring, and (3) an estimate for *r*. The method is quick and generally yields a reasonable estimate when the "window of data" is approximately square.

Note: This estimation technique does not replace the calculation of *r*. It is very sensitive to the "spread" of the diagram. However, if the "window of data" is approximately square, this approximation will be useful as a mental estimate or check.

FIGURE 3.10
Focusing on Pattern

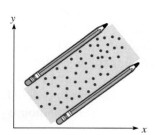

Procedure

1. Construct a scatter diagram of your data, being sure to scale the axes so that the resulting graph has an approximately square "window of data," as demonstrated in Figure 3.9 by the light green frame. The window may not be the same region as determined by the bounds of the two scales, shown as a green rectangle on Figure 3.9.

2. Lay two pencils on your scatter diagram. Keeping them parallel, move them to a position so that they are as close together as possible yet have all the points on the scatter diagram between them. (See Figure 3.10.)

3. Visualize a rectangular region that is bounded by the two pencils and that ends just beyond the points on the scatter diagram. (See the shaded portion of Figure 3.10.)

FIGURE 3.11

Finding *k*

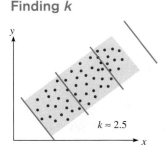

$k \approx 2.5$

4. Estimate the number of times longer the rectangle is than it is wide. An easy way to do this is to mentally mark off squares in the rectangle. (See Figure 3.11.) Call this number of multiples *k*.

5. The value of *r* may be estimated as $\pm \left(1 - \dfrac{1}{k}\right)$.

6. The sign assigned to *r* is determined by the general position of the length of the rectangular region. If it lies in an increasing position, *r* will be positive; if it lies in a decreasing position, *r* will be negative (see Figure 3.12). If the rectangle is in either a horizontal or a vertical position, then *r* will be zero, regardless of the length–width ratio.

FIGURE 3.12

(a) Increasing Position;

(b) Decreasing Position

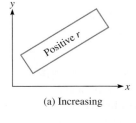

(a) Increasing

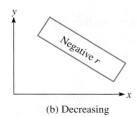

(b) Decreasing

FIGURE 3.13

Push-ups versus Sit-ups for 10 Students

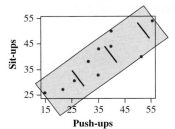

Push-ups

Let's use this method to estimate the value of the linear correlation coefficient for the relationship between the number of push-ups and sit-ups. As shown in Figure 3.13, we find that the rectangle is approximately 3.5 times longer than it is wide—that is, $k \approx 3.5$—and the rectangle lies in an increasing position. Therefore, our estimate for *r* is

$$r \approx + \left(1 - \frac{1}{3.5}\right) \approx +0.70$$

Causation and Lurking Variables

As we try to explain the past, understand the present, and estimate the future, judgments about cause and effect are necessary because of our desire to impose order on our environment.

The *cause-and-effect relationship* is fairly straightforward. You may focus on a situation, the *effect* (e.g., a disease or social problem), and try to determine its *cause(s)*, or you may begin with a *cause* (unsanitary conditions or poverty) and discuss its *effect(s)*. To determine the cause of something, ask yourself **why** it happened. To determine the effect, ask yourself **what** happened.

> **Lurking variable:** A variable that is not included in a study but has an effect on the variables of the study and makes it appear that those variables are related.

A good example is the strong positive relationship shown between the amount of damage caused by a fire and the number of firefighters who work the fire. The "size" of the fire is the lurking variable; it "causes" both the "amount" of damage and the "number" of firefighters.

If there is a strong linear correlation between two variables, then one of the following situations may be true about the relationship between the two variables:

1. There is a direct cause-and-effect relationship between the two variables.
2. There is a reverse cause-and-effect relationship between the two variables.
3. Their relationship may be caused by a third variable.
4. Their relationship may be caused by the interactions of several other variables.
5. The apparent relationship may be strictly a coincidence.

Remember that a strong correlation does not necessarily imply causation.

Here are some pitfalls to avoid:

1. In a direct cause-and-effect relationship, an increase (or decrease) in one variable causes an increase (or decrease) in another. Suppose there is a strong positive correlation between weight and height. Does an increase in weight *cause* an increase in height? Not necessarily. Or to put it another way, does a decrease in weight *cause* a decrease in height? Many other possible variables are involved, such as gender, age, and body type. These other variables are called *lurking variables.*

2. In Applied Example 3.4 (p. 154), a negative correlation existed between the percent of students who received free or reduced-price lunches and the percent of students who passed the mathematics proficiency test. Shall we hold back on the free lunches so that more students pass the mathematics test? A third variable is the motivation for this relationship, namely, poverty level.

3. Don't reason from *correlation* to *cause:* Just because all people who move to the city get old doesn't mean that the city *causes* aging. The city may be a factor, but you can't base your argument on the correlation.

APPLIED EXAMPLE 3.6

Life Insurance Rates

Does a high linear correlation coefficient, r, imply that the data are linear in nature? The issue age of the insured and the monthly life insurance rate for nontobacco users appears highly correlated looking at the chart presented here. As the issue age increases, the monthly rate for insurance increases for each of the genders.

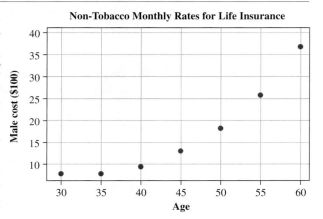

Non-Tobacco Monthly Rates for Life Insurance

TABLE 3.12

Nontobacco Monthly Rates for Life Insurance [TA03-012]

Issue Age	$100,000 Male ($)	$100,000 Female ($)	$250,000 Male ($)	$250,000 Female ($)	$500,000 Male ($)	$500,000 Female ($)
30	7.96	6.59	11.96	9.13	19.25	12.46
35	8.05	6.56	11.96	9.13	19.57	12.46
40	9.63	7.79	15.22	10.89	23.19	16.47
45	13.14	9.80	22.40	15.44	35.87	24.03
50	18.44	12.42	33.69	21.10	53.81	33.38
55	26.01	15.75	49.22	29.37	87.59	48.06
60	37.10	20.83	74.59	42.05	137.38	69.87

Source: http://www.reliaquote.com/termlife/default.asp; accessed March 11, 2005
All of the rates listed are for each carrier's best nontobacco classifications.

Let's consider the issue age of the insured and the male monthly rate for a $100,000 policy. The calculated correlation coefficient for this specific class of insurance results in a value of $r = 0.932$. Typically, a value of r this close to 1.0 would indicate a fairly strong straight-line relationship; but wait. Do we have a linear relationship? Only a scatter diagram can tell us that.

The scatter diagram clearly shows a non-straight-line pattern. Yet, the correlation coefficient was so high. It is the elongated pattern in the data that produces a calculated r so large. The lesson from this example is that one should always begin with a scatter diagram when considering linear correlation. The correlation coefficient only tells one side of the story!

SECTION 3.3 EXERCISES

Statistics⟨△⟩Now™

3.29 Skillbuilder Applet Exercise provides scatter diagrams for various correlation coefficients.

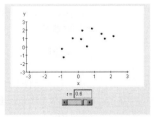

a. Starting at $r = 0$, move the slider to the right until $r = 1$. Explain what is happening to the corresponding scatter diagrams.

b. Starting at $r = 0$, move the slider to the left until $r = -1$. Explain what is happening to the corresponding scatter diagrams.

3.30 How would you interpret the findings of a correlation study that reported a linear correlation coefficient of -1.34?

3.31 How would you interpret the findings of a correlation study that reported a linear correlation coefficient of $+0.3$?

3.32 Explain why it makes sense for a set of data to have a correlation coefficient of zero when the data show a very definite pattern, as in Figure 3.9 (p. 163).

3.33 Does studying for an exam pay off? The number of hours studied, x, is compared with the exam grade received, y:

x	2	5	1	4	2
y	80	80	70	90	60

a. Complete the preliminary calculations: extensions, five sums, $SS(x)$, $SS(y)$, and $SS(xy)$.

b. Find r.

3.34 [EX03-034] The following table lists the percents of students who receive free or reduced-price lunches compared with the percents who passed the reading portion of the state exam. The results are for Sandusky County, Ohio, and were reported in *The Blade,* a Toledo newspaper, on August 5, 2001. Sandusky County is a combination of 13 rural and urban schools. (Same data as in Exercise 3.21, p. 159.)

School	% Free/ Reduced- Price Lunches	% Passing Reading	School	% Free/ Reduced- Price Lunches	% Passing Reading
1	29	66	8	47	58
2	29	59	9	29	88
3	23	62	10	17	68
4	60	53	11	22	60
5	57	53	12	38	47
6	50	57	13	15	62
7	49	54			

Find: a. $SS(x)$ b. $SS(y)$ c. $SS(xy)$ d. r

3.35 [EX03-035] Many organizations offer "special" magazine subscription rates to their members. The American Federation of Teachers is no different, and here are a few of the rates they offer their members.

Magazine	Usual Rate	Your Price
Cosmopolitan	$29.97	$18.00
Sports Illustrated	$78.97	$39.75
Ebony	$20.00	$14.97
Rolling Stone	$23.94	$11.97
Martha Stewart Living	$24.95	$20.00

Source: American Federation of Teachers

a. Construct a scatter diagram with "Your Price" as the dependent variable, y, and "Usual Rate" as the independent variable, x.

Find:

b. $SS(x)$ c. $SS(y)$ d. $SS(xy)$

e.   Pearson's product moment, r

3.36 Estimate the correlation coefficient for each of the following:

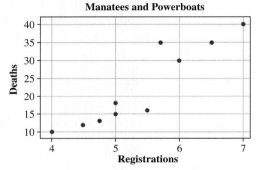

3.37 Manatees swim near the surface of the water. They often run into trouble with the many powerboats in Florida. Consider the graph that follows.

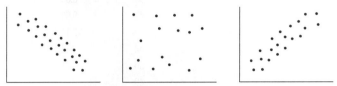

a. What two groups of subjects are being compared?

b. What two variables are being used to make the comparison?

c. What conclusion can one make based on this scatterplot?

d. What might you do if you were a wildlife official in Florida?

3.38 a. Use the scatter diagram you drew in Exercise 3.19 (p. 159) to estimate r for the sample data on the number of hours studied and the exam grade.

b. Calculate r.

3.39 a. Use the scatter diagram you drew in Exercise 3.20 (p. 159) to estimate r for the sample data on the number of irrelevant answers and the child's age.

b. Calculate r.

FYI Have you tried to use the correlation commands on your computer or calculator?

3.40 [EX03-040] A marketing firm wished to determine whether the number of television commercials broadcast were linearly correlated with the sales of its product. The data, obtained from each of several cities, are shown in the following table.

City	A	B	C	D	E	F	G	H	I	J
Commercials, x	12	6	9	15	11	15	8	16	12	6
Sales Units, y	7	5	10	14	12	9	6	11	11	8

a. Draw a scatter diagram. b. Estimate r.

c. Calculate r.

3.41 Skillbuilder Applet Exercise matches correlation coefficients with their scatter diagrams. After several practice rounds using "New Plots," explain your method of matching.

3.42 Skillbuilder Applet Exercise provide practice in constructing scatter diagrams to match given correlation coefficients.

a. After placing just 2 points, what is the calculated r value for each scatter diagram? Why?

b. Which scatter diagram did you find easier to construct?

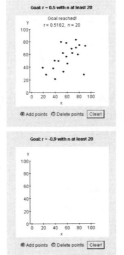

3.43 [EX03-043] Sports drinks are very popular in today's culture around the world. The following table lists 10 different products you can buy in England and the values for three variables: cost per serving (in pence), energy per serving (in kilocalories), and carbohydrates per serving (in grams).

Sports Drink	Cost	Energy	Carbs
Lucozade Sport RTD 330ml pouch/can	72	92	21.1
Lucozade Sport RTD 500ml bot.	79	140	32
Lucozade Sport RTD 650ml sports bot.	119	182	41.6
POWERade 500ml bot.	119	120	30
Gatorade Sports 750ml	89	188	45
Science in Sport Go Electrolyte (500ml)	99	160	40
High Five Isotonic electrolyte (750ml)	99	220	55
Isostar powder (per litre) 5l tub	126	320	77
Isostar RTD 500ml bot.	99	150	35
Maxim Electrolyte (per litre) 2kg bag	66	296	75

Note: Cost is in pence (p), 0.01 of a British pound, worth $0.0187 on March 28, 2005

Energy is measured in kilocalories; carbs (carbohydrates) are measured in grams.

Source: http://www.simplyrunning.net/articles/sports_drinks.htm

a. Draw a scatter diagram using x = carbs/serving and y = energy/serving.

b. Does there appear to be a linear relationship?

c. Calculate the linear correlation coefficient, r.

d. What does this value of correlation seem to be telling us? Explain.

e. Repeat parts a through d using x = cost/serving and y = energy/serving. (Retain these solutions to use in Exercise 3.59, p. 185.)

3.44 [EX03-044] The National Adoption Information Clearinghouse tracks and posts information about adoptions in the United States. The following table lists the number of children adopted in each of 16 randomly identified states for 1991 and 2001.

State	1991	2001	State	1991	2001	State	1991	2001
DE	190	225	IA	1518	1116	WY	425	412
NV	779	764	NJ	2382	2384	AL	1939	1857
MI	4758	6274	AR	1678	1698	ID	879	1048
SC	1471	1648	HI	592	766	WA	2603	2748
GA	2330	3499	TN	751	2633	VT	350	407
AK	898	616						

Source: Children's Bureau, Administration for Children and Families, U.S. Department of Health and Human Services, 2004, http://naic.acf.hhs.gov/pubs/s_adopted/index.cfm

Is there a linear relationship between the 1991 and 2001 data? Use graphic and numerical statistics to support your answer.

3.45 [EX03-045] Interstate 95, the longest of the north–south U.S. interstate highways, is 1907 miles long, stretching from Houlton, Maine, at the Canadian border on the northern end to Miami, Florida, at US 1 on the southern end. It travels across 15 east–coast states; the number of miles and the number of intersections in each of those states are listed here.

State	FL	GA	SC	NC	VA	MD	DE
Intersections	73	19	39	44	51	38	9
Miles	381	112	201	183	178	110	26

State	PA	NJ	NY	CT	RI	MA	NH	ME
Intersections	16	28	12	68	26	44	4	52
Miles	58	44	29	118	47	97	17	306

Source: Rand McNally and http://www.ihoz.com/I90.html

Using all 15 data:

a. Construct a scatter diagram with number of intersections as the dependent variable, y, and miles as the independent variable, x.

b. Does there seem to be a linear pattern to the data? Does the pattern seem reasonable for the variables? Explain why or why not.

c. Calculate the linear correlation coefficient, r.

d. Does the value of r seem reasonable compared with the pattern demonstrated on the scatter diagram? Explain.

Remove the CT (118, 68) from the data, then:

e. What is there about the Connecticut data point that makes it different? Is it understandable why it is different?

f. What effect did the removal of the Connecticut data point seem to have on the pattern?

g. Calculate the linear correlation coefficient, r.

h. What effect did the removal of the Connecticut point have on the value of r? How does this compare with the effect you anticipated? Explain.

3.46 [EX03-046] Movie production companies spend millions of dollars to produce movies with the great hope of attracting millions of people to the theater. The success of a movie can be measured in many ways, two of which are the box office receipts and the number of Oscar nominations received.

Following is a list of ten 2005 movies and their corresponding "report cards." Each movie is measured based on its budget cost (in millions of dollars), its box office receipts (in millions of dollars), and the number of Oscar nominations it received.

Movie	Budget	Box Office	Nominations
The Aviator	110	82.3	11
Finding Neverland	24	42.5	7
Million Dollar Baby	30	44.9	7
Ray	35	74.7	6
Sideways	16	52.8	5
Hotel Rwanda	17	14.2	3
Vera Drake	8.5	2.8	3
Eternal Sunshine of the Spotless Mind	20	34.1	2
Being Julia	10	5.1	1
Maria Full of Grace	3	6.5	1

Source: USA Today, February 8, 2005, "A quick guide to movies up for top awards"

a. Draw a scatter diagram using x = budget and y = box office.

b. Does there appear to be a linear relationship?

c. Calculate the linear correlation coefficient, r.

d. What does this value of correlation seem to be telling us? Explain.

e. Repeat parts a through d using x = box office and y = nominations.

3.47 [EX03-047] The national highway system is made up of interstate highways and noninterstate highways. Listed here are 15 randomly selected U.S. states and their corresponding number of miles of interstate and noninterstate highway.

National Highway System, Number of Miles–October 2005

State	Interstate	Noninterstate	State	Interstate	Noninterstate
AL	905	2715	NE	482	2496
VT	320	373	UT	940	1253
NH	235	589	TX	3233	10157
RI	71	197	OK	930	2431
AZ	1167	1565	WV	549	1195
IA	782	2433	AK	1082	1030
WI	745	3404	GA	1245	3384
NY	1674	3476			

Source: U.S. Department of Transportation

a. Construct a scatter diagram using $x =$ interstate and $y =$ noninterstate miles.

b. Describe the pattern displayed, including any unusual characteristics.

c. Calculate the correlation coefficient.

d. Remove Texas from the data and repeat parts a through c.

e. Compare the answers found in part d with the answers found in parts a and c, including comments about what effect removing Texas from the data had on the correlation coefficient.

3.48 [EX03-048] NBA players, teams, and fans are interested in seeing their leading scorer score lots of points, yet at the same time, the number of personal fouls they commit tends to limit their playing time. For each team, the following table lists the number of minutes played per game (Min/G) and the number of personal fouls committed per game (PF/G) by the leading scorer during the 2003–2004 season.

Team	Min/G	PF/G
Bulls	35.14	2.01
Lakers	37.65	2.71

••• Remainder of data on Student's Suite CD-ROM

Source: NBA.com

a. Construct a scatter diagram.

b. Describe the resulting pattern. Are there any unusual characteristics displayed?

c. Calculate the correlation coefficient, r.

d. Does the value of the correlation coefficient seem reasonable?

3.49 [TA03-012] By looking at the insurance rates on the table in Applied Example 3.6, we can see that as a person ages, the insurance rate increases. You probably anticipated that, but let's take a closer look at one of specific situations listed.

a. Calculate the correlation coefficient, r, for the variables issue age (x) and monthly rate for $250,000 for males.

b. Draw a scatter diagram of the insurance data for males at the $250,000 based on age ($x$).

c. Do the data appear to have a linear pattern? Explain.

d. Explain how a nonlinear data pattern can have a high linear correlation coefficient.

e. Explain why you should have anticipated this nonlinear pattern.

f. (Optional) Investigate one or more of the other five columns of insurance rates answering parts a through e for each.

3.50 In many communities there is a strong positive correlation between the amount of ice cream sold in a given month and the number of drownings that occur in that month. Does this mean that ice cream causes drowning? If not, can you think of an alternative explanation for the strong association? Write a few sentences addressing these questions.

3.51 Explain why one would expect to find a positive correlation between the number of fire engines that respond to a fire and the amount of damage done in the fire. Does this mean that the damage would be less extensive if fewer fire engines were dispatched? Explain.

3.4 Linear Regression

Although the correlation coefficient measures the strength of a linear relationship, it does not tell us about the mathematical relationship between the two variables. In Section 3.3, the correlation coefficient for the push-up/sit-up data was found to be 0.84 (see pp. 164–165). This along with the pattern on the scatter diagram imply that there is a linear relationship between the number of push-ups and the number of sit-ups a student does. However, the correlation coefficient does not help us predict the number of sit-ups a person can do based on knowing he or she can do 28 push-ups. **Regression analysis** finds the equation of the line that best describes the relationship between the two variables. One use of this equation is to make predictions. We make use of these predictions regularly—for example, predicting the success a student will have in college based on high school results and predicting the distance required to stop a car based on its speed. Generally, the exact value of *y* is not predictable, and we are usually satisfied if the predictions are reasonably close.

The relationship between two variables will be an algebraic expression describing the mathematical relationship between *x* and *y*. Here are some examples of various possible relationships, called *models* or **prediction equations:**

Linear (straight-line):	$\hat{y} = b_0 + b_1 x$
Quadratic:	$\hat{y} = a + bx + cx^2$
Exponential:	$\hat{y} = a(b^x)$
Logarithmic:	$\hat{y} = a \log_b x$

Figures 3.14, 3.15, and 3.16 show patterns of bivariate data that appear to have a relationship, whereas in Figure 3.17 the variables do not seem to be related.

FIGURE 3.14
Linear Regression with Positive Slope

FIGURE 3.15
Linear Regression with Negative Slope

FIGURE 3.16
Curvilinear Regression (Quadratic)

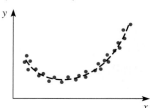

FIGURE 3.17
No Relationship

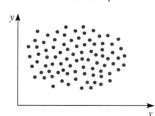

If a straight-line model seems appropriate, the best-fitting straight line is found by using the **method of least squares.** Suppose that $\hat{y} = b_0 + b_1 x$ is the equation of a straight line, where $\hat{y}$ (read "y-hat") represents the **predicted value of y** that corresponds to a particular value of *x*. The **least squares criterion** requires that we find the constants b_0 and b_1 such that $\sum (y - \hat{y})^2$ is as small as possible.

Figure 3.18 shows the distance of an observed value of *y* from a **predicted value of ŷ.** The length of this distance represents the value $(y - \hat{y})$ (shown as the red line segment in Figure 3.18). Note that $(y - \hat{y})$ is positive when the point (*x*, *y*) is above the line and negative when (*x*, *y*) is below the line.

Figure 3.19 shows a scatter diagram with what appears to be the **line of best fit,** along with 10 individual $(y - \hat{y})$ values. (Positive values are shown in red; negative, in green.) The sum of the squares of these differences is minimized (made as small as possible) if the line is indeed the line of best fit.

Figure 3.20 shows the same data points as Figure 3.19. The 10 individual values of $(y - \hat{y})$ are plotted with a line that is definitely not the line of best fit. [The value of $\sum (y - \hat{y})^2$ is 149, much larger than 23 from Figure 3.19.] Every different line drawn through this set of 10 points will result in a different value for $\sum (y - \hat{y})^2$. Our job is to find the one line that will make $\sum (y - \hat{y})^2$ the smallest possible value.

FIGURE 3.18 Observed and Predicted Values of *y*

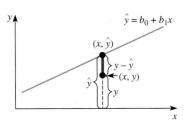

FIGURE 3.19 The Line of Best Fit

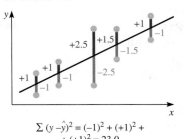

$$\Sigma\,(y-\hat{y})^2 = (-1)^2 + (+1)^2 + \ldots + (+1)^2 = 23.0$$

FIGURE 3.20 Not the Line of Best Fit

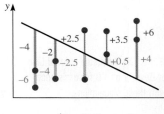

$$\Sigma\,(y-\hat{y})^2 = (-6)^2 + (-4)^2 + \ldots + (+6)^2 = 149.0$$

The equation of the line of best fit is determined by its **slope (b_1)** and its **y-intercept (b_0).** (See the *Students Solution Manual* for a review of the concepts of slope and intercept of a straight line.) The values of the constants—slope and y-intercept—that satisfy the least squares criterion are found by using the formulas presented next:

Definition Formula

slope: $b_1 = \dfrac{\Sigma(x-\bar{x})(y-\bar{y})}{\Sigma(x-\bar{x})^2}$ 　　　(3.5)

We will use a mathematical equivalent of formula (3.5) for the slope, b_1, that uses the sums of squares found in the preliminary calculations for correlation:

Computational Formula

slope: $b_1 = \dfrac{SS(xy)}{SS(x)}$ 　　　(3.6)

Notice that the numerator of formula (3.6) is the SS(xy) formula (3.4) (p. 164) and the denominator is formula (2.9) (p. 87) from the correlation coefficient calculations. Thus, if you have previously calculated the linear correlation coefficient using the procedure outlined on pages 164–165, you can easily find the slope of the line of best fit. If you did not previously calculate r, set up a table similar to Table 3.11 (p. 165) and complete the necessary preliminary calculations.

For the y-intercept, we have:

Computational Formula

y-intercept $= \dfrac{(sum\ of\ y) - [(slope)(sum\ of\ x)]}{number}$

$$b_0 = \dfrac{\Sigma y - (b_1 \cdot \Sigma x)}{n}$$ 　　　(3.7)

Alternative Computational Formula

y-intercept $= y\text{-}bar - (slope \cdot x\text{-}bar)$

$$b_0 = \bar{y} - (b_1 \cdot \bar{x})$$ 　　　(3.7a)

Now let's consider the data in Example 3.3 (p. 153) and the question of predicting a student's number of sit-ups based on the number of push-ups. We want to find the line of best fit, $\hat{y} = b_0 + b_1 x$. The preliminary calculations have already been completed in Table 3.11 (p. 165). To calculate the slope, b_1, using formula (3.6), recall that $SS(xy) = 919.0$ and $SS(x) = 1396.9$. Therefore,

FYI See Animated Tutorial Line of Best Fit on CD for assistance with these calculations.

$$\text{slope:} \quad b_1 = \frac{SS(xy)}{SS(x)} = \frac{919.0}{1396.9} = 0.6579 = \mathbf{0.66}$$

To calculate the y-intercept, b_0, using formula (3.7), recall that $\sum x = 351$ and $\sum y = 380$ from the extensions table. We have

$$\text{y-intercept:} \quad b_0 = \frac{\sum y - (b_1 \cdot \sum x)}{n} = \frac{380 - (0.6579)(351)}{10}$$

$$= \frac{380 - 230.9229}{10} = 14.9077 = \mathbf{14.9}$$

By placing the two values just found into the model $\hat{y} = b_0 + b_1 x$, we get the equation of the line of best fit:

$$\hat{y} = \mathbf{14.9 + 0.66}x$$

Notes:

1. Remember to keep at least three extra decimal places while doing the calculations to ensure an accurate answer.

2. When rounding off the calculated values of b_0 and b_1, always keep at least two significant digits in the final answer.

Now that we know the equation for the line of best fit, let's draw the line on the scatter diagram so that we can see the relationship between the line and the data. We need two points in order to draw the line on the diagram. Select two convenient x values, one near each extreme of the domain ($x = 10$ and $x = 60$ are good choices for this illustration), and find their corresponding y values.

For $x = 10$: $\hat{y} = 14.9 + 0.66x = 14.9 + 0.66(10) = 21.5$; **(10, 21.5)**

For $x = 60$: $\hat{y} = 14.9 + 0.66x = 14.9 + 0.66(60) = 54.5$; **(60, 54.5)**

These two points, (10, 21.5) and (60, 54.5), are then located on the scatter diagram (we use a purple + to distinguish them from data points) and the line of best fit is drawn (shown in red in Figure 3.21).

There are some additional facts about the least squares method that we need to discuss.

1. The slope, b_1, represents the predicted change in y per unit increase in x. In our example, where $b_1 = 0.66$, if a student can do an additional 10 push-ups (x), we predict that he or she would be able to do approximately 7 (0.66 × 10) additional sit-ups (y).

2. The y-intercept is the value of y where the line of best fit intersects the y-axis. (When the vertical scale is located above $x = 0$, the y-intercept is easily seen on the scatter diagram, shown as a green + in Figure 3.21.)

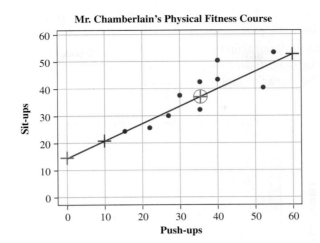

FIGURE 3.21
Line of Best Fit for
Push-ups versus
Sit-ups

First, however, in interpreting b_0, you must consider whether $x = 0$ is a realistic x value before you conclude that you would predict $\hat{y} = b_0$ if $x = 0$. To predict that if a student did no push-ups, he or she would still do approximately 15 sit-ups ($b_0 = 14.9$) is probably incorrect. Second, the x value of zero may be outside the domain of the data on which the regression line is based. In predicting y based on an x value, check to be sure that the x value is within the domain of the x values observed.

3. The line of best fit will always pass through the *centroid*, the point $(\bar{x}, \bar{y})$. When drawing the line of best fit on your scatter diagram, use this point as a check. For our illustration,

$$\bar{x} = \frac{\sum x}{n} = \frac{351}{10} = 35.1, \qquad \bar{y} = \frac{\sum y}{n} = \frac{380}{10} = 38.0$$

We see that the line of best fit does pass through $(\bar{x}, \bar{y}) = (35.1, 38.0)$, as shown in green $\oplus$ in Figure 3.21.

Let's work through another example to clarify the steps involved in regression analysis.

EXAMPLE 3.7

Calculating the Line of Best Fit Equation

In a random sample of eight college women, each woman was asked her height (to the nearest inch) and her weight (to the nearest 5 pounds). The data obtained are shown in Table 3.13. Find an equation to predict the weight of a college woman based on her height (the equation of the line of best fit), and draw it on the scatter diagram in Figure 3.22.

TABLE 3.13

College Women's Heights and Weights [TA03-13]

	1	2	3	4	5	6	7	8
Height, x	65	65	62	67	69	65	61	67
Weight, y	105	125	110	120	140	135	95	130

SOLUTION Before we start to find the equation for the line of best fit, it is often helpful to draw the scatter diagram, which provides visual insight into the relationship between the two variables. The scatter diagram for the data on the heights and weights of college women, shown in Figure 3.22, indicates that the linear model is appropriate.

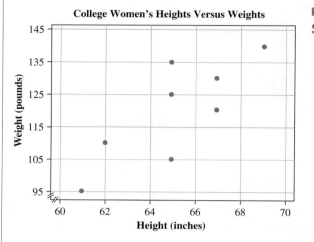

College Women's Heights Versus Weights

FIGURE 3.22
Scatter Diagram

To find the equation for the line of best fit, we first need to complete the preliminary calculations, as shown in Table 3.14. The other preliminary calculations include finding SS(x) from formula (2.9) and SS(xy) from formula (3.4):

TABLE 3.14

Preliminary Calculations Needed to Find b_1 and b_0

Student	Height, x	x^2	Weight, y	xy
1	65	4225	105	6825
2	65	4225	125	8125
3	62	3844	110	6820
4	67	4489	120	8040
5	69	4761	140	9660
6	65	4225	135	8775
7	61	3721	95	5795
8	67	4489	130	8710
	$\sum x = 521$	$\sum x^2 = 33{,}979$	$\sum y = 960$	$\sum xy = 62{,}750$

$$SS(x) = \sum x^2 - \frac{(\sum x)^2}{n} = 33{,}979 - \frac{(521)^2}{8} = 48.875$$

$$SS(xy) = \sum xy - \frac{\sum x \sum y}{n} = 62{,}750 - \frac{(521)(960)}{8} = 230.0$$

Second, we need to find the slope and the *y*-intercept using formulas (3.6) and (3.7):

slope: $b_1 = \dfrac{SS(xy)}{SS(x)} = \dfrac{230.0}{48.875} = 4.706 = \mathbf{4.71}$

y-intercept: $b_0 = \dfrac{\sum y - (b_1 \cdot \sum x)}{n} = \dfrac{960 - (4.706)(521)}{8} = -186.478 = \mathbf{-186.5}$

Thus, the equation of the line of best fit is $\hat{y} = \mathbf{-186.5 + 4.71}x$.

To draw the line of best fit on the scatter diagram, we need to locate two points. Substitute two values for *x*—for example, 60 and 70—into the equation for the line of best fit and obtain two corresponding values for $\hat{y}$:

$\hat{y} = -186.5 + 4.71x = -186.5 + (4.71)(60) = -186.5 + 282.6 = 96.1 \approx 96$

$\hat{y} = -186.5 + 4.71x = -186.5 + (4.71)(70) = -186.5 + 329.7 = 143.2 \approx 143$

The values (60, 96) and (70, 143) represent two points (designated by a red + in Figure 3.23) that enable us to draw the line of best fit.

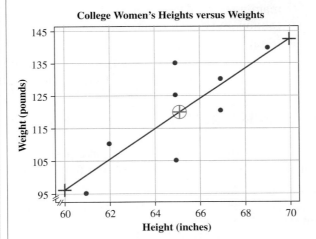

College Women's Heights versus Weights

FIGURE 3.23
Scatter Diagram with Line of Best Fit

Note: In Figure 3.23, $(\bar{x}, \bar{y}) = (65.1, 120)$ is also on the line of best fit. It is the green $\oplus$. Use $(\bar{x}, \bar{y})$ as a check on your work.

Making Predictions

One of the main reasons for finding a regression equation is to make predictions. Once a linear relationship has been established and the value of the input variable *x* is known, we can predict a value of *y*, $\hat{y}$. Consider the equation $\hat{y} = -186.5 + 4.71x$ relating the height and weight of college women. If a particular female college student is 66 inches tall, what do you predict her weight to be? The predicted value is

$\hat{y} = -186.5 + 4.71x = -186.5 + (4.71)(66) = -186.5 + 310.86$

$= 124.36 \approx 124$ lb

You should not expect this predicted value to occur exactly; rather, it is the average weight you would expect for all female college students who are 66 inches tall.

When you make predictions based on the line of best fit, observe the following restrictions:

1. The equation should be used to make predictions only about the population from which the sample was drawn. For example, using our relationship between the height and the weight of college women to predict the weight of professional athletes given their height would be questionable.

2. The equation should be used only within the sample domain of the input variable. We know the data demonstrate a linear trend within the domain of the x data, but we do not know what the trend is outside this interval. Hence, predictions can be very dangerous outside the domain of the x data. For instance, in Example 3.7 it is nonsense to predict that a college woman of height zero will weigh -186.5 pounds. Do not use a height outside the sample domain of 61 to 69 inches to predict weight. On occasion you might wish to use the line of best fit to estimate values outside the domain interval of the sample. This can be done, but you should do it with caution and only for values close to the domain interval.

3. If the sample was taken in 2006, do not expect the results to have been valid in 1929 or to hold in 2010. The women of today may be different from the women of 1929 and the women in 2010.

TECHNOLOGY INSTRUCTIONS: LINE OF BEST FIT

MINITAB (Release 14)   Input the x values into C1 and the corresponding y values into C2; then to obtain the equation for the line of best fit, continue with:

Method 1—
Choose: **Stat > Regression > Regression . . .**
Enter: Response (*y*): **C2**
               Predictors (*x*): **C1 > OK**

To draw the scatter diagram with the line of best fit superimposed on the data points, FITS must have been selected previously; then continue with:

Choose: **Graph > Scatterplot**
Select: **With Regression > OK**
Enter: Y variable: **C2** X variable: **C1**
Select: **Labels > Titles/Footnotes**
Enter: Title: **your title > OK > OK**

OR
Method 2—

Choose: **Stat > Regression > Fitted Line Plot**
Enter: Response (*Y*): **C2**
               Response (*X*): **C1**
Select: **Linear**
Select: **Options**
Enter: Title: **your title > OK > OK**

Excel Input the *x*-variable data into column A and the corresponding *y*-variable data into column B; then continue with:

Choose: **Tools > Data Analysis > Regression > OK**
Enter: Input Y Range: **(B1:B10 or select cells)**
                 Input X Range: **(A1:A10 or select cells)**
Select: **Labels** (if necessary)
               Output Range
               Enter: **(C1 or select cell)**
               Line Fits Plots > OK

To make the output readable; continue with:

Choose: **Format > Column > Autofit Selection**

To form the regression equation, the *y*-intercept is located at the intersection of the intercept and coefficients columns, whereas the slope is located at the intersection of the *x* variable and the coefficients columns.

To draw the line of best fit on the scatter diagram, activate the chart; then continue with:

Choose: **Chart > Add Trendline > Linear > OK**
(This command also works with the scatter diagram Excel commands on p. 155.)

TI-83/84 Plus Input the *x*-variable data into L1 and the corresponding *y*-variable data into L2; then continue with:

     If just the equation is desired:

Choose: **STAT > CALC > 8:LinReg(a + bx)**
Enter: **L1, L2***
*If the equation and graph on the scatter diagram are desired, use:

Enter: **L1, L2, Y1†**

then continue with the same commands for a scatter diagram as shown on page 155.
†To enter Y1, use:
Choose: **VARS > Y-VARS > 1:Function > 1:Y1 > ENTER**

Understanding the Line of Best Fit

The following method will create (1) a visual meaning for the line of best fit, (2) a visual meaning for what the line of best fit is describing, and (3) an estimate for the slope and *y*-intercept of the line of best fit. As with the approximation of *r*, estimations of the slope and *y*-intercept of the line of best fit should be used only as a mental estimate or check.

Note: This estimation technique *does not* replace the calculations for b_1 and b_0.

Procedure
  1. On the scatter diagram of the data, draw the straight line that appears to be the line of best fit. (*Hint:* If you draw a line parallel to and halfway between the two pencils described in Section 3.3 on page 166 [Figure 3.10], you will have a reasonable estimate for the line of best fit.) The two pencils border the "path" demonstrated by the ordered pairs, and the line

down the center of this path approximates the line of best fit. Figure 3.24 shows the pencils and the resulting estimated line for Example 3.7.

FIGURE 3.24

Estimate the Line of Best Fit for the College Woman Data

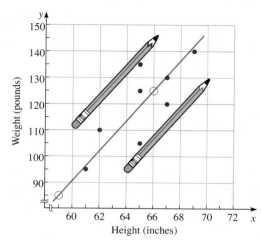

2. This line can now be used to approximate the equation. First, locate any two points (x_1, y_1) and (x_2, y_2) along the line and determine their coordinates. Two such points, circled in Figure 3.24, have the coordinates (59, 85) and (66, 125). These two pairs of coordinates can now be used in the following formula to estimate the slope b_1:

estimate of the slope, b_1: $\qquad b_1 \approx \dfrac{y_2 - y_1}{x_2 - x_1} = \dfrac{125 - 85}{66 - 59} = \dfrac{40}{7} = 5.7$

3. Using this result, the coordinates of one of the points, and the following formula, we can determine an estimate for the y-intercept, b_0:

estimate of the y-intercept, b_0:

$$b_0 \approx y - b_1 \cdot x = 85 - (5.7)(59) = 85 - 336.3 = -251.3$$

Thus, b_0 is approximately -250.

4. We now can write the estimated equation for the line of best fit:

$$\hat{y} = -250 + 5.7x$$

This should serve as a crude estimate. The actual equation calculated using all of the ordered pairs was $\hat{y} = -186.5 + 4.71x$.

APPLIED EXAMPLE 3.8

Concrete Shrinkage

DRYING SHRINKAGE

Drying shrinkage is defined as the contracting of a hardened concrete mixture due to the loss of capillary water. This shrinkage causes an increase in tensile stress, which may lead to cracking, internal warping, and external deflection, before the concrete is subjected to any kind of loading. All portland cement concrete undergoes drying shrinkage, or hydral volume change, as the concrete ages. The hydral volume change in concrete is very important to the engineer in the design of a structure. . . .

Drying shrinkage is dependent upon several factors. These factors include the properties of the components, proportions of the components, mixing manner, amount of moisture while curing, dry environment, and member size. . . . Drying shrinkage happens mostly because of the reduction of capillary water by evaporation and the water in the cement paste. The higher amount of water in the fresh concrete, the greater the drying shrinkage affects. . . .

The concrete properties influence on drying shrinkage depends on the ratio of water to cementitious materials content, aggregate content, and total water content. The total water content is the most important of these. The relationship be-

tween the amount of water content of fresh concrete and the drying shrinkage is linear. Increase of the water content by one percent will approximately increase the drying shrinkage by three percent.

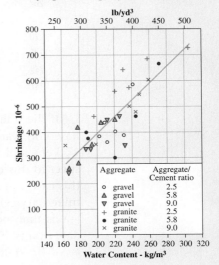

Source: http://www.engr.psu.edu/ce/concrete_clinic/expansionscontractions/ dryshrinkage/dryingshrinkage.htm

The following data were obtained from the website where the preceding article was posted.

Because these are bivariate data, our first consideration is to sketch a scatter diagram. Water content will be the independent variable and plotted along the *x*-axis; shrinkage will be the dependent variable and plotted along the *y*-axis.

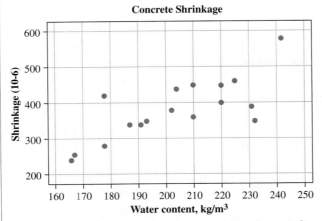

Note that the two variables appear to have a linear relationship as claimed in the article. There is an elongated pattern from the lower left corner to the upper right corner of the scatter diagram. As the water content increased, the shrinkage also increased.

TABLE 3.15

Relationship between Water Content and Drying Shrinkage [TA03-015]

Water Content (kg/m³)	Shrinkage (10⁻⁶)	Water Content (kg/m³)	Shrinkage (10⁻⁶)	Water Content (kg/m³)	Shrinkage (10⁻⁶)
202	380	178	280	187	340
210	360	178	420	191	340
220	400	193	350	210	450
231	390	204	440	225	460
242	580	220	450	232	350
167	255	166	240		

If the line of best fit is calculated, predictions of shrinkage could be made based on water content. The line of best fit is $y = -166.4 + 2.69x$.

Let's look at the y-intercept and slope and see what each means with respect to this concrete shrinkage data. The y-intercept corresponds to $x = 0$. In this case, $x = 0$ means there is no water content and the "concrete" is still unmixed dry cement. Think of the y-intercept, -166.4, as being the value on the y-axis that locates (vertically) the line of best fit so that it passes through the data points. The slope, 2.69, in this example is the amount of shrinkage, $2.69 \times 10^{-6} = 0.00000269$, per one unit of increase in water content. This indicates that for each additional 1 kilogram of water per cubic meter of concrete, there is an increase of 0.00000269 units of shrinkage. To see this on the scatter diagram, use the same units used for calculations [ignoring the (10^{-6}) factor]; then, for every one unit of increase horizontally, you should see 2.69 units of change vertically. This might be easier to see if the ratio 2.69 to 1 is rewritten as 26.9/10. This now says that for every 10 additional kilograms of water/cubic meter of mix, there is an increase of almost 27 units of shrinkage.

SECTION 3.4 EXERCISES

Statistics⊜Now™

Datasets can be found on your Student's Suite CD-ROM or at the StatisticsNow website at **http://1pass.thomson.com**.

3.52 Draw a scatter diagram for these data:

x	1	2.5	3	4	5	1.5
y	1.5	2.2	3.5	3	4	2.5

Would you be justified in using the techniques of linear regression on these data to find the line of best fit? Explain.

3.53 [EX03-053] Draw a scatter diagram for these data:

x	2	12	4	6	9	4	11	3	10	11	3	1	13	12	14	7	2	8
y	4	8	10	9	10	8	8	5	10	9	8	3	9	8	8	11	6	9

Would you be justified in using the techniques of linear regression on these data to find the line of best fit? Explain.

3.54 [EX03-054] Twenty-four countries were randomly selected from *The World Factbook*'s 2004 list of world countries. Data on the percentage of each country below the poverty line and the countries' life expectancies were collected.

Country	Below Poverty	Life Expectancy
Afghanistan	23.0	42.46
Albania	30.0	77.06

••• Remainder of data on Student's Suite CD-ROM

Source: The World Factbook, 2004, http://www.cia.gov/cia/publications/factbook/index.html

a.   Construct a scatter diagram of the percentage below the poverty line, *x*, and the life expectancy, *y*.

b.   Does it appear that these two variables are correlated?

c.   Are we justified in using the techniques of linear regression on these data? Explain.

3.55 The formulas for finding the slope and the *y*-intercept of the line of best fit use both summations, $\sum$'s, and sums of squares, SS( )'s. It is important to know the difference. In reference to Example 3.5 (p. 164):

a.   Find three pairs of values: $\sum x^2$, SS(*x*); $\sum y^2$, SS(*y*); and $\sum xy$, SS(*xy*).

b.   Explain the difference between the numbers for each pair of numbers.

3.56 Show that formula (3.7a) is equivalent to formula (3.7) (p. 175).

3.57 The values of *x* used to find points for graphing the line $\hat{y} = 14.9 + 0.66x$ in Figure 3.21 (p. 177) are arbitrary. Suppose you choose to use *x* = 20 and *x* = 50.

a.   What are the corresponding $\hat{y}$ values?

b.   Locate these two points on Figure 3.21. Are these points on the line of best fit? Explain why or why not.

3.58 Does it pay to study for an exam? The number of hours studied, *x*, is compared to the exam grade received, *y*:

x	2	5	1	4	2
y	80	80	70	90	60

a.   Find the equation for the line of best fit.

b.   Draw the line of best fit on the scatter diagram of the data drawn in Exercise 3.15 (p. 158).

c.   Based on what you see in your answers to parts a and b, does it pay to study for an exam? Explain.

3.59 [EX03-059] What is the relationship between carbohydrates consumed and energy released in a sports drink? Did you ever wonder if there was a relationship? Let's use the sports drink data listed in Exercise 3.43 on page 171 to investigate the relationship.

a.   In Exercise 3.43 a scatter diagram was drawn using *x* = carbs/serving and *y* = energy/serving. Review the scatter diagram (if you did not draw it before, do so now), and describe why you believe there is or is not a linear relationship.

b.   Find the equation for the line of best fit.

c.   Using the equation found in part b, estimate the amount of energy that one can expect to gain from consuming 40 grams of carbohydrates.

d.   Using the equation found in part b, estimate the amount of energy that one can expect to gain from consuming 65 grams of carbohydrates.

3.60 AJ used linear regression to help him understand his monthly telephone bill. The line of best fit was $\hat{y} = 23.65 + 1.28x$, where *x* is the number of long-distance calls made during a month, and *y* is the total telephone cost for a month. In terms of number of long-distance calls and cost:

a.   Explain the meaning of the *y*-intercept, 23.65.

b.   Explain the meaning of the slope, 1.28.

3.61 For Example 3.7 (p. 177) and the scatter diagram in Figure 3.23 on page 179:

a.   Explain how the slope of 4.71 can be seen.

b.   Explain why the *y*-intercept of -186.5 cannot be seen.

3.62 If all students from Mr. Chamberlain's physical fitness course on pages 153 and 164 who can do 40 push-ups are asked to do as many sit-ups as possible:

a.   How many sit-ups do you expect each can do?

b   Will they all be able to do the same number?

c.   Explain the meaning of the answer to part a.

3.63 A study was conducted to investigate the relationship between the resale price, y (in hundreds of dollars), and the age, x (in years), of midsize luxury American automobiles. The equation of the line of best fit was determined to be $\hat{y} = 185.7 - 21.52x$.

a. Find the resale value of such a car when it is 3 years old.

b. Find the resale value of such a car when it is 6 years old.

c. What is the average annual decrease in the resale price of these cars?

3.64 A study was conducted to investigate the relationship between the cost, y (in tens of thousands of dollars), per unit of equipment manufactured and the number of units produced per run, x. The resulting equation for the line of best fit was $\hat{y} = 7.31 - 0.01x$, with x being observed for values between 10 and 200. If a production run was scheduled to produce 50 units, what would you predict the cost per unit to be?

3.65 The Federal Highway Administration annually reports on state motor-fuel taxes. Based on the latest report, in thousands of dollars, the amount of receipts can be estimated using the equation: Receipts = $-5359 + 0.9956$ Collections.

a. If a state collected $500,000, what would you estimate the receipts to be?

b. If a state collected $1,000,000, what would you estimate the receipts to be?

c. If a state collected $1,500,000, what would you estimate the receipts to be?

3.66 A study of the tipping habits of restaurant-goers was completed. The data for two of the variables—x, the amount of the restaurant check, and y, the amount left as a tip for the servers—were used to construct a scatter diagram. What do you expect the scatter diagram to reveal?

a. Do you expect the two variables will show a linear relationship? Explain.

b. What will the scatter diagram suggest about linear correlation? Explain.

c. What value do you expect for the slope of the line of best fit? Explain.

d. What value do you expect for the y-intercept of the line of best fit? Explain.

The data are used to determine the equation for the line of best fit: $\hat{y} = 0.02 + 0.177x$.

e. What does the slope of this line represent as applied to the actual situation? Does the value 0.177 make sense? Explain.

f. What does the y-intercept of this line represent as applied to the actual situation? Does the value 0.02 make sense? Explain.

g. If the next restaurant check was for $30, what would the line of best fit predict for the tip?

h. Using the line of best fit, predict the tip for a check of $31. What is the difference between this amount and the amount in part g for a $30 check? Does this difference make sense? Where do you see it in the equation for the line of best fit?

3.67 Consider Figure 3.24 on page 182. The graph's y-intercept is -250, not approximately 80, as might be read from the figure. Explain why?

3.68 [EX03-068] Stride rate (number of steps per second) is important to the serious runner. Stride rate is closely related to speed, and a runner's goal is to achieve the optimal stride rate. As part of a study, researchers measured the stride rate at seven different speeds for 21 top female runners. The average stride rates for these women and the test speeds are listed in the accompanying table.

Speed, x (ft/sec)	15.86	16.88	17.50	18.62	19.97	21.06	22.11
Stride Rate, y	3.05	3.12	3.17	3.25	3.36	3.46	3.55

Source: R. C. Nelson, C. M. Brooks, and N. L. Pike, Biomechanical comparison of male and female runners, in P. Milvy (ed.), The Marathon: Physiological, Medical, Epistemological, and Psychological Studies (New York Academy of Sciences, 1977), pp. 793–807

a. Construct a scatter diagram.

b. Does the relationship of the two variables appear to be linear?

c. Find the equation of the line of best fit.

d. Interpret the slope of the equation in part c. In other words, what are the "units" of the slope?

e. Plot the line of best fit on the scatter diagram.

f. Using the line drawn in part e, predict the average stride rate if the speed is 19 feet per second.

g. What is the stride rate if the speed is zero? Interpret your result. Do the results make sense? Explain.

FYI Have you tried to use the computer or calculator commands yet?

3.69 Consider the College Women's data presented in Example 3.7 and the line of best fit. When estimating the line of best fit from a scatter diagram, the choice for the two points (x_1, y_1) and (x_2, y_2) to be used is somewhat arbitrary. When different points are used, slightly different values for b_0 and b_1 will result, but they should be approximately the same.

a. What points on the scatter diagram (Figure 3.24, p. 182) were used to estimate the slope and y-intercept in the example on page 182? What were the resulting estimates?

b. Use points (61, 95) and (67, 130) and find the approximate slope and y-intercept values.

c. Compare the values found in part b with those described in part a. How similar in value are they?

d. Compare both sets of estimates with the actual values of slope and y-intercept found in Example 3.7 on pages 177–179. Draw both estimated lines of best fit on scatter diagram shown in Figure 3.23. How useful do you think estimated values might be? Explain.

3.70 [EX03-070] Luxury cars are sweet if you can afford them. If not, you may want to consider a more affordable "feel alike" model, according to Mitch McCullough, who evaluates 60 to 70 vehicles annually as editor of New Car Test Drive (http://www.nctd.com).

Luxury Models	Cost ($1000)
Mercedes-Benz SLK320	46
Chevrolet Corvette	45
BMW 330i	35
Lexus ES 330	32
Lincoln Town Car	42
Lexus RX 330 SUV	36
Lincoln Aviator SUV	41
Porsche Cayenne S SUV	56
Land Rover Range Rover SUV	73
Cadillac Escalade SUV	53

"Feel Alike" Alternative	Cost ($1000)
Chrysler Crossfire	34
Nissan 350Z	27
Infiniti G35	30
Hyundai XG350L	26
Mercury Grand Marquis LS	30
Nissan Murano	28
Ford Explorer Eddie Bauer	34
Infiniti FX35	34
Volkswagen Touareg V8	43
GMC Yukon	36

Source: Reader's Digest, June 2004

a. Do you expect that the two variables will show a linear relationship? Explain.

b. Construct the scatter diagram using x = luxury cost and y = "feel alike cost."

c. Does there appear to be a linear pattern? Explain.

d. Calculate the equation for the line of best fit.

e. Use the equation found in part d to estimate the cost of a "feel alike" vehicle comparable to a $40,000 luxury vehicle. Explain the meaning of your answer.

f. Use the equation found in part d to estimate the cost of a "feel alike" vehicle comparable to a $60,000 luxury vehicle. Explain the meaning of your answer.

3.71 [EX03-071] Professional golfers have a classic golf quandary: "drive for show, putt for dough." It is often their short game (on the putting green) that determines whether they win a tournament. In the January 7, 2005, *USA Today* article titled "In short, Durant aims to improve," a chart was provided showing the success rates for the 2004 season PGA Tour players for hitting the greens from various distances.

Yards	Mid-yard	Success Rate (%)
More than 200	213	44
176–200	188	53
151–175	163	61
126–150	138	68
101–125	113	72
76–100	88	78
75 or less	63	85

Source: PGA Tour Shotlink

Using the mid-yardage as the independent variable, x, and success rate as the dependent variable, y:

a. Construct a scatter diagram.

b. Does there appear to be a linear correlation? Justify your answer.

c. Calculate the linear correlation coefficient, r.

d. Interpret the correlation coefficient found in part c. Comment on its direction and strength.

e. Does there appear to be a linear relationship? Justify your answer.

f. Calculate the equation of the line of best fit.

g. Graph the line of best fit on the scatter diagram.

h. Predict the average success rate for a professional golfer if he made it to the green from a distance of 90 yards.

3.72 [EX03-072] The following data are a sample of the ages, x (years), and the asking prices, y (×$1000), for used Honda Accords that were listed on AutoTrader.com on March 10, 2005:

x	y	x	y	x	y
3	24.9	7	11.9	6	16.4
7	9.0	6	15.2	4	21.2
5	17.8	2	25.9	3	24.9
4	29.2	2	26.9	5	20.0
6	15.7	4	23.8	7	13.6
3	24.9	5	19.3	5	18.8
2	25.7	4	21.9		

Source: http://autotrader.com

a. Draw a scatter diagram.

b. Calculate the equation of the line of best fit.

c. Graph the line of best fit on the scatter diagram.

d. Predict the average asking price for all Honda Accords that are 5 years old. Obtain this answer in two ways: using the equation from part b and using the line drawn in part c.

e. Can you think of any potential lurking variables for this situation? Explain any possible role they might play.

3.73 [EX03-073] Baseball teams win and lose games. Many fans believe that a team's earned run average (ERA) has a major effect on winning. During the 2004 season, the 30 major league baseball teams recorded the following number of wins while generating these ERAs.

Wins	ERA	Wins	ERA	Wins	ERA
96	3.74	98	4.18	63	4.76
105	3.75	67	4.24	80	4.81
89	3.81	92	4.28	70	4.81
93	4.01	72	4.29	83	4.91
92	4.03	91	4.29	67	4.91
87	4.03	67	4.33	72	4.93
92	4.05	86	4.45	51	4.98
71	4.09	89	4.53	58	5.15
83	4.10	101	4.69	76	5.19
91	4.17	78	4.70	68	5.54

Source: http://mlb.mlb.com

a. Do you think the teams with the better ERAs have the most wins? (The lower the ERA, the fewer earned runs the other team scored.)

b. If this is true, what will the pattern on the scatter diagram look like? Be specific.

c. Construct a scatter diagram of these data.

d. Does the scatter diagram suggest that teams tend to win more games when their team ERA is lower? Explain.

e. Calculate the line of best fit using x = ERA and y = number of wins.

f. On the average, how is the number of wins affected by an increase of 1 in the ERA? Explain how you determined this number.

g. Do your findings seem to support the idea that the teams with the better ERAs will have the most wins? Justify your response.

3.74 [EX03-074] Interstate 90, the longest of the east–west U.S. interstate highways, is 3112 miles long, stretching from Boston, Massachusetts, at I-93 on the eastern end to Seattle, Washington, at the Kingdome on the western end. It travels across 13 northern states; the number of miles and number of intersections in each of these states are listed below.

State	WA	ID	MT	WY	SD	MN	WI
No. of Inter.	57	15	83	23	61	52	40
Miles	298	73	558	207	412	275	188

State	IL	IN	OH	PA	NY	MA
No. of Inter.	19	21	40	14	48	18
Miles	103	157	244	47	391	159

Source: Rand McNally and http://www.ihoz.com/I90.html

a. Construct a scatter diagram.

b. Find the equation for the line of best fit using x = miles and y = intersections.

c. Using the equation found in part b, estimate the average number of interchanges per mile along I-90.

3.75 [TA03-12] By looking at the insurance rates shown in Applied Example 3.6, one can easily see that males pay higher rates for insurance than females of the same age. Is there a consistent pattern to this higher rate? To uncover any such pattern in the male/female $250,000 insurance rates from Applied Example 3.6:

a. Draw a scatter diagram of the insurance rates for males (y) versus females (x). Does the diagram show a linear relationship? Explain.

b. Calculate the linear correlation coefficient, r, for the variables. Is this a strong linear relationship? Why?

c. Calculate the equation of the line of best fit.

d. Predict the monthly rate for a male who is the same age as a female whose monthly rate is $15.00.

e. Based on the preceding answers, what conclusion can you make about the relationship between these male and female insurance rates? What role does the slope of the line of best fit serve in describing the relationship?

3.76 [EX03-076] The success of a professional golfer can be measured along a number of dimensions. The bottom line is probably how much money a golfer earns in a given year, but golfers are also given a world ranking by points for each event that they enter. Following is a combined table extracted from the PGA Tour website (http://www.pgatour.com) that shows the 2004 season's 20 money leaders as well as their world ranking at the end of 2004, number of events played, money earned, and average points per event.

Rank	Player	2004 Events	2004 Money	Avg. Points*
1	Vijay Singh	29	10,905,166	12.97
2	Tiger Woods	19	5,365,472	11.90

••• Remainder of data on Student's Suite CD-ROM

*The Official World Golf Ranking. This statistic is the average number of points earned per event in the last 104 weeks. These points are awarded based on finish position as well as the strength of the field. The points are initially worth double their original value and decline gradually over this 2-year period. There are eight 13-week periods, and points decline by 0.25x their value each period.
Source: PGA TOUR, Inc.

a. Draw a scatter diagram with "2004 money" as the dependent variable, y, and "rank" as the predictor variable, x.

b. Does the scatter diagram in part a suggest that a linear regression will be useful? Explain.

c. Calculate the equation of best fit.

d. Draw the line of best fit on the scatter diagram you obtained in part a. Explain the role of a negative slope for this pair of variables.

e. Do you see a potential lurking variable? Explain its possible role.

f. Draw a scatter diagram with "2004 money" as the dependent variable, y, and "average points" as the predictor variable, x.

g. Does the scatter diagram in part f suggest that a linear regression will be useful? Explain.

h. Calculate the equation of best fit.

i.   Draw the line of best fit on the scatter diagram you obtained in part f.

j.   Do you see a potential lurking variable? Explain its possible role.

k.   Draw a scatter diagram with "2004 money" as the dependent variable, *y*, and "2004 events" as the predictor variable, *x*.

l.   Does the scatter diagram in part k suggest that a linear regression will be useful? Explain.

m.   Calculate the equation of best fit.

n.   Draw the line of best fit on the scatter diagram you obtained in part k.

o.   Is the line of best fit useful in predicting 2004 money based on number of 2004 events played? Explain.

3.77 [EX03-077] The Office of Aviation Enforcement & Proceedings, U.S. Department of Transportation, reported the number of mishandled baggage reports filed by airline passengers (in thousands) during October 2004. The industry average was 4.02 reports per 1000 passengers.

Airline	Reports	Passengers
AirTran	2,084	1148.8
JetBlue	2,295	1057.5
••• Remainder of data on Student's Suite CD-ROM		

Source: Office of Aviation Enforcement & Proceedings, U.S. Department of Transportation

a.   Draw a scatter diagram with number of reports as the dependent variable, *y*, and number of passengers (in thousands) as the predictor variable, *x*.

b.   Does the scatter diagram in part a suggest that a linear regression will be useful? Explain.

c.   Calculate the equation of best fit.

d.   Draw the line of best fit on the scatter diagram you obtained in part a. How well does the line fit the pattern of data? Explain.

3.78 Nielsen Ratings are often published in national newspapers. The number of viewers (in millions) for each program is given with the corresponding Nielsen Rating. With the number of viewers as *x* and the ratings as *y*, the data published

in *USA Today* (February 7, 2002) for the 7–10 PM time slot resulted in a correlation coefficient of 0.99 and the regression equation $\hat{y} = 0.12 + 0.6x$. Explain how the slope and *y* intercept demonstrate that the number of viewers, *x*, must have the largest impact on predicting the Nielsen Rating.

3.79 The following graph shows the relationship between three variables: number of licensed drivers, number of registered vehicles, and the size of the resident population for the United States from 1961 to 2003. Study the graph and answer these questions:

a.   Does it seem reasonable that the Population line and the Drivers line run basically parallel to each other and that the Population line is above the Drivers line? Explain what it means for them to be parallel. What would it mean if they were not parallel?

b.   The Drivers and Motor Vehicles lines cross. What does this mean? When do the lines intersect, and what does the point of intersection represent?

c.   Explain the relationship between motor vehicles and drivers before 1973.

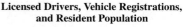

Licensed Drivers, Vehicle Registrations, and Resident Population

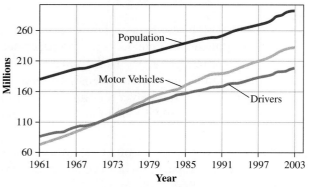

Source: U.S. Dept. of Transportation–Federal Highway Administration

d.   Explain the relationship between motor vehicles and drivers after 1973.

e.   Do you predict drivers will ever surpass motor vehicles after 2003? Why or why not?

f. Using the years 1982 and 2000, estimate the slopes of the Motor Vehicles line and the Drivers line. Compare and contrast the slopes found.

3.80 The correlation coefficient and slope of the line of best fit are related by definition.

a. Verify this statement.

b. Describe how the relationship between correlation coefficient and slope can be seen in the statistics that describe a particular set of data.

c. Show that $b_1 = r(s_y/s_x)$. Comment on this relationship.

CHAPTER REVIEW

In Retrospect

To sum up what we have just learned: There is a distinct difference between the purpose of regression analysis and the purpose of correlation. In regression analysis, we seek a relationship between the variables. The equation that represents this relationship may be the answer that is desired, or it may be the means to the prediction that is desired. In correlation analysis, we measure the strength of the linear relationship between the two variables.

The Applied Examples in the text show a variety of uses for the techniques of correlation and regression. These examples are worth reading again. When bivariate data appear to fall along a straight line on the scatter diagram, they suggest a linear relationship. But this is not proof of cause and effect. Clearly, if a basketball player commits too many fouls, he or she will not be scoring more points. Players in foul trouble are "riding the pine" with no chance to score. It also seems reasonable that the more game time they have, the more points they will score and the more fouls they will commit. Thus, a positive correlation and a positive regression relationship will exist between these two variables. Time is a lurking variable here.

The bivariate linear methods we have studied thus far have been presented as a first, descriptive look. More details must, by necessity, wait until additional developmental work has been done. After completing this chapter, you should have a basic understanding of bivariate data, how they are different from just two sets of data, how to present them, what correlation and regression analysis are, and how each is used.

Vocabulary and Key Concepts

bivariate data (p. 146)
cause-and-effect relationship (p. 167)
coefficient of linear correlation (p. 163)
contingency table (p. 147)
correlation (p. 162)
correlation analysis (p. 162)
cross-tabulation (p. 147)
dependent variable (pp. 152, 162)

independent variable (pp. 152, 162)
input variable (p. 152)
least squares criterion (p. 174)
line of best fit (p. 174)
linear correlation (p. 162)
linear regression (p. 173)
lurking variable (p. 167)
method of least squares (p. 174)
negative correlation (p. 163)
ordered pair (p. 152)

output variable (p. 152)
Pearson's product moment, r (p. 163)
positive correlation (p. 163)
predicted value (p. 174)
prediction equation (p. 174)
regression (p. 173)
regression analysis (p. 174)
scatter diagram (p. 153)
slope, b_1 (p. 175)
y-intercept, b_0 (p. 175)

Learning Outcomes

✓ Understand and be able to present and describe data in the form of two qualitative variables, both in contingency table format and appropriate graphs.
EXP 3.1, pp. 146–149, Ex. 3.83

✓ Understand and be able to present and describe data in the form of one qualitative variable and one quantitative variable, both in table format and appropriate graphs.
EXP 3.2, pp. 151–152, Ex. 3.09, 3.10

✓ Understand and be able to present and describe the relationship between two quantitative variables using a scatter diagram.
EXP 3.3, APP EXP 3.4, pp. 152–155, Ex. 3.15

✓ Understand and be able to explain a linear relationship.
pp. 162–163

✓ Compute, describe, and interpret a correlation coefficient.
pp. 162–164, EXP 3.5, Ex. 3.33

✓ Compute, describe, and interpret a line of best fit.
EXP 3.7

✓ Define and understand the difference between correlation and causation.
pp. 167–168, Ex. 3.50, 3.51

✓ Determine and explain possible lurking variables and their effects on a linear relationship.
pp. 167–168, Ex. 3.50, 3.51

✓ Understand and be able to explain the slope of the line of best fit with respect to the context it is presented in.
Ex. 3.60, 3.66

✓ Understand and be able to explain the y-intercept of the line of best fit with respect to the context it is presented in.
Ex. 3.60, 3.66

✓ Create a scatter diagram with the line of best fit drawn on it.
Ex. 3.58

✓ Compute prediction values based on the line of best fit.
pp. 179–180, Ex. 3.63

✓ Understand and be able to explain what prediction values are.
pp. 179–180, 173–175

✓ Understand that predictions should be made only for values within the sample domain and that caution must be exercised for values outside that domain.
pp. 179–180

Chapter Exercises

Statistics⏂Now™

Go to the StatisticsNow website **http://1pass.thomson.com** to

• Assess your understanding of this chapter

• Check your readiness for an exam by taking the Pre-Test quiz and exploring the resources in the Personalized Learning Plan

Datasets can be found on your Student's Suite CD-ROM or at the StatisticsNow website at **http://1pass.thomson.com**.

3.81 [EX03-81] Fear of the dentist (or the dentist's chair) is an emotion felt by many people of all ages. A survey of 100 individuals in five age groups was conducted about this fear, and these were the results:

	Elementary	Jr. High	Sr. High	College	Adult
Fear	37	28	25	27	21
Do Not Fear	63	72	75	73	79

a. Find the marginal totals.

b. Express the frequencies as percentages of the grand total.

c. Express the frequencies as percentages of each age group's marginal totals.

d. Express the frequencies as percentages of those who fear and those who do not fear.

e. Draw a bar graph based on age groups.

3.82 The accompanying "Emergency cash stash" graphic lists in percentages the distributions for the amount both genders have saved for emergencies.

a. Identify the population, the variables, and the type of variables.

b. Construct a bar graph showing the two distributions side by side.

c. Do the distributions seem to differ for the genders? Explain.

"EMERGENCY CASH STASH"

Among workers ages 25–64, 62% of men and 53% of women have savings set aside for emergencies.

	Men	Women
Less than a months income	12%	18%
1 to less than 3 months	31%	24%
3 to less than 6 months	21%	29%
6 or more months income	36%	26%
Don't know	0%	3%

Data from Anne R. Carey and Grant Jerding, USA TODAY; Source: Merrill Lynch. © 1998 USA TODAY, reprinted by permission.

3.83 [EX03-083] Six breeds of dogs have been rather popular in the United States over the past few years. The following table lists the breeds coupled with the number of registrations filed with the American Kennel Club in 2003 and 2004.

Breeds	2003	2004
Retrievers (Labrador)	144,896	146,692
Retrievers (Golden)	52,520	52,550
German Shepherds	43,938	46,046
Beagles	45,021	44,555
Yorkshire Terriers	38,246	43,522
Dachshunds	39,468	40,770

Source: American Kennel Club, http://www.akc.org/reg/dogreg_stats.cfm

a. A cross-tabulation of the two variables, year (columns) and dog breed (rows), is given. Determine the marginal totals.

b. Express the contingency table in part a in percentages based on the grand total.

c. Draw a bar graph showing the results from part b.

d. Express the contingency table in part a in percentages based on the marginal total for the year.

e. Draw a bar graph showing the results from part d.

3.84 [EX03-84] When was the last time you saw your doctor? That question was asked for the survey summarized in the following table.

		Time Since Last Consultation		
		Less Than 6 Months	6 Months to Less Than 1 Year	1 Year or More
Age	Younger than 28 years	413	192	295
	28–40	574	208	218
	Older than 40	653	288	259

a. Find the marginal totals.

b. Express the frequencies as percentages of the grand total.

c. Express the frequencies as percentages of each age group's marginal totals.

d. Express the frequencies as percentages of each time period.

e. Draw a bar graph based on the grand total.

3.85 [EX03-85] Part of quality control is keeping track of what is occurring. The following contingency table shows the number of rejected castings last month categorized by their cause and the work shift during which they occurred.

	1st Shift	2nd Shift	3rd Shift
Sand	87	110	72
Shift	16	17	4
Drop	12	17	16
Corebreak	18	16	33
Broken	17	12	20
Other	8	18	22

a. Find the marginal totals.

b. Express the numbers as percentages of the grand total.

c. Express the numbers as percentages of each shift's marginal total.

d. Express the numbers as percentages of each type of rejection.

e. Draw a bar graph based on the shifts.

3.86 Determine whether each of the following questions requires correlation analysis or regression analysis to obtain an answer.

a. Is there a correlation between the grades a student attained in high school and the grades he or she attained in college?

b. What is the relationship between the weight of a package and the cost of mailing it first class?

c. Is there a linear relationship between a person's height and shoe size?

d. What is the relationship between the number of worker-hours and the number of units of production completed?

e. Is the score obtained on a certain aptitude test linearly related to a person's ability to perform a certain job?

3.87 An automobile owner records the number of gallons of gasoline, x, required to fill the gasoline tank and the number of miles traveled, y, between fill-ups.

a. If she does a correlation analysis on the data, what would be her purpose and what would be the nature of her results?

b. If she does a regression analysis on the data, what would be her purpose and what would be the nature of her results?

3.88 These data were generated using the equation $y = 2x + 1$.

x	0	1	2	3	4
y	1	3	5	7	9

A scatter diagram of the data results in five points that fall perfectly on a straight line. Find the correlation coefficient and the equation of the line of best fit.

3.89 Consider this set of bivariate data:

x	1	1	3	3
y	1	3	1	3

a. Draw a scatter diagram.

b. Calculate the correlation coefficient.

c. Calculate the line of best fit.

3.90 Start with the point (5, 5) and add at least four ordered pairs, (x, y), to make a set of ordered pairs that display the following properties. Show that your sample satisfies the requirements.

a. The correlation of x and y is 0.0.

b. The correlation of x and y is +1.0.

c. The correlation of x and y is −1.0.

d. The correlation of x and y is between −0.2 and 0.0.

e. The correlation of x and y is between +0.5 and +0.7.

3.91 A scatter diagram is drawn showing the data for x and y, two normally distributed variables. The data fall within the intervals $20 \leq x \leq 40$ and $60 \leq y \leq 100$. Where would you expect to find the data on the scatter diagram, if:

a. The correlation coefficient is 0.0

b. The correlation coefficient is 0.3

c. The correlation coefficient is 0.8

d. The correlation coefficient is −0.3

e. The correlation coefficient is −0.8

3.92 Start with the point (5, 5) and add at least four ordered pairs, (x, y), to make a set of ordered pairs that display the following properties. Show that your sample satisfies the requirements.

a. The correlation of x and y is between +0.9 and +1.0, and the slope of the line of best fit is 0.5.

b. The correlation of x and y is between +0.5 and +0.7, and the slope of the line of best fit is 0.5.

c. The correlation of x and y is between −0.7 and −0.9, and the slope of the line of best fit is −0.5.

d. The correlation of x and y is between +0.5 and +0.7, and the slope of the line of best fit is −1.0.

3.93 [EX03-093] Major league baseball teams often say they sign players with skills that match the confines of their home stadium, thinking that because half of their games are at home, this will work to their advantage. If this were the case, then it would seem that teams would tend to score more runs in their home stadium than in their opponents'. Following is a list of the average number of runs scored at home (Av Runs H) and the average number of runs scored at opponents' stadiums (Av Runs A).

Av Runs H	Av Runs A
4.83	5.49
6.38	5.33

••• Remainder of data on Student's Suite CD-ROM

Source: http://mlb.mlb.com

a. Do teams score, on the average, more runs in their home stadium than at opponents' stadiums? What do you think?

b. If there is not a relationship between *x*, average runs at home, and *y*, average runs away, what do you expect the pattern on a scatter diagram to look like?

c. If they do have a relationship, what do you expect the pattern on a scatter diagram to look like?

d. Construct the scatter diagram.

e. Does the scatter diagram seem to support your answers in parts b and c. Explain why or why not.

3.94 [EX03-094] A biological study of a minnow called the blacknose dace* was conducted. The length, *y* (in millimeters), and the age, *x* (to the nearest year), were recorded.
*Visit: http://www.dnr.state.oh.us/dnap/rivfish/bndace.html

x	0	3	2	2	1	3	2	4	1	1
y	25	80	45	40	36	75	50	95	30	15

a. Draw a scatter diagram of these data.

b. Calculate the correlation coefficient.

c. Find the equation of the line of best fit.

d. Explain the meaning of the answers to parts a through c.

3.95 [EX03-095] Twenty-four countries were randomly selected from *The World Factbook*'s 2004 list of world countries. The life expectancies for males and females were recorded for those countries.

Country	Male Life Expectancy	Female Life Expectancy
Albania	74.37	80.02
American Samoa	72.05	79.41

••• Remainder of data on Student's Suite CD-ROM

Source: The World Factbook, 2004, http://www.cia.gov/cia/publications/factbook/geos/ve.html

a. Construct a scatter diagram of life expectancy for males, *x*, and life expectancy for females, *y*.

b. Does it appear that these two variables are correlated?

c. Find the equation of the line of best fit.

d. What does the numerical value of the slope represent?

3.96 [EX03-96] The sound of crickets chirping is a welcome summer night's sound. In fact, those chirping crickets may be telling you the temperature. In the book *The Song of Insects*, George W. Pierce, a Harvard physics professor, presented real data relating the number of chirps per second, *x*, for striped ground crickets to the temperature in °F, *y*. The following table gives real cricket and temperature data. It appears that the number of chirps represents an average, because it is given to the nearest tenth.

x	*y*	*x*	*y*	*x*	*y*
20.0	88.6	15.5	75.2	15.0	79.6
16.0	71.6	14.7	69.7	17.2	82.6
19.8	93.3	17.1	82.0	16.0	80.6
18.4	84.3	15.4	69.4	17.0	83.5
17.1	80.6	16.2	83.3	14.4	76.3

Source: George W. Pierce, *The Song of Insects*, Harvard University Press, 1948

a. Draw a scatter diagram of the number of chirps per second, *x*, and the air temperature, *y*.

b. Describe the pattern displayed.

c. Find the equation for the line of best fit.

d. Using the equation from part c, find the temperatures that correspond to 14 and 20 chirps, the approximate bounds for the domain of the study.

e. Does the range of temperature values bounded by the temperature values found in part d seem reasonable for this study? Explain.

f. The next time you are out where crickets chirp on a summer night and you find yourself without a thermometer, just count the chirps and you will be able to tell the temperature. If the count is 16, what temperature would you suspect it is?

3.97 [EX03-97] Lakes are bodies of water surrounded by land and may include seas. The accompanying table lists the areas and maximum depths of 32 lakes throughout the world.

a. Draw a scatter diagram showing area, *x*, and maximum depth, *y*, for the lakes.

b. Find the linear correlation coefficient between area and maximum depth. What does the value of this linear correlation imply?

Lake	Area (sq mi)	Max. Depth (ft)
Caspian Sea	143,244	3,363
Superior	31,700	1,330
••• Remainder of data on Student's Suite CD-ROM		

Source: Geological Survey, U.S. Department of the Interior

3.98 [EX03-98] Wildlife populations are monitored with aerial photographs. The number of animals and their locations relative to areas inhabited by the human population are useful information. Sometimes it is possible to monitor the physical characteristics of the animals. The length of an alligator can be estimated quite accurately from aerial photographs, but its weight cannot. The following data are the lengths, *x* (in inches), and weights, *y* (in pounds), of alligators captured in central Florida and can be used to predict the weight of an alligator based on its length.

Weight	Length	Weight	Length	Weight	Length
130	94	38	72	44	61
51	74	366	128	106	90
640	147	84	85	84	89
28	58	80	82	39	68
80	86	83	86	42	76
110	94	70	88	197	114
33	63	61	72	102	90
90	86	54	74	57	78
36	69				

Source: http://exploringdata.cqu.edu.au/stories.htm alligatr

a. Construct a scatter diagram for length, *x*, and weight, *y*.

b. Does it appear that the weight of an alligator is predictable from its length? Explain.

c. Is the relationship linear?

d. Explain why the line of best fit, as described in this chapter, is not adequate for estimating weight based on length.

e. Find the value of the linear correlation coefficient.

f. Explain why the value of *r* can be so high for a set of data that is so obviously not linear in nature.

3.99 [EX03-99] Sugar cane growers are interested in the relationship between the acres of crop harvested and the total sugar cane production (tons) of those acres. The data listed here are for the 2001 crop from 14 randomly selected sugar cane–producing counties in Louisiana.

Acres	Production	Acres	Production
33,700	940,000	20,200	590,000
15,200	460,000	33,800	1,020,000
14,400	440,000	20,500	585,000
2,300	65,000	33,100	1,020,000
30,200	830,000	8,000	200,000
13,100	380,000	41,100	1,130,000
29,600	860,000	17,900	570,000

Source: http://www.usda.gov/nass/graphics/county01/data/

a. These data values have many zeros that will be in the way. Change acres harvested to 100s of acres and production to 1000s of tons of production before continuing.

b. Construct a scatter diagram of acres harvested, *x*, and tons of production, *y*.

c. Does the relationship between the variables appear to be linear? Explain.

d. Find the equation for the line of best fit.

e. What is the slope for the line of best fit? What does the slope represent? Explain what it means to the sugar cane grower.

3.100 [EX03-100] Relatively few business travelers use mass transit systems when visiting large cities. The payoff could be substantial—both in time and money—if they learned how to use the systems, as noted in the December 28, 2004, *USA Today* article "Mass transit could save business travelers big bucks." *USA Today* gathered the following information on the busiest U.S. rail systems.

City	Stations	Vehicles	Track (miles)
Atlanta	38	252	193
Baltimore	14	100	34
Boston	53	408	108
Chicago	144	1190	288
Cleveland	18	60	42
Los Angeles	16	102	34
Miami	22	136	57
New York	468	6333	835
Philadelphia	53	371	102
San Francisco	43	669	246
Washington	86	950	226

Source: USA Today, December 28, 2004

Suppose a mass transit system is being proposed for a city and you have been put in charge of preparing statistical information (both graphic and numerical) about the relationship between the following three variables: the number of stations, the number of cars, and the number of miles of rail. You were provided with the preceding data.

a. Start by inspecting the data given. Do you notice any thing unusual about the data? Are there any values that seem quite different from the rest? Explain.

b. Your supervisor suggests that you remove the data for New York. Make a case for that being acceptable. Include some preliminary graphs and calculated statistics to justify removing these values.

Using the data from the other 10 cities:

c. Construct a scatter diagram using miles of track as the independent variable, *x*, and the number of stations as the dependent variable, *y*.

d. Is there evidence of a linear relationship between these two variables? Justify your answer.

e. Find the equation of the line of best fit for part c.

f. Interpret the meaning of the equation for the line of best fit. What does it tell you?

g. Construct a scatter diagram using miles of track as the independent variable, *x*, and the number of vehicles as the dependent variable, *y*.

h. Is there evidence of a linear relationship between these two variables? Justify your answer.

i. Find the equation of the line of best fit for part g.

j. Interpret the meaning of the equation for the line of best fit. What does it tell you?

k. Construct a scatter diagram using number of stations as the independent variable, *x*, and the number of vehicles as the dependent variable, *y*.

l. Is there evidence of a linear relationship between these two variables? Justify your answer.

m. Find the equation of the line of best fit for part k.

n. Interpret the meaning of the equation for the line of best fit. What does it tell you?

o. The city is entertaining initial proposals for a mass transit system of 50 miles of track. Based on the answers found in parts c through n, how many stations and how many vehicles will be needed for the system? Justify your answers.

p. If someone wants an estimate for the number of stations and vehicles needed for a 100-mile system, they should not just double the results found in part o. Explain why not.

q. Based on the answers found in parts c through n, how many stations and how many vehicles will be needed for a 100-mile system? Justify your answers.

3.101 [EX03-101] Cicadas are flying, plant-eating insects. One particular species, 13-year cicadas *(Magicicada)*, spends five juvenile stages in underground burrows. During the 13 years underground, the cicadas grow from approximately the size of a small ant to nearly the size of an adult cicada. Every 13 years, the animals then emerge from their burrows as adults. The following table presents three different species of these 13-year cicadas and their corresponding adult body weight

(BW), in grams, and wing length (WL), in millimeters.

Species	BW	WL	Species	BW	WL
tredecula	0.15	28	tredecula	0.18	29
tredecim	0.29	32	tredecassini	0.21	27
tredecim	0.17	27	tredecula	0.15	30
tredecula	0.18	30	tredecula	0.17	27
tredecim	0.39	35	tredecassini	0.13	27
tredecim	0.26	31	tredecassini	0.17	29
tredecassini	0.17	29	tredecassini	0.23	30
tredecassini	0.16	28	tredecim	0.12	22
tredecassini	0.14	25	tredecula	0.26	30
tredecassini	0.14	28	tredecula	0.19	30
tredecassini	0.28	25	tredecassini	0.20	30
tredecim	0.12	28	tredecula	0.14	23

Source: http://insects/ummz.lsa.umich.edu

a. Construct a scatter diagram of the body weights, x, and the corresponding wing lengths, y. Use a different symbol to represent the ordered pairs for each species.

b. Describe what the scatter diagram displays with respect to relationship and species.

c. Calculate the correlation coefficient, r.

d. Find the equation for the line of best fit.

e. Suppose the body weight of a cicada is 0.20 gram. What wing length would you predict? Which species do you think this cicada might be?

3.102 [EX03-102] Yellowstone National Park's Old Faithful has been a major tourist attraction for a long time. Understanding the duration of eruptions and the time between eruptions is necessary to predict the timing of the next eruption. The Old Faithful data set variables are as follows: date: an index of the date the observation was taken (days 1, 2, and 3 are given here—all 16 days are on your CD); duration: the duration of an eruption of the geyser, in minutes; and intereruption: the time until the next eruption, in minutes.

Day 1		Day 2		Day 3	
Duration	Intereruption	Duration	Intereruption	Duration	Intereruption
4.4	78	4.3	80	4.5	76
3.9	74	1.7	56	3.9	82
4.0	68	3.9	80	4.3	84
4.0	76	3.7	69	2.3	53
3.5	80	3.1	57	3.8	86
4.1	84	4.0	90	1.9	51
2.3	50	1.8	42	4.6	85
4.7	93	4.1	91	1.8	45
1.7	55	1.8	51	4.7	88
4.9	76	3.2	79	1.8	51
1.7	58	1.9	53	4.6	80
4.6	74	4.6	82	1.9	49
3.4	75	2.0	51	3.5	82

Source: http://comp.uark.edu/~jtubbs/Biostat/Labs/Oldfaithful/oldfaithful.html

a. Construct a scatter diagram of the 39 durations, x, and intereruptions, y. Use a different symbol to represent the ordered pairs for each day.

b. Describe the pattern displayed by all 39 ordered pairs.

c. Do the data for the individual days show the same pattern as one another and as the total data set?

d. Based on information in the scatter diagram, if Old Faithful's last eruption lasted 4 minutes, how long would you predict we will need to wait until the next eruption starts?

e. Find the line of best fit for the data listed in the table.

f. Based on the line of best fit, if Old Faithful's last eruption lasted 4 minutes, how long would you predict we will need to wait until the next eruption starts?

g. What effect do you think the distinctive pattern shown on the scatter diagram has on the line of best fit? Explain.

h. Repeat parts a through g using the data set for 16 days of observations. The data are on your CD. [DS03-102]

i. Compare the results found in part h to the results in parts a through g. Discuss your conclusions.

3.103 a. Verify, algebraically, that formula (3.2) for calculating r is equivalent to the definition formula (3.1).

b. Verify, algebraically, that formula (3.6) is equivalent to formula (3.5).

3.104 This equation gives a relationship that exists between b_1 and r:

$$r = b_1 \sqrt{\frac{SS(x)}{SS(y)}}$$

a. Verify the equation for these data:

x	4	3	2	3	0
y	11	8	6	7	4

b. Verify this equation using formulas (3.2) and (3.6).

Chapter Project

The Kid Is All Grown Up

As a way of assessing the statistical techniques for bivariate data that we have learned in this chapter, let's return to Section 3.1, "The Kid Is All Grown Up," on page 145. For any basketball player, the number of points scored per game and the number of personal fouls committed per game are of interest. Could a clear and definite relationship exist between these two variables, and if so, why?

Putting Chapter 3 to Work

3.105 [EX03-001]

Minnesota Timberwolves, 2003–2004 Regular Season

Player	Personal Fouls per Game	Points per Game	Player	Personal Fouls per Game	Points per Game
Garnett	2.5	24.2	Madsen	2.4	3.6
Cassell	3	19.8	Martin	1.4	3.4
Sprewell	1.2	16.8	McLeod	1.2	2.7
Szczerbiak	1.5	10.2	Goldwire	1.0	2.6
Hudson	1.1	7.5	Miller	1.9	2.5
Hoiberg	1.7	6.7	Johnson	2.4	1.9
Olowokandi	3.2	6.5	Lewis	0.7	1.1
Trent	1.9	5.6	Ebi	0.4	0.8
Hassell	2.5	5			

Source: http://sports.espn.go.com/nba/teams

a. Construct a scatter diagram, using points scored per game, y, and number of personal fouls committed per game, x. Explain why you believe there is or is not a relationship.

b. Are the two variables points scored per game and number of personal fouls committed per game correlated? Use the correlation coefficient to justify your answer.

c. Express the relationship between the two variables total points scored, y, and number of personal fouls committed, x, as a linear equation.

d. Using the results from part c, if a Minnesota Timberwolves player committed two fouls in a game, how many points would you expect him to score?

e. If the player in part d committed a third personal foul, how many extra points would you expect him to score?

f. How does the slope for the line of best fit relate to the number of additional points expected when the player commits one extra personal foul?

g. Do the preceding results show a cause-and-effect relationship between total points scored and number of personal fouls committed? Explain.

h. Should the team coach instruct a player to commit an extra personal foul so that he will score more points? Explain.

i. Name at least one possible lurking variable for the preceding situation.

Suppose the preceding investigation is to be expanded to include an additional variable, "minutes played per game."

j. Describe the relationship you think might exist between the variables "minutes played per game" and "points scored per game." Explain why.

k. Describe the relationship you think might exist between the variables "minutes played per game" and "number of personal fouls committed per game." Explain why.

l. Could "minutes played per game" be a lurking variable for the work completed in parts a–h? Explain.

3.106 a. The situation described in Exercise 3.105 only occurred with the Minnesota Timberwolves during the 2003–2004 regular season. Use the Internet (search by the team name) to obtain the season team statistics for your favorite professional or intercollegiate basketball team, or see the coach of a local high school or college team.

b. Answer the same questions asked in Exercise 3.105 for your selected team.

c. Discuss the differences and similarities between the Minnesota Timberwolves and your selected team. Consider other lurking variables.

Chapter Practice Test

PART I: Knowing the Definitions

Answer "True" if the statement is always true. If the statement is not always true, replace the words shown in bold with words that make the statement always true.

3.1 **Correlation** analysis is a method of obtaining the equation that represents the relationship between two variables.

3.2 The linear correlation coefficient is used to determine the **equation that represents** the relationship between two variables.

3.3 A correlation coefficient of **zero** means that the two variables are perfectly correlated.

3.4 Whenever the slope of the regression line is zero, the **correlation coefficient** will also be zero.

3.5 When r is positive, b_1 will always be **negative.**

3.6 The **slope** of the regression line represents the amount of change expected to take place in y when x increases by one unit.

3.7 When the calculated value of r is positive, the calculated value of b_1 will be **negative.**

3.8 Correlation coefficients range between **0 and +1.**

3.9 The value being predicted is called the **input variable.**

3.10 The line of best fit is used to predict the **average value of** y that can be expected to occur at a given value of x.

PART II: Applying the Concepts

3.11 Refer to the scatter diagram that follows.

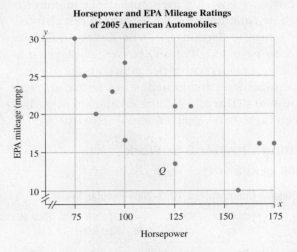

Horsepower and EPA Mileage Ratings of 2005 American Automobiles

a. Match the descriptions in column 2 with the terms in column 1.

____ population   (a) the horsepower rating for an automobile

____ sample   (b) all 2005 American-made automobiles

____ input variable   (c) the EPA mileage rating for an automobile

____ output variable   (d) the 2005 automobiles with ratings shown on the scatter diagram

b. Find the sample size.

c. What is the smallest value reported for the output variable?

d. What is the largest value reported for the input variable?

e. Does the scatter diagram suggest a positive, negative, or zero linear correlation coefficient?

f. What are the coordinates of point Q?

g. Will the slope for the line of best fit be positive, negative, or zero?

h. Will the intercept for the line of best fit be positive, negative, or zero?

3.12 A research group reports a 2.3 correlation coefficient for two variables. What can you conclude from this information?

3.13 For the bivariate data, the extensions, and the totals shown on the table, find the following:

a. $SS(x)$

b. $SS(y)$

c. $SS(xy)$

d. The linear correlation coefficient, r

e. The slope, b_1

f. The y-intercept, b_0

g. The equation of the line of best fit

x	y	x^2	xy	y^2
2	6	4	12	36
3	5	9	15	25
3	7	9	21	49
4	7	16	28	49
5	7	25	35	49
5	9	25	45	81
6	8	36	48	64
28	49	124	204	353

PART III: Understanding the Concepts

3.14 A test was administered to measure the mathematics ability of the people in a certain town. Some of the townspeople were surprised to find out that their test results and their shoe sizes correlated strongly. Explain why a strong positive correlation should not have been a surprise.

3.15 Student A collected a set of bivariate data and calculated r, the linear correlation coefficient. Its value was -1.78. Student A proclaimed that there was no correlation between the two variables because the value of r was not between -1.0 and $+1.0$. Student B argued that -1.78 was impossible and that only values of r near zero implied no correlation. Who is correct? Justify your answer.

3.16 The linear correlation coefficient, r, is a numerical value that ranges from -1.0 to $+1.0$. Write a sentence or two describing the meaning of r for each of these values:

a. -0.93 d. $+0.08$

b. $+0.89$ e. -2.3

c. -0.03

3.17 Make up a set of three or more ordered pairs such that:

a. $r = 0.0$ c. $r = -1.0$

b. $r = +1.0$ d. $b_1 = 0.0$

Working with Your Own Data

Each semester, new students enter your college environment. You may have wondered, What will the student body be like this semester? As a beginning statistics student, you have just finished studying three chapters of basic descriptive statistical techniques. You can use some of these techniques to describe some characteristics of your college's student body.

A Single-Variable Data

1. Define the population to be studied.
2. Choose a variable to define. (You may define your own variable, or you may use one of the variables in the accompanying table* if you are not able to collect your own data. Ask your instructor for guidance.)
3. Collect 35 pieces of data for your variable.
4. Construct a stem-and-leaf display of your data. (Be sure to label it.)
5. Calculate the value of the measure of central tendency that you believe best answers the question: What is the average value of your variable? Explain why you chose this measure.
6. Calculate the sample mean for your data (unless you used the mean in question 5).
7. Calculate the sample standard deviation for your data.
8. Find the value of the 85th percentile, P_{85}.
9. Construct a graphic display (other than a stem-and-leaf) that you believe "best" displays your data. Explain why the graph best presents your data.
10. Write a summary paragraph describing your findings.

B Bivariate Data

1. Define the population to be studied.
2. Choose and define two quantitative variables that will produce bivariate data. (You may define your own variables, or you may use two of the variables in the accompanying table if you are not able to collect your own data. Ask your instructor for guidance.)
3. Collect 15 ordered pairs of data.
4. Construct a scatter diagram of your data. (Be sure to label it completely.)
5. Using a table to assist with the organization, calculate the extensions x^2, xy, and y^2, and the summations of x, y, x^2, xy, and y^2.
6. Calculate the linear correlation coefficient, r.
7. Calculate the equation of the line of best fit.
8. Draw the line of best fit on your scatter diagram.
9. Write a summary paragraph describing your findings.

*The table of data on page 203 was collected on the first day of class last semester. You may use it as a source for your data if you are not able to collect your own.

Variable *A:* student's gender (male/female)

Variable *B:* student's age at last birthday

Variable *C:* number of completed credit hours toward degree

Variable *D:* "Do you have a job (full/part time)?" (yes/no)

Variable *E:* number of hours worked last week, if D = yes

Variable *F:* wages (before taxes) earned last week, if D = yes

FYI The computer will select your random sample (see p. 101).

[DS-1]

Student	A	B	C	D	E	F	Student	A	B	C	D	E	F
1	M	21	16	No			51	F	42	34	Yes	40	244
2	M	18	0	Yes	10	34	52	M	25	60	Yes	60	503
3	F	23	18	Yes	46	206	53	M	39	32	Yes	40	500
4	M	17	0	No			54	M	29	13	Yes	39	375
5	M	17	0	Yes	40	157	55	M	19	18	Yes	51	201
6	M	40	17	No			56	M	25	0	Yes	48	500
7	M	20	16	Yes	40	300	57	F	18	0	No		
8	M	18	0	No			58	M	32	68	Yes	44	473
9	F	18	0	Yes	20	70	59	F	21	0	No		
10	M	29	9	Yes	8	32	60	F	26	0	Yes	40	320
11	M	20	22	Yes	38	146	61	M	24	11	Yes	45	330
12	M	34	0	Yes	40	340	62	F	19	0	Yes	40	220
13	M	19	31	Yes	29	105	63	M	19	0	Yes	10	33
14	M	18	0	No			64	F	35	59	Yes	25	88
15	M	20	0	Yes	48	350	65	F	24	6	Yes	40	300
16	F	27	3	Yes	40	130	66	F	20	33	Yes	40	170
17	M	19	10	Yes	40	202	67	F	26	0	Yes	52	300
18	F	18	16	Yes	40	140	68	F	17	0	Yes	27	100
19	M	19	4	Yes	6	22	69	M	25	18	Yes	41	355
20	F	29	9	No			70	M	24	0	No		
21	F	21	0	Yes	20	80	71	M	21	0	Yes	30	150
22	F	39	6	No			72	M	30	12	Yes	48	555
23	M	23	34	Yes	42	415	73	F	19	0	Yes	38	169
24	F	31	0	Yes	48	325	74	M	32	45	Yes	40	385
25	F	22	7	Yes	40	195	75	M	26	90	Yes	40	340
26	F	27	75	Yes	20	130	76	M	20	64	Yes	10	45
27	F	19	0	No			77	M	24	0	Yes	30	150
28	M	22	20	Yes	40	470	78	M	20	14	No		
29	F	60	0	Yes	40	390	79	M	21	70	Yes	40	340
30	M	25	14	No			80	F	20	13	Yes	40	206
31	F	24	45	No			81	F	33	3	Yes	32	246
32	M	34	4	No			82	F	25	68	Yes	40	330
33	M	29	48	No			83	F	29	48	Yes	40	525
34	M	22	80	Yes	40	336	84	F	40	0	Yes	40	400
35	M	21	12	Yes	26	143	85	F	36	3	Yes	40	300
36	F	18	0	No			86	F	35	0	Yes	40	280
37	M	18	0	Yes	13	65	87	F	28	0	Yes	40	350
38	M	40	64	Yes	40	390	88	F	27	9	Yes	40	260
39	F	31	0	Yes	40	200	89	F	26	3	Yes	40	240
40	F	32	0	Yes	40	270	90	F	23	9	Yes	40	330
41	F	37	0	Yes	24	150	91	M	41	3	Yes	23	253
42	F	35	0	Yes	40	350	92	M	39	0	Yes	40	110
43	M	21	72	Yes	45	470	93	M	21	0	Yes	40	246
44	F	27	0	Yes	40	550	94	F	32	0	Yes	40	350
45	F	42	47	Yes	37	300	95	F	48	58	Yes	40	714
46	F	41	21	Yes	40	250	96	F	26	0	Yes	32	200
47	M	36	0	Yes	40	400	97	F	27	0	Yes	40	350
48	M	25	16	Yes	40	480	98	F	52	56	Yes	40	390
49	F	18	0	Yes	45	189	99	F	34	27	Yes	8	77
50	M	22	0	Yes	40	385	100	F	49	3	Yes	24	260

© Rachel Epstein/The Image Works

4.1 Sweet Statistics

Where did all these colorful candies come from?

Did you know they have 21 different colors?

Did you know that the idea for "M&M's" Plain Chocolate Candies was born in the backdrop of the Spanish Civil War? Legend has it that on a trip to Spain, Forrest Mars Sr. encountered soldiers who were eating pellets of chocolate that were encased in a hard sugary coating to prevent them from melting. Mr. Mars was inspired by this concept and returned home and invented the recipe for "M&M's" Plain Chocolate Candies.

Statistics class had started and the teacher was discussing percentages, proportions, and probabilities—how the three are similar and yet different. All of a sudden a student mentioned that she heard that the previous semester's class did a lesson using, and eating, M&M's; she asked if this year's class would be doing something similar. The conversation was soon focused entirely on M&M's—their color combinations and the percentage of each color. The 24 class members were each asked to guess the percentage of each color they believed were contained in those little brown bags of M&M's Plain Chocolate Candies. They were told there would be a prize for the person whose guess was the closest to the actual number. Each student wrote down the percentages and turned them in; in return, the students received a little brown bag. "Oh, this is that lesson!" "Yes!" the teacher replied, "and before you open those bags, we *must* have a plan." Each student was to count the number of M&M's of each color in his or her bag and record the six counts; then the class totals would be determined. The resulting distribution of counts is shown in Table 4.1.

The class totals were converted to percentages (Table 4.2), and each student was asked to determine the six percentages they observed for their own bag of M&M's.

The discussion that followed centered on the variation that occurred from bag to bag, with some students being quite surprised to see so much variation. Several bags either had none or only one of a color, and a few bags had a fairly

M&M Colors by Count

Color	Count
Brown	91
Yellow	112
Red	102
Blue	151
Orange	137
Green	99
	692

TABLE 4.2

M&M Colors by Percentages

Color	Percent
Brown	13.2
Yellow	16.2
Red	14.7
Blue	21.8
Orange	19.8
Green	14.3
	100.0

large proportion of just one or two colors. Have you ever noticed either of these extremes when you opened a bag of M&M's?

The percents reported on Table 4.2 are the percentages for each color found in this sample of 692 M&M's. Percentages behave very much like probability numbers, but the question being asked in probability is quite different. In the preceding illustration, we are treating the information as sample data and describing the results found. If we now think in terms of a probability, we will turn the orientation around and treat the complete set of 692 M&M's as the complete list of possibilities and ask questions about the likeliness of certain events when one M&M is randomly selected from the entire collection of 692.

For example, suppose we were to dump all 692 M&M's into a large bowl and thoroughly mix them. Now consider the question, "If one M&M is selected at random from the bowl, what is the probability that it will be orange?" We hope that your thinking is along the following lines: selected randomly means each M&M has the same chance of being selected, and because there are 137 orange M&M's in the bowl, the probability of selecting an orange M&M is 137/692, or 0.198.

You have seen the number 0.198 before, only it was expressed as 19.8%. Percentages and probability numbers are "the same, but different." (You have probably heard that before, somewhere.) The numbers have the same value and behave with the same properties; however, the orientation of the situation and the questions asked are different, as you will see in Section 4.2.

After completing Chapter 4, you will have an opportunity to further investigate "Sweet Statistics" in the Chapter 4 Project section.

SECTION 4.1 EXERCISES

4.1 a. If you bought a bag of M&M's, what color M&M would you expect to see the most? What color the least? Why?

b. If you bought a bag of M&M's, would you expect to find the percentages listed previously in Table 4.2? If not, why and what would you expect?

4.2 a. Construct a bar graph showing the Table 4.2 percentages obtained from the 692 M&M's.

b. Based on your graph, which color M&M occurred most often? How does this show on your graph?

c. Based on your graph, which color M&M occurred the least? How does this show on your graph?

4.3 If you were given a small bag of M&M's with 40 candies in it, using the percentages in Table 4.2, how many of each color would you "expect" to find?

4.4 *Bad charts?* Just like there are bad graphs (as seen in Section 2.8), there are bad charts—misleading and hard-to-read charts. Mothers Against Drunk Driving (MADD) presented the following chart regarding the 6764 holiday traffic fatalities that occurred in 2002.

Holiday 2002	Total Traffic Fatalities	Total Fatalities Alcohol-Related
New Year's Eve (2001)	118	45
New Year's Day	165	94
New Year's Holiday	575	301
Super Bowl Sunday	147	86
St. Patrick's Day	158	72
Memorial Day	491	237
Fourth of July	683	330
Labor Day weekend	541	300

Halloween	268	109
Thanksgiving	543	255
Thanksgiving–New Year's	4019	1561
Christmas	130	68
New Year's Eve (2002)	123	57

Source: Mothers Against Drunk Driving (MADD), http://www.infoplease.com/ipa/A0777960.html

a. The column totals are not included because they would be meaningless values. Examine the table and explain why.

b. Select the appropriate nonoverlapping holidays (column 1) and verify the 6764 total number of traffic fatalities for 2002.

c. Using the holidays selected in part b, find the total number of holiday alcohol-related traffic fatalities for 2002?

d. Describe how you would organize this chart to make it more meaningful.

4.5 Use either the random-number table (Appendix B), a calculator, or a computer (see p. 101) to simulate the following:

a. The rolling of a die 50 times; express your results as relative frequencies.

b. The tossing of a coin 100 times; express your results as relative frequencies.

4.6 Use either the random-number table (Appendix B), a calculator, or a computer (see p. 101) to simulate the random selection of 100 single-digit numbers, 0 through 9.

a. List the 100 digits.

b. Prepare a relative frequency distribution of the 100 digits.

c. Prepare a relative frequency histogram of the distribution in part b.

4.2	# Probability of Events

We are now ready to define what is meant by *probability.* Specifically, we talk about "the probability that a certain event will occur."

> **Probability an event will occur:** The relative frequency with which that event can be expected to occur.

The probability of an event may be obtained in three different ways: (1) *empirically*, (2) *theoretically*, and (3) *subjectively*.

The **empirical** method was illustrated by the M&M's and their percentages in Section 4.1 and might be called *experimental* or *empirical probability*. This probability is the *observed relative frequency* with which an event occurs. In our M&M example, we observed that 137 of the 692 M&M's were orange. The observed empirical probability for the occurrence of orange was 137/692, or 0.198.

The value assigned to the probability of event A as a result of experimentation can be found by means of the formula:

> **Empirical (Observed) Probability: $P'(A)$**
>
> In words: $empirical\ probability\ of\ A = \dfrac{number\ of\ times\ A\ occurred}{number\ of\ trials}$
>
> In algebra: $P'(A) = \dfrac{n(A)}{n}$ (4.1)

Notation for empirical probability: When the value assigned to the probability of an event results from experimental or empirical data, we will identify the probability of the event with the symbol $P'(\ )$.

The **theoretical** method for obtaining the probability of an event uses a sample space. A sample space is a listing of all possible outcomes from the experiment being considered. When this method is used, the sample space must contain equally likely sample points. For example, the sample space for the rolling of one die is {1, 2, 3, 4, 5, 6}. Each outcome (i.e., number) is equally likely. An event is a subset of the sample space. Therefore, the probability of an event A, $P(A)$, is the ratio of the number of points that satisfy the definition of event A, $n(A)$, to the number of sample points in the entire sample space, $n(S)$. That is,

---

Theoretical (Expected) Probability: $P(A)$

In words: theoretical probability of $A = \dfrac{\text{number of times A occurs in sample space}}{\text{number of elements in sample space}}$

In algebra: $P(A) = \dfrac{n(A)}{n(S)}$ (4.2)

Notes:

1. When the value assigned to the probability of an event results from a theoretical source, we will identify the probability of the event with the symbol $P(\ )$.

2. The prime symbol is *not used* with theoretical probabilities; it is used only for empirical probabilities.

EXAMPLE 4.1

One Die

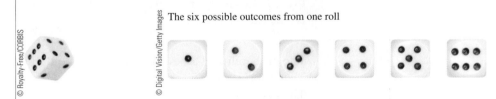

The six possible outcomes from one roll

© Royalty-Free/CORBIS

© Digital Vision/Getty Images

Consider one rolling of one die. Define event A as the occurrence of a number "greater than 4." In a single roll of a die, there are six possible outcomes, making $n(S) = 6$. The event "greater than 4" is satisfied by the occurrence of either a 5 or a 6; thus, $n(A) = 2$. Assuming that the die is symmetrical and each number has an equal likelihood of occurring, the probability of A is $\dfrac{2}{6}$, or $\dfrac{1}{3}$.

EXAMPLE 4.2

A Pair of Dice

A pair of dice (one white, one black) is rolled one time, and the number of dots showing on each die is observed. The sample space is shown in a chart format:

Milk in Your Tea?

In the late 1920s at a summer afternoon tea party in Cambridge, England, a guest proclaimed that tea tastes different depending on whether the tea is poured into the milk or the milk is poured into the tea. Her claim was met with much ridicule. After much bantering, one man, Ronald A. Fisher, proposed a scientific way to test her hypothesis: combine the milk and tea in both manners, then offer her one of each, two at a time in random order, for identification. Others quickly joined him and assisted with the testing–she correctly identified 10 in a row. What do you think? Could she tell the difference?

Chart Representation

$n(S) = 36$

The sum of their dots is to be considered. A listing of the possible "sums" forms a sample space, S = {2, 3, 4, 5, 6, 7, 8, 9, 10, 11, 12} and $n(S) = 11$. However the elements of this sample space are not equally likely; therefore, this sample space cannot be used to find theoretical probabilities—we must use the 36-point sample space shown in the preceding chart. By using the 36-point sample space, the sample space is entirely made up of equally likely sample points and the probabilities for the sums of 2, 3, 4, and so on, can be found quite easily. The sum of 2 represents {(1, 1)}, where the first element of the ordered pair is the outcome for the white die and the second element of the ordered pair is the outcome for the black die. The sum of 3 represents {(2, 1), (1, 2)}; and the sum of 4 represents {(1, 3), (3, 1), (2, 2)}; and so on. Thus, we can use formula (4.2) and the 36-point sample space to obtain the probabilities for the 11 sums.

$$P(2) = \frac{n(2)}{n(S)} = \frac{1}{36}, \ P(3) = \frac{n(3)}{n(S)} = \frac{2}{36}, \ P(4) = \frac{n(4)}{n(S)} = \frac{3}{36}$$

and so forth.

When a probability experiment can be thought of as a sequence of events, a *tree diagram* often is a very helpful way to picture the sample space.

EXAMPLE 4.3

Using Tree Diagrams

A family with two children is to be selected at random, and we want to find the probability that the family selected has one child of each gender. Because there will always be a firstborn and a second-born child, we will use a tree diagram to show the possible arrangements of gender, thus making it possible for us to determine the probability. Start by determining the sequence of events involved—firstborn and second-born in this case. Use the tree to show the possible outcomes of the first event (shown in brown on Figure 4.1) and then add branch segments to show the possible outcomes for the second event (shown in orange in Figure 4.1).

Notes:

1. The two branch segments representing B and G for the second-born child must be drawn from each outcome for the firstborn child, thus creating the "tree" appearance.

FIGURE 4.1 Tree Diagram Representation* of Family with Two Children

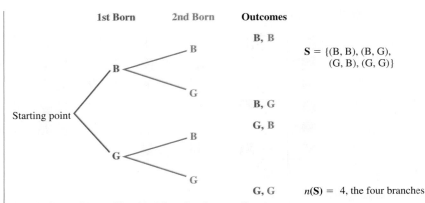

*See the *Student Solutions Manual* for information about tree diagrams.

2. There are four branches; each branch starts at the "tree root" and continues to an "end" (made up of two branch segments each), showing a possible outcome.

Because the branch segments are equally likely, assuming equal likeliness of gender, the four branches are then equally likely. This means we need only the count of branches to use formula 4.2 to find the probability of the family having one child of each gender. The two middle branches, (B,G) and (G,B), represent the event of interest, so $n(A) = n$(one of each) $= 2$, whereas $n(S) = 4$ because there are a total of four branches. Thus,

$$P(\text{one of each gender in family of two children}) = \frac{2}{4} = \frac{1}{2} = 0.5$$

Now let's consider selecting a family of three children and finding the probability of "at least one boy" in that family. Again the family can be thought of as a sequence of three events—firstborn, second-born, and third-born. To create a tree diagram of this family, we need to add a third set of branch segments to our two-child family tree diagram. The green branch segments represent the third child (see Figure 4.2).

FIGURE 4.2 Tree Diagram Representation* of Family with Three Children

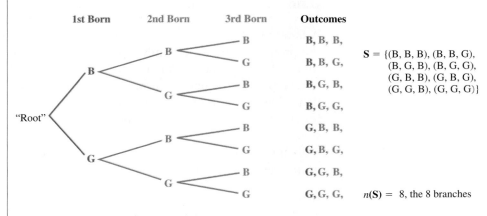

*See the *Student Solutions Manual* for information about tree diagrams.

Again, because the branch segments are equally likely, assuming equal likeliness of gender, the eight branches are then equally likely. This means we need only the count of branches to use formula 4.2 to find the probability of the family having at least one boy. The top seven branches all have one or more boys, the equivalent of "at least one."

$$P(\text{at least one boy in a family of three children}) = \frac{7}{8} = 0.875$$

Let's consider one other question before we leave this example. What is the probability that the third child in this family of three children is a girl? The question is actually an easy one; the answer is 0.5, because we have assumed equal likelihood of either gender. However, if we look at the tree diagram in Figure 4.2, there are two ways to view the answer. First, if you look at only the branch segments for the third-born child, you see one of two is for a girl in each set, thus $\frac{1}{2}$, or 0.5. Also, if you look at the entire tree diagram, the last child is a girl on four of the eight branches, thus $\frac{4}{8}$, or 0.5.

When a probability question provides information about the events in the form of the number of items per set, the percentage of each set, or the probability of the various events, a *Venn diagram* often is a very helpful way to picture the sample space.

EXAMPLE 4.4

Using Venn Diagrams

A lucky customer at Used Car Charlie's lot will get to randomly select one key from a barrel of keys. The barrel contains the keys to all of the cars on Charlie's lot. Charlie's inventory lists 80 cars, of which 38 are foreign models, 50 are compact models, and 22 are foreign compact models. The Venn diagram shown in Figure 4.3 summarizes Charlie's inventory. Notice that some of the 38 foreign models are compact and some are not. The same is true with the compact models; some are foreign, and some are not. Therefore, when breaking this kind of information down, you must start with the most specific. In this case, 22 cars are foreign and compact; they are represented by the center region of the Venn diagram. From there, you can determine how many cars are foreign but not compact and how many are compact but not foreign. Refer to Figure 4.3.

FIGURE 4.3
Venn Diagram Representation* of Used Car Charlie's Inventory

*See the *Student Solutions Manual* for information about Venn diagrams.

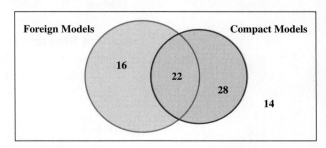

You are the lucky customer who has won the chance to draw for a free car at Used Car Charlie's, and you are about to draw 1 of the 80 keys. What is the probability that you win a nonforeign compact car? Looking at the Venn diagram, foreign cars are inside the blue circle; therefore, nonforeign are outside the blue circle. The event of interest along with nonforeign is compact (inside the red circle), which, based on Figure 4.3, we can determine to be 28 of these cars. Using formula (4.2), we find

$$P(\text{nonforeign compact}) = \frac{28}{80} = 0.35$$

Conveniently, the Venn diagram works equally well if the information had been given in percentages or probabilities. The diagram looks the same except the values become either probabilities or percentages. To be sure that the entire sample space has been covered, the sum for all regions must be exactly 1.0 in order for the labeling to be correct.

Note: Sometimes it is helpful to place a coin on the circle representing an event when you are looking at an event that did "not" occur. In the Venn diagram shown in Figure 4.3, a quarter placed on the "Foreign Models" circle leaves all nonforeign models visible.

Special attention should always be given to the sample space. Like the statistical population, the sample space must be well defined. Once the sample space is defined, you will find the remaining work much easier.

A **subjective** probability generally results from personal judgment. Your local weather forecaster often assigns a probability to the event "precipitation." For example, "there is a 20% chance of rain today," or "there is a 70% chance of snow tomorrow." In such cases, the only method available for assigning probabilities is personal judgment. These probability assignments are called *subjective probabilities.* The accuracy of subjective probabilities depends on the individual's ability to correctly assess the situation.

Properties of Probability Numbers

Whether the probability is empirical, theoretical, or subjective, the following properties must hold.

Property 1

In words: "A probability is always a numerical value between zero and one."

In algebra: $0 \leq$ each $P(A) \leq 1$

Notes about Property 1:
1. The probability is 0 if the event cannot occur.
2. The probability is 1 if the event occurs every time.
3. Otherwise, the probability is a fractional number between 0 and 1.

Property 2

In words: "The sum of the probabilities for all outcomes of an experiment is equal to exactly one."

In algebra: $\sum_{\text{all outcomes}} P(A) = 1$

Note about Property 2: The list of "all outcomes" must be a nonoverlapping (mutually exclusive) set of events that includes all the possibilities (all-inclusive).

Notes about probability numbers:

1. Probability represents a relative frequency.
2. $P(A)$ is the ratio of the number of times an event can be expected to occur divided by the number of trials.
3. The numerator of the probability ratio must be a positive number or zero.
4. The denominator of the probability ratio must be a positive number (greater than zero).
5. The number of times an event can be expected to occur in n trials is always less than or equal to the total number of trials, n.

How Are Empirical and Theoretical Probabilities Related?

Consider one rolling of one die and define event A as the occurrence of a "1." An ordinary die has six equally likely sides, so the theoretical probability of event A is $P(A) = \dfrac{1}{6}$.

What does this mean?

Do you expect to see one "1" in each trial of six rolls? Explain. If not, what results do you expect? If we were to roll the die several times and keep track of the proportion of the time event A occurs, we would observe an empirical probability for event A. What value would you expect to observe for $P'(A)$? Explain. How are the two probabilities $P(A)$ and $P'(A)$ related? Explain.

To gain some insight into this relationship, let's perform an experiment.

EXAMPLE 4.5	**Demonstration–Law of Large Numbers**

The experiment will consist of 20 trials. Each trial of the experiment will consist of rolling a die six times and recording the number of times the "1" occurs. Perform 20 trials.

Each row of Table 4.3 shows the results of one trial; we conduct 20 trials, so there are 20 rows. Column 1 lists the number of 1s observed in each trial (set of six rolls); column 2, the observed relative frequency for each trial; and column 3, the cumulative relative frequency as each trial was completed.

Figure 4.4a shows the fluctuation (above and below) of the observed probability, $P'(A)$ (Table 4.3, column 2), about the theoretical probability, $P(A) = \dfrac{1}{6}$, whereas Figure 4.4b shows the fluctuation of the cumulative relative frequency (Table 4.3, column 3) and how it becomes more stable. In fact, the cumulative relative frequency becomes relatively close to the theoretical or expected probability $\dfrac{1}{6}$, or $0.166\overline{6} = 0.167$.

TABLE 4.3

Experimental Results of Rolling a Die Six Times in Each Trial

Trial	Column 1: Number of 1s Observed	Column 2: Relative Frequency	Column 3: Cumulative Relative Frequency	Trial	Column 1: Number of 1s Observed	Column 2: Relative Frequency	Column 3: Cumulative Relative Frequency
1	1	1/6	1/6 = 0.17	11	1	1/6	10/66 = 0.15
2	2	2/6	3/12 = 0.25	12	0	0/6	10/72 = 0.14
3	0	0/6	3/18 = 0.17	13	2	2/6	12/78 = 0.15
4	1	1/6	4/24 = 0.17	14	1	1/6	13/84 = 0.15
5	0	0/6	4/30 = 0.13	15	1	1/6	14/90 = 0.16
6	1	1/6	5/36 = 0.14	16	3	3/6	17/96 = 0.18
7	2	2/6	7/42 = 0.17	17	0	0/6	17/102 = 0.17
8	2	2/6	9/48 = 0.19	18	1	1/6	18/108 = 0.17
9	0	0/6	9/54 = 0.17	19	0	0/6	18/114 = 0.16
10	0	0/6	9/60 = 0.15	20	1	1/6	19/120 = 0.16

FIGURE 4.4
Fluctuations Found in the Die-Tossing Experiment

(a) Relative Frequency

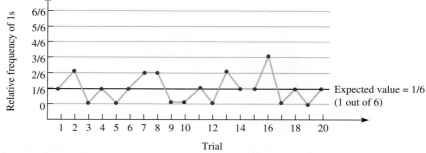

(b) Cumulative Relative Frequency

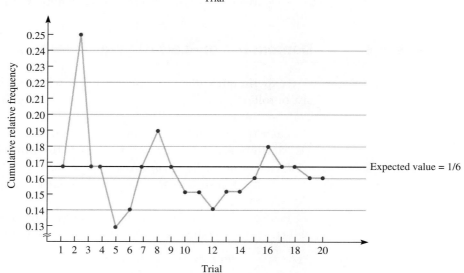

A cumulative graph such as that shown in Figure 4.4b demonstrates the idea of a long-term average and is often referred to as the *law of large numbers*.

> **Law of large numbers:** As the number of times an experiment is repeated increases, the ratio of the number of successful occurrences to the number of trials will tend to approach the theoretical probability of the outcome for an individual trial.

The law of large numbers is telling us that the larger the number of experimental trials, n, the closer the empirical probability, $P'(A)$, is expected to be to the true or theoretical probability, $P(A)$. This concept has many applications. The preceding die-tossing experiment is an example in which we can easily compare actual results against what we expected to happen; it gave us a chance to verify the claim of the law of large numbers.

Example 4.6 is an illustration in which we live with the results obtained from large sets of data when the theoretical expectation is unknown.

EXAMPLE 4.6

Uses of Empirical Probabilities

The key to establishing proper life insurance rates is using the probability that the insureds' will live 1, 2, or 3 years, and so forth, from the time they purchase their policies. These probabilities are derived from actual life and death statistics and hence are empirical probabilities. They are published by the government and are extremely important to the life insurance industry.

Probabilities as Odds

Probabilities can be and are expressed in many ways; we see and hear many of them in the news nearly every day. Odds are a way of expressing probabilities by expressing the number of ways an event can happen compared with the number of ways it cannot happen. The statement "It is four times more likely to rain tomorrow than not rain" is a probability statement and is expressed as odds: "The odds are 4 to 1 in favor of rain tomorrow" (also written 4:1).

The relationship between odds and probability is shown here.

> If the odds in favor of an event A are **a to b** (or **$a:b$**), then
> 1. The odds against event A are **b to a** (or **$b:a$**).
> 2. The probability of event A is $P(A) = \dfrac{a}{a+b}$.
> 3. The probability that event A will not occur is $P(\text{not A}) = \dfrac{a}{a+b}$.

To illustrate this relationship, consider the statement "The odds favoring rain tomorrow are 4 to 1." Using the preceding notation, $a = 4$ and $b = 1$.

Therefore, the probability of rain tomorrow is $\frac{4}{4+1}$, or $\frac{4}{5} = 0.8$. The odds against rain tomorrow are 1 to 4 (or 1:4), and the probability that there is no rain tomorrow is $\frac{1}{4+1}$, or $\frac{1}{5} = 0.2$.

APPLIED EXAMPLE 4.7	**Trying to Beat the Odds**

Many young men aspire to become professional athletes. Only a few make it to the big time, as indicated in the following graph. For every 2400 college senior basketball players, only 64 make a professional team; that translates to a probability of only 0.027 (64/2400).

There are many other interesting specifics hidden in this information. For example, many high school boys dream of becoming a professional basketball player, but according to these numbers, the probability of their dream being realized is only 0.000427 (64/150000).

Once a player has made a college basketball team, he might be very interested in the odds that he will play as a senior. Of the 3800 players making a college team, 2400 play as seniors, whereas 1400 do not. Thus, if a player has made a college team, the odds he will play as a senior are 2400 to 1400, which reduces to 12 to 7.

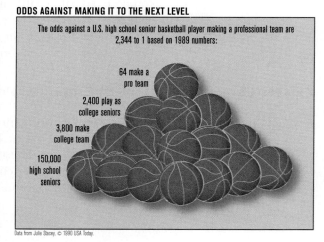

ODDS AGAINST MAKING IT TO THE NEXT LEVEL

The odds against a U.S. high school senior basketball player making a professional team are 2,344 to 1 based on 1989 numbers:

64 make a pro team

2,400 play as college seniors

3,800 make college team

150,000 high school seniors

Data from Julie Stacey, © 1990 USA Today.

The college senior who is playing is interested in his chances of making it to the next level. We see that of the 2400 college seniors, only 64 make the pros, whereas 2336 do not; thus, the odds against him making it to the next level are 2336 to 64, which reduces to 73 to 2. Odds are strongly against him making it.

SECTION 4.2 EXERCISES

Statistics Now™

Skillbuilder Applet Exercises must be worked using an accompanying applet found on your Student's Suite CD-ROM or at the StatisticsNow website at **http://1pass.thomson.com**.

4.7 If you roll a die 40 times and 9 of the rolls result in a "5," what empirical probability was observed for the event "5"?

4.8 Explain why an empirical probability, an observed proportion, and a relative frequency are actually three different names for the same thing.

4.9 Millions of people ride railroads every year. The National Association of Railroad Passengers provides the following figures for railroad ridership in 2004.

Rail System	Riders (millions)
Amtrak system	25.0
Northeast Corridor	14.2
Intercity + West	10.8

Source: National Association of Railroad Passengers, http://www.infoplease.com/ipa/A0855824.html

a. What percentage of the railroad riders rode the Amtrak system in 2004?

b. If one of these riders is to be interviewed, what is the probability that the rider rode the Amtrak system in 2004 if he or she is selected at random?

c. Explain the difference and the relationship between questions and answers in parts a and b.

4.10 Webster Aquatic Center offers various levels of swimming lessons year-round. The March 2005 Monday and Wednesday evening lessons included instructions from Water Babies through Adults. The number in each classification is given in the table that follows.

Swim Lesson Types	No. of Participants
Water Babies	15
Tiny Tots	12
Tadpoles	12
Level 2	15
Level 3	10
Level 4	6
Level 5	2
Level 6	1
Adults	4
Total	77

If one participant is selected at random, find the probability of the following:

a. The participant is in Tiny Tots.

b. The participant is in the Adults lesson.

c. The participant is in a Level 2 to Level 6 lesson.

4.11 In September 2004, the American Payroll Association reported the results of their National Payroll Week 2004 Survey. One of the questions asked about annual household income.

Annual Household Income	Number	Percentage
Less than $15,000	423	1.9%
$15,001–$30,000	2225	9.8%
$30,001–$50,000	5394	23.9%
$50,001–$75,000	5772	25.5%
$75,001–$100,000	4730	20.9%
$100,001–$150,000	3065	13.6%
More than $150,000	984	4.4%

Source: American Payroll Association, http://www.AmericanPayroll.org

Suppose one of the respondents from the survey is to be selected at random for a follow-up interview. Find the probability of the following events.

a. The respondent's family income is $50,000 or less.

b. The respondent's family income is $75,001 or more.

c. The respondent's family income is between $30,000 and $100,000.

d. The respondent's family income is at least $100,001.

4.12 The U.S. Department of Transportation annually reports the number of consumer complaints against the top U.S. airlines by category. Following are the figures for 2002.

Complaint Category	Number of Complaints	Complaint Category	Number of Complaints
Flight problems	2031	Oversales	454
Customer service	1715	Fares	523
Baggage	1421	Disability	477
Reservations/ticketing/ boarding	1159	Advertising	68
		Other	322
Refunds	1106		

Source: Office of Aviation Enforcement & Proceedings, U.S. Department of Transportation, Air Travel Consumer Report, http://www.infoplease.com/ipa/A0198353.html

If one of these complaints is selected at random for follow-up evaluation, what is the probability that the complaint is:

a. About flight problems?

b. About customer service or baggage?

c. About reservations/ticketing/boarding or refunds or oversales?

d. Not about baggage?

4.13 The Weather Underground, Inc., provides a trip planner on the Internet. The weather statistics they post are based on searched dates from 1975 to present. You are planning a trip from March 1 to March 15 and obtained the following weather forecast information from the trip planner.

Seattle, WA
Average High Temperature is 52°F ~ historical range 33°F to 70°F
Average Low Temperature is 39°F ~ historical range 20°F to 53°F
  0% chance of a Hot Day (temp. over 90°F) (0 days out of 390 in historical record).
  9% chance of a Warm Day (temp. over 60°F) (37 days out of 390 in historical record).
Average Daily Precipitation is 0.23 ~ historical range 0.00 to 1.47
  48% chance of a Precipitation Day (188 days out of 390 in historical record).
Average Cloud Cover is mostly cloudy
  71% chance of a Cloudy Day (75 days out of 105 in historical record).

San Diego, CA
Average High Temperature is 66°F ~ historical range 56°F to 85°F
Average Low Temperature is 53°F ~ historical range 44°F to 62°F
  0% chance of a Hot Day (temp. over 90°F) (0 days out of 390 in historical record).
  95% chance of a Warm Day (temp. over 60°F) (371 days out of 390 in historical record).
Average Daily Precipitation is 0.12 ~ historical range 0.00 to 1.95
  27% chance of a Precipitation Day (104 days out of 390 in historical record).
Average Cloud Cover is partly cloudy
  35% chance of a Cloudy Day (37 days out of 105 in historical record).

Source: The Weather Underground, Inc., http://www.wunderground.com/tripplanner/index.asp

a. The probabilities reported are the relative frequencies of the event based on historical records. Verify the probability (chance) of a warm day in Seattle. In San Diego.

b. Verify the probability (chance) of a precipitation day in Seattle. In San Diego.

c. While you are on your trip, you plan to use half of your days just to "relax and get some rays." Based on the preceding information, which city is the better choice for warm (but not hot) and sunny (not cloudy) days? Make a case for your answer.

4.14 The two professional football coaches who won the most games during their careers were Don Shula and George Halas. Shula's teams (Colts and Dolphins) won 347 games (a 347-173-6 record) and tied 6 of the 526 games that he coached, whereas Halas's team (Bears) won 324 games (a 324-151-31 record) and tied 31 of the 506 games that he coached.

Source: Pro Football Hall of Fame

Suppose one filmstrip from every game each man coached is thrown into a bin and mixed. You select one filmstrip randomly from the bin and load it into a projector. What is the probability that the film you select shows the following:

a. A tie game   b. A losing game

c. One of Shula's teams winning a game

d. Halas's team winning a game

e. One of Shula's teams losing a game

f. Halas's team losing a game

g. One of Shula's teams playing to a tie

h. Halas's team playing to a tie

i. A game coached by Halas

j. A game coached by Shula

4.15 One single-digit number is to be selected randomly. List the sample space.

4.16 A single die is rolled. What is the probability that the number on top is the following:

a. A 3   b. An odd number

c. A number less than 5

d. A number no greater than 3

4.17 A pair of dice is to be rolled. In Example 4.2, the probability for each of the possible sums was discussed and three of the probabilities, $P(2)$, $P(3)$, and $P(4)$, were found. Find the probability for each of the remaining sums of two dice: $P(5)$, $P(6)$, $P(7)$, $P(8)$, $P(9)$, $P(10)$, $P(11)$, and $P(12)$.

4.18 Two dice are rolled. Find the probabilities in parts b–e. Use the sample space given in Example 4.2 (pp. 208–209).

a. Why is the set {2, 3, 4, . . . , 12} not a useful sample space?

b. P(white die is an odd number)

c. P(sum is 6)

d. P(both dice show odd numbers)

e. P(number on black die is larger than number on white die)

4.19 Take two dice (one white and one colored) and roll them 50 times, recording the results as ordered pairs [(white, color); for example, (3, 5) represents 3 on the white die and 5 on the colored die]. (You could simulate these 50 rolls using a random-number table or a computer.) Then calculate each observed probability:

a. P'(white die is an odd number)

b. P'(sum is 6)

c. P'(both dice show odd number)

d. P'(number on color die is larger than number on white die)

e. Explain why these answers and the answers found in Exercise 4.18 above are not exactly the same.

4.20 Use a random-number table or a computer to simulate rolling a pair of dice 100 times.

a. List the results of each roll as an ordered pair and the sum.

b. Prepare an ungrouped frequency distribution and a histogram of the sums.

c. Describe how these results compare with what you expect to occur when two dice are rolled.

TECHNOLOGY INSTRUCTIONS: SIMULATE DICE

MINITAB (Release 14)

```
Choose:   Calc > Random Data > Integer
Enter:    Generate: 100
          Store in column(s): C1 C2
          Minimum value: 1 Maximum value: 6 > OK
Choose:   Calc > Calculator
Enter:    Store result in variable: C3
          Expression: C1 + C2 > OK
Choose:   Stat > Tables > Tally Individual Variables
Enter:    Variable: C3
Select:   Counts > OK
```

Use the MINITAB commands on page 61 to construct a frequency histogram of the data in C3. (Use Binning > midpoint and midpoint positions 2:12/1 if necessary.)

Excel

Enter 1, 2, 3, 4, 5, 6 into column A, label C1: **Die1;** D1: **Die2;** E1: **Dice,** and activate B1.

```
Choose:   Format > Cells > Number > Number
Enter:    Decimal places: 8 > OK
Enter:    1/6 in B1
Drag:     Bottom right corner of B1 down for 6 entries
Choose:   Tools > Data Analysis > Random Number Genera-
          tion > OK
Enter:    Number of Variables: 2
          Number of Random Numbers: 100
          Distribution: Discrete
          Value and Probability Input Range:  (A1:B6 or
          select cells)
Select:   Output Range
Enter:    (C2 or select cells) > OK
```

Activate the **E2** cell.

```
Enter:    = C2 + D2  > Enter
Drag:     Bottom right corner of E2 down for 100 entries
Choose:   Data > Pivot and PivotChart Table Report . . .
Select:   Microsoft Excel list or database > Next
Enter:    Range: (E1:E101 or select cells) > Next
Select:   Existing Worksheet
Enter:    (F1 or select cell)
Choose:   Layout
Drag:     "Dice" heading into both row & data areas
```

Double click the "sum of dice" in data area box; then continue with:

```
Choose:   Summarize by: Count > OK > OK > Finish
```

Label column J "sums" and input the numbers 2, 3, 4, . . . , 12 into it. Use the Excel histogram commands on pages 61–62 with column E as the input range and column J as the bin range.

TI-83/84 Plus

```
Choose:   MATH > PRB > 5:randInt(
Enter:    1,6,100)
Choose:   STO→ > 2nd L1
```

Repeat preceding for L2.

```
Choose:   STAT > EDIT > 1:Edit
Highlight:           L3
Enter:    L3 =L1 + L2
Choose:   2nd > STAT PLOT
          > 1:Plot1
Choose:   WINDOW
Enter:    −.5, 12.5, 1,
          −10, 40, 10,1
Choose:   TRACE > > >
```

4.21 The 12 face cards (4 jacks, 4 queens, and 4 kings) are removed from a regular deck of playing cards, and then one card is selected from this set of face cards. List the sample space for this experiment.

4.22 An experiment consists of drawing one marble from a box that contains a mixture of red, yellow, and green marbles. There are at least two marbles of each color.

a. List the sample space.

b. Can we be sure that each outcome in the sample space in part a is equally likely? Explain.

c. If two marbles are drawn from the box, list the sample space.

d. Are the outcomes of the sample space in part c equally likely? Explain.

4.23 A box contains one each of $1, $5, $10, and $20 bills.

a. One is selected at random; list the sample space.

b. Two bills are drawn at random (without replacement); list the sample space as a tree diagram.

4.24 Three coins are tossed, and the number of heads observed is recorded. Find the probability for each of the possible results: 0H, 1H, 2H, and 3H.

4.25 A group of files in a medical clinic classifies the patients by gender and by type of diabetes (type 1 or type 2). The groupings may be shown as follows. The table gives the number in each classification.

Gender	Type of Diabetes	
	1	2
Male	30	15
Female	35	20

a. Display the information on this 2 × 2 table as a Venn diagram using "type 1" and "male" as the two events displayed as circles. Explain how the Venn diagram and the given 2 × 2 table display the same information.

If one file is selected at random, find the probability of the following:

b. The selected individual is female.

c. The selected individual has type 2 diabetes.

4.26 Researchers have for a long time been interested in the relationship between cigarette smoking and lung cancer. The following table shows the percentages of adult females observed in a recent study.

	Smokes	Does Not Smoke
Gets Cancer	0.06	0.03
Does Not Get Cancer	0.15	0.76

a. Display the information on this 2 × 2 table as a Venn diagram using "smokes" and "gets cancer" as the two events displayed as circles. Explain how the Venn diagram and the given 2 × 2 table display the same information.

Suppose an adult female is randomly selected from this particular population. What is the probability of the following:

b.   She smokes and gets cancer.

c.   She smokes.

d.   She does not get cancer.

e.   She does not smoke or does not get cancer.

f.   She gets cancer if she smokes.

g.   She does not get cancer, knowing she does not smoke.

4.27 A parts store sells both new and used parts. Sixty percent of the parts in stock are used. Sixty-one percent are used or defective. If 5% of the store's parts are defective, what percentage is both used and defective? Solve using a Venn diagram.

4.28 Union officials report that 60% of the workers at a large factory belong to the union, 90% make more than $12 per hour, and 40% belong to the union and make more than $12 per hour. Do you believe these percentages? Explain. Solve using a Venn diagram.

4.29 Let x be the success rating of a new television show. The following table lists the subjective probabilities assigned to each x for a particular new show by three different media critics. Which of these sets of probabilities are inappropriate because they violate a basic rule of probability? Explain.

	Judge		
Success Rating, x	A	B	C
Highly Successful	0.5	0.6	0.3
Successful	0.4	0.5	0.3
Not Successful	0.3	−0.1	0.3

4.30 A transportation engineer in charge of a new traffic-control system expresses the subjective probability that the system functions correctly 99 times as often as it malfunctions.

a.   Based on this belief, what is the probability that the system functions properly?

b.   Based on this belief, what is the probability that the system malfunctions?

4.31 a.   Explain what is meant by the statement: "When a single die is rolled, the probability of a 1 is $\frac{1}{6}$."

b.   Explain what is meant by the statement: "When one coin is tossed one time, there is a 50-50 chance of getting a tail."

4.32 Skillbuilder Applet Exercise demonstrates the law of large numbers and also allows you to see if you have psychic powers. Repeat the simulations at least 50 times, guessing between picking either a red card or a black card from a deck of cards.

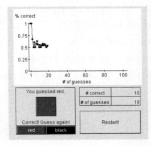

a.   What proportion of time did you guess correctly?

b.   As more and more guesses were made, were your proportions starting to stabilize? If so, at what value? Does this value make sense for the experiment? Why?

c.   How might you know if you have ESP?

4.33 A computer generates (in random fashion) pairs of integers. The first integer is between 1 and 5, inclusive, and the second is between 1 and 4, inclusive.

a.   Represent the sample space as a tree diagram.

b.   List your outcomes as ordered pairs, with x as the first integer and y as the second integer.

4.34 Use a computer (or a random-number table) to simulate the experiment described in Exercise 4.33; x is an integer 1 to 5, and y is an integer 1 to 4. Generate a list of 100 random x values and 100 y values. List the resulting 100 ordered pairs of integers.

a.   Find the relative frequency for $x = 2$.

b.   Find the relative frequency for $y = 3$.

c.   Find the relative frequency for the ordered pair (2, 3).

4.35 An experiment consists of two trials. The first is tossing a penny and observing whether it lands with heads or tails facing up; the second is rolling a die and observing a 1, 2, 3, 4, 5, or 6.

a. Construct the sample space using a tree diagram.

b. List your outcomes as ordered pairs, with the first element representing the coin and the second, the die.

4.36 Use a computer (or a random-number table) to simulate 200 trials of the experiment described in Exercise 4.35: the tossing of a penny and the rolling of a die. Let $1 = H$ and $2 = T$ for the penny, and 1, 2, 3, 4, 5, 6 for the die. Report your results using a cross-tabulated table showing the frequency of each outcome.

a. Find the relative frequency for heads.

b. Find the relative frequency for 3.

c. Find the relative frequency for (H, 3).

4.37 Using a coin, perform the experiment discussed on pages 213–214. Toss a coin 10 times, observe the number of heads (or put 10 coins in a cup, shake and dump them into a box, and use each toss for a block of 10), and record the results. Repeat until you have 200 tosses. Chart and graph the data as individual sets of 10 and as cumulative relative frequencies. Do your data tend to support the claim that $P(\text{head}) = \frac{1}{2}$? Explain.

4.38 A chocolate kiss is to be tossed into the air and will be landing on a smooth hard surface. (Similar to tossing a coin or rolling dice.)

a. What proportion of the time do you believe the kiss will land "point up" ⌣ (as opposed to "point down" ⌢)?

b. Let's estimate the probability that a chocolate kiss lands "point up" when it lands on a smooth hard surface after being tossed. Using a chocolate kiss, with the wrapper still on, perform the die experiment discussed on pages 213–214. Toss the kiss 10 times, record the number of "point up" (or put 10 kisses in a

cup, shake and dump them onto a hard smooth surface, and use each toss for a block of 10), and record the results. Repeat until you have 200 tosses. Chart and graph the data as individual sets of 10 and as cumulative relative frequencies.

c. What is your best estimate for the true $P(\text{⌣})$? Explain.

d. If unwrapped kisses were to be tossed, what do you think the probability of "point up" would be? Will it be the different? Explain.

e. Unwrap the chocolate kisses used in part b and repeat the experiment.

f. Are the results in part e what you anticipated? Explain.

4.39 A box contains marbles of five different colors: red, green, blue, yellow, and purple. There is an equal number of each color. Assign probabilities to each color in the sample space.

4.40 Suppose a box of marbles contains equal numbers of red marbles and yellow marbles but twice as many green marbles as red marbles. Draw one marble from the box and observe its color. Assign probabilities to the elements in the sample space.

4.41 If four times as many students pass a statistics course as those who fail and one statistics student is selected at random, what is the probability that the student will pass statistics?

4.42 Events A, B, and C are defined on sample space S. Their corresponding sets of sample points do not intersect, and their union is S. Furthermore, event B is twice as likely to occur as event A, and event C is twice as likely to occur as event B. Determine the probability of each of the three events.

4.43 The odds of the Patriots winning next year's Super Bowl are 1 to 12.

a. What is the probability the Patriots will win next year's Super Bowl?

b. What are the odds against the Patriots winning next year's Super Bowl?

4.44 The NCAA men's basketball season starts with 327 college teams all dreaming of making it to "the big dance" and attaining the National Championship. Sixty-four teams are selected for the tournament, and only one wins it all.

a. What are the odds against a team being selected for the tournament?

b. What are the odds of a team that is in the tournament winning the National Championship?

c. Now wait a minute! What assumption did you make in order to answer parts a and b? Does this seem realistic?

4.45 The odds against being dealt a contract bridge hand containing 13 cards of the same suit are 158,753,389,899 to 1. The odds against being dealt a royal flush while playing poker are 649,739 to 1.

a. What is the probability of being dealt a contract bridge hand containing 13 cards all of the same suit?

b. What is the probability of being dealt a royal flush poker hand?

c. Express the answers to parts a and b in scientific notation (powers of 10).

4.46 Worldwide the rate of maternal deaths (a woman's risk of dying from pregnancy and childbirth) is 1 in 233. By regions around the world this rate is as follows: North America—1 in 3700, Northern Europe—1 in 4000, Africa—1 in 16, Asia—1 in 65, and Latin America/Caribbean—1 in 130. Express the risk of maternal death as (i) odds in favor of dying, (ii) odds against dying, and (iii) probability of dying for each of the following:

a. Worldwide   b. North America

c. Northern Europe   d. Africa

e. Asia   f. Latin America/Caribbean

4.47 a. A balanced coin is tossed twice. List a sample space showing the possible outcomes.

     b. A biased coin (it favors heads in a ratio of 3 to 1) is tossed twice. List a sample space showing the possible outcomes.

4.48 A box stored in a warehouse contains 100 units of a specific part, of which 10 are defective and 90 are nondefective. Three parts are selected without replacement. Construct a tree diagram representing the sample space.

4.3   Conditional Probability of Events

Many of the probabilities that we see or hear being used on a daily basis are the result of conditions existing at the time. In this section, we will learn about conditional probabilities.

> **Conditional probability an event will occur:** A conditional probability is the relative frequency with which an event can be expected to occur under the condition that additional preexisting information is known about some other event.
> $P(A \mid B)$ is used to symbolize the probability of event A occurring under the condition that event B is known to already exist.

Some ways to say or express the conditional probability, $P(A \mid B)$, are:

The *"probability of A, given B."*

The *"probability of A, knowing B."*

The *"probability of A happening, knowing B has already occurred."*

The concept of conditional probability is actually very familiar and occurs frequently without us even being aware of it. The news media often report many conditional probability values. However, no one makes the point that it is a conditional probability, and it passes for simply everyday arithmetic as illustrated in the following example.

EXAMPLE 4.8

Finding Probabilities from Table of Percentages

From a National Election Pool exit poll of 13,660 voters in 250 precincts across the country on November 2, 2004, we have the following:

Gender	Percentage of Voters	Percent for Bush	Percent for Kerry	Percent for Others
Men	46	55	44	1
Women	54	48	51	1
Age				
18–29	17	45	54	1
30–44	29	53	46	1
45–59	30	51	48	1
60 and older	24	54	46	0

All of the percentages listed are to the nearest integer.

One person is to be selected at random from the sample of 13,600 voters. Using the table, find the answer to the following probability questions.

1. What is the probability the person selected is a man?   You answer: 0.46.

   Expressed in equation form: P(voter selected is a man) = 0.46.

2. What is the probability the person selected is from age 18 to 29?   You answer: 0.17.

   Expressed in equation form: P(voter selected is of age 18 to 29) = 0.17.

3. Knowing the voter selected was a woman, what is the probability she voted for Kerry?   You answer: 0.51.

   Expressed in equation form: P(Kerry $\mid$ woman) = 0.51.

4. What is the probability the person selected voted for Bush if the voter was 60 or older?   Answer: 0.54.

   Expressed in equation form: P(Bush $\mid$ 60 and older) = 0.54.

Note: The first two are simple probabilities, whereas the last two are conditional probabilities.

EXAMPLE 4.9 | **Finding Conditional Probabilities from Table of Count Data**

From a nationwide exit poll of 1000 voters in 25 precincts across the country on November 2, 2004, we have the following.

Education	Number for Bush	Number for Kerry	Number for Others	Number of Voters
No high school	19	20	1	40
High school graduate	114	103	3	220
Some college	172	147	1	320
College graduate	135	119	6	260
Postgraduate	70	88	2	160
Total	510	477	13	1000

One person is to be selected at random from the preceding sample of 1000 voters. Using the table, find the answer to the following probability questions.

1. Knowing the voter selected was a high school graduate, what is the probability the person voted for Kerry?   Answer: $103/220 = 0.46818 = 0.47$.

   Expressed in equation form:
   P (Kerry $\mid$ high school graduate) = $103/220 = 0.46818 = \underline{0.47}$.

2. Knowing the voter selected had some college education, what is the probability the person voted for Bush?   Answer: $172/320 = 0.5375 = 0.54$.

   Expressed in equation form: P(Bush $\mid$ some college) = $172/320 = 0.5375 = \underline{0.54}$.

3. Knowing the selected person voted for Kerry, what is the probability the voter has a postgraduate education?   Answer: $88/477 = 0.1844 = 0.18$.

   Expressed in equation form: P(postgraduate $\mid$ Kerry) = $88/477 = 0.1844 = \underline{0.18}$.

4. Knowing the selected person voted for Bush, what is the probability the voter does not have a high school education? Answer: $19/510 = 0.0372 = 0.04$.

   Expressed in equation form:
   P(no high school $\mid$ Bush) = $19/510 = 0.0372 = \underline{0.04}$.

Notes:

1. The conditional probability notation is very informative and useful. When you express a conditional probability in equation form, it is to your advantage to use the most complete notation. That way, when you read the information back, all the information is there.

2. When finding a conditional probability, some outcomes from the list of possible outcomes will be eliminated as possibilities as soon as the condition is known. Consider question 4 in Example 4.9. As soon as the conditional stated "knowing the selected person voted for Bush," the 477 who voted for Kerry and the 13 voting for Others were eliminated, leaving the 510 possible outcomes.

SECTION 4.3 EXERCISES

4.49 Three hundred viewers were asked if they were satisfied with TV coverage of a recent disaster.

	Gender	
	Female	Male
Satisfied	80	55
Not Satisfied	120	45

One viewer is to be randomly selected from those surveyed.

a. Find P(satisfied)

b. Find P(satisfied | female)

c. Find P(satisfied | male)

4.50 Saturday mornings are busy times at the Webster Aquatic Center. Swim lessons ranging from the Red Cross Level 2, Fundamental Aquatic Skills, through the Red Cross Level 6, Swimming and Skill Proficiency, are offered during two sessions.

Level	Number of People in 10:00 AM Class	Number of People in 11:00 AM Class
2	16	16
3	15	11
4	9	7
5	8	3
6	0	3

Lauren, the program coordinator, is going to randomly select one swimmer to be interviewed for a local television spot on the center and its swim program. What is the probability that the selected swimmer is in the following:

a. A Level 4 class

b. The 10:00 AM class

c. A Level 3 class given it is the 10:00 AM session

d. The 11:00 AM class given it is the Level 5 class

4.51 *The World Factbook,* 2004, reports that U.S. airports have the following number of meters of runways that are either paved or unpaved.

	Number of Airports	
Total Runway (meters)	Paved	Unpaved
More than 3047	188	1
2438–3047	221	7
1524–2437	1375	160
914–1523	2383	1718
Less than 914	961	7843
Total	5128	9729

Source: *The World Factbook,* January 2004, http://www.cia.gov/cia/publications/factbook/geos/us.html#People

If one of these airports is selected at random for inspection, what is the probability that it will have the following:

a. Paved runways

b. 914 to 2437 meters of runway

c. Less than 1524 meters of runway and unpaved

d. More than 2437 meters of runway and paved

e. Paved runway, given that it has more than 1523 meters of runway

f. Unpaved, knowing that it has less than 1524 meters of runway

g. Less than 1524 meters of runway, given that it is unpaved

4.52 During the month of August in 2002, the faculty and staff at Boise State University were asked to participate in a survey to identify the general level of satisfaction with the newly modified workweek in the summer. The following table lists how 620 respondents answered the question: "How satisfied are you with the Boise State University Summer 2002 Schedule?"

Group	Very Satisfied	Somewhat Satisfied	Neither Satisfied Nor Dissatisfied	Somewhat Dissatisfied	Very Dissatisfied	Total
Faculty	65	24	21	13	9	132
Classified staff	190	61	16	15	2	284
Professional staff	139	38	7	12	8	204
All respondents	394	123	44	40	19	620

Source: Boise State University, http://www2.boisestate.edu/iassess/summer_schedule_survey.htm

Find the probability of the following for a randomly selected respondent.

a. Was "somewhat satisfied" with the summer 2002 schedule

b. Was a member of the "professional staff"

c. Was "very satisfied" with the summer 2002 schedule given the respondent was a faculty member

d. Was a member of the "classified staff" given that the respondent was "very dissatisfied" with the summer 2002 schedule

4.53 A *USA Today* article titled "Yum Brands builds dynasty in China" (February 7, 2005) reports on how Yum Brands, the world's largest restaurant company, is bringing the fast-food industry to China, India, and other big countries. Yum Brands, a spin-off from PepsiCo, has been delivering double-digit earnings growth in the past year.

Location and Number of Yum Brands Fast-Food Stores

Store	USA	Abroad	Total
KFC	5,450	7,676	13,126
Pizza Hut	6,306	4,680	10,986
Taco Bell	5,030	193	5,223
Long John Silver's	1,200	33	1,233
A&W All-American	485	209	694
Total	18,471	12,791	31,262

Source: USA Today, February 7, 2005, and Yum Brands

Suppose when the CEO of Yum Brands was interviewed for this article, he was asked the following questions. How might he have answered based on the accompanying table?

a. What percent of your locations are in the United States?

b. What percent of your locations are abroad?

c. What percent of your stores are Pizza Huts?

d. What percent of your stores are Taco Bell given that the location in the United States?

e. What percent of your locations are abroad given that the store is an A&W All-American?

f. What percent of your stores are KFCs given that the location is abroad?

g. What percent of your abroad locations are KFCs?

Review your answers to parts f and g to answer the following:

h. What do you notice about these two answers? Why is this happening?

4.54 In a 2000 census, The National Highway Traffic Safety Administration reported that, nationally, 2% of all traffic fatalities are bicycle deaths. The California Highway Patrol's Statewide Integrated Traffic Records System reports that bicycle deaths account for 4% of the state's traffic fatalities. Information from that report is outlined in the following table.

California Bicycle Fatalities and Injuries by Age Group, 2000

Age (yr)	Bicycle Fatalities	Bicycle Injuries	Total Collisions
0 to 4	0	14	14
5 to 14	21	3,210	3,231
15 to 24	9	2,945	2,954
25 to 34	9	1,907	1,916
35 to 44	23	1,904	1,927
45 to 54	22	1,212	1,234
55 to 64	8	505	513
65 to 74	10	207	217
75 to 84	8	117	125
85 or over	3	22	25
Not stated	3	102	105
Total	116	12,145	12,261

Source: 2000 Statewide Integrated Traffic Records System

a. What percent of collisions were bicycle fatalities in California in 2000?

b. What percent of collisions resulted in bicycle injuries in California in 2000?

c. What percent of collisions involved someone from the 5- to 14-year age group?

d. What percent of bicycle injuries were incurred given there was interest in only the 35- to 44-year age bracket?

e. What percent of bicycle fatalities involved someone from the 75- to 84-year age group?

f. What percent of the 15- to 24-year age group were involved in bicycle injuries?

4.55 The American Housing Survey reported its findings about the principal means of transportation to work by worker in Washington, DC, during the year of 2001.

Means of Transportation	Number (thousands)
All workers	120,191
Automobile	105,586
Drives self	93,942
Carpool	11,644
2-person	9,036
3-person	1,635
4+ person	973
Public transportation[1]	5,627
Taxicab	133
Bicycle or motorcycle	847
Walks only	3,408
Other means[2]	1,049
Works at home	3,401

NOTE: Principal means of transportation refers to the mode used most often by individual.
1. Public transportation refers to bus, streetcar, subway, or elevated trains.
2. Other means include ferryboats, surface trains, and van service.
Source: U.S. Department of Housing and Urban Development, American Housing Survey, Washington, DC, 2001, http://www.infoplease.com/ipa/A0908113.html

a. The column total is not included because they would be meaningless values. Examine the table and explain why.

One person is to be selected and asked additional questions as part of this survey. If that person is selected at random, find the probability for each of the following events.

b. Person selected is a member of a two-person carpool.

c. Person selected is a member of a two-person carpool given that the person carpools.

d. Person selected does not arrive by car.

e. Person selected uses public transportation knowing that person does not use an automobile.

4.56 The five most popular colors for luxury cars manufactured during the 2003 model year in North America are reported here in percentages.

Luxury Car	Percentage	Luxury Car	Percentage
1. Med./Dk. Gray	23.30	4. White	12.6
2. Silver	18.8	5. Black	10.9
3. White Met.	17.8		

Source: DuPont Herberts Automotive Systems, Troy, Michigan, 2003 DuPont Automotive Color Popularity Survey Results, http://www.infoplease.com/ipa/A0855652.html

a. Why does the column of percentages not total 100%?

b. Why are all probabilities based on this table conditional? What is that condition?

c. Did your favorite color make the list?

If one 2003 luxury car was picked at random from all luxury cars manufactured in the United States in 2003, what is the probability that its color is the following:

d. Black, silver, gray, or white

e. Not white

f. Black, knowing that the luxury car has one of the five most popular colors

g. Black, knowing that the luxury car has one of the five most popular colors but is not white

4.4   Rules of Probability

Often, one wants to know the probability of a compound event and the only data available are the probabilities of the related simple events. (Compound events are combinations of more than one simple event.) In the next few paragraphs, the relationship between these probabilities is summarized.

Finding the Probability of "Not A"

The concept of complementary events is fundamental to finding the probability of "not A."

> **Complementary events:** The *complement of an event A, $\overline{A}$,* is the set of all sample points in the sample space that do not belong to event A.

Note: The complement of event A is denoted by $\overline{A}$ (read "A complement").

A few examples of complementary events are (1) the complement of the event "success" is "failure," (2) the complement of "selected voter is Republican" is "selected voter is not Republican," and (3) the complement of "no heads" on 10 tosses of a coin is "at least one head."

By combining the information in the definition of complement with Property 2 (p. 212), we can say that

$$P(A) + P(\overline{A}) = 1.0 \text{ for any event A}$$

As a result of this relationship, we have the complement rule:

> **Complement Rule**
> In words: *probability of A complement = one − probability of A*
> In algebra: $P(\overline{A}) = 1 - P(A)$
>
> $\hspace{11cm}$ (4.3)

Note: Every event A has a complementary event $\overline{A}$. Complementary probabilities are very useful when the question asks for the probability of "at least one." Generally, this represents a combination of several events, but the complementary event "none" is a single outcome. It is easier to solve for the complementary event and get the answer by using formula (4.3).

EXAMPLE 4.10 | ## Using Complements to Find Probabilities

Two dice are rolled. What is the probability that the sum is at least 3 (that is, 3, 4, 5, . . . , 12)?

SOLUTION Suppose one of the dice is black and the other is white. (See the chart in Example 4.2 on pages 208–209; it shows all 36 possible pairs of results when rolling a pair of dice.)

Rather than finding the probability for each of the sums 3, 4, 5, . . . , 12 separately and adding, it is much simpler to find the probability that the sum is 2 ("less than 3") and then use formula (4.3) to find the probability of "at least 3," because "less than 3" and "at least 3" are complementary events.

$P(\text{sum of 2}) = P(A) = \dfrac{1}{36}$ ("2" occurs only once in the 36-point sample space)

$P(\text{sum is at least 3}) = P(\overline{A}) = 1 - P(A) = 1 - \dfrac{1}{36} = \dfrac{35}{36}$ [using formula (4.3)]

Finding the Probability of "A or B"

An hourly wage earner wants to estimate the chances of "receiving a promotion or getting a pay raise." The worker would be happy with either outcome. Historical information is available that will allow the worker to estimate the probability of "receiving a promotion" and "getting a pay raise" separately. In this section we will learn how to apply the addition rule to find the compound probability of interest.

General Addition Rule

Let A and B be two events defined in a sample space, *S*.

In words: *probability of A or B = probability of A + probability of B − probability of A and B*

In algebra: $P(A \text{ or } B) = P(A) + P(B) - P(A \text{ and } B)$

$$(4.4)$$

To see if the relationship expressed by the general addition rule works, let's look at Example 4.11.

EXAMPLE 4.11

Understanding the Addition Rule

A statewide poll of 800 registered voters in 25 precincts from across New York State was taken. Each voter was identified as being registered as Republican, Democrat, or other and then asked, "Are you are in favor of or against the current budget proposal awaiting the governor's signature?" The resulting tallies are shown here.

	Number in Favor	Number Against	Number of Voters
Republican	136	88	224
Democrat	314	212	526
Other	14	36	50
Totals	464	336	800

Suppose one voter is to be selected at random from the 800 voters summarized in the preceding table. Let's consider the two events: "The voter selected is in favor" and "The voter is a Republican." Suppose one voter is picked at random from these 800 voters; find the four probabilities: P(in favor), P(Republican), P(in favor or Republican), and P(in favor and Republican). Then use the results to check the truth of the addition rule.

SOLUTION

Probability the voter selected is "in favor" = P(in favor) = 464/800 = <u>0.58</u>.

Probability the voter selected is "Republican" = P(Republican) = 224/800 = <u>0.28</u>.

Probability the voter selected is "in favor or Republican" = P(in favor or Republican) = $(136 + 314 + 14 + 88)/800 = 552/800 = $ <u>0.69</u>.

Probability the voter selected is "in favor" and "Republican" = P(in favor and Republican) = 136/800 = 0.17.

Notes about finding the preceding probabilities:
1. The connective "or" means "one or the other or both"; thus, "in favor or Republican" means all voters who satisfy either event.
2. The connective "and" means "both" or "in common"; thus, "in favor and Republican" means all voters who satisfy both events.

Now let's use the preceding probabilities to demonstrate the truth of the addition rule.

Let A = "in favor" and B = "Republican." The general addition rule then becomes:

P(in favor or Republican) = P(in favor) + P(Republican) − P(in favor and Republican)

Remember: Previously we found: P(in favor or Republican) = 0.69.
   Using the other three probabilities, we see:

P (in favor) + P (Republican) − P (in Favor and Republican) = 0.58 + 0.28 − 0.17 = 0.69

Thus, we obtain identical answers by applying the addition rule and by referring to the relevant cells in the table. You typically do not have the option of finding the P(A or B) two ways, as we did here. You will be asked to find P(A or B) starting with the P(A) and P(B). However, you will need a third piece of information. In the previous situation, we needed P(A and B). We will either need to know P(A and B) or some information that allows us to find it.

Finding the Probability of "A and B"

Suppose a criminal justice professor wants his class to determine the likeliness of the event "a driver is ticketed for a speeding violation and the driver had previously attended a defensive driving class." The students are confident they can find the probabilities of "a driver being ticketed for speeding" and "a driver has attended a defensive driving class" separately. In this section we will learn how to apply the multiplication rule to find the compound probability of interest.

General Multiplication Rule

Let A and B be two events defined in sample space S.

In words: *probability of A and B = probability of A × probability of B, knowing A*

In algebra: $P(A \text{ and } B) = P(A) \cdot P(B \mid A)$

(4.5)

Note: When two events are involved, either event can be identified as A, with the other identified as B. The general multiplication rule could also be written as $P(B \text{ and } A) = P(B) \cdot P(A \mid B)$.

EXAMPLE 4.12   **Understanding the Multiplication Rule**

A statewide poll of 800 registered voters in 25 precincts from across New York State was taken. Each voter was identified as being registered as Republican, Democrat, or other and then asked, "Are you are in favor of or against the current budget proposal awaiting the governor's signature?" The resulting tallies are shown here.

	Number in Favor	Number Against	Number of Voters
Republican	136	88	224
Democrat	314	212	526
Other	14	36	50
Totals	464	336	800

Suppose one voter is to be selected at random from the 800 voters summarized in the preceding table. Let's consider the two events: "The voter selected is in favor" and "The voter is a Republican." Suppose one voter is picked at random from these 800 voters; find the three probabilities: P(in favor), P(Republican | in favor), and P(in favor and Republican). Then use the results to check the truth of the multiplication rule.

SOLUTION

Probability the voter selected is "in favor" = P(in favor) = $464/800 = \dfrac{464}{800}$.

Probability the voter selected is "Republican, given in favor"

$= P$ (Republican | in favor) = $136/464 = \dfrac{136}{464}$.

Probability the voter selected is "in favor" and "Republican"

$= P$(in favor and Republican) = $136/800 = \dfrac{136}{800} = \underline{0.17}$.

Notes about finding the preceding probabilities:
1. The conditional "given" means there is a restriction; thus, "Republican | in favor" means we start with only those voters who are "in favor." In this case, this means we are looking only at 464 voters when determining this probability.
2. The connective "and" means "both" or "in common"; thus, "in favor and Republican" means all voters who satisfy both events.

Now let's use the previous probabilities to demonstrate the truth of the multiplication rule.

Let A = "in favor" and B = "Republican." The general multiplication rule then becomes:

$$P(\text{in favor and Republican}) = P(\text{in favor}) \cdot P(\text{Republican} \mid \text{in favor})$$

Previously we found: $P(\text{in favor and Republican}) = \dfrac{136}{800} = \underline{0.17}$.
Using the other two probabilities, we see:

$$P(\text{in favor}) \cdot P(\text{Republican} \mid \text{in favor}) = \frac{464}{800} \cdot \frac{136}{464} = \frac{136}{800} = \underline{0.17}.$$

You typically do not have the option of finding the $P(A \text{ and } B)$ two ways, as we did here. When you are asked to find $P(A \text{ and } B)$, you will often be given the $P(A)$ and $P(B)$. However, you will not always get the correct answer by just multiplying those two probabilities together. You will need a third piece of information. You will need the conditional probability of one of the two events or information that will allow you to find it.

EXAMPLE 4.13

Drawing without Replacement

At a carnival game, the player blindly draws one colored marble at a time from a box containing two red and four blue marbles. The chosen marble is not returned to the box after being selected; that is, each drawing is done without replacement. The marbles are mixed before each drawing. It costs $1 to play, and if the first two marbles drawn are red, the player receives a $2 prize. If the first four marbles drawn are all blue, the player receives a $5 prize. Otherwise, no prize is awarded. To find the probability of winning a prize, let's look first at the probability of drawing red or blue on consecutive drawings and organize the information on a tree diagram.

On the first draw (represented by the purple branch segments in Figure 4.5), the probability of red is two chances out of six, 2/6 or 1/3, whereas the probability of blue is 4/6, or 2/3. Because the marble is not replaced, only five marbles are left in the box; the number of each color remaining depends on the color of the first marble drawn. If the first marble was red, then the probabilities are 1/5 and 4/5 as shown on the tree diagram (green branch segments in Figure 4.5). If the first marble was blue, then the probabilities are 2/5 and 3/5 as shown on the tree diagram (orange branch segments in Figure 4.5). The probabilities change with each drawing, because the number of marbles available keeps decreasing as each drawing takes place. The tree diagram is a marvelous pictorial aid in following the progression.

FIGURE 4.5 Tree Diagram—First Two Drawings, Carnival Game

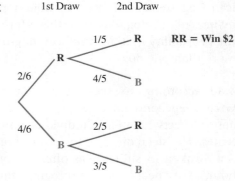

The probability of winning the $2 prize can now be found using formula (4.5):

$$P(\text{A and B}) = P(\text{A}) \cdot P(\text{B} \mid \text{A})$$

$$P(\text{winning } \$2) = P(\text{R}_1 \text{ and } \text{R}_2) = P(\text{R}_1) \cdot P(\text{R}_2 \mid \text{R}_1) = \frac{2}{6} \cdot \frac{1}{5} = \frac{1}{15} = 0.067$$

(Winning the $5 prize is left as Exercise 4.77.)

Note: The tree diagram, when labeled, has the probabilities needed for multiplying listed along the branch representing the winning effort.

SECTION 4.4 EXERCISES

4.57 a. If the probability that event A occurs during an experiment is 0.7, what is the probability that event A does not occur during that experiment?

b. If the results of a probability experiment can be any integer from 16 to 28 and the probability that the integer is less than 20 is 0.78, what is the probability the integer will be 20 or more?

4.58 a. If the probability that you pass the next exam in statistics is accurately assessed at 0.75, what is the probability that you will not pass the next statistics exam?

b. The weather forecaster predicts that there is a "70 percent" chance of less than 1 inch of rain during the next 30-day period. What is the probability of at least 1 inch of rain in the next 30 days?

4.59 According to the U.S. Pet Ownership & Demographic Sourcebook, from Pedigree Food for Dogs (USA Snapshot "Most give puppy love to just one," February 22, 2005), about 66% of all American dog owners—some 60 million people—are owners of one dog. Based on this information, find the probability that an American dog-owner owns more than one dog.

4.60 According to the Sleep Channel (http://www.sleepchannel.net, September 2002), sleep apnea affects 12 million individuals in the United States. The sleep disorder interrupts breathing and can awaken its sufferers as often as five times an hour. Many people do not recognize the condition even though it causes loud snoring. Assuming there are 275 million people in the United States, what is the probability that an individual chosen at random will not be affected by sleep apnea?

4.61 If $P(\text{A}) = 0.4$, $P(\text{B}) = 0.5$, and $P(\text{A and B}) = 0.1$, find $P(\text{A or B})$.

4.62 If $P(\text{A}) = 0.5$, $P(\text{B}) = 0.3$, and $P(\text{A and B}) = 0.2$, find $P(\text{A or B})$.

4.63 If $P(\text{A}) = 0.4$, $P(\text{B}) = 0.5$, and $P(\text{A or B}) = 0.7$, find $P(\text{A and B})$.

4.64 If $P(\text{A}) = 0.4$, $P(\text{A or B}) = 0.9$, and $P(\text{A and B}) = 0.1$, find $P(\text{B})$.

4.65 The entertainment sports industry employs athletes, coaches, referees, and related workers. Of these, 0.37 work part time and 0.50 earn more than $20,540 annually. If 0.32 of these employees work full time and earn more than $20,540, what proportion of the industry's employees are full time or earn more than $20,540?

4.66 Jason attends his high school reunion. Of the attendees, 50% are female. Common knowledge has it that 88% of people are right-handed. Being a left-handed male, Jason knows that of a given crowd, only approximately 6% are left-handed males. If Jason talks to the first person he meets at the reunion, what is the probability that the person is a male or left-handed?

4.67 A parts store sells both new and used parts. Sixty percent of the parts in stock are used. Sixty-one percent are used or defective. If 5% of the store's parts are defective, what percentage is both used and defective? Solve using formulas. Compare your solution to your answer to Exercise 4.27.

4.68 Union officials report that 60% of the workers at a large factory belong to the union, 90% make more than $12 per hour, and 40% belong to the union and make more than $12 per hour. Do you believe these percentages? Explain. Solve using formulas. Compare your solution to your answer to Exercise 4.28.

4.69 A and B are events defined on a sample space, with $P(A) = 0.7$ and $P(B \mid A) = 0.4$. Find $P(A \text{ and } B)$.

4.70 A and B are events defined on a sample space, with $P(A \mid B) = 0.5$ and $P(B) = 0.8$. Find $P(A \text{ and } B)$.

4.71 A and B are events defined on a sample space, with $P(A) = 0.6$ and $P(A \text{ and } B) = 0.3$. Find $P(B \mid A)$.

4.72 A and B are events defined on a sample space, with $P(B) = 0.4$ and $P(A \text{ and } B) = 0.5$. Find $P(A \mid B)$.

4.73 It is known that steroids give users an advantage in athletic contests, but it is also known that steroid use is banned in athletes. As a result, a steroid testing program has been instituted and athletes are randomly tested. The test procedures are believed to be equally effective on both users and nonusers and claim to be 98% accurate. If 90% of the athletes affected by this testing program are clean, what is the probability that the next athlete tested will be a user and fail the test?

4.74 Juan lives in a large city and commutes to work daily by subway or by taxi. He takes the subway 80% of the time because it costs less, and he takes a taxi the other 20% of the time. When taking the subway, he arrives at work on time 70% of

the time, whereas he makes it on time 90% of the time when traveling by taxi.

a. What is the probability that Juan took the subway and is at work on time on any given day?

b. What is the probability that Juan took a taxi and is at work on time on any given day?

4.75 Nobody likes paying taxes, but this is not the way to get out of it! It is believed that 10% of all taxpayers intentionally claim some deductions to which they are not entitled. If 9% of all taxpayers both intentionally claim extra deductions and deny such when audited, find the probability that a taxpayer who does take extra deductions intentionally will deny it.

4.76 Casey loves his mid-morning coffee and always stops by one of his favorite coffeehouses for a cup. When he gets take-out, there is a 0.6 chance that he will also have a pastry. He gets both coffee and a pastry as take-out with a probability of 0.48. What is the probability that he does take-out?

4.77 Find the probability of winning $5 if you play the carnival game described in Example 4.13.

a. Complete the branches of the tree diagram started in Figure 4.5, listing the probabilities for all possible drawings.

b. What is the probability of drawing a red marble on the second drawing? What additional information is needed to find the probability? What "conditions" could exist?

c. Calculate the probability of winning the $5 prize.

d. Is the $2 prize or the $5 prize harder to win? Which is more likely? Justify your answer.

4.78 Suppose the rules for the carnival game in Example 4.13 were modified so that the marble drawn each time is returned to the box before the next drawing.

a. Redraw the tree diagram drawn for Exercise 4.77, listing the probabilities for the game when played "with replacement."

b. What is the probability of drawing a red marble on the second drawing? What additional information is needed to find the probability? What effect does this have on P(red)?

c. Calculate the probability of winning the $2 prize.

d. Calculate the probability of winning the $5 prize.

e. When the game is played with replacement, is the $2 or the $5 prize harder to win? Which is more likely? Justify your answer.

4.79 Suppose that A and B are events defined on a common sample space and that the following probabilities are known: $P(A) = 0.3$, $P(B) = 0.4$, and $P(A \mid B) = 0.2$. Find P(A or B).

4.80 Suppose that A and B are events defined on a common sample space and that the following probabilities are known: $P(A \text{ or } B) = 0.7$, $P(B) = 0.5$, and $P(A \mid B) = 0.2$. Find $P(A)$.

4.81 Suppose that A and B are events defined on a common sample space and that the following probabilities are known: $P(A) = 0.4$, $P(B) = 0.3$, and $P(A \text{ or } B) = 0.66$. Find $P(A \mid B)$.

4.82 Suppose that A and B are events defined on a common sample space and that the following

probabilities are known: $P(A) = 0.5$, $P(A \text{ and } B) = 0.24$, and $P(A \mid B) = 0.4$. Find P(A or B).

4.83 Given $P(A \text{ or } B) = 1.0$, $P(\overline{A \text{ and } B}) = 0.7$, and $P(\overline{B}) = 0.4$, find:

a. $P(B)$ b. $P(A)$ c. $P(A \mid B)$

4.84 Given $P(A \text{ or } B) = 1.0$, $P(\overline{A \text{ and } B}) = 0.3$, and $P(\overline{B}) = 0.4$, find:

a. $P(B)$ b. $P(A)$ c. $P(A \mid B)$

4.85 The probability of A is 0.5. The conditional probability that A occurs given that B occurs is 0.25. The conditional probability that B occurs given that A occurs is 0.2.

a. What is the probability that B occurs?

b. What is the conditional probability that B does not occur given that A does not occur?

4.86 The probability of C is 0.4. The conditional probability that C occurs given that D occurs is 0.5. The conditional probability that C occurs given that D does not occur is 0.25.

a. What is the probability that D occurs?

b. What is the conditional probability that D occurs given that C occurs?

4.5   Mutually Exclusive Events

To further our discussion of compound events, the concept of "mutually exclusive" must be introduced.

> **Mutually exclusive events:** Nonempty events defined on the same sample space with each event excluding the occurrence of the other. In other words, they are events that share no common elements.
>
> In algebra: P(A and B) = 0
>
> In words: There are several equivalent ways to express the concept of mutually exclusive:
>
> 1. If you know that either one of the events has occurred, then the other event is excluded or cannot have occurred.
> 2. If you are looking at the lists of the elements making up each event, none of the elements listed for either event will appear on the other event's list; there are "no shared elements."

3. If you are looking at a Venn diagram, the closed areas representing each event "do not intersect"—that is, there are "no shared elements," or as another way to say it, "they are disjoint."
4. The equation says, "the intersection of the two events has a probability of zero," meaning "the intersection is an empty set" or "there is no intersection."

Note: The concept of mutually exclusive events is based on the relationship between the sets of elements that satisfy the events. Mutually exclusive is not a probability concept by definition; it just happens to be easy to express the concept using a probability statement.

Let's look at some examples.

EXAMPLE 4.14 | **Understanding Mutually Exclusive Events**

From a nationwide exit poll of 1000 voters in 25 precincts across the country on November 2, 2004, we have the following.

Education	Number for Bush	Number for Kerry	Number for Other	Number of Voters
No high school	19	20	1	40
High school graduate	114	103	3	220
Some college	172	147	1	320
College graduate	135	119	6	260
Postgraduate	70	88	2	160
Total	510	477	13	1000

Consider the two events the voter selected "voted for Bush" and the voter selected "voted for Kerry." Suppose one voter is selected at random from the 1000 voters summarized in the table. In order for the event the voter selected "voted for Bush" to occur, the voter selected must be 1 of the 510 voters listed in "Number for Bush" column. In order for the event the voter selected "voted for Kerry" to occur, the voter selected must be 1 of the 477 voters listed in "Number for Kerry" column. Because no voter listed in the Bush column is also listed in the Kerry column and because no voter listed in the Kerry column is also listed in the Bush column, these two events are mutually exclusive.

In equation form: P(voted for Bush and voted for Kerry) = 0.

EXAMPLE 4.15 | **Understanding Not Mutually Exclusive Events**

From a nationwide exit poll of 1000 voters in 25 precincts across the country on November 2, 2004, we have the following.

Education	Number for Bush	Number for Kerry	Number for Other	Number of Voters
No high school	19	20	1	40
High school graduate	114	103	3	220
Some college	172	147	1	320
College graduate	135	119	6	260
Postgraduate	70	88	2	160
Total	510	477	13	1000

Consider the two events the voter selected "voted for Bush" and the voter selected had "some college education." Suppose one voter is selected at random from the 1000 voters summarized in the table. In order for the event the voter selected "voted for Bush" to occur, the voter selected must be 1 of the 510 voters listed in "Number for Bush" column. In order for the event the voter selected had "some college education" to occur, the voter selected must be 1 of the 320 voters listed in the "Some college" row. Because the 172 voters shown in the intersection of the "Number for Bush" column and the "Some college" row belong to both of the events (the voter selected "voted for Bush" and the voter selected had "some college education"), these two events are NOT mutually exclusive.

In equation form: P(voted for Bush and some college education) $= 172/1000 = 0.172$; which is not equal to zero.

EXAMPLE 4.16 **Mutually Exclusive Card Events**

Consider a regular deck of playing cards and the two events "card drawn is a queen" and "card drawn is an ace." The deck is to be shuffled and one card randomly drawn. In order for the event "card drawn is a queen" to occur, the card drawn must be one of the four queens: queen of hearts, queen of diamonds, queen of spades, or queen of clubs. In order for the event "card drawn is an ace" to occur, the card drawn must be one of the four aces: ace of hearts, ace of diamonds, ace of spades, or ace of clubs. Notice that there is no card that is both a queen and an ace. Therefore, these two events, "card drawn is a queen" and "card drawn is an ace," are mutually exclusive events.

In equation form: P(queen and ace) $= 0$.

EXAMPLE 4.17 **Not Mutually Exclusive Card Events**

Consider a regular deck of playing cards and the two events "card drawn is a queen" and "card drawn is a heart." The deck is to be shuffled and one card randomly drawn. Are the events "queen" and "heart" mutually exclusive? The event "card drawn is a queen" is made up of the four queens: queen of hearts, queen of diamonds, queen of spades, and queen of clubs. The event "card drawn is a heart" is made up of the 13 hearts: ace of hearts, king of hearts, queen of hearts, jack of hearts, and the other nine hearts. Notice that the "queen of hearts" is on both lists, thereby making it possible for both events "card drawn is a queen" and "card drawn is a heart" to occur simultaneously. That means, when one of these two events occurs, it does not exclude the possibility of the other's occurrence. These events are not mutually exclusive events.

In equation form: P(queen and heart) $= 1/52$; which is not equal to zero.

EXAMPLE 4.18 **Visual Display and Understanding
of Mutually Exclusive Events**

Statistics ⬡ Now™

Watch a video example at
http://1pass.thomson.com
or on your CD.

Consider an experiment in which two dice are rolled. Three events are defined as follows:

A: The sum of the numbers on the two dice is 7.

B: The sum of the numbers on the two dice is 10.

C: Each of the two dice shows the same number.

Let's determine whether these three events are mutually exclusive.

We can show three events are mutually exclusive by showing that each pair of events is mutually exclusive. Are events A and B mutually exclusive? Yes, they are, because the sum on the two dice cannot be both 7 and 10 at the same time. If a sum of 7 occurs, it is impossible for the sum to be 10.

Figure 4.6 presents the sample space for this experiment. This is the same sample space shown in Example 4.2, except that ordered pairs are used in place of the pictures. The ovals, diamonds, and rectangles show the ordered pairs that are in events A, B, and C, respectively. We can see that events A and B do not intersect. Therefore, they are mutually exclusive. Point (5, 5) in Figure 4.6 satisfies both events B and C. Therefore, B and C are not mutually exclusive. Two dice can each show a 5, which satisfies C, and the total satisfies B. Since we found one pair of events that are not mutually exclusive, events A, B, and C are not mutually exclusive.

FIGURE 4.6
Sample Space for the Roll of Two Dice

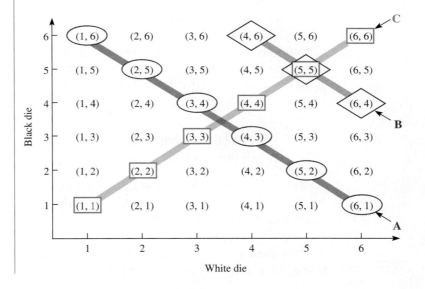

Special Addition Rule

The addition rule simplifies when the events involved are mutually exclusive.

If we know two events are mutually exclusive, then by applying $P(A \text{ and } B) = 0$ to the addition rule for probabilities, it follows that $P(A \text{ or } B) = P(A) + P(B) - P(A \text{ and } B)$ becomes $P(A \text{ or } B) = P(A) + P(B)$.

> **Special Addition Rule**
>
> Let A and B be two mutually exclusive events defined in a sample space *S*.
>
> In words: *probability of A or B = probability of A + probability of B*
>
> In algebra: $P(A \text{ or } B) = P(A) + P(B)$
>
> (4.6)

This formula can be expanded to consider more than two mutually exclusive events:

$$P(A \text{ or } B \text{ or } C \text{ or } \dots \text{ or } E) = P(A) + P(B) + P(C) + \dots + P(E)$$

This equation is often convenient for calculating probabilities, but it does not help us understand the relationship between the events A and B. It is the *definition* that tells us how we should think about mutually exclusive events. Students who understand mutual exclusiveness this way gain insight into what mutual exclusiveness is all about. This should lead you to think more clearly about situations dealing with mutually exclusive events, thereby making you less likely to confuse the concept of mutually exclusive events with independent events (to be defined in Section 4.6) or to make other common mistakes regarding the concept of mutually exclusive.

Notes:
1. Define mutually exclusive events in terms of the sets of elements satisfying the events and test for mutual exclusiveness in that manner.
2. Do not use $P(A \text{ and } B) = 0$ as the definition of mutually exclusive events. It is a property that results from the definition. It can be used as a test for mutually exclusive events; however, as a statement, it shows no meaning or insight into the concept of mutually exclusive events.
3. In equation form, the *definition* of mutually exclusive events states:

$$P(A \text{ and } B) = 0 \text{ (Both cannot happen at same time.)}$$

$$P(A \mid B) = 0 \text{ and } P(B \mid A) = 0$$
(If one is known to have occurred, then the other has not.)

Reconsider Example 4.16, with the two events "card drawn is a queen" and "card drawn is an ace" when drawing exactly one card from a deck of regular playing cards. The one card drawn is a queen, or the one card drawn is an ace. That one card cannot be both a queen and an ace at the same time, thereby making these two events mutually exclusive. The special addition rule therefore applies to the situation of finding $P(\text{queen or ace})$.

$$P(\text{queen or ace}) = P(\text{queen}) + P(\text{ace}) = \frac{4}{52} + \frac{4}{52} = \frac{8}{52} = \frac{2}{13}$$

SECTION 4.5 EXERCISES

4.87 Determine whether each of the following pairs of events is mutually exclusive.

a. Five coins are tossed: "one head is observed," "at least one head is observed."

b. A salesperson calls on a client and makes a sale: "the sale exceeds \$100," "the sale exceeds \$1000."

c. One student is selected at random from a student body: the person selected is "male," the person selected is "older than 21 years of age."

d. Two dice are rolled: the total showing is "less than 7," the total showing is "more than 9."

4.88 Determine whether each of the following sets of events is mutually exclusive.

a. Five coins are tossed: "no more than one head is observed," "two heads are observed," "three or more heads are observed."

b. A salesperson calls on a client and makes a sale: the amount of the sale is "less than $100," is "between $100 and $1000," is "more than $500."

c. One student is selected at random from the student body: the person selected is "female," is "male," is "older than 21."

d. Two dice are rolled: the numbers of dots showing on the dice are "both odd," "both even," "total 7," "total 11."

4.89 Explain why $P(A \text{ and } B) = 0$ when events A and B are mutually exclusive.

4.90 Explain why $P(A \text{ occurring when } B \text{ has occurred}) = 0$ when events A and B are mutually exclusive.

4.91 If $P(A) = 0.3$ and $P(B) = 0.4$, and if A and B are mutually exclusive events, find:

a. $P(\overline{A})$ b. $P(\overline{B})$ c. $P(A \text{ or } B)$ d. $P(A \text{ and } B)$

4.92 If $P(A) = 0.4$ and $P(B) = 0.5$, and if A and B are mutually exclusive events, find $P(A \text{ or } B)$.

4.93 One student is selected from the student body of your college. Define the following events: M—the student selected is male, F—the student selected is female, S—the student selected is registered for statistics.

a. Are events M and F mutually exclusive? Explain.

b. Are events M and S mutually exclusive? Explain.

c. Are events F and S mutually exclusive? Explain.

d. Are events M and F complementary? Explain.

e. Are events M and S complementary? Explain.

f. Are complementary events also mutually exclusive events? Explain.

g. Are mutually exclusive events also complementary events? Explain.

4.94 One student is selected at random from a student body. Suppose the probability that this student is female is 0.5 and the probability that this student works part time is 0.6. Are the two events "female" and "working" mutually exclusive? Explain.

4.95 Two dice are rolled. Define events as follows: A—sum of 7, C—doubles, E—sum of 8.

a. Which pairs of events, A and C, A and E, or C and E, are mutually exclusive? Explain.

b. Find the probabilities $P(A \text{ or } C)$, $P(A \text{ or } E)$, and $P(C \text{ or } E)$.

4.96 An aquarium at a pet store contains 40 orange swordfish (22 females and 18 males) and 28 green swordtails (12 females and 16 males). You randomly net one of the fish.

a. What is the probability that it is an orange swordfish?

b. What is the probability that it is a male fish?

c. What is the probability that it is an orange female swordfish?

d. What is the probability that it is a female or a green swordtail?

e. Are the events "male" and "female" mutually exclusive? Explain.

f. Are the events "male" and "swordfish" mutually exclusive? Explain.

4.97 Do people take indoor swimming lessons in the middle of the hot summer? They sure do at the Webster Aquatic Center. During the month of July 2004 alone, 179 people participated in various forms of lessons.

Swim Categories	Daytime	Evenings
Preschool	26	29
Levels	75	39
Adult and diving	4	6
Total	105	74

If one swimmer was selected at random from the July participants:

a. Are the events the selected participant is "daytime" and "evening" mutually exclusive? Explain.

b. Are the events the selected participant is "preschool" and "levels" mutually exclusive? Explain.

c. Are the events the selected participant is "daytime" and "preschool" mutually exclusive? Explain.

d. Find P(preschool).

e. Find P(daytime).

f. Find P(not levels).

g. Find P(preschool or evening).

h. Find P(preschool and daytime).

i. Find P(daytime | levels).

j. Find P(adult and diving | evening).

4.98 Injuries are unfortunately part of every sport. High school basketball is no exception, as the following table shows. The percentages listed are the percent of reported injuries that occur to high school male and female basketball players and the location on their body that was injured.

Injury Location	Males	Females
Ankle/foot	38.3%	36.0%
Hip/thigh/leg	14.7%	16.6%
Knee	10.3%	13.0%
Forearm/wrist/hand	11.5%	11.2%
Face/scalp	12.2%	8.8%
Other	13.0%	14.4%
Total	100.0%	100.0%

If one player is selected at random from those included in the table:

a. Are the events the selected player was "male" and "female" mutually exclusive? Explain.

b. Are the events the selected player's injury was "ankle/foot" and "knee" mutually exclusive? Explain.

c. Are the events "female" and "face/scalp" mutually exclusive? Explain.

d. Find P(ankle/foot | male).

e. Find P(ankle/foot | female).

f. Find P(not leg related | male).

g. Find P(knee or face/scalp | male).

h. Find P(knee or face/scalp | female).

i. Explain why P(knee) for all high school basketball players cannot be found using the information in the table. What additional information is needed?

4.99 Most Americans, 70% in fact, say frequent hand washing is the best way to fend off the flu. Despite that, when using public restrooms, women wash their hands only 62% of the time and men wash only 43% of the time. Of the adults using the public restroom at a large grocery chain store, 58% are women. What is the probability that the next person to enter the restroom in this store washes his or her hands?

4.100 He is the last guy you want to see in your rearview mirror when you are speeding down the highway, but research shows that a traffic ticket reduces a driver's chance of being involved in a fatal accident, at least for a few weeks. By age group, 13.3% of all drivers are younger than age 25, 58.6% are between ages 25 and 54, and 28.1% are 55 or older. Statistics show that 1.6% of the drivers younger than age 25, 2.2% of those ages 25 to 54, and 0.5% of those 55 or older will have an accident in the next month. What is the probability that a randomly identified driver will have an accident in the next month?

| | 4.6 | | **Independent Events** |

The concept of independent events is necessary to continue our discussion on compound events.

> **Independent events:** Two events are *independent* if the occurrence (or nonoccurrence) of one gives us no information about the likeliness of occurrence for the other. In other words, if the probability of A remains unchanged after we know that B has happened (or has not happened), the events are independent.
>
> In algebra: $P(A) = P(A \mid B) = P(A \mid \text{not } B)$
>
> In words: There are several equivalent ways to express the concept of independence:
>
> 1. The probability of event A is unaffected by knowledge that a second event, B, has occurred, knowledge that B has not occurred, or no knowledge about event B whatsoever.
> 2. The probability of event A is unaffected by knowledge, or no knowledge, about a second event, B, having occurred or not occurred.
> 3. The probability of event A (with no knowledge about event B) is the same as the probability of event A, knowing event B has occurred, and both are the same as the probability of event A, knowing event B has not occurred.

Not all events are independent.

> **Dependent events:** Events that are not independent. That is, the occurrence of one event does have an effect on the probability for occurrence of the other event.

Let's look at some examples.

EXAMPLE 4.19   **Understanding Independent Events**

A statewide poll of 750 registered Republicans and Democrats in 25 precincts from across New York State was taken. Each voter was identified as being registered as a Republican or a Democrat and then asked, "Are you are in favor of or against the current budget proposal awaiting the governor's signature?" The resulting tallies are shown here.

	Number in Favor	Number Against	Number of Voters
Republican	135	90	225
Democrat	315	210	525
Totals	450	300	750

Suppose one voter is to be selected at random from the 750 voters summarized in the preceding table. Let's consider the two events, "The voter selected is in favor" and "The voter is a Republican." Are these two events independent?

To answer this, consider the following three probabilities: (1) probability the voter selected is in favor; (2) probability the voter selected is in favor, knowing the voter is a Republican; and (3) probability the voter selected is in favor, knowing the voter is not a Republican.

Probability the voter selected is in favor = P(in favor) = 450/750 = <u>0.60</u>.

Probability the voter selected is in favor, knowing voter is a Republican = P(in favor | Republican) = 135/225 = <u>0.60</u>.

Probability the voter selected is in favor, knowing voter is not a Republican = Probability the voter selected is in favor, knowing voter is a Democrat = P(in favor | not Republican) = P(in favor | Democrat) = 315/525 = <u>0.60</u>.

Does knowing the voter's political affiliation have an influencing effect on the probability that the voter is in favor of the budget proposal? With no information about political affiliation, the probability of being in favor is 0.60. Information about the event "Republican" does not alter the probability of "in favor." They are all the value 0.60. Therefore, these two events are said to be *independent events*.

When checking the three probabilities, $P(A)$, $P(A | B)$, and $P(A | \text{not } B)$, we need to compare only two of them. If any two of the three probabilities are equal, the third will be the same value. Furthermore, if any two of the three probabilities are unequal, then all three will be different in value.

Note: Determine all three values, using the third as a check. All will be the same, or all will be different—there is no other possible outcome.

EXAMPLE 4.20

Understanding Not Independent Events

From a National Election Pool exit poll of 13,660 voters in 250 precincts across the country on November 2, 2004, we have the following.

	Percentage of Voters	Percent for Bush	Percent for Kerry	Percent for Other
Men	46	55	44	1
Women	54	48	51	1

Suppose one voter is selected at random from the 13,660 voters summarized in the preceding table. Let's consider the two events: "The voter is a woman" and "The voter voted for Bush." Are these two events independent? To answer this, consider this question, "Does knowing the voter is a woman have an influencing effect on the probability that the voter voted for Bush?" What is the probability of voting for Bush, if the voter is a woman? You say, "0.48." Now compare this to the probability of voting for Bush, if the voter is not a woman. You say that probability is 0.55. So I ask you, "Did knowing the voter was a woman influence the probability of voting for Bush?" Yes, it did; it is 0.48 when the voter is a woman and 0.55 when it is not a woman. Information about the event "woman" does alter the probability of "voted for

Bush." Therefore, these two events are *not independent* and said to be *dependent events*.

In equation form:

$$P(\text{voted for Bush} \mid \text{voter is a woman}) = P(B \mid W) = 0.48$$

$$P(\text{voted for Bush} \mid \text{voter is not a woman}) = P(B \mid \text{not W}) = 0.51$$

Therefore, $P(B \mid W) \neq P(B \mid \text{not W})$, and the two events are not independent.

EXAMPLE 4.21 **Independent Card Events**

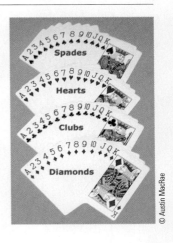
© Austin MacRae

Consider a regular deck of playing cards and the two events "card drawn is a queen" and "card drawn is a heart." Suppose that I shuffle the deck, randomly draw one card, and, before looking at the card, ask you the probability that it is a "queen." You say 4/52, or 1/13. Then I peek at the card and tell you that it is a "heart." Now, what is the probability that the card is a "queen"? You say it is 1/13, the same as before knowing the card was a "heart."

The hint that the card was a heart provided you with additional information, but that information did not change the probability that it was a queen. Therefore, "queen" and "heart" are independent. Furthermore, suppose that after I drew the card and looked at it, I had told you the card was "not a heart." What would be the probability the card is a "queen"? You say 3/39, or 1/13. Again, notice that knowing the card was "not a heart" did provide additional information, but that information did not change the probability that it was a "queen." This is what it means for the two events "card is a queen" and "card is a heart" to be independent.

In equation form:

$$P(\text{queen} \mid \text{card is heart}) = P(Q \mid H) = P(Q)$$

$$P(\text{queen} \mid \text{card is not heart}) = P(Q \mid \text{not H}) = P(Q)$$

Therefore, $P(Q) = P(Q \mid H) = P(Q \mid \text{not H})$, and the two events are independent.

EXAMPLE 4.22 **Not Independent Card Events**

Now, let's consider the two events "card drawn is a heart" and "card drawn is red." Are the events "heart" and "red" independent? Following the same scenario as in Example 4.21, I shuffle the deck of 52 cards, randomly draw one card, and before looking at it, you say the probability that the unknown card is "red" is 26/52 = 1/2. However, when told the additional information that the card is a "heart," you change your probability that the card is "red" to 13/13, or 1. This additional information results in a different probability of "red."

$P(\text{red} \mid \text{card is heart}) = P(R \mid H) = 13/13 = 1$, and $P(\text{red}) = P(\text{red} \mid \text{having no additional information}) = 26/52 = 1/2$. Therefore, the additional information did change the probability of the event "red." These two events are not independent and therefore said to be dependent events.

In equation form, the definition states:

A and B are independent if and only if $P(A \mid B) = P(A)$

Note: Define *independence* in terms of conditional probability, and test for independence in that manner.

Special Multiplication Rule

The multiplication rule simplifies when the events involved are independent.

If we know two events are independent, then by applying the definition of independence, $P(B \mid A) = P(B)$, to the multiplication rule, it follows that:

$$P(A \text{ and } B) = P(A) \cdot P(B \mid A) \quad \text{becomes} \quad P(A \text{ and } B) = P(A) \cdot P(B)$$

Special Multiplication Rule

Let A and B be two independent events defined in a sample space S.

In words: *probability of A and B = probability of A × probability of B*

In algebra: $P(A \text{ and } B) = P(A) \cdot P(B)$ (4.7)

This formula can be expanded to consider more than two independent events:

$$P(A \text{ and } B \text{ and } C \text{ and } \ldots \text{ and } E) = P(A) \cdot P(B) \cdot P(C) \cdot \ldots \cdot P(E)$$

This equation is often convenient for calculating probabilities, but it does not help us understand the independence relationship between the events A and B. It is *the definition* that tells us how we should think about independent events. Students who understand independence this way gain insight into what independence is all about. This should lead you to think more clearly about situations dealing with independent events, thereby making you less likely to confuse the concept of independent events with mutually exclusive events or to make other common mistakes regarding independence.

Note: Do not use $P(A \text{ and } B) = P(A) \cdot P(B)$ as the definition of independence. It is a property that results from the definition. It can be used as a test for independence, but as a statement, it shows no meaning or insight into the concept of independent events.

SECTION 4.6 EXERCISES

4.101 Determine whether each of the following pairs of events is independent:

a. Rolling a pair of dice and observing a "1" on the first die and a "1" on the second die

b. Drawing a "spade" from a regular deck of playing cards and then drawing another "spade" from the same deck without replacing the first card

c. Same as part b except the first card is returned to the deck before the second drawing

d. Owning a red automobile and having blonde hair

e. Owning a red automobile and having a flat tire today

f. Studying for an exam and passing the exam

4.102 Determine whether each of the following pairs of events is independent:

a. Rolling a pair of dice and observing a "2" on one of the dice and having a "total of 10"

b. Drawing one card from a regular deck of playing cards and having a "red" card and having an "ace"

c. Raining today and passing today's exam

d. Raining today and playing golf today

e. Completing today's homework assignment and being on time for class

4.103 A and B are independent events, and $P(A) = 0.7$ and $P(B) = 0.4$. Find $P(A \text{ and } B)$.

4.104 A and B are independent events, and $P(A) = 0.5$ and $P(B) = 0.8$. Find $P(A \text{ and } B)$.

4.105 A and B are independent events, and $P(A) = 0.6$ and $P(A \text{ and } B) = 0.3$. Find $P(B)$.

4.106 A and B are independent events, and $P(A) = 0.4$ and $P(A \text{ and } B) = 0.5$. Find $P(B)$.

4.107 If $P(A) = 0.3$ and $P(B) = 0.4$ and A and B are independent events, what is the probability of each of the following:

a. $P(A \text{ and } B)$ b. $P(B \mid A)$ c. $P(A \mid B)$

4.108 Suppose that $P(A) = 0.3$, $P(B) = 0.4$, and $P(A \text{ and } B) = 0.12$.

a. What is $P(A \mid B)$?

b. What is $P(B \mid A)$?

c. Are A and B independent?

4.109 Suppose that $P(A) = 0.3$, $P(B) = 0.4$, and $P(A \text{ and } B) = 0.20$.

a. What is $P(A \mid B)$?

b. What is $P(B \mid A)$?

c. Are A and B independent?

4.110 One student is selected at random from a group of 200 students known to consist of 140 full-time (80 female and 60 male) students and 60 part-time (40 female and 20 male) students. Event A is "the student selected is full time," and event C is "the student selected is female."

a. Are events A and C independent? Justify your answer.

b. Find the probability $P(A \text{ and } C)$.

4.111 A single card is drawn from a standard deck. Let A be the event that "the card is a face card" (a jack, a queen, or a king), B is a "red card," and C is "the card is a heart." Determine whether the following pairs of events are independent or dependent:

a. A and B   b. A and C   c. B and C

4.112 A box contains four red and three blue poker chips. Three poker chips are to be randomly selected, one at a time.

a. What is the probability that all three chips will be red if the selection is done with replacement?

b. What is the probability that all three chips will be red if the selection is done without replacement?

c. Are the drawings independent in either part a or b? Justify your answer.

4.113 Excluding job benefit coverage, approximately 49% of adults have purchased life insurance. The likelihood that those aged 18 to 24 without life insurance will purchase life insurance in the next year is 15%, and for those aged 25 to 34, it is 26%. (Opinion Research)

a. Find the probability that a randomly selected adult has not purchased life insurance.

b. What is the probability that an adult aged 18 to 24 will purchase life insurance within the next year?

c. Find the probability that a randomly selected adult will be 25 to 34 years old, does not currently have life insurance, and will purchase it within the next year?

4.114 The U.S. space program has a history made up of many successes and some failures. Space flight reliability is of the utmost importance in the launching of space shuttles. The reliability of the complete mission is based on all of its components. Each of the six joints in the *Challenger* space shuttle's booster rocket had a 0.977 reliability. The six joints worked independently.

a. What does it mean to say that the six joints work independently?

b. What was the reliability (probability) for all six of the joints working together?

4.115 Of households in the United States, 18 million, or 17%, have three or more vehicles, as stated in *USA Today* (June 12, 2002), quoting the Census Bureau as the source.

a. If two U.S. households are randomly selected, find the probability that both will have three or more vehicles.

b. If two U.S. households are randomly selected, find the probability that neither of the two has three or more vehicles.

c. If four households are selected, what is the probability that all four have three or more vehicles?

4.116 A *USA Today* article titled "Survey: Records tainted—Fans want drug tests for baseball players" (June 12, 2002) quotes a *USA Today*/CNN Gallup Poll as finding 86% of baseball fans saying they favor testing ballplayers for steroids and other performance-enhancing drugs. If five baseball fans are randomly selected, what is the probability that all five will be in favor of drug testing?

4.117 The July 8, 2002, issue of *Democrat & Chronicle* gave the results from the 2000 census that 42% of grandparents are responsible for "most of the basic needs" of a grandchild in the home. If three American grandparents are contacted, what is the probability that all three are the primary caregiver for their grandchildren?

4.118 You have applied for two scholarships: a merit scholarship (M) and an athletic scholarship (A). Assume the probability that you receive the athletic scholarship is 0.25, the probability you receive both scholarships is 0.15, and the probability you get at least one of the scholarships is 0.37. Use a Venn diagram to answer these questions:

a. What is the probability you receive the merit scholarship?

b. What is the probability you do not receive either of the two scholarships?

c. What is the probability you receive the merit scholarship given that you have been awarded the athletic scholarship?

d. What is the probability you receive the athletic scholarship given that you have been awarded the merit scholarship?

e. Are the events "receiving an athletic scholarship" and "receiving a merit scholarship" independent events? Explain.

4.119 The owners of a two-person business make their decisions independently of each other and then compare their decisions. If they agree, the decision is made; if they do not agree, then further consideration is necessary before a decision is reached. If each has a history of making the right decision 60% of the time, what is the probability that together they:

a. Make the right decision on the first try

b. Make the wrong decision on the first try

c. Delay the decision for further study

4.120 The odds against throwing a pair of dice and getting a total of 5 are 8 to 1. The odds against throwing a pair of dice and getting a total of 10 are 11 to 1. What is the probability of throwing the dice twice and getting a total of 5 on the first throw and 10 on the second throw?

4.121 Consider the set of integers 1, 2, 3, 4, and 5.

a. One integer is selected at random. What is the probability that it is odd?

b. Two integers are selected at random (one at a time with replacement so that each of the five is available for a second selection). Find the probability that neither is odd; exactly one of them is odd; both are odd.

4.122 A box contains 25 parts, of which 3 are defective and 22 are nondefective. If 2 parts are selected without replacement, find the following probabilities:

a. P(both are defective)

b. P(exactly one is defective)

c. P(neither is defective)

4.123 Graduation rates reached a record low in 2001. The percentage of students who graduate within 5 years was 41.9% for public and 55.1% for private colleges. One of the reasons for this might be that 42% of the students attend only part time. (ACT)

a. What additional information do you need to determine the probability that a student selected at random is part time and will graduate within 5 years?

b. Is it likely that these two events have the needed property? Explain.

c. If appropriate, find the probability that a student selected at random is part time and will graduate within 5 years.

4.124 From a survey of adults, 48% plan to buy candy this year at Easter. The types of candy they will buy are described in the following table.

Chocolate	Nonchocolate	Jellybeans	Cream-Filled	Marshmallow	Malted	Don't Know
30%	25%	13%	11%	8%	7%	6%

Source: International Mass Retail Association

a. What additional information do you need to determine the probability that a customer selected at random will buy candy and it will be chocolate?

b. Is it likely that these two events have the needed property? Explain.

c. If appropriate, find the probability that a customer selected at random will buy candy and it will be chocolate.

4.7 Are Mutual Exclusiveness and Independence Related?

Mutually exclusive events and independent events are two very different concepts based on definitions that start from very different orientations. The two concepts can easily become confused because these two concepts interact with each other and are intertwined by the probability statements we use in describing these concepts.

To describe these two concepts and eventually understand the distinction between them as well as the relationship between them, we need to agree that the events being considered are two nonempty events defined on the same sample space and therefore each has nonzero probabilities.

Note: Students often have a hard time realizing that when we say, "event A is a nonempty event" and write, "$P(A) > 0$" that we are describing the same situation. The words and the algebra often do not seem to have the same meaning. In this case the words and the probability statement both tell us that event A exists within the sample space.

Mutually Exclusive

Mutually exclusive events are two nonempty events defined on the same sample space that share no common elements.

This means:

1. In words: If you are looking at a Venn diagram, the closed areas representing each event "do not intersect"; in other words, they are disjoint sets, or no intersection occurs between their respective sets.

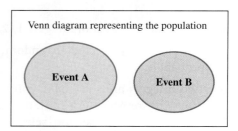

Venn diagram representing the population

Event A

Event B

2. In algebra: $P(A \text{ and } B) = 0$, which says, "the intersection of the two events is an empty set"; in other words, there is no intersection between their respective sets.

Notice that the concept of mutually exclusive is based on the relationship of the elements that satisfy the events. Mutually exclusive is not a probability concept by definition—it just happens to be easy to express the concept using a probability statement.

Independence

Independent events are two nonempty events defined on the same sample space that are related in such a way that the occurrence of either event does not affect the probability of the other event.

This means that:

1. In words: If event A has (or is known to have) already occurred, the probability of event B is unaffected (that is the probability of B after knowing event A had occurred remains the same as it was before knowing event A had occurred).

   In addition, it is also the case when A and B interchange roles that if event B has (or is known to have) already occurred, the probability of event A is unaffected (i.e., the probability of A is still the same after knowing event B had occurred as it was before).

   This is a "mutual relationship"; it works both ways.

2. In algebra: $P(B|A) = P(B|\text{not } A) = P(B)$ and $P(A|B) = P(A|\text{not } B) = P(A)$
   Or with a few words to help read the algebra, $P(B$, knowing A has occurred$) = P(B$, knowing A has not occurred$) = P(B)$ and $P(A$, knowing B has occurred$) = P(A$, knowing B has not occurred$) = P(A)$.
   Notice that the concept of independence is based on the effect one event (in this case, the lack of effect) has on the probability of the other event.

Let's look at the following four demonstrations relating to mutually exclusive and independent events:

Demonstration A

Given: $P(A) = 0.4$, $P(B) = 0.5$, and A and B are mutually exclusive; are they independent?

Answer: If A and B are mutually exclusive events, $P(A|B) = 0.0$, and because we are given $P(A) = 0.4$, we see that the occurrence of B has an effect on the probability of A. Therefore, A and B are not independent events.

Conclusion A: If the events are mutually exclusive, they are NOT independent.

Demonstration B

Given: $P(A) = 0.4$, $P(B) = 0.5$, and A and B are independent; are events A and B mutually exclusive?

Answer: If A and B are independent events, then the $P(A \text{ and } B) = P(A) \cdot P(B) = 0.4 \cdot 0.5 = 0.20$, and because the $P(A \text{ and } B)$ is greater than zero, events A and B must intersect, meaning the events are not mutually exclusive.

Conclusion B: If the events are independent, they are NOT mutually exclusive.

Demonstration C

Given: $P(A) = 0.4$, $P(B) = 0.5$, and A and B are not mutually exclusive; are events A and B independent?

Answer: Because A and B are not mutually exclusive events, it must be that $P(A \text{ and } B)$ is greater than zero. Now, if the $P(A \text{ and } B)$ happens to be exactly 0.20, then events A and B are independent [$P(A) \cdot P(B) = 0.4 \cdot 0.5 = 0.20$], but if the $P(A \text{ and } B)$ is any other positive value, say 0.1, then events A and B are not independent. Therefore, events A and B could be either independent or dependent; some other information is needed to make that determination.

Conclusion C: If the events are not mutually exclusive, they MAY be either independent or dependent; additional information is needed to determine which.

Demonstration D

Given: $P(A) = 0.4$, $P(B) = 0.5$, and A and B are not independent; are events A and B mutually exclusive?

Answer: Because A and B are not independent events, it must be that $P(A$ and $B)$ is different than 0.20, the value it would be if they were independent [$P(A) \cdot P(B) = 0.4 \cdot 0.5 = 0.20$]. Now, if $P(A$ and $B)$ happens to be exactly 0.00, then events A and B are mutually exclusive, but if $P(A$ and $B)$ is any other positive value, say 0.1, then events A and B are not mutually exclusive. Therefore, events A and B could be either mutually exclusive or not; some other information is needed to make that determination.

Conclusion D: If the events are NOT independent, they MAY be either mutually exclusive or not mutually exclusive; additional information is needed to determine which.

Advice

Work very carefully, starting with the information you are given and the definitions of the concepts involved.

What not to do:

Do not rely on the first "off-the-top" example you can think of to lead you to the correct answer. It typically will not!

The following examples give further practice with these probability concepts.

EXAMPLE 4.23

Calculating Probabilities and the Addition Rule

A pair of dice is rolled. Event T is defined as the occurrence of a "total of 10 or 11," and event D is the occurrence of "doubles." Find the probability $P(T$ or $D)$.

SOLUTION Look at the sample space of 36 ordered pairs for the rolling of two dice in Figure 4.6 (p. 239). Event T occurs if any one of 5 ordered pairs occurs:(4, 6), (5, 5), (6, 4), (5, 6), (6, 5). Therefore, $P(T) = \dfrac{5}{36}$. Event D occurs if any one of 6 ordered pairs occurs: (1, 1), (2, 2), (3, 3), (4, 4), (5, 5), (6, 6). Therefore, $P(D) = \dfrac{6}{36}$. Notice, however, that these two events are not mutually exclusive.

The two events "share" the point (5, 5). Thus, the probability $P(T$ and $D) = \dfrac{1}{36}$. As a result, the probability $P(T$ or $D)$ will be found using formula (4.4).

$$P(T \text{ or } D) = P(T) + P(D) - P(T \text{ and } D)$$

$$= \frac{5}{36} + \frac{6}{36} - \frac{1}{36} = \frac{10}{36} = \mathbf{\frac{5}{18}}$$

(Look at the sample space in Figure 4.6 and verify $P(T$ or $D) = \dfrac{5}{18}$.)

EXAMPLE 4.24

Using Conditional Probabilities to Determine Independence

In a sample of 150 residents, each person was asked if he or she favored the concept of having a single countywide police agency. The county is composed of one large city and many suburban townships. The residence (city or outside the city) and the responses of the residents are summarized in Table 4.4. If one of these residents was to be selected at random, what is the probability that the person will (a) favor the concept? (b) favor the concept if the person selected is a city resident? (c) favor the concept if the person selected is a resident from outside the city? (d) Are the events F (favor the concept) and C (reside in city) independent?

TABLE 4.4

Sample Results for Example 4.24

Residence	Favor (F)	Oppose ($\bar{F}$)	Total
In city (C)	80	40	120
Outside of city ($\bar{C}$)	20	10	30
Total	100	50	150

SOLUTION

(a) $P(F)$ is the proportion of the total sample that favors the concept. Therefore,

$$P(F) = \frac{n(F)}{n(S)} = \frac{100}{150} = \frac{2}{3}$$

(b) $P(F \mid C)$ is the probability that the person selected favors the concept given that he or she lives in the city. The condition, is city resident, reduces the sample space to the 120 city residents in the sample. Of these, 80 favored the concept; therefore,

$$P(F \mid C) = \frac{n(F \text{ and } C)}{n(C)} = \frac{80}{120} = \frac{2}{3}$$

(c) $P(F \mid \bar{C})$ is the probability that the person selected favors the concept, knowing that the person lives outside the city. The condition, lives outside the city, reduces the sample space to the 30 noncity residents; therefore,

$$P(F \mid \bar{C}) = \frac{n(F \text{ and } \bar{C})}{n(\bar{C})} = \frac{20}{30} = \frac{2}{3}$$

(d) All three probabilities have the same value, $\frac{2}{3}$. Therefore, we can say that the events F (favor) and C (reside in city) are independent. The location of residence did not affect $P(F)$.

EXAMPLE 4.25 Determining Independence and Using the Multiplication Rule

One student is selected at random from a group of 200 known to consist of 140 full-time (80 female and 60 male) students and 60 part-time (40 female and 20 male) students. Event A is "the student selected is full-time," and event C is "the student selected is female."

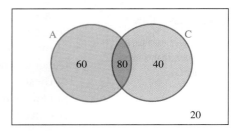

(a) Are events A and C independent?

(b) Find the probability $P(A \text{ and } C)$ using the multiplication rule.

SOLUTION 1

(a) First find the probabilities $P(A)$, $P(C)$, and $P(A \mid C)$:

$$P(A) = \frac{n(A)}{n(S)} = \frac{140}{200} = 0.7$$

$$P(C) = \frac{n(C)}{n(S)} = \frac{120}{200} = 0.6$$

$$P(A \mid C) = \frac{n(A \text{ and } C)}{n(C)} = \frac{80}{120} = 0.67$$

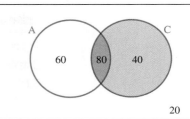

A and C are dependent events because $P(A) \neq P(A \mid C)$.

(b) $P(A \text{ and } C) = P(C) \cdot P(A \mid C) = \frac{120}{200} \cdot \frac{80}{120} = \frac{80}{200} = \mathbf{0.4}$

SOLUTION 2

(a) First find the probabilities $P(A)$, $P(C)$, and $P(C \mid A)$:

$$P(A) = \frac{n(A)}{n(S)} = \frac{140}{200} = 0.7$$

$$P(C) = \frac{n(C)}{n(S)} = \frac{120}{200} = 0.6$$

$$P(C \mid A) = \frac{n(C \text{ and } A)}{n(A)} = \frac{80}{140} = 0.57$$

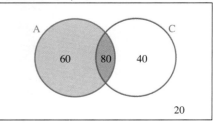

A and C are dependent events because $P(C) \neq P(C \mid A)$.

(b) $P(C \text{ and } A) = P(A) \cdot P(C \mid A) = \frac{140}{200} \cdot \frac{80}{140} = \frac{80}{200} = \mathbf{0.4}$

EXAMPLE 4.26 Using Several Probability Rules

A production process produces thousands of items. On the average, 20% of all items produced are defective. Each item is inspected before it is shipped. The inspector misclassifies an item 10% of the time; that is,

$P(\text{classified good} \mid \text{defective item}) = P(\text{classified defective} \mid \text{good item})$
$$= 0.10$$

Statistics ⟨△⟩ Now™

Watch a video example at
http://1pass.thomson.com
or on your CD.

What proportion of the items will be "classified good"?

SOLUTION What do we mean by the event "classified good"?

G: The item is good.

D: The item is defective.

CG: The item is classified good by the inspector.

CD: The item is classified defective by the inspector.

FIGURE 4.7
Using Several
Probability Rules

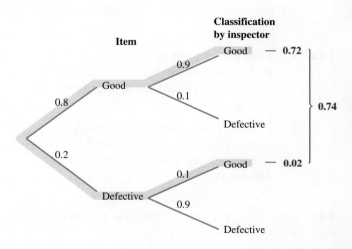

CG consists of two possibilities: "the item is good and is correctly classified good" and "the item is defective and is misclassified good." Thus,

$$P(CG) = P[(CG \text{ and } G) \text{ or } (CG \text{ and } D)]$$

Since the two possibilities are mutually exclusive, we can start by using the addition rule, formula (4.6):

$$P(CG) = P(CG \text{ and } G) + P(CG \text{ and } D)$$

The condition of an item and its classification by the inspector are not independent. The multiplication rule for dependent events must be used. Therefore,

$$P(CG) = [P(G) \cdot P(CG \mid G)] + [P(D) \cdot P(CG \mid D)]$$

Substituting the known probabilities in Figure 4.7, we get

$$P(CG) = [(0.8)(0.9)] + [(0.2)(0.1)]$$
$$= 0.72 + 0.02$$
$$= \mathbf{0.74}$$

That is, 74% of the items are classified good.

SECTION 4.7 EXERCISES

4.125 a. Describe in your own words what it means for two events to be mutually exclusive.

b. Describe in your own words what it means for two events to be independent.

c. Explain how mutually exclusive and independent are two very different properties.

4.126 a. Describe in your own words why two events cannot be independent if they are already known to be mutually exclusive.

b. Describe in your own words why two events cannot be mutually exclusive if they are already known to be independent.

4.127 $P(G) = 0.5$, $P(H) = 0.4$, and $P(G \text{ and } H) = 0.1$ (see the diagram).

a. Find $P(G \mid H)$.

b. Find $P(H \mid G)$.

c. Find $P(\overline{H})$.

d. Find $P(G \text{ or } H)$.

e. Find $P(G \text{ or } \overline{H})$.

f. Are events G and H mutually exclusive? Explain.

g. Are events G and H independent? Explain.

4.128 $P(R) = 0.5$, $P(S) = 0.3$, and events R and S are independent.

a. Find $P(R \text{ and } S)$.
b. Find $P(R \text{ or } S)$.

c. Find $P(\overline{S})$.
d. Find $P(R \mid S)$.

e. Find $P(\overline{S} \mid R)$.

f. Are events R and S mutually exclusive? Explain.

4.129 $P(M) = 0.3$, $P(N) = 0.4$, and events M and N are mutually exclusive.

a. Find $P(M \text{ and } N)$.
b. Find $P(M \text{ or } N)$.

c. Find $P(M \text{ or } \overline{N})$.
d. Find $P(M \mid N)$.

e. Find $P(M \mid \overline{N})$.

f. Are events M and N independent? Explain.

4.130 Two flower seeds are randomly selected from a package that contains five seeds for red flowers and three seeds for white flowers.

a. What is the probability that both seeds will result in red flowers?

b. What is the probability that one of each color is selected?

c. What is the probability that both seeds are for white flowers?

FYI Draw a tree diagram.

4.131 One thousand employees at the Russell Microprocessor Company were polled about worker satisfaction. One employee is selected at random.

	Male		Female		
	Skilled	Unskilled	Skilled	Unskilled	Total
Satisfied	350	150	25	100	625
Unsatisfied	150	100	75	50	375
Total	500	250	100	150	1000

a. Find the probability that an unskilled worker is satisfied with work.

b. Find the probability that a skilled female employee is satisfied with work.

c. Is satisfaction for female employees independent of their being skilled or unskilled?

4.132 A company that manufactures shoes has three factories. Factory 1 produces 25% of the company's shoes, Factory 2 produces 60%, and Factory 3 produces 15%. One percent of the shoes produced by Factory 1 are mislabeled, 0.5% of those produced by Factory 2 are mislabeled, and 2% of those produced by Factory 3 are mislabeled. If you purchase one pair of shoes manufactured by this company, what is the probability that the shoes are mislabeled?

CHAPTER REVIEW

In Retrospect

You have been studying the basic concepts of probability. These fundamentals need to be mastered before we continue with our study of statistics. Probability is the vehicle of statistics, and we have begun to see how probabilistic events occur. We have explored theoretical and experimental probabilities for the same event. Does the experimental probability turn out to have the same value as the theoretical? Not exactly, but we have seen that over the long run it does have approximately the same value.

Upon completion of this chapter, you should understand the properties of mutual exclusiveness

and independence and be able to apply the multiplication and addition rules to "and" and "or" compound events. You should also be able to calculate conditional probabilities.

In the next three chapters we will look at distributions associated with probabilistic events. This will prepare us for the statistics that follow. We must be able to predict the variability that the sample will show with respect to the population before we can be successful at "inferential statistics," in which we describe the population based on the sample statistics available.

Vocabulary and Key Concepts

addition rule (pp. 230, 239)
all-inclusive events (p. 213)
complementary event (p. 229)
compound event (p. 228)
conditional probability (p. 223)
dependent events (p. 243)
empirical probability (p. 207)
equally likely events (p. 208)
event (p. 208)
experimental probability (p. 207)
general addition rule (p. 230)

general multiplication rule (p. 231)
independence (p. 243)
independent events (pp. 240, 243)
intersection (p. 237)
law of large numbers (pp. 213, 215)
long-term average (p. 215)
multiplication rule (pp. 231, 246)
mutually exclusive events (p. 236)
odds (p. 215)

ordered pair (p. 209)
outcome (p. 208)
probability of an event (p. 207)
relative frequency (p. 207)
sample point (p. 208)
sample space (p. 208)
special addition rule (p. 239)
special multiplication rule (p. 246)
subjective probability (p. 207)
theoretical probability (p. 207)
tree diagram (p. 209)
Venn diagram (p. 211)

Learning Outcomes

✓ Understand and be able to describe the basic concept of probability. — pp. 205–207
✓ Understand and describe a simple event. — EXP. 4.1
✓ Understand and be able to describe the differences between empirical, theoretical, and subjective probabilities. — pp. 208–209, 212
✓ Compute and interpret relative frequencies. — Ex. 4.7, 4.10, 4.11, 4.133
✓ Identify and describe a sample space for an experiment. — pp. 208–209, Ex. 4.15, 4.21, 4.22
✓ Construct tables, tree diagrams, and/or Venn diagrams to aid in computing and interpreting probabilities. — EXP. 4.2, 4.3, 4.4, Ex. 4.23, 4.25
✓ Understand the properties of probability numbers: — pp. 212–213, Ex. 4.29, 4.41
  1. $0 \leq$ Each $P(A) \leq 1$
  2. $\sum_{\text{all outcomes}} P(A) = 1$

✓ Understand, describe, and use the law of large numbers to determine probabilities.	EXP. 4.5, p. 215, Ex. 4.32, 4.171
✓ Understand, compute, and interpret odds of an event.	EXP. 4.6, Ex. 4.43, 4.46, 4.120
✓ Understand that compound events involve the occurrence of more than one event.	Ex. 4.35, 4.51
✓ Construct, describe, compute, and interpret a conditional probability.	EXP. 4.9, Ex. 4.49, 4.53, 4.141
✓ Understand and be able to utilize the complement rule.	EXP. 4.10, Ex. 4.59, 4.60
✓ Compute probabilities of compound events using the addition rule.	EXP. 4.11, Ex. 4.65, EXP. 4.23
✓ Compute probabilities of compound events using the multiplication rule.	EXP. 4.12, Ex. 4.74
✓ Understand, describe, and determine mutually exclusive events.	p. 236, EXP. 4.14, 4.15, Ex. 4.87, 4.93
✓ Compute probabilities of compound events using the addition rule for mutually exclusive events.	EXP. 4.18, Ex. 4.97
✓ Understand, describe, and determine independent events.	p. 243, EXP 4.19, 4.20, Ex. 4.101
✓ Compute probabilities of compound events using the multiplication rule for independent events.	EXP. 4.24, 4.25, Ex. 4.11, 4.115
✓ Recognize and compare the differences between mutually exclusive events and independent events.	pp. 250–252, Ex. 4.127, 4.147, 4.155

Chapter Exercises

Statistics⊙Now™

Go to the StatisticsNow website **http://1pass.thomson.com** to

- Assess your understanding of this chapter
- Check your readiness for an exam by taking the Pre-Test quiz and exploring the resources in the Personalized Learning Plan

Skillbuilder Applet Exercises must be worked using an accompanying applet found on your Student's Suite CD-ROM or at the StatisticsNow website at **http://1pass.thomson.com.**

Datasets can be found on your Student's Suite CD-ROM or at the StatisticsNow website at **http://1pass.thomson.com.**

4.133 The Federal Railroad Administration provided the top five categories of violations for the CSX railroad for the years 1999–2003 in the following table. There were a total of 1897 violations. The information was contained in the December 29, 2004, *Democrat and Chronicle* article titled "Rail cop lacks a 'big stick'."

Category	Number
Track safety	485
Train safety equipment	324
Employee work hours	323
Freight car safety	289
Locomotives	248
All others	228
Total	1897

If one violation is selected at random for review, what is the probability that the violation for CSX is due to the following:

a. Train safety equipment

b. Employee work hours

c. Freight car safety or track safety

What if two violations were selected?

d. Would this be an example of sampling with or without replacement? Explain why.

4.134 [EX04-134] The number of people living in the 50 U.S. states and the District of Columbia in September 2004 is reported by age groups in the following table.

Age Group	Percentage	Number (1000s)
0–17	25%	73,447.7
18–24	10%	28,855.7
25–34	13%	39,892.5
35–49	23%	66,620.3
50+	29%	84,119.8

Source: Sales & Marketing Management Survey of Buying Power, September 2004, for the 50 U.S. states and the District of Columbia

a. Verify the percentages reported in the table.

If one person is picked at random from all the people represented in the table, what is the probability of the following events:

b. "Between 18 and 24." How is this related to the 10% listed on the table?

c. "Older than 17"

d. "Between 18 and 24" and "older than 17"

e. "Between 18 and 24" or "older than 17"

f. "At least 25"

g. "No more than 24"

4.135 One thousand persons screened for a certain disease are given a clinical exam. As a result of the exam, the sample of 1000 persons is classified according to height and disease status.

	Disease Status				
Height	None	Mild	Moderate	Severe	Total
Tall	122	78	139	61	400
Medium	74	51	90	35	250
Short	104	71	121	54	350
Total	300	200	350	150	1000

Use the information in the table to estimate the probability of being medium or short and of having moderate or severe disease status.

4.136 [EX04-136] The Federal Highway Administration periodically tracks the number of licensed vehicle drivers by gender and by age. The following table shows the results of the administration's findings in 2002.

Age Group (years)	Male	Female
19 and under	4,772,152	4,526,106
20–24	8,424,540	8,115,247
25–29	8,727,305	8,372,379
30–34	9,737,052	9,378,312
35–39	10,189,184	9,936,933
40–44	10,614,344	10,584,498
45–49	9,941,582	9,997,864
50–54	8,735,627	8,788,501
55–59	7,148,429	7,141,534
60–64	5,371,340	5,377,859
65–69	4,253,857	4,284,304
70–74	3,647,137	3,788,721
75–79	2,936,969	3,173,171
80–84	1,849,298	2,079,929
85 and older	1,112,647	1,288,812
Total	97,461,463	96,834,170

Source: U.S. Department of Transportation, Federal Highway Administration, *Highway Statistics 2002*

Suppose you encountered a driver of a vehicle at random. Find the probabilities of the following events:

a. The driver is a male and older than age 59.

b. The driver is a female or younger than age 30.

c. The driver is younger than age 25.

d. The driver is a female.

e. The driver is a male between the ages of 35 and 49.

f. The driver is older than age 69.

g. The driver is a female, given driver is between the ages of 25 and 44.

h. The driver is between the ages of 25 and 44, given the driver is female.

4.137 Let's assume there are three traffic lights between your house and a friend's house. As you arrive at each light, it may be red (R) or green (G).

a. List the sample space showing all possible sequences of red and green lights that could occur on a trip from your house to your friend's. (RGG represents red at the first light and green at the other two.)

Assume that each element of the sample space is equally likely to occur.

b. What is the probability that on your next trip to your friend's house you will have to stop for exactly one red light?

c. What is the probability that you will have to stop for at least one red light?

4.138 Assuming that a woman is equally likely to bear a boy as a girl, use a tree diagram to compute the probability that a four-child family consists of one boy and three girls.

4.139 Skillbuilder Applet Exercise simulates generating a family. The "family" will stop having children when they have a boy or three girls, whichever comes first. Assuming that a woman is equally likely to bear a boy or a girl, perform the simulation 24 times. What is the probability that the family will have a boy?

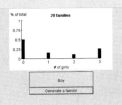

4.140 A coin is flipped three times.

a. Draw a tree diagram that represents all possible outcomes.

b. Identify all branches that represent the event "exactly one head occurred."

c. Find the probability of "exactly one head occurred."

4.141 A recent survey of New York State families asked about their vacation habits. The accompanying two-way table shows the number of families according to where they live (rural, suburban, urban) and the length of their last vacation (1–7 days, 8 days or more).

	Rural	Suburban	Urban	Total
1–7 Days	90	57	52	199
8 Days or More	74	38	21	133
Total	164	95	73	332

If one family is selected at random from these 332 families, what is the probability of the following:

a. They spent 8 days or more on vacation.

b. They were a rural family.

c. They were an urban family and spent 8 days or more on vacation.

d. They were a rural family or spent 1 to 7 days on vacation.

e. They spent 8 days or more on vacation, given they were a suburban family.

f. They were a rural family, given they spent 1 to 7 days on vacation.

4.142 The age and gender demographics for the 2004 Monroe Community College students are outlined in the following table.

	19 and Younger	20–24	25–29	30 and Older
Female	3136	2736	1067	2648
Male	2877	2757	779	1502
Total	6013	5493	1846	4150

If one of these students is selected at random, what is the probability that the student is the following:

a. Male

b. Between 20 to 24 years of age

c. Female and 30 or older

d. Male or 19 or younger

e. Between 25 to 29 years of age, given she was a female student

f. Male, given the student was 20 or older

4.143 This bar graph shows the number of registered automobiles in each of several countries.

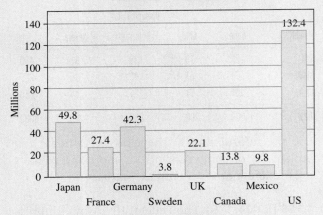

Number of Automobiles

a. Name at least two countries not included in the information.

b. Why are all probabilities resulting from these information conditional probabilities?

Based on the information in the accompanying graph:

c. What percentage of all cars in these countries is registered in the United States?

d. If one registered car was selected at random from all of these cars, what is the probability that it is registered in the United States?

e. Explain the relationship between your answers to parts c and d.

4.144 Probabilities for events A, B, and C are distributed as shown in the figure. Find:

a. $P(A \text{ and } B)$ b. $P(A \text{ or } C)$ c. $P(A \mid C)$

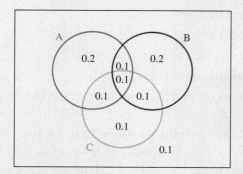

4.145 Show that if event A is a subset of event B, then $P(A \text{ or } B) = P(B)$.

4.146 Explain why these probabilities cannot be legitimate: $P(A) = 0.6$, $P(B) = 0.4$, $P(A \text{ and } B) = 0.7$.

4.147 A shipment of grapefruit arrived containing the following proportions of types: 10% pink seedless, 20% white seedless, 30% pink with seeds, and 40% white with seeds. A grapefruit is selected at random from the shipment. Find the probability of these events:

a. It is seedless.

b. It is white.

c. It is pink and seedless.

d. It is pink or seedless.

e. It is pink, given that it is seedless.

f. It is seedless, given that it is pink.

4.148 A traffic analysis at a busy traffic circle in Washington, DC, showed that 0.8 of the autos using the circle entered from Connecticut Avenue. Of those entering the traffic circle from Connecticut Avenue, 0.7 continued on Connecticut Avenue at the opposite side of the circle. What is the probability that a randomly selected auto observed in the traffic circle entered from Connecticut and will continue on Connecticut?

4.149 Suppose that when a job candidate comes to interview for a job at RJB Enterprises, the probability that he or she will want the job (A) after the interview is 0.68. Also, the probability that RJB wants the candidate (B) is 0.36. The probability $P(A \mid B)$ is 0.88.

a. Find $P(A \text{ and } B)$.   b. Find $P(B \mid A)$.

c. Are events A and B independent? Explain.

d. Are events A and B mutually exclusive? Explain.

e. What would it mean to say A and B are mutually exclusive events in this exercise?

4.150 The probability that thunderstorms are in the vicinity of a particular midwestern airport on an August day is 0.70. When thunderstorms are in the vicinity, the probability that an airplane lands on time is 0.80. Find the probability that thunderstorms are in the vicinity and the plane lands on time.

4.151 Tires salvaged from a train wreck are on sale at the Getrich Tire Company. Of the 15 tires offered in the sale, 5 have suffered internal damage and the remaining 10 are damage free. You randomly selected and purchased two of these tires.

a. What is the probability that the tires you purchased are both damage free?

b. What is the probability that exactly one of the tires you purchased is damage free?

c. What is the probability that at least one of the tires you purchased is damage free?

4.152 According to automobile accident statistics, one out of every six accidents results in an insurance claim of $100 or less in property damage. Three cars insured by an insurance company are involved in different accidents. Consider these two events:

A: The majority of claims exceed $100.

B: Exactly two claims are $100 or less.

a. List the sample points for this experiment.

b. Are the sample points equally likely?

c. Find $P(A)$ and $P(B)$.

d. Are events A and B independent? Justify your answer.

4.153 A testing organization wishes to rate a particular brand of television. Six TVs are selected at random from stock. If nothing is found wrong with any of the six, the brand is judged satisfactory.

a. What is the probability that the brand will be rated satisfactory if 10% of the TVs actually are defective?

b. What is the probability that the brand will be rated satisfactory if 20% of the TVs actually are defective?

c. What is the probability that the brand will be rated satisfactory if 40% of the TVs actually are defective?

4.154 Suppose a certain ophthalmic trait is associated with eye color. Three hundred randomly selected individuals are studied, with results given in the following table.

Trait	Eye Color			
	Blue	Brown	Other	Total
Yes	70	30	20	120
No	20	110	50	180
Total	90	140	70	300

a. What is the probability that a person selected at random has blue eyes?

b. What is the probability that a person selected at random has the trait?

c. Are events A (has blue eyes) and B (has the trait) independent? Justify your answer.

d. How are the two events A (has blue eyes) and C (has brown eyes) related—independent, mutually exclusive, complementary, or all inclusive? Explain why or why not each term applies.

4.155 As listed in *The World Factbook,* 2004, the age structure of the U.S. population is as follows.

	Male	Female
0–14 years	31,122,974	29,713,748
15–64 years	97,756,380	98,183,309
65 years and over	15,078,204	21,172,956

Source: *The World Factbook*, January 2004, http://www.cia.gov/cia/publications/factbook/geos/us.html#Geo

If one U.S. citizen were to be selected at random, what is the probability that the person selected from this population is the following:

a. Female

b. 0 to 14 years old

c. Male and 15 to 64 years old

d. Female or 65 years or older

e. Younger than age 15, knowing the person is female

f. Male, given the person is 15 to 64 years old

The events "person selected is male" and "person selected is female" are not independent events.

g. Is this statement correct? Justify your answer. What is the relationship between female and male in this situation?

4.156 The following table shows the sentiments of 2500 wage-earning employees at the Spruce Company on a proposal to emphasize fringe benefits rather than wage increases during their impending contract discussions.

Employee	Opinion			
	Favor	Neutral	Opposed	Total
Male	800	200	500	1500
Female	400	100	500	1000
Total	1200	300	1000	2500

a. Calculate the probability that an employee selected at random from this group will be opposed.

b. Calculate the probability that an employee selected at random from this group will be female.

c. Calculate the probability that an employee selected at random from this group will be opposed, given that the person is male.

d. Are the events "opposed" and "female" independent? Explain.

4.157 Events R and S are defined on a sample space. If $P(R) = 0.2$ and $P(S) = 0.5$, explain why each of the following statements is either true or false:

a. If R and S are mutually exclusive, then $P(R \text{ or } S) = 0.10$.

b. If R and S are independent, then $P(R \text{ or } S) = 0.6$.

c. If R and S are mutually exclusive, then $P(R \text{ and } S) = 0.7$.

d. If R and S are mutually exclusive, then $P(R \text{ or } S) = 0.6$.

4.158 It is believed that 3% of a clinic's patients have cancer. A particular blood test yields a positive result for 98% of patients with cancer, but it also shows positive for 4% of patients who do not have cancer. One patient is chosen at random from the clinic's patient list and is tested. What is the probability that if the test result is positive, the person actually has cancer?

4.159 Box 1 contains two red balls and three green balls, and Box 2 contains four red balls and one green ball. One ball is randomly selected from Box 1 and placed in Box 2. Then one ball is randomly selected from Box 2. What is the probability that the ball selected from Box 2 is green?

4.160 Salespersons Adams and Jones call on three and four customers, respectively, on a given day. Adams could make 0, 1, 2, or 3 sales, whereas Jones could make 0, 1, 2, 3, or 4 sales. The sample space listing the number of possible sales for each person on a given day is shown in the table. (3, 1 stands for 3 sales by Jones and 1 sale by Adams.)

	Jones				
Adams	0	1	2	3	4
0	0, 0	1, 0	2, 0	3, 0	4, 0
1	0, 1	1, 1	2, 1	3, 1	4, 1
2	0, 2	1, 2	2, 2	3, 2	4, 2
3	0, 3	1, 3	2, 3	3, 3	4, 3

Assume that each sample point is equally likely. Consider these events:

A: At least one of the salespersons made no sales.

B: Together they made exactly three sales.

C: Each made the same number of sales.

D: Adams made exactly one sale.

Find the probabilities by counting sample points:

a. $P(A)$ b. $P(B)$ c. $P(C)$
d. $P(D)$ e. $P(A \text{ and } B)$ f. $P(B \text{ and } C)$
g. $P(A \text{ or } B)$ h. $P(B \text{ or } C)$ i. $P(A \mid B)$
j. $P(B \mid D)$ k. $P(C \mid B)$ l. $P(B \mid \overline{A})$
m. $P(C \mid \overline{A})$ n. $P(A \text{ or } B \text{ or } C)$

Are the following pairs of events mutually exclusive? Explain.

o. A and B p. B and C q. B and D

Are the following pairs of events independent? Explain.

r. A and B s. B and C t. B and D

4.161 Alex, Bill, and Chen each, in turn, toss a balanced coin. The first one to throw a head wins.

a. What are their respective chances of winning if each tosses only one time?

b. What are their respective chances of winning if they continue, given a maximum of two tosses each?

FYI Draw a tree diagram.

4.162 Coin A is loaded in such a way that P (heads) is 0.6. Coin B is a balanced coin. Both coins are tossed. Find:

a. The sample space that represents this experiment; assign a probability measure to each outcome

b. $P(\text{both show heads})$

c. $P(\text{exactly one head shows})$

d. $P(\text{neither coin shows a head})$

e. $P(\text{both show heads} \mid \text{coin A shows a head})$

f. $P(\text{both show heads} \mid \text{coin B shows a head})$

g. $P(\text{heads on coin A} \mid \text{exactly one head shows})$

4.163 Professor French forgets to set his alarm with a probability of 0.3. If he sets the alarm, it rings with a probability of 0.8. If the alarm rings, it will wake him on time to make his first class with a probability of 0.9. If the alarm does not ring, he wakes in time for his first class with a probability of 0.2. What is the probability that Professor French wakes in time to make his first class tomorrow?

4.164 The probability that a certain door is locked is 0.6. The key to the door is one of five unidentified keys hanging on a key rack. You randomly select two keys before approaching the door. What is the probability that you can open the door without returning for another key?

4.165 Your local art museum has planned next year's 52-week calendar by scheduling a mixture of 1-week and 2-week shows that feature the works of 22 painters and 20 sculptors. There is a showing scheduled for every week of the year, and only one artist is featured at a time. There are 42 different shows scheduled for next year. You have randomly selected one week to attend and have been told the probability of it being a 2-week show of sculpture is 3/13.

a. What is the probability that the show you have selected is a painter's showing?

b. What is the probability that the show you have selected is a sculptor's showing?

c. What is the probability that the show you have selected is a 1-week show?

d. What is the probability that the show you have selected is a 2-week show?

4.166 A two-page typed report contains an error on one of the pages. Two proofreaders review the copy. Each has an 80% chance of catching the error. What is the probability that the error will be identified in the following cases:

a. Each reads a different page.

b. They each read both pages.

c. The first proofreader randomly selects a page to read and then the second proofreader randomly selects a page unaware of which page the first selected.

4.167 In sports, championships are often decided by two teams playing in a championship series. Often the fans of the losing team claim they were unlucky and their team is actually the better team. Suppose Team A is the better team, and the probability it will defeat Team B in any one game is 0.6.

a. What is the probability that the better team, Team A, will win the series if it is a one-game series?

b. What is the probability that the better team, Team A, will win the series if it is a best out of three series?

c. What is the probability that the better team, Team A, will win the series if it is a best out of seven series?

d. Suppose the probability that A would beat B in any given game were actually 0.7. Recompute parts a–c.

e. Suppose the probability that A would beat B in any given game were actually 0.9. Recompute parts a–c.

f. What is the relationship between the "best" team winning and the number of games played? The best team winning and the probabilities that each will win?

4.168 A woman and a man (unrelated) each have two children. At least one of the woman's children is a boy, and the man's older child is a boy. Is the probability that the woman has two boys greater than, equal to, or less than the probability that the man has two boys?

a. Demonstrate the truth of your answer using a simple sample to represent each family.

b. Demonstrate the truth of your answer by taking two samples, one from men with two-children families and one from women with two-children families.

c. Demonstrate the truth of your answer using computer simulation. Using the Bernoulli probability function with $p = 0.5$ (let 0 = girl and 1 = boy), generate 500 "families of two children" for the man and the woman. Determine which of the 500 satisfy the condition for each and determine the observed proportion with two boys.

d. Demonstrate the truth of your answer by repeating the computer simulation several times. Repeat the simulation in part c several times.

e. Do the preceding procedures seem to yield the same results? Explain.

4.169 Three balanced coins are tossed simultaneously. Find the probability of obtaining three heads, given that at least one of the coins shows heads.

a. Solve using an equally likely sample space.

b. Solve using the formula for conditional probability.

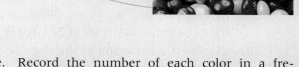

Chapter Project

Sweet Statistics

The chapter project takes us back to Section 4.1, "Sweet Statistics" (p. 205), as a way to assess what we have learned in this chapter. And what a better way to do that than with some candy! We can explore the differences between theoretical and experimental probabilities as well as see the law of large numbers in action—all with M&M's. Now that is "Sweet Statistics." Let's begin.

Putting Chapter 4 to Work

4.170 Let's take a theoretical look at the expected. Mars, Inc., currently uses the following percentages to mix the colors for M&M's Milk Chocolate Candies: 13% brown, 13% red, 14% yellow, 16% green, 20% orange, 24% blue.

a. Construct a bar graph showing the expected (theoretical) proportion of M&M's for each color.

b. Theoretically, what percentage of red M&M's should you expect in a bag of M&M's?

c. If you opened a bag of M&M's right now, would you be surprised to find color percentages different from those given by Mars? Explain.

An empirical (experimental) look at what happened.

d. Obtain a pack of M&M's (at least a 1.69 oz. size—approximately $0.50 in cost).

e. Record the number of each color in a frequency distribution with the headings "Color" and "Frequency."

f. Verify the total number of M&M's with the sum of the Frequency column.

g. Now you may snack! ☺

h. Present the frequency distribution as a relative frequency distribution, using the heading "Empirical Probability."

i. Verify that the sum of the Empirical Probability column is equal to 1. Explain the meaning of this sum.

j. Construct a bar graph showing the relative frequency for each color. Use the same color order as in part a.

k. Empirically, what percentage of red M&M's should you expect in a bag of M&M's?

l. What other statistical displays could you use to present the data from the bag of M&M's? Present them.

m. Compare your empirical (experimental) findings to the expectations (theoretical) expressed in part a.

Your Study

4.171 a. Use a computer (or random-number table) to generate a random sample of 56 M&M's, using the corresponding theoretical probabilities for each color.

b. Form a frequency distribution of the random data.

c. Construct a bar graph showing the relative frequencies for each color. Use the same color order as in part a of Exercise 4.170.

d. Compare your experimental findings with the theoretical expectations.

e. Repeat parts a–d three more times.

f. Describe the variability you observe between the samples.

g. Consolidate your four frequency distributions into one frequency distribution having a frequency total of 224 M&M's.

h. Construct a bar graph of the consolidation showing relative frequencies for each color. Use the same color order as in part a of Exercise 4.170.

i. Compare these experimental findings with the theoretical expectations.

j. Compare the consolidated findings with the four previous individual findings.

k. How does the law of large numbers impact this mini study?

MINITAB and Excel can only generate random numbers. Therefore, it is common practice to use numbers in place of the colors (words). Use the numbers 1, 2, 3, 4, 5, 6 to correspond to brown, red, . . . , blue, respectively.

MINITAB Release 14

a. Input the numbers 1-6 into C1 and their corresponding probabilities in C2; then continue with:

```
Choose:   Calc > Random Data > Discrete
Enter:    Generate: 56 (# of M&M's® in a pack)
          Store in column(s): C3
          Values in: C1 (color numbers)
          Probabilities in: C2 > OK
```

b. To obtain the frequency distribution, continue with:

```
Choose:   Stat > Tables > Cross Tabulation & Chi Square
Enter:    Categorical variables: For rows: C3
Select:   Display: Counts and Column percents > OK
```

c. To construct a bar graph enter the actual colors in C4 and the corresponding probabilities (%) found in step b in C5:

```
Choose:   Graph > Bar Chart > Bar represent: Values from a
          table > One Column of values: Simple > OK
Enter:    Graph variables: C5  Categorical variables: C4
Select:   Labels > Data Labels > Label Type: Use y-value
          labels > OK
Select:   Data View > Data Display: Bars > OK > OK
```

Excel

a. Input the numbers 1-6 in column A and their corresponding probabilities in column B; then continue with:

```
Choose:   Tools > Data Analysis > Random Number Generation
          > OK
Enter:    Number of Variables: 1
          Number of Random Numbers: 56 (# of M&M's® in a
          pack)
          Distribution: Discrete
          Value & Prob. Input Range: (A1:B7 select data
          cells)
Select:   Output range
Enter:    (C1 or select cell) > OK
```

b. The frequency distribution is given with the histogram of the generated data. Use the histogram Excel commands on page 61 using the data in column C and the bin range in column A.

c. Divide the frequencies by 56 to obtain the corresponding probabilities. Enter the actual colors in column D (ex. D13:D18) and the corresponding probabilities in column E (ex. E13:E18). To construct a bar graph, continue with:

```
Choose:   Chart Wizard > Column > 1st picture(usually)
          > Next
Enter:    Data range: (D13:E18 or select cells) > Next
Enter:    Chart and axes titles > Finish (Edit as
          needed)
```

Chapter Practice Test

PART I: Knowing the Definitions

Answer "True" if the statement is always true. If the statement is not always true, replace the words shown in bold with words that make the statement always true.

4.1 The probability of an event is a **whole number.**

4.2 The concepts of probability and relative frequency as related to an event are **very similar.**

4.3 The **sample space** is the theoretical population for probability problems.

4.4 The sample points of a sample space are **equally likely** events.

4.5 The value found for experimental probability will **always be** exactly equal to the theoretical probability assigned to the same event.

4.6 The probabilities of complementary events always **are equal.**

4.7 If two events are mutually exclusive, they are also **independent.**

4.8 If events A and B are **mutually exclusive,** the sum of their probabilities must be exactly 1.

4.9 If the sets of sample points that belong to two different events do not intersect, the events are **independent.**

4.10 A compound event formed with the word "and" requires the use of the **addition rule.**

PART II: Applying the Concepts

4.11 A computer is programmed to generate the eight single-digit integers 1, 2, 3, 4, 5, 6, 7, and 8 with equal frequency. Consider the experiment "the next integer generated" and these events:

  A: odd number, {1, 3, 5, 7}

  B: number greater than 4, {5, 6, 7, 8}

  C: 1 or 2, {1, 2}

  a. Find $P(A)$.   b. Find $P(B)$.
  c. Find $P(C)$.   d. Find $P(\overline{C})$.
  e. Find $P(A \text{ and } B)$. f. Find $P(A \text{ or } B)$.
  g. Find $P(B \text{ and } C)$. h. Find $P(B \text{ or } C)$.
  i. Find $P(A \text{ and } C)$. j. Find $P(A \text{ or } C)$.
  k. Find $P(A \mid B)$.   l. Find $P(B \mid C)$.
  m. Find $P(A \mid C)$.

  n. Are events A and B mutually exclusive? Explain.

  o. Are events B and C mutually exclusive? Explain.

  p. Are events A and C mutually exclusive? Explain.

  q. Are events A and B independent? Explain.

  r. Are events B and C independent? Explain.

  s. Are events A and C independent? Explain.

4.12 Events A and B are mutually exclusive and $P(A) = 0.4$ and $P(B) = 0.3$.

  a. Find $P(A \text{ and } B)$.

  b. Find $P(A \text{ or } B)$.

  c. Find $P(A \mid B)$.

  d. Are events A and B independent? Explain.

4.13 Events E and F have probabilities $P(E) = 0.5$, $P(F) = 0.4$, and $P(E \text{ and } F) = 0.2$.

  a. Find $P(E \text{ or } F)$.

  b. Find $P(E \mid F)$.

  c. Are E and F mutually exclusive? Explain.

  d. Are E and F independent? Explain.

  d. Are G and H independent? Explain.

4.14 Janice wants to become a police officer. She must pass a physical exam and then a written exam. Records show that the probability of passing the physical exam is 0.85 and that once the physical is passed, the probability of passing the written exam is 0.60. What is the probability that Janice passes both exams?

PART III: Understanding the Concepts

4.15 Student A says that independence and mutually exclusive are basically the same thing; namely, both mean neither event has anything to do with the other one. Student B argues that although Student A's statement has some truth in it, Student A has missed the point of these two properties. Student B is correct. Carefully explain why.

4.16 Using complete sentences, describe the following in your own words:

  a. Mutually exclusive events

  b. Independent events

  c. The probability of an event

  d. A conditional probability

Statistics ⌂ Now™ Preparing for an exam? Assess your progress by taking the post-test at **http://1pass.thomson.com**.

VMentor Do you need a live tutor for homework problems? Access vMentor on the StatisticsNow website at **http://1pass.thomson.com** for one-on-one tutoring from a statistics expert.

CHAPTER

5

Probability Distributions (Discrete Variables)

© Digital Vision/Getty Images

5.1 Caffeine Drinking

Are Starbucks and other coffee purveyors taking over the country? It appears that way. One of the most common sights is a person on a cell phone with a cup of coffee. Think about it. How many people fitting that description have you seen today? Maybe you are one of them!

Consider the graphic "Americans like their coffee!" It displays the number of cups or cans of caffeinated beverages adult Americans say they drink daily. The number of daily cups ranges from zero cups to four or more cups. Can you find yourself on the graphic?

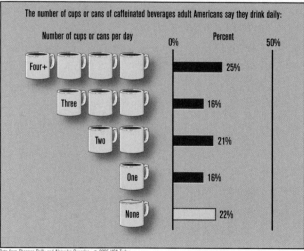

AMERICANS LIKE THEIR COFFEE!

The number of cups or cans of caffeinated beverages adult Americans say they drink daily:

Number of cups or cans per day	Percent
Four+	25%
Three	16%
Two	21%
One	16%
None	22%

Data from Shannon Reilly and Alejandro Gonzalez, © 2005 USA Today.

Who else might be interested in this information besides Starbucks? Apparently, the National Sleep Foundation is. Their mission and goals statement, as noted on their web page, is:

> The National Sleep Foundation (NSF) is an independent nonprofit organization dedicated to improving public health and safety by achieving understanding of sleep and sleep disorders, and by supporting education, sleep-related research, and advocacy.
>
> *Source:* http://www.sleepfoundation.org

Based on the poll of 1506 adults and the common belief that caffeine consumption affects sleep, does it appear that caffeine should be a concern of NSF? Probably not.

As you work through Chapter 5, you will be combining the fundamentals of frequency distributions of Chapter 2 with the fundamentals of probability from Chapter 4. This combination will be referred to as *probability distributions,* which in reality are much like relative frequency distributions. The basic difference between probability distributions and relative frequency distributions is that probability distributions are theoretical probabilities (populations), whereas relative frequency distributions are empirical probabilities (samples). You will be able to further investigate "Americans like their coffee!" in the Chapter Project section with Exercises 5.128 and 5.129 (p. 310).

SECTION 5.1 EXERCISES

5.1 Refer to the graphic "Americans like their coffee!" to answer the following questions:

a. What percentage of adults do not drink any caffeinated beverages?

b. What number of cups or cans of caffeinated beverages has the highest likelihood?

c. What variable could be used to describe all five of the events shown on the graph?

d. Are the events mutually exclusive? Explain.

5.2 Refer to the graphic "Americans like their coffee!" to answer the following questions:

a. What other statistical graph could be used to picture or display this information? Draw it.

b. What other statistical methods could be used to describe this information?

5.2   Random Variables

If each outcome of a probability **experiment** is assigned a numerical value, then as we observe the results of the experiment we are observing the values of a random variable. This numerical value is the *random variable value.*

> **Random variable:** A variable that assumes a unique numerical value for each of the outcomes in the sample space of a probability experiment.

In other words, a random variable is used to denote the outcomes of a probability experiment. The random variable can take on any numerical value that belongs to the set of all possible outcomes of the experiment. (It is called "random" because the value it assumes is the result of a chance, or random, event.) Each event in a probability experiment must also be defined in such a way that only one value of the random variable is assigned to it **(mutually exclusive events)**, and every event must have a value assigned to it **(all-inclusive events).**

The following example demonstrates random variables.

EXAMPLE 5.1   **Random Variables**

a. We toss five coins and observe the "number of heads" visible. The random variable x is the number of heads observed and may take on integer values from 0 to 5.
b. Let the "number of phone calls received" per day by a company be the random variable. Integer values ranging from zero to some very large number are possible values.
c. Let the "length of the cord" on an electrical appliance be a random variable. The random variable is a numerical value between 12 and 72 inches for most appliances.
d. Let the "qualifying speed" for racecars trying to qualify for the Indianapolis 500 be a random variable. Depending on how fast the driver can go, the speeds are approximately 220 and faster and are measured in miles per hour (to the nearest thousandth).

Numerical random variables can be subdivided into two classifications: *discrete random variables* and *continuous random variables.*

FYI Discrete and continuous variables were defined on page 11.

Discrete random variable: A quantitative random variable that can assume a countable number of values.

Continuous random variable: A quantitative random variable that can assume an uncountable number of values.

The random variables "number of heads" and "number of phone calls received" in Example 5.1 parts a and b are discrete. They each represent a count, and therefore there is a countable number of possible values. The random variables "length of the cord" and "qualifying speed" in Example 5.1 parts c and d are continuous. They each represent measurements that can assume any value along an interval, and therefore there is an infinite number of possible values.

SECTION 5.2 EXERCISES

Statistics⬡Now™

Datasets can be found on your Student's Suite CD-ROM or at the StatisticsNow website at **http://1pass.thomson.com**.

5.3 Survey your classmates about the number of siblings they have and the length of the last conversation they had with their mother. Identify the two random variables of interest and list their possible values.

5.4 a. Explain why the variable "number of saved telephone numbers on a person's cell phone" is discrete.

b. Explain why the variable "weight of a statistics textbook" is continuous.

5.5 a. The variables in Exercise 5.3 are either discrete or continuous. Which are they and why?

b. Explain why the variable "number of dinner guests for Thanksgiving dinner" is discrete.

c. Explain why the variable "number of miles to your grandmother's house" is continuous.

5.6 A social worker is involved in a study about family structure. She obtains information regarding the number of children per family for a certain community from the census data. Identify the random variable of interest, determine whether it is discrete or continuous, and list its possible values.

5.7 The staff at *Fortune* recently isolated what they considered to be the 100 best companies in America to work for. Many companies on the list were hiring last year. Those adding the most employees are listed here.

Company	New Jobs	Company	New Jobs
Marriott International	3679	Booz Allen Hamilton	2463
Whole Foods Market	3569		

Source: Fortune, "The 100 Best Companies to Work for 2005"

a.   What is the random variable involved in this study?

b.   Is the random variable discrete or continuous? Explain.

5.8 Above-average warmth extended over the east and southeast on January 13, 2005. The day's forecasted high temperatures in four cities in the affected area were as follows.

City	Temperature	City	Temperature
Burlington, VT	55°F	Durham, NC	74°F
Williamsburg, VA	74°F	Augusta, GA	75°F

a.   What is the random variable involved in this study?

b.   Is the random variable discrete or continuous? Explain.

5.9 An archer shoots arrows at a bull's-eye of a target and measures the distance from the center of the target to the arrow. Identify the random variable of interest, determine whether it is discrete or continuous, and list its possible values.

5.10 A USA Snapshot titled "Are you getting a summer job?" (July 8, 2002) reported that 49% of high school students said, "Getting? I already have one"; 26% said, "Maybe. Depends on my cash situation"; and 25% said, "No! Nothing interferes with my beach time."

a.   What is the variable involved, and what are the possible values?

b.   Why is this variable not a random variable?

5.11 A *USA Today* article titled "Electronic world swallows up kids' time, study finds" (March 10, 2005) presented the following chart depicting the average amount of time 8- to 18-year-olds spend daily on various activities. The Kaiser Family Foundation had conducted the study of 2000 children in grades 3 through 12.

Activity	Average Amount of Time
Watching TV	3 hours, 51 minutes
Listening to music	1 hour, 44 minutes
Using a computer	1 hour, 2 minutes
Playing video games	49 minutes
Reading	43 minutes
Watching movies	25 minutes

a.   What is the random variable involved in this study?

b.   Is the random variable discrete or continuous? Explain.

5.12 [EX05-012] If you could stop time and live forever in good health, what age would you pick? Answers to this question were reported in a USA Snapshot. The average ideal age for each age group is listed in the following table; the average ideal age for all adults was found to be 41. Interestingly, those younger than 30 years want to be older, whereas those older than 30 years want to be younger.

Age Group	18–24	25–29	30–39	40–49	50–64	65+
Ideal Age	27	31	37	40	44	59

Age is used as a variable twice in this application.

a.   The age of the person being interviewed is not the random variable in this situation. Explain why and describe how "age" is used with regard to age group.

b.   What is the random variable involved in this study? Describe its role in this situation.

c.   Is the random variable discrete or continuous? Explain.

5.3 Probability Distributions of a Discrete Random Variable

Consider a coin-tossing experiment where two coins are tossed and no heads, one head, or two heads are observed. If we define the random variable x to be the number of heads observed when two coins are tossed, x can take on the value 0, 1, or 2. The probability of each of these three events can be calculated using techniques from Chapter 4:

$$P(x = 0) = P(\text{0H}) = P(\text{TT}) = \frac{1}{2} \cdot \frac{1}{2} = \frac{1}{4} = 0.25$$

$$P(x = 1) = P(\text{1H}) = P(\text{HT or TH}) = \frac{1}{2} \cdot \frac{1}{2} + \frac{1}{2} \cdot \frac{1}{2} = \frac{1}{2} = 0.50$$

$$P(x = 2) = P(\text{2H}) = P(\text{HH}) = \frac{1}{2} \cdot \frac{1}{2} = \frac{1}{4} = 0.25$$

These probabilities can be listed in any number of ways. One of the most convenient is a table format known as a *probability distribution* (see Table 5.1).

Probability distribution: A distribution of the probabilities associated with each of the values of a random variable. The probability distribution is a theoretical distribution; it is used to represent populations.

TABLE 5.1

Probability Distribution: Tossing Two Coins

x	$P(x)$
0	0.25
1	0.50
2	0.25

FYI Can you see why the name "probability distribution" is used?

In an experiment in which a single die is rolled and the number of dots on the top surface is observed, the random variable is the number observed. The probability distribution for this random variable is shown in Table 5.2.

TABLE 5.2

Probability Distribution: Rolling a Die

x	1	2	3	4	5	6
$P(x)$	$\frac{1}{6}$	$\frac{1}{6}$	$\frac{1}{6}$	$\frac{1}{6}$	$\frac{1}{6}$	$\frac{1}{6}$

Sometimes it is convenient to write a rule that algebraically expresses the probability of an event in terms of the value of the random variable. This expression is typically written in formula form and is called a *probability function*.

Probability function: A rule that assigns probabilities to the values of the random variables.

A probability function can be as simple as a list that pairs the values of a random variable with their probabilities. Tables 5.1 and 5.2 show two such listings. However, a probability function is most often expressed in formula form.

Consider a die that has been modified so that it has one face with one dot, two faces with two dots, and three faces with three dots. Let x be the number of dots observed when this die is rolled. The probability distribution for this experiment is presented in Table 5.3.

Each of the probabilities can be represented by the value of x divided by 6; that is, each $P(x)$ is equal to the value of x divided by 6, where $x = 1, 2$, or 3. Thus,

$$P(x) = \frac{x}{6} \quad \text{for} \quad x = 1, 2, 3$$

is the formula for the probability function of this experiment.

The probability function for the experiment of rolling one ordinary die is

$$P(x) = \frac{1}{6} \quad \text{for} \quad x = 1, 2, 3, 4, 5, 6$$

This particular function is called a **constant function** because the value of $P(x)$ does not change as x changes.

Every probability function must display the two basic properties of probability (see p. 212). These two properties are (1) the probability assigned to each value of the random variable must be between zero and one, inclusive and (2) the sum of the probabilities assigned to all the values of the random variable must equal one—that is,

FYI These properties were presented in Chapter 4.

Property 1 $0 \le$ each $P(x) \le 1$

Property 2 $\sum_{\text{all } x} P(x) = 1$

TABLE 5.3

Probability Distribution: Rolling the Modified Die

x	$P(x)$
1	$\dfrac{1}{6}$
2	$\dfrac{2}{6}$
3	$\dfrac{3}{6}$

EXAMPLE 5.2

Determining a Probability Function

TABLE 5.4

Probability Distribution for $P(x) = \dfrac{x}{10}$ **for** $x = 1, 2, 3, 4$

x	$P(x)$
1	$\dfrac{1}{10} = 0.1$ ✓
2	$\dfrac{2}{10} = 0.2$ ✓
3	$\dfrac{3}{10} = 0.3$ ✓
4	$\dfrac{4}{10} = 0.4$ ✓
	$\dfrac{10}{10} = 1.0$ ⓒⓚ

Is $P(x) = \dfrac{x}{10}$ for $x = 1, 2, 3, 4$ a probability function?

SOLUTION To answer this question we need only test the function in terms of the two basic properties. The probability distribution is shown in Table 5.4.

Property 1 is satisfied because 0.1, 0.2, 0.3, and 0.4 are all numerical values between zero and one. (See the ✓ showing each value was checked.) Property 2 is also satisfied because the sum of all four probabilities is exactly one. (See the ⓒⓚ showing the sum was checked.) Since both properties are satisfied, we can conclude that $P(x) = \dfrac{x}{10}$ for $x = 1, 2, 3, 4$ is a probability function.

What about $P(x = 5)$ (or any value other than $x = 1, 2, 3,$ or 4) for the function $P(x) = \dfrac{x}{10}$ for $x = 1, 2, 3, 4$? $P(x = 5)$ is considered to be zero. That is, the probability function provides a probability of zero for all values of x other than the values specified as part of the domain.

Probability distributions can be presented graphically. Regardless of the specific graphic representation used, the values of the random variable are plotted on the horizontal scale, and the probability associated with each value of the random variable is plotted on the vertical scale. The probability distribution of a discrete random variable could be presented by a set of line segments drawn at the values of x with lengths that represent the probability of each x. Figure 5.1 shows the probability distribution of $P(x) = \dfrac{x}{10}$ for $x = 1, 2, 3, 4$.

Statistics ☁ Now™

Watch a video example at
http://1pass.thomson.com
or on your CD.

FIGURE 5.1   Line Representation: Probability Distribution for

$P(x) = \dfrac{x}{10}$ for $x = 1, 2, 3, 4$

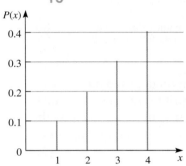

FIGURE 5.2   Histogram: Probability Distribution for

$P(x) = \dfrac{x}{10}$ for $x = 1, 2, 3, 4$

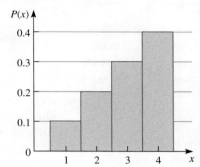

FYI The graph in Figure 5.1 is sometimes called a needle graph.

A regular histogram is used more frequently to present probability distributions. Figure 5.2 shows the probability distribution of Figure 5.1 as a **probability histogram.** The histogram of a probability distribution uses the physical area of each bar to represent its assigned probability. The bar for $x = 2$ is 1 unit wide (from 1.5 to 2.5) and 0.2 unit high. Therefore, its area (length × width) is $(1)(0.2) = 0.2$, the probability assigned to $x = 2$. The areas of the other bars can be determined in similar fashion. This area representation will be an important concept in Chapter 6 when we begin to work with continuous random variables.

TECHNOLOGY INSTRUCTIONS: GENERATE RANDOM DATA

MINITAB (Release 14)   Input the possible values of the random variable into C1 and the corresponding probabilities into C2; then continue with:

Choose: **Calc > Random Data > Discrete**
Enter: Generate: **25** (number wanted)
             Store in column(s): **C3**
             Values (of x) in: **C1**
             Probabilities in: **C2 > OK**

Excel   Input the possible values of the random variable into column A and the corresponding probabilities into column B; then continue with:

Choose: **Tools > Data Analysis >Random Number Generation > OK**
Enter: Number of Variables: **1**
             Number of Random Numbers: **25** (# wanted)
             Distribution: **Discrete**
             Value & Prob. Input Range: **(A2:B5 select data cells, not labels)**
Select: **Output Range**
Enter   **(C1 or select cell)**

APPLIED
EXAMPLE 5.3   Applying for Admission

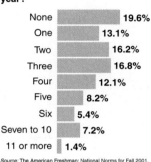

Students hedge their bets

Most students apply to more than one school, making it difficult for colleges to predict how many will actually enroll. Last fall's freshman class was asked:

To how many colleges, other than the one where you enrolled, did you apply for admission this year?

None	19.6%
One	13.1%
Two	16.2%
Three	16.8%
Four	12.1%
Five	8.2%
Six	5.4%
Seven to 10	7.2%
11 or more	1.4%

Source: The American Freshman: National Norms for Fall 2001, survey of 281,064 freshmen entering 421 four-year colleges and universities.

Data from Julie Snider, © 2002 USA Today.

COLLEGES STRIVE TO FILL DORMS

By Mary Beth Marklein, *USA Today*

Colleges and universities will mail their last batch of admission offers in the next few days, but the process is far from over.

Now, students have until May 1 to decide where they'll go this fall. And with lingering concerns about the economy and residual fears about travel and security since Sept. 11, many admissions officials are less able this year to predict how students will respond.

Note the distribution depicted on the bar graph. It has the makings of a discrete probability distribution. The random variable, "number of colleges applied to," is a discrete random variable with values from zero to 11 or more. Each of the values has a corresponding probability, and the sum of the probabilities is equal to 1.

SECTION 5.3 EXERCISES

5.13 Express the tossing of one coin as a probability distribution of x, the number of heads occurring (that is, $x = 1$ if a head occurs and $x = 0$ if a tail occurs).

5.14 a. Express $P(x) = \dfrac{1}{6}$; for $x = 1, 2, 3, 4, 5, 6$, in distribution form.

b. Construct a histogram of the probability distribution $P(x) = \dfrac{1}{6}$; for $x = 1, 2, 3, 4, 5, 6$.

c. Describe the shape of the histogram in part b.

5.15 a. Explain how the various values of x in a probability distribution form a set of mutually exclusive events.

b. Explain how the various values of x in a probability distribution form a set of "all inclusive" events.

5.16 Test the following function to determine whether it is a probability function. If it is not, try to make it into a probability function.

$$R(x) = 0.2 \text{ for } x = 0, 1, 2, 3, 4.$$

a. List the distribution of probabilities.

b. Sketch a histogram.

5.17 Test the following function to determine whether it is a probability function.

$$P(x) = \frac{x^2 + 5}{50}, \text{ for } x = 1, 2, 3, 4$$

a. List the probability distribution.

b. Sketch a histogram.

5.18 Test the following function to determine whether it is a probability function. If it is not, try to make it into a probability function.

$$S(x) = \frac{6 - |x - 7|}{36}, \text{ for } x = 2, 3, 4, 5, 6, 7, \ldots, 11, 12$$

a. List the distribution of probabilities and sketch a histogram.

b. Do you recognize $S(x)$? If so, identify it.

5.19 Census data are often used to obtain probability distributions for various random variables. Census data for families in a particular state with a combined income of $50,000 or more show that 20% of these families have no children, 30% have one child, 40% have two children, and 10% have three children. From this information, construct the probability distribution for x, where x represents the number of children per family for this income group.

5.20 Is "a dog, man's best friend"? One would think so with 60 million pet dogs nationwide. But how many friends are needed? In the USA Snapshot (February 22, 2005), the following statistics were reported.

Number of Pet Dogs	Percentage	Number of Pet Dogs	Percentage
One	66	Four	3
Two	24	Five or more	2
Three	5		

Source: U.S. Pet Ownership & Demographics Sourcebook, Pedigree Food for Dogs

a. Is this a probability distribution? Explain.

b. Draw a relative frequency histogram to depict the results shown in the table.

5.21 How many colleges did you apply to other than the one in which are enrolled? This was just the question asked and illustrated in Applied Example 5.3, "Applying for Admission," on page 276.

a. Using the variable x, number of additional applications for admission completed, express the information on the bar graph "Students hedge their bets" as a discrete probability distribution.

b. Explain how the distribution supports the article's opening statement, ". . . but the process is far from over."

5.22 In February 2004, the Oregon's Medically Needy Program Survey reported the following statistics pertaining to the medically needy population in their state. Due to budget cuts, the Medically Needy Program was eliminated. It had provided Medicare assistance to certain groups who were not eligible for Medicaid but had significant health needs.

Number of Chronic Conditions	Oregon's Medically Needy Population	Number of Chronic Conditions	Oregon's Medically Needy Population
0	2%	3	21%
1	12%	4–5	31%
2	23%		

Source: http://www.ohpr.state.or.us/OHRECwelcome2_files/ReportsandBriefs/MedicallyNeedyFINAL.pdf

a. Is this a probability distribution? Explain.

b. What information could you add to make it into a probability distribution?

c. Draw a relative frequency histogram to depict the results shown in the table plus part b.

5.23 As part of a 2003 Consumer Preferences Report, the following information was gathered. It indicates the percentage, as a relative frequency, of new home customers who desired each outdoor feature as part of their new home. Is this a probability distribution? Explain.

Outdoor Feature	Percentage (Rel. Freq.)	Outdoor Feature	Percentage (Rel. Freq.)
Front porch	0.56	Fencing	0.23
Patio	0.49	Landscape wall	0.14
Deck	0.35		

Source: NAHB Research Center

5.24 A USA Snapshot (March 10, 2005) presented a bar graph depicting business travelers' impression of wait times in airport security lines over the past 12 months. Statistics were derived from a Travel Industry Association of America Business Traveler Survey of 2034 respondents. Is this a probability distribution? Explain.

Impression	Percentage	Impression	Percentage	Impression	Percentage
Worse	49	Same	40	Better	11

5.25 a. Use a computer (or random-number table) to generate a random sample of 25 observations drawn from the discrete probability distribution.

x	1	2	3	4	5
$P(x)$	0.2	0.3	0.3	0.1	0.1

Compare the resulting data with your expectations.

b. Form a relative frequency distribution of the random data.

c. Construct a probability histogram of the given distribution and a relative frequency histogram of the observed data using class midpoints of 1, 2, 3, 4, and 5.

d. Compare the observed data with the theoretical distribution. Describe your conclusions.

e. Repeat parts a–d several times with $n = 25$. Describe the variability you observe between samples.

f. Repeat parts a–d several times with $n = 250$. Describe the variability you see between samples of this much larger size.

MINITAB (Release 14)

a. Input the x values of the random variable into C1 and their corresponding probabilities, $P(x)$, into C2; then continue with the generating random data MINITAB commands on page 275.

b. To obtain the frequency distribution, continue with:

```
Choose:   Stat > Tables > Cross Tabulation
Enter:    Categorical variables: For rows: C3
Select:   Display: Total percents > OK
```

c. To construct the histogram of the generated data in C3, continue with the histogram MINITAB commands on page 61, selecting scale > Y-Scale Type > Percent. (Use Binning followed by midpoint and midpoint positions 1:5/1 if necessary.)

To construct a bar graph of the given distribution, continue with the bar graph MINITAB commands on page 266 using C2 as the Graph variable and C1 as the Categorical variable.

Excel

a. Input the x values of the random variable in column A and their corresponding probabili-

ties, $P(x)$, in column B; then continue with the generating random data Excel commands on page 275 for $n = 25$.

b. & c. The frequency distribution is given with the histogram of the generated data. Use the histogram Excel commands on pages 61–62 using the data in column C and the bin range in column A.

To construct a histogram of the given distribution, continue with:

```
Choose:   Chart Wizard > Column > 1ˢᵗ picture(usually) >
          Next
Enter:    Data range: (A1:B6 or select cells)
Choose:   Series > Remove (Series 1: x column) > Next >
          Titles
Enter:    Chart and axes titles > Finish   (Edit as needed)
```

5.26 a. Use a computer (or random-number table) and generate a random sample of 100 observations drawn from the discrete probability population $P(x) = \dfrac{5 - x}{10}$, for $x = 1, 2, 3, 4$. List the resulting sample. (Use the computer commands in Exercise 5.25; just change the arguments.)

b. Form a relative frequency distribution of the random data.

c. Form a probability distribution of the expected probability distribution. Compare the resulting data with your expectations.

d. Construct a probability histogram of the given distribution and a relative frequency histogram of the observed data using class midpoints of 1, 2, 3, and 4.

e. Compare the observed data with the theoretical distribution. Describe your conclusions.

f. Repeat parts a–d several times with $n = 100$. Describe the variability you observe between samples.

5.4 Mean and Variance of a Discrete Probability Distribution

Recall that in Chapter 2 we calculated several numerical sample statistics (mean, variance, standard deviation, and others) to describe empirical sets of data. Probability distributions may be used to represent theoretical populations, the counterpart to samples. We use **population parameters** (mean, variance, and standard deviation) to describe these probability distributions just as we use **sample statistics** to describe samples.

Notes:
1. $\bar{x}$ is the mean of the sample.
2. s^2 and s are the variance and standard deviation of the sample, respectively.
3. $\bar{x}$, s^2, and s are called *sample statistics*.
4. μ (lowercase Greek letter mu) is the mean of the population.
5. σ^2 (sigma squared) is the variance of the population.
6. σ (lowercase Greek letter sigma) is the standard deviation of the population.
7. μ, σ^2, and σ are called *population parameters.* (A parameter is a constant; μ, σ^2, and σ are typically unknown values in real statistics problems. About the only time they are known is in a textbook problem setting for the purpose of learning and understanding.)

The *mean of the probability distribution* of a discrete random variable, or the *mean of a discrete random variable,* is found in a manner somewhat similar to that used to find the mean of a frequency distribution. The mean of a discrete random variable is often referred to as its *expected value.*

> **Mean of a discrete random variable (expected value):** The mean, μ, of a discrete random variable x is found by multiplying each possible value of x by its own probability and then adding all the products together:
>
> *mean of x:*   *mu = sum of (each x multiplied by its own probability)*
>
> $$\mu = \sum [xP(x)] \tag{5.1}$$

The variance of a discrete random variable is defined in much the same way as the variance of sample data, the mean of the squared deviations from the mean.

> **Variance of a discrete random variable:** The variance, σ^2, of a discrete random variable x is found by multiplying each possible value of the squared deviation from the mean, $(x - \mu)^2$, by its own probability and then adding all the products together:
>
> *variance:*   *sigma squared = sum of (squared deviation times probability)*
>
> $$\sigma^2 = \sum [(x - \mu)^2 P(x)] \tag{5.2}$$

Formula (5.2) is often inconvenient to use; it can be reworked into the following form(s):

variance: *sigma squared = sum of (x^2 times probability) − [sum of (x times probability)]²*

$$\sigma^2 = \sum [x^2 P(x)] - \left\{ \sum [xP(x)] \right\}^2 \tag{5.3a}$$

or

$$\sigma^2 = \sum [x^2 P(x)] - \mu^2 \tag{5.3b}$$

Likewise, standard deviation of a random variable is calculated in the same manner as the standard deviation of sample data.

> **Standard deviation of a discrete random variable:** The positive square root of variance.
>
> $$\text{standard deviation: } \sigma = \sqrt{\sigma^2} \qquad (5.4)$$

EXAMPLE 5.4

Statistics for a Probability Function (Distribution)

Find the mean, variance, and standard deviation of the probability function

$$P(x) = \frac{x}{10} \qquad \text{for } x = 1, 2, 3, 4$$

SOLUTION We will find the mean using formula (5.1), the variance using formula (5.3a), and the standard deviation using formula (5.4). The most convenient way to organize the products and find the totals we need is to expand the probability distribution into an extensions table (see Table 5.5).

TABLE 5.5

Extensions Table: Probability Distribution, $P(x) = \dfrac{x}{10}$ for $x = 1, 2, 3, 4$

x	$P(x)$	$xP(x)$	x^2	$x^2P(x)$
1	$\dfrac{1}{10} = 0.1$ ✓	0.1	1	0.1
2	$\dfrac{2}{10} = 0.2$ ✓	0.4	4	0.8
3	$\dfrac{3}{10} = 0.3$ ✓	0.9	9	2.7
4	$\dfrac{4}{10} = 0.4$ ✓	1.6	16	6.4
	$\dfrac{10}{10} = 1.0$ ⓒⓚ	$\sum[xP(x)] = 3.0$		$\sum[x^2P(x)] = 10.0$

Find the mean of x: The $xP(x)$ column contains each value of x multiplied by its corresponding probability, and the sum at the bottom is the value needed in formula (5.1):

$$\mu = \sum[xP(x)] = \mathbf{3.0}$$

Find the variance of x: The totals at the bottom of the $xP(x)$ and $x^2P(x)$ columns are substituted into formula (5.3a):

$$\sigma^2 = \sum[x^2P(x)] - \{\sum[xP(x)]\}^2$$
$$= 10.0 - \{3.0\}^2 = \mathbf{1.0}$$

Find the standard deviation of x: Use formula (5.4):

$$\sigma = \sqrt{\sigma^2} = \sqrt{1.0} = \mathbf{1.0}$$

Notes:
1. The purpose of the extensions table is to organize the process of finding the three column totals: $\sum[P(x)]$, $\sum[xP(x)]$, and $\sum[x^2P(x)]$.
2. The other columns, x and x^2, should not be totaled; they are not used.
3. $\sum[P(x)]$ will always be 1.0; use this only as a check.
4. $\sum[xP(x)]$ and $\sum[x^2P(x)]$ are used to find the mean and variance of x.

EXAMPLE 5.5

Mean, Variance, and Standard Deviation of a Discrete Random Variable

Statistics⟨△⟩Now™

Watch a video example at
http://1pass.thomson.com
or on your CD.

A coin is tossed three times. Let the "number of heads" that occur in those three tosses be the random variable, x. Find the mean, variance, and standard deviation of x.

SOLUTION There are eight possible outcomes (all equally likely) to this experiment: {HHH, HHT, HTH, HTT, THH, THT, TTH, TTT}. One outcome results in $x = 0$, three in $x = 1$, three in $x = 2$, and one in $x = 3$. Therefore, the probabilities for this random variable are $\dfrac{1}{8}$, $\dfrac{3}{8}$, $\dfrac{3}{8}$, and $\dfrac{1}{8}$. The probability distribution associated with this experiment is shown in Figure 5.3 and in Table 5.6. The necessary extensions and summations for the calculation of the mean, variance, and standard deviation are also shown in Table 5.6.

FIGURE 5.3
Probability Distribution: Number of Heads in Three Tosses of Coin

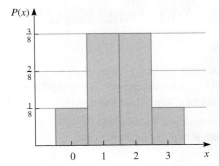

TABLE 5.6

Extensions Table of Probability Distribution of Number of Heads in Three Coin Tosses

x	$P(x)$	$xP(x)$	x^2	$x^2P(x)$
0	$\dfrac{1}{8}$ ✓	$\dfrac{0}{8}$	0	$\dfrac{0}{8}$
1	$\dfrac{3}{8}$ ✓	$\dfrac{3}{8}$	1	$\dfrac{3}{8}$
2	$\dfrac{3}{8}$ ✓	$\dfrac{6}{8}$	4	$\dfrac{12}{8}$
3	$\dfrac{1}{8}$ ✓	$\dfrac{3}{8}$	9	$\dfrac{9}{8}$

$$\sum[P(x)] = \frac{8}{8} = 1.0 \ \text{(ck)} \qquad \sum[xP(x)] = \frac{12}{8} = 1.5 \qquad \sum[x^2P(x)] = \frac{24}{8} = 3.0$$

The mean is found using formula (5.1):

$$\mu = \sum [xP(x)] = \mathbf{1.5}$$

This result, 1.5, is the mean of the theoretical distribution for the random variable "number of heads" observed per set of three coin tosses. It is expected that the mean for many observed values of the random variable would also be approximately equal to this value.

The variance is found using formula (5.3a):

$$\sigma^2 = \sum [x^2 P(x)] - \{\sum [xP(x)]\}^2$$
$$= 3.0 - \{1.5\}^2 = 3.0 - 2.25 = \mathbf{0.75}$$

The standard deviation is found using formula (5.4):

$$\sigma = \sqrt{\sigma^2} = \sqrt{0.75} = 0.866 = \mathbf{0.87}$$

That is, 0.87 is the standard deviation of the theoretical distribution for the random variable "number of heads" observed per set of three coin tosses. It is expected that the standard deviation for many observed values of the random variable would also be approximately equal to this value.

SECTION 5.4 EXERCISES

Statistics⊖Now™

Skillbuilder Applet Exercises must be worked using an accompanying applet found on your Student's Suite CD-ROM or at the StatisticsNow website at **http://1pass.thomson.com**.

Datasets can be found on your Student's Suite CD-ROM or at the StatisticsNow website at **http://1pass.thomson.com**.

5.27 Verify that formulas (5.3a) and (5.3b) are equivalent to formula (5.2).

5.28 a. Form the probability distribution table for $P(x) = \dfrac{x}{6}$, for $x = 1, 2, 3$.

 b. Find the extensions $xP(x)$ and $x^2 P(x)$ for each x.

 c. Find $\sum [xP(x)]$ and $\sum [x^2 P(x)]$.

 d. Find the mean for $P(x) = \dfrac{x}{6}$, for $x = 1, 2, 3$.

 e. Find the variance for $P(x) = \dfrac{x}{6}$, for $x = 1, 2, 3$.

 f. Find the standard deviation for $P(x) = \dfrac{x}{6}$, for $x = 1, 2, 3$.

5.29 If you find the sum of the x and the x^2 columns on the extensions table, exactly what have you found?

5.30 Given the probability function $P(x) = \dfrac{5 - x}{10}$ for $x = 1, 2, 3, 4$, find the mean and standard deviation.

5.31 Given the probability function $R(x) = 0.2$ for $x = 0, 1, 2, 3, 4$, find the mean and standard deviation.

5.32 a. Draw a histogram of the probability distribution for the single-digit random numbers 0, 1, 2, . . . , 9.

 b. Calculate the mean and standard deviation associated with the population of single-digit random numbers.

 c. Represent (1) the location of the mean on the histogram with a vertical line and (2) the magnitude of the standard deviation with a line segment.

d. How much of this probability distribution is within 2 standard deviations of the mean?

5.33 Forecasting hurricanes has become a fine art in Florida. It takes a combination of meteorology and statistics to build forecast models. The following probability distribution was reported in "What Seasonal Hurricane Forecasts Mean to the Residents of Florida" in April 2003.

Number of Florida Hurricanes	Annual Probability	Number of Florida Hurricanes	Annual Probability
0	0.60	3	0.02
1	0.30	4	0.01
2	0.07		

Source: http://garnet.acns.fsu.edu/~jelsner/PDF/Research/Floridafcsts.pdf

a. Build an extensions table of the probability distribution and use it to find the mean and standard deviation of the number of hurricanes experienced annually in Florida.

b. Draw the histogram of the relative frequencies.

5.34 [EX05-034] In a USA Snapshot (June 12, 2002), the U.S. Census Bureau describes the number of vehicles per household in the United States as follows.

Number	Percent	Number (million)	Number	Percent	Number (million)
0	10.3%	10.9	2	38.4%	40.5
1	34.2%	36.1	3 or more	17.1%	18.0

a. Replacing the category "3 or more" with exactly "3," find the mean and standard deviation of the number of vehicles per household in the United States.

b. Explain the effect that replacing the category "3 or more" with "3" had on the mean and standard deviation.

5.35 The number of ships to arrive at a harbor on any given day is a random variable represented by *x*. The probability distribution for *x* is as follows:

x	10	11	12	13	14
P(x)	0.4	0.2	0.2	0.1	0.1

Find the mean and standard deviation of the number of ships that arrive at a harbor on a given day.

5.36 In a USA Today Snapshot (February 22, 2005), the following statistics were reported on pet ownership per household.

Number of Pet Dogs	Percentage	Number of Pet Dogs	Percentage
One	66	Four	3
Two	24	Five or more	2
Three	5		

Source: U.S. Pet Ownership & Demographics Sourcebook, Pedigree Food for Dogs

a. Replacing the category "five or more" with exactly "five," find the mean and standard deviation of the number of pet dogs per household.

b. How do you interpret the mean?

c. Explain the effect that replacing the category "five or more" with "five" had on the mean and standard deviation.

5.37 The random variable *A* has the following probability distribution:

A	1	2	3	4	5
P(A)	0.6	0.1	0.1	0.1	0.1

a. Find the mean and standard deviation of *A*.

b. How much of the probability distribution is within 2 standard deviations of the mean?

c. What is the probability that *A* is between $\mu - 2\sigma$ and $\mu + 2\sigma$?

5.38 The random variable $\bar{x}$ has the following probability distribution:

$\bar{x}$	1	2	3	4	5
$P(\bar{x})$	0.6	0.1	0.1	0.1	0.1

a. Find the mean and standard deviation of $\bar{x}$.

b. What is the probability that $\bar{x}$ is between $\mu - \sigma$ and $\mu + \sigma$?

5.39 Skillbuilder Applet Exercise simulates playing a game where a player has a 0.2 probability of winning $3 and a 0.8 probability of losing $1. Repeat the simulations for several sets of 100 plays using the "Play 25 times" button.

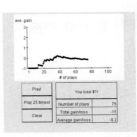

a. What would you estimate for your expected value (average gain or loss) from the results?

b. Using the following probability distribution, calculate the mean.

x	$P(x)$
$3	0.2
−$1	0.8

c. How do your answers to parts a and b compare? Would you consider this a fair game? Why?

5.40 Every Tuesday, Jason's Video has "roll-the-dice" day. A customer may roll two fair dice and rent a second movie for an amount (in cents) determined by the numbers showing on the dice, the larger number first. For example, if the customer rolls a one and a five, a second movie may be rented for $0.51. Let x represent the amount paid for a second movie on roll-the-dice Tuesday.

a. Use the sample space for the rolling of a pair of dice and express the rental cost of the second movie, x, as a probability distribution.

b. What is the expected mean rental cost (mean of x) of the second movie on roll-the-dice Tuesday?

c. What is the standard deviation of x?

d. Using a computer and the probability distribution found in part a, generate a random sample of 30 values for x and determine the total cost of renting the second movie for 30 rentals.

e. Using a computer, obtain an estimate for the probability that the total amount paid for 30 second movies will exceed $15.00 by repeating part d 500 times and using the 500 results.

5.5 The Binomial Probability Distribution

Consider the following probability experiment. Your instructor gives the class a surprise four-question multiple-choice quiz. You have not studied the material, and therefore you decide to answer the four questions by randomly guessing the answers without reading the questions or the answers.

Answer Page to Quiz

Directions: Circle the best answer to each question.

1. a   b   c
2. a   b   c
3 a   b   c
4. a   b   c

FYI That's right, guess!

Circle your answers before continuing.

Before we look at the correct answers to the quiz and find out how you did, let's think about some of the things that might happen if you answer a quiz this way.

1. How many of the four questions are you likely to have answered correctly?

2. How likely are you to have more than half of the answers correct?

3. What is the probability that you selected the correct answers to all four questions?

4. What is the probability that you selected wrong answers for all four questions?

5. If an entire class answers the quiz by guessing, what do you think the class "average" number of correct answers will be?

To find the answers to these questions, let's start with a tree diagram of the sample space, showing all 16 possible ways to answer the four-question quiz. Each of the four questions is answered with the correct answer (C) or with a wrong answer (W). See Figure 5.4.

FIGURE 5.4 Tree Diagram: Possible Answers to a Four-Question Quiz

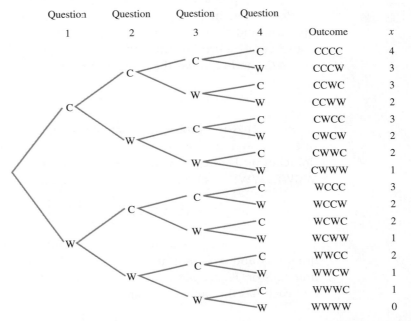

Question 1	Question 2	Question 3	Question 4	Outcome	x
			C	CCCC	4
		C	W	CCCW	3
	C		C	CCWC	3
		W	W	CCWW	2
C			C	CWCC	3
		C	W	CWCW	2
	W		C	CWWC	2
		W	W	CWWW	1
			C	WCCC	3
		C	W	WCCW	2
	C		C	WCWC	2
		W	W	WCWW	1
W			C	WWCC	2
		C	W	WWCW	1
	W		C	WWWC	1
		W	W	WWWW	0

We can convert the information on the tree diagram into a probability distribution. Let x be the "number of correct answers" on one person's quiz when the quiz was taken by randomly guessing. The random variable x may take on any one of the values 0, 1, 2, 3, or 4 for each quiz. Figure 5.4 shows 16 branches representing five different values of x. Notice that the event $x = 4$, "four correct answers," is represented by the top branch of the tree diagram, and the event $x = 0$, "zero correct answers," is shown on the bottom branch. The other events, "one correct answer," "two correct answers," and "three correct answers," are each represented by several branches of the tree. We find that the event $x = 1$ occurs on four different branches, event $x = 2$ occurs on six branches, and event $x = 3$ occurs on four branches.

Each individual question has only one correct answer among the three possible answers, so the probability of selecting the correct answer to an individual question is $\frac{1}{3}$. The probability that a wrong answer is selected on an individual question is $\frac{2}{3}$. The probability of each value of x can be found by calculating the probabilities of all the branches and then combining the probabilities

TABLE 5.7	

Probability Distribution for the Four-Question Quiz

x	$P(x)$
0	0.198
1	0.395
2	0.296
3	0.099
4	0.012
	1.000 (ck)

for branches that have the same x values. The calculations follow, and the resulting probability distribution appears in Table 5.7.

$P(x = 0)$ is the probability that the correct answers are given for zero questions and the wrong answers given for four questions (there is only one branch on Figure 5.4 where all four are wrong—WWWW):

$$P(x = 0) = \frac{2}{3} \times \frac{2}{3} \times \frac{2}{3} \times \frac{2}{3} = \left(\frac{2}{3}\right)^4 = \frac{16}{81} = \textbf{0.198}$$

Note: Answering each individual question is a separate and independent event, thereby allowing us to use formula (4.7), which states that we should multiply the probabilities.

$P(x = 1)$ is the probability that the correct answer is given for exactly one question and wrong answers are given for the other three (there are four branches on Figure 5.4 where this occurs—namely, CWWW, WCWW, WWCW, WWWC—and each has the same probability):

$$P(x = 1) = (4) \times \frac{1}{3} \times \frac{2}{3} \times \frac{2}{3} \times \frac{2}{3} = (4)\left(\frac{1}{3}\right)^1\left(\frac{2}{3}\right)^3 = \textbf{0.395}$$

$P(x = 2)$ is the probability that correct answers are given for exactly two questions and wrong answers are given for the other two (there are six branches on Figure 5.4 where this occurs—CCWW, CWCW, CWWC, WCCW, WCWC, WWCC—and each has the same probability):

$$P(x = 2) = (6) \times \frac{1}{3} \times \frac{1}{3} \times \frac{2}{3} \times \frac{2}{3} = (6)\left(\frac{1}{3}\right)^2\left(\frac{2}{3}\right)^2 = \textbf{0.296}$$

$P(x = 3)$ is the probability that correct answers are given for exactly three questions and a wrong answer is given for the other one (there are four branches on Figure 5.4 where this occurs—CCCW, CCWC, CWCC, WCCC—and each has the same probability):

$$P(x = 3) = (4) \times \frac{1}{3} \times \frac{1}{3} \times \frac{1}{3} \times \frac{2}{3} = (4)\left(\frac{1}{3}\right)^3\left(\frac{2}{3}\right)^1 = \textbf{0.099}$$

$P(x = 4)$ is the probability that correct answers are given for all four questions (there is only one branch on Figure 5.4 where all four are correct—CCCC):

$$P(x = 4) = \frac{1}{3} \times \frac{1}{3} \times \frac{1}{3} \times \frac{1}{3} = \left(\frac{1}{3}\right)^4 = \frac{1}{81} = \textbf{0.012}$$

Now we can answer the five questions that were asked about the four-question quiz (pp. 284–285).

Answer 1: The most likely occurrence would be to get one answer correct; it has a probability of 0.395. Zero, one, or two correct answers are expected to result approximately 89% of the time (0.198 + 0.395 + 0.296 = 0.889).

Answer 2: To have more than half correct is represented by x = 3 or 4; their total probability is 0.099 + 0.012 = 0.111. (You will pass this quiz only 11% of the time by random guessing.)

Answer 3: P(all four correct) = $P(x = 4)$ = 0.012. (All correct occurs only 1% of the time.)

Answer 4: P(all four wrong) $= P(x = 0) = 0.198$. (That's almost 20% of the time.)

Answer 5: The class average is expected to be $\frac{1}{3}$ of 4, or 1.33 correct answers.

The correct answers to the quiz are b, c, b, a. How many correct answers did you have? Which branch of the tree in Figure 5.4 represents your quiz results? You might ask several people to answer this same quiz by guessing the answers. Then construct an observed relative frequency distribution and compare it with the distribution shown in Table 5.7.

Many experiments are composed of repeated trials whose outcomes can be classified into one of two categories: **success** or **failure.** Examples of such experiments are coin tosses, right/wrong quiz answers, and other more practical experiments such as determining whether a product did or did not do its prescribed job and whether a candidate gets elected or not. There are experiments in which the trials have many outcomes that, under the right conditions, may fit this general description of being classified in one of two categories. For example, when we roll a single die, we usually consider six possible outcomes. However, if we are interested only in knowing whether a "one" shows or not, there are really only two outcomes: the "one" shows or "something else" shows. The experiments just described are called *binomial probability experiments.*

Binomial probability experiment: An experiment that is made up of repeated trials that possess the following properties:

1. There are n repeated identical independent trials.
2. Each trial has two possible outcomes (success or failure).
3. P(success) $= p$, P(failure) $= q$, and $p + q = 1$.
4. The binomial random variable x is the count of the number of successful trials that occur; x may take on any integer value from zero to n.

Notes:

1. Properties 1 and 2 describe the two basic characteristics of any binomial experiment.
2. *Independent trials* means that the result of one trial does not affect the probability of success on any other trial in the experiment. In other words, the probability of success remains constant throughout the entire experiment.
3. Property 3 gives the algebraic notation for each trial.
4. Property 4 concerns the algebraic notation for the complete experiment.
5. It is of utmost importance that both x and p be associated with "success."

The four-question quiz qualifies as a binomial experiment made up of four trials when all four of the answers are obtained by random guessing.

Property 1: A <u>trial</u> is the <u>answering of one question</u>, and it is repeated <u>$n = 4$</u> times. The trials are <u>independent</u> because the probability of a correct answer on any one question is not affected by the answers on other questions.

Property 2: The two possible outcomes on each trial are <u>success = C</u>, correct answer, and <u>failure = W</u>, wrong answer.

Property 3: For each trial (each question): $\underline{p = P(\text{correct}) = \frac{1}{3}}$ and $\underline{q = P(\text{wrong}) = \frac{2}{3}}$. $[p + q = 1 \; \text{ck}]$

Property 4: For the total experiment (the quiz): $\underline{x = \text{number of correct answers}}$ and can be any integer value from zero to $n = 4$.

EXAMPLE 5.6

Demonstrating the Properties of a Binomial Probability Experiment

Consider the experiment of rolling a die 12 times and observing a "one" or "something else." At the end of all 12 rolls, the number of "ones" is reported. The random variable x is the number of times that a "one" is observed in the $n = 12$ trials. Since "one" is the outcome of concern, it is considered "success"; therefore, $p = P(\text{one}) = \frac{1}{6}$ and $q = P(\text{not one}) = \frac{5}{6}$. This experiment is binomial.

EXAMPLE 5.7

Demonstrating the Properties of a Binomial Probability Experiment

If you were an inspector on a production line in a plant where television sets are manufactured, you would be concerned with identifying the number of defective television sets. You probably would define "success" as the occurrence of a defective television. This is not what we normally think of as success, but if we count "defective" sets in a binomial experiment, we must define "success" as a "defective." The random variable x indicates the number of defective sets found per lot of n sets; $p = P(\text{television is defective})$ and $q = P(\text{television is good})$.

The key to working with any probability experiment is its probability distribution. All binomial probability experiments have the same properties, and therefore the same organization scheme can be used to represent all of them. The *binomial probability function* allows us to find the probability for each possible value of x.

> **Binomial probability function:** For a binomial experiment, let p represent the probability of a "success" and q represent the probability of a "failure" on a single trial. Then $P(x)$, the probability that there will be exactly x successes in n trials, is
>
> $$P(x) = \binom{n}{x}(p^x)(q^{n-x}) \qquad \text{for } x = 0, 1, 2, \ldots, n \qquad (5.5)$$

When you look at the probability function, you notice that it is the product of three basic factors:

1. The number of ways that exactly x successes can occur in n trials, $\binom{n}{x}$

2. The probability of exactly x successes, p^x

3. The probability that failure will occur on the remaining $(n - x)$ trials, q^{n-x}

The number of ways that exactly x successes can occur in a set of n trials is represented by the symbol $\binom{n}{x}$, which must always be a positive integer. This term is called the **binomial coefficient** and is found by using the formula

$$\binom{n}{x} = \frac{n!}{x!(n-x)!} \tag{5.6}$$

Notes:

1. $n!$ ("*n factorial*") is an abbreviation for the product of the sequence of integers starting with n and ending with one. For example, $3! = 3 \cdot 2 \cdot 1 = 6$ and $5! = 5 \cdot 4 \cdot 3 \cdot 2 \cdot 1 = 120$. There is one special case, $0!$, that is defined to be 1. For more information about **factorial notation,** see the *Student Solutions Manual.*

2. The values for $n!$ and $\binom{n}{x}$ can be readily found using most scientific calculators.

3. The binomial coefficient $\binom{n}{x}$ is equivalent to the number of combinations $_nC_x$, the symbol most likely on your calculator.

4. See the *Student Solutions Manual* for general information on the binomial coefficient.

Let's reconsider Example 5.5 (pp. 281–282): A coin is tossed three times and we observe the number of heads that occur in the three tosses. This is a binomial experiment because it displays all the properties of a binomial experiment:

1. There are $\underline{n = 3}$ repeated <u>independent</u> trials (each coin toss is a separate trial, and the outcome of any one trial has no effect on the probability of another).

2. Each trial (each toss of the coin) results in one of two possible outcomes: success = <u>heads</u> (what we are counting) or failure = <u>tails</u>.

3. The probability of success is $p = P(\text{H}) = \underline{0.5}$, and the probability of failure is $q = P(\text{T}) = \underline{0.5}$. [$p + q = 0.5 + 0.5 = 1$ ⓒⓚ]

4. The random variable x is the <u>number of heads </u>that occur in the three trials. x will assume exactly one of the values <u>0, 1, 2, or 3</u> when the experiment is complete.

The binomial probability function for the tossing of three coins is

$$P(x) = \binom{n}{x}(p^x)\,(q^{n-x}) = \binom{3}{x}(0.5)^x(0.5)^{n-x} \qquad \text{for } x = 0, 1, 2, 3$$

Let's find the probability of $x = 1$ using the preceding binomial probability function:

FYI In Table 5.6 (p. 281), $P(1) = \frac{3}{8}$. Here, $P(1) = 0.375$ and $\frac{3}{8} = 0.375$.

$$P(x = 1) = \binom{3}{1}(0.5)^1(0.5)^2 = 3(0.5)(0.25) = \mathbf{0.375}$$

Note that this is the same value found in Example 5.5 (p. 281).

EXAMPLE 5.8   **Determining a Binomial Experiment and Its Probabilities**

Statistics⬡Now™

Watch a video example at
http://1pass.thomson.com
or on your CD.

Consider an experiment that calls for drawing five cards, one at a time with replacement, from a well-shuffled deck of playing cards. The drawn card is identified as a spade or not a spade, it is returned to the deck, the deck is reshuffled, and so on. The random variable x is the number of spades observed in the set of five drawings. Is this a binomial experiment? Let's identify the four properties.

1. There are <u>five repeated drawings</u>; $\underline{n = 5}$. These individual trials are <u>independent</u> because the drawn card is returned to the deck and the deck is reshuffled before the next drawing.

2. Each drawing is a trial, and each drawing has two outcomes: <u>spade</u> or <u>not spade</u>.

3. $p = P(\underline{\text{spade}}) = \dfrac{13}{52}$ and $q = P(\underline{\text{not spade}}) = \dfrac{39}{52}$. $[p + q = 1 \text{ ⓒₖ}]$

4. x is the <u>number of spades</u> recorded upon completion of the five trials; the possible values are <u>0, 1, 2, . . . , 5</u>.

The binomial probability function is

$$P(x) = \binom{5}{x}\left(\frac{13}{52}\right)^x\left(\frac{39}{52}\right)^{5-x} = \binom{5}{x}\left(\frac{1}{4}\right)^x\left(\frac{3}{4}\right)^{5-x} = \binom{5}{x}(0.25)^x(0.75)^{5-x}$$

$$\text{for } x = 0, 1, \dots, 5$$

$$P(0) = \binom{5}{0}(0.25)^0(0.75)^5 = (1)(1)(0.2373) = \mathbf{0.2373}$$

$$P(1) = \binom{5}{1}(0.25)^1(0.75)^4 = (5)(0.25)(0.3164) = \mathbf{0.3955}$$

$$P(2) = \binom{5}{2}(0.25)^2(0.75)^3 = (10)(0.0625)(0.421875) = \mathbf{0.2637}$$

$$P(3) = \binom{5}{3}(0.25)^3(0.75)^2 = (10)(0.015625)(0.5625) = \mathbf{0.0879}$$

The two remaining probabilities are left for you to compute in Exercise 5.54.

The preceding distribution of probabilities indicates that the single most likely value of x is one, the event of observing exactly one spade in a hand of five cards. What is the least likely number of spades that would be observed?

FYI Answer: five

EXAMPLE 5.9   **Binomial Probability of "Bad Eggs"**

Statistics⬡Now™

Watch a video example at
http://1pass.thomson.com
or on your CD.

The manager of Steve's Food Market guarantees that none of his cartons of a dozen eggs will contain more than one bad egg. If a carton contains more than one bad egg, he will replace the whole dozen and allow the customer to keep the original eggs. If the probability that an individual egg is bad is 0.05, what is the probability that the manager will have to replace a given carton of eggs?

SOLUTION   At first glance, the manager's situation appears to fit the properties of a binomial experiment if we let x be the number of bad eggs found in a carton of a dozen eggs, let $p = P(\text{bad}) = 0.05$, and let the inspection of each egg be a trial that results in finding a "bad" or "not bad" egg. There will be $n = 12$ trials to account for the 12 eggs in a carton. However, trials of a binomial experiment must be independent; therefore, we will assume the quality of one egg in a carton is independent of the quality of any of the other eggs. (This may be a big assumption! But with this assumption, we will be able to use the binomial probability distribution as our model.) Now, based on this assumption, we will be able to find/estimate the probability that the manager will have to make good on his guarantee. The probability function associated with this experiment will be:

$$P(x) = \binom{12}{x}(0.05)^x(0.95)^{12-x} \quad \text{for } x = 0, 1, 2, \dots, 12$$

The probability that the manager will replace a dozen eggs is the probability that $x = 2, 3, 4, \dots, 12$. Recall that $\sum P(x) = 1$; that is,

$$\boldsymbol{P(0) + P(1)} + P(2) + \cdots + P(12) = 1$$
$$P(\text{replacement}) = P(2) + P(3) + \cdots + P(12) = 1 - \boldsymbol{[P(0) + P(1)]}$$

It is easier to find the probability of replacement by finding $P(x = 0)$ and $P(x = 1)$ and subtracting their total from 1 than by finding all of the other probabilities. We have

$$P(x) = \binom{12}{x}(0.05)^x(0.95)^{12-x}$$

$$P(0) = \binom{12}{0}(0.05)^0(0.95)^{12} = \boldsymbol{0.540}$$

$$P(1) = \binom{12}{1}(0.05)^1(0.95)^{11} = \boldsymbol{0.341}$$

$$P(\text{replacement}) = 1 - (0.540 + 0.341) = \boldsymbol{0.119}$$

If $p = 0.05$ is correct, then the manager will be busy replacing cartons of eggs. If he replaces 11.9% of all the cartons of eggs he sells, he certainly will be giving away a substantial proportion of his eggs. This suggests that he should adjust his guarantee (or market better eggs). For example, if he were to replace a carton of eggs only when four or more were found to be bad, he would expect to replace only 3 out of 1000 cartons [$1.0 - (0.540 + 0.341 + 0.099 + 0.017)$], or 0.3% of the cartons sold. Notice that the manager will be able to control his "risk" (probability of replacement) if he adjusts the value of the random variable stated in his guarantee.

Note: The value of many binomial probabilities for values of $n \le 15$ and common values of p are found in Table 2 of Appendix B. In this example, we have $n = 12$ and $p = 0.05$, and we want the probabilities for $x = 0$ and 1. We need to locate the section of Table 2 where $n = 12$, find the column headed $p = 0.05$, and read the numbers across from $x = 0$ and $x = 1$. We find .540 and .341, as shown in Table 5.8. (Look up these values in Table 2 in Appendix B.)

TABLE 5.8

Excerpt of Table 2 in Appendix B, Binomial Probabilities

		p													
n	*x*	0.01	0.05	0.10	0.20	0.30	0.40	0.50	0.60	0.70	0.80	0.90	0.95	0.99	*x*
	⋮														
12	0	.886	.540	.282	.069	.014	.002	0+	0+	0+	0+	0+	0+	0+	0
	1	.107	.341	.377	.206	.071	.017	.003	0+	0+	0+	0+	0+	0+	1
	2	.006	.099	.230	.283	.168	.064	.016	.002	0+	0+	0+	0+	0+	2
	3	0+	.017	.085	.236	.240	.142	.054	.012	.001	0+	0+	0+	0+	3
	4	0+	.002	.021	.133	.231	.213	.121	.042	.008	.001	0+	0+	0+	4
	⋮														

Note: A convenient notation to identify the binomial probability distribution for a binomial experiment with $n = 12$ and $p = 0.05$ is $B(12, 0.05)$. $B(12, 0.05)$, read "*binomial distribution for* $n = 12$ and $p = 0.05$," represents the entire distribution or "block" of probabilities shown in purple in Table 5.8. When used in combination with the $P(x)$ notation, $P(x = 1 \mid B(12, 0.05))$ indicates the probability of $x = 1$ from this distribution, or 0.341 as shown on Table 5.8.

TECHNOLOGY INSTRUCTIONS: BINOMIAL AND CUMULATIVE BINOMIAL PROBABILITIES

MINITAB (Release 14)

For binomial probabilities, input *x* values into C1; then continue with:

```
Choose:     Calc > Probability Distributions > Binomial
Select:     Probability *
Enter:      Number of trials: n
            Probability of success: p
Select:     Input column
Enter:      C1
            Optional Storage: C2 (not necessary) > OK
Or
Select:     Input constant
Enter:      One single x value > OK
```

*For cumulative binomial probabilities, repeat the preceding commands but replace the probability selection with:

```
Select:     Cumulative Probability
```

Excel

For binomial probabilities, input *x* values into column A and activate the column B cell across from the first *x* value; then continue with:

```
Choose:     Insert function, fx > Statistical > BINOMDIST > OK
Enter:      Number_s: (A1:A4 or select 'x value' cells)
            Trials: n
            Probability_s: p
            Cumulative: false* (gives individual probabilities) > OK
```

Drag: **Bottom right corner of probability value cell in column B down to give other probabilities**

*For cumulative binomial probabilities, repeat the preceding commands but replace the false cumulative with:

Cumulative: **true (gives cumulative probabilities) > OK**

TI-83/84 Plus   To obtain a complete list of probabilities for a particular n and p, continue with:

Choose: **2nd > DISTR > 0:binompdf(**
Enter: **n, p)**

Use the right arrow key to scroll through the probabilities.
To scroll through a vertical list in L1:

Choose: **STO→ > L1 > ENTER**
                 STAT > EDIT > 1:Edit

To obtain individual probabilities for a particular n, p, and x, continue with:

Choose: **2nd > DISTR > 0:binompdf(**
Enter: **n, p, x)**

To obtain cumulative probabilities for $x = 0$ to $x = n$ for a particular n and p, continue with:

Choose: **2nd > DISTR > A:binomcdf(**
Enter: **n, p)***   (see previous for scrolling through probabilities)

*To obtain individual cumulative probabilities for a particular n, p, and x, repeat the preceding commands but replace the enter with:

Enter: **n, p, x)**

APPLIED EXAMPLE 5.10   **Living with the Law**

WHAT IS AN AFFIRMATIVE ACTION PROGRAM (AAP)?

As a condition of doing business with the federal government, federal contractors meeting certain contract and employee population levels agree to prepare, in accordance with federal regulations at 41 CFR 60-1, 60-2, etc., an Affirmative Action Program (AAP). A contractor's AAP is a combination of numerical reports, commitments of action and description of policies. A quick overview of an AAP based on the federal regulations (41 CFR 60-2.10), is as follows:

AAPs must be developed for
- Minorities and women (41 CFR 60-1 and 60-2)
- Special disabled veterans, Vietnam era veterans, and other covered veterans (41 CFR 60-250)
- Individuals with disabilities (41 CFR 60-741)

Source: http://eeosource.peopleclick.com/maintopic/default.asp?MainTopicID=1

The AAP regulations do not endorse the use of a specific test for determining whether the percentage of minorities or women is less than would be reasonably expected. However, several tests are commonly used. One of the tests is called the *exact binomial test* as defined here.

EXACT BINOMIAL TEST

The variables used are:
T = The total number of employees in the job group
M = The number of females or minorities in the job group
A = The availability percentage of females or minorities for the job group
This test involves the calculation of a probability, denoted as P, and the comparison of that probability to 0.05. If P is less than or equal to 0.05, the percentage

of minorities or women is considered to be "less than would be reasonably expected." The formula for calculating P is as follows:
1. Calculate probability, Q, the cumulative binomial probability for the binomial probability distribution with $n = T$, $x = M$, and $p = A/100$.
2. If Q is less than or equal to 0.5, then $P = 2Q$; otherwise, $P = Q$.

For example, if T = 50 employees and M = 2 females, A = 6% female availability.

Using a computer, the value Q is found: $Q = 0.41625$. Since Q is less than 0.5, $P = 2Q = 0.8325$. P, 0.8325, is greater than 0.05, so the percentage of women is found to be "**not** less than would be reasonably expected."

SECTION 5.5 EXERCISES

Statistics⬡Now™

Skillbuilder Applet Exercises must be worked using an accompanying applet found on your Student's Suite CD-ROM or at the StatisticsNow website at **http://1pass.thomson.com**.

5.41 Consider the four-question multiple-choice quiz presented at the beginning of this section (pp. 284–287).

a. Explain why the four questions represent four independent trials.

b. Explain why the number 4 is multiplied into the $P(x = 1)$.

c. In Answer 5 on page 287, where did $\frac{1}{3}$ and 4 come from? Why multiply them to find an expected average?

5.42 Identify the properties that make flipping a coin 50 times and keeping track of heads a binomial experiment.

5.43 State a very practical reason why the defective item in an industrial situation might be defined to be the "success" in a binomial experiment.

5.44 What does it mean for the trials to be independent in a binomial experiment?

5.45 Evaluate each of the following.

a. 4!   b. 7!   c. 0!   d. $\dfrac{6!}{2!}$

e. $\dfrac{5!}{2!3!}$ f. $\dfrac{6!}{4!(6-4)!}$ g. $(0.3)^4$ h. $\dbinom{7}{3}$

i. $\dbinom{5}{2}$ j. $\dbinom{3}{0}$ k. $\dbinom{4}{1}(0.2)^1(0.8)^3$

l. $\dbinom{5}{0}(0.3)^0(0.7)^5$

5.46 Show that each of the following is true for any values of n and k. Use two specific sets of values for n and k to show that each is true.

a. $\dbinom{n}{0} = 1$ and $\dbinom{n}{n} = 1$

b. $\dbinom{n}{1} = n$ and $\dbinom{n}{n-1} = n$ c. $\dbinom{n}{k} = \dbinom{n}{n-k}$

5.47 A carton containing 100 T-shirts is inspected. Each T-shirt is rated "first quality" or "irregular." After all 100 T-shirts have been inspected, the number of irregulars is reported as a random variable. Explain why x is a binomial random variable.

5.48 A die is rolled 20 times, and the number of "fives" that occurred is reported as being the random variable. Explain why x is a binomial random variable.

5.49 Four cards are selected, one at a time, from a standard deck of 52 playing cards. Let x represent the number of aces drawn in the set of four cards.

a. If this experiment is completed without replacement, explain why x is not a binomial random variable.

b. If this experiment is completed with replacement, explain why x is a binomial random variable.

5.50 The employees at a General Motors assembly plant are polled as they leave work. Each is asked, "What brand of automobile are you riding home in?" The random variable to be reported is the number of each brand mentioned. Is x a binomial random variable? Justify your answer.

5.51 Consider a binomial experiment made up of three trials with outcomes of success, S, and failure, F, where $P(S) = p$ and $P(F) = q$.

a. Complete the accompanying tree diagram. Label all branches completely.

	Trial	Trial	Trial	(b)	(c)
	1	2	3	Probability	x
			p S	p^3	3
		p S	q		
			F	p^2q	2
	S	q			
p		F $\cdots$		$\vdots$	$\vdots$
Start					
q					
F $\cdots$					

b. In column (b) of the tree diagram, express the probability of each outcome represented by the branches as a product of powers of p and q.

c. Let x be the random variable, the number of successes observed. In column (c), identify the value of x for each branch of the tree diagram.

d. Notice that all the products in column (b) are made up of three factors and that the value of the random variable is the same as the exponent for the number p.

e. Write the equation for the binomial probability function for this situation.

5.52 Draw a tree diagram picturing a binomial experiment of four trials.

5.53 Use the probability function for three coin tosses as demonstrated on page 289 and verify the probabilities for $x = 0$, 2, and 3.

5.54 a. Calculate $P(4)$ and $P(5)$ for Example 5.8 on page 290.

b. Verify that the six probabilities $P(0)$, $P(1)$, $P(2)$, . . . , $P(5)$ form a probability distribution.

5.55 Skillbuilder Applet Exercise demonstrates calculating a binomial probability along with a visual interpretation. Suppose that you buy 20 plants from a nursery and the nursery claims that 95% of its plants survive when planted. Inputting $n = 20$ and $p = 0.95$, compute the following: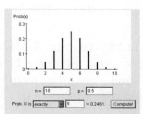

a. The probability that all 20 will survive

b. The probability that at most 16 survive

c. The probability that at least 18 survive

5.56 Skillbuilder Applet Exercise demonstrates calculating a binomial probability along with a visual interpretation. Suppose that you are in a class of 30 stu-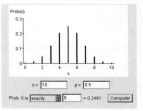

dents and it is assumed that approximately 11% of the population is left-handed. Inputting $n = 30$ and $p = 0.11$, compute the following:

a. The probability that exactly five students are left-handed

b. The probability that at most four students are left-handed

c. The probability that at least six students are left-handed

5.57 If x is a binomial random variable, calculate the probability of x for each case.

a. $n = 4$, $x = 1$, $p = 0.3$ b. $n = 3$, $x = 2$, $p = 0.8$

c. $n = 2$, $x = 0$, $p = \dfrac{1}{4}$ d. $n = 5$, $x = 2$, $p = \dfrac{1}{3}$

e. $n = 4$, $x = 2$, $p = 0.5$ f. $n = 3$, $x = 3$, $p = \dfrac{1}{6}$

5.58 If x is a binomial random variable, use Table 2 in Appendix B to determine the probability of x for each of the following:

a. $n = 10$, $x = 8$, $p = 0.3$ b. $n = 8$, $x = 7$, $p = 0.95$

c. $n = 15$, $x = 3$, $p = 0.05$ d. $n = 12$, $x = 12$, $p = 0.99$

e. $n = 9$, $x = 0$, $p = 0.5$ f. $n = 6$, $x = 1$, $p = 0.01$

g. Explain the meaning of the symbol 0+ that appears in Table 2.

5.59 Test the following function to determine whether or not it is a binomial probability function. List the distribution of probabilities and sketch a histogram.

$$T(x) = \binom{5}{x}\left(\frac{1}{2}\right)^x\left(\frac{1}{2}\right)^{5-x} \qquad \text{for } x = 0, 1, 2, 3, 4, 5$$

5.60 Let x be a random variable with the following probability distribution:

x	0	1	2	3
$P(x)$	0.4	0.3	0.2	0.1

Does x have a binomial distribution? Justify your answer.

5.61 Of all the trees planted by a landscaping firm, 90% survive. What is the probability that 8 or more of the 10 trees they just planted will survive? (Find the answer by using a table.)

5.62 According to the Federal Trade Commission, more than half of the 358,603 consumer fraud complaints in 2004 were Internet related—53% to be more accurate. In a group of 20 people who have filed a fraud complaint, what is the probability that exactly half are Internet related?

5.63 First graders who misbehave in school may be more likely to be regular smokers as young adults according to a new study presented in the July 2004 issue of the *American Journal of Epidemiology*. After following a group of U.S. first graders for 15 years, it was found that among those kids who had tried smoking and misbehaved, 66% were daily smokers.

Source: http://preventdisease.com/news/articles/troubled_kids_more_likely_smokers.shtml

a. What is the probability that exactly two of the next three randomly selected young adults who misbehaved in early grades and have tried smoking are daily smokers?

b. What is the probability that exactly 8 of the next 12 randomly selected young adults who misbehaved in early grades and have tried smoking are daily smokers?

c. What is the probability that exactly 20 of the next 30 randomly selected young adults who misbehaved in early grades and have tried smoking are daily smokers?

5.64 The Pew Internet & American Life Project produces reports that explore the impact of the Internet on many facets of our daily life, whether at home, school, or the office. In its January 2005 Tracking Survey, it found that more than 80% of 18- to 29-year-olds use the Internet. Consider a randomly selected group of ten 18- to 29-year-olds.

Source: http://www.pewinternet.org/trends/User_Demo_03.07.05.htm

a. What is the probability that exactly eight use the Internet?

b. What is the probability that at least five use the Internet?

5.65 In the biathlon event of the Olympic Games, a participant skis cross-country and on four intermittent occasions stops at a rifle range and shoots a set of five shots. If the center of the target is hit, no penalty points are assessed. If a particular man has a history of hitting the center of the target with 90% of his shots, what is the probability of the following:

a. He will hit the center of the target with all five of his next set of five shots.

b. He will hit the center of the target with at least four of his next set of five shots. (Assume independence.)

5.66 The survival rate during a risky operation for patients with no other hope of survival is 80%. What is the probability that exactly four of the next five patients survive this operation?

5.67 Of the parts produced by a particular machine, 0.5% are defective. If a random sample of 10 parts produced by this machine contains 2 or more defective parts, the machine is shut down for repairs. Find the probability that the machine will be shut down for repairs based on this sampling plan.

5.68 A January 2005 survey of bikers, commissioned by the Progressive Group of Insurance Companies, showed that 40% of bikers have body art, such as tattoos and piercings. A group of 10 bikers are in the process of buying motorcycle insurance.

Source: http://www.syracuse.com/business/poststandard/index.ssf?/base/business-1/

a. What is the probability that none of the 10 have any body art?

b. What is the probability that exactly 3 have any body art?

c. What is the probability that at least 4 have any body art?

d. What is the probability that no more than 2 have any body art?

5.69 If boys and girls are equally likely to be born, what is the probability that in a randomly selected family of six children, there will be at least one boy? (Find the answer using a formula.)

5.70 One-fourth of a certain breed of rabbits are born with long hair. What is the probability that in a litter of six rabbits, exactly three will have long hair? (Find the answer by using a formula.)

5.71 St. Louis Cardinal baseball player Albert Pujols has a 3-year batting average (ratio of hits to at bats) of 0.334 for the 2002–2004 seasons. Suppose Pujols has five official times at bat during his next game. Assuming no extenuating circumstances and that the binomial model will produce reasonable approximations, what is the probability of the following:

a. Pujols gets less than two hits.

b. Pujols gets more than three hits.

c. Pujols goes five-for-five (all hits).

5.72 As a quality-control inspector for toy trucks, you have observed that 3% of the time, the wooden wheels are bored off-center. If six wooden wheels are used on each toy truck produced, what is the probability that a randomly selected toy truck has no off-center wheels?

5.73 Consider the manager of Steve's Food Market as illustrated in Example 5.9. What would be the manager's "risk" if he bought "better" eggs, say with $P(\text{bad}) = 0.01$ using the "more than one" guarantee?

5.74 According to the USA Snapshot "Knowing drug addicts," 45% of Americans know somebody who became addicted to a drug other than alcohol. Assuming this to be true, what is the probability of the following:

a. Exactly 3 people of a random sample of 5 know someone who became addicted. Calculate the value.

b. Exactly 7 people of a random sample of 15 know someone who became addicted. Estimate using Table 2 in Appendix B.

c. At least 7 people of a random sample of 15 know someone who became addicted. Estimate using Table 2.

d. No more than 7 people of a random sample of 15 know someone who became addicted. Estimate using Table 2.

5.75 Of all mortgage foreclosures in the United States, 48% are caused by disability. People who are injured or ill cannot work—they then lose their jobs and thus their incomes. With no income, they cannot make their mortgage payment and the bank forecloses.

Source: http://www.ricedelman.com 06.11.02

Given that 20 mortgage foreclosures are audited by a large lending institution, find the probability of the following:

a. Five or fewer of the foreclosures are due to a disability.

b. At least three foreclosures are due to a disability.

5.76 a. Use a calculator or computer to find the probability that $x = 3$ in a binomial experiment where $n = 12$ and $p = 0.30$: $P(x = 3 \mid B(12, 0.30))$. (See Note about this notation on p. 292.)

    b. Use Table 5.8 to verify the answer in part a.

5.77 Use a computer to find the probabilities for all possible x values for a binomial experiment where $n = 30$ and $p = 0.35$.

MINITAB (Release 14)

```
Choose:  Calc > Make Patterned Data > Simple Set of
         Numbers
Enter:   Store patterned data in: C1
         From first value: 0
         To last value: 30
         In steps of: 1 > OK
```

Continue with the binomial probability MINITAB commands on page 292, using $n = 30$, $p = 0.35$, and C2 for optional storage.

Excel

```
Enter:   0,1,2, . . . , 30 into column A
```

Continue with the binomial probability Excel commands on pages 292–293, using $n = 30$ and $p = 0.35$.

TI-83/84 Plus

```
Use the binomial probability TI-83/84 commands on
pages 293, using n = 30 and p = 0.35.
```

5.78 Use a computer to find the cumulative probabilities for all possible x values for a binomial experiment where $n = 45$ and $p = 0.125$.

a. Explain why there are so many 1.000s listed.

b. Explain what is represented by each number listed.

MINITAB (Release 14)

```
Choose:  Calc > Make Patterned Data > Simple Set of
         Numbers . . .
Enter:   Store patterned data in: C1
         From first value: 0
         To last value: 45
         In steps of: 1 > OK
```

Continue with the <u>cumulative</u> binomial probability MINITAB commands on page 292, using $n = 45$, $p = 0.125$, and C2 as optional storage.

Excel

```
Enter:   0,1,2, . . . , 45 into column A
```

Continue with the <u>cumulative</u> binomial probability Excel commands on pages 292–293, using $n = 45$ and $p = 0.125$.

TI-83/84 Plus

Use the <u>cumulative</u> binomial probability TI-83/84 commands on page 293, using $n = 45$ and $p = 0.125$.

5.79 The increase in Internet usage over the past few years has been phenomenal, as demonstrated by the February 2004 report from Pew Internet & American Life Project. The survey of Americans 65 or older (about 8 million adults) reported that 22% have access to the Internet. By contrast, 58% of 50- to 64-year-olds, 75% of 30- to 49-year-olds, and 77% of 18- to 29-year-olds currently go online.

Source: http://www.suddenlysenior.com/ maturemarketstatsmore.html

Suppose that 50 adults in each age group are to be interviewed.

a. What is the probability that "have Internet access" is the response for 10 to 20 adults in the 65 or older group?

b. What is the probability that "have Internet access" is the response for 30 to 40 adults in the 50- to 64-year-old group?

c. What is the probability that "have Internet access" is the response for 30 to 40 adults in the 30- to 49-year-old group?

d. What is the probability that "have Internet access" is the response for 30 to 40 adults in the 18- to 29-year-old group?

e. Why are the answers for parts a and d nearly the same? Explain.

f. What effect did the various values of p have on the probabilities? Explain.

5.80 Where does all that Halloween candy go? The October 2004 issue of *Readers' Digest* quoted that "90% of parents admit taking Halloween candy from their children's trick-or-treat bags." The source of information was the National Confectioners Association. Suppose that 25 parents are interviewed. What is the probability that 20 or more took Halloween candy from their children's trick-or-treat bags?

5.81 Harris Interactive conducted a survey for Tylenol PM asking U.S. drivers what they do if they are driving while drowsy. The results were reported in a USA Snapshot on January 18, 2005, with 40% of the respondents saying they "open the windows" to fight off sleep. Suppose that 35 U.S. drivers are interviewed. What is the probability that between 10 and 20 of the drivers will say they "open the windows" to fight off sleep?

5.82 a. When using the exact binomial test (Applied Example 5.10, pp. 293–294), what is the interpretation of the situation when the calculated value of P is less than or equal to 0.05?

b. When using the exact binomial test, what is the interpretation of the situation when the calculated value of P is larger than 0.05?

c. An employer has 15 employees in a very specialized job group, of which 2 are minorities. Based on 2000 census information, the proportion of minorities available for this type of work is 5%. Using the binomial test, is the percentage of minorities what would be reasonably expected?

d. For this same employer and the same job group, there are three female employees. The percentage of female availability for this position is 50%. Does it appear that the percentage of females is what would be reasonably expected?

5.83 Extended to overtime in a game 7 on the road in the 2002 NBA play-offs, the two-time defending champion Los Angeles Lakers did what they do best—thrived when the pressure was at its highest. Both of the Lakers' star players had their chance at the foul line late in overtime.

a. With 1:27 minutes left in overtime and the game tied at 106–106, Shaquille (Shaq) O'Neal was at the line for two free-throw attempts. He has a history of making 0.555 of his free-throw attempts, and during this game, prior to these two shots, he had made 9 of his 13 attempts. Justify the statement, "The law of averages is working against him."

b. With 0:06 seconds left in overtime and the game score standing at 110–106, Kobe Bryant was at the line for two free-throw shots. He has a history of making 0.829 of his free throws, and during this game, prior to these two shots, he had made 6 of his 8 attempts. Justify the statement, "The law of averages is working for him."

Both players made both shots, and the series with the Sacramento Kings was over.

5.84 If the binomial $(q + p)$ is squared, the result is $(q + p)^2 = q^2 + 2qp + p^2$. For the binomial experiment with $n = 2$, the probability of no successes in two trials is q^2 (the first term in the expansion), the probability of one success in two trials is $2qp$ (the second term in the expansion), and the probability of two successes in two trials is p^2 (the third term in the expansion). Find $(q + p)^3$ and compare its terms to the binomial probabilities for $n = 3$ trials.

Mean and Standard Deviation of the Binomial Distribution

The mean and standard deviation of a theoretical binomial probability distribution can be found by using these two formulas:

Mean of Binomial Distribution

$$\mu = np \tag{5.7}$$

and

Standard Deviation of Binomial Distribution

$$\sigma = \sqrt{npq} \tag{5.8}$$

The formula for the mean, μ, seems appropriate: the number of trials multiplied by the probability of "success." [Recall that the mean number of correct answers on the binomial quiz (Answer 5, p. 287) was expected to be $\frac{1}{3}$ of 4, $4(\frac{1}{3})$, or np.] The formula for the standard deviation, σ, is not as easily understood. Thus, at this point it is appropriate to look at an example, which demonstrates that formulas (5.7) and (5.8) yield the same results as formulas (5.1), (5.3a), and (5.4).

In Example 5.5 (pp. 281–282), x is the number of heads in three coin tosses, $n = 3$, and $p = \frac{1}{2} = 0.5$. Using formula (5.7), we find the mean of x to be

$$\mu = np = (3)(0.5) = \mathbf{1.5}$$

Using formula (5.8), we find the standard deviation of x to be

$$\sigma = \sqrt{npq} = \sqrt{(3)(0.5)(0.5)} = \sqrt{0.75} = 0.866 = \mathbf{0.87}$$

Now look back at the solution for Example 5.5 (p. 282). Note that the results are the same, regardless of which formula you use. However, formulas (5.7) and (5.8) are much easier to use when x is a binomial random variable.

EXAMPLE 5.11

Calculating the Mean and Standard Deviation of a Binomial Distribution

Find the mean and standard deviation of the binomial distribution when $n = 20$ and $p = \frac{1}{5}$ (or 0.2, in decimal form). Recall that the "binomial distribution where $n = 20$ and $p = 0.2$" has the probability function

$$P(x) = \binom{20}{x}(0.2)^x(0.8)^{20-x} \qquad \text{for } x = 0, 1, 2, \ldots, 20$$

and a corresponding distribution with 21 x values and 21 probabilities, as shown in the distribution chart, Table 5.9, and on the histogram in Figure 5.5.

TABLE 5.9	
Binomial Distribution:	
$n = 20$, $p = 0.2$	
x	$P(x)$
0	0.012
1	0.058
2	0.137
3	0.205
4	0.218
5	0.175
6	0.109
7	0.055
8	0.022
9	0.007
10	0.002
11	0+
12	0+
13	0+
·	·
:	:
20	0+

FIGURE 5.5   Histogram of Binomial Distribution $B(20, 0.2)$

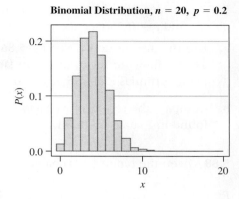

Let's find the mean and the standard deviation of this distribution of x using formulas (5.7) and (5.8):

$$\mu = np = (20)(0.2) = \mathbf{4.0}$$
$$\sigma = \sqrt{npq} = \sqrt{(20)(0.2)(0.8)} = \sqrt{3.2} = \mathbf{1.79}$$

FIGURE 5.6   Histogram of Binomial Distribution $B(20, 0.2)$

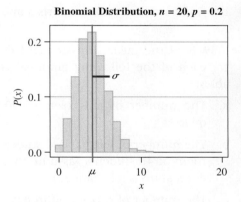

Figure 5.6 shows the mean, $\mu = 4$, (shown by the location of the vertical blue line along the x-axis) relative to the variable x. This 4.0 is the mean value expected for x, the number of successes in each random sample of size 20 drawn from a population with $p = 0.2$. Figure 5.6 also shows the size of the standard deviation, $\sigma = 1.79$ (as shown by the length of the horizontal red line segment). It is the expected standard deviation for the values of the random variable x that occur in samples of size 20 drawn from this same population.

SECTION 5.6 EXERCISES

5.85 Find the mean and standard deviation for the binomial random variable x with $n = 30$ and $p = 0.6$, using formulas (5.7) and (5.8).

5.86 Consider the binomial distribution where $n = 11$ and $p = 0.05$.

a.   Find the mean and standard deviation using formulas (5.7) and (5.8).

b.   Using Table 2 in Appendix B, list the probability distribution and draw a histogram.

c.   Locate μ and σ on the histogram.

5.87 Consider the binomial distribution where $n = 11$ and $p = 0.05$ (see Exercise 5.86).

a. Use the distribution [Exercise 5.86(b) or Table 2] and find the mean and standard deviation using formulas (5.1), (5.3a), and (5.4).

b. Compare the results of part a with the answers found in Exercise 5.86(a).

5.88 Given the binomial probability function

$$P(x) = \binom{5}{x} \cdot \left(\frac{1}{2}\right)^x \cdot \left(\frac{1}{2}\right)^{5-x} \quad \text{for } x = 0, 1, 2, 3, 4, 5$$

a. Calculate the mean and standard deviation of the random variable by using formulas (5.1), (5.3a), and (5.4).

b. Calculate the mean and standard deviation using formulas (5.7) and (5.8).

c. Compare the results of parts a and b.

5.89 Find the mean and standard deviation of x for each of the following binomial random variables:

a. The number of tails seen in 50 tosses of a quarter

b. The number of left-handed students in a classroom of 40 students (assume that 11% of the population is left-handed)

c. The number of cars found to have unsafe tires among the 400 cars stopped at a roadblock for inspection (assume that 6% of all cars have one or more unsafe tires)

d. The number of melon seeds that germinate when a package of 50 seeds is planted (the package states that the probability of germination is 0.88)

5.90 Find the mean and standard deviation for each of the following binomial random variables in parts a–c:

a. The number of sixes seen in 50 rolls of a die

b. The number of defective televisions in a shipment of 125 (The manufacturer claimed that 98% of the sets were operative.)

c. The number of operative televisions in a shipment of 125 (the manufacturer claimed that 98% of the sets were operative)

d. How are parts b and c related? Explain.

5.91 According to United Mileage Plus Visa (November 22, 2004), 41% of passengers say they "put on the earphones" to avoid being bothered by their seat mates during flights. To show how important, or not important, the earphones are to people, consider the variable x to be the number of people in a sample of 12 who say they "put on the earphones" to avoid their seat mates. Assume the 41% is true for the whole population of airline travelers and that a random sample is selected.

a. Is x a binomial random variable? Justify your answer.

b. Find probability that $x = 4$ or 5?

c. Find the mean and standard deviation of x.

d. Draw a histogram of the distribution of x: label it completely, highlight the area representing $x = 4$ and $x = 5$, draw a vertical line at the value of the mean, and mark the location of x that is 1 standard deviation larger than the mean.

5.92 According to the American Payroll Association (September 10, 2004), when asked, "Do you support the use of biometric technology (fingerprints or hand recognition) to record employee time and attendance?" the majority of workers say they wouldn't mind. However, 28% object to it, believing it is an invasion of personal privacy. To better understand the extent of this objection, consider the variable x to be the number of people in a sample of 25 who say they "object." Assume the 28% is true for the whole population of workers and that a random sample is selected.

a. Is x a binomial random variable? Justify your answer.

b. Find the probability that x has a value from 4 to 10.

c. Find the mean and standard deviation of x.

d. Draw a histogram of the distribution of x: label it completely, highlight the area representing $x = 4$ to $x = 10$, draw a vertical line at the value of the mean, and mark the location of *x* that is 1 standard deviation larger than the mean.

5.93 A binomial random variable has a mean equal to 200 and a standard deviation of 10. Find the values of *n* and *p*.

5.94 The probability of success on a single trial of a binomial experiment is known to be $\frac{1}{4}$. The random variable *x*, number of successes, has a mean value of 80. Find the number of trials involved in this experiment and the standard deviation of *x*.

5.95 A binomial random variable *x* is based on 15 trials with the probability of success equal to 0.4. Find the probability that this variable will take on a value more than 2 standard deviations above the mean.

5.96 A binomial random variable *x* is based on 15 trials with the probability of success equal to 0.2. Find the probability that this variable will take on a value more than 2 standard deviations from the mean.

5.97 Imprints Galore buys T-shirts (to be imprinted with an item of the customer's choice) from a manufacturer who guarantees that the shirts have been inspected and that no more than 1% are imperfect in any way. The shirts arrive in boxes of 12. Let *x* be the number of imperfect shirts found in any one box.

a. List the probability distribution and draw the histogram of *x*.

b. What is the probability that any one box has no imperfect shirts?

c. What is the probability that any one box has no more than one imperfect shirt?

d. Find the mean and standard deviation of *x*.

e. What proportion of the distribution is between $\mu - \sigma$ and $\mu + \sigma$?

f. What proportion of the distribution is between $\mu - 2\sigma$ and $\mu + 2\sigma$?

g. How does this information relate to the empirical rule and Chebyshev's theorem? Explain.

h. Use a computer to simulate Imprints Galore's buying 200 boxes of shirts and observing *x*, the number of imperfect shirts per box of 12. Describe how the information from the simulation compares to what was expected (answers to parts a–g describe the expected results).

i. Repeat part h several times. Describe how these results compare with those of parts a–g and with part h.

MINITAB (Release 14)

a.
```
Choose:  Calc > Make Patterned Data > Simple Set of
         Numbers . . .
Enter:   Store patterned data in: C1
         From first value: -1 (see note)
         To last value: 12
         In steps of: 1    > OK
```

c. Continue with the binomial probability MINITAB commands on page 292, using *n* = 12, *p* = 0.01, and C2 for optional storage.

```
Choose:  Graph > Scatterplot > Simple > OK
Enter:   Y variables: C2 X variables: C1
Select:  Data view: Data Display: Area > OK
```
The graph is not a histogram, but can be converted to a histogram by double clicking on 'area' of graph.
```
Select:  Options   Select: Step > OK > OK
```

h. Continue with the cumulative binomial probability MINITAB commands on page 292, using *n* = 12, *p* = 0.01, and C3 for optional storage.

```
Choose:  Calc > Random Data > Binomial
Enter:   Generate: 200 rows of data
         Store in column C4
         Number of trials: 12
         Probability: .01    > OK
Choose:  Stat > Tables > Cross Tabulation
Enter:   Categorical variables: For rows: C4
Select:  Display: Total percents    >OK
Choose:  Calc > Column Statistics
Select:  Statistic: Mean
Enter:   Input variable: C4    > OK
Choose:  Calc > Column Statistics
Select:  Statistic: Standard deviation
Enter:   Input variable: C4    > OK
```

Continue with the histogram MINITAB commands on page 61, using the data in C4 and selecting the options: percent and midpoint with intervals 0:12/1.

Note: The binomial variable *x* cannot take on the value −1. The use of −1 (the next would-be class midpoint to left of 0) allows MINITAB to draw the histogram of a probability distribution. Without −1, PLOT will draw only half of the bar representing *x* = 0.

Excel

a.

Enter: `0,1,2, . . . ,12` into column A

Continue with the binomial probability Excel commands on pages 292–293, using $n = 12$ and $p = 0.01$. Activate columns A and B; then continue with:

Choose: **Chart Wizard > Column > 1ˢᵗ picture**(usually) >
           Next > Series

Choose: **Series 1 > Remove**

Enter: Category (x)axis labels: **(A1:A13 or select 'x
           value' cells)**

Choose: **Next > Finish**

Click on: **Anywhere clear on the chart**
           **—use handles to size so x values fall under
           corresponding bars**

Continue with the <u>cumulative</u> binomial probability Excel commands on pages 292–293, using $n = 12$, $p = 0.01$, and column C for the activated cell.

h.

Choose: **Tools > Data analysis > Random Number Genera-
           tion > OK**

Enter: Number of Variables: **1**
           Number of Random Numbers: **200**
           Distribution: **Binomial**
           p Value = **0.01**
           Number of Trials = **12**

Select: Output Options: **Output Range**

Enter   **(D1 or select cell) > OK**
           Activate the E1 cell, then:

Choose: **Insert function, fₓ > Statistical > AVERAGE > OK**

Enter: Number 1: **D1:D200 > OK**

Activate the E2 cell, then:

Choose: **Insert function, fₓ > Statistical > STDEV > OK**

Enter: Number 1: **D1:D200 > OK**

Continue with the histogram Excel commands on pages 61–62, using the data in column D and the bin range in column A.

TI-83/84 Plus

a.

Choose: **STAT > EDIT > 1:Edit**

Enter: L1: **0,1,2,3,4,5,6,7,8,9,10,11,12**

Choose: **2ⁿᵈ QUIT > 2ⁿᵈ DISTR > 0:binompdf(**

Enter: **12, 0.01) > ENTER**

Choose: **STO→ > L2 > ENTER**

Choose: **2ⁿᵈ > STAT PLOT > 1:Plot1**

Screen capture 5.5A

Choose: **WINDOW**

Enter: **0, 13, 1, −.1, .9, .1, 1**

Choose: **TRACE > > >**

c.

Choose: **2ⁿᵈ > DISTR > A:binomcdf(**

Enter: **12, 0.01)**

Choose: **STO→ > L3 > ENTER**
           STAT > EDIT > 1:Edit

h.

Choose: **MATH > PRB > 7:randBin(**

Enter: **12, .01, 200)** (takes a while to process)

Choose: **STO→ > L4 > ENTER**

Choose: **2ⁿᵈ LIST > Math > 3:mean(**

Enter: **L4**

Choose: **2ⁿᵈ LIST > Math > 7:StdDev(**

Enter: **L4**

Continue with the histogram TI-83/84 commands on pages 62–63, using the data in column L4 and adjusting the window after the initial look using ZoomStat.

5.98 Did you ever buy an incandescent light bulb that failed (either burned out or did not work) the first time you turned the switch on? When you put a new bulb into a light fixture, you expect it to light, and most of the time it does. Consider 8-packs of 60-watt bulbs and let x be the number of bulbs in a pack that "fail" the first time they are used. If 0.02 of all bulbs of this type fail on their first use and each 8-pack is considered a random sample,

a.   List the probability distribution and draw the histogram of x.

b.   What is the probability that any one 8-pack has no bulbs that fail on first use?

c.   What is the probability that any one 8-pack has no more than one bulb that fails on first use?

d.   Find the mean and standard deviation of x.

e.   What proportion of the distribution is between $\mu - \sigma$ and $\mu + \sigma$?

f.   What proportion of the distribution is between $\mu - 2\sigma$ and $\mu + 2\sigma$?

g.   How does this information relate to the empirical rule and Chebyshev's theorem? Explain.

h.   Use a computer to simulate testing 100 8-packs of bulbs and observing x, the number of failures per 8-pack. Describe how the information from the simulation compares with what was expected (answers to parts a–g describe the expected results).

i.   Repeat part h several times. Describe how these results compare with those of parts a–g and with part h.

CHAPTER REVIEW

In Retrospect

In this chapter we combined concepts of probability with some of the ideas presented in Chapter 2. We now are able to deal with distributions of probability values and find means, standard deviations, and other statistics.

In Chapter 4 we explored the concepts of mutually exclusive events and independent events. We used the addition and multiplication rules on several occasions in this chapter, but very little was said about mutual exclusiveness or independence. Recall that every time we add probabilities, as we did in each of the probability distributions, we need to know that the associated events are mutually exclusive. If you look back over the chapter, you will notice that the random variable actually requires events to be mutually exclusive; therefore, no real emphasis was placed on this concept. The same basic comment can be made in reference to the multiplication of probabilities and the concept of independent events. Throughout this chapter, probabilities were multiplied and occasionally independence was mentioned. Independence, of course, is necessary to be able to multiply probabilities.

Now, after completing Chapter 5, if we were to take a close look at some of the sets of data in Chapter 2, we would see that several problems could be reorganized to form probability distributions. Here are some examples: (1) Let x be the number of credit hours for which a student is registered this semester, paired with the percentage of the entire student body reported for each value of x. (2) Let x be the number of correct passageways through which an experimental laboratory animal passes before taking a wrong one, paired with the probability of each x value. (3) Let x be the number of college applications made other than the one where you enrolled (Applied Example 5.3), paired with the probability of each x value. The list of examples is endless.

We are now ready to extend these concepts to continuous random variables in Chapter 6.

Vocabulary List and Key Concepts

binomial coefficient (p. 289)
binomial experiment (p. 287)
binomial probability function (p. 288)
binomial random variable (p. 287)
constant function (p. 274)
continuous random variable (p. 271)
discrete random variable (p. 271)

experiment (p. 270)
failure (p. 287)
independent trials (p. 287)
mean of discrete random variable (p. 279)
mutually exclusive events (p. 270)
population parameter (p. 278)
probability distribution (p. 273)
probability function (p. 273)

probability histogram (p. 275)
random variable (p. 270)
sample statistic (p. 278)
standard deviation of discrete random variable (p. 280)
success (p. 287)
trial (p. 287)
variance of discrete random variable (p. 279)

Learning Outcomes

✓ Understand that a random variable is a numerical quantity whose value depends on the conditions and probabilities associated with an experiment.   pp. 270–271, EXP. 5.1

✓ Understand the difference between a discrete and a continuous random variable.   Ex. 5.4, 5.5, 5.9

✓ Be able to construct a discrete probability distribution based on an experiment or given function.

pp. 273–274, Ex. 5.13, 5.19

✓ Understand the terms *mutually exclusive* and *all inclusive* as they apply to the variables for probability distributions.

p. 270, Ex. 5.15

✓ Understand the similarities and differences between frequency distributions and probability distributions.

p. 270, Ex. 5.100

✓ Understand and be able to utilize the two main properties of probability distributions to verify compliance.

p. 274, EXP. 5.2, Ex. 5.17, 5.99, 5.101

✓ Understand that a probability distribution is a theoretical probability distribution and that the mean and standard deviation (μ and σ, respectively) are parameters.

pp. 278–280, Ex. 5.100

✓ Compute, describe, and interpret the mean and standard deviation of a probability distribution.

EXP. 5.5, Ex. 5.33, 5.35

✓ Understand the key elements of a binomial experiment and be able to define x, n, p, and q.

p. 287, EXP. 5.6, 5.7

✓ Know and be able to calculate binomial probabilities using the binomial probability function.

EXP. 5.8, Ex. 5.57, 5.63

✓ Understand and be able to use Table 2 in Appendix B, Binomial Probabilities, to determine binomial probabilities.

p. 292, Ex. 5.58, 5.111

✓ Compute, describe, and interpret the mean and standard deviation of a binomial probability distribution.

EXP. 5.11, Ex. 5.89, 5.91

Chapter Exercises

Statistics ⊘Now™

Go to the StatisticsNow website **http://1pass.thomson.com** to
- Assess your understanding of this chapter
- Check your readiness for an exam by taking the Pre-Test quiz and exploring the resources in the Personalized Learning Plan

Datasets can be found on your Student's Suite CD-ROM or at the StatisticsNow website at **http://1pass.thomson.com.**

5.99 What are the two basic properties of every probability distribution?

5.100 a. Explain the difference and the relationship between a probability distribution and a probability function.

  b. Explain the difference and the relationship between a probability distribution and a frequency distribution, and explain how they relate to a population and a sample.

5.101 Verify whether or not each of the following is a probability function. State your conclusion and explain.

a. $f(x) = \dfrac{\frac{3}{4}}{x!(3-x)!}$ for $x = 0, 1, 2, 3$

b. $f(x) = 0.25$ for $x = 9, 10, 11, 12$

c. $f(x) = (3-x)/2$ for $x = 1, 2, 3, 4$

d. $f(x) = (x^2 + x + 1)/25$ for $x = 0, 1, 2, 3$

5.102 Verify whether or not each of the following is a probability function. State your conclusion and explain.

a. $f(x) = \dfrac{3x}{8x!}$ for $x = 1, 2, 3, 4$

b. $f(x) = 0.125$ for $x = 0, 1, 2, 3$ and $f(x) = 0.25$ for $x = 4, 5$

c. $f(x) = (7-x)/28$ for $x = 0, 1, 2, 3, 4, 5, 6, 7$

d. $f(x) = (x^2 + 1)/60$ for $x = 0, 1, 2, 3, 4, 5$

5.103 The number of ships to arrive at a harbor on any given day is a random variable represented by x. The probability distribution for x is as follows:

x	10	11	12	13	14
$P(x)$	0.4	0.2	0.2	0.1	0.1

Find the probability of the following for any a given day:

a. Exactly 14 ships arrive.

b. At least 12 ships arrive.

c. At most 11 ships arrive.

5.104 "How many TV's are there in your household?" was one of the questions on a questionnaire sent to 5000 people in Japan. The collected data resulted in the following distribution:

Number of TV's/Household	0	1	2	3	4	5 or more
Percentage	1.9	31.4	23.0	24.4	13.0	6.3

Source: http://www.japan-guide.com/topic/0107.html

One of these households is selected at random.

a. What percentage of the households have at least one television?

b. What percentage of the households have at most three televisions?

c. What percentage of the households have three or more televisions?

d. Is this a binomial probability experiment? Justify your answer.

e. Let x be the number of televisions per household. Is this a probability distribution? Explain.

f. Assign $x = 5$ for "5 or more" and find the mean and standard deviation of x.

5.105 Patients who have artificial hip-replacement surgery experience pain the first day after surgery. Typically, the pain is measured on a subjective scale using values of 1 to 5. Let x represent the random variable, the pain score as determined by the patient. The probability distribution for x is believed to be:

x	1	2	3	4	5
$P(x)$	0.10	0.15	0.25	0.35	0.15

a. Find the mean of x.

b. Find the standard deviation of x.

5.106 The 2000 census produced the following figures for the city of Loveland, Colorado, with respect to the number of available vehicles per household:

x	0	1	2	3 or more
Percent	4.6	30.0	43.3	22.1

Source:
http://www.co.larimer.co.us/compass/vehicleperhousehold_cd_trans.htm#chart2

Replacing the category of "3 or more" with "3":

a. Find the mean of x.

b. Find the standard deviation of x.

5.107 A doctor knows from experience that 10% of the patients to whom she gives a certain drug will have undesirable side effects. Find the probabilities that among the 10 patients to whom she gives the drug:

a. At most two will have undesirable side effects.

b. At least two will have undesirable side effects.

5.108 In a recent survey of women, 90% admitted that they had never looked at a copy of *Vogue* magazine. Assuming that this is accurate information, what is the probability that a random sample of three women will show that fewer than two have read the magazine?

5.109 Of those seeking a driver's license, 70% admitted that they would not report someone if he or she copied some answers during the written exam. You have just entered the room and see 10 people waiting to take the written exam. What is the probability that if the incident happened, 5 of the 10 would not report what they saw?

5.110 The engines on an airliner operate independently. The probability that an individual engine operates for a given trip is 0.95. A plane will be able to complete a trip successfully if at least one-half of its engines operate for the entire trip. Determine whether a four-engine or a two-engine plane has the higher probability of a successful trip.

5.111 The Pew Internet & American Life Project found that nearly 70% of "wired" senior citizens go online every day. In a randomly selected group of 15 "wired" senior citizens:

a. What is the probability that more than four will say they go online every day?

b. What is the probability that exactly 10 will say that they go online every day?

c. What is the probability that fewer than 10 will say that they go online every day?

5.112 R&B/hip-hop songs accounted for more than 60% of radio's top 100 hits in 2004, according to data from Nielsen BDS and Arbitron as reported in *USA Today* on January 5, 2005. A new radio station, appropriately named Fickle, plays all types of music from the list of the top 100 hits including R&B/hip-hop, pop, rock, and country. In the next randomly selected group of 14 songs being played on Fickle and using the 60% for R&B/hip-hop songs:

a. What is the probability that more than seven songs are R&B/hip-hop songs?

b. What is the probability that exactly 10 songs are R&B/hip-hop songs?

c. What is the probability that fewer than five songs are R&B/hip-hop songs?

5.113 Imagine that you are in the midst of purchasing a lottery ticket and the person behind the counter prints too many tickets with your numbers. What would you do? The results of an online survey were as follows:

Let them keep the tickets?	30.77%
Trust the person to delete them?	15.38%
Buy the extra ones and hope they win?	30.77%
Other	23.08%

Is this a probability distribution? Explain.

5.114 Learning is a lifetime activity. For some, it means learning from everyday experiences; for others, it means taking classes in a more traditional atmosphere. The percentage of people participating in organized learning situations during 2002 for each age group is reported here by NIACE.

Age group	17–19	20–24	25–35	35–44	45–54	55–64	65–74	75+	All
Percentage	78	72	51	49	44	30	20	10	42

Source: NIACE Adult Participation in Learning Surveys

Is this a probability distribution? Explain.

5.115 The town council has nine members. A proposal to establish a new industry in this town has been tabled, and all proposals must have at least two-thirds of the votes to be accepted. If we know that two members of the town council are opposed and that the others randomly vote "in favor" and

"against," what is the probability that the proposal will be accepted?

5.116 There are 750 players on the active rosters of the 30 major league baseball teams. A random sample of 15 players is to be selected and tested for use of illegal drugs.

a. If 5% of all the players are using illegal drugs at the time of the test, what is the probability that 1 or more players test positive and fail the test?

b. If 10% of all the players are using illegal drugs at the time of the test, what is the probability that 1 or more players test positive and fail the test?

c. If 20% of all the players are using illegal drugs at the time of the test, what is the probability that 1 or more players test positive and fail the test?

5.117 A box contains 10 items, of which 3 are defective and 7 are nondefective. Two items are selected without replacement, and x is the number of defective items in the sample of two. Explain why x is not a binomial random variable.

5.118 A box contains 10 items, of which 3 are defective and 7 are nondefective. Two items are randomly selected, one at a time, with replacement, and x is the number of defectives in the sample of two. Explain why x is a binomial random variable.

5.119 A large shipment of radios is accepted upon delivery if an inspection of 10 randomly selected radios yields no more than 1 defective radio.

a. Find the probability that this shipment is accepted if 5% of the total shipment is defective.

b. Find the probability that this shipment is not accepted if 20% of this shipment is defective.

c. The binomial probability distribution is often used in situations similar to this one, namely, large populations sampled without replacement. Explain why the binomial yields a good estimate.

5.120 The state bridge design engineer has devised a plan to repair North Carolina's 4706 bridges that are currently listed as being in either poor or fair condition. The state has a total of 13,268 bridges. Before the governor will include the cost of this plan in his budget, he has decided to personally visit and inspect five bridges that are to be randomly selected. What is the probability that in the sample of five bridges, the governor will visit the following:

a. No bridges rated as poor or fair

b. One or two bridges rated as poor or fair

c. Five bridges rated as poor or fair

5.121 A discrete random variable has a standard deviation equal to 10 and a mean equal to 50. Find $\sum x^2 P(x)$.

5.122 A binomial random variable is based on $n = 20$ and $p = 0.4$. Find $\sum x^2 P(x)$.

5.123 [EX05-123] In a germination trial, 50 seeds were planted in each of 40 rows. The number of seeds germinating in each row was recorded as listed in the following table.

Number Germinated	Number of Rows	Number Germinated	Number of Rows
39	1	45	8
40	2	46	4
41	3	47	3
42	4	48	1
43	6	49	1
44	7		

a. Use the preceding frequency distribution table to determine the observed rate of germination for these seeds.

b. The binomial probability experiment with its corresponding probability distribution can be used with the variable "number of seeds germinating per row" when 50 seeds are planted in every row. Identify the specific binomial function and list its distribution using the germination rate found in part a. Justify your answer.

c. Suppose you are planning to repeat this experiment by planting 40 rows of these seeds, with 50 seeds in each row. Use your probability

model from part b to find the frequency distribution for x that you would expect to result from your planned experiment.

d. Compare your answer in part c with the results that were given in the preceding table. Describe any similarities and differences.

5.124 In another germination experiment involving old seed, 50 rows of seeds were planted. The number of seeds germinating in each row were recorded in the following table (each row contained the same number of seeds).

Number Germinating	Number of Rows	Number Germinating	Number of Rows
0	17	3	2
1	20	4	1
2	10	5 or more	0

a. What probability distribution (or function) would be helpful in modeling the variable "number of seeds germinating per row"? Justify your choice.

b. What information is missing in order to apply the probability distribution you chose in part a?

c. Based on the information you do have, what is the highest or lowest rate of germination that you can estimate for these seeds? Explain.

5.125 A business firm is considering two investments. It will choose the one that promises the greater payoff. Which of the investments should it accept? (Let the mean profit measure the payoff.)

Invest in Tool Shop		**Invest in Book Store**	
Profit	Probability	Profit	Probability
$100,000	0.10	$400,000	0.20
50,000	0.30	90,000	0.10
20,000	0.30	−20,000	0.40
−80,000	0.30	−250,000	0.30
Total	1.00	Total	1.00

5.126 Bill has completed a 10-question multiple-choice test on which he answered 7 questions correctly. Each question had one correct answer to be chosen from five alternatives. Bill says that he answered the test by randomly guessing the answers without reading the questions or answers.

a. Define the random variable x to be the number of correct answers on this test, and construct the probability distribution if the answers were obtained by random guessing.

b. What is the probability that Bill guessed 7 of the 10 answers correctly?

c. What is the probability that anybody can guess six or more answers correctly?

d. Do you believe that Bill actually randomly guessed as he claims? Explain.

5.127 A random variable that can assume any one of the integer values 1, 2, ..., n with equal probabilities of $\frac{1}{n}$ is said to have a uniform distribution. The probability function is written $P(x) = \frac{1}{n}$, for $x = 1, \ 2, \ 3, \ldots, \ n$. Show that $\mu = \frac{(n + 1)}{2}$.

(*Hint:* $1 + 2 + 3 + \ldots + n = [n(n + 1)]/2$.)

Chapter Project

Caffeine Drinking

Let's take a second look at Section 5.1, "Caffeine Drinking" (p. 269), and test our knowledge of the material presented in this chapter. Based on the USA Snapshot, we have the number of cups or cans of caffeinated beverages adult Americans say they drink daily and their corresponding probabilities. Consider where you might fit into this situation.

Putting Chapter 5 to Work

5.128 a. What variable could be used to describe all five of the events shown in the "Americans like their coffee!" graphic (p. 269)?

b. Is the variable in part a discrete or continuous? Why?

c. Are events $x = 0, 1, 2, \ldots$ mutually exclusive? Explain why or why not.

d. What characteristics of a circle graph make it appropriate for use with a probability distribution? Be specific.

e. Construct a circle graph depicting the information described in the graphic.

f. Express the information in the circle graph as a probability distribution.

g. Assuming the information on the circle graph represents the population, find the mean and standard deviation of the variable described in part a.

h. Draw a histogram to display the information in the graphic. Describe the histogram. Is it a normal distribution? Explain.

i. Locate the mean and standard deviation found in part g on the histogram drawn in part h.

j. Do the empirical and Chebyshev's rules apply? Justify your answer.

Your Study

5.129 Design your own study of caffeine drinking.

a. Define a specific population that you will sample, describe your sampling plan, and collect your data.

b. Express your sample as a relative frequency distribution and draw a histogram.

c. Express your sample as a frequency distribution and find the sample mean and sample standard deviation.

d. Discuss the differences and similarities between your sample and the distribution shown in the graphic "Americans like their coffee!"

Chapter Practice Test

PART I: Knowing the Definitions

Answer "True" if the statement is always true. If the statement is not always true, replace the words shown in bold with words that make the statement always true.

5.1 The number of hours you waited in line to register this semester is an example of a **discrete** random variable.

5.2 The number of automobile accidents you were involved in as a driver last year is an example of a **discrete** random variable.

5.3 The sum of all the probabilities in any probability distribution is always exactly **two.**

5.4 The various values of a random variable form a list of **mutually exclusive events.**

5.5 A binomial experiment always has **three or more** possible outcomes to each trial.

5.6 The formula $\mu = np$ may be used to compute the mean of a **discrete** population.

5.7 The binomial parameter p is the probability of **one success occurring in n trials** when a binomial experiment is performed.

5.8 A parameter is a statistical measure of some aspect of a **sample.**

5.9 **Sample statistics** are represented by letters from the Greek alphabet.

5.10 The probability of event A or B is equal to the sum of the probability of event A and the probability of event B when A and B are **mutually exclusive events.**

PART II: Applying the Concepts

5.11 a. Show that the following is a probability distribution:

x	1	3	4	5
$P(x)$	0.2	0.3	0.4	0.1

   b. Find $P(x = 1)$.
   c. Find $P(x = 2)$.
   d. Find $P(x > 2)$.
   e. Find the mean of x.
   f. Find the standard deviation of x.

5.12 A T-shirt manufacturing company advertises that the probability of an individual T-shirt being irregular is 0.1. A box of 12 such T-shirts is randomly selected and inspected.

a. What is the probability that exactly 2 of these 12 T-shirts are irregular?

b. What is the probability that exactly 9 of these 12 T-shirts are not irregular?

Let x be the number of T-shirts that are irregular in all such boxes of 12 T-shirts.

c. Find the mean of x.

d. Find the standard deviation of x.

PART III: Understanding the Concepts

5.13 What properties must an experiment possess in order for it to be a binomial probability experiment?

5.14 Student A uses a relative frequency distribution for a set of sample data and calculates the mean and standard deviation using formulas from Chapter 5. Student A justifies her choice of formulas by saying that since relative frequencies are empirical probabilities, her sample is represented by a probability distribution and therefore her choice of formulas was correct. Student B argues that since the distribution represented a sample, the mean and standard deviation involved are known as $\bar{x}$ and s and must be calculated using the corresponding frequency distribution and formulas from Chapter 2. Who is correct, A or B? Justify your choice.

5.15 Student A and Student B were discussing one entry in a probability distribution chart:

x	$P(x)$
-2	0.1

Student B thought this entry was okay because $P(x)$ was a value between 0.0 and 1.0. Student A argued that this entry was impossible for a probability distribution because x was -2 and negatives are not possible. Who is correct, A or B? Justify your choice.

CHAPTER

6

Normal Probability Distributions

6.1 Intelligence Scores

MEASURES OF INTELLIGENCE
Aptitude Tests and Their Interpretation

There are many kinds of aptitude tests. Some are for specific purposes, such as measurement of finger dexterity, something that might be important on a particular job. Others are of more general aptitudes. So-called intelligence tests are examples of general aptitude tests.

The Binet Intelligence Scale. Alfred Binet, who devised the first general aptitude test at the beginning of the 20th century, defined intelligence as *the ability to make adaptations.* The general purpose of the test was to determine which children in Paris could benefit from school. Binet's test, like its subsequent revisions, consists of a series of progressively more difficult tasks that children of different ages can successfully complete. A child who can solve problems typically solved by children at a particular age level is said to have that mental age. For example, if a child can successfully do the same tasks that an average eight-year-old can do, he or she is said to have a mental age of eight. The *intelligence quotient,* or IQ, is defined by the formula:

intelligence quotient =
    $100 \times$ (mental age/chronological age)

There has been a great deal of controversy in recent years over what intelligence tests measure. Many of the test items depend on either language or other specific cultural experiences for correct answers. Nevertheless, such tests can rather effectively predict school success. If school requires language and the tests measure language ability at a particular point of time in a child's life, then the test is a better-than-chance predictor of school performance.

Deviation IQ Scores. Present-day tests of intelligence or other abilities use *deviation scores.* These scores represent the deviation of a particular person from the average score for similar persons. Suppose you take a "general aptitude test" and get a score of 115. This does not mean that your mental age is greater than your chronological age; it means that you are "above average" in some degree. Because we have become accustomed to thinking of an IQ score of 100 as average, most general aptitude tests are scored in such a way that 100 is average. A person scoring 115 would generally have a score higher than the scores of about 85 percent of people who take the test; a score of 84 would be better than about 16 percent. The exact interpretation of a test score depends on the particular test, but Figure 2.2 (p. 314) shows how the scores on a number of commonly used aptitude tests are interpreted in terms of how an individual compares with a group.

Figure 2.2 pictures the comparison of several deviation scores and the normal distribution: Standard scores have a mean of zero and a standard deviation of 1.0. Scholastic Aptitude Test scores have a mean of 500 and a standard deviation of 100.

Binet Intelligence Scale scores have a mean of 100 and a standard deviation of 16. In each case there are 34 percent of the scores between the mean and one standard deviation, 14 percent between one and two standard deviations, and 2 percent beyond two standard deviations.

FIGURE 2.2

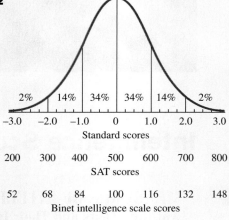

| 2% | 14% | 34% | 34% | 14% | 2% |

| −3.0 | −2.0 | −1.0 | 0 | 1.0 | 2.0 | 3.0 |

Standard scores

| 200 | 300 | 400 | 500 | 600 | 700 | 800 |

SAT scores

| 52 | 68 | 84 | 100 | 116 | 132 | 148 |

Binet intelligence scale scores

Source: Robert C. Beck, *Applying Psychology, Critical and Creative Thinking,* 3rd ed. (Englewood Cliffs, NJ: Prentice Hall, 1992)

After completing Chapter 6, further investigate the intelligence scores in the Chapter Project with Exercises 6.137 and 6.138 (p. 356).

SECTION 6.1 EXERCISES

6.1 a. Explain why the IQ score is a continuous variable.

b. What are the mean and the standard deviation for the distribution of IQ scores? SAT scores? Standard scores?

c. Express, algebraically or as an equation, the relationship between standard scores and IQ scores and between standard scores and SAT scores.

d. What standard score is 2 standard deviations above the mean? What IQ score is 2 standard deviations above the mean? What SAT score is 2 standard deviations above the mean?

e. Compare the information about percentage of distribution shown in Figure 2.2 above with the empirical rule studied in Chapter 2. Explain the similarities.

6.2 Examine the intelligence quotient, or IQ, as it is defined by the formula:

intelligence quotient = 100 × (mental age/chronological age)

Justify why it is reasonable for the mean to be 100.

6.2 Normal Probability Distributions

The **normal probability distribution** is considered the single most important probability distribution. An unlimited number of **continuous random variables** have either a normal or an approximately normal distribution. Several other probability distributions of both discrete and continuous random variables are also approximately normal under certain conditions.

Recall that in Chapter 5 we learned how to use a probability function to calculate the probabilities associated with **discrete random variables.** The normal probability distribution has a continuous random variable and uses two functions: one function to determine the ordinates (*y* values) of the graph picturing the distribution and a second to determine the probabilities. Formula (6.1) expresses the ordinate (*y* value) that corresponds to each abscissa (*x* value).

Normal Probability Distribution Function

$$y = f(x) = \frac{e^{-\frac{1}{2}\left(\frac{x-\mu}{\sigma}\right)^2}}{\sigma\sqrt{2\pi}} \quad \text{for all real } x \tag{6.1}$$

FIGURE 6.1 The Normal Probability Distribution

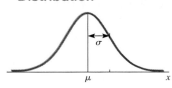

When a graph of all such points is drawn, the **normal (bell-shaped) curve** will appear as shown in Figure 6.1.

Note: Each different pair of values for the mean, μ, and standard deviation, σ, will result in a different normal probability distribution function.

Formula (6.2) yields the probability associated with the interval from $x = a$ to $x = b$:

$$P(a \le x \le b) = \int_a^b f(x)\ dx \tag{6.2}$$

FIGURE 6.2 Shaded Area: $P(a \le x \le b)$

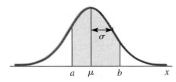

The probability that *x* is within the interval from $x = a$ to $x = b$ is shown as the shaded area in Figure 6.2.

We will not be using the preceding formulas to calculate probabilities for normal distributions. The definite integral of formula (6.2) is a calculus topic and is mathematically beyond what is expected in elementary statistics. (These formulas often appear at the top of normal probability tables as identification.) Instead of using formulas (6.1) and (6.2), we will use a table to find probabilities for normal distributions. Formulas 6.1 and 6.2 were used to generate that table. Before we learn to use the table, however, it must be pointed out that the table is expressed in "standardized" form. It is standardized so that this one table can be used to find probabilities for all combinations of mean, μ, and standard deviation, σ, values. That is, the normal probability distribution with mean 38 and standard deviation 7 is similar to the normal probability distribution with mean 123 and standard deviation 32. Recall the empirical rule and the percentages of the distribution that fall within certain intervals of the mean (p. 116). The same three percentages hold true for all normal distributions.

Note: Percentage, proportion, and **probability** are basically the same concepts. Percentage (25%) or proportion ($\frac{1}{4}$) is used when talking about part of a population, percentage being the more common. Probability is usually used when talking about the chance that the next individual item will possess a certain property. Area is the graphic representation of all three when we draw a picture to illustrate the situation. The empirical rule is a fairly crude measuring device; with it we are able to find probabilities associated only with whole-number multiples of the standard deviation (within 1, 2, or 3 standard deviations of the mean). We will often be interested in the probabilities associated with fractional parts of the standard deviation. For example, we might want to know the probability that *x* is within 1.37 standard deviations of the mean. Therefore, we must refine the empirical rule so that we can deal with more precise measurements. This refinement is discussed in the next section.

SECTION 6.2 EXERCISES

6.3 Percentage, proportion, or probability—identify which is illustrated by each of the following statements.

a. One-third of the crowd had a clear view of the event.

b. Fifteen percent of the voters were polled as they left the voting precinct.

c. The chance of rain during the day tomorrow is 0.2.

6.4 Percentage, proportion, or probability—in your own words, using between 25 and 50 words for each, describe the following:

a. How percentage is different than the other two

b. How proportion is different than the other two

c. How probability is different than the other two

d. How all three are basically the same thing

6.3 | The Standard Normal Distribution

There are an unlimited number of normal probability distributions, but fortunately they are all related to one distribution: the **standard normal distribution.** The standard normal distribution is the normal distribution of the standard variable *z* (called **"standard score"** or **"z-score"**).

Properties of the Standard Normal Distribution:

1. The total area under the normal curve is equal to 1.

2. The distribution is mounded and symmetrical; it extends indefinitely in both directions, approaching but never touching the horizontal axis.

3. The distribution has a mean of 0 and a standard deviation of 1.

4. The mean divides the area in half—0.50 on each side.

5. Nearly all the area is between $z = -3.00$ and $z = 3.00$.

Table 3 in Appendix B lists the probabilities associated with the intervals from the mean (located at $z = 0.00$) to a specific value of z. Probabilities of other intervals may be found by using the table entries and the operations of addition and subtraction, in accordance with the preceding properties. Let's look at several illustrations demonstrating how to use Table 3 to find probabilities of the standard normal score, z.

EXAMPLE 6.1

Statistics⬡Now™

Watch a video example at **http://1pass.thomson.com** or on your CD.

DID YOU KNOW

The Bell-Shaped Curve

In the 18th and 19th centuries, astronomers and physicists described their observations using precise mathematical formulas. They then explained that the difference between the observed and predicted values was a result of their instruments' lack of precision and was therefore unimportant. As instruments became more precise, it became apparent that this error was inherent randomness of the observations. In 1820, Laplace described this with his error function. This error distribution gained popularity and is now known as the normal probability distribution and is often called a *bell-shaped curve.*

Finding Area to the Right of $z = 0$

Find the area under the standard normal curve between $z = 0$ and $z = 1.52$ (see Figure 6.3).

FIGURE 6.3
Area from $z = 0$ to $z = 1.52$

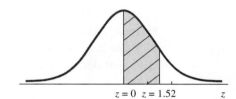

$z = 0$ $z = 1.52$ z

SOLUTION Table 3 is designed to give the area between $z = 0$ and $z = 1.52$ directly. The z-score is located on the margins, with the units and tenths digit along the left side and the hundredths digit across the top. For $z = 1.52$, locate the row labeled 1.5 and the column labeled 0.02; at their intersection you will find 0.4357, the measure of the area or the probability for the interval $z = 0.00$ to $z = 1.52$ (see Table 6.1). Expressed as a probability: $P(0.00 < z < 1.52) = \mathbf{0.4357}.$

TABLE 6.1

A Portion of Table 3

z	0.00	0.01	0.02	...
⋮				
1.5			0.4357	...
⋮				

Recall that one of the basic properties of probability is that the sum of all probabilities is exactly 1.0. Since the area under the normal curve represents the measure of probability, the total area under the bell-shaped curve is exactly 1. This distribution is also symmetrical with respect to the vertical line drawn through $z = 0$, which cuts the area in half at the mean. Can you verify this fact by inspecting formula (6.1)? That is, the area under the curve to the right of the mean is exactly one-half, 0.5, and the area to the left is also one-half, 0.5. Areas (probabilities) not given directly in the table can be found by relying on these facts.

Now let's look at some examples.

EXAMPLE 6.2

Finding Area in the Right Tail of a Normal Curve

Find the area under the normal curve to the right of $z = 1.52$: $P(z > 1.52)$.

SOLUTION The area to the right of the mean (all the shading in the figure) is exactly 0.5000. The problem asks for the shaded area that is not included in the 0.4357. Therefore, we subtract 0.4357 from 0.5000:

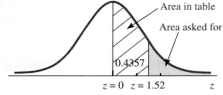

$$P(z > 1.52) = 0.5000 - 0.4357 = \mathbf{0.0643}$$

Notes: 1. As we have done here, always draw and label a sketch. It is most helpful. 2. Make it a habit to write z with two decimal places and areas and probabilities with four decimal places, as done in Table 3.

EXAMPLE 6.3

Finding Area to the Left of a Positive z Value

Find the area to the left of $z = 1.52$: $P(z < 1.52)$.

SOLUTION The total shaded area is made up of 0.4357 found in the table and the 0.5000 that is to the left of the mean. Therefore, we add 0.4357 to 0.5000:

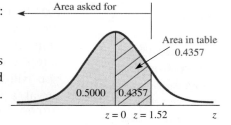

$$P(z \leq 1.52) = P(z < 0) + P(0 < z < 1.52)$$
$$= 0.5000 + 0.4357 = \mathbf{0.9357}$$

Note: The addition and subtraction done in Examples 6.2 and 6.3 are correct because the "areas" represent mutually exclusive events (discussed in Section 4.5).

The symmetry of the normal distribution is a key factor in determining probabilities associated with values below (to the left of) the mean. The area between the mean and $z = -1.52$ is exactly the same as the area between the mean and $z = +1.52$. This fact allows us to find values related to the left side of the distribution, as illustrated in the next two examples.

EXAMPLE 6.4

Finding Area from a Negative z to z = 0

The area between the mean ($z = 0$) and $z = -2.1$ is the same as the area between $z = 0$ and $z = +2.1$; that is,

$$P(-2.1 < z < 0) = P(0 < z < 2.1)$$

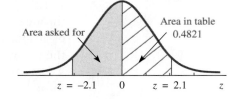

Thus, we have

$$P(-2.1 < z < 0) = P(0 < z < 2.1) = \mathbf{0.4821}$$

EXAMPLE 6.5   **Finding Area in the Left Tail of a Normal Curve**

The area to the left of $z = -1.35$ is found by subtracting 0.4115 from 0.5000. Therefore, we obtain

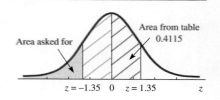

$P(z < -1.35) = P(z < 0) - P(-1.35 < z < 0)$
$= 0.5000 - 0.4115 = \mathbf{0.0885}$

EXAMPLE 6.6   **Finding Area from a Negative z to a Positive z**

The area between $z = -1.5$ and $z = 2.1$, $P(-1.5 < z < 2.1)$, is found by adding two areas together. Both required probabilities are read directly from Table 3.
      Therefore, we obtain

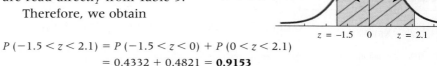

$P(-1.5 < z < 2.1) = P(-1.5 < z < 0) + P(0 < z < 2.1)$
$= 0.4332 + 0.4821 = \mathbf{0.9153}$

EXAMPLE 6.7   **Finding Area between Two z Values of the Same Sign**

Statistics △ Now™

Watch a video example at
http://1pass.thomson.com
or on your CD.

The area between $z = 0.7$ and $z = 2.1$, $P(0.7 < z < 2.1)$, is found by subtracting. The area between $z = 0$ and $z = 2.1$ includes all the area between $z = 0$ and $z = 0.7$. Therefore, we subtract the area between $z = 0$ and $z = 0.7$ from the area between $z = 0$ and $z = 2.1$.
      Thus, we have

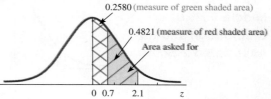

$$P(0.7 < z < 2.1) = P(0 < z < 2.1) - P(0 < z < 0.7)$$
$$= 0.4821 - 0.2580 = \mathbf{0.2241}$$

The standard normal distribution table can also be used to find a *z*-score when we are given an area. The next example considers this idea.

EXAMPLE 6.8   **Finding z-scores Associated with a Percentile**

What is the *z*-score associated with the 75th percentile of a normal distribution? See Figure 6.4.

SOLUTION

FIGURE 6.4 P_{75}
and Its Associated
z-Score

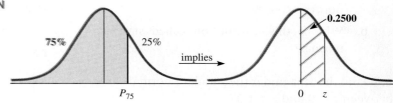

To find this *z*-score, look in Table 3 in Appendix B and find the "area" entry that is closest to 0.2500; this area entry is 0.2486. Now read the *z*-score that corresponds to this area.

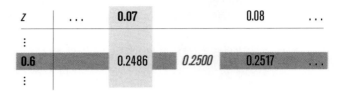

From the table, the *z*-score is found to be $z = \mathbf{0.67}$. This says that the 75th percentile in a normal distribution is 0.67 (approximately $\frac{2}{3}$) standard deviation above the mean.

EXAMPLE 6.9

Finding *z*-Scores That Bound an Area

FIGURE 6.5
Middle 95% of
Distribution and Its
Associated *z*-Score

What *z*-scores bound the middle 95% of a normal distribution?

SOLUTION The 95% is split into two equal parts by the mean, so 0.4750 is the area (percentage) between $z = 0$, the mean, and the *z*-score at the right boundary. See Figure 6.5.

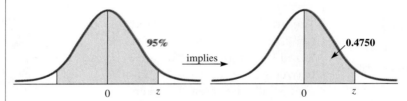

Since we have the area, we look for the entry in Table 3 closest to 0.4750 (it happens to be exactly 0.4750) and read the *z*-score. We obtain $z = 1.96$.

Therefore, $z = \mathbf{-1.96}$ and $z = \mathbf{1.96}$ bound the middle 95% of a normal distribution.

SECTION 6.3 EXERCISES

6.5 a. Describe the distribution of the standard normal score *z*.

b. Why is this distribution called standard normal?

6.6 Find the area under the standard normal curve between $z = 0$ and $z = 1.37$.

6.7 Find the area under the normal curve that lies between the following pairs of *z*-values.

a. $z = 0$ to $z = 1.30$ b. $z = 0$ to $z = 1.28$

c. $z = 0$ to $z = -3.20$ d. $z = 0$ to $z = -1.98$

6.8 Find the probability that a data value picked at random from a normal population will have a standard score (z) that lies between the following pairs of z-values.

a. $z = 0$ to $z = 2.10$ b. $z = 0$ to $z = 2.57$

c. $z = 0$ to $z = -1.20$ d. $z = 0$ to $z = -1.57$

6.9 Find the area under the standard normal curve to the right of $z = 2.03$, $P(z > 2.03)$.

6.10 Find the area under the standard normal curve to the left of $z = 1.73$, $P(z < 1.73)$.

6.11 Find the area under the standard normal curve between -1.39 and the mean, $P(-1.39 < z < 0.00)$.

6.12 Find the area under the standard normal curve to the left of $z = -1.53$, $P(z < -1.53)$.

6.13 Find the area under the standard normal curve between $z = -1.83$ and $z = 1.23$, $P(-1.83 < z < 1.23)$.

6.14 Find the area under the standard normal curve between $z = -2.46$ and $z = 1.46$, $P(-2.46 < z < 1.46)$.

6.15 Find the area under the standard normal curve that corresponds to the following z-values:

a. Between 0 and 1.55   b. To the right of 1.55

c. To the left of 1.55   d. Between -1.55 and 1.55

6.16 Find the probability that a data value picked at random from a normally distributed population will have a standard score (z) that corresponds to the following:

a. Between 0 and 0.84   b. To the right of 0.84

c. To the left of 0.84   d. Between -0.84 and 0.84

6.17 Find the following areas under the normal curve.

a. To the right of $z = 0.00$

b. To the right of $z = 1.05$

c. To the right of $z = -2.3$

d. To the left of $z = 1.60$

e. To the left of $z = -1.60$

6.18 Find the probability that a data value picked at random from a normally distributed population will have a standard score that corresponds to the following:

a. Less than 3.00   b. Greater than -1.55

c. Less than -0.75 d. Less than 1.25

e. Greater than -1.25

6.19 Find the following:

a. $P(0.00 < z < 2.35)$ b. $P(-2.10 < z < 2.34)$

c. $P(z > 0.13)$ d. $P(z < 1.48)$

6.20 Find the following:

a. $P(-2.05 < z < 0.00)$ b. $P(-1.83 < z < 2.07)$

c. $P(z < -1.52)$ d. $P(z < -0.43)$

6.21 Find the following:

a. $P(0.00 < z < 0.74)$ b. $P(-1.17 < z < 1.94)$

c. $P(z > 1.25)$ d. $P(z < 1.75)$

6.22 Find the following:

a. $P(-3.05 < z < 0.00)$ b. $P(-2.43 < z < 1.37)$

c. $P(z < -2.17)$ d. $P(z > 2.43)$

6.23 Find the area under the standard normal curve between $z = 0.75$ and $z = 2.25$, $P(0.75 < z < 2.25)$.

6.24 Find the area under the standard normal curve between $z = -2.75$ and $z = -1.28$, $P(-2.75 < z < -1.28)$.

6.25 Find the area under the normal curve that lies between the following pairs of z-values:

a. $z = -1.20$ to $z = 1.22$ b. $z = -1.75$ to $z = 1.54$

c. $z = -1.30$ to $z = 2.58$ d. $z = -3.5$ to $z = -0.35$

6.26 Find the probability that a data value picked at random from a normally distributed population will have a standard score (z) that lies between the following pairs of z-values:

a. $z = -2.75$ to $z = 1.38$ b. $z = 0.67$ to $z = 2.95$

c. $z = -2.95$ to $z = -1.18$

6.27 Find the *z*-score for the standard normal distribution shown on each of the following diagrams.

a.
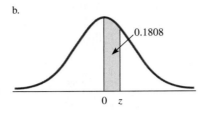

b.
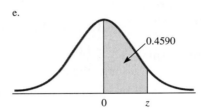

c.
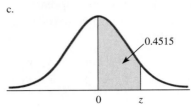

d.

e.

f.
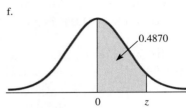

6.28 Find the *z*-score for the standard normal distribution shown in each of the following diagrams:

a.
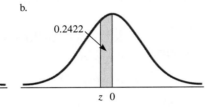

b.

c.
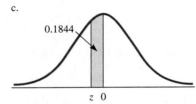

d.

e.
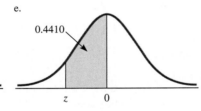

f.
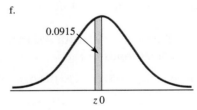

6.29 Find the standard score (*z*) shown on each of the following diagrams.

a.
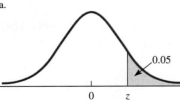

b.
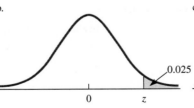

c.
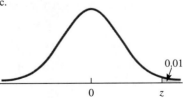

6.30 Find the standard score (*z*) shown on each of the following diagrams.

a.
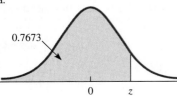

b.
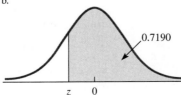

c.
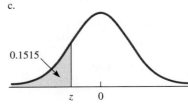

6.31 Find a value of z such that 40% of the distribution lies between it and the mean. (There are two possible answers.)

6.32 Find the standard z-score that corresponds to the following:

a. Eighty percent of the distribution is below (to the left of) this value.

b. The area to the right of this value is 0.15.

6.33 Find the two z-scores that bound the middle 50% of a normal distribution.

6.34 Find the two standard scores (z) that correspond to the following:

a. The middle 90% of a normal distribution is bounded by them.

b. The middle 98% of a normal distribution is bounded by them.

6.35 a. Find the z-score for the 80th percentile of the standard normal distribution.

b. Find the z-scores that bound the middle 75% of the standard normal distribution.

6.36 a. Find the z-score for the 33rd percentile of the standard normal distribution.

b. Find the z-scores that bound the middle 40% of the standard normal distribution.

6.37 Assuming a normal distribution, find the z-score associated with the following:

a. The 90th percentile

b. The 95th percentile

c. The 99th percentile

6.38 Assuming a normal distribution, what is the z-score associated with the following:

a. 1st quartile

b. 2nd quartile

c. 3rd quartile

6.4 Applications of Normal Distributions

In Section 6.3 we learned how to use Table 3 in Appendix B to convert information about the standard normal variable z into probability and vice versa—how to convert probability information about the standard normal distribution into z-scores. Now we are ready to apply this methodology to all normal distributions. The key is the standard score, z. The information associated with a normal distribution will be in terms of x values or probabilities. We will use the z-score and Table 3 as the tools to "go between" the given information and the desired answer.

Recall that the standard score, z, was defined in Chapter 2.

Standard Score

In words: $z = \dfrac{x - (\text{mean of } x)}{\text{standard deviation of } x}$

In algebra: $z = \dfrac{x - \mu}{\sigma}$ (6.3)

(Note that when $x = \mu$, the standard score $z = 0$.)

EXAMPLE 6.10 — Converting to a Standard Normal Curve to Find Probabilities

Statistics⌾Now™

Watch a video example at
http://1pass.thomson.com
or on your CD.

Consider IQ scores. IQ scores are normally distributed with a mean of 100 and a standard deviation of 16. If a person is picked at random, what is the probability that his or her IQ is between 100 and 115; that is, what is $P(100 < x < 115)$?

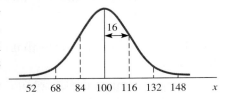

SOLUTION $P(100 < x < 115)$ is represented by the shaded area in the figure.

The variable x must be standardized using formula (6.3). The z values are shown on the next figure.

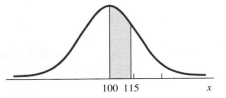

$$z = \frac{x - \mu}{\sigma}$$

when $x = 100$: $z = \dfrac{100 - 100}{16} = \mathbf{0.00}$

when $x = 115$: $z = \dfrac{115 - 100}{16} = \mathbf{0.94}$

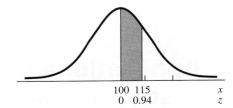

Therefore,

FYI The value 0.3264 is found by using Table 3 in Appendix B.

$$P(100 < x < 115) = P(0.00 < z < 0.94) = \mathbf{0.3264}$$

Thus, the probability is 0.3264 that a person picked at random has an IQ between 100 and 115.

EXAMPLE 6.11 — Calculating Probability under "Any" Normal Curve

Find the probability that a person selected at random will have an IQ greater than 90.

SOLUTION

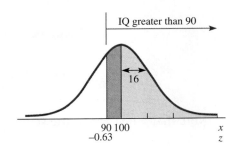

$$z = \frac{x - \mu}{\sigma} = \frac{90 - 100}{16} = \frac{-10}{16} = -0.625 = -0.63$$

$$P(x > 90) = P(z > -0.63)$$

$$= 0.2357 + 0.5000 = \mathbf{0.7357}$$

Thus, the probability is 0.7357 that a person selected at random will have an IQ greater than 90.

The normal table can be used to answer many kinds of questions that involve a normal distribution. Many times a problem will call for the location of a "cutoff point"—that is, a particular value of x such that exactly a certain percentage is in a specified area. The following examples concern some of these problems.

EXAMPLE 6.12

Using the Normal Curve and z to Determine Data Values

In a large class, suppose your instructor tells you that you need to obtain a grade in the top 10% of your class to get an A on a particular exam. From past experience, she is able to estimate that the mean and standard deviation on this exam will be 72 and 13, respectively. What will be the minimum grade needed to obtain an A? (Assume that the grades will be approximately normally distributed.)

SOLUTION Start by converting the 10% to information that is compatible with Table 3 by subtracting:

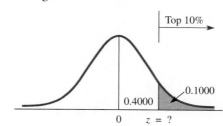

FYI Why is 0.5000 used?

$$10\% = 0.1000; \quad 0.5000 - 0.1000 = 0.4000$$

Look in Table 3 to find the value of z associated with the area entry closest to 0.4000; it is $z = 1.28$. Thus,

$$P(z > 1.28) = 0.10$$

Now find the x value that corresponds to $z = 1.28$ by using formula (6.3):

$$z = \frac{x - \mu}{\sigma}: \quad 1.28 = \frac{x - 72}{13}$$

$$x - 72 = (13)(1.28)$$

$$x = 72 + (13)(1.28) = 72 + 16.64 = 88.64, \text{ or } \mathbf{89}$$

Thus, if you receive an 89 or higher, you can expect to be in the top 10% (which means an A).

EXAMPLE 6.13   Using the Normal Curve and *z* to Determine Percentiles

Statistics⬡Now™

Watch a video example at
http://1pass.thomson.com
or on your CD.

Find the 33rd percentile for IQ scores ($\mu = 100$ and $\sigma = 16$ from Example 6.10, p. 324).

SOLUTION

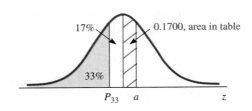

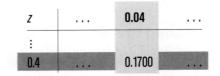

$$P(0 < z < a) = 0.17$$
$$a = \mathbf{0.44} \quad \text{(cutoff value of } z \text{ from Table 3)}$$
$$\text{33rd percentile of } z = -0.44 \text{ (below mean)}$$

Now we convert the 33rd percentile of the *z*-scores, -0.44, to an *x*-score using formula (6.3):

$$z = \frac{x - \mu}{\sigma}: \quad -0.44 = \frac{x - 100}{16}$$

$$x - 100 = 16(-0.44)$$

$$x = 100 - 7.04 = \mathbf{92.96}$$

Thus, 92.96 is the 33rd percentile for IQ scores.

Example 6.14 concerns a situation in which you are asked to find the mean, μ, when given related information.

EXAMPLE 6.14   Using the Normal Curve and *z* to Determine Population Parameters

The incomes of junior executives in a large corporation are normally distributed with a standard deviation of $1200. A cutback is pending, at which time those who earn less than $28,000 will be discharged. If such a cut represents 10% of the junior executives, what is the current mean salary of the group of junior executives?

SOLUTION If 10% of the salaries are less than $28,000, then 40% (or 0.4000) are between $28,000 and the mean, μ. Table 3 indicates that $z = -1.28$ is the standard score that occurs at $x = \$28,000$.

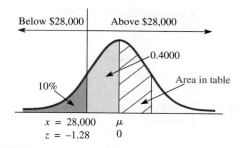

Using formula (6.3), we can find the value of μ:

$$z = \frac{x - \mu}{\sigma}: \quad -1.28 = \frac{28,000 - \mu}{1,200}$$

$$-1,536 = 28,000 - \mu$$

$$\mu = 28,000 + 1,536 = \mathbf{\$29,536}$$

That is, the current mean salary of junior executives is $29,536.

Referring again to IQ scores, what is the probability that a person picked at random has an IQ of 125: $P(x = 125)$? (IQ scores are normally distributed with a mean of 100 and a standard deviation of 16.) This situation has two interpretations: theoretical and practical. Let's look at the theoretical interpretation first. Recall that the probability associated with an interval for a continuous random variable is represented by the area under the curve; that is, $P(a \le x \le b)$ is equal to the area between a and b under the curve. $P(x = 125)$ (that is, x is exactly 125) is then $P(125 \le x \le 125)$, or the area of the vertical line segment at $x = 125$. This area is zero. However, this is not the practical meaning of $x = 125$. It generally means 125 to the nearest integer value. Thus, $P(x = 125)$ would most likely be interpreted as

$$P(124.5 < x < 125.5)$$

The interval from 124.5 to 125.5 under the curve has a measurable area and is then nonzero. In situations of this nature, you must be sure what meaning is being used.

Note: A standard notation used to abbreviate "normal distribution with mean μ and standard deviation σ" is $N(\mu, \sigma)$. That is, $N(58, 7)$ represents "a normal distribution with mean = 58 and standard deviation = 7."

TECHNOLOGY INSTRUCTIONS: GENERATE RANDOM DATA FROM A NORMAL DISTRIBUTION

MINITAB (Release 14)

```
Choose:      Calc > Random Data > Normal
Enter:       Generate:  n  rows of data
             Store in column(s):  C1
             Mean:  μ
             Stand. dev.:  σ > OK
```

If multiple samples (say, 12), all of the same size, are wanted, modify the preceding commands: Store in column(s): C1–C12.

Note: To find descriptive statistics for each of these samples, use the commands: Stat > Basic Statistics > Display Descriptive Statistics for C1–C12.

Excel	Choose:	**Tools > Data Analysis > Random Number Generation > OK**
	Enter:	Number of Variables: **1**
		Number of Random Numbers: **n**
		Distribution: **Normal**
		Mean = : μ
		Standard Deviation = : σ
	Select:	Output Options: **Output Range**
	Enter:	**(A1 or select cell) > OK**

If multiple samples (say, 12), all of the same size, are wanted, modify the preceding commands: Number of variables: 12.

Note: To find descriptive statistics for each of these samples, use the commands: Tools > Data Analysis > Descriptive Statistics for columns A through L.

TI-83/84 Plus	Choose:	**MATH > PRB > 6:randNorm(**
	Enter:	μ, σ, **# of trials)**
	Choose:	**STO→ > L1 > ENTER**

If multiple samples (say, six), all of the same size, are wanted, repeat the preceding commands six times and store in L1–L6.

Note: To find descriptive statistics for each of these samples, use the commands: STAT > CALC > 1:1-Var Stats for L1–L6.

TECHNOLOGY INSTRUCTIONS: CALCULATING ORDINATE (y) VALUES FOR A NORMAL DISTRIBUTION CURVE

MINITAB (Release 14) Input the desired abscissas (x values) into C1; then continue with:

Choose:	**Calc > Probability Distributions > Normal**
Select:	**Probability Density**
Enter:	Mean: μ
	Stand. dev.: σ
	Input column: **C1**
	Optional Storage: **C2 > OK**

To draw the graph of a normal probability curve with the x values in C1 and the y values in C2, continue with:

Choose:	**Graph > Scatterplot**
Select:	**With Connect Line > OK**
Enter:	Y variables: **C2** X variables: **C1 > OK**

Excel Input the desired abscissas (x values) into column A and activate B1; then continue with:

Choose:	**Insert function f$_x$ > Statistical > NORMDIST > OK**
Enter:	X: **(A1:A100 or select 'x value' cells)**

```
                Mean: μ
                Standard dev.: σ
                Cumulative: False > OK
Drag:           Bottom right corner of the ordinate value box down to give
                other ordinates
```

To draw the graph of a normal probability curve with the *x* values in column A and the *y* values in column B, continue with:

```
Choose:    Chart Wizard > XY(Scatter) > 1st picture > Next > Data Range
Enter:     Data range: (A1:B100 or select x & y cells)
Choose:    Next > Finish
```

TI-83/84 Plus The ordinate values can be calculated for individual abscissa values, *x*:

```
Choose:    2nd > DISTR > 1:normalpdf(
Enter:     x, μ, σ)
```

To draw the graph of the normal probability curve for a particular μ and σ, continue with:

```
Choose:    WINDOW
Enter:     μ − 3σ, μ + 3σ, σ, − .05, 1, .1, 0)
Choose:    Y = > 2nd > DISTR > 1:normalpdf(
Enter:     x, μ, σ)
```

After an initial graph, adjust with 0:ZoomFit from the ZOOM menu.

TECHNOLOGY INSTRUCTIONS: CUMULATIVE PROBABILITY FOR NORMAL DISTRIBUTIONS

MINITAB (Release 14) Input the desired abscissas (*x* values) into C1; then continue with:

```
Choose:    Calc > Probability Distributions > Normal
Select:    Cumulative probability
Enter:     Mean: μ
           Stand. dev.: σ
           Input column: C1
           Optional Storage: C3 > OK
```

Notes:
1. To find the probability between two *x* values, enter the two values into C1, use the preceding commands, and subtract using the numbers in C3.
2. To draw a graph of the cumulative probability distribution (ogive), use the Scatterplot commands on page 328 with C3 as the *y* variable.

Excel Input the desired abscissas (*x* values) into column A and activate C1; then continue with:

```
Choose:    Insert function f_x > Statistical > NORMDIST > OK
Enter:     X: (A1:A100 or select 'x value' cells)
           Mean: μ
           Standard dev.: σ
           Cumulative: True > OK
Drag:      Bottom right corner of the cumulative probability box down to
           give other cumulative probabilities
```

Notes:

1. To find the probability between two *x* values, enter the two values into column A, use the preceding commands, and subtract using the numbers in column C.

2. To draw a graph of the cumulative probability distribution (ogive), use the Chart Wizard commands on page 329, choosing the subcommand Series with column C as the *y* values and column A as the *x* values.

TI-83/84 Plus The cumulative probabilities can be calculated for individual abscissa values, *x*:

Choose: **2nd > DISTR > 2:normalcdf(**
Enter: **−1 EE 99, x, μ, σ)**

Notes:

1. To find the probability between two *x* values, enter the two values in place of −1 EE 99 and the *x*.

2. To draw a graph of the cumulative probability distribution (ogive), use either the Scatter command under STATPLOTS, with the *x* values and their cumulative probabilities in a pair of lists, or normalcdf(−1EE99, *x*, μ, σ) in the Y = editor.

**APPLIED
EXAMPLE 6.15**

Bottle Stoppers–Corks

You're probably aware of that seemingly insignificant little cylinder of squeezable woody material called a bottle cork. But are you aware that the process by which raw bark from the Oak cork tree becomes a cork is anything but simple? The cork industry has very high standards, and there are very strict international laws covering everything from the harvesting of the cork to the delivery of the corks to the user.

Corks start as bark of the *Quercus suber* tree, and after being peeled from the tree, the bark goes through a series of storage and cooking processes to stabilize, clean, and increase the elasticity of the cork. Next it is cut into strips, and the corks are punched out. This is followed by a series of washing, bleaching, disinfecting, and coloring processes, while all the time being inspected and sorted. The finishing processes include inspections, coatings, printing, cessation of moisture, surface treating, sterilization, packing, and quality control certification.

The standard no. 9 size cork is 24 mm in diameter by 1.75 inches (45 mm) in length. Some of its characteristics (and specifications used) that must pass inspection are as follows:

- Defects/faults (e.g., worm holes, cracks, pores, green wood)
- Length (45.0 + 1.0 mm/−0.5 mm)
- Average diameter (24 mm + 0.6mm/−0.4 mm)
- Ovality (out of round, <1.0 mm)
- Weight (grams)
- Specific weight (g/cc)
- Humidity (customer's requirements ± 1.5%)

- Residual peroxide (<0.2 ppm)
- Extraction force (300 N + 100 N/−150 N)

Length is the one variable that is not very important in evaluating the quality of corks because it has little to do with the effectiveness of a cork in preserving wine. Long corks are chosen over shorter corks largely because of their aesthetic appeal—the loud pop when you uncork the bottle is appealing.

Some of the aforementioned variables have normal distributions; others do not. Two of them with normal distributions are the average diameter of the cork and the extraction force. The diameter of each cork is measured in several places, and an average diameter is reported for the cork. It has a normal distribution with a mean of 24.0 mm and standard deviation of 0.13 mm. A sample of 250 corks produced the following summary.

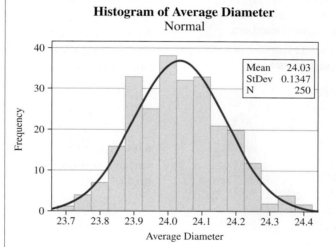

Histogram of Average Diameter
Normal

To obtain the extraction force, each bottle is filled, corked, and allowed to sit for 24 hours. It is then placed on a machine that removes the cork and records the force required to extract it from the bottle. This force has a normal distribution with a mean of 310 Newtons and a standard deviation of 36 Newtons. (A Newton is a unit of force; 1 N = 1 Newton = 1 kilogram meter/sec^2.) A sample of 400 corks produced this summary.

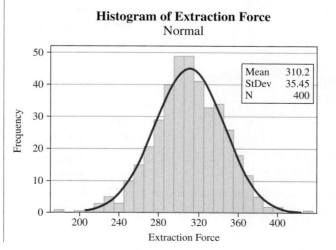

Histogram of Extraction Force
Normal

Ovality (the measure of out-of-round) is the difference between a cork's maximum diameter and minimum diameter. As you might expect, ovality does not have a normal distribution. Its lowest possible value is 0, and it increases from there. It does have a mounded but skewed right distribution.

What kind of distribution do you anticipate for the variables length, weight, and specific weight?

Source: Courtesy of Gültig GmbH

SECTION 6.4 EXERCISES

Statistics ◯ Now ™

Skillbuilder Applet Exercises must be worked using an accompanying applet found on your Student's Suite CD-ROM or at the StatisticsNow website at **http://1pass.thomson.com**.

Datasets can be found on your Student's Suite CD-ROM or at the StatisticsNow website at **http://1pass.thomson.com**.

6.39 Skillbuilder Applet Exercise demonstrates that probability is equal to the area under a curve. Given that college students sleep an average of 7 hours per night with a standard deviation equal to 1.7 hours, use the scroll bar in the applet to find the following:

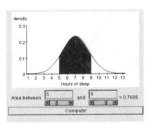

a. P(a student sleeps between 5 and 9 hours)

b. P(a student sleeps less than 4 hours)

c. P(a student sleeps between 8 and 11 hours)

6.40 Skillbuilder Applet Exercise demonstrates the effects that the mean and standard deviation have on a normal curve.

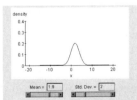

a. Leaving the standard deviation at 1, increase the mean to 3. What happens to the curve?

b. Reset the mean to 0 and increase the standard deviation to 2. What happens to the curve?

c. If you could decrease the standard deviation to 0.5, what do you think would happen to the normal curve?

6.41 Given $x = 58$, $\mu = 43$, and $\sigma = 5.2$, find z.

6.42 Given $x = 237$, $\mu = 220$, and $\sigma = 12.3$, find z.

6.43 Given that x is a normally distributed random variable with a mean of 60 and a standard deviation of 10, find the following probabilities:

a. $P(x > 60)$ b. $P(60 < x < 72)$ c. $P(57 < x < 83)$

d. $P(65 < x < 82)$ e. $P(38 < x < 78)$ f. $P(x < 38)$

6.44 Given that x is a normally distributed random variable with a mean of 28 and a standard deviation of 7, find the following probabilities:

a. $P(x < 28)$ b. $P(28 < x < 38)$ c. $P(24 < x < 40)$

d. $P(30 < x < 45)$ e. $P(19 < x < 35)$ f. $P(x < 48)$

6.45 Using the information given in Example 6.10 (p. 324):

a. Find the probability that a randomly selected person will have an IQ score between 100 and 120.

b. Find the probability that a randomly selected person will have an IQ score above 80.

6.46 Using the information given in Example 6.14 on page 326:

a. Find the probability that a randomly selected junior executive has a salary between $27,000 and $31,000.

b. Find the probability that a randomly selected junior executive has a salary greater than $33,000.

6.47 Depending on where you live and on the quality of the day care, costs of day care can range from $3000 to $15,000 a year (or $250 to $1250 monthly) for one child, according to the Baby Center. Day-care centers in large cities such as New York and San Francisco are notoriously expensive.

Source: http://www.babycenter.com/refcap/baby/ babychildcare/6056.html#0 4/16/2005

Suppose that annual day-care costs are normally distributed with a mean equal to $9000 and a standard deviation equal to $1800.

a. What percentage of day-care centers will cost between $7200 and $10,800 annually?

b. What percentage of day-care centers will cost between $5400 and $12,600 annually?

c. What percentage of day-care centers will cost between $3600 and $14,400 annually?

d. Compare the results for parts a–c with the empirical rule. Explain the relationship.

6.48 According to Wageweb.com (http://www .wageweb.com/hr1.htm), the national average salary as of October 2003 for a human resources clerk was $29,932. If we assume that the annual salaries for clerks are normally distributed with a standard deviation of $1850, find the following:

a. The percentage who earn less than $27,000

b. The percentage who earn more than $32,000

6.49 According to the Federal Highway Administration's 2003 highway statistics (http://www .fhwa.dot.gov), the distribution of ages for licensed drivers has a mean of 44.5 years and a standard deviation of 17.1 years. Assuming the distribution of ages is normally distributed, what percentage of the drivers are:

a. Between the ages of 17 and 22

b. Younger than 25 years of age

c. Older than 21 years of age

d. Between the ages of 45 and 65

e. Older than 75 years of age

6.50 There is a new working class with money to burn according to the *USA Today* article titled "New 'gold-collar' young workers gain clout" (March 1, 2005). "Gold-collar" is a subset of blue-collar workers defined by researchers as those working in fast-food and retail jobs or as security guards, office workers, or hairdressers. These 18- to 25-year-old "gold-collar" workers are spending an average of $729 a month on themselves (versus $267 for college students and $609 for blue-collar workers). Assuming this spending is normally distributed with a standard deviation of $92.00, what percentage of gold-collar workers spend:

a. Between $600 and $900 a month on themselves

b. Between $400 and $1000 a month on themselves

c. More than $1050 a month on themselves

d. Less than $500 a month on themselves

6.51 The International Bottled Water Association says that Americans on the average drink 4.6 (8-oz.) servings of water a day (http://www.bottledwater .org/public/summary.htm). Assuming that the number of 8-oz. servings of water is approximately normally distributed with a standard deviation of 1.4 servings, what proportion of Americans drink:

a. More than the recommended 8 servings

b. Less than half the recommended 8 servings

6.52 According to the American College Test (ACT), results from the 2004 ACT testing found that students had a mean reading score of 21.3 with a standard deviation of 6.0. Assuming that the scores are normally distributed:

a. Find the probability that a randomly selected student has a reading ACT score less than 20.

b. Find the probability that a randomly selected student has a reading ACT score between 18 and 24.

c. Find the probability that a randomly selected student has a reading ACT score greater than 30.

d. Find the value of the 75th percentile for ACT scores.

6.53 A brewery's filling machine is adjusted to fill quart bottles with a mean of 32.0 oz. of ale and a variance of 0.003. Periodically, a bottle is checked and the amount of ale noted.

a. Assuming the amount of fill is normally distributed, what is the probability that the next randomly checked bottle contains more than 32.02 oz.?

b. Let's say you buy 100 quart bottles of this ale for a party. How many bottles would you expect to find containing more than 32.02 oz. of ale?

6.54 The extraction force required to remove a cork from a bottle of wine has a normal distribution with a mean of 310 Newtons and a standard deviation of 36 Newtons.

a. The specs for this variable, given in Applied Example 6.15, were "300 N + 100 N/−150 N." Express these specs as an interval.

b. What percentage of the corks is expected to fall within the specs?

c. What percentage of the tested corks will have an extraction force of more than 250 Newtons?

d. What percentage of the tested corks will have an extraction force within 50 Newtons of 310?

6.55 The diameter of each cork, as described in Applied Example 6.15, is measured in several places, and an average diameter is reported for the cork. The average diameter has a normal distribution with a mean of 24.0 mm and standard deviation of 0.13 mm.

a. The specs for this variable, given in Applied Example 6.15, were "24 mm + 0.6 mm/−0.4 mm." Express these specs as an interval.

b. What percentage of the corks is expected to fall within the specs?

c. What percentage of the tested corks will have an average diameter of more than 24.5 mm?

d. What percentage of the tested corks will have an average diameter within 0.35 mm of 24?

6.56 Using the standard normal curve and z:

a. Find the minimum score needed to receive an A if the instructor in Example 6.12 (p. 325) said the top 15% were to get As.

b. Find the 25th percentile for IQ scores in Example 6.10 (p. 324).

c. If 20% of the salaries in Example 6.14 (p. 326) are below $28,000, find the current mean salary.

6.57 Final averages are typically approximately normally distributed with a mean of 72 and a standard deviation of 12.5. Your professor says that the top 8% of the class will receive an A; the next 20%, a B; the next 42%, a C; the next 18%, a D; and the bottom 12%, an F.

a. What average must you exceed to obtain an A?

b. What average must you exceed to receive a grade better than a C?

c. What average must you obtain to pass the course? (You'll need a D or better.)

6.58 A radar unit is used to measure the speed of automobiles on an expressway during rush-hour traffic. The speeds of individual automobiles are normally distributed with a mean of 62 mph.

a. Find the standard deviation of all speeds if 3% of the automobiles travel faster than 72 mph.

b. Using the standard deviation found in part a, find the percentage of these cars that are traveling less than 55 mph.

c. Using the standard deviation found in part a, find the 95th percentile for the variable "speed."

6.59 The weights of ripe watermelons grown at Mr. Smith's farm are normally distributed with a standard deviation of 2.8 lb. Find the mean weight of Mr. Smith's ripe watermelons if only 3% weigh less than 15 lb.

6.60 A machine fills containers with a mean weight per container of 16.0 oz. If no more than 5% of the containers are to weigh less than 15.8 oz., what must the standard deviation of the weights equal? (Assume normality.)

6.61 "On hold" times for callers to a local cable television company are known to be normally distributed with a standard deviation of 1.3 minutes. Find the average caller "on hold" time if the company maintains that no more than 10% of callers wait more than 6 minutes.

6.62 On a given day, the number of square feet of office space available for lease in a small city is a normally distributed random variable with a mean of 750,000 square feet and a standard deviation of 60,000 square feet. The number of square feet available in a second small city is normally distributed with a mean of 800,000 square feet and a standard deviation of 60,000 square feet.

a. Sketch the distribution of leasable office space for both cities on the same graph.

b. What is the probability that the number of square feet available in the first city is less than 800,000?

c. What is the probability that the number of square feet available in the second city is more than 750,000?

6.63 [EX06-063] The data are the net weights (in grams) for a sample of 30 bags of M&M's. The advertised net weight is 47.9 grams per bag.

46.22	46.72	46.94	47.61	47.67	47.70
47.98	48.28	48.33	48.45	48.49	48.72
48.74	48.95	48.98	49.16	49.40	49.69
49.79	49.80	49.80	50.01	50.23	50.40
50.43	50.97	51.53	51.68	51.71	52.06

Source: http://www.math.uah.edu/stat/, Christine Nickel and Jason York, ST 687 project, fall 1998

The FDA requires that (nearly) every bag contain the advertised weight; otherwise, violations (less than 47.9 grams per bag) will bring about mandated fines. (M&M's are manufactured and distributed by Mars Inc.)

a. What percentage of the bags in the sample are in violation?

b. If the weight of all filled bags is normally distributed with a mean weight of 47.9 g, what percentage of the bags will be in violation?

c. Assuming the bag weights are normally distributed with a standard deviation of 1.5 g, what mean value would leave 5% of the weights below 47.9 g?

d. Assuming the bag weights are normally distributed with a standard deviation of 1.0 g, what mean value would leave 5% of the weights below 47.9 g?

e. Assuming the bag weights are normally distributed with a standard deviation of 1.5 g, what mean value would leave 1% of the weights below 47.9 g?

f. Why is it important for Mars to keep the percentage of violations low?

g. It is important for Mars to keep the standard deviation as small as possible so that in turn the mean can be as small as possible to maintain net weight. Explain the relationship between the standard deviation and the mean. Explain why this is important to Mars.

6.64 a. Generate a random sample of 100 simulated values from a normal distribution with a mean of 50 and a standard deviation of 12.

b. Using the random sample of 100 simulated values found in part a and the technology commands for calculating ordinate values on page 328, find the 100 corresponding *y* values for the normal distribution curve with a mean of 50 and a standard deviation of 12.

c. Use the 100 ordered pairs found in part b and draw the curve for the normal distribution with a mean of 50 and a standard deviation of 12. (Technology commands are included with the part b commands on pages 328–329.)

d. Using the technology commands for cumulative probability on page 329, find the probability that a randomly selected value from a normal distribution with a mean of 50 and a standard deviation of 12 will be between 55 and 65. Verify your results by using Table 3 in Appendix B.

6.65 Use a computer or calculator to find the probability that one randomly selected value of x from a normal distribution (mean of 584.2 and standard deviation of 37.3) will have a value that corresponds to the following:

a. Less than 525   b. Between 525 and 590

c. At least 590

d. Verify the results of parts a–c using Table 3.

e. Explain any differences you may find between answers in part d and those in parts a–c.

MINITAB

Input 525 and 590 into C1; then continue with the cumulative probability commands on page 329, using 584.2 as μ, 37.3 as σ, and C2 as optional storage.

Excel

Input 525 and 590 into column A and activate the B1 cell; then continue with the cumulative probability commands on page 329, using 584.2 as μ and 37.3 as σ.

TI-83/84

Input 525 and 590 into L1; then continue with the cumulative probability commands on page 330 in L2, using 584.2 as μ and 37.3 as σ.

6.66 a. Use a computer to generate your own abbreviated standard normal probability table (a short version of Table 3). Use z-values of 0.0 to 5.0 in intervals of 0.1.

b. How are the values obtained related to Table 3 entries? Make the necessary adjustment and store the results in a column.

c. Compare your results in part b with the first column of Table 3. Comment on any differences you see.

MINITAB (Release 14)

a.
```
Choose:   Calc > Make Patterned Data > Simple Set of
          Numbers
Enter:    Store patterned data in: C1
From first value: 0
To last value: 5
In steps of: 0.1 > OK
```

Continue with the cumulative probability commands on page 329, using 0 as μ, 1 as σ, and C2 as optional storage.

b.
```
Choose:   Calc > Calculator
Enter:    Store result in variable: C3
          Expression: C2 − 0.5 > OK
Choose:   Data > Display Data
Enter:    Columns to display: C1 C3 > OK
```

Excel

a.
```
Choose:   Tools > Data Analysis > Random Number Genera-
          tion > OK
Enter:    Number of variables: 1
          Distribution: Patterned
          From: 0 to 5.0 in steps of 0.1
          Repeat each number: 1 times
Select:   Output Range
Enter:    (A1 or select cell)
```

Continue with the cumulative probability commands on page 329, activating cell B1 and using 0 as μ and 1 as σ.

b. Activate cell C1; then continue with:
```
Enter:    = B1 − 0.5 > Enter
Drag:     Bottom right corner of the C1 box down to give
          probabilities for the x values
```

6.67 Use a computer to compare a random sample to the population from which the sample was drawn. Consider the normal population with mean 100 and standard deviation 16.

a. List values of x from $\mu - 4\sigma$ to $\mu + 4\sigma$ in increments of half standard deviations and store them in a column.

b. Find the ordinate (y value) corresponding to each abscissa (x value) for the normal distribution curve for $N(100, 16)$ and store them in a column.

c. Graph the normal probability distribution curve for $N(100, 16)$.

d. Generate a random sample of 100 simulated values from the $N(100, 16)$ distribution and store them in a column.

e. Graph the histogram of the 100 values obtained in part d using the numbers listed in part a as class boundaries.

f. Calculate other helpful descriptive statistics of the 100 values and compare the data with the expected distribution. Comment on the similarities and the differences you see.

MINITAB (Release 14)

a. Use the Make Patterned Data commands in Exercise 6.66, replacing the first value with 36, the last value with 164, and the steps with 8.

b.

```
Choose:  Calc > Prob. Dist. > Normal
Select:  Probability density
Enter:   Mean:     100
         Stand. dev.:  16
         Input column:  C1
         Optional Storage:  C2 > OK
```

c. Use the Scatterplot commands on page 328 for the data in C1 and C2.

d. Use the Calculate RANDOM DATA commands on page 327, replacing n with 100, store in with C3, mean with 100, and standard deviation with 16.

e. Use the HISTOGRAM with Fits commands on page 61 for the data in C3. To adjust histogram, select Binning with cutpoint and cutpoint positions 36:148/8.

f. Use the MEAN and STANDARD DEVIATION commands on pages 74 and 88 for the data in C3.

Excel

a. Use the RANDOM NUMBER GENERATION Patterned Distribution commands in Exercise 6.66, replacing the first value with 36, the last value with 172, and the steps with 8.

b. Activate B1; then continue with:

```
Choose:  Insert function fₓ > Statistical > NORMDIST >
         OK
Enter:   X: (A1:A? or select 'x value' cells)
         Mean: 100
         Standard dev.: 16
         Cumulative: False > OK
Drag:    Bottom right corner of the ordinate value box
         down to give other ordinates
```

c. Use the CHART WIZARD XY(Scatter) commands on page 329 for the data in columns A and B.

d. Activate cell C1; then use the Normal RANDOM NUMBER GENERATION commands on page 328, replacing number of random numbers with 100, mean with 100, and standard deviation with 16.

e. Use the HISTOGRAM commands on page 61 with column C as the input range and column A as the bin range

f. Use the MEAN and STANDARD DEVIATION commands on pages 74 and 88 for the data in column C.

6.68 Use a computer to compare a random sample to the population from which the sample was drawn. Consider the normal population with mean 75 and standard deviation 14. Answer questions a–f of Exercise 6.67 using $N(75, 14)$.

6.69 Suppose you were to generate several random samples, all the same size, all from the same normal probability distribution. Will they all be the same? How will they differ? By how much will they differ?

a. Use a computer or calculator to generate 10 different samples, all of size 100, all from the normal probability distribution of mean 200 and standard deviation 25.

b. Draw histograms of all 10 samples using the same class boundaries.

c. Calculate several descriptive statistics for all 10 samples, separately.

d. Comment on the similarities and the differences you see.

MINITAB (Release 14)

a. Use the generate RANDOM DATA commands on page 327, replacing n with 100, store in with C1–C10, mean with 200, and standard deviation with 25.

b. Use the HISTOGRAM commands on page 61 for the data in C1–C10. To adjust histogram, select Binning with cutpoint and cutpoint positions 36:148/8.

c. Use the DISPLAY DESCRIPTIVE STATISTICS command on page 98 for the data in C1–C10.

Excel

a. Use the Normal RANDOM NUMBER GENERATION commands on page 328, replacing number of variables with 10, number of random numbers with 100, mean with 200, and standard deviation with 25.

b. Use the RANDOM NUMBER GENERATION Patterned Distribution commands in Exercise 6.66, replacing the first value with 100, the last value with 300, the steps with 25, and the output range with K1. Use the HISTOGRAM commands on page 61 for each of the columns A through J (input range) with column K as the bin range.

c. Use the DESCRIPTIVE STATISTICS commands on page 98 for the data in columns A through J.

TI-83/84 Plus

a. Use the 6:randNorm commands on page 328, replacing the mean with 200, the standard de-

viation with 25, and the number of trials with 100. Repeat six times, using L1–L6 for storage.

b. Use the HISTOGRAM commands on page 62 for the data in L1–L6, entering WINDOW values 100, 300, 25, −10, 60, 10, and 1. Adjust with ZoomStat.

c. Use the 1-Var Stats command on page 98 for the data in L1–L6.

6.70 Generate 10 random samples, each of size 25, from a normal distribution with mean 75 and standard deviation 14. Answer questions parts b–d of Exercise 6.69.

6.5 Notation

The *z*-score is used throughout statistics in a variety of ways; however, the relationship between the numerical value of *z* and the area under the **standard normal distribution** curve does not change. Since *z* will be used with great frequency, we want a convenient notation to identify the necessary information. The convention that we will use as an "algebraic name" for a specific *z*-score is $z(\alpha)$, where α represents the "area to the right" of the *z* being named.

EXAMPLE 6.16

Visual Interpretation of $z(\alpha)$

a. $z(0.05)$ (read "*z* of 0.05") is the algebraic name for *z* such that the area to the right and under the standard normal curve is exactly 0.05, as shown in Figure 6.6.

FIGURE 6.6 Area Associated with **z(0.05)**

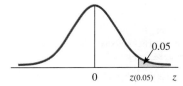

b. $z(0.60)$ (read "*z* of 0.60") is that value of *z* such that 0.60 of the area lies to its right, as shown in Figure 6.7.

FIGURE 6.7 Area Associated with **z(0.60)**

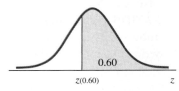

Now let's find the numerical values of $z(0.05)$, $z(0.60)$, and $z(0.95)$.

EXAMPLE 6.17   **Determining Corresponding *z* Values for *z*(α)**

a. Find the numerical value of $z(0.05)$.

SOLUTION
We must convert the area information in the notation into information that we can use with Table 3 in Appendix B. See the areas shown in Figure 6.8.

FIGURE 6.8 Find the value of *z*(0.05)

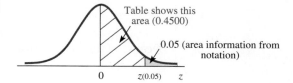

When we look in Table 3, we look for an area as close as possible to 0.4500.

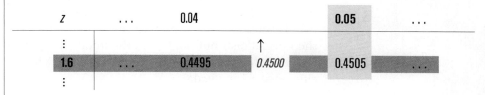

Therefore, $z(0.05) = \mathbf{1.65.}$

Note: We will use the *z* corresponding to the area closest in value. If the value is exactly halfway between the table entries, always use the larger value of *z*.

b. Find the numerical value of $z(0.60)$.

SOLUTION The value 0.60 is related to Table 3 by use of the area 0.1000, as shown in the diagram.

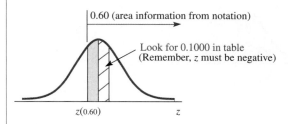

The closest values in Table 3 are 0.0987 and 0.1026.

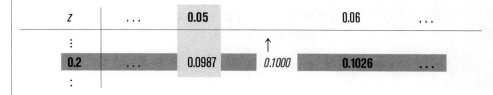

Therefore, $z(0.60)$ is related to 0.25. Since $z(0.60)$ is below the mean, we conclude that $z(0.60) = \mathbf{-0.25.}$

Statistics ⟡ Now™

Watch a video example at
http://1pass.thomson.com
or on your CD.

c. Find $z(0.95)$.

SOLUTION $z(0.95)$ is located on the left-hand side of the normal distribution because the area to the right is 0.95. The area in the tail to the left then contains the other 0.05, as shown in Figure 6.9.

FIGURE 6.9   Area Associated with z(0.95)

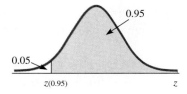

Because of the symmetrical nature of the normal distribution, $z(0.95)$ is $-z(0.05)$—that is, $z(0.05)$ with its sign changed. Thus, $z(0.95) = -z(0.05) = \mathbf{-1.65.}$

In the following chapters we will use this notation on a regular basis. The values of z that will be used regularly come from one of the following situations: (1) the z-score such that there is a specified area in one tail of the normal distribution or (2) the z-scores that bound a specified middle proportion of the normal distribution. When the middle proportion of a normal distribution is specified, we can still use the "area to the right" notation to identify the specific z-score involved.

EXAMPLE 6.18

Determining z-Scores for Bounded Areas

Find the z-scores that bound the middle 0.95 of the normal distribution.

SOLUTION   Given 0.95 as the area in the middle (see Figure 6.10), the two tails must contain a total of 0.05. Therefore, each tail contains $\frac{1}{2}$ of 0.05, or 0.025, as shown in Figure 6.11.

FIGURE 6.10   Area Associated with Middle 0.95

FIGURE 6.11   Finding z-Scores for Middle 0.95

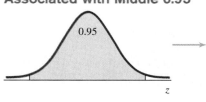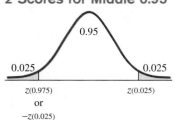

In order to find $z(0.025)$ in Table 3, we must determine the area between the mean and $z(0.025)$. It is $0.5000 - 0.0250 = 0.4750$, as shown in Figure 6.12.

FIGURE 6.12   Finding the Value of z(0.025)

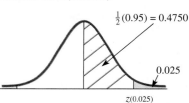

Table 3 shows us:

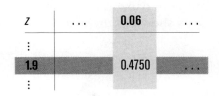

Therefore, $z(0.025) = 1.96$ and $z(0.975) = -z(0.025) = -1.96$. The middle 0.95 of the normal distribution is bounded by **−1.96** and **1.96.**

SECTION 6.5 EXERCISES

6.71 Using the $z(\alpha)$ notation (identify the value of α used within the parentheses), name each of the standard normal variable z's shown in the following diagrams.

a.

b.

c.

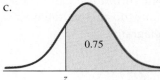

d.

e.

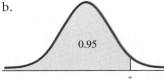

f.

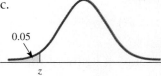

6.72 Using the $z(\alpha)$ notation (identify the value of α used within the parentheses), name each of the standard normal variable z's shown in the following diagrams.

a.

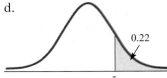

b.

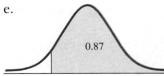

c.

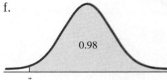

d.

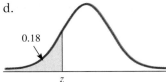

e.

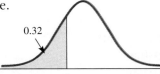

f.

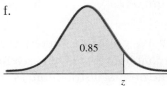

6.73 Using the $z(\alpha)$ notation (identify the value of α used within the parentheses), name each of the standard normal variable z's shown in the following diagrams.

a.   b.

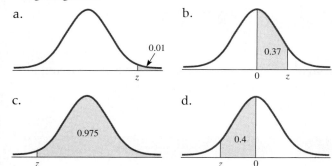

c.   d.

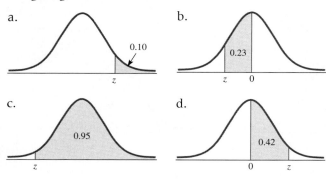

6.74 Using the $z(\alpha)$ notation (identify the value of α used within the parentheses), name each of the standard normal variable z's shown in the following diagrams.

a.   b.

c.   d.

6.75 Draw a figure of the standard normal curve showing:

a. $z(0.15)$ b. $z(0.82)$

6.76 Draw a figure of the standard normal curve showing:

a. $z(0.04)$ b. $z(0.94)$

6.77 We are often interested in finding the value of z that bounds a given area in the right-hand tail of the normal distribution, as shown in the accompanying figure. The notation $z(\alpha)$ represents the value of z such that $P(z > z(\alpha)) = \alpha$.

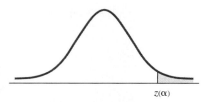

Find the following:

a. $z(0.025)$ b. $z(0.05)$ c. $z(0.01)$

6.78 Find the value of the following:

a. $z(0.15)$ b. $z(0.82)$

6.79 Find the value of the following:

a. $z(0.08)$ b. $z(0.92)$

6.80 Use Table 3 in Appendix B to find the following values of z.

a. $z(0.05)$ b. $z(0.01)$ c. $z(0.025)$

d. $z(0.975)$ e. $z(0.98)$

6.81 Complete the following charts of z-scores. The area A given in the tables is the area to the right under the normal distribution in the figures.

a. z-scores associated with the right-hand tail: Given the area A, find $z(A)$.

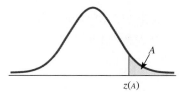

A	0.10	0.05	0.025	0.02	0.01	0.005
$z(A)$						

b. z-scores associated with the left-hand tail: Given the area B, find $z(B)$.

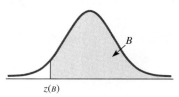

B	0.995	0.99	0.98	0.975	0.95	0.90
$z(B)$						

6.82 a. Find the area under the normal curve for z between $z(0.95)$ and $z(0.025)$.

  b. Find $z(0.025) - z(0.95)$.

6.83 The z notation, $z(\alpha)$, combines two related concepts—the z-score and the area to the right—into a mathematical symbol. Identify the letter in each of the following as being a z-score or being an area; then, with the aid of a diagram, explain what both the given number and the letter represent on the standard normal curve.

a. $z(A) = 0.10$ **b.** $z(0.10) = B$

c. $z(C) = -0.05$ **d.** $-z(0.05) = D$

6.84 Understanding the z notation, $z(\alpha)$, requires us to know whether we have a z-score or an area. Each of the following expressions use the z notation in a variety of ways, some typical and some not so typical. Find the value asked for in each of the following; then, with the aid of a diagram, explain what your answer represents.

a. $z(0.08)$

b. The area between $z(0.98)$ and $z(0.02)$

c. $z(1.00 - 0.01)$ **d.** $z(0.025) - z(0.975)$

6.6   Normal Approximation of the Binomial

In Chapter 5 we introduced the **binomial distribution.** Recall that the binomial distribution is a probability distribution of the discrete random variable x, the number of successes observed in n repeated independent trials. We will now see how **binomial probabilities**—that is, probabilities associated with a binomial distribution—can be reasonably approximated by using the normal probability distribution.

Let's look first at a few specific binomial distributions. Figure 6.13 shows the probabilities of x for 0 to n for three situations: $n = 4$, $n = 8$, and $n = 24$. For each of these distributions, the probability of success for one trial is 0.5. Notice that as n becomes larger, the distribution appears more and more like the normal distribution.

FIGURE 6.13
Binomial Distributions

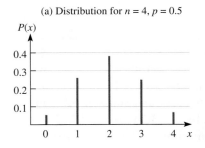

(a) Distribution for $n = 4$, $p = 0.5$

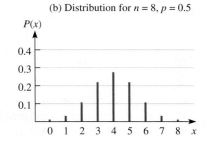

(b) Distribution for $n = 8$, $p = 0.5$

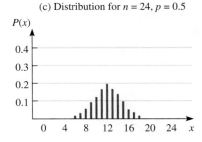

(c) Distribution for $n = 24$, $p = 0.5$

To make the desired approximation, we need to take into account one major difference between the binomial and the normal probability distribution. The binomial random variable is **discrete,** whereas the normal random variable is **continuous.** Recall that Chapter 5 demonstrated that the probability assigned to a particular value of x should be shown on a diagram by means of a straight-line segment whose length represents the probability (as in Figure 6.13). Chapter 5 suggested, however, that we can also use a histogram in which the area of each bar is equal to the probability of x.

Let's look at the distribution of the binomial variable x, when $n = 14$ and $p = 0.5$. The probabilities for each x value can be obtained from Table 2 in Appendix B. This distribution of x is shown in Figure 6.14. We see the very same distribution in Figure 6.15 in histogram form.

FIGURE 6.14   The Distribution of x when $n = 14, p = 0.5$

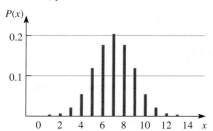

FIGURE 6.15   Histogram for the Distribution of x when $n = 14, p = 0.5$

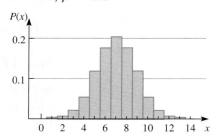

Let's examine $P(x = 4)$ for $n = 14$ and $p = 0.5$ to study the approximation technique. $P(x = 4)$ is equal to 0.061 (see Table 2 in Appendix B), the area of the bar (rectangle) above $x = 4$ in Figure 6.16.

FIGURE 6.16   Area of Bar above $x = 4$ Is 0.061, for $B(n = 14, p = 0.5)$

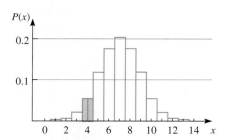

The area of a rectangle is the product of its width and height. In this case the height is 0.061 and the width is 1.0, so the area is 0.061. Let's take a closer look at the width. For $x = 4$, the bar starts at 3.5 and ends at 4.5, so we are looking at an area bounded by $x = 3.5$ and $x = 4.5$. The addition and subtraction of 0.5 to the x value is commonly called the **continuity correction factor.** It is our method of converting a discrete variable into a continuous variable.

Now let's look at the normal distribution related to this situation. We will first need a normal distribution with a mean and a standard deviation equal to those of the binomial distribution we are discussing. Formulas (5.7) and (5.8) give us these values:

$$\mu = np = (14)(0.5) = \mathbf{7.0}$$
$$\sigma = \sqrt{npq} = \sqrt{(14)(0.5)(0.5)} = \sqrt{3.5} = \mathbf{1.87}$$

The probability that $x = 4$ is approximated by the area under the normal curve between $x = 3.5$ and $x = 4.5$, as shown in Figure 6.17. Figure 6.18 shows the entire distribution of the binomial variable x with a normal distribution of the same mean and standard deviation superimposed. Notice that the bars and the interval areas under the curve cover nearly the same area.

FIGURE 6.17 Probability That $x = 4$ Is Approximated by Shaded Area

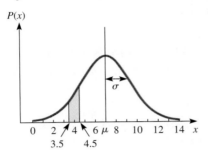

FIGURE 6.18 Normal Distribution Superimposed over Distribution for Binomial Variable x

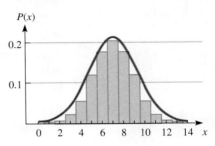

The probability that x is between 3.5 and 4.5 under this normal curve is found by using formula (6.3), Table 3 and the methods outlined in Section 6.4:

$$z = \frac{x - \mu}{\sigma}: \qquad P(3.5 < x < 4.5) = P\left(\frac{3.5 - 7.0}{1.87} < z < \frac{4.5 - 7.0}{1.87}\right)$$

$$= P(-1.87 < z < -1.34)$$

$$= 0.4693 - 0.4099 = \mathbf{0.0594}$$

Since the binomial probability of 0.061 and the normal probability of 0.0594 are reasonably close, the normal probability distribution seems to be a reasonable approximation of the binomial distribution.

The normal approximation of the binomial distribution is also useful for values of p that are not close to 0.5. The binomial probability distributions shown in Figures 6.19 and 6.20 suggest that binomial probabilities can be approximated using the normal distribution. Notice that as n increases, the binomial distribution begins to look like the normal distribution. As the value of

FIGURE 6.19 Binomial Distributions

(a) Distribution for $n = 4$, $p = 0.3$

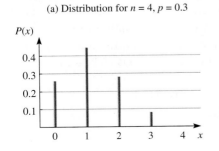

(b) Distribution for $n = 8$, $p = 0.3$

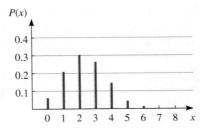

(c) Distribution for $n = 24$, $p = 0.3$

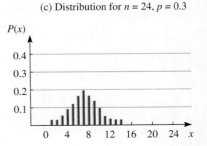

FIGURE 6.20 Binomial Distributions

(a) Distribution for $n = 4$, $p = 0.1$

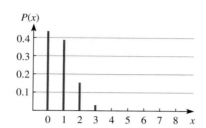

(b) Distribution for $n = 8$, $p = 0.1$

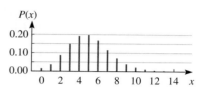

(c) Distribution for $n = 50$, $p = 0.1$

p moves away from 0.5, a larger n is needed in order for the normal approximation to be reasonable. The following *rule of thumb* is generally used as a guideline:

> **Rule:** The normal distribution provides a reasonable approximation to a binomial probability distribution whenever the values of np and $n(1 - p)$ both equal or exceed 5.

By now you may be thinking, "So what? I will just use the binomial table and find the probabilities directly and avoid all the extra work." But consider for a moment the situation presented in Example 6.19.

EXAMPLE 6.19

Solving a Binomial Probability Problem with the Normal Distribution

An unnoticed mechanical failure has caused $\frac{1}{3}$ of a machine shop's production of 5000 rifle firing pins to be defective. What is the probability that an inspector will find no more than 3 defective firing pins in a random sample of 25?

SOLUTION In this example of a binomial experiment, x is the number of defectives found in the sample, $n = 25$, and $p = P(\text{defective}) = \frac{1}{3}$. To answer the question using the binomial distribution, we will need to use the binomial probability function, formula (5.5):

$$P(x) = \binom{25}{x}\left(\frac{1}{3}\right)^x\left(\frac{2}{3}\right)^{25-x} \quad \text{for } x = 0, 1, 2, \dots, 25$$

We must calculate the values for $P(0)$, $P(1)$, $P(2)$, and $P(3)$, because they do not appear in Table 2. This is a very tedious job because of the size of the exponent. In situations such as this, we can use the normal approximation method.

Now let's find $P(x \le 3)$ by using the normal approximation method. We first need to find the mean and standard deviation of x, formulas (5.7) and (5.8):

$$\mu = np = (25)\left(\frac{1}{3}\right) = \mathbf{8.333}$$

$$\sigma = \sqrt{npq} = \sqrt{(25)\left(\frac{1}{3}\right)\left(\frac{2}{3}\right)} = \sqrt{5.55556} = \mathbf{2.357}$$

These values are shown in the figure. The area of the shaded region ($x < 3.5$) represents the probability of $x = 0$, 1, 2, or 3. Remember that $x = 3$, the discrete binomial variable, covers the continuous interval from 2.5 to 3.5.

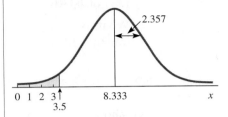

$$P(x \text{ is no more than } 3) = P(x \le 3) \quad \text{(for a discrete variable } x\text{)}$$

$$= P(x < 3.5) \quad \text{(for a continuous variable } x\text{)}$$

$$z = \frac{x - \mu}{\sigma}: \qquad P(x < 3.5) = P\left(z < \frac{3.5 - 8.333}{2.357}\right) = P(z < -2.05)$$

$$= 0.5000 - 0.4798 = \mathbf{0.0202}$$

Thus, $P(\text{no more than three defectives})$ is approximately 0.02.

SECTION 6.6 EXERCISES

6.85 Find the values np and nq (recall: $q = 1 - p$) for a binomial experiment with $n = 100$ and $p = 0.02$. Does this binomial distribution satisfy the rule for normal approximation? Explain.

6.86 In which of the following binomial distributions does the normal distribution provide a reasonable approximation? Use computer commands to generate a graph of the distribution and compare the results to the "rule of thumb." State your conclusions.

a. $n = 10$, $p = 0.3$ b. $n = 100$, $p = 0.005$

c. $n = 500$, $p = 0.1$ d. $n = 50$, $p = 0.2$

MINITAB (Release 14)

Insert the specific n and p as needed in the following procedure.
Use the Make Patterned Data commands in Exercise 6.66, replacing the first value with 0, the last value with n, and the steps with 1.
Use the Binomial Probability Distribution commands on page 292, using C2 as optional storage.

Use the Scatterplot Simple commands for the data in C1 and C2. Select Data View, Data Display, Project Lines to complete the graph.

Excel

Insert the specific n and p as needed in the following procedure.
Use the RANDOM NUMBER GENERATION Patterned Distribution commands in Exercise 6.66, replacing the first value with 0, the last value with n, the steps with 1, and the output range with A1.
Activate cell B1; then use the Binomial Probability Distribution commands on page 292.
Use the Chart Wizard Column commands for the data in columns A and B. Choosing the Series subcommand, input column B for the y values and column A for the category (x) axis labels.

6.87 In order to see what happens when the normal approximation is improperly used, consider the binomial distribution with $n = 15$ and $p = 0.05$. Since $np = 0.75$, the rule of thumb ($np > 5$ and $nq > 5$) is not satisfied. Using the binomial tables, find the probability of one or fewer successes and compare this with the normal approximation.

6.88 Find the normal approximation for the binomial probability $P(x = 6)$, where $n = 12$ and $p = 0.6$. Compare this to the value of $P(x = 6)$ obtained from Table 2 in Appendix B.

6.89 Find the normal approximation for the binomial probability $P(x = 4, 5)$, where $n = 14$ and $p = 0.5$. Compare this to the value of $P(x = 4, 5)$ obtained from Table 2 in Appendix B.

6.90 Find the normal approximation for the binomial probability $P(x \leq 8)$, where $n = 14$ and $p = 0.4$. Compare this to the value of $P(x \leq 8)$ obtained from Table 2 in Appendix B.

6.91 Find the normal approximation for the binomial probability $P(x \geq 9)$, where $n = 13$ and $p = 0.7$. Compare this to the value of $P(x \geq 9)$ obtained from Table 2 in Appendix B.

6.92 Referring to Example 6.19 (p. 346):

a. Calculate $P(x \leq 3 \mid B(25, \frac{1}{3}))$. (*Hint:* If you use a computer or calculator, use the commands on p. 292.)

b. How good was the normal approximation? Explain.

6.93 Melanoma is the most serious form of skin cancer and is increasing at a rate higher than any other cancer in the United States. If it is caught in its early stage, the survival rate for patients is almost 90% in the United States. What is the probability that 200 or more of some group of 250 early-stage patients will survive melanoma?

Source: http://www.channelonenews.com/articles/2002/05/31/skin.cancer/

6.94 If 30% of all students entering a certain university drop out during or at the end of their first year, what is the probability that more than 600 of this year's entering class of 1800 will drop out during or at the end of their first year?

6.95 According to the Federal Highway Administration, nearly 50% of U.S. drivers are females. As-

sume a random sample of 50 drivers is to be selected for survey.

a. What is the probability that no more than half (25) of the drivers will be female?

b. What is the probability that at least three-fourths (38) of the drivers will be female?

6.96 According to a February 2005 survey completed by the Pew Internet & American Life Project, about 36 million Americans, or 27% of Internet users, say they download either music or video files. Assuming the percentage is correct, use the normal approximation to the binomial to find the probability of the following in a survey of 2000 American Internet users.

Source: http://www.pewinternet.org/PPF/r/153/report_display.asp

a. At least 500 download music or video files

b. At least 575 download music or video files

c. At most 520 download music or video files

d. At most 555 download music or video files

6.97 Not all NBA coaches who enjoyed lengthy careers were consistently putting together winning seasons with the teams they coached. For example, Bill Fitch, who coached for 25 seasons of professional basketball after starting his coaching career at the University of Minnesota, won 944 games but lost 1106 while working with the Cavaliers, Celtics, Rockets, Nets, and Clippers. If you were to randomly select 60 box scores from the historical records of games in which Bill Fitch coached one of the teams, what is the probability that less than half of them show his team winning? To obtain your answer, use the normal approximation to the binomial distribution.

Source: http://www.basketball-reference.com

6.98 One poll found that more than 6 in 10 voters say they believe the United States is ready for a female president. The poll was conducted in February 2005 by the Siena College Research Institute and was sponsored by Hearst Newspapers. Assuming that the proportion is 0.6, what is the probability that another poll of 1125 registered voters conducted randomly will result in the following:

a. More than two-thirds believe the United States is ready for a female president.

b. Less than 58% believe the United States is ready for a female president.

6.99 According to an April 2005 report from the Substance Abuse and Mental Health Services Administration (SAMHSA), 35% of people involved in treatment began drinking heavily between ages 15 and 17 (http://www.jointogether.org). Use the normal approximation to the binomial distribution to find the probability that in a poll of 1200 people in treatment, between 450 and 500 inclusive will have begun drinking heavily between ages 15 and 17.

a. Solve using normal approximation and Table 3 in Appendix B.

b. Solve using a computer or calculator and the normal approximation method.

c. Solve using a computer or calculator and the binomial probability function.

6.100 In 2003, of the nearly 105 million native wage and salary workers in the United States, approximately 15.4 million were affiliated with a union. Use the normal approximation to the binomial distribution to find the probability that in a national survey of 2500 workers, at most 400 will be union members.

a. Solve using normal approximation and Table 3 in Appendix B.

b. Solve using a computer or calculator and the normal approximation method.

CHAPTER REVIEW

In Retrospect

We have learned about the standard normal probability distribution, the most important family of continuous random variables. We have learned to apply it to all other normal probability distributions and how to use it to estimate probabilities of binomial distributions. We have seen a wide variety of variables that have this normal distribution or are reasonably well approximated by it.

In the next chapter we will examine sampling distributions and learn how to use the standard normal probability to solve additional applications.

Vocabulary and Key Concepts

area representation for probability (p. 316)
bell-shaped curve (p. 315)
binomial distribution (p. 343)
binomial probability (p. 343)
continuity correction factor (p. 344)

continuous random variable (pp. 315, 344)
discrete random variable (pp. 315, 344)
normal approximation of binomial (p. 343)
normal curve (p. 316)
normal distribution (p. 315)

percentage (p. 316)
probability (p. 316)
proportion (p. 316)
random variable (p. 315)
standard normal distribution (pp. 316, 323, 338)
standard score (pp. 316, 323)
z-score (pp. 316, 323)

Learning Outcomes

Chapter Exercises

Statistics ⬡ Now™

Go to the StatisticsNow website **http://1pass.thomson.com** to

• Assess your understanding of this chapter

• Check your readiness for an exam by taking the Pre-Test quiz and exploring the resources in the Personalized Learning Plan

Datasets can be found on your Student's Suite CD-ROM or at the StatisticsNow website at **http://1pass.thomson.com**.

6.101 According to Chebyshev's theorem, at least how much area is there under the standard normal distribution between $z = -2$ and $z = +2$? What is the actual area under the standard normal distribution between $z = -2$ and $z = +2$?

6.102 The middle 60% of a normally distributed population lies between what two standard scores?

6.103 Find the standard score (z) such that the area above the mean and below z under the normal curve is:

a. 0.3962   b. 0.4846   c. 0.3712

6.104 Find the standard score (z) such that the area below the mean and above z under the normal curve is:

a. 0.3212   b. 0.4788   c. 0.2700

6.105 Given that z is the standard normal variable, find the value of k such that:

a. $P(|z| > 1.68) = k$ b. $P(|z| < 2.15) = k$

6.106 Given that z is the standard normal variable, find the value of c such that:

a. $P(|z| > c) = 0.0384$ b. $P(|z| < c) = 0.8740$

6.107 Find the following values of z:

a. $z(0.12)$ b. $z(0.28)$ c. $z(0.85)$ d. $z(0.99)$

6.108 Find the area under the normal curve that lies between the following pairs of z-values:

a. $z = -3.00$ and $z = 3.00$

b. $z(0.975)$ and $z(0.025)$

c. $z(0.10)$ and $z(0.01)$

6.109 Based on data from ACT in 2004, the average science reasoning test score was 20.9, with a standard deviation of 4.6. Assuming that the scores are normally distributed:

a. Find the probability that a randomly selected student has a science reasoning ACT score of least 25.

b. Find the probability that a randomly selected student has a science reasoning ACT score between 20 and 26.

c. Find the probability that a randomly selected student has a science reasoning ACT score less than 16.

6.110 The 70-year long-term record for weather shows that for New York State, the annual precipitation has a mean of 39.67 inches and a standard deviation of 4.38 inches.

Source: Department of Commerce; State, Regional and National Monthly Precipitation Report

If the annual precipitation amount has a normal distribution, what is the probability that next year the total precipitation for New York State is:

a. More than 50.0 inches

b. Between 42.0 and 48.0 inches

c. Between 30.0 and 37.5 inches

d. More than 35.0 inches

e. Less than 45.0 inches

f. Less than 32.0 inches

6.111 American Express charges merchants higher fees than any other credit or debit card, according to the *USA Today* article "American Express fees take flak" (December 23, 2004). The company believes they can do this because they claim the customers using the American Express card spend more. The average annual charges per card in 2003 were $9600 according to data from American Express and *The Neilson Report*. Assuming that the annual charges per card are approximately normally distributed with a standard deviation of $2100, what is the probability that an American Express customer's annual charges are:

a. Less than $4000

b. Between $5000 and $10,000

c. Greater than $16,000

6.112 A company that produces rivets used by commercial aircraft manufacturers knows that the shearing strength (force required to break) of its rivets is of major concern. The company believes the shearing strength of its rivets is normally distributed with a mean of 925 pounds and a standard deviation of 18 pounds.

a. If the company is correct, what percentage of its rivets have a shearing strength greater than 900 pounds?

b. What is the upper bound for the shearing strength of the weakest 1% of the rivets?

c. If one rivet is randomly selected from all of the rivets, what is the probability that it will require a force of at least 920 pounds to break it?

d. Using the probability found in part c, rounded to nearest tenth, what is the probability that 3 rivets in a random sample of 10 will break at a force less than 920 pounds?

6.113 In a study of the length of time it takes to play major league baseball games during the early 2005 season, the variable "time of game" appears to be normally distributed with a mean of 2 hours, 50.1 minutes and a standard deviation of 20.99 minutes.

Source: MLB.com

a. Some fans describe a game as "unmanageably long" if it takes more than 3 hours. What is the probability that a randomly identified game was unmanageably long?

b. Many fans describe a game lasting less than 2 hours, 30 minutes as "quick." What is the probability that a randomly selected game was quick?

c. What are the bounds of the interquartile range for the variable "time of game"?

d. What are the bounds for the middle 90% of the variable "time of game"?

6.114 A certain type of refrigerator has a length of life that is approximately normally distributed with a mean of 4.8 years and a standard deviation of 1.3 years.

a. If this machine is guaranteed for 2 years, what is the probability that the machine you purchased will require replacement under the guarantee?

b. What period of time should the manufacturer give as a guarantee if it is willing to replace only 0.5% of the machines?

6.115 A machine is programmed to fill 10-oz. containers with a cleanser. However, the variability inherent in any machine causes the actual amounts of fill to vary. The distribution is normal with a standard deviation of 0.02 oz. What must the mean amount μ be in order that only 5% of the containers receive less than 10 oz.?

6.116 In a large industrial complex, the maintenance department has been instructed to replace light bulbs before they burn out. It is known that the life of light bulbs is normally distributed with a mean life of 900 hours of use and a standard deviation of 75 hours. When should the light bulbs be replaced so that no more than 10% of them will burn out while in use?

6.117 The grades on an examination whose mean is 525 and whose standard deviation is 80 are normally distributed.

a. Anyone who scores below 350 will be retested. What percentage does this represent?

b. The top 12% are to receive a special commendation. What score must be surpassed to receive this special commendation?

c. The interquartile range of a distribution is the difference between Q_1 and Q_3 (that is, $Q_3 - Q_1$). Find the interquartile range for the grades on this examination.

d. Find the grade such that only 1 out of 500 will score above it.

6.118 A soft-drink vending machine can be regulated to ensure that it dispenses an average of μ oz. of soft drink per glass.

a. If the ounces dispensed per glass are normally distributed with a standard deviation of 0.2 oz., find the setting for μ that will allow a 6-oz. glass to hold (without overflowing) the amount dispensed 99% of the time.

b. Use a computer or calculator to simulate drawing a sample of 40 glasses of soft drink from the machine (set using your answer to part a).

MINITAB (Release 14)

Use the Calculate RANDOM DATA commands on page 327, replacing *n* with 40, store in with C1, mean with the value calculated in part a, and standard deviation with 0.2.

Use the HISTOGRAM commands on page 61 for the data in C1. To adjust the histogram, select Binning with cutpoint and cutpoint positions 5:6.2/0.05.

Excel

Use the Normal RANDOM NUMBER GENERATION commands on page 328, replacing *n* with 40, the mean with the value calculated in part a, the standard deviation with 0.2, and the output range with A1.

Use the RANDOM NUMBER GENERATION Patterned Distribution on page 336, replacing the first value with 5, the last value with 6.2, the steps with 0.05, and the output range with B1.

Use the HISTOGRAM commands on page 61 with column A as the input range and column B as the bin range.

TI-83/84 Plus

Use the 6:randNorm commands on page 328, replacing the mean with the value calculated in part a, the standard deviation with 0.2, and the number of trials with 40. Store in with L1.

Use the HISTOGRAM commands on page 62 for the data in L1, entering the following WINDOW VALUES: 5, 6.2, 0.05, −1, 10, 1, 1.

c. What percentage of your sample would have overflowed the cup?

d. Does your sample seem to indicate the setting for μ is going to work? Explain.

FYI Repeat part b a few times. Try a different value for the mean amount dispensed and repeat part b. Observe how many would overflow in each set of 40.

6.119 Suppose that x has a binomial distribution with $n = 25$ and $p = 0.3$.

a. Explain why the normal approximation is reasonable.

b. Find the mean and standard deviation of the normal distribution that is used in the approximation.

6.120 Let x be a binomial random variable for $n = 30$ and $p = 0.1$.

a. Explain why the normal approximation is not reasonable.

b. Find the function used to calculate the probability of any x from $x = 0$ to $x = 30$.

c. Use a computer or calculator to list the probability distribution.

6.121 a. Use a computer or calculator to list the binomial probabilities for the distribution where $n = 50$ and $p = 0.1$.

b. Use the results from part a and find $P(x \le 6)$.

c. Find the normal approximation for $P(x \le 6)$, and compare the results with those in part b.

6.122 a. Use a computer or calculator to list both the probability distribution and the cumulative probability distribution for the binomial probability experiment with $n = 40$ and $p = 0.4$.

b. Explain the relationship between the two distributions found in part a.

c. If you could use only one of these lists when solving problems, which one would you prefer and why?

6.123 Consider the binomial experiment with $n = 300$ and $p = 0.2$.

a. Set up, but do not evaluate, the probability expression for 75 or fewer successes in the 300 trials.

b. Use a computer or calculator to find $P(x \le 75)$ using the binomial probability function.

c. Use a computer or calculator to find $P(x \le 75)$ using the normal approximation.

d. Compare the answers in parts b and c.

FYI Use the cumulative probability commands.

6.124 A test-scoring machine is known to record an incorrect grade on 5% of the exams it grades. Use the appropriate method to find the probability that the machine records the following:

a. Exactly 3 wrong grades in a set of 5 exams

b. No more than 3 wrong grades in a set of 5 exams

c. No more than 3 wrong grades in a set of 15 exams

d. No more than 3 wrong grades in a set of 150 exams

6.125 A company asserts that 80% of the customers who purchase its special lawn mower will require no repairs during the first 2 years of ownership. Your personal study has shown that only 70 of the 100 in your sample lasted the 2 years without incurring repair expenses. What is the probability of your sample outcome or less if the actual repair expense–free percentage is 80%?

6.126 It is believed that 58% of married couples with children agree on methods of disciplining their children. Assuming this to be the case, what is the probability that in a random survey of 200 married couples, we would find:

a. Exactly 110 couples who agree

b. Fewer than 110 couples who agree

c. More than 100 couples who agree

6.127 In a February 2005 poll conducted by Salary.com, firefighters hosed down the competition and won the title of "sexiest job," with 16% of the votes. Suppose you randomly selected 50 adults. Use the normal approximation to the binomial distribution to find the probability that from within your collection:

Source: http://salary.com/careers/layoutscripts/crel_display .asp?tab=cre&cat=nocat&ser=Ser348&part=Par516

a. More than 12 of the adults pick firefighter as the sexiest job.

b. Less than 8 of the adults pick firefighter as the sexiest job.

c. From 7 to 14 of the adults pick firefighter as the sexiest job.

6.128 The 2004 Pew Internet & American Life Project survey revealed that 4 in 10 online Americans—about 53 millions American adults—use instant messaging (IM) software.

Source: http://www.pewinternet.org/PPF/r/133/report_ display.asp

Use the normal approximation to the binomial to find the probability that in a random sample of 100 Internet users, no more than 50 use IM programs.

6.129 The National Coffee Drinking Trends is "the publication" in the coffee industry. For more than five decades, it has tracked annual consumption patterns in a wide variety of situations and categories. The 2004 edition says that 39% of the total coffee drinkers 18 years and older purchased shade-grown coffee in 2004.

Source: http://www.ncausa.org/public/pages/index .cfm?pageid=38

If this percentage is true for coffee drinkers at Crimson Light's coffeehouse, what is the probability of the following for the next 50 customers purchasing coffee at Crimson Light's:

a. More than 20 have purchased a shade-grown variety.

b. Fewer than 15 have purchased a shade-grown variety.

6.130 Apparently, playing video games, watching TV, and instant messaging friends isn't relaxing enough. In a February 2005 poll from Yesawich, Pepperdine, Brown and Russell found that one-third of the children polled said they helped research some aspect of their family's vacation on the Internet. If a follow-up survey of 100 of these children is taken, what is the probability of the following:

a. Less than 25% of the new sample will say they helped research the family vacation on the Internet.

b. More than 40% of the new sample will say they helped research the family vacation on the Internet.

6.131 The U.S. civilian labor force of 148,157,000 workers was 94.8% employed in March 2005. If a random sample of 2500 is taken from the civilian labor force, what is the probability of the following:

a. More than 6% of the sample will be unemployed.

b. Less than 5% of the sample will be unemployed.

6.132 During the first 2 months of 2005, there were 1,140,256 commercial airline flights in and out of U.S. airports. Of these, 74.35% were on-time arrivals and 18.96% were late departures. Three hundred flights are to be randomly identified from all flights and their flight logs examined closely. What is the probability of the following:

a. More than 80% of the sample will be an on-time arrival.

b. Less than 15% of the sample will have departed late.

6.133 Infant mortality rates are often used to assess quality of life and adequacy of health care. The rate is based on the number of deaths of infants younger than 1 year old in a given year per 1000 live births in the same year. Listed here are the infant mortality rates, to the nearest integer, for eight nations throughout the world, as found in *The World Factbook*, 2004.

Nation	Infant Mortality (per 1000 live births)	Nation	Infant Mortality (per 1000 live births)
China	25	Mexico	22
Germany	4	Russia	17
India	58	S. Africa	62
Japan	3	United States	7

Source: http://www.cia.gov/cia/publications/factbook/docs/notesanddefs.html

Suppose the next 2000 births within each nation are tracked for the occurrence of infant deaths.

a. Construct a table showing the mean and standard deviation of the associated binomial distributions.

b. In the final column of the table, find the probability that at least 70 infants from the samples within each nation will become casualties that contribute to the nation's mortality rate. Show all work.

c. Explain what caused the answers to vary so much.

6.134 [EX06-134] A large sample was randomly selected from a competitive product and evaluated for a particular lens dimension. It was then compared with its specification range of Nominal (0.000) ± 0.030 unit. A total of 110 lenses were evaluated. The data were coded in two ways and shown here:

−0.020	−0.043	−0.002	0.002	−0.018
−0.016	−0.051	0.024	−0.024	−0.032

••• Remainder of data on Student's Suite CD-ROM

Source: Courtesy of Bausch & Lomb (Variable not named and data coded at B&L's request.)

a. Calculate the mean and standard deviation of the data.

b. Create a histogram and comment on the pattern of variability of the data.

c. Use tests for normality and/or the empirical rule as confirmation of the normal appearance. Explain your findings.

d. Determine the observed percentage of conformance to specification. That is, what percentage of the measurements did fall within the specification range of 0.000 ± 0.030 unit?

6.135 Assume the distribution of data in Exercise 6.134 was exactly normally distributed with a mean of 0.00 and standard deviation of 0.020.

a. Find the bounds of the middle 95% of the distribution?

b. What percent of the data actually is within the interval found in part a?

c. Using *z*-scores, determine the percentage of estimated conformance to specification. That is, what percentage of the measurements would be expected to fall within the specification range of 0.000 ± 0.030 unit?

6.136 The following triangular distribution provides an approximation to the normal distribution. Line segment l_1 has the equation $y = x/9 + 1/3$, and segment l_2 has the equation $y = -x/9 + 1/3$.

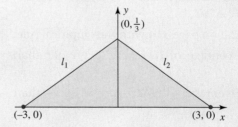

a. Find the area under the entire triangular distribution.

b. Find the area under the triangular distribution between 0 and 2.

c. Find the area under the standard normal distribution between 0 and 2.

d. Discuss the effectiveness of this "triangular" approximation.

Chapter Project

Intelligence Scores

All normal probability distributions have the same shape and distribution relative to the mean and standard deviation. In this chapter we learned how to use the standard normal probability distribution to answer questions about all normal distributions. Let's return to distribution of IQ scores discussed in the Section 6.1, "Intelligence Scores" (p. 313), and try out some of our new knowledge.

Putting Chapter 6 to Work

6.137 Let's take a second look at the normally distributed IQ scores illustrated in Section 6.1, "Intelligence Scores" (p. 313). If completed, use your Exercise 6.1 solutions as a basis.

a. How is an IQ score converted to a standard score?

b. What is the standard score for an IQ score of 90? 110? 120?

c. What is the standard score for an SAT score of 465? 575? 650?

Using Figure 2.2 on page 314 with the empirical rule:

d. What percentage of IQ scores is greater than 132?

e. What percentage of SAT scores is less than 700?

Using Table 3 in Appendix B:

f. What is the probability that an IQ score is greater than 132?

g. What is the probability that an SAT score is less than 700?

h. Compare your answers to parts f and g with your answers to parts d and e that used the empirical rule and Figure 2.2 on page 314. Explain any similarities.

i. What proportion of the IQ scores fall within the range of 80 to 120?

j. What proportion of the IQ scores exceed 125?

k. What percentage of the SAT scores are below 450?

l. What percentage of the SAT scores are above 575?

m. What SAT score is at the 95th percentile? Explain what this means.

Your Study

6.138 Intelligence Tests

The Wechsler Tests, Wechsler Adult Intelligence Scale-Revised, WAIS-R, and Wechsler Intelligence Scale for Children, WISC-III, are a widely used alternative to the Stanford-Binet. The Wechsler tests rates performance (non-verbal) intelligence in addition to verbal intelligence, and can be broken down to reveal strengths and weaknesses in various areas.

Based on scores from a large number of randomly selected people, IQ ranges have been classified as shown in Table 10-4. A look at the percentages reveals a definite pattern. The distribution of IQ' approximates a normal curve, in which the majority of scores fall close to the average, with fewer at the extremes.

TABLE 10-4

Distribution of Adult IQ Scores on WAIS-R

IQ	Description	Percent
Above 130	Very superior	2.2
120–129	Superior	6.7
110–119	Bright normal	16.1
90–109	Average	50.0
80–89	Dull normal	16.1
70–79	Borderline	6.7
Below 70	Mentally retarded	2.2

Source: Dennis Coon, *Essentials of Psychology, Exploration and Application,* 8th ed. (Belmont, CA: Wadsworth, 1999)

a. Use the information in Table 10-4 on page 356 and estimate the standard deviation for adult WAIS-R scores. Use at least two different pieces of information to obtain two separate estimates. Determine your answer.

b. Does the IQ score discussed here seem to have a normal distribution? Give reasons to support your answer.

c. What percentage of the adult population has "superior" intelligence?

d. What is the probability of randomly selecting one person from this population who is classified below "average"?

e. What IQ score is at the 95th percentile? Explain what this means.

Chapter Practice Test

PART I: Knowing the Definitions

Answer "True" if the statement is always true. If the statement is not always true, replace the words shown in bold with words that make the statement always true.

6.1 The normal probability distribution is symmetric about **zero.**

6.2 The total area under the curve of any normal distribution is **1.0.**

6.3 The theoretical probability that a particular value of a **continuous** random variable will occur is exactly zero.

6.4 The unit of measure for the standard score is the **same as the unit of measure of the data.**

6.5 All **normal** distributions have the same general probability function and distribution.

6.6 In the notation $z(0.05)$, the number in parentheses is the measure of the area to the **left** of the z-score.

6.7 Standard normal scores have a mean of **one** and a standard deviation of **zero.**

6.8 Probability distributions of **all** continuous random variables are normally distributed.

6.9 We are able to add and subtract the areas under the curve of a continuous distribution because these areas represent probabilities of **independent** events.

6.10 The most common distribution of a continuous random variable is the **binomial** probability.

PART II: Applying the Concepts

6.11 Find the following probabilities for z, the standard normal score:

a. $P(0 < z < 2.42)$ b. $P(z < 1.38)$

c. $P(z < -1.27)$ d. $P(-1.35 < z < 2.72)$

6.12 Find the value of each z-score:

a. $P(z > ?) = 0.2643$ b. $P(z < ?) = 0.17$

c. $z(0.04)$

6.13 Use the symbolic notation $z(\alpha)$ to give the symbolic name for each z-score shown in the figure at the bottom of the page.

6.14 The lifetimes of flashlight batteries are normally distributed about a mean of 35.6 hr with a standard deviation of 5.4 hr. Kevin selected one of these batteries at random and tested it. What is the probability that this one battery will last less than 40.0 hr?

6.15 The lengths of time, x, spent commuting daily, one-way, to college by students are believed to have a mean of 22 min with a standard deviation of 9 min. If the lengths of time spent commuting are approximately normally distributed, find the time, x, that separates the 25% who spend the most time commuting from the rest of the commuters.

Figure for 6.13

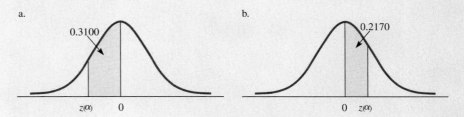

a.

0.3100

$z(\alpha)$ 0

b.

0.2170

0    $z(\alpha)$

6.16 Thousands of high school students take the SAT each year. The scores attained by the students in a certain city are approximately normally distributed with a mean of 490 and a standard deviation of 70. Find:

    a. the percentage of students who score between 600 and 700

    b. the percentage of students who score less than 650

    c. the third quartile

    d. the 15th percentile, P_{15}

    e. the 95th percentile, P_{95}

PART III: Understanding the Concepts

6.17 In 50 words, describe the standard normal distribution.

6.18 Describe the meaning of the symbol $z(\alpha)$.

6.19 Explain why the standard normal distribution, as computed in Table 3 in Appendix B, can be used to find probabilities for all normal distributions.

Statistics⊘Now™   Preparing for an exam? Assess your progress by taking the post-test at **http://1pass.thomson.com**.

⊘Mentor™   Do you need a live tutor for homework problems? Access vMentor on the StatisticsNow website at **http://1pass.thomson.com** for one-on-one tutoring from a statistics expert.

CHAPTER

7 Sample Variability

© Spencer Grant/PhotoEdit

Recall our primary question, "What can be deduced about the statistical population from which the sample is taken?" The objective of this chapter is to study the measures and the patterns of variability for the distribution formed by repeatedly observed values of a sample mean.

275 Million Americans

The U.S. Census and Sampling It

Statistics⬡Now™
Throughout the chapter, this icon introduces a list of resources on the StatisticsNow website at
http://1pass.thomson.com
that will:

- Help you evaluate your knowledge of the material
- Allow you to take an exam-prep quiz
- Provide a Personalized Learning Plan targeting resources that address areas you should study

According to the 2000 census, the U.S. population consists of more than 275 million people. We read and hear about this population often; the news media reports on results of samples nearly every day. One of the variables of interest to many is the "age" of Americans.

According to the 2000 census, the approximately 275 million Americans have a mean age of 36.5 years

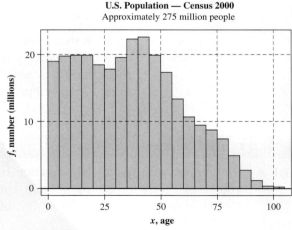

U.S. Population — Census 2000
Approximately 275 million people

f, number (millions) vs. *x*, age

and a standard deviation of 22.5 years. The ages are distributed as shown in the accompanying histogram.

A census in the United States is done only every 10 years. It is an enormous and overwhelming job, but the information that is obtained is vital to our country's organization and structure. Issues come up and times change; information is needed, and a census is impractical. This is where a representative sample comes in.

THE SAMPLING ISSUE

The fundamental goal of a survey is to come up with the same results that would have been obtained had every single member of a population been interviewed. For national Gallup polls, in other words, the objective is to present the opinions of a sample of people which are exactly the same opinions that would have been obtained had it been possible to interview all adult Americans in the country.

The key to reaching this goal is a fundamental principle called *equal proba-*

bility of selection, which states that if every member of a population has an equal probability of being selected in a sample, then that sample will be representative of the population. It's that straightforward.

Thus, it is Gallup's goal in selecting samples to allow every adult American an equal chance of falling into the sample. How that is done, of course, is the key to the success or failure of the process.

Source: http://www.gallup.com/help/FAQs/poll1.asp

Suppose a random sample of 100 ages was taken from 2000 census distribution. [EX07-01]

45	78	55	15	47	85	93	46	13	41
87	78	7	7	94	48	11	41	81	32
59	8	15	20	49	66	11	61	16	19
39	74	34	6	46	8	46	21	44	41
52	84	27	53	33	48	80	6	62	21
47	11	17	3	31	43	46	23	52	20
35	24	30	37	54	90	26	55	89	2
58	44	30	45	15	25	47	13	28	10
80	41	30	57	63	79	75	7	26	4
2	10	21	19	5	62	32	59	40	16

How well does this sample represent the population? What should we look at? How should we compare? After completing Chapter 7, further investigate these questions concerning the ages of Americans based on the 2000 Census in the Chapter Project on page 389.

SECTION 7.1 EXERCISES

7.1 a. How would you graphically describe the 100 "ages" in the preceding random sample taken from the 2000 census distribution? Construct the graph.

b. Using the graph that you constructed in part a, describe the shape of the distribution of sample data.

c. How well did the sample describe the population of ages from the 2000 census? Explain using the graphical displays.

d. If another sample was collected, would you expect the same results? Explain.

7.2 a. How would you numerically describe the 100 "ages" in the preceding random sample taken from the 2000 census distribution? Calculate the statistics.

b. How well do the statistics calculated in part a compare with the parameters from the 2000 census? Be specific.

c. If another sample was collected, would you expect the same results? Explain.

Sampling Distributions

To make inferences about a population, we need to discuss sample results a little more. A sample mean, $\bar{x}$, is obtained from a sample. Do you expect that this value, $\bar{x}$, is exactly equal to the value of the population mean, μ? Your answer should be no. We do not expect the means to be identical, but we will be satisfied with our sample results if the sample mean is "close" to the value of the population mean. Let's consider a second question: If a second sample is taken, will the second sample have a mean equal to the population mean? Equal to the first sample mean? Again, no, we do not expect the sample mean to be equal to the population mean, nor do we expect the second sample mean to be a repeat of the first one. We do, however, again expect the values to be "close." (This argument should hold for any other sample statistic and its corresponding population value.)

The next questions should already have come to mind: What is "close"? How do we determine (and measure) this closeness? Just how will **repeated sample statistics** be distributed? To answer these questions we must look at a *sampling distribution.*

> **Sampling distribution of a sample statistic:** The distribution of values for a sample statistic obtained from repeated samples, all of the same size and all drawn from the same population.

Let's start by investigating two different small theoretical sampling distributions.

EXAMPLE 7.1

Forming a Sampling Distribution of Means and Ranges

FYI Samples are drawn with replacement.

Consider as a population the set of single-digit even integers, {0, 2, 4, 6, 8}. In addition, consider all possible samples of size 2. We will look at two different sampling distributions that might be formed: the sampling distribution of sample means and the sampling distribution of sample ranges.

First we need to list all possible samples of size 2; there are 25 possible samples:

{0, 0}	{2, 0}	{4, 0}	{6, 0}	{8, 0}
{0, 2}	{2, 2}	{4, 2}	{6, 2}	{8, 2}
{0, 4}	{2, 4}	{4, 4}	{6, 4}	{8, 4}
{0, 6}	{2, 6}	{4, 6}	{6, 6}	{8, 6}
{0, 8}	{2, 8}	{4, 8}	{6, 8}	{8, 8}

Each of these samples has a mean $\bar{x}$. These means are, respectively:

0	1	2	3	4
1	2	3	4	5
2	3	4	5	6
3	4	5	6	7
4	5	6	7	8

TABLE 7.1

**Probability Distribution:
Sampling Distribution
of Sample Means**

$\bar{x}$	$P(\bar{x})$
0	0.04
1	0.08
2	0.12
3	0.16
4	0.20
5	0.16
6	0.12
7	0.08
8	0.04

Each of these samples is equally likely, and thus each of the 25 sample means can be assigned a probability of $\frac{1}{25} = 0.04$. The **sampling distribution of sample means** is shown in Table 7.1 as a **probability distribution** and shown in Figure 7.1 as a histogram.

FIGURE 7.1
**Histogram: Sampling
Distribution of Sample
Means**

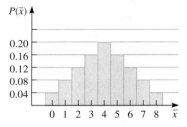

For the same set of all possible samples of size 2, let's find the sampling distribution of sample ranges. Each sample has a range R. The ranges are:

0	2	4	6	8
2	0	2	4	6
4	2	0	2	4
6	4	2	0	2
8	6	4	2	0

Again, each of these 25 sample ranges has a probability of 0.04. Table 7.2 shows the sampling distribution of sample ranges as a probability distribution, and Figure 7.2 shows the sampling distribution as a histogram.

TABLE 7.2

**Probability Distribution:
Sampling Distribution of
Sample Ranges**

R	$P(R)$
0	0.20
2	0.32
4	0.24
6	0.16
8	0.08

FIGURE 7.2
**Histogram: Sampling
Distribution of Sample
Ranges**

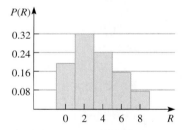

Example 7.1 is theoretical in nature and therefore expressed in probabilities. Since this population is small, it is easy to list all 25 possible samples of size 2 (a sample space) and assign probabilities. However, it is not always possible to do this.

Now, let's empirically (that is, by experimentation) investigate another sampling distribution.

EXAMPLE 7.2

Creating a Sampling Distribution of Sample Means

Let's consider a population that consists of five equally likely integers: 1, 2, 3, 4, and 5. Figure 7.3 shows a histogram representation of the population. We can observe a portion of the sampling distribution of sample means when 30 samples of size 5 are randomly selected.

Table 7.3 shows 30 samples and their means. The resulting sampling distribution, a **frequency distribution,** of sample means is shown in Figure 7.4. Notice that this distribution of sample means does not look like the population. Rather, it seems to display the characteristics of a normal distribution; it is mounded and nearly symmetrical about its mean (approximately 3.0).

FIGURE 7.3 The Population: Theoretical Probability Distribution

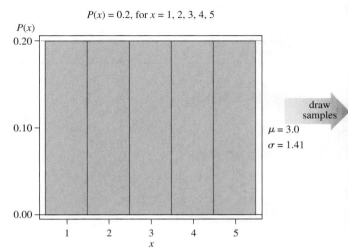

$P(x) = 0.2$, for $x = 1, 2, 3, 4, 5$

draw
samples

$\mu = 3.0$
$\sigma = 1.41$

TABLE 7.3

30 Samples of Size 5 [TA07-03]

No.	Sample	$\bar{x}$	No.	Sample	$\bar{x}$
1	4,5,1,4,5	3.8	16	4,5,5,3,5	4.4
2	1,1,3,5,1	2.2	17	3,3,1,2,1	2.0
3	2,5,1,5,1	2.8	18	2,1,3,2,2	2.0
4	4,3,3,1,1	2.4	19	4,3,4,2,1	2.8
5	1,2,5,2,4	2.8	20	5,3,1,4,2	3.0
6	4,2,2,5,4	3.4	21	4,4,2,2,5	3.4
7	1,4,5,5,2	3.4	22	3,3,5,3,5	3.8
8	4,5,3,1,2	3.0	23	3,4,4,2,2	3.0
9	5,3,3,3,5	3.8	24	3,3,4,5,3	3.6
10	5,2,1,1,2	2.2	25	5,1,5,2,3	3.2
11	2,1,4,1,3	2.2	26	3,3,3,5,2	3.2
12	5,4,3,1,1	2.8	27	3,4,4,4,4	3.8
13	1,3,1,5,5	3.0	28	2,3,2,4,1	2.4
14	3,4,5,1,1	2.8	29	2,1,1,2,4	2.0
15	3,1,5,3,1	2.6	30	5,3,3,2,5	3.6

using
the
30
means

FIGURE 7.4
Empirical
Distribution
of Sample Means

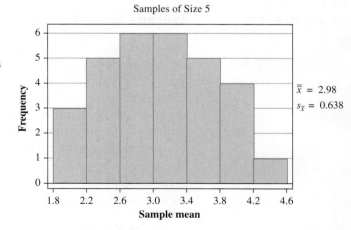

Samples of Size 5

$\bar{\bar{x}} = 2.98$
$s_{\bar{x}} = 0.638$

Note: The variable for the sampling distribution is $\bar{x}$; therefore, the mean of the $\bar{x}$'s is $\bar{\bar{x}}$ and the standard deviation of $\bar{x}$ is $s_{\bar{x}}$.

The theory involved with sampling distributions that will be described in the remainder of this chapter requires *random sampling*.

Random sample: A sample obtained in such a way that each possible sample of fixed size n has an equal probability of being selected (see p. 22).

Figure 7.5 shows how the sampling distribution of sample means is formed.

FIGURE 7.5 The Sampling Distribution of Sample Means

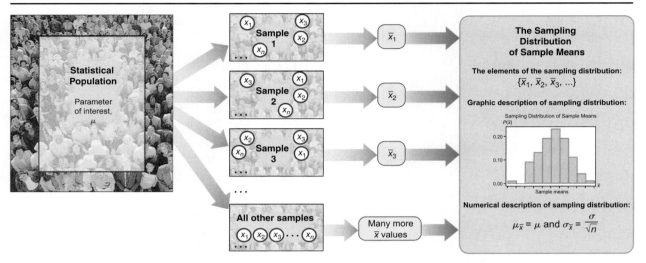

APPLIED EXAMPLE 7.3

Average Age of Urban Transit Rail Vehicles

There are many reasons for collecting data repeatedly. Not all repeated data collections are performed to form a sampling distribution. Consider the "Average Age of Urban Transit Rail Vehicles (Years)" statistics from the U.S. Department of Transportation that follows. The table shows the average age for four different classifications of transit rail vehicles tracked over several years. By studying the pattern of change in the average age for each class of vehicle, a person can draw conclusions about what has been happening to the fleet over several years. Chances are the people involved in maintaining each fleet can also detect when a change in policies regarding replacement of older vehicles is needed. However useful this information is, there is no sampling distribution involved here.

Average Age of Urban Transit Rail Vehicles (Years)

	1985	1990	1995	2000	2003
Transit rail					
Commuter rail locomotives[a]	16.3	15.7	15.9	13.4	16.6
Commuter rail passenger coaches	19.1	17.6	21.4	16.9	20.5
Heavy-rail passenger cars	17.1	16.2	19.3	22.9	19.0
Light-rail vehicles (streetcars)	20.6	15.2	16.8	16.1	15.6

[a]Locomotives used in Amtrak inter-city passenger services are not included.

Source: U.S. Department of Transportation, Federal Transit Administration

7.3 Manufacturers use random samples to test whether or not their product is meeting specifications. These samples could be people, manufactured parts, or even samples during the manufacturing of potato chips.

a. Do you think that all random samples taken from the same population will lead to the same result?

b. What characteristic (or property) of random samples could be observed during the sampling process?

7.4 Refer to Table 7.1 in Example 7.1 (p. 363) and explain why the samples are equally likely; that is, why $P(0) = 0.04$, and why $P(2) = 0.12$.

7.5 a. What is a sampling distribution of sample means?

b. A sample of size 3 is taken from a population, and the sample mean is found. Describe how this sample mean is related to the sampling distribution of sample means.

7.6 Consider the set of odd single-digit integers {1, 3, 5, 7, 9}.

a. Make a list of all samples of size 2 that can be drawn from this set of integers. (Sample with replacement; that is, the first number is drawn, observed, and then replaced [returned to the sample set] before the next drawing.)

b. Construct the sampling distribution of sample means for samples of size 2 selected from this set.

c. Construct the sampling distributions of sample ranges for samples of size 2.

7.7 Consider the set of even single-digit integers {0, 2, 4, 6, 8}.

a. Make a list of all the possible samples of size 3 that can be drawn from this set of integers. (Sample with replacement; that is, the first number is drawn, observed, and then replaced [returned to the sample set] before the next drawing.)

b. Construct the sampling distribution of the sample medians for samples of size 3.

c. Construct the sampling distribution of the sample means for samples of size 3.

7.8 Using the telephone numbers listed in your local directory as your population, randomly obtain 20 samples of size 3. From each telephone number identified as a source, take the fourth, fifth, and sixth digits. (For example, for 245-8269, you would take the 8, the 2, and the 6 as your sample of size 3.)

a. Calculate the mean of the 20 samples.

b. Draw a histogram showing the 20 sample means. (Use classes −0.5 to 0.5, 0.5 to 1.5, 1.5 to 2.5, and so on.)

c. Describe the distribution of $\bar{x}$'s that you see in part b (shape of distribution, center, and amount of dispersion).

d. Draw 20 more samples and add the 20 new $\bar{x}$'s to the histogram in part b. Describe the distribution that seems to be developing.

7.9 Using a set of five dice, roll the dice and determine the mean number of dots showing on the five dice. Repeat the experiment until you have 25 sample means.

a. Draw a dotplot showing the distribution of the 25 sample means. (See Example 7.2, p. 364.)

b. Describe the distribution of $\bar{x}$'s in part a.

c. Repeat the experiment to obtain 25 more sample means and add these 25 $\bar{x}$'s to your dotplot. Describe the distribution of 50 means.

7.10 Considering the population of five equally likely integers in Example 7.2:

a. Verify μ and σ for the population in Example 7.2.

b. Table 7.3 lists 30 $\bar{x}$ values. Construct a grouped frequency distribution to verify the frequency distribution shown in Figure 7.4.

c. Find the mean and standard deviation of the 30 $\bar{x}$ values in Table 7.3 to verify the values for $\bar{\bar{x}}$ and $s_{\bar{x}}$. Explain the meaning of the two symbols $\bar{\bar{x}}$ and $s_{\bar{x}}$.

7.11 In reference to Applied Example 7.3 on page 366:

a. Explain why the numerical values on this table do not form a sampling distribution.

b. Explain how this repeated gathering of data differs from the idea of repeated sampling to gather information about a sampling distribution?

7.12 From the table of random numbers in Table 1 in Appendix B, construct another table showing 20 sets of 5 randomly selected single-digit integers. Find the mean of each set (the grand mean) and compare this value with the theoretical population mean, μ, using the absolute difference and the % error. Show all work.

7.13 a. Using a computer or a random-numbers table, simulate the drawing of 100 samples, each of size 5, from the uniform probability distribution of single-digit integers, 0 to 9.

   b. Find the mean for each sample.

   c. Construct a histogram of the sample means. (Use integer values as class midpoints.)

   d. Describe the sampling distribution shown in the histogram in part c.

MINITAB (Release 14)

a. Use the Integer RANDOM DATA commands on page 101, replacing generate with 100, store in with C1–C5, minimum value with 0, and maximum value with 9.

b. Choose: **Calc > Row Statistics**
   Select: **Mean**
   Enter: Input variables: **C1–C5**
               Store result in: **C6 > OK**

c. Use the HISTOGRAM commands on page 61 for the data in C6. To adjust the histogram, select Binning with midpoint and midpoint positions 0:9/1.

Excel

a. Input 0 through 9 into column A and corresponding 0.1's into column B; then continue with:

Choose: **Tools > Data Analysis > Random Number Generation > OK**

Enter: Number of Variables: **5**
         Number of Random Numbers: **100**
         Distribution: **Discrete**
         Value and Probability Input Range: **(A1:B10 or select cells)**
Select: **Output Range:**
Enter: **(C1 or select cell) > OK**

b. Activate cell H1.

Choose: **Insert function, f_x > Statistical > AVERAGE > OK**
Enter: Number1: **(C1:G1 or select cells)**
Drag: **Bottom right corner of average value box down to give other averages**

c. Use the HISTOGRAM commands on pages 61–62 with column H as the input range and column A as the bin range.

TI-83/84 Plus

a. Use the Integer RANDOM DATA and STO commands on page 101, replacing the Enter with 0,9,100). Repeat preceding commands four more times, storing data in L2, L3, L4, and L5, respectively.

b. Choose: **STAT > EDIT > 1:Edit**
   Highlight: **L6 (column heading)**
   Enter: **(L1+L2+L3+L4+L5)/5**
c. Choose: **2nd > STAT PLOT > 1:Plot1**
   Choose: **Window**
   Enter: **0, 9, 1, 0,**
               30, 5, 1
   Choose: **Trace > > >**

7.14 a. Using a computer or a random-numbers table, simulate the drawing of 250 samples, each of size 18, from the uniform probability distribution of single-digit integers, 0 to 9.

   b. Find the mean for each sample.

   c. Construct a histogram of the sample means.

   d. Describe the sampling distribution shown in the histogram in part c.

7.15 a. Use a computer to draw 200 random samples, each of size 10, from the normal probability distribution with mean 100 and standard deviation 20.

b. Find the mean for each sample.

c. Construct a frequency histogram of the 200 sample means.

d. Describe the sampling distribution shown in the histogram in part c.

MINITAB (Release 14)

a. Use the Normal RANDOM DATA commands on page 101, replacing generate with 200, store in with C1–C10, mean with 100, and standard deviation with 20.

b. Choose: `Calc > Row Statistics`
   Select: `Mean`
   Enter: `Input variables: C1–C10`
               `Store result in: C11 > OK`

c. Use the HISTOGRAM commands on page 61 for the data in C11. To adjust the histogram, select Binning with midpoint and midpoint positions 74.8:125.2/6.3.

Excel

a. Use the Normal RANDOM NUMBER GENERATION commands on page 101, replacing number of variables with 10, number of random numbers with 200, mean with 100, and standard deviation with 20.

b. Activate cell K1.
   Choose: `Insert function, fₓ > Statistical >`
               `AVERAGE > OK`
   Enter: `Number1: (A1:J1 or select cells)`
   Drag: `Bottom right corner of average value`
               `box down to give other averages`

c. Use the RANDOM NUMBER GENERATION Patterned Distribution commands in Exercise 6.66 on page 336, replacing the first value with 74.8, the last value with 125.2, the steps with 6.3, and the output range with L1. Use the HISTOGRAM commands on pages 61–62 with column K as the input range and column L as the bin range.

7.16 a. Use a computer to draw 500 random samples, each of size 20, from the normal probability distribution with mean 80 and standard deviation 15.

b. Find the mean for each sample.

c. Construct a frequency histogram of the 500 sample means.

d. Describe the sampling distribution shown in the histogram in part c, including the mean and standard deviation.

7.3

The Sampling Distribution of Sample Means

On the preceding pages we discussed the sampling distributions of two statistics: sample means and sample ranges. Many others could be discussed; however, the only sampling distribution of concern to us at this time is the sampling distribution of sample means.

FYI This is very useful information!

Sampling distribution of sample means (SDSM): If all possible random samples, each of size n, are taken from any population with mean μ and standard deviation σ, then the sampling distribution of sample means will have the following:

1. A mean $\mu_{\bar{x}}$ equal to μ
2. A standard deviation $\sigma_{\bar{x}}$ equal to $\dfrac{\sigma}{\sqrt{n}}$

Furthermore, if the sampled population has a normal distribution, then the sampling distribution of $\bar{x}$ will also be normal for samples of all sizes.

This is a very interesting two-part statement. The first part tells us about the relationship between the population mean and standard deviation, and the sampling distribution mean and standard deviation for all sampling distributions of sample means. The standard deviation of the sampling distribution is denoted by $\sigma_{\bar{x}}$ and given a specific name to avoid confusion with the population standard deviation, σ.

> **Standard error of the mean ($\sigma_{\bar{x}}$):** The standard deviation of the sampling distribution of sample means.

The second part indicates that this information is not always useful. Stated differently, it says that the mean value of only a few observations will be normally distributed when samples are drawn from a normally distributed population, but it will not be normally distributed when the sampled population is uniform, skewed, or otherwise not normal. However, the *central limit theorem* gives us some additional and very important information about the sampling distribution of sample means.

> **Central limit theorem (CLT):** The sampling distribution of sample means will more closely resemble the normal distribution as the sample size increases.

If the sampled distribution is normal, then the sampling distribution of sample means (SDSM) is normal, as stated previously, and the central limit theorem (CLT) is not needed. But, if the sampled population is not normal, the CLT tells us that the sampling distribution will still be approximately normally distributed under the right conditions. If the sampled population distribution is nearly normal, the $\bar{x}$ distribution is approximately normal for fairly small n (possibly as small as 15). When the sampled population distribution lacks symmetry, n may have to be quite large (maybe 50 or more) before the normal distribution provides a satisfactory approximation.

By combining the preceding information, we can describe the sampling distribution of $\bar{x}$ completely: (1) the location of the center (mean), (2) a measure of spread indicating how widely the distribution is dispersed (standard error of the mean), and (3) an indication of how it is distributed.

1. $\mu_{\bar{x}} = \mu$; the mean of the sampling distribution ($\mu_{\bar{x}}$) is equal to the mean of the population (μ).

2. $\sigma_{\bar{x}} = \dfrac{\sigma}{\sqrt{n}}$; the standard error of the mean ($\sigma_{\bar{x}}$) is equal to the standard deviation of the population (σ) divided by the square root of the sample size, n.

3. The distribution of sample means is normal when the parent population is normally distributed, and the CLT tells us that the distribution of sample means becomes approximately normal (regardless of the shape of the parent population) when the sample size is large enough.

FYI Truly amazing: $\bar{x}$ is normally distributed when n is large enough, no matter what shape the population is!

DID YOU KNOW

Central Limit Theorem

Abraham de Moivre was a pioneer in the theory of probability and published *The Doctrine of Chance*, first in Latin in 1711 and then in expanded editions in 1718, 1738, and 1756. The 1756 edition contained his most important contribution—the approximation of the binomial distributions for a large number of trials using the normal distribution. The definition of statistical independence also made its debut along with many dice and other games. de Moivre proved the central limit theorem holds for numbers resulting from games of chance. With the use of mathematics, he also successfully predicted the date of his own death.

Note: The n referred to is the size of each sample in the sampling distribution. (The number of repeated samples used in an empirical situation has no effect on the standard error.)

We do not show the proof for the preceding three facts in this text; however, their validity will be demonstrated by examining two examples. For the first example, let's consider a population for which we can construct the theoretical sampling distribution of all possible samples.

EXAMPLE 7.4

Constructing a Sampling Distribution of Sample Means

Let's consider all possible samples of size 2 that could be drawn from a population that contains the three numbers 2, 4, and 6. First let's look at the population itself. Construct a histogram to picture its distribution, Figure 7.6; calculate the mean, μ, and the standard deviation, σ, Table 7.4. (Remember: We must use the techniques from Chapter 5 for discrete probability distributions.)

FIGURE 7.6 Population

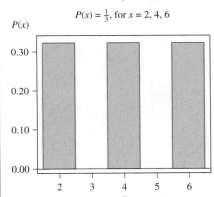

$P(x) = \frac{1}{3}$, for $x = 2, 4, 6$

TABLE 7.4

Extensions Table for x

x	$P(x)$	$xP(x)$	$x^2P(x)$
2	$\frac{1}{3}$	$\frac{2}{3}$	$\frac{4}{3}$
4	$\frac{1}{3}$	$\frac{4}{3}$	$\frac{16}{3}$
6	$\frac{1}{3}$	$\frac{6}{3}$	$\frac{36}{3}$
Σ	$\frac{3}{3}$	$\frac{12}{3}$	$\frac{56}{3}$
	1.0	4.0	18.6$\overline{6}$

$\mu = \mathbf{4.0}$

$\sigma = \sqrt{18.6\overline{6} - (4.0)^2} = \sqrt{2.6\overline{6}} = \mathbf{1.63}$

Table 7.5 (see p. 372) lists all the possible samples of size 2 that can be drawn from this population. (One number is drawn, observed, and then returned to the population before the second number is drawn.) Table 7.5 also lists the means of these samples. The sample means are then collected to form the sampling distribution. The distribution for these means and the extensions are given in Table 7.6 (p. 372), along with the calculation of the mean and the standard error of the mean for the sampling distribution. The histogram for the sampling distribution of sample means is shown in Figure 7.7 (p. 372).

Let's now check the truth of the three facts about the sampling distribution of sample means:

1. The mean $\mu_{\bar{x}}$ of the sampling distribution will equal the mean μ of the population: both μ and $\mu_{\bar{x}}$ have the value **4.0.**
2. The standard error of the mean $\sigma_{\bar{x}}$ for the sampling distribution will equal the standard deviation σ of the population divided by the square root of the sample size, n: $\sigma_{\bar{x}} = \mathbf{1.15}$ and $\sigma = 1.63$, $n = 2$, $\dfrac{\sigma}{\sqrt{n}} = \dfrac{1.63}{\sqrt{2}} = \mathbf{1.15}$; they are equal: $\sigma_{\bar{x}} = \dfrac{\sigma}{\sqrt{n}}$.
3. The distribution will become approximately normally distributed: the histogram in Figure 7.7 very strongly suggests normality.

TABLE 7.5					
All Nine Possible Samples of Size 2					
Sample	$\bar{x}$	Sample	$\bar{x}$	Sample	$\bar{x}$
2, 2	2	4, 2	3	6, 2	4
2, 4	3	4, 4	4	6, 4	5
2, 6	4	4, 6	5	6, 6	6

TABLE 7.6			
Extensions Table for $\bar{x}$			
$\bar{x}$	$P(\bar{x})$	$\bar{x}P(\bar{x})$	$\bar{x}^2P(\bar{x})$
2	$\frac{1}{9}$	$\frac{2}{9}$	$\frac{4}{9}$
3	$\frac{2}{9}$	$\frac{6}{9}$	$\frac{18}{9}$
4	$\frac{3}{9}$	$\frac{12}{9}$	$\frac{48}{9}$
5	$\frac{2}{9}$	$\frac{10}{9}$	$\frac{50}{9}$
6	$\frac{1}{9}$	$\frac{6}{9}$	$\frac{36}{9}$
Σ	$\frac{9}{9}$	$\frac{36}{9}$	$\frac{156}{9}$
	1.0	4.0	$17.3\overline{3}$

$$\mu_{\bar{x}} = \mathbf{4.0}$$

$$\sigma_{\bar{x}} = \sqrt{17.3\overline{3} - (4.0)^2} = \sqrt{1.3\overline{3}} = \mathbf{1.15}$$

FIGURE 7.7 Sampling Distribution of Sample Means

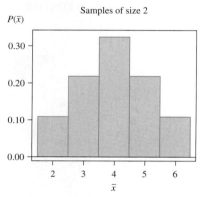

Example 7.4, a theoretical situation, suggests that all three facts appear to hold true. Do these three facts hold when actual data are collected? Let's look back at Example 7.2 (p. 364) and see if all three facts are supported by the empirical sampling distribution there.

First, let's look at the population—the theoretical probability distribution from which the samples in Example 7.2 were taken. Figure 7.3 is a histogram showing the probability distribution for randomly selected data from the population of equally likely integers 1, 2, 3, 4, 5. The population mean μ equals 3.0. The population standard deviation σ is $\sqrt{2}$, or 1.41. The population has a uniform distribution.

Now let's look at the empirical distribution of the 30 sample means found in Example 7.2. From the 30 values of $\bar{x}$ in Table 7.3, the observed mean of the $\bar{x}$'s, $\bar{\bar{x}}$, is 2.98 and the observed standard error of the mean, $s_{\bar{x}}$, is 0.638. The histogram of the sampling distribution in Figure 7.4 appears to be mounded, approximately symmetrical, and centered near the value 3.0.

Now let's check the truth of the three specific properties:

1. $\mu_{\bar{x}}$ and μ will be equal. The mean of the population μ is 3.0, and the observed sampling distribution mean $\bar{\bar{x}}$ is 2.98; they are very close in value.

2. $\sigma_{\bar{x}}$ will equal $\frac{\sigma}{\sqrt{n}}$. $\sigma = 1.41$ and $n = 5$; therefore, $\frac{\sigma}{\sqrt{n}} = \frac{1.41}{\sqrt{5}} = \mathbf{0.632}$, and $s_{\bar{x}} = \mathbf{0.638}$; they are very close in value. (Remember that we have taken only 30 samples, not all possible samples, of size 5.)

3. The sampling distribution of $\bar{x}$ will be approximately normally distributed. Even though the population has a rectangular distribution, the histogram in Figure 7.4 suggests that the x distribution has some of the properties of normality (mounded, symmetrical).

Although Examples 7.2 and 7.4 do not constitute a proof, the evidence seems to strongly suggest that both statements, the sampling distribution of sample means and the CLT, are true.

Having taken a look at these two specific examples, let's now look at four graphic illustrations that present the sampling distribution information and the CLT in a slightly different form. Each of these illustrations has four distributions. The first graph shows the distribution of the parent population, the distribution of the individual x values. Each of the other three graphs shows a sampling distribution of sample means, $\bar{x}$'s, using three different sample sizes.

In Figure 7.8 we have a uniform distribution, much like Figure 7.3 for the integer illustration, and the resulting distributions of sample means for samples of sizes 2, 5, and 30.

FIGURE 7.8
Uniform Distribution

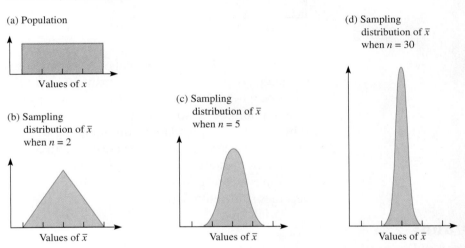

Figure 7.9 shows a U-shaped population and the three sampling distributions.

FIGURE 7.9
U-Shaped Distribution

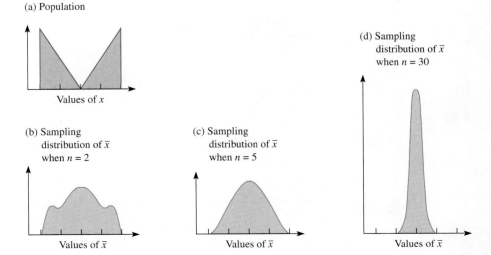

Figure 7.10 shows a J-shaped population and the three sampling distributions.

FIGURE 7.10
J-Shaped
Distribution

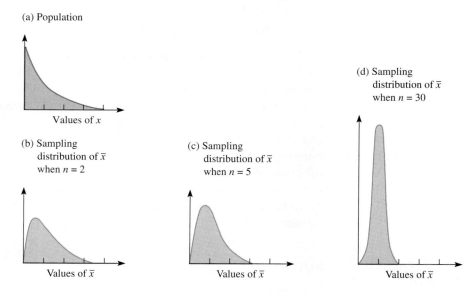

All three nonnormal population distributions seem to verify the CLT; the sampling distributions of sample means appear to be approximately normal for all three when samples of size 30 were used. Now consider Figure 7.11, which shows a normally distributed population and the three sampling distributions. With the normal population, the sampling distributions of the sample means for all sample sizes appear to be normal. Thus, you have seen an amazing phenomenon: no matter what the shape of a population, the sampling distribution of sample means either is normal or becomes approximately normal when n becomes sufficiently large.

FIGURE 7.11
Normal Distribution

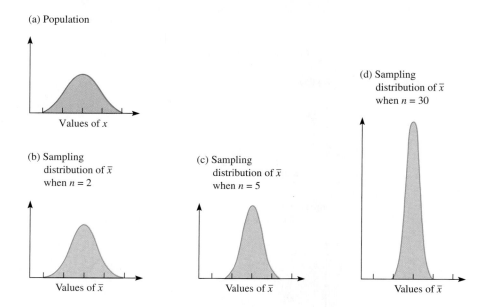

You should notice one other point: the sample mean becomes less variable as the sample size increases. Notice that as *n* increases from 2 to 30, all the distributions become narrower and taller.

SECTION 7.3 EXERCISES

Statistics☯Now™

Skillbuilder Applet Exercises must be worked using an accompanying applet found on your Student's Suite CD-ROM or at the StatisticsNow website at **http://1pass.thomson.com**

7.17 Skillbuilder Applet Exercise simulates taking samples of size 4 from an approximately normal population, where $\mu = 65.15$ and $\sigma = 2.754$.

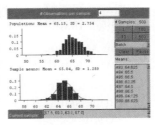

a. Click "1" for "# Samples." Note the four data values and their mean. Change "slow" to "batch" and take at least 1000 samples using the "500" for "# Samples."

b. What is the mean for the 1001 sample means? How close is it to the population mean, μ?

c. Compare the sample standard deviation to the population standard deviation, σ. What is happening to the sample standard deviation? Compare it with $\sigma/\sqrt{n}$, which is $2.754/\sqrt{4}$.

d. Does the histogram of sample means have an approximately normal shape?

e. Relate your findings to the SDSM.

7.18 Skillbuilder Applet Exercise simulates sampling from a skewed population, where $\mu = 6.029$ and $\sigma = 10.79$.

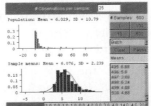

a. Change the "# Observations per sample" to "4." Using batch and 500, take 1000 samples of size 4.

b. Compare the mean and standard deviation for the sample means with μ and σ. Compare the

sample standard deviation with $\sigma/\sqrt{n}$, which is $10.79/\sqrt{4}$. Does the histogram have an approximately normal shape? If not, what shape is it?

c. Using the "clear" button each time, repeat the directions in parts a and b for samples of size 25, 100, and 1000. Table your findings for each sample size.

d. Relate your findings to the SDSM and the CLT.

7.19 a. What is the total measure of the area for any probability distribution?

    b. Justify the statement "$\bar{x}$ becomes less variable as *n* increases."

7.20 If a population has a standard deviation σ of 25 units, what is the standard error of the mean if samples of size 16 are selected? Samples of size 36? Samples of size 100?

7.21 A certain population has a mean of 500 and a standard deviation of 30. Many samples of size 36 are randomly selected and the means calculated.

a. What value would you expect to find for the mean of all these sample means?

b. What value would you expect to find for the standard deviation of all these sample means?

c. What shape would you expect the distribution of all these sample means to have?

7.22 An April 2004 article on HearTheIssues.com stated that Americans have an average of 2.24 televisions per household (*source:* Nielsen Media Research). If the standard deviation for the number of televisions in a U.S. household is 1.2 and a random sample of 80 American households is selected, the mean of this sample belongs to a sampling distribution.

a. What is the shape of this sampling distribution?

b. What is the mean of this sampling distribution?

c. What is the standard deviation of this sampling distribution?

7.23 An April 2004 article on HearTheIssues.com stated that Americans watch an average of 4.0 hours of television per person per day (*source:* Nielsen Media Research). If the standard deviation for the number of hours of television watched per day is 2.1 and a random sample of 250 Americans is selected, the mean of this sample belongs to a sampling distribution.

a. What is the shape of this sampling distribution?

b. What is the mean of this sampling distribution?

c. What is the standard deviation of this sampling distribution?

7.24 According to *The World Factbook,* 2004, the total fertility rate (estimated mean number of children born per woman) for Madagascar is 5.7. Suppose that the standard deviation of the total fertility rate is 2.6. The mean number of children for a sample of 200 randomly selected women is one value of many that form the SDSM.

a. What is the mean value for this sampling distribution?

b. What is the standard deviation of this sampling distribution?

c. Describe the shape of this sampling distribution.

7.25 The USDA Economics and Statistics System at Cornell University maintains a *Poultry Yearbook* in which they list monthly, quarterly, and annual facts about the poultry industry. The 2004 yearbook lists the annual consumption of turkey meat as 17.71 pounds per person. Suppose the standard deviation for the consumption of turkey per person is 6.3 pounds. The mean weight of turkey consumed for a sample of 150 randomly selected people is one value of many that form the SDSM.

a. What is the mean value for this sampling distribution?

b. What is the standard deviation of this sampling distribution?

c. Describe the shape of this sampling distribution.

7.26 A researcher wants to take a simple random sample of about 5% of the student body at each of two schools. The university has approximately 20,000 students, and the college has about 5000 students. Identify each of the following as true or false and justify your answer.

a. The sampling variability is the same for both schools.

b. The sampling variability for the university is higher than that for the college.

c. The sampling variability for the university is lower than that for the college.

d. No conclusion about the sampling variability can be stated without knowing the results of the study.

7.27 a. Use a computer to randomly select 100 samples of size 6 from a normal population with mean $\mu = 20$ and standard deviation $\sigma = 4.5$.

b. Find mean $\bar{x}$ for each of the 100 samples.

c. Using the 100 sample means, construct a histogram, find mean $\bar{\bar{x}}$, and find the standard deviation $s_{\bar{x}}$.

d. Compare the results of part c with the three statements made in the SDSM.

MINITAB (Release 14)

a. Use the Normal RANDOM DATA commands on page 101, replacing generate with 100, store in with C1–C6, mean with 20, and standard deviation with 4.5.

b. Use the ROW STATISTICS commands on page 368, replacing input variables with C1–C6 and store result in with C7.

c. Use the HISTOGRAM commands on page 61 for the data in C7. To adjust the histogram, select Binning with midpoint and midpoint positions 12.8:27.2/1.8. Use the MEAN and

STANDARD DEVIATION commands on pages 74 and 88 for the data in C7.

Excel

a. Use the Normal RANDOM NUMBER GENERATION commands on page 110, replacing number of variables with 6, number of random numbers with 100, mean with 20, and standard deviation with 4.5.

b. Activate cell G1.

Choose:	**Insert function, f$_x$ > Statistical > AVERAGE > OK**
Enter:	**Number1: (A1:F1 or select cells)**
Drag:	**Bottom right corner of average value box down to give other averages**

c. Use the RANDOM NUMBER GENERATION Patterned Distribution commands in Exercise 6.66 on page 336, replacing the first value with 12.8, the last value with 27.2, the steps with 1.8, and the output range with H1. Use the HISTOGRAM commands on page 61 with column G as the input range and column H as the bin range. Use the MEAN and STANDARD DEVIATION commands on pages 74 and 88 for the data in column G.

TI-83/84 Plus

a. Use the Integer RANDOM DATA and STO commands on page 101, replacing Enter with 20,4.5,100). Repeat the preceding commands five more times, storing data in L2, L3, L4, L5, and L6, respectively.

b. Enter:	**(L1 + L2 + L3 + L4 + L5 + L6)/6**
Choose:	**STO→ L7 (use ALPHA key for the 'L' or use 'MEAN')**
c. Choose:	**2nd > STAT**
	PLOT > 1:Plot1
Choose:	**Window**
Enter:	**12.8, 27.2, 1.8, 0, 40, 5, 1**
Choose:	**Trace > > >**
Choose:	**STAT > CALC > 1:1-VAR STATS > 2nd > LIST**
Select:	**L7**

7.28 a. Use a computer to randomly select 200 samples of size 24 from a normal population with mean $\mu = 20$ and standard deviation $\sigma = 4.5$.

b. Find mean $\bar{x}$ for each of the 200 samples.

c. Using the 200 sample means, construct a histogram, find mean $\bar{\bar{x}}$, and find the standard deviation $s_{\bar{x}}$.

d. Compare the results of part c with the three statements made for the SDSM and CLT on page 370.

e. Compare these results with the results obtained in Exercise 7.27. Specifically, what effect did the increase in sample size from 6 to 24 have? What effect did the increase from 100 samples to 200 samples have?

FYI If you use a computer, see Exercise 7.27.

7.4 Application of the Sampling Distribution of Sample Means

When the sampling distribution of sample means is normally distributed, or approximately normally distributed, we will be able to answer probability questions with the aid of the standard normal distribution (Table 3 of Appendix B).

EXAMPLE 7.5   ### Converting $\bar{x}$ Information into z-Scores

Consider a normal population with $\mu = 100$ and $\sigma = 20$. If a random sample of size 16 is selected, what is the probability that this sample will have a mean value between 90 and 110? That is, what is $P(90 < \bar{x} < 110)$?

SOLUTION   Since the population is normally distributed, the sampling distribution of $\bar{x}$'s is normally distributed. To determine probabilities associated with a normal distribution, we will need to convert the statement $P(90 < \bar{x} < 110)$ to a probability statement involving the **z-score.** This will allow us to use Table 3 in Appendix B, the standard normal distribution table. The sampling distribution is shown in the figure, where the shaded area represents $P(90 < \bar{x} < 110)$.

The formula for finding the z-score corresponding to a known value of $\bar{x}$ is

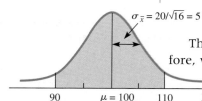

$$z = \frac{\bar{x} - \mu_{\bar{x}}}{\sigma_{\bar{x}}} \tag{7.1}$$

The mean and standard error of the mean are $\mu_{\bar{x}} = \mu$ and $\sigma_{\bar{x}} = \dfrac{\sigma}{\sqrt{n}}$. Therefore, we will rewrite formula (7.1) in terms of μ, σ, and n:

$$z = \frac{\bar{x} - \mu}{\sigma/\sqrt{n}} \tag{7.2}$$

Returning to the example and applying formula (7.2), we find:

z-score for $\bar{x} = 90$: $z = \dfrac{\bar{x} - \mu}{\sigma/\sqrt{n}} = \dfrac{90 - 100}{20/\sqrt{16}} = \dfrac{-10}{5} = \mathbf{-2.00}$

z-score for $\bar{x} = 110$: $z = \dfrac{\bar{x} - \mu}{\sigma/\sqrt{n}} = \dfrac{110 - 100}{20/\sqrt{16}} = \dfrac{10}{5} = \mathbf{2.00}$

Therefore,

$$P(90 < \bar{x} < 110) = P(-2.00 < z < 2.00) = 2(0.4772) = \mathbf{0.9544}$$

Before we look at additional examples, let's consider what is implied by $\sigma_{\bar{x}} = \dfrac{\sigma}{\sqrt{n}}$. To demonstrate, let's suppose that $\sigma = 20$ and let's use a sampling distribution of samples of size 4. Now $\sigma_{\bar{x}}$ is $20/\sqrt{4}$, or 10, and approximately 95% (0.9544) of all such sample means should be within the interval from 20 below to 20 above the population mean (within 2 standard deviations of the population mean). However, if the sample size is increased to 16, $\sigma_{\bar{x}}$ becomes $20/\sqrt{16} = 5$ and approximately 95% of the sampling distribution should be within 10 units of the mean, and so on. As the sample size increases, the size of $\sigma_{\bar{x}}$ becomes smaller so that the distribution of sample means becomes much narrower. Figure 7.12 illustrates what happens to the distribution of $\bar{x}$'s as the size of the individual samples increases.

FIGURE 7.12
Distributions of Sample Means

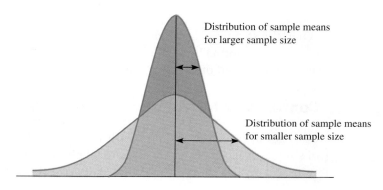

Distribution of sample means for larger sample size

Distribution of sample means for smaller sample size

Recall that the area (probability) under the normal curve is always exactly 1. So as the width of the curve narrows, the height has to increase to maintain this area.

EXAMPLE 7.6

Calculating Probabilities for the Mean Height of Kindergarten Children

Statistics⌂Now™

Watch a video example at
http://1pass.thomson.com
or on your CD.

Kindergarten children have heights that are approximately normally distributed about a mean of 39 inches and a standard deviation of 2 inches. A random sample of size 25 is taken, and the mean $\bar{x}$ is calculated. What is the probability that this mean value will be between 38.5 and 40.0 inches?

SOLUTION We want to find $P(38.5 < \bar{x} < 40.0)$. The values of $\bar{x}$, 38.5 and 40.0, must be converted to z-scores (necessary for use of Table 3 in Appendix B) using

$$z = \frac{\bar{x} - \mu}{\sigma/\sqrt{n}}:$$

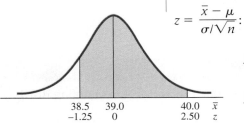

$$\bar{x} = 38.5: \quad z = \frac{\bar{x} - \mu}{\sigma/\sqrt{n}} = \frac{38.5 - 39.0}{2/\sqrt{25}} = \frac{-0.5}{0.4} = \mathbf{-1.25}$$

$$\bar{x} = 40.0: \quad z = \frac{\bar{x} - \mu}{\sigma/\sqrt{n}} = \frac{40.0 - 39.0}{2/\sqrt{25}} = \frac{1.0}{0.4} = \mathbf{2.50}$$

Therefore,

$$P(38.5 < \bar{x} < 40.0) = P(-1.25 < z < 2.50) = 0.3944 + 0.4938 = \mathbf{0.8882}$$

EXAMPLE 7.7

Calculating Mean Height Limits for the Middle 90% of Kindergarten Children

Use the heights of kindergarten children given in Example 7.6. Within what limits does the middle 90% of the sampling distribution of sample means for samples of size 100 fall?

SOLUTION The two tools we have to work with are formula (7.2) and Table 3 in Appendix B. The formula relates the key values of the population to the key values of the sampling distribution, and Table 3 relates areas to z-scores. First, using Table 3, we find that the middle 0.9000 is bounded by $z = \pm 1.65$.

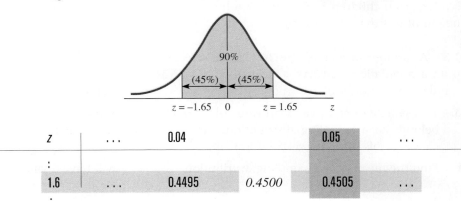

FYI Remember: If value is exactly halfway, use the larger z.

z	...	0.04		0.05	...
:					
1.6	...	0.4495	*0.4500*	0.4505	...
:					

Second, we use formula (7.2), $z = \dfrac{\bar{x} - \mu}{\sigma/\sqrt{n}}$:

$z = -1.65:$ $\quad -1.65 = \dfrac{\bar{x} - 39.0}{2/\sqrt{100}}$ $\qquad z = 1.65:$ $\quad 1.65 = \dfrac{\bar{x} - 39.0}{2/\sqrt{100}}$

$$\bar{x} - 39 = (-1.65)(0.2) \qquad\qquad \bar{x} - 39 = (1.65)(0.2)$$
$$\bar{x} = 39 - 0.33 \qquad\qquad\qquad \bar{x} = 39 + 0.33$$
$$= 38.67 \qquad\qquad\qquad\qquad = 39.33$$

Thus,

$$P(38.67 < \bar{x} < 39.33) = 0.90$$

Therefore, 38.67 inches and 39.33 inches are the limits that capture the middle 90% of the sample means.

SECTION 7.4 EXERCISES

7.29 Consider a normal population with $\mu = 43$ and $\sigma = 5.2$. Calculate the z-score for an $\bar{x}$ of 46.5 from a sample of size 16.

7.30 Consider a population with $\mu = 43$ and $\sigma = 5.2$.

a. Calculate the z-score for an $\bar{x}$ of 46.5 from a sample of size 35.

b. Could this z-score be used in calculating probabilities using Table 3 in Appendix B? Why or why not?

7.31 In Example 7.5, explain how 0.4772 was obtained and what it is.

7.32 What is the probability that the sample of kindergarten children in Example 7.6 has a mean height of less than 39.75 inches?

7.33 A random sample of size 36 is to be selected from a population that has a mean $\mu = 50$ and a standard deviation σ of 10.

a. This sample of 36 has a mean value of $\bar{x}$, which belongs to a sampling distribution. Find the shape of this sampling distribution.

b. Find the mean of this sampling distribution.

c. Find the standard error of this sampling distribution.

d. What is the probability that this sample mean will be between 45 and 55?

e. What is the probability that the sample mean will have a value greater than 48?

f. What is the probability that the sample mean will be within 3 units of the mean?

7.34 The local bakery bakes more than a thousand 1-pound loaves of bread daily, and the weights of these loaves varies. The mean weight is 1 lb. and 1 oz., or 482 grams. Assume the standard deviation of the weights is 18 grams and a sample of 40 loaves is to be randomly selected.

a. This sample of 40 has a mean value of $\bar{x}$, which belongs to a sampling distribution. Find the shape of this sampling distribution.

b. Find the mean of this sampling distribution.

c. Find the standard error of this sampling distribution.

d. What is the probability that this sample mean will be between 475 and 495?

e. What is the probability that the sample mean will have a value less than 478?

f. What is the probability that the sample mean will be within 5 grams of the mean?

7.35 Consider the approximately normal population of heights of male college students with mean $\mu = 69$ inches and standard deviation $\sigma = 4$ inches. A random sample of 16 heights is obtained.

a. Describe the distribution of x, height of male college students.

b. Find the proportion of male college students whose height is greater than 70 inches.

c. Describe the distribution of $\bar{x}$, the mean of samples of size 16.

d. Find the mean and standard error of the $\bar{x}$ distribution.

e. Find $P(\bar{x} > 70)$.   f.   Find $P(\bar{x} < 67)$.

7.36 The amount of fill (weight of contents) put into a glass jar of spaghetti sauce is normally distributed with mean $\mu = 850$ grams and standard deviation $\sigma = 8$ grams.

a. Describe the distribution of x, the amount of fill per jar.

b. Find the probability that one jar selected at random contains between 848 and 855 grams.

c. Describe the distribution of $\bar{x}$, the mean weight for a sample of 24 such jars of sauce.

d. Find the probability that a random sample of 24 jars has a mean weight between 848 and 855 grams.

7.37 The heights of the kindergarten children mentioned in Example 7.6 (p. 379) are approximately normally distributed with $\mu = 39$ and $\sigma = 2$.

a. If an individual kindergarten child is selected at random, what is the probability that he or she has a height between 38 and 40 inches?

b. A classroom of 30 of these children is used as a sample. What is the probability that the class mean $\bar{x}$ is between 38 and 40 inches?

c. If an individual kindergarten child is selected at random, what is the probability that he or she is taller than 40 inches?

d. A classroom of 30 of these kindergarten children is used as a sample. What is the probability that the class mean $\bar{x}$ is greater than 40 inches?

7.38 WageWeb (http://www.wageweb.com/health1 .htm) is a service of HRPDI and provides compensation information on more than 170 benchmark positions in human resources. The October 2003 posting indicated that labor relation managers earn a mean annual salary of $86,700. Assume that annual salaries are normally distributed and have a standard deviation of $8850.

a. What is the probability that a randomly selected labor relation manager earned more than $100,000 in 2003?

b. A sample of 20 labor relation managers is taken, and annual salaries are reported. What is the probability that the sample mean annual salary falls between $80,000 and $90,000?

7.39 Based on 53 years of data compiled by the National Climatic Data Center (http://lwf.ncdc.noaa .gov/oa/climate/online/ccd/avgwind.html), the average speed of winds in Honolulu, Hawaii, equals 11.3 mph, as of June 2004. Assume that wind speeds are approximately normally distributed with a standard deviation of 3.5 mph.

a. Find the probability that the wind speed on any one reading will exceed 13.5 mph.

b. Find the probability that the mean of a random sample of nine readings will exceed 13.5 mph.

c. Do you think the assumption of normality is reasonable? Explain.

d. What effect do you think the assumption of normality had on the answers to parts a and b? Explain.

7.40 TIMSS 2003 (Trends in International Mathematics and Science Study) focused on the mathematics and science achievement of eighth-graders throughout the world. A total of 45 countries (including the United States) participated in the study. The mean math exam score for U.S. students was 504 with a standard deviation of 88.

Source: http://nces.ed.gov/timss/TIMSS03Tables

Assume the scores are normally distributed and a sample of 150 students is taken.

a. Find the probability that the mean TIMSS score for a randomly selected group of eighth-graders would be between 495 and 510.

b. Find the probability that the mean TIMSS score for a randomly selected group of eighth-graders would be less than 520.

c. Do you think the assumption of normality is reasonable? Explain.

7.41 According to the June 2004 *Readers' Digest* article "Only in America," the average amount that a 17-year-old spends on his or her high school prom is $638. Assume that the amounts spent are normally distributed with a standard deviation of $175.

a. Find the probability that the mean cost to attend a high school prom for 36 randomly selected high school 17-year-olds is between $550 and $700.

b. Find the probability that the mean cost to attend a high school prom for 36 randomly selected high school 17-year-olds is greater than $750.

c. Do you think the assumption of normality is reasonable? Explain.

7.42 WageWeb (http://www.wageweb.com/health1.htm) provides compensation information and services on more than 160 positions. As of October 1, 2003, the national average salary for a registered nurse (RN) was $47,858. Suppose the standard deviation is $7750.

a. Find the probability that the mean of a sample of 100 such nurses is less than $45,000.

b. Find the probability that the sample mean of a sample of 100 such nurses is between $46,000 and $48,000.

c. Find the probability that the sample mean of a sample of 100 such nurses is greater than $50,000.

d. Explain why the assumption of normality about the distribution of wages was not involved in the solution to parts a–c.

7.43 Referring to Example 7.6 (p. 379), what height would bound the lower 25% of all samples of size 25?

7.44 A popular flashlight that uses two D-size batteries was selected, and several of the same models were purchased to test the "continuous-use life" of D batteries. As fresh batteries were installed, each flashlight was turned on and the time noted. When the flashlight no longer produced light, the time was again noted. The resulting "life" data from Rayovac batteries had a mean of 21.0 hours (*source:* http://www.rayovac.com). Assume these values have a normal distribution with a standard deviation of 1.38 hours.

a. What is the probability that one randomly selected Rayovac battery will have a test life of between 20.5 and 21.5 hours?

b. What is the probability that a randomly selected sample of 4 Rayovac batteries will have a mean test life of between 20.5 and 21.5 hours?

c. What is the probability that a randomly selected sample of 16 Rayovac batteries will have a mean test life of between 20.5 and 21.5 hours?

d. What is the probability that a randomly selected sample of 64 Rayovac batteries will have a mean test life of between 20.5 and 21.5 hours?

e. Describe the effect that the increase in sample size had on the answers for parts b–d.

7.45 a. Find $P(4 < \bar{x} < 6)$ for a random sample of size 4 drawn from a normal population with $\mu = 5$ and $\sigma = 2$.

b. Use a computer to randomly generate 100 samples, each of size 4, from a normal probability distribution with $\mu = 5$ and $\sigma = 2$. Calculate the mean, $\bar{x}$, for each sample.

c. How many of the sample means in part b have values between 4 and 6? What percentage is that?

d. Compare the answers to parts a and c, and explain any differences that occurred.

MINITAB (Release 14)

a. Input the numbers 4 and 6 into C1. Use the CUMULATIVE NORMAL PROBABILITY DISTRIBUTION commands on page 329, replacing the mean with 5, the standard deviation with 1 $(2/\sqrt{4})$, the input column with C1, and the optional storage with C2. Find $CDF(6) - CDF(4)$.

b. Use the Normal RANDOM DATA commands on page 327, replacing generate with 100, store in with C3–C6, mean with 5, and standard deviation with 2. Use the ROW STATISTICS commands on page 368, replacing input variables with C3–C6 and store result in with C7.

c. Use the HISTOGRAM commands on page 61 for the data in C7. Select Labels, Data Labels, Label Type; use *y*-value levels. To adjust the histogram, select Binning with midpoint and midpoint positions 0:10/1.

Excel

a. Input the numbers 4 and 6 into column A. Activate cell B1. Use the CUMULATIVE NORMAL DISTRIBUTION commands on page 329, replacing X with A1:A2. Find $CDF(6)-CDF(4)$.

b. Use the Normal RANDOM NUMBER GENERATION commands on page 328, replacing number of variables with 4, number of random numbers with 100, mean with 5, standard deviation with 2, and output range with C1. Activate cell G1. Use the AVERAGE INSERT FUNCTION commands in Exercise 7.13b on page 368, replacing Number1 with C1:F1.

c. Use the RANDOM NUMBER GENERATION Patterned Distribution commands in Exercise 6.66 on page 336, replacing the first value with

0, the last value with 9, the steps with 1, and the output range with H1. Use the HISTOGRAM commands on page 61 with column G as the input range, column H as the bin range, and column I as the output range.

TI-83/84 Plus

a. Use the CUMULATIVE NORMAL PROBABILITY commands on page 330, replacing the Enter with 4,6,5,1). (The standard deviation is 1; from $2/\sqrt{4}$.)

b. Use the Normal RANDOM DATA and STO commands on page 328, replacing the Enter with 5,2,100). Repeat these commands three more times, storing data in L2, L3, and L4, respectively.

```
Choose:      STAT > EDIT > 1:Edit
Highlight:   L5 (column heading)
Enter:       (L1+L2+L3+L4)/4
```

c. Use the HISTOGRAM and TRACE commands on page 62 to count. Enter 0,9,1,0,45,1 for the Window.

7.46 a. Find $P(46 < \bar{x} < 55)$ for a random sample size 16 drawn from a normal population with mean $\mu = 50$ and standard deviation $\sigma = 10$.

b. Use a computer to randomly generate 200 samples, each of size 16, from a normal probability distribution with mean $\mu = 50$ and standard deviation $\sigma = 10$. Calculate the mean, $\bar{x}$, for each sample.

c. How many of the sample means in part b have values between 46 and 55? What percentage is that?

d. Compare the answers to parts a and c, and explain any differences that occurred.

FYI If you use a computer, see Exercise 7.45.

CHAPTER REVIEW

In Retrospect

In Chapters 6 and 7 we have learned to use the standard normal probability distribution. We now have two formulas for calculating a *z*-score:

$$z = \frac{x - \mu}{\sigma} \quad \text{and} \quad z = \frac{\bar{x} - \mu}{\sigma/\sqrt{n}}$$

You must be careful to distinguish between these two formulas. The first gives the standard score when we have individual values from a normal distribution (*x* values). The second formula deals with a sample mean ($\bar{x}$ value). The key to distinguishing between the formulas is to decide whether the problem deals with an individual *x* or a sample mean $\bar{x}$. If it deals with the individual values of *x*, we use the first formula, as presented in Chapter 6. If the problem deals with a sample mean, $\bar{x}$, we use the second formula and proceed as illustrated in this chapter.

The basic purpose for considering what happens when a population is repeatedly sampled, as discussed in this chapter, is to form sampling distributions. The sampling distribution is then used to describe the variability that occurs from one sample to the next. Once this pattern of variability is known and understood for a specific sample statistic, we are able to make predictions about the corresponding population parameter with a measure of how accurate the prediction is. The SDSM and the central limit theorem help describe the distribution for sample means. We will begin to make inferences about population means in Chapter 8.

There are other reasons for repeated sampling. Repeated samples are commonly used in the field of production control, in which samples are taken to determine whether a product is of the proper size or quantity. When the sample statistic does not fit the standards, a mechanical adjustment of the machinery is necessary. The adjustment is then followed by another sampling to be sure the production process is in control.

The "standard error of the _____" is the name used for the standard deviation of the sampling distribution for whatever statistic is named in the blank. In this chapter we have been concerned with the standard error of the mean. However, we could also work with the standard error of the proportion, median, or any other statistic.

You should now be familiar with the concept of a sampling distribution and, in particular, with the sampling distribution of sample means. In Chapter 8 we will begin to make predictions about the values of population parameters.

Vocabulary and Key Concepts

central limit theorem (p. 370)
frequency distribution (p. 364)
probability distribution (p. 364)
random sample (p. 365)

repeated sampling (p. 366)
sampling distribution (p. 373)
sampling distribution of sample
   means (pp. 363, 369)

standard error of the mean
  (p. 370)
z-score (p. 377)

Learning Outcomes

✓ Understand what a sampling distribution of a sample statistic is and that the distribution is obtained from repeated samples, all of the same size.   pp. 363–364, EXP 7.1

✓ Be able to form a sampling distribution for a mean, median, or range based on a small, finite population.   EXP 7.1, Ex. 7.6, 7.7

✓ Understand that a sampling distribution is a probability distribution for a sample statistic.   EXP 7.2

✓ Understand and be able to present and describe the sampling distribution of sample means and the central limit theorem.　　pp. 369–371, EXP 7.4

✓ Understand and be able to explain the relationship between the sampling distribution of sample means and the central limit theorem.　　pp. 369–371, Ex. 7.17, 7.18, 7.21

✓ Determine and be able to explain the effect of sample size on the standard error of the mean.　　pp. 373–375, Ex. 7.20, 7.26, 7.47

✓ Understand when and how the normal distribution can be used to find probabilities corresponding to sample means.　　EXP 7.5

✓ Compute, describe, and interpret z-scores corresponding to known values of $\bar{x}$.　　EXP 7.6, EXP 7.7, Ex. 7.29, 7.30, 7.48

✓ Compute z-scores and probabilities for applications of the sampling distribution of sample means.　　Ex. 7.33, 7.35

Chapter Exercises

Statistics⌂Now™

Go to the StatisticsNow website **http://1pass.thomson.com** to

- Assess your understanding of this chapter
- Check your readiness for an exam by taking the Pre-Test quiz and exploring the resources in the Personalized Learning Plan

7.47 If a population has a standard deviation σ of 18.2 units, what is the standard error of the mean if samples of size 9 are selected? Samples of size 25? Samples of size 49? Samples of size 100?

7.48 Consider a normal population with $\mu = 24.7$ and $\sigma = 4.5$.

a. Calculate the z-score for an x of 21.5.

b. Calculate the z-score for an $\bar{x}$ of 21.5 from a sample of size 25.

c. Explain how 21.5 can have such different z-scores.

7.49 The Dean of Nursing tells students being recruited for the incoming class that 1 year after graduation, the university's graduates can expect to be earning a mean weekly income of $675. Assume that the dean's statement is true and that the weekly salaries 1 year after graduation are normally distributed with a standard deviation of $85. If one graduate is randomly selected:

a. Describe the distribution of the weekly salaries being earned 1 year after graduation.

b. What is the probability that the selected graduate is making between $550 and $825?

If a random sample of 25 graduates is selected:

c. Describe the distribution of mean weekly salaries being earned 1 year after graduation.

d. What is the probability that the sample mean is between $650 and $705?

e. Why is the z-score used in answering parts b and d?

f. Why is the formula for z-score used in part d different from that used in part b?

7.50 The diameters of Red Delicious apples in a certain orchard are normally distributed with a mean of 2.63 inches and a standard deviation of 0.25 inch.

a. What percentage of the apples in this orchard have diameters less than 2.25 inches?

b. What percentage of the apples in this orchard are larger than 2.56 inches in diameter?

A random sample of 100 apples is gathered, and the mean diameter obtained is $\bar{x} = 2.56$.

c. If another sample of size 100 is taken, what is the probability that its sample mean will be greater than 2.56 inches?

d. Why is the z-score used in answering parts a–c?

e. Why is the formula for z-score used in part c different from that used in parts a and b?

7.51 a. Find a value for e such that 95% of the apples in Exercise 7.50 are within e units of the mean, 2.63. That is, find e such that $P(2.63 - e < x < 2.63 + e) = 0.95$.

b. Find a value for E such that 95% of the samples of 100 apples taken from the orchard in Exercise 7.50 will have mean values within E units of the mean, 2.63. That is, find E such that $P(2.63 - E < \bar{x} < 2.63 + E) = 0.95$.

7.52 Americans spend billions of dollars on veterinary care each year, predicted to hit $31 billion this year. The health care services offered to animals rival those provided to humans, with the typical surgery costing from $1700 to $3000, or even more. In 2003, on average, dog owners spent $196 on veterinary-related expenses in the prior 12 months.

Source: American Pet Products Manufacturers Association

Assume that annual dog owner expenditure on health care is normally distributed with a mean of $196 and a standard deviation of $95.

a. What is the probability that a dog owner, randomly selected from the population, spent more than $300 for dog health care in 2003?

b. Suppose a survey of 300 dog owners is conducted, and each person is asked to report the total of their vet care bills for 2003. What is the probability that the mean annual expenditure of this sample falls between $200 and $225?

c. The assumption of a normal distribution in this situation is likely misguided. Explain why and what effect this had on the answers.

7.53 The statistics-conscious store manager at Marketview records the number of customers who walk through the door each day. Years of records show the mean number of customers per day to be 586 with a standard deviation of 165. Assume the number of customers is normally distributed.

a. What is the probability that on any given day, the number of customers exceeds 1000?

b. If 20 days are randomly selected, what is the probability that the mean of this sample is less than 550?

c. The assumption of normality allowed you to calculate the probabilities; however, this may not be a reasonable assumption. Explain why and how that affects the probabilities found in parts a and b.

7.54 A study from the University of Michigan, as noted in *Newsweek* (March 25, 2002), stated that men average 16 hours of housework each week (up from an average of 12 hours in 1965). If we assume that the number of hours in which men engage in housework each week is normally distributed with a standard deviation of 5.4 hours, what is the probability that the mean number of housework hours for a sample of 20 randomly selected men is between 15 to 18 hours?

7.55 A shipment of steel bars will be accepted if the mean breaking strength of a random sample of 10 steel bars is greater than 250 pounds per square inch. In the past, the breaking strength of such bars has had a mean of 235 and a variance of 400.

a. Assuming that the breaking strengths are normally distributed, what is the probability that one randomly selected steel bar will have a breaking strength in the range from 245 to 255 pounds per square inch?

b. What is the probability that the shipment will be accepted?

7.56 An April 15, 2002, report in *Time* magazine stated that the average age for women to marry in the United States is now 25 years of age. If the standard deviation is assumed to be 3.2 years, find the probability that a random sample of 40 U.S. women would show a mean age at marriage of less than or equal to 24 years.

7.57 A manufacturer of light bulbs claims that its light bulbs have a mean life of 700 hours and a standard deviation of 120 hours. You purchased 144 of these bulbs and decided that you would purchase more if the mean life of your current sample exceeded 680 hours. What is the probability that you will not buy again from this manufacturer?

7.58 A tire manufacturer claims (based on years of experience with its tires) that the mean mileage is 35,000 miles and the standard deviation is 5000 miles. A consumer agency randomly selects 100 of these tires and finds a sample mean of 31,000. Should the consumer agency doubt the manufacturer's claim?

7.59 For large samples, the sample sum $(\sum x)$ has an approximately normal distribution. The mean of the sample sum is $n \cdot \mu$ and the standard deviation is $\sqrt{n} \cdot \sigma$. The distribution of savings per account for a savings and loan institution has a mean equal to $750 and a standard deviation equal to $25. For a sample of 50 such accounts, find the probability that the sum in the 50 accounts exceeds $38,000.

7.60 The baggage weights for passengers using a particular airline are normally distributed with a mean of 20 lb. and a standard deviation of 4 lb. If the limit on total luggage weight is 2125 lb., what is the probability that the limit will be exceeded for 100 passengers?

7.61 A trucking firm delivers appliances for a large retail operation. The packages (or crates) have a mean weight of 300 lb. and a variance of 2500.

a. If a truck can carry 4000 lb. and 25 appliances need to be picked up, what is the probability that the 25 appliances will have an aggregate weight greater than the truck's capacity? Assume that the 25 appliances represent a random sample.

b. If the truck has a capacity of 8000 lb., what is the probability that it will be able to carry the entire lot of 25 appliances?

7.62 A pop-music record firm wants the distribution of lengths of cuts on its records to have an average of 2 minutes and 15 seconds (135 seconds) and a standard deviation of 10 seconds so that disc jockeys will have plenty of time for commercials within each 5-minute period. The population of times for cuts is approximately normally distributed with only a negligible skew to the right. You have just timed the cuts on a new release and have found that the 10 cuts average 140 seconds.

a. What percentage of the time will the average be 140 seconds or longer if the new release is randomly selected?

b. If the music firm wants 10 cuts to average 140 seconds less than 5% of the time, what must the population mean be, given that the standard deviation remains at 10 seconds?

7.63 Let's simulate the sampling distribution related to the disc jockey's concern for "length of cut" in Exercise 7.62.

a. Use a computer to randomly generate 50 samples, each of size 10, from a normal distribution with mean 135 and standard deviation 10. Find the "sample total" and the sample mean for each sample.

b. Using the 50 sample means, construct a histogram and find their mean and standard deviation.

c. Using the 50 sample "totals," construct a histogram and find their mean and standard deviation.

d. Compare the results obtained in parts b and c. Explain any similarities and any differences observed.

MINITAB (Release 14)

a. Use the Normal RANDOM DATA commands on page 327, replacing generate with 50, store in with C1–C10, mean with 135, and standard deviation with 10. Use the ROW STATISTICS commands on page 368, selecting Sum and replacing input variables with C1–C10 and store result in with C11. Use the ROW STATISTICS commands, again selecting Mean and then replacing input variables with C1–C10 and store result in with C12.

b. Use the HISTOGRAM commands on page 61 for the data in C12. To adjust the histogram, select Binning with midpoint. Use the MEAN and STANDARD DEVIATION commands on pages 74 and 88 for the data in C12.

c. Use the HISTOGRAM commands on page 61 for the data in C11. To adjust the histogram, select Binning with midpoints. Use the MEAN and STANDARD DEVIATION commands on pages 74 and 88 for the data in C11.

d. Use the DISPLAY DESCRIPTIVE STATISTICS commands on page 98 for the data in C11 and C12.

Excel

a. Use the Normal RANDOM NUMBER GENERATION commands on page 328, replacing number of variables with 10, number of random numbers with 50, mean with 135, and standard deviation with 10.

Activate cell K1.

```
Choose:  Insert function, fₓ > All > SUM > OK
Enter:   Number1: (A1:J1 or select cells)
Drag:    Bottom right corner of sum value box down to
         give other sums
```

Activate cell L1. Use the AVERAGE INSERT FUNCTION commands in Exercise 7.13b on page 368, replacing Number1 with A1:J1.

b. Use the RANDOM NUMBER GENERATION Patterned Distribution commands in Exercise 6.66 on page 336, replacing the first value with 125.4, the last value with 144.6, the steps with 3.2, and the output range with M1. Use the HISTOGRAM commands on page 61 with column L as the input range and column M as the bin range. Use the MEAN and STANDARD DEVIATION commands on pages 74 and 88 for the data in column L.

c. Use the RANDOM NUMBER GENERATION Patterned Distribution commands in Exercise 6.66 on page 336, replacing the first value with 1254, the last value with 1446, the steps with 32, and the output range with M20. Use the HISTOGRAM commands on page 61 with column L as the input range and cells M20–? as the bin range. Use the MEAN and STANDARD DEVIATION commands on pages 74 and 88 for the data in column K.

d. Use the DESCRIPTIVE STATISTICS commands on page 98 for the data in columns K and L.

7.64 a. Find the mean and standard deviation of x for a binomial probability distribution with $n = 16$ and $p = 0.5$.

b. Use a computer to construct the probability distribution and histogram for the binomial probability experiment with $n = 16$ and $p = 0.5$.

c. Use a computer to randomly generate 200 samples of size 25 from a binomial probability distribution with $n = 16$ and $p = 0.5$. Calculate the mean of each sample.

d. Construct a histogram and find the mean and standard deviation of the 200 sample means.

e. Compare the probability distribution of x found in part b and the frequency distribution of $\bar{x}$ in part d. Does your information support the CLT? Explain.

MINITAB (Release 14)

a. Use the MAKE PATTERNED DATA commands in Exercise 6.66 on page 336, replacing the first value with 0, the last value with 16, and the steps with 1. Use the BINOMIAL PROBABILITY DISTRIBUTIONS commands on page 292, replacing n with 16, p with 0.5, input column with C1, and optional storage with C2. Use the Scatterplot with Connect Line commands on page 155, replacing Y with C2 and X with C1.

b. Use the BINOMIAL RANDOM DATA commands on page 303, replacing generate with 200, store in with C3–C27, number of trials with 16, and probability with 0.5. Use the ROW STATISTICS commands for a mean on page 368, replacing input variables with C3–C27 and store result in with C28. Use the HISTOGRAM commands on page 61 for the data in C28. To adjust histogram, select Binning with midpoints. Use the MEAN and STANDARD DEVIATION commands on pages 74 and 88 for the data in C28.

Excel

a. Input 0 through 16 into column A. Continue with the binomial probability commands on page 292, using $n = 16$ and $p = 0.5$. Activate columns A and B; then continue with:

```
Choose:  Chart Wizard > Column > 1st picture > Next >
         Series
Choose:  Series 1 > Remove
Enter:   Category (x)axis labels: (A1:A17 or select 'x
         value' cells)
Choose:  Next > Finish
```

b. Use the Binomial RANDOM NUMBER GEN-ERATION commands from Exercise 5.97 on page 304, replacing number of variables with 25, number of random numbers with 200, p value with 0.5, number of trials with 16, and output range with C1. Activate cell BB1. Use the AVERAGE INSERT FUNCTION commands in Exercise 7.13b on page 368, replacing Number1 with C1:AA1.

c. Use the RANDOM NUMBER GENERATION Patterned Distribution commands in Exercise 6.66 on page 336, replacing the first value with 6.8, the last value with 9.2, the steps with 0.4, and the output range with CC1. Use the HIS-TOGRAM commands on page 61 with column BB as the input range and column CC as the bin range. Use the MEAN and STANDARD DE-VIATION commands on pages 74 and 88 for the data in column BB.

7.65 a. Find the mean and standard deviation of x for a binomial probability distribution with $n = 200$ and $p = 0.3$.

b. Use a computer to construct the probability distribution and histogram for the random variable x of the binomial probability experiment with $n = 200$ and $p = 0.3$.

c. Use a computer to randomly generate 200 samples of size 25 from a binomial probability distribution with $n = 200$ and $p = 0.3$. Calculate the mean $\bar{x}$ of each sample.

d. Construct a histogram and find the mean and standard deviation of the 200 sample means.

e. Compare the probability distribution of x found in part b and the frequency distribution of $\bar{x}$ in part d. Does your information support the CLT? Explain.

FYI Use the commands in Exercise 7.64, making the necessary adjustments.

7.66 A sample of 144 values is randomly selected from a population with mean, μ, equal to 45 and standard deviation, σ, equal to 18.

a. Determine the interval (smallest value to largest value) within which you would expect a sample mean to lie.

b. What is the amount of deviation from the mean for a sample mean of 45.3?

c. What is the maximum deviation you have allowed for in your answer to part a?

d. How is this maximum deviation related to the standard error of the mean?

Chapter Project

275 Million Americans

As noted in the "The Sampling Issue" of Section 7.1, "275 Million Americans" (p. 361), the fundamental goal of a survey is to come up with the same results that we would have obtained had we interviewed every person of the population. Knowing that interviewing every person of a population is nearly impossible for most populations promotes the importance of a good representative sample. In addition, we now have the sampling distribution of sample means and the central limit theorem to help us make predictions about the population by using the sample. Putting Chapter 7 to Work will help us put these new concepts together.

Putting Chapter 7 to Work

7.67 A second sample of 100 ages as been collected from the U.S. 2000 census and is listed here. [EX07-67]

14	6	59	64	39	12	8	34	27	4
16	18	17	33	56	60	65	73	53	43
26	42	60	87	58	42	82	21	35	64
58	53	36	66	63	66	39	62	58	49
31	27	39	35	12	28	28	20	3	54
41	41	63	39	37	23	79	43	28	17
12	45	52	10	11	32	32	23	86	61
50	27	19	15	3	51	5	36	83	39
35	44	59	30	31	69	40	16	40	66
15	55	32	4	43	41	23	46	61	30

a. How would you describe the preceding "ages" sample data graphically? Construct the graph.

b. Using the graph that you constructed in part a, describe the shape of the distribution of sample data.

c. How well did the sample describe the population of ages from the 2000 census shown in Section 7.1? Explain using the graphical displays.

d. How would you describe the preceding "ages" sample data numerically? Calculate the statistics.

e. How well do the statistics calculated in part d compare with the parameters from the 2000 census given in Section 7.1?

f. (optional) If you completed Exercises 7.1 and 7.2, how does your graphical display and statistics compare with those constructed and calculated in Exercises 7.1 and 7.2 using a different sample of 100 ages?

g. Is the distribution of ages for the population of Americans in Section 7.1 normal? Is it approximately normal?

h. Will the SDSM apply to samples taken from this population? Explain.

i. Will the CLT apply to samples taken from this population? Explain.

j. Describe the SDSM for samples of size 100. Be sure to include center, spread, and shape.

k. Compare your results in parts a and d with the theoretical answers in part j. Be sure to include center, spread, and shape.

l. Describe the SDSM for samples of size 30. Be sure to include center, spread, and shape.

m. Describe the sampling SDSM for samples of size 1000. Be sure to include center, spread, and shape.

n. Relate your findings in parts j, l, and m to the SDSM and the CLT.

Your Study

7.68 Skillbuilder Applet Exercise simulates taking samples of size 50 from the population of American ages from the 2000 census, where $\mu = 36.5$ and $\sigma = 22.5$ and the shape is skewed right.

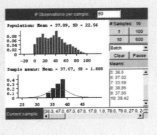

a. Click "1" for "# Samples." Note the 50 data values and their mean. Change "slow" to "batch" and take at least 1000 samples of size 50.

b. What is the mean of the sample means? How close is it to the population mean?

c. What is the standard deviation of the sample means?

d. Based on the SDSM (as described in Section 7.3), what should you expect for the standard deviation of sample means? How close was your standard deviation from part c?

e. What shape is the histogram of the 1000 means?

f. Relate your findings to the SDSM and the CLT.

Chapter Practice Test

PART I: Knowing the Definitions

Answer "True" if the statement is always true. If the statement is not always true, replace the words shown in bold with words that make the statement always true.

7.1 A sampling distribution **is** a distribution listing all the sample statistics that describe a particular sample.

7.2 The histograms of **all** sampling distributions are symmetrical.

7.3 The mean of the sampling distribution of $\bar{x}$'s is equal to the mean of the **sample.**

7.4 The standard error of the mean is the standard deviation of the population **from which the samples have been taken.**

7.5 The standard error of the mean **increases** as the sample size increases.

7.6 The shape of the distribution of sample means is always that of a **normal** distribution.

7.7 A **probability** distribution of a sample statistic is a distribution of all the values of that statistic that were obtained from all possible samples.

7.8 The sampling distribution of sample means provides us with a description of the three characteristics of a sampling distribution of sample **medians.**

7.9 A **frequency** sample is obtained in such a way that all possible samples of a given size have an equal chance of being selected.

7.10 We **do not need** to take repeated samples in order to use the concept of the sampling distribution.

PART II: Applying the Concepts

7.11 The lengths of the lake trout in Conesus Lake are believed to have a normal distribution with a mean of 15.6 inches and a standard deviation of 3.8 inches.

    a. Kevin is going fishing at Conesus Lake tomorrow. If he catches one lake trout, what is the probability that it is less than 15.0 inches long?

    b. If Captain Brian's fishing boat takes 10 people fishing on Conesus Lake tomorrow and they catch a random sample of 16 lake trout, what is the probability that the mean length of their total catch is less than 15 inches?

7.12 Cigarette lighters manufactured by EasyVice Company are claimed to have a mean lifetime of 20 months with a standard deviation of 6 months. The money-back guarantee allows you to return the lighter if it does not last at least 12 months from the date of purchase.

    a. If the lifetimes of these lighters are normally distributed, what percentage of the lighters will be returned to the company?

    b. If a random sample of 25 lighters is tested, what is the probability the sample mean lifetime will be more than 18 months?

7.13 Aluminum rivets produced by Rivets Forever, Inc., are believed to have shearing strengths that are distributed about a mean of 13.75 with a standard deviation of 2.4. If this information is true and a sample of 64 such rivets is tested for shear strength, what is the probability that the mean strength will be between 13.6 and 14.2?

PART III: Understanding the Concepts

7.14 "Two heads are better than one." If that's true, then how good would several heads be? To find out, a statistics instructor drew a line across the chalkboard and asked her class to estimate its length to the nearest inch. She collected their estimates, which ranged from 33 to 61 inches, and calculated the mean value. She reported that the mean was 42.25 inches. She then measured the line and found it to be 41.75 inches long. Does this show that "several heads are better than one"? What statistical theory supports this occurrence? Explain how.

7.15 The sampling distribution of sample means is more than just a distribution of the mean values that occur from many repeated samples taken from the same population. Describe what other specific condition must be met in order to have a sampling distribution of sample means.

7.16 Student A states, "A sampling distribution of the standard deviations tells you how the standard deviation varies from sample to sample." Student B argues, "A population distribution tells you that." Who is right? Justify your answer.

7.17 Student A says it is the "size of each sample used" and Student B says it is the "number of samples used" that determines the spread of an empirical sampling distribution. Who is right? Justify your choice.

Working with Your Own Data

Putting Probability to Work

The sampling distribution of sample means and the central limit theorem are very important to the development of the rest of this course. The proof, which requires the use of calculus, is not included in this textbook. However, the truth of the SDSM and the CLT can be demonstrated both theoretically and by experimentation. The following activities will help to verify both statements.

A The Population

Consider the theoretical population that contains the three numbers 0, 3, and 6 in equal proportions.

1. a. Construct the theoretical probability distribution for the drawing of a single number, with replacement, from this population.

   b. Draw a histogram of this probability distribution.

   c. Calculate the mean, μ, and the standard deviation, σ, for this population.

B The Sampling Distribution, Theoretically

Let's study the theoretical sampling distribution formed by the means of all possible samples of size 3 that can be drawn from the given population.

2. Construct a list showing all the possible samples of size 3 that could be drawn from this population. (There are 27 possibilities.)

3. Find the mean for each of the 27 possible samples listed in answer to question 2.

4. Construct the probability distribution (the theoretical sampling distribution of sample means) for these 27 sample means.

5. Construct a histogram for this sampling distribution of sample means.

6. Calculate the mean $\mu_{\bar{x}}$ and the standard error of the mean $\sigma_{\bar{x}}$ using the probability distribution found in question 4.

7. Show that the results found in questions 1c, 5, and 6 support the three claims made by the sampling distribution of sample means and the central limit theorem. Cite specific values to support your conclusions.

C The Sampling Distribution, Empirically

Let's now see whether the sampling distribution of sample means and the central limit theorem can be verified empirically; that is, does it hold when the sampling distribution is formed by the sample means that result from several random samples?

8. Draw a random sample of size 3 from the given population. List your sample of three numbers and calculate the mean for this sample.

You may use a computer to generate your samples. You may take three identical "tags" numbered 0, 3, and 6, put them in a "hat," and draw your sample using replacement between each drawing. Or you may use dice; let 0 be represented by 1 and 2; 3, by 3 and 4; and 6, by 5 and 6. You may also use random numbers to simulate the drawing of your samples. Or you may draw your sample from the list of random samples at the end of this section. Describe the method you decide to use. (Ask your instructor for guidance.)

9. Repeat question 8 forty-nine more times so that you have a total of 50 sample means that have resulted from samples of size 3.

10. Construct a frequency distribution of the 50 sample means found in questions 8 and 9.

11. Construct a histogram of the frequency distribution of observed sample means.

12. Calculate the mean $\bar{\bar{x}}$ and standard deviation $s_{\bar{x}}$ of the frequency distribution formed by the 50 sample means.

13. Compare the observed values of $\bar{\bar{x}}$ and $s_{\bar{x}}$ with the values of $\mu_{\bar{x}}$ and $\sigma_{\bar{x}}$. Do they agree? Does the empirical distribution of $\bar{x}$ look like the theoretical one?

Here are 100 random samples of size 3 that were generated by computer:

6 3 0	0 3 0	6 6 0	3 3 6	6 6 3	6 3 3
0 0 3	3 0 6	3 3 0	3 6 6	0 3 0	6 6 3
6 6 6	0 3 0	6 3 6	0 6 3	6 0 3	6 3 3
6 0 0	3 0 6	6 3 3	3 3 0	3 3 0	3 3 3
3 3 3	3 0 0	6 6 6	3 3 6	0 0 6	0 6 3
6 6 6	0 0 6	3 3 0	0 6 6	0 0 3	6 6 3
0 0 6	0 0 6	6 6 6	6 3 6	6 6 0	3 0 0
3 6 6	6 3 0	3 6 3	3 0 0	3 3 6	0 6 0
3 0 0	0 3 6	6 3 3	6 0 6	3 3 6	6 0 3
0 3 6	3 6 3	6 6 3	6 6 0	3 3 3	3 0 0
6 3 0	6 6 0	0 3 0	6 6 0	3 6 6	0 3 6
6 3 3	0 3 0	6 6 0	6 6 3	6 6 0	3 0 3
3 6 3	3 6 0	0 0 6	0 3 3	3 6 6	0 3 6
0 6 0	6 0 0	0 6 0	0 6 6	0 3 3	0 3 6
3 3 6	3 3 3	3 3 6	6 3 6	3 3 3	3 6 6
6 3 3	3 0 0	3 0 6	6 0 3	3 6 6	6 0 3
0 3 3	6 3 0	0 3 6	0 3 6		

CHAPTER

8

Introduction to Statistical Inferences

© Christa Renee/Getty Images

Were They Shorter Back Then?

WERE THEY SHORTER BACK THEN?

The average height for an early 17th-century English man was approximately 5' 6". For 17th-century English women, it was about 5' 1/2". While average heights in England remained virtually unchanged in the 17th and 18th centuries, American colonists grew taller. Averages for modern Americans are just over 5' 9" for men, and about 5' 3 3/4" for women. The main reasons for this difference are improved nutrition, notably increased consumption of meat and milk, and antibiotics.

Source: http://www.plimoth.org/Library/l-short.htm

The National Center for Health Statistics (NCHS) provides statistical information that guides actions and policies to improve the health of the American people. Recent data from NCHS give the average height of females in the United States to be 63.7 inches, with a standard deviation of 2.75 inches.

A random sample of 50 females from the health profession yielded the following height data. [EX08-001]

65.0	66.0	64.0	67.0	59.0	69.0	66.0	69.0	64.0	61.5
63.0	62.0	63.0	64.0	72.0	66.0	65.0	64.0	67.0	68.0
70.0	63.0	63.0	68.0	58.0	60.0	63.5	66.0	64.0	62.0
64.5	69.0	63.5	69.0	62.0	58.0	66.0	68.0	59.0	56.0
64.0	66.0	65.0	69.0	67.0	66.5	67.5	62.0	70.0	62.0

Do you expect the mean of this random sample of 50 females to be exactly equal to the population mean of 63.7 inches given by NCHS? If the sample mean is greater than 63.7 inches, does it mean that we are even taller today? After completing Chapter 8, we will know how to answer these questions. We can then further investigate "Were They Shorter Back Then?" in the Chapter Project on page 469.

SECTION 8.1 EXERCISES

Statistics⏃Now™

Datasets can be found on your Student's Suite CD-ROM or at the StatisticsNow website at **http://1pass.thomson.com.**

8.1 [EX08-001] a. What population was sampled to obtain the height data listed in Section 8.1?

b. Describe the sample data using the mean and standard deviation, plus any other statistics that help describe the sample. Construct a histogram and comment on the shape of the distribution.

8.2 a. How is the distribution of the sample height data in Section 8.1 related to the distribution of the population and the sampling distribution of sample means?

b. Using the techniques in Chapter 7, find the limits that would bound the middle 90% of the sampling distribution of sam-

ple means for samples of size 50 randomly selected from the population of female heights with a known mean of 63.7 inches and a standard deviation of 2.75 inches.

c. On the histogram drawn in Exercise 8.1, draw a vertical line at the population mean of 63.7 and draw a horizontal line segment showing the interval found in part b. Does the sample mean found in part b of Exercise 8.1 fall in the interval? Explain what this means.

d. Using the techniques of Chapter 7, find $P(\bar{x} \geq 64.7)$ for a random sample of 50 drawn from a population with a known mean of 63.7 inches and a standard deviation of 2.75 inches. Explain what the resulting value means.

e. Does the sample of 50 height data values appear to belong to the population described by the NCHS? Explain.

The objective of inferential statistics is to use the information contained in the sample data to increase our knowledge of the sampled population. We will learn about making two types of inferences: (1) estimating the value of a population parameter and (2) testing a hypothesis. The sampling distribution of sample means (SDSM) is the key to making these inferences, as shown in Figure 8.1.

In this chapter, we deal with questions about the population mean using two methods that assume the value of the population standard deviation is a known quantity. This assumption is seldom realized in real-life problems, but it will make our first look at the techniques of inference much simpler.

FIGURE 8.1 Where the Sampling Distribution Fits into the Statistical Process

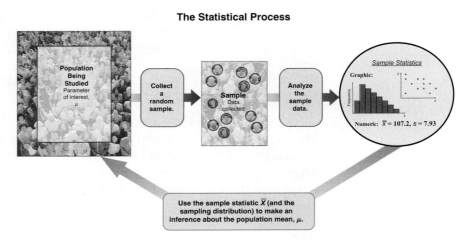

8.2 The Nature of Estimation

A company manufactures rivets for use in building aircraft. One characteristic of extreme importance is the "shearing strength" of each rivet. The company's engineers must monitor production to be certain that the shearing strength of the rivets meets the required specs. To accomplish this, they take a sample and determine the mean shearing strength of the sample. Based on this sample information, the company can estimate the mean shearing strength for all the rivets it is manufacturing.

A random sample of 36 rivets is selected, and each rivet is tested for shearing strength. The resulting sample mean is $\bar{x} = 924.23$ lb. Based on this sample, we say, "We believe the mean shearing strength of all such rivets is 924.23 lb."

Notes:

1. Shearing strength is the force required to break a material in a "cutting" action. Obviously, the manufacturer is not going to test all rivets because the test destroys each rivet tested. Therefore, samples are tested and the information about the sample must be used to make inferences about the population of all such rivets.

2. Throughout Chapter 8 we will treat the standard deviation, σ, as a known, or given, quantity and concentrate on learning the procedures for making statistical inferences about the population mean, μ. Therefore, to continue the explanation of statistical inferences, we will assume $\sigma = 18$ for the specific rivets described in our example.

> **Point estimate for a parameter:** A single number designed to estimate a quantitative parameter of a population, usually the value of the corresponding **sample statistic**.

That is, the sample mean, $\bar{x}$, is the point estimate (single number value) for the mean, μ, of the sampled population. For our rivet example, 924.23 is the point estimate for μ, the mean shearing strength of all rivets.

The quality of this point estimate should be questioned. Is the estimate exact? Is the estimate likely to be high? Or low? Would another sample yield the same result? Would another sample yield an estimate of nearly the same value? Or a value that is very different? How is "nearly the same" or "very different" measured? The quality of an estimation procedure (or method) is greatly enhanced if the sample statistic is both *less variable* and *unbiased*. The variability of a statistic is measured by the standard error of its sampling distribution. The sample mean can be made less variable by reducing its standard error, $\sigma/\sqrt{n}$. That requires using a larger sample because as n increases, the standard error decreases.

> **Unbiased statistic:** A sample statistic whose sampling distribution has a mean value equal to the value of the population parameter being estimated. A statistic that is not unbiased is a **biased statistic**.

Figure 8.2 illustrates the concept of being unbiased and the effect of variability on the point estimate. The value A is the parameter being estimated, and the dots represent possible sample statistic values from the sampling distribution of the statistic. If A represents the true population mean, μ, then the dots represent possible sample means from the $\bar{x}$ sampling distribution.

FIGURE 8.2   Effects of Variability and Bias

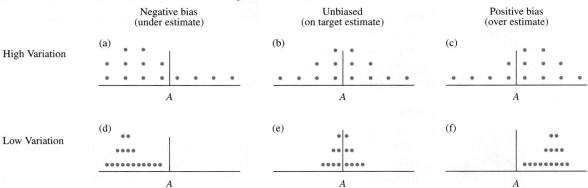

Figure 8.2(a), (c), (d), and (f) show biased statistics; (a) and (d) show sampling distributions whose mean values are less than the value of the parameter, whereas (c) and (f) show sampling distributions whose mean values are greater than the parameter. Figure 8.2(b) and (e) show sampling distributions that appear to have a mean value equal to the value of the parameter; therefore, they are unbiased. Figure 8.2(a), (b), and (c) show more variability, whereas (d), (e), and (f) show less variability in the sampling distributions. Diagram (e) represents the best situation, an estimator that is unbiased (on target) and has low variability (all values close to the target).

The sample mean, $\bar{x}$, is an unbiased statistic because the mean value of the sampling distribution of sample means, $\mu_{\bar{x}}$, is equal to the population mean, μ. (Recall that the sampling distribution of sample means has a mean $\mu_{\bar{x}} = \mu$.) Therefore, the sample statistic $\bar{x} = 924.23$ is an unbiased point estimate for the mean strength of all rivets being manufactured in our example.

Sample means vary in value and form a sampling distribution in which not all samples result in $\bar{x}$ values equal to the population mean. Therefore, we should not expect this sample of 36 rivets to produce a point estimate (sample mean) that is exactly equal to the mean μ of the sampled population. We should, however, expect the point estimate to be fairly close in value to the population mean. The sampling distribution of sample means (SDSM) and the central limit theorem (CLT) provide the information needed to describe how close the point estimate, $\bar{x}$, is expected to be to the population mean, μ.

Recall that approximately 95% of a normal distribution is within 2 standard deviations of the mean and that the CLT describes the sampling distribution of sample means as being nearly normal when samples are large enough. Samples of size 36 from populations of variables like rivet strength are generally considered large enough. Therefore, we should anticipate that 95% of all random samples selected from a population with unknown mean μ and standard deviation $\sigma = 18$ will have means $\bar{x}$ between

$$\mu - 2(\sigma_{\bar{x}}) \quad \text{and} \quad \mu + 2(\sigma_{\bar{x}})$$

$$\mu - 2\left(\frac{\sigma}{\sqrt{n}}\right) \quad \text{and} \quad \mu + \left(\frac{\sigma}{\sqrt{n}}\right)$$

$$\mu - 2\left(\frac{18}{\sqrt{36}}\right) \quad \text{and} \quad \mu + 2\left(\frac{18}{\sqrt{36}}\right)$$

$$\mu - 6 \quad \text{and} \quad \mu + 6$$

This suggests that 95% of all random samples of size 36 selected from the population of rivets should have a mean $\bar{x}$ between $\mu - 6$ and $\mu + 6$. Figure 8.3 shows the middle 95% of the distribution, the bounds of the interval covering the 95%, and the mean μ.

FIGURE 8.3
Sampling Distribution
of $\bar{x}$'s, Unknown μ

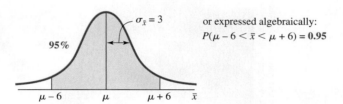

or expressed algebraically:
$P(\mu - 6 < \bar{x} < \mu + 6) = 0.95$

Now let's put all of this information together in the form of a *confidence interval.*

> **Interval Estimate:** An interval bounded by two values and used to estimate the value of a population parameter. The values that bound this interval are statistics calculated from the sample that is being used as the basis for the estimation.
>
> **Level of confidence 1 − α:** The portion of all interval estimates that include the parameter being estimated.
>
> **Confidence interval:** An interval estimate with a specified level of confidence.

To construct the confidence interval, we will use the point estimate $\bar{x}$ as the central value of an interval in much the same way as we used the mean μ as the central value to find the interval that captures the middle 95% of the $\bar{x}$ distribution in Figure 8.3.

For our rivet example, we can find the bounds to an interval centered at $\bar{x}$:

$$\bar{x} - 2(\sigma_{\bar{x}}) \quad \text{to} \quad \bar{x} + 2(\sigma_{\bar{x}})$$

$$924.23 - 6 \quad \text{to} \quad 924.23 + 6$$

The resulting interval is 918.23 to 930.23

The level of confidence assigned to this interval is approximately 95%, or 0.95. The bounds of the interval are two multiples ($z = 2.0$) of the standard error from the sample mean, and by looking at Table 3 in Appendix B, we can more accurately determine the level of confidence as 0.9544. Putting all of this information together, we express the estimate as a confidence interval: **918.23 to 930.23** *is the 95.44% confidence interval for the mean shear strength of the rivets.* Or in an abbreviated form: **918.23 to 930.23,** *the 95.44% confidence interval for* μ.

**APPLIED
EXAMPLE 8.1**

Yellowstone Park's Old Faithful

FYI
Visit the Old Faithful WebCam. When is the next eruption predicted to occur?

Next Prediction: 11:27 AM +/− 10 min.   08/20/2005 11:20:48 AM

Courtesy National Park Service, Yellowstone National Park, http://www.nps.gov/yell.oldfaithfulcam.htm

Welcome to the Old Faithful WebCam.

Predictions for the time of the next eruption of Old Faithful are made by the rangers using a formula that takes into account the length of the previous eruption. The formula has proved to be accurate, plus or minus 10 minutes, 90% of the time. At 10:35 AM on August 20, 2005, the posted prediction time of the next eruption was:

Predicted Eruption: 11:27 AM +/− 10 min.

Note the time at which the picture was recorded: 11:20:48 AM. (*Source:* http://www.nps.gov/yell/oldfaithfulcam.htm)

SECTION 8.2 EXERCISES

Statistics⬡Now™

Datasets can be found on your Student's Suite CD-ROM or at the StatisticsNow website at **http://1pass.thomson.com**.

8.3 Explain the difference between a point estimate and an interval estimate.

8.4 Identify each numerical value by "name" (e.g., mean, variance) and by symbol (e.g., $\bar{x}$):

a. The mean height of 24 junior high school girls is 4′11″.

b. The standard deviation for IQ scores is 16.

c. The variance among the test scores on last week's exam was 190.

d. The mean height of all cadets who have ever entered West Point is 69 inches.

8.5 [EX08-005] A random sample of the amount paid (in dollars) for taxi fare from downtown to the airport was obtained:

15 19 17 23 21 17 16 18 12 18 20 22 15 18 20

Use the data to find a point estimate for each of the following parameters.

a. Mean   b. Variance   c. Standard deviation

8.6 [EX08-006] The number of engines owned per fire department was obtained from a random sample taken from the profiles of fire departments from across the United States (*Firehouse*/June 2003).

29 8 7 33 21 26 6 11 4 54 7 4

Use the data to find a point estimate for each of the following parameters:

a. Mean   b. Variance   c. Standard deviation

8.7 In each diagram at the bottom of the page, I and II represent sampling distributions of two statistics that might be used to estimate a parameter. In each case, identify the statistic that you think would be the better estimator and describe why it is your choice.

8.8 Suppose that there are two statistics that will serve as an estimator for the same parameter. One of them is biased, and the other is unbiased.

a. Everything else being equal, explain why you usually would prefer an unbiased estimator to a biased estimator.

b. If a statistic is unbiased, does that ensure that it is a good estimator? Why or why not? What other considerations must be taken into account?

c. Describe a situation that might occur in which the biased statistic might be a better choice as an estimator than the unbiased statistic.

8.9 The use of a tremendously large sample does not solve the question of quality for an estimator. What problems do you anticipate with very large samples?

8.10 Being unbiased and having a small variability are two desirable characteristics of a statistic if it is going to be used as an estimator. Describe how the SDSM addresses both of these properties when estimating the mean of a population.

8.11 Explain why the standard error of sample means is 3 for the rivet example on page 399.

8.12 a. Verify that a 95% level of confidence requires a 1.96-standard-deviation interval.

   b. Verify that the level of confidence for a 2-standard-deviation interval is 95.44%.

8.13 Find the level of confidence assigned to an interval estimate of the mean formed using the following intervals:

a. $\bar{x} - 1.28 \cdot \sigma_{\bar{x}}$ to $\bar{x} + 1.28 \cdot \sigma_{\bar{x}}$

b. $\bar{x} - 1.44 \cdot \sigma_{\bar{x}}$ to $\bar{x} + 1.44 \cdot \sigma_{\bar{x}}$

c. $\bar{x} - 1.96 \cdot \sigma_{\bar{x}}$ to $\bar{x} + 1.96 \cdot \sigma_{\bar{x}}$

d. $\bar{x} - 2.33 \cdot \sigma_{\bar{x}}$ to $\bar{x} + 2.33 \cdot \sigma_{\bar{x}}$

8.14 Find the level of confidence assigned to an interval estimate of the mean formed using the following intervals:

a. $\bar{x} - 1.15 \cdot \sigma_{\bar{x}}$ to $\bar{x} + 1.15 \cdot \sigma_{\bar{x}}$

b. $\bar{x} - 1.65 \cdot \sigma_{\bar{x}}$ to $\bar{x} + 1.65 \cdot \sigma_{\bar{x}}$

c. $\bar{x} - 2.17 \cdot \sigma_{\bar{x}}$ to $\bar{x} + 2.17 \cdot \sigma_{\bar{x}}$

d. $\bar{x} - 2.58 \cdot \sigma_{\bar{x}}$ to $\bar{x} + 2.58 \cdot \sigma_{\bar{x}}$

Figure for Exercise 8.7

a.
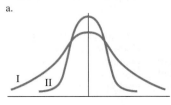

b.
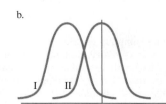

c.
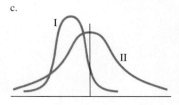

8.15 A sample of 25 of 174 funded projects revealed that 19 were valued at $17,320 each and 6 were valued at $20,200 each. From the sample data, estimate the total value of the funding for all the projects.

8.16 A stamp dealer wishes to purchase a stamp collection that is believed to contain approximately 7000 individual stamps and approximately 4000 first-day covers. Devise a plan that might be used to estimate the collection's worth.

8.17 Using the Old Faithful eruption information in Applied Example 8.1 on page 400:

a. What does "11:27 AM +/− 10 min." mean? Explain.

b. Did this eruption occur during the predicted time interval?

c. What does "90% of the time" mean? Explain.

8.18 A recruiter estimates that if you are hired to work for her company and you put in a full week at the commissioned sales representative position she is offering, you will make "$525 plus or minus $250, 80% of the time." She adds, "It all depends on you!"

a. What does the "$525 plus or minus $250" mean?

b. What does the "80% of the time" mean?

c. If you make $300 to the nearest $10 most weeks, will she have told you the truth? Explain.

8.3   Estimation of Mean μ (σ Known)

In Section 8.2 we surveyed the basic ideas of estimation: point estimate, interval estimate, level of confidence, and confidence interval. These basic ideas are interrelated and used throughout statistics when an inference calls for an estimate. In this section we formalize the interval estimation process as it applies to estimating the population mean μ based on a random sample under the restriction that the population standard deviation σ is a known value.

The sampling distribution of sample means and the CLT provide us with the information we need to ensure that the necessary *assumptions* for estimating a population mean are satisfied.

> **The assumption for estimating mean μ using a known σ:** The sampling distribution of $\bar{x}$ has a normal distribution.

Note: The word *assumptions* is somewhat of a misnomer. It does not mean that we "assume" something to be the situation and continue but rather that we must be sure the conditions expressed by the assumptions do exist before we apply a particular statistical method.

The information needed to ensure that this assumption (or condition) is satisfied is contained in the SDSM and in the CLT (see Chapter 7, pp. 369–370):

> The sampling distribution of sample means $\bar{x}$ is distributed about a mean equal to μ with a standard error equal to $\sigma/\sqrt{n}$; and (1) if the randomly sampled population is normally distributed, then $\bar{x}$ is normally distributed for all sample sizes, or (2) if the randomly sampled population is not normally distributed, then $\bar{x}$ is approximately normally distributed for sufficiently large sample sizes.

Therefore, we can satisfy the required assumption by either (1) knowing that the sampled population is normally distributed or (2) using a random sample that contains a sufficiently large amount of data. The first possibility is obvious. We either know enough about the population to know that it is normally distributed or we don't. The second way to satisfy the assumption is by applying the CLT. Inspection of various graphic displays of the sample data should yield an indication of the type of distribution the population possesses. The CLT can be applied to smaller samples (say, $n = 15$ or larger) when the data provide a strong indication of a unimodal distribution that is approximately symmetrical. If there is evidence of some skewness in the data, then the sample size needs to be much larger (perhaps $n \geq 50$). If the data provide evidence of an extremely skewed or J-shaped distribution, the CLT will still apply if the sample is large enough. In extreme cases, "large enough" may be unrealistically or impracticably large.

FYI The help of a professional statistician should be sought when treating extremely skewed data.

Note: There is no hard and fast rule defining "large enough"; the sample size that is "large enough" varies greatly according to the distribution of the population.

The $1 - \alpha$ confidence interval for the estimation of mean μ is found using formula (8.1).

Confidence Interval for Mean

$$\bar{x} - z(\alpha/2)\left(\frac{\sigma}{\sqrt{n}}\right) \quad \text{to} \quad \bar{x} + z(\alpha/2)\left(\frac{\sigma}{\sqrt{n}}\right) \tag{8.1}$$

Here are the parts of the confidence interval formula:

1. $\bar{x}$ is the point estimate and the center point of the confidence interval.

2. $z(\alpha/2)$ is the **confidence coefficient.** It is the number of multiples of the standard error needed to formulate an interval estimate of the correct width to have a level of confidence of $1 - \alpha$. Figure 8.4 shows the relationship among the level of confidence $1 - \alpha$ (the middle portion of the distribution), $\alpha/2$ (the "area to the right" used with the critical-value notation), and the confidence coefficient $z(\alpha/2)$ (whose value is found using Table 4B of Appendix B). Alpha, α, is the first letter of the Greek alphabet and represents the portion associated with the tails of the distribution.

3. $\sigma/\sqrt{n}$ is the **standard error of the mean,** or the standard deviation of the sampling distribution of sample means.

4. $z(\alpha/2)\left(\dfrac{\sigma}{\sqrt{n}}\right)$ is one-half the width of the confidence interval (the product of the confidence coefficient and the standard error) and is called the **maximum error of estimate, *E*.**

5. $\bar{x} - z(\alpha/2)\left(\dfrac{\sigma}{\sqrt{n}}\right)$ is called the **lower confidence limit** (LCL), and
$\bar{x} + z(\alpha/2)\left(\dfrac{\sigma}{\sqrt{n}}\right)$ is called the **upper confidence limit** (UCL) for the confidence interval.

FIGURE 8.4
Confidence Coefficient $z(\alpha/2)$

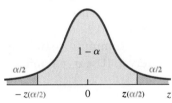

FYI Basically, the confidence interval is "point estimate ± maximum error."

The estimation procedure is organized into a five-step process that will take into account all of the preceding information and produce both the point estimate and the confidence interval.

THE CONFIDENCE INTERVAL: A FIVE-STEP PROCEDURE

Step 1 **The Set-Up:**
Describe the population parameter of interest.

Step 2 **The Confidence Interval Criteria:**
a. Check the assumptions.
b. Identify the probability distribution and the formula to be used.
c. State the level of confidence, $1 - \alpha$.

Step 3 **The Sample Evidence:**
Collect the sample information.

Step 4 **The Confidence Interval:**
a. Determine the confidence coefficient.
b. Find the maximum error of estimate.
c. Find the lower and upper confidence limits.

Step 5 **The Results:**
State the confidence interval.

Example 8.2 will illustrate this five-step confidence interval procedure.

EXAMPLE 8.2

Constructing a Confidence Interval for the Mean One-Way Commute Distance

The student body at many community colleges is considered a "commuter population." The student activities office wishes to obtain an answer to the question, How far (one way) does the average community college student commute to college each day? (Typically the "average student's commute distance" is meant to be the "mean distance" commuted by all students who commute.) A random sample of 100 commuting students was identified, and the one-way distance each commuted was obtained. The resulting sample mean distance was 10.22 miles.

Estimate the mean one-way distance commuted by all commuting students using (a) a point estimate and (b) a 95% confidence interval. (Use $\sigma = 6$ miles.)

SOLUTION

(a) The point estimate for the mean one-way distance is **10.22** miles (the sample mean).

(b) We use the five-step procedure to find the 95% confidence interval.

STEP 1 **The Set-Up:**
Describe the population parameter of interest.
The mean μ of the one-way distances commuted by all commuting community college students is the parameter of interest.

STEP 2 **The Confidence Interval Criteria:**
a. Check the assumptions.
σ is known. The variable "distance commuted" most likely has a skewed distribution because the vast majority of the students will commute between 0 and 25 miles, with fewer commuting more than 25 miles. A sample size of 100 should be large

enough for the CLT to satisfy the assumption; the $\bar{x}$ sampling distribution is approximately normal.

b. **Identify the probability distribution and the formula to be used.**

The standard normal distribution, z, will be used to determine the confidence coefficient, and formula (8.1) with $\sigma = 6$.

c. **State the level of confidence, $1 - \alpha$.**

The question asks for 95% confidence, or $1 - \alpha = 0.95$.

STEP 3   The Sample Evidence:

Collect the sample information.

The sample information is given in the statement of the problem: $n = 100, \bar{x} = 10.22$.

STEP 4   The Confidence Interval:

a. **Determine the confidence coefficient.**

The confidence coefficient is found using Table 4B:

A Portion of Table 4B

Level of confidence:	α	$\ldots$	0.05
	$z(\alpha/2)$	$\ldots$	1.96 $\rightarrow$ Confidence coefficient:
$1 - \alpha = 0.95 \rightarrow$	$1 - \alpha$	$\ldots$	0.95 $z(\alpha/2) = 1.96$

b. **Find the maximum error of estimate.**

Use the maximum error part of formula (8.1):

$$E = z(\alpha/2)\left(\frac{\sigma}{\sqrt{n}}\right) = 1.96\left(\frac{6}{\sqrt{100}}\right) = (1.96)(0.6) = 1.176$$

c. **Find the lower and upper confidence limits.**

Using the point estimate, $\bar{x}$, from Step 3 and the maximum error, E, from Step 4b, we find the confidence interval limits:

$$\bar{x} - z(\alpha/2)\left(\frac{\sigma}{\sqrt{n}}\right) \quad \text{to} \quad \bar{x} + z(\alpha/2)\left(\frac{\sigma}{\sqrt{n}}\right)$$

$$10.22 - 1.176 \quad \text{to} \quad 10.22 + 1.176$$

$$9.044 \quad \text{to} \quad 11.396$$

$$9.04 \quad \text{to} \quad 11.40$$

STEP 5   The Results:

State the confidence interval.

9.04 to 11.40, the 95% confidence interval for μ. That is, with 95% confidence we can say, "The mean one-way distance is between 9.04 and 11.40 miles."

Let's look at another example of the estimation procedure.

Constructing a Confidence Interval for the Mean Particle Size

"Particle size" is an important property of latex paint and is monitored during production as part of the quality-control process. Thirteen particle-size measurements were taken using the Dwight P. Joyce Disc, and the sample mean was 3978.1 angstroms (where 1 angstrom [1 Å] = 10^{-8} cm). The particle size, x, is normally distributed with a standard deviation $\sigma = 200$ angstroms. Find the 98% confidence interval for the mean particle size for this batch of paint.

SOLUTION

STEP 1   The Set-Up:
Describe the population parameter of interest.
The mean particle size, μ, for the batch of paint from which the sample was drawn.

STEP 2   The Confidence Interval Criteria:
a. **Check the assumptions.**
σ is known. The variable "particle size" is normally distributed; therefore, the sampling distribution of sample means is normal for all sample sizes.
b. **Identify the probability distribution and the formula to be used.**
The standard normal variable z, and formula (8.1) with $\sigma = 200$.
c. **State the level of confidence, $1 - \alpha$.**
98%, or $1 - \alpha = 0.98$.

STEP 3   The Sample Evidence:
Collect the sample information: $n = 13$ and $\bar{x} = 3978.1$.

STEP 4   The Confidence Interval:
a. **Determine the confidence coefficient.**
The confidence coefficient is found using Table 4B: $z(\alpha/2) = z(0.01) = 2.33$.

A Portion of Table 4B

	α	$\ldots$	0.02
Level of confidence:	$z(\alpha/2)$	$\ldots$	**2.33** → Confidence coefficient:
$1 - \alpha = 0.98$ →	$1 - \alpha$	$\ldots$	0.98 $z(\alpha/2) = $ **2.33**

b. **Find the maximum error of estimate.**
$$E = z(\alpha/2)\left(\frac{\sigma}{\sqrt{n}}\right) = 2.33\left(\frac{200}{\sqrt{13}}\right) = (2.33)(55.47) = 129.2$$

c. **Find the lower and upper confidence limits.**
Using the point estimate, $\bar{x}$, from Step 3 and the maximum error, E, from Step 4b, we find the confidence interval limits:
$$\bar{x} - z(\alpha/2)\left(\frac{\sigma}{\sqrt{n}}\right) \quad \text{to} \quad \bar{x} + z(\alpha/2)\left(\frac{\sigma}{\sqrt{n}}\right)$$

$$3978.1 - 129.2 = 3848.9 \quad \text{to} \quad 3978.1 + 129.2 = 4107.3$$

STEP 5   The Results:
State the confidence interval.

3848.9 to 4107.3, the 98% confidence interval for μ. With 98% confidence we can say, "The mean particle size is between 3848.9 and 4107.3 angstroms."

Let's take another look at the concept "level of confidence." It was defined to be the probability that the sample to be selected will produce interval bounds that contain the parameter.

EXAMPLE 8.4

Demonstrating the Meaning of a Confidence Interval

Single-digit random numbers, like the ones in Table 1 in Appendix B, have a mean value μ = 4.5 and a standard deviation σ = 2.87 (see Exercise 5.32, p. 282). Draw a sample of 40 single-digit numbers from Table 1 and construct the 90% confidence interval for the mean. Does the resulting interval contain the expected value of μ, 4.5? If we were to select another sample of 40 single-digit numbers from Table 1, would we get the same result? What might happen if we selected a total of 15 different samples and constructed the 90% confidence interval for each? Would the expected value for μ—namely, 4.5—be contained in all of them? Should we expect all 15 confidence intervals to contain 4.5? Think about the definition of "level of confidence"; it says that in the long run, 90% of the samples will result in bounds that contain μ. In other words, 10% of the samples will not contain μ. Let's see what happens.

First we need to address the assumptions; if the assumptions are not satisfied, we cannot expect the 90% and the 10% to occur. We know: (1) the distribution of single-digit random numbers is rectangular (definitely not normal), (2) the distribution of single-digit random numbers is symmetrical about their mean, (3) the $\bar{x}$ distribution for very small samples ($n = 5$) in Example 7.2 (pp. 364–365) displayed a distribution that appeared to be approximately normal, and (4) there should be no skewness involved. Therefore, it seems reasonable to assume that $n = 40$ is large enough for the CLT to apply.

The first random sample was drawn from Table 1 in Appendix B:

TABLE 8.1

Random Sample of Single-Digit Numbers [TA08-01]

2	8	2	1	5	5	4	0	9	1
0	4	6	1	5	1	1	3	8	0
3	6	8	4	8	6	8	9	5	0
1	4	1	2	1	7	1	7	9	3

The sample statistics are $n = 40$, $\Sigma x = 159$, and $\bar{x} = 3.98$. Here is the resulting 90% confidence interval:

$$\bar{x} \pm z(\alpha/2)\left(\frac{\sigma}{\sqrt{n}}\right): \quad 3.98 \pm 1.65\left(\frac{2.87}{\sqrt{40}}\right)$$

$$3.98 \pm (1.65)(0.454)$$

$$3.98 \pm 0.75$$

$$3.98 - 0.75 = 3.23 \quad \text{to} \quad 3.98 + 0.75 = 4.73$$

3.23 to 4.73, the 90% confidence interval for μ

Figure 8.5 shows this confidence interval, its bounds, and the expected mean μ.

FIGURE 8.5   **The 90% Confidence Interval**

With 90% confidence, we think μ is somewhere within this interval

3.23 $\mu = 4.50$ 4.73 $\bar{x}$

The expected value for the mean, 4.5, does fall within the bounds of the confidence interval for this sample. Let's now select 14 more random samples from Table 1 in Appendix B, each of size 40.

Table 8.2 lists the mean from the first sample and the means obtained from the 14 additional random samples of size 40. The 90% confidence intervals for the estimation of μ based on each of the 15 samples are listed in Table 8.2 and shown in Figure 8.6.

TABLE 8.2

Fifteen Samples of Size 40 [TA08-02]

Sample Number	Sample Mean, $\bar{x}$	90% Confidence Interval Estimate for μ	Sample Number	Sample Mean, $\bar{x}$	90% Confidence Interval Estimate for μ
1	3.98	3.23 to 4.73	9	4.08	3.33 to 4.83
2	4.64	3.89 to 5.39	10	5.20	4.45 to 5.95
3	4.56	3.81 to 5.31	11	4.88	4.13 to 5.63
4	3.96	3.21 to 4.71	12	5.36	4.61 to 6.11
5	5.12	4.37 to 5.87	13	4.18	3.43 to 4.93
6	4.24	3.49 to 4.99	14	4.90	4.15 to 5.65
7	3.44	2.69 to 4.19	15	4.48	3.73 to 5.23
8	4.60	3.85 to 5.35			

FIGURE 8.6
Confidence Intervals from Table 8.2

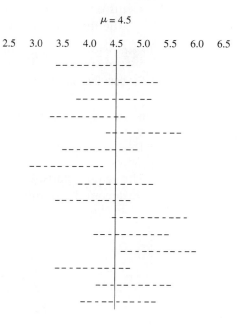

We see that 86.7% (13 of the 15) of the intervals contain μ and 2 of the 15 samples (sample 7 and sample 12) do not contain μ. The results here are "typical"; repeated experimentation might result in any number of intervals that contain 4.5. However, in the long run, we should expect approximately $1 - \alpha = 0.90$ (or 90%) of the samples to result in bounds that contain 4.5 and approximately 10% that do not contain 4.5.

APPLIED EXAMPLE 8.5

Rockies' Snow Melt Produces Less Water

When snow melts it becomes water, sometimes more water than at other times. The accompanying graphic compares the water content of snow from two areas in the United States that typically get about the same amount of snow annually. However, the water content is very different. There are several point estimates for the average included in the graphic.

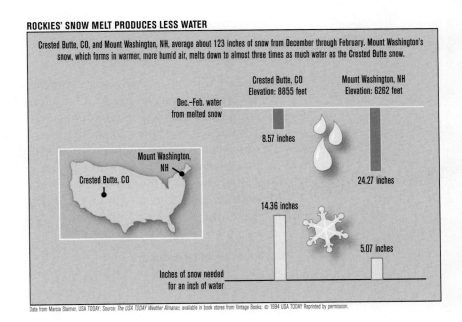

ROCKIES' SNOW MELT PRODUCES LESS WATER

Crested Butte, CO, and Mount Washington, NH, average about 123 inches of snow from December through February. Mount Washington's snow, which forms in warmer, more humid air, melts down to almost three times as much water as the Crested Butte snow.

Crested Butte, CO
Elevation: 8855 feet

Mount Washington, NH
Elevation: 6262 feet

Dec.–Feb. water from melted snow

8.57 inches

24.27 inches

14.36 inches

5.07 inches

Mount Washington, NH

Crested Butte, CO

Inches of snow needed for an inch of water

Data from Marcia Stainer, USA TODAY; Source: *The USA TODAY Weather Almanac*, available in book stores from Vintage Books; © 1994 USA TODAY Reprinted by permission.

TECHNOLOGY INSTRUCTIONS: CONFIDENCE INTERVAL FOR MEAN μ WITH A GIVEN σ

MINITAB (Release 14)   Input the data into C1; then continue with:

Choose:	**Stat** > **Basic Statistics** > **1-Sample Z**
Enter:	Variables: **C1**
	Standard deviation: **σ**
Select:	**Options**
Enter:	Confidence Level: **1 − α** (ex.: 0.95 or 95.0)
Select:	Alternative: **not equal** > **OK** > **OK**

Excel Input the data into column A; then continue with:

Choose: **Tools > Data Analysis Plus > Z-Test: Mean > OK**
Enter: **Input Range: (A1:A20 or select cells) > OK**
   Standard Deviation (SIGMA): σ > OK
   Alpha: α (ex.: 0.05) > OK

TI-83/84 PLUS Input the data into L1; then continue with the following, entering the appropriate values and highlighting Calculate:

Choose: **STAT > TESTS > 7:Zinterval**

```
ZInterval
 Inpt:DATA Stats
 σ:0
 List:L₁
 Freq:1
 C-Level:.95
 Calculate
```

Sample Size

The confidence interval has two basic characteristics that determine its quality: its level of confidence and its width. It is preferred that the interval have a high level of confidence and be precise (narrow) at the same time. The higher the level of confidence, the more likely the interval is to contain the parameter, and the narrower the interval, the more precise the estimation. However, these two properties seem to work against each other, because it would seem that a narrower interval would tend to have a lower probability and a wider interval would be less precise. The maximum error part of the confidence interval formula specifies the relationship involved.

Maximum Error of Estimate

$$E = z(\alpha/2)\left(\frac{\sigma}{\sqrt{n}}\right)$$
(8.2)

This formula has four components: (1) the maximum error E, half of the width of the confidence interval; (2) the confidence coefficient, $z(\alpha/2)$, which is determined by the level of confidence; (3) the sample size, n; and (4) the standard deviation, σ. The standard deviation σ is not a concern in this discussion because it is a constant (the standard deviation of a population does not change in value). That leaves three factors. Inspection of formula (8.2) indicates the following: increasing the level of confidence will make the confidence coefficient larger and thereby require either the maximum error to increase or the sample size to increase; decreasing the maximum error will require the level of confidence to decrease or the sample size to increase; and decreasing the sample size will force the maximum error to become larger or the level of confidence to decrease. We have a "three-way tug-of-war," as pictured in Figure 8.7.

FIGURE 8.7   The "Three-Way Tug-of-War" between $1 - \alpha$, n, and E

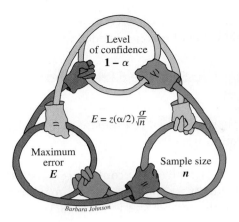

Barbara Johnson

FYI When the denominator increases, the value of the fraction decreases.

An increase or decrease to any one of the three factors has an effect on one or both of the other two factors. The statistician's job is to "balance" the level of confidence, the sample size, and the maximum error so that an acceptable interval results.

Let's look at an example of this relationship in action.

EXAMPLE 8.6

Statistics Now™

Watch a video example at http://1pass.thomson.com or on your CD.

FYI Instructions for using Table 4B are given on page 403.

Determining the Sample Size for a Confidence Interval

Determine the size sample needed to estimate the mean weight of all second-grade boys if we want to be accurate within 1 lb with 95% confidence. Assume a normal distribution and that the standard deviation of the boys' weights is 3 lb.

SOLUTION   The desired level of confidence determines the confidence coefficient: the confidence coefficient is found using Table 4B: $z(\alpha/2) = z(0.025) = \mathbf{1.96}$.

The desired maximum error is $E = 1.0$. Now we are ready to use the maximum error formula:

$$E = z(\alpha/2)\left(\frac{\sigma}{\sqrt{n}}\right): \qquad 1.0 = 1.96\left(\frac{3}{\sqrt{n}}\right)$$

$$\text{Solve for } n: \qquad 1.0 = \frac{5.88}{\sqrt{n}}$$

$$\sqrt{n} = 5.88$$

$$n = (5.88)^2 = 34.57 = \mathbf{35}$$

Therefore, $n = \mathbf{35}$ is the sample size needed if you want a 95% confidence interval with a maximum error no greater than 1 lb.

Note: When we solve for the sample size n, it is customary to round up to the next larger integer, no matter what fraction (or decimal) results.

Using the maximum error formula (8.2) can be made a little easier by rewriting the formula in a form that expresses n in terms of the other values.

Sample Size

$$n = \left(\frac{z(\alpha/2) \cdot \sigma}{E}\right)^2 \qquad\qquad (8.3)$$

If the maximum error is expressed as a multiple of the standard deviation σ, then the actual value of σ is not needed in order to calculate the sample size.

EXAMPLE 8.7

Determining the Sample Size without a Known Value of Sigma (σ)

Find the sample size needed to estimate the population mean to within $\frac{1}{5}$ of a standard deviation with 99% confidence.

SOLUTION Determine the confidence coefficient (using Table 4B): $1 - \alpha = 0.99$, $z(\alpha/2) = 2.58$. The desired maximum error is $E = \frac{\sigma}{5}$. Now we are ready to use the sample size formula (8.3):

$$n = \left(\frac{z(\alpha/2) \cdot \sigma}{E}\right)^2: \quad n = \left(\frac{(2.58) \cdot \sigma}{\sigma/5}\right)^2 = \left(\frac{(2.58\sigma)(5)}{\sigma}\right)^2 = [(2.58)(5)]^2$$

$$= (12.90)^2 = 166.41 = \mathbf{167}$$

SECTION 8.3 EXERCISES

8.19 Discuss the conditions that must exist before we can estimate the population mean using the interval techniques of formula (8.1).

8.20 Determine the value of the confidence coefficient $z(\alpha/2)$ for each situation described:

a. $1 - \alpha = 0.90$ b. $1 - \alpha = 0.95$

8.21 Determine the value of the confidence coefficient $z(\alpha/2)$ for each situation described:

a. 98% confidence   b. 99% confidence

8.22 Determine the level of the confidence given the confidence coefficient $z(\alpha/2)$ for each situation:

a. $z(\alpha/2) = 1.645$ b. $z(\alpha/2) = 1.96$

c. $z(\alpha/2) = 2.575$ d. $z(\alpha/2) = 2.05$

8.23 Given the information, the sampled population is normally distributed, $n = 16$, $\bar{x} = 28.7$, and $\sigma = 6$:

a. Find the 0.95 confidence interval for μ.

b. Are the assumptions satisfied? Explain.

8.24 Given the information, the sampled population is normally distributed, $n = 55$, $\bar{x} = 78.2$, and $\sigma = 12$:

a. Find the 0.98 confidence interval for μ.

b. Are the assumptions satisfied? Explain.

8.25 Given the information, $n = 86$, $\bar{x} = 128.5$, and $\sigma = 16.4$:

a. Find the 0.90 confidence interval for μ.

b. Are the assumptions satisfied? Explain.

8.26 Given the information, $n = 22$, $\bar{x} = 72.3$, and $\sigma = 6.4$:

a. Find the 0.99 confidence interval for μ.

b. Are the assumptions satisfied? Explain.

8.27 Based on the confidence interval formed in Exercise 8.25, give the value for each of the following:

a. Point estimate   b. Confidence coefficient

c. Standard error of the mean

d. Maximum error of estimate, *E*

e. Lower confidence limit

f. Upper confidence limit

8.28 Based on the confidence interval formed in Exercise 8.24, give the value for each of the following:

a. Point estimate   b. Confidence coefficient

c. Standard error of the mean

d. Maximum error of estimate, *E*

e. Lower confidence limit

f. Upper confidence limit

8.29 In your own words, describe the relationship between the following:

a. Sample mean and point estimate

b. Sample size, sample standard deviation, and standard error

c. Standard error and maximum error

8.30 In your own words, describe the relationship between the point estimate, the level of confidence, the maximum error, and the confidence interval.

8.31 Skillbuilder Applet Exercise demonstrates the effect that the level of confidence $(1 - \alpha)$ has on the width of a confidence interval. Consider sampling from a population where $\mu = 300$ and $\sigma = 80$.

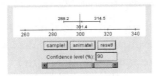

a. Set the slider for level of confidence to 68%. Click "sample!" to construct one 68% confidence interval. Note the upper and lower confidence limits and calculate the width of the interval. Using "animate!" construct many samples and note the percent of intervals containing the true mean of 300. Click "stop" and "reset."

b. Set the slider for level of confidence to 95%. Click "sample!" to construct one 95% confidence interval. Note the upper and lower confidence limits and calculate the width of the interval. Using "animate!" construct many samples and note the percent of intervals containing the true mean of 300. Click "stop" and "reset."

c. Set the slider for level of confidence to 99%. Click "sample!" to construct one 99% confidence interval. Note the upper and lower confidence limits and calculate the width of the interval. Using "animate!" construct many samples and note the percent of intervals containing the true mean of 300. Click "stop."

d. Using the information collected in parts a–c, what effect does the level of confidence have on the width of the interval. Why is this happening?

8.32 Discuss the effect that each of the following have on the confidence interval:

a. Point estimate   b. Level of confidence

c. Sample size

d. Variability of the characteristic being measured

8.33 A machine produces parts with lengths that are normally distributed with $\sigma = 0.5$. A sample of 10 parts has a mean length of 75.92.

a. Find the point estimate for μ.

b. Find the 98% confidence maximum error of estimate for μ.

c. Find the 98% confidence interval for μ.

8.34 A sample of 60 night-school students' ages is obtained in order to estimate the mean age of night-school students. $\bar{x} = 25.3$ years. The population variance is 16.

a. Give a point estimate for μ.

b. Find the 95% confidence interval for μ.

c. Find the 99% confidence interval for μ.

8.35 Two hundred fish caught in Cayuga Lake had a mean length of 14.3 inches. The population standard deviation is 2.5 inches.

a. Find the 90% confidence interval for the population mean length.

b. Find the 98% confidence interval for the population mean length.

8.36 The Eurostar was Europe's first international train, designed to take advantage of the Channel Tunnel that connects England with Continental Europe. It carries nearly 800 passengers and occasionally reaches a peak speed of more than 190 mph (http://www.o-keating.com/hsr/eurostar.htm). Assume the standard deviation of train speed is 19 mph in the course of all the journeys back and forth and that the train's speed is normally distributed. Suppose speed readings are made during the next 20 trips of the Eurostar and the mean speed of these measurements is 184 mph.

a. What is the variable being studied?

b. Find the 90% confidence interval estimate for the mean speed.

c. Find the 95% confidence interval estimate for the mean speed.

8.37 In 2003, the Trends International Mathematics and Science Study (TIMSS) examined eighth-graders' proficiency in math and science. The mean mathematics scale score for the sample of eighth-grade students in the United States was 504, with a standard error of 8.4. Construct a 95% confidence interval for the mean mathematics score for all eighth-grade students in the United States.

8.38 About 67% of married adults say they consult with their spouse before spending $352, the average of the amount that married adults say they consult with each other before spending.

Source: Yankelovich Partner for Lutheran Brotherhood

a. Based on the preceding information, what can you conclude about the variable, dollar amount, that requires consultation with a spouse before spending? What is the $352?

A survey of 500 married adults was taken from a nearby neighborhood and gave a sample mean of $289.75.

b. Construct a 0.98 confidence interval for the mean amount for all married adults. Use $\sigma = \$600$.

c. Based on the answers to parts a and b, what can you conclude about the mean dollar amount married adults discuss with their spouses before spending in the sampled neighborhood compared with the general population?

8.39 [EX08-039] A certain adjustment to a machine will change the length of the parts it makes but will not affect the standard deviation. The length of the parts is normally distributed, and the standard deviation is 0.5 mm. After an adjustment is made, a random sample is taken to determine the mean length of the parts now being produced. The resulting lengths are as follows:

| 75.3 | 76.0 | 75.0 | 77.0 | 75.4 | 76.3 | 77.0 | 74.9 | 76.5 | 75.8 |

a. What is the parameter of interest?

b. Find the point estimate for the mean length of all parts now being produced.

c. Find the 0.99 confidence interval for μ.

8.40 [EX08-040] The atomic weight of a reference sample of silver was measured at the National Institute of Standards and Technology (NIST) using two nearly identical mass spectrometers. This project was undertaken in conjunction with the redetermination of the Faraday constant. Following are 48 observations:

| 107.8681568 | 107.8681465 | 107.8681572 |

••• Remainder of data on Student's Suite CD-ROM

Source: StatLib, http://lib.stat.cmu.edu/datasets/

Notice that the data differ only in the fifth, sixth, and seventh decimal places. Most computers will round the data and their calculated results; thus, the variation is seemingly lost. The statistics can be computed using just the last three digits of each data value (i.e., 107.8681568 will become 568). Algebraically this coding looks like this:

Atom Wt Coded = (Atomic weight − 107.8681000) × 10,000,000.

The data are listed in both the original and coded formats on the Student Suite CD.

a. Construct a graph of the coded data. How does the coding show on the graph?

b. Find the mean and standard deviation of the coded data.

c. Convert the answers found in part b to original units.

d. Determine whether the data has an approximately normal distribution. Present your case.

e. Do the SDSM and CLT apply? Explain.

f. Is sigma known?

g. If the goal is to find the 95% confidence interval for the mean value of all observations, what would you do?

h. Find the 95% confidence interval for the mean value of all such observations. Justify your method.

8.41 [EX08-041] The force required to extract a cork from a wine bottle is an important property of the cork. If the force is too little, the cork probably is not a good protector of the wine inside. If the force is too great, it will be difficult to remove. Neither is desirable. The no. 9 corks in Applied Example 6.15 (p. 330) are believed to have an extraction force that is normally distributed with a standard deviation of 36 Newtons.

a. A sample of 20 randomly chosen bottles is selected for testing.

Extraction Force in Newtons

296	338	341	261	250	347	336	297	279	297
259	334	281	284	279	266	300	305	310	253

Find the 98% confidence interval for the mean extraction force.

b. During a different testing, a sample of eight bottles was randomly selected and tested.

Extraction Force in Newtons

331.9	312.0	289.4	303.6	346.9	308.1	346.9	276.0

Find the 98% confidence interval for the mean extraction force.

c. What effect did the two different sample means have on the answers in parts a and b? Explain.

d. What effect did the two different sample sizes have on the answers in parts a and b? Explain.

e. The mean extraction force was claimed to be 310 Newtons. Does either sample show sufficient reason to doubt the truthfulness of the claim? Explain.

8.42 "College costs spike again" (October 19, 2005), an article that appeared on the CNN Money website, gave the latest figures from the College Board on annual tuition, fees, and room and board. The average total figure for private colleges is $27,516 and $11,354 for public colleges.

Source: http://money.cnn.com/2004/10/18/pf/college/college_costs/

In an effort to compare those same costs in New York State, a sample of 32 students in their junior year of college is randomly selected statewide from the private colleges and 32 more students are selected from public colleges. The private college sample resulted in a mean of $27,436, and the public college sample mean was $11,147.

a. Assume the annual college fees for private colleges have a mounded distribution and the standard deviation is $1800. Find the 95% confidence interval for the mean college costs in New York State.

b. Assume the annual college fees for public colleges have a mounded distribution and the standard deviation is $1200. Find the 95% confidence interval for the mean college costs in New York State.

c. How do New York State college costs compare with the College Board's values? Explain how they differ.

d. Compare the confidence intervals found in parts a and b and describe the effect the two different sample means had on the resulting answers.

e. Compare the confidence intervals found in parts a and b and describe the effect the two different standard deviations had on the resulting answers.

8.43 "Rockies' snow melt produces less water" (Applied Example 8.5) lists "14.36 inches" and "5.07 inches" as statistics and uses them as point estimates. Describe why these numbers are statistics and why they are also point estimates.

8.44 Using a computer or calculator, randomly select a sample of 40 single-digit numbers and find the 90% confidence interval for μ. Repeat several times, observing whether or not 4.5 is in the interval each time. Refer to Example 8.4, page 407. Describe your results.

FYI Use commands for generating integer data on page 407; then continue with confidence interval commands on pages 409–410.

8.45 Find the sample size needed to estimate μ of a normal population with $\sigma = 3$ to within 1 unit at the 98% level of confidence.

8.46 How large a sample should be taken if the population mean is to be estimated with 99% confidence to within $75? The population has a standard deviation of $900.

8.47 A high-tech company wants to estimate the mean number of years of college education its employees have completed. A good estimate of the standard deviation for the number of years of college is 1.0. How large a sample needs to be taken to estimate μ to within 0.5 of a year with 99% confidence?

8.48 By measuring the amount of time it takes a component of a product to move from one workstation to the next, an engineer has estimated that the standard deviation is 5 seconds.

a. How many measurements should be made to be 95% certain that the maximum error of estimation will not exceed 1 second?

b. What sample size is required for a maximum error of 2 seconds?

8.49 The new mini-laptop computers can deliver as much computing power as machines several times their size, but they weigh in at less than 3 lb. How large a sample would be needed to estimate the population mean weight if the maximum error of estimate is to be 0.4 of 1 standard deviation with 95% confidence?

8.50 The image of the public library is constantly changing, and their online services continue to grow. Usage of the library's home page grew by 17% during the past 12 months. It has been estimated that the current average length of a visit to the library's home page is approximately 20 minutes. The library wants to take a sample to statistically estimate this mean. How large will the sample need to be to estimate the mean within 0.3 of 1 standard deviation with 0.98 confidence?

Source: http://library.loganutah.org/library/annual04/annualreport2004.html

| 8.4 | # The Nature of Hypothesis Testing |

We make decisions every day of our lives. Some of these decisions are of major importance; others are seemingly insignificant. All decisions follow the same basic pattern. We weigh the alternatives; then, based on our beliefs and preferences and whatever evidence is available, we arrive at a decision and take the appropriate action. The statistical hypothesis test follows much the same process, except that it involves statistical information. In this section we develop many of the concepts and attitudes of the hypothesis test while looking at several decision-making situations without using any statistics.

A friend is having a party (Super Bowl party, home-from-college party—you know the situation, any excuse will do), and you have been invited. You must make a decision: attend or not attend. That's simple; well maybe, except that you want to go only if you can be convinced the party is going to be more fun than your friend's typical party. Furthermore, you definitely do not want to go if the party is going to be just another dud. You have taken the position that "the party will be a dud" and you will not go unless you become convinced otherwise. Your friend assures you, "Guaranteed, the party will be a great time!" Do you go or not?

The decision-making process starts by identifying **something of concern** and then formulating **two hypotheses** about it.

> **Hypothesis:** A statement that something is true.

Your friend's statement, "The party will be a great time," is a hypothesis. Your position, "The party will be a dud," is also a hypothesis.

> **Statistical hypothesis test:** A process by which a decision is made between two opposing hypotheses. The two opposing hypotheses are formulated so that each hypothesis is the negation of the other. (That way one of them is always true, and the other one is always false.) Then one hypothesis is tested in hopes that it can be shown to be a very improbable occurrence, thereby implying the other hypothesis is likely the truth.

The two hypotheses involved in making a decision are known as the *null hypothesis* and the *alternative hypothesis*.

> **Null hypothesis,* H_o:** The hypothesis we will test. Generally, this is a statement that a population parameter has a specific value. The null hypothesis is so named because it is the "starting point" for the investigation. (The phrase "there is no difference" is often used in its interpretation.)
>
> **Alternative hypothesis, H_a:** A statement about the same population parameter that is used in the null hypothesis. Generally, this is a statement that specifies the population parameter has a value different, in some way, from the value given in the null hypothesis. The rejection of the null hypothesis will imply the likely truth of this alternative hypothesis.

With regard to your friend's party, the two opposing viewpoints or hypotheses are "The party will be a great time" and "The party will be a dud." Which statement becomes the null hypothesis, and which becomes the alternative hypothesis?

Determining the statement of the null hypothesis and the statement of the alternative hypothesis is a very important step. The *basic idea* of the hypothesis test is for the evidence to have a chance to "disprove" the null hypothesis. The null hypothesis is the statement that the evidence might disprove. *Your concern* (belief or desired outcome), as the person doing the testing, is expressed in the alternative hypothesis. As the person making the decision, you believe that the evidence will demonstrate the feasibility of your "theory" by demon-

---

*We use the notation H_o for the null hypothesis to contrast it with H_a for the alternative hypothesis. Other texts may use H_0 (subscript zero) in place of H_o and H_1 in place of H_a.

strating the *unlikeliness* of the truth of the null hypothesis. The alternative hypothesis is sometimes referred to as the *research hypothesis,* because it represents what the researcher hopes will be found to be "true."

Because the "evidence" (who's going to the party, what is going to be served, and so on) can demonstrate only the unlikeliness of the party being a dud, your initial position, "The party will be a dud," becomes the null hypothesis. Your friend's claim, "The party will be a great time," then becomes the alternative hypothesis.

H_o: "Party will be a dud"   vs. H_a: "Party will be a great time"

The following examples will illustrate the formation of and the relationship between null and alternative hypotheses.

EXAMPLE 8.8 — Writing Hypotheses

You are testing a new design for air bags used in automobiles, and you are concerned that they might not open properly. State the null and alternative hypotheses.

SOLUTION The two opposing possibilities are "Bags open properly" and "Bags do not open properly." Testing could produce evidence that discredits the hypothesis "Bags open properly" plus your concern is that the "Bags do not open properly." Therefore, "Bags do not open properly" would become the alternative hypothesis and "Bags open properly" would be the null hypothesis.

The alternative hypothesis can be the statement the experimenter wants to show to be true.

EXAMPLE 8.9 — Writing Hypotheses

An engineer wishes to show that the new formula that was just developed results in a quicker-drying paint. State the null and alternative hypotheses.

SOLUTION The two opposing possibilities are "does dry quicker" and "does not dry quicker." Because the engineer wishes to show "does dry quicker," the alternative hypothesis is "Paint made with the new formula does dry quicker" and the null hypothesis is "Paint made with the new formula does not dry quicker."

Occasionally it might be reasonable to hope that the evidence does not lead to a rejection of the null hypothesis. Such is the case in Example 8.10.

EXAMPLE 8.10 — Writing Hypotheses

Statistics ⬡ Now™

Watch a video example at
http://1pass.thomson.com
or on your CD.

You suspect that a brand-name detergent outperforms the store's brand of detergent, and you wish to test the two detergents because you would prefer to buy the cheaper store brand. State the null and alternative hypotheses.

SOLUTION Your suspicion, "The brand-name detergent outperforms the store brand," is the reason for the test and therefore becomes the alternative hypothesis.

H_o: "There is no difference in detergent performance."

H_a: "The brand-name detergent performs better than the store brand."

However, as a consumer, you are hoping not to reject the null hypothesis for budgetary reasons.

**APPLIED
EXAMPLE 8.11**

Evaluation of Teaching Techniques

ABSTRACT: THIS STUDY TESTS THE EFFECT OF HOMEWORK COLLECTION AND QUIZZES ON EXAM SCORES.

The hypothesis for this study is that an instructor can improve a student's performance (exam scores) through influencing the student's perceived effort-reward probability. An instructor accomplishes this by assigning tasks (teaching techniques) which are a part of a student's grade and are perceived by the student as a means of improving his or her grade in the class. The student is motivated to increase effort to complete those tasks which should also improve understanding of course material. The expected final result is improved exam scores. The null hypothesis for this study is:

H_o: Teaching techniques have no significant effect on students' exam scores. . . .

Source: David R. Vruwink and Janon R. Otto, *The Accounting Review*, Vol. LXII, No. 2, April 1987. Reprinted by permission.

Before returning to our example about the party, we need to look at the four possible outcomes that could result from the null hypothesis being either true or false and the decision being either to "reject H_o" or to "fail to reject H_o." Table 8.3 shows these four possible outcomes.

A **type A correct decision** occurs when the null hypothesis is true and we decide in its favor. A **type B correct decision** occurs when the null hypothesis is false and the decision is in opposition to the null hypothesis. A **type I error** is committed when a true null hypothesis is rejected—that is, when the null hypothesis is true but we decide against it. A **type II error** is committed when we decide in favor of a null hypothesis that is actually false.

TABLE 8.3

Four Possible Outcomes in a Hypothesis Test

Decision	Null Hypothesis	
	True	False
Fail to reject H_o	Type A correct decision	Type II error
Reject H_o	Type I error	Type B correct decision

EXAMPLE 8.12

Describing the Possible Outcomes and Resulting Actions (on Hypothesis Tests)

Describe the four possible outcomes and the resulting actions that would occur for the hypothesis test in Example 8.10.

SOLUTION
Recall: H_o: "There is no difference in detergent performance."
H_a: "The brand-name detergent performs better than the store brand."

	Null Hypothesis Is True	Null Hypothesis Is False
Fail to Reject H_o	**Type A Correct Decision** **Truth of situation:** There is no difference between the detergents. **Conclusion:** It was determined that there was no difference. **Action:** The consumer bought the cheaper detergent, saving money and getting the same results.	**Type II Error** **Truth of situation:** The brand-name detergent is better. **Conclusion:** It was determined that there was no difference. **Action:** The consumer bought the cheaper detergent, saving money and getting inferior results.
Reject H_o	**Type I Error** **Truth of situation:** There is no difference between the detergents. **Conclusion:** It was determined that the brand-name detergent was better. **Action:** The consumer bought the brand-name detergent, spending extra money to attain no better results.	**Type B Correct Decision** **Truth of situation:** The brand-name detergent is better. **Conclusion:** It was determined that the brand-name detergent was better. **Action:** The consumer bought the brand-name detergent, spending more and getting better results.

Notes:
1. The truth of the situation is not known before the decision is made, the conclusion reached, and the resulting actions take place. The truth of H_o may never be known.
2. The type II error often results in what represents a "lost opportunity"; lost in this situation is the chance to use a product that yields better results.

When a decision is made, it would be nice to always make the correct decision. This, however, is not possible in statistics because we make our decisions on the basis of sample information. The best we can hope for is to control the probability with which an error occurs. The probability assigned to the type I error is **α** (called **"alpha"**). The probability of the type II error is **β** (called **"beta"**; β is the second letter of the Greek alphabet). See Table 8.4.

To control these errors we assign a small probability to each of them. The most frequently used probability values for α and β are 0.01 and 0.05. The probability assigned to each error depends on its seriousness. The more serious the error, the less willing we are to have it occur; therefore, a smaller probabil-

Probability with Which Decisions Occur

Error in Decision	Type	Probability	Correct Decision	Type	Probability
Rejection of a true H_o	I	α	Failure to reject a true H_o	A	$1 - \alpha$
Failure to reject a false H_o	II	β	Rejection of a false H_o	B	$1 - \beta$

ity will be assigned. α and β are probabilities of errors, each under separate conditions, and they cannot be combined. Therefore, we cannot determine a single probability for making an incorrect decision. Likewise, the two correct decisions are distinctly separate, and each has its own probability; $1 - \alpha$ is the probability of a correct decision when the null hypothesis is true, and $1 - \beta$ is the probability of a correct decision when the null hypothesis is false. $1 - \beta$ is called the *power of the statistical test,* because it is the measure of the ability of a hypothesis test to reject a false null hypothesis, a very important characteristic.

Note: Regardless of the outcome of a hypothesis test, you are never certain that a correct decision has been reached.

Let's look back at the two possible errors in decision that could occur in Example 8.10. Most people would become upset if they found out they were spending extra money for a detergent that performed no better than the cheaper brand. Likewise, many people would become upset if they found out they could have been buying a better detergent. Evaluating the relative seriousness of these errors requires knowing whether this is your personal laundry or a professional laundry business, how much extra the brand-name detergent costs, and so on.

There is an interrelationship among the probability of the type I error (α), the probability of the type II error (β), and the sample size (n). This is very much like the interrelationship among level of confidence, maximum error, and sample size discussed on pages 410–411. Figure 8.8 shows the "three-way tug-of-war" among α, β, and n. If any one of the three is increased or decreased, it has an effect on one or both of the others. The statistician's job is to "balance" the three values of α, β, and n to achieve an acceptable testing situation.

FIGURE 8.8 The "Three-Way Tug-of-War" between α, β, and n

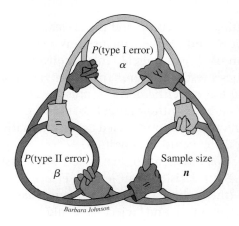

Barbara Johnson

If α is reduced, then either β must increase or n must be increased; if β is decreased, then either α increases or n must be increased; if n is decreased, then either α increases or β increases. The choices for α, β, and n are definitely not arbitrary. At this time in our study of statistics, only the sample size, n, and α, P(type I error), will be given and used to complete a hypothesis test. β, P(type II error), is further investigated in the section exercises but will not be utilized in this introduction to hypothesis testing.

Can the Lady Taste the Difference?

In the late 1920s, the question was asked, Can the lady taste the difference between milk poured into a cup of tea versus tea poured into a cup of milk? The lady was randomly presented with two cups, one of each in pairs, and she correctly identified all of them. If she guessed, the probability of her guessing correctly was 0.5. So we hypothesize that she guessed and look at the sample evidence. She correctly identified the 10 cups offered. What is the probability of guessing correctly 10 times in a row? This is Fisher's *p*-value. Is it likely that she was guessing and identified them correctly 10 times in a row? (1 in 1024 = 0.00098.)

Sample size, n, is self-explanatory, so let's look at the role of α.

> **Level of significance α:** The probability of committing the type I error.

Establishing the level of significance can be thought of as a "managerial decision." Typically, someone in charge determines the level of probability with which he or she is willing to risk a type I error.

At this point in the hypothesis test procedure, the evidence is collected and summarized and the value of a *test statistic* is calculated.

> **Test statistic:** A random variable whose value is calculated from the sample data and is used in making the decision "reject H_o" or "fail to reject H_o."

The value of the calculated test statistic is used in conjunction with a decision rule to determine either "reject H_o" or "fail to reject H_o." This **decision rule** must be established prior to collecting the data; it specifies how you will reach the decision.

Back to your friend's party: You have to weigh the history of your friend's parties, the time and place, others going, and so on, against your own criteria and then make your decision. As a result of the decision about the null hypothesis ("The party will be a dud"), you will take the appropriate action; you will either go to or not go to the party.

To complete a hypothesis test, you will need to write a conclusion that carefully describes the meaning of the decision relative to the intent of the hypothesis test.

> **The Conclusion:**
> a. If the decision is "reject H_o," then the conclusion should be worded something like, "There is sufficient evidence at the α level of significance to show that . . . (the meaning of the alternative hypothesis)."
> b. If the decision is "fail to reject H_o," then the conclusion should be worded something like, "There is not sufficient evidence at the α level of significance to show that . . . (the meaning of the alternative hypothesis)."

When writing the decision and the conclusion, remember that (1) the decision is about H_o and (2) the conclusion is a statement about whether or not the contention of H_a was upheld. This is consistent with the "attitude" of the whole hypothesis test procedure. The null hypothesis is the statement that is "on trial," and therefore the decision must be about it. The contention of the alternative hypothesis is the thought that brought about the need for a decision. Therefore, the question that led to the alternative hypothesis must be answered when the conclusion is written.

We must always remember that when the decision is made, nothing has been proved. Both decisions can lead to errors: "fail to reject H_o" could be a type II error (the lack of sufficient evidence has led to great parties being missed more than once), and "reject H_o" could be a type I error (more than one person has decided to go to a party that was a dud).

SECTION 8.4 EXERCISES

Statistics⬡Now™

Datasets can be found on your Student's Suite CD-ROM or at the StatisticsNow website at **http://1pass.thomson.com**.

8.51 You are testing a new detonating system for explosives and are concerned that the system is not reliable. State the null and alternative hypotheses.

8.52 Referring to Applied Example 8.11, state the instructor's hypothesis, the alternative hypothesis.

8.53 State the null and alternative hypotheses for each of the following:

a. You are investigating a complaint that "special delivery mail takes too much time" to be delivered.

b. You want to show that people find the new design for a recliner chair more comfortable than the old design.

c. You are trying to show that cigarette smoke affects the quality of a person's life.

d. You are testing a new formula for hair conditioner and hope to show that it is effective on "split ends."

8.54 State the null and alternative hypotheses for each of the following:

a. You want to show an increase in buying and selling of single-family homes this year when compared with last year's rate.

b. You are testing a new recipe for "low-fat" cheesecake and expect to find that its taste is not as good as traditional cheesecake.

c. You are trying to show that music lessons have a positive effect on a child's self-esteem.

d. You are investigating the relationship between a person's gender and the automobile he or she drives—specifically you want to show that males tend to drive truck-type vehicles more than females do.

8.55 Using the example of your friend's party (pp. 417 and 422) with H_o: "Party will be a dud" versus H_a: "The party will be a great time," describe the four possible decisions and the resulting actions as described in Example 8.12.

8.56 When a parachute is inspected, the inspector is looking for anything that might indicate the parachute might not open.

a. State the null and alternative hypotheses.

b. Describe the four possible outcomes that can result depending on the truth of the null hypothesis and the decision reached.

c. Describe the seriousness of the two possible errors.

8.57 When a medic at the scene of a serious accident inspects each victim, she administers the appropriate medical assistance to all victims, unless she is certain the victim is dead.

a. State the null and alternative hypotheses.

b. Describe the four possible outcomes that can result depending on the truth of the null hypothesis and the decision reached.

c. Describe the seriousness of the two possible errors.

8.58 A supplier of highway construction materials claims he can supply an asphalt mixture that will make roads that are paved with his materials less slippery when wet. A general contractor who builds roads wishes to test the supplier's claim. The null hypothesis is "Roads paved with this asphalt mixture are no less slippery than roads paved with other asphalt." The alternative hypothesis is "Roads paved with this asphalt mixture are less slippery than roads paved with other asphalt."

a. Describe the meaning of the two possible types of errors that can occur in the decision when this hypothesis test is completed.

b. Describe how the null hypothesis, as stated previously, is a "starting point" for the decision to be made about the asphalt.

8.59 Using the information from Exercise 8.55, describe how the type II error in the party example represents a "lost opportunity."

8.60 Describe the actions that would result in a type I error and a type II error if each of the following null hypotheses were tested. (Remember, the alternative hypothesis is the negation of the null hypothesis.)

a. H_o: The majority of Americans favor laws against assault weapons.

b. H_o: The choices on the fast-food menu are not low in salt.

c. H_o: This building must not be demolished.

d. H_o: There is no waste in government spending.

8.61 Describe the action that would result in a correct decision type A and a correct decision type B if each of the null hypotheses in Exercise 8.60 were tested.

8.62 Describe the action that would result in a correct decision type A and a correct decision type B if the hypotheses for the new detonating system for explosives of Exercise 8.51 were tested.

8.63 Consider the null hypothesis in Applied Example 8.11, "H_o: Teaching techniques have no significant effect on students' exam scores." Describe the actions that would result in a type I and a type II error if H_o were tested.

8.64 Consider the null hypothesis in Applied Example 8.11, "H_o: Teaching techniques have no significant effect on students' exam scores." Describe the actions that would result in a correct decision type A and correct decision type B if H_o were tested.

8.65 a.   If the null hypothesis is true, what decision error could be made?

   b.   If the null hypothesis is false, what decision error could be made?

   c.   If the decision "reject H_o" is made, what decision error could have been made?

   d.   If the decision "fail to reject H_o" is made, what decision error could have been made?

8.66 The director of an advertising agency is concerned with the effectiveness of a television commercial.

a.   What null hypothesis is she testing if she commits a type I error when she erroneously says that the commercial is effective?

b.   What null hypothesis is she testing if she commits a type II error when she erroneously says that the commercial is effective?

8.67 The director of an advertising agency is concerned with the effectiveness of a television commercial.

a.   What null hypothesis is she testing if she makes a correct decision type A when she correctly says that the commercial is not effective?

b.   What null hypothesis is she testing if she makes a correct decision type B when she correctly says that the commercial is not effective?

8.68 A politician is concerned with winning an upcoming election.

a.   What null hypothesis is he testing if he commits a type I error when he erroneously says that he will win the election?

b.   What null hypothesis is he testing if he commits a type II error when he erroneously says that he will win the election?

8.69 a.   If α is assigned the value 0.001, what are we saying about the type I error?

   b.   If α is assigned the value 0.05, what are we saying about the type I error?

   c.   If α is assigned the value 0.10, what are we saying about the type I error?

8.70 a.   If β is assigned the value 0.001, what are we saying about the type II error?

   b.   If β is assigned the value 0.05, what are we saying about the type II error?

   c.   If β is assigned the value 0.10, what are we saying about the type II error?

8.71 a. If the null hypothesis is true, the probability of a decision error is identified by what name?

b. If the null hypothesis is false, the probability of a decision error is identified by what name?

8.72 Suppose that a hypothesis test is to be carried out by using $\alpha = 0.05$. What is the probability of committing a type I error?

8.73 Explain why α is not always the probability of rejecting the null hypothesis.

8.74 Explain how assigning a small probability to an error controls the likelihood of its occurrence.

8.75 The conclusion is part of the hypothesis test that communicates the findings of the test to the reader. As such, it needs special attention so that the reader receives an accurate picture of the findings.

a. Carefully describe the "attitude" of the statistician and the statement of the conclusion when the decision is "reject H_o."

b. Carefully describe the "attitude" and the statement of the conclusion when the decision is "fail to reject H_o."

8.76 Find the power of a test when the probability of the type II error is:

a. 0.01   b. 0.05   c. 0.10

8.77 A normally distributed population is known to have a standard deviation of 5, but its mean is in question. It has been argued to be either $\mu = 80$ or $\mu = 90$, and the following hypothesis test has been devised to settle the argument. The null hypothesis, H_o: $\mu = 80$, will be tested using one randomly selected data value and comparing it with the critical value of 86. If the data value is greater than or equal to 86, the null hypothesis will be rejected.

a. Find α, the probability of the type I error.

b. Find β, the probability of the type II error.

8.78 Suppose the argument in Exercise 8.77 was to be settled using a sample of size 4; find α and β.

8.79 [EX08-079] You are a quality-control inspector and are in a position to make the decision as to whether a large shipment of cork stoppers for use in bottling still (versus bubbly) wine passes inspection. Once you inspect the mandatory number in the approved manner you will make a decision to accept or reject the lot.

Part 1 of the inspection requires you to randomly select 32 corks and measure three physical dimensions of the cylindrical stopper according to defined procedures.

Specification Limits

Diameter	24 mm ± 0.5 mm
Ovalization	≤0.7 mm
Length	45 mm ± 0.7 mm

Acceptance Quality Levels (AQL)

The batch is accepted if no more than two corks present an inferior or superior result to the limits of specification.

The batch may be refused if three corks or more present an inferior or superior result to the limits of specification.

Source: http://www.codiliege.org

The results of inspecting the mandated sample follow. (All measurements are in millimeters.)

Cork	1	2	3	4	5	6	7	8
Diameter	24.51	24.13	24.28	24.27	23.79	24.11	24.08	23.66
Ovalization	0.20	0.88	0.38	0.20	0.29	0.14	0.20	0.32
Length	44.89	44.69	45.36	44.94	44.65	45.50	44.86	44.67

Cork	9	10	11	12	13	14	15	16
Diameter	24.41	24.08	24.02	23.94	23.71	24.18	24.13	24.30
Ovalization	0.03	0.43	0.50	0.43	0.51	0.46	0.53	0.14
Length	45.13	44.92	44.88	45.14	44.87	44.67	45.01	44.86

Cork	17	18	19	20	21	22	23	24
Diameter	23.78	24.01	24.03	24.10	23.77	24.28	23.85	24.39
Ovalization	0.07	0.32	0.34	0.23	0.76	0.39	0.47	0.43
Length	45.12	45.21	45.70	44.95	44.27	45.23	45.29	44.98

Cork	25	26	27	28	29	30	31	32
Diameter	24.27	23.92	24.23	24.17	23.77	24.40	24.31	23.85
Ovalization	0.20	0.47	0.23	0.23	0.28	0.34	0.56	0.05
Length	44.80	45.06	45.38	45.11	44.75	45.42	45.04	44.53

a. Determine the number of corks that pass part 1 of the inspection.

b. State the decision and explain how it was reached.

c. Prepare a short written report summarizing the requirements and your findings and decision.

8.80 [EX08-080] As the quality-control inspector in Exercise 8.79 you are ready for the second phase of the inspection.

Part 2 requires that the humidity percentage of 20 cork stoppers be determined while following the prescribed procedure.

Specification Limits

Nominal value: 6%

Limits of specification: ±2% (i.e., from 4% to 8%)

Acceptance Quality Levels (AQL)

The batch is accepted if no more than two corks present an inferior or superior result to the limits of specification.

The batch may be refused if three corks or more present an inferior or superior result to the limits of specification.

Source: http://www.codiliege.org

Listed are three different samples, each taken from different lots. Review the sample results and answer these questions for each sample separately.

Sample 1:	5	5	6	3	7	6	6	7	8	6
	6	7	5	7	6	6	7	6	4	5
Sample 2:	1	6	6	8	6	5	7	6	10	6
	7	5	7	6	5	6	6	8	5	9
Sample 3:	5	7	3	5	5	5	6	5	9	3
	5	7	7	9	7	8	5	10	8	9

a. Construct a dotplot of the data.

b. Completely label the dotplot and circle the dots representing cork percents inferior or superior to the limits of specification.

c. State the decision and explain how it was reached.

d. Prepare a short written report summarizing the requirements and your findings and decision for each sample.

Hypothesis Test of Mean μ (σ Known): A Probability-Value Approach

In Section 8.4 we surveyed the concepts and much of the reasoning behind a hypothesis test while looking at nonstatistical examples. In this section we are going to formalize the hypothesis test procedure as it applies to statements concerning the mean μ of a population under the restriction that σ, the population standard deviation, is a known value.

The assumption for hypothesis tests about mean μ using a known σ: The sampling distribution of $\bar{x}$ has a normal distribution.

The information we need to ensure that this assumption is satisfied is contained in the sampling distribution of sample means and in the CLT (see Chapter 7, pp. 369–370):

The sampling distribution of sample means $\bar{x}$ is distributed about a mean equal to μ with a standard error equal to $\sigma/\sqrt{n}$; and (1) if the randomly sampled population is normally distributed, then $\bar{x}$ is normally distributed for all sample sizes, or (2) if the randomly sampled population is not normally distributed, then $\bar{x}$ is approximately normally distributed for sufficiently large sample sizes.

The hypothesis test is a well-organized, step-by-step procedure used to make a decision. Two different formats are commonly used for hypothesis testing. The *probability-value approach,* or simply *p-value approach,* is the hypothesis test process that has gained popularity in recent years, largely as a result of the convenience and the "number crunching" ability of the computer. This approach is organized as a five-step procedure.

THE PROBABILITY-VALUE HYPOTHESIS TEST: A FIVE-STEP PROCEDURE

Step 1   The Set-Up:
   a. Describe the population parameter of interest.
   b. State the null hypothesis (H_o) and the alternative hypothesis (H_a).

Step 2   The Hypothesis Test Criteria:
   a. Check the assumptions.
   b. Identify the probability distribution and the test statistic to be used.
   c. Determine the level of significance, α.

Step 3   The Sample Evidence:
   a. Collect the sample information.
   b. Calculate the value of the test statistic.

Step 4   The Probability Distribution:
   a. Calculate the *p*-value for the test statistic.
   b. Determine whether or not the *p*-value is smaller than α.

Step 5   The Results:
   a. State the decision about H_o.
   b. State the conclusion about H_a.

A commercial aircraft manufacturer buys rivets to use in assembling airliners. Each rivet supplier that wants to sell rivets to the aircraft manufacturer must demonstrate that its rivets meet the required specifications. One of the specs is, "The mean shearing strength of all such rivets, μ, is at least 925 lb." Each time the aircraft manufacturer buys rivets, it is concerned that the mean strength might be less than the 925-lb specification.

FYI Think about the consequences of using weak rivets.

Note: Each individual rivet has a shearing strength, which is determined by measuring the force required to shear ("break") the rivet. Clearly, not all the rivets can be tested. Therefore, a sample of rivets will be tested, and a decision about the mean strength of all the untested rivets will be based on the mean from those sampled and tested.

STEP 1   The Set-Up:
   a. **Describe the population parameter of interest.**
      The population parameter of interest is the mean μ, the mean shearing strength (or mean force required to shear) of the rivets being considered for purchase.

b. State the null hypothesis (H_o) and the alternative hypothesis (H_a).

The null hypothesis and the alternative hypothesis are formulated by inspecting the problem or statement to be investigated and first formulating two opposing statements about the mean μ. For our example, these two opposing statements are (A) "The mean shearing strength is less than 925" ($\mu < 925$, the aircraft manufacturer's concern), and (B) "The mean shearing strength is at least 925" ($\mu = 925$, the rivet supplier's claim and the aircraft manufacturer's spec).

FYI More specific instruction are given on pages 417–418.

Note: The trichotomy law from algebra states that two numerical values must be related in exactly one of three possible relationships: $<$, $=$, or $>$. All three of these possibilities must be accounted for in the two opposing hypotheses in order for the two hypotheses to be negations of each other. The three possible combinations of signs and hypotheses are shown in Table 8.5. Recall that the null hypothesis assigns a specific value to the parameter in question, and therefore "equals" will always be part of the null hypothesis.

TABLE 8.5

The Three Possible Statements of Null and Alternative Hypotheses

Null Hypothesis	Alternative Hypothesis
1. Greater than or equal to ($\geq$)	Less than ($<$)
2. Less than or equal to ($\leq$)	Greater than ($>$)
3. Equal to ($=$)	Not equal to ($\neq$)

The parameter of interest, the population mean μ, is related to the value 925. Statement (A) becomes the alternative hypothesis:

$$H_a: \ \mu < 925 \text{ (the mean is less than 925)}$$

This statement represents the aircraft manufacturer's concern and says, "The rivets do not meet the required specs." Statement (B) becomes the null hypothesis:

$$H_o: \ \mu = 925 \ (\geq) \text{ (the mean is at least 925)}$$

This hypothesis represents the negation of the aircraft manufacturer's concern and says, "The rivets do meet the required specs."

Note: We will write the null hypothesis with just the equal sign, thereby stating the exact value assigned. When "equal" is paired with "less than" or paired with "greater than," the combined symbol is written beside the null hypothesis as a reminder that all three signs have been accounted for in these two opposing statements.

Before continuing with our example, let's look at three examples that demonstrate formulating the statistical null and alternative hypotheses involving the population mean μ. Examples 8.13 and 8.14 each demonstrate a "one-tailed" alternative hypothesis.

EXAMPLE 8.13

Writing Null and Alternative Hypotheses (One-Tailed Situation)

Statistics◯Now™

Watch a video example at
http://1pass.thomson.com
or on your CD.

Suppose the Environmental Protection Agency was suing the city of Rochester for noncompliance with carbon monoxide standards. Specifically, the EPA would want to show that the mean level of carbon monoxide in downtown Rochester's air is dangerously high, higher than 4.9 parts per million. State the null and alternative hypotheses.

SOLUTION To state the two hypotheses, we first need to identify the population parameter in question: the "mean level of carbon monoxide in Rochester." The parameter μ is being compared with the value 4.9 parts per million, the specific value of interest. The EPA is questioning the value of μ and wishes to show it is higher than 4.9 (i.e., $\mu > 4.9$). The three possible relationships—(1) $\mu < 4.9$, (2) $\mu = 4.9$, and (3) $\mu > 4.9$—must be arranged to form two opposing statements: one states the EPA's position, "The mean level is higher than 4.9 ($\mu > 4.9$)," and the other states the negation, "The mean level is not higher than 4.9 ($\mu \leq 4.9$)." One of these two statements will become the null hypothesis, H_o, and the other will become the alternative hypothesis, H_a.

Recall that there are two rules for forming the hypotheses: (1) the null hypothesis states that the parameter in question has a specified value ("H_o must contain the equal sign"), and (2) the EPA's contention becomes the alternative hypothesis ("higher than"). Both rules indicate:

$$H_o: \mu = 4.9 \ (\leq) \quad \text{and} \quad H_a: \mu > 4.9$$

EXAMPLE 8.14

Writing Null and Alternative Hypotheses (One-Tailed Situation)

Statistics◯Now™

Watch a video example at
http://1pass.thomson.com
or on your CD.

An engineer wants to show that applications of paint made with the new formula dry and are ready for the next coat in a mean time of less than 30 minutes. State the null and alternative hypotheses for this test situation.

SOLUTION The parameter of interest is the mean drying time per application, and 30 minutes is the specified value. $\mu < 30$ corresponds to "The mean time is less than 30," whereas $\mu \geq 30$ corresponds to the negation, "The mean time is not less than 30." Therefore, the hypotheses are

$$H_o: \mu = 30 \ (\geq) \quad \text{and} \quad H_a: \mu < 30$$

Example 8.15 demonstrates a "two-tailed" alternative hypothesis.

EXAMPLE 8.15

Writing Null and Alternative Hypotheses (Two-Tailed Situation)

Statistics◯Now™

Watch a video example at
http://1pass.thomson.com
or on your CD.

Job satisfaction is very important to worker productivity. A standard job-satisfaction questionnaire was administered by union officers to a sample of assembly-line workers in a large plant in hopes of showing that the assembly workers' mean score on this questionnaire would be different from the established mean of 68. State the null and alternative hypotheses.

SOLUTION Either the mean job satisfaction score is different from 68 ($\mu \neq 68$) or the mean is equal to 68 ($\mu = 68$). Therefore,

$$H_o: \mu = 68 \quad \text{and} \quad H_a: \mu \neq 68$$

Notes:

1. The alternative hypothesis is referred to as being "two-tailed" when H_a is "not equal."
2. When "less than" is combined with "greater than," they become "not equal to."

The viewpoint of the experimenter greatly affects the way the hypotheses are formed. Generally, the experimenter is trying to show that the parameter value is different from the value specified. Thus, the experimenter is often hoping to be able to reject the null hypothesis so that the experimenter's theory has been substantiated. Examples 8.13, 8.14, and 8.15 also represent the three possible arrangements for the $<$, $=$, and $>$ relationships between the parameter μ and a specified value.

Table 8.6 lists some additional common phrases used in claims and indicates their negations and the hypothesis in which each phrase will be used. Again, notice that "equals" is always in the null hypothesis. Also notice that the negation of "less than" is "greater than or equal to." Think of negation as "all the others" from the set of three signs.

TABLE 8.6

Common Phrases and Their Negations

$H_o: (\geq)$ vs. $H_a: (<)$		$H_o: (\leq)$ vs. $H_a: (>)$		$H_o: (=)$ vs. $H_a: (\neq)$	
At least	Less than	At most	More than	Is	Is not
No less than	Less than	No more than	More than	Not different from	Different from
Not less than	Less than	Not greater than	Greater than	Same as	Not same as

After the null and alternative hypotheses are established, we will work under the assumption that the null hypothesis is a true statement until there is sufficient evidence to reject it. This situation might be compared with a courtroom trial, where the accused is assumed to be innocent (H_o: Defendant is innocent vs. H_a: Defendant is not innocent) until sufficient evidence has been presented to show that innocence is totally unbelievable ("beyond reasonable doubt"). At the conclusion of the hypothesis test, we will make one of two possible decisions. We will decide in opposition to the null hypothesis and say that we "reject H_o" (this corresponds to "conviction" of the accused in a trial), or we will decide in agreement with the null hypothesis and say that we "fail to reject H_o" (this corresponds to "fail to convict" or an "acquittal" of the accused in a trial).

Let's return to the rivet example we interrupted on page 428 and continue with Step 2. Recall that

$$H_o: \mu = 925 \; (\geq) \; \text{(at least 925)} \quad H_a: \mu < 925 \; \text{(less than 925)}$$

STEP 2 **The Hypothesis Test Criteria:**
   a. **Check the assumptions.**

   Assume that the standard deviation of the shearing strength of rivets is known from past experience to be $\sigma = 18$. Variables like shearing strength typically have a mounded distribution; therefore, a sample of size 50 should be large enough for the CLT to apply and ensure that the SDSM will be normally distributed.

   b. **Identify the probability distribution and the test statistic to be used.**

   The standard normal probability distribution is used because $\bar{x}$ is expected to have a normal distribution.

For a hypothesis test of μ, we want to compare the value of the sample mean with the value of the population mean as stated in the null hypothesis. This comparison is accomplished using the test statistic in formula (8.4):

Test Statistic for Mean

$$z\star = \frac{\bar{x} - \mu}{\sigma/\sqrt{n}} \qquad (8.4)$$

The resulting calculated value is identified as $z\star$ ("z star") because it is expected to have a standard normal distribution when the null hypothesis is true and the assumptions have been satisfied. The $\star$ ("star") is to remind us that this is the calculated value of the test statistic.

The test statistic to be used is $z\star = \dfrac{\bar{x} - \mu}{\sigma/\sqrt{n}}$ with $\sigma = 18$.

   c. **Determine the level of significance, α.**

Setting α was described as a managerial decision in Section 8.4. To see what is involved in determining α, the probability of the type I error, for our rivet example, we start by identifying the four possible outcomes, their meaning, and the action related to each.

The type I error occurs when a true null hypothesis is rejected. This would occur when the manufacturer tested rivets that did meet the specs and rejected them. Undoubtedly this would lead to the rivets not being purchased even though they did meet the specs. In order for the manager to set a level of significance, related information is needed—namely, how soon is the new supply of rivets needed? If they are needed tomorrow and this is the only vendor with an available supply, waiting a week to find acceptable rivets could be very expensive; therefore, rejecting good rivets could be considered a serious error. On the other hand, if the rivets are not needed until next month, then this error may not be very serious. Only the manager will know all the ramifications, and therefore the manager's input is important here.

After much consideration, the manager assigns the level of significance: $\alpha = 0.05$.

STEP 3 **The Sample Evidence:**
   a. **Collect the sample information.**

   The sample must be a random sample drawn from the population whose mean μ is being questioned. A random sample of 50 rivets is selected, each rivet is tested, and the sample mean shearing strength is calculated: $\bar{x} = 921.18$ and $n = 50$.

b.   Calculate the value of the test statistic.

The sample evidence ($\bar{x}$ and n found in Step 3a) is next converted into the **calculated value of the test statistic, $z\star$,** using formula (8.4). (μ is 925 from H_o, and $\sigma = 18$ is a known quantity.) We have

$$z\star = \frac{\bar{x} - \mu}{\sigma/\sqrt{n}}: \quad z\star = \frac{921.18 - 925.0}{18/\sqrt{50}} = \frac{-3.82}{2.5456} = -1.50$$

STEP 4   The Probability Distribution:

   a.   Calculate the p-value for the test statistic.

> **Probability value, or p-value:** The probability that the test statistic could be the value it is or a more extreme value (in the direction of the alternative hypothesis) when the null hypothesis is true. (*Note:* The symbol **P** will be used to represent the p-value, especially in algebraic situations.)

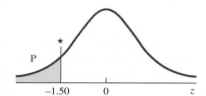

Draw a sketch of the standard normal distribution and locate $z\star$ (found in Step 3b) on it. To identify the area that represents the p-value, look at the sign in the alternative hypothesis. For this test, the alternative hypothesis indicates that we are interested in that part of the sampling distribution that is *"less than"* $z\star$. Therefore, the p-value is the area that lies to the *left* of $z\star$. Shade this area.

To find the p-value, you may use any one of the three methods outlined here. The method you use is not the important thing, because each of them is just the tool of choice to help you find the p-value.

FYI Complete instructions for using Table 3 are given on pages 317–320.

Method 1: Use Table 3 in Appendix B to determine the tabled area related to $z = 1.50$; then calculate the p-value by subtracting from 0.5000:

$$p\text{-value} = P(z < z\star) = P(z < -1.50) = P(z > 1.50) = 0.5000 - 0.4332 = \mathbf{0.0668}$$

Method 2: Use Table 5 in Appendix B and the symmetry property: Table 5 is set up to allow you to read the p-value directly from the table. Since $P(z < -1.50) = P(z > 1.50)$, simply locate $z\star = 1.50$ on Table 5 and read the p-value:

FYI You will use only one of these three equivalent methods.

$$P(z < -1.50) = \mathbf{0.0668}$$

FYI Instructions for using this computer command are given on pages 329–330. Try it! See if you get the same answer.

Method 3:   Use the cumulative probability function on a computer or calculator to find the p-value:

$$P(z < -1.50) = \mathbf{0.0668}$$

b.   Determine whether or not the p-value is smaller than α.

The p-value (0.0668) is not smaller than α (0.05).

STEP 5   The Results:

   a.   State the decision about H_o.

Is the p-value small enough to indicate that the sample evidence is highly unlikely in the event that the null hypothesis is true? In order to make the decision, we need to know the *decision rule*.

> **Decision rule:**
> a. If the *p*-value is *less than or equal to* the level of significance α, then the decision must be **reject H_o.**
> b. If the *p*-value is *greater than* the level of significance α, then the decision must be **fail to reject H_o.**

FYI Specific information about writing the conclusion is given on page 422.

Decision about H_o: Fail to reject H_o.

b. State the conclusion about H_a.

There is not sufficient evidence at the 0.05 level of significance to show that the mean shearing strength of the rivets is less than 925. "We failed to convict" the null hypothesis. In other words, a sample mean as small as 921.18 is likely to occur (as defined by α) when the true population mean value is 925.0 and $\bar{x}$ is normally distributed. The resulting action by the manager would be to buy the rivets.

Note: When the decision reached is "fail to reject H_o," it simply means "for the lack of better information, act as if the null hypothesis is true." ("Accept H_o" is a misnomer.)

Before looking at another example, let's look at the procedures for finding the *p*-value. The *p*-value is represented by the area under the curve of the probability distribution for the test statistic that is more extreme than the calculated value of the test statistic. There are three separate cases, and the direction (or sign) of the alternative hypothesis is the key. Table 8.7 outlines the procedure for all three cases.

TABLE 8.7

Finding *p*-Values

Case 1 H_a contains ">" "Right tail"	*p*-value is the *area to right of z★* *p*-value $= P(z > z★)$	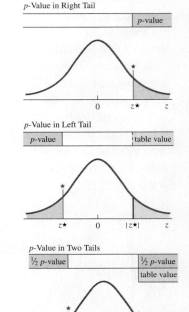
Case 2 H_a contains "<" "Left tail"	*p*-value is the *area to left of z★* the area of the left tail is the same as the area in the right tail bounded by the positive z★; therefore, *p*-value $= P(z < z★) = P(z > |z★|)$	
Case 3 H_a contains "≠" "Two-tailed"	*p*-value is the *total area of both tails* *p*-value $= P(z < -|z★|) + P(z > |z★|)$ z★ may be in either tail, and since both areas are equal, find the probability of one tail and double it. Thus, *p*-value $= 2 \times P(z > |z★|)$	

Let's look at an example involving the two-tailed procedure.

DID YOU KNOW

The Fisher and Neyman Conflict

Their differences centered on their approaches to hypothesis testing. Both methods start with a null hypothesis and use the same test statistic; however, Neyman and Pearson use a set risk of error and Fisher does not. The Neyman/Pearson approach follows a basic deductive logic method of assuming the hypothesis is true, then looking for evidence that contradicts the assumptions. The Fisher approach determines the probability of the occurrence for the data that result and uses that probability value to assess the data. A small probability shows "unlikeliness" for the data to have occurred under a true null hypothesis. The probability that the null hypothesis is correct is another story—that requires a Bayesian approach and leads to yet another academic dispute.

EXAMPLE 8.16

Two-Tailed Hypothesis Test

For years, many large companies in a certain city have used the Kelley Employment Agency for testing prospective employees. The employment selection test used has historically resulted in scores normally distributed about a mean of 82 and a standard deviation of 8. The Brown Agency has developed a new test that is quicker and easier to administer and therefore less expensive. Brown claims that its test results are the same as those obtained on the Kelley test. Many of the companies are considering a change from the Kelley Agency to the Brown Agency to cut costs. However, they are unwilling to make the change if the Brown test results have a different mean value. An independent testing firm tested 36 prospective employees with the Brown test. A sample mean of 79 resulted. Determine the *p*-value associated with this hypothesis test. (Assume $\sigma = 8$.)

SOLUTION

STEP 1   **The Set-Up:**
  a. **Describe the population parameter of interest.**
     The population mean μ, the mean of all test scores using the Brown Agency test.
  b. **State the null hypothesis (H_o) and the alternative hypothesis (H_a).**
     The Brown Agency's test results "will be different" (the concern) if the mean test score is not equal to 82. They "will be the same" if the mean is equal to 82. Therefore,

     H_o: $\mu = 82$ (test results have the same mean)

     H_a: $\mu \neq 82$ (test results have a different mean)

STEP 2   **The Hypothesis Test Criteria:**
  a. **Check the assumptions.**
     σ is known. If the Brown test scores are distributed the same as the Kelley test scores, they will be normally distributed and the sampling distribution will be normal for all sample sizes.
  b. **Identify the probability distribution and the test statistic to be used.**
     The standard normal probability distribution and the test statistic

     $z\bigstar = \dfrac{\bar{x} - \mu}{\sigma/\sqrt{n}}$ will be used with $\sigma = 8$.

  c. **Determine the level of significance, α.**
     The level of significance is omitted because the question asks for the *p*-value and not a decision.

STEP 3   **The Sample Evidence:**
  a. **Collect the sample information:** $n = 36, \bar{x} = 79$.

b. Calculate the value of the test statistic.

μ is 82 from H_o; $\sigma = 8$ is a known quantity. We have

$$z\star = \frac{\bar{x} - \mu}{\sigma/\sqrt{n}}: \quad z\star = \frac{79 - 82}{8/\sqrt{36}} = \frac{-3}{1.3333} = -2.25$$

STEP 4   The Probability Distribution:

a. Calculate the *p*-value for the test statistic.

Because the alternative hypothesis indicates a two-tailed test, we must find the probability associated with both tails. The *p*-value is found by doubling the area of one tail (see Table 8.7, p. 433). Since $z\star = -2.25$, the value of $|z\star| = 2.25$. The *p*-value $= 2 \times P(z > |z\star|) = 2 \times P(z > 2.25)$.

From Table 3: *p*-value $= 2 \times P(z > 2.25) = 2 \times (0.5000 - 0.4878) = 2(0.0122) = 0.0244$.

or

From Table 5: *p*-value $= 2 \times P(z > 2.25) = 2(0.0122) = 0.0244$.

or

Use the cumulative probability function on a computer or calculator: *p*-value $= 2 \times P(z < -2.25) = 0.0244$.

b. Determine whether or not the *p*-value is smaller than α.

A comparison is not possible; no α value was given in the statement of the question.

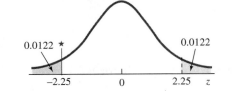

½ *p*-value   table value

0.0122 ★   0.0122

−2.25   0   2.25   *z*

FYI

See the instructions on page 432.

STEP 5   The Results:

The *p*-value for this hypothesis test is 0.0244. Each individual company now will decide whether to continue to use the Kelley Agency's services or change to the Brown Agency. Each will need to establish the level of significance that best fits its own situation and then make a decision using the decision rule described previously.

The *fundamental idea of the p-value* is to express the degree of belief in the null hypothesis:

- When the *p*-value is minuscule (something like 0.0003), the null hypothesis would be rejected by everybody because the sample results are very unlikely for a true H_o.

- When the *p*-value is fairly small (like 0.012), the evidence against H_o is quite strong and H_o will be rejected by many.

- When the *p*-value begins to get larger (say, 0.02 to 0.08), there is too much probability that data like the sample involved could have occurred even if H_o were true and the rejection of H_o is not an easy decision.

- When the *p*-value gets large (like 0.15 or more), the data are not at all unlikely if the H_o is true and no one will reject H_o.

The *advantages of the p-value approach* are as follows: (1) The results of the test procedure are expressed in terms of a continuous probability scale from 0.0 to 1.0, rather than simply on a "reject" or "fail to reject" basis. (2) A *p*-value can be reported and the user of the information can decide on the strength of the evidence as it applies to his or her own situation. (3) Computers can do all the calculations and report the *p*-value, thus eliminating the need for tables.

The *disadvantage of the p-value approach* is the tendency for people to put off determining the level of significance. This should not be allowed to happen, because it is then possible for someone to set the level of significance after the fact, leaving open the possibility that the "preferred" decision will result. This is probably important only when the reported *p*-value falls in the "hard choice" range (say, 0.02 to 0.08), as described previously.

FYI
Do your opponents show you their poker hands before you bet?

EXAMPLE 8.17   **Two-Tailed Hypothesis Test with Sample Data**

According to the results of Exercise 5.32 (p. 282), the mean of single-digit random numbers is 4.5 and the standard deviation is $\sigma = 2.87$. Draw a random sample of 40 single-digit numbers from Table 1 in Appendix B and test the hypothesis, "The mean of the single-digit numbers in Table 1 is 4.5." Use $\alpha = 0.10$.

SOLUTION

STEP 1   The Set-Up:
    a.   Describe the population parameter of interest.
        The population parameter of interest is the mean μ of the population of single-digit numbers in Table 1 of Appendix B.
    b.   State the null hypothesis (H_o) and the alternative hypothesis (H_a).

$$H_o: \mu = 4.5 \text{ (mean is 4.5)}$$
$$H_a: \mu \neq 4.5 \text{ (mean is not 4.5)}$$

STEP 2   The Hypothesis Test Criteria:
    a.   Check the assumptions.
        σ is known. Samples of size 40 should be large enough to satisfy the CLT; see the discussion of this issue on page 426.
    b.   Identify the probability distribution and the test statistic to be used.
        We use the standard normal probability distribution, and the test statistic is $z\star = \dfrac{\bar{x} - \mu}{\sigma/\sqrt{n}}$; $\sigma = 2.87$.
    c.   Determine the level of significance, α.
        $\alpha = 0.10$ (given in the statement of the problem).

STEP 3   The Sample Evidence:
    a.   Collect the sample information.
        This random sample was drawn from Table 1 in Appendix B [TA08-01]:

2	8	2	1	5	5	4	0	9	1	0	4	6	1
5	1	1	3	8	0	3	6	8	4	8	6	8	
9	5	0	1	4	1	2	1	7	1	7	9	3	

From the sample: $\bar{x} = 3.975$ and $n = 40$.

b. Calculate the value of the test statistic.

We use formula (8.4), and μ is 4.5 from H_o, and $\sigma = 2.87$:

$$z\star = \frac{\bar{x} - \mu}{\sigma/\sqrt{n}}: \quad z\star = \frac{3.975 - 4.50}{2.87/\sqrt{40}} = \frac{-0.525}{0.454} = -1.156 = -1.16$$

STEP 4   The Probability Distribution:

a. Calculate the *p*-value for the test statistic.

Because the alternative hypothesis indicates a two-tailed test, we must find the probability associated with both tails. The *p*-value is found by doubling the area of one tail. Since $z\star = -1.16$, the value of $|z\star| = 1.16$.

The *p*-value $= 2 \times P(z > |z\star|)$:

$$\mathbf{P} = 2 \times P(z > 1.16) = 2 \times (0.5000 - 0.3770)$$
$$= 2(0.1230) = 0.2460$$

b. Determine whether or not the *p*-value is smaller than α.

The *p*-value (0.2460) is greater than α (0.10).

STEP 5   The Results:

a. State the decision about H_o: Fail to reject H_o.

b. State the conclusion about H_a.

The observed sample mean is not significantly different from 4.5 at the 0.10 level of significance.

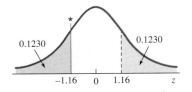

Suppose we were to take another sample of size 40 from Table 1. Would we obtain the same results? Suppose we took a third sample and a fourth. What results might we expect? What does the *p*-value in Example 8.17 measure? Table 8.8 lists (1) the means obtained from 50 different random samples of size

TABLE 8.8

a. The Means of 50 Random Samples Taken from Table 1 in Appendix B [TA08-08]

3.850	5.075	4.375	4.675	5.200	4.250	3.775	4.075	5.800	4.975
4.225	4.125	4.350	4.925	5.100	4.175	4.300	4.400	4.775	4.525
4.225	5.075	4.325	5.025	4.725	4.600	4.525	4.800	4.550	3.875
4.750	4.675	4.700	4.400	5.150	4.725	4.350	3.950	4.300	4.725
4.975	4.325	4.700	4.325	4.175	3.800	3.775	4.525	5.375	4.225

b. The *z*★ Values Corresponding to the 50 Means

−1.432	1.267	−0.275	0.386	1.543	−0.551	−1.598	−0.937	2.865	1.047
−0.606	−0.826	−0.331	0.937	1.322	−0.716	−0.441	−0.220	0.606	0.055
−0.606	1.267	−0.386	1.157	0.496	0.220	0.055	0.661	0.110	−1.377
0.551	0.386	0.441	−0.220	1.432	0.496	−0.331	−1.212	−0.441	0.496
1.047	−0.386	0.441	−0.386	−0.716	−1.543	−1.598	0.055	1.928	−0.606

c. The *p*-Values Corresponding to the 50 Means

0.152	0.205	0.783	0.700	0.123	0.582	0.110	0.349	0.004	0.295
0.545	0.409	0.741	0.349	0.186	0.474	0.659	0.826	0.545	0.956
0.545	0.205	0.700	0.247	0.620	0.826	0.956	0.509	0.912	0.168
0.582	0.700	0.659	0.826	0.152	0.620	0.741	0.226	0.659	0.620
0.295	0.700	0.659	0.700	0.474	0.123	0.110	0.956	0.054	0.545

40 that were taken from Table 1 in Appendix B, (2) the 50 values of $z\star$ corresponding to the 50 $\bar{x}$'s, and (3) their 50 corresponding p-values. Figure 8.9 shows a histogram of the 50 $z\star$ values.

FIGURE 8.9   The 50 Values of $z\star$ from Table 8.8

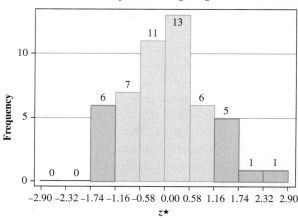

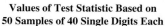

The histogram shows that six values of $z\star$ were less than -1.16 and seven values were greater than 1.16. That means 13 of the 50 samples, or 26%, have mean values more extreme than the mean ($\bar{x} = 3.975$) in Example 8.17. This observed relative frequency of 0.26 represents an empirical look at the p-value. Notice that the empirical value for the p-value (0.26) is very similar to the calculated p-value of 0.2460. Check the list of p-values; do you find that 13 of the 50 p-values are less than 0.2460? Which samples resulted in $|z\star| > 1.16$? Which samples resulted in a p-value greater than 0.2460? How do they compare?

TECHNOLOGY INSTRUCTIONS: HYPOTHESIS TEST FOR MEAN μ WITH A GIVEN σ

MINITAB (Release 14)   Input the data into C1; then continue with:

```
Choose:      Stat > Basic Statistics > 1-Sample Z
Enter:       Samples in columns: C1
             Standard deviation: σ
             Test mean: μ
Select:      Options
Select:      Alternative: less than or not equal to or greater than
             > OK > OK
```

Excel   Input the data into column A; then continue with:

```
Choose:      Tools > Data Analysis Plus > Z-Test: Mean > OK
Enter:       Input Range: (A1:A20 or select cells)
             Hypothesized Mean: μ
             Standard Deviation (SIGMA): σ > OK
Gives p-values for both one-tailed and two-tailed tests.
```

TI-83/84 Plus   Input the data into L1; then continue with the following, entering the appropriate values and highlighting Calculate:

Choose: **STAT > TESTS > 1:Z-Test**

```
Z-Test
 Inpt:DATA Stats
 µ0:0
 σ:0
 List:L1
 Freq:1
 µ:≠µ0 <µ0 >µ0
Calculate Draw
```

FYI The *p*-value approach was "made" for the computer!

The MINITAB solution to the rivet example, used in this section (pp. 427–428, 430–433), is shown here:

```
One-sample Z: C1
Test of mu = 925.00 vs < 925.00
The assumed standard deviation = 18.0
N        Mean        StDev        SE Mean        Z           P
50       921.18      17.58        2.546          −1.50       0.0668
```

When the computer is used, all that is left is for you to do is to make the decision and to write the conclusion.

SECTION 8.5 EXERCISES

Statistics Now™

Skillbuilder Applet Exercises must be worked using an accompanying applet found on your Student's Suite CD-ROM or at the StatisticsNow website at **http://1pass.thomson.com**.

Datasets can be found on your Student's Suite CD-ROM or at the StatisticsNow website at **http://1pass.thomson.com**.

8.81 In the example starting on page 427, the aircraft builder who is buying the rivets is concerned that the rivets might not meet the mean-strength spec. State the aircraft manufacturer's null and alternative hypotheses.

8.82 Professor Hart does not believe a statement he heard: "The mean weight of college women is 54.4 kg." State the null and alternative hypotheses he would use to challenge this statement.

8.83 State the null and alternative hypotheses used to test each of the following claims:

a. The mean reaction time is greater than 1.25 seconds.

b. The mean score on that qualifying exam is less than 335.

c. The mean selling price of homes in the area is not $230,000.

d. The mean weight of college football players is no more than 210 lb.

e. The mean hourly wage for a childcare giver is at most $9.

8.84 State the null hypothesis H_o and the alternative hypothesis H_a that would be used for a hypothesis test related to each of the following statements:

a. The mean age of the students enrolled in evening classes at a certain college is greater than 26 years.

b. The mean weight of packages shipped on Air Express during the past month was less than 36.7 lb.

c. The mean life of fluorescent light bulbs is at least 1600 hours.

d. The mean strength of welds by a new process is different from 570 lb per unit area, the mean strength of welds by the old process.

8.85 Identify the four possible outcomes and describe the situation involved with each outcome with regard to the aircraft manufacturer's testing and buying of rivets. Which is the more serious error: the type I or type II error? Explain.

8.86 A manufacturer wishes to test the hypothesis that "by changing the formula of its toothpaste it will give its users improved protection." The null hypothesis represents the idea that "the change will not improve the protection," and the alternative hypothesis is "the change will improve the protection." Describe the meaning of the two possible types of errors that can occur in the decision when the test of the hypothesis is conducted.

8.87 Suppose we want to test the hypothesis that the mean hourly charge for automobile repairs is at least $60 per hour at the repair shops in a nearby city. Explain the conditions that would exist if we make an error in decision by committing a type I error. What about a type II error?

8.88 Describe how the null hypothesis, as stated in Example 8.14 (p. 429), is a "starting point" for the decision to be made about the drying time for paint made with the new formula.

8.89 Assume that z is the test statistic and calculate the value of $z\star$ for each of the following:

a. H_o: $\mu = 10$, $\sigma = 3$, $n = 40$, $\bar{x} = 10.6$

b. H_o: $\mu = 120$, $\sigma = 23$, $n = 25$, $\bar{x} = 126.2$

c. H_o: $\mu = 18.2$, $\sigma = 3.7$, $n = 140$, $\bar{x} = 18.93$

d. H_o: $\mu = 81$, $\sigma = 13.3$, $n = 50$, $\bar{x} = 79.6$

8.90 Assume that z is the test statistic and calculate the value of $z\star$ for each of the following:

a. H_o: $\mu = 51$, $\sigma = 4.5$, $n = 40$, $\bar{x} = 49.6$

b. H_o: $\mu = 20$, $\sigma = 4.3$, $n = 75$, $\bar{x} = 21.2$

c. H_o: $\mu = 138.5$, $\sigma = 3.7$, $n = 14$, $\bar{x} = 142.93$

d. H_o: $\mu = 815$, $\sigma = 43.3$, $n = 60$, $\bar{x} = 799.6$

8.91 There are only two possible decisions that can result from a hypothesis test.

a. State the two possible decisions.

b. Describe the conditions that will lead to each of the two decisions identified in part a.

8.92 a. What decision is reached when the p-value is greater than α?

b. What decision is reached when α is greater than the p-value?

8.93 For each of the following pairs of values, state the decision that will occur and why.

a. p-value = 0.014, $\alpha = 0.02$

b. p-value = 0.118, $\alpha = 0.05$

c. p-value = 0.048, $\alpha = 0.05$

d. p-value = 0.064, $\alpha = 0.10$

8.94 For each of the following pairs of values, state the decision that will occur and why.

a. p-value = 0.018, $\alpha = 0.01$

b. p-value = 0.033, $\alpha = 0.05$

c. p-value = 0.078, $\alpha = 0.05$

d. p-value = 0.235, $\alpha = 0.10$

8.95 The calculated p-value for a hypothesis test is 0.084. What decision about the null hypothesis would occur in the following:

a. The hypothesis test is completed at the 0.05 level of significance.

b. The hypothesis test is completed at the 0.10 level of significance.

8.96 a. A one-tailed hypothesis test is to be completed at the 0.05 level of significance. What calculated values of p will cause a rejection of H_o?

b. A two-tailed hypothesis test is to be completed at the 0.02 level of significance. What calculated values of p will cause a "fail to reject H_o" decision?

8.97 Skillbuilder Applet Exercise estimates the p-value for a one-tailed hypothesis test by simulating the taking of many samples. The given hypothesis test is for an H_o: $\mu = 1500$ versus H_a: $\mu < 1500$. A

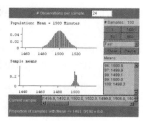

sample of 24 has been taken and the sample mean is 1451.

a. Click "10" for "# of samples." Note the sample means and the probability of being less than 1451 if the true mean is really 1500.

b. Change to "Batch" and simulate 1000 more samples. What is the probability of being less than 1451? This is your estimated p-value.

c. How does your estimated p-value show on the histogram formed from the taking of many samples? Explain what this p-value means with respect to the test.

d. If the level of significance were 0.01, what would your decision be?

8.98 Skillbuilder Applet Exercise estimates the p-value for a two-tailed hypothesis test by simulating the taking of many samples. The given hypothesis test is for H_o: $\mu = 4$ versus H_a: $\mu \neq 4$. A sample of 100 has been taken and the sample mean is 3.6.

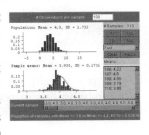

a. Click "10" for "# of samples." Note the sample means and the probability of being less than 3.6 or greater than 4.4. Why are we including the greater than 4.4?

b. Change to "Batch" and simulate 1000 more samples. What is the probability of being less than 3.6 or greater than 4.4? This is your estimated p-value.

c. How does your estimated p-value show on the histogram formed from the taking of many samples? Explain what this p-value means with respect to the test.

d. If the level of significance were 0.05, what would your decision be?

8.99 Describe in your own words what the p-value measures.

8.100 a. Calculate the p-value, given H_a: $\mu < 45$ and $z\star = -2.3$.

b. Calculate the p-value, given H_a: $\mu > 58$ and $z\star = 1.8$.

8.101 Calculate the p-value, given H_a: $\mu \neq 245$ and $z\star = 1.1$.

8.102 Find the test statistic $z\star$ and the p-value for each of the following situations.

a. H_o: $\mu = 22.5$, H_a: $\mu > 22.5$; $\bar{x} = 24.5$, $\sigma = 6$, $n = 36$

b. H_o: $\mu = 200$, H_a: $\mu < 200$; $\bar{x} = 192.5$, $\sigma = 40$, $n = 50$

c. H_o: $\mu = 12.4$, H_a: $\mu \neq 2.4$; $\bar{x} = 11.52$, $\sigma = 2.2$, $n = 16$

8.103 Calculate the p-value for each of the following:

a. H_o: $\mu = 10$, H_a: $\mu > 10$, $z\star = 1.48$

b. H_o: $\mu = 105$, H_a: $\mu < 105$, $z\star = -0.85$

c. H_o: $\mu = 13.4$, H_a: $\mu \neq 13.4$, $z\star = 1.17$

d. H_o: $\mu = 8.56$, H_a: $\mu < 8.56$, $z\star = -2.11$

e. H_o: $\mu = 110$, H_a: $\mu \neq 110$, $z\star = -0.93$

8.104 Calculate the p-value for each of the following:

a. H_o: $\mu = 20$, H_a: $\mu < 20$; $\bar{x} = 17.8$, $\sigma = 9$, $n = 36$

b. H_o: $\mu = 78.5$, H_a: $\mu > 78.5$; $\bar{x} = 79.8$, $\sigma = 15$, $n = 100$

c. H_o: $\mu = 1.587$, H_a: $\mu \neq 1.587$; $\bar{x} = 1.602$, $\sigma = 0.15$, $n = 50$

8.105 Find the value of $z\star$ for each of the following:

a. H_o: $\mu = 35$ versus H_a: $\mu > 35$ when p-value $= 0.0582$

b. H_o: $\mu = 35$ versus H_a: $\mu < 35$ when p-value $= 0.0166$

c. H_a: $\mu = 35$ versus H_a: $\mu \neq 35$ when p-value $= 0.0042$

8.106 The null hypothesis, H_o: $\mu = 48$, was tested against the alternative hypothesis, H_a: $\mu > 48$. A sample of 75 resulted in a calculated p-value of 0.102. If $\sigma = 3.5$, find the value of the sample mean, $\bar{x}$.

8.107 The null hypothesis, H_o: $\mu = 16$, was tested against the alternative hypothesis, H_a: $\mu < 16$. A sample of 50 resulted in a calculated p-value of 0.017. If $\bar{x} = 14$, find the value of the population standard deviation.

8.108 Using the MINITAB solution to the rivet example as shown on page 439, describe how MINITAB found each of the six numerical values it reported as results.

8.109 The following computer output was used to complete a hypothesis test.

```
TEST OF MU = 525.00 VS MU < 525.00
THE ASSUMED SIGMA = 60.0
N     MEAN     STDEV    SE MEAN    Z       P VALUE
38    512.14   64.78    9.733     -1.32    0.093
```

a. State the null and alternative hypotheses.

b. If the test is completed using $\alpha = 0.05$, what decision and conclusion are reached?

c. Verify the value of the standard error of the mean.

8.110 Using the computer output and information in Exercise 8.109, determine the value of the following:

a. Hypothesized value of population mean

b. Sample mean

c. Population standard deviation

d. Test statistic

8.111 The following computer output was used to complete a hypothesis test.

```
TEST OF MU = 6.250 VS MU not = 6.250
THE ASSUMED SIGMA = 1.40
N     MEAN    STDEV    SE MEAN    Z      P VALUE
78    6.596   1.273    0.1585    2.18    0.029
```

a. State the null and alternative hypotheses.

b. If the test is completed using $\alpha = 0.05$, what decision and conclusion are reached?

c. Verify the value of the standard error of the mean.

d. Find the values for Σx and Σx^2.

8.112 Using the computer output and information in Exercise 8.111, determine the value of the following:

a. Hypothesized value of population mean

b. Sample mean

c. Population standard deviation

d. Test statistic

8.113 According to the *USA Today* article "Laptops inching closer to PCs in popularity" (February 6, 2005), the desktop PC era is ending and the age of laptops has begun. A study by researcher Current Analysis found that laptop prices fell to an average of $1211 during the 2004 December holidays. The mean purchase price was thought to be $1240. The researcher's finding is based on the purchase price for a sample of 35 recently purchased laptops. Assuming that $\sigma = \$66.75$, does the researcher have sufficient evidence to make his claim or could his sample results have realistically occurred by chance?

a. Describe the parameter of interest.

b. State the null and alternative hypotheses.

c. Calculate the value for $z\star$ and find the *p*-value.

d. State your decision and conclusion using $\alpha = 0.001$.

8.114 One of the best indicators of a baby's health is his or her weight at birth. In the United States, mothers who live in poverty generally have babies with lower birth weights than those who do not live in poverty. Although the average birth weight for babies born in the United States is approximately 3300 grams, the birth weight for babies of women living in poverty is 2800 grams with a standard deviation of 500 grams. Recently, a local hospital introduced an innovative new prenatal care program to reduce the number of low-birth-weight babies born in the hospital. At the end of the first year, the birth weights of 25 randomly selected babies were collected; all of the babies were born to women who lived in poverty and participated in the program. Their mean birth weight was 3075 grams. The question posed to you, the researcher, is, "Has there been a significant improvement in the birth weights of babies born to poor women?" Use $\alpha = 0.02$.

Source: http://www.ccnmtl.columbia.edu/projects/qmss/ t_one.html

a. Define the parameter.

b. State the null and alternative hypotheses.

c. Specify the hypothesis test criteria.

d. Present the sample evidence.

e. Find the probability distribution information.

f. Determine the results.

8.115 The owner of a local chain of grocery stores is always trying to minimize the time it takes her customers to check out. In the past, she has conducted many studies of the checkout times, and they have displayed a normal distribution with a mean time of 12 minutes and a standard deviation of 2.3 minutes. She has implemented a new schedule for cashiers in hopes of reducing the mean checkout time. A random sample of 28 customers visiting her store this week resulted in a mean of 10.9 minutes. Does she have sufficient evidence to claim the mean checkout time this week was less than 12 minutes? Use $\alpha = 0.02$.

8.116 The average size of a home in 2003 was 2320 square feet according to the National Association of Home Builders and reported in *USA Today* Snapshots (November 27, 2004). The homebuilders of a northeastern city believe that the average size of homes continues to increase each year. To test their claim, a random sample of 45 new homes was selected; they revealed an average of 2490 square feet. Assuming that the population standard deviation is approximately 450 square feet, is there evidence that the average size has increased since the 2003 figure? Use a 0.05 level of significance.

8.117 From candy to jewelry to flowers, the average consumer was expected to spend $104.63 for Mother's Day in 2005, according to the *Democrat & Chronicle* article "Mom's getting more this year" (May 7, 2005). Local merchants thought this average was too high for their area. They contracted an agency to conduct a study. A random sample of 60 consumers was taken at a local shopping mall the Saturday before Mother's Day and produced a sample mean amount of $94.27. If $\sigma = \$29.50$, does the sample provide sufficient evidence to support the merchants' claim at the 0.05 level of significance?

8.118 Imagine that you are a customer living in the shopping area described in Exercise 8.117 and you need to buy a Mother's Day gift. Identify the four possible outcomes and describe the situation involved with each outcome with regard to the average amount spent on a Mother's Day gift. Which is the more serious error: the type I or type II error? Explain.

8.119 Who says that the more you spend on a wristwatch, the more accurately the watch will keep time? Some say that you can now buy a quartz watch for less than $25 and that it keeps time just as accurately as watches that cost four times as much. Suppose the average accuracy for all watches being sold today, regardless of price, is within 19.8 seconds per month with a standard deviation of 9.1 seconds. A random sample of 36 quartz watches priced less than $25 is taken, and their accuracy check reveals a sample mean error of 22.7 seconds per month. Based on this evidence, complete the hypothesis test of H_o: $\mu = 20$ vs. H_a: $\mu > 20$ at the 0.05 level of significance using the probability-value approach.

a. Define the parameter.

b. State the null and alternative hypotheses.

c. Specify the hypothesis test criteria.

d. Present the sample evidence.

e. Find the probability distribution information.

f. Determine the results.

8.120 [EX08-120] The National Health and Nutrition Examination Survey (NHANES) indicates that more U.S. adults are becoming either overweight or obese, which is defined as having a body mass index (BMI) of 25 or more. Data from the Centers for Disease Control and Prevention (CDC) indicate that for females aged 35 to 55, the mean BMI is 25.12 with a standard deviation of 5.3. In a similar study that examined female cardiovascular technologists who were registered in the United States and were within the same age range, the following BMI scores resulted:

22	28	26	19

••• Remainder of data on Student's Suite CD-ROM

Source: "An Assessment of Cardiovascular Risk Behaviors of Registered Cardiovascular Technologists," Dissertation by Dr. Susan Wambold, University of Toledo, 2002

Test the claim that the cardiovascular technologists have a lower average BMI than the general population. Use $\alpha = 0.05$.

a. Describe the parameter of interest.

b. State the null and alternative hypotheses.

c. Calculate the value for $z\star$ and find the *p*-value.

d. State your decision and conclusion using $\alpha = 0.05$.

8.121 Use a computer or calculator to select 40 random single-digit numbers. Find the sample mean, $z\star$, and p-value for testing H_o: $\mu = 4.5$ against a two-tailed alternative. Repeat several times as in Table 8.8. Describe your findings.

FYI Use commands for generating integer data on page 101; then continue with hypothesis test commands on pages 438–439.

8.122 Use a computer or calculator to select 36 random numbers from a normal distribution with mean 100 and standard deviation 15. Find the sample mean, $z\star$, and p-value for testing a two-tailed hypothesis test of $\mu = 100$. Repeat several times as in Table 8.8. Describe your findings.

FYI Use commands for generating data on pages 327–328, then continue with hypothesis test commands on pages 438–439.

8.6 Hypothesis Test of Mean μ (σ Known): A Classical Approach

In Section 8.4 we surveyed the concepts and much of the reasoning behind a hypothesis test while looking at nonstatistical examples. In this section we are going to formalize the hypothesis test procedure as it applies to statements concerning the mean μ of a population under the restriction that σ, the population standard deviation, is a known value.

> **The assumption for hypothesis tests about mean μ using a known σ:** The sampling distribution of $\bar{x}$ has a normal distribution.

The information we need to ensure that this assumption is satisfied is contained in the sampling distribution of sample means and in the central limit theorem (see Chapter 7, pp. 369–370).

> The sampling distribution of sample means $\bar{x}$ is distributed about a mean equal to μ with a standard error equal to $\sigma/\sqrt{n}$; and (1) if the randomly sampled population is normally distributed, then $\bar{x}$ is normally distributed for all sample sizes, or (2) if the randomly sampled population is not normally distributed, then $\bar{x}$ is approximately normally distributed for sufficiently large sample sizes.

The hypothesis test is a well-organized, step-by-step procedure used to make a decision. Two different formats are commonly used for hypothesis testing. The *classical approach* is the hypothesis test process that has enjoyed popularity for many years. This approach is organized as a five-step procedure.

> **THE CLASSICAL HYPOTHESIS TEST: A FIVE-STEP PROCEDURE**
>
> **Step 1   The Set-Up:**
>   a. Describe the population parameter of interest.
>   b. State the null hypothesis (H_o) and the alternative hypothesis (H_a).

Step 2 **The Hypothesis Test Criteria:**
   a. Check the assumptions.
   b. Identify the probability distribution and the test statistic to be used.
   c. Determine the level of significance, α.

Step 3 **The Sample Evidence:**
   a. Collect the sample information.
   b. Calculate the value of the test statistic.

Step 4 **The Probability Distribution:**
   a. Determine the critical region and critical value(s).
   b. Determine whether or not the calculated test statistic is in the critical region.

Step 5 **The Results:**
   a. State the decision about H_o.
   b. State the conclusion about H_a.

A commercial aircraft manufacturer buys rivets to use in assembling airliners. Each rivet supplier that wants to sell rivets to the aircraft manufacturer must demonstrate that its rivets meet the required specifications. One of the specs is "The mean shearing strength of all such rivets, μ, is at least 925 lb." Each time the aircraft manufacturer buys rivets, it is concerned that the mean strength might be less than the 925-lb specification.

Note: Each individual rivet has a shearing strength, which is determined by measuring the force required to shear ("break") the rivet. Clearly, not all the rivets can be tested. Therefore, a sample of rivets will be tested, and a decision about the mean strength of all the untested rivets will be based on the mean from those sampled and tested.

STEP 1 **The Set-Up:**
   a. **Describe the population parameter of interest.**
       The population parameter of interest is the mean μ, the mean shearing strength (or mean force required to shear) of the rivets being considered for purchase.
   b. **State the null hypothesis (H_o) and the alternative hypothesis (H_a).**
       The null hypothesis and the alternative hypothesis are formulated by inspecting the problem or statement to be investigated and first formulating two opposing statements about the mean μ. For our example, these two opposing statements are: (A) "The mean shearing strength is less than 925" ($\mu < 925$, the aircraft manufacturer's concern), and (B) "The mean shearing strength is at least 925" ($\mu = 925$, the rivet supplier's claim and the aircraft manufacturer's spec).

Note: The trichotomy law from algebra states that two numerical values must be related in exactly one of three possible relationships: $<$, $=$, or $>$. All three

FYI More specific instructions are given on pages 417–418.

of these possibilities must be accounted for in the two opposing hypotheses in order for the two hypotheses in order to be negations of each other. The three possible combinations of signs and hypotheses are shown in Table 8.9. Recall that the null hypothesis assigns a specific value to the parameter in question, and therefore "equals" will always be part of the null hypothesis.

TABLE 8.9

The Three Possible Statements of Null and Alternative Hypotheses

Null Hypothesis	Alternative Hypothesis
1. Greater than or equal to ($\geq$)	Less than ($<$)
2. Less than or equal to ($\leq$)	Greater than ($>$)
3. Equal to ($=$)	Not equal to ($\neq$)

The parameter of interest, the population mean μ, is related to the value 925. Statement (A) becomes the alternative hypothesis:

$$H_a: \mu < 925 \text{ (the mean is less than 925)}$$

This statement represents the aircraft manufacturer's concern and says, "The rivets do not meet the required specs." Statement (B) becomes the null hypothesis:

$$H_o: \mu = 925 \ (\geq) \text{ (the mean is at least 925)}$$

This hypothesis represents the negation of the aircraft manufacturer's concern and says, "The rivets do meet the required specs."

Note: We will write the null hypothesis with just the equal sign, thereby stating the exact value assigned. When "equal" is paired with "less than" or paired with "greater than," the combined symbol is written beside the null hypothesis as a reminder that all three signs have been accounted for in these two opposing statements.

Before continuing with our example, let's look at three examples that demonstrate formulating the statistical null and alternative hypotheses involving population mean μ. Examples 8.18 and 8.19 each demonstrate a "one-tailed" alternative hypothesis.

EXAMPLE 8.18

Writing Null and Alternative Hypotheses (One-Tailed Situation)

A consumer advocate group would like to disprove a car manufacturer's claim that a specific model will average 24 miles per gallon of gasoline. Specifically, the group would like to show that the mean miles per gallon is considerably less than 24. State the null and alternative hypotheses.

SOLUTION   To state the two hypotheses, we first need to identify the population parameter in question: the "mean mileage attained by this car model." The parameter μ is being compared with the value 24 miles per gallon, the specific value of interest. The advocates are questioning the value of μ and wish to show it to be less than 24 (i.e., $\mu < 24$). There are three possible relationships: (1) $\mu < 24$, (2) $\mu = 24$, and (3) $\mu > 24$. These three cases must be arranged to form two opposing statements: one states what the advocates are trying to show, "The mean level is less than 24 ($\mu < 24$)," whereas the "negation" is "The mean level is not less than 24 ($\mu \geq 24$)." One of these two statements will become the null hypothesis H_o, and the other will become the alternative hypothesis H_a.

Note: Recall that there are two rules for forming the hypotheses: (1) the null hypothesis states that the parameter in question has a specified value ("H_o must contain the equal sign"), and (2) the consumer advocate group's contention becomes the alternative hypothesis ("less than"). Both rules indicate:

$$H_o: \mu = 24 \ (\geq) \quad \text{and} \quad H_a: \mu < 24$$

EXAMPLE 8.19   **Writing Null and Alternative Hypotheses (One-Tailed Situation)**

Suppose the EPA is suing a large manufacturing company for not meeting federal emissions guidelines. Specifically, the EPA is claiming that the mean amount of sulfur dioxide in the air is dangerously high, higher than 0.09 part per million. State the null and alternative hypotheses for this test situation.

SOLUTION   The parameter of interest is the mean amount of sulfur dioxide in the air, and 0.09 part per million is the specified value. $\mu > 0.09$ corresponds to "The mean amount is greater than 0.09," whereas $\mu \leq 0.09$ corresponds to the negation, "The mean amount is not greater than 0.09." Therefore, the hypotheses are

$$H_o: \mu = 0.09 \ (\leq) \quad \text{and} \quad H_a: \mu > 0.09$$

Example 8.20 demonstrates a "two-tailed" alternative hypothesis.

EXAMPLE 8.20   **Writing Null and Alternative Hypotheses (Two-Tailed Situation)**

Job satisfaction is very important to worker productivity. A standard job-satisfaction questionnaire was administered by union officers to a sample of assembly-line workers in a large plant in hopes of showing that the assembly workers' mean score on this questionnaire would be different from the established mean of 68. State the null and alternative hypotheses.

SOLUTION   Either the mean job-satisfaction score is different from 68 ($\mu \neq 68$) or the mean score is equal to 68 ($\mu = 68$). Therefore,

$$H_o: \mu = 68 \quad \text{and} \quad H_a: \mu \neq 68$$

Notes:

1. The alternative hypothesis is referred to as being "two-tailed" when H_a is "not equal."
2. When "less than" is combined with "greater than," they become "not equal to."

The viewpoint of the experimenter greatly affects the way the hypotheses are formed. Generally, the experimenter is trying to show that the parameter value is different from the value specified. Thus, the experimenter is often hoping to be able to reject the null hypothesis so that the experimenter's theory has been substantiated. Examples 8.18, 8.19, and 8.20 also represent the three possible arrangements for the $<$, $=$, and $>$ relationships between the parameter μ and a specified value.

Table 8.10 lists some additional common phrases used in claims and indicates the phrase of its negation and the hypothesis in which each phrase will be used. Again, notice that "equals" is always in the null hypothesis. Also notice that the negation of "less than" is "not less than," which is equivalent to "greater than or equal to." Think of negation of one sign as the other two signs combined.

TABLE 8.10

Common Phrases and Their Negations

$H_o: (\geq)$	vs.	$H_a: (<)$	$H_o: (\leq)$	vs.	$H_a: (>)$	$H_o: (=)$	vs.	$H_a: (\neq)$
At least		Less than	At most		More than	Is		Is not
No less than		Less than	No more than		More than	Not different from		Different from
Not less than		Less than	Not greater than		Greater than	Same as		Not same as

After the null and alternative hypotheses are established, we will work under the assumption that the null hypothesis is a true statement until there is sufficient evidence to reject it. This situation might be compared with a courtroom trial, where the accused is assumed to be innocent (H_o: Defendant is innocent vs. H_a: Defendant is not innocent) until sufficient evidence has been presented to show that innocence is totally unbelievable ("beyond reasonable doubt"). At the conclusion of the hypothesis test, we will make one of two possible decisions. We will decide in opposition to the null hypothesis and say that we "reject H_o" (this corresponds to "conviction" of the accused in a trial), or we will decide in agreement with the null hypothesis and say that we "fail to reject H_o" (this corresponds to "fail to convict" or an "acquittal" of the accused in a trial).

Let's return to the rivet example we interrupted on page 446 and continue with Step 2. Recall that

$$H_o: \mu = 925 \ (\geq) \ \text{(at least 925)} \quad H_a: \mu < 925 \ \text{(less than 925)}$$

STEP 2 **The Hypothesis Test Criteria:**
  a. Check the assumptions.
    Assume that the standard deviation of the shearing strength of rivets is known from past experience to be $\sigma = 18$. Variables like

shearing strength typically have a mounded distribution; therefore, a sample of size 50 should be large enough for the CLT to satisfy the assumption; the SDSM is normally distributed.

b. Identify the probability distribution and the test statistic to be used.

The standard normal probability distribution is used because $\bar{x}$ is expected to have a normal or approximately normal distribution.

For a hypothesis test of μ, we want to compare the value of the sample mean with the value of the population mean as stated in the null hypothesis. This comparison is accomplished using the test statistic in formula (8.4):

Test Statistic for Mean

$$z\star = \frac{\bar{x} - \mu}{\sigma/\sqrt{n}} \tag{8.4}$$

The resulting calculated value is identified as $z\star$ ("z star") because it is expected to have a standard normal distribution when the null hypothesis is true and the assumptions have been satisfied. The $\star$ ("star") is to remind us that this is the calculated value of the test statistic.

The test statistic to be used is $z\star = \dfrac{\bar{x} - \mu}{\sigma/\sqrt{n}}$.

c. Determine the level of significance, α.

Setting α was described as a managerial decision in Section 8.4. To see what is involved in determining α, the probability of the type I error, for our rivet example, we start by identifying the four possible outcomes, their meaning, and the action related to each.

The type I error occurs when a true null hypothesis is rejected. This would occur when the manufacturer tested rivets that did meet the specs and rejected them. Undoubtedly this would lead to the rivets not being purchased even though they did meet the specs. In order for the manager to set a level of significance, related information is needed—namely, how soon is the new supply of rivets needed? If they are needed tomorrow and this is the only vendor with an available supply, waiting a week to find acceptable rivets could be very expensive; therefore, rejecting good rivets could be considered a serious error. On the other hand, if the rivets are not needed until next month, then this error may not be very serious. Only the manager will know all the ramifications, and therefore the manager's input is important here.

FYI There is more to this scenario, but we hope you get the idea.

FYI α will be assigned in the statement of exercises.

After much consideration, the manager assigns the level of significance: $\alpha = 0.05$.

STEP 3 The Sample Evidence:

a. Collect the sample information.

We are ready for the data. The sample must be a random sample drawn from the population whose mean μ is being questioned. A random sample of 50 rivets is selected, each rivet is tested, and the sample mean shearing strength is calculated: $\bar{x} = 921.18$ and $n = 50$.

b. **Calculate the value of the test statistic.**

The sample evidence ($\bar{x}$ and n found in Step 3a) is next converted into the calculated value of the test statistic, $z\star$, using formula (8.4). (μ is 925 from H_o, and $\sigma = 18$ is the known quantity). We have

$$z\star = \frac{\bar{x} - \mu}{\sigma/\sqrt{n}}; \quad z\star = \frac{921.18 - 925.0}{18/\sqrt{50}} = \frac{-3.82}{2.5456} = -1.50$$

STEP 4   The Probability Distribution:

a. **Determine the critical region and critical value(s).**

The standard normal variable z is our test statistic for this hypothesis test.

> **Critical region:** The set of values for the test statistic that will cause us to reject the null hypothesis. The set of values that are not in the critical region is called the **noncritical region** (sometimes called the *acceptance region*).

Recall that we are working under the assumption that the null hypothesis is true. Thus, we are assuming that the mean shearing strength of all rivets in the sampled population is 925. If this is the case, then when we select a random sample of 50 rivets, we can expect this sample mean, $\bar{x}$, to be part of a normal distribution that is centered at 925 and to have a standard error of $\sigma/\sqrt{n} = 18/\sqrt{50}$, or approximately 2.55. Approximately 95% of the sample mean values will be greater than 920.8 (a value 1.65 standard errors below the mean: $925 - (1.65)(2.55) = 920.8$). Thus, if H_o is true and $\mu = 925$, then we expect $\bar{x}$ to be greater than 920.8 approximately 95% of the time and less than 920.8 only 5% of the time.

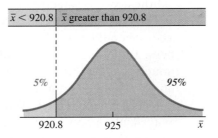

If, however, the value of $\bar{x}$ that we obtain from our sample is less than 920.8—say, 919.5—we will have to make a choice. It could be that either: (A) such an $\bar{x}$ value (919.5) is a member of the sampling distribution with mean 925 although it has a very low probability of occurrence (less than 0.05), or (B) $\bar{x} = 919.5$ is a member of a sampling distribution whose mean is less than 925, which would make it a value that is more likely to occur.

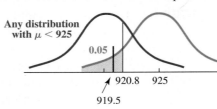

In statistics, we "bet" on the "more probable to occur" and consider the second choice (B) to be the right one. Thus, the left-hand tail of the z-distribution becomes the critical region, and the level of significance α becomes the measure of its area.

FYI Information about the critical value notation, $z(\alpha)$, is given on pages 338–340.

FYI Shading will be used to identify the critical region.

> **Critical value(s):** The "first" or "boundary" value(s) of the critical region(s).

The critical value for our example is $-z(0.05)$ and has the value of -1.65, as found in Table 4A in Appendix B.

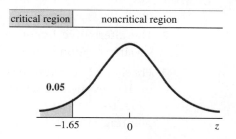

b. **Determine whether or not the calculated test statistic is in the critical region.**

Graphically this determination is shown by locating the value for $z\star$ on the sketch in Step 4a.

The calculated value of z, $z\star = -1.50$, is **not in the critical region** (it is in the unshaded portion of the figure).

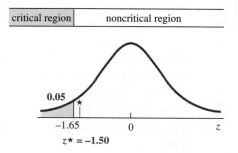

STEP 5 **The Result:**

a. **State the decision about H_o.**

In order to make the decision, we need to know the *decision rule*.

> **Decision rule:**
> a. If the test statistic falls *within the critical region*, then the decision must be **reject H_o.** (The critical value is part of the critical region.)
> b. If the test statistic is *not in the critical region*, then the decision must be **fail to reject H_o.**

The decision is: Fail to reject H_o.

FYI Specific information about writing the conclusion is given on page 422.

b. **State the conclusion about H_a.**

There is not sufficient evidence at the 0.05 level of significance to show that the rivets have a mean shearing strength less than 925. "We failed to convict" the null hypothesis. In other words, a sample mean as small as 921.18 is not unlikely to occur (as defined by α) when the true population mean value is 925.0. Therefore, the resulting action would be to buy the rivets.

Before we look at another example, let's summarize briefly some of the details we have seen thus far:

1. The null hypothesis specifies a particular value of a population parameter.

2. The alternative hypothesis can take three forms. Each form dictates a specific location of the critical region(s), as shown in the following table.

3. For many hypothesis tests, the sign in the alternative hypothesis "points" in the direction in which the critical region is located. (Think of the not equal to sign [$\neq$] as being both less than [$<$] and greater than [$>$], thus pointing in both directions.)

	Sign in the Alternative Hypothesis		
	$<$	$\neq$	$>$
Critical Region	One region Left side **One-tailed test**	Two regions Half on each side **Two-tailed test**	One region Right side **One-tailed test**

The value assigned to α is called the *significance level* of the hypothesis test. Alpha cannot be interpreted to be anything other than the risk (or probability) of rejecting the null hypothesis when it is actually true. We will seldom be able to determine whether the null hypothesis is true or false; we will decide only to "reject H_o" or to "fail to reject H_o." The relative frequency with which we reject a true hypothesis is α, but we will never know the relative frequency with which we make an error in decision. The two ideas are quite different; that is, a type I error and an error in decision are two different things altogether. Remember that there are two types of errors: type I and type II.

Let's look at another hypothesis test, one involving the two-tailed procedure.

EXAMPLE 8.21   **Two-Tailed Hypothesis Test**

Statistics⬡Now™

Watch a video example at
http://1pass.thomson.com
or on your CD.

It has been claimed that the mean weight of female students at a college is 54.4 kg. Professor Hart does not believe the claim and sets out to show that the mean weight is not 54.4 kg. To test the claim he collects a random sample of 100 weights from among the female students. A sample mean of 53.75 kg results. Is this sufficient evidence for Professor Hart to reject the statement? Use $\alpha = 0.05$ and $\sigma = 5.4$ kg.

SOLUTION

STEP 1 **The Set-Up:**

a. **Describe the population parameter of interest.**

The population parameter of interest is the mean μ, the mean weight of all female students at the college.

b. **State the null hypothesis (H_o) and the alternative hypothesis (H_a).**

The mean weight is equal to 54.4 kg, or the mean weight is not equal to 54.4 kg.

H_o: $\mu = 54.4$ (mean weight is 54.4)

H_a: $\mu \neq 54.4$ (mean weight is not 54.4)

(Remember: $\neq$ is $<$ and $>$ together.)

STEP 2 **The Hypothesis Test Criteria:**

a. **Check the assumptions.**

σ is known. The weights of an adult group of women are generally approximately normally distributed; therefore, a sample of $n = 100$ is large enough to allow the CLT to apply.

b. **Identify the probability distribution and the test statistic to be used.**

The standard normal probability distribution and the test statistic $z\star = \dfrac{\bar{x} - \mu}{\sigma/\sqrt{n}}$ will be used; $\sigma = 5.4$.

c. **Determine the level of significance, α.**

$\alpha = 0.05$ (given in the statement of problem).

STEP 3 **The Sample Evidence:**

a. **Collect the sample information:** $\bar{x} = 53.75$ and $n = 100$.

b. **Calculate the value of the test statistic.**

Use formula (8.4), information from H_o: $\mu = 54.4$, and $\sigma = 5.4$ (known):

$$z\star = \frac{\bar{x} - \mu}{\sigma/\sqrt{n}}: \quad z\star = \frac{53.75 - 54.4}{5.4/\sqrt{100}} = \frac{-0.65}{0.54} = -1.204 = -1.20$$

STEP 4 **The Probability Distribution:**

a. **Determine the critical region and critical value(s).**

The critical region is both the left tail and the right tail because both smaller and larger values of the sample mean suggest that the null hypothesis is wrong. The level of significance will be split in half, with 0.025 being the measure of each tail. The critical values are found in Table 4B in Appendix B: $\pm z(0.025) = \pm 1.96$. (Table 4B instructions are on page 403.)

b. **Determine whether or not the calculated test statistic is in the critical region.**

The calculated value of z, $z\star = -1.20$, is not in the critical region (shown in red on the adjacent figure).

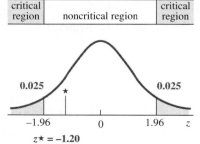

STEP 5 **The Results:**

a. **State the decision about H_o:** Fail to reject H_o.

b. State the conclusion about H_a.

There is not sufficient evidence at the 0.05 level of significance to show that the female students have a mean weight different from the 54.4 kg claimed. In other words, there is no statistical evidence to support Professor Hart's contentions.

EXAMPLE 8.22 | **Two-Tailed Hypothesis Test with Sample Data**

According to the results of Exercise 5.32 (p. 282), the mean of single-digit random numbers is 4.5 and the standard deviation is $\sigma = 2.87$. Draw a random sample of 40 single-digit numbers from Table 1 in Appendix B and test the hypothesis, "The mean of the single-digit numbers in Table 1 is 4.5." Use $\alpha = 0.10$.

SOLUTION

STEP 1   The Set-Up:
    a. Describe the population parameter of interest.

The parameter of interest is the mean μ of the population of single-digit numbers in Table 1 of Appendix B.

    b. State the null hypothesis (H_o) and the alternative hypothesis (H_a).

$$H_o: \mu = 4.5 \text{ (mean is 4.5)}$$
$$H_a: \mu \neq 4.5 \text{ (mean is not 4.5)}$$

STEP 2   The Hypothesis Test Criteria:
    a. Check the assumptions.

σ is known. Samples of size 40 should be large enough to satisfy the CLT; see the discussion of this issue on page 426.

    b. Identify the probability distribution and the test statistic to be used.

We use the standard normal probability distribution and the test statistic $z\star = \dfrac{\bar{x} - \mu}{\sigma/\sqrt{n}}$; $\sigma = 2.87$.

    c. Determine the level of significance, α.

$\alpha = 0.10$ (given in the statement of problem).

STEP 3   The Sample Evidence:
    a. Collect the sample information.

This random sample was drawn from Table 1 in Appendix B.

TABLE 8.11

Random Sample of Single-Digit Numbers [TA08-01]

2	8	2	1	5	5	4	0	9	1
0	4	6	1	5	1	1	3	8	0
3	6	8	4	8	6	8	9	5	0
1	4	1	2	1	7	1	7	9	3

The sample statistics are $\bar{x} = 3.975$ and $n = 40$.

b. Calculate the value of the test statistic.

Use formula (8.4), information from H_o: $\mu = 4.5$, and $\sigma = 2.87$:

$$z\star = \frac{\bar{x} - \mu}{\sigma/\sqrt{n}}: \quad z\star = \frac{3.975 - 4.50}{2.87/\sqrt{40}} = \frac{-0.525}{0.454} = -1.156 = -1.16$$

STEP 4   The Probability Distribution:

a. Determine the critical region and critical value(s).

A two-tailed critical region will be used, and 0.05 will be the area in each tail. The critical values are $\pm z(0.05) = \pm 1.65$.

b. Determine whether or not the calculated test statistic is in the critical region.

The calculated value of z, $z\star = -1.16$, is not in the critical region (shown in red on the figure).

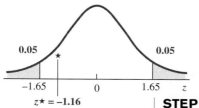

STEP 5   The Result:

a. State the decision about H_o: Fail to reject H_o.

b. State the conclusion about H_a.

The observed sample mean is not significantly different from 4.5 at the 0.10 significance level.

Suppose we were to take another sample of size 40 from Table 1. Would we obtain the same results? Suppose we took a third sample and a fourth. What results might we expect? What is the level of significance? Yes, its value is 0.10, but what does it measure? Table 8.12 lists the means obtained from 20 different random samples of size 40 that were taken from Table 1 in Appendix B. The calculated value of $z\star$ that corresponds to each $\bar{x}$ and the decision each would dictate are also listed. The 20 calculated z-scores are shown in Figure 8.10. Note that 3 of the 20 samples (or 15%) caused us to reject the null hypothesis, even though we know the null hypothesis is true for this situation. Can you explain this?

TABLE 8.12

Twenty Random Samples of Size 40 Taken from Table 1 in Appendix B [TA08-12]

Sample Number	Sample Mean, $\bar{x}$	Calculated z, $z\star$	Decision Reached	Sample Number	Sample Mean, $\bar{x}$	Calculated z, $z\star$	Decision Reached
1	4.62	+0.26	Fail to reject H_o	11	4.70	+0.44	Fail to reject H_o
2	4.55	+0.11	Fail to reject H_o	12	4.88	+0.83	Fail to reject H_o
3	4.08	−0.93	Fail to reject H_o	13	4.45	−0.11	Fail to reject H_o
4	5.00	+1.10	Fail to reject H_o	14	3.93	−1.27	Fail to reject H_o
5	4.30	−0.44	Fail to reject H_o	15	5.28	+1.71	Reject H_o
6	3.65	−1.87	Reject H_o	16	4.20	−0.66	Fail to reject H_o
7	4.60	+0.22	Fail to reject H_o	17	3.48	−2.26	Reject H_o
8	4.15	−0.77	Fail to reject H_o	18	4.78	+0.61	Fail to reject H_o
9	5.05	+1.21	Fail to reject H_o	19	4.28	−0.50	Fail to reject H_o
10	4.80	+0.66	Fail to reject H_o	20	4.23	−0.61	Fail to reject H_o

FIGURE 8.10
z-Scores from Table
8.12

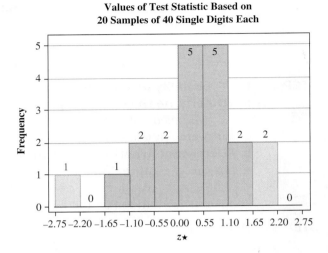

**Values of Test Statistic Based on
20 Samples of 40 Single Digits Each**

Note: Remember that α is the probability that we "reject H_o" when it is actually a true statement. Therefore, we can anticipate that a type I error will occur α of the time when testing a true null hypothesis. In the preceding empirical situation, we observed a 15% rejection rate. If we were to repeat this experiment many times, the proportion of samples that would lead to a rejection would vary, but the observed relative frequency of rejection should be approximately α or 10%.

SECTION 8.6 EXERCISES

8.123 In the example on page 445, the aircraft builder who is buying the rivets is concerned that the rivets might not meet the mean-strength spec. State the aircraft manufacturer's null and alternative hypotheses.

8.124 Professor Hart does not believe the statement "the mean distance commuted daily by the nonresident students at our college is no more than 9 miles." State the null and alternative hypotheses he would use to challenge this statement.

8.125 State the null and alternative hypotheses used to test each of the following claims:

a. The mean reaction time is less than 1.25 seconds.

b. The mean score on that qualifying exam is different from 335.

c. The mean selling price of homes in the area is no more than $230,000.

8.126 State the null hypothesis, H_o, and the alternative hypothesis, H_a, that would be used for a hypothesis test for each of the following statements:

a. The mean age of the youths who hang out at the mall is less than 16 years.

b. The mean height of professional basketball players is greater than 6'6''.

c. The mean elevation drop for ski trails at eastern ski centers is at least 285 feet.

d. The mean diameter of the rivets is no more than 0.375 inches.

e. The mean cholesterol level of male college students is different from 200 mg/dL.

8.127 Suppose you want to test the hypothesis that "the mean salt content of frozen 'lite' dinners is more than 350 mg per serving." An average of 350 mg is an acceptable amount of salt per serving; therefore, you use it as the standard. The null hypothesis is "The average content is not more than 350 mg" ($\mu = 350$). The alternative hypothesis is "The average content is more than 350 mg" ($\mu > 350$).

a. Describe the conditions that would exist if your decision results in a type I error.

b. Describe the conditions that would exist if your decision results in a type II error.

8.128 Identify the four possible outcomes and describe the situation involved with each outcome with regard to the aircraft manufacturer's testing and buying of rivets. Which is the more serious error: the type I or type II error? Explain.

8.129 Suppose you wanted to test the hypothesis that the mean minimum charge in your area for a home service call by a plumber is at most $85. Explain the conditions that would exist if you make an error in decision by committing a type I error. What about a type II error?

8.130 Describe how the null hypothesis in Example 8.21 is a "starting point" for the decision to be made about the mean weight of all female students at the college.

8.131 a. What is the critical region?

b. What is the critical value?

8.132 a. What decision is reached when the test statistic falls in the critical region?

b. What decision is reached when the test statistic falls in the noncritical region?

8.133 Because the size of the type I error can always be made smaller by reducing the size of the critical region, why don't we always choose critical regions that make α extremely small?

8.134 Calculate the test statistic $z\star$, given H_o: $\mu = 356$, $\sigma = 17$, $\bar{x} = 354.3$, and $n = 120$.

8.135 Find the critical region and value(s) for H_a: $\mu < 19$ and $\alpha = 0.01$.

8.136 Find the critical region and value(s) for H_a: $\mu > 34$ and $\alpha = 0.02$.

8.137 Determine the critical region and critical values for z that would be used to test the null hypothesis at the given level of significance, as described in each of the following:

a. H_o: $\mu = 20$, H_a: $\mu \neq 20$, $\alpha = 0.10$

b. H_o: $\mu = 24$ ($\leq$), H_a: $\mu > 24$, $\alpha = 0.01$

c. H_o: $\mu = 10.5$ ($\geq$), H_a: $\mu < 10.5$, $\alpha = 0.05$

d. H_o: $\mu = 35$, H_a: $\mu \neq 35$, $\alpha = 0.01$

8.138 Determine the critical region and the critical values used to test the following null hypotheses:

a. H_o: $\mu = 55$ ($\geq$), H_a: $\mu < 55$, $\alpha = 0.02$

b. H_o: $\mu = -86$ ($\geq$), H_a: $\mu < -86$, $\alpha = 0.01$

c. H_o: $\mu = 107$, H_a: $\mu \neq 107$, $\alpha = 0.05$

d. H_o: $\mu = 17.4$ ($\leq$), H_a: $\mu > 17.4$, $\alpha = 0.10$

8.139 The null hypothesis, H_o: $\mu = 250$, was tested against the alternative hypothesis, H_a: $\mu < 250$. A sample of $n = 85$ resulted in a calculated test statistic of $z\star = -1.18$. If $\sigma = 22.6$, find the value of the sample mean, $\bar{x}$. Find the sum of the sample data, Σx.

8.140 Find the value of $\bar{x}$ for each of the following:

a. H_o: $\mu = 580$, $z\star = 2.10$, $\sigma = 26$, $n = 55$

b. H_o: $\mu = 75$, $z\star = -0.87$, $\sigma = 9.2$, $n = 35$

8.141 The calculated value of the test statistic is actually the number of standard errors that the sample mean differs from the hypothesized value of μ in the null hypothesis. Suppose that the null hypothesis is H_o: $\mu = 4.5$, σ is known to be 1.0, and a sample of size 100 results in $\bar{x} = 4.8$.

a. How many standard errors is $\bar{x}$ above 4.5?

b. If the alternative hypothesis is H_a: $\mu > 4.5$ and $\alpha = 0.01$, would you reject H_o?

8.142 Consider the hypothesis test where the hypotheses are $H_o: \mu = 26.4$ and $H_a: \mu < 26.4$. A sample of size 64 is randomly selected and yields a sample mean of 23.6.

a. If it is known that $\sigma = 12$, how many standard errors below $\mu = 26.4$ is the sample mean, $\bar{x} = 23.6$?

b. If $\alpha = 0.05$, would you reject H_o? Explain.

8.143 There are only two possible decisions as a result of a hypothesis test.

a. State the two possible decisions.

b. Describe the conditions that will lead to each of the two decisions identified in part a.

8.144 a. What proportion of the probability distribution is in the critical region, provided the null hypothesis is correct?

b. What error could be made if the test statistic falls in the critical region?

c. What proportion of the probability distribution is in the noncritical region, provided the null hypothesis is not correct?

d. What error could be made if the test statistic falls in the noncritical region?

8.145 The following computer output was used to complete a hypothesis test.

```
TEST OF MU = 15.0000 VS MU not = 15.0000
THE ASSUMED SIGMA = 0.50
N       MEAN      STDEV     SE MEAN      Z
30      15.6333   0.4270    0.0913       6.94
```

a. State the null and alternative hypotheses.

b. If the test is completed using $\alpha = 0.01$, what decision and conclusion are reached?

c. Verify the value of the standard error of the mean.

8.146 Using the computer output and information in Exercise 8.145, determine the value of the following:

a. Hypothesized value of population mean

b. Sample mean

c. Population standard deviation

d. Test statistic

8.147 The following computer output was used to complete a hypothesis test.

```
TEST OF MU = 72.00 VS MU > 72.00
THE ASSUMED SIGMA = 12.0
N       MEAN      STDEV     SE MEAN      Z
36      75.2      11.87     2.00         1.60
```

a. State the null and alternative hypotheses.

b. If the test is completed using $\alpha = 0.05$, what decision and conclusion are reached?

c. Verify the value of the standard error of the mean.

8.148 Using the computer output and information in Exercise 8.147, determine the value of the following:

a. Hypothesized value of population mean

b. Sample mean

c. Population standard deviation

d. Test statistic

8.149 The Texas Department of Health published the statewide results for the Emergency Medical Services Certification Examination. Data for those taking the paramedic exam for the first time gave an average score of 79.68 (out of a possible 100) with a standard deviation of 9.06. Suppose a random sample of 50 individuals taking the exam yielded a mean score of 81.05. Is there sufficient evidence to conclude that "the population from which this random sample was taken, on the average, scored higher than the state"? Use $\alpha = 0.05$.

8.150 According to the *Democrat & Chronicle* article "Millions forfeited in health accounts" (December 26, 2004), an estimated $210 million is forfeited each year because employees do not use all of the money that they contributed to their medical flexible spending accounts. The average amount put into these accounts annually is about $1000.

Suppose a random sample of 150 employees at a local business was taken in 2005 and an average amount of $925 was deposited into their flexible spending accounts. Test the hypothesis that there was no significant change in average flexible spending amounts despite the millions of dollars forfeited. Assume that $\sigma = \$307$ per year. Use $\alpha = 0.05$.

a. Define the parameter.

b. State the null and alternative hypotheses.

c. Specify the hypothesis test criteria.

d. Present the sample evidence.

e. Find the probability distribution information.

f. Determine the results.

8.151 The National Thoroughbred Racing Association's figures show that the average age of its 3 million committed fans, those who attend multiple races every year, is now 55 years old. Despite a $30 million a year promotional campaign by the association and its member tracks aimed at younger and female fans, their guess is that 10 years from now the average age of fans will be close to 65.

Source: http://money.cnn.com/2002/05/03/commentary/column_sportsbiz/horse_racing/

Suppose a random sample of 35 patrons at Finger Lakes Race Track is taken and the sample mean is 52.7 years. If $\sigma = 8$ years, does the sample provide sufficient evidence to support the racing association's current average age figure? Use a 0.10 level of significance.

8.152 A fire insurance company thought that the mean distance from a home to the nearest fire department in a suburb of Chicago was at least 4.7 miles. It set its fire insurance rates accordingly. Members of the community set out to show that the mean distance was less than 4.7 miles. This, they thought, would convince the insurance company to lower its rates. They randomly identified 64 homes and measured the distance to the nearest fire department from each. The resulting sample mean was 4.4. If $\sigma = 2.4$ miles, does the sample show sufficient evidence to support the community's claim at the $\alpha = 0.05$ level of significance?

8.153 [EX08-153] The length of major league baseball games are approximately normally distributed and average 2 hours and 50.1 minutes, with

a standard deviation of 21.0 minutes. It has been claimed that New York Yankee baseball games last, on the average, longer than the games of the other major league teams. To test the truth of this statement, a sample of eight Yankee games was randomly identified and the "time of game" (in minutes) for each obtained:

| 199 | 196 | 202 | 213 | 187 | 169 | 169 | 188 |

Source: MLB.com

At the 0.05 level of significance, does this data show sufficient evidence to conclude that the mean time of Yankee baseball games is longer than that of other major league baseball teams?

8.154 [EX08-154] The manager at Air Express believes that the weights of packages shipped recently are less than those in the past. Records show that in the past, packages have had a mean weight of 36.5 lb and a standard deviation of 14.2 lb. A random sample of last month's shipping records yielded the following 64 data values:

32.1	41.5	16.1	8.9	36.2	12.3	28.4	40.4
45.5	15.2	26.5	13.3	23.5	33.7	18.3	16.3
15.4	39.7	50.3	14.8	44.4	47.7	45.8	52.3
48.4	10.4	59.9	5.5	6.7	17.1	20.0	28.1
48.1	29.5	22.9	47.8	24.8	20.1	40.1	12.6
24.3	43.3	32.4	57.7	42.9	36.7	15.5	46.4
51.3	38.6	39.4	27.1	55.7	37.7	39.4	55.5
26.9	15.7	32.3	47.8	33.2	29.1	31.1	34.5

Is this sufficient evidence to reject the null hypothesis in favor of the manager's claim? Use $\alpha = 0.01$.

8.155 Do you drink the recommended amount of water each day? Most Americans don't! On average, Americans drink 4.6 eight-oz servings of water a day.

Source: http://www.bottledwater.org/public/summary.htm

A sample of 42 education professionals was randomly selected and their water consumption for a 24-hour period was monitored; the mean amount consumed was 39.3 oz. Assuming the amount of water consumed daily by adults is normally distributed and the standard deviation is 11.2 oz, is there sufficient evidence to show that education professionals consume, on average, more water daily than the national average? Use $\alpha = 0.05$.

8.156 The recommended amount of water a person should drink is eight 8-oz servings per day.

a. Does the sample of educational professionals in Exercise 8.155 show sufficient evidence that the education professionals consume, on average, significantly less water daily than the recommended amount? Use $\alpha = 0.05$.

b. The value of the calculated z-score in part a is unusual. In what way is it unusual, and what does that mean?

8.157 Use a computer or calculator to select 40 random single-digit numbers. Find the sample mean and $z\star$. Using $\alpha = 0.05$, state the decision for testing $H_o: \mu = 4.5$ against a two-tailed alterna-tive. Repeat it several times as in Table 8.12. Describe your findings after several tries.

FYI Use commands for generating integer data on page 101; then continue with the hypothesis test commands on pages 438–439.

8.158 Use a computer or calculator to select 36 random numbers from a normal distribution with mean 100 and standard deviation 15. Find the sample mean and $z\star$ for testing a two-tailed hypothesis test of $\mu = 100$. Using $\alpha = 0.05$, state the decision. Repeat several times as in Table 8.12. Describe your findings.

FYI Use commands for generating data on pages 327–328; then continue with hypothesis test commands on pages 438–439.

CHAPTER REVIEW

In Retrospect

Two forms of inference were presented in this chapter: estimation and hypothesis testing. They may be, and often are, used separately. It seems natural, however, for the rejection of a null hypothesis to be followed by a confidence interval. (If the value claimed is wrong, we often want an estimate for the true value.)

These two forms of inference are quite different, but they are related. There is a certain amount of crossover between the use of the two inferences. For example, suppose that you had sampled and calculated a 90% confidence interval for the mean of a population. The interval was 10.5 to 15.6. Then someone claims that the true mean is 15.2. Your confidence interval can be compared with this claim. If the claimed value falls within your interval estimate, you would fail to reject the null hypothesis that $\mu = 15.2$ at a 10% level of significance in a two-tailed test. If the claimed value (say, 16.0) falls outside the interval, you would reject the null hypothesis that $\mu = 16.0$ at $\alpha = 0.10$ in a two-tailed test. If a one-tailed test is required, or if

you prefer a different value of α, a separate hypothesis test must be used.

Many users of statistics (especially those marketing a product) will claim that their statistical results prove that their product is superior. But remember, the hypothesis test does not *prove* or *disprove* anything. The decision reached in a hypothesis test has probabilities associated with the four various situations. If "fail to reject H_o" is the decision, it is possible that an error has occurred. Furthermore, if "reject H_o" is the decision reached, it is possible for this to be an error. Both errors have probabilities greater than zero.

In this chapter we have restricted our discussion of inferences to the mean of a population for which the standard deviation is known. In Chapters 9 and 10 we will discuss inferences about the population mean and remove the restriction about the known value for standard deviation. We will also look at inferences about the parameters proportion, variance and standard deviation.

Vocabulary and Key Concepts

alpha (α) (pp. 403, 420)

alternative hypothesis (pp. 417, 428, 445)

assumptions (pp. 402, 426, 444)

beta (β) (p. 420)

biased statistics (p. 397)

calculated value ($z\star$) (pp. 432, 450)

conclusion (pp. 422, 433, 452)

confidence coefficient (p. 403)

confidence interval (p. 399)

confidence interval procedure (p. 404)

critical region (p. 450)

critical value (p. 451)

decision rule (pp. 422, 433, 451)

estimation (p. 397)

hypothesis (p. 417)

hypothesis test (p. 417)

hypothesis test, classical procedure (p. 444)

hypothesis test, p-value procedure (p. 427)

interval estimate (p. 399)

level of confidence (p. 399)

level of significance (pp. 422, 431, 449)

lower confidence limit (p. 403)

maximum error of estimate (pp. 403, 410)

noncritical region (p. 450)

null hypothesis (pp. 417, 428, 445)

parameter (p. 397)

point estimate (p. 397)

p-value (p. 432)

sample size (p. 410)

sample statistic (p. 397)

standard error of mean (p. 403)

test criteria (pp. 431, 448)

test statistic (pp. 422, 431, 449)

type A correct decision (p. 419)

type B correct decision (p. 419)

type I error (p. 419)

type II error (p. 419)

unbiased statistic (p. 397)

upper confidence limit (p. 403)

$z(\alpha)$ (pp. 402, 451)

Learning Outcomes

✔ Understand the difference between descriptive statistics and inferential statistics.   p. 4, Ex. 1.6, p. 396

✔ Understand that an unbiased statistic has a sampling distribution with a mean that is equal to the population parameter being estimated.   pp. 397–398

With respect to confidence intervals:

✔ Understand that a confidence interval is an interval estimate of a population parameter, with a degree of certainty, used when the population parameter is unknown.   p. 399

✔ Understand that a point estimate for a population parameter is the value of the corresponding sample statistic.   p. 397, Ex. 8.5

✔ Understand that the level of confidence is the long-run proportion of the intervals, which will contain the true population parameters, based on repeated sampling.   EXP 8.4

✔ Understand and be able to describe the key components for a confidence interval: point estimate, level of confidence, confidence coefficient, maximum error of estimate, lower confidence limit, and upper confidence limit.   p. 403, Ex. 8.25, 8.27, 8.163

✔ Understand that the assumption for a confidence interval for μ using a known σ is that the sampling distribution of $\bar{x}$ has a normal distribution. Based on this assumption, the standard normal z distribution will be utilized.   pp. 402–403

✔ Compute, describe, and interpret a confidence interval for the population mean, μ.

EXP 8.2, Ex. 8.33

✔ Compute sample sizes required for constructing confidence intervals with varying levels of confidence and acceptable errors.

pp. 410–412, Ex. 8.45, 8.49

With respect to hypothesis tests:

✔ Understand that a hypothesis test is used to make a decision about the value of a population parameter.

p. 417

✔ Understand and be able to define null and alternative hypotheses.

p. 417

✔ Understand and be able to describe the two types of error in a hypothesis test, type I and type II. Understand that the probability of these errors are α and β, respectively.

pp. 419–421, Ex. 8.57

✔ Understand and be able to describe the two types of correct decisions in a hypothesis test, type A and type B.

pp. 419–421, Ex. 8.57

✔ Understand and be able to describe the relationship between the four possible outcomes of a hypothesis test— the two types of errors and the two types of correct decisions.

pp. 419–421, Ex. 8.57

✔ Demonstrate and understand the three possible combinations for the null and alternative hypotheses.

pp. 428–430, Ex. 8.83, pp. 446–448, Ex. 8.125

✔ Understand that the assumption for a hypothesis test for μ using a known σ is that the sampling distribution of $\bar{x}$ has a normal distribution. Based on this assumption, the standard normal z distribution will be utilized.

pp. 426, 444

✔ Compute and understand the value of the test statistic. Compute the p-value for the test statistic and/or determine the critical region and critical value(s).

pp. 432–433, Ex. 8.102, pp. 450–452, Ex. 8.134, 8.137

✔ Understand and be able to describe what a p-value and/or a critical region is with respect to a hypothesis test.

pp. 432–433, 435, 450–452

✔ Determine and know the proper format for stating a decision in a hypothesis test.

pp. 422, 433, 451

✔ Understand and be able to state the conclusion for a hypothesis test.

pp. 422, 433, 452, Ex. 8.179, 8.185, 8.186

Chapter Exercises

8.159 A sample of 64 measurements is taken from a continuous population, and the sample mean is found to be 32.0. The standard deviation of the population is known to be 2.4. An interval estimation is to be made of the mean with a level of confidence of 90%. State or calculate the following items.

a. $\bar{x}$ b. σ c. n d. $1 - \alpha$ e. $z(\alpha/2)$ f. $\sigma_{\bar{x}}$

g. E (maximum error of estimate)

h. Upper confidence limit

i. Lower confidence limit

8.160 Suppose that a confidence interval is assigned a level of confidence of $1 - \alpha = 95\%$. How is the 95% used in constructing the confidence interval? If $1 - \alpha$ was changed to 90%, what effect would this have on the confidence interval?

8.161 The average volunteer ambulance member is 45 years old and has 8 years of service according to the *Democrat & Chronicle* article "Unpaid ambulance workers could get 'pension'" (January 23, 2005). The quoted statistics were based on the Penfield Volunteer Ambulance Squad of 80 members. If the Penfield Volunteer Ambulance Squad is considered representative of all upstate New York volunteer ambulance squads, determine a 95% confidence interval for the mean age of all volunteer ambulance members in upstate New York. Assume the population standard deviation is 7.8 years.

8.162 The standard deviation of a normally distributed population is equal to 10. A sample size of 25 is selected, and its mean is found to be 95.

a. Find an 80% confidence interval for μ.

b. What would the 80% confidence interval be for a sample of size 100?

c. What would be the 80% confidence interval for a sample size of 25 with a standard deviation of 5 (instead of 10)?

8.163 The weights of full boxes of a certain kind of cereal are normally distributed with a standard deviation of 0.27 oz. A sample of 18 randomly selected boxes produced a mean weight of 9.87 oz.

a. Find the 95% confidence interval for the true mean weight of a box of this cereal.

b. Find the 99% confidence interval for the true mean weight of a box of this cereal.

c. What effect did the increase in the level of confidence have on the width of the confidence interval?

8.164 Waiting times (in hours) at a popular restaurant are believed to be approximately normally distributed with a variance of 2.25 during busy periods.

a. A sample of 20 customers revealed a mean waiting time of 1.52 hours. Construct the 95% confidence interval for the population mean.

b. Suppose that the mean of 1.52 hours had resulted from a sample of 32 customers. Find the 95% confidence interval.

c. What effect does a larger sample size have on the confidence interval?

8.165 A random sample of the scores of 100 applicants for clerk-typist positions at a large insurance company showed a mean score of 72.6. The preparer of the test maintained that qualified applicants should average 75.0.

a. Determine the 99% confidence interval for the mean score of all applicants at the insurance company. Assume that the standard deviation of test scores is 10.5.

b. Can the insurance company conclude that it is getting qualified applicants (as measured by this test)?

8.166 Are you worried about identity theft? If you do business mostly online, you may be in a safer position as stated in the January 30, 2005, *USA Today* article, "Odds low for online theft." In a telephone survey done by the Better Business Bureau and Javelin Strategy & Research, the average loss for online identity theft was $551 versus $4543 for paper identity theft. If the online average was based on a survey of 60 selected randomly from those who had experienced online identity theft, find the 95% confidence interval for the mean online identity theft amount. Assume the standard deviation is $180.

8.167 The length of time it takes to play a major league baseball game is of interest to many fans. To estimate the mean "time of game," a random sample of 48 National League games was identified and the "time of game" (in minutes) obtained for each. The resulting sample mean was 2 hours and

49.1 minutes, and the history of baseball indicates the time of game variable has a standard deviation of 21 minutes. Construct the 98% confidence interval for the mean time for all National League games.

8.168 [EX08-168] A large order of the no. 9 corks described in Applied Example 6.15 (p. 330) is about to be shipped. The final quality-control inspection includes an estimation of the mean ovality (ovalization; out-of-roundness) of the corks. The diameter of each cork is measured in several places, and the difference between the maximum and minimum diameters is the measure of ovality for each cork. After years of measuring corks, the manufacturer is sure that ovality has a mounded distribution with a standard deviation of 0.10 mm. A random sample of 36 corks is taken from the batch and the ovality is determined for each.

0.32	0.27	0.24	0.31	0.20	0.38	0.32	0.11	0.25
0.22	0.35	0.20	0.28	0.17	0.36	0.28	0.38	0.17
0.34	0.06	0.43	0.13	0.39	0.15	0.18	0.13	0.25
0.20	0.16	0.26	0.47	0.21	0.19	0.34	0.24	0.20

a. The out-of-round spec is "less than 1.0 mm." Does it appear this order meets the spec on an individual cork basis? Explain.

b. The certification sheet that accompanies the shipment includes a 95% confidence interval for the mean ovality. Construct the confidence interval.

c. Explain what the confidence interval found in part b tells about this shipment of corks.

8.169 [EX08-169] This computer output shows a simulated sample of size 25 randomly generated from a normal population with $\mu = 130$ and $\sigma = 10$. A confidence interval command was then used to set a 95% confidence interval for μ.

```
116.187 119.832 121.782 122.320 141.436 129.197 119.172
120.713 135.765 131.153 122.307 126.155 137.545 141.154
123.405 143.331 121.767 109.742 140.524 150.600 121.655
127.992 136.434 139.768 125.594
```

N	MEAN	STDEV	SE MEAN	95.0 PERCENT C.I.
25	129.02	10.18	2.00	(125.10, 132.95)

a. State the confidence interval that resulted.

b. Verify the values reported for the standard error of mean and the interval bounds.

8.170 Use a computer and generate 50 random samples, each of size $n = 25$, from a normal probability distribution with $\mu = 130$ and $\sigma = 10$.

a. Calculate the 95% confidence interval based on each sample mean.

b. What proportion of these confidence intervals contain $\mu = 130$?

c. Explain what the proportion found in part b represents.

8.171 A pharmaceutical company wants to estimate the mean response time for a supplement to reduce blood pressure. How large of a sample should they take to estimate the mean response time to within 1 week at 99% confidence. Assume $\sigma = 3.7$ weeks.

8.172 An automobile manufacturer wants to estimate the mean gasoline mileage of its new compact model. How many sample runs must be performed to ensure that the estimate is accurate to within 0.3 mpg at 95% confidence? (Assume $\sigma = 1.5$.)

8.173 A fish hatchery manager wants to estimate the mean length of her 3-year-old hatchery-raised trout. She wants to make a 99% confidence interval accurate to within $\frac{1}{3}$ of a standard deviation. How large a sample does she need to take?

8.174 We are interested in estimating the mean life of a new product. How large a sample do we need to take to estimate the mean to within $\frac{1}{10}$ of a standard deviation with 90% confidence?

8.175 Suppose a hypothesis test is conducted using the p-value approach and assigned a level of significance of $\alpha = 0.01$.

a. How is the 0.01 used in completing the hypothesis test?

b. If α is changed to 0.05, what effect would this have on the test procedure?

8.176 Suppose a hypothesis test is conducted using the classical approach and assigned a level of significance of $\alpha = 0.01$.

a. How is the 0.01 used in completing the hypothesis test?

b. If α is changed to 0.05, what effect would this have on the test procedure?

8.177 The expected mean of a continuous population is 100, and its standard deviation is 12. A sample of 50 measurements gives a sample mean of 96. Using a 0.01 level of significance, a test is to be made to decide between "the population mean is 100" or "the population mean is different from 100." State or find each of the following:

a. H_o b. H_a c. α d. μ (based on H_o)

e. $\bar{x}$ f. σ g. $\sigma_{\bar{x}}$ h. $z\star$, z-score for $\bar{x}$

i. p-value   j. Decision

k. Sketch the standard normal curve and locate $z\star$ and p-value.

8.178 The expected mean of a continuous population is 200, and its standard deviation is 15. A sample of 80 measurements gives a sample mean of 205. Using a 0.01 level of significance, a test is to be made to decide between "the population mean is 200" or "the population mean is different from 200." State or find each of the following:

a. H_o b. H_a c. α d. $z_{(\alpha/2)}$

e. μ (based on H_o)   f. $\bar{x}$ g. σ h. $\sigma_{\bar{x}}$

i. $z\star$, z-score for $\bar{x}$ j. decision

k. Sketch the standard normal curve and locate $\alpha/2$, $z_{(\alpha/2)}$, the critical region, and $z\star$.

8.179 A lawn and garden sprinkler system is designed to have a delayed start; that is, there is a delay from the moment it is turned on until the water starts. The delay times form a normal distribution with mean 45 seconds and standard deviation 8 seconds. Several customers have complained that the delay time is considerably longer than claimed. The system engineer has selected a random sample of 15 installed systems and has obtained one delay time from each system. The sample mean is 50.1 seconds. Using $\alpha = 0.02$, is there significant evidence to show that the customers might be correct that the mean delay time is more than 45 seconds?

a. Solve using the p-value approach.

b. Solve using the classical approach.

8.180 The college bookstore tells prospective students that the average cost of its textbooks is $90 per book with a standard deviation of $15. The engineering science students think that the average cost of their books is higher than the average for all students. To test the bookstore's claim against their alternative, the engineering students collect a random sample of size 45.

a. If they use $\alpha = 0.05$, what is the critical value of the test statistic?

b. The engineering students' sample data are summarized by $n = 45$ and $\Sigma x = 4380.30$. Is this sufficient evidence to support their contention?

8.181 A manufacturing process produces ball bearings with diameters having a normal distribution and a standard deviation of $\sigma = 0.04$ cm. Ball bearings that have diameters that are too small or too large are undesirable. To test the null hypothesis that $\mu = 0.50$ cm, a sample of 25 is randomly selected and the sample mean is found to be 0.51.

a. Design null and alternative hypotheses such that rejection of the null hypothesis will imply that the ball bearings are undesirable.

b. Using the decision rule established in part a, what is the p-value for the sample results?

c. If the decision rule in part a is used with $\alpha = 0.02$, what is the critical value for the test statistic?

8.182 After conducting a large number of tests over a long period, a rope manufacturer has found that its rope has a mean breaking strength of 300 lb and a standard deviation of 24 lb. Assume that these values are μ and σ. It is believed that by using a recently developed high-speed process, the mean breaking strength has been decreased.

a. Design null and alternative hypotheses such that rejection of the null hypothesis will imply that the mean breaking strength has decreased.

b. Using the decision rule established in part a, what is the *p-value* associated with rejecting the null hypothesis when 45 tests result in a sample mean of 295?

c. If the decision rule in part a is used with $\alpha = 0.01$, what is the critical value for the test statistic and what value of $\bar{x}$ corresponds to it if a sample of size 45 is used?

8.183 A worker honeybee leaves the hive on a regular basis and travels to flowers and other sources of pollen and nectar before returning to the hive to deliver its cargo. The process is repeated several times each day in order to feed younger bees and support the hive's production of honey and wax. The worker bee can carry an average of 0.0113 gram of pollen and nectar per trip, with a standard deviation of 0.0063 gram. Fuzzy Drone is entering the honey and beeswax business with a new strain of Italian bees that are reportedly capable of carrying larger loads of pollen and nectar than the typical honeybee. After installing three hives, Fuzzy isolated 200 bees before and after their return trip and carefully weighed their cargoes. The sample mean weight of the pollen and nectar was 0.0124 gram. Can Fuzzy's bees carry a greater load of pollen and nectar than the rest of the honeybee population? Complete the appropriate hypothesis test at the 0.01 level of significance.

a. Solve using the *p-value* approach.

b. Solve using the classical approach.

8.184 The average weight for a 10-year-old girl in 2002 was 88.0 lb according to the National Center for Health Statistics. This average exceeded the 1966 figure by more than 10 lb. Suppose a random sample of thirty-five 10-year-old girls is taken and the sample mean is 90.5 lb. Assuming that the population standard deviation is 10.7 lb, is there evidence from this sample that the average weight for 10-year-old girls has increased since 2002?

Complete the appropriate hypothesis test at the 0.05 level of significance.

a. Solve using the *p-value* approach.

b. Solve using the classical approach.

8.185 In a large supermarket the customer's waiting time to check out is approximately normally distributed with a standard deviation of 2.5 minutes. A sample of 24 customer waiting times produced a mean of 10.6 minutes. Is this evidence sufficient to reject the supermarket's claim that its customer checkout time averages no more than 9 minutes? Complete this hypothesis test using the 0.02 level of significance.

a. Solve using the *p-value* approach.

b. Solve using the classical approach.

8.186 At a very large firm, the clerk-typists were sampled to see whether salaries differed among departments for workers in similar categories. In a sample of 50 of the firm's accounting clerks, the average annual salary was $16,010. The firm's personnel office insists that the average salary paid to all clerk-typists in the firm is $15,650 and that the standard deviation is $1800. At the 0.05 level of significance, can we conclude that the accounting clerks receive, on average, a different salary from that of the clerk-typists?

a. Solve using the *p-value* approach.

b. Solve using the classical approach.

8.187 Jack Williams is vice president of marketing for one of the largest natural gas companies in the nation. During the past 4 years, he has watched two major factors erode the profits and sales of the company. First, the average price of crude oil has been virtually flat, and many of his industrial customers are burning heavy oil rather than natural gas to fire their furnaces, regardless of added smokestack emissions. Second, both residential and commercial customers are still pursuing energy-conservation techniques (e.g., adding extra insulation, installing clock-drive thermostats, and sealing cracks around doors and windows to eliminate cold air infiltration). In previous years, residential customers bought an average of 129.2 mcf of natural gas from Jack's company ($\sigma = 18$ mcf), based on internal company billing records, but environmentalists have claimed that conservation is cutting fuel consumption up to 3% per year. Jack has commissioned you to conduct a spot check to see if any change in annual usage has transpired before his next meeting with the officers of the corporation. A sample of 300 customers selected randomly from the billing records reveals an average of 127.1 mcf during the past 12 months. Is there a significant decline in consumption?

a. Complete the appropriate hypothesis test at the 0.01 level of significance using the p-value approach so that you can properly advise Jack before his meeting.

b. Because you are Jack's assistant, why is it best for you to use the p-value approach?

8.188 With a nationwide average drive time of about 24.3 minutes, Americans now spend more than 100 hours a year commuting to work, according to the U.S. Census Bureau's American Community Survey. Yes, that's more than the average 2 weeks of vacation time (80 hours) taken by many workers during a year.

Source: http://usgovinfo.about.com/od/censusandstatistics/a/commutetimes.htm

A random sample of 150 workers at a large nearby industry was polled about their commute time. If the standard deviation is known to be 10.7 minutes, is the resulting sample mean of 21.7 minutes significantly lower than the nationwide average? Use $\alpha = 0.01$.

a. Solve using the p-value approach.

b. Solve using the classical approach.

8.189 [EX08-189] This computer output shows a simulated sample of size 28 randomly generated from a normal population with $\mu = 18$ and $\sigma = 4$. Computer commands were then used to complete a hypothesis test for $\mu = 18$ against a two-tailed alternative.

a. State the alternative hypothesis, the decision, and the conclusion that resulted.

b. Verify the values reported for the standard error of mean, $z\star$, and the p-value.

```
18.7734  21.4352  15.5438  20.2764  23.2434  15.7222  13.9368
14.4112  15.7403  19.0970  19.0032  20.0688  12.2466  10.4158
 8.9755  18.0094  20.0112  23.2721  16.6458  24.6146  17.8078
16.5922  16.1385  12.3115  12.5674  18.9141  22.9315  13.3658
```

```
TEST OF  MU = 18.000  VS  MU not = 18.000
THE ASSUMED STANDARD DEVIATION = 4.00
```

N	MEAN	STDEV	SE MEAN	Z	P VALUE
28	17.217	4.053	0.756	−1.04	0.30

8.190 A manufacturer of automobile tires believes it has developed a new rubber compound that has superior antiwearing qualities. It produced a test run of tires made with this new compound and had them road tested. The data values recorded were the amount of tread wear per 10,000 miles. In the past, the mean amount of tread wear per 10,000 miles, for tires of this quality, has been 0.0625 inches.

The null hypothesis to be tested here is "The mean amount of wear on the tires made with the new compound is the same mean amount of wear with the old compound, 0.0625 inches per 10,000 miles," H_o: $\mu = 0.0625$. Three possible alternative hypotheses could be used: (1) H_a: $\mu < 0.0625$, (2) H_a: $\mu \neq 0.0625$, (3) H_a: $\mu > 0.0625$.

a. Explain the meaning of each of these three alternatives.

b. Which one of the possible alternative hypotheses should the manufacturer use if it hopes to conclude that "use of the new compound does yield superior wear"?

8.191 From a population of unknown mean μ and a standard deviation $\sigma = 5.0$, a sample of $n = 100$ is selected and the sample mean 40.6 is found. Compare the concepts of estimation and hypothesis testing by completing the following:

a. Determine the 95% confidence interval for μ.

b. Complete the hypothesis test involving H_a: $\mu \neq 40$ using the p-value approach and $\alpha = 0.05$.

c. Complete the hypothesis test involving H_a: $\mu \neq 40$ using the classical approach and $\alpha = 0.05$.

d. On one sketch of the standard normal curve, locate the interval representing the confidence interval from part a; the $z\star$, p-value, and α from part b; and the $z\star$ and critical regions from part c. Describe the relationship between these three separate procedures.

8.192 From a population of unknown mean μ and a standard deviation $\sigma = 5.0$, a sample of $n = 100$ is selected and the sample mean 41.5 is found. Compare the concepts of estimation and hypothesis testing by completing the following:

a. Determine the 95% confidence interval for μ.

b. Complete the hypothesis test involving H_a: $\mu \neq 40$ using the p-value approach and $\alpha = 0.05$.

c. Complete the hypothesis test involving H_a: $\mu \neq 40$ using the classical approach and $\alpha = 0.05$.

d. On one sketch of the standard normal curve, locate the interval representing the confidence interval from part a; the $z\star$, p-value, and α from part b; and the $z\star$ and critical regions from part c. Describe the relationship between these three separate procedures.

8.193 From a population of unknown mean μ and a standard deviation $\sigma = 5.0$, a sample of $n = 100$ is selected and the sample mean 40.9 is found. Compare the concepts of estimation and hypothesis testing by completing the following:

a. Determine the 95% confidence interval for μ.

b. Complete the hypothesis test involving H_a: $\mu > 40$ using the p-value approach and $\alpha = 0.05$.

c. Complete the hypothesis test involving H_a: $\mu > 40$ using the classical approach and $\alpha = 0.05$.

d. On one sketch of the standard normal curve, locate the interval representing the confidence interval from part a; the $z\star$, p-value, and α from part b; and the $z\star$ and critical regions from part c. Describe the relationship between these three separate procedures.

8.194 A manufacturer of stone-ground, deli-style mustard uses a high-speed machine to fill jars. The amount of mustard dispensed into the jars forms a normal distribution with a mean 290 grams and a standard deviation 4 grams. Each hour a random sample of 12 jars is taken from that hour's production. If the sample mean is between 287.74 and 292.26, that hour's production is accepted; other-

wise, it is rejected and the machine is recalibrated before continuing.

a. What is the probability of the type I error by rejecting the previous hour's production when the mean jar weight is 290 grams?

b. What is the probability of the type II error by accepting the previous hour's production when the mean jar weight is actually 288 grams?

8.195 All drugs must be approved by the U.S. Food and Drug Administration (FDA) before they can be marketed by a drug company. The FDA must weigh the error of marketing an ineffective drug, with the usual risks of side effects, against the consequences of not allowing an effective drug to be sold. Suppose, using standard medical treatment, that the mortality rate (r) of a certain disease is known to be A. A manufacturer submits for approval a drug that is supposed to treat this disease. The FDA sets up the hypothesis to test the mortality rate for the drug as (1) H_o: $r = A$, H_a: $r < A$, $\alpha = 0.005$ or (2) H_o: $r = A$, H_a: $r > A$, $\alpha = 0.005$.

a. If $A = 0.95$, which test do you think the FDA should use? Explain.

b. If $A = 0.05$, which test do you think the FDA should use? Explain.

8.196 The drug manufacturer in Exercise 8.195 has a different viewpoint on the matter. It wants to market the new drug starting as soon as possible so that it can beat its competitors to the marketplace and make lots of money. Its position is, "Market the drug unless the drug is totally ineffective."

a. How would the drug company set up the alternative hypothesis if it were doing the testing? H_a: $r < A$, H_a: $r \neq A$, or H_a: $r > A$. Explain.

b. Does the mortality rate ($A = 0.95$ or $A = 0.05$) of the existing treatment affect the alternative? Explain.

8.197 Use a computer and generate 50 random samples, each of size $n = 28$, from a normal probability distribution with $\mu = 18$ and $\sigma = 4$.

a. Calculate the $z\star$ corresponding to each sample mean.

b. In regard to the *p*-value approach, find the proportion of 50 $z\star$-values that are "more extreme" than the $z = -1.04$ that occurred in Exercise 8.189 (H_a: $\mu \neq 18$). Explain what this proportion represents.

c. In regard to the classical approach, find the critical values for a two-tailed test using $\alpha = 0.01$; find the proportion of 50 $z\star$-values that fall in the critical region. Explain what this proportion represents.

8.198 Use a computer and generate 50 random samples, each of size $n = 28$, from a normal probability distribution with $\mu = 19$ and $\sigma = 4$.

a. Calculate the $z\star$ corresponding to each sample mean that would result when testing the null hypothesis $\mu = 18$.

b. In regard to the *p*-value approach, find the proportion of 50 $z\star$-values that are "more extreme" than the $z = -1.04$ that occurred in Exercise 8.189 (H_a: $\mu \neq 18$). Explain what this proportion represents.

c. In regard to the classical approach, find the critical values for a two-tailed test using $\alpha = 0.01$; find the proportion of 50 $z\star$-values that fall in the noncritical region. Explain what this proportion represents.

Chapter Project

Were They Shorter Back Then?

Data from the National Center for Health Statistics indicate that the average height of a female in the United States is 63.7 inches with a standard deviation of 2.75 inches. Use the data on heights of females in the health profession from Section 8.1, "Were They Shorter Back Then?" (p. 395), to answer the following questions.   [EX08-001]

65.0	66.0	64.0	67.0	59.0	69.0	66.0	69.0	64.0	61.5
63.0	62.0	63.0	64.0	72.0	66.0	65.0	64.0	67.0	68.0
70.0	63.0	63.0	68.0	58.0	60.0	63.5	66.0	64.0	62.0
64.5	69.0	63.5	69.0	62.0	58.0	66.0	68.0	59.0	56.0
64.0	66.0	65.0	69.0	67.0	66.5	67.5	62.0	70.0	62.0

Putting Chapter 8 to Work

8.199 a. Are the assumptions of the confidence interval and hypothesis test methods of this chapter satisfied? Explain.

b. Using the sample data and a 95% level of confidence, estimate the mean height of females in the health profession. Use the given population standard deviation of 2.75 inches.

c. Test the claim that the mean height of females in the health profession is different from 63.7 inches, the mean height for all females in the United States. Use a 0.05 level of significance.

d. On the same histogram used in part b of Exercise 8.1 on page 396:

(i) Draw a vertical line at the hypothesized population mean value, 63.7.

(ii) Draw a horizontal line segment showing the 95% confidence interval found in part b.

e. Does the mean $\mu = 63.7$ fall in the interval? Explain what this means.

f. Describe the relationship between the two lines drawn on your graph for part c of Exercise 8.2 on page 396 and the two lines drawn for part d of this exercise.

g. On the basis of the results obtained earlier, does it appear that the females in this study, on average, are the same height as all females in the United States as reported by the NCHS? Explain.

Your Study

8.200 Design your own study of female heights. Define a specific population that you will sample, describe your sampling plan, collect your data, and answer part b of Exercise 8.1 (p. 396) and parts a, b, c, and g of Exercise 8.199, replacing health profession with your particular population. Discuss the differences and similarities between your sample and the population and between your sample and the sample of 50 female health professionals.

Chapter Practice Test

PART I: Knowing the Definitions

Answer "True" if the statement is always true. If the statement is not always true, replace the words shown in bold with words that make the statement always true.

8.1 **Beta** is the probability of a type I error.

8.2 $1 - \alpha$ is known as the level of significance of a hypothesis test.

8.3 The standard error of the mean is the standard deviation of the **sample selected.**

8.4 The maximum error of estimate is controlled by three factors: **level of confidence, sample size,** and **standard deviation.**

8.5 Alpha is the measure of the area under the curve of the standard score that lies in the **rejection region** for H_o.

8.6 The risk of making a **type I error** is directly controlled in a hypothesis test by establishing a level for α.

8.7 Failing to reject the null hypothesis when it is false is a **correct decision.**

8.8 If the noncritical region in a hypothesis test is made wider (assuming σ and n remain fixed), α becomes larger.

8.9 Rejection of a null hypothesis that is false is a **type II error.**

8.10 To conclude that the mean is greater (or less) than a claimed value, the value of the test statistic must fall in the **acceptance region.**

PART II: Applying the Concepts

Answer all questions, showing all formulas, substitutions, and work.

8.11 An unhappy post office customer is frustrated with the waiting time to buy stamps. Upon registering his complaint, he was told, "The average waiting time in the past has been about 4 minutes with a standard deviation of 2 minutes." The customer collected a sample of $n = 45$ customers and found the mean wait was 5.3 minutes. Find the 95% confidence interval for the mean waiting time.

8.12 State the null (H_o) and the alternative (H_a) hypotheses that would be used to test each of these claims:

   a. The mean weight of professional football players is more than 245 lb.

   b. The mean monthly amount of rainfall in Monroe County is less than 4.5 inches.

   c. The mean weight of the baseball bats used by major league players is not equal to 35 oz.

8.13 Determine the level of significance, test statistic, critical region, and critical value(s) that would be used in completing each hypothesis test using $\alpha = 0.05$:

a. H_o: $\mu = 43$ b. H_o: $\mu = 0.80$ c. H_o: $\mu = 95$

   H_a: $\mu < 43$ H_a: $\mu > 0.80$ H_a: $\mu \neq 95$

   (given $\sigma = 6$)   (given $\sigma = 0.13$)   (given $\sigma = 12$)

8.14 Find each value:

   a. $z(0.05)$ b. $z(0.01)$ c. $z(0.12)$

8.15 In the past, the grapefruits grown in a particular orchard have had a mean diameter of 5.50 inches. and a standard deviation of 0.6 inches. The owner believes this year's crop is larger than those in the past. He collected a random sample of 100 grapefruits and found a sample mean diameter of 5.65 inches.

   a. Find the value of the test statistic, $z\star$, that corresponds to $\bar{x} = 5.65$.

   b. Calculate the p-value for the owner's hypothesis.

8.16 A manufacturer claims that its light bulbs have a mean lifetime of 1520 hours with a standard deviation of 85 hours. A random sample of 40 such bulbs is selected for testing. If the sample produces a mean value of 1498.3 hours, is there sufficient evidence to claim that the mean lifetime is less than the manufacturer claimed? Use $\alpha = 0.01$.

PART III: Understanding the Concepts

8.17 Sugar Creek Convenience Stores has commissioned a statistics firm to survey its customers in order to estimate the mean amount spent per customer. From previous records the standard deviation is believed to be $\sigma = \$5$. In its proposal to Sugar Creek, the statistics firm states that it plans to base the estimate for the mean amount spent on a sample of size 100 and use the 95% confidence level. Sugar Creek's president has suggested that the sample size be increased to 400. If nothing else changes, what effect will this increase in the sample size have on the following:

a. The point estimate for the mean

b. The maximum error of estimation

c. The confidence interval

The CEO wants the level of confidence increased to 99%. If nothing else changes, what effect will this increase in level of confidence have on the following:

d. The point estimate for the mean

e. The maximum error of estimation

f. The confidence interval

8.18 The noise level in a hospital may be a critical factor influencing a patient's speed of recovery. Suppose for the sake of discussion that a research commission has recommended a maximum mean noise level of 30 decibels (db) with a standard deviation of 10 db. The staff of a hospital intend to sample one of its wards to determine whether the noise level is significantly higher than the recommended level. The following hypothesis test will be completed:

$$H_o: \mu = 30 \; (\leq) \quad \text{vs.} \quad H_a: \mu > 30, \quad \alpha = 0.05$$

a. Identify the correct interpretation for each hypothesis with regard to the recommendation and justify your choice.

H_o: (1) Noise level is not significantly higher than the recommended level, or (2) Noise level is significantly higher than the recommended level.

H_a: (1) Noise level is not significantly higher than the recommended level, or (2) Noise level is significantly higher than the recommended level.

b. Which statement best describes the type I error?

(i) Decision reached was that noise level is within the recommended level when in fact it actually was within.

(ii) Decision reached was that noise level is within the recommended level when in fact it actually exceeded it.

(iii) Decision reached was that noise level exceeds the recommended level when in fact it actually was within.

(iv) Decision reached was that noise level is within the recommended level when in fact it actually exceeded it.

c. Which statement in part b best describes the type II error?

d. If α were changed from 0.05 to 0.01, identify and justify the effect (increases, decreases, or remains the same) on $P(\text{type I error})$ and on $P(\text{type II error})$.

8.19 The alternative hypothesis is sometimes called the *research hypothesis*. The conclusion is a statement written about the alternative hypothesis. Explain why these two statements are compatible.

CHAPTER
9

Inferences Involving One Population

© Thinkstock/Alamy

Get Enough Daily Exercise?

Statistics⬣Now™

Throughout the chapter, this icon introduces a list of resources on the StatisticsNow website at **http://1pass.thomson.com** that will:

- Help you evaluate your knowledge of the material

- Allow you to take an exam-prep quiz

- Provide a Personalized Learning Plan targeting resources that address areas you should study

GOOD NEWS FOR WOMEN

The National Women's Health Information Center reports, "It's never too late to start an active lifestyle." No matter how old you are, how unfit you feel, or how long you have been inactive, research shows that starting a more active lifestyle now through regular, moderate-intensity activity can make you healthier and improve your quality of life. Here's what you should do:

If:	Then:
You do not currently engage in regular physical activity,	you should begin by incorporating a few minutes of physical activity into each day, gradually building up to 30 minutes or more of moderate-intensity activities.
You are now active, but at less than the recommended levels,	you should strive to adopt more consistent activity: moderate-intensity physical activity for 30 minutes or more on 5 days or more of the week *or* vigorous-intensity physical activity for 20 minutes or more on 3 days or more of the week.
You currently engage in moderate-intensity activities for at least 30 minutes on 5 days or more of the week, You currently regularly engage in vigorous-intensity activities 20 minutes or more on 3 days or more of the week,	you may achieve even greater health benefits by increasing the time spent or intensity of those activities. you should continue to do so.

Scientific evidence to date supports the statements in this table.
Source: Centers for Disease Control and Prevention,
http://www.4woman.gov/pub/steps/Physical%20Activity.htm

The article recommends different amounts and different levels of activity depending on a woman's current level of activity. Some of the recommendations require as little as 60 minutes of exercise a week. The data values that follow are from a study surveying cardiovascular technicians (individuals who perform various cardiovascular diagnostic procedures) as to their own physical exercise per week, measured in minutes. [EX09-001]

60	40	50	30	60	50	90	30	60	60
60	80	90	90	60	30	20	120	60	50
20	60	30	120	50	30	90	20	30	40
50	40	30	40	20	30	60	50	60	80

Do the technicians, on average, appear to exercise at least 60 minutes per week?

After completing Chapter 9, further investigate the preceding data values and "Get Enough Daily Exercise?" in the Chapter Project on page 540.

SECTION 9.1 EXERCISES

9.1 Consider the preceding sample data.

a. What is the population parameter of interest?

b. Construct a histogram of the data.

c. What name would you give to the shape of the histogram in part b?

d. Would you say the histogram in part b suggests the variable, amount of time, does not have a normal distribution?

9.2 Based on the data presented in Section 9.1 for the cardiovascular technicians:

a. Find the mean and standard deviation for the amount of time the cardiovascular technicians exercised per week.

b. How would you estimate the mean amount of time exercised per week by all cardiovascular technicians?

c. Does it appear that cardiovascular technicians exercise "at least 60 minutes per week"? Justify your answer.

9.2 Inferences about the Mean μ (σ Unknown)

Inferences about the population mean μ are based on the sample mean $\bar{x}$ and information obtained from the sampling distribution of sample means. Recall that the sampling distribution of sample means has a mean μ and a **standard error** of $\sigma/\sqrt{n}$ for all samples of size n, and it is normally distributed when the sampled population has a normal distribution or approximately normally distributed when the **sample size** is sufficiently large. This means the test statistic $z\bigstar = \dfrac{\bar{x} - \mu}{\sigma/\sqrt{n}}$ has a standard normal distribution. However, when σ **is unknown,** the standard error $\sigma/\sqrt{n}$ is also unknown. Therefore, the sample standard deviation s will be used as the point estimate for σ. As a result, an estimated standard error of the mean, $s/\sqrt{n}$, will be used and our test statistic will become $\dfrac{\bar{x} - \mu}{s/\sqrt{n}}$.

When a **known σ** is being used to make an inference about the mean μ, a sample provides one value for use in the formulas; that one value is $\bar{x}$. When the sample standard deviation s is also used, the sample provides two values: the sample mean $\bar{x}$ and the estimated standard error $s/\sqrt{n}$. As a result, the z-statistic will be replaced with a statistic that accounts for the use of an estimated standard error. This new statistic is known as **Student's t-statistic.**

In 1908, W. S. Gosset, an Irish brewery employee, published a paper about this t-distribution under the pseudonym "Student." In deriving the t-distribution, Gosset assumed that the samples were taken from normal populations. Although this might seem to be restrictive, satisfactory results are obtained when large samples are selected from many nonnormal populations.

FIGURE 9.1 Do I Use the *z*-Statistic or the *t*-Statistic

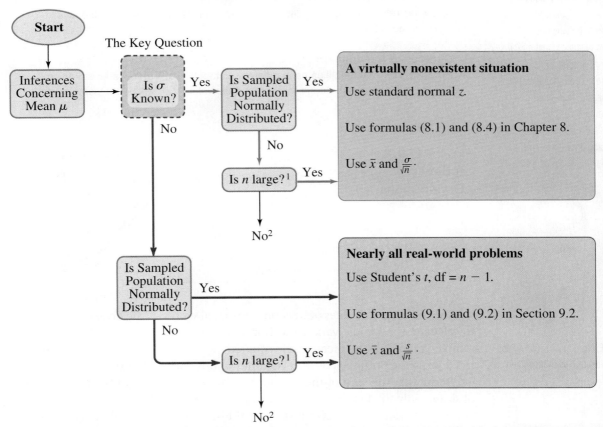

1. Is *n* large? Samples as small as *n* = 15 or 20 may be considered large enough for the central limit theorem to hold if the sample data are unimodal, nearly symmetrical, short-tailed, and without outliers. Samples that are not symmetrical require larger sample sizes, with 50 sufficing except for extremely skewed samples. See the discussion on page 403.

2. Requires the use of a nonparametric technique; see Chapter 14.

Figure 9.1 presents a diagrammatic organization for the inferences about the population mean as discussed in Chapter 8 and in this second section of Chap-

William Gosset ("Student")

William Gosset studied mathematics and chemistry at Oxford University and upon graduation took a position with Guinness Brewery in Dublin, where he found a mass of collected data related to the brewing process. In 1905, he met with Karl Pearson to discuss his statistical problems, and a year later, with Guinness' approval, he went to work at Pearson's Biometric Laboratory. Upon returning to Guinness, he was put in charge of their Experimental Brewery. During these years he wrote several papers, which Guinness agreed to let him publish, provided he used a pseudonym and did not include company data; he used the pseudonym "A Student."

FYI Explore Skillbuilder Applet "Properties of *t*-distribution" on your CD.

ter 9. Two situations exist: σ is known, or σ is unknown. As stated before, σ is almost never a known quantity in real-world problems; therefore, the standard error will almost always be estimated by $s/\sqrt{n}$. The use of an estimated standard error of the mean requires the use of the *t*-distribution. Almost all real-world inferences about the population mean will be made with Student's *t*-statistic.

The *t*-distribution has the following properties (see also Figure 9.2):

Properties of the *t*-distribution (df > 2)*
1. *t* is distributed with a mean of zero.
2. *t* is distributed symmetrically about its mean.
3. *t* is distributed so as to form a family of distributions, a separate distribution for each different number of degrees of freedom (df ≥ 1).
4. The *t*-distribution approaches the **standard normal distribution** as the number of degrees of freedom increases.
5. *t* is distributed with a variance greater than 1, but as the degrees of freedom increases, the variance approaches 1.
6. *t* is distributed so as to be less peaked at the mean and thicker at the tails than is the normal distribution.

Degrees of freedom, df:
A **parameter** that identifies each different distribution of Student's *t*-distribution. For the methods presented in this chapter, the value of df will be the sample size minus 1: df = $n - 1$.

FIGURE 9.2
Student's *t*-Distributions

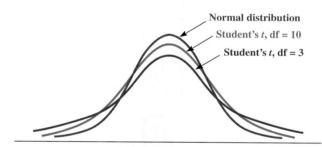

The number of degrees of freedom associated with s^2 is the divisor $(n - 1)$ used to calculate the sample variance s^2 [formula (2.6), p. 94]; that is, df = $n - 1$. The sample variance is the mean of the squared deviations. The number of degrees of freedom is the "number of unrelated deviations" available for use in estimating σ^2. Recall that the sum of the deviations, $\sum(x - \bar{x})$, must be zero. From a sample of size *n*, only the first $n - 1$ of these deviations has freedom of value. That is, the last, or *n*th, value of $(x - \bar{x})$ must make the sum of the *n* deviations total exactly zero. As a result, variance is said to average $n - 1$ unrelated squared deviation values, and this number, $n - 1$, was named "degrees of freedom."

Although there is a separate *t*-distribution for each degrees of freedom, df = 1, df = 2, . . . , df = 20, . . . , df = 40, and so on, only certain key **critical val-**

---

*Not all of the properties hold for df = 1 and df = 2. Since we will not encounter situations where df = 1 or df = 2, these special cases are not discussed further.

ues of *t* will be necessary for our work. Consequently, the table for Student's *t*-distribution (Table 6 in Appendix B) is a table of critical values rather than a complete table, such as Table 3 is for the standard normal distribution for *z*. As you look at Table 6, you will note that the left side of the table is identified by "df," degrees of freedom. This left-hand column starts at 3 at the top and lists consecutive df values to 30, then jumps to 35, . . . , to "df > 100" at the bottom. As we stated, as the degrees of freedom increases, the *t*-distribution approaches the characteristics of the standard normal *z*-distribution. Once df is "greater than 100," the critical values of the *t*-distribution are the same as the corresponding critical values of the standard normal distribution as given in Table 4A in Appendix B.

Using the *t*-Distribution Table (Table 6, Appendix B)

The critical values of Student's *t*-distribution that are to be used both for constructing a confidence interval and for hypothesis testing will be obtained from Table 6 in Appendix B. To find the value of *t*, you will need to know two identifying values: (1) df, the number of degrees of freedom (identifying the distribution of interest), and (2) α, the area under the curve to the right of the right-hand critical value. A notation much like that used with *z* will be used to identify a critical value. $t_{(df, \alpha)}$, read as "*t* of df, α," is the symbol for the value of *t* with df degrees of freedom and an area of α in the right-hand tail, as shown in Figure 9.3.

FIGURE 9.3
***t*-Distribution
Showing $t_{(df, \alpha)}$**

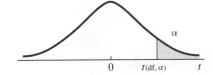

EXAMPLE 9.1

t on the Right Side of the Mean

Find the value of $t_{(10, 0.05)}$ (see the diagram).

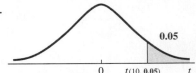

SOLUTION There are 10 degrees of freedom, and 0.05 is to be the area to the right of the critical value. In Table 6 of Appendix B, we look for the row df = 10 and the column marked "Amount of α in One Tail," $\alpha = 0.05$. At their intersection, we see that $t_{(10, 0.05)} = $ **1.81.**

Portion of Table 6

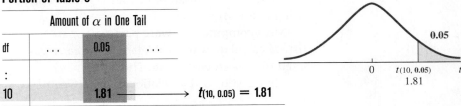

For the values of *t* on the left side of the mean, we can use one of two notations. The *t*-value shown in Figure 9.4 could be named $t_{(df, 0.95)}$, because the area to the right of it is 0.95, or it could be identified by $-t_{(df, 0.05)}$, because the *t*-distribution is symmetrical about its mean, zero.

FIGURE 9.4
t-Value on Left Side

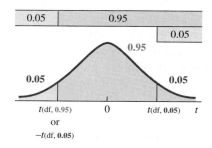

EXAMPLE 9.2   *t* on the Left Side of the Mean

Find the value of $t_{(15, 0.95)}$.

SOLUTION   There are 15 degrees of freedom. In Table 6 we look for the column marked $\alpha = 0.05$ (one tail) and its intersection with the row df = 15. The table gives us $t_{(15, 0.05)} = 1.75$; therefore, $t_{(15, 0.95)} = -t_{(15, 0.05)} = -1.75$. The value is negative because it is to the left of the mean; see the figure.

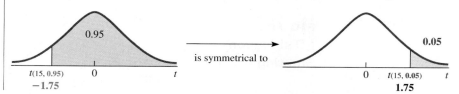

Let's look at another example that connects the *t*-distribution to percentiles.

EXAMPLE 9.3   *t*-Values That Bound a Middle Percentage

Find the values of the *t*-distribution that bound the middle 0.90 of the area under the curve for the distribution with df = 17.

SOLUTION   The middle 0.90 leaves 0.05 for the area of each tail. The value of *t* that bounds the right-hand tail is $t_{(17, 0.05)} = $ **1.74,** as found in Table 6. The value that bounds the left-hand tail is **−1.74** because the *t*-distribution is symmetrical about its mean, zero.

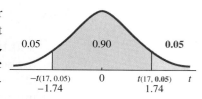

If the df needed is not listed in the left-hand column of Table 6, then use the next smaller value of df that is listed. For example, $t_{(72, 0.05)}$ is estimated using $t_{(70, 0.05)} = 1.67$.

Most computer software packages or statistical calculators will calculate the area related to a specified *t*-value. The accompanying figure shows the relationship between

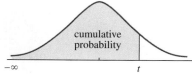

the cumulative probability and a specific *t*-value for a *t*-distribution with df degrees of freedom.

TECHNOLOGY INSTRUCTIONS: PROBABILITY ASSOCIATED WITH A SPECIFIED VALUE OF *t*

MINITAB (Release 14) Cumulative probability for a specified value of *t*:

Choose:	**Calc > Probability Distribution > t**
Select:	**Cumulative Probability**
	Noncentrality parameter: 0.0
Enter:	Degrees of freedom: **df**
Select:	**Input constant***
Enter:	**t-value** (ex. 1.74) > **OK**

*Select Input column if several *t*-values are stored in C1. Use C2 for optional storage. If the area in the right tail is needed, subtract the calculated probability from 1.

Excel Probability in one or two tails for a given *t*-value:

If several *t*-values (nonnegative) are to be used, input the values into column A and activate B1; then continue with:

Choose:	**Insert function f_x > Statistical > TDIST > OK**
Enter:	X: **individual t-value** or **(A1:A5 or select 't-value' cells)***
	Deg_freedom: **df**
	Tails: **1** or **2** (one or two-tailed distributions) > **OK**
Drag*:	**Bottom right corner of the B1 cell down to give other probabilities**

To find the probability within the two tails or the cumulative probability for one tail, subtract the calculated probability from 1.

TI-83/84 Plus Cumulative probability for a specified value of *t*:

Choose:	**2nd > DISTR > 5:tcdf(**†
Enter:	**−1EE99, t-value, df)**

†To find the probability between two *t*-values, enter the two values in place of −1EE99 and *t*-value.

If the area in the right tail is needed, subtract the calculated probability from 1.

Confidence Interval Procedure

We are now ready to make inferences about the population mean μ using the sample standard deviation. As we mentioned earlier, use of the *t*-distribution has a condition.

> **The assumption for inferences about the mean μ when σ is unknown:** The sampled population is normally distributed.

The procedure to make confidence intervals using the sample standard deviation is very similar to that used when σ is known (see pp. 402–407). The difference is the use of Student's t in place of the standard normal z and the use of s, the sample standard deviation, as an estimate of σ. The central limit theorem (CLT) implies that this technique can also be applied to nonnormal populations when the sample size is sufficiently large.

> **Confidence Interval for Mean**
>
> $$\bar{x} - t(\text{df}, \alpha/2)\left(\frac{s}{\sqrt{n}}\right) \quad \text{to} \quad \bar{x} + t(\text{df}, \alpha/2)\left(\frac{s}{\sqrt{n}}\right), \quad \text{with df} = n - 1 \qquad (9.1)$$

Example 9.4 will illustrate the formation of a confidence interval utilizing the t-distribution.

EXAMPLE 9.4 — Confidence Interval for μ with σ Unknown

Statistics⊜Now™

Watch a video example at
http://1pass.thomson.com
or on your CD.

A random sample of 20 weights is taken from babies born at Northside Hospital. A mean of 6.87 lb and a standard deviation of 1.76 lb were found for the sample. Estimate, with 95% confidence, the mean weight of all babies born in this hospital. Based on past information, it is assumed that weights of newborns are normally distributed.

SOLUTION

FYI The five-step confidence interval procedure is given on page 404.

Step 1   The Set-Up:
Describe the population parameter of interest.
μ, the mean weight of newborns at Northside Hospital.

Step 2   The Confidence Interval Criteria:
a. **Check the assumptions.**
Past information indicates that the sampled population is normal.
b. **Identify the probability distribution and the formula to be used. The value of the population standard deviation, σ, is unknown.**
Student's t-distribution will be used with formula (9.1).
c. **State the level of confidence:** $1 - \alpha = 0.95$.

FYI Recall that confidence intervals are two-tailed situations.

Step 3   The Sample Evidence:
Collect the sample information: $n = 20$, $\bar{x} = 6.87$, and $s = 1.76$.

Step 4   The Confidence Interval:
a. **Determine the confidence coefficients.**

FYI df is used to find the confidence coefficient in Table 6; n is used in the formula.

Since $1 - \alpha = 0.95$, $\alpha = 0.05$, and therefore $\alpha/2 = 0.025$. Also, since $n = 20$, df = 19. At the intersection of row df = 19 and the one-tailed column $\alpha = 0.025$ in Table 6, we find $t(\text{df}, \alpha/2) = t(19, 0.025) = 2.09$. See the figure.
Information about the confidence coefficient and using Table 6 is on pages 477–479.

b. **Find the maximum error of estimate.**

$$E = t(\text{df}, \alpha/2)\left(\frac{s}{\sqrt{n}}\right): \quad E = t(19, 0.025)\left(\frac{s}{\sqrt{n}}\right)$$

$$= 2.09\left(\frac{1.76}{\sqrt{20}}\right) = (2.09)(0.394) = 0.82$$

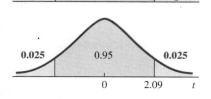

| $\frac{\alpha}{2}$ | $1 - \alpha$ | $\frac{\alpha}{2}$ |

0.025   0.95   0.025

0   2.09 t

c. **Find the lower and upper confidence limits.**

$$\bar{x} - E \quad \text{to} \quad \bar{x} + E$$

$$6.87 - 0.82 \quad \text{to} \quad 6.87 + 0.82$$

$$6.05 \quad \text{to} \quad 7.69$$

Step 5   **The Results:**

State the confidence interval.

6.05 to 7.69, the 95% confidence interval for μ. That is, with 95% confidence we estimate the mean weight of babies born at Northside Hospital to be between 6.05 and 7.69 lb.

TECHNOLOGY INSTRUCTIONS: 1 − α CONFIDENCE INTERVAL FOR MEAN μ WITH σ UNKNOWN

MINITAB (Release 14)   Input the data into C1; then continue with:

Choose:	**Stat > Basic Statistics > 1-Sample t**
Enter:	Samples in columns: **C1**
Select:	**Options**
Enter:	Confidence level: **1 − α** (ex. 95.0)
Select:	Alternative: **not equal > OK > OK**

Excel   Input the data into column A; then continue with:

Choose:	**Tools > Data Analysis Plus > t-Estimate: Mean > OK**
Enter:	Input Range: **(A1:A20 or select cells)**
Enter:	Alpha: **α** (ex. 0.05) **> OK**

TI-83/84 Plus   Input the data into L1; then continue with the following, entering the appropriate values and highlighting Calculate:

Choose:	**STAT > TESTS > 8:Tinterval**

```
TInterval
Inpt:DATA Stats
List:L₁
Freq:1
C-Level:.95
Calculate
```

The MINITAB solution to Example 9.4 looks like this:

```
One-Sample T: C1
Variable    N    Mean    StDev    SE Mean    95% CI
C1          20   6.870   1.760    0.394      (6.047, 7.693)
```

Hypothesis-Testing Procedure

The *t*-statistic is used to complete a hypothesis test about the population mean μ in much the same manner z was used in Chapter 8. In hypothesis-testing situations, we use formula (9.2) to calculate the value of the **test statistic $t\star$:**

Test Statistic for Mean

$$t\star = \frac{\bar{x} - \mu}{s/\sqrt{n}} \quad \text{with } df = n - 1 \tag{9.2}$$

The **calculated** t is the number of estimated standard errors $\bar{x}$ is from the hypothesized mean μ. As with confidence intervals, the CLT indicates that the t-distribution can also be applied to nonnormal populations when the **sample size** is sufficiently large.

EXAMPLE 9.5   ## One-Tailed Hypothesis Test for μ with σ Unknown

Let's return to the hypothesis of Example 8.13 (p. 429) where the Environmental Protection Agency (EPA) wanted to show that the mean carbon monoxide level is higher than 4.9 parts per million. Does a random sample of 22 readings (sample results: $\bar{x} = 5.1$ and $s = 1.17$) present sufficient evidence to support the EPA's claim? Use $\alpha = 0.05$. Previous studies have indicated that such readings have an approximately normal distribution.

SOLUTION

FYI The five-step p-value hypothesis test procedure is given on page 404.

Step 1   **The Set-Up:**
   a. **Describe the population parameter of interest.**
       μ, the mean carbon monoxide level of air in downtown Rochester.
   b. **State the null hypothesis (H_o) and the alternative hypothesis (H_a).**
       H_o: $\mu = 4.9$ ($\leq$) (no higher than)
       H_a: $\mu > 4.9$ (higher than)

FYI Procedures for writing H_a and H_a are discussed on pages 428–430.

Step 2   **The Hypothesis Test Criteria:**
   a. **Check the assumptions.**
       The assumptions are satisfied because the sampled population is approximately normal and the sample size is large enough for the CLT to apply (see p. 475).
   b. **Identify the probability distribution and the test statistic to be used.**
       σ is unknown; therefore, the t-distribution with df $= n - 1 = 21$ will be used, and the test statistic is $t\star$, formula (9.2).
   c. **Determine the level of significance:** $\alpha = 0.05$.

Step 3   **The Sample Evidence:**
   a. **Collect the sample information:** $n = 22$, $\bar{x} = 5.1$, and $s = 1.17$.
   b. **Calculate the value of the test statistic.**
       Use formula (9.2):

$$t\star = \frac{\bar{x} - \mu}{s/\sqrt{n}} : \quad t\star = \frac{5.1 - 4.9}{1.17/\sqrt{22}} = \frac{0.20}{0.2494} = 0.8018 = 0.80$$

Step 4   **The Probability Distribution:**

Using the p-value procedure: **OR**   **Using the classical procedure:**

a. **Calculate the p-value for the test statistic.**
Use the right-hand tail because H_a expresses concern for values related to "higher than."

a. **Determine the critical region and critical value(s).**
The critical region is the right-hand tail because H_a expresses concern for values related to "higher than." The

$\mathbf{P} = P(t \star > 0.80$, with df $= 21$) as shown on the figure.

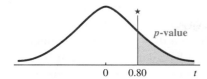

p-value

0   0.80   *t*

To find the *p*-value, use one of three methods:

1. Use Table 6 in Appendix B to place bounds on the *p*-value: $0.10 < \mathbf{P} < 0.25$.

2. Use Table 7 in Appendix B to read the value directly: $\mathbf{P} = 0.216$.

3. Use a computer or calculator to calculate the *p*-value: $\mathbf{P} = 0.2163$.

Specific details follow this example.

b. Determine whether or not the *p*-value is smaller than α.

The *p*-value is not smaller than α, the level of significance.

critical value is found at the intersection of the df $= 21$ row and the one-tailed 0.05 column of Table 6: $t(21, 0.05) = 1.72$.

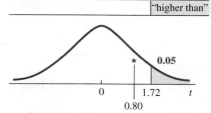

"higher than"

0.05

0   1.72   *t*
0.80

Specific instructions are given on pages 477–479.

b. Determine whether or not the calculated test statistic is in the critical region.

$t \star$ is not in the critical region, as shown in **red** in the figure above.

Step 5   The Results:

    a. State the decision about H_o: Fail to reject H_o.

    b. State the conclusion about H_a.

        At the 0.05 level of significance, the EPA does not have sufficient evidence to show that the mean carbon monoxide level is higher than 4.9.

Calculating the *p*-value when using the *t*-distribution

Method 1: Use Table 6 in Appendix B to place bounds on the p-value. By inspecting the df $= 21$ row of Table 6, you can determine an interval within which the *p*-value lies. Locate $t \star$ along the row labeled df $= 21$. If $t \star$ is not listed, locate the two table values it falls between, and read the bounds for the *p*-value from the top of the table. In this case, $t \star = 0.80$ is between 0.686 and 1.32; therefore, **P** is between 0.10 and 0.25. Use the one-tailed heading, since H_a is one-tailed in this illustration. (Use the two-tailed heading when H_a is two-tailed.)

Finding $\mathbf{P} = P(t \star > 0.80$, with df $= 21)$

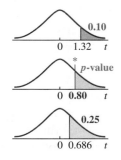

0.10

0   1.32   *t*

p-value

0   0.80   *t*

0.25

0   0.686   *t*

Portion of Table 6			
	Amount of α in One-Tail		
df	0.25	P	0.10
⋮	⋮	↑	⋮
⋮	⋮		⋮
21	0.686	0.80	1.32

$$0.10 < \mathbf{P} < 0.25$$

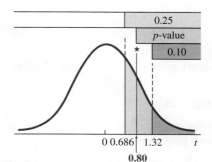

0.25

p-value

0.10

0 0.686 1.32   *t*
0.80

    The 0.686 entry in the table tells us that $P(t > 0.686) = 0.25$, as shown on the figure in purple. The 1.32 entry in the table tells us that $P(t > 1.32) = 0.10$, as shown in green. You can see that the *p*-value **P** (shown in blue) is between 0.10 and 0.25. Therefore, $0.10 < \mathbf{P} < 0.25$, and we say that 0.10 and 0.25 are the "bounds" for the *p*-value.

Method 2: Use Table 7 in Appendix B to read the p-value or to "place bounds" on the p-value. Table 7 is designed to yields *p*-values given the *t*★ and df values or produce bounds on **P** that are narrower than Table 6 produces.

In the preceding example, *t*★ = 0.80 and df = 21. These happen to be row and column headings, so the *p*-value can be read directly from the table. Locate the *p*-value at the intersection of the *t*★ = 0.80 row and the df = 21 column. The *p*-value for *t*★ = 0.80 with df = 21 is **0.216.**

Portion of Table 7

t★	df	. . .	21
⋮			
0.80			0.216

⟶ **P** = *P*(*t*★ > 0.80, with df = 21) = **0.216**

To illustrate how to place bounds on the *p*-value when *t*★ and df are not the heading values, let's consider the situation where *t*★ = 2.43 with df = 16. The *t*★ = 2.43 is between rows *t* = 2.4 and *t* = 2.5, while df = 16 is between columns df = 15 and df = 18. These two rows and two columns intersect a total of four times, namely at 0.015 and 0.014 in the row *t*★ = 2.4 and at 0.012 and 0.011 in the row *t*★ = 2.5. The *p*-value we are looking for is bounded by the smallest and largest of these four values, namely, 0.011 (lower right) and 0.015 (upper left). Therefore, the bounds for the *p*-value are 0.011 < **P** < 0.015.

Portion of Table 7

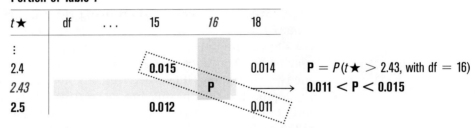

t★	df	. . .	15	16	18
⋮					
2.4			0.015		0.014
2.43				P	
2.5			0.012		0.011

P = *P*(*t*★ > 2.43, with df = 16)
0.011 < **P** < 0.015

Method 3: If you are doing the hypothesis test with the aid of a computer or calculator, most likely it will calculate the *p*-value for you, or you may use the cumulative probability distribution commands described on page 479.

Let's look at a two-tailed hypothesis-testing situation.

EXAMPLE 9.6

Two-Tailed Hypothesis Test for *μ* with *σ* Unknown

On a popular self-image test that results in normally distributed scores, the mean score for public-assistance recipients is expected to be 65. A random sample of 28 public-assistance recipients in Emerson County is given the test. They achieve a mean score of 62.1, and their scores have a standard deviation of 5.83. Do the Emerson County public-assistance recipients test differently, on average, than what is expected, at the 0.02 level of significance?

SOLUTION

Step 1   **The Set-Up:**
      a. Describe the population parameter of interest.

μ, the mean self-image test score for all Emerson County pub-lic-assistance recipients.

b. State the null hypothesis (H_o) and the alternative hypothesis (H_a).

H_o: $\mu = 65$ (mean is 65)

H_a: $\mu \neq 65$ (mean is different from 65)

Step 2   **The Hypothesis Test Criteria:**

a. Check the assumptions.

The test is expected to produce normally distributed scores; therefore, the assumption has been satisfied; σ is unknown.

b. Identify the probability distribution and the test statistic to be used.

The *t*-distribution with df $= n - 1 = 27$, and the test statistic is $t\star$, formula (9.2).

c. Determine the level of significance: $\alpha = 0.02$ (given in statement of problem).

Step 3   **The Sample Evidence:**

a. Collect the sample information: $n = 28$, $\bar{x} = 62.1$, and $s = 5.83$.

b. Calculate the value of the test statistic.

Use formula (9.2):

$$t\star = \frac{\bar{x} - \mu}{s/\sqrt{n}} \; : \; t\star = \frac{62.1 - 65.0}{5.83/\sqrt{28}} = \frac{-2.9}{1.1018} = -2.632 = -2.63$$

Step 4   **The Probability Distribution:**

OR

Using the *p*-value procedure:

a. Calculate the *p*-value for the test statistic.

Use both tails because H_a expresses concern for values related to "different from."

$\mathbf{P} = P(t < -2.63) + P(t > 2.63) = 2 \cdot P(t > 2.63)$, with df $= 27$ as shown in the figure.

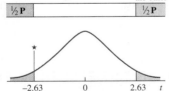

To find the *p*-value, use one of three methods:

1. Use Table 6 in Appendix B to place bounds on the *p*-value: $0.01 < \mathbf{P} < 0.02$.

2. Use Table 7 in Appendix B to place bounds on the *p*-value: $0.012 < \mathbf{P} < 0.016$.

3. Use a computer or calculator to calculate the *p*-value: $\mathbf{P} = 0.0140$.

Specific details follow this example:

b. Determine whether or not the *p*-value is smaller than α.

The *p*-value is smaller than the level of significance, α.

Using the classical procedure:

a. Determine the critical region and critical value(s).

The critical region is both tails because H_a expresses concern for values related to "different from." The critical value is found at the intersection of the df $= 27$ row and the one-tailed 0.01 column of Table 6: $t(27, 0.01) = 2.47$.

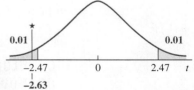

b. Determine whether or not the calculated test statistic is in the critical region.

$t\star$ is in the critical region, as shown in **red** in the preceding figure.

Step 5 **The Results:**
   a. **State the decision about H_o:** Reject H_o.
   b. **State the conclusion about H_a.**
      At the 0.02 level of significance, we do have sufficient evidence to conclude that the Emerson County assistance recipients test significantly different, on average, from the expected 65.

Calculating the *p*-value when using the *t*-distribution

Method 1: Using Table 6, find 2.63 between two entries in the df = 27 row and read the bounds for **P** from the two-tailed heading at the top of the table:

$$0.01 < P < 0.02.$$

Method 2: Generally, bounds found using Table 7 will be narrower than bounds found using Table 6. The following table shows you how to read the bounds from Table 7; find $t\star = 2.63$ between two rows and df = 27 between two columns, and locate the four intersections of these columns and rows. The value of $\frac{1}{2}$ **P** is bounded by the upper left and the lower right of these table entries.

Portion of Table 7

		Degrees of Freedom		
$t\star$	25	27	29	$P = 2P(t\star > 2.63,$ with df $= 27)$
⋮		↓	⋮	
2.6	0.008		0.007	
2.63		½ P		$0.006 < \frac{1}{2}P < 0.008$
2.7	0.006		0.006	$0.012 < P < 0.016$

Method 3: If you are doing the hypothesis test with the aid of a computer or calculator, most likely it will calculate the *p*-value for you (do not double it). Or you may use the cumulative probability distribution commands described on page 479.

TECHNOLOGY INSTRUCTIONS: HYPOTHESIS TEST FOR MEAN μ WHEN σ UNKNOWN

MINITAB (Release 14)   Input the data into C1; then continue with:

```
Choose:    Stat > Basic Statistics > 1-Sample t
Enter:     Samples in columns: C1
           Test mean: μ (value in H₀)
Select:    Options
Select:    Alternative: less than or not equal or greater than > OK >
           OK
```

Excel Input the data into column A; then continue with:

Choose: **Tools > Data Analysis Plus > t-Test: Mean > OK**
Enter: Input Range: **(A1:A20 or select cells) > OK**
         Hypothesized Mean: **μ**
         Alpha: **α** (ex. 0.05) > OK
         Gives *p*-values and critical values for both one-tailed and two-tailed tests.

TI-83/84 Plus Input the data into L1; then continue with the following, entering the appropriate values and highlighting Calculate:

Choose: **STAT > TESTS > 2:T-Test**

```
T-Test
 Inpt:DATA Stats
 μ0:0
 List:L1
 Freq:1
 μ:≠μ0 <μ0 >μ0
 Calculate Draw
```

FYI Compare the MINITAB results to the solution found in Example 9.6.

Here is the MINITAB solution to Example 9.6:

```
One-Sample T: C1
Test of mu 5 65 vs not 5 65
```

Variable	N	Mean	StDev	SE Mean	T	P
C1	28	62.1	5.83	1.102	22.63	0.0140

APPLIED EXAMPLE 9.7

Mothers' Use of Personal Pronouns When Talking with Toddlers

The calculated *t*-value and the probability value for five different hypothesis tests are given in the following article. The expression $t(44) = 1.92$ means $t\star = 1.92$ with df = 44 and is significant with *p*-value < 0.05. Can you verify the *p*-values? Explain.

ABSTRACT

The verbal interaction of 2-year-old children ($N = 46$; 16 girls, 30 boys) and their mothers was audiotaped, transcribed, and analyzed for the use of personal pronouns, the total number of utterances, the child's mean length of utterance, and the mother's responsiveness to her child's utterances. Mothers' use of the personal pronoun "we" was significantly related to their children's performance on the Stanford-Binet at age 5 and the Wechsler Intelligence Scale for Children at age 8. Mothers' use of "we" in social-vocal interchange, indicating a system for establishing a shared relationship with the child, was closely connected with their verbal responsiveness to their children. The total amount of ma-

ternal talking, the number of personal pronouns used by mothers, and their verbal responsiveness to their children were not related to mothers' social class or years of education.

Mothers tended to use more first person singular pronouns (I and me), $t(44) = 1.81, p < .10$, and used significantly more first person plural pronouns (we), $t(44) = 1.92$, $p < .05$, with female children than with male children. The mothers also were more verbally responsive to their female children, $t(44) = 2.0, p < .06$.

In general, mothers talked more to their first born children, $t(44) = 3.41$, $p < .001$, and were more responsive to their first born children, $t(44) = 3.71$, $p < .001$. Yet, the proportion of personal pronouns used when speaking to first born children was not different from that used when speaking to later born children.

Source: Dan R. Laks, Leila Beckwith, and Sarale E. Cohen, *The Journal of Genetic Psychology*, 151(1), 25–32, 1990. Reprinted with permission of the Helen Dwight Reid Educational Foundation. Published by Heldref Publications, 1319 Eighteenth St. N.W., Washington, D.C., 20036-1802. Copyright © 1990.

SECTION 9.2 EXERCISES

Statistics⊖Now™

Datasets can be found on your Student's Suite CD-ROM or at the StatisticsNow website at **http://1pass.thomson.com**.

9.3 Make a list of four numbers that total "zero." How many numbers were you able to pick without restriction? Explain how this demonstrates degrees of freedom.

9.4 Explain the relationship between the critical values found in the bottom row of Table 6 and the critical values of z given in Table 4A.

9.5 Find:

a. $t(12, 0.01)$ b. $t(22, 0.025)$

c. $t(50, 0.10)$ d. $t(8, 0.005)$

9.6 Find these critical values using Table 6 in Appendix B:

a. $t(25, 0.05)$ b. $t(10, 0.10)$

c. $t(15, 0.01)$ d. $t(21, 0.025)$

9.7 Find:

a. $t(18, 0.90)$ b. $t(9, 0.99)$

c. $t(35, 0.975)$ d. $t(14, 0.98)$

9.8 Find these critical values using Table 6 in Appendix B:

a. $t(21, 0.95)$ b. $t(26, 0.975)$

c. $t(27, 0.99)$ d. $t(60, 0.025)$

9.9 Using the notation of Exercise 9.8, name and find the following critical values of t:

a.
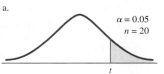

b.
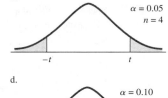

c.
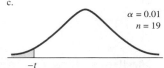

d.

9.10 Using the notation of Exercise 9.8, name and find the following critical values of t:

a.
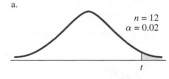

b.

c.

d.

9.11 Find the values of t that bound the middle 0.95 of the distribution for df $= 12$.

9.12 Find the values of t that bound the middle 0.80 of the distribution for df $= 26$.

9.13 a.   Find the first percentile of Student's t-distribution with 24 degrees of freedom.

b.   Find the 95th percentile of Student's t-distribution with 24 degrees of freedom.

c. Find the first quartile of Student's *t*-distribution with 24 degrees of freedom.

9.14 Find the percent of the Student's *t*-distribution that lies between the following values:

a. df = 12 and *t* ranges from −1.36 to 2.68

b. df = 15 and *t* ranges from −1.75 to 2.95

9.15 Ninety percent of Student's *t*-distribution lies between $t = -1.89$ and $t = 1.89$ for how many degrees of freedom?

9.16 Ninety percent of Student's *t*-distribution lies to the right of $t = -1.37$ for how many degrees of freedom?

9.17 Use a computer or calculator to find the area to the left of $t = -2.12$ with df = 18. Draw a sketch showing the question with the answer.

9.18 Use a computer or calculator to find the area to the right of $t = 1.12$ with df = 15. Draw a sketch showing the question with the answer.

9.19 a. State two ways in which the standard normal distribution and Student's *t*-distribution are alike.

b. State two ways in which they are different.

9.20 The variance for each of Student's *t*-distributions is equal to $df/(df - 2)$. Find the standard deviation for a Student's *t*-distribution with each of the following degrees of freedom:

a. 10   b. 20   c. 30

In summary:

d. Explain how this verifies Property 5 of the *t*-distributions listed on p. 476.

9.21 Construct a 95% confidence interval estimate for the mean μ using the sample information $n = 24$, $\bar{x} = 16.7$, and $s = 2.6$.

9.22 In a study of 25 criminals convicted of antitrust offenses, the average age was 54 years, with a standard deviation of 7.5 years. Construct a 90% confidence interval on the true mean age.

9.23 A survey of 3000 randomly selected Minnesotans aged 65 and older revealed that, on average, they spent $85 per month on prescription drugs, with a standard deviation of $50.35 per month. Construct a 99% confidence interval for the true mean amount spent per month.

9.24 The Robertson square drive screw was invented in 1908, but it has gained in popularity with American woodworkers and home craftspeople only within the last 10 years. The advantages of square drives over conventional screws is indeed remarkable—most notably greater strength, increased holding power, and reduced driving resistance and "cam-out." Strength test results published in McFeely's 2005 catalog revealed that the no. 8 Robertson square drive flat head steel screws fail only after an average of 46 inch-pounds of torque is applied, a strength nearly 50% greater than that of the more common slotted- or Phillips-head wood screw.

Source: McFeely's Square Drive Screws, 2005.

Suppose an independent testing laboratory randomly selects 22 square drive flat head steel screws from a box of 1000 screws and obtains a mean failure torque of 45.2 inch-pounds and a standard deviation of 5.1 inch-pounds. Estimate with 95% confidence the mean failure torque of the no. 8 wood screws based on the study by the independent laboratory. Specify the population parameter of interest, the criteria, the sample evidence, and the interval limits.

9.25 While writing an article on the high cost of college education, a reporter took a random sample of the cost of new textbooks for a semester. The random variable *x* is the cost of one book. Her sample data can be summarized by $n = 41$, $\sum x = 3582.17$, and $\sum (x - \bar{x})^2 = 9960.336$.

a. Find the sample mean, $\bar{x}$.

b. Find the sample standard deviation, *s*.

c. Find the 90% confidence interval to estimate the true mean textbook cost for the semester based on this sample.

9.26 [EX09-026] Lunch breaks are often considered too short, and employees frequently develop a habit of "stretching" them. The manager at Giant Mart randomly identified 22 employees and observed the length of their lunch breaks (in minutes) for one randomly selected day during the week:

30	24	38	35	27	35	23	28	28	22	26
34	29	25	28	34	24	26	28	32	29	40

a. Show evidence that the normality assumptions are satisfied.

b. Find the 95% confidence interval for "mean length of lunch breaks" at Giant Mart.

9.27 [EX09-027] Our modern meat or chicken "broiler" can weigh 5.5 lb at 49 days, a common time to sell the broiler to a meat processor. Before a poultry farmer sells her broilers, she wants to estimate the mean weight of the flock. She selects a random sample of 15 from the current 7-week old broilers and obtains their weights (in ounces):

74.9	74.2	73.9	72.6	70.4	66.9	76.6	73.3
75.4	78.9	76.6	75.2	73.6	78.6	70.4	

a. How many ounces is 5.5 lb? How many pounds is 74.9 oz?

b. Show evidence that the normality assumptions are satisfied.

c. Calculate the sample statistics mean and standard deviation.

d. Estimate the mean weight per broiler using a 98% confidence interval.

e. If the farmer sells 1000 broilers, construct a 98% confidence interval for total weight, in pounds.

9.28 [EX09-028] The addition of a new accelerator is claimed to decrease the drying time of latex paint by more than 4%. Several test samples were conducted with the following percentage decrease in drying time.

5.2	6.4	3.8	6.3	4.1	2.8	3.2	4.7

Assume that the percentage decrease in drying time is normally distributed.

a. Find the 95% confidence interval for the true mean decrease in the drying time based on this sample. (The sample mean and standard deviation were found in answering Exercise 2.181, p. 129.)

b. Did the interval estimate reached in part a result in the same conclusion as you expressed in answering part c of Exercise 2.181 for these same data?

9.29 [EX09-029] The pulse rates for 13 adult women were as follows:

83	58	70	56	76	64	80	76	70	97	68	78	108

Verify the results shown on the last line of the MINITAB output:

```
MTB > TINTERVAL 90 PERCENT CONFIDENCE INTERVAL FOR DATA
IN C1
        N    MEAN    STDEV   SE MEAN    90% CI
C1     13   75.69   14.54    4.03    (68.50, 82.88)
```

9.30 Using the computer output in Exercise 9.29, determine the value for each of the following:

a. Point estimate   b. Confidence coefficient

c. Standard error of the mean

d. Maximum error of estimate, E

e. Lower confidence limit

f. Upper confidence limit

9.31 Use a computer or calculator to construct a 0.98 confidence interval using the sample data:

6	7	12	9	10	8	5	9	7	9	6	5

9.32 [EX09-032] James Short (1708–1768), a Scottish optician, constructed the highest-quality reflectors of his time. It was with these reflectors that Short obtained the following measurements of the parallax of the sun (in seconds of a degree), based on the 1761 transit of Venus. The parallax of the sun is the angle α subtended by the earth, as seen from the surface of the sun. (See accompanying diagram.)

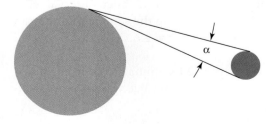

8.50	8.50	7.33	8.64	9.27	9.06	9.25	9.09	8.50
8.06	8.43	8.44	8.14	7.68	10.34	8.07	8.36	9.71
8.65	8.35	8.71	8.31	8.36	8.58	7.80	7.71	8.30
9.71	8.50	8.28	9.87	8.86	5.76	8.44	8.23	8.50
8.80	8.40	8.82	9.02	10.57	9.11	8.66	8.34	8.60
7.99	8.58	8.34	9.64	8.34	8.55	9.54	9.07	

Source: The data and descriptive information are based on material from Stephen M. Stigler. (1977). Do robust estimators work with real data? *Annals of Statistics, 5,* 1055–1098.

a. Determine whether an assumption of normality is reasonable. Explain.

b. Construct a 95% confidence interval for the estimate of the mean parallax of the sun.

c. If the true value is 8.798 seconds of a degree, what does the confidence interval suggest about Short's measurements?

9.33 [EX09-033] The fuel economy information on a new SUV window sticker indicates that its new owner can expect 16 mpg (miles per gallon) in city driving and 20 mpg for highway driving and 18 mpg overall. Accurate gasoline records for one such vehicle were kept, and a random sample of mileage per tank of gasoline was collected:

17.6	17.7	18.1	22.0	17.0	19.4	18.9	17.4	21.0	19.2
18.3	19.1	20.7	16.7	19.4	18.2	18.4	17.1	17.4	15.8
17.9	18.0	16.3	17.5	17.3	20.4	19.1	21.0	18.1	19.0
19.6	18.9	16.8	18.2	17.6	19.1	18.0	16.8	20.9	17.9
17.7	20.3	18.6	19.0	16.5	19.4	18.6	18.6	17.3	18.7

a. Determine whether an assumption of normality is reasonable. Explain.

b. Construct a 95% confidence interval for the estimate of the mean mileage per gallon.

c. What does the confidence interval suggest about SUVs' fuel economy expectations as expressed on the window sticker?

9.34 [EX09-034] College students throw out an average of 640 lb of trash each year, 30% of that in the month before graduation according to the June 2004 *Readers Digest* article "Campus Castoffs." Last year, 20 universities raised more than $100,000 for local charities by selling the "junk." To estimate the amount of trash discarded by the students at State University, 18 students were randomly selected and carefully monitored for 1 year. The amounts of trash discarded (in pounds) were as follows:

| 692 | 563 | 482 | 413 | 437 | 456 | 509 | 347 | 676 |
| 944 | 678 | 392 | 399 | 742 | 584 | 492 | 514 | 758 |

a. Describe the data using a histogram and its mean and standard deviation.

b. Determine whether an assumption of normality is reasonable. Explain.

c. Construct a 95% confidence interval for the mean amount of trash.

d. What does the confidence interval suggest about the mean amount of trash discarded by the students at State University compared with that discarded by all college students.

9.35 State the null hypothesis, H_o, and the alternative hypothesis, H_a, that would be used to test each of the following claims:

a. The mean weight of honeybees is at least 11 grams.

b. The mean age of patients at Memorial Hospital is no more than 54 years.

c. The mean amount of salt in granola snack bars is different from 75 mg.

9.36 State the null hypothesis, H_o, and the alternative hypothesis, H_a, that would be used to test each of the following claims:

a. A chicken farmer at Best Broilers claims that his chickens have a mean weight of 56 oz.

b. The mean age of U.S. commercial jets is less than 18 years.

c. The mean monthly unpaid balance on credit card accounts is more than $400.

9.37 Calculate the value of $t\star$ for the hypothesis test: H_o: $\mu = 32$, H_a: $\mu > 32$, $n = 16$, $\bar{x} = 32.93$, $s = 3.1$.

9.38 Calculate the value of $t\star$ for the following hypothesis test: H_o: $\mu = 73$, H_a: $\mu \neq 73$, $n = 12$, $\bar{x} = 71.46$, $s = 4.1$.

9.39 Determine the *p*-value for the following hypothesis tests involving Student's *t*-distribution with 10 degrees of freedom.

a. $H_o: \mu = 15.5$, $H_a: \mu < 15.5$, $t\star = -2.01$

b. $H_o: \mu = 15.5$, $H_a: \mu > 15.5$, $t\star = 2.01$

c. $H_o: \mu = 15.5$, $H_a: \mu \neq 15.5$, $t\star = 2.01$

d. $H_o: \mu = 15.5$, $H_a: \mu \neq 15.5$, $t\star = -2.01$

9.40 Determine the critical region and critical value(s) that would be used in the classical approach to test the following null hypotheses:

a. $H_o: \mu = 10$, $H_a: \mu \neq 10$ ($\alpha = 0.05$, $n = 15$)

b. $H_o: \mu = 37.2$, $H_a: \mu > 37.2$ ($\alpha = 0.01$, $n = 25$)

c. $H_o: \mu = -20.5$, $H_a: \mu < -20.5$ ($\alpha = 0.05$, $n = 18$)

d. $H_o: \mu = 32.0$, $H_a: \mu > 32.0$ ($\alpha = 0.01$, $n = 42$)

9.41 a.   Find the value of **P** and state the decision for the hypothesis test in Exercise 9.37 using $\alpha = 0.05$.

  b.   Find the critical region and critical value and state the decision for the hypothesis test in Exercise 9.37 using $\alpha = 0.05$.

9.42 a.   Use Table 6 or Table 7 in Appendix B to find the value of **P** for the hypothesis test in Exercise 9.38; state the decision using $\alpha = 0.05$.

  b.   Find the critical region and critical values for the hypothesis test in Exercise 9.38; state the decision using $\alpha = 0.05$.

9.43 Use a computer or calculator to find the *p*-value for the hypothesis test: $H_o: \mu = 32$, $H_a: \mu > 32$, $n = 16$, $\bar{x} = 32.93$, $s = 3.1$.

9.44 Use a computer or calculator to find the *p*-value for the following hypothesis test: $H_o: \mu = 73$, $H_a: \mu \neq 73$, $n = 12$, $\bar{x} = 71.46$, $s = 4.1$.

9.45 Use both the *p*-value and classical approaches to hypothesis testing to reach a decision for each of the following situations. Use $\alpha = 0.05$.

a. $H_o: \mu = 128$, $H_a: \mu \neq 128$, $n = 15$, $t\star = 1.60$

b. $H_o: \mu = 18$, $H_a: \mu > 18$, $n = 25$, $t\star = 2.16$

c. $H_o: \mu = 38$, $H_a: \mu < 38$, $n = 45$, $t\star = -1.73$

d.   Compare the results of the two techniques for each case.

9.46 In reference to Applied Example 9.7 (p. 487):

a.   Verify that $t(44) = 1.92$ is significant at the 0.05 level.

b.   Verify that $t(44) = 3.41$ is significant at the 0.01 level.

c.   Explain why $t(44) = 1.81$, $p < .10$, makes sense only if the hypothesis test is two-tailed.

d.   If the test is one-tailed, what level would be reported?

9.47 A student group maintains that each day, the average student must travel for at least 25 minutes one way to reach college. The college admissions office obtained a random sample of 31 one-way travel times from students. The sample had a mean of 19.4 minutes and a standard deviation of 9.6 minutes. Does the admissions office have sufficient evidence to reject the students' claim? Use $\alpha = 0.01$.

a.   Solve using the *p*-value approach.

b.   Solve using the classical approach.

9.48 Homes in a nearby college town have a mean value of $88,950. It is assumed that homes in the vicinity of the college have a higher mean value. To test this theory, a random sample of 12 homes is chosen from the college area. Their mean valuation is $92,460, and the standard deviation is $5200. Complete a hypothesis test using $\alpha = 0.05$. Assume prices are normally distributed.

a.   Solve using the *p*-value approach.

b.   Solve using the classical approach.

9.49 Consumers enjoy the deep selection of merchandise made possible by specialty stores that sacrifice breadth for greater depth. Consider stores that carry only Levi Strauss pants. The company reports that a fully stocked Levi's store carries 130 ready-to-wear pairs of jeans for any given waist and inseam, and the company is phasing in two more lines of pants (Personal Pair and Original Spin) that it claims will eventually quadruple that number.

Source: Fortune, "The Customized, Digitized, Have-It-Your-Way Economy"

Suppose a random sample of 24 Levi stores is sampled 2 months after the phase-in process has been launched and inventories are taken at each of the stores in the sample for all sizes of jeans. The sample mean number of choices for any given size is 141.3, and the standard deviation is 36.2. Does this sample of stores carry a greater selection of jeans, on average, than what is expected at the 0.01 level of significance?

a. Solve using the *p*-value approach.

b. Solve using the classical approach.

9.50 Up all night? Caffeine crave may cause long-term health problems. Homework, jobs, and studying all may be causes for teens to consume too much coffee in their everyday lives. Health officials warn that high caffeine intake is not good for anyone, but coffee drinking continues to become more and more popular. It is not the coffee that is of concern; it is the amount of caffeine. A moderate amount of caffeine is nothing to worry about, health experts say. There are no health risks drinking three 8-oz cups of regular coffee, which is about 250 mg of caffeine each day, according to the Henry Ford Health System.

Source: New Expressions, http://www.newexpression.org/main/cover/

A nationwide random sample of college students revealed that 24 students consumed a total of 5428 mg of caffeine each day, with a standard deviation of 48 mg. Assuming that the amount of caffeine consumed per person daily is normally distributed, is there sufficient evidence to conclude that the mean amount of caffeine consumed daily by college students is less than 250 mg, using $\alpha = 0.05$?

a. Complete the test using the *p*-value approach. Include $t\star$, *p*-value, and your conclusion.

b. Complete the test using the classical approach. Include the critical values, $t\star$, and your conclusion.

9.51 [EX09-051] To test the null hypothesis "the mean weight for adult males equals 160 lb" against the alternative "the mean weight for adult males exceeds 160 lb," the weights of 16 males were obtained:

173	178	145	146	157	175	173	137
152	171	163	170	135	159	199	131

Assume normality and verify the results shown on the following MINITAB analysis by calculating the values yourself.

```
TEST OF MU = 160.00 VS MU > 160.00
          N      MEAN    STDEV   SE MEAN     T       P
C1       16    160.25    18.49    4.62     0.05    0.48
```

9.52 Using the computer output in Exercise 9.51, determine the values of the following terms:

a. Hypothesized value of population mean

b. Sample mean

c. Population standard deviation

d. Sample standard deviation

e. Test statistic

9.53 [EX09-053] Use a computer or calculator to complete the hypothesis test: H_o: $\mu = 52$, H_a: $\mu < 52$, $\alpha = 0.01$ using the data:

45	47	46	58	59	49	46	54	53	52	47	41

9.54 [EX09-054] The recommended number of hours of sleep per night is 8 hours, but everybody "knows" that the average college student sleeps less than 7 hours. The number of hours slept last night by 10 randomly selected college students is listed here:

5.2	6.8	6.2	5.5	7.8	5.8	7.1	8.1	6.9	5.6

Use a computer or calculator to complete the hypothesis test: H_o: $\mu = 7$, H_a: $\mu < 7$, $\alpha = 0.05$.

9.55 It is claimed that the students at a certain university will score an average of 35 on a given test. Is the claim reasonable if a random sample of test scores from this university yields 33, 42, 38, 37, 30, 42? Complete a hypothesis test using $\alpha = 0.05$. Assume test results are normally distributed.

a. Solve using the *p*-value approach.

b. Solve using the classical approach.

9.56 [EX02-182] Gasoline pumped from a supplier's pipeline is supposed to have an octane rating of 87.5. On 13 consecutive days, a sample was taken and analyzed with the following results.

88.6 86.4 87.2 88.4 87.2 87.6 86.8 86.1 87.4 87.3 86.4 86.6 87.1

a. If the octane ratings have a normal distribution, is there sufficient evidence to show that these octane readings were taken from gasoline with a mean octane significantly less than 87.5 at the 0.05 level? (The sample mean and standard deviation were found in answering Exercise 2.182, p. 129.)

b. Did the statistical decision reached in part a result in the same conclusion as you expressed in answering part c of Exercise 2.182 for these same data?

9.57 [EX09-057] The density of the earth relative to the density of water is known to be 5.517 g/cm³. Henry Cavendish, an English chemist and physicist (1731–1810), was the first scientist to accurately measure the density of the earth. Following are 29 measurements taken by Cavendish in 1798 using a torsion balance.

5.50 5.61 4.88 5.07 5.26 5.55 5.36 5.29 5.58 5.65 5.57
5.53 5.62 5.29 5.44 5.34 5.79 5.10 5.27 5.39 5.42 5.47
5.63 5.34 5.46 5.30 5.75 5.68 5.85

Source: The data and descriptive information are based on material from "Do robust estimators work with real data?" by Stephen M. Stigler, *Annals of Statistics,* *5* (1977), 1055–1098.

a. What evidence do you have that the assumption of normality is reasonable? Explain.

b. Is the mean of Cavendish's data significantly less than today's recognized standard? Use a 0.05 level of significance.

9.58 [EX09-058] Use a computer or calculator to complete the calculations and the hypothesis test for this exercise. Delco Products, a division of General Motors, produces commutators designed to be 18.810 mm in overall length. (A commutator is a device used in the electrical system of an automobile.) The following data are the lengths of a sample of 35 commutators taken while monitoring the manufacturing process:

18.802	18.810	18.780	18.757	18.824	18.827	18.825
18.809	18.794	18.787	18.844	18.824	18.829	18.817
18.785	18.747	18.802	18.826	18.810	18.802	18.780
18.830	18.874	18.836	18.758	18.813	18.844	18.861
18.824	18.835	18.794	18.853	18.823	18.863	18.808

Source: With permission of Delco Products Division, GMC

Is there sufficient evidence to reject the claim that these parts meet the design requirements "mean length is 18.810" at the $\alpha = 0.01$ level of significance?

9.59 Acetaminophen is an active ingredient found in more than 600 over-the-counter and prescription medicines, such as pain relievers, cough suppressants, and cold medications. It is safe and effective when used correctly, but taking too much can lead to liver damage.

Source: http://www.keepkidshealthy.com/medicine_cabinet/acetaminophen.html

A researcher believes the amount of acetaminophen in a particular brand of cold tablets contains a mean amount of acetaminophen per tablet different from the 600 mg claimed by the manufacturer. A random sample of 30 tablets had a mean acetaminophen content of 596.3 mg with a standard deviation of 4.7 mg.

a. Is the assumption of normality reasonable? Explain.

b. Construct a 99% confidence interval for the estimate of the mean acetaminophen content.

c. What does the confidence interval found in part b suggest about the mean acetaminophen content of one pill? Do you believe there is 600 mg per tablet? Explain.

9.60 [EX09-060] A winemaker has placed a large order for the no. 9 corks described in Applied Example 6.15 (p. 330) and is concerned about the number of corks that might have smaller diameters. During the corking process, the corks are squeezed down to 16 to 17 mm in diameter for insertion into bottles with an 18 mm opening. The cork then expands to make the seal. The wine-maker wants the corks to be as tight as possible and is therefore concerned about any that might be undersized. The diameter of each cork is measured in several places, and an average diameter is reported for each cork. The cork manufacturer has assured the winemaker that each cork has an average diameter within the specs and that all average diameters have a normal distribution with a mean of 24.0 mm.

a. Why does it make sense for the diameter of the cork to be assigned the average of several different diameter measurements?

A random sample of 18 corks is taken from the batch to be shipped and the diameters (in millimeters) obtained:

| 23.93 | 23.91 | 23.82 | 24.02 | 23.93 | 24.17 | 23.93 | 23.84 | 24.13 |
| 24.01 | 23.83 | 23.74 | 23.73 | 24.10 | 23.86 | 23.90 | 24.32 | 23.83 |

b. The average diameter spec is "24 mm + 0.6mm/−0.4 mm." Does it appear this order meets the spec on an individual cork basis? Explain.

c. Does the sample in part a show sufficient reason to doubt the truthfulness of the claim, the mean average diameter is 24.0 mm, at the 0.02 level of significance?

A different sample of 18 corks was randomly selected and the diameters (in millimeters) obtained:

| 23.90 | 23.98 | 24.28 | 24.22 | 24.07 | 23.87 | 24.05 | 24.06 | 23.82 |
| 24.03 | 23.87 | 24.08 | 23.98 | 24.21 | 24.08 | 24.06 | 23.87 | 23.95 |

d. Does the preceding sample show sufficient reason to doubt the truthfulness of the claim, the mean average diameter is 24.0 mm, at the 0.02 level of significance?

e. What effect did the two different sample means have on the calculated test statistic in parts c and d? Explain.

f. What effect did the two different sample standard deviations have on the calculated test statistic in parts c and d? Explain.

9.61 [EX09-061] Length is not very important in evaluating the quality of corks because it has little to do with the effectiveness of a cork in preserving wine. Winemakers have several lengths to choose from and order the length cork they prefer (long corks tend to make a louder pop when the bottle is uncorked). Length is monitored very closely, though, because it is a specified quality of the cork. The lengths of no. 9 natural corks (24 mm diameter by 45 mm length) have a normal distribution. Twelve randomly selected corks were measured to the nearest hundredth of a millimeter.

| 44.95 | 44.95 | 44.80 | 44.93 | 45.22 | 44.82 |
| 45.12 | 44.62 | 45.17 | 44.60 | 44.60 | 44.75 |

a. Does the preceding sample give sufficient reason to show that the mean length is different than 45.0 mm, at the 0.02 level of significance?

A different random sample of 18 corks is taken from the same batch.

| 45.17 | 45.02 | 45.30 | 45.14 | 45.35 | 45.50 | 45.26 | 44.88 | 44.71 |
| 44.07 | 45.10 | 45.01 | 44.83 | 45.13 | 44.69 | 44.89 | 45.15 | 45.13 |

b. Does the preceding sample give sufficient reason to show that the mean length is different than 45.0 mm, at the 0.02 level of significance?

c. What effect did the two different sample means have on the calculated test statistic in parts a and b? Explain.

d. What effect did the two different sample sizes have on the calculated test statistic in parts a and b? Explain.

e. What effect did the two different sample standard deviations have on the calculated test statistic in parts a and b? Explain.

9.62 How important is the assumption, "the sampled population is normally distributed," to the use of Student's t-distribution? Using a computer, simulate drawing 100 samples of size 10 from each of three different types of population distributions, namely, a normal, a uniform, and an exponential. First generate 1000 data values from the population and construct a histogram to see what the population looks like. Then generate 100 samples of size 10 from the same population; each row represents a sample. Calculate the mean and standard deviation for each of the 100 samples. Calculate $t\star$ for each of the 100 samples. Construct histograms of the 100 sample means and the 100 $t\star$ values. (Additional details can be found in the *Student Solutions Manual*.)

For the samples from the normal population:

a. Does the $\bar{x}$ distribution appear to be normal? Find percentages for intervals and compare with the normal distribution.

b. Does the distribution of $t\star$ appear to have a t-distribution with df = 9? Find percentages for intervals and compare them with the t-distribution.

For the samples from the rectangular or uniform population:

c. Does the $\bar{x}$ distribution appear to be normal? Find percentages for intervals and compare them with the normal distribution.

d. Does the distribution of $t\star$ appear to have a t-distribution with df = 9? Find percentages for intervals and compare them with the t-distribution.

For the samples from the skewed (exponential) population:

e. Does the $\bar{x}$ distribution appear to be normal? Find percentages for intervals and compare them with the normal distribution.

f. Does the distribution of $t\star$ appear to have a t-distribution with df = 9? Find percentages for intervals and compare them with the t-distribution.

In summary:

g. In each of the preceding three situations, the sampling distribution for $\bar{x}$ appears to be slightly different than the distribution of $t\star$. Explain why.

h. Does the normality condition appear to be necessary in order for the calculated test statistic $t\star$ to have a Student's t-distribution? Explain.

9.3 Inferences about the Binomial Probability of Success

Perhaps the most common inference involves the **binomial parameter p,** the "probability of success." Yes, every one of us uses this inference, even if only casually. In thousands of situations we are concerned about something either "happening" or "not happening." There are only two possible outcomes of concern, and that is the fundamental property of a **binomial experiment.** The other necessary ingredient is multiple independent trials. Asking five people whether they are "for" or "against" some issue can create five independent trials; if 200 people are asked the same question, 200 independent trials may be involved; if 30 items are inspected to see if each "exhibits a particular property" or "not," there will be 30 repeated trials; these are the makings of a binomial inference.

The binomial parameter p is defined to be the probability of success on a single trial in a binomial experiment.

Sample Binomial Probability

$$p' = \frac{x}{n} \tag{9.3}$$

where the **random variable x** represents the number of successes that occur in a sample consisting of n trials

FYI Complete details about binomial experimentation can be found on pages 287–290.

Recall that the mean and standard deviation of the binomial random variable x are found by using formula (5.7), $\mu = np$, and formula (5.8), $\sigma = \sqrt{npq}$, where $q = 1 - p$. The distribution of x is considered to be approximately normal if n is greater than 20 and if np and nq are both greater than 5. This commonly accepted *rule of thumb* allows us to use the **standard normal distribution** to estimate probabilities for the binomial random variable x, the number of successes in n trials, and to make inferences concerning the binomial parameter p, the probability of success on an individual trial.

Generally, it is easier and more meaningful to work with the distribution of p' (the observed probability of occurrence) than with x (the number of occurrences). Consequently, we will convert formulas (5.7) and (5.8) from units of x (integers) to units of proportions (percentages expressed as decimals) by dividing each formula by n, as shown in Table 9.1.

TABLE 9.1

Formulas (9.4) and (9.5)

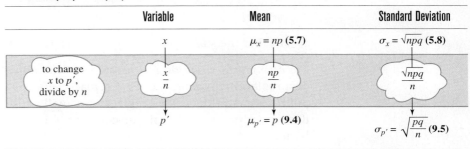

	Variable	Mean	Standard Deviation
	x	$\mu_x = np$ **(5.7)**	$\sigma_x = \sqrt{npq}$ **(5.8)**
to change x to p', divide by n	$\dfrac{x}{n}$	$\dfrac{np}{n}$	$\dfrac{\sqrt{npq}}{n}$
	p'	$\mu_{p'} = p$ **(9.4)**	$\sigma_{p'} = \sqrt{\dfrac{pq}{n}}$ **(9.5)**

Recall that $\mu_{p'} = p$ and that the *sample statistic p'* is an **unbiased estimator for p.** Therefore, the information about the sampling distribution of p' is summarized as follows:

> If a random sample of size n is selected from a large population with $p = P(\text{success})$, then the sampling distribution of p' has:
>
> 1. A mean $\mu_{p'}$ equal to p
> 2. A standard error $\sigma_{p'}$ equal to $\sqrt{\dfrac{pq}{n}}$
> 3. An approximately normal distribution if n is sufficiently large

In practice, using these guidelines will ensure normality:

1. The sample size is greater than 20.
2. The products np and nq are both greater than 5.
3. The sample consists of less than 10% of the population.

We are now ready to make inferences about the population parameter p. Use of the z-distribution involves an assumption.

> **The assumptions for inferences about the binomial parameter p:** The n random observations that form the sample are selected independently from a population that is not changing during the sampling.

Confidence Interval Procedure

FYI The standard deviation of a sampling distribution is called the "standard error."

Inferences concerning the population binomial parameter p, P(success), are made using procedures that closely parallel the inference procedures used for the population mean μ. When we estimate the **population proportion p,** we will base our estimations on the **unbiased estimator p'.** The point estimate, the sample statistic p', becomes the center of the confidence interval, and the maximum error of estimate is a multiple of the **standard error.** The **level of confidence** determines the confidence coefficient, the number of multiples of the standard error.

Confidence Interval for a Proportion

$$p' - z(\alpha/2)\left(\sqrt{\frac{p'q'}{n}}\right) \quad \text{to} \quad p' + z(\alpha/2)\ \left(\sqrt{\frac{p'q'}{n}}\right) \qquad (9.6)$$

where $p' = \dfrac{x}{n}$ and $q' = 1 - p'$

Notice that the standard error, $\sqrt{\dfrac{pq}{n}}$, has been replaced by $\sqrt{\dfrac{p'q'}{n}}$. Since we are estimating p, we do not know its value and therefore we must use the best replacement available. That replacement is p', the observed value or the point estimate for p. This replacement will cause little change in the standard error or the width of our confidence interval provided n is sufficiently large.

    Example 9.8 will illustrate the formation of a confidence interval for the binomial parameter, p.

EXAMPLE 9.8   Confidence Interval for p

Statistics ⬡ Now™

Watch a video example at http://1pass.thomson.com or on your CD.

In a discussion about the cars that fellow students drive, several statements were made about types, ages, makes, colors, and so on. Dana decided he wanted to estimate the proportion of cars students drive that are convertibles, so he randomly identified 200 cars in the student parking lot and found 17 to be convertibles. Find the 90% confidence interval for the proportion of cars driven by students that are convertibles.

SOLUTION

FYI The five-step confidence interval procedure is given on page 404.

Step 1   **The Set-Up:**
       Describe the population parameter of interest.
       p, the proportion (percentage) of students' cars that are convertibles.

Step 2   **The Confidence Interval Criteria:**
       a.   **Check the assumptions.**
           The sample was randomly selected, and each student's response is independent of those of the others surveyed.
       b.   **Identify the probability distribution and the formula to be used.**
           The standard normal distribution will be used with formula (9.6) as the test statistic. p' is expected to be approximately normal because:
           (1) $n = 200$ is greater than 20, and

(2) both np [approximated by $np' = 200(17/200) = 17$] and nq [approximated by $nq' = 200(183/200) = 183$] are greater than 5.

c. **State the level of confidence:** $1 - \alpha = 0.90$.

Step 3 **The Sample Evidence:**
Collect the sample information.
$n = 200$ cars were identified, and $x = 17$ were convertibles:

$$p' = \frac{x}{n} = \frac{17}{200} = 0.085$$

Step 4 **The Confidence Interval:**

a. **Determine the confidence coefficient.**
This is the z-score [$z(\alpha/2)$, "z of one-half of alpha"] identifying the number of standard errors needed to attain the level of confidence and is found using Table 4 in Appendix B; $z(\alpha/2) = z(0.05) = 1.65$ (see the diagram).

b. **Find the maximum error of estimate.**
Use the maximum error part of formula (9.6):

$$E = z(\alpha/2)\left(\sqrt{\frac{p'q'}{n}}\right) = 1.65\left(\sqrt{\frac{(0.085)(0.915)}{200}}\right)$$
$$= (1.65)\sqrt{0.000389} = (1.65)(0.020) = \mathbf{0.033}$$

c. **Find the lower and upper confidence limits.**

$$p' - E \quad \text{to} \quad p' + E$$
$$0.085 - 0.033 \quad \text{to} \quad 0.085 + 0.033$$
$$0.052 \quad \text{to} \quad 0.118$$

Step 5 **The Results:**
State the confidence interval.
0.052 to 0.118 is the 90% confidence interval for $p = P(\text{drives convertible})$.
That is, the true proportion of students who drive convertibles is between 0.052 and 0.118, with 90% confidence.

FYI Specific instructions are given on pages 403–405.

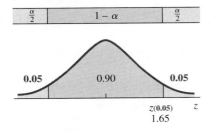

FYI Explore Skillbuilder Applet "$z\star$ & Confidence Level" on your CD.

TECHNOLOGY INSTRUCTIONS: $1 - \alpha$ CONFIDENCE INTERVAL FOR A PROPORTION p

MINITAB (Release 14)

Choose: `Stat > Basic Statistics > 1 Proportion`
Select: `Summarized Data`
Enter: `Number of trials: n`
          `Number of events: x`
Select: `Options`
Enter: `Confidence level: 1 - α (ex. 95.0)`
Select: `Alternative: not equal`
          `Use test and interval based on normal distribution. > OK > OK`

Excel   Input the data into column A using 0's for failures (or no's) and 1's for successes (or yes's); then continue with:

Choose: **Tools > Data Analysis Plus > Z-Estimate: Proportion**
Enter: Input Range: **(A2:A20 or select cells) > OK**
          Code for success: **1**
          Alpha: α (ex. 0.05) **> OK**

TI-83/84 Plus   Choose: **STAT > TESTS > A:1-PropZint**
                     Enter the appropriate values
                     and highlight Calculate.

```
1-PropZInt
x:0
n:0
C-Level:.95
Calculate
```

**APPLIED
EXAMPLE 9.9**   # Myth and Reality in Reporting Sampling Error

On almost every occasion when we release a new survey, someone in the media will ask, "What is the margin of error for this survey?" When the media print sentences such as "the margin of error is plus or minus three percentage points," they strongly suggest that the results are accurate to within the percentage stated. They want to warn people about sampling error. But they might be better off assuming that all surveys, all opinion polls are estimates, which may be wrong.

In the real world, "random sampling error"—or the likelihood that a pure probability sample would produce replies within a certain band of percentages only because of the sample size—is one of the least of our measurement problems.

For this reason, we (Harris) include a strong warning in all of the surveys that we publish. Typically, it goes as follows: In theory, with a sample of this size, one can say with 95 percent certainty that the results have a statistical precision of plus or minus _ percentage points of what they would be if the entire adult population had been polled with complete accuracy. Unfortunately, there are several other possible sources of error in all polls or surveys that are probably more serious than theoretical calculations of sampling error. They include refusals to be interviewed (non-response), question wording and question order, interviewer bias, weighting by demographic control data, and screening. It is difficult or impossible to quantify the errors that may result from these factors.

If journalists are the least bit interested in all of this they may well ask, "If there are so many sources of error in surveys, why should we bother to read or report any poll results?" To which I normally give two replies:

1. Well-designed, well-conducted surveys work. Their record overall is pretty good. Most social, and marketing, researchers would be very happy with the average forecasting errors of the polls. However, there are enough disasters in the history of election predictions for readers to be cautious about interpreting the results.

2. (And this is more effective.) I re-word Winston Churchill's famous remarks about democracy and say, "Polls are the worst way of measuring public opinion and public behavior, or of predicting elections—except for all of the others."

Source: The Polling Report, May 4, 1998, by Humphrey Taylor, Chairman, Louis Harris & Assoc., Inc. http://www.pollingreport.com/sampling.htm

Determining the Sample Size

By using the maximum error part of the confidence interval formula, it is possible to determine the **size of the sample** that must be taken in order to estimate p with a desired accuracy. Here is the formula for the **maximum error of estimate for a proportion**:

$$E = z_{(\alpha/2)}\left(\sqrt{\frac{pq}{n}}\right) \tag{9.7}$$

To determine the sample size from this formula, we must decide on the quality we want for our final confidence interval. This quality is measured in two ways: the level of confidence and the preciseness (narrowness) of the interval. The level of confidence we establish will in turn determine the confidence coefficient, $z_{(\alpha/2)}$. The desired preciseness will determine the maximum error of estimate, E. (Remember that we are estimating p, the binomial probability; therefore, E will typically be expressed in hundredths.)

For ease of use, we can solve formula (9.7) for n as follows:

Sample Size for $1 - \alpha$ Confidence Interval of p

$$n = \frac{[z_{(\alpha/2)}]^2 \cdot p^* \cdot q^*}{E^2} \tag{9.8}$$

where p^* and q^* are provisional values of p and q used for planning

---

FYI Remember that $q = 1 - p$.

By inspecting formula (9.8), we can observe that three components determine the sample size:

1. The level of confidence [$1 - \alpha$, which determines the confidence coefficient, $z_{(\alpha/2)}$]
2. The provisional value of p (p^* determines the value of q^*)
3. The maximum error, E

An increase or decrease in one of these three components affects the sample size. If the level of confidence is increased or decreased (while the other components are held constant), then the sample size will increase or decrease, respectively. If the product of p^* and q^* is increased or decreased (with other components held constant), then the sample size will increase or decrease, respectively. (The product $p^* \cdot q^*$ is largest when $p^* = 0.5$ and decreases as the value of p^* becomes further from 0.5.) An increase or decrease in the desired maximum error will have the opposite effect on the sample size, since E appears in the denominator of the formula. If no provisional values for p and q are available, then use $p^* = 0.5$ and $q^* = 0.5$. Using $p^* = 0.5$ is safe because it gives the largest sample size of any possible value of p. Using $p^* = 0.5$ works reasonably well when the true value is "near 0.5" (say, between 0.3 and 0.7); however, as p gets nearer to either 0 or 1, a sizable overestimate in sample size will occur.

EXAMPLE 9.10   ### Sample Size for Estimating p (No Prior Information)

Determine the sample size that is required to estimate the true proportion of community college students who are blue-eyed if you want your estimate to be within 0.02 with 90% confidence.

SOLUTION

STEP 1   The level of confidence is $1 - \alpha = 0.90$; therefore, the confidence coefficient is $z(\alpha/2) = z(0.05) = 1.65$ from Table 4 in Appendix B; see the diagram.

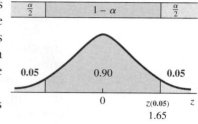

STEP 2   The desired maximum error is $E = 0.02$.

STEP 3   No estimate was given for p, so use $p^* = 0.5$ and $q^* = 1 - p^* = 0.5$.

STEP 4   Use formula (9.8) to find n:

$$n = \frac{[z(\alpha/2)]^2 \cdot p^* \cdot q^*}{E^2}: \quad n = \frac{(1.65)^2 \cdot 0.5 \cdot 0.5}{(0.02)^2} = \frac{0.680625}{0.0004} = 1701.56 = \mathbf{1702}$$

EXAMPLE 9.11   ### Sample Size for Estimating p (Prior Information)

An automobile manufacturer purchases bolts from a supplier who claims the bolts are approximately 5% defective. Determine the sample size that will be required to estimate the true proportion of defective bolts if we want our estimate to be within ±0.02 with 90% confidence.

SOLUTION

STEP 1   The level of confidence is $1 - \alpha = 0.90$; the confidence coefficient is $z(\alpha/2) = z(0.05) = 1.65$.

STEP 2   The desired maximum error is $E = 0.02$.

STEP 3   There is an estimate for p (supplier's claim is "5% defective"), so use $p^* = 0.05$ and $q^* = 1 - p^* = 0.95$.

STEP 4   Use formula (9.8) to find n:

$$n = \frac{[z(\alpha/2)]^2 \cdot p^* \cdot q^*}{E^2}: \quad n = \frac{(1.65)^2 \cdot 0.05 \cdot 0.95}{(0.02)^2} = \frac{0.12931875}{0.0004} = 323.3 = \mathbf{324}$$

Notice the difference in the sample sizes required in Examples 9.10 and 9.11. The only mathematical difference between the problems is the value used for p^*. In Example 9.10 we used $p^* = 0.5$, and in Example 9.11 we used $p^* = 0.05$. Recall that the use of the provisional value $p^* = 0.5$ gives the maximum sample size. As you can see, it will be an advantage to have some indication of the value expected for p, especially as p becomes increasingly further from 0.5.

Hypothesis-Testing Procedure

When the binomial parameter p is to be tested using a hypothesis-testing procedure, we will use a test statistic that represents the difference between the observed proportion and the hypothesized proportion, divided by the standard error. This test statistic is assumed to be normally distributed when the null hypothesis is true, when the assumptions for the test have been satisfied, and when n is sufficiently large ($n > 20$, $np > 5$, and $nq > 5$).

FYI p' is from the sample, p is from H_o, and $q = 1 - p$.

Test Statistic for a Proportion

$$z\star = \frac{p' - p}{\sqrt{\dfrac{pq}{n}}} \quad \text{with } p' = \frac{x}{n} \tag{9.9}$$

EXAMPLE 9.12

One-Tailed Hypothesis Test for Proportion p

Statistics Now™

Watch a video example at
http://1pass.thomson.com
or on your CD.

Many people sleep late on the weekends to make up for "short nights" during the workweek. The Better Sleep Council reports that 61% of us get more than 7 hours of sleep per night on the weekend. A random sample of 350 adults found that 235 had more than 7 hours of sleep each night last weekend. At the 0.05 level of significance, does this evidence show that more than 61% sleep 7 hours or more per night on the weekend?

SOLUTION

Step 1   **The Set-Up:**
a. **Describe the population parameter of interest.**
p, the proportion of adults who get more than 7 hours of sleep per night on weekends.
b. **State the null hypothesis (H_o) and the alternative hypothesis (H_a).**
$H_o: p = P(7+ \text{ hours of sleep}) = 0.61$ ($\leq$) (no more than 61%)
$H_a: p > 0.61$ (more than 61%)

Step 2   **The Hypothesis Test Criteria:**
a. **Check the assumptions.**
The random sample of 350 adults was independently surveyed.
b. **Identify the probability distribution and the test statistic to be used.**
The standard normal z will be used with formula (9.9). Since $n = 350$ is greater than 20 and both $np = (350)(0.61) = 213.5$ and $nq = (350)(0.39) = 136.5$ are greater than 5, p' is expected to be approximately normally distributed.
c. **Determine the level of significance:** $\alpha = 0.05$.

Step 3   **The Sample Evidence:**
a. **Collect the sample information:** $n = 350$ and $x = 235$:

$$p' = \frac{x}{n} = \frac{235}{350} = 0.671$$

b. **Calculate the value of the test statistic.**
Use formula (9.9):

$$z\star = \frac{p' - p}{\sqrt{\dfrac{pq}{n}}} : \quad z\star = \frac{0.671 - 0.61}{\sqrt{\dfrac{(0.61)(0.39)}{350}}} = \frac{0.061}{\sqrt{0.0006797}} = \frac{0.061}{0.0261} = 2.34$$

Step 4   The Probability Distribution:

Using the *p*-value procedure: OR   Using the classical procedure:

a. Calculate the *p*-value for the test statistic.

Use the right-hand tail because H_a expresses concern for values related to "more than."

P = *p*-value = $P(z > 2.34)$ as shown in the figure.

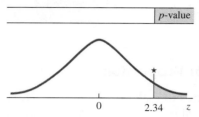

0   2.34 z

To find the *p*-value, use one of three methods:

1. Use Table 3 in Appendix B to calculate the *p*-value: $P = 0.5000 - 0.4904 = \mathbf{0.0096}.$

2. Use Table 5 in Appendix B to place bounds on the *p*-value: $0.0094 < P < 0.0107.$

3. Use a computer or calculator to calculate the *p*-value: $P = 0.0096.$

For specific instructions, see Method 3 below.

b. Determine whether or not the *p*-value is smaller than α.

The *p*-value is smaller than α.

a. Determine the critical region and critical value(s).

The critical region is the right-hand tail because H_a expresses concern for values related to "more than." The critical value is obtained from Table 4A: $z(0.05) = \mathbf{1.65}.$

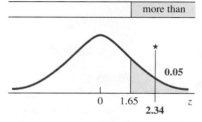

0   1.65 z
     2.34

Specific instructions for finding critical values are given on pages 450–451.

b. Determine whether or not the calculated test statistic is in the critical region.

$z \star$ is in the critical region, as shown in **red** in the accompanying figure.

Step 5   The Results:

a. State the decision about H_o: Reject H_o.

b. State the conclusion about H_a.

There is sufficient reason to conclude that the proportion of adults in the sampled population who are getting more than 7 hours of sleep nightly on weekends is significantly higher than 61% at the 0.05 level of significance.

Method 3: If you are doing the hypothesis test with the aid of a computer or calculator, most likely it will calculate the *p*-value for you, or you may use the cumulative probability distribution commands described on pages 329–330.

EXAMPLE 9.13   **Two-Tailed Hypothesis Test for Proportion *p***

While talking about the cars that fellow students drive (see Example 9.8, p. 498), Tom claimed that 15% of the students drive convertibles. Jody finds this hard to believe, and she wants to check the validity of Tom's claim using Dana's random sample. At a level of significance of 0.10, is there sufficient evidence to reject Tom's claim if there were 17 convertibles in his sample of 200 cars?

SOLUTION

Step 1   The Set-Up:

a. Describe the population parameter of interest.

$p = P$(student drives convertible).

b. **State the null hypothesis (H_o) and the alternative hypothesis (H_a).**

H_o: $p = 0.15$ (15% do drive convertibles)
H_a: $p \neq 0.15$ (the percentage is different from 15%)

Step 2 **The Hypothesis Test Criteria:**

a. **Check the assumptions.**
The sample was randomly selected, and each subject's response is independent of other responses.

b. **Identify the probability distribution and the test statistic to be used.**
The standard normal z and formula (9.9) will be used. Since $n = 200$ is greater than 20 and both np and nq are greater than 5, p' is expected to be approximately normally distributed.

c. **Determine the level of significance: $\alpha = 0.10$.**

Step 3 **The Sample Evidence:**

a. **Collect the sample information:** $n = 200$ and $x = 17$:

$$p' = \frac{x}{n} = \frac{17}{200} = 0.085$$

b. **Calculate the value of the test statistic.**
Use formula (9.9):

$$z\star = \frac{p' - p}{\sqrt{\dfrac{pq}{n}}}: \quad z\star = \frac{0.085 - 0.150}{\sqrt{\dfrac{(0.15)(0.85)}{200}}}$$

$$= \frac{-0.065}{\sqrt{0.00064}} = \frac{-0.065}{0.022525} = -2.57$$

Step 4 **The Probability Distribution:**

Using the *p*-value procedure:	OR	Using the classical procedure:

a. Calculate the *p*-value for the test statistic.
Use both tails because H_a expresses concern for values related to "different from."

$\mathbf{P} = p\text{-value} = P(z < -2.57) + P(z > 2.57)$
$= 2 \times P(|z| > 2.57)$ as shown in the figure.

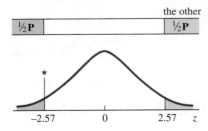

To find the *p*-value, use one of three methods:
1. Use Table 3 in Appendix B to calculate the *p*-value:
   $\mathbf{P} = 2 \times (0.5000 - 0.4949) = 0.0102$.
2. Use Table 5 in Appendix B to place bounds on the *p*-value: $0.0094 < \mathbf{P} < 0.0108$.

a. Determine the critical region and critical value(s).
The critical region is two-tailed because H_a expresses concern for values related to "different from." The critical value is obtained from Table 4B: $z(0.05) = 1.65$.

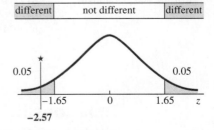

For specific instructions, see pages 453–454.

3. Use a computer or calculator to calculate the *p*-value: **P** = 0.0102.

For specific instructions, see page 432.

b. Determine whether or not the *p*-value is smaller than α.

The *p*-value is smaller than *α*.

b. Determine whether or not the calculated test statistic is in the critical region.

z ★ is in the critical region, as shown in **red** in the accompanying figure.

Step 5 **The Results:**

    a. **State the decision about H_o:** Reject H_o.

    b. **State the conclusion about H_a.**

        There is sufficient evidence to reject Tom's claim and conclude that the percentage of students who drive convertibles is different from 15% at the 0.10 level of significance.

TECHNOLOGY INSTRUCTIONS: HYPOTHESIS TEST FOR A PROPORTION *p*

MINITAB (Release 14)

```
Choose:   Stat > Basic Statistics > 1 Proportion
Select:   Summarized Data
Enter:    Number of trials: n
          Number of events: x
Select:   Options
Enter:    Test proportion: p
Select:   Alternative: less than or not equal or greater than
          Use test and interval based on normal distribution. > OK > OK
```

Excel

Input the data into column A using 0's for failures (or no's) and 1's for successes (or yes's); then continue with:

```
Choose:   Tools > Data Analysis Plus > Z-Test: Proportion
Enter:    Input Range: (A2:A20 or select cells) > OK
          Code for success: 1
          Hypothesized Proportion: p
          Alpha: α (ex. 0.05)
Choose:   Alternative: less than or not equal or greater than > OK
          Gives p-values and critical values for both one-tailed and
          two-tailed tests.
```

TI-83/84 Plus

```
Choose:   STAT > TESTS > 5:1-PropZTest
          Enter the appropriate values and
          highlight Calculate.
```

```
1-PropZTest
 p₀:0
 x:0
 n:0
 prop≠p₀ <p₀ >p₀
 Calculate Draw
```

Relationship between Confidence Intervals and Hypothesis Tests

There is a relationship between confidence intervals and two-tailed hypothesis tests when the level of confidence and the level of significance add up to 1. The confidence coefficients and the critical values are the same, which means the width of the confidence interval and the width of the noncritical region are the same. The point estimate is the center of the confidence interval, and the hypothesized mean is the center of the noncritical region. Therefore, if the hypothesized value of p is contained in the confidence interval, then the test statistic will be in the noncritical region (see Figure 9.5).

FIGURE 9.5
Confidence Interval
Contains p

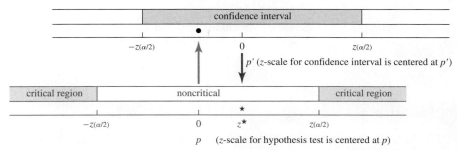

Furthermore, if the hypothesized probability p does not fall within the confidence interval, then the test statistic will be in the critical region (see Figure 9.6).

FIGURE 9.6
Confidence Interval
Does Not Contain p

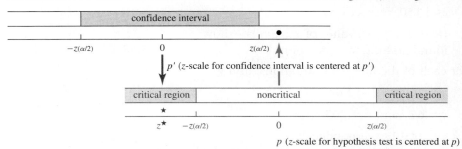

FYI Explore Skillbuilder Applet "$z\star$ & Confidence Level" on your CD.

This comparison should be used only when the hypothesis test is two-tailed and when the same value of α is used in both procedures.

APPLIED
EXAMPLE 9.14   **Heads or Tails?**

HEADS, BELGIUM WINS—AND WINS

Memo to all teams playing Belgium in the World Cup this year: "Don't let them use their own coins for the toss."

Mathematicians say the coins issued in the eurozone's administrative heartland are more likely to land heads up than down. While the notes which began circulating in the 12 members of the eurozone on January 1 are all the same, the coins show national symbols on one side and a map of Europe on the other. King Albert, who appears on Belgian coins, appears to be a bit of a lightweight, according to Polish mathematicians Tomasz Gliszczynski and Waclaw Zawadowski. The two professors and their

students at the Podlaska Academy in Siedlce spun a Belgian one euro coin 250 times, and found it landed heads up 140 times.

"The euro is struck asymmetrically," Prof Gliszczynski, who teaches statistics, told Germany's *Die Welt* newspaper. The head of the mint said yesterday that the Polish mathematicians' findings were "just luck." "When the coins were made they were struck in exactly the same way on all sides and the metal was evenly distributed," said Romain Coenen. "I haven't heard of any problems with the coins." But a variation of the experiment at the Guardian office suggested that the

> Polish mathematicians may be right. When tossed 250 times, the one euro coin came up heads 139 times and tails 111. "It looks very suspicious to me," said Barry Blight, a statistics lecturer at the London School of Economics. "If the coin were unbiased the chance of getting a result as extreme as that would be less than 7%."
>
> [See Ex. 9.111]
>
> *Source:* Charlotte Denny and Sarah Dennis, *The Guardian*, Friday, January 4, 2002, http://www.guardian.co.uk/euro/story/0,11306,627496,00.html

SECTION 9.3 EXERCISES

9.63 Of the 150 elements in a random sample, 45 are classified as "success."

a. Explain why x and n are assigned the values 45 and 150, respectively.

b. Determine the value of p'. Explain how p' is found and the meaning of p'.

For each of the following situations, find p'.

c. $x = 24$ and $n = 250$

d. $x = 640$ and $n = 2050$

e. 892 of 1280 responded "yes"

9.64 a. What is the relationship between $p = P$ (success) and $q = P$(failure)? Explain.

b. Explain why the relationship between p and q can be expressed by the formula $q = 1 - p$.

c. If $p = 0.6$, what is the value of q?

d. If the value of $q' = 0.273$, what is the value of p'?

9.65 a. Does it seem reasonable that the mean of the sampling distribution of observed values of p' should be p, the true proportion? Explain.

b. Explain why p' is an unbiased estimator for the population p.

9.66 Show that $\dfrac{\sqrt{npq}}{n}$ simplifies to $\sqrt{\dfrac{pq}{n}}$.

9.67 Find α, the area of one tail, and the confidence coefficients of z that are used with each of the following levels of confidence.

a. $1 - \alpha = 0.90$ b. $1 - \alpha = 0.95$ c. $1 - \alpha = 0.99$

9.68 Find α, the area of one tail, and the confidence coefficients of z that are used with each of the following levels of confidence.

a. $1 - \alpha = 0.80$ b. $1 - \alpha = 0.98$ c. $1 - \alpha = 0.75$

9.69 Refer back to Example 9.8. Another sample is taken to estimate the proportion of students who drive convertibles. Results are $n = 400$ and $x = 92$. Find:

a. The estimate for the standard error

b. The 95% confidence interval

9.70 "You say tomato, burger lovers say ketchup!" According to a recent T.G.I. Friday's restaurants' random survey of 1027 Americans, approximately half (47%) say that ketchup is their preferred burger condiment. The survey quoted a margin of error of plus or minus 3.1%.

Source: Harris Interactive/Yankelovich Partners for T.G.I. Friday's restaurants, http://www.knoxville3.com/fridays/News/burger.htm

a. Describe how this survey of 1027 Americans fits the properties of a binomial experiment. Specifically identify n, a trial, success, p, and x.

b. What is the point estimate for the proportion of all Americans who prefer ketchup on their burger? Is it a parameter or a statistic?

c. Calculate the 95% confidence maximum error of estimate for a binomial experiment of 1027 trials that result in an observed proportion of 0.47.

d. How is the maximum error, found in part c, related to the 3.1% margin of error quoted in the survey report?

e. Find the 95% confidence interval for the true proportion p based on a binomial experiment of 1027 trials that result in an observed proportion of 0.47.

9.71 Although most people are aware of minor dehydration symptoms such as dry skin and headaches, many are less knowledgeable about the causes of dehydration. According to a poll done for the Nutrition Information Center, the results of a random sample of 3003 American adults showed that 20% did not know that caffeine dehydrates. The survey listed a margin of error of plus or minus 1.8%.

Source: Yankelovich Partners for the Nutrition Information Center of the New York Hospital–Cornell Medical Center and the International Bottled Water Association

a. Describe how this survey of 3003 American adults fits the properties of a binomial experiment. Specifically identify n, a trial, success, p, and x.

b. What is the point estimate for the proportion of all Americans who do not know that caffeine dehydrates? Is it a parameter or a statistic?

c. Calculate the 95% confidence maximum error of estimate for a binomial experiment of 3003 trials that result in an observed proportion of 0.20.

d. How is the maximum error, found in part c, related to the 1.8% margin of error quoted in the survey report?

e. Find the 95% confidence interval for the true proportion p based on a binomial experiment of 3003 trials that result in an observed proportion of 0.20.

9.72 A bank randomly selected 250 checking account customers and found that 110 of them also

had savings accounts at this same bank. Construct a 95% confidence interval for the true proportion of checking account customers who also have savings accounts.

9.73 In a sample of 60 randomly selected students, only 22 favored the amount being budgeted for next year's intramural and interscholastic sports. Construct the 99% confidence interval for the proportion of all students who support the proposed budget amount.

9.74 The December 21, 2004, *USA Today* article "On the Path to Problem Gambling: National Survey Shows Casinos, Slots and Lotteries Attract Youth Into Potentially Addictive Habit" reported on a poll of 200 male and 199 female teenagers, aged 14 to 17. The findings reported were that 66% of the male teenagers have "ever gambled" and that 37% of the female teenagers have "ever gambled." The poll was sponsored by the 2003 Annenberg National Risk Survey of Youth.

a. Find the point estimate, the maximum error of estimate, and the 95% confidence interval that results from a binomial experiment with $n = 200$ and observed success proportion 0.66.

b. Find the point estimate, the maximum error of estimate, and the 95% confidence interval that results from a binomial experiment with $n = 199$ and observed success proportion 0.37.

c. Compare the resulting confidence intervals and make a statement based on the two intervals comparing male and female teenage gambling habits.

9.75 Of the 1742 managers and professionals polled by Management Recruiters International, May 2002, 27.8% work late 5 days a week on average. Using a 99% confidence interval for the true binomial proportion based on a random sample of 1742 binomial trials and an observed proportion of 0.278, estimate the proportion of managers and professionals who work late 5 days a week.

9.76 New research suggests that children often learn coping skills from imaginary friends. In a study of 100 children done at the University of

Oregon and reported in *Developmental Psychology* (*USA Today,* December 20, 2004, "'Pretend' friends, actual benefits"), 33% of the children still had an imaginary friend at the age of 7. Find the 90% confidence interval for the true proportion of "success" for a binomial experiment with $n = 100$ and an observed proportion of 0.33.

9.77 A Cambridge Consumer Credit Index nationwide telephone survey of 1000 people found that most Americans are not easily swayed by the lure of reward points or rebates when deciding to use a credit card or pay by cash or check. The survey found that 2 out of 3 consumers do not even have credit cards offering reward points or rebates. Explain why you would be reluctant to use this information to construct a confidence interval estimating the true proportion of consumers who do not have credit cards offering reward points or rebates.

9.78 Construct 90% confidence intervals for the binomial parameter p for each of the following pairs of values. Write your answers on the chart.

Observed Proportion $p' = x/n$		Sample Size	Lower Limit	Upper Limit
a.	$p' = 0.3$	$n = 30$		
b.	$p' = 0.7$	$n = 30$		
c.	$p' = 0.5$	$n = 10$		
d.	$p' = 0.5$	$n = 100$		
e.	$p' = 0.5$	$n = 1000$		

f. Explain the relationship between the answers to parts a and b.

g. Explain the relationship between the answers to parts c–e.

9.79 Descriptions of three nationwide poll results follow.

> CNN/*USA Today*/Gallup Poll, June 21–23, 2002; $N = 1020$ adults nationwide; MoE ± 3. (MoE is margin of error.)
>
> "Do you think school districts should or should not be allowed to test public school students for illegal drugs before those students can participate in non-athletic activities?" Should be allowed—70%, Should not be allowed—29%, No opinion—1%

> CNN/*USA Today*/Gallup Poll, June 28–30, 2002; $N = 1019$ adults nationwide; MoE ± 3.
>
> "A proposal has been made that would allow people to put a portion of their Social Security payroll taxes into personal retirement accounts that would be invested in private stocks and bonds. Do you favor or oppose this proposal?"
>
> Favor—57%, Oppose—39%, No—4%

> *International Communications Research* (ICR), December 2004; $N = 1021$ adults; MoE ± 3.1.
>
> The Soap and Detergent Association (SDA) survey reported that 15% said that the most important factor in choosing a cleaning product was the brand name.

Each of the polls is based on approximately 1020 randomly selected adults.

a. Calculate the 95% confidence maximum error of estimate for the true binomial proportion based on binomial experiments with the same sample size and observed proportion as listed first in each article.

b. Explain what caused the values of the maximum errors to vary.

c. The margin of error being reported is typically the value of the maximum error rounded to the next larger whole percentage. Do your results in part a verify this?

d. Explain why the round-up practice is considered "conservative"?

e. What value of p should be used to calculate the standard error if the most conservative margin of error is desired?

9.80 a. If x successes result from a binomial experiment with $n = 1000$ and $p = P(\text{success})$, and the 95% confidence interval for the true probability of success is determined, what is the maximum value possible for the "maximum error of estimate"?

b. Compare the numerical value of the "maximum error of estimate" found in part a with "the margin of error" discussed in the Applied Example 9.9.

c. Under what conditions are they the same? Not the same?

d. Explain how the results of national polls, like those of Harris and Gallup, are related (similarities and differences) to the confidence interval technique studied in this section.

e. The theoretical sampling error with a level of confidence can be calculated, but the polls typically report only a "margin of error" with no probability (level of confidence). Why is that?

9.81 Adverse drug reactions to legally prescribed medicines are among the leading causes of drug-related death in the United States. Suppose you investigate drug-related deaths in your city and find that 223 out of 250 incidences were caused by legally prescribed drugs and the rest were the result of illicit drug use. MINITAB was then used to form the 98% confidence interval for the proportion of drug-related deaths that are caused by legally prescribed drugs. Verify the MINITAB results that follow.

```
CI for One Proportion
Sample    X      N      Sample p           98% CI
1        223    250     0.892000     (0.846333, 0.937667)
```

9.82 Using the MINITAB output and information in Exercise 9.81, determine the values of the following terms:

a. Point estimate   b. Confidence coefficient

c. Standard error of the mean

d. Maximum error of estimate, E

e. Lower confidence limit

f. Upper confidence limit

9.83 Karl Pearson once tossed a coin 24,000 times and recorded 12,012 heads.

a. Calculate the point estimate for $p = P(\text{head})$ based on Pearson's results.

b. Determine the standard error of proportion.

c. Determine the 95% confidence interval estimate for $p = P(\text{head})$.

d. It must have taken Mr. Pearson many hours to toss a coin 24,000 times. You can simulate 24,000 coin tosses using the computer and calculator commands that follow. (*Note:* A Bernoulli experiment is like a "single" trial bi-

nomial experiment. That is, one toss of a coin is one Bernoulli experiment with $p = 0.5$; and 24,000 tosses of a coin either is a binomial experiment with $n = 24,000$ or is 24,000 Bernoulli experiments. Code: $0 = \text{tail}$, $1 = \text{head}$. The sum of the 1s will be the number of heads in the 24,000 tosses.)

MINITAB

```
Choose Calc > Random Data > Bernoulli, entering 24000
for generate, C1 for Store in column(s) and 0.5 for
Probability of success. Sum the data and divide by
24,000.
```

Excel

```
Choose Tools > Data Analysis > Random Number Generation
> Bernoulli, entering 1 for Number of Variables, 24000
for Number of Random Numbers and 0.5 for p Value. Sum
the data and divide by 24,000.
```

TI-83/84 PLUS

```
Choose MATH > PRB > 5:randInt, then enter 0, 1, number
of trials. The maximum number of elements (trials) in a
list is 999. (slow process for large n's) Sum the data
and divide by n.
```

e. How do your simulated results compare with Pearson's?

f. Use the commands (part d) and generate another set of 24,000 coin tosses. Compare these results to those obtained by Pearson. Also, compare the two simulated samples to each other. Explain what you can conclude from these results.

9.84 When a single die is rolled, the probability of a "one" is 1/6, or 0.167. Let's simulate 3000 rolls of a die. (*Note:* A Bernoulli experiment is like a "single" trial binomial experiment. That is, one roll of a die is one Bernoulli experiment with $p = 1/6$; and 3000 rolls of a die either is a binomial experiment with $n = 3000$ or is 3000 Bernoulli experiments. Code: $0 = 2$, 3, 4, 5, or 6, and $1 = 1$. The sum of the 1s will be the number of ones in the 3000 tosses.)

a. Use the commands given in Exercise 9.83 and a calculator or computer to simulate the rolling of a single die 3000 times.

Using the results from the simulation:

b. Sum the data and divide by 3000. Explain what this value represents.

c. Determine the standard error of proportion.

d. Determine the 95% confidence interval for $p = P(\text{one})$.

e. How do the results from the simulation compare with your expectations? Explain.

9.85 The "rule of thumb," stated on page 497, indicated that we would expect the sampling distribution of p' to be approximately normal when "$n > 20$ and both np and nq were greater than 5." What happens when these guidelines are not followed?

a. Use the following set of computer or calculator commands to show you what happens. Try $n = 15$ and $p = 0.1$ ($K1 = n$ and $K2 = p$). Do the distributions look normal? Explain what causes the "gaps." Why do the histograms look alike? Try some different combinations of n (K1) and p (K2):

MINITAB

Choose Calc > Random Data > Binomial to simulate a 1000 trials for an n of 15 and a p of 0.5. Divide each generated value by n, forming a column of sample p's. Calculate a z value for each sample p by using $z = (p' - p) / \sqrt{p(1-p)/n}$. Construct a histogram for the sample p's and another histogram for the z's.

Excel

Choose Tools > Data Analysis > Random Number Generation > Binomial to simulate a 1000 trials for an n of 15 and a p of 0.5. Divide each generated value by n, forming a column of sample p's. Calculate a z value for each sample p by using $z = (p' - p) / \sqrt{p(1-p)/n}$. Construct a histogram for the sample p's and another histogram for the z's.

TI-83/84 Plus

Choose MATH > PRB > 7:randBin, then enter n, p, number of trials. The maximum number of elements (trials) in a list is 999. (slow process for large n's) Divide each generated value by n, forming a list of sample p's. Calculate a z value for each sample p by using $z = (p' - p) / \sqrt{p(1-p)/n}$. Construct a histogram for the sample p's and another histogram for the z's.

b. Try $n = 15$ and $p = 0.01$.

c. Try $n = 50$ and $p = 0.03$.

d. Try $n = 20$ and $p = 0.2$.

e. Try $n = 20$ and $p = 0.8$.

f. What happens when the rule of thumb is not followed?

9.86 Has the law requiring bike helmet use failed? Yankelovich Partners conducted a survey of bicycle riders in the United States. Only 60% of the nationally representative sample of 1020 bike riders reported owning a bike helmet.

Source: http://www.cpsc.gov/library/helmet.html

a. Find the 95% confidence interval for the true proportion p for a binomial experiment of 1020 trials that resulted in an observed proportion of 0.60. Use this to estimate the percentage of bike riders who report owning a helmet.

b. Based on the survey results, would you say there is compliance with the law requiring bike helmet use? Explain.

Suppose you wish to conduct a survey in your city to determine what percent of bicyclists own helmets. Use the national figure of 60% for your initial estimate of p.

c. Find the sample size if you want your estimate to be within 0.02 with 95% confidence.

d. Find the sample size if you want your estimate to be within 0.04 with 95% confidence.

e. Find the sample size if you want your estimate to be within 0.02 with 90% confidence.

f. What effect does changing the maximum error have on the sample size? Explain.

g. What effect does changing the level of confidence have on the sample size? Explain.

9.87 Find the sample size n needed for a 95% interval estimate in Example 9.10.

9.88 Find n for a 90% confidence interval for p with $E = 0.02$ using an estimate of $p = 0.25$.

9.89 According to a May 14, 2002, USA Snapshot, 81% of all drivers use their seat belts. You wish to conduct a survey in your city to determine what percent of the drivers use seat belts. Use the national figure of 81% for your initial estimate of p.

a. Find the sample size if you want your estimate to be within 0.02 with 90% confidence.

b. Find the sample size if you want your estimate to be within 0.04 with 90% confidence.

c. Find the sample size if you want your estimate to be within 0.02 with 98% confidence.

d. What effect does changing the maximum error have on the sample size? Explain.

e. What effect does changing the level of confidence have on the sample size? Explain.

9.90 Lung cancer is the leading cause of cancer deaths in both women and men in the United States, Canada, and China. In several other countries, lung cancer is the number one cause of cancer deaths in men and the second or third cause among women. Only about 14% of all people who develop lung cancer survive for 5 years.

Source: eMedicine Consumer Journal, June 20, 2002, Volume 3, Number 6

Suppose you wanted to see if this survival rate were still true. How large a sample would you need to take to estimate the true proportion surviving for 5 years after diagnosis to within 1% with 95% confidence? (Use the 14% as the value of p.)

9.91 State the null hypothesis, H_o, and the alternative hypothesis, H_a, that would be used to test these claims:

a. More than 60% of all students at our college work part-time jobs during the academic year.

b. No more than one-third of cigarette smokers are interested in quitting.

c. A majority of the voters will vote for the school budget this year.

d. At least three-fourths of the trees in our county were seriously damaged by the storm.

e. The results show the coin was not tossed fairly.

9.92 State the null hypothesis, H_o, and the alternative hypothesis, H_a, that would be used to test these claims:

a. The probability of our team winning tonight is less than 0.50.

b. At least 50% of all parents believe in spanking their children when appropriate.

c. At most, 80% of the invited guests will attend the wedding.

d. The single-digit numbers generated by the computer do not seem to be equally likely with regard to being odd or even.

e. Less than half of the customers like the new pizza.

9.93 Calculate the test statistic $z\star$ used in testing the following:

a. H_o: $p = 0.70$ vs. H_a: $p > 0.70$, with the sample $n = 300$ and $x = 224$

b. H_o: $p = 0.50$ vs. H_a: $p < 0.50$, with the sample $n = 450$ and $x = 207$

c. H_o: $p = 0.35$ vs. H_a: $p \neq 0.35$, with the sample $n = 280$ and $x = 94$

d. H_o: $p = 0.90$ vs. H_a: $p > 0.90$, with the sample $n = 550$ and $x = 508$

9.94 Find the value of **P** for each of the hypothesis tests in Exercise 9.93; state the decision using $\alpha = 0.05$.

9.95 Determine the p-value for each of the following hypothesis-testing situations.

a. H_o: $p = 0.5$, H_a: $p \neq 0.5$, $z\star = 1.48$

b. H_o: $p = 0.7$, H_a: $p \neq 0.7$, $z\star = -2.26$

c. H_o: $p = 0.4$, H_a: $p > 0.4$, $z\star = 0.98$

d. H_o: $p = 0.2$, H_a: $p < 0.2$, $z\star = -1.59$

9.96 Find the critical region, critical values, for each of the hypothesis tests in Exercise 9.93; state the decision using $\alpha = 0.05$.

9.97 Determine the test criteria that would be used to test the following hypotheses when z is used as the test statistic and the classical approach is used.

a. $H_o: p = 0.5$ and $H_a: p > 0.5$, with $\alpha = 0.05$

b. $H_o: p = 0.5$ and $H_a: p \neq 0.5$, with $\alpha = 0.05$

c. $H_o: p = 0.4$ and $H_a: p < 0.4$, with $\alpha = 0.10$

d. $H_o: p = 0.7$ and $H_a: p > 0.7$, with $\alpha = 0.01$

9.98 The binomial random variable, x, may be used as the test statistic when testing hypotheses about the binomial parameter, p, when n is small (say, 15 or less). Use Table 2 in Appendix B and determine the p-value for each of the following situations.

a. $H_o: p = 0.5$, $H_a: p \neq 0.5$, where $n = 15$ and $x = 12$

b. $H_o: p = 0.8$, $H_a: p \neq 0.8$, where $n = 12$ and $x = 4$

c. $H_o: p = 0.3$, $H_a: p > 0.3$, where $n = 14$ and $x = 7$

d. $H_o: p = 0.9$, $H_a: p < 0.9$, where $n = 13$ and $x = 9$

9.99 The binomial random variable, x, may be used as the test statistic when testing hypotheses about the binomial parameter, p. When n is small (say, 15 or less), Table 2 in Appendix B provides the probabilities for each value of x separately, thereby making it unnecessary to estimate probabilities of the discrete binomial random variable with the continuous standard normal variable z. Use Table 2 to determine the value of α for each of the following:

a. $H_o: p = 0.5$ and $H_a: p > 0.5$, where $n = 15$ and the critical region is $x = 12, 13, 14, 15$

b. $H_o: p = 0.3$ and $H_a: p < 0.3$, where $n = 12$ and the critical region is $x = 0, 1$

c. $H_o: p = 0.6$ and $H_a: p \neq 0.6$, where $n = 10$ and the critical region is $x = 0, 1, 2, 3, 9, 10$

d. $H_o: p = 0.05$ and $H_a: p > 0.05$, where $n = 14$ and the critical region is $x = 4, 5, 6, 7, \ldots , 14$

9.100 Use Table 2 in Appendix B and determine the critical region used in testing each of the following hypotheses. (*Note:* Since x is discrete, choose critical regions that do not exceed the value of α given.)

a. $H_o: p = 0.5$ and $H_a: p > 0.5$, where $n = 15$ and $\alpha = 0.05$

b. $H_o: p = 0.5$ and $H_a: p \neq 0.5$, where $n = 14$ and $\alpha = 0.05$

c. $H_o: p = 0.4$ and $H_a: p < 0.4$, where $n = 10$ and $\alpha = 0.10$

d. $H_o: p = 0.7$ and $H_a: p > 0.7$, where $n = 13$ and $\alpha = 0.01$

9.101 You are testing the hypothesis $p = 0.7$ and have decided to reject this hypothesis if after 15 trials you observe 14 or more successes.

a. If the null hypothesis is true and you observe 13 successes, which of the following will you do? (1) Correctly fail to reject H_o. (2) Correctly reject H_o. (3) Commit a type I error. (4) Commit a type II error.

b. Find the significance level of your test.

c. If the true probability of success is $1/2$ and you observe 13 successes, which of the following will you do? (1) Correctly fail to reject H_o. (2) Correctly reject H_o. (3) Commit a type I error. (4) Commit a type II error.

d. Calculate the p-value for your hypothesis test after 13 successes are observed.

9.102 You are testing the null hypothesis $p = 0.4$ and will reject this hypothesis if $z\star$ is less than -2.05.

a. If the null hypothesis is true and you observe $z\star$ equal to -2.12, which of the following will you do? (1) Correctly fail to reject H_o. (2) Correctly reject H_o. (3) Commit a type I error. (4) Commit a type II error.

b. What is the significance level for this test?

c. What is the p-value for $z\star = -2.12$?

9.103 An insurance company states that 90% of its claims are settled within 30 days. A consumer group selected a random sample of 75 of the company's claims to test this statement. If the consumer group found that 55 of the claims were settled within 30 days, do they have sufficient reason to support their contention that less than 90% of the claims are settled within 30 days? Use $\alpha = 0.05$.

a. Solve using the *p*-value approach.

b. Solve using the classical approach.

9.104 A recent survey conducted by ZOOM and Applied Research & Consulting LLC reported that the events of September 11 have motivated kids to volunteer and that more than 80% volunteer. A disbeliever of this information took his own random sample of 500 kids in an attempt to show that the true percentage of kids who volunteer is less than 80%.

a. Find the *p*-value if 384 of the surveyed kids said they do community volunteer work.

b. Explain why it is important for the level of significance to be established before the sample results are known.

9.105 A politician claims that she will receive 60% of the vote in an upcoming election. The results of a properly designed random sample of 100 voters showed that 50 of those sampled will vote for her. Is it likely that her assertion is correct at the 0.05 level of significance?

a. Solve using the *p*-value approach.

b. Solve using the classical approach.

9.106 The full-time student body of a college is composed of 50% males and 50% females. Does a random sample of students (30 male, 20 female) from an introductory chemistry course show sufficient evidence to reject the hypothesis that the proportion of male and of female students who take this course is the same as that of the whole student body? Use $\alpha = 0.05$.

a. Solve using the *p*-value approach.

b. Solve using the classical approach.

9.107 In a poll conducted by the American Association of Retired Persons (AARP) of 1706 adults aged 45 and older, 72% agreed with the statement that "adults should be allowed to legally use mari-

juana for medical purposes if a physician recommends it."

Source: D&C, December 19, 2004, "Most older Americans OK with medical pot"

Suppose a recent study of 200 adults in the Midwest showed 134 in favor of legally using marijuana for medical purposes. Do these results show a lower proportion for the Midwest with respect to the rest of the country? Use a 0.05 level of significance.

a. Solve using the *p*-value approach.

b. Solve using the classical approach.

9.108 The popularity of personal watercraft (PWCs, also known as jet skis) continues to increase, despite the apparent danger associated with their use. In fact, a sample of 54 reported watercraft accidents to the Game and Parks Commission in the state of Nebraska revealed that 85% of them involved PWCs, even though only 8% of the motorized boats registered in the state are PWCs.

Source: Nebraskaland, "Officer's Notebook: The Personal Problem"

Suppose the national average proportion of watercraft accidents involving PWCs was 78%. Does the watercraft accident rate for PWCs in the state of Nebraska exceed the nation as a whole? Use a 0.01 level of significance.

a. Solve using the *p*-value approach.

b. Solve using the classical approach.

9.109 The USA Snapshot "Facing a crowd isn't easy" (May 30, 2002) reported that 35% of U.S. professional women fear public speaking. Suppose you conduct a survey of 1000 randomly chosen professional women to test H_o: $p = 0.35$ versus H_a: $p < 0.35$, where *p* represents the proportion who fear public speaking. Of the 1000 sampled, 324 feared public speaking. Use $\alpha = 0.01$.

a. Calculate the value of the test statistic.

b. Solve using the *p*-value approach.

c. Solve using the classical approach.

9.110 Show that the hypothesis test completed as Example 9.13 was unnecessary because the confidence interval had already been completed in Example 9.8.

9.111 Refer to Applied Example 9.14.

a. Find the probability that 139 or more heads result when a balanced coin is tossed fairly 250 times.

b. Find the probability that 110 or fewer heads result when a balanced coin is tossed fairly 250 times.

c. Find the probability that the results of fairly tossing a balanced coin 250 times is as extreme as the 139 heads.

d. What is Romain Coenen's claim? What is it that Barry Blight says is "suspicious"? How do these two statements form the opposing sides of a hypothesis test? State the null and alternative hypotheses.

e. Barry Blight's statement, "If the coin were unbiased the chance of getting a result as extreme as that would be less than 7%," is a statement of a *p*-value for a two-tailed hypothesis test. Explain why.

f. If the sample results had been used to estimate the probability that the euro coin lands heads up, what would have been the confidence interval?

g. If the media had reported these results as 56% with a ±6% margin of error, how would this be similar to and different than a national opinion poll that resulted in 56% ± 6%?

9.112 Reliable Equipment has developed a machine, *The Flipper,* that will flip a coin with predictable results. They claim that a coin flipped by The Flipper will land heads up at least 88% of the time. What conclusion would result in a hypothesis test, using $\alpha = 0.05$, when 200 coins are flipped and the following results are achieved:

a. 181 heads   b. 172 heads

c. 168 heads   d. 153 heads

9.113 The following computer output was used to complete a hypothesis test.

Test for One Proportion

Test of p = 0.225 vs p > 0.225

Sample	X	N	Sample p	95% Lower Bound	Z-Value	P-Value
1	61	200	0.305000	0.251451	2.71	0.003

a. State the null and alternative hypotheses.

b. If the test is completed using $\alpha = 0.05$, what decision and conclusion are reached?

c. Verify the "Sample *p*."

9.114 Using the computer output and information in Exercise 9.113, determine the value of the following:

a. Hypothesized value of population proportion

b. Sample proportion   c. Test statistic

9.4 Inferences about the Variance and Standard Deviation

Problems often arise that require us to make inferences about variability. For example, a soft-drink bottling company has a machine that fills 16-oz bottles. The company needs to control the standard deviation σ (or variance σ^2) in the amount of soft drink, *x*, put into each bottle. The mean amount placed in each bottle is important, but a correct mean amount does not ensure that the filling machine is working correctly. If the variance is too large, many bottles will be overfilled and many underfilled. Thus, the bottling company wants to maintain as small a standard deviation (or variance) as possible.

When discussing inferences about the spread of data, we usually talk about variance instead of standard deviation because the techniques (the formulas used) employ the sample variance rather than the standard deviation. However, remember that the standard deviation is the positive square root of the variance; thus, talking about the variance of a population is comparable to talking about the standard deviation.

Inferences about the variance of a normally distributed population use the **chi-square, χ^2,** distributions (*"ki-square"*: that's *"ki"* as in *"kite"* and χ is the Greek lowercase letter chi). The chi-square distributions, like Student's t-distributions, are a family of probability distributions, each one identified by the **parameter** number of **degrees of freedom.** To use the chi-square distribution, we must be aware of its properties (also, see Figure 9.7).

Properties of the chi-square distribution

1. χ^2 is nonnegative in value; it is zero or positively valued.
2. χ^2 is not symmetrical; it is skewed to the right.
3. χ^2 is distributed so as to form a family of distributions, a separate distribution for each different number of degrees of freedom.

FIGURE 9.7
Various Chi-Square Distributions

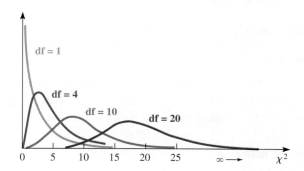

Note: When df > 2, the mean value of the chi-square distribution is df. The mean is located to the right of the mode (the value where the curve reaches its high point) and just to the right of the median (the value that splits the distribution, 50% on each side). By locating zero at the left extreme and the value of df on your sketch of the χ^2 distribution, you will establish an approximate scale so that other values can be located in their respective positions. See Figure 9.9.

FIGURE 9.9
Location of Mean, Median, and Mode for χ^2 Distribution

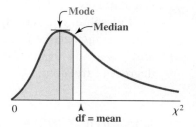

For values of χ^2 on the left side of the median, the area to the right will be greater than 0.50.

The **critical values for chi-square** are obtained from Table 8 in Appendix B. Each critical value is identified by two pieces of information: df and area under the curve to the right of the critical value being sought. Thus, $\chi^2(\text{df}, \alpha)$ (read *"chi-square of df, alpha"*) is the symbol used to identify the critical value of chi-

FIGURE 9.8 Chi-Square Distribution Showing $\chi^2(\text{df}, \alpha)$

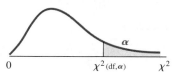

square with df degrees of freedom and with α area to the right, as shown in Figure 9.8. Since the chi-square distribution is not symmetrical, the critical values associated with the right and left tails are given separately in Table 8.

EXAMPLE 9.15

χ^2 Associated with the Right Tail

Find $\chi^2(20, 0.05)$.

SOLUTION See the figure. Use Table 8 in Appendix B to find the value of $\chi^2(20, 0.05)$ at the intersection of row df = 20 and column $\alpha = 0.05$, as shown in the portion of the table that follows:

Portion of Table 8

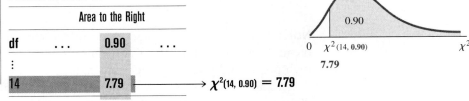

df		Area to the Right 0.05	
⋮	...		...
20		31.4	

$\chi^2(20, 0.05) = 31.4$

EXAMPLE 9.16

χ^2 Associated with the Left Tail

Find $\chi^2(14, 0.90)$.

SOLUTION See the figure that follows. Use Table 8 in Appendix B to find the value of $\chi^2(14, 0.90)$ at the intersection of row df = 14 and column $\alpha = 0.90$, as shown in the portion of the table that follows:

Portion of Table 8

df		Area to the Right 0.90	
⋮	...		...
14		7.79	

$\chi^2(14, 0.90) = 7.79$

FYI Explore Skillbuilder Applet "Chi-Square Probabilities" on your CD.

Most computer software packages or statistical calculators will calculate the area related to a specified χ^2-value. The accompanying figure shows the relationship between the cumulative probability and a specific χ^2-value for a χ^2-distribution with df degrees of freedom.

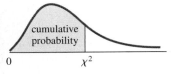

TECHNOLOGY INSTRUCTIONS: CUMULATIVE PROBABILITIES FOR χ^2

MINITAB (Release 14) Input the data into C1; then continue with:

```
Choose:     Calc > Probability Distributions > Chi-Square
Select:     Cumulative Probability
            Noncentrality Parameter: 0.0
```

```
Enter:        Degrees of freedom: df
Select:       Input constant*
Enter:        χ²-value (ex. 47.25) > OK
```

*Select Input column if several χ²-values are stored in C1. Use C2 for optional storage.
If the area in the right tail is needed, subtract the calculated probability from one.

Excel   If several χ²-values are to be used, input the values into column A and activate B1; then continue with:

```
Choose:       Insert function fₓ > Statistical > CHIDIST > OK
Enter:        X: individual χ²-value or (A1:A5 or select "χ²-value" cells)*
              Deg_freedom: df > OK
Drag*:        Bottom right corner of the B1 cell down to give other proba-
              bilities
```

TI-83/84 Plus
```
Choose:       2nd > DISTR > 7: χ²cdf(
Enter:        0, χ²-value, df)
```

If the area in the right tail is needed, subtract the calculated probability from one.

We are now ready to use chi-square to make inferences about the population variance or standard deviation.

> **The assumptions for inferences about the variance σ^2 or standard deviation σ:** The sampled population is normally distributed.

The t procedures for inferences about the mean (see Section 9.2) were based on the assumption of normality, but the t procedures are generally useful even when the sampled population is nonnormal, especially for larger samples. However, the same is not true about the inference procedures for the standard deviation. The statistical procedures for the standard deviation are very sensitive to nonnormal distributions (skewness, in particular), and this makes it difficult to determine whether an apparent significant result is the result of the sample evidence or a violation of the assumptions. Therefore, the only inference procedure to be presented here is the hypothesis test for the standard deviation of a normal population.

The **test statistic** that will be used in testing hypotheses about the population variance or standard deviation is obtained by using the formula

> **Test Statistic for Variance and Standard Deviation**
>
> $$\chi^2\star = \frac{(n-1)s^2}{\sigma^2}, \quad \text{with df} = n-1 \tag{9.10}$$

When random samples are drawn from a normal population with a known variance σ^2, the quantity $\dfrac{(n-1)s^2}{\sigma^2}$ possesses a probability distribution that is known as the chi-square distribution with $n-1$ degrees of freedom.

Hypothesis-Testing Procedure

Let's return to the example about the bottling company that wishes to detect when the variability in the amount of soft drink placed into each bottle gets out of control. A variance of 0.0004 is considered acceptable, and the company wants to adjust the bottle-filling machine when the variance, σ^2, becomes larger than this value. The decision will be made using the hypothesis-testing procedure.

EXAMPLE 9.17

One-Tailed Hypothesis Test for Variance, σ^2

Statistics△Now™

Watch a video example at
http://1pass.thomson.com
or on your CD.

The soft-drink bottling company wants to control the variability in the amount of fill by not allowing the variance to exceed 0.0004. Does a sample of size 28 with a variance of 0.0007 indicate that the bottling process is out of control (with regard to variance) at the 0.05 level of significance?

SOLUTION

Step 1 **The Set-Up:**
   a. **Describe the population parameter of interest.**
      σ^2, the variance in the amount of fill of a soft drink during a bottling process.
   b. **State the null hypothesis (H_o) and the alternative hypothesis (H_a).**
      H_o: $\sigma^2 = 0.0004$ ($\leq$) (variance is not larger than 0.0004)
      H_a: $\sigma^2 > 0.0004$ (variance is larger than 0.0004)

Step 2 **The Hypothesis Test Criteria:**
   a. **Check the assumptions.**
      The amount of fill put into a bottle is generally normally distributed. By checking the distribution of the sample, we could verify this.
   b. **Identify the probability distribution and the test statistic to be used.**
      The chi-square distribution will be used and formula (9.10), with df $= n - 1 = 28 - 1 = 27$.
   c. **Determine the level of significance:** $\alpha = 0.05$.

Step 3 **The Sample Evidence:**
   a. **Collect the sample information:** $n = 28$ and $s^2 = 0.0007$.
   b. **Calculate the value of the test statistic.**
      Use formula (9.10):

$$\chi^2\bigstar = \frac{(n-1)s^2}{\sigma^2}: \quad \chi^2\bigstar = \frac{(28-1)(0.0007)}{0.0004} = \frac{(27)(0.0007)}{0.0004} = 47.25$$

Step 4 **The Probability Distribution:**

Using the *p*-value procedure:	**OR**	Using the classical procedure:
a. **Calculate the *p*-value for the test statistic.**		a. **Determine the critical region and critical value(s).**
Use the right-hand tail because H_a expresses concern for values related to "larger than."		The critical region is the right-hand tail because H_a expresses concern for values related to "larger than." The

$\mathbf{P} = P\,(\chi^2\star > 47.25,$ with df $= 27)$ as shown in the figure.

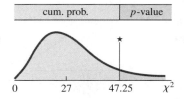

To find the *p*-value, use one of two methods:

1. Use Table 8 in Appendix B to place bounds on the *p*-value: $0.005 < \mathbf{P} < 0.01$.
2. Use a computer or calculator to calculate the *p*-value: $\mathbf{P} = 0.0093$.

Specific instructions follow this illustration.

b. Determine whether or not the *p*-value is smaller than α.

The *p*-value is smaller than the level of significance, α (0.05).

critical value is obtained from Table 8, at the intersection of row df $= 27$ and column $\alpha = 0.05$: $\chi^2(27, 0.05) = 40.1$.

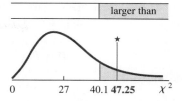

For specific instructions, see page 518.

b. Determine whether or not the calculated test statistic is in the critical region.

$\chi^2\star$ is in the critical region, as shown in **red** in the accompanying figure.

Step 5   The Results:

a. State the decision about H_o: Reject H_o.

b. State the conclusion about H_a.

At the 0.05 level of significance, we conclude that the bottling process is out of control with regard to the variance.

Calculating the *p*-value when using the χ^2-distribution

Method 1: Use Table 8 in Appendix B to place bounds on the p-value. By inspecting the df $= 27$ row of Table 8, you can determine an interval within which the *p*-value lies. Locate $\chi^2\star$ along the row labeled df $= 27$. If $\chi^2\star$ is not listed, locate the two values that $\chi^2\star$ falls between, and then read the bounds for the *p*-value from the top of the table. In this case, $\chi^2\star = 47.25$ is between 47.0 and 49.6; therefore, **P** is between 0.005 and 0.01.

Portion of Table 8   Finding $\mathbf{P} = P(\chi^2\star > 47.25,$ with df $= 27)$

df	...	0.01	P	0.005	
⋮		↑		↑	
27		47.0	47.25	49.6	$\longrightarrow 0.005 < P < 0.01$

Area in Right-Hand Tail

Method 2: Use a computer or calculator. Use the χ^2 probability distribution commands on pages 518–519 to find the *p*-value associated with $\chi^2\star = 47.25$.

EXAMPLE 9.18   **One-Tailed *p*-Value Hypothesis Test for Variance, σ^2**

Find the *p*-value for this hypothesis test:

$H_o\colon \sigma^2 = 12$

$H_a\colon \sigma^2 < 12$ with df $= 15$ and $\chi^2\star = 7.88$

SOLUTION Since the concern is for "smaller" values (the alternative hypothesis is "less than"), the *p*-value is the area to the left of $\chi^2\bigstar = 7.88$, as shown in the figure:

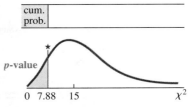

$$\mathbf{P} = P(\chi^2\bigstar < 7.88 \text{ with df} = 15)$$

To find the *p*-value, use one of two methods:

Method 1: Use Table 8 in Appendix B to place bounds on the p-value. Inspect the df = 15 row to find $\chi^2\bigstar = 7.88$. The $\chi^2\bigstar$ value is between entries, so the interval that bounds **P** is read from the Left-Hand Tail heading at the top of the table.

Portion of Table 8   Finding $\mathbf{P} = P(\chi^2\bigstar < 7.88,$ with df $= 15)$

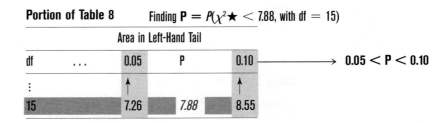

Method 2: Use a computer or calculator. Use the χ^2 probability distribution commands on pages 518–519 to find the *p*-value associated with $\chi^2\bigstar = 7.88$.

EXAMPLE 9.19   # Two-Tailed Hypothesis Test for Standard Deviation, σ

The manufacturer claims that a photographic chemical has a shelf life that is normally distributed about a mean of 180 days with a standard deviation of no more than 10 days. As a user of this chemical, Fast Photo is concerned that the standard deviation might be different from 10 days; otherwise, it will buy a larger quantity while the chemical is part of a special promotion. Twelve random samples were selected and tested, with a standard deviation of 14 days resulting. At the 0.05 level of significance, does this sample present sufficient evidence to show that the standard deviation is different from 10 days?

SOLUTION

Step 1   **The Set-Up:**
a. **Describe the population parameter of interest.**
   σ, the standard deviation for the shelf life of the chemical.
b. **State the null hypothesis (H_o) and the alternative hypothesis (H_a).**
   H_o: $\sigma = 10$ (standard deviation is 10 days)
   H_a: $\sigma \neq 10$ (standard deviation is different from 10 days)

Step 2   **The Hypothesis Test Criteria:**
a. **Check the assumptions.**
   The manufacturer claims shelf life is normally distributed; this could be verified by checking the distribution of the sample.

b. Identify the probability distribution and the test statistic to be used.

The chi-square distribution will be used and formula (9.10), with df $= n - 1 = 12 - 1 = 11$.

c. Determine the level of significance: $\alpha = 0.05$.

Step 3   The Sample Evidence:

a. Collect the sample information: $n = 12$ and $s = 14$.

b. Calculate the value of the test statistic.

Use formula (9.10):

$$\chi^2\star = \frac{(n-1)s^2}{\sigma^2}: \quad \chi^2\star = \frac{(12-1)(14)^2}{(10)^2} = \frac{2156}{100} = 21.56$$

Step 4   The Probability Distribution:

Using the p-value procedure: OR   **Using the classical procedure:**

a. Calculate the p-value for the test statistic.

Since the concern is for values "different from" 10, the p-value is the area of both tails. The area of each tail will represent $\frac{1}{2}$ **P**. Since $\chi^2\star = 21.56$ is in the right tail, the area of the right tail is $\frac{1}{2}$**P**:

$\frac{1}{2}$**P** $= P(\chi^2 > 21.56$, with df $= 11)$ as shown in the figure.

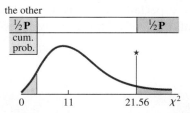

To find $\frac{1}{2}$**P**, use one of two methods:

1. Use Table 8 in Appendix B to place bounds on $\frac{1}{2}$**P**: $0.025 < \frac{1}{2}$**P** < 0.05. Double both bounds to find the bounds for **P**: $2 \times (0.025 < \frac{1}{2}$**P** $< 0.05)$ becomes $0.05 < $ **P** < 0.10.

2. Use a computer or calculator to find $\frac{1}{2}$**P**: $\frac{1}{2}$**P** $= 0.0280$; therefore, **P** $= 0.0560$.

Specific instructions follow this illustration.

b. Determine whether or not the p-value is smaller than α.

The p-value is not smaller than the level of significance, α (0.05).

a. Determine the critical region and critical value(s).

The critical region is split into two equal parts because H_a expresses concern for values related to "different from." The critical values are obtained from Table 8 at the intersections of row df $= 11$ with columns $\alpha = 0.975$ and 0.025 (area to right): $\chi^2(11, 0.975) = 3.82$ and $\chi^2(11, 0.025) = 21.9$.

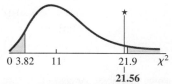

For specific instructions, see page 518.

b. Determine whether or not the calculated test statistic is in the critical region.

$\chi^2\star$ is not in the critical region; see the accompanying figure.

Step 5   The Results:

a. State the decision about H_o: Fail to reject H_o.

b. State the conclusion about H_a.

There is not sufficient evidence at the 0.05 significance level to conclude that the shelf life of this chemical has a standard deviation different from 10 days. Therefore, Fast Photo should purchase the chemical accordingly.

Calculating the *p*-value when using the χ^2-distribution

Method 1: Use Table 8 in Appendix B to place bounds on the p-value. Inspect the df = 11 row to locate $\chi^2\star = 21.56$. Notice that 21.56 is between two table entries. The bounds for $^1/_2\mathbf{P}$ are read from the Right-Hand Tail heading at the top of the table.

Portion of Table 8   Finding $\mathbf{P} = 2 \cdot P(\chi^2\star > 21.56$, with df $= 11)$

Area in Right-Hand Tail

df	...	0.05	$^1/_2$P	0.025
⋮		↑		↑
11		19.7	*21.56*	21.9

⟶ $0.025 < {}^1/_2\mathbf{P} < 0.05$
$0.05 < \mathbf{P} < 0.10$

Double both bounds to find the bounds for **P**: $2 \times (0.025 < {}^1/_2\mathbf{P} < 0.05)$ becomes $\mathbf{0.05 < P < 0.10}$.

Method 2: Use a computer or calculator. Use the χ^2 probability distribution commands on pages 518–519 to find the *p*-value associated with $\chi^2\star = 21.56$. Remember to double the probability.

Note: When sample data are skewed, just one outlier can greatly affect the standard deviation. It is very important, especially when using small samples, that the sampled population be normal; otherwise, these procedures are not reliable.

SECTION 9.4 EXERCISES

9.115 a. Calculate the standard deviation for each set.

   A: 5, 6, 7, 7, 8, 10   B: 5, 6, 7, 7, 8, 15

   b. What effect did the largest value changing from 10 to 15 have on the standard deviation?

   c. Why do you think 15 might be called an outlier?

9.116 The variance of shoe sizes for all manufacturers is 0.1024. What is the standard deviation?

9.117 Find:

   a. $\chi^2(10, 0.01)$ b. $\chi^2(12, 0.025)$

   c. $\chi^2(10, 0.95)$ d. $\chi^2(22, 0.995)$

9.118 Find these critical values by using Table 8 of Appendix B.

   a. $\chi^2(18, 0.01)$ b. $\chi^2(16, 0.025)$ c. $\chi^2(8, 0.10)$

   d. $\chi^2(28, 0.01)$ e. $\chi^2(22, 0.95)$ f. $\chi^2(10, 0.975)$

   g. $\chi^2(50, 0.90)$ h. $\chi^2(24, 0.99)$

9.119 Using the notation of Exercise 9.118, name and find the critical values of χ^2.

a.

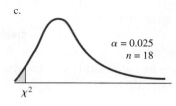

$\alpha = 0.05$
$n = 20$

χ^2

b.

$\alpha = 0.01$
$n = 5$

χ^2

c.

$\alpha = 0.025$
$n = 18$

χ^2

d.

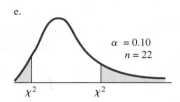

$\alpha = 0.05$
$n = 61$

χ^2

e.

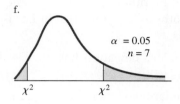

$\alpha = 0.10$
$n = 22$

χ^2 χ^2

f.

$\alpha = 0.05$
$n = 7$

χ^2 χ^2

9.120 Using the notation of Exercise 9.118, name and find the critical values of χ^2.

a.

$n = 14$
$\alpha = 0.005$

χ^2

b.

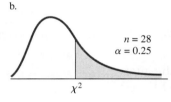

$n = 28$
$\alpha = 0.25$

χ^2

c.

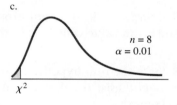

$n = 8$
$\alpha = 0.01$

χ^2

d.

$n = 16$
$\alpha = 0.025$

χ^2

e.

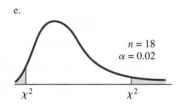

$n = 18$
$\alpha = 0.02$

χ^2 χ^2

f.

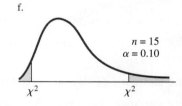

$n = 15$
$\alpha = 0.10$

χ^2 χ^2

9.121 a. What value of chi-square for 5 degrees of freedom subdivides the area under the distribution curve such that 5% is to the right and 95% is to the left?

b. What is the value of the 95th percentile for the chi-square distribution with 5 degrees of freedom?

c. What is the value of the 90th percentile for the chi-square distribution with 5 degrees of freedom?

9.122 a. The central 90% of the chi-square distribution with 11 degrees of freedom lies between what values?

b. The central 95% of the chi-square distribution with 11 degrees of freedom lies between what values?

c. The central 99% of the chi-square distribution with 11 degrees of freedom lies between what values?

9.123 For a chi-square distribution having 12 degrees of freedom, find the area under the curve for chi-square values ranging from 3.57 to 21.0.

9.124 For a chi-square distribution having 35 degrees of freedom, find the area under the curve between $\chi^2(35, 0.96)$ and $\chi^2(35, 0.15)$.

9.125 Use a computer or calculator to find the area (a) to the left, and (b) to the right of $\chi^2\star = 20.2$ with df $= 15$.

9.126 Use a computer or calculator to find the area (a) to the left, and (b) to the right of $\chi^2\star = 14.7$ with df $= 24$.

9.127 State the null hypothesis, H_o, and the alternative hypothesis, H_a, that would be used to test these claims:

a. The standard deviation has increased from its previous value of 24.

b. The standard deviation is no larger than 0.5 oz.

c. The standard deviation is not equal to 10.

d. The variance is no less than 18.

e. The variance is different from the value of 0.025, the value called for in the specs.

9.128 State the null hypothesis, H_o, and the alternative hypothesis, H_a, that would be used to test these claims:

a. The variance has decreased from 34.5.

b. The standard deviation of shoe size is more than 0.32.

c. The standard deviation is at least 5.5.

d. The variance is at most 35.

e. The variance has shrunk from the value of 0.34 since the assembly lines were retooled.

9.129 Find the test statistic for the hypothesis test:

a. $H_o: \sigma^2 = 532$ vs. $H_a: \sigma^2 > 532$ using sample information $n = 18$ and $s^2 = 785$.

b. $H_o: \sigma^2 = 52$ vs. $H_a: \sigma^2 \neq 52$ using sample information $n = 41$ and $s^2 = 78.2$.

9.130 Calculate the value for the test statistic, $\chi^2\star$, for each of these situations:

a. $H_o: \sigma^2 = 20$, $n = 15$, $s^2 = 17.8$

b. $H_o: \sigma^2 = 30$, $n = 18$, $s = 5.7$

c. $H_o: \sigma = 42$, $n = 25$, $s = 37.8$

d. $H_o: \sigma = 12$, $n = 37$, $s^2 = 163$

9.131 Calculate the p-value for each of the following hypothesis tests.

a. $H_a: \sigma^2 \neq 20$, $n = 15$, $\chi^2\star = 27.8$

b. $H_a: \sigma^2 > 30$, $n = 18$, $\chi^2\star = 33.4$

c. $H_a: \sigma^2 \neq 42$, df $= 25$, $\chi^2\star = 37.9$

d. $H_a: \sigma^2 < 12$, df $= 40$, $\chi^2\star = 26.3$

9.132 Determine the critical region and critical value(s) that would be used to test the following using the classical approach:

a. $H_o: \sigma = 0.5$ and $H_a: \sigma > 0.5$, with $n = 18$ and $\alpha = 0.05$

b. $H_o: \sigma^2 = 8.5$ and $H_a: \sigma^2 < 8.5$, with $n = 15$ and $\alpha = 0.01$

c. $H_o: \sigma = 20.3$ and $H_a: \sigma \neq 20.3$, with $n = 10$ and $\alpha = 0.10$

d. $H_o: \sigma^2 = 0.05$ and $H_a: \sigma^2 \neq 0.05$, with $n = 8$ and $\alpha = 0.02$

e. $H_o: \sigma = 0.5$ and $H_a: \sigma < 0.5$, with $n = 12$ and $\alpha = 0.10$

9.133 Complete the hypothesis test in Exercise 9.129a using the following:

a. The p-value method and $\alpha = 0.01$

b. The classical method and $\alpha = 0.01$

9.134 Complete the hypothesis test in Exercise 9.129b using the following:

a. The p-value method and $\alpha = 0.05$

b. The classical method and $\alpha = 0.05$

9.135 A random sample of 51 observations was selected from a normally distributed population. The sample mean was $\bar{x} = 98.2$, and the sample variance was $s^2 = 37.5$. Does this sample show sufficient reason to conclude that the population standard deviation is not equal to 8 at the 0.05 level of significance?

a. Solve using the p-value approach.

b. Solve using the classical approach.

9.136 In the past the standard deviation of weights of certain 32.0-oz packages filled by a machine was 0.25 oz. A random sample of 20 packages showed a standard deviation of 0.35 oz. Is the apparent increase in variability significant at the 0.10 level of significance? Assume package weight is normally distributed.

a. Solve using the *p*-value approach.

b. Solve using the classical approach.

9.137 [EX09-137] Maybe even more important than how much they weigh, it is very important that the plates used in weightlifting be the same weight. When one of each weight is hanging on opposite ends of a bar, they need to balance. A random sample of twenty-four 25-lb weights used for weightlifting were randomly selected and their weights (in pounds) determined:

25.3	22.1	25.7	24.2	25.7	23.9	23.1	21.9
24.7	26.3	26.5	22.2	25.9	23.5	25.8	27.1
25.4	22.0	25.2	21.1	27.9	22.9	27.3	25.7

There have been complaints about the excessive variability in the weights of these 25-lb plates. Does the sample show sufficient evidence to conclude that the variability in the weights is greater than the acceptable 1-lb standard deviation? Use $\alpha = 0.01$.

a. What role does the assumption of normality play in this solution? Explain.

b. What evidence do you have that the assumption of normality is reasonable? Explain.

c. Solve using the *p*-value approach.

d. Solve using the classical approach.

9.138 A commercial farmer harvests his entire field of a vegetable crop at one time. Therefore, he would like to plant a variety of green beans that mature all at one time (small standard deviation between maturity times of individual plants). A seed company has developed a new hybrid strain of green beans that it believes to be better for the commercial farmer. The maturity time of the standard variety has an average of 50 days and a standard deviation of 2.1 days. A random sample of 30 plants of the new hybrid showed a standard deviation of 1.65 days. Does this sample show a significant lowering of the standard deviation at the 0.05

level of significance? Assume that maturity time is normally distributed.

a. Solve using the *p*-value approach.

b. Solve using the classical approach.

9.139 Farm real estate values in rural America fluctuate substantially from state to state and county to county, thus making it difficult for buyers purchasing land or landowners to know precisely what their property is actually worth. For example, the average value of ranch land in Missouri was $548 per acre, whereas the same average in three nearby states (Kansas, Nebraska, and Oklahoma) was more than $200 less.

Source: Regional Economic Digest, "Survey of Agricultural Credit Conditions"

This discrepancy could be caused by an exaggerated variability in the value of ranch land acreage in Missouri. Assume that the combined four-state region yielded a standard deviation of $85 per acre. Suppose a sample of 31 landowners in Missouri who recently sold their property was taken and a sample standard deviation of $125 per acre resulted. Is the variability in ranch land value in Missouri, at the 0.05 level of significance, greater than the variability for the region as a whole?

a. Solve using the *p*-value approach.

b. Solve using the classical approach.

9.140 [EX09-140] A car manufacturer claims that the miles per gallon for a certain model has a mean equal to 40.5 miles with a standard deviation equal to 3.5 miles. Use the following data, obtained from a random sample of 15 such cars, to test the hypothesis that the standard deviation differs from 3.5. Use $\alpha = 0.05$. Assume normality.

37.0	38.0	42.5	45.0	34.0	32.0	36.0	35.5
38.0	42.5	40.0	42.5	35.0	30.0	37.5	

a. Solve using the *p*-value approach.

b. Solve using the classical approach.

9.141 [EX09-141] The dry weight of a cork is another quality that does not affect the ability of the cork to seal the bottle, but it is a variable that is monitored regularly. The weights of the no. 9 natural corks (24 mm in diameter by 45 mm in length) have a normal distribution. Ten randomly selected corks were weighed to the nearest hundredth of a gram.

Dry Weight (in grams)

3.26	3.58	3.07	3.09	3.16	3.02	3.64	3.61	3.02	2.79

a. Does the preceding sample show sufficient reason to show the standard deviation of the dry weights is different from 0.3275 gram at the 0.02 level of significance?

A different random sample of 20 is taken from the same batch.

Dry Weight (in grams)

3.53	3.77	3.49	3.24	3.00	3.41	3.33	3.51	3.02	3.46
2.80	3.58	3.05	3.51	3.61	2.90	3.69	3.62	3.26	3.58

b. Does the preceding sample show sufficient reason to show the standard deviation of the dry weights is different from 0.3275 gram at the 0.02 level of significance?

c. What effect did the two different sample standard deviations have on the calculated test statistic in parts a and b? What effect did they have on the *p*-value or critical value? Explain.

d. What effect did the two different sample sizes have on the calculated test statistic in parts a and b? What effect did they have on the *p*-value or critical value? Explain.

9.142 Use a computer or calculator to find the *p*-value for the following hypothesis test: $H_o: \sigma^2 = 7$ vs. $H_a: \sigma^2 \neq 7$, if $\chi^2\star = 6.87$ for a sample of $n = 15$.

9.143 Use a computer or calculator to find the *p*-value for the following hypothesis test: $H_o: \sigma = 12.4$ vs. $H_a: \sigma > 12.4$, if $\chi^2\star = 36.59$ for a sample of $n = 24$.

9.144 The chi-square distribution was described on page 517 as a family of distributions. Let's investigate these distributions and observe some of their properties.

a. Use the MINITAB commands that follow and generate several large random samples of data

from various chi-square distributions. Use df values of 1, 2, 3, 5, 10, 20, and 80 (and others if you wish).

```
Choose:  Calc > Random Data > ChiSquare
Enter:   Generate: 1000 rows of data
         Store in column(s): C1
         Degrees of freedom: df
Use Stat > Basic Statistics > Display Descriptive Sta-
tistics to calculate the mean and median of the data in
C1. Use Graph > Histogram to construct a histogram of
the data in C1.
```

b. What appears to be the relationship between the mean of the sample and the number of degrees of freedom?

c. How do the values of the mean, median, and mode appear to be related? Do your results agree with the information on page 517?

d. Have the computer generate samples for two additional degrees of freedom df = 120 and 150. Describe how these distributions seem to be changing as df increases.

9.145 How important is the assumption "the sampled population is normally distributed" for the use of the chi-square distributions? Use a computer and the two sets of MINITAB commands that can be found in the *Student Solutions Manual* to simulate drawing 200 samples of size 10 from each of two different types of population distributions. The first commands will generate 2000 data values and construct a histogram so that you can see what the population looks like. The next commands will generate 200 samples of size 10 from the same population; each row represents a sample. The following commands will calculate the standard deviation and $\chi^2\star$ for each of the 200 samples. The last commands will construct histograms of the 200 sample standard deviations and the 200 $\chi^2\star$-values. (Additional details can be found in the *Student Solutions Manual.*)

For the samples from the normal population:

a. Does the sampling distribution of sample standard deviations appear to be normal? Describe the distribution.

b. Does the χ^2-distribution appear to have a chi-square distribution with df = 9? Find percentages for intervals (less than 2, less than 4, ..., more than 15, more than 20, etc.), and compare them with the percentages expected as estimated using Table 8 in Appendix B.

For the samples from the skewed population:

c. Does the sampling distribution of sample standard deviations appear to be normal? Describe the distribution.

d. Does the χ^2-distribution appear to have a chi-square distribution with df = 9? Find percent-

ages for intervals (less than 2, less than 4, . . . , more than 15, more than 20, etc.), and compare them with the percentages expected as estimated using Table 8.

In summary:

e. Does the normality condition appear to be necessary in order for the calculated test statistic $\chi^2\star$ to have a χ^2-distribution? Explain.

CHAPTER REVIEW

In Retrospect

We have been studying inferences, both confidence intervals and hypothesis tests, for the three basic population parameters (mean μ, proportion p, and standard deviation σ) of a single population. Most inferences about a single population are concerned with one of these three parameters. Figure 9.10 (p. 530) presents a visual organization of the techniques presented in Chapters 8 and 9 along with the key questions that you must ask as you are deciding which test statistic and formula to use.

In this chapter we also used the maximum error of estimate, formula (9.7), to determine the size of the sample required to make estimations

about the population proportion with the desired accuracy. In Applied Example 9.9 the margin of error reported by the media is described, and its relationship to the maximum error of estimate, as presented in this chapter, is discussed. By combining the reported point estimate and the sample size, we can determine the corresponding binomial proportion maximum error of estimate. Most polls and surveys use the 95% confidence level and then use the maximum error as an estimate for margin of error and do not report a level of confidence, as Humphrey Taylor explained.

In the next chapter we will discuss inferences about two populations whose respective means, proportions, and standard deviations are to be compared.

Vocabulary and Key Concepts

assumptions (pp. 479, 497, 519)
binomial experiment (p. 496)
calculated value
　(pp. 482, 503, 520)
chi-square (p. 517)
conclusion (pp. 483, 504, 521)
confidence interval
　(pp. 480, 498)
critical region
　(pp. 482, 504, 520)
critical value (pp. 477, 504, 517)
decision (pp. 483, 504, 521)
degrees of freedom
　(pp. 475, 517)

hypothesis test
　(pp. 481, 502, 520)
inference (pp. 474, 496, 516)
level of confidence
　(pp. 480, 498, 506)
level of significance
　(pp. 482, 503, 520)
maximum error of estimate
　(pp. 480, 498, 501)
observed binomial probability,
　p' (p. 496)
p-value (pp. 482, 504, 520)
parameter (pp. 475, 480, 496,
　498, 503, 517, 520)

proportion (p. 497)
random variable (p. 496)
rule of thumb (p. 497)
sample size (pp. 474, 501)
sample statistic (p. 497)
σ known (p. 475)
σ unknown (p. 474)
standard error (pp. 474, 498)
standard normal, z
　(pp. 474, 497)
Student's t (p. 475)
test statistic (pp. 474, 481,
　503, 519)
unbiased estimator (p. 497)

FIGURE 9.10 Choosing the Right Inference Technique

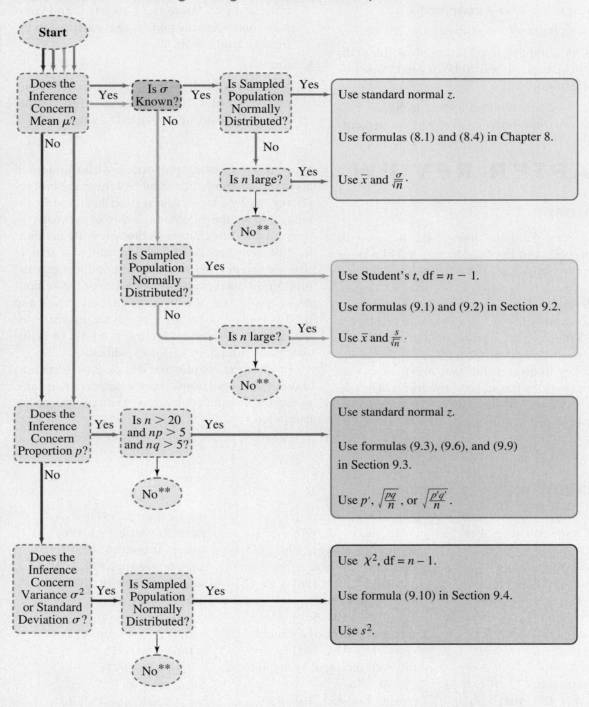

No** means that a nonparametric technique (normal distribution not required) is used: see Chapter 14.

Learning Outcomes

✓ Understand that s, the sample standard deviation, is a point estimate for σ, the population standard deviation. pp. 474–475

✓ Understand that in most real-life cases, σ is unknown and s is used as its best estimate. pp. 474–475

✓ Understand that when σ is unknown, the z-statistic is replaced with Student's t-statistic. pp. 474–475

✓ Understand the properties of the t-distribution, how it is a series of distributions based on sample size (using degrees of freedom as the index), and how it approaches the standard normal distribution as the sample size increases. pp. 475–476, EXP 9.1, 9.2, 9.3, Ex. 9.4, 9.19

✓ Understand that the assumption for inferences about the mean μ when σ is unknown is that the sampled population is normally distributed. p. 479, Ex. 9.27b

✓ Compute, describe, and interpret a confidence interval for the population mean, μ, using the t-distribution. p. 480, EXP 9.4, Ex. 9.23, 9.27, 9.147

✓ Perform, describe, and interpret a hypothesis test for the population mean, μ, using the t-distribution with the p-value approach and classical approach. pp. 481–482, EXP 9.5, 9.6, Ex. 9.47, 9.154

✓ Understand the fundamental properties of a binomial experiment and the binomial parameter, p. pp. 496–497, Ex. 9.63, 9.64

✓ Understand that p', the sample proportion, is an unbiased estimator of the population proportion, p. pp. 496–498, Ex. 9.65

✓ Understand that the sampling distribution of p' has an approximately normal distribution if n is sufficiently large and therefore the standard normal distribution can be used for inferences. pp. 496–498, Ex. 9.85

✓ Understand that the assumption for inferences about the binomial parameter, p, is that the n random observations forming the sample are selected independently from a population that is not changing during the sampling. p. 497

✓ Compute, describe, and interpret a confidence interval for the population proportion, p, using the z-distribution. EXP 9.8, Ex. 9.69, 9.73, 9.161

✓ Compute and describe the required sample size for a confidence interval of p, the population proportion. p. 501, EXP 9.10, 9.11, Ex. 9.86

✓ Perform, describe, and interpret a hypothesis test for the population proportion, p, using the z-distribution with the p-value approach and classical approach. EXP 9.12, 9.13, Ex. 9.103, 9.108

✓ Understand the properties of the chi-square distribution and how it is a series of distributions based on sample size (using degrees of freedom as the index). p. 517

✓ Understand that the assumption for inferences about the variance, σ^2, or standard deviation, σ, is that the sampled population is normally distributed. p. 519, Ex. 9.145

✓ Perform, describe, and interpret a hypothesis test for the population variance, σ^2, or standard deviation, σ, using the χ^2-distribution with the p-value approach and classical approach. EXP 9.17, 9.19, Ex. 9.136, 9.177

Chapter Exercises

9.146 You hurry to the local emergency department in hopes of immediate urgent care only to find yourself waiting for what seems like hours. The manager of a large emergency department believes that his new procedures have substantially reduced the wait time for the average urgent care patient. He initiates a study to evaluate the wait time. The records of 18 randomly selected patients seen since the new procedures have been in place are checked, and the time between entering the emergency department and being seen by urgent care personnel was observed. The mean wait time was 17.82 minutes with a standard deviation of 5.68 minutes. Estimate the mean wait time using a 99% confidence interval. Assume that wait times are normally distributed.

9.147 A natural-gas utility is considering a contract for purchasing tires for its fleet of service trucks. The decision will be based on expected mileage. For a sample of 100 tires tested, the mean mileage was 36,000 and the standard deviation was 2000 miles. Estimate the mean mileage that the utility should expect from these tires using a 98% confidence interval.

9.148 One of the objectives of a large medical study was to estimate the mean physician fee for cataract removal. For 25 randomly selected cases, the mean fee was found to be $1550 with a standard deviation of $125. Set a 99% confidence interval on μ, the mean fee for all physicians. Assume fees are normally distributed.

9.149 Oranges are selected at random from a large shipment that just arrived. The sample was taken to estimate the size (circumference, in inches) of the oranges. The sample data are summarized as follows: $n = 100$, $\sum x = 878.2$, and $\sum (x - \bar{x})^2 = 49.91$.

a. Determine the sample mean and standard deviation.

b. What is the point estimate for μ, the mean circumference of all oranges in the shipment?

c. Find the 95% confidence interval for μ.

9.150 [EX09-150] Molds are used in the manufacture of contact lenses so that the lens material on proper preparation and curing will be consistent and meet designated dimensional criteria. Molds were fabricated and a critical dimension measured for 15 randomly selected molds. (Data have been doubly coded to insure propriety.)

140	130	15	180	95	135	220	105
195	110	150	150	130	120	120	

Courtesy of Bausch & Lomb

a. Construct a histogram and find the mean and standard deviation.

b. Demonstrate how this set of data satisfies the assumptions for inference.

c. Find the 95% confidence interval for μ.

d. Interpret the meaning of the confidence interval.

9.151 A company claims that its battery lasts no less than 42.5 hours in continuous use in a specified toy. A simple random sample of batteries yields a sample mean life of 41.89 hours with a standard deviation of 4.75 hours. A computer calculates a test statistic of $t = -1.09$ and a p-value of 0.139. If the test uses df = 71, what is the best estimate of the sample size?

9.152 [EX09-152] Getting a college education today is almost as important as breathing and it's expensive! It is not just the tuition, room, and board; textbooks are expensive too. It is very important for students, and their parents, to have an accurate estimate of total textbook costs. The total cost of required textbooks for nine freshman- or sophomore-level classes at 10 randomly selected New York public colleges was collected:

582.19	806.40	913.44	915.75	932.35
957.45	960.92	996.24	1070.44	1223.44

a. Construct a histogram and find the mean and standard deviation.

b. Demonstrate how this set of data satisfies the assumptions for inference.

c. Find the 95% confidence interval for μ, the mean total cost of required textbooks.

d. Interpret the meaning of the confidence interval.

9.153 [EX09-153] The total cost of required textbooks for nine freshman- or sophomore-level classes at 10 randomly selected New York private colleges was collected:

639.00	865.75	868.20	874.25	887.06
890.50	970.13	1013.22	1026.00	1048.96

a. Construct a histogram and find the mean and standard deviation.

b. Demonstrate how this set of data satisfies the assumptions for inference.

c. Find the 95% confidence interval for μ, the mean total cost of required textbooks.

d. Interpret the meaning of the confidence interval.

e. Is there a difference in the mean total cost of the nine required textbooks between the public colleges in Exercise 9.152 and in the private colleges in this exercise? Explain.

f. Explain why the confidence interval for the public colleges is so much wider than the corresponding interval for the private colleges. Be exact and detailed.

9.154 A manufacturer of television sets claims that the maintenance expenditures for its product will average no more than $50 during the first year following the expiration of the warranty. A consumer group has asked you to substantiate or discredit the claim. The results of a random sample of 50 owners of such television sets showed that the mean expenditure was $61.60 with a standard deviation of $32.46. At the 0.01 level of significance,

should you conclude that the producer's claim is true or not likely to be true?

9.155 [EX09-155] In a large cherry orchard the average yield has been 4.35 tons per acre for the last several years. A new fertilizer was tested on 15 randomly selected 1-acre plots. The yields (in tons) from these plots follow:

3.56	5.00	4.88	4.93	3.92	4.25	5.12	5.13
4.79	4.45	5.35	4.81	3.48	4.45	4.72	

At the 0.05 level of significance, do we have sufficient evidence to claim that there was a significant increase in production? Assume yield per acre is normally distributed.

9.156 [EX09-156] The water pollution readings at State Park Beach seem to be lower than that of the prior year. A sample of 12 readings (measured in coliform/100 mL) was randomly selected from the records of this year's daily readings:

3.5	3.9	2.8	3.1	3.1	3.4	4.8	3.2	2.5	3.5	4.4	3.1

Does this sample provide sufficient evidence to conclude that the mean of this year's pollution readings is significantly lower than last year's mean of 3.8 at the 0.05 level? Assume that all such readings have a normal distribution.

9.157 [EX09-157] It has been suggested that abnormal male children tend to occur more in children born to older-than-average parents. Case histories of 20 abnormal males were obtained, and the ages of the 20 mothers were as follows:

31	21	29	28	34	45	21	41	27	31
43	21	39	38	32	28	37	28	16	39

The mean age at which mothers in the general population give birth is 28.0 years.

a. Calculate the sample mean and standard deviation.

b. Does the sample give sufficient evidence to support the claim that abnormal male children have older-than-average mothers? Use $\alpha = 0.05$. Assume ages have a normal distribution.

9.158 [EX09-158] Twenty-four oat-producing counties in the United States were randomly identified for the purpose of testing the claim, "the mean oat crop yield rate is less than 60 bushels per acre." For each county identified, the yield rate, in bushels of oats per harvested acre, was obtained:

44.0 65.0 78.5 50.0 52.5 51.0 47.5 67.5 76.7 33.3 73.6 20.0
57.0 52.0 68.6 42.9 63.0 80.0 37.0 70.0 43.0 30.0 60.0 67.5

Source: http://www.usda.gov/nass/graphics/county01/data/ot01.csv

a. Are the test assumptions satisfied? Explain.

b. Complete the test using $\alpha = 0.05$.

9.159 [EX09-159] Presented here are 100 measurements of the velocity of light in air (km/sec) recorded by Albert Michelson, an American physicist, from June 5 to July 2, 1879. The measurements have had 299,000 subtracted from them and then adjusted for corrections used by Michelson. In this form, the true constant value for the velocity of light in air becomes 734.5 km/sec. Do Michelson's measurements support the true value that he was trying to measure? Use a 0.01 level of significance.

850	740	900	1070	930	850	950	980	980	880	1000
980	930	650	760	810	1000	1000	960	960	960	940
960	940	880	800	850	880	900	840	830	790	810
880	880	830	800	790	760	800	880	880	880	860
720	720	620	860	970	950	880	910	850	870	840
840	850	840	840	840	890	810	810	820	800	770
760	740	750	760	910	920	890	860	880	720	840
850	850	780	890	840	780	810	760	810	790	810
820	850	870	870	810	740	810	940	950	800	810
870										

Source: http://lib.stat.cmu.edu/DASL/Stories/SpeedofLight.html

Note: The currently accepted "true" value is 299,792.5 km/sec (with no adjustments).

9.160 Even with a heightened awareness of beef quality, 82% of Americans indicated their recent burger-eating behavior has remained the same, according to a recent T.G.I. Friday's restaurants random survey of 1027 Americans. In fact, half of Americans eat at least one beef burger each week. That's a minimum of 52 burgers each year.

Source: Harris Interactive/Yankelovich Partners for T.G.I. Friday's restaurants, http://www.knoxville3.com/fridays/News/burger.htm

a. What is the point estimate for the proportion of all Americans who eat at least one beef burger per week?

b. Find the 98% confidence interval for the true proportion p in the binomial situation where $n = 1027$ and the observed proportion is one-half.

c. Use the results of part b to estimate the percentage of all Americans who eat at least one beef burger per week.

9.161 The marketing research department of an instant-coffee company conducted a survey of married men to determine the proportion of married men who preferred their brand. Of the 100 men in the random sample, 20 preferred the company's brand. Use a 95% confidence interval to estimate the proportion of all married men who prefer this company's brand of instant coffee. Interpret your answer.

9.162 A company is drafting an advertising campaign that will involve endorsements by noted athletes. For the campaign to succeed, the endorser must be both highly respected and easily recognized. A random sample of 100 prospective customers is shown photos of various athletes. If the customer recognizes an athlete, then the customer is asked whether he or she respects the athlete. In the case of a top woman golfer, 16 of the 100 respondents recognized her picture and indicated that they also respected her. At the 95% level of confidence, what is the true proportion with which this woman golfer is both recognized and respected?

9.163 A local auto dealership advertises that 90% of customers whose autos were serviced by their service department are pleased with the results. As a researcher, you take exception to this statement because you are aware that many people are reluctant to express dissatisfaction even if they are not pleased. A research experiment was set up in which those in the sample had received service by this dealer within the past 2 weeks. During the interview, the individuals were led to believe that the interviewer was new in town and was considering taking his car to this dealer's service department. Of the 60 sampled, 14 said that they were dissatisfied and would not recommend the department.

a. Estimate the proportion of dissatisfied customers using a 95% confidence interval.

b. Given your answer to part a, what can be concluded about the dealer's claim?

9.164 In obtaining the sample size to estimate a proportion, the formula $n = [z(\alpha/2)]^2 \, pq/E^2$ is used. If a reasonable estimate of p is not available, it is suggested that $p = 0.5$ be used because this will give the maximum value for n. Calculate the value of $pq = p(1 - p)$ for $p = 0.1, 0.2, 0.3, \ldots, 0.8, 0.9$ in order to obtain some idea about the behavior of the quantity pq.

9.165 A Pew Internet & American Life Project study based on 1100 random telephone calls with children ages 12 to 17 and their parents was conducted from October 26 to November 28, 2004. The survey found that 13% of American teens do not use the Internet. The findings have a margin of sampling error of plus or minus 3 percentage points.

Source: http://www.usatoday.com/news/bythenumbers/2005-03-18-teen-net-use_x.htm

You wish to conduct a study to estimate the percentage of children 12 to 17 years old who do not use the Internet and live in your state. Assume the population proportion is 13% as reported by Pew Internet & American Life Project. What sample size must you use if you want your estimate to be within:

a. 0.03 with 90% confidence

b. 0.06 with 95% confidence

c. 0.09 with 99% confidence

9.166 The chief executive officer (CEO) of a small business wishes to hire your consulting firm to conduct a simple random sample of its customers. She wants to determine the proportion of her customers that consider her company the primary source of their products. She requests the margin of error in the proportion be no more than 3%

with 95% confidence. Earlier studies have indicated that the approximate proportion is 37%.

a. What is the minimum size of the sample that you would recommend to meet the requirements of your client if you use the earlier results?

b. What is the minimum size of the sample that you would recommend to meet the requirements of your client if you ignore the earlier results?

c. Is the approximate proportion of value in conducting the survey? Explain.

9.167 The March 25, 2002, *Newsweek* article "Bringing Up Adultolescents" quoted an online survey by Monster-TRAK.com. The survey found that 60% of college students planned to live at home after graduation. How large of a sample size would you need to estimate the true proportion of students who plan on living at home after graduation to within 2% with 98% confidence.

9.168 A machine is considered to be operating in an acceptable manner if it produces 0.5% or fewer defective parts. It is not performing in an acceptable manner if more than 0.5% of its production is defective. The hypothesis H_o: $p = 0.005$ is tested against the hypothesis H_a: $p > 0.005$ by taking a random sample of 50 parts produced by the machine. The null hypothesis is rejected if two or more defective parts are found in the sample. Find the probability of the type I error.

9.169 You are interested in comparing the null hypothesis $p = 0.8$ against the alternative hypothesis $p < 0.8$. In 100 trials you observe 73 successes. Calculate the p-value associated with this result.

9.170 The Kaiser Family Foundation conducted a national survey in 2003 of 17,685 seniors. The purpose of the survey was to capture detailed information about seniors' prescription drug use, coverage, and experiences.

Source: http://www.kff.org/medicare/med041905nr.cfm

a. If this were a random sample that satisfied all the requirements for an inference about p, what would be the standard error?

b. What would be the maximum error of estimate for a 95% confidence interval?

c. Is a sample this size worthwhile? Give reasons to support your answer.

9.171 The Pizza Shack has been experimenting with different recipes for their pizza crust, thinking they might replace their current recipe. They are planning to sample pizza made with the new crust. Before sampling, a strategy is needed so that after the tasting results are in, Pizza Shack will know how to interpret their customers' preferences. The decision is not being taken lightly because there is much to be gained or lost depending on whether or not the decision is a popular one. A one-tailed hypothesis test of $p = P(\text{prefer new crust}) = 0.50$ is being planned.

a. If H_a: $p > 0.50$ is used, explain the meaning of the four possible outcomes and their resulting actions.

b. If H_a: $p < 0.50$ is used, explain the meaning of the four possible outcomes and their resulting actions.

c. Which alternative hypothesis do you recommend be used, $p > 0.5$ or $p < 0.5$? Explain.

9.172 The Pizza Shack in Exercise 9.171 has completed its sampling and the results are in! On Tuesday afternoon, they sampled 15 customers and 9 preferred the new pizza crust. On Friday evening, they sampled 200 customers and 120 preferred the new pizza crust. Help the manager interpret the meaning of these results. Use a one-tailed test with H_a: $p > 0.50$ and $\alpha = 0.02$. Use z as the test statistic.

a. Is there sufficient evidence to conclude a significant preference for the new crust based on Tuesday's customers?

b. Is there sufficient evidence to conclude a significant preference for the new crust based on Friday's customers?

c. Since the percentage of customers preferring the new crust was the same, $p' = 0.60$, in both samplings, explain why the answers in parts a and b are not the same.

9.173 The owner of the Pizza Shack in Exercises 9.171 and 9.172 does not understand the use of the normal distribution and z in Exercise 9.172. Help the manager interpret the meaning of the results by redoing both hypothesis tests using $x =$ number of customers preferring the new crust as the test statistic and its binomial probability distribution. Use a one-tailed test with H_a: $p > 0.50$ and $\alpha = 0.02$.

The results were as follows: on Tuesday afternoon, they sampled 15 customers and 9 preferred the new pizza crust; on Friday evening, they sampled 200 customers and found 120 preferred the new pizza crust.

a. Is there sufficient evidence to conclude a significant preference for the new crust based on Tuesday's customers?

b. Is there sufficient evidence to conclude a significant preference for the new crust based on Friday's customers?

c. Explain the relationship between the solutions obtained in Exercise 9.172 and here.

9.174 An instructor asks each of the 54 members of his class to write down "at random" one of the numbers 1, 2, 3, . . . , 13, 14, 15. Since the instructor believes that students like gambling, he considers that 7 and 11 are lucky numbers. He counts the number of students, x, who selected 7 or 11. How large must x be before the hypothesis of randomness can be rejected at the 0.05 level?

9.175 Today's newspapers and magazines often report the findings of survey polls about various aspects of life. The Pew Internet & American Life Project (January 13–February 9, 2005) found that "63% of cell phone users ages 18-27 have used text messaging within the past month." Other information obtained from the project included "random telephone survey of 1,460 cell phone users" and "has a margin of sampling error of plus or minus 3 percentage points." Relate this information to the statistical inferences you have been studying in this chapter.

a. Is a percentage of people a population parameter, and if so, how is it related to any of the parameters that we have studied?

b. Based on the information given, find the 95% confidence interval for the true proportion of cell phone users who have used text messaging.

c. Explain how the terms "point estimate," "level of confidence," "maximum error of estimate," and "confidence interval" relate to the values reported in the article and to your answers in part b.

9.176 To test the hypothesis that the standard deviation on a standard test is 12, a sample of 40 randomly selected students was tested. The sample variance was found to be 155. Does this sample provide sufficient evidence to show that the standard deviation differs from 12 at the 0.05 level of significance?

9.177 Bright-Lite claims that its 60-watt light bulb burns with a length of life that is approximately normally distributed with a standard deviation of 81 hours. A sample of 101 bulbs had a variance of 8075. Is this sufficient evidence to reject Bright-Lite's claim in favor of the alternative, "the standard deviation is larger than 81 hours," at the 0.05 level of significance?

9.178 A production process is considered out of control if the produced parts have a mean length different from 27.5 mm or a standard deviation that is greater than 0.5 mm. A sample of 30 parts yields a sample mean of 27.63 mm and a sample standard deviation of 0.87 mm. If we assume part length is a normally distributed variable, does this sample indicate that the process should be adjusted to correct the standard deviation of the product? Use $\alpha = 0.05$.

9.179 Julia Jackson operates a franchised restaurant that specializes in soft ice cream cones and sundaes. Recently she received a letter from corporate headquarters warning her that her shop was in danger of losing its franchise because the average sales per customer had dropped "substantially below the average for the rest of the corporation." The statement may be true, but Julie is convinced that such a statement is completely invalid to justify threatening a closing. The variation in sales at her restaurant is bound to be larger than most, primarily because she serves more children, elderly, and single adults rather than large families who run up big bills at the other restaurants. Therefore, her average ticket is likely to be smaller and exhibit greater variability. To prove her point, Julie obtained the sales records from the whole company and found that the standard deviation was $2.45 per sales ticket. She then conducted a study of the last 71 sales tickets at her store and found a standard deviation of $2.95 per ticket. Is the variability in sales at Julie's franchise, at the 0.05 level of significance, greater than the variability for the company?

9.180 All tomatoes that a certain supermarket buys from growers must meet the store's specifications of a mean diameter of 6.0 cm and a standard deviation of no more than 0.2 cm. The supermarket's buyer visits a potential new supplier and selects a random sample of 36 tomatoes from the grower's greenhouse. The diameter of each tomato is measured, and the mean is found to be 5.94 and the standard deviation is 0.24. Do the tomatoes meet the supermarket's specs?

a. Determine whether an assumption of normality is reasonable. Explain.

b. Does the sample evidence show sufficient evidence to conclude the tomatoes do not meet the specs with regard to the mean diameter? Use $\alpha = 0.05$.

c. Does the sample evidence show sufficient evidence to conclude the tomatoes do not meet the specs with regard to the standard deviation? Use $\alpha = 0.05$.

d. Write a short report for the buyer outlining the findings and recommendations as to whether or not to use this tomato grower to supply tomatoes for sale in the supermarket.

9.181 The uniform length of nails is very important to a carpenter—the length of the nails being used are matched to the materials being fastened together, thereby making a small standard deviation an important property of the nails. A sample of 35 randomly selected 2-inch nails is taken from a large quantity of Nails, Inc.'s, recent production run. The resulting length measurements have a mean length of 2.025 inches and a standard deviation of 0.048 inch.

a. Determine whether an assumption of normality is reasonable. Explain.

b. Does the sample evidence show sufficient evidence to reject the idea that the nails have a mean length of 2 inches? Use $\alpha = 0.05$.

c. Is there sufficient evidence, at the 0.05 level, to show that the length of nails from this production run have a standard deviation greater than the advertised 0.040 inch?

d. Write a short report outlining the findings and recommendations as to whether or not the carpenter should use these nails for an application that requires 2-inch nails.

9.182 [EX09-182] It is important that the force required to extract a cork from a wine bottle not have a large standard deviation. Years of production and testing indicate that the no. 9 corks in Applied Example 6.15 (p. 330) have an extraction force that is normally distributed with a standard deviation of 36 Newtons. Recent changes in the manufacturing process are thought to have reduced the standard deviation.

a. What would be the problem with the standard deviation being relatively large? What would be the advantage of a smaller standard deviation?

A sample of 20 randomly selected bottles is used for testing.

Extraction Force in Newtons

296	338	341	261	250	347	336	297	279	297
259	334	281	284	279	266	300	305	310	253

b. Does the preceding sample show sufficient reason to show that the standard deviation of extraction force is less than 36.0 Newtons, at the 0.02 level of significance?

During a different testing, a sample of eight bottles was randomly selected and tested.

Extraction Force in Newtons

331.9	312.0	289.4	303.6	346.9	308.1	346.9	276.0

c. Does the preceding sample show sufficient reason to show that the standard deviation of ex-

traction force is less than 36.0 Newtons, at the 0.02 level of significance?

d. What effect did the two different sample sizes have on the calculated test statistic in parts b and c? What effect did they have on the *p*-value or critical value? Explain.

e. What effect did the two different sample standard deviations have on the answers in parts b and c? What effect did they have on the *p*-value or critical value? Explain.

9.183 [EX09-183] A box of Corn Flakes that is labeled "NET WT. 14 OZ." should have 14 oz or more of cereal inside. Twenty of these boxes were randomly selected and the weight of the contents (in ounces) determined.

14.52	14.47	14.80	14.60	14.45	14.25	14.15	14.12	14.36	14.39
14.50	14.29	14.28	14.60	13.85	14.18	14.39	14.45	14.69	14.38

a. Draw a histogram of the weight of cereal per box.

b. Find the sample statistics mean and standard deviation.

c. What percent of the sample is below the 14.0 oz weight?

The plant foreman is studying the filling process and needs to estimate the mean weight of all boxes being filled.

d. Determine whether an assumption of normality is reasonable. Explain.

e. Find the 95% confidence interval for the mean weight

f. The filling process is believed to be running with a standard deviation of fill of no more than 0.2 oz. Test this hypothesis at the 0.01 level.

9.184 [EX09-184] The foreman in Exercise 9.183 believes that the cereal-filling machine used for Corn Flakes needs to be replaced and that the new one he is considering will pay for the upgrade within a short time, mainly due to less variability in the fill amount. The new machine is started, and a test run is made. Twenty of these boxes were randomly selected from the run and the contents weighed (in ounces).

14.17 14.25 14.17 14.16 14.18 14.09 14.19 14.17 14.16 14.06
14.11 14.15 14.12 14.19 14.14 14.19 14.13 14.12 14.16 14.15

a. Draw a histogram of the weight of cereal per box.

b. Find the sample statistics mean and standard deviation.

c. What percent of the sample from the new machine is below the 14.0 oz weight?

The foreman needs to estimate the mean weight and test the standard deviation of all boxes being filled.

d. Determine whether an assumption of normality is reasonable. Explain.

e. Find the 95% confidence interval for the mean weight.

f. The filling process for the new machine is claimed to be running with a standard deviation of fill of less than 0.1 oz. Test this hypothesis at the 0.01 level.

9.185 The boxes of Corn Flakes in Exercises 9.183 and 9.184 that have more than 14.2 oz of cereal are being considered "too full." Since the weights appear to be normally distributed for both filling-machines, use the normal distribution and find the following information for the foreman.

a. What proportion of the boxes filled by the current machine fill the boxes with too much cereal?

b. What proportion of the boxes filled by the new machine fill the boxes with too much cereal?

c. For every 1000 boxes of cereal filled by the current machine, how many boxes can be filled by the new machine using the same total amount of cereal?

d. Summarize what you believe should be the foreman's pitch to the company for getting the new filling-machine.

9.186 [EX09-186] Traffic gridlock is no small problem! In 2003, congestion-delayed travelers wasted 3.7 billion hours and 2.3 billion gallons of fuel for a total cost of more than $63 billion. It is estimated that half of all traffic delays are caused by car crashes. One variable often included in studies of traffic delays is the travel time index (TTI). The TTI is a ratio of peak period to free-flow travel time. A value of 1.30 indicates a free-flow trip of 20 minutes takes 26 minutes in the peak as a result of heavy traffic demand and incidents. The national big-city mean TTI is 1.37.

Source: http://mobility.tamu.edu/ums/report/

In May 2005, 25 commuters who typically commuted during peak rush hour were randomly identified. They were monitored while driving their regular route to work both during a free-flow period and during a peak period.

Free-Flow	Peak	Free-Flow	Peak	Free-Flow	Peak
50.7	64.9	21.0	30.5	24.1	34.7
27.4	32.1	51.5	66.4	48.3	64.2
47.9	69.5	50.1	66.6	34.2	40.7
22.9	30.0	48.9	72.4	27.5	35.8
29.0	45.8	36.6	42.1	44.6	57.1
29.7	38.3	35.9	43.1	33.5	44.9
34.4	40.6	38.1	54.9	22.2	34.4
34.4	46.4	31.2	44.9	26.1	32.6
38.7	55.3				

a. Calculate the TTI for each of the commuters.

b. Construct a histogram and find the mean and standard deviation of the TTI.

c. Demonstrate how this set of data satisfies the assumptions for inference.

d. Find the 95% confidence interval for μ, the mean TTI.

e. Test the hypothesis that the mean TTI for this urban/suburban area is different from the big-city mean of 1.37. Use $\alpha = 0.05$.

f. Compare the answers found in parts d and e. Do you think these commuters have a traffic problem? Why? Explain why part e is a duplication in this case.

Chapter Project

Get Enough Daily Exercise?

Many studies have proved that we need to exercise to lower various health risks, such as high blood pressure, heart disease, and high cholesterol. But knowing and doing are not the same thing. People in the health profession should be even more aware of the need for exercise. The following data values are from Section 9.1, "Get Enough Daily Exercise?" a study surveying cardiovascular technicians (individuals who perform various cardiovascular diagnostic procedures) as to their own physical exercise per week (measured in minutes).
[EX09-001]

60	40	50	30	60	50	90	30	60	60
60	80	90	90	60	30	20	120	60	50
20	60	30	120	50	30	90	20	30	40
50	40	30	40	20	30	60	50	60	80

Putting Chapter 9 to Work

9.187 a. What evidence do you have to show that the assumption of normality is reasonable? Explain.

b. Estimate the mean amount of weekly exercise time for all cardiovascular technicians using a point estimate and a 95% confidence interval.

c. The "Good News for Women" article in Section 9.1, "Get Enough Daily Exercise?" says that people should exercise at least 60 minutes a week. Based on the data from the study, determine whether the technicians exercise at least 60 minutes a week. Use a 0.05 level of significance.

Your Study

9.188 a. Define the population whose amount of exercise time per week you would be interested in investigating.

b. Collect the "times" from a sample of 40 members of your population.

c. Find the mean and standard deviation for the amount of time exercised per week by the members of your sample.

d. Construct a graph displaying the distribution of your data.

e. Estimate the mean amount of weekly exercise time for your population using a point estimate and a 95% confidence interval.

f. The "Good News for Women" article says that people (can?) should exercise at least 60 minutes a week. Does it appear that the members of your sample exercise "at least 60 minutes per week"? Use a 0.05 level of significance. Justify your answer.

g. Did your data satisfy the assumptions? Explain.

Chapter Practice Test

PART I: Knowing the Definitions

Answer "True" if the statement is always true. If the statement is not always true, replace the words shown in bold with words that make the statement always true.

9.1 Student's t-distributions have an approximately normal distribution but are **more** dispersed than the standard normal distribution.

9.2 The **chi-square** distribution is used for inferences about the mean when σ is unknown.

9.3 **Student's t**-distribution is used for all inferences about a population's variance.

9.4 If the test statistic falls in the critical region, the null hypothesis has **been proved true.**

9.5 When the test statistic is t and the number of degrees of freedom gets very large, the critical value of t is very close to that of the **standard normal z.**

9.6 When making inferences about one mean when the value of σ is not known, the **z-score** is the test statistic.

9.7 The chi-square distribution is a skewed distribution whose mean value is **2** for df > 2.

9.8 Often, the concern with testing the variance (or standard deviation) is to keep its size under control or relatively small. Therefore, many of the hypothesis tests with chi-square are **one-tailed.**

9.9 $\sqrt{npq}$ is the standard error of proportion.

9.10 The sampling distribution of p' is distributed approximately as a **Student's t**-distribution.

PART II: Applying the Concepts

Answer all questions, showing all formulas, substitutions, and work.

9.11 Find each value:

a. $z(0.02)$ b. $t(18, 0.95)$ c. $\chi^2(25, 0.95)$

9.12 A random sample of 25 data values was selected from a normally distributed population for the purpose of estimating the population mean, μ. The sample statistics are $n = 25$, $\bar{x} = 28.6$, and $s = 3.50$.

a. Find the point estimate for μ.

b. Find the maximum error of estimate for the 0.95 confidence interval estimate.

c. Find the lower confidence limit (LCL) and the upper confidence limit (UCL) for the 0.95 confidence interval estimate for μ.

9.13 Thousands of area elementary school students were recently given a nationwide standardized exam to test their composition skills. If 64 of a random sample of 100 students passed this exam, construct the 0.98 confidence interval estimate for the true proportion of all area students who passed the exam.

9.14 State the null (H_o) and the alternative (H_a) hypotheses that would be used to test each of these claims:

a. The mean weight of professional basketball players is no more than 225 lb.

b. Approximately 40% of daytime students own their own car.

c. The standard deviation for the monthly amounts of rainfall in Monroe County is less than 3.7 inches.

9.15 Determine the level of significance, test statistic, critical region, and critical values(s) that would be used in completing each hypothesis test using the classical approach with $\alpha = 0.05$.

a. H_o: $\mu = 43$ vs. H_a: $\mu < 43$, $\sigma = 6$

b. H_o: $\mu = 95$ vs. H_a: $\mu \neq 95$, σ unknown, $n = 22$

c. H_o: $p = 0.80$ vs. H_a: $p > 0.80$

d. H_o: $\sigma = 12$ vs. H_a: $\sigma \neq 12$, $n = 28$

9.16 The automobile manufacturer of the Alero claims that the typical Alero will average 32 mpg of gasoline. An independent consumer group is somewhat skeptical of this claim and thinks the mean gas mileage is less than the 32 claimed. A sample of 24 randomly selected Aleros produced these sample statistics: mean 30.15 and standard deviation 4.87. At the 0.05 level of significance, does the consumer group have sufficient evidence to refute the manufacturer's claim?

9.17 A coffee machine is supposed to dispense 6 fluid ounces of coffee into a paper cup. In reality, the amount dispensed varies from cup to cup. However, if the machine is operating properly, the standard deviation of the amounts dispensed should be 0.1 oz or less. A random sample of 15 cups produced a standard deviation of 0.13 oz. Does this represent sufficient evidence, at the 0.10 level of significance, to conclude that the machine is not operating properly?

9.18 An unhappy customer is frustrated with the waiting time at the post office when buying stamps. Upon registering his complaint, he was told, "You wait more than 1 minute for service no more than half of the time when you only buy stamps." Not believing this to be the case, the customer collected some data from people who had just purchased stamps only. The sample statistics are $n = 60$ and $x = n$ (wait more than 1 minute) $= 35$. At the 0.02 level of significance, does our unhappy customer have sufficient evidence to refute the post office's claim?

PART III: Understanding the Concepts

9.19 Student B says the range of a set of data may be used to obtain a crude estimate for the standard deviation of a population. Student A is not sure. How will student B correctly explain how and under what circumstances his statement is true?

9.20 Is it the null hypothesis or the alternative hypothesis that the researcher usually believes to be true? Explain.

9.21 When you reject a null hypothesis, student A says that you are expressing disbelief in the value of the parameter as claimed in the null hypothesis. Student B says that instead you are expressing the belief that the sample statistic came from a population other than the one related to the parameter claimed in the null hypothesis. Who is correct? Explain.

9.22 "Student's *t*-distribution must be used when making inferences about the population mean, μ, when the population standard deviation, σ, is not known" is a true statement. Student A states that the *z*-score sometimes plays a role when the *t*-distribution is used. Explain the conditions that exist and the role played by *z* that make student A's statement correct.

9.23 Student A says that the percentage of the sample means that fall outside the critical values of the sampling distribution determined by a true null hypothesis is the *p*-value for the test. Student B says that the percentage student A is describing is the level of significance. Who is correct? Explain.

9.24 Student A carries out a study in which she is willing to run a 1% risk of making a type I error. She rejects the null hypothesis and claims that her statistic is significant at the 99% level of confidence. Student B argues that student A's claim is not properly worded. Who is correct? Explain.

9.25 Student A claims that when you employ a 95% confidence interval to determine an estimation, you do not know for sure whether or not your inference is correct (the parameter is contained within the interval). Student B claims that you do know; you have shown that the parameter cannot be less than the lower limit or greater than the upper limit of the interval. Who is right? Explain.

9.26 Student A says that the best way to improve a confidence interval estimate is to increase the level of confidence. Student B argues that using a high confidence level does not really improve the resulting interval estimate. Who is right? Explain.

Statistics ⊜ Now™ Preparing for an exam? Assess your progress by taking the post-test at **http://1pass.thomson.com**.

⊙**Mentor** Do you need a live tutor for homework problems? Access vMentor on the StatisticsNow website at **http://1pass.thomson.com** for one-on-one tutoring from a statistics expert.

Inferences Involving Two Populations

© David Young-Wolff/Getty Images

10.1 · Students, Credit Cards, and Debt

We all know, "College is expensive" and "Credit cards are readily available." We also know and believe that young adults need experience at handling their own finances. But two old adages, "Buyers beware" and "Know the facts" probably should each play a role in a college student's approach to credit card use. Below is a portion of the report Nellie Mae published in 2002.

CREDIT CARDS AND DEBT

Not surprisingly, the freshman population has a lower overall percentage of credit cards and lower debt levels on their cards than students in upper classes. However, more than half of all freshmen (54%) had at least one credit card, with the average number of cards being 2.5; among those who have credit cards, 26% have four or more. Freshman debt levels are also lower than the over-all counts in all categories. Their median debt amount is $901, lower than the overall median of $1770; their average balance is $1533 versus $2327 overall; those with balances exceeding $7000 account for only 4% of freshmen as opposed to 6% overall; and those with high-level balances between $3000 and $7000 account for 8% of freshmen compared with 21% overall.

Credit Card Usage by Grade Level	01Fresh	02Soph	03Jr	04/05Sr/+
Percentage Who Have Credit Cards	54%	92%	87%	96%
Average Number of Credit Cards	2.5	3.67	4.5	6.13
Percentage Who Have 4 or More Cards	26%	44%	50%	66%
Average Credit Card Debt	$1533	$1825	$2705	$3262
Median Credit Card Debt	$901	$1564	$1872	$2185
Percentage with Balances between $3000 and $7000	8%	18%	24%	31%
Percentage with Balances Exceeding $7000	4%	4%	7%	9%

As students progress through their 4 years (or more) in college, there is a steady increase in credit card usage rates and balances each year. By gradu-ation, most students have more than doubled their average debt and almost tripled the number of cards they hold. Most dramatic, however, is the 70% jump occurring between freshman and sophomore year in the percentage of students with at least one card—from 54% to 92% of the total population.

Once freshmen arrive on campus, there are many tempting incentives to sign up for new credit cards and many opportunities to use them. The fact that the average number of cards per student continues to increase is not surprising. The proliferation of on-campus, mail, and Internet offers of free gifts, bonus airline miles, and low introductory rates for each new card is difficult for students to resist.

Source: Undergraduate Students and Credit Cards, published April 2002, Nellie Mae, 50 Braintree Hill Park, Suite 300, Braintree, MA 02184; 781-849-1325; http://www.nelliemae.com

The data that follow are from random samples of 200 freshman and 200 sophomore college students who were asked, "Do you have your own credit card?" A total of 97 freshman and 187 sophomores answered that they had one or more credit cards with their name on it. The first 40 freshmen and the first 44 sophomores answering "yes" were then asked for their "current total credit card debt balance." The total credit card debt balances are listed here.

Freshman Debt Balance, *n* = 40 [EX10-001]

1011.97	3998.72	2447.93	2457.39	855.63	1602.74	912.39	2478.49	1014.39	444.48	1293.36
1065.82	989.56	412.53	321.85	2578.39	2103.35	2917.65	3218.54	1384.34	4368.28	244.33
190.24	2778.17	1702.65	616.31	491.73	2205.95	1130.09	2402.92	767.42	657.83	1150.78
1102.28	154.11	1494.48	1324.01	2054.76	1762.31	644.31				

Sophomore Debt Balance, *n* = 44

690.08	595.04	2983.50	1761.21	1020.91	2143.18	3048.87	1314.36	1378.99	1456.10	1893.37	1287.47
284.93	7135.64	3194.07	2565.71	3298.15	2747.14	839.57	393.20	1422.73	1652.03	2214.77	1126.76
3433.80	3962.25	1849.23	3037.52	328.29	3074.19	1194.87	889.40	1480.94	486.22	1688.81	1317.27
2624.01	2286.74	5341.94	633.37	873.18	3601.18	2023.29	4898.46				

After completing Chapter 10, further investigate the preceding data and "Students, Credit Cards, and Debt" in the Chapter Project on page 613.

SECTION 10.1 EXERCISES

Statistics⌂Now™

Datasets can be found on your Student's Suite CD-ROM or at the StatisticsNow website at **http://1pass.thomson.com**.

10.1 [EX10-001] Consider the sample data above in Section 10.1.

a. What is the population of interest?

b. What percentage of each group has their own credit card? How does this compare with the findings reported by the Nellie Mae organization?

c. Describe the shape of the distribution you believe total credit card debt will display. Explain.

d. Construct a histogram of total credit card debt for each class. Use the same class intervals for both histograms. Compare your findings to your thoughts in part c.

e.   Find the mean and standard deviation for each data set. Compare the two samples using these findings.

f.   Compare the distribution of credit card debts for freshmen to that for sophomores. Describe ways in which they appear to be similar. Describe ways in which they appear to be quite different.

10.2 a.   Estimate the mean credit card debt for freshmen with a 95% confidence interval.

   b.   Does the sampled population of sophomores have a significantly higher credit card debt than the national average reported by Nellie Mae? Use $\alpha = 0.05$.

FIGURE 10.1   "Road Map" to Two Population Inferences

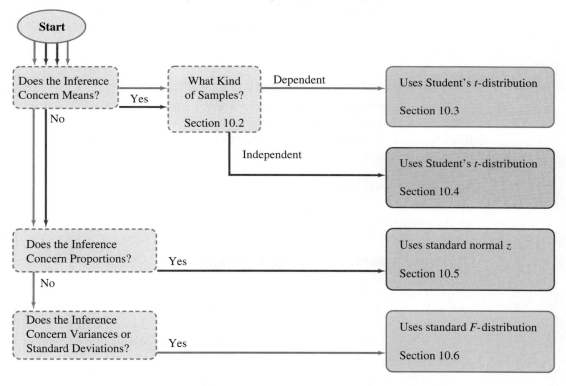

| **10.2** | # Dependent and Independent Samples |

In this chapter we are going to study the procedures for making inferences about two populations. When comparing two populations, we need two samples, one from each population. Two basic kinds of samples can be used: independent and dependent. The dependence or independence of two samples is determined by the sources of the data. A **source** can be a person, an object, or anything that yields a data value. If the same set of sources or related sets are used to obtain the data representing both populations, we have **dependent samples.** If two unrelated sets of sources are used, one set from each population, we have **independent samples.** The following examples should clarify these ideas.

EXAMPLE 10.1 Dependent versus Independent Samples

A test will be conducted to see whether the participants in a physical fitness class actually improve in their level of fitness. It is anticipated that approximately 500 people will sign up for this course. The instructor decides that she will give 50 of the participants a set of tests before the course begins (a pretest), and then she will give another set of tests to 50 participants at the end of the course (a posttest). Two sampling procedures are proposed:

Plan A: Randomly select 50 participants from the list of those enrolled and give them the pretest. At the end of the course, make a second random selection of size 50 and give them the posttest.

Plan B: Randomly select 50 participants and give them the pretest; give the same set of 50 the posttest when they complete the course.

Plan A illustrates independent sampling; the sources (the class participants) used for each sample (pretest and posttest) were selected separately. Plan B illustrates dependent sampling; the sources used for both samples (pretest and posttest) are the same.

Typically, when both a pretest and a posttest are used, the same subjects participate in the study. Thus, pretest versus posttest (before versus after) studies usually use dependent samples.

EXAMPLE 10.2 Dependent versus Independent Samples

A test is being designed to compare the wearing quality of two brands of automobile tires. The automobiles will be selected and equipped with the new tires and then driven under "normal" conditions for 1 month. Then a measurement will be taken to determine how much wear took place. Two plans are proposed:

Plan C: A sample of cars will be selected randomly, equipped with brand A tires, and driven for 1 month. Another sample of cars will be selected, equipped with brand B tires, and driven for 1 month.

Plan D: A sample of cars will be selected randomly, equipped with one tire of brand A and one tire of brand B (the other two tires are not part of the test), and driven for 1 month.

We suspect that many other factors must be taken into account when testing automobile tires—such as age, weight, and mechanical condition of the car; driving habits of drivers; location of the tire on the car; and where and how much the car is driven. However, at this time we are trying only to illustrate dependent and independent samples. Plan C is independent (unrelated sources), and plan D is dependent (common sources).

APPLIED
EXAMPLE 10.3   **Exploring the Traits of Twins**

Studies that involve identical twins are a natural for the dependent sampling technique discussed in this section.

A NEW STUDY SHOWS THAT KEY CHARACTERISTICS MAY BE INHERITED

Like many identical twins reared apart, Jim Lewis and Jim Springer found they had been leading eerily similar lives. Separated four weeks after birth in 1940, the Jim twins grew up 45 miles apart in Ohio and were reunited in 1979. Eventually they discovered that both drove the same model blue Chevrolet, chain-smoked Salems, chewed their fingernails, and owned dogs named Toy. Each had spent a good deal of time vacationing at the same three-block strip of beach in Florida. More important, when tested for such personality traits as flexibility, self-control, and sociability, the twins responded almost exactly alike.

The project is considered the most comprehensive of its kind. The Minnesota researchers report the results of six-day tests of their subjects, including 44 pairs of identical twins who were brought up apart. Well-being, alienation, aggression, and the shunning of risk or danger were found to owe as much or more to nature as to nurture. Of eleven key traits or clusters of traits analyzed in the study, researchers estimated that a high of 61 percent of what they call "society potency" (a tendency toward leadership or dominance) is inherited, while "social closeness" (the need for intimacy, comfort, and help) was lowest, at 33 percent.

Source: Time Inc. All rights reserved. Reprinted by permission of *TIME*.

Independent and dependent samples each have their advantages; these will be emphasized later. Both methods of sampling are often used.

SECTION 10.2 EXERCISES

10.3 Explain why studies involving identical twins, as in Applied Example 10.3, result in dependent samples of data.

10.4 a. Describe how you could select two independent samples from among your classmates to compare the heights of female and male students.

b. Describe how you could select two dependent samples from among your classmates to compare their heights as they entered high school with their heights when they entered college.

10.5 The students at a local high school were assigned to do a project for their statistics class. The project involved having sophomores take a timed test on geometric concepts. The statistics students would then use these data to determine whether there was a difference between male and female performances. Would the resulting sets of data represent dependent or independent samples? Explain.

10.6 In trying to estimate the amount of growth that took place in the trees recently planted by the County Parks Commission, 36 trees were randomly selected from the 4000 planted. The heights of these trees were measured and recorded. One year later, another set of 42 trees was randomly selected and measured. Do the two sets of data (36 heights, 42 heights) represent dependent or independent samples? Explain.

10.7 Twenty people were selected to participate in a psychology experiment. They answered a short multiple-choice quiz about their attitudes on a particular subject and then viewed a 45-minute film.

The following day the same 20 people were asked to answer a follow-up questionnaire about their attitudes. At the completion of the experiment, the experimenter will have two sets of scores. Do these two samples represent dependent or independent samples? Explain.

10.8 An experiment is designed to study the effect diet has on the uric acid level. The study includes 20 white rats. Ten rats are randomly selected and given a junk-food diet; the other 10 rats receive a high-fiber, low-fat diet. Uric acid levels of the two groups are determined. Do the resulting sets of data represent dependent or independent samples? Explain.

10.9 Two different types of disc centrifuges are used to measure the particle size in latex paint. A gallon of paint is randomly selected, and 10 specimens are taken from it for testing on each of the centrifuges. There will be two sets of data, 10 data values each, as a result of the testing. Do the two sets of data represent dependent or independent samples? Explain.

10.10 An insurance company is concerned that garage A charges more for repair work than garage B charges. It plans to send 25 cars to each garage and obtain separate estimates for the repairs needed for each car.

a. How can the company do this and obtain independent samples? Explain in detail.

b. How can the company do this and obtain dependent samples? Explain in detail.

10.11 A study is being designed to determine the reasons why adults choose to follow a healthy diet plan. The study will survey 1000 men and 1000 women. Upon completion of the study, the reasons men choose a healthy diet will be compared with the reasons women choose a healthy diet.

a. How can the data be collected if independent samples are to be obtained? Explain in detail.

b. How can the data be collected if dependent samples are to be obtained? Explain in detail.

10.12 Suppose that 400 students in a certain college are taking elementary statistics this semester. Two samples of size 25 are needed to test some precourse skill against the same skill after the students complete the course.

a. Describe how you would obtain your samples if you were to use dependent samples.

b. Describe how you would obtain your samples if you were to use independent samples.

10.3 Inferences Concerning the Mean Difference Using Two Dependent Samples

The procedures for comparing two population means are based on the relationship between two sets of sample data, one sample from each population. When dependent samples are involved, the data are thought of as "paired data." The data may be paired as a result of being obtained from "before" and "after" studies; from pairs of identical twins as in Applied Example 10.3; from a "common" source, as with the amounts of tire wear for each brand in plan D of Example 10.2; or from matching two subjects with similar traits to form "matched pairs." The pairs of data values are compared directly to each other by using the difference in their numerical values. The resulting difference is called a **paired difference.**

Paired Difference

$$d = x_1 - x_2 \tag{10.1}$$

Using paired data this way has a built-in ability to remove the effect of otherwise uncontrolled factors. The tire-wear problem in Example 10.2 is an excellent example of such additional factors. The wearing ability of a tire is greatly affected by a multitude of factors: the size, weight, age, and condition of the car; the driving habits of the driver; the number of miles driven; the condition and types of roads driven on; the quality of the material used to make the tire; and so on. We create paired data by mounting one tire from each brand on the same car. Since one tire of each brand will be tested under the same conditions, using the same car, same driver, and so on, the extraneous causes of wear are neutralized.

Procedures and Assumptions for Inferences Involving Paired Data

A test was conducted to compare the wearing quality of the tires produced by two tire companies using plan D, as described in Example 10.2. All the aforementioned factors will have an equal effect on both brands of tires, car by car. One tire of each brand was placed on each of six test cars. The position (left or right side, front or back) was determined with the aid of a random-number table. Table 10.1 lists the amounts of wear (in thousandths of an inch) that resulted from the test.

TABLE 10.1

Amount of Tire Wear [TA10-01]

Car	1	2	3	4	5	6
Brand A	125	64	94	38	90	106
Brand B	133	65	103	37	102	115

Since the various cars, drivers, and conditions are the same for each tire of a paired set of data, it makes sense to use a third variable, the paired difference *d*. Our two dependent samples of data may be combined into one set of *d* values, where $d = B - A$.

Car	1	2	3	4	5	6
$d = B - A$	8	1	9	−1	12	9

The difference between the two population means, when dependent samples are used (often called **dependent means**), is equivalent to the **mean of the paired differences.** Therefore, when an inference is to be made about the difference of two means and paired differences are used, the inference will in fact be about the mean of the paired differences. The sample mean of the paired differences will be used as the point estimate for these inferences.

In order to make inferences about the mean of all possible paired differences μ_d, we need to know about the *sampling distribution* of $\bar{d}$.

> When paired observations are randomly selected from normal populations, the paired difference, $d = x_1 - x_2$, will be approximately normally distributed about a mean μ_d with a standard deviation of σ_d.

This is another situation in which the *t*-test for one mean is applied; namely, we wish to make inferences about an unknown mean (μ_d) where the random variable (d) involved has an approximately normal distribution with an unknown standard deviation (σ_d).

Inferences about the mean of all possible paired differences μ_d are based on samples of n dependent pairs of data and the **t-distribution** with $n - 1$ degrees of freedom (df), under the following assumption:

> **Assumption for inferences about the mean of paired differences μ_d:** The paired data are randomly selected from normally distributed populations.

Confidence Interval Procedure

The $1 - \alpha$ **confidence interval for estimating the mean difference μ_d** is found using this formula:

FYI Formula (10.2) is an adaptation of formula (9.1).

> **Confidence Interval for Mean Difference (Dependent Samples)**
>
> $$\bar{d} - t(\text{df}, \alpha/2) \cdot \frac{s_d}{\sqrt{n}} \quad \text{to} \quad \bar{d} + t(\text{df}, \alpha/2) \cdot \frac{s_d}{\sqrt{n}}, \quad \text{where df} = n - 1 \qquad (10.2)$$

Where $\bar{d}$ is the mean of the sample differences:

$$\bar{d} = \frac{\sum d}{n} \qquad (10.3)$$

FYI Formulas (10.3) and (10.4) are adaptations of formulas (2.1) and (2.10).

and s_d is the standard deviation of the sample differences:

$$s_d = \sqrt{\frac{\sum d^2 - \left[\dfrac{(\sum d)^2}{n}\right]}{n - 1}} \qquad (10.4)$$

EXAMPLE 10.4

Constructing a Confidence Interval for μ_d

Statistics⟁Now™

Watch a video example at
http://1pass.thomson.com
or on your CD.

Construct the 95% confidence interval for the mean difference in the paired data on tire wear, as reported in Table 10.1. The sample information is $n = 6$

pieces of paired data, $\bar{d} = 6.3$, and $s_d = 5.1$. Assume the amounts of wear are approximately normally distributed for both brands of tires.

SOLUTION

Step 1 **Parameter of interest:** μ_d, the mean difference in the amounts of wear between the two brands of tires.

Step 2 a. **Assumptions:** Both sampled populations are approximately normal.

    b. **Probability distribution:** The t-distribution with df $= 6 - 1 = 5$ and formula (10.2) will be used.

    c. **Level of confidence:** $1 - \alpha = 0.95$.

Step 3 **Sample information:** $n = 6$, $\bar{d} = 6.3$, and $s_d = 5.1$.

The mean:

$$\bar{d} = \frac{\sum d}{n} \; : \; \bar{d} = \frac{38}{6} = 6.333 = \mathbf{6.3}$$

The standard deviation:

$$s_d = \sqrt{\frac{\sum d^2 - \left[\dfrac{(\sum d)^2}{n}\right]}{n-1}} : \quad s_d = \sqrt{\frac{372 - \left[\dfrac{(38)^2}{6}\right]}{6-1}} = \sqrt{26.27} = 5.13 = \mathbf{5.1}$$

Step 4 a. **Confidence coefficient:**
This is a two-tailed situation with $\alpha/2 = 0.025$ in one tail. From Table 6 in Appendix B, $t(\mathrm{df}, \alpha/2) = t(5, 0.025) = \mathbf{2.57}$.

    b. **Maximum error of estimate:** Using the maximum error part of formula (10.2), we have

$$E = t(\mathrm{df}, \alpha/2) \cdot \frac{s_d}{\sqrt{n}} : \quad E = 2.57 \cdot \left(\frac{5.1}{\sqrt{6}}\right) = (2.57)(2.082) = 5.351 = \mathbf{5.4}$$

    c. **Lower/upper confidence limits:**

$$\bar{d} \pm E$$

$$6.3 \pm 5.4$$

$$6.3 - 5.4 = \mathbf{0.9} \quad \text{to} \quad 6.3 + 5.4 = \mathbf{11.7}$$

Step 5 a. **Confidence interval:** 0.9 to 11.7 is the 95% confidence interval for μ_d.

    b. That is, with 95% confidence we can say that the mean difference in the amounts of wear is between 0.9 and 11.7 thousandths of an inch. Or, in other words, the population mean tire wear from Brand B is between 0.9 and 11.7 thousandths of an inch greater than the population mean tire wear for Brand A.

Note: This confidence interval is quite wide, in part because of the small sample size. Recall from the central limit theorem that as the sample size increases, the standard error (estimated by $s_d/\sqrt{n}$) decreases.

For specific instructions about confidence coefficients and Table 6, see pages 477–478.

Statistics⌂Now™

Watch the supplemental video example "Calculating the mean and the standard deviation for paired differences" at http://1pass.thomson.com or on your CD.

TECHNOLOGY INSTRUCTIONS: $1 - \alpha$ CONFIDENCE INTERVAL FOR MEAN μ_d WITH UNKNOWN STANDARD DEVIATION FOR TWO DEPENDENT SETS OF SAMPLE DATA

MINITAB (Release 14) Input the paired data into C1 and C2; then continue with:

Choose:	**Stat > Basic Statistics > Paired t**
Select:	Samples in columns
Enter:	First sample: **C1***
	Second sample: **C2**
Select:	**Options**
Enter:	Confidence level: **1 − α** (ex. 0.95 or 95.0)
Select:	Alternative: **not equal > OK > OK**

*Paired *t* evaluates the first sample minus the second sample.

Excel Input the paired data into columns A and B; activate C1 or C2 (depending on whether column headings are used or not); then continue with:

Enter:	= **A2 − B2*** (if column headings are used)
Drag:	Bottom right corner of C2 down to give other differences
Choose:	**Tools > Data Analysis Plus > t-Estimate: Mean**
Enter:	Input range: **(C2:C20 or select cells)**
Select:	**Labels** (if necessary)
Enter:	Alpha: *α* (ex. 0.05) > **OK**

*Enter the expression in the order that is needed: A2 − B2 or B2 − A2.

TI-83/84 Plus Input the paired data into L1 and L2; then continue with the following, entering the appropriate values and highlighting Calculate:

Highlight:	**L3**
Enter:	L3 = **L1 − L2***
Choose:	**STAT > TESTS > 8:Tinterval**

```
TInterval
 Inpt:Data Stats
 List:L3
 Freq:1
 C-Level:.95
 Calculate
```

*Enter the expression in the order that is needed: L1 − L2 or L2 − L1.

The solution to Example 10.4 looks like this when solved in MINITAB:

```
Paired T for Brand B − Brand A
             N      Mean     StDev    SE Mean
Brand B      6      92.5     35.2     14.4
Brand A      6      86.2     30.9     12.6
Difference   6      6.33     5.13     2.09
95% CI for mean difference: (0.95, 11.71)
```

Hypothesis-Testing Procedure

When we test a null **hypothesis about the mean difference,** the test statistic used will be the difference between the sample mean $\bar{d}$ and the hypothesized value of μ_d, divided by the estimated **standard error.** This statistic is assumed to have a *t*-distribution when the null hypothesis is true and the assumptions for the test are satisfied. The value of the **test statistic** *t*★ is calculated as follows:

FYI Formula (10.5) is an adaptation of formula (9.2).

Test Statistic for Mean Difference (Dependent Samples)

$$t \star = \frac{\bar{d} - \mu_d}{s_d / \sqrt{n}}, \quad \text{where df} = n - 1 \tag{10.5}$$

Note: A hypothesized mean difference, μ_d, can be any specified value. The most common value specified is zero; however, the difference can be nonzero.

EXAMPLE 10.5	**One-Tailed Hypothesis Test for μ_d**

Statistics⟁Now™

Watch a video example at
http://1pass.thomson.com
or on your CD.

In a study on high blood pressure and the drugs used to control it, the effect of calcium channel blockers on pulse rate was one of many specific concerns. Twenty-six patients were randomly selected from a large pool of potential subjects, and their pulse rates were recorded. A calcium channel blocker was administered to each patient for a fixed period of time, and then each patient's pulse rate was again determined. The two resulting sets of data appeared to have approximately normal distributions, and the statistics were $\bar{d} = 1.07$ and $s_d = 1.74$ ($d = $ before $-$ after). Does the sample information provide sufficient evidence to show that the pulse rate is lower after taking the medication? Use $\alpha = 0.05$.

FYI "Lower rate" means "after" is less than "before" and "before $-$ after" is positive.

SOLUTION

Step 1 a. **Parameter of interest:** μ_d, the mean difference (reduction) in pulse rate from before to after using the calcium channel blocker for the time period of the test.

b. **Statement of hypotheses:**

$H_o: \mu_d = 0$ ($\leq$) (did not lower rate)   Remember: $d = $ before $-$ after.
$H_a: \mu_d > 0$ (did lower rate)

Step 2 a. **Assumptions:** Since the data in both sets are approximately normal, it seems reasonable to assume that the two populations are approximately normally distributed.

b. **Test statistic:** The *t*-distribution with df $= n - 1 = 25$, and the test statistic is *t*★ from formula (10.5).

c. **Level of significance:** $\alpha = 0.05$.

Step 3 a. **Sample information:** $n = 26$, $\bar{d} = 1.07$, and $s_d = 1.74$.

b. **Calculated test statistic:**

$$t \star = \frac{\bar{d} - \mu_d}{s_d / \sqrt{n}}: \quad t \star = \frac{1.07 - 0.0}{1.74 / \sqrt{26}} = \frac{1.07}{0.34} = \mathbf{3.14}$$

Step 4   **The Probability Distribution:**

p-value: 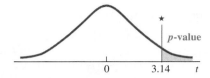 Classical:

a. Use the right-hand tail because H_a expresses concern for values related to "greater than." $\mathbf{P} = P(t\star > 3.14$, with df = 25) as shown in the figure.

a. The critical region is the right-hand tail because H_a expresses concern for values related to "greater than." The critical value is obtained from Table 6: $t(25, 0.05) = \mathbf{1.71}$.

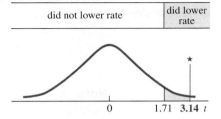

To find the *p*-value, you have three options:
1. Use Table 6 (Appendix B): **P < 0.005.**
2. Use Table 7 (Appendix B) to read the value directly: **P = 0.002.**
3. Use a computer or calculator to find the *p*-value: **P = 0.0022.**
Specific instructions are on pages 483–484.
b. The *p*-value is smaller than the level of significance, α.

Specific instructions are on pages 477–478.
b. $t\star$ is in the critical region, as shown in **red** in the figure.

Step 5   **a.** **Decision:** Reject H_o.

  b. **Conclusion:** At the 0.05 level of significance, we can conclude that the average pulse rate is lower after the administration of the calcium channel blocker.

Statistical significance does not always have the same meaning when the "practical" application of the results is considered. In the preceding detailed hypothesis test, the results showed a statistical significance with a *p*-value of 0.002—that is, 2 chances in 1000. However, a more practical question might be: Is lowering the pulse rate by this small average amount, estimated to be 1.07 beats per minute, worth the risks of possible side effects of this medication? Actually the whole issue is much broader than just this one issue of pulse rate.

TECHNOLOGY INSTRUCTIONS: HYPOTHESIS TEST FOR THE MEAN μ_d WITH UNKNOWN STANDARD DEVIATION FOR TWO DEPENDENT SETS OF SAMPLE DATA.

MINITAB (Release 14)   Input the paired data into C1 and C2; then continue with:

```
Choose:      Stat > Basic Statistics > Paired t
Select:      Samples in columns
Enter:       First sample: C1*
             Second sample: C2
Select:      Options
Enter:       Test mean: 0.0 or μd
Select:      Alternative: less than or not equal or greater than > OK > OK
```

*Paired *t* evaluates the first sample minus the second sample.

Excel Input the paired data into columns A and B; then continue with:

Choose: **Tools > Data Analysis > t-Test: Paired Two Sample for Means**

Enter: **Variable 1 Range: (A1:A20 or select cells)**

                  Variable 2 Range: (B1:B20 or select cells)

                            (subtracts: Var1 − Var2)

                  Hypothesized Mean Difference: μ_d (usually 0)

Select: **Labels** (if necessary)

Enter: α (ex. 0.05)

Select: **Output Range**

Enter: **(C1 or select cell) > OK**

Use Format > Column > AutoFit Selection to make the output more readable. The output shows *p*-values and critical values for one- and two-tailed tests. The hypothesis test may also be done by first subtracting the two columns and then using the inference about a mean (sigma unknown) commands on page 487 on the differences.

TI-83/84 Plus Input the paired data into L1 and L2; then continue with the following, entering the appropriate values, and highlighting Calculate:

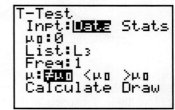

Highlight: **L3**

Enter: **L3 = L1 − L2***

Choose: **STAT > TESTS > 2:T-Test . . .**

*Enter the expression in the order that is needed: L1 − L2 or L2 − L1.

The solution to Example 10.5 looks like this when solved in MINITAB:

```
Paired T for Before − After
                  N      Mean     StDev     SE Mean
Difference       26      1.07      1.74       0.34
T-Test of mean difference = 0 (vs > 0): T-Value = 3.14
P-Value = 0.002
```

EXAMPLE 10.6   **Two-Tailed Hypothesis Test for μ_d**

Suppose the sample data in Table 10.1 (p. 551) were collected with the hope of showing that the two tire brands do not wear equally. Do the data provide sufficient evidence for us to conclude that the two brands show unequal wear, at the 0.05 level of significance? Assume the amounts of wear are approximately normally distributed for both brands of tires.

SOLUTION

Step 1   a. **Parameter of interest:** μ_d, the mean difference in the amounts of wear between the two brands.

          b. **Statement of hypotheses:**

             H_o: $\mu_d = 0$ (no difference)   Remember: $d = B - A$.

             H_a: $\mu_d \neq 0$ (difference)

Step 2 a. **Assumptions:** The assumption of normality is included in the statement of this problem.
  b. **Test statistic:** The *t*-distribution with df $= n - 1 = 6 - 1 = 5$, and $t\star = (\bar{d} - \mu_d)/(s_d/\sqrt{n})$.
  c. **Level of significance:** $\alpha = 0.05$.

Step 3 a. **Sample information:** $n = 6$, $\bar{d} = 6.3$, and $s_d = 5.1$.
  b. **Calculated test statistic:**

$$t\star = \frac{\bar{d} - \mu_d}{s_d/\sqrt{n}}: \quad t\star = \frac{6.3 - 0.0}{5.1/\sqrt{6}} = \frac{6.3}{2.08} = \mathbf{3.03}$$

Step 4   **Probability Distribution:**

p-**Value:** **OR**   **Classical:**

a. Use both tails because H_a expresses concern for values related to "different from."

$\mathbf{P} = p\text{-value} = P\,(t\star < -3.03) + P\,(t\star > 3.03)$

$= 2 \times P(\,|t\star|\, > 3.03)$ as shown in the figure

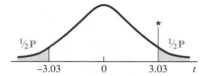

To find the *p*-value, you have three options:
1. Use Table 6 (Appendix B): $\mathbf{0.02 < P < 0.05}$.
2. Use Table 7 (Appendix B) to place bounds on the *p*-value: $\mathbf{0.026 < P < 0.030}$.
3. Use a computer or calculator to find the *p*-value: $\mathbf{P = 2 \times 0.0145 = 0.0290}$.
For specific instructions, see page 486.
b. The *p*-value is smaller than α.

a. The critical region is two-tailed because H_a expresses concern for values related to "different than." The critical value is obtained from Table 6: $t(5, 0.025) = \mathbf{2.57}$.

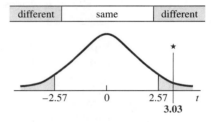

For specific instructions, see pages 477–478.
b. $t\star$ is in the critical region, as shown in **red** in the figure.

Step 5 a. **Decision:** Reject H_o.
  b. **Conclusion:** There is a significant difference in the mean amounts of wear at the 0.05 level of significance.

APPLIED
EXAMPLE 10.7   **Testing Asphalt Sampling Procedures**

This Application is an excerpt from a Florida Department of Transportation research report.

COMPARISON OF THE SCOOPING VS. QUARTERING METHODS FOR OBTAINING ASPHALT MIXTURE SAMPLES
Research Report FL/DOT/SMO/00-441
Gregory A. Sholar James A. Musselman Gale C. Page
State Materials Office

ABSTRACT - The standard method of quartering plant produced asphalt mix to obtain samples for maximum specific gravity, gradation and asphalt binder content has been used by the Florida Department of Transportation (FDOT), contractors and independent testing laboratories for many years with great success. This report examines an alternative method for obtaining samples that is somewhat easier and less time consuming than the traditional quartering method. This method, hereafter referred to as the "scooping" method, involves some of the same procedures and techniques that are used with the quartering method. The principle difference is that samples are scooped from the pile of asphalt mix until the desired sample weight is obtained instead of quartering the pile down until the desired sample weight is obtained. Twelve different mixtures were sampled for this study and the following mixture properties were compared for the two different sampling methods: bulk density, maximum specific gravity, % air voids, asphalt binder content and gradation. Analysis of the data indicates that the two sampling methods provide statistically equivalent results for the aforementioned mixture properties. Included in this report is a new version of FM 1-T 168, "Sampling Bituminous Paving Mixtures," which encompasses this new method for sampling asphalt mixtures.

DATA ANALYSIS - Theoretically, if the two sampling methods were identical then the average difference between values obtained for any asphalt property (ex., asphalt binder content) for a particular mix would be zero. A paired difference analysis was performed for each property measured. A paired difference analysis is a t-test performed on the differences between each sampling method. A 95% confidence interval was used, i.e. $\alpha = 0.05$, to calculate the two-sided t-critical value. The null hypothesis is that the average difference is zero. If t-calculated is less than t-critical, then the null hypothesis cannot be rejected. In the t-test summaries, the important values are the "t-calculated" and the "t-critical" values. For simplicity, all of these "t" values have been summarized in Table 14. Examination of the statistical results indicate that for all of the properties measured, except for % passing the No. 4 sieve, the null hypothesis cannot be rejected. This indicates that the two methods are statistically equivalent. The one exception is for the % passing the No. 4 sieve. The t-calculated and t-critical values were nearly identical (2.224 vs. 2.228).

TABLE 14

Summary of Paired Difference Analysis

Asphalt Mixture Property	Absolute Value		
	t-Calculated	t-Critical	t-calc. $<$ t-crit. ?
Gmb (Nmax)	1.442	2.306	YES
Gmm	0.802	2.201	YES
% Air Voids	1.719	2.306	YES
% AC (ignition)	0.534	2.201	YES
Sieve Size			
1/2″	0.672	2.228	YES
3/8″	0.783	2.228	YES
No. 4	2.224	2.228	Equal
No. 8	1.819	2.228	YES
No. 16	1.047	2.228	YES
No. 30	0.814	2.228	YES
No. 50	0.753	2.228	YES
No. 100	0.387	2.228	YES
No. 200	0.305	2.228	YES

CONCLUSION-Based on the statistical analysis of the data, the two methods of sampling are equivalent with respect to Gmb, Gmm, asphalt binder content and gradation. Since the scooping method is easier and faster it is recommended that the revised Florida method for sampling (FM 1-T 168) be accepted and implemented statewide.

SECTION 10.3 EXERCISES

Statistics⬡Now™

Datasets can be found on your Student's Suite CD-ROM or at the StatisticsNow website at **http://1pass.thomson.com**.

10.13 Given this set of paired data:

Pairs	1	2	3	4	5
Sample A	3	6	1	4	7
Sample B	2	5	1	2	8

Find:

a. The paired differences, $d = A - B$, for this set of data

b. The mean $\bar{d}$ of the paired differences

c. The standard deviation s_d of the paired differences

10.14 Find $t(15, 0.025)$. Describe the role this number plays when forming a confidence interval for the mean difference.

10.15 a. Find the 95% confidence interval for μ_d given $n = 26$, $\bar{d} = 6.3$, and $s_d = 5.1$. Assume the data are randomly selected from a normal population.

    b. Compare your interval to the interval found in Example 10.4 (p. 552).

10.16 [EX10-016] All students who enroll in a certain memory course are given a pretest before the course begins. At the completion of the course, 10 students are selected at random and given a posttest; their scores are listed here.

Student	1	2	3	4	5	6	7	8	9	10
Before	93	86	72	54	92	65	80	81	62	73
After	98	92	80	62	91	78	89	78	71	80

MINITAB was used to find the 95% confidence interval for the mean improvement in memory resulting from taking the memory course, as measured by the difference in test scores (d = after − before). Verify the results shown on the output by calculating the values yourself. Assume normality.

```
Confidence Intervals
Variable   N    Mean   StDev   SE Mean    95% C.I.
C3         10   6.10   4.79    1.52      (2.67, 9.53)
```

10.17 [EX10-017] Ten subjects with borderline-high cholesterol levels were randomly recruited for a study. The study involved taking a nutrition education class. Cholesterol readings were taken before the class and 3 months after the class.

Subject	1	2	3	4	5	6	7	8	9	10
Preclass	295	279	250	235	255	290	310	260	275	240
Postclass	265	266	245	240	230	230	235	250	250	215

Let d = preclass cholesterol − postclass cholesterol. Excel was used to find the 95% confidence interval for the mean amount of reduction in cholesterol readings after taking the nutrition education class. Verify the results shown on the output by calculating the values yourself. Assume normality.

t-Estimate: Mean

	$d = pre - post$
Mean	26.3
Standard deviation	24.4997
LCL	8.773924024
UCL	43.82607598

10.18 [EX10-018] Use a computer or calculator to find the 95% confidence interval for estimating μ_d based on these paired data and assuming normality:

Before	75	68	40	30	43	65
After	70	69	32	30	39	63

10.19 [EX10-019] Salt-free diets are often prescribed to people with high blood pressure. The following data values were obtained from an experiment designed to estimate the reduction in diastolic blood pressure as a result of consuming a salt-free diet for 2 weeks. Assume diastolic readings to be normally distributed.

Before	93	106	87	92	102	95	88	110
After	92	102	89	92	101	96	88	105

a. What is the point estimate for the mean reduction in the diastolic reading after 2 weeks on this diet?

b. Find the 98% confidence interval for the mean reduction.

10.20 [EX10-020] An experiment was designed to estimate the mean difference in weight gain for pigs fed ration A as compared with those fed ration B. Eight pairs of pigs were used. The pigs within each pair were littermates. The rations were assigned at random to the two animals within each pair. The gains (in pounds) after 45 days are shown in the following table.

Litter	1	2	3	4	5	6	7	8
Ration A	65	37	40	47	49	65	53	59
Ration B	58	39	31	45	47	55	59	51

Assuming weight gain is normal, find the 95% confidence interval estimate for the mean of the differences μ_d, where d = ration A − ration B.

10.21 [EX10-021] Two men, A and B, who usually commute to work together, decide to conduct an experiment to see whether one route is faster than the other. The men believe that their driving habits are approximately the same, and therefore they decide on the following procedure. Each morning for 2 weeks, A will drive to work on one route and B will use the other route. On the first morning, A will toss a coin. If heads appear, he will use route I; if tails appear, he will use route II. On the second morning, B will toss the coin: heads, route I; tails, route II. The times, recorded to the nearest minute, are shown in the following table. Assume commute times are normal and estimate the population mean difference with a 95% confidence interval.

				Day						
Route	M	Tu	W	Th	F	M	Tu	W	Th	F
I	29	26	25	25	25	24	26	26	30	31
II	25	26	25	25	24	23	27	25	29	30

10.22 [EX10-022] In evaluating different measuring instruments, one must first determine whether there is a systematic difference between the instruments. Lenses from several different powers were measured once each by two different instruments. The measurement differences (Instrument A − Instrument B) were recorded. The measurement units have been coded for proprietary reasons.

```
 4    5  -2  -3   -7  10    11  -1  3  7  -5    3  -4
-5   -7   4  -1  -18   0  -17  12  9  4   17  -2
```

Does there appear to be a systematic difference between the two instruments?

a. Describe the data using a histogram and one other graph.

b. Find the mean and the standard deviation.

c. Are the assumptions required for making inferences satisfied? Explain.

d. Using a 95% confidence interval, estimate the population mean of the differences.

e. Is there any evidence of a difference? Explain.

10.23 State the null hypothesis, H_o, and the alternative hypothesis, H_a, that would be used to test these claims:

a. There is an increase in the mean difference between posttest and pretest scores.

b. Following a special training session, it is believed that the mean of the difference in performance scores will not be zero.

c. On average, there is no difference between the readings from two inspectors on each of the selected parts.

d. The mean of the differences between pre–self-esteem and post–self-esteem scores showed improvement after involvement in a college learning community.

10.24 State the null hypothesis, H_o, and the alternative hypothesis, H_a, that would be used to test these claims:

a. The mean of the differences between the posttest and the pretest scores is greater than 15.

b. The mean weight gain, after the change in diet for the laboratory animals, is at least 10 oz.

c. The mean weight loss experienced by people on a new diet plan was no less than 12 lb.

d. The mean difference in the home reassessments from the two town assessors was no more than $200.

10.25 Determine the p-value for each hypothesis test for the mean difference.

a. H_o: $\mu_d = 0$ and H_a: $\mu_d > 0$, with $n = 20$ and $t\bigstar = 1.86$

b. H_o: $\mu_d = 0$ and H_a: $\mu_d \neq 0$, with $n = 20$ and $t\bigstar = -1.86$

c. H_o: $\mu_d = 0$ and H_a: $\mu_d < 0$, with $n = 29$ and $t\bigstar = -2.63$

d. H_o: $\mu_d = 0.75$ and H_a: $\mu_d > 0.75$, with $n = 10$ and $t\bigstar = 3.57$

10.26 Determine the test criteria that would be used with the classical approach to test the following hypotheses when t is used as the test statistic.

a. H_o: $\mu_d = 0$ and H_a: $\mu_d > 0$, with $n = 15$ and $\alpha = 0.05$

b. H_o: $\mu_d = 0$ and H_a: $\mu_d \neq 0$, with $n = 25$ and $\alpha = 0.05$

c. H_o: $\mu_d = 0$ and H_a: $\mu_d < 0$, with $n = 12$ and $\alpha = 0.10$

d. H_o: $\mu_d = 0.75$ and H_a: $\mu_d > 0.75$, with $n = 18$ and $\alpha = 0.01$

10.27 The corrosive effects of various soils on coated and uncoated steel pipe was tested by using a dependent sampling plan. The data collected are summarized by $n = 40$, $\sum d = 220$, $\sum d^2 = 6222$, where d is the amount of corrosion on the coated portion subtracted from the amount of corrosion on the uncoated portion. Does this random sample provide sufficient reason to conclude that the coating is beneficial? Use $\alpha = 0.01$ and assume normality.

a. Solve using the p-value approach.

b. Solve using the classical approach.

10.28 Does a content title help a reader comprehend a piece of writing? Twenty-six participants were given an article to read without a title. They then rated themselves on comprehension of the information on a scale from 1 to 10, where 10 was complete comprehension. The same 26 participants were then given the article again, this time with an appropriate title, and asked to rate their comprehension. The resulting summarized data was given as $\bar{d} = 4.76$ and $s_d = 2.33$, where d = rating with title $-$ rating without title. Comprehension is generally higher on the second reading than on the first by an average of 3.2 on this scale. Does this sample provide sufficient evidence that a content title does make a difference with respect to comprehension? Use $\alpha = 0.05$.

10.29 [EX10-029] Ten people recently diagnosed with diabetes were tested to determine whether an educational program was effective in increasing their knowledge of diabetes. They were given a test, before and after the educational program, concerning self-care aspects of diabetes. The scores on the test were as follows:

Patient	1	2	3	4	5	6	7	8	9	10
Before	75	62	67	70	55	59	60	64	72	59
After	77	65	68	72	62	61	60	67	75	68

The following MINITAB output may be used to determine whether the population mean difference is greater than zero after the program. Verify the values shown on the output (mean difference [MEAN], standard deviation [STDEV], standard error of the difference [SE MEAN], $t\bigstar$ [T], and p-value) by calculating the values yourself.

```
TEST OF MU = 0.000 VS MU G.T. 0.000
      N     MEAN    STDEV   SE MEAN    T     P VALUE
C3   10    3.200    2.741    0.867    3.69    0.0025
```

10.30 [EX10-030] Ten subjects with borderline-high cholesterol levels were recruited for a study. The study involved taking a nutrition education class. Cholesterol readings were taken before the class and 3 months after the class.

Subject	1	2	3	4	5	6	7	8	9	10
Preclass	295	279	250	235	255	290	310	260	275	240
Postclass	265	266	245	240	230	230	235	250	250	215

Let d = preclass cholesterol $-$ postclass cholesterol. Use the following Excel output to test the null hypothesis that the population mean difference equals zero versus the alternative hypothesis that the population mean difference is positive at $\alpha = 0.05$. Rejection of the null hypothesis would indicate that the (population) average cholesterol level after the class is lower than the average level before the class. Assume normality.

```
t-Test: Paired Two Sample for Means
                        Pretest         Posttest

Mean                    268.9           242.6
Variance                618.7666667     256.4888889
Observations            10              10
Hypothesized mean       0
  difference
df                      9
t Stat                  3.394655392
P(T≤t) one-tail         0.003970146
t Critical one-tail     1.833113856
```

10.31 Complete the hypothesis test with alternative hypothesis $\mu_d > 0$ based on the paired data that follow and $d = B - A$. Use $\alpha = 0.05$. Assume normality.

A	700	830	860	1080	930
B	720	820	890	1100	960

a. Solve using the *p*-value approach.

b. Solve using the classical approach.

10.32 Use a computer or calculator to complete the hypothesis test with alternative hypothesis $\mu_d < 0$ based on the paired data that follow and $d = M - N$. Use $\alpha = 0.02$. Assume normality.

M	58	78	45	38	49	62
N	62	86	42	39	47	68

10.33 Complete the hypothesis test with alternative hypothesis $\mu_d \neq 0$ based on the paired data that follows and $d = O - Y$. Use $\alpha = 0.01$. Assume normality.

Oldest	199	162	174	159	173
Youngest	194	162	167	156	176

a. Solve using the *p*-value approach.

b. Solve using the classical approach.

10.34 [EX10-034] Ten randomly selected college students, who participated in a learning community, were given pre–self-esteem and post–self-

esteem surveys. A learning community is a group of students who take two or more courses together. Typically, each learning community has a theme, and the faculty involved coordinate assignments linking the courses. Research has shown that the benefits of higher self-esteem, higher grade point averages (GPAs), and improved satisfaction in courses, as well as better retention rates, result from involvement in a learning community. The scores on the surveys are as follows:

Student	1	2	3	4	5	6	7	8	9	10
Prescore	18	14	11	23	19	21	21	21	11	22
Postscore	17	17	10	25	20	10	24	22	10	24

Does this sample of students show sufficient evidence that self-esteem scores were higher after participation in a learning community? Lower scores indicate higher self-esteem. Use the 0.05 level of significance and assume normality of scores.

10.35 [EX10-035] In reference to the college students who participated in a learning community in Exercise 10.34, a control group of students was also formed for testing and comparison. Ten randomly selected college students, who were not involved in the learning community, were given pre–self-esteem and post–self-esteem surveys. The scores on the surveys for the control group are as follows:

Student	1	2	3	4	5	6	7	8	9	10
Prescore	19	23	12	20	26	20	15	10	22	12
Postscore	19	21	9	10	23	20	19	10	21	19

Does this sample of students show sufficient evidence that self-esteem scores were higher after participation in college courses? Lower scores indicate higher self-esteem. Use the 0.05 level of significance and assume normality of scores.

10.36 [EX10-036] To test the effect of a physical fitness course on one's physical ability, the number of sit-ups that a person could do in 1 minute, both before and after the course, was recorded. Ten randomly selected participants scored as shown in the following table. Can you conclude that a significant amount of improvement took place? Use $\alpha = 0.01$ and assume normality.

Before	29	22	25	29	26	24	31	46	34	28
After	30	26	25	35	33	36	32	54	50	43

a. Solve using the *p*-value approach.

b. Solve using the classical approach.

10.37 Referring to Applied Example 10.7:

a. What null hypothesis is being tested in each of these 13 tests?

b. Why are the "*t*-calculated" and the "*t*-critical" values the important values?

c. Why is it correct to report their absolute values for both *t*-values in Table 14?

d. What decision is reached for each of these 13 hypotheses tests?

e. What conclusion is reached as a result of these tests?

f. What action is recommended to the State of Florida as a result of the conclusion?

10.38 **[EX10-038]** A research project was undertaken to evaluate two focimeters. Each of 20 lenses of varying powers was read once each on each focimeter. The measurement differences were then calculated, where each difference is Focimeter A − Focimeter B. Assume readings are normally distributed.

−0.016 0.013 0.009 0.000 −0.005 −0.015 −0.006 −0.016 −0.022 −0.006
−0.020 0.015 −0.017 −0.010 −0.003 0.011 −0.012 0.008 −0.005 −0.009

Courtesy: Bausch & Lomb

a. Using a *t*-test on these paired differences and an $\alpha = 0.01$, determine whether the corresponding population mean difference is significantly different from zero.

b. Construct a 99% confidence interval for the mean difference in focimeter readings.

c. Explain what both inferential procedures indicate about the differences?

d. If an enterprising experimenter performed this same test using $\alpha = 10\%$, what would the outcome be? Offer comments about proceeding using these ground rules.

10.4 | Inferences Concerning the Difference between Means Using Two Independent Samples

When comparing the means of two populations, we typically consider the difference between their means, $\mu_1 - \mu_2$ (often called **"independent means"**). The inferences about $\mu_1 - \mu_2$ will be based on the difference between the observed sample means, $\bar{x}_1 - \bar{x}_2$. This observed difference, $\bar{x}_1 - \bar{x}_2$, belongs to a sampling distribution with the characteristics described in the following statement.

FYI Why is $\bar{x}_1 - \bar{x}_2$ an *unbiased* estimator of $\mu_1 - \mu_2$?

If independent samples of sizes n_1 and n_2 are drawn randomly from large populations with means μ_1 and μ_2 and variances σ_1^2 and σ_2^2, respectively, then the sampling distribution of $\bar{x}_1 - \bar{x}_2$, the difference between the sample means, has

1. mean $\mu_{\bar{x}_1 - \bar{x}_2} = \mu_1 - \mu_2$ and

2. standard error $\sigma_{\bar{x}_1 - \bar{x}_2} = \sqrt{\left(\dfrac{\sigma_1^2}{n_1}\right) + \left(\dfrac{\sigma_2^2}{n_2}\right)}.$ (10.6)

If both populations have normal distributions, then the sampling distribution of $\bar{x}_1 - \bar{x}_2$ will also be normally distributed.

The preceding statement is true for all sample sizes given that the populations involved are normal and the population variances σ_1^2 and σ_2^2 are known quantities. However, as with inferences about one mean, the variance of a

The "*t*-Distribution"

As head brewer at Guinness Brewing Company, William Gosset was faced with many small sets of data–small by necessity because a 24-hour period often resulted in only one data value. Thus, he developed the *t*-test to handle these small samples for quality control in brewing. In his paper *The Probable Error of a Mean*, he set out to find the distribution of the amount of error in the sample mean, $(x - \mu)$ divided by s, where s was from a sample of any known size. He then found the probable error of a mean, $\overline{x}$, for any size sample, by using the distribution of $(x - \mu)/(s/\sqrt{n})$. Student's *t*-distribution did not immediately gain popularity, and in 1922, even 14 years after its publication, Student wrote to Fisher: "I am sending you a copy of Student's Tables as you are the only man that's ever likely to use them!" Today, Student's *t*-distribution is widely used and respected in statistical research.

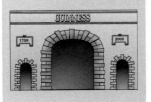

population is generally an unknown quantity. Therefore, it will be necessary to estimate the standard error by replacing the variances, σ_1^2 and σ_2^2, in formula (10.6) with the best estimates available—namely, the sample variances, s_1^2 and s_2^2. The *estimated standard error* will be found using the following formula:

$$\text{estimated standard error} = \sqrt{\left(\frac{s_1^2}{n_1}\right) + \left(\frac{s_2^2}{n_2}\right)} \qquad (10.7)$$

Inferences about the difference between two population means, $\mu_1 - \mu_2$, will be based on the following assumptions.

> **Assumptions for inferences about the difference between two means,**
> $\mu_1 - \mu_2$: The samples are randomly selected from normally distributed populations, and the samples are selected in an independent manner.
>
> NO ASSUMPTIONS ARE MADE ABOUT THE POPULATION VARIANCES.

Since the samples provide the information for determining the standard error, the **t-distribution** will be used as the test statistic. The inferences are divided into two cases.

Case 1: The *t*-distribution will be used, and the number of degrees of freedom will be calculated.

Case 2: The *t*-distribution will be used, and the number of degrees of freedom will be approximated.

Case 1 will occur when you are completing the inference *using a computer or statistical calculator and the statistical software or program calculates the number of degrees of freedom* for you. The calculated value for df is a function of both sample sizes and their relative sizes, and both sample variances and their relative sizes. The value of df will be a number between the smaller of $df_1 = n_1 - 1$ or $df_2 = n_2 - 1$, and the sum of the degrees of freedom, $df_1 + df_2 = [(n_1 - 1) + (n_2 - 1)] = n_1 + n_2 - 2$.

Case 2 will occur when you are completing the inference *without the aid of a computer or calculator and its statistical software package.* Use of the *t*-distribution with the smaller of $df_1 = n_1 - 1$ or $df_2 = n_2 - 1$ will give *conservative* results. Because of this approximation, the true level of confidence for an interval estimate will be slightly higher than the reported level of confidence; or the true *p*-value and the true level of significance for a hypothesis test will be slightly less than reported. The gap between these reported values and the true values will be quite small, unless the sample sizes are quite small and unequal or the sample variances are very different. The gap will decrease as the samples increase in size or as the sample variances are more alike.

Since the only difference between the two cases is the number of degrees of freedom used to identify the *t*-distribution involved, we will study case 2 first.

Note: $A > B$ ("A is greater than B") is equivalent to $B < A$ ("B is less than A"). When the difference between A and B is being discussed, it is customary to express the difference as "larger − smaller" so that the resulting difference is pos-

FYI Would you say the difference between 5 and 8 is −3? How would you express the difference? Explain.

itive: $A - B > 0$. To express the difference as "smaller − larger" results in $B - A < 0$ (the difference is negative) and is usually unnecessarily confusing. Therefore, it is recommended that the difference be expressed as "larger − smaller."

Confidence Interval Procedure

We will use the following formula for calculating the endpoints of the $1 - \alpha$ confidence interval.

Confidence Interval for the Difference between Two Means (Independent Samples)

$$(\bar{x}_1 - \bar{x}_2) - t(df, \alpha/2) \cdot \sqrt{\left(\frac{s_1^2}{n_1}\right) + \left(\frac{s_2^2}{n_2}\right)} \text{ to } (\bar{x}_1 - \bar{x}_2) + t(df, \alpha/2) \cdot \sqrt{\left(\frac{s_1^2}{n_1}\right) + \left(\frac{s_2^2}{n_2}\right)}$$

where df is either calculated or is the smaller of df_1 or df_2 (see p. 565)

(10.8)

EXAMPLE 10.8

Constructing a Confidence Interval for the Difference between Two Means

Statistics⬥Now™

Watch a video example at http://1pass.thomson.com or on your CD.

The heights (in inches) of 20 randomly selected women and 30 randomly selected men were independently obtained from the student body of a certain college in order to estimate the difference in their mean heights. The sample information is given in Table 10.2. Assume that heights are approximately normally distributed for both populations.

TABLE 10.2

Sample Information on Student Heights

Sample	Number	Mean	Standard Deviation
Female (*f*)	20	63.8	2.18
Male (*m*)	30	69.8	1.92

Find the 95% confidence interval for the difference between the mean heights, $\mu_m - \mu_f$.

SOLUTION

Step 1 **Parameter of interest:** $\mu_m - \mu_f$, the difference between the mean height of male students and the mean height of female students.

Step 2 **Assumptions:** Both populations are approximately normal, and the samples were random and independently selected.

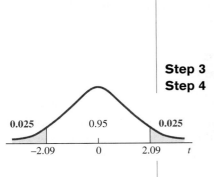

b. **Probability distribution:** The t-distribution with df = 19, the smaller of $n_m - 1 = 30 - 1 = 29$ or $n_f - 1 = 20 - 1 = 19$, and formula (10.8).

c. **Level of confidence:** $1 - \alpha = 0.95$.

Step 3 **Sample information:** See Table 10.2.

Step 4 **Confidence coefficient:** We have a two-tailed situation with $\alpha/2 = 0.025$ in one tail and df = 19. From Table 6 in Appendix B, $t(\mathrm{df},\alpha/2) = t(19,0.025) = 2.09$. See the figure. See pages 477–479 for instructions on using Table 6.

b. **Maximum error of estimate:** Using the maximum error part of formula (10.8), we have

$$E = t_{(\mathrm{df},\,\alpha/2)} \cdot \sqrt{\left(\frac{s_1^2}{n_1}\right) + \left(\frac{s_2^2}{n_2}\right)}: \qquad E = 2.09 \cdot \sqrt{\left(\frac{1.92^2}{30}\right) + \left(\frac{2.18^2}{20}\right)}$$

$$= (2.09)(0.60) = \mathbf{1.25}$$

c. **Lower and upper confidence limits:**

$$(\bar{x}_1 - \bar{x}_2) \pm E$$

$$6.00 \pm 1.25$$

$$6.00 - 1.25 = \mathbf{4.75} \quad \text{to} \quad 6.00 + 1.25 = \mathbf{7.25}$$

Step 5 a. **Confidence interval.**
4.75 to 7.25 is the 95% confidence interval for $\mu_m - \mu_f$.

b. That is, with 95% confidence, we can say that the difference between the mean heights of the male and female students is between 4.75 and 7.25 inches; that is, the mean height of male students is between 4.75 and 7.25 inches greater than the mean height of female students.

Hypothesis-Testing Procedure

When we test a null **hypothesis about the difference between two population means,** the test statistic used will be the difference between the observed difference of the sample means and the hypothesized difference of the population means, divided by the estimated standard error. The test statistic is assumed to have approximately a t-distribution when the null hypothesis is true and the normality assumption has been satisfied. The calculated value of the **test statistic** is found using this formula:

Test Statistic for the Difference between Two Means (Independent Samples)

$$t\bigstar = \frac{(\bar{x}_1 - \bar{x}_2) - (\mu_1 - \mu_2)}{\sqrt{\left(\frac{s_1^2}{n_1}\right) + \left(\frac{s_2^2}{n_2}\right)}} \tag{10.9}$$

where df is either calculated or is the smaller of df_1 or df_2 (see p. 565)

Note: A hypothesized difference between the two population means, $\mu_1 - \mu_2$, can be any specified value. The most common value specified is zero; however, the difference can be nonzero.

EXAMPLE 10.9

One-Tailed Hypothesis Test for the Difference between Two Means

Suppose that we are interested in comparing the academic success of college students who belong to fraternal organizations with the academic success of those who do not belong to fraternal organizations. The reason for the comparison is the recent concern that fraternity members, on average, are achieving at a lower academic level than nonfraternal students achieve. (Cumulative GPA is used to measure academic success.) Random samples of size 40 are taken from each population. The sample results are listed in Table 10.3.

TABLE 10.3

Sample Information on Academic Success

Sample	Number	Mean	Standard Deviation
Fraternity members (*f*)	40	2.03	0.68
Nonmembers (*n*)	40	2.21	0.59

Complete a hypothesis test using $\alpha = 0.05$. Assume that the GPAs for both groups are approximately normally distributed.

SOLUTION

Step 1 a. **Parameter of interest:** $\mu_n - \mu_f$, the difference between the mean GPAs for the nonfraternity members and the fraternity members.

b. **Statement of hypotheses:**

$H_o: \mu_n - \mu_f = 0$ ($\leq$) (fraternity averages are no lower)
$H_a: \mu_n - \mu_f > 0$ (fraternity averages are lower)

Step 2 a. **Assumptions:** Both populations are approximately normal, and random samples were selected. Since the two populations are separate, the samples are independent.

b. **Test statistic:** The *t*-distribution with df = the smaller of df_n or df_f; since both *n*'s are 40, df = 40 − 1 = **39;** and $t\star$ is calculated using formula (10.9).

c. **Level of significance:** $\alpha = 0.05$.

Step 3 a. **Sample information:** See Table 10.3.

b. **Calculated test statistic:**

$$t\star = \frac{(\bar{x}_1 - \bar{x}_2) - (\mu_1 - \mu_2)}{\sqrt{\left(\frac{s_1^2}{n_1}\right) + \left(\frac{s_2^2}{n_2}\right)}} : \qquad t\star = \frac{(2.21 - 2.03) - (0.00)}{\sqrt{\left(\frac{0.59^2}{40}\right) + \left(\frac{0.68^2}{40}\right)}}$$

$$= \frac{0.18}{\sqrt{0.00870 + 0.01156}} = \frac{0.18}{0.1423} = \mathbf{1.26}$$

Step 4 **Probability Distribution:**

p-value: **OR** **Classical:**

a. Use the right-hand tail because H_a expresses concern for values related to "greater than." **P** = $P(t\star > 1.26$, with df = 39) as shown on the figure.

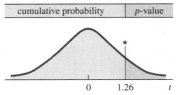

To find the *p*-value, use one of three methods:
1. Use Table 6 (Appendix B) to place bounds on the *p*-value: **0.10 < P < 0.25.**
2. Use Table 7 (Appendix B) to place bounds on the *p*-value: **0.100 < P < 0.119.**
3. Use a computer or calculator to find the *p*-value: **P = 0.1076.**

Specific details follow this illustration.
b. The *p*-value is not smaller than α.

a. The critical region is the right-hand tail because H_a expresses concern for values related to "greater than." The critical value is obtained from Table 6: t (39, 0.05) = **1.69.**

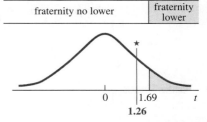

See pages 477–478 for information about critical values.
b. $t\star$ is not in the critical region, as shown in **red** on the figure.

Step 5 **a.** **Decision:** Fail to reject H_o.
b. **Conclusion:** At the 0.05 level of significance, the claim that the fraternity members achieve at a lower level than nonmembers is not supported by the sample data.

To find the *p*-value for Example 10.9, use one of three methods:

Method 1: Use Table 6. Find 1.26 between two entries in the df = 39 (use df = 35) row and read the bounds for **P** from the one-tail heading at the top of the table: **0.10 < P < 0.25.**

Method 2: Use Table 7. Find $t\star$ = 1.26 between two rows and df = 39 between two columns; read the bounds for $P(t\star > 1.26 \,|\, df = 39)$; **0.100 < P < 0.119.**

Method 3: If you are doing the hypothesis test with the aid of a computer or calculator, most likely it will calculate the *p*-value for you (see pp. 486–487), or you may use the cumulative probability distribution commands described in Chapter 9 (p. 479).

EXAMPLE 10.10 **Two-Tailed Hypothesis for the Difference between Two Means**

Many students have complained that the soft-drink vending machine in the student recreation room (A) dispenses a different amount of drink than machine in the faculty lounge (B). To test this belief, a student randomly sampled several servings from each machine and carefully measured them, with the results shown in Table 10.4.

TABLE 10.4

Sample Information on Vending Machines

Machine	Number	Mean	Standard Deviation
A	10	5.38	1.59
B	12	5.92	0.83

Does this evidence support the hypothesis that the mean amount dispensed by machine A is different from the amount dispensed by machine B? Assume the amounts dispensed by both machines are normally distributed, and complete the test using $\alpha = 0.10$.

SOLUTION

Step 1   **a.** **Parameter of interest:** $\mu_B - \mu_A$, the difference between the mean amount dispensed by machine B and the mean amount dispensed by machine A.

   b. **Statement of hypotheses:**

FYI "Larger − smaller" results in a positive difference.

$H_o: \mu_B - \mu_A = 0$ (A dispenses the same average amount as B)

$H_a: \mu_B - \mu_A \neq 0$ (A dispenses a different average amount than B)

Step 2   **Assumptions:** Both populations are assumed to be approximately normal, and the samples were random and independently selected.

   b. **Test statistic:** The t-distribution with df = the smaller of $n_A - 1 = 10 - 1 = 9$ or $n_B - 1 = 12 - 1 = 11$, df = 9, and $t\star$ calculated using formula (10.9).

   c. **Level of significance:** $\alpha = 0.10$.

Step 3   **a.** **Sample information:** See Table 10.4.

   b. **Calculated test statistic:**

$$t\star = \frac{(\bar{x}_B - \bar{x}_A) - (\mu_B - \mu_A)}{\sqrt{\left(\frac{s_B^2}{n_B}\right) + \left(\frac{s_A^2}{n_A}\right)}} \quad : \quad t\star = \frac{(5.92 - 5.38) - (0.00)}{\sqrt{\left(\frac{0.83^2}{12}\right) + \left(\frac{1.59^2}{10}\right)}}$$

$$= \frac{0.54}{\sqrt{0.0574 + 0.2528}} = \frac{0.54}{0.557} = \mathbf{0.97}$$

Step 4   **Probability Distribution:**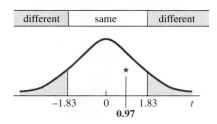

p-value: **OR**   Classical:

a. Use both tails because H_a expresses concern for values related to "different than."

$\mathbf{P} = p\text{-value} = P(t\star < -0.97) + P(t\star > 0.97)$
$= 2 \times P(|t\star| > 0.97 \mid \text{df} = 9)$ as in the figure.

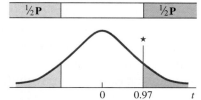

To find the *p*-value, you have three options:
1. Use Table 6 (Appendix B) **0.20 < P < 0.50.**
2. Use Table 7 (Appendix B) to place bounds on the *p*-value: **0.340 < P < 0.394.**
3. Use a computer or calculator to find the *p*-value: **P = 2 × 0.1787 = 0.3574.**

Specific instructions follow this illustration.
b. The *p*-value is not smaller than α.

a. The critical region is two-tailed because H_a expresses concern for values related to "different than." The right-hand critical value is obtained from Table 6: $t(9, 0.05) = \mathbf{1.83}$. See the figure.

For specific instructions, see pages 477–478.
b. $t\star$ is not in the critical region as shown in **red** on the figure.

Step 5   a.   **Decision:** Fail to reject H_o.

b.   **Conclusion:** The evidence is not sufficient to show that machine A dispenses a different average amount of soft drink than machine B, at the 0.10 level of significance. Thus, for lack of evidence we will proceed as though the two machines dispense, on average, the same amount.

To find the *p*-value for Example 10.10, use one of three methods:

Method 1: Use Table 6. Find 0.97 between two entries in the df = 9 row and read the bounds for **P** from the two-tail heading at the top of the table: **0.20 < P < 0.50.**

Method 2: Use Table 7. Find $t\star = 0.97$ between two rows and df = 9 between two columns; read the bounds for $P(t\star > 0.97 \mid \text{df} = 9)$: $0.170 < \frac{1}{2}P < 0.197$; therefore, **0.340 < P < 0.394.**

Method 3: If you are doing the hypothesis test with the aid of a computer or calculator, most likely it will calculate the *p*-value (do not double) for you (see pp. 486–487), or you may use the cumulative probability distribution commands described in Chapter 9 (p. 479).

Most computer or calculator statistical packages will complete the inferences for the difference between two means by calculating the number of degrees of freedom.

TECHNOLOGY INSTRUCTIONS: HYPOTHESIS TEST FOR THE DIFFERENCE BETWEEN TWO POPULATION MEANS WITH UNKNOWN STANDARD DEVIATION GIVEN TWO INDEPENDENT SETS OF SAMPLE DATA

MINITAB (Release 14)   MINITAB's 2-Sample *t* (Test and Confidence Interval) command performs both the confidence interval and the hypothesis test at the same time.
Input the two independent sets of data into C1 and C2; then continue with:

Choose: **Stat > Basic Statistics > 2-Sample t**
Select: **Samples in different columns***
Enter: First: **C1** Second: **C2**
Select: **Assume equal variances** (if known)
Select: **Options**
Enter: Confidence level: **1 − α** (ex. 0.95 or 95.0)
             Test mean: **0.0**
Choose: Alternative: **less than** or **not equal** or **greater than** > **OK** > **OK**

*Note the other possible data formats.

Excel   Input the two independent sets of data into columns A and B; then continue with:

Choose: **Tools > Data Analysis > t-Test: Two-Sample Assuming Unequal Variances**
Enter: Variable 1 Range: **(A1:A20 or select cells)**
             Variable 2 Range: **(B1:B20 or select cells)**
             Hypothesized Mean Difference: $\mu_B - \mu_A$ (usually 0)
Select: **Labels** (if necessary)

```
Enter:        α (ex. 0.05)
Select:       Output Range
Enter:        (C1 or select cell) > OK
```

Use Format > Column > AutoFit Selection to make the output more readable. The output shows *p*-values and critical values for one- and two-tailed tests.

TI-83/84 Plus   Input the two independent sets of data into L1 and L2.*

To construct a $1 - \alpha$ confidence interval for the mean difference, continue with the following, entering the appropriate values and highlighting Calculate:

```
Choose:       STAT > TESTS > 0:2-SampTInt ...
```

 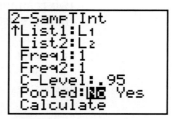

To complete a hypothesis test for the mean difference, continue with the following, entering the appropriate values and highlighting Calculate:

```
Choose:       STAT > TESTS > 4:2-SampTTest ...
```

*Enter the data in the order that is needed; the program subtracts as L1 − L2.

Highlight No for Pooled if there are no assumptions about the equality of variances.

Example 10.9 was solved using MINITAB. With 40 cumulative GPAs for nonmembers in C1 and 40 averages for fraternity members in C2, the preceding commands resulted in the output shown here. Compare these results to the solution of Example 10.9. Notice the difference in **P** and df values. Explain.

Two-Sample T-Test and CI

Sample	N	Mean	StDev	SE Mean
1	40	2.210	0.590	0.093
2	40	2.030	0.680	0.11

Difference = mu (1) − mu (2) Est. diff.: 0.180
95% CI for difference: (−0.10, 0.46)
T-Test diff. = 0 (vs >): T = 1.26 P = 0.105 DF = 76

APPLIED

EXAMPLE 10.11   ## Sectional Anatomy: Strategy for Mastery

A strategy for mastery of sectional anatomy must contain research to determine whether or not the methodology is sound. The research-based approach that follows demonstrates the effectiveness of a specific strategy. One application that deals with a comprehensive understanding of human anatomical features and their adjacent structures is a prescribed, sequenced labeling method (PSLM). The method contrasts the modern convention of random labeled human anatomical sections. The PSLM is based on studies utilizing images acquired from The Visible Human Project and was performed at Triton College. These studies have shown a significant impact on the learning rate and the comprehension of structures and adjacent anatomical relationships.

The initial component of this research consisted of a group of 28 students who were divided into two groups: Group A: 15, Group B: 13. Both groups were given a pretest, study period, and a singular Posttest. Both images presented for study were identical cross sections of the brain. The Group A image for study was labeled in accordance with PSLM protocol. The Group B image for study was randomly labeled. Both groups were instructed to "list and recognize the parts/layers of a transverse brain section, from superficial to deep" by writing their answers in spaces provided as the only posttest in this preliminary study.

First Set of PSLM Studies

Variable	Number of Cases	Mean	SD	SE of Mean
Group A	15	9.6667	1.589	0.410
Group B	13	3.1538	3.023	0.839

Variances	t-value	df	2-Tail Sig	SE of Difference	95% CI of Diff
Unequal	6.98	17.57	.000	0.933	(4.548, 8.477)

Group A's average score was 9.6667 of a total possible right of 11 with a standard deviation of 1.589. Group B scored on average 3.1538 out of 11 possible points.

The mean difference between Group A and Group B is 6.5129, which is highly significant.

Source: Alexander Lane, Ph.D., Triton College, River Grove, Illinois, http://www.nlm.nih.gov/research/visible/vhpconf98/AUTHORS/LANE/LANE.HTM

SECTION 10.4 EXERCISES

Statistics⬙Now™

Datasets can be found on your Student's Suite CD-ROM or at the StatisticsNow website at **http://1pass.thomson.com**.

10.39 Two independent random samples resulted in the following:

Sample 1: $n_1 = 12$, $s_1^2 = 190$
Sample 2: $n_2 = 18$, $s_2^2 = 150$

Find the estimate for the standard error for the difference between two means.

10.40 Two independent random samples resulted in the following:

Sample A: $n_A = 24$, $s_A = 8.5$
Sample B: $n_B = 21$, $s_B = 11.3$

Find the estimate for the standard error for the difference between two means.

10.41 Two independent random samples of sizes 18 and 24 were obtained to make inferences about the difference between two means. What is the number of degrees of freedom? Discuss both cases.

10.42 Find the confidence coefficient, $t_{(df, \frac{\alpha}{2})}$, that would be used to find the maximum error for each of the following situations when estimating the difference between two means, $\mu_1 - \mu_2$.

a. $1 - \alpha = 0.95$, $n_1 = 25$, $n_2 = 15$

b. $1 - \alpha = 0.98$, $n_1 = 43$, $n_2 = 32$

c. $1 - \alpha = 0.99$, $n_1 = 19$, $n_2 = 45$

10.43 Find the 90% confidence interval for the difference between two means based on this information about two samples. Assume independent samples from normal populations.

Sample	Number	Mean	Std. Dev.
1	20	35	22
2	15	30	16

10.44 A study comparing attitudes toward death was conducted in which organ donors (individuals who had signed organ donor cards) were compared with nondonors. The study is reported in the journal *Death Studies*. Templer's Death Anxiety Scale (DAS) was administered to both groups. On this scale, high scores indicate high anxiety concerning death. The results were reported as follows.

	n	Mean	Std. Dev.
Organ Donors	25	5.36	2.91
Nonorgan Donors	69	7.62	3.45

Construct the 95% confidence interval for the difference between the means, $\mu_{non} - \mu_{donor}$.

FYI Results obtained may be noticeably different depending on use of the calculated df or use of the df for a smaller sample.

10.45 "Car renters come across steeper rates" (*USA Today*, May 17, 2005) reported that rates for car rentals are rising but have not surpassed the national average peak of $35.98 a day in 2001. The average national daily rate for the April–June quarter of 2005 was $30.15, yet in some cities, car rentals

can cost more than $100 a day. A similar study of two major cities found the following results:

City	n	Average Daily Rate	Std. Dev.
Boston	10	128.25	7.50
New York City	15	116.60	8.90

Set a 95% confidence interval on the difference in average daily rates between the two major East Coast cities of Boston and New York City. Assume normality for the sampled populations and that the samples were selected randomly.

10.46 Part of a testing program on a family of dual bell rockets designed to compensate for low-altitude performance led to two slightly different new designs. The two sets of test flow rates, measured in pounds per second, resulted from testing these two new designs.

	n	$\sum x$	$\sum x^2$
Design 1	36	278.4	2163.76
Design 2	42	310.8	2332.26

Determine the 99% confidence interval for the difference between the two mean test flow rates. Assume normality.

10.47 [EX10-047] A study was designed to estimate the difference in diastolic blood pressure readings between men and women. MINITAB was used to construct a 99% confidence interval for the difference between the means based on the following sample data.

Males	76	76	74	70	80	68	90	70
	90	72	76	80	68	72	96	80
Females	76	70	82	90	68	60	62	68
	80	74	60	62	72			

```
Two-sample T for Males vs Females
             N      Mean     StDev    SE Mean
Males       16      77.37    8.35      2.1
Females     13      71.08    9.22      2.6
99% C.I. for mu males - mu females: ( -2.9, 15.5)
```

Verify the results (the two sample means and standard deviations, and the confidence interval bounds) by calculating the values yourself. Assume normality of blood pressure readings.

10.48 [EX10-048] "Is the length of a steel bar affected by the heat treatment technique used?" This was the question being tested when the following data were collected.

Heat Treatment	Lengths (to the nearest inch)										
1	156	159	151	153	157	159	155	155	151	152	158
	154	156	156	157	155	156	159	153	157	157	159
	158	155	159	152	150	154	156	156	157	160	
2	154	156	150	151	156	155	153	154	149	150	150
	151	154	155	155	154	154	156	150	151	156	154
	153	154	149	150	150	151	154	148	155	158	

a. Find the means and standard deviations for the two sets of data.

b. Find evidence about the sample data (both graphic and numeric) that supports the assumption of normality for the two sampled populations.

c. Find the 95% confidence interval for $\mu_1 - \mu_2$.

10.49 [EX10-049] Approximately 95% of the sunflowers raised in the United States are grown in North Dakota, South Dakota, and Minnesota. To compare yield rates between North and South Dakota, 11 sunflower-producing counties were randomly selected from North Dakota and 14 sunflower-producing counties were randomly selected from South Dakota. Their 2004 yields (in pounds per acre) follow.

N. Dakota

1012	780	711	805	1294	666	779	1099	1314	1358	920

S. Dakota

1101	1347	1520	1025	1462	1650	1903	1652	486	1144	1684
1530	1800	876								

Source: http://www.usda.gov/nass/graphics/county04/data/sf04.csv

Find the 95% confidence interval for the difference between the mean sunflower yield for all South Dakota sunflower-producing counties and all North Dakota sunflower-producing counties. Assume normality.

10.50 [EX10-050] At a large university, a mathematics placement exam is administered to all students. Samples of 36 male and 30 female students

are randomly selected from this year's student body and the following scores recorded:

Male	72	68	75	82	81	60	75	85	80	70
	71	84	68	85	82	80	54	81	86	79
	99	90	68	82	60	63	67	72	77	51
	61	71	81	74	79	76				
Female	81	76	94	89	83	78	85	91	83	83
	84	80	84	88	77	74	63	69	80	82
	89	69	74	97	73	79	55	76	78	81

a. Describe each set of data with a histogram (use the same class intervals on both histograms), the mean, and standard deviation.

b. Construct 95% confidence interval for the mean score for all male students. Do the same for all female students.

c. Do the results found in part b show that the mean scores for males and females could be the same? Justify your answer. Be careful!

d. Construct the 95% confidence interval for the difference between the mean scores for male and female students.

e. Do the results found in part d show the mean scores for male and female students could be the same? Explain.

f. Explain why the results in part b cannot be used to draw conclusions about the difference between the two means.

10.51 State the null and alternative hypotheses that would be used to test the following claims:

a. There is a difference between the mean age of employees at two different large companies.

b. The mean of population 1 is greater than the mean of population 2.

c. The mean yield per county of sunflower seeds in North Dakota is less than the mean yield per county in South Dakota.

d. There is no difference in the mean number of hours spent studying per week between male and female college students.

10.52 State the null and alternative hypotheses that would be used to test the following claims:

a. The difference between the means of the two populations is more than 20 lb.

b. The mean of population A is less than 50 more than the mean of population B.

c. The difference between the two populations is at least $500.

d. The average size yard for neighborhood A is no more than 30 square yards greater than the average yard in neighborhood B.

10.53 Calculate the estimate for the standard error of difference between two independent means for each of the following cases:

a. $s_1^2 = 12$, $s_2^2 = 15$, $n_1 = 16$, and $n_2 = 21$

b. $s_1^2 = 0.054$, $s_2^2 = 0.087$, $n_1 = 8$, and $n_2 = 10$

c. $s_1 = 2.8$, $s_2 = 6.4$, $n_1 = 16$, and $n_2 = 21$

10.54 Find the value of $t\star$ for the difference between two means based on an assumption of normality and this information about two samples:

Sample	Number	Mean	Std. Dev.
1	18	38.2	14.2
2	25	43.1	10.6

10.55 Find the value of $t\star$ for the difference between two means based on an assumption of normality and this information about two samples:

Sample	Number	Mean	Std. Dev.
1	21	1.66	0.29
2	9	1.43	0.18

10.56 Determine the p-value for the following hypothesis tests for the difference between two means with population variances unknown.

a. $H_a: \mu_1 - \mu_2 > 0$, $n_1 = 6$, $n_2 = 10$, $t\star = 1.3$

b. $H_a: \mu_1 - \mu_2 < 0$, $n_1 = 16$, $n_2 = 9$, $t\star = -2.8$

c. $H_a: \mu_1 - \mu_2 \neq 0$, $n_1 = 26$, $n_2 = 16$, $t\star = 1.8$

d. $H_a: \mu_1 - \mu_2 \neq 5$, $n_1 = 26$, $n_2 = 35$, $t\star = -1.8$

10.57 Determine the critical values that would be used for the following hypothesis tests (using the classical approach) about the difference between two means with population variances unknown.

a. $H_a: \mu_1 - \mu_2 \neq 0$, $n_1 = 26$, $n_2 = 16$, $\alpha = 0.05$

b. $H_a: \mu_1 - \mu_2 < 0$, $n_1 = 36$, $n_2 = 27$, $\alpha = 0.01$

c. $H_a: \mu_1 - \mu_2 > 0$, $n_1 = 8$, $n_2 = 11$, $\alpha = 0.10$

d. $H_a: \mu_1 - \mu_2 \neq 10$, $n_1 = 14$, $n_2 = 15$, $\alpha = 0.05$

10.58 For the hypothesis test involving $H_a: \mu_B - \mu_A \neq 0$ with df $= 18$ and $t\star = 1.3$.

a. Find the p-value.

b. Find the critical values given $\alpha = 0.05$.

10.59 Suppose the calculated $t\star$ had been 1.80 in Example 10.10 (pp. 569–570). Using df $= 9$ or using df $= 20$ results in different answers. Explain how the word *conservative* (p. 565) applies here.

10.60 "In a typical month, men spend $178 and women spend $96 on leisure activities," according to the results of an International Communications Research (ICR) for American Express poll, as reported in USA Snapshot found on the Internet June 25, 2005.

Suppose random samples are taken from the population of college male and female students. Each student is asked to determine his or her expenditures for leisure activities in the prior month. The sample data results have a standard deviation of $75 for the men and a standard deviation of $50 for the women.

a. If both samples are of size 20, what is the standard error for the difference of two means?

b. Assuming normality in leisure activity expenditures, is the difference found in the ICR poll significant at $\alpha = 0.05$ if the samples in part a are used? Explain.

10.61 Many cheeses are produced in the shape of a wheel. Because of the differences in consistency between these different types of cheese, the amount of cheese, measured by weight, varies from wheel to wheel. Heidi Cembert wishes to determine whether there is a significant difference, at the 10% level, between the weight per wheel of Gouda and Brie cheese. She randomly samples 16 wheels of Gouda and finds the mean is 1.2 lb with a standard deviation of 0.32 lb; she then randomly samples 14 wheels of Brie and finds a mean of 1.05 lb and a standard deviation of 0.25 lb. What is the *p*-value for Heidi's hypothesis of equality? Assume normality.

10.62 If a random sample of 18 homes south of Center Street in Provo has a mean selling price of $145,200 and a standard deviation of $4700, and a random sample of 18 homes north of Center Street has a mean selling price of $148,600 and a standard deviation of $5800, can you conclude that there is a significant difference between the selling price of homes in these two areas of Provo at the 0.05 level? Assume normality.

a.   Solve using the *p*-value approach.

b.   Solve using the classical approach.

10.63 The computer age has allowed teachers to use electronic tutorials to motivate their students to learn. *Issues in Accounting Education* published the results of a study that showed that an electronic tutorial, along with intentionally induced peer pressure, was effective in enhancing preclass preparations and in improving class attendance, test scores, and course evaluations when used by students studying tax accounting.

Suppose a similar study is conducted at your school using an electronic study guide (ESG) as a tutor for students of accounting principles. For one course section, the students were required to use a new ESG computer program that generated and scored chapter review quizzes and practice exams, presented textbook chapter reviews, and tracked progress. Students could use the computer to build, take, and score their own simulated tests and review materials at their own pace before they took their formal in-class quizzes and exams composed of different questions. The same instructor taught the other course section, used the same textbook, and gave the same daily assignments, but he did not require the students to use the ESG. Identical tests were administered to both sections, and the mean scores of all tests and assignments at the end of the year were tabulated:

Section	*n*	Mean Score	Std. Dev.
ESG (1)	38	79.6	6.9
No ESG (2)	36	72.8	7.6

Do these results show that the mean scores of tests and assignments for students taking accounting principles with an ESG to help them is significantly greater than those not using an ESG? Use a 0.01 level of significance.

a.   Solve using the *p*-value approach.

b.   Solve using the classical approach.

10.64 The purchasing department for a regional supermarket chain is considering two sources from which to purchase 10-lb bags of potatoes. A random sample taken from each source shows the following results.

	Idaho Supers	Idaho Best
Number of Bags Weighed	100	100
Mean Weight	10.2 lb	10.4 lb
Sample Variance	0.36	0.25

At the 0.05 level of significance, is there a difference between the mean weights of the 10-lb bags of potatoes?

a.   Solve using the *p*-value approach.

b.   Solve using the classical approach.

10.65 Lauren, a brunette, was tired of hearing "blondes have more fun." She set out to "prove" that "brunettes are more intelligent." Lauren randomly (as best she could) selected 40 blondes and 40 brunettes at her high school. The following overall grade statistics were calculated:

Blondes	$n_{Bl} = 40$	$\bar{x}_{Bl} = 88.375$	$s_{Bl} = 6.134$
Brunettes	$n_{Br} = 40$	$\bar{x}_{Br} = 87.600$	$s_{Br} = 6.640$

Upon seeing the sample results, does Lauren have support for her claim that "brunettes are more intelligent than blondes"? Explain. What could Lauren say about blondes' and brunettes' intelligence?

10.66 One could reason that high school seniors would seem to have more money issues than high school juniors. Seniors foresee expenses for college as well as their senior trip and prom. So does this mean that they work more than their junior classmates? Christine, a senior at HFL High School, randomly collected the following data (recorded in hours/week) from students that work:

Seniors $n_S = 17$ $\bar{x}_S = 16.4$ $s_S = 10.48$
Juniors $n_J = 20$ $\bar{x}_J = 18.405$ $s_J = 9.69$

Assuming that work hours are normally distributed, do these data suggest that there is a significant difference between the average number of hours that HFL seniors and juniors work per week? Use $\alpha = 0.10$.

10.67 Referring to Applied Example 10.11:

a. The "Variances − Unequal" is equivalent to making no assumption about the variances. Verify the value reported for the estimated standard error, SE of Difference.

b. Verify the *t*-value.

c. What is the range of possible values for df for this study?

d. Explain how df = 17.57 was obtained.

e. What df value would you use based on the material presented in Section 10.4?

f. Explain why the *t*-value of 6.98 is said to be "highly significant."

10.68 A study was conducted to assess the safety and efficiency of receiving nitroglycerin from a transdermal system (i.e., a patch worn on the skin), which intermittently delivers the medication, versus taking oral medication (pills). Twenty patients who suffer from angina (chest pain) due to physical effort were enrolled in trials. All received patches, some ($n = 8$) contained nitroglycerin; the others ($n = 12$) contained a placebo. Suppose the resulting "time to angina" data were summarized:

Mean Time to Angina (sec)

	Active	Placebo	Difference	SE	*p*-Value[a]
Day 1 AM	320.00	287.00	33.00	9.68	0.0029
Day 7 PM	314.00	285.25	28.75	13.74	0.0500

[a]For treatment difference.

a. Determine the value of *t* for the difference between two independent means given the dif-

ference and the standard error (SE) for the day 1 AM data. Assume normality.

b. Verify the *p*-value.

c. Determine the value of *t* for the difference between two independent means given the difference and the standard error (SE) for the day 7 PM data.

d. Verify the *p*-value.

10.69 [EX10-069] MINITAB was used to complete a *t*-test of the difference between the two means using the following two independent samples.

Sample 1	33.7	21.6	32.1	38.2	33.2	35.9	34.1	39.8
	23.5	21.2	23.3	18.9	30.3			
Sample 2	28.0	59.9	22.3	43.3	43.6	24.1	6.9	14.1
	30.2	3.1	13.9	19.7	16.6	13.8	62.1	28.1

```
Two-sample T for sample 1 vs sample 2
              N        Mean      StDev     SE Mean
sample1       13       29.68     7.07      2.0
sample2       16       26.9      17.4      4.4
T-Test mu sample1 = mu sample2 (vs not =): T=0.59
P=0.56 DF=20
```

a. Assuming normality, verify the results (two sample means and standard deviations, and the calculated *t*★) by calculating the values yourself.

b. Use Table 7 in Appendix B to verify the *p*-value based on the calculated df.

c. Find the *p*-value using the smaller number of degrees of freedom. Compare the two *p*-values.

10.70 [EX10-070] According to the College Board (http://www.collegeboard.com/press/article/ 0,3,8993,00.html), the average 2004–2005 cost (tuition, fees, and room and board) for a public college is $11,354 versus $26,057 for a private college. Is there also a difference in the average cost of required textbooks between public and private colleges? The following samples of size 10 were taken.

Public	Private	Public	Private	Public	Private
64.69	71.00	103.59	98.56	110.69	112.58
89.60	96.19	106.38	98.94	118.94	114.00
101.49	96.47	106.77	107.79	135.94	116.55
101.75	97.14				

Using the Excel output that follows and $\alpha = 0.05$, determine whether the average cost of required textbooks per class is different between public and private colleges. Assume normality.

a. Solve using the p-value approach.

b. Solve using the classical approach.

```
t-Test: Two-Sample Assuming Unequal Variances
                        Public         Private
Mean                    103.984        100.922
Variance                340.6249822    173.2995511
Observations            10             10
Hypothesized Mean
  Difference            0
df                      16
t Stat                  0.427125511
P(T≤t) two-tail         0.674980208
  t Critical two-tail   2.119904821
```

10.71 [EX10-071] Twenty laboratory mice were randomly divided into two groups of 10. Each group was fed according to a prescribed diet. At the end of 3 weeks, the weight gained by each animal was recorded. Do the data in the following table justify the conclusion that the mean weight gained on diet B was greater than the mean weight gained on diet A, at the $\alpha = 0.05$ level of significance? Assume normality.

Diet A	5	14	7	9	11	7	13	14	12	8
Diet B	5	21	16	23	4	16	13	19	9	21

a. Solve using the p-value approach.

b. Solve using the classical approach.

10.72 [EX10-072] Many people who are involved with major league baseball (MLB) believe that games played by the New York Yankees tend to last longer than games played by other teams. To test this theory, one other MLB team was picked at random, namely, the St. Louis Cardinals. The time of game (minutes) for seven randomly selected 2005 Cardinals' games and eight randomly selected 2005 Yankees' games were obtained:

Cardinals	Yankees	Cardinals	Yankees
164	199	163	187
160	196	175	169
196	202	190	169
171	213		188

Source: MLB.com

a. Assuming normality, do these samples provide significant evidence to conclude the mean time of all Yankees' baseball games is significantly greater than the mean time of all Cardinals' games? Use $\alpha = 0.05$.

b. What decision will result if $\alpha = 0.01$? Explain the change.

10.73 [EX10-073] Penfield and Perinton are two adjacent eastside suburbs of Rochester, New York. In previous years, they have always been considered on equal ground with respect to quality of life, housing, and education. Many new housing developments are currently going up in Penfield, and it appears that the average home value in Penfield is higher than that in Perinton. To test this theory, random samples of real estate transactions were taken in each suburb during the week of June 12, 2005. Do the data support the theory for this time frame? Use $\alpha = 0.10$ and assume normality.

Penfield	Perinton	Penfield	Perinton	Penfield	Perinton
$164,500	$127,500	$204,000	$300,000		$140,000
$134,900	$189,500		$212,500		$114,000
$295,000	$106,000		$210,000		$106,000
$235,000	$83,900				

10.74 [EX10-074] When evaluating different measuring instruments, one must first determine whether there is a systematic difference between the instruments. Lenses from two different groups (1 and 2) were measured once each by two different instruments. The measurement differences (Instrument A − Instrument B) were recorded. Measurement units have been coded for proprietary reasons.

```
Group 1    4   5 −2 −3 −7  10   11 −1   3   7 −5   3 −4
          −5 −7   4 −1 −18   0 −17  12   9   4  17 −2
Group 2 −13 −12 −5  11  15   7 −33 −10 −6 −2 −16   2
           0 −19   6 −17 −4 −19 −22 −4   8  10 −6
```

Does there appear to be a systematic difference between the two instruments?

a. Describe each set of data separately using a histogram and comparatively using one side-by-side graph.

b. Find the mean and the standard deviation for each set of data.

c. Are the assumptions satisfied? Explain.

d. Test the hypothesis that there is no difference between the means of the two differences. Use $\alpha = 0.05$.

e. Is there any evidence of a difference between the two instruments? Explain.

10.75 Use a computer to demonstrate the truth of the statement describing the sampling distribution of $\bar{x}_1 - \bar{x}_2$. Use two theoretical normal populations: $N_1(100, 20)$ and $N_2(120, 20)$.

a. To get acquainted with the two theoretical populations, randomly select a very large sample from each. Generate 2000 data values, calculate mean and standard deviation, and construct a histogram using class boundaries that are multiples of one-half of a standard deviation (10) starting at the mean for each population.

b. If samples of size eight are randomly selected from each population, what do you expect the distribution of $\bar{x}_1 - \bar{x}_2$ to be like (shape of distribution, mean, standard error)?

c. Randomly draw a sample of size eight from each population, and find the mean of each sample. Find the difference between the sample means. Repeat 99 more times.

d. The set of 100 $(\bar{x}_1 - \bar{x}_2)$ values forms an empirical sampling distribution of $\bar{x}_1 - \bar{x}_2$. Describe the empirical distribution: shape (histogram), mean, and standard error. (Use class boundaries that are multiples of standard error from mean for easy comparison to the expected.)

e. Using the information found in parts a–d, verify the statement about the $\bar{x}_1 - \bar{x}_2$ sampling distribution made on page 564.

f. Repeat the experiment a few times and compare the results.

10.76 One reason for being conservative when determining the number of degrees of freedom to use with the t-distribution is the possibility that the population variances might be unequal. Extremely different values cause a lowering in the number of df used. Repeat Exercise 10.75 using theoretical normal distributions of $N(100, 9)$ and $N(120, 27)$ and both sample sizes of eight. Check all three properties of the sampling distribution: normality, its mean value, and its standard error. Describe in detail what you discover. Do you think we should be concerned about the choice of df? Explain.

10.77 Unbalanced sample sizes is a factor in determining the number of degrees of freedom for inferences about the difference between two means. Repeat Exercise 10.75 using theoretical normal distributions of $N(100, 20)$ and $N(120, 20)$ and sample sizes of 5 and 20. Check all three properties of the sampling distribution: normality, its mean value, and its standard error. Describe in detail what you discover. Do you think we should be concerned when using unbalanced sample sizes? Explain.

10.78 One assumption for the two-sample t-test is the "sampled populations are to be normally distributed." What happens when they are not normally distributed? Repeat Exercise 10.75 using two theoretical populations that are not normal and using samples of size 10. The exponential distribution uses a continuous random variable, it has a J-shaped distribution, and its mean and standard deviation are the same value. Use two exponential distributions with means of 50 and 80: Exp(50) and Exp(80). Check all three properties of the sampling distribution: normality, its mean value, and its standard error. Describe in detail what you discover. Do you think we should be concerned when sampling nonnormal populations? Explain.

FYI See the *Student Solutions Manual* for additional information about commands.

10.5 Inferences Concerning the Difference between Proportions Using Two Independent Samples

We are often interested in making statistical comparisons between the **proportions, percentages,** or **probabilities** associated with two populations. These questions ask for such comparisons: Is the proportion of homeowners who favor a certain tax proposal different from the proportion of renters who favor it? Did a larger percentage of this semester's class than of last semester's class pass statistics? Is the probability of a Democratic candidate winning in New York greater than the probability of a Republican candidate winning in Texas? Do students' opinions about the new code of conduct differ from those of the faculty? You have probably asked similar questions.

FYI The 3 "p" words (*proportion, percentage, probability*) are all the binomial parameter p, P(success).

FYI Binomial experiments are defined in more detail on page 287.

Note: These are the properties of a **binomial experiment:**

1. The observed probability is $p' = x/n$, where x is the number of observed successes in n trials.
2. $q' = 1 - p'$.
3. p is the probability of success on an individual trial in a binomial probability experiment of n repeated independent trials.

In this section, we will compare two population proportions by using the difference between the observed proportions, $p'_1 - p'_2$, of two independent samples. The observed difference, $p'_1 - p'_2$, belongs to a sampling distribution with the characteristics described in the following statement.

If independent samples of sizes n_1 and n_2 are drawn randomly from large populations with $p_1 = P_1$(success) and $p_2 = P_2$(success), respectively, then the sampling distribution of $p'_1 - p'_2$ has these properties:

1. mean $\mu_{p'_1 - p'_2} = p_1 - p_2$,

2. standard error $\sigma_{p'_1 - p'_2} = \sqrt{\left(\dfrac{p_1 q_1}{n_1}\right) + \left(\dfrac{p_2 q_2}{n_2}\right)}$ $\hspace{2em}$ (10.10)

3. an approximately normal distribution if n_1 and n_2 are sufficiently large

---

In practice, we use the following *guidelines to ensure normality:*

1. The sample sizes are both larger than 20.
2. The products $n_1 p_1$, $n_1 q_1$, $n_2 p_2$, and $n_2 q_2$ are all larger than 5.
3. The samples consist of less than 10% of their respective populations.

Note: p_1 and p_2 are unknown; therefore, the products mentioned in guideline 2 will be estimated by $n_1 p'_1$, $n_1 q'_1$, $n_2 p'_2$, and $n_2 q'_2$.

Inferences about the difference between two population proportions, $p_1 - p_2$, will be based on the following assumptions.

Assumptions for inferences about the difference between two proportions $p_1 - p_2$: The n_1 random observations and the n_2 random observations that form the two samples are selected independently from two populations that are not changing during the sampling.

Confidence Interval Procedure

When we estimate the **difference between two proportions,** $p_1 - p_2$, we will base our estimates on the **unbiased sample statistic** $p_1' - p_2'$. The point estimate, $p_1' - p_2'$, becomes the center of the confidence interval and the confidence interval limits are found using the following formula:

Confidence Interval for the Difference between Two Proportions

$$(p_1' - p_2') - z(\alpha/2) \cdot \sqrt{\left(\frac{p_1' q_1'}{n_1}\right) + \left(\frac{p_2' q_2'}{n_2}\right)} \quad \text{to} \quad (p_1' - p_2') + z(\alpha/2) \cdot \sqrt{\left(\frac{p_1' q_1'}{n_1}\right) + \left(\frac{p_2' q_2'}{n_2}\right)}$$

$$(10.11)$$

EXAMPLE 10.12

Constructing a Confidence Interval for the Difference between Two Proportions

Statistics⬥Now™

Watch a video example at
http://1pass.thomson.com
or on your CD.

In studying his campaign plans, Mr. Morris wishes to estimate the difference between men's and women's views regarding his appeal as a candidate. He asks his campaign manager to take two random independent samples and find the 99% confidence interval for the difference between the proportions of women and men voters who plan to vote for Morris. A sample of 1000 voters was taken from each population, with 388 men and 459 women favoring Mr. Morris.

SOLUTION

FYI It is customary to place the larger value first, that way, the point estimate for the difference is a positive value.

Step 1 **Parameter of interest:** $p_w - p_m$, the difference between the proportion of women voters and the proportion of men voters who plan to vote for Mr. Morris.

Step 2 a. **Assumptions:** The samples are randomly and independently selected.

b. **Probability distribution:** The standard normal distribution. The populations are large (all voters); the sample sizes are larger than 20; and the estimated values for $n_m p_m$, $n_m q_m$, $n_w p_w$, and $n_w q_w$ are all larger than 5. Therefore, the sampling distribution of $p_w' - p_m'$ should have an approximately normal distribution. $z\star$ will be calculated using formula (10.11).

c. **Level of confidence:** $1 - \alpha = 0.99$.

Step 3 **Sample information:**
We have $n_m = 1000$, $x_m = 388$, $n_w = 1000$, and $x_w = 459$.

$$p'_m = \frac{x_m}{n_m} = \frac{388}{1000} = \mathbf{0.388} \qquad q'_m = 1 - 0.388 = \mathbf{0.612}$$

$$p'_w = \frac{x_w}{n_w} = \frac{459}{1000} = \mathbf{0.459} \qquad q'_w = 1 - 0.459 = \mathbf{0.541}$$

Step 4 **a.** **Confidence coefficient:** This is a two-tailed situation, with $\alpha/2$ in each tail. From Table 4B, $z_{(\alpha/2)} = z_{(0.005)} = 2.58$. Instructions for using Table 4B are on page 405.

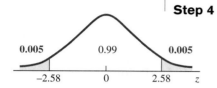

0.005   0.99   0.005

−2.58   0   2.58 z

b. **Maximum error of estimate:** using the maximum error part of formula (10.11), we have

$$E = z_{(\alpha/2)} \cdot \sqrt{\left(\frac{p'_w q'_w}{n_w}\right) + \left(\frac{p'_m q'_m}{n_m}\right)}:$$

$$E = 2.58 \cdot \sqrt{\left(\frac{(0.459)(0.541)}{1000}\right) + \left(\frac{(0.388)(0.612)}{1000}\right)}$$

$$= 2.58\sqrt{0.000248 + 0.000237} = (2.58)(0.022) = \mathbf{0.057}$$

c. **Lower/upper confidence limits:**

$$(p'_w - p'_m) \pm E$$

$$0.071 \pm 0.057$$

$$0.071 - 0.057 = \mathbf{0.014} \quad \text{to} \quad 0.071 + 0.057 = \mathbf{0.128}$$

Step 5 **a.** **Confidence interval:** 0.014 to 0.128 is the 99% confidence interval for $p_w - p_m$. With 99% confidence, we can say that there is a difference of from 1.4% to 12.8% in Mr. Morris's voter appeal.

b. That is, a larger proportion of women than men favor Mr. Morris, and the difference in the proportions is between 1.4% and 12.8%.

Confidence intervals and hypothesis tests can sometimes be interchanged; that is, a confidence interval can be used in place of a hypothesis test. For example, Example 10.12 called for a confidence interval. Now suppose that Mr. Morris asked, "Is there a difference in my voter appeal to men voters as opposed to women voters?" To answer his question, you would not need to complete a hypothesis test if you chose to test at $\alpha = 0.01$ using a two-tailed test. "No difference" would mean a difference of zero, which is not included in the interval from 0.014 to 0.128 (the interval determined in Example 10.12). Therefore, a null hypothesis of "no difference" would be rejected, thereby substantiating the conclusion that a significant difference exists in voter appeal between the two groups.

TECHNOLOGY INSTRUCTIONS: CONFIDENCE INTERVALS FOR THE DIFFERENCE BETWEEN TWO PROPORTIONS GIVEN TWO INDEPENDENT SETS OF SAMPLE DATA

MINITAB (Release 14)

Choose:	**Stat > Basic Statistics > 2 Proportions**
Select:	**Summarized data:**
Enter:	First: **n** (trials) **x** (events)
	Second: **n** (trials) **x** (events)
Select:	**Options**
Enter:	Confidence level: **1 − α** (ex. 0.95 or 95.0)
Select:	Alternative: **not equal > OK > OK**

Excel

Input the data for the first sample into column A using 0s for failures (or no's) and 1s for successes (or yes's); then repeat the same procedure for the second sample in column B; then continue with:

Choose:	**Tools > Data Analysis Plus > Z-Estimate: Two Proportions**
Enter:	Variable 1 Range: **(A2:A20 or select cells)**
	Variable 2 Range: **(B1:B20 or select cells)**
	Code for success: **1**
Select:	**Labels** (if necessary)
Enter:	Alpha: **α** (ex. 0.05) **> OK**

TI-83/84 Plus

Choose:	**STAT >TESTS > B:2-PropZint**

Enter the appropriate values and highlight Calculate.

```
2-PropZInt
 x1:0
 n1:0
 x2:0
 n2:0
 C-Level:.95
 Calculate
```

Hypothesis-Testing Procedure

When the null **hypothesis, there is no difference between two proportions,** is being tested, the **test statistic** will be the difference between the observed proportions divided by the **standard error;** it is found with the following formula:

Test Statistic for the Difference between Two Proportions–Population Proportion Known

$$z\bigstar = \frac{p'_1 - p'_2}{\sqrt{pq\left[\left(\dfrac{1}{n_1}\right) + \left(\dfrac{1}{n_2}\right)\right]}}$$

(10.12)

Notes:

1. The null hypothesis is $p_1 = p_2$ or $p_1 - p_2 = 0$ (the difference is zero).
2. Nonzero differences between proportions are not discussed in this section.

3. The numerator of formula (10.12) could be written as $(p_1' - p_2') - (p_1 - p_2)$, but since the null hypothesis is assumed to be true during the test, $p_1 - p_2 = 0$. By substitution, the numerator becomes simply $p_1' - p_2'$.

4. Since the null hypothesis is $p_1 = p_2$, the standard error of $p_1' - p_2'$, $\sqrt{\left(\dfrac{p_1 q_1}{n_1}\right) + \left(\dfrac{p_2 q_2}{n_2}\right)}$, can be written as $\sqrt{pq\left[\left(\dfrac{1}{n_1}\right) + \left(\dfrac{1}{n_2}\right)\right]}$, where $p = p_1 = p_2$ and $q = 1 - p$.

5. When the null hypothesis states $p_1 = p_2$ and does not specify the value of either p_1 or p_2, the two sets of sample data will be pooled to obtain the estimate for p. This pooled probability (known as p_p') is the total number of successes divided by the total number of observations with the two samples combined; it is found using the next formula:

$$p_p' = \frac{x_1 + x_2}{n_1 + n_2} \tag{10.13}$$

and q_p' is its complement,

$$q_p' = 1 - p_p' \tag{10.14}$$

When the pooled estimate, p_p', is being used, formula (10.12) becomes formula (10.15):

> **Test Statistic for the Difference between Two Proportions–Population Proportion Unknown**
>
> $$z\bigstar = \frac{p_1' - p_2'}{\sqrt{(p_p')(q_p')\left[\left(\dfrac{1}{n_1}\right) + \left(\dfrac{1}{n_2}\right)\right]}} \tag{10.15}$$

EXAMPLE 10.13

One-Tailed Hypothesis Test for the Difference between Two Proportions

Statistics◁▷Now™

Watch a video example at
http://1pass.thomson.com
or on your CD.

A salesperson for a new manufacturer of cellular phones claims not only that they cost the retailer less but also that the percentage of defective cellular phones found among her products will be no higher than the percentage of defectives found in a competitor's line. To test this statement, the retailer took random samples of each manufacturer's product. The sample summaries are given in Table 10.5. Can we reject the salesperson's claim at the 0.05 level of significance?

TABLE 10.5

Cellular Phone Sample Information

Product	Number Defective	Number Checked
Salesperson's	15	150
Competitor's	6	150

SOLUTION

Step 1 **a. Parameter of interest:** $p_s - p_c$, the difference between the proportion of defectives in the salesperson's product and the proportion of defectives in the competitor's product.

  b. Statement of hypotheses: The concern of the retailer is that the salesperson's less expensive product may be of a poorer quality, meaning a greater proportion of defectives. If we use the difference "suspected larger proportion − smaller proportion," then the alternative hypothesis is "The difference is positive (greater than zero)."

$H_o: p_s - p_c = 0$ ($\leq$) (salesperson's defective rate is no higher than competitor's)
$H_a: p_s - p_c > 0$ (salesperson's defective rate is higher than competitor's)

Step 2 **a. Assumptions:** Random samples were selected from the products of two different manufacturers.

  b. The test statistic to be used: The standard normal distribution. Populations are very large (all cellular phones produced); the samples are larger than 20; and the estimated products $n_s p'_s$, $n_s q'_s$, $n_c p'_c$, and $n_c q'_c$ are all larger than 5. Therefore, the sampling distribution should have an approximately normal distribution. $z\star$ will be calculated using formula (10.15).

  c. Level of significance: $\alpha = 0.05$.

Step 3 **a. Sample information:**

$$p'_s = \frac{x_s}{n_s} = \frac{15}{150} = \mathbf{0.10} \qquad\qquad p'_c = \frac{x_c}{n_c} = \frac{6}{150} = \mathbf{0.04}$$

$$p'_p = \frac{x_1 + x_2}{n_1 + n_2} = \frac{15 + 6}{150 + 150} = \frac{21}{300} = \mathbf{0.07} \quad q'_p = 1 - p'_p = 1 - 0.07 = \mathbf{0.93}$$

  b. Calculated test statistic:

$$z\star = \frac{p'_s - p'_c}{\sqrt{(p'_p)(q'_p)\left[\left(\frac{1}{n_s}\right) + \left(\frac{1}{n_c}\right)\right]}} : \quad z\star = \frac{0.10 - 0.04}{\sqrt{(0.07)(0.93)\left[\left(\frac{1}{150}\right) + \left(\frac{1}{150}\right)\right]}}$$

$$= \frac{0.06}{\sqrt{0.000868}} = \frac{0.06}{0.02946} = \mathbf{2.04}$$

Step 4 **Probability Distribution:**

OR

p-value:

a. Use the right-hand tail because H_a expresses concern for values related to "higher than." **P** = p-value = $P(z\star > 2.04)$ as shown in the figure.

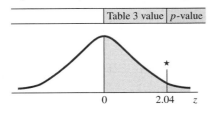

Table 3 value	p-value

Classical:

a. The critical region is the right-hand tail because H_a expresses concern for values related to "higher than." The critical value is obtained from Table 4A: $z(0.05) = \mathbf{1.65}$.

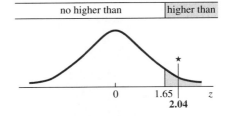

no higher than	higher than

To find the *p*-value, you have three options:
1. Use Table 3 (Appendix B) to calculate the *p*-value: **P = 0.5000 − 0.4793 = 0.0207.**
2. Use Table 5 (Appendix B) to place bounds on the *p*-value: **0.0202 < P < 0.0228.**
3. Use a computer or calculator: **P = 0.0207.**
For specific instructions, see page 432.
b. The *p*-value is smaller than α.

For specific instructions, see pages 451.
b. $z\star$ is in the critical region, as shown in **red** in the figure.

Step 5   **a. Decision:** Reject H_o.

b. Conclusion: At the 0.05 level of significance, there is sufficient evidence to reject the salesperson's claim; the proportion of his company's cellular phones that are defective is higher than the proportion of her competitor's cellular phones that are defective.

TECHNOLOGY INSTRUCTIONS: HYPOTHESIS TEST FOR THE DIFFERENCE BETWEEN TWO PROPORTIONS, $p_1 - p_2$, FOR TWO INDEPENDENT SETS OF SAMPLE DATA

MINITAB (Release 14)

Choose:	**Stat > Basic Statistics > 2 Proportions**
Select:	**Summarized data:**
Enter:	First: **n** (trials) **x** (events)
	Second: **n** (trials) **x** (events)
Select:	**Options**
Enter:	Test difference: **0.0**
Select:	Alternative: **less than** or **not equal** or **greater than**
Select:	**Use pooled estimate of p for test > OK > OK**

Excel

Input the data for the first sample into column A using 0s for failures (or no's) and 1s for successes (or yes's); then repeat the same procedure for the second sample in column B; then continue with:

Choose:	**Tools > Data Analysis Plus > Z-Test: Two Proportions**
Enter:	Variable 1 Range: **(A1:A20 or select cells)**
	Variable 2 Range: **(B1:B20 or select cells)**
	Code for success: **1**
	Hypothesized difference: **0**
Select:	**Labels** (if necessary)
Enter:	Alpha: α (ex. 0.05)

TI-83/84 Plus

Choose:	**STAT > TESTS > 6:2-PropZTest . . .**

Enter the appropriate values and highlight Calculate.

```
2-PropZTest
 x1:0
 n1:0
 x2:0
 n2:0
 p1:≠p2 <p2 >p2
Calculate Draw
```

APPLIED
EXAMPLE 10.14 **Cadaver Kidneys Are Good for Transplants**

In a discovery that could ease the severe shortage of donor organs, Swiss researchers found that kidneys transplanted from cadavers keep working just as long as those from a patient whose heart is still beating. Most transplant organs are taken from brain-dead patients whose hearts have not stopped because doctors have long believed that if they wait until the heart stops, the organs will become damaged from lack of oxygen.

But in the first long-term study comparing the two approaches, doctors at University Hospital Zurich followed nearly 250 transplant patients for up to 15 years and found nearly identical survival rates. At 10 years, 79 percent of patients whose kidney came from a donor with no heartbeat were alive, as were 77 percent of patients whose organ came from a brain-dead donor whose heart was beating. The study, published in Thursday's *New England Journal of Medicine,* could prove especially influential because it was a head-to-head comparison of the two approaches and was the first to follow patients for many years.

Doctors believe similar results may be found for transplants of the liver, pancreas, and lungs. By using organs from "cardiac death" donors, the number of kidneys available could increase up to 30 percent, meaning some 1000 or more extra U.S. donors a year, experts estimate.

Source: Reprinted with permission of The Associated Press.

SECTION 10.5 EXERCISES

10.79 Only 75 of the 250 people interviewed were able to name the vice president of the United States. Find the values for x, n, p', and q'.

10.80 If $n_1 = 40$, $p'_1 = 0.9$, $n_2 = 50$, and $p'_2 = 0.9$:

a. Find the estimated values for both np's and both nq's.

b. Would this situation satisfy the guidelines for approximately normal? Explain.

10.81 Calculate the estimate for the standard error of the difference between two proportions for each of the following cases:

a. $n_1 = 40$, $p'_1 = 0.8$, $n_2 = 50$, and $p'_2 = 0.8$

b. $n_1 = 33$, $p'_1 = 0.6$, $n_2 = 38$, and $p'_2 = 0.65$

10.82 Calculate the maximum error of estimate for a 90% confidence interval for the difference between two proportions for the following cases:

a. $n_1 = 40$, $p'_1 = 0.7$, $n_2 = 44$, and $p'_2 = 0.75$

b. $n_1 = 36$, $p'_1 = 0.33$, $n_2 = 38$, and $p'_2 = 0.42$

10.83 A *Nursing Economics* article titled "Nurse Executive Turnover" compared two groups of nurse executives. One group had participated in a unique program for nurse executives called the Wharton Fellows Program, and the other group had not participated in the program. Of 341 Wharton Fellows, 87 had experienced one change in position; of 40 non-Wharton Fellows, 9 had experienced one change in position. MINITAB was used to construct a 99% confidence interval for the difference in population proportions. Verify the results that follow by calculating them yourself.

```
Test and CI for Two Proportions
Sample   X        N        Sample p
1        87       341      0.255132
2        9        40       0.225000
Difference = p (1) - p (2)
Estimate for difference: 0.0301320
99% CI for difference: (-0.150483, 0.210747)
```

10.84 Find the 95% confidence interval for $p_A - p_B$.

Sample	n	X
A	125	45
B	150	48

10.85 "The game's appeal for teenagers 'is the same as for adults'," an article in *USA Today* (December 21, 2004), reported on a poll of 200 male and 199 female teenagers aged 14 to 17. The findings reported were that 66% of the male teenagers have "ever gambled" and 37% of the female teenagers have "ever gambled." The poll was sponsored by the 2003 Annenberg National Risk Survey of Youth.

Use a 95% confidence interval to estimate the difference in the proportion of male and female teenagers who have ever gambled. Compare your answer to the confidence intervals formed in Exercise 9.74 on page 509. Comment on what these two approaches are telling you. Include a revisit with the comments written in answer to Exercise 9.74 part c.

10.86 In a random sample of 40 brown-haired individuals, 22 indicated that they used hair coloring. In another random sample of 40 blonde individuals, 26 indicated that they used hair coloring. Use a 92% confidence interval to estimate the difference in the population proportions of brunettes and blondes that use hair coloring.

10.87 The proportions of defective parts produced by two machines were compared, and the following data were collected:

Machine 1: $n = 150$; number of defective parts = 12
Machine 2: $n = 150$: number of defective parts = 6

Determine a 90% confidence interval for $p_1 - p_2$.

10.88 In a survey of 300 people from city A, 128 preferred New Spring soap to all other brands of deodorant soap. In city B, 149 of 400 people preferred New Spring soap. Find the 98% confidence interval for the difference in the proportions of people from the two cities who prefer New Spring soap.

10.89 State the null hypothesis, H_o, and the alternative hypothesis, H_a, that would be used to test these claims:

a. There is no difference between the proportions of men and women who will vote for the incumbent in next month's election.

b. The percentage of boys who cut classes is greater than the percentage of girls who cut classes.

c. The percentage of college students who drive old cars is higher than the percentage of non-college people of the same age who drive old cars.

10.90 Show that the standard error of $p_1' - p_2'$, which is $\sqrt{\left(\dfrac{p_1 q_1}{n_1}\right) + \left(\dfrac{p_2 q_2}{n_2}\right)}$, reduces to $\sqrt{pq\left[\left(\dfrac{1}{n_1}\right) + \left(\dfrac{1}{n_2}\right)\right]}$ when $p_1 = p_2 = p$.

10.91 Find the values of p_p' and q_p' for these samples:

Sample	X	n
E	15	250
R	25	275

10.92 Find the value of $z\bigstar$ that would be used to test the difference between the proportions, given the following:

Sample	n	X
G	380	323
H	420	332

10.93 Find the p-value for the test with alternative hypothesis $p_E < p_R$ using the data in Exercise 10.91.

10.94 Determine the p-value that would be used to test the following hypotheses when z is used as the test statistic.

a. $H_o: p_1 = p_2$ vs. $H_a: p_1 > p_2$, with $z\bigstar = 2.47$

b. $H_o: p_A = p_B$ vs. $H_a: p_A \neq p_B$, with $z\bigstar = -1.33$

c. $H_o: p_1 - p_2 = 0$ vs. $H_a: p_1 - p_2 < 0$, with $z\bigstar = -0.85$

d. $H_o: p_m - p_f = 0$ vs. $H_a: p_m - p_f > 0$, with $z\bigstar = 3.04$

10.95 Determine the critical region and critical value(s) that would be used to test (classical procedure) the following hypotheses when z is used as the test statistic.

a. $H_o: p_1 = p_2$ vs. $H_a: p_1 > p_2$, with $\alpha = 0.05$

b. $H_o: p_A = p_B$ vs. $H_a: p_A \neq p_B$, with $\alpha = 0.05$

c. $H_o: p_1 - p_2 = 0$ vs. $H_a: p_1 - p_2 < 0$,   with $\alpha = 0.04$

d. $H_o: p_m - p_f = 0$ vs. $H_a: p_m - p_f > 0$,   with $\alpha = 0.01$

10.96 PC users are often victimized by hardware problems. A study revealed that hardware problems reported to manufacturers could not be fixed by one in three owners of personal computers. Home PC owners fared even worse than those with work PCs, facing longer waits for service and getting even fewer problems resolved. Relatively few owners gave service technicians high marks for having adequate knowledge or for exerting sincere efforts to help solve the problems with the hardware.

Source: PC World, "Which PC Makers Can You Trust?"

Suppose a study is conducted to compare the service provided by manufacturers to both home PC owners and work PC owners. Of 220 home PC owners who had trouble, 98 reported that their problem was not resolved satisfactorily. When the same question was asked of 180 work PC owners who experienced difficulty, 52 reported that the problem was not resolved. Did the home PC owners experience a greater proportion of problems that could not be solved with help from the manufacturer? Use the 0.05 level of significance and the MINITAB output that follows to answer the question.

a. Solve using the *p*-value approach.

b. Solve using the classical approach.

```
Test and CI for Two Proportions
Sample    X        N        Sample p
1         98       220      0.445455
2         52       180      0.288889
Difference = p (1) - p (2)
Estimate for difference: 0.156566
Test for difference = 0 (vs > 0): Z = 3.22
P-Value = 0.001
```

10.97 Two randomly selected groups of citizens were exposed to different media campaigns that dealt with the image of a political candidate. One week later, the citizen groups were surveyed to see whether they would vote for the candidate. The results were as follows:

	Exposed to Conservative Image	Exposed to Moderate Image
Number in Sample	100	100
Proportion for the Candidate	0.40	0.50

Is there sufficient evidence to show a difference in the effectiveness of the two image campaigns at the 0.05 level of significance?

a. Solve using the *p*-value approach.

b. Solve using the classical approach.

10.98 In a survey of families in which both parents work, one of the questions asked was, "Have you refused a job, promotion, or transfer because it would mean less time with your family?" A total of 200 men and 200 women were asked this question. "Yes" was the response given by 29% of the men and 24% of the women. Based on this survey, can we conclude that there is a difference in the proportion of men and women responding "yes" at the 0.05 level of significance?

10.99 The Committee of 200, a professional organization of preeminent women entrepreneurs and corporate leaders, reported the following: 60% of women MBA students say "businesses pay their executives too much money" and 50% of the men MBA students agreed.

a. Does there appear to be a difference in the proportion of women and men who say, "Executives are paid too much"? Explain the meaning of your answer.

b. If the preceding percentages resulted from two samples of size 20 each, is the difference statistically significant at a 0.05 level of significance? Justify your answer.

c. If the preceding percentages resulted from two samples of size 500 each, is the difference statistically significant at a 0.05 level of significance? Justify your answer.

d. Explain how answers to parts b and c affect your thoughts about your answer to part a.

10.100 Both parents and students have many concerns when considering colleges. One of the top three concerns according to the Fall 2004 College Partnership study is "Choosing best major/career." Nineteen percent of the parents reported "Choosing best major/career" as a major concern, whereas 15% of students reported it as a major concern.

Source: http://www.collegepartnership.com/pdf/
Fall%202004%20Study%20Charts%20&%20Analysis%
20_1_.pdf

If the study was conducted with a sample of 1750 students and their parents, test the hypothesis that "Choosing best major/career" is a bigger concern for the parents, at the 0.05 level of significance.

10.101 Forty-one small lots of experimental product were manufactured and tested for the occurrence of a particular indication that is attribute in nature yet causes rejection of the part. Thirty-one lots were made using one particular processing method, and ten lots were made using yet a second processing method. Each lot was equally sampled ($n = 32$) for the presence of this indication. In practice, optimal processing conditions show little or no occurrence of the indication. Method 1, involving the ten lots, was run before Method 2.

Methods	n	Number of Rejects
Method 1	320	4
Method 2	992	26

Courtesy of Bausch & Lomb

Determine, at the 0.05 level of significance, whether there is a difference in the proportion of reject product between the two methods. (Save your answer for comparison with Exercises 11.44 on p. 645.)

10.102 Adverse side effects are always a concern when testing and trying new medicines. Placebo-controlled clinical studies were conducted in patients 12 years of age and older who were receiving "once-a-day" doses of Allegra, a seasonal allergy drug. The following results were published in the April 2005 edition of *Readers' Digest*.

	Allegra (dose once a day)	Placebo (dose once a day)
Side effects	$n = 283$	$n = 293$
Number reporting headaches	30	22

Determine, at the 0.05 level of significance, whether there is a difference in the proportion of patients reporting headaches between the two groups.

10.103 A 2005 Harris Interactive for Korbel poll found that 63% of men and 55% of women think that it is okay for women to make marriage proposals to men. An 8% difference may or may not be statistically significant. What size sample is required to make this difference significant?

Source: USA Snapshot found on Internet, June 25, 2005

a. If the preceding sample statistics had resulted from a sample of 250 men and a sample of 250 women, would the difference be significant, using $\alpha = 0.05$? Explain.

b. If the samples had each been of size 500, would the difference be significant, using $\alpha = 0.05$? Explain.

c. Determine the size sample that would have the difference of 0.08 corresponding to $p = 0.05$.

10.104 The guidelines to ensure the sampling distribution of $p_1' - p_2'$ is normal include several conditions about the size of several values. The two binomial distributions $B(100, 0.3)$ and $B(100, 0.4)$ satisfy all of those guidelines.

a. Verify that $B(100, 0.3)$ and $B(100, 0.4)$ satisfy all guidelines.

b. Use a computer to randomly generate 200 random samples from each of the binomial populations. Find the observed proportion for each sample and the value of the 200 differences between two proportions.

c. Describe the observed sampling distribution using both graphic and numerical statistics.

d. Does the empirical sampling distribution appear to have an approximately normal distribution? Explain

FYI See the *Student Solutions Manual* for additional information about commands.

10.6 Inferences Concerning the Ratio of Variances Using Two Independent Samples

When comparing two populations, we naturally compare their two most fundamental distribution characteristics, their "center" and their "spread," by comparing their means and standard deviations. We have learned, in two of the previous sections, how to use the *t*-distribution to make inferences comparing two population means with either dependent or independent samples. These procedures were intended to be used with normal populations, but they work quite well even when the populations are not exactly normally distributed.

The next logical step in comparing two populations is to compare their standard deviations, the most often used measure of spread. However, sampling distributions that deal with sample standard deviations (or variances) are very sensitive to slight departures from the assumptions. Therefore, the only inference procedure to be presented here will be the **hypothesis test for the equality of standard deviations (or variances)** for two normal populations.

The soft-drink bottling company discussed in Section 9.4 (pp. 516, 520) is trying to decide whether to install a modern, high-speed bottling machine. There are, of course, many concerns in making this decision, and one of them is that the increased speed may result in increased variability in the amount of fill placed in each bottle; such an increase would not be acceptable. To this concern, the manufacturer of the new system responded that the variance in fills will be no greater with the new machine than with the old. (The new system will fill several bottles in the same amount of time as the old system fills one bottle; this is the reason the change is being considered.) A test is set up to statistically test the bottling company's concern, "Standard deviation of new machine is greater than standard deviation of old," against the manufacturer's claim, "Standard deviation of new is no greater than standard deviation of old."

EXAMPLE 10.15

Writing Hypotheses for the Equality of Variances

State the null and alternative hypotheses to be used for comparing the variances of the two soft-drink bottling machines.

SOLUTION There are several equivalent ways to express the null and alternative hypotheses, but because the test procedure uses the ratio of variances, the recommended convention is to express the null and alternative hypotheses as ratios of the population variances. Furthermore, it is recommended that the "larger" or "expected to be larger" variance be the numerator. The concern of the soft-drink company is that the new modern machine (*m*) will result in a larger standard deviation in the amounts of fill than its present machine (p); $\sigma_m > \sigma_p$ or equivalently $\sigma_m^2 > \sigma_p^2$, which becomes $\dfrac{\sigma_m^2}{\sigma_p^2} > 1$.

We want to test the manufacturer's claim (the null hypothesis) against the company's concern (the alternative hypothesis):

$$H_o: \frac{\sigma_m^2}{\sigma_p^2} = 1 \quad (m \text{ is no more variable})$$

$$H_a: \frac{\sigma_m^2}{\sigma_p^2} > 1 \quad (m \text{ is more variable})$$

Inferences about the ratio of variances for two normally distributed populations use the **F-distribution.** The *F*-distribution, similar to Student's *t*-distribution and the χ^2-distribution, is a family of probability distributions. Each *F*-distribution is identified by two numbers of degrees of freedom, one for each of the two samples involved.

Before continuing with the details of the hypothesis-testing procedure, let's learn about the *F*-distribution.

Properties of the *F*-distribution:

1. *F* is nonnegative; it is zero or positive.
2. *F* is nonsymmetrical; it is skewed to the right.
3. *F* is distributed so as to form a family of distributions; there is a separate distribution for each pair of numbers of degrees of freedom.

For inferences discussed in this section, the number of degrees of freedom for each sample is $df_1 = n_1 - 1$ and $df_2 = n_2 - 1$. Each different combination of degrees of freedom results in a different *F*-distribution, and each *F*-distribution looks approximately like the distribution shown in Figure 10.2.

FIGURE 10.2
F-Distribution

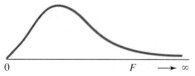

$0 \qquad\qquad F \longrightarrow \infty$

The critical values for the *F*-distribution are identified using three values:

FYI Explore Skillbuilder Applet "Properties of *F*-distribution" on your CD.

df_n, the degrees of freedom associated with the sample whose variance is in the numerator of the calculated *F*

df_d, the degrees of freedom associated with the sample whose variance is in the denominator

α, the area under the distribution curve to the right of the critical value being sought

Therefore, the symbolic name for a critical value of *F* will be $F(df_n, df_d, \alpha)$, as shown in Figure 10.3 (see p. 594).

Because it takes three values to identify a single critical value of *F*, making tables for *F* is not as simple as with previously studied distributions. The tables presented in this textbook are organized so as to have a different table for each different value of α, the "area to the right." Table 9A in Appendix B shows the

FIGURE 10.3

A Critical Value of *F*

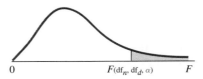

$0 \qquad\qquad F_{(df_n,\, df_d,\, \alpha)} \qquad F$

critical values for $F_{(df_n,\, df_d,\, \alpha)}$, when $\alpha = 0.05$; Table 9B gives the critical values when $\alpha = 0.025$; Table 9C gives the values when $\alpha = 0.01$.

EXAMPLE 10.16

Finding Critical *F*-Values

Find $F_{(5,\, 8,\, 0.05)}$, the critical *F*-value for samples of size 6 and size 9 with 5% of the area in the right-hand tail.

SOLUTION Using Table 9A ($\alpha = 0.05$), find the intersection of column df = 5 (for the numerator) and row df = 8 (for the denominator) and read the value: $F_{(5,\, 8,\, 0.05)} = $ **3.69.** See the accompanying partial table.

Portion of Table 9A ($\alpha = 0.05$)

			df for Numerator				
		...	5	...	8	...	
df		⋮					
for	5				**4.82** ←	$F_{(8,\, 5,\, 0.05)} = $ **4.82**	
Denom-		⋮					
inator	8		**3.69** ←			$F_{(5,\, 8,\, 0.05)} = $ **3.69**	
		⋮					

FYI Explore Skillbuilder Applet "*F*-distribution Probabilities" on your CD.

Notice that $F_{(8,\, 5,\, 0.05)}$ is **4.82.** The degrees of freedom associated with the numerator and with the denominator must be kept in the correct order; 3.69 is different from 4.82. Check some other pairs to verify that interchanging the degrees of freedom numbers will result in different *F*-values.

TECHNOLOGY INSTRUCTIONS: CUMULATIVE PROBABILITY ASSOCIATED WITH A SPECIFIED VALUE OF *F*

MINITAB (Release 14)

```
Choose:     Calc > Probability Distributions > F
Select:     Cumulative Probability Noncentrality parameter: 0.0
Enter:      Numerator degrees of freedom: dfₙ
            Denominator degrees of freedom: df_d
Select:     Input constant*
Enter:      F-value (ex. 1.74) > OK
```

*Select the Input column if several *F*-values are stored in C1. Use C2 for optional storage. If the area in the right tail is needed, subtract the calculated probability from one.

Excel If several *F*-values are to be used, input the values into column A and activate B1; then continue with:

Choose: **Insert function fₓ > Statistical > FDIST > OK**
Enter: **X: individual F-value or (A1:A5 or select 'F-value' cells)***
             Deg_freedom 1: df_n
             Deg_freedom 2: df_d > OK
*Drag: **Bottom right corner of the B1 cell down to give other proba-**
             bilities

To find the probability for the left tail (the cumulative probability up to the *F*-value), subtract the calculated probability from one.

TI-83/84 Plus Choose: **2nd > DISTR > 9:Fcdf(**
Enter: **0, F-value, df_n, df_d)**

Note: To find the probability between two *F*-values, enter the two values in place of 0 and the *F*-value.

If the area in the right tail is needed, subtract the calculated probability from one.

Use of the *F*-distribution has a condition.

> **Assumptions for inferences about the ratio of two variances:** The samples are randomly selected from normally distributed populations, and the two samples are selected in an independent manner.

> **Test Statistic for Equality of Variances**
>
> $$F\star = \frac{s_m^2}{s_p^2}, \qquad \text{with } df_m = n_m - 1 \text{ and } df_p = n_p - 1 \qquad (10.16)$$

The sample variances are assigned to the numerator and denominator in the order established by the null and alternative hypotheses for one-tailed tests. The calculated ratio, $F\star$, will have an *F*-distribution with $df_n = n_n - 1$ (numerator) and $df_d = n_d - 1$ (denominator) when the assumptions are met and the null hypothesis is true.

We are ready to use *F* to complete a hypothesis test about the ratio of two population variances.

EXAMPLE 10.17

One-Tailed Hypothesis Test for the Equality of Variances

Recall that our soft-drink bottling company was to make a decision about the equality of the variances of amounts of fill between its present machine and a modern high-speed outfit. Does the sample information in Table 10.6 (p. 596) present sufficient evidence to reject the null hypothesis (the manufacturer's claim) that the modern high-speed bottle-filling machine fills bottles with no greater variance than the company's present machine? Assume the amounts of fill are normally distributed for both machines, and complete the test using $\alpha = 0.01$.

TABLE 10.6

Sample Information on Variances of Fills

Sample	n	s^2
Present machine (p)	22	0.0008
Modern high-speed machine (m)	25	0.0018

SOLUTION

Step 1   **a.** **Parameter of interest:** $\dfrac{\sigma_m^2}{\sigma_p^2}$, the ratio of the variances in the amounts of fill placed in bottles for the modern machine versus the company's present machine.

b. **Statement of hypotheses:** The hypotheses were established in Example 10.15 (pp. 592–593):

$$H_o: \frac{\sigma_m^2}{\sigma_p^2} = 1 \;\; (\le) \;\; (m \text{ is no more variable})$$

$$H_a: \frac{\sigma_m^2}{\sigma_p^2} > 1 \qquad (m \text{ is more variable})$$

Note: When the "expected to be larger" variance is in the numerator for a one-tailed test, the alternative hypothesis states "The ratio of the variances is greater than one."

Step 2   **a.** **Assumptions:** The sampled populations are normally distributed (given in the statement of the problem), and the samples are independently selected (drawn from two separate populations).

b. **Test statistic:** The F-distribution with the ratio of the sample variances and formula (10.16):

c. **Level of significance:** $\alpha = 0.01$.

Step 3   **a.** **Sample information:** See Table 10.6.

b. **Calculated test statistic:**
Using formula (10.16), we have

$$F\star = \frac{s_m^2}{s_p^2}: \qquad F\star = \frac{0.0018}{0.0008} = \mathbf{2.25}$$

The number of degrees of freedom for the numerator is $\text{df}_n = 24$ (or $25 - 1$) because the sample from the modern high-speed machine is associated with the numerator, as specified by the null hypothesis. Also, $\text{df}_d = 21$ because the sample associated with the denominator has size 22.

Step 4   **Probability Distribution:**

OR

p-value:

a. Use the right-hand tail because H_a expresses concern for values related to "more than." **P** $= P(F\star > 2.25$, with $\text{df}_n = 24$ and $\text{df}_d = 21$) as shown in the figure.

Classical:

a. The critical region is the right-hand tail becaue H_a expresses concern for values related to "more than." $\text{df}_n = 24$ and $\text{df}_d = 21$. The critical value is obtained from Table 9C: $F(24, 21, 0.01) = \mathbf{2.80}$.

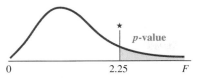

To find the *p*-value, you have two options:
1. Use Tables 9A and 9B (Appendix B) to place bounds on the *p*-value: **0.025 < P < 0.05.**
2. Use a computer or calculator to find the *p*-value: **P = 0.0323.**

Specific instructions follow this illustration.

b. The *p*-value is not smaller than the level of significance, α (0.01).

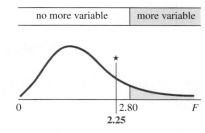

For additional instructions, see page 594.

b. $F\star$ is not in the critical region, as shown in **red** in the figure.

Step 5   **a.**   **Decision:** Fail to reject H_o.

b.   **Conclusion:** At the 0.01 level of significance, the samples do not present sufficient evidence to indicate an increase in variance with the new machine.

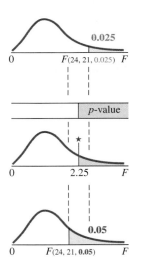

FYI α must still be split between the two tails for a two-tailed H_a.

Calculating the *p*-value when using the *F*-distribution

Method 1: Use Table 9 in Appendix B to place bounds on the p-value. Using Tables 9A, 9B, and 9C in Appendix B to estimate the *p*-value is very limited. However, for Example 10.17, the *p*-value can be estimated. By inspecting Tables 9A and 9B, you will find that $F_{(24,21,0.025)} = 2.37$ and $F_{(24,21,0.05)} = 2.05$. $F\star = 2.25$ is between the values 2.37 and 2.05; therefore, the *p*-value is between 0.025 and 0.05: **0.025 < P < 0.05.** (See figure in margin.)

Method 2: If you are doing the hypothesis test with the aid of a computer or calculator, most likely it will calculate the *p*-value for you, or you may use the cumulative probability distribution commands described on pages 594–595.

Critical *F*-Values for One- and Two-Tailed Tests

The tables of critical values for the *F*-distribution give only the right-hand critical values. This will not be a problem because the right-hand critical value is the only critical value that will be needed. You can adjust the numerator–denominator order so that all the "activity" is in the right-hand tail. There are two cases: one-tailed tests and two-tailed tests.

One-tailed tests: Arrange the null and alternative hypotheses so that the alternative is always "greater than." The $F\star$-value is calculated using the same order as specified in the null hypothesis (as in Example 10.17; also see Example 10.18).

Two-tailed tests: When the value of $F\star$ is calculated, always use the sample with the larger variance for the numerator; this will make $F\star$ greater than one and place it in the right-hand tail of the distribution. Thus, you will need only the critical value for the right-hand tail (see Example 10.19).

All hypothesis tests about two variances can be formulated and completed in a way that both the critical value of F and calculated value of $F\star$ will be in the right-hand tail of the distribution. Since Tables 9A, 9B, and 9C contain only critical values for the right-hand tail, this will be convenient and you will never need critical values for the left-hand tail. The following two examples will demonstrate how this is accomplished.

EXAMPLE 10.18 Format for Writing Hypotheses for the Equality of Variances

Reorganize the alternative hypothesis so that the critical region will be the right-hand tail:

$$H_a: \sigma_1^2 < \sigma_2^2 \quad \text{or} \quad \frac{\sigma_1^2}{\sigma_2^2} < 1 \quad \text{(population 1 is less variable)}$$

SOLUTION Reverse the direction of the inequality, and reverse the roles of the numerator and denominator.

$$H_a: \sigma_2^2 > \sigma_1^2 \quad \text{or} \quad \frac{\sigma_2^2}{\sigma_1^2} > 1 \quad \text{(population 2 is more variable)}$$

The calculated test statistic $F\bigstar$ will be $\dfrac{s_2^2}{s_1^2}$.

EXAMPLE 10.19 Two-Tailed Hypothesis Test for the Equality of Variances

Statistics ⬙ Now™

Watch a video example at
http://1pass.thomson.com
or on your CD.

Find $F\bigstar$ and the critical values for the following hypothesis test so that only the right-hand critical value is needed. Use $\alpha = 0.05$ and the sample information $n_1 = 10$, $n_2 = 8$, $s_1 = 5.4$, and $s_2 = 3.8$.

$$H_o: \sigma_2^2 = \sigma_1^2 \quad \text{or} \quad \frac{\sigma_2^2}{\sigma_1^2} = 1$$

$$H_a: \sigma_2^2 \neq \sigma_1^2 \quad \text{or} \quad \frac{\sigma_2^2}{\sigma_1^2} \neq 1$$

SOLUTION When the alternative hypothesis is two-tailed ($\neq$), the calculated $F\bigstar$ can be either $F\bigstar = \dfrac{s_1^2}{s_2^2}$ or $F\bigstar = \dfrac{s_2^2}{s_1^2}$. The choice is ours; we only need to make sure that we keep df_n and df_d in the correct order. We make the choice by looking at the sample information and using the sample with the larger standard deviation or variance as the numerator. Therefore, in this illustration,

$$F\bigstar = \frac{s_1^2}{s_2^2} = \frac{5.4^2}{3.8^2} = \frac{29.16}{14.44} = \textbf{2.02}$$

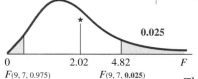

0.025

| 0 | 2.02 | 4.82 | F |

$F(9, 7, 0.975)$ $F(9, 7, 0.025)$

The critical values for this test are left tail, $F(9, 7, 0.975)$, and right tail, $F(9, 7, 0.025)$, as shown in the figure.

Since we chose the sample with the larger standard deviation (or variance) for the numerator, the value of $F\bigstar$ will be greater than 1 and will be in the right-hand tail; therefore, only the right-hand critical value is needed. (All critical values for left-hand tails will be values between 0 and 1.)

TECHNOLOGY INSTRUCTIONS: HYPOTHESIS TEST FOR THE RATIO BETWEEN TWO POPULATION VARIANCES, σ_1^2/σ_2^2, FOR TWO INDEPENDENT SETS OF SAMPLE DATA

MINITAB (Release 14)

Choose:	**Stat > Basic Statistics > 2 Variances**
Select:	**Samples in one column:**
Enter:	Samples: **C1** Subscripts: **C2**
Or	Select: **Samples in different columns:**
	Enter: First: **C1** Second: **C2**
Or	Select: **Summarized data**
	Enter: Sample size and Variance for each sample:
Select:	Storage: **Standard Deviations > OK > OK**

*The 2 Variances procedure evaluates the first sample divided by the second sample.

Excel

Input the data for the numerator (larger spread) into column A and the data for the denominator (smaller spread) into column B; then continue with:

Choose:	**Tools > Data Analysis > F-Test: Two-Sample for Variances**
Enter:	Variable 1 Range: **(A1:A20 or select cells)**
	Variable 2 Range: **(B1:B20 or select cells)**
Select:	**Labels** (if necessary)
Enter:	α (ex. 0.05)
Select:	**Output Range**
Enter:	**(C1 or select cell) > OK**

Use Format > Column > AutoFit Selection to make the output more readable. The output shows the *p*-value and the critical value for a one-tailed test.

TI-83/84 Plus

Input the data for the numerator (larger spread) into L1 and the data for the denominator (smaller spread) into L2; then continue with the following, entering the appropriate values and highlighting Calculate:

Choose:	**STAT > TESTS > D:2-SampFTest . . .**

```
2-SampFTest
 Inpt:DATA Stats
 List1:L1
 List2:L2
 Freq1:1
 Freq2:1
 σ1:≠σ2 <σ2 >σ2
 Calculate Draw
```

APPLIED EXAMPLE 10.20

Personality Characteristics of Police Academy Applicants

Bruce N. Carpenter and Susan M. Raza concluded that "police applicants are somewhat more like each other than are those in the normative population" when the *F*-test of homogeneity of variance resulted in a *p*-value of less than 0.005. *Homogeneity* means that the group's scores are less variable than the scores for the normative population.

COMPARISONS ACROSS SUBGROUPS AND WITH OTHER POPULATIONS

To determine whether police applicants are a more homogeneous group than the normative population, the *F*-test of homogeneity of variance was used. With the exception of scales *F*, *K*, and 6, where the differences are nonsignificant, the results indicate that the police appli-cants form a somewhat more homogeneous group than the normative population [$F(237, 305) = 1.36$, $p < 0.005$]. Thus, police applicants are somewhat more like each other than are individuals in the normative population.

Source: Reproduced from the *Journal of Police Science and Administration*, Vol. 15, no. 1, pp. 10–17, with permission of the International Association of Chiefs of Police, PO Box 6010, 13 Firstfield Road, Gaithersburg, MD 20878.

SECTION 10.6 EXERCISES

Statistics⬯Now™

Datasets can be found on your Student's Suite CD-ROM or at the StatisticsNow website at **http://1pass.thomson.com**.

10.105 State the null hypothesis, H_o, and the alternative hypothesis, H_a, that would be used to test the following claims:

a. The variances of populations A and B are not equal.

b. The standard deviation of population I is larger than the standard deviation of population II.

c. The ratio of the variances for populations A and B is different from 1.

d. The variability within population C is less than the variability within population D.

10.106 State the null hypothesis, H_o, and the alternative hypothesis, H_a, that would be used to test the following claims:

a. The standard deviation of population X is smaller than the standard deviation of population Y.

b. The ratio of the variances of population A over population B is greater than 1.

c. The standard deviation of population Q_1 is at most that of population Q_2.

d. The variability within population I is more than the variability within population II.

10.107 Explain why the inequality $\sigma_m^2 > \sigma_p^2$ is equivalent to $\dfrac{\sigma_m^2}{\sigma_p^2} > 1$.

10.108 Express the H_o and H_a of Example 10.18 (p. 598) equivalently in terms of standard deviations.

10.109 Using the $F_{(df_1, df_2, \alpha)}$ notation, name each of the critical values shown on the following figures.

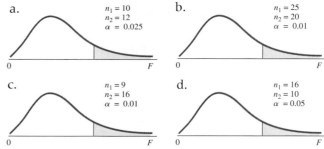

a. $n_1 = 10$
     $n_2 = 12$
     $\alpha = 0.025$

b. $n_1 = 25$
     $n_2 = 20$
     $\alpha = 0.01$

c. $n_1 = 9$
     $n_2 = 16$
     $\alpha = 0.01$

d. $n_1 = 16$
     $n_2 = 10$
     $\alpha = 0.05$

10.110 Find the values of $F(12, 24, 0.01)$ and $F(24, 12, 0.01)$.

10.111 Find the following critical values for F from Tables 9A, 9B, and 9C in Appendix B.

a. $F(24, 12, 0.05)$ b. $F(30, 40, 0.01)$

c. $F(12, 10, 0.05)$ d. $F(5, 20, 0.01)$

e. $F(15, 18, 0.025)$ f. $F(15, 9, 0.025)$

g. $F(40, 30, 0.05)$ h. $F(8, 40, 0.01)$

10.112 Determine the *p*-value that would be used to test the following hypotheses when *F* is used as the test statistic:

a. $H_o: \sigma_1 = \sigma_2$ vs. $H_a: \sigma_1 > \sigma_2$,
   with $n_1 = 10$, $n_2 = 16$, and $F\star = 2.47$

b. $H_o: \sigma_1^2 = \sigma_2^2$ vs. $H_a: \sigma_1^2 > \sigma_2^2$,
   with $n_1 = 25$, $n_2 = 21$, and $F\star = 2.31$

c. $H_o: \dfrac{\sigma_1^2}{\sigma_2^2} = 1$ vs. $H_a: \dfrac{\sigma_1^2}{\sigma_2^2} \neq 1$,

   with $n_1 = 41$, $n_2 = 61$, and $F\star = 4.78$

d. $H_o: \sigma_1 = \sigma_2$ vs. $H_a: \sigma_1 < \sigma_2$,
   with $n_1 = 10$, $n_2 = 16$, and $F\star = 2.47$

10.113 Find the critical value for the hypothesis test with $H_a: \sigma_1 > \sigma_2$, with $n_1 = 7$, $n_2 = 10$, and $\alpha = 0.05$.

10.114 Determine the critical region and critical value(s) that would be used to test the following hypotheses using the classical approach when $F\star$ is used as the test statistic.

a. $H_o: \sigma_1^2 = \sigma_2^2$ vs. $H_a: \sigma_1^2 > \sigma_2^2$,
   with $n_1 = 10$, $n_2 = 16$, and $\alpha = 0.05$

b. $H_o: \dfrac{\sigma_1^2}{\sigma_2^2} = 1$ vs. $H_a: \dfrac{\sigma_1^2}{\sigma_2^2} \neq 1$,

   with $n_1 = 25$, $n_2 = 31$, and $\alpha = 0.05$

c. $H_o: \dfrac{\sigma_1^2}{\sigma_2^2} = 1$ vs. $H_a: \dfrac{\sigma_1^2}{\sigma_2^2} > 1$,

   with $n_1 = 10$, $n_2 = 10$, $\alpha = 0.01$

d. $H_o: \sigma_1 = \sigma_2$ vs. $H_a: \sigma_1 < \sigma_2$,
   with $n_1 = 25$, $n_2 = 16$, and $\alpha = 0.01$

10.115 Calculate $F\star$ given $s_1 = 3.2$ and $s_2 = 2.6$.

10.116 Calculate $F\star$ given $s_1^2 = 3.2$ and $s_2^2 = 2.6$.

10.117 What would be the value of $F\star$ in Example 10.20 if $F\star = \dfrac{s_2^2}{s_1^2}$ were used? Why is it less than 1?

10.118 a. Two independent samples, each of size 3, are drawn from a normally distributed population. Find the probability that one of the sample variances is at least 19 times larger than the other one.

b. Two independent samples, each of size 6, are drawn from a normally distributed population. Find the probability that one of the sample variances is no more than 11 times larger than the other one.

10.119 A bakery is considering buying one of two gas ovens. The bakery requires that the temperature remain constant during a baking operation. A study was conducted to measure the variance in temperature of the ovens during the baking process. The variance in temperature before the thermostat restarted the flame for the Monarch oven was 2.4 for 16 measurements. The variance for the Kraft oven was 3.2 for 12 measurements. Does this information provide sufficient reason to conclude that there is a difference in the variances for the two ovens? Assume measurements are normally distributed and use a 0.02 level of significance.

10.120 A study in *Pediatric Emergency Care* compared the injury severity between younger and older children. One measure reported was the Injury Severity Score (ISS). The standard deviation of ISS scores for 37 children 8 years or younger was 23.9, and the standard deviation for 36 children older than 8 years was 6.8. Assume that ISS scores are normally distributed for both age groups. At the 0.01 level of significance, is there sufficient reason to conclude that the standard deviation of ISS scores for younger children is larger than the standard deviation of ISS scores for older children?

10.121 Sucrose, ordinary table sugar, is probably the single most abundant pure organic chemical in the world and the one most widely known to nonchemists. Whether from sugar cane (20% by weight) or sugar beets (15% by weight), and whether raw or refined, common sugar is still sucrose. Fifteen U.S. sugar beet–producing counties were randomly selected, and their 2001 sucrose percentages gave a standard deviation of 0.862. Similarly, twelve U.S. sugar cane–producing counties were randomly selected, and their 2001 sucrose percentages recorded a standard deviation of 0.912. At the 0.05 level, is there significant difference between the standard deviations for sugar beet and sugar cane sucrose percentages?

10.122 [EX10-122] A study was conducted to determine whether or not there was equal variability in male and female systolic blood pressure readings. Random samples of 16 men and 13 women were used to test the experimenter's claim that the variances were unequal. MINITAB was used to calculate the standard deviations, $F\star$, and the p-value. Assume normality.

Men	120	120	118	112	120	114	130	114	124	125
	130	100	120	108	112	122				
Women	122	102	118	126	108	130	104	116	102	122
	120	118	130							

```
Standard deviation of Men = 7.8864
Standard deviation of Women = 9.9176
F-Test (normal distribution)
Test Statistic: 1.581
P-Value: 0.398
```

Verify these results by calculating the values yourself.

10.123 When a hypothesis test is two-tailed and Excel is used to calculate the p-value, what additional step must be taken?

10.124 Referring to Applied Example 10.20 (p. 599):

a. What null and alternative hypotheses did Carpenter and Raza test?

b. What does "$p < 0.005$" mean?

c. Use a computer or calculator to calculate the p-value for $F_{(237, 305)} = 1.36$.

10.125 [EX10-125] The quality of the end product is somewhat determined by the quality of the materials used. Textile mills monitor the tensile strength of the fibers used in weaving their yard goods. The following independent random samples

are tensile strengths of cotton fibers from two suppliers.

Supplier A	78	82	85	83	77	84	90	82	93	82
	80	82	77	80	80					
Supplier B	76	79	83	78	72	73	69	80	74	77
	78	78	73	76	78	79				

Calculate the observed value of F, $F\star$, for comparing the variances of these two sets of data.

10.126 [EX10-126] Several counties in Minnesota and Wisconsin were randomly selected, and information about the 2001 sweet corn crop was collected. The following yield rates in tons of sweet corn per acre harvested resulted.

Yield, MN 5.90 6.50 6.20 5.90 6.20 5.70 5.91 5.90 6.00 5.60 6.51 6.30

Yield, WI 7.80 6.80 7.00 5.30 6.50 6.90 6.60 5.70 6.60 6.00 6.20

Source: http://www.usda.gov/nass/graphics/county01/data/vsc01.csv

a. Is there a difference in the variability of county yields as measured by standard deviation of yield rate? Use $\alpha = 0.05$.

b. Assuming normality in yield rates, is the mean yield rate in Wisconsin significantly higher than the mean yield rate in Minnesota? Use $\alpha = 0.05$.

10.127 [EX10-127] A constant objective in the manufacture of contact lenses is to improve the level and variation for those features that affect lens power and visual acuity. One such feature involves the tooling from which lenses are ultimately manufactured.

The results of two initial process development runs were examined for Critical Feature A. Two distinct product lots were manufactured with slight differences designed to affect the feature in question. Each lot was then sampled. Lot 1 had a smaller run time than Lot 2, and therefore more samples were taken for Lot 2.

a. Calculate the mean and standard deviation of Critical Feature A for Lots 1 and 2.

b. Is there evidence of a difference in the variability of Critical Feature A between Lots 1 and 2? Use an alpha value of 5% to make a determination.

c. Is there evidence of a difference in the mean levels of Critical Feature A between Lots 1 and

2. Use an alpha value of 5% to make a determination.

Lot 1 Sample	Critical Feature A	Lot 2 Sample	Critical Feature A	Lot 2 Sample	Critical Feature A
1	0.017	1	0.026	14	0.041
2	0.021	2	0.027	15	0.021
3	0.006	3	0.024	16	0.022
4	0.009	4	0.023	17	0.027
5	0.018	5	0.034	18	0.032
6	0.021	6	0.035	19	0.023
7	0.013	7	0.035	20	0.023
8	0.017	8	0.033	21	0.024
		9	0.034	22	0.017
		10	0.033	23	0.023
		11	0.032	24	0.019
		12	0.038	25	0.027
		13	0.041		

10.128 [EX10-128] Americans snooze on the weekends, according to a poll of 1506 adults for the National Sleep Foundation and reported in a USA Snapshot during April 2005.

Hours of Sleep	Weekdays	Weekends
Less than 6	0.16	0.10
6–6.9	0.24	0.15
7–7.9	0.31	0.24
8 or more	0.26	0.49

Two independent random samples were taken at a large industrial complex. The workers selected in one sample were asked, "How many hours, to the nearest 15 minutes, did you sleep on Tuesday night this week?" The workers selected for the second sample were asked, "How many hours, to nearest 15 minutes, did you sleep on Saturday night last weekend?"

Weekday			Weekend			
5.00	7.75	7.25	9.00	7.25	8.75	7.50
9.25	7.25	8.75	6.25	5.25	9.25	9.25
7.00	7.75	6.75	7.50	8.50	8.75	6.50
9.25	7.00	7.75	8.00	8.75	9.50	8.00
9.25	9.25	6.00	8.75	7.75	8.75	7.50

a. Construct a histogram and find the mean and standard deviation for each set of data.

b. Do the distributions of "hours of sleep on weekday" and "hours of sleep on weekend" resulting from the poll appear to be similar in shape? center? spread? Discuss your responses.

c. Is it possible that both of the samples were drawn from normal populations? Justify your answer.

d. Is the mean number of hours slept on the weekend statistically greater than the mean number of hours slept on the weekday? Use $\alpha = 0.05$.

e. Is there sufficient evidence to show that the standard deviation of these two samples are statistically different? Use $\alpha = 0.05$.

f. Explain how the answers to parts b–e now affect your thoughts about your answer to part a.

10.129 [EX10-129] What dollar amount should someone spend on a Valentine's Day gift for you? The results found by a Greenfield Online survey of 653 respondents was pictured in the accompanying graphic.

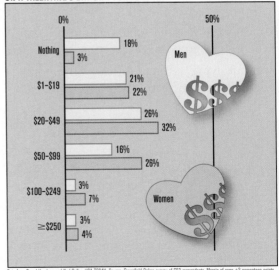

WHAT DOLLAR AMOUNT SHOULD SOMEONE SPEND ON A VALENTINE'S DAY GIFT FOR YOU?

Data from Darryl Haralson and Karl Gelles, USA TODAY; Source: Greenfield Online survey of 653 respondents. Margin of error ±3 percentage points.

Random samples were selected in central New York State with the following results.

Men	103	100	100	67	77	63	55	43	139	2	51	5	100	52	139
	86	23	56	40	84	15	32	157	35	4	24	102	52	43	75
	128	206	16	13	98										
Women	36	5	77	97	25	62	91	170	108	112	198	161	54	40	111
	107	241	89	37	10	175	10	84	102	17	32	25	1	38	126
	121	30	147	135	45	230	29	88							

a. Construct a histogram and find the mean and standard deviation for each set of central New York State data.

b. Do the shapes of "amounts spent" suggested by males and females in central New York State appear to be similar in shape? center? spread? Discuss your responses.

c. Is it possible that both of the samples were drawn from normal populations? Justify your answer.

d. Is the mean amount stated by women statistically greater than the mean amount stated by men? Use $\alpha = 0.05$.

e. Is there sufficient evidence to show that the standard deviations of these two samples are statistically different? Use $\alpha = 0.05$.

f. Explain how your answers to parts b–e now affect your thoughts about your answer to part a.

10.130 Use a computer to demonstrate the truth of the theory presented in this section.

a. The underlying assumptions are "the populations are normally distributed," and while conducting a hypothesis test for the equality of two standard deviations, it is assumed that the standard deviations are equal. Generate very large samples of two theoretical populations: $N(100, 20)$ and $N(120, 20)$. Find graphic and numerical evidence that the populations satisfy the assumptions.

b. Randomly select 100 samples, each of size eight, from both populations and find the standard deviation of each sample.

c. Using the first sample drawn from each population as a pair, calculate the $F\star$-statistic. Repeat for all samples. Describe the sampling distribution of the 100 $F\star$-values using both graphic and numerical statistics.

d. Generate the probability distribution for $F_{(7, 7)}$, and compare it with the observed distribution of $F\star$. Do the two graphs agree? Explain.

FYI See the *Student Solutions Manual* for additional information about commands.

10.131 It was stated in this section that the F-test is very sensitive to minor departures from the assumptions. Repeat Exercise 10.130 using $N(100, 20)$ and $N(120, 30)$. Notice that the only change from Exercise 10.130 is the seemingly slight increase in the standard deviation of the second population. Answer the same questions using the same kind of information and you will see very different results.

CHAPTER REVIEW

In Retrospect

In this chapter we began the comparisons of two populations by distinguishing between independent and dependent samples, which are statistically important and useful sampling procedures. We then proceeded to examine the inferences concerning the comparison of means, proportions, and variances for two populations.

We are always making comparisons between two groups. We compare means and we compare proportions. In this chapter we have learned how to statistically compare two populations by making inferences about their means, proportions, or variances. For convenience, Table 10.7 identifies the formulas to use when making in-

ferences about comparisons between two populations.

In Chapters 8, 9, and 10 we have learned how
to use confidence intervals and hypothesis tests to
answer questions about means, proportions, and

standard deviations for one or two populations.
From here we can expand our techniques to include inferences about more than two populations
as well as inferences of different types.

Formulas to Use for Inferences Involving Two Populations

Situations	Test Statistic	Formula to Be Used	
		Confidence Interval	Hypothesis Test
Difference between two means			
Dependent samples	t	Formula (10.2) (p. 552)	Formula (10.5) (p. 555)
Independent samples	t	Formula (10.8) (p. 566)	Formula (10.9) (p. 567)
Difference between two proportions	z	Formula (10.11) (p. 582)	Formula (10.15) (p. 585)
Difference between two variances	F		Formula (10.16) (p. 595)

Vocabulary and Key Concepts

assumptions (pp. 551, 565, 582, 595)
binomial experiment (p. 581)
binomial p (p. 581)
confidence interval (pp. 552, 566, 582)
dependent means (p. 551)
dependent samples (pp. 547, 550)
F-distribution (p. 593)
F-statistic (p. 595)

hypothesis test (pp. 554, 567, 584, 592)
independent means (p. 564)
independent samples (pp. 548, 564, 581, 592)
mean difference (pp. 552, 554)
paired difference (p. 550)
percentage (p. 581)
pooled observed probability (p. 585)
probability (p. 581)

proportion (p. 581)
p-value (pp. 556, 569, 586, 596)
source (of data) (p. 547)
standard error (pp. 555, 564, 581)
t-distribution (pp. 552, 565)
test statistic (pp. 555, 567, 584, 595)
t-statistic (pp. 555, 567)
z-statistic (p. 584)

Learning Outcomes

✓ Understand the difference between dependent and independent samples.　　EXP 10.1, 10.2, Ex. 10.5, 10.7

✓ Understand that the mean difference (mean of the paired differences) should be used to analyze dependent samples.　　pp. 550–551

✓ Compute and/or understand how to compute the mean difference and standard deviation for paired data.　　pp. 551–552, Ex. 10.13

✓ Compute, describe, and interpret a confidence interval for the population mean difference.　　EXP 10.4, Ex. 10.19

✓ Perform, describe, and interpret a hypothesis test for the population mean difference, μ_d, using the p-value approach and classical approach.　　EXP 10.5, 10.6, Ex. 10.35, 10.138

✓ Understand that the difference between two means should be used to analyze independent samples.　　pp. 564–565

✓ Understand how to determine the *t*-distribution's degrees of freedom for the difference between means using two independent samples.

p. 565, Ex. 10.14

✓ Compute, describe, and interpret a confidence interval for the difference between two means using independent samples.

EXP 10.8, Ex. 10.43, 10.45

✓ Perform, describe, and interpret a hypothesis test for the difference between two population means, $\mu_1 - \mu_2$, using the *p*-value approach and classical approach.

EXP 10.9, 10.10, Ex. 10.62, 10.71

✓ Understand that the *z*-distribution will be used to analyze the difference between two proportions using independent samples, provided the guidelines to ensure normality are met.

pp. 581–582

✓ Compute sample proportions based on sample size and number of successes.

Ex. 10.79

✓ Compute, describe, and interpret a confidence interval for the difference between two proportions using independent samples.

EXP 10.12, Ex. 10.87, 10.153

✓ Perform, describe, and interpret a hypothesis test for the difference between two population proportions, $p_1 - p_2$, using the *p*-value approach and classical approach.

EXP 10.13, Ex. 10.101, 10.154

✓ Understand the properties of the *F*-distribution and how it is a series of distributions based on sample sizes (using pairs of numbers for degrees of freedom as the index).

pp. 593–594, EXP 10.16, 10.18, 10.19, Ex. 10.110

✓ Understand that the assumption for inferences about the ratio of two variances is that the sampled populations are normally distributed and the two samples are selected independently.

p. 595

✓ Perform, describe, and interpret a hypothesis test for the ratio of two population variances, $\dfrac{\sigma_1^2}{\sigma_2^2}$, using the *F*-distribution with the *p*-value approach and classical approach.

EXP 10.17, Ex. 10.119, 10.159

Chapter Exercises

Statistics ⊘ Now™

Go to the StatisticsNow website **http://1pass.thomson.com** to

• Assess your understanding of this chapter

• Check your readiness for an exam by taking the Pre-Test quiz and exploring the resources in the Personalized Learning Plan

Datasets can be found on your Student's Suite CD-ROM or at the StatisticsNow website at **http://1pass.thomson.com**.

10.132 A chemist is testing a newly proposed analytical method and decides to use the currently accepted method for comparison. She takes 12 specimens of unknown concentrate and determines the concentration of each specimen, using both the proposed method and the current method. Do these two samples represent dependent or independent samples? Explain.

10.133 [EX10-133] Using a 95% confidence interval, estimate the mean difference in IQ between the oldest and the youngest members (brothers and sisters) of a family based on the following random sample of IQs. Assume normality.

Oldest	145	133	116	128	85	100	105	150	97	110	120	130
Youngest	131	119	103	93	108	100	111	130	135	113	108	125

10.134 [EX10-134] The diastolic blood pressure readings for 15 patients were determined using two techniques: the standard method used by medical personnel and a method using an electronic device with a digital readout. The results were as follows:

Patient	1	2	3	4	5	6	7	8	9	10	11	12	13	14	15
Standard method	72	80	88	80	80	75	92	77	80	65	69	96	77	75	60
Digital method	70	76	87	77	81	75	90	75	82	64	72	95	80	70	61

Assuming blood pressure is normally distributed, determine the 90% confidence interval for the mean difference in the two readings, where d = standard method − digital readout.

10.135 [EX10-135] We want to know which of two types of filters should be used. A test was designed in which the strength of a signal could be varied from zero to the point where the operator first detects the image. At this point, the intensity setting is recorded. Lower settings are better. Twenty operators were asked to make one reading for each filter.

Operator	1	2	3	4	5	6	7	8	9	10
Filter1	96	83	97	93	99	95	97	91	100	92
Filter2	92	84	92	90	93	91	92	90	93	90

Operator	11	12	13	14	15	16	17	18	19	20
Filter1	88	89	85	94	90	92	91	78	77	93
Filter2	88	89	86	91	89	90	90	80	80	90

Assuming the intensity readings are normally distributed, estimate the mean difference between the two readings using a 90% confidence interval.

10.136 [EX10-136] At the end of their first day at training camp, 10 new recruits participated in a rifle-shooting competition. The same 10 competed again at the end of a full week of training and practice. Their resulting scores are shown in the following table.

	Recruit									
Time of Competition	1	2	3	4	5	6	7	8	9	10
First day	72	29	62	60	68	59	61	73	38	48
One week later	75	43	63	63	61	72	73	82	47	43

Does this set of 10 pairs of data show that there was a significant amount of improvement in the recruits' shooting abilities during the week? Use $\alpha = 0.05$ and assume normality.

10.137 [EX10-137] The amount of general anesthetic a patient should receive prior to surgery has received considerable public attention. According to the American Society of Anesthesiologists, every year, about 40,000 (some researchers have put the figure closer to 200,000) of the 28 million patients who undergo general anesthesia experience limited awareness during surgery because of resistance to the medication or from too little dosage. Patients commonly report overhearing doctors conversing with nurses and assistants during operations.

Source: People, "Wake-Up Call"

Suppose a study is conducted using 20 patients who are having eye surgery performed on both eyes, with 2 weeks separating the treatments on each eye. Ten of the patients are given a lighter dose of general anesthetic prior to surgery on the first eye, and the other ten patients are given a heavier dose. The following week, the procedure is reversed. Two days after each surgery is performed, the patients are asked to rate the amount of pain and discomfort they experienced on a scale from 0 (none) to 10 (unbearable). Results follow:

Subject	Light Dosage	Heavy Dosage	Subject	Light Dosage	Heavy Dosage
1	4	3	11	6	7
2	6	5	12	7	5
3	5	6	13	10	7
4	8	4	14	3	2
5	4	5	15	1	0
6	9	6	16	5	6
7	3	2	17	6	3
8	7	8	18	8	5
9	8	5	19	4	2
10	9	7	20	2	0

Can you conclude that the heavier dose of anesthetic resulted in the patients experiencing lower pain and discomfort after the eye surgery? Use the 0.01 level of significance.

a. Solve using the p-value approach.

b. Solve using the classical approach.

10.138 [EX10-138] Immediate-release medications quickly deliver their drug content, with the maximum concentration reached in a short time; sustained-release medications, on the other hand, take longer to reach maximum concentration. As part of a study, immediate-release codeine (irc) was compared with sustained-release codeine (src) using 13 healthy patients. The patients were randomly assigned to one of the two types of codeine and treated for 2.5 days; after a 7-day wash-out period, each patient was given the other type of codeine. Thus, each patient received both types. The total amount (A) of drug available over the life of the treatment in (ng · mL)/hr follow:

Patient	1	2	3	4	5	6	7
Airc	1091.3	1064.5	1281.1	1921.4	1649.9	1423.6	1308.4
Asrc	1308.5	1494.2	1382.2	1978.3	2004.6	*	1211.1

Patient	8	9	10	11	12	13
Airc	1192.1	766.2	978.6	1618.9	582.9	972.1
Asrc	1002.4	866.6	1345.8	979.2	576.3	999.1

Source: http://exploringdata.cqu.edu.au/ws_coedn.htm

a. Explain why this is a paired-difference design.

b. What adjustment is needed since there is no Asrc for patient 6?

Is there is a significant difference in the total amount of drug available over the life of the treatment?

c. Check the test assumptions and describe your findings.

d. Test the claim using $\alpha = 0.05$.

10.139 A test that measures math anxiety was given to 50 male and 50 female students. The results were as follows:

  Males: $\bar{x} = 70.5$, $s = 13.2$
  Females: $\bar{x} = 75.7$, $s = 13.6$

Construct a 95% confidence interval for the difference between the mean anxiety scores.

10.140 The same achievement test is given to soldiers selected at random from two units. The scores they attained are summarized as follows:

  Unit 1: $n_1 = 70$, $\bar{x}_1 = 73.2$, $s_1 = 6.1$
  Unit 2: $n_2 = 60$, $\bar{x}_2 = 70.5$, $s_2 = 5.5$

Construct a 90% confidence interval for the difference in the mean level of the two units.

10.141 [EX10-141] Ten soldiers were selected at random from each of two companies to participate in a rifle-shooting competition. Their scores are shown in the following table.

Company A	72	29	62	60	68	59	61	73	38	48
Company B	75	43	63	63	61	72	73	82	47	43

Construct a 95% confidence interval for the difference between the mean scores for the two companies. Assume normality.

10.142 [EX10-142] An achievement test in a beginning computer science course was administered to two groups. One group had a previous computer science course in high school; the other group did not. The test results follow. Assuming test scores are normal, construct a 98% confidence interval for the difference between the two population means.

Group 1 (had high school course)	17	18	27	19	24	36	27	26	35	22	18
	29	29	26	33							

Group 2 (no high school course)	19	25	28	27	21	24	18	14	28	21	22
	20	21	14	29	28	25	17	20	28	31	27

10.143 [EX10-143] Two methods were used to study the latent heat of ice fusion. Both method A (an electrical method) and method B (a method of mixtures) were conducted with the specimens cooled to $-0.72°C$. The data in the following table represent the change in total heat from $0.72°C$ to water at $0°C$ in calories per gram of mass.

Method A	79.98	80.04	80.02	80.04	80.03	80.03	80.04
	79.97	80.05	80.03	80.02	80.00	80.02	
Method B	80.02	79.94	79.98	79.97	79.97	80.03	79.95
	79.97						

Assuming normality, construct a 95% confidence interval for the difference between the means.

10.144 [EX10-144] Sucrose, ordinary table sugar, is probably the single most abundant pure organic chemical in the world and the one most widely known to nonchemists. Whether from sugar cane (20% by weight) or sugar beets (15% by weight), and whether raw or refined, common sugar is still sucrose. Fifteen U.S. sugar beet–producing counties were randomly selected and their 2001 sucrose percentages recorded. Similarly, twelve U.S. sugar cane–producing counties were randomly selected and their 2001 sucrose percentages recorded.

SBsucrose	17.30	16.46	16.20	17.53	17.00	18.53	16.77	16.11
	15.30	17.90	15.98	17.30	17.94	17.30	16.60	
SCsucrose	14.1	13.5	15.2	15.0	13.6	13.6	11.7	14.3
	13.8	13.8	14.8	13.7				

Source: http://www.usda.gov/nass/graphics/county01/data/

Find the 95% confidence interval for the difference between the mean sucrose percentage for all U.S. sugar beet–producing counties and all U.S. sugar cane–producing counties. Assume normality.

10.145 George Johnson is the head coach of a college football team that trains and competes at home on artificial turf. George is concerned that the 40-yard sprint time recorded by his players and others increases substantially when running on natural turf as opposed to artificial turf. If so, there is little comparison between his players' speed and those of his opponents whenever his team plays on grass. George's next opponent plays on grass, so he surveyed all the starters in the next game and obtained their best 40-yard sprint times. He then compared them with the best times turned in by his own players. The results are shown in the table that follows:

Player Group	n	Mean (sec)	Std. Dev.
Artificial turf	22	4.85	0.31
Grass	22	4.96	0.42

Do Coach Johnson's players have a lower mean sprint time? Assuming normality, test at the 0.05 level of significance to advise Coach Johnson.

a. Solve using the *p*-value approach.

b. Solve using the classical approach.

10.146 A test concerning some of the fundamental facts about acquired immunodeficiency syndrome (AIDS) was administered to two groups, one consisting of college graduates and the other consisting of high school graduates. A summary of the test results follows:

College graduates: $n = 75, \bar{x} = 77.5, s = 6.2$
High school graduates: $n = 75, \bar{x} = 50.4, s = 9.4$

Do these data show that the college graduates, on average, score significantly higher on the test? Use $\alpha = 0.05$.

10.147 Some 20 million Americans visit chiropractors annually, and the number of practitioners in the United States is 55,000, nearly double the number two decades ago, according to the American Chiropractic Association. The *New England Journal of Medicine* released a report showing the results of a study that compared chiropractic spinal manipulation (CSM) with physical therapy for treatment of acute lower back pain. After 2 years of treatment, CSM was found to be no more effective at either reducing missed work or preventing a relapse.

Suppose a similar study of 60 patients is made by dividing the sample into two groups. For 1 year, one group is given CSM and the other, physical therapy. During the 1-year period, the number of missed days at work as a result of lower back pain is measured:

Group	n	Mean	Std. Dev.
CSM (1)	32	10.6	4.8
Therapy (2)	28	12.5	6.3

Do these results show that the mean number of missed days of work for people suffering from acute back pain is significantly less for those receiving CSM than for those undergoing physical therapy? Assume normality and use a 0.01 level of significance.

a. Solve using the *p*-value approach.

b. Solve using the classical approach.

10.148 To compare the merits of two short-range rockets, 8 of the first kind and 10 of the second kind are fired at a target. If the first kind has a mean target error of 36 feet and a standard deviation of 15 feet and the second kind has a mean target error of 52 feet and a standard deviation of 18 feet, does this indicate that the second kind of rocket is less accurate than the first? Use $\alpha = 0.01$ and assume normal distribution for target error.

10.149 [EX10-149] The material used in making parts affects not only how long the part lasts but also how difficult it is to repair. The following measurements are for screw torque removal for a specific screw after several operations of use. The first row lists the part number, the second row lists the screw torque removal measurements for assemblies made with material A, and the third row lists the screw torque removal measurements for assemblies made with material B. Assume torque measurements are normally distributed.

Removal Torque (NM, Newton-meters)

Part Number	1	2	3	4	5	6	7	8	9	10	11	12	13	14	15
Material A	16	14	13	17	18	15	17	16	14	16	15	17	14	16	15
Material B	11	14	13	13	10	15	14	12	11	14	13	12	11	13	12

Source: Problem data provided by AC Rochester Division, General Motors, Rochester, NY

a. Find the sample mean, variance, and standard deviation for the material A data.

b. Find the sample mean, variance, and standard deviation for the material B data.

c. At the 0.01 level, do these data show a significant difference in the mean torque required to remove the screws from the two different materials?

10.150 [EX10-150] A group of 17 students participated in an evaluation of a special training session that claimed to improve memory. The students were randomly assigned to two groups: group A, the test group, and group B, the control group. All 17 students were tested for the ability to remember certain material. Group A was given the special training; group B was not. After 1 month, both groups were tested again, with the results as shown in the following table. Do these data support the alternative hypothesis that the special training is effective at the $\alpha = 0.01$ level of significance? Assume normality.

Time of Test	Group A Students									Group B Students							
	1	2	3	4	5	6	7	8	9	10	11	12	13	14	15	16	17
Before	23	22	20	21	23	18	17	20	23	22	20	23	17	21	19	20	20
After	28	29	26	23	31	25	22	26	26	23	25	26	18	21	17	18	20

10.151 [EX10-151] At a large university, a mathematics placement exam is administered to all students. This exam has a history of producing scores with a mean of 77. Samples of 36 male students and 30 females students are randomly selected from this year's student body and the following scores recorded.

Male	72	68	75	82	81	60	75	85	80	70
	71	84	68	85	82	80	54	81	86	79
	99	90	68	82	60	63	67	72	77	51
	61	71	81	74	79	76				
Female	81	76	94	89	83	78	85	91	83	83
	84	80	84	88	77	74	63	69	80	82
	89	69	74	97	73	79	55	76	78	81

a. Describe each set of data with a histogram (use the same class intervals on both histograms), mean, and standard deviation.

b. Test the hypotheses, "Mean score for all males is 77" and "Mean score for all females is 77," using $\alpha = 0.05$.

c. Do the preceding results show that the mean scores for males and females are the same? Justify your answer. Be careful!

d. Test the hypothesis, "There is no difference between the mean scores for male and female students," using $\alpha = 0.05$.

e. Do the results found in part d show that the mean scores for male and females are the same? Explain.

f. Explain why the results found in part b cannot be used to conclude, "The two means are the same."

10.152 A survey was conducted to determine the proportion of Democrats as well as Republicans who support a "get tough" policy in South America. The results of the survey were as follows:

Democrats: $n = 250$, number in support = 120
Republicans: $n = 200$, number in support = 105

Construct the 98% confidence interval for the difference between the proportions of support.

10.153 A consumer group compared the reliability of two comparable microcomputers from two different manufacturers. The proportion requiring service within the first year after purchase was determined for samples from each of two manufacturers.

Manufacturer	Sample Size	Proportion Needing Service
1	75	0.15
2	75	0.09

Find a 0.95 confidence interval for $p_1 - p_2$.

10.154 In determining the "goodness" of a test question, a teacher will often compare the percentage of better students who answer it correctly with the percentage of poorer students who answer it correctly. One expects that the proportion of better students who will answer the question correctly is greater than the proportion of poorer students who will answer it correctly. On the last test, 35 of the students with the top 60 grades and 27 of the students with the bottom 60 grades answered a certain question correctly. Did the students with the top grades do significantly better on this question? Use $\alpha = 0.05$.

a. Solve using the *p*-value approach.

b. Solve using the classical approach.

10.155 According to "Venus vs. Mars" in the May/June 2005 issue of *Arthritis Today*, men and women may be more alike than we think. The Boomers Wellness Lifestyle Survey of men and women aged 35 to 65 found that 88% of women considered "managing stress" important for maintaining overall well-being. The same survey found that 75% of men consider "managing stress" important. Answer the following and give details to support each of your answers.

a. If these statistics came from samples of 100 men and 100 women, is the difference significant?

b. If these statistics came from samples of 150 men and 150 women, is the difference significant?

c. If these statistics came from samples of 200 men and 200 women, is the difference significant?

d. What effects did the increase in sample size have on the solutions in parts a–c?

10.156 A study in the *New England Journal of Medicine* reported that based on 987 deaths in southern California, right-handers died at an average age of 75 and left-handers died at an average age of 66. In addition, it was found that 7.9% of the lefties died from accident-related injuries, excluding vehicles, versus 1.5% for the right-handers; and 5.3% of the left-handers died while driving vehicles versus 1.4% of the right-handers.

Suppose you examine 1000 randomly selected death certificates, of which 100 were left-handers and 900 were right-handers. If you found that 5 of the left-handers and 18 of the right-handers died while driving a vehicle, would you have evidence to show that the proportion of left-handers who die at the wheel is significantly higher than the proportion of right-handers who die while driving? Calculate the *p*-value and interpret its meaning.

10.157 Who wins disputed cases whenever there is a change made to the tax laws, the taxpayer or the Internal Revenue Service (IRS)? The latest trend indicates that the burden of proof in all court cases has shifted from the taxpayers to the IRS, which tax experts predict could set off more intrusive questioning. Of the accountants, lawyers, and other tax professionals surveyed by RIA Group, a tax-information publisher, 55% expect at least a slight increase in taxpayer wins.

Source: Fortune, "Tax Reform?"

Suppose samples of 175 accountants and 165 lawyers are asked, "Do you expect taxpayers to win more court cases because of the new burden of proof rules?" Of those surveyed, 101 accountants replied "yes" and 84 lawyers said "yes." Do the two expert groups differ in their opinions? Use a 0.01 level of significance to answer the question.

a. Solve using the *p*-value approach.

b. Solve using the classical approach.

10.158 "It's a draw," according to two Australian researchers. "By age 25, up to 29% of all men and up to 34% of all women have some gray hair, but this difference is so small that it's considered insignificant."

Source: "Silver Threads Among the Gold: Who'll Find Them First, a Man or a Woman?" *Family Circle*

If 1000 men and 1000 women were involved in this research, would the 5% difference mentioned be significant at the 0.01 level? Explain, include details to support your answer.

10.159 A manufacturer designed an experiment to compare the difference between men and women with respect to the times they require to assemble a product. A total of 15 men and 15 women were tested to determine the time they required, on average, to assemble the product. The time required by the men had a standard deviation of 4.5 minutes, and the time required by the women had a standard deviation of 2.8 minutes. Do these data show that the amount of time needed by men is more variable than the amount of time needed by women? Use $\alpha = 0.05$ and assume the times are approximately normally distributed.

a. Solve using the p-value approach.

b. Solve using the classical approach.

10.160 A soft-drink distributor is considering two new models of dispensing machines. Both the Harvard Company machine and the Fizzit machine can be adjusted to fill the cups to a certain mean amount. However, the variation in the amount dispensed from cup to cup is a primary concern. Ten cups dispensed from the Harvard machine showed a variance of 0.065, whereas 15 cups dispensed from the Fizzit machine showed a variance of 0.033. The factory representative from the Harvard Company maintains that his machine had no more variability than the Fizzit machine. Assume the amount dispensed is normally distributed.

At the 0.05 level of significance, does the sample refute the representative's assertion?

a. Solve using the p-value approach.

b. Solve using the classical approach.

10.161 Mindy Fernandez is in charge of production at the new sport utility vehicle (SUV) assembly plant that just opened in her town. Lately she has been concerned that the wheel lug bolts do not match the chrome lug nuts close enough to keep the assembly of the wheels operating smoothly. Workers are complaining that cross-threading is happening so often that threads are being stripped by the air wrenches and that torque settings also have to be adjusted downward to prevent stripped threads even if the parts match up. In an effort to determine whether the fault lies with the lug nuts

or the studs, Mindy has decided to ask the quality-control department to test a random sample of 60 lug nuts and 40 studs to see if the variances in threads are the same for both parts. The report from the technician indicated that the thread variance of the sampled lug nuts was 0.00213 and that the thread variance for the sampled studs was 0.00166. What can Mindy conclude about the equality of the variances at the 0.05 level of significance?

a. Solve using the p-value approach.

b. Solve using the classical approach.

10.162 [EX10-162] Random samples of counties in North Dakota and South Dakota were selected from the USDA-NASS website for the purpose of estimating the difference between the mean 2001 yield rates for oat production for the two states.

Yield, ND	66.0	73.8	51.9	61.3	67.4	54.0	71.4	58.0	56.2	40.0
	66.7	64.4	75.7	74.4						
Yield, SD	65.0	62.5	30.0	90.0	70.0	62.7	42.4	47.1	62.2	76.2
	56.1	50.0	65.6	45.5	79.2	59.3				

Source: http://www.usda.gov/nass/graphics/county01/data/ot01.csv

a. Are the assumptions satisfied? Explain.

b. Is there sufficient evidence to reject the hypothesis of equal variances for the oat production yield rates for these two states? Use $\alpha = 0.05$.

c. Is there sufficient evidence to reject the hypothesis that there is no difference between the mean oat production yield rates for these two states? Use $\alpha = 0.05$.

FYI When using the two-sample t-test, select "assume equal variances" according to result (b).

10.163 [EX10-163] A research project was undertaken to evaluate the amount of force needed to elicit a designated response on equipment made from two distinct designs: the existing design and an improved design. The expectation was that new design equipment would require less force than the current equipment. Fifty units of each design were tested and the required force recorded. A lower force level and reduced variability are both considered desirable.

Control or Existing	Test or New Design
0.003562	0.002477
0.005216	0.002725

••• Remainder of data on Student's Suite CD-ROM

a. Describe both sets of data using means, standard deviations, and histograms.

b. Check the assumptions for comparing the variances and means of two independent samples. Describe your findings.

c. Will one- or two-tailed tests be appropriate for testing the expectations for the new design? Why?

d. Is there significant evidence to show that the new design has reduced the variability in the required force? Use $\alpha = 0.05$.

e. Is there significant evidence to show that the new design has reduced the mean amount of force? Use $\alpha = 0.05$.

f. Did the new design live up to expectations? Explain.

Chapter Project

Students, Credit Cards, and Debt

As a way of assessing the statistical techniques for two populations that we have learned in this chapter, let's return to Section 10.1, "Students, Credit Cards, and Debt" (p. 545). Credit card companies are notorious for enticing college students to sign up for new credit cards. The convenience of the credit cards along with money-handling inexperience lead to huge debts for many college students. Is there a significant difference between freshmen and sophomore college students with respect to credit cards and debt? Let's investigate.

Putting Chapter 10 to Work

10.164 [EX10-001] How do credit card debts for freshmen and sophomores compare? Using the two sets of sample data on page 546:

a. Find the proportion of each sample that have at least one credit card.

b. Find the point estimate for the difference between the two proportions.

c. Are the assumptions for making inferences about the difference between two proportions satisfied? Explain.

d. Find the 95% confidence interval for the difference between proportion of sophomore and freshmen students having their own credit cards.

e. Find the mean credit card debt for the freshmen and sophomores in the samples.

f. Draw dotplots for the amount of debt for both groups using a common scale. Interpret what the dotplots are showing you, including shape, center, and spread.

g. Find the point estimate for difference between two means.

h. Check the assumptions for normality for both sets of credit card debt. Verify.

i. What effect does your answer in part h have on answering the question: "Is the difference between the mean credit care debt for sophomores and freshmen in part g significantly greater than the $292 difference reported by Nellie Mae (pages 545–546)?

j. Based on your finding in parts a–i, compare and contrast the spending habits of freshman and sophomore college students.

Your Study

10.165 Design your own study involving two populations.

a. Determine a set of questions that compare the means, proportions, or variances of two populations of interest to you. You might consider two different class levels as the two populations and questions similar to the following: Is there is a difference between their mean cost of books and supplies for a semester? Is there is a difference between the proportion of those with credit cards of their own? Is there is a difference between the proportion of those with four or more credit cards of their own?

b. Define two specific populations that you will sample, describe your sampling plan, and collect the data needed to answer your questions.

c. Discuss any differences and similarities between your study and the Chapter Case Study.

Chapter Practice Test

PART I: Knowing the Definitions

Answer "True" if the statement is always true. If the statement is not always true, replace the words shown in bold with words that make the statement always true.

10.1 When the means of two unrelated samples are used to compare two populations, we are dealing with **two dependent means.**

10.2 The use of **paired data (dependent means)** often allows for the control of unmeasurable or confounding variables because each pair is subjected to these confounding effects equally.

10.3 The **chi-square distribution** is used for making inferences about the ratio of the variances of two populations.

10.4 The **z-distribution** is used when two dependent means are to be compared.

10.5 In comparing two independent means when the σ's are unknown, we need to use the **standard normal** distribution.

10.6 The **standard normal score** is used for all inferences concerning population proportions.

10.7 The F-distribution is a **symmetrical** distribution.

10.8 The number of degrees of freedom for the critical value of t is equal to **the smaller**

of $n_1 - 1$ or $n_2 - 1$ when inferences are made about the difference between two independent means in the case when the degrees of freedom are estimated.

10.9 In a confidence interval for the mean difference in paired data, the interval **increases** in width when the sample size is increased.

10.10 A **pooled estimate** for any statistic in a problem dealing with two populations is a value arrived at by combining the two separate sample statistics so as to achieve the best possible point estimate.

PART II: Applying the Concepts

Answer all questions, showing all formulas, substitutions, and work.

10.11 State the null (H_o) and the alternative (H_a) hypotheses that would be used to test each of these claims:

a. There is no significant difference in the mean batting averages for the baseball players of the two major leagues.

b. The standard deviation for the monthly amounts of rainfall in Monroe County is less than the standard deviation for the monthly amounts of rainfall in Orange County.

c. There is a significant difference between the percentages of male and female college students who own their own car.

10.12 Determine the test statistic, critical region, and critical value(s) that would be used in completing each hypothesis test using the classical procedure with $\alpha = 0.05$.

a. $H_o: p_1 - p_2 = 0$ b. $H_o: \mu_d = 12$
$H_a: p_1 - p_2 \neq 0$ $H_a: \mu_d \neq 12$
                            $(n = 28)$

c. $H_o: \mu_1 - \mu_2 = 17$ d. $H_o: \mu_1 - \mu_2 = 37$
$H_a: \mu_1 - \mu_2 > 17$ $H_a: \mu_1 - \mu_2 < 37$
$(n_1 = 8, n_2 = 10)$ $(n_1 = 38, n_2 = 50)$

e. $H_o: \sigma_m^2 = \sigma_p^2$
$H_a: \sigma_m^2 > \sigma_p^2$
$(n_m = 16, n_p = 25)$

10.13 Find each of the following:

a. $z(0.02)$ b. $t(15, 0.025)$ c. $F(24, 12, 0.05)$

d. $F(12, 24, 0.05)$ e. $z(0.04)$

f. $t(38, 0.05)$ g. $t(23, 0.99)$ h. $z(0.90)$

10.14 [PT10-14] Twenty college freshmen were randomly divided into two groups. Members of one group were assigned to a statistics section that used programmed materials only. Members of the other group were assigned to a section in which the professor lectured. At the end of the semester, all were given the same final exam. Here are the results:

Programmed	76	60	85	58	91
	44	82	64	79	88
Lecture	81	62	87	70	86
	77	90	63	85	83

At the 5% level of significance, do these data provide sufficient evidence to conclude that on the average the students in the lecture sections performed significantly better on the final exam? Assume normality.

10.15 [PT10-15] The weights of eight people before they stopped smoking and 5 weeks after they stopped smoking are as follows:

	1	2	3	4	5	6	7	8
Before	148	176	153	116	129	128	120	132
After	154	179	151	121	130	136	125	128

At the 0.05 level of significance, does this sample present enough evidence to justify the conclusion that weight increases if one quits smoking? Assume normality.

10.16 In a nationwide sample of 600 school-age boys and 500 school-age girls, 288 boys and 175 girls admitted to having committed a destruction-of-property offense. Use these sample data to construct a 95% confidence interval for the difference between the proportions of boys and girls who have committed this offense.

PART III: Understanding The Concepts

10.17 To compare the accuracy of two short-range missiles, 8 of the first kind and 10 of the second kind are fired at a target. Let x be the distance by which the missile missed the target. Do these two sets of data (8 distances and 10 distances) represent dependent or independent samples? Explain.

10.18 Let's assume that 400 students in our college are taking elementary statistics this semester. Describe how you could obtain two dependent samples of size 20 from these students to test some precourse skill against the same skill after completing the course. Be very specific.

10.19 Student A says, "I don't see what all the fuss is about the difference between independent and dependent means; the results are almost the same regardless of the method used." Professor C suggests student A should compare the procedures a bit more carefully. Help student A discover that there is a substantial difference between the procedures.

10.20 Suppose you are testing H_o: $\mu_d = 0$ versus H_a: $\mu_d < 0$ and the sample paired differences are all negative. Does this mean there is sufficient evidence to reject the null hypothesis? How can it not be significant? Explain.

10.21 Truancy is very disruptive to the educational system. A group of high school teachers and counselors have developed a group counseling program that they hope will help improve the truancy situation in their school. They have selected the 80 students in their school with the worst truancy records and have randomly assigned half of them to the group counseling program. At the end of the school year, the 80 students will be rated with regard to their truancy. When the scores have been collected, they will be turned over to you for evaluation. Explain what you will do to complete the study.

10.22 You wish to estimate and compare the proportion of Catholic families whose children attend a private school to the proportion of non-Catholic families whose children attend private schools. How would you go about estimating the two proportions and the difference between them?

Working with Your Own Data

History contains many stories about consumers and the various products they purchase. An exhibit at the Boston Museum of Science tells such a mathematician–baker story. A man named Poincaré bought one loaf of bread daily from his local baker, a loaf that was supposed to weigh 1 kilogram. After a year of weighing and recording the weight of each loaf, Poincaré found a normal distribution with a mean of 950 grams. The police were called and the baker was told to behave himself; however, a year later Poincaré reported that the baker had not reformed and the police confronted the baker again. The baker questioned, "How could Poincaré have known that we always gave him the largest loaf?" Poincaré then showed the police the second year of his record, a bell-shaped curve with a mean of 950 grams but truncated on the left side.

As consumers, we all purchase many bottled, boxed, canned, and packaged products. Seldom, if ever, do any of us question whether or not the content is really the amount stated on the container. Here are some content listings found on containers we purchase:

28 FL OZ (1 PT 12 OZ)	750 ml
5 FL OZ (148 ml)	32 FL OZ (1 QT) 0.95 l
NET WT 10 OZ	NET WT $3^3/_4$ OZ
283 GRAMS	106g—48 tea bags
140 1-PLY NAPKINS	77 SQ FT—92
	TWO-PLY SHEETS—
	11 × 11 IN.

Have you ever wondered, "Am I getting the amount that I am paying for?" And if this thought did cross your mind, did you attempt to check the validity of the content claim? The following article appeared in the *Times Union* of Rochester, New York, in 1972.

MILK FIRM ACCUSED OF SHORT MEASURE

The processing manager of Dairylea Cooperative, Inc., has been named in a warrant charging that the cooperative is distributing cartons of milk in the Rochester area containing less than the quantity represented.

. . . an investigator found shortages in four quarts of Dairylea milk purchased Friday.

Asst. Dist. Atty. Howard R. Relin, who issued the warrant, said the shortages ranged from $1\frac{1}{8}$ to $1\frac{1}{4}$ ounces per quart. A quart of milk contains 32 fluid ounces.

. . . the state Agriculture and Markets Law . . . provides that a seller of a commodity shall not sell or deliver less of the commodity than the quantity represented to be sold.

. . . the purpose of the law under which . . . the dairy is charged is to ensure honest, accurate, and fair dealing with the public. There is no requirement that intent to violate the law be proved, he said.

From The Times-Union, Rochester, NY, February 16, 1972

This situation poses a very interesting legal problem: there is no need to show intent to "short the customer." If caught, violators are fined automatically and the fines are often quite severe.

A   A High-Speed Filling Operation

A high-speed piston-type machine used to fill cans with hot tomato juice was sold to a canning company. The guarantee stated that the machine would fill 48-oz cans with a mean amount of 49.5 oz, a standard deviation of 0.072 oz, and a maximum spread of 0.282 oz while operating at a rate of filling 150 to 170 cans per minute. On August 12, 1994, a sample of 42 cans was gathered and the following weights were recorded. The weights, measured to the nearest $\frac{1}{8}$ oz, are recorded as variations from 49.5 oz.

[DS-3]

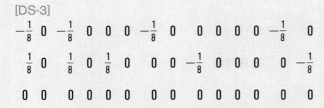

1. Calculate the mean $\bar{x}$, the standard deviation s, and the range of the sample data.

2. Construct a histogram picturing the sample data.

3. Does the amount of fill differ from the prescribed 49.5 oz at the $\alpha = 0.05$ level? Test the hypothesis that $\mu = 49.5$ against an appropriate alternative.

4. Does the amount of variation, as measured by the range, satisfy the guarantee?

5. Assuming that the filling machine continues to fill cans with an amount of tomato juice that is distributed normally and the mean and standard deviation are equal to the values found in question 1, what is the probability that a randomly selected can will contain less than the 48 oz claimed on the label?

6. If the amount of fill per can is normally distributed and the standard deviation can be maintained, find the setting for the mean value that would allow only 1 can in every 10,000 to contain less than 48 oz.

B   Your Own Investigation

Select a packaged product that has a quantity of fill per package that you can and would like to investigate.

1. Describe your selected product, including the quantity per package, and describe how you plan to obtain your data.

2. Collect your sample of data. (Consult your instructor for advice on size of sample.)

3. Calculate the mean $\bar{x}$ and the standard deviation s for your sample data.

4. Construct a histogram or stem-and-leaf diagram picturing the sample data.

5. Does the mean amount of fill agree with the amount given on the label? Test using $\alpha = 0.05$.

6. Assume that the item you selected is filled continually. The amount of fill is normally distributed, and the mean and standard deviation are equal to the values found in question 3. What is the probability that one randomly selected package contains less than the prescribed amount?

CHAPTER

11

Applications of Chi-Square

© Bonnie Kamin/Index Stock Imagery

11.1 — Cooling a Great Hot Taste

If you like hot foods, you probably have a preferred way to "cool" your mouth after eating a delicious spicy favorite. Some of the more common methods used by people are drinking water, milk, soda, or beer or eating bread or other food. There are even a few people who prefer not to cool their mouth on such occasions and therefore do nothing. The graphic shown here lists the top six ways adults say they cool their mouths after eating hot sauce.

PUTTING OUT THE FIRE

Top six ways American adults say they cool their mouths after eating hot sauce:

| Water 43% | Bread 19% | Milk 15% | Beer 7% | Soda 7% | Don't 6% |

Data from Anne R. Carey and Suzy Parker. © 1995 USA Today.

Two hundred adults professing to love hot spicy food were asked to name their favorite way to cool their mouth after eating food with hot sauce. Following is the summary of the resulting sample. [EX11-01]

Method	Water	Milk	Soda	Beer	Bread	Other	Nothing
Number	73	35	20	19	29	11	13

After completing Chapter 11, further investigate the preceding data and "Cooling a Great Hot Taste" in the Chapter Project on page 652.

SECTION 11.1 EXERCISES

11.1 [EX11-01] Referring to the sample of 200 adults collected in Section 11.1:

a. What information was collected from each adult in the sample?

b. Define the population and the variable involved in the sample.

c. Using the sample data, calculate percentages for the various methods of cooling one's mouth.

d. How do the sample percentages compare with the graphic percentages?

11.2 [EX11-01] Referring to the sample of 200 adults collected in Section 11.1, how similar is the distribution in the sample to the distribution of percentages in the graphic?

a. Construct a horizontal bar graph of the 200 adults using relative frequency for the horizontal scale.

b. Superimpose the bar graph from "Putting out the fire" on the bar graph in part a.

c. Would you say the sample's distribution looks "similar to" or "quite different than" the distribution shown in the "Putting out the fire" graph? Explain your answer.

11.2 Chi-Square Statistic

There are many problems for which **enumerative** data are categorized and the results shown by way of counts. For example, a set of final exam scores can be displayed as a frequency distribution. These frequency numbers are counts, the number of data that fall in each cell. A survey asks voters whether they are registered as Republican, Democrat, or other, and whether or not they support a particular candidate. The results are usually displayed on a chart that shows the number of voters in each possible category. Numerous illustrations of this way of presenting data have been given throughout the previous 10 chapters.

Data Setup

Suppose that we have a number of **cells** into which n observations have been sorted. (The term *cell* is synonymous with the term *class;* the terms *class* and *frequency* were defined and first used in earlier chapters. Before you continue, a brief review of Sections 2.2, 2.3, and 3.2 might be beneficial.) The **observed frequencies** in each cell are denoted by $O_1, O_2, O_3, \ldots, O_k$ (see Table 11.1). Note that the sum of all the observed frequencies is

$$O_1 + O_2 + \cdots + O_k = n$$

where n is the sample size. What we would like to do is compare the observed frequencies with some **expected,** or theoretical **frequencies,** denoted by $E_1, E_2, E_3, \ldots, E_k$ (see Table 11.1), for each of these cells. Again, the sum of these expected frequencies must be exactly n:

$$E_1 + E_2 + \cdots + E_k = n$$

TABLE 11.1

Observed Frequencies

	k Categories					
	1st	2nd	3rd	...	kth	Total
Observed frequencies	O_1	O_2	O_3	...	O_k	n
Expected frequencies	E_1	E_2	E_3	...	E_k	n

We will then decide whether the observed frequencies seem to agree or disagree with the expected frequencies. We will do this by using a **hypothesis test** with **chi-square, χ^2** ("ki-square"; that's "ki" as in *kite*; χ is the Greek lowercase letter chi).

Outline of Test Procedure

Test Statistic for Chi-Square:

$$\chi^2 \bigstar = \sum_{\text{all cells}} \frac{(O - E)^2}{E} \qquad (11.1)$$

This calculated value for chi-square is the sum of several nonnegative numbers, one from each cell (or category). The numerator of each term in the formula for $\chi^2 \bigstar$ is the square of the difference between the values of the observed and the expected frequencies. The closer together these values are, the smaller the value of $(O - E)^2$; the farther apart, the larger the value of $(O - E)^2$. The denominator for each cell puts the size of the numerator into perspective; that is, a difference $(O - E)$ of 10 resulting from frequencies of 110 (O) and 100 (E) is quite different from a difference of 10 resulting from 15 (O) and 5 (E).

These ideas suggest that small values of chi-square indicate agreement between the two sets of frequencies, whereas larger values indicate disagreement. Therefore, it is customary for these tests to be one-tailed, with the critical region on the right.

In repeated sampling, the calculated value of $\chi^2 \bigstar$ in formula (11.1) will have a sampling distribution that can be approximated by the chi-square probability distribution when n is large. This approximation is generally considered adequate when all the expected frequencies are equal to or greater than 5. Recall that the chi-square distributions, like Student's t-distributions, are a family of probability distributions, each one being identified by the parameter number of **degrees of freedom,** df. The appropriate value of df will be described with each specific test. In order to use the chi-square distribution, we must be aware of its properties, which were listed in Section 9.4 on page 517. (Also see Figure 9.7.) The critical values for chi-square are obtained from Table 8 in Appendix B. (Specific instructions were given in Section 9.4; see pp. 517–518.)

> **Assumption for using chi-square to make inferences based on enumerative data:** The sample information is obtained using a random sample drawn from a population in which each individual is classified according to the categorical variable(s) involved in the test.

A *categorical variable* is a variable that classifies or categorizes each individual into exactly one of several cells or classes; these cells or classes are all-inclusive and mutually exclusive. The side facing up on a rolled die is a categorical variable: the list of outcomes {1, 2, 3, 4, 5, 6} is a set of all-inclusive and mutually exclusive categories.

In this chapter we permit a certain amount of "liberalization" with respect to the null hypothesis and its testing. In previous chapters the null hypothesis was always a statement about a population parameter (μ, σ, or p). However, there are other types of hypotheses that can be tested, such as "This die is fair" or "The height and weight of individuals are independent." Notice that these hypotheses are not claims about a parameter, although sometimes they could be stated with parameter values specified.

Suppose that I claim "This die is fair," $p = P(\text{any one number}) = \frac{1}{6}$, and you want to test the claim. What would you do? Was your answer something like: Roll this die many times and record the results? Suppose that you decide to roll the die 60 times. If the die is fair, what do you expect will happen? Each number (1, 2, . . . , 6) should appear approximately $\frac{1}{6}$ of the time (that is, 10 times). If it happens that approximately 10 of each number occur, you will certainly accept the claim of fairness ($p = \frac{1}{6}$ for each value). If it happens that the die seems to favor some particular numbers, you will reject the claim. (The test statistic $\chi^2\star$ will have a large value in this case, as we will soon see.)

SECTION 11.2 EXERCISES

11.3 Using Table 8 of Appendix B, find the following:

a. $\chi^2(10, 0.01)$ b. $\chi^2(12, 0.025)$

c. $\chi^2(10, 0.95)$ d. $\chi^2(22, 0.995)$

11.4 Find these critical values by using Table 8 of Appendix B.

a. $\chi^2(18, 0.01)$ b. $\chi^2(16, 0.025)$

c. $\chi^2(40, 0.10)$ d. $\chi^2(45, 0.01)$

11.5 Using the notation seen in Exercise 11.4, name and find the critical values of χ^2.

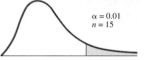

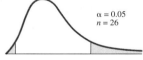

11.6 Using the notation seen in Exercise 11.4, name and find the critical values of χ^2.

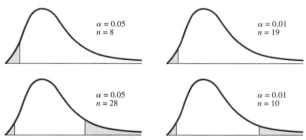

11.3   Inferences Concerning Multinomial Experiments

The preceding die problem is a good illustration of a **multinomial experiment.** Let's consider this problem again. Suppose that we want to test this die (at $\alpha = 0.05$) and decide whether to fail to reject or reject the claim "This die is fair." (The probability of each number is $\frac{1}{6}$.) The die is rolled from a cup onto a smooth, flat surface 60 times, with the following observed frequencies:

Number	1	2	3	4	5	6
Observed frequency	7	12	10	12	8	11

The null hypothesis that the die is fair is assumed to be true. This allows us to calculate the expected frequencies. If the die is fair, we certainly expect 10 occurrences of each number.

Now let's calculate an observed value of χ^2. These calculations are shown in Table 11.2. The calculated value is $\chi^2\star = 2.2$.

TABLE 11.2

Computations for Calculating χ^2

Number	Observed (O)	Expected (E)	$O - E$	$(O - E)^2$	$\dfrac{(O - E)^2}{E}$
1	7	10	-3	9	0.9
2	12	10	2	4	0.4
3	10	10	0	0	0.0
4	12	10	2	4	0.4
5	8	10	-2	4	0.4
6	11	10	1	1	0.1
Total	60	60	0 ⓒⓚ		2.2

Note: $\sum(O - E)$ must equal zero because $\sum O = \sum E = n$. You can use this fact as a check, as shown in Table 11.2.

Now let's use our familiar hypothesis-testing format.

Step 1 **a. Parameter of interest:** The probability with which each side faces up: $P(1)$, $P(2)$, $P(3)$, $P(4)$, $P(5)$, $P(6)$.

   b. Statement of hypotheses:

   H_o: The die is fair   (each $p = \dfrac{1}{6}$).

   H_a: The die is not fair (at least one p is different from the others).

Step 2 **a. Assumptions:** The data were collected in a random manner, and each outcome is one of the six numbers.

   b. Test statistic: The chi-square distribution and formula (11.1), with df $= k - 1 = 6 - 1 = 5$.

In a multinomial experiment, df $= k - 1$, where k is the number of cells.

   c. Level of significance: $\alpha = 0.05$.

Step 3 **a. Sample information:** See Table 11.2.

   b. Calculated test statistic: Using formula (11.1), we have

$$\chi^2\star = \sum_{\text{all cells}} \frac{(O - E)^2}{E}: \quad \chi^2\star = 2.2$$
(calculations are shown in Table 11.2)

Step 4   **Probability Distribution:**

p-value: **OR**   **Classical:**

a. Use the right-hand tail because "larger" values of chi-square disagree with the null hypothesis:

$P = P(\chi^2\star > 2.2 \,|\, \text{df} = 5)$ as shown in the figure.

a. The critical region is the right-hand tail because "larger" values of chi-square disagree with the null hypothesis. The critical value is obtained from

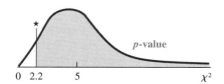

To find the *p*-value, you have two options:
1. Use Table 8 (Appendix B) to place bounds on the *p*-value: **0.75 < P < 0.90.**
2. Use a computer or calculator to find the *p*-value: **P = 0.821.**

For specific instructions, see page 518.

b. The *p*-value is not smaller than the level of significance, α.

Table 8, at the intersection of row df = 5 and column α = 0.05:

$$\chi^2(5, 0.05) = \mathbf{11.1}$$

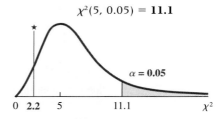

For specific instructions, see page 518.

b. $\chi^2\star$ is not in the critical region, as shown in **red** in the figure.

FYI Computer and calculator commands to find the probability associated with a specified chi-square value can be found in Chapter 9 (pp. 518–519).

Step 5   **a.**   **Decision:** Fail to reject H_o.

            b.   **Conclusion:** At the 0.05 level of significance, the observed frequencies are not significantly different from those expected of a fair die.

Before we look at other examples, we must define the term *multinomial experiment* and state the guidelines for completing the chi-square test for it.

> **Multinomial experiment:** A multinomial experiment has the following characteristics:
> 1. It consists of *n* identical independent trials.
> 2. The outcome of each trial fits into exactly one of *k* possible cells.
> 3. There is a probability associated with each particular cell, and these individual probabilities remain constant during the experiment. (It must be the case that $p_1 + p_2 + \cdots + p_k = 1$.)
> 4. The experiment will result in a set of *k* observed frequencies, $O_1, O_2, \ldots, O_k$, where each O_i is the number of times a trial outcome falls into that particular cell. (It must be the case that $O_1 + O_2 + \cdots + O_k = n$.)

The die example meets the definition of a multinomial experiment because it has all four of the characteristics described in the definition.

1. The die was rolled *n* (60) times in an identical fashion, and these trials were independent of each other. (The result of each trial was unaffected by the results of other trials.)

2. Each time the die was rolled, one of six numbers resulted, and each number was associated with a cell.

3. The probability associated with each cell was $\frac{1}{6}$, and this was constant from trial to trial. (Six values of $\frac{1}{6}$ sum to 1.0.)

4. When the experiment was complete, we had a list of six frequencies (7, 12, 10, 12, 8, and 11) that summed to 60, indicating that each of the outcomes was taken into account.

The testing procedure for multinomial experiments is very similar to the testing procedure described in previous chapters. The biggest change comes with the statement of the null hypothesis. It may be a verbal statement, such as in the die example: "This die is fair." Often the alternative to the null hypothesis is not stated. However, in this book the alternative hypothesis will be shown, because it aids in organizing and understanding the problem. It will not be used to determine the location of the critical region, though, as was the case in previous chapters. For multinomial experiments we will always use a one-tailed critical region, and it will be the right-hand tail of the χ^2-distribution because larger deviations (positive or negative) from the expected values lead to an increase in the calculated $\chi^2\star$ value.

The critical value will be determined by the level of significance assigned (α) and the number of degrees of freedom. The number of degrees of freedom (df) will be 1 less than the number of cells (k) into which the data are divided:

> **Degrees of Freedom for Multinomial Experiments**
>
> $$df = k - 1 \qquad (11.2)$$

Each expected frequency, E_i, will be determined by multiplying the total number of trials n by the corresponding probability (p_i) for that cell; that is,

> **Expected Value for Multinomial Experiments**
>
> $$E_i = n \cdot p_i \qquad (11.3)$$

One guideline should be met to ensure a good approximation to the chi-square distribution: each expected frequency should be at least 5 (i.e., each $E_i \geq 5$). Sometimes it is possible to combine "smaller" cells to meet this guideline. If this guideline cannot be met, then corrective measures to ensure a good approximation should be used. These corrective measures are not covered in this book but are discussed in many other sources.

EXAMPLE 11.1

A Multinomial Hypothesis Test with Equal Expected Frequencies

College students have regularly insisted on freedom of choice when they register for courses. This semester there were seven sections of a particular mathematics course. The sections were scheduled to meet at various times with a variety of instructors. Table 11.3 shows the number of students who selected each of the seven sections. Do the data indicate that the students had a preference for certain sections, or do they indicate that each section was equally likely to be chosen?

TABLE 11.3

Data on Section Enrollments

	Section							Total
	1	2	3	4	5	6	7	
Number of students	18	12	25	23	8	19	14	119

SOLUTION If no preference were shown in the selection of sections, then we would expect the 119 students to be equally distributed among the seven classes. We would expect 17 students to register for each section. The hypothesis test is completed at the 5% level of significance.

Step 1 a. **Parameter of interest:** Preference for each section, the probability that a particular section is selected at registration.

  b. **Statement of hypotheses:**
  H_o: There was no preference shown (equally distributed).
  H_a: There was a preference shown (not equally distributed).

Step 2 a. **Assumptions:** The 119 students represent a random sample of the population of all students who register for this particular course. Since no new regulations were introduced in the selection of courses and registration seemed to proceed in its usual pattern, there is no reason to believe this is other than a random sample.

  b. **Test statistic:** The chi-square distribution and formula (11.1), with df = 6.

  c. **Level of significance:** $\alpha = 0.05$.

Step 3 a. **Sample information:** See Table 11.3 (p. 625).

  b. **Calculated test statistic:** Using formula (11.1), we have

$$\chi^2\star = \sum_{\text{all cells}} \frac{(O-E)^2}{E}: \quad \chi^2\star = \frac{(18-17)^2}{17} + \frac{(12-17)^2}{17} + \frac{(25-17)^2}{17}$$

$$+ \frac{(23-17)^2}{17} + \frac{(8-17)^2}{17} + \frac{(19-17)^2}{17} + \frac{(14-17)^2}{17}$$

$$= \frac{(1)^2 + (-5)^2 + (8)^2 + (6)^2 + (-9)^2 + (2)^2 + (-3)^2}{17}$$

$$= \frac{1 + 25 + 64 + 36 + 81 + 4 + 9}{17} = \frac{220}{17} = 12.9411$$

$$= \mathbf{12.94}$$

Step 4 **Probability Distribution:**

OR

p-value:

a. Use the right-hand tail because "larger" values of chi-square disagree with the null hypothesis:

$P = P(\chi^2\star > 12.94 \mid df = 6)$ as shown in the figure:

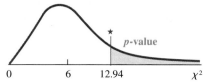

To find the *p*-value, you have two options:
1. Use Table 8 (Appendix B) to place bounds on the *p*-value: **0.025 < P < 0.05.**
2. Use a computer or calculator to find the *p*-value: **P = 0.044.**

For specific instructions, see page 518.

b. The *p*-value is smaller than the level of significance, α.

Classical:

a. The critical region is the right-hand tail because "larger" values of chi-square disagree with the null hypothesis. The critical value is obtained from Table 8, at the intersection of row df = 6 and column $\alpha = 0.05$:

$$\chi^2(6, 0.05) = \mathbf{12.6}$$

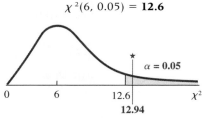

For specific instructions, see page 518.

b. $\chi^2\star$ is in the critical region, as shown in **red** in the figure.

Step 5   a. **Decision:** Reject H_o.

   b. **Conclusion:** At the 0.05 level of significance, there does seem to be a preference shown. We cannot determine from the given information what the preference is. It could be teacher preference, time preference, or a schedule conflict.

Conclusions must be worded carefully to avoid suggesting conclusions that the data cannot support.

Not all multinomial experiments result in equal expected frequencies, as we will see in Example 11.2.

EXAMPLE 11.2

A Multinomial Hypothesis Test with Unequal Expected Frequencies

Statistics△Now™

Watch a video example at
http://1pass.thomson.com
or on your CD.

The Mendelian theory of inheritance claims that the frequencies of round and yellow, wrinkled and yellow, round and green, and wrinkled and green peas will occur in the ratio $9:3:3:1$ when two specific varieties of peas are crossed. In testing this theory, Mendel obtained frequencies of 315, 101, 108, and 32, respectively. Do these sample data provide sufficient evidence to reject the theory at the 0.05 level of significance?

SOLUTION

Step 1   a. **Parameter of interest:** The proportions: P(round and yellow), P(wrinkled and yellow), P(round and green), P(wrinkled and green).

   b. **Statement of hypotheses:**

   H_o: $9:3:3:1$ is the ratio of inheritance.

   H_a: $9:3:3:1$ is not the ratio of inheritance.

Step 2   a. **Assumptions:** We will assume that Mendel's results form a random sample.

   b. **Test statistic:** The chi-square distribution and formula (11.1), with df = 3.

   c. **Level of significance:** $\alpha = 0.05$.

Step 3   a. **Sample information:** The observed frequencies were: 315, 101, 108, and 32.

   b. **Calculated test statistic:** The ratio $9:3:3:1$ indicates probabilities of $\dfrac{9}{16}$, $\dfrac{3}{16}$, $\dfrac{3}{16}$, and $\dfrac{1}{16}$.

Therefore, the expected frequencies are $\dfrac{9n}{16}$, $\dfrac{3n}{16}$, $\dfrac{3n}{16}$, and $\dfrac{1n}{16}$. We have

$$n = \sum O_i = 315 + 101 + 108 + 32 = 556$$

The computations for calculating $\chi^2\star$ are shown in Table 11.4.

TABLE 11.4

Computations Needed to Calculate $\chi^2\star$

O	E	$O - E$	$\dfrac{(O - E)^2}{E}$
315	312.75	2.25	0.0162
101	104.25	−3.25	0.1013
108	104.25	3.75	0.1349
32	34.75	−2.75	0.2176
556	556.00	0 (ck)	0.4700

$$\longrightarrow \chi^2\star = \sum_{\text{all cells}} \frac{(O - E)^2}{E} = 0.47$$

Step 4   **Probability Distribution:**

p-value: **OR**   Classical:

a. Use the right-hand tail because "larger" values of chi-square disagree with the null hypothesis:

$\mathbf{P} = P(\chi^2\star > 0.47 \mid \mathbf{df = 3})$ as shown in the figure.

To find the *p*-value, you have two options:

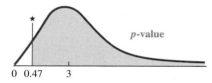

1. Use Table 8 (Appendix B) to place bounds on the *p*-value: **0.90 < P < 0.95.**
2. Use a computer or calculator to find the *p*-value: **P = 0.925.**

For specific instructions, see page 518.

b. The *p*-value is not smaller than the level of significance, α.

a. The critical region is the right-hand tail because "larger" values of chi-square disagree with the null hypothesis. The critical value is obtained from Table 8, at the intersection of row df = 3 and column $\alpha = 0.05$:

$$\chi^2(3, 0.05) = \mathbf{7.82}$$

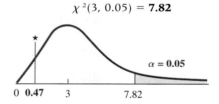

For specific instructions, see page 518.

b. $\chi^2\star$ is not in the critical region, as shown in **red** in the figure.

Step 5   **a.** **Decision:** Fail to reject H_o.

b. **Conclusion:** At the 0.05 level of significance, there is not sufficient evidence to reject Mendel's theory.

APPLIED EXAMPLE 11.3

Birth Days

The Census Bureau collects data for many variables. The information exhibited by the following graphic is based on the 2003 U.S. census and fits the format of a multinomial experiment. Verify that these data qualify as a multinomial experiment (see Exercise 11.7).

BOUNTIFUL BABY BIRTH DATES

More babies are born on Tuesdays than on any other day of the week. In the United States, an average 13,000 babies are born each Tuesday. Slowest day for new births: Sunday.

Tuesday	Thursday	Friday	Wednesday
happy birthday! :)			
1	2	3	4

Data from Anne R. Carey and Ron Coddington. © 2003 USA Today.

APPLIED
EXAMPLE 11.4

Downloading What?

The graphic "Teens and downloading" displays the results of surveying 8- to 18-year-olds about what they download using their cell phones. This information does not qualify as a multinomial experiment. What property is violated? (See Exercise 11.8.)

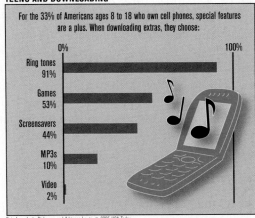

TEENS AND DOWNLOADING

For the 33% of Americans ages 8 to 18 who own cell phones, special features are a plus. When downloading extras, they choose:

Ring tones	91%
Games	53%
Screensavers	44%
MP3s	10%
Video	2%

Data from Justin Dickerson and Adrienne Lewis, © 2005 USA Today.

SECTION 11.3 EXERCISES

Statistics⊖Now™

Datasets can be found on your Student's Suite CD-ROM or at the StatisticsNow website at **http://1pass.thomson.com**.

11.7 Verify that Applied Example 11.3 (p. 628) is a multinomial experiment. Be specific.

a. What is one trial?

b. What is the variable?

c. What are the possible levels of results from each trial?

11.8 Why is the information shown in Applied Example 11.4 (above) not that of a multinomial experiment? Be specific.

11.9 State the null hypothesis, H_o, and the alternative hypothesis, H_a, that would be used to test the following statements:

a. The five numbers 1, 2, 3, 4, and 5 are equally likely to be drawn.

b. The multiple-choice question has a history of students selecting answers in the ratio of 2:3:2:1.

c. The poll will show a distribution of 16%, 38%, 41%, and 5% for the possible ratings of excellent, good, fair, and poor on that issue.

11.10 Determine the p-value for the following hypotheses tests involving the χ^2-distribution.

a. H_o: $P(1) = P(2) = P(3) = P(4) = 0.25$, with $\chi^2\star = 12.25$

b. H_o: $P(I) = 0.25$, $P(II) = 0.40$, $P(III) = 0.35$, with $\chi^2\star = 5.98$

11.11 Determine the critical value and critical region that would be used in the classical approach to test the null hypothesis for each of the following multinomial experiments.

a. H_o: $P(1) = P(2) = P(3) = P(4) = 0.25$, with $\alpha = 0.05$

b. H_o: $P(I) = 0.25$, $P(II) = 0.40$, $P(III) = 0.35$, with $\alpha = 0.01$

11.12 Explain how $9:3:3:1$ becomes $\dfrac{9}{16}$, $\dfrac{3}{16}$, $\dfrac{3}{16}$, and $\dfrac{1}{16}$ in Example 11.2 on page 627.

11.13 Explain how 312.75, 2.25, and 0.0162 were obtained in the first row of Table 11.4 on page 628.

11.14 A manufacturer of floor polish conducted a consumer-preference experiment to determine which of five different floor polishes was the most appealing in appearance. A sample of 100 consumers viewed five patches of flooring that had each received one of the five polishes. Each consumer indicated the patch he or she preferred. The lighting and background were approximately the same for all patches. The results were as follows:

Polish	A	B	C	D	E	Total
Frequency	27	17	15	22	19	100

Solve the following using the *p*-value approach and solve using the classical approach:

a. State the hypothesis for "no preference" in statistical terminology.

b. What test statistic will be used in testing this null hypothesis?

c. Complete the hypothesis test using $\alpha = 0.10$.

11.15 A certain type of flower seed will produce magenta, chartreuse, and ochre flowers in the ratio 6:3:1 (one flower per seed). A total of 100 seeds are planted and all germinate, yielding the following results.

Magenta	Chartreuse	Ochre
52	36	12

Solve the following using the *p*-value approach and solve using the classical approach:

a. If the null hypothesis (6:3:1) is true, what is the expected number of magenta flowers?

b. How many degrees of freedom are associated with chi-square?

c. Complete the hypothesis test using $\alpha = 0.10$.

11.16 [EX11-016] Over the years, African American actors in major cinema releases are more likely than white actors to have major roles in comedies. The following table shows the percent of all roles by type of picture.

Type of Picture	Percent of Roles	Type of Picture	Percent of Roles
Action and adventure	13.2	Horror and suspense	12.5
Comedy	31.9	Romantic comedy	8.2
Drama	23.0	Other	11.2

The next table shows the number of leading roles played by African Americans for each type of film for the last 89 films released.

Type of Picture	Number of Roles	Type of Picture	Number of Roles
Action and adventure	9	Horror and suspense	11
Comedy	40	Romantic comedy	5
Drama	17	Other	7

Does the distribution of African American roles differ from the overall distribution of roles in major cinema releases?

a. Solve using the *p*-value approach.

b. Solve using the classical approach.

11.17 A large supermarket carries four qualities of ground beef. Customers are believed to purchase these four varieties with probabilities of 0.10, 0.30, 0.35, and 0.25, respectively, from the least to most expensive variety. A sample of 500 purchases resulted in sales of 46, 162, 191, and 101 of the respective qualities. Does this sample contradict the expected proportions? Use $\alpha = 0.05$.

a. Solve using the *p*-value approach.

b. Solve using the classical approach.

11.18 [EX11-18] One of the major benefits of e-mail is that it makes it possible to communicate rapidly without getting a busy signal or no answer, two major criticisms of telephone calls. But does e-mail succeed in helping solve the problems people have trying to run computer software? A study polled the opinions of consumers who tried to use e-mail to obtain help by posting a message online to their PC manufacturer or authorized representative. Results are shown in the following table.

Result of Online Query	Percent
Never got a response	14
Got a response, but it didn't help	30
Response helped, but didn't solve problem	34
Response solved problem	22

Source: PC World, "PC World's Reliability and Service Survey"

As marketing manager for a large PC manufacturer, you decide to conduct a survey of your customers using your e-mail records to compare against the published results. To ensure a fair comparison, you elect to use the same questionnaire and examine returns from 500 customers who attempted to use e-mail for help from your technical support staff. The results follow:

Result of Online Query	Number Responding
Never got a response	35
Got a response, but it didn't help	102
Response helped, but didn't solve problem	125
Response solved problem	238
Total	500

Does the distribution of responses differ from the distribution obtained from the published survey? Test at the 0.01 level of significance.

a. Solve using the *p*-value approach.

b. Solve using the classical approach.

11.19 [EX11-19] *Nursing Magazine* reported results of a survey of more than 1800 nurses across the country concerning job satisfaction and retention. Nurses from magnet hospitals (hospitals that successfully attract and retain nurses) describe the staffing situation in their units as follows:

Staffing Situation	Percent
1. Desperately short of help–patient care has suffered	12
2. Short, but patient care hasn't suffered	32
3. Adequate	38
4. More than adequate	12
5. Excellent	6

A survey of 500 nurses from nonmagnet hospitals gave the following responses to the staffing situation.

Staffing Situation	1	2	3	4	5
Number	165	140	125	50	20

Do the data indicate that the nurses from the nonmagnet hospitals have a different distribution of opinions? Use $\alpha = 0.05$.

a. Solve using the *p*-value approach.

b. Solve using the classical approach.

11.20 [EX11-20] A program for generating random numbers on a computer is to be tested. The program is instructed to generate 100 single-digit integers between 0 and 9. The frequencies of the observed integers were as follows:

Integer	0	1	2	3	4	5	6	7	8	9
Frequency	11	8	7	7	10	10	8	11	14	14

At the 0.05 level of significance, is there sufficient reason to believe that the integers are not being generated uniformly?

a. Solve using the *p*-value approach.

b. Solve using the classical approach.

11.21 [EX11-21] "Climbing out of debt, step by step," an article in the April 29, 2005, *USA Today*, reported results of a survey of 260 members of the Financial Planning Association. Financial planners each reported what they consider to be the one most valuable step people can take to improve their financial life.

Most Valuable Step	Percent
1. Establish goals	30
2. Pay yourself first	21
3. Create and stick to a budget	17
4. Save on a regular basis	12
5. Pay down credit card debt	7
6. Invest the maximum in 401(k)	5
7. Other	8

A survey of 60 financial planners from an upstate metropolitan area gave the following responses to the "one most valuable goal" question.

Answer to Question	1	2	3	4	5	6	7
Number	10	13	13	8	9	3	4

Do the data indicate that the financial planners from the upstate metropolitan area have a different distribution of opinions? Use $\alpha = 0.05$.

a. Solve using the *p*-value approach.

b. Solve using the classical approach.

11.22 [EX11-22] The 2003 U.S. census found that babies entered the world on the days of the week in the proportions that follow.

Weekday	P(Day)	Weekday	P(Day)
Sunday	0.098	Thursday	0.160
Monday	0.149	Friday	0.159
Tuesday	0.166	Saturday	0.111
Wednesday	0.157		

Source: U.S. Census Bureau

A random sample selected from the birth records for a large metropolitan area resulted in the data:

Day	Su	M	Tu	W	Th	F	Sa
Observed	10	6	9	13	9	17	11

a. Do these data provide sufficient evidence to reject the claim, "births occur in this metropolitan area in the same daily proportions," as reported by the U.S. Census Bureau? Use $\alpha = 0.05$.

b. Do these data provide sufficient evidence to reject the claim, "births occur in this metropolitan area on all days with the same likeliness"? Use $\alpha = 0.05$.

c. Compare the results obtained in parts a and b. State your conclusions.

11.23 Skittles Original Fruit bite-size candies are multicolored candies in a bag, and you can "Taste the Rainbow" with their five colors and flavors: green, lime; purple, grape; yellow, lemon; orange, orange; and red, strawberry. Unlike some of the other multicolored candies available, Skittles claims that their five colors are equally likely. In an attempt to reject this claim, a 4-oz bag of Skittles was purchased and the colors counted:

Red	Orange	Yellow	Green	Purple
18	21	23	17	27

Does this sample contradict Skittles's claim at the 0.05 level?

a. Solve using the *p*-value approach.

b. Solve using the classical approach.

11.24 To demonstrate/explore the effect increased sample size has on the calculated chi-square value, let's consider the Skittles candies in Exercise 11.23 and sample some larger bags of the candy.

a. Suppose we purchase a 16-oz bag of Skittles, count the colors, and observe exactly the same proportion of colors as found in Exercise 11.23:

Red	Orange	Yellow	Green	Purple
72	84	92	68	108

Calculate the value of chi-square for these data. How is the new chi-square value related to the one found in Exercise 11.23? What effect does this new value have on the test results? Explain.

b. To continue this demonstration/exploration, suppose we purchase a 48-oz bag, count the colors, and observe exactly the same proportion of colors as found in Exercise 11.23 and part b of this exercise.

Red	Orange	Yellow	Green	Purple
216	252	276	204	324

Calculate the value of chi-square for these data. How is the new chi-square value related to the one found in Exercise 11.23? Explain.

c. What effect does the size of the sample have on the calculated chi-square value when the proportion of observed frequencies stays the same as the sample size increases?

d. Explain in what way this indicates that if a large enough sample is taken, the hypothesis test will eventually result in a rejection.

11.25 [EX11-25] According to The Harris Poll, the proportion of all adults who live in households with rifles (29%), shotguns (29%), or pistols (23%) has not changed significantly since 1996. However, today more people live in households with no guns (61%). The 1014 adults surveyed gave the following results.

	All Adults (%)	All Gun Owners (%)
Have rifle, shotgun, and pistol (3 out of 3)	16	41
Have 2 out of 3 (rifle, shotgun, or pistol)	11	27
Have 1 out of 3 (rifle, shotgun, or pistol)	11	29
Decline to answer/Not sure	1	3
TOTAL	39%	100%

In a survey of 2000 adults in Memphis who said they own guns, 780 said they own all three types, 550 said they owned 2 of the 3, 560 said they owned 1 of the 3 types, and 110 declined to specify what types of guns they owned.

a. Test the null hypothesis that the distribution of number of types owned is the same in Memphis as it is nationally as reported by The Harris Poll. Use a level of significance equal to 0.05.

b. What caused the calculated value of $\chi^2\star$ to be so large? Does it seem right that one cell should have this much effect on the results? How could this test be completed differently (hopefully, more meaningfully) so that the results might not be affected as they were in part a? Be specific.

11.26 Why is the chi-square test typically a one-tail test with the critical region in the right tail?

a. What kind of value would result if the observed frequencies and the expected frequencies were very close in value? Explain how you would interpret this situation.

b. Suppose you had to roll a die 60 times as an experiment to test the fairness of the die as discussed in the example on pages 622–624; but instead of rolling the die yourself, you paid your little brother $1 to roll it 60 times and keep a tally of the numbers. He agreed to perform this deed for you and ran off to his room with the die, returning in a few minutes with his resulting frequencies. He demanded his $1. You, of course, pay him before he hands over his results, which are as follows: 10, 10, 10, 10, 10, and 10. The observed results are exactly what you had "expected," right? Explain your reactions. What value of $\chi^2\star$ will result? What do you think happened? What do you demand of your little brother and why? What possible role might the left tail have in the hypothesis test?

c. Why is the left tail not typically of concern?

11.4 Inferences Concerning Contingency Tables

A **contingency table** is an arrangement of data in a two-way classification. The data are sorted into cells, and the count for each cell is reported. The contingency table involves two factors (or variables), and a common question concerning such tables is whether the data indicate that the two variables are independent or dependent (see pp. 147–149, 243–245).

Two different tests use the contingency table format. The first one we will look at is the *test of independence*.

Test of Independence

To illustrate a test of independence, let's consider a random sample that shows the gender of liberal arts college students and their favorite academic area.

EXAMPLE 11.5

Hypothesis Test for Independence

Each person in a group of 300 students was identified as male or female and then asked whether he or she preferred taking liberal arts courses in the area of math–science, social science, or humanities. Table 11.5 (p. 634) is a contingency table that shows the frequencies found for these categories. Does this sample present sufficient evidence to reject the null hypothesis: "Preference for math–science, social science, or humanities is independent of the gender of a college student"? Complete the **hypothesis test** using the 0.05 level of significance.

TABLE 11.5

Sample Results for Gender and Subject Preference

Gender	Favorite Subject Area			Total
	Math–Science (MS)	Social Science (SS)	Humanities (H)	
Male (M)	37	41	44	122
Female (F)	35	72	71	178
Total	72	113	115	300

SOLUTION

Step 1   a.   **Parameter of interest:** Determining the independence of the variables "gender" and "favorite subject area" requires us to discuss the probability of the various cases and the effect that answers about one variable have on the probability of answers about the other variable. Independence, as defined in Chapter 4, requires $P(\text{MS} \mid \text{M}) = P(\text{MS} \mid \text{F}) = P(\text{MS})$; that is, gender has no effect on the probability of a person's choice of subject area.

b.   **Statement of hypotheses:**

H_o: Preference for math–science, social science, or humanities is independent of the gender of a college student.

H_a: Subject area preference is not independent of the gender of the student.

Step 2   a.   **Assumptions:** The sample information is obtained using one random sample drawn from one population, with each individual then classified according to gender and favorite subject area.

b.   **Test statistic.**

In the case of contingency tables, the number of degrees of freedom is exactly the same as the number of cells in the table that may be filled in freely when you are given the *marginal totals*. The totals in this example are shown in the following table:

Given these totals, you can fill in only two cells before the others are all determined. (The totals must, of course, remain the same.) For example, once we pick two arbitrary values (say, 50 and 60) for the first two cells of the first row, the other four cell values are fixed (see the following table):

50	60	C	122
D	E	F	178
72	113	115	300

The values have to be $C = 12$, $D = 22$, $E = 53$, and $F = 103$. Otherwise, the totals will not be correct. Therefore, for this problem there are two free choices. Each free choice corresponds to 1 degree of freedom. Hence, the number of degrees of freedom for our example is 2 (df = 2).

> The chi-square distribution will be used along with formula (11.1), with df = 2.

c. Level of significance: $\alpha = 0.05$.

Step 3   **a. Sample information:** See Table 11.5.
   b. Calculated test statistic.

Before we can calculate the value of chi-square, we need to determine the expected values, E, for each cell. To do this we must recall the null hypothesis, which asserts that these factors are independent. Therefore, we would expect the values to be distributed in proportion to the marginal totals. There are 122 males; we would expect them to be distributed among MS, SS, and H proportionally to the 72, 113, and 115 totals. Thus, the expected cell counts for males are

$$\frac{72}{300} \cdot 122 \qquad \frac{113}{300} \cdot 122 \qquad \frac{115}{300} \cdot 122$$

Similarly, we would expect for the females

$$\frac{72}{300} \cdot 178 \qquad \frac{113}{300} \cdot 178 \qquad \frac{115}{300} \cdot 178$$

Thus, the expected values are as shown in Table 11.6. Always check the marginal totals for the expected values against the marginal totals for the observed values.

TABLE 11.6

Expected Values

	MS	SS	H	Total
Male	29.28	45.95	46.77	122.00
Female	42.72	67.05	68.23	178.00
Total	72.00	113.00	115.00	300.00

Note: We can think of the computation of the expected values in a second way. Recall that we assume the null hypothesis to be true until there is evidence to reject it. Having made this assumption in our example, we are saying in effect that the event that a student picked at random is male and the event that a student picked at random prefers math–science courses are independent. Our point estimate for the probability that a student is male is $\frac{122}{300}$, and the point estimate for the probability that the student prefers math–science courses is $\frac{72}{300}$. Therefore, the probability that both events occur is the product of the probabilities. [Refer to formula (4.7), p. 246.] Thus, $\left(\frac{122}{300}\right)\left(\frac{72}{300}\right)$ is the probability of a selected student being male and preferring math–science. The number of students out of 300 who are expected to be male and prefer math–science is found by multiplying the probability (or proportion) by the total number of students (300). Thus, the expected number of males who prefer math–science is $\left(\frac{122}{300}\right)\left(\frac{72}{300}\right)(300) = \left(\frac{122}{300}\right)(72) = 29.28$. The other expected values can be determined in the same manner.

Typically, the contingency table is written so that it contains all this information (see Table 11.7).

TABLE 11.7

Contingency Table Showing Sample Results and Expected Values

| Gender | Favorite Subject Area | | | Total |
	MS	SS	H	
Male	37 (29.28)	41 (45.95)	44 (46.77)	122
Female	35 (42.72)	72 (67.05)	71 (68.23)	178
Total	72	113	115	300

The calculated chi-square is

$$\chi^2\star = \sum_{\text{all cells}} \frac{(O - E)^2}{E} : \chi^2\star = \frac{(37 - 29.28)^2}{29.28} + \frac{(41 - 45.95)^2}{45.95} + \frac{(44 - 46.77)^2}{46.77}$$

$$+ \frac{(35 - 42.72)^2}{42.72} + \frac{(72 - 67.05)^2}{67.05} + \frac{(71 - 68.23)^2}{68.23}$$

$$= 2.035 + 0.533 + 0.164 + 1.395 + 0.365 + 0.112$$

$$= \mathbf{4.604}$$

Step 4   **Probability Distribution:**

OR

p-value:

a. Use the right-hand tail because "larger" values of chi-square disagree with the null hypothesis:

P = P ($\chi^2\star > 4.604$ | df = 2) as shown in the figure.

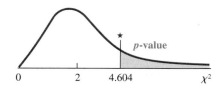

Classical:

a. The critical region is the right-hand tail because "larger" values of chi-square disagree with the null hypothesis. The critical value is obtained from Table 8, at the intersection of row df = 2 and column $\alpha = 0.05$:

$$\chi^2(2, 0.05) = \mathbf{5.99}$$

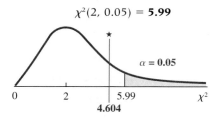

To find the *p*-value, you have two options:
1. Use Table 8 (Appendix B) to place bounds on the *p*-value: **0.10 < P < 0.25.**
2. Use a computer or calculator to find the *p*-value: **P = 0.1001.**

For specific instructions, see page 518.
b. The *p*-value is not smaller than α.

For specific instructions, see page 518.
b. $\chi^2\bigstar$ is not in the critical region, as shown in **red** in the figure.

Step 5 **a.** **Decision:** Fail to reject H_o.

b. **Conclusion:** At the 0.05 level of significance, the evidence does not allow us to reject independence between the gender of a student and the student's preferred academic subject area.

In general, the **$r \times c$ contingency table** (*r* is the number of **rows;** *c* is the number of **columns**) is used to test the independence of the row factor and the column factor. The number of **degrees of freedom** is determined by

---

Degrees of Freedom for Contingency Tables

$$df = (r - 1) \cdot (c - 1) \qquad (11.4)$$

where *r* and *c* are both greater than 1

(This value for df should agree with the number of cells counted according to the general description on pages 634–635.)

The **expected frequencies** for an $r \times c$ contingency table are found by means of the formulas given in each cell in Table 11.8, where *n* = grand total. In general, the expected frequency at the intersection of the *i*th row and the *j*th column is given by

---

Expected Frequencies for Contingency Tables

$$E_{i,j} = \frac{\text{row total} \times \text{column total}}{\text{grand total}} = \frac{R_i \times C_j}{n} \qquad (11.5)$$

TABLE 11.8

Expected Frequencies for an $r \times c$ Contingency Table

Row	1	2	...	*j*th column	...	*c*	Total
				Column			
1	$\dfrac{R_1 \times C_1}{n}$	$\dfrac{R_1 \times C_2}{n}$	...	$\dfrac{R_1 \times C_j}{n}$	...	$\dfrac{R_1 \times C_c}{n}$	R_1
2	$\dfrac{R_2 \times C_1}{n}$						R_2
⋮	⋮			⋮			⋮
*i*th row	$\dfrac{R_i \times C_1}{n}$			$\dfrac{R_i \times C_j}{n}$	...		R_i
⋮	⋮			⋮			⋮
r	$\dfrac{R_r \times C_1}{n}$						
Total	C_1	C_2	...	C_j	...	...	*n*

We should again observe the previously mentioned guideline: Each $E_{i,j}$ should be at least 5.

Note: The notation used in Table 11.8 and formula (11.5) may be unfamiliar to you. For convenience in referring to cells or entries in a table, we use $E_{i,j}$ to denote the entry in the ith row and the jth column. That is, the first letter in the subscript corresponds to the row number and the second letter corresponds to the column number. Thus, $E_{1,2}$ is the entry in the first row, second column, and $E_{2,1}$ is the entry in the second row, first column. In Table 11.6 (p. 635), $E_{1,2}$ is 45.95 and $E_{2,1}$ is 42.72. The notation used in Table 11.8 is interpreted in a similar manner; that is, R_1 corresponds to the total from row 1, and C_1 corresponds to the total from column 1.

Test of Homogeneity

The second type of contingency table problem is called a *test of homogeneity*. This test is used when one of the two variables is controlled by the experimenter so that the row or column totals are predetermined.

For example, suppose that we want to poll registered voters about a piece of legislation proposed by the governor. In the poll, 200 urban, 200 suburban, and 100 rural residents are randomly selected and asked whether they favor or oppose the governor's proposal. That is, a simple random sample is taken for each of these three groups. A total of 500 voters are polled. But notice that it has been predetermined (before the sample is taken) just how many are to fall within each row category, as shown in Table 11.9, and each category is sampled separately.

TABLE 11.9

Registered Voter Poll with Predetermined Row Totals

Residence	Governor's Proposal		Total
	Favor	Oppose	
Urban			200
Suburban			200
Rural			100
Total			500

In a test of this nature, we are actually testing the hypothesis: The distribution of proportions within the rows is the same for all rows. That is, the distribution of proportions in row 1 is the same as in row 2, is the same as in row 3, and so on. The alternative is: The distribution of proportions within the rows is not the same for all rows. This type of example may be thought of as a comparison of several multinomial experiments.

Beyond this conceptual difference, the actual testing for independence and homogeneity with contingency tables is the same. Let's demonstrate this **hypothesis test** by completing the polling illustration.

EXAMPLE 11.6

Hypothesis Test for Homogeneity

Each person in a random sample of 500 registered voters (200 urban, 200 suburban, and 100 rural residents) was asked his or her opinion about the governor's proposed legislation. Does the sample evidence shown in Table 11.10 support the hypothesis: "Voters within the different residence groups have different opinions about the governor's proposal"? Use $\alpha = 0.05$.

TABLE 11.10

Sample Results for Residence and Opinion

Residence	Governor's Proposal		Total
	Favor	Oppose	
Urban	143	57	200
Suburban	98	102	200
Rural	13	87	100
Total	254	246	500

SOLUTION

Step 1 a. **Parameter of interest:** The proportion of voters who favor or oppose (i.e., the proportion of urban voters who favor, the proportion of suburban voters who favor, the proportion of rural voters who favor, and the proportion of all three groups, separately, who oppose).

b. **Statement of hypotheses:**

H_o: The proportion of voters who favor the proposed legislation is the same in all three residence groups.

H_a: The proportion of voters who favor the proposed legislation is not the same in all three groups. (That is, in at least one group the proportion is different from the others.)

Step 2 a. **Assumptions:** The sample information is obtained using three random samples drawn from three separate populations in which each individual is classified according to his or her opinion.

b. **Test statistic:** The chi-square distribution and formula (11.1), with df $= (r - 1)(c - 1) = (3 - 1)(2 - 1) = 2$

c. **Level of significance:** $\alpha = 0.05$.

Step 3 a. **Sample information:** See Table 11.10.

b. **Calculated test statistic:** The expected values are found by using formula (11.5) (p. 637) and are given in Table 11.11.

TABLE 11.11

Sample Results and Expected Values

Residence	Governor's Proposal		Total
	Favor	Oppose	
Urban	143 (101.6)	57 (98.4)	200
Suburban	98 (101.6)	102 (98.4)	200
Rural	13 (50.8)	87 (49.2)	100
Total	254	246	500

Note: Each expected value is used twice in the calculation of $\chi^2\star$; therefore, it is a good idea to keep extra decimal places while doing the calculations.

The calculated chi-square is

$$\chi^2\star = \sum_{\text{all cells}} \frac{(O - E)^2}{E} : \chi^2\star = \frac{(143 - 101.6)^2}{101.6} + \frac{(57 - 98.4)^2}{98.4} + \frac{(98 - 101.6)^2}{101.6}$$

$$+ \frac{(102 - 98.4)^2}{98.4} + \frac{(13 - 50.8)^2}{50.8} + \frac{(87 - 49.2)^2}{49.2}$$

$$= 16.87 + 17.42 + 0.13 + 0.13 + 28.13 + 29.04$$

$$= \mathbf{91.72}$$

Step 4   **Probability Distribution:**

OR

p-value:

a. Use the right-hand tail because "larger" values of chi-square disagree with the null hypothesis:
$\mathbf{P} = P(\chi^2\star > 91.72 \mid \mathbf{df = 2})$ as shown in the figure.

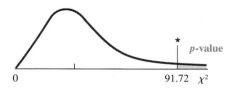

To find the *p*-value, you have two options:
1. Use Table 8 (Appendix B) to place bounds on the *p*-value: **P < 0.005.**
2. Use a computer or calculator to find the *p*-value: **P = 0.000+.**
For specific instructions, see page 518.
b. The *p*-value is smaller than α.

Classical:

a. The critical region is the right-hand tail because "larger" values of chi-square disagree with the null hypothesis. The critical value is obtained from Table 8, at the intersection of row df = 2 and column $\alpha = 0.05$:

$$\chi^2(2, 0.05) = \mathbf{5.99}$$

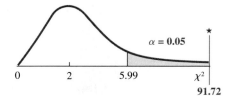

For specific instructions, see page 518.
b. $\chi^2\star$ is in the critical region, as shown in **red** in the figure.

Step 5   **a.** **Decision:** Reject H_o.
           b. **Conclusion:** The three groups of voters do not all have the same proportions favoring the proposed legislation.

TECHNOLOGY INSTRUCTIONS: HYPOTHESIS TEST OF INDEPENDENCE OR HOMOGENEITY

MINITAB (Release 14)   Input each column of observed frequencies from the contingency table into C1, C2, . . . ; then continue with:

Choose: **Stat > Tables > Chi-Square Test (Table in Worksheet)**
Enter: Columns containing the table: **C1 C2 > OK**

COMPUTER SOLUTION MINITAB Printout for Example 11.6:
Chi-square Test: C1, C2
Expected counts are printed below observed counts
Chi-Square contributions are printed below expected counts

```
              C1       C2      Total
     1    143      57       200
          101.60   98.40
          16.870   17.418
     2    98       102      200
          101.60   98.40
          0.128    0.132
     3    13       87       100
          50.80    49.20
          28.127   29.041
Total     254      246      500
Chi-Sq = 91.715, DF = 2, P-Value = 0.000
```

Excel   Input each column of observed frequencies from the contingency table into columns A, B, . . .; then continue with:

Choose: **Tools > Data Analysis Plus > Contingency Table > OK**
Enter: Input range: **(A1:B4 or select cells)**
Select: **Labels** (if necessary)
Enter: Alpha: α (ex. 0.05)

TI-83/84 Plus   Input the observed frequencies from the $r \times c$ contingency table into an $r \times c$ matrix A. Set up matrix B as an empty $r \times c$ matrix for the expected frequencies.

Choose: **MATRX > EDIT > 1:[A]**
Enter: **r > ENTER > c > ENTER**
            Each observed frequency with an ENTER afterward

Then continue with:

Choose: **MATRX > EDIT > 2[B]**
Enter: **r > ENTER > c > ENTER**
Choose: **STAT > TESTS > C:χ^2–Test...**
Enter: Observed: **[A]** or wherever the contingency table is located
            Expected: **[B]** place for expected frequencies
Highlight: **Calculate > ENTER**

APPLIED
EXAMPLE 11.7 Baked Potatoes Rule for Westerners

The graphic "Baked potatoes rule for Westerners" reports the percentage of American's who prefer to eat baked potatoes by region as well as for the whole country. If the actual number of people in each category were given, we would have a contingency table and we would be able to complete a hypothesis test about the homogeneity of the four regions. (See Exercises 11.38 and 11.45.)

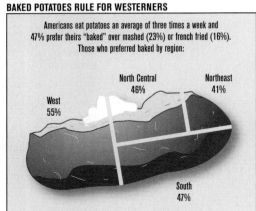

BAKED POTATOES RULE FOR WESTERNERS

Americans eat potatoes an average of three times a week and 47% prefer theirs "baked" over mashed (23%) or french fried (16%). Those who preferred baked by region:

West 55%
North Central 46%
Northeast 41%
South 47%

Data from Anne R. Carey and Sam Ward, © 1998 USA Today.

SECTION 11.4 EXERCISES

Statistics△Now™

Datasets can be found on your Student's Suite CD-ROM or at the StatisticsNow website at **http://1pass.thomson.com**.

11.27 State the null hypothesis, H_o, and the alternative hypothesis, H_a, that would be used to test the following statements:

a. The voters expressed preferences that were not independent of their party affiliations.

b. The distribution of opinions is the same for all three communities.

c. The proportion of "yes" responses was the same for all categories surveyed.

11.28 The "test of independence" and the "test of homogeneity" are completed in identical fashion, using the contingency table to display and organize the calculations. Explain how these two hypothesis tests differ.

11.29 Find the expected value for the cell shown.

□	...	50
⋮		
40		200

11.30 Identify these values from Table 11.7:

a. C_2 b. R_1 c. n d. $E_{2,3}$

11.31 MINITAB was used to complete a chi-square test of independence between the number of boat-related manatee deaths and two Florida counties.

County	Boat-Related Deaths	Non–Boat-Related Deaths	Total Deaths
Lee County	23	25	48
Collier County	8	23	31

Chi-Square Test: Boat-Related Deaths, Non-boat-Related Deaths

Expected counts are printed below observed counts
Chi-Square contributions are printed below expected counts

	Boat-Related Deaths	Non-boat-Related Deaths	Total
1	23	25	48
	18.84	29.16	
	0.921	0.595	
2	8	23	31
	12.16	18.84	
	1.426	0.921	
Total	31	48	79

Chi-Sq = 3.862, DF = 1, P-Value = 0.049

a. Verify the results (expected values and the calculated $\chi^2\star$) by calculating the values yourself.

b. Use Table 8 to verify the p-value based on the calculated df.

c. Is the proportion of boat-related deaths independent of the county? Use $\alpha = 0.05$.

11.32 Results on seat belt usage from the 2003 Youth Risk Behavior Survey were published in a USA Snapshot on January 13, 2005. The following table outlines the results from the high school students who were surveyed in the state of Nebraska. They were asked whether or not they rarely or never wear seat belts when riding in someone else's car.

	Female	Male
Rarely or never use seat belt	208	324
Uses seat belt	1217	1184

Source: http://www.cdc.gov/mmwr/preview/mmwrhtml/ss5302a1.htm#top

Using $\alpha = 0.05$, does this sample present sufficient evidence to reject the hypothesis that gender is independent of seat belt usage?

a. Solve using the *p*-value approach.

b. Solve using the classical approach.

11.33 A random sample of 500 married men was taken; each person was cross-classified as to the size community that he was presently residing in and the size community that he was reared in. The results are shown in the following table.

Size of Community Reared In	Size of Community Residing In			
	Less Than 10,000	10,000 to 49,999	50,000 or Over	Total
Less than 10,000	24	45	45	114
10,000 to 49,999	18	64	70	152
50,000 or over	21	54	159	234
Total	63	163	274	500

Does this sample contradict the claim of independence, at the 0.01 level of significance?

a. Solve using the *p*-value approach.

b. Solve using the classical approach.

11.34 A survey of randomly selected travelers who visited the service station restrooms of a large

U.S. petroleum distributor showed the following results:

Gender of Respondent	Quality of Restroom Facilities			
	Above Average	Average	Below Average	Totals
Female	7	24	28	59
Male	8	26	7	41
Totals	15	50	35	100

Using $\alpha = 0.05$, does the sample present sufficient evidence to reject the hypothesis: "Quality of responses is independent of the gender of the respondent"?

a. Solve using the *p*-value approach.

b. Solve using the classical approach.

11.35 A survey of employees at an insurance firm was concerned with worker–supervisor relationships. One statement for evaluation was, "I am not sure what my supervisor expects." The results of the survey are presented in the following contingency table.

Years of Employment	I Am Not Sure What My Supervisor Expects		
	True	Not True	Totals
Less than 1 year	18	13	31
1 to 3 years	20	8	28
3 to 10 years	28	9	37
10 years or more	26	8	34
Totals	92	38	130

Can we reject the hypothesis that "The responses to the statement and the years of employment are independent" at the 0.10 level of significance?

a. Solve using the *p*-value approach.

b. Solve using the classical approach.

11.36 [EX11-36] The following table is from the publication of *Vital and Health Statistics* from the Centers for Disease Control and Prevention/National Center for Health Statistics. The individuals in the following table have an eye irritation, a nose irritation, or a throat irritation. They have only one of the three.

Type of Irritation	Age (years)			
	18-29	30-44	45-64	65 and Older
Eye	440	567	349	59
Nose	924	1311	794	102
Throat	253	311	157	19

Is there sufficient evidence to reject the hypothesis that the type of ear, nose, or throat irritation is independent of the age group at a level of significance equal to 0.05.

a. Solve using the *p*-value approach.

b. Solve using the classical approach.

11.37 The manager of an assembly process wants to determine whether or not the number of defective articles manufactured depends on the day of the week the articles are produced. She collected the following information.

Day of Week	M	Tu	W	Th	F
Nondefective	85	90	95	95	90
Defective	15	10	5	5	10

Is there sufficient evidence to reject the hypothesis that the number of defective articles is independent of the day of the week on which they are produced? Use $\alpha = 0.05$.

a. Solve using the *p*-value approach.

b. Solve using the classical approach.

11.38 Referring to Applied Example 11.7 (p. 642):

a. Express the percentage of Americans who "prefer baked" to "other" by region as a 2 × 4 contingency table.

b. Explain why the following question could be tested using the chi-square statistic: "Is the preference for baked the same in all four regions of the United States?"

c. Explain why this is a test of homogeneity.

11.39 Blogging is a hot topic nowadays. A "blog" is an Internet log. Blogs are created for personal or professional use. According to the Xtreme Recruiting website (http://www.xtremerecruiting.org/blog/archives/news/cat_news.html), there is a new blog born every 7 seconds—and quite a few

people are reading these blogs. The table that follows shows the number of new blog readers for each of the months listed. Is the distribution of blog creators and readers the same for the months listed? Use $\alpha = 0.05$.

	Blog Creators	Blog Readers
March 2003	74	205
February 2004	93	316
November 2004	130	502

Source: USA Today, "Warning: Your clever little blog could get you fired," June 15, 2005

11.40 Students use many kinds of criteria when selecting courses. "Teacher who is a very easy grader" is often one criterion. Three teachers are scheduled to teach statistics next semester. A sample of previous grade distributions for these three teachers is shown here.

Grades	Professor		
	#1	#2	#3
A	12	11	27
B	16	29	25
C	35	30	15
Other	27	40	23

At the 0.01 level of significance, is there sufficient evidence to conclude "The distribution of grades is not the same for all three professors."

a. Solve using the *p*-value approach.

b. Solve using the classical approach.

c. Which professor is the easiest grader? Explain, citing specific supporting evidence.

11.41 Fear of darkness is a common emotion. The following data were obtained by asking 200 individuals in each age group whether they had serious fears of darkness. At $\alpha = 0.01$, do we have sufficient evidence to reject the hypothesis that "the same proportion of each age group has serious fears of darkness"? (*Hint:* The contingency table must account for all 1000 people.)

Age Group	Elementary	Jr. High	Sr. High	College	Adult
No. Who Fear Darkness	83	72	49	36	114

a. Solve using the *p*-value approach.

b. Solve using the classical approach.

11.42 On May 21, 2004, the National Center for Chronic Disease Prevention and Health Promotion, the Centers for Disease Control and Prevention, reported the results of the Youth Risk Behavior Surveillance—United States, 2003. The report split the sample of 15,184 American teenagers into grade levels as noted in the table that follows. The students admitted to carrying a weapon within the 30 days preceding the survey and to being in a physical fight during the past year. The following table summarizes two portions of the results.

	At Least Once	Never	Total
Carried a weapon			
Grades 9 and 10	1,436	7,008	8,444
Grades 11 and 12	1,140	5,600	6,740
Total	2,576	12,608	15,184
In a physical fight			
Grades 9 and 10	3,057	5,387	8,444
Grades 11 and 12	1,942	4,798	6,740
Total	4,999	10,185	15,184

Source: Data from http://www.cdc.gov/mmwr/preview/mmwrhtml/ss5302a1.htm#top

Does the sample evidence show that students in grades 9 and 10 and grades 11 and 12 have a different tendency to carry weapons to school? get into a physical fight? Use the 0.01 level of significance in each case.

a. Solve using the *p*-value approach.

b. Solve using the classical approach.

11.43 [EX11-43] All new drugs must go through a drug study before being approved by the U.S. Food and Drug Administration (FDA). A drug study typically includes clinical trials whereby participants are randomized to receive different dosages as well as a placebo but are unaware of which group they are in. To control as many factors as possible, it is best to assign participants randomly yet homogeneously across the treatments. Consider the following arrangement for homogeneity with respect to gender and dosages.

Gender	10-mg Drug	20-mg Drug	Placebo
Female	54	56	60
Male	32	27	26

At the 0.01 level of significance, is the distribution of drug the same for both genders?

Considering the same study, homogeneity of ages would also be an important feature. At the 0.01 level of significance, is the distribution of drug that follows the same for all age groups?

Age	10-mg Drug	20-mg Drug	Placebo
40–49	18	20	19
50–59	48	41	57
60–69	20	22	10

11.44 Forty-one small lots of experimental product were manufactured and tested for the occurrence of a particular indication that is attributed in nature yet causes rejection of the part. Ten lots were made using one particular processing method, and thirty-one lots were made using yet a second processing method. Each lot was equally sampled ($n = 32$) for presence of this indication. In practice, optimal processing conditions show little or no occurrence of the indication. Method 1, involving the ten lots, was run before Method 2.

Source: Courtesy of Bausch & Lomb

Methods	n	Number of Rejects
Method 1	320	4
Method 2	992	26

a. Determine, at the 0.05 level of significance, if there is a difference in the proportion of reject product between the two methods.

b. Compare your finding in part a with the findings in Exercise 10.101 (p. 591). Include all parts of the hypothesis tests in your comparison.

c. Would you say these two different testing procedures are equivalent? Give specific evidence supporting your answer.

11.45 Applied Example 11.7 (p. 642) reports percentages describing people's preferences with regard to how potatoes are prepared. Do you believe there is a significant difference between the four regions of America with regard to the percentage who prefer baked? Notice that the article does not mention the sample size.

a. Assume the percentages reported were based on four samples of size 100 from each region and calculate $\chi^2\star$ and its *p*-value.

b. Repeat part a using sample sizes of 200 and 300.

c. Are the four percentages, reported in the graphic, who prefer baked potatoes significantly different? Describe in detail the circumstances for which they are significantly different.

CHAPTER REVIEW

In Retrospect

In this chapter we have been concerned with tests of hypotheses using chi-square, with the cell probabilities associated with the multinomial experiment, and with the simple contingency table. In each case the basic assumptions are that a large number of observations have been made and that the resulting test statistic, $\sum \frac{(O - E)^2}{E}$, is approximately distributed as chi-square. In general, if *n* is large and the minimum allowable expected cell size is 5, then this assumption is satisfied.

The contingency table can be used to test independence and homogeneity. The test for homogeneity and the test for independence look very similar and, in fact, are carried out in exactly the same way. The concepts being tested, however—

same distributions and independence—are quite different. The two tests are easily distinguished because the test of homogeneity has predetermined marginal totals in one direction in the table. That is, before the data are collected, the experimenter determines how many subjects will be observed in each category. The only predetermined number in the test of independence is the grand total.

A few words of caution: the correct number of degrees of freedom is critical if the test results are to be meaningful. The degrees of freedom determine, in part, the critical region, and its size is important. As in other tests of hypothesis, failure to reject H_o does not mean outright acceptance of the null hypothesis.

Vocabulary and Key Concepts

assumptions (p. 621)
cell (p. 620)
chi-square (p. 621)
column (p. 637)
contingency table (pp. 633, 637)
degrees of freedom (pp. 625, 634)
enumerative data (p. 620)

expected frequency (pp. 625, 637)
homogeneity (p. 638)
hypothesis test (pp. 625, 627, 633, 639)
independence (p. 633)
marginal totals (p. 634)

multinomial experiment (pp. 622, 624)
observed frequency (p. 620)
$r \times c$ contingency table (p. 633)
rows (p. 637)
test statistic (p. 621)

Learning Outcomes

✓ Understand that enumerative data are data that can be counted and placed into categories. — pp. 620–622

✓ Understand that the chi-square distribution will be used to test hypotheses involving enumerative data. — pp. 620–621

✓ Understand the properties of the chi-square distribution and how it is a series of distributions based on sample size (using degrees of freedom as the index). — pp. 517, 621, Ex. 11.3, 11.5

✓ Understand the key elements of a multinomial experiment and be able to define n, k, O_i, and P_i. — pp. 622–624

✓ Know and be able to calculate expected values using $E = np$. — EXP 11.2, Ex. 11.13

✓ Know and be able to calculate a chi-square statistic: $\chi^2 = \sum\limits_{\text{all cells}} \dfrac{(O - E)^2}{E}$. — EXP 11.1

✓ Know and be able to calculate the degrees of freedom for a multinomial experiment (df $= k - 1$). — EXP 11.1, 11.2

✓ Perform, describe, and interpret a hypothesis test for a multinomial experiment, using the chi-square distribution with the p-value approach and classical approach. — Ex. 11.14, 11.15, 11.17

✓ Understand and know the definition of independence of two events. — pp. 635–636

✓ Know and be able to calculate expected values using $E_{ij} = \dfrac{R_i \cdot C_j}{n}$. — pp. 635, 637–638, Ex. 11.29

✓ Know and be able to calculate the degrees of freedom for a test of independence or homogeneity [df $= (r - 1)(c - 1)$]. — p. 637

✓ Perform, describe, and interpret a hypothesis test for a test of independence or homogeneity, using the chi-square distribution with the p-value approach and classical approach. — EXP 11.5, 11.6, Ex. 11.34, 11.41

✓ Understand the differences and similarities between tests of independence and tests of homogeneity — p. 646

Chapter Exercises

11.46 The psychology department at a certain college claims that the grades in its introductory course are distributed as follows: 10% As, 20% Bs, 40% Cs, 20% Ds, and 10% Fs. In a poll of 200 randomly selected students who had completed this course, it was found that 16 had received As; 43, Bs; 65, Cs; 48, Ds; and 28, Fs. Does this sample contradict the department's claim at the 0.05 level?

a. Solve using the p-value approach.

b. Solve using the classical approach.

11.47 When interbreeding two strains of roses, we expect the hybrid to appear in three genetic classes in the ratio $1:3:4$. If the results of an experiment yield 80 hybrids of the first type, 340 of the second type, and 380 of the third type, do we have sufficient evidence to reject the hypothesized genetic ratio at the 0.05 level of significance?

a. Solve using the *p*-value approach.

b. Solve using the classical approach.

11.48 A sample of 200 individuals are tested for their blood type, and the results are used to test the hypothesized distribution of blood types:

Blood Type	A	B	O	AB
Percent	0.41	0.09	0.46	0.04

The observed results were as follows:

Blood Type	A	B	O	AB
Number	75	20	95	10

At the 0.05 level of significance, is there sufficient evidence to show that the stated distribution is incorrect?

11.49 [EX11-49] As reported in *USA Today*, about 8.9 million families sent students to college this year and more than half live away from home. Where students live:

Parents' or guardian's home	46%
Campus housing	26%
Off-campus rental	18%
Own off-campus housing	9%
Other arrangements	2%

Note: Exceeds 100% due to rounding error.

A random sample of 1000 college students resulted in the following information:

Parents' or guardian's home	484
Campus housing	230
Off-campus rental	168
Own off-campus housing	96
Other arrangements	22

Is the distribution of this sample significantly different than the distribution reported in the newspaper? Use $\alpha = 0.05$. (To adjust for the rounding error, subtract 2 from each expected frequency.)

a. Solve using the *p*-value approach.

b. Solve using the classical approach.

11.50 [EX11-50] How often do you review your pay stub to check that the correct taxes are being withheld? *USA Today* reported that American adults check as follows:

Always	53%
Most of the time	12%
Occasionally	14%
Never	10%
Don't get a paycheck	10%
Not sure	1%

A random sample of 650 workers resulted in the following frequencies:

Always	342
Most of the time	94
Occasionally	68
Never	73
Don't get a paycheck	60
Not sure	13

Is the distribution of this sample significantly different than the distribution reported in the newspaper? Use $\alpha = 0.05$.

a. Solve using the *p*-value approach.

b. Solve using the classical approach.

11.51 [EX11-51] Most golfers are probably happy to play 18 holes of golf whenever they get a chance to play. Ben Winter, a club professional, played 306 holes in 1 day at a charity golf marathon in Stevens, Pennsylvania. A nationwide survey conducted by *Golf* magazine over the Internet revealed the following frequency distribution of the most number of holes ever played by the respondents in 1 day:

Most Holes Played in 1 Day	Percent	Most Holes Played in 1 Day	Percent
18	5	37 to 45	20
19 to 27	12	46 to 54	18
28 to 36	28	55 or more	17

Source: Golf, "18 Is Not Enough"

Suppose one of your local public golf courses asks the next 200 golfers who tee off to answer the same question. The following table summarizes their responses:

Most Holes Played in 1 Day	Number	Most Holes Played in 1 Day	Number
18	12	37 to 45	44
19 to 27	35	46 to 54	35
28 to 36	60	55 or more	14

Does the distribution of largest number of holes played by "marathon golfers" at your public course differ from the distribution compiled by *Golf* magazine using responses polled on the Internet? Test at the 0.01 level of significance.

a. Solve using the *p*-value approach.

b. Solve using the classical approach.

11.52 [EX11-52] The 2003 U.S. census found that babies entered the world during the various months in the proportions that follow.

Month	P(Month)	Month	P(Month)	Month	P(Month)
January	0.082	May	0.084	September	0.087
February	0.076	June	0.081	October	0.086
March	0.082	July	0.089	November	0.079
April	0.081	August	0.089	December	0.083

Source: U.S. Census Bureau

A random sample selected from the birth records for a large metropolitan area resulted in the following data:

Month	Jan	Feb	Mar	Apr	May	Jun	Jul	Aug	Sep	Oct	Nov	Dec
Observed	14	12	12	10	16	9	16	11	17	7	17	9

a. Do these data provide sufficient evidence to reject the claim, "Births occur in this metropolitan area in the same monthly proportions," as reported by the U.S. Census Bureau? Use $\alpha = 0.05$.

b. Do these data provide sufficient evidence to reject the claim, "Births occur in this metropolitan area in all months with the same likeliness"? Use $\alpha = 0.05$.

c. Compare the results obtained in parts a and b. State your conclusions.

11.53 [EX11-53] The weights (*x*) of 300 adult males were determined and used to test the hypothesis that the weights were normally distributed with a mean of 160 lb and a standard devia-

tion of 15 lb. The data were grouped into the following classes.

Weight (*x*)	Observed Frequency	Weight (*x*)	Observed Frequency
$x < 130$	7	$160 \leq x < 175$	102
$130 \leq x < 145$	38	$175 \leq x < 190$	40
$145 \leq x < 160$	100	190 and over	13

Using the normal tables, the percentages for the classes are 2.28%, 13.59%, 34.13%, 34.13%, 13.59%, and 2.28%, respectively. Do the observed data show significant reason to discredit the hypothesis that the weights are normally distributed with a mean of 160 lb and a standard deviation of 15 lb? Use $\alpha = 0.05$.

a. Verify the percentages for the classes.

b. Solve using the *p*-value approach.

c. Solve using the classical approach.

11.54 Do you have a favorite "comfort food"? How do you obtain it? The following graphic lists three methods used by Americans and the percentage each method is used. Which category do you belong to? A random sample of 120 Americans living on the East Coast was asked, "How do you obtain your favorite comfort food?"

OBTAINING OUR FAVORITE COMFORT FOOD

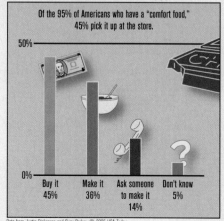

Of the 95% of Americans who have a "comfort food," 45% pick it up at the store.

| | Buy it 45% | Make it 36% | Ask someone to make it 14% | Don't know 5% |

Data from Justin Dickerson and Suzy Parker. © 2005 USA Today.

Comfort Food	Buy it	Make it	Ask someone to make it	Don't know
East Coast	57	44	12	7

Using the percentages given in the graphic as the national "standard," does the evidence indicate that the East Coast responses are different than those of the nation as a whole? Use $\alpha = 0.05$.

a. Solve using the *p*-value approach.

b. Solve using the classical approach.

11.55 A random sample of 120 Americans living in the Midwest was asked, "How do you obtain your favorite comfort food?"

Comfort Food	Buy it	Make it	Ask someone to make it	Don't know
Midwest	54	48	17	1

Using the percentages given in the graphic (p. 649) as the national "standard," does the evidence indicate that the Midwest responses are different than those of the nation as a whole? Use $\alpha = 0.05$.

a. Solve using the *p*-value approach.

b. Solve using the classical approach.

c. Explain the similarities and differences found in Exercises 11.54 and 11.55.

11.56 A random sample of 120 Americans living along the West Coast was asked, "How do you obtain your favorite comfort food?"

Comfort Food	Buy it	Make it	Ask someone to make it	Don't know
West Coast	73	26	18	3

Using the percentages given in the graphic (p. 649) as the national "standard," does the evidence indicate that the West Coast responses are different than those of the nation as a whole? Use $\alpha = 0.05$.

a. Solve using the *p*-value approach.

b. Solve using the classical approach.

c. Explain the similarities and differences found in Exercises 11.54, 11.55, and 11.56.

11.57 [EX11-57] The following table gives the color counts for a sample of 30 bags (47.9 gram size) of M&M's.

Case	Red	Gr	Blue	Or	Yel	Br
1	15	9	3	3	9	19
2	9	17	19	3	3	8
••• Remainder of data on Student's Suite CD-ROM

Source: http://www.math.uah.edu/stat/, Christine Nickel and Jason York, ST 687 project, Fall 1998

Before the Global Color Vote (GCV) of 2002, the target percentage for each color in the six-color mix was as follows: brown, 30%; red and yellow, 20%; blue, green, and orange, 10%.

a. Does case 1 show that bag 1 has a significantly different distribution of colors than the target distribution? Use $\alpha = 0.05$.

b. Combine cases 1 and 2. Does the total of bags 1 and 2 show a significantly different distribution of colors than the target distribution?

c. Combine the results of all 30 cases. Does the total of all 30 bags show a significantly different distribution of colors than the target distribution?

d. Discuss the findings of parts a–c.

11.58 The January 5, 2005, *USA Today* article "Hip-hop rules radio" gave the following results from Nielsen BDS and Arbitron on the top 100 hits of 2003 and 2004.

	R&B/Hip-Hop	Pop	Rock	Country	Total
2003	53	25	8	14	100
2004	61	16	12	11	100

Does this sample information show that the distribution of radio's top 100 hits changed from 2003 to 2004? Use $\alpha = 0.05$.

11.59 [EX11-59] Based on data from the U.S. Census Bureau, the National Association of Home Builders forecasts a rise in homeownership rates for this decade. Part of the forecast is to predict new housing starts by region. The following table shows what they are forecasting.

	Average Housing Starts		
Region	1996–2000	2001–2005	2006–2010
Northeast	145	161	170
South	710	687	688
Midwest	331	314	313
West	382	385	373

Do the data present sufficient evidence to reject the hypothesis that the distribution of housing starts across the regions is the same for all years?

a. Solve using the *p*-value approach.

b. Solve using the classical approach.

11.60 [EX11-60] The following table shows the number of reported crimes committed last year in the inner part of a large city. The crimes were classified according to the type of crime and district of the inner city where it occurred. Do these data show sufficient evidence to reject the hypothesis that the type of crime and the district in which it occurred are independent? Use $\alpha = 0.01$.

District	Crime				
	Robbery	Assault	Burglary	Larceny	Stolen Vehicle
1	54	331	227	1090	41
2	42	274	220	488	71
3	50	306	206	422	83
4	48	184	148	480	42
5	31	102	94	596	56
6	10	53	92	236	45

a. Solve using the *p*-value approach.

b. Solve using the classical approach.

11.61 [EX11-61] Based on the results of a survey questionnaire, 400 individuals were classified as either politically conservative, moderate, or liberal. In addition, each person was classified by age, as shown in the following table.

	Age Group			
	20–35	36–50	Older Than 50	Totals
Conservative	20	40	20	80
Moderate	80	85	45	210
Liberal	40	25	45	110
Totals	140	150	110	400

Is there sufficient evidence to reject the hypothesis that "political preference is independent of age"? Use $\alpha = 0.01$.

a. Solve using the *p*-value approach.

b. Solve using the classical approach.

11.62 [EX11-62] "Cramped quarters" is a common complaint by airline travelers. A random sample of 150 business travelers and 150 leisure travelers were asked where they would "most like more space." The resulting answers are summarized as follows.

Place	Business	Leisure
Overhead space on plane	15	9
Hotel room	29	49
Leg room on plane	91	66
Rental car size	10	20
Other	5	6

Does this sample information present sufficient evidence to conclude that the business traveler and the leisure traveler differ in where they would most like additional space? Use $\alpha = 0.05$.

a. Solve using the *p*-value approach.

b. Solve using the classical approach.

11.63 Four brands of popcorn were tested for popping. One hundred kernels of each brand were popped, and the number of kernels not popped was recorded in each test (see the following table). Can we reject the null hypothesis that all four brands pop equally? Test at $\alpha = 0.05$.

Brand	A	B	C	D
No. Not Popped	14	8	11	15

a. Solve using the *p*-value approach.

b. Solve using the classical approach.

11.64 [EX11-64] An average of two players per boys' or girls' high school basketball team is injured during a season. The following table shows the distribution of injuries for a random sample of 1000 girls and 1000 boys taken from the season records of all reported injuries.

Injury	Girls	Boys
Ankle/foot	360	383
Hip/thigh/leg	166	147
Knee	130	103
Forearm/wrist/hand	112	115
Face/scalp	88	122
All others	144	130

Does this sample information present sufficient evidence to conclude that the distribution of injuries is different for girls than for boys? Use $\alpha = 0.05$.

a. Solve using the *p*-value approach.

b. Solve using the classical approach.

11.65 [EX11-65] "Have you designated on your driving license that you are an organ donor?" We have all heard this question, and the word must be getting out according to the March 30, 2005, article "Organ Transplants Reach New High of Almost 27,000 in 2004." The exact results for types of organ donations are as follows.

	From a Deceased Donor	From a Living Donor
2003	18,650	6,812
2004	20,018	6,966

Source: http://www.seniorjournal.com/NEWS/Features/5-03-30DonateLife.htm

a. What percent of organ donors were deceased for each year? Do you view these percentages as significantly different? Explain.

b. At the 0.05 level of significance, did the rates of deceased donor to living donor change significantly between 2003 and 2004?

c. Compare the decision reached in part b to your answer in part a. Describe any differences and explain what caused them.

11.66 [EX11-66] Last year's work record for absenteeism in each of four categories for 100 randomly selected employees is compiled in the following table. Do these data provide sufficient evidence to reject the hypothesis that the rate of absenteeism is the same for all categories of employees? Use $\alpha = 0.01$ and 240 workdays for the year.

	Married Male	Single Male	Married Female	Single Female
Number of Employees	40	14	16	30
Days Absent	180	110	75	135

a. Solve using the *p*-value approach.

b. Solve using the classical approach.

11.67 If you were to roll a die 600 times, how different from 100 could the observed frequencies for each face be before the results would become significantly different from equally likely at the 0.05 level?

11.68 Consider the following set of data.

	Response		Total
	Yes	No	
Group 1	75	25	100
Group 2	70	30	100
Total	145	55	200

a. Compute the value of the test statistic $z\star$ that would be used to test the null hypothesis that $p_1 = p_2$, where p_1 and p_2 are the proportions of "yes" responses in the respective groups.

b. Compute the value of the test statistic $\chi^2\star$ that would be used to test the hypothesis that "response is independent of group."

c. Show that $\chi^2\star = (z\star)^2$.

Chapter Project

Cooling a Great Hot Taste

The graphic shown in Section 11.1, "Cooling a Great Hot Taste" (p. 619), shows the top six ways adults say they cool their mouths after eating food with hot sauce. Do these variables and percents seem reasonable? They may to some people and may not to others. That's where a multinomial experiment comes into play. A sample can be tested against the stated statistics to determine whether the sample comes from the same given distribution or if it contradicts the given distribution.

Suppose 200 adults professing a love for hot, spicy food were asked to name their favorite way to cool their mouth after eating food with hot sauce. Following is the summary of the resulting sample.

Method	Water	Milk	Soda	Beer	Bread	Other	Nothing
Number	73	35	20	19	29	11	13

Let's investigate this sample with respect to the distribution given in the graphic on page 619.

Putting Chapter 11 to Work

11.69 [EX11-01] a. Does the sample show a distribution that is significantly different from the distribution shown in the "Putting out the fire" graph (p. 619)? Use $\alpha = 0.05$.

b. Write a paragraph (50+ words) describing why the statistical method used in part a is appropriate for this set of data.

c. Write a paragraph (50+ words) describing the meaning of the assumptions and the results of the statistical procedure chosen in part b.

Your Study

11.70 Design your own study for "favorite way people cool their mouth after eating something hot."

a. Define a specific population that you will sample, describe your sampling plan, and collect a random sample of at least 100 observations.

b. Make a descriptive presentation of your sample data, using a chart, at least one graph, and a descriptive paragraph.

c. Does your sample show a distribution that is significantly different from the distribution shown in the graphic on page 619? Use $\alpha = 0.05$.

d. Discuss the differences and similarities between your sample and the distribution shown in the graphic "Putting out the fire."

Chapter Practice Test

Part I: Knowing the Definitions

Answer "True" if the statement is always true. If the statement is not always true, replace the words shown in bold with words that make the statement always true.

11.1 The number of degrees of freedom for a test of a multinomial experiment is **equal to** the number of cells in the experimental data.

11.2 The **expected frequency** in a chi-square test is found by multiplying the hypothe-sized probability of a cell by the total number of observations in the sample.

11.3 The **observed** frequency of a cell should not be allowed to be smaller than 5 when a chi-square test is being conducted.

11.4 In a **multinomial experiment** we have $(r - 1)(c - 1)$ degrees of freedom (r is the number of rows, and c is the number of columns).

11.5 A multinomial experiment consists of n **identical independent trials.**

11.6 A **multinomial experiment** arranges the data in a two-way classification such that the totals in one direction are predetermined.

11.7 The charts for both the multinomial experiment and the contingency table **must** be set in such a way that each piece of data will fall into exactly one of the categories.

11.8 The test statistic $\sum \frac{(O - E)^2}{E}$ has a distribution that is **approximately normal.**

11.9 The data used in a chi-square multinomial test are always **enumerative.**

11.10 The null hypothesis being tested by a test of **homogeneity** is that the distribution of proportions is the same for each of the subpopulations.

Part II: Applying the Concepts

Answer all questions. Show formulas, substitutions, and work.

11.11 State the null and alternative hypotheses that would be used to test each of these claims:

a. The single-digit numerals generated by a certain random-number generator were not equally likely.

b. The results of the last election in our city suggest that the votes cast were not independent of the voter's registered party.

c. The distributions of types of crimes committed against society are the same in the four largest U.S. cities.

11.12 Find each value:

a. $\chi^2(12, 0.975)$ b. $\chi^2(17, 0.005)$

11.13 Three hundred consumers were asked to identify which one of three different items they found to be the most appealing. The table shows the number that preferred each item.

Item	1	2	3
Number	85	103	112

Do these data present sufficient evidence at the 0.05 level of significance to indicate that the three items are not equally preferred?

11.14 To study the effect of the type of soil on the amount of growth attained by a new hybrid plant, saplings were planted in three different types of soil and their subsequent amounts of growth classified into three categories:

	Soil Type		
Growth	Clay	Sand	Loam
Poor	16	8	14
Average	31	16	21
Good	18	36	25
Total	65	60	60

Does the quality of growth appear to be distributed differently for the tested soil types at the 0.05 level?

a. State the null and alternative hypotheses.

b. Find the expected value for the cell containing 36.

c. Calculate the value of chi-square for these data.

d. Find the *p*-value.

e. Find the test criteria [level of significance, test statistic, its distribution, critical region, and critical value(s)].

f. State the decision and the conclusion for this hypothesis test.

Part III: Understanding the Concepts

11.15 Explain how a multinomial experiment and a binomial experiment are similar and also how they are different.

11.16 Explain the distinction between a test for independence and a test for homogeneity.

11.17 Student A says that tests for independence and homogeneity are the same, and student B says that they are not at all alike because they are tests of different concepts. Both students are partially right and partially wrong. Explain.

11.18 You are interpreting the results of an opinion poll on the role of recycling in your town. A random sample of 400 people was asked to respond strongly in favor, slightly in favor, neutral, slightly against, or strongly against on each of several questions. There are four key questions that concern you, and you plan to analyze their results.

a. How do you calculate the expected probabilities for each answer?

b. How would you decide whether the four questions were answered the same?

CHAPTER

12

Analysis of Variance

© Derek Trask/CORBIS

12.1   Time Spent Commuting to Work

Statistics ◯ Now™

Throughout the chapter, this icon introduces a list of resources on the StatisticsNow website at **http://1pass.thomson.com** that will:

- Help you evaluate your knowledge of the material
- Allow you to take an exam-prep quiz
- Provide a Personalized Learning Plan targeting resources that address areas you should study

How much time did America's workforce spend getting to their jobs this morning? How much time did your parents spend commuting to work this morning? Does everybody spend the same amount of time? What was the mean amount of time spent commuting to work this morning by people in Boston? What was the mean amount of time spent commuting to work this morning by people in Dallas? Do you think that the city will have any effect on the amount of time spent commuting to work this morning? The graphic "Longest commute to work" seems to suggest that some cities have longer commuting times than others. From studying previous chapters, we know that the statistics from different samples, even if drawn from the same population, vary. The question that might be asked here is, "Is the variation between the samples greater than would be expected if the samples were all drawn from one population?"

To compare the commuting time in various locations, independent and random samples were obtained in each of six different U.S. cities from workers who commute to work during the 8:00 AM rush hour.

LONGEST COMMUTE TO WORK

Among U.S. cities with populations of 250,000 or more, New York City has the longest commute. The average one-way commuting time for all large U.S. cities is 24.3 minutes.

Riverside 31.2 min.   Newark 31.5 min.   Chicago 33.2 min.   New York 38.3 min.

Data from Anne R. Carey and Juan Thomassie, © 2005 USA TODAY.

TABLE 12.1

One-Way Travel to Work (in minutes) [EX12-01]

Atlanta	Boston	Dallas	Philadelphia	Seattle	St. Louis	Atlanta	Boston	Dallas	Philadelphia	Seattle	St. Louis
29	18	42	29	30	15	37	32	20	42	30	33
21	37	25	20	19	54	26	34	26			35
20	37	36	33	31	42		48	35			
15	25	32	37	39	23						

Does there seem to be a difference between the mean one-way commute time for these six cities? How would you measure the variability in the six means? How would you describe the six samples individually so that you could make a comparison? How would you measure the variability within the six samples? Some of the techniques studied in previous chapters help, but in this chapter we will learn some new techniques designed specifically to make comparisons similar to the ones proposed here. After completing Chapter 12, you will have an opportunity to further investigate the preceding situation in the Chapter Project section on page 690.

SECTION 12.1 EXERCISES

Statistics⊖Now™

Datasets can be found on your Student's Suite CD-ROM or at the StatisticsNow website at **http://1pass.thomson.com**.

12.1 [EX12-01] a. Construct a graphic representation of the Table 12.1 data using six side-by-side dotplots.

b. Visually estimate the mean commute time for each city and locate it with an X.

c. Does it appear that the city has an effect on the average amount of time spent by workers who commute to work during the 8:00 AM rush hour? Explain.

d. Does it visually appear that the city has an effect on the variation in the amount of time spent by workers who commute to work during the 8:00 AM rush hour? Explain.

12.2 a. Calculate the mean commute time for each city in Table 12.1.

b. Does there seem to be a difference among the mean one-way commute times for these six cities?

c. Calculate the standard deviation for each of the six cities' commute times.

d. Does there seem to be a difference among the standard deviations of the one-way commute times for these six cities?

12.2 Introduction to the Analysis of Variance Technique

Previously, we have tested hypotheses about two means. In this chapter we are concerned with testing a hypothesis about several means. The analysis of variance technique (ANOVA), which we are about to explore, will be used to test a null hypothesis about several means, for example,

$$H_o: \mu_1 = \mu_2 = \mu_3 = \mu_4 = \mu_5$$

By using our former technique for hypotheses about two means, we could test several hypotheses if each stated a comparison of two means. For example, we could test

$$H_1: \mu_1 = \mu_2 \quad H_2: \mu_1 = \mu_3 \quad H_3: \mu_1 = \mu_4 \quad H_4: \mu_1 = \mu_5 \quad H_5: \mu_2 = \mu_3$$
$$H_6: \mu_2 = \mu_4 \quad H_7: \mu_2 = \mu_5 \quad H_8: \mu_3 = \mu_4 \quad H_9: \mu_3 = \mu_5 \quad H_{10}: \mu_4 = \mu_5$$

To test the null hypothesis, H_o, that all five means are equal, we would have to test each of these 10 hypotheses using our former technique. Rejection of any one of the 10 hypotheses about two means would cause us to reject the null hypothesis that all five means are equal. If we failed to reject all 10 hypotheses, we would fail to reject the main null hypothesis. By testing in this manner, the overall type I error rate would become much larger than the value of α associated with a single test. The ANOVA techniques allow us to test the null hypothesis (all means are equal) against the alternative hypothesis (at least one mean value is different) with a specified value of α.

In this chapter we introduce ANOVA. ANOVA experiments can be very complex, depending on the situation. We will restrict our discussion to the most basic experimental design—the single-factor ANOVA. We will begin our discussion of the analysis of variance technique by looking at an example.

EXAMPLE 12.1

Hypothesis Test for Several Means

The temperature at which a manufacturing plant is maintained is believed to affect the rate of production in the plant. The data in Table 12.2 are the number, x, of units produced in 1 hour for randomly selected 1-hour periods when the production process in the plant was operating at each of three temperature *levels*. The data values from repeated samplings are called **replicates.** Four replicates, or data values, were obtained for two of the temperatures and five were obtained for the third temperature. Do these data suggest that temperature has a significant effect on the production level at $\alpha = 0.05$?

TABLE 12.2

Sample Results on Temperature and Production

	Temperature Levels		
	Sample from 68°F ($i = 1$)	Sample from 72°F ($i = 2$)	Sample from 76°F ($i = 3$)
	10	7	3
	12	6	3
	10	7	5
	9	8	4
		7	
Column totals	$C_1 = 41$ $\bar{x}_1 = 10.25$	$C_2 = 35$ $\bar{x}_2 = 7.0$	$C_3 = 15$ $\bar{x}_3 = 3.75$

The level of production is measured by the mean value; $\bar{x}_i$ indicates the observed production mean at level i, where $i = 1$, 2, and 3 corresponds to temperatures of 68°F, 72°F, and 76°F, respectively. There is a certain amount of variation among these means. Since sample means are not necessarily the same when repeated samples are taken from a population, some variation can be expected, even if all three population means are equal. We will next pursue the question: Is this variation among the $\bar{x}$'s due to chance, or is it due to the effect that temperature has on the production rate?

SOLUTION

STEP 1 a. **Parameter of interest:** The "mean" at each *level of the test factor* is of interest: the mean production rate at 68°F, μ_{68}; the mean production rate at 72°F, μ_{72}; and the mean production rate at 76°F, μ_{76}. The factor being tested, plant temperature, has three levels: 68°F, 72°F, or 76°F.

b. **Statement of hypotheses:**

$$H_o: \mu_{68} = \mu_{72} = \mu_{76}$$

That is, the true production mean is the same at each temperature level tested. In other words, the temperature does not have a significant effect on the production rate. The alternative to the null hypothesis is

$$H_a: \text{Not all temperature level means are equal.}$$

Thus, we will want to reject the null hypothesis if the data show that one or more of the means are significantly different from the others.

STEP 2 a. **Assumptions:** The data were randomly collected and are independent of each other. The effects due to chance and untested factors are assumed to be normally distributed. (See p. 668 for further discussion.)

b. **Test statistic:** We will make the decision to reject H_o or fail to reject H_o by using the F-distribution and an F-test statistic.

c. **Level of significance:** $\alpha = 0.05$ (given in the statement of the problem).

STEP 3 a. **Sample information:** See Table 12.2.

b. **Calculated test statistic.**

Recall from Chapter 10 that the calculated value of F is the ratio of two variances. The analysis of variance procedure will separate the variation among the entire set of data into two categories. To accomplish this separation, we first work with the numerator of the fraction used to define **sample variance,** formula (2.6) (p. 86):

$$s^2 = \frac{\sum (x - \bar{x})^2}{n - 1}$$

The numerator of this fraction is called the **sum of squares:**

Total Sum of Squares

$$\text{sum of squares} = \sum (x - \bar{x})^2 \qquad (12.1)$$

We calculate the **total sum of squares, SS(total),** for the total set of data by using a formula that is equivalent to formula (12.1) but does not require the use of $\bar{x}$. This equivalent formula is

Shortcut for Total Sum of Squares

$$SS(total) = \sum(x^2) - \frac{(\sum x)^2}{n} \qquad (12.2)$$

Now we can find SS(total) for our example by using formula (12.2). First,

$$\sum(x^2) = 10^2 + 12^2 + 10^2 + 9^2 + 7^2 + 6^2 + 7^2 + 8^2 + 7^2 + 3^2 + 3^2 + 5^2 + 4^2 = 731$$

$$\sum x = 10 + 12 + 10 + 9 + 7 + 6 + 7 + 8 + 7 + 3 + 3 + 5 + 4 = 91$$

Then, using formula (12.2), we have

$$SS(total) = \sum(x^2) - \frac{(\sum x)^2}{n}: \qquad SS(total) = 731 - \frac{(91)^2}{13} = 731 - 637 = \mathbf{94}$$

Next, 94, SS(total), must be separated into two parts: the sum of squares due to temperature levels, SS(temperature), and the sum of squares due to experimental error of replication, SS(error). This splitting is often called **partitioning,** since SS(temperature) + SS(error) = SS(total); that is, in our example SS(temperature) + SS(error) = 94. The sum of squares, **SS(factor)** [SS(temperature) for our example], that measures the **variation between the factor levels** (temperatures) is found by using formula (12.3):

Sum of Squares Due to Factor

$$SS(factor) = \left(\frac{C_1^2}{k_1} + \frac{C_2^2}{k_2} + \frac{C_3^2}{k_3} + \cdots \right) - \frac{(\sum x)^2}{n} \qquad (12.3)$$

where C_i represents the column total, k_i represents the number of replicates at each level of the factor, and n represents the total sample size ($n = \sum k_i$)

Note: The data have been arranged so that each column represents a different level of the factor being tested.

Now we can find SS(temperature) for our example by using formula (12.3):

$$SS(factor) = \left(\frac{C_1^2}{k_1} + \frac{C_2^2}{k_2} + \frac{C_3^2}{k_3} + \cdots \right) - \frac{(\sum x)^2}{n}:$$

$$SS(temperature) = \left(\frac{41^2}{4} + \frac{35^2}{5} + \frac{15^2}{4} \right) - \frac{(91)^2}{13}$$

$$= (420.25 + 245.00 + 56.25) - 637.0 = 721.5 - 637.0 = \mathbf{84.5}$$

The sum of squares, **SS(error),** that measures the **variation within the rows** is found by using formula (12.4):

Sum of Squares Due to Error

$$SS(\text{error}) = \sum (x^2) - \left(\frac{C_1^2}{k_1} + \frac{C_2^2}{k_2} + \frac{C_3^2}{k_3} + \cdots \right) \qquad (12.4)$$

The SS(error) for our example can now be found. First,

$$\sum (x^2) = 731 \qquad \text{(found previously)}$$

$$\left(\frac{C_1^2}{k_1} + \frac{C_2^2}{k_2} + \frac{C_3^2}{k_3} + \cdots \right) = 721.5 \qquad \text{(found previously)}$$

Then, using formula (12.4), we have

$$SS(\text{error}) = \sum (x^2) - \left(\frac{C_1^2}{k_1} + \frac{C_2^2}{k_2} + \frac{C_3^2}{k_3} + \cdots \right): \quad SS(\text{error}) = 731.0 - 721.5 = \textbf{9.5}$$

Note: SS(total) = SS(factor) + SS(error). Inspection of formulas (12.2), (12.3), and (12.4) will verify this.

For convenience we will use an ANOVA table to record the sums of squares and to organize the rest of the calculations. The format of an ANOVA table is shown in Table 12.3.

TABLE 12.3

Format for ANOVA Table

Source	df	SS	MS
Factor		84.5	
Error		9.5	
Total		94.0	

We have calculated the three sums of squares for our example. The degrees of freedom, df, associated with each of the three sources are determined as follows:

1. df(factor) is 1 less than the number of levels (columns) for which the factor is tested:

Degrees of Freedom for Factor

$$df(\text{factor}) = c - 1 \qquad (12.5)$$

where c is the number of *levels for which the factor is being tested* (number of columns on the data table)

2. df(total) is 1 less than the total number of data:

Degrees of Freedom for Total

$$\text{df(total)} = n - 1 \qquad (12.6)$$

where n is the number of data in the total sample (i.e., $n = k_1 + k_2 + k_3 + \cdots$, where k_i is the number of replicates at each level tested)

3. df(error) is the sum of the degrees of freedom for all the levels tested (columns in the data table). Each column has $k_i - 1$ degrees of freedom; therefore,

$$\text{df(error)} = (k_1 - 1) + (k_2 - 1) + (k_3 - 1) + \cdots$$

or

Degrees of Freedom for Error

$$\text{df(error)} = n - c \qquad (12.7)$$

The degrees of freedom for our illustration are

$$\text{df(temperature)} = c - 1 = 3 - 1 = \mathbf{2}$$

$$\text{df(total)} = n - 1 = 13 - 1 = \mathbf{12}$$

$$\text{df(error)} = n - c = 13 - 3 = \mathbf{10}$$

The sums of squares and the degrees of freedom must check; that is,

$$\text{SS(factor)} + \text{SS(error)} = \text{SS(total)} \qquad (12.8)$$

and

$$\text{df(factor)} + \text{df(error)} = \text{df(total)} \qquad (12.9)$$

The **mean square** for the factor being tested, **MS(factor),** and for error, **MS(error),** are obtained by dividing the sum-of-squares value by the corresponding number of degrees of freedom:

Mean Square for Factor

$$\text{MS(factor)} = \frac{\text{SS(factor)}}{\text{df(factor)}} \qquad (12.10)$$

Mean Square for Error

$$\text{MS(error)} = \frac{\text{SS(error)}}{\text{df(error)}} \qquad (12.11)$$

The mean squares for our example are

$$MS(\text{temperature}) = \frac{SS(\text{temperature})}{df(\text{temperature})} = \frac{84.5}{2} = \mathbf{42.25}$$

$$MS(\text{error}) = \frac{SS(\text{error})}{df(\text{error})} = \frac{9.5}{10} = \mathbf{0.95}$$

The complete ANOVA table appears in Table 12.4.

TABLE 12.4

ANOVA Table for Example 12.1

Source	df	SS	MS
Temperature	2	84.5	42.25
Error	10	9.5	0.95
Total	12	94.0	

The hypothesis test is now completed using the two mean squares as the measures of variance. The calculated value of the test statistic, $F\star$, is found by dividing the MS(factor) by the MS(error):

Test Statistic for ANOVA

$$F\star = \frac{MS(\text{factor})}{MS(\text{error})} \tag{12.12}$$

The calculated value of F for our example is found by using formula (12.12):

$$F\star = \frac{MS(\text{factor})}{MS(\text{error})}: \qquad F\star = \frac{MS(\text{temperature})}{MS(\text{error})} = \frac{42.25}{0.95} = \mathbf{44.47}$$

Note: Since the calculated value of F, $F\star$, is found by dividing MS(temperature) by MS(error), the number of degrees of freedom for the numerator is df(temperature) = 2 and the number of degrees of freedom for the denominator is df(error) = 10.

Step 4   **Probability Distribution:**

OR

p-**Value:**

a. Use the right-hand tail because larger values of $F\star$ indicate "not all equal" as expressed by H_a, $\mathbf{P} = P(F\star > 44.47 \mid df_n = 2, df_d = 10)$ as shown in the figure.

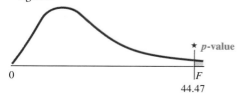

Classical:

a. The critical region is the right-hand tail because larger values of $F\star$ indicate "not all equal" as expressed by H_a, $df_n = 2$ and $df_d = 10$. The critical value is obtained from Table 9A:

$$F(2, 10, 0.05) = 4.10$$

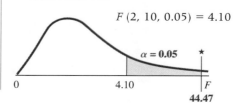

To find the *p*-value, you have two options:
1. Use Table 9C (Appendix B) to place bounds on the *p*-value: **P < 0.01.**
2. Use a computer or calculator to find the *p*-value: **P = 0.00001.**
For additional instructions, see page 597.
b. The *p*-value is smaller than the level of significance, $\alpha(0.05)$.

For additional instructions, see page 594.
b. $F\star$ is in the critical region, as shown in **red** in the figure.

Step 5   **a.** **Decision:** Reject H_o.
    b. **Conclusion:** At least one of the room temperatures does have a significant effect on the production rate. The differences in the mean production rates at the tested temperature levels were found to be significant.

The mean at 68°F is certainly different from the mean at 75°F because the sample means for these levels are the largest and the smallest, respectively. Whether any other pairs of means are significantly different cannot be determined from the ANOVA procedure alone.

In this section we have seen how the ANOVA technique separated the variance among the sample data into two measures of variance: (1) MS(factor), the measure of variance between the levels being tested, and (2) MS(error), the measure of variance within the levels being tested. Then these measures of variance can be compared. For our example, the between-level variance was found to be significantly greater than the within-level variance (experimental error). This led us to the conclusion that temperature did have a significant effect on the variable *x*, the number of units of production completed per hour.

In the next section we will demonstrate the logic of the analysis of variance technique.

SECTION 12.2 EXERCISES

12.3 Draw a dotplot of the data in Table 12.2 (p. 659). Represent the data using the integers 1, 2, and 3, indicating the level of test factor the data are from. Do you see a "difference" between the levels?

12.4 Each department at a large industrial plant is rated weekly. State the hypotheses used to test that "the mean weekly ratings are the same in three departments."

12.5 Refer to the following ANOVA table.

Source	df	SS	MS
Factor	3		
Error		40.4	
Total	20	164.2	

a. Find the four missing values.
b. Find the calculated value for F, $F\star$.

12.6 An analysis of variance experiment with level A containing 10 data values; level B, 12 data values; level C, 10 values; level D, 12 values; level E, 9 values; and level F, 10 values was analyzed using MINITAB.

```
One-way ANOVA: Level A, Level B, Level C, Level D, Level
E, Level F
Source      DF      SS      MS      F       P
Factor      5       6355    1271    3.15
0.014
Error       57      22964   403
Total       62      29319
```

a. Verify the three values for df shown on the printout. Also verify the relationship between the three numbers.

b. Verify the two MS values reported on the printout.

c. Verify the *F*-value.   d. Verify the *p*-value.

12.3 The Logic behind ANOVA

Many experiments are conducted to determine the effect that different levels of some test factor have on a **response variable.** The test factor may be temperature (as in Example 12.1), the manufacturer of a product, the day of the week, or any number of other things. In this chapter we are investigating the single-factor analysis of variance. Basically, the design for the single-factor ANOVA is to obtain independent random samples at each of the several *levels of the factor being tested.* We then make a statistical decision concerning the effect that the levels of the test factors have on the response (observed) variable.

Examples 12.2 and 12.3 demonstrate the logic of the analysis of variance technique. Briefly, the reasoning behind the technique proceeds like this: to compare the means of the levels of the test factor, a measure of the **variation between the levels** (between the columns on the data table), the **MS(factor),** will be compared with a measure of the **variation within the levels** (within the columns on the data table), the **MS(error).** If MS(factor) is significantly larger than MS(error), we will conclude that the means for the factor levels being tested are not all the same. This implies that the factor being tested does have a significant effect on the response variable. If, however, MS(factor) is not significantly larger than MS(error), we will not be able to reject the null hypothesis that all means are equal.

EXAMPLE 12.2

Visualizing the Difference among Several Means

Do the data in Table 12.5 provide sufficient evidence to conclude that there is a difference in the three population means μ_F, μ_G, and μ_H?

TABLE 12.5

Sample Results

	Factor Levels	
Sample from Level *F*	Sample from Level *G*	Sample from Level *H*
3	5	8
2	6	7
3	5	7
4	5	8
$C_F = 12$	$C_G = 21$	$C_H = 30$
$\bar{x}_F = 3.00$	$\bar{x}_G = 5.25$	$\bar{x}_H = 7.50$

SOLUTION Figure 12.1 shows the relative relationship among the three samples. A quick look at the figure suggests that the three sample means are different from each other, implying that the sampled populations have different mean values. These three samples demonstrate relatively little within-sample variation, although there is a relatively large amount of between-sample variation.

FIGURE 12.1 Data from Table 12.5

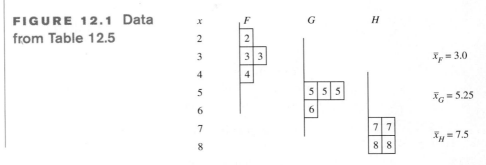

Let's look at another example.

EXAMPLE 12.3

Visualizing the Equality of Several Means

Do the data in Table 12.6 provide sufficient evidence to conclude that there is a difference in the three population means μ_J, μ_K, and μ_L?

TABLE 12.6

Sample Results

	Factor Levels	
Sample from Level J	Sample from Level K	Sample from Level L
3	5	6
8	4	2
6	3	7
4	7	5
$C_J = 21$	$C_K = 19$	$C_L = 20$
$\bar{x}_J = 5.25$	$\bar{x}_K = 4.75$	$\bar{x}_L = 5.00$

SOLUTION Figure 12.2 (p. 668) shows the relative relationship among the three samples. A quick look at the figure does not suggest that the three sample means are different from each other. There is little between-sample variation for these three samples (i.e., the sample means are relatively close in value), whereas the within-sample variation is relatively large (i.e., the data values within each sample cover a relatively wide range of values).

FIGURE 12.2 Data
from Table 12.6

x	J	K	L
2			2
3	3	3	
4	4	4	
5		5	5
6	6		6
7		7	7
8	8		

$\bar{x}_K = 4.75$
$\bar{x}_L = 5.00$
$\bar{x}_J = 5.25$

To complete a hypothesis test for analysis of variance, we must agree on some ground rules, or **assumptions.** In this chapter we will use the following three basic assumptions:

1. Our goal is to investigate the effect that various levels of the factor being tested have on the response variable. Typically, we want to find the level that yields the most advantageous values of the response variable. This, of course, means that we probably will want to reject the null hypothesis in favor of the alternative. Then a follow-up study could determine the "best" level of the factor.

2. We must assume that the effects due to chance and due to untested factors are normally distributed and that the variance caused by these effects is constant throughout the experiment.

3. We must assume independence among all observations of the experiment. (Recall that independence means that the results of one observation of the experiment do not affect the results of any other observation.) We will usually conduct the tests in a **randomized** order to ensure independence. This technique also helps avoid data contamination.

**APPLIED
EXAMPLE 12.4**

PC Access for Pupils

This graphic reports that in 2004, the average number of "students per computer" in U.S. schools decreases as the student progresses through school. Does it appear that the age (grade) category to which the student belongs has an effect on the average number of students per school computer? (See Exercise 12.9.)

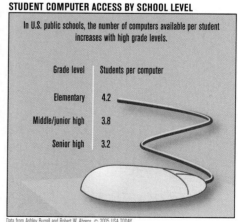

STUDENT COMPUTER ACCESS BY SCHOOL LEVEL

In U.S. public schools, the number of computers available per student increases with high grade levels.

Grade level	Students per computer
Elementary	4.2
Middle/junior high	3.8
Senior high	3.2

Data from Ashley Burrell and Robert W. Ahrens, © 2005 USA TODAY.

APPLIED
EXAMPLE 12.5

April Not the Wettest Month for All

The average amount of rainfall varies by month and by location. This graphic reports the average amount of rainfall for April and for the wettest month of the year for each of six U.S. cities. Does it appear that the city and the month have an effect on the average monthly rainfall? (See Exercise 12.10.)

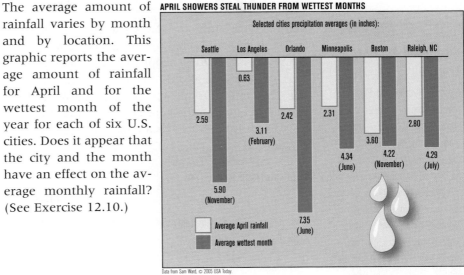

SECTION 12.3 EXERCISES

12.7 Do the data shown in the boxplot have a greater amount of variability within levels A, B, C, and D or between the four levels? Explain.

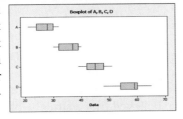

12.8 Do the data shown in the boxplot have a greater amount of variability within levels A, B, C, and D or between the four levels? Explain.

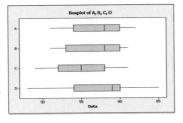

12.9 Referring to Applied Example 12.4:

a. Does the category appear to have an effect on the average number of students per computer? Explain.

b. The graphic shows three categories. Explain how these three categories could be used to organize data for one-way ANOVA. What would be used as the levels? What would be used as the data (replicates)? How would the data be related to the 4.2, 3.8, and 3.2 values given in the graphic?

12.10 The amount of monthly rainfall varies from month to month and from city to city. Applied Example 12.5 suggests that the average monthly rainfall is affected by both the month and the location.

a. What variable was used to collect the data used to find the monthly averages shown in Applied Example 12.5?

b. Explain what data would be needed and how it would be arranged to analyze the location (represented by cities) effect on the amount of rainfall during the month of April.

12.4 Applications of Single-Factor ANOVA

Before continuing our ANOVA discussion, let's identify the notation, particularly the subscripts that are used; see Table 12.7. Notice that each data value (i.e., $x_{2,3}$) has two subscripts; the first subscript indicates the column number (test factor level), and the second subscript identifies the replicate (row) number. The column totals, C_i, are listed across the bottom of the table. The grand total, T, is equal to the sum of all x's and is found by adding the column totals. Row totals can be used as a cross-check but serve no other purpose.

TABLE 12.7

Notation Used in ANOVA

Replicates	Factor Levels				
	Sample from Level 1	Sample from Level 2	Sample from Level 3	...	Sample from Level c
$k = 1$	$x_{1,1}$	$x_{2,1}$	$x_{3,1}$		$x_{c,1}$
$k = 2$	$x_{1,2}$	$x_{2,2}$	$x_{3,2}$		$x_{c,2}$
$k = 3$	$x_{1,3}$	$x_{2,3}$	$x_{3,3}$		$x_{c,3}$
$\vdots$					
Column Totals	C_1	C_2	C_3	...	C_c | T

$$T = \text{grand total} = \text{sum of all } x\text{'s} = \sum x = \sum C_i$$

A **mathematical model** (equation) is often used to express a particular situation. In Chapter 3 we used a mathematical model to help explain the relationship between the values of bivariate data. The equation $\hat{y} = b_0 + b_1x$ served as the model when we believed that a straight-line relationship existed. The probability functions studied in Chapter 5 are also examples of mathematical models. For the single-factor ANOVA, the mathematical model, formula (12.13), is an expression of the composition of each data value, $x_{c,k}$ entered in our data table:

Mathematical Model for Single-Factor ANOVA

$$x_{c,k} = \mu + F_c + \epsilon_{k(c)} \tag{12.13}$$

We interpret each term of this model as follows:

$x_{c,k}$, is the value of the variable at the kth replicate of level c.

μ is the mean value for all the data without respect to the test factor.

F_c is the effect that the factor being tested has on the response variable at each different level c.

$\epsilon_{k(c)}$ (ϵ is the lowercase Greek letter epsilon) is the *experimental error* that occurs among the k replicates in each of the c columns.

Let's look at another hypothesis test using an analysis of variance.

EXAMPLE 12.6

Hypothesis Test for the Equality of Several Means

TABLE 12.8

Sample Results on Target Shooting [TA12-8]

Method of Sighting		
Right Eye	Left Eye	Both Eyes
12	10	16
10	17	14
18	16	16
12	13	11
14		20
		21

A rifle club performed an experiment on a randomly selected group of first-time shooters. The purpose of the experiment was to determine whether shooting accuracy is affected by the method of sighting used: only the right eye open, only the left eye open, or both eyes open. Fifteen first-time shooters were selected and divided into three groups. Each group experienced the same training and practicing procedures with one exception—the method of sighting used. After completing training, each student was given the same number of rounds and asked to shoot at a target. Their scores are listed in Table 12.8.

At the 0.05 level of significance, is there sufficient evidence to reject the claim that the three methods of sighting are equally effective?

SOLUTION In this experiment the factor is method of sighting and the levels are the three different methods of sighting (right eye, left eye, and both eyes open). The replicates are the scores received by the students in each group. The null hypothesis to be tested is: The three methods of sighting are equally effective, or the mean scores attained using each of the three methods are the same.

Step 1 a. **Parameter of interest:** The "mean" at each level of the test factor is of interest: the mean score using the right eye, μ_R; the mean score using the left eye, μ_L; and the mean score using both eyes, μ_B. The factor being tested, "method of sighting," has three levels: right, left, and both.

 b. **Statement of hypotheses:**

H_o: $\mu_R = \mu_L = \mu_B$

H_a: The means are not all equal (i.e., at least one mean is different).

Step 2 a. **Assumptions:** The shooters were randomly assigned to the method, and their scores are independent of each other. The effects due to chance and untested factors are assumed to be normally distributed.

 b. **Test statistic:** The F-distribution and formula (12.12) will be used with df(numerator) = df(method) = 2 and df(denominator) = df(error) = 12.

 c. **Level of significance:** $\alpha = 0.05$.

Step 3 a. **Sample information:** See Table 12.8.

 b. **Calculated test statistic:** The test statistic is $F\star$: Table 12.9 is used to find the column totals.

First, the summations $\sum x$ and $\sum x^2$ need to be calculated:

$$\sum x = 12 + 10 + 18 + 12 + 14 + 10 + 17 + \cdots + 21 = \mathbf{220}$$
$$(\text{Or } 66 + 56 + 98 = 220 \ \text{ⓒⓚ})$$

$$\sum x^2 = 12^2 + 10^2 + 18^2 + 12^2 + 14^2 + 10^2 + \cdots + 21^2 = \mathbf{3392}$$

TABLE 12.9

Sample Results for Target Shooting

Replicates	Factor Levels: Method of Sighting		
	Right Eye	Left Eye	Both Eyes
$k = 1$	12	10	16
$k = 2$	10	17	14
$k = 3$	18	16	16
$k = 4$	12	13	11
$k = 5$	14		20
$k = 6$			21
Totals	$C_R = 66$	$C_L = 56$	$C_B = 98$

Using formula (12.2), we find

$$\text{SS(total)} = \sum (x^2) - \frac{(\sum x)^2}{n}: \qquad \text{SS(total)} = 3392 - \frac{(220)^2}{15}$$

$$= 3392 - 3226.67 = \mathbf{165.33}$$

Using formula (12.3), we find

$$\text{SS(method)} = \left(\frac{C_1^2}{k_1} + \frac{C_2^2}{k_2} + \frac{C_3^2}{k_3} + \cdots \right) - \frac{(\sum x)^2}{n}:$$

$$\text{SS(method)} = \left(\frac{66^2}{5} + \frac{56^2}{4} + \frac{98^2}{6} \right) - \frac{(220)^2}{15}$$

$$= (871.2 + 784 + 1600.67) - 3226.67 = 3255.87 - 3226.67 = \mathbf{29.20}$$

To find SS(error) we need first:

$$\sum (x^2) = 3392 \quad \text{(found previously)}$$

$$\left(\frac{C_1^2}{k_1} + \frac{C_2^2}{k_2} + \frac{C_3^2}{k_3} + \cdots \right) = 3255.87 \quad \text{(found previously)}$$

Then using formula (12.4), we have

$$\text{SS(error)} = \sum (x^2) - \left(\frac{C_1^2}{k_1} + \frac{C_2^2}{k_2} + \frac{C_3^2}{k_3} + \cdots \right):$$

$$\text{SS(error)} = 3392 - 3255.87 = \mathbf{136.13}$$

We use formula (12.8) to check the sum of squares:

$$\text{SS(method)} + \text{SS(error)} = \text{SS(total)}: \qquad 29.20 + 136.13 = 165.33$$

The degrees of freedom are found using formulas (12.5), (12.6), and (12.7):

$$\text{df(method)} = c - 1 = 3 - 1 = \mathbf{2}$$

$$\text{df(total)} = n - 1 = 15 - 1 = \mathbf{14}$$

$$\text{df(error)} = n - c = 15 - 3 = \mathbf{12}$$

Using formulas (12.10) and (12.11), we find

$$\text{MS(method)} = \frac{\text{SS(method)}}{\text{df(method)}}: \quad \text{MS(method)} = \frac{29.20}{2} = \mathbf{14.60}$$

$$\text{MS(error)} = \frac{\text{SS(error)}}{\text{df(error)}}: \quad \text{MS(error)} = \frac{136.13}{12} = \mathbf{11.34}$$

The results of these computations are recorded in the ANOVA table in Table 12.10.

TABLE 12.10

ANOVA Table for Example 12.6

Source	df	SS	MS
Method	2	29.20	14.60
Error	12	136.13	11.34
Total	14	165.33	

The calculated value of the test statistic is then found using formula (12.12):

$$F\star = \frac{\text{MS(factor)}}{\text{MS(error)}}: \quad F\star = \frac{\text{MS(method)}}{\text{MS(error)}} = \frac{14.60}{11.34} = \mathbf{1.287}$$

Step 4   **Probability Distribution:**

OR

p-Value:

a. Use the right-hand tail: $\mathbf{P} = P(F\star > 1.287$, with $\text{df}_n = 2$ and $\text{df}_d = 12$) as shown on the figure.

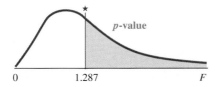

To find the *p*-value, you have two options:
1. Use Table 9A (Appendix B) to place bounds on the *p*-value: **P > 0.05.**
2. Use a computer or calculator to find the *p*-value: **P = 0.312.**
For additional instructions, see page 597.
b. The *p*-value is not smaller than the level of significance, $\alpha(0.05)$.

Classical:

a. The critical region is the right-hand tail; the critical value is obtained from Table 9A:

$$F\,(2, 12, 0.05) = 3.89$$

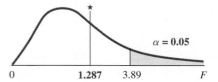

For additional instructions, see page 594.
b. $F\star$ is not in the critical region, as shown in **red** in the figure.

Step 5 a. **Decision:** Fail to reject H_o.

b. **Conclusion:** The data show no evidence to reject the null hypothesis that the three methods are equally effective.

TECHNOLOGY INSTRUCTIONS: ONE-WAY ANALYSIS OF VARIANCE

MINITAB (Release 14) Input the data for each level into columns C1, C2, . . . ; then continue with:

```
Choose:     Stat > ANVOA > One-Way (Unstacked)
Enter:      Responses: C1 C2 ...* >OK
```

OR

Input all of the data into C1 with the corresponding levels of factors into C2; then continue with:

```
Choose:     Stat > ANOVA > One-Way
Enter:      Response: C1
            Factor: C2* > OK
```

*Optional for either method:
```
Choose      Graphs...
Select:     Individual value plot and/or Boxplots of data > OK > OK
```

Excel Input the data for each level into columns A, B, . . . ; then continue with:

```
Choose:     Tools > Data Analysis > Anova: Single Factor
Enter:      Input Range: (A1:C4 or select cells)
Select:     Grouped By: Columns
            Labels in First Row (if necessary)
Enter:      Alpha: α
Select:     Output Range:
Enter:      (D1 or select cell)
```

To make the output more readable, continue with: Format > Column > Autofit Selection.

TI-83/84 Plus Input the data for each level into lists L1, L2, . . . ; then continue with:

```
Choose:     STAT > TESTS > F: ANOVA(
Enter:      L1, L2, ...)
```

Note: Side-by-side dotplots are very useful for visualizing the within-sample variation, the between-sample variation, and the relationship between them. Commands for side-by-side dotplots can be found in Chapter 2, pp. 49–50, 152.

Computer Solution   MINITAB Printout for Example 12.6:

Information given to computer →

Row	Right eye	Left eye	Both eyes
1	12	10	16
2	10	17	14
3	18	16	16
4	12	13	11
5	14		20
6			21

ANALYSIS OF VARIANCE

The ANOVA table ——→
compare with Table 12.10

SOURCE	DF	SS	MS	F	P
FACTOR	2	29.2	14.6	1.29	0.312
ERROR	12	136.1	11.3		
TOTAL	14	165.3			

The calculated value of
F, $F\bigstar$

Sample statistics for
each factor level ——→

The calculated p-value ——

LEVEL	N	MEAN	ST. DEV.
1	5	13.200	3.033
2	4	14.000	3.162
3	6	16.333	3.724

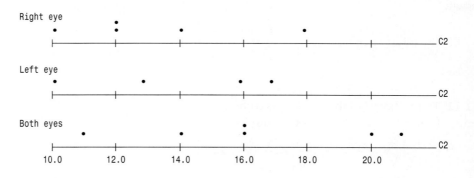

Recall the null hypothesis: There is no difference between the levels of the factor being tested. A "fail to reject H_o" decision must be interpreted as the conclusion that there is no evidence of a difference due to the levels of the tested factor, whereas the rejection of H_o implies that there is a difference between the levels. That is, at least one level is different from the others. If there is a difference, the next problem is to locate the level or levels that are different. Locating this difference may be the main objective of the analysis. To find the difference, the only method that is appropriate at this stage is to inspect the data. It may be obvious which level(s) caused the rejection of H_o. In Example 12.1 it seems quite obvious that at least one of the levels [level 1 (68°F) or level 3

(76°F) because they have the largest and the smallest sample means] is different from the other two. If the higher values are more desirable for finding the "best" level to use, we would choose that corresponding level of the factor.

Thus far we have discussed analysis of variance for data dealing with one factor. It is not unusual for problems to have several factors of interest. The ANOVA techniques presented in this chapter can be developed further and applied to more complex cases.

SECTION 12.4 EXERCISES

Statistics⬯Now™

Datasets can be found on your Student's Suite CD-ROM or at the StatisticsNow website at **http://1pass.thomson.com.**

12.11 Consider the following table for a single-factor ANOVA. Find the following:

a. $x_{1,2}$ b. $x_{2,1}$ c. C_1 d. $\sum x$ e. $\sum (C_i)^2$

	Level of Factor		
Replicates	1	2	3
1	3	2	7
2	0	5	4
3	1	4	5

12.12 The following table of data is to be used for single-factor ANOVA. Find each of the following:

a. $x_{3,2}$ b. $x_{4,3}$ c. C_3 d. $\sum x$ e. $\sum (C_i)^2$

	Level of Factor			
Replicates	1	2	3	4
1	13	12	16	14
2	17	8	18	11
3	9	15	10	19

12.13 State the null hypothesis, H_o, and the alternative hypothesis, H_a, that would be used to test the following statements:

a. The mean value of x is the same at all five levels of the experiment.

b. The scores are the same at all four locations.

c. The four levels of the test factor do not significantly affect the data.

d. The three different methods of treatment do affect the variable.

12.14 Find the p-value for each of the following situations:

a. $F\star = 3.852$, df(factor) $= 3$, df(error) $= 12$

b. $F\star = 4.152$, df(factor) $= 5$, df(error) $= 18$

c. $F\star = 4.572$, df(factor) $= 5$, df(error) $= 22$

12.15 For the following ANOVA experiments, determine the critical region(s) and critical value(s) that are used in the classical approach for testing the null hypothesis.

a. $H_o: \mu_1 = \mu_2 = \mu_3 = \mu_4$,

with $n = 18$ and $\alpha = 0.05$

b. $H_o: \mu_1 = \mu_2 = \mu_3 = \mu_4, = \mu_5$

with $n = 15$ and $\alpha = 0.01$

c. $H_o: \mu_1 = \mu_2 = \mu_3$,

with $n = 25$ and $\alpha = 0.05$

12.16 Why does df(factor), the number of degrees of freedom associated with the factor, always appear first in the critical value notation $F[\text{df(factor)}, \text{df(error)}, \alpha]$?

12.17 Suppose that an F test (as described in this chapter using the p-value approach) has a p-value of 0.04.

a. What is the interpretation of p-value $= 0.04$?

b. What is the interpretation of the situation if you had previously decided on a 0.05 level of significance?

c. What is the interpretation of the situation if you had previously decided on a 0.02 level of significance?

12.18 Suppose that an *F* test (as described in this chapter using the classical approach) has a critical value of 2.2, as shown in this figure:

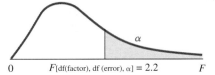

a. What is the interpretation of a calculated value of *F* larger than 2.2?

b. What is the interpretation of a calculated value of *F* smaller than 2.2?

c. What is the interpretation if the calculated *F* were 0.1? 0.01?

12.19 a. State the null hypothesis, in a general form, for the one-way ANOVA.

b. State the alternative hypothesis, in a general form, for the one-way ANOVA.

c. What must happen in order to "reject H_o"? Answer for both the *p*-value approach and the classical approach.

d. How would a decision of "reject H_o" be interpreted?

e. What must happen in order to "fail to reject H_o"? Answer for both the *p*-value approach and the classical approach.

f. How would a decision of "fail to reject H_o" be interpreted?

12.20 The following two excerpts were taken from "Documentation of Structured Analysis for Reviewing Scientifically-Based Research: Instructional Strategies and Programs," reviewed for NCTM on August 28, 2004.

I. A statistically significant main effect was obtained for group only, $F_{(1, 31)} = 6.23$, $p = 0.02$, favoring the schema condition.

II. No significant differences between the two conditions on test time, $F_{(1, 31)} = 1.8$, $p = 0.19$, were found.

a. Verify the *p*-value stated in I, and explain why they concluded a significant effect. [*Note:* Use the $F_{(1, 31)}$ distribution, with $F\star = 6.23$.]

b. Verify the *p*-value stated in II, and explain why they concluded no significant effect.

12.21 Two new drugs are being tested for their effect on the number of days a patient must remain hospitalized after surgery. A control group is receiving a placebo, and two treatment groups are each receiving one of two new drugs, both developed to promote recovery. The null hypothesis is that there is no difference between the means. The results of an analysis of variance used to analyze the data are shown here.

One-way ANOVA: Days versus Group

Source	DF	SS	MS	F	P
Group	2	11.00	5.50	2.11	0.159
Error	14	36.53	2.61		
Total	16	47.53			

a. How many patients were there?

b. How does the printout verify that there was one control and two test groups?

c. Using the SS values, verify the two mean square values.

d. Using the MS values, verify the *F*-value.

e. Verify the *p*-value.

f. State the decision and conclusion reached as a result of this analysis.

12.22 The article "An Investigation of High School Preparation as Predictors of the Cultural Literacy of Developmental, Nondevelopmental and ESL College Students" (RTDE) reported on a study that examined the cultural literacy of developmental, nondevelopmental, and English as a Second Language (ESL) college freshmen.

Analysis of Variance by Group for Total Score

Source	df	SS	MS	F	P
Group	2	4062.06	2031.03	14.49	0.0001
Error	117	16394.53	140.12		
Total	119	20456.59			

Analysis of Variance by Group for Foreign Language Preparation

Source	df	SS	MS	F	P
Group	2	0.95	0.475	1.93	0.1493
Error	117	28.75	0.246		
Total	119	29.70			

a. How many student scores were in the samples?

b. The students were divided into how many groups?

c. Given the SS and df values, verify the MS, the calculated *F*-value, and the *p*-value for each figure.

d. Do the statistics in the first table show that the total scores were different for the groups involved? Explain.

e. Do the statistics in the second table show that the foreign-language preparation scores were different for the groups involved? Explain.

12.23 An article titled "The Effectiveness of Biofeedback and Home Relaxation Training on Reduction of Borderline Hypertension" *(Health Education)* compared different methods of reducing blood pressure. Biofeedback ($n = 13$ subjects), biofeedback/relaxation ($n = 15$), and relaxation ($n = 14$) were the three methods compared. There were no differences among the three groups on pretest diastolic or systolic blood pressure readings. There was a significant posttest difference between groups on the systolic measure, $F_{(2, 39)} = 4.14$, $p < 0.025$, and diastolic measure, $F_{(2, 39)} = 5.56$, $p < 0.008$.

a. Verify that df(method) = 2 and df(error) = 39.

b. Use Tables 9A, 9B, and 9C in Appendix B to verify that for systolic, $p < 0.025$, and that for diastolic, $p < 0.008$.

12.24 [EX12-24] "The NBA is a big man's game. Average height for the league is about 6 feet 7 inches, with guards averaging 6 ft 4 in, forwards averaging 6 ft 9 in and centers 7 feet," as reported in a USA Snapshot in April 2004.

Basketball players are typically divided into three groups: guards, forwards, and centers. For many reasons, as most know and understand, "the guards are usually the shortest and the centers are typically the tallest." A random sample of 2005 NBA players was selected, and each player's height was recorded to nearest inch.

Guards	Forwards	Centers	Guards	Forwards	Centers
78	81	84	73	81	83
74	84	90	72	82	87
78	80	83	80	80	84
74	84	83		80	
77	82	85			

Source: NBA.com

a. Do you expect to find the mean heights for the three positions to be different from each other? Do you expect to find more variation between the positions or within the positions? Explain.

b. Construct a side-by-side graph (dotplot, box-plot, other) of your choice. Describe the data pattern shown.

c. Does the graph in part b show a relative large amount of variability between the positions? Explain, in detail, what you can determine from the graph.

d. Is there a significant difference in the heights of NBA players by position? Use $\alpha = 0.05$.

e. Do the results found in part d confirm your answer to part c? Explain.

f. Are the results what you anticipated they would be? Explain why or why not.

12.25 [EX12-25] A new operator was recently assigned to a crew of workers who perform a certain job. From the records of the number of units of work completed by each worker each day last month, a sample of size five was randomly selected for each of the two experienced workers and the new worker. At the 0.05 level of significance, does the evidence provide sufficient reason to reject the claim that there is no difference in the amount of work done by the three workers?

	Workers		
	New	A	B
Units of work (replicates)	8	11	10
	10	12	13
	9	10	9
	11	12	12
	8	13	13

a. Solve using the *p*-value approach.

b. Solve using the classical approach.

12.26 [EX12-26] An employment agency wants to see which of three types of ads in the help-wanted section of local newspapers is the most effective. Three types of ads (big headline, straightforward, and bold print) were randomly alternated over a period of weeks, and the number of people responding to the ads was noted each week. Do these data support the null hypothesis that there is no difference in the effectiveness of the ads, as measured by the mean number responding, at the 0.01 level of significance?

	Type of Advertisement		
	Big Headline	Straightforward	Bold Print
Number of responses (replicates)	23	19	28
	42	31	33
	36	18	46
	48	24	29
	33	26	34
	26		34

a. Solve using the *p*-value approach.

b. Solve using the classical approach.

12.27 [EX12-27] Cities across the United States have restaurants that offer themes associated with foreign countries. German-style food and drink have become popular since many communities started hosting Oktoberfests, but authentic German food restaurants offer the fare all year long. The following ratings, based on three categorical judgments of food quality, décor, and service, were assembled from different German restaurants located in various cities. The ratings were made on the same scale from 0 to 30 (the higher the better).

Restaurant Rating Category			Restaurant Rating Category		
Food Quality	Décor	Service	Food Quality	Décor	Service
19	19	18	21	16	18
17	15	14	19	15	18
19	17	16			

Source: Newsweek, "Meal Ticket Oktoberfest," September 27, 2004

Is there any significant difference in the ratings given to the German restaurants in each category? Construct a one-way ANOVA table and test for the difference at the 0.05 level of significance.

a. Solve using the *p*-value approach.

b. Solve using the classical approach.

12.28 [EX12-28] Thirty-nine counties from the six-state upper Midwest area of the United States were randomly selected from the USDA-NASS website and the following data on oat-production yield per acre were obtained.

County	IA	MN	ND	NE	SD	WI
1	76.2	53.0	71.4	60.0	76.5	52.0
2	65.3	70.0	64.3	37.0	50.0	53.0
3	86.0	71.0	66.7	53.0	42.0	72.0
4	73.6	54.0	61.4	50.0	62.5	81.0
5	61.3	64.0	66.0	56.0	55.7	57.0
6	74.3	40.0		58.0	59.1	64.0
7	58.3				59.3	
8	56.0					
9	61.4					

Source: http://www.usda.gov/nass/graphics/county01/data/ot01.csv

a. Do these data show a significant difference in the mean yield rates for the six states? Use $\alpha = 0.05$.

b. Draw a graph that demonstrates the results found in part a.

c. Explain the meaning of the results, including an explanation of how the graph portrays the results.

12.29 [EX12-29] A number of sports enthusiasts have argued that major league baseball players from teams in the Central Division have an unfair advantage over coastal players in the Western and Eastern Divisions. When playing games on the road (i.e., away from home), players from the Western and Eastern Divisions could gain (going west) or lose (going east) up to 3 hours as a result of the time zone differences; Central Division players, on the other hand, would seldom gain or lose more than 1 hour. The following data show the win/loss percentages by division for games played on the road by all three divisions of major league baseball teams for the 2004 season.

East	Central	West	East	Central	West
54.3	53.1	58.0	54.3	54.3	55.6
53.1	45.7	48.1	50.0	55.0	37.0
49.4	44.4	46.9	40.7	44.4	27.2
35.8	42.0	31.3	39.0	40.7	
33.8	30.5	54.3		38.8	
58.0	64.2	55.0			

Source: MLB.com

Complete an ANOVA table for won/loss percentages by teams, representing each division. Test the null hypothesis that when teams play on the road, the mean won/loss percentage is the same for each of the three divisions. Use the 0.05 level of significance.

a. Solve using the *p*-value approach.

b. Solve using the classical approach.

12.30 [EX12-30] Does the position of a suburb with respect to a city have an effect on school population? The following table lists 2004–2005 student enrollment in the suburbs of Rochester as shown in the May 9, 2005, *Democrat & Chronicle.*

East	West	South	East	West	South
3728	4367	2667	4986	4354	4471
3604	4956	5944	6029		
1277	13589	3725	9031		
7251	4481	896	3949		

Complete an ANOVA for the data, and test the hypothesis that the mean student enrollment per suburb is not the same for each of the geographical areas around Rochester, New York. Use a 0.05 level of significance. What data value(s) is influencing the results the most in this hypothesis test?

12.31 [EX12-31] A study was conducted to assess the effectiveness of treating vertigo (motion sickness) with the transdermal therapeutic system (TTS; a patch worn on the skin). Two other treatments, both oral (one pill containing a drug and one a placebo), were used. The age and the gender of the patients for each treatment are listed here.

TTS		Antivert		Placebo	
47-f	53-m	51-f	43-f	67-f	38-m
41-f	58-f	53-f	56-f	52-m	59-m
63-m	62-f	27-m	48-m	47-m	33-f
59-f	34-f	29-f	52-f	35-f	32-f
62-f	47-f	31-f	19-f	37-f	26-f
24-m	35-f	25-f	31-f	40-f	37-m
43-m	34-f	52-f	48-f	31-f	49-f
20-m	63-m	55-f	53-m	45-f	49-m
55-f	46-f	32-f	63-m	41-f	38-f
21-f		51-f	54-m	49-m	

Is there a significant difference between the mean age of the three test groups? Use $\alpha = 0.05$. Use a computer or calculator to complete this exercise.

a. Solve using the *p*-value approach.

b. Solve using the classical approach.

12.32 [EX12-32] Tracy works in a drugstore photo-developing department. Over time, Tracy felt that the number of rolls of film brought in for development varied among the days of the week. Were the weekends having an effect one way or another? Tracy decided to keep a daily record of the number of rolls of film dropped off for development for 4 weeks. Her results follow.

M	Tu	W	Th	F	Sa	Su
33	43	27	28	26	40	18
21	23	29	39	32	21	7
47	36	14	21	17	12	5
38	41	47	14	33	8	11

a. Using a one-way ANOVA, test the claim that the mean number of rolls of film dropped off is not the same for all 7 days of the week. Use a level of significance of 0.05.

b. Explain the meaning of the conclusion in part a. Does the conclusion tell you which days are different? which days have the larger means?

c. Construct a side-by-side dotplot of the data. Explain how the multiple dotplot coupled with the hypothesis test in part a help identify the difference between days.

d. How might the drugstore use this information?

e. Could there be other factors affecting these results? If so, name a few.

12.33 [EX12-33] Let x = a person's "ideal age" in years. Independent and random samples were obtained from U.S. adults in each of six different age groups.

18-24	25-29	30-39	40-49	50-64	65+
21	28	30	38	45	54
24	29	35	40	51	48
28	31	37	45	39	59
30	25	32	39	45	60
32	27	39	35	42	65
28	35	37			60
	32	40			

a. Construct a side-by-side boxplot showing the "ideal age" for each of the six age groups. What does this graph suggest?

b. Using a one-way ANOVA, test the claim that the "ideal age" is not the same for all age groups. Use a level of significance of 0.05.

c. What conclusions might you draw from the results of the hypothesis test?

d. Explain how the boxplots drawn in part a demonstrate the results found in part b.

12.34 [EX12-34] Albert Michelson, the first American citizen to be awarded a Nobel Prize in Physics, conducted many experiments in determining the velocity of light in air. Following is an excerpt of five trials of 20 measurements each taken by Michelson from June 5 to July 2, 1879. The measurements have had 299,000 subtracted from them.

Trial 1	Trial 2	Trial 3	Trial 4	Trial 5
850	960	880	890	890
740	940	880	810	840

••• Remainder of data on Student's Suite CD-ROM

Source: http://lib.stat.cmu.edu/DASL/Stories/SpeedofLight.html

a. Construct a boxplot showing the five trials side by side. What does this graph suggest?

b. Using a one-way ANOVA, test the claim that not all trial results were the same. Use a level of significance of 0.05.

c. What conclusions might you draw from the results of the hypothesis test?

12.35 [EX12-35] The U.S. Department of Labor Bureau of Labor Statistics posts a table of average hourly wages paid to production workers on private nonfarm payrolls by major industry. The following data were taken from their website on May 18, 2005.

Average Hourly Earnings of Production Workers

Year	Jan	Feb	Mar	Apr	to	Dec
1995 to	14.37	14.59	14.62	14.66		14.76
2005	19.24	19.31	19.35	19.38		

••• Remainder of data on Student's Suite CD-ROM

Source: http://data.bls.gov/cgi-bin/surveymost

a. At the 0.05 level of significance, does the evidence provide sufficient reason to reject the claim that there is no difference in the mean hourly wages paid by month? Show graphical evidence to visually support your conclusion.

b. At the 0.05 level of significance, does the evidence provide sufficient reason to reject the claim that there is no difference in the mean hourly wages paid each year? Show graphical evidence to visually support your conclusion.

12.36 [EX12-36] Mr. B, the manager at a large retail store, is investigating several variables while measuring the level of his business. His store is open every day during the year except for New Year's Day, Christmas, and all Sundays. From his records, which cover several years prior, Mr. B has randomly identified 62 days and collected data for the daily total for three variables: number of paying customers, number of items purchased, and total cost of items purchased.

Day	Month	Customers	Items	Sales
2	1	425	1311	$12,707.00
1	1	412	1123	$11,467.50

••• Remainder of data on Student's Suite CD-ROM

Data are actual values; store name withheld for privacy reasons.
Day Code: 1 = M, 2 = Tu, 3 = W, 4 = Th, 5 = F, 6 = Sa
Month Code: 1 = Jan, 2 = Feb, 3 = Mar, . . ., 12 = Dec

Is the mean number of customers per day affected by the month? Or equivalently, "Is the mean number of customers per day the same for all months?" versus "Is there at least 1 month when the mean number of customers per day is significantly different from the others?" The following computer output resulted from analysis of the data.

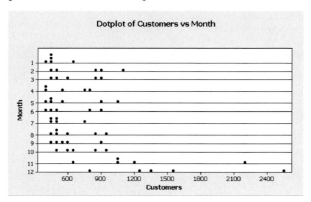

One-way ANOVA: Customers versus Month

Source	DF	SS	MS	F	P
Month	11	5224286	474935	5.03	0.000
Error	50	4724554	94491		
Total	61	9948840			

Inspect the preceding dotplot for number of customers per day for the 12 months and the ANOVA output for number of customers versus months. Look for evidence that leads to the conclusion, "Not all months have the same mean number of customers per day."

a. Describe the graphical evidence found and discuss how it shows that not all months are the same. Which month or months appear to be different from the others?

b. Describe the numerical evidence found and discuss how it shows that not all months are the same.

c. Can you tell which months are different based on the numerical evidence? Explain.

d. Does there appear to be any support to the idea that the Thanksgiving to New Year's holiday season period is the most important sales time of the year? Explain.

e. Use your calculator or computer to perform the ANOVA shown and to verify the results.

12.37 [EX12-36] Mr. B, the manager of the retail store in Exercise 12.36, has seen that the mean number of customers per day varies by month. and he is now wondering if month has a similar effect on number of items purchased.

a. Do you think "month" has an effect on the mean number of items purchased per day? Equivalently, do you think the mean number of items purchased per day is the same for all months? If not, which months do you believe will be different? Explain.

b. Construct a dotplot for the number of items purchased per day for each different month.

c. Does the dotplot in part b support your conjectures in part a? Explain.

d. Use the ANOVA technique to answer the question, "Does month affect the mean number of items purchased per day?" Use $\alpha = 0.05$.

e. Explain any differences and similarities between your answer in part a and the answer found in d.

12.38 [EX12-36] Mr. B, the manager of the retail store in Exercise 12.36, has seen that the mean number of customers per day and the mean number of items purchased (Exercise 12.37) are affected by the month, and he is now wondering if month has a similar effect on the total cost of items purchased.

a. Do you think "month" has an effect on the mean total cost of items purchased per day? Equivalently, do you think the mean total cost of items purchased per day is the same for all months? If not, which months do you believe will be different? Explain.

b. Construct a dotplot for the total cost of items purchased per day for each different month.

c. Does the dotplot in part b support your conjectures in part a? Explain.

d. Use the ANOVA technique to answer the question, "Does month affect the mean total cost of items purchased per day?" Use $\alpha = 0.05$.

e. Explain any differences and similarities between your answer in part a and the answer found in part d.

f. Explain any differences and similarities between the answers found for Exercises 12.36, 12.37, and 12.38. Do the similarities seem reasonable? What does this imply about these variables?

CHAPTER REVIEW

In Retrospect

In this chapter we have presented an introduction to the statistical techniques known as analysis of variance. The techniques studied here were restricted to the test of a hypothesis that dealt with questions about the means from several populations. We were restricted to normal populations and populations with homogeneous (equal) variances. The test of multiple means is done by partitioning the sum of squares into two segments: (1) the sum of squares due to variation between

the levels of the factor being tested and (2) the sum of squares due to variation between the replicates within each level. The null hypothesis about means is then tested by using the appropriate variance measurements.

Note that we restricted our development to one-factor experiments. This one-factor technique represents only a beginning to the study of analysis of variance techniques.

Vocabulary and Key Concepts

analysis of variance (ANOVA) (pp. 658, 666)
assumptions (p. 668)
between-sample variation (p. 667)
degrees of freedom (p. 662)
experimental error (p. 661)
levels of the tested factor (pp. 659, 662, 666)
mathematical model (p. 670)

mean square, MS(factor), MS(error) (p. 663)
partitioning (p. 661)
randomize (p. 668)
replicate (pp. 659, 670)
response variable (pp. 666, 668, 670)
sum of squares (p. 660)
test statistic, $F\star$ (p. 664)

total sum of squares, SS(total) (p. 660)
variance (p. 660)
variation between levels, MS(factor) (pp. 663, 667, 674)
variation within a level, MS(error) (pp. 663, 667, 674)
within-sample variation (pp. 667, 674)

Learning Outcomes

✓ Understand that analysis of variance techniques (ANOVA) are used to test differences among more than two means.	pp. 658–659
✓ Understand that ANOVA uses variances to complete the testing of several means.	EXP 12.1
✓ Understand that the *F*-distribution is used to test the ratio of the variation between the means being tested to the variation within the samples being tested.	EXP 12.1, Ex. 12.18
✓ Understand that if the variation between the means is significantly more than the variation within the samples, then the means are considered unequal.	EXP 12.2, 12.3, Ex. 12.7, 12.8
✓ Compute, describe, and interpret a hypothesis test for the differences among several means, using the *F*-distribution with the *p*-value approach and classical approach.	EXP 12.6, Ex. 12.25, 12.27

Chapter Exercises

12.39 [EX12-39] Samples of peanut butter produced by three different manufacturers were tested for salt content (in milligrams), with the following results:

Brand 1	2.5	8.3	3.1	4.7	7.5	6.3
Brand 2	4.5	3.8	5.6	7.2	3.2	2.7
Brand 3	5.3	3.5	2.4	6.8	4.2	3.0

Is there a significant difference in the mean amount of salt in these samples? Use $\alpha = 0.05$.

a. State the null and alternative hypotheses.

b. Determine the test criteria: assumptions, level of significance, test statistic.

c. Using the information on the computer printout that follows, state the decision and conclusion to the hypothesis test.

d. What does the *p*-value tell you? Explain.

Hint: Each level of data is entered into a separate column.

```
Analysis of Variance
Source    DF      SS       MS      F       P
Factor    2       4.68     2.34    0.64    0.541
Error     15      54.88    3.66
Total     17      59.56
                                 Individual 95% CIs For
                                 Mean
                                 Based on Pooled StDev
Level    N     Mean    StDev   --+------+------+------+--
Brand1   6     5.400   2.359            (-----*-----)
Brand2   6     4.500   1.669        (-----*-----)
Brand3   6     4.200   1.621      (-----*-----)
                                 --+------+------+------+--
Pooled StDev = 1.913             3.0   4.5   6.0   7.5
```

12.40 [EX12-40] A new all-purpose cleaner is being test-marketed by placing sales displays in three different locations within various supermarkets. The number of bottles sold from each location

within each of the supermarkets being tested is reported here:

	I	40	35	44	38
Locations	II	32	38	30	35
	III	45	48	50	52

a. State the null and alternative hypotheses for testing "the location of the sales display had no effect on the number of bottles sold."

b. Using $\alpha = 0.01$, determine the test criteria: assumptions, level of significance, test statistic.

c. Using the information on the computer printout that follows, state the decision and conclusion to the hypothesis test.

d. What does the *p*-value tell you? Explain.

Hint: Each level of data is entered into a separate column.

```
Analysis of Variance
Source    DF      SS       MS      F       P
Factor    2       460.7    230.3   19.51   0.001
Error     9       106.2    11.8
Total     11      566.9

                                 Individual 95% CIs For
                                 Mean
                                 Based on Pooled StDev
Level      N    Mean     StDev   ----+------+------+----
Location   4    39.250   3.775        (---*---)
Location   4    33.750   3.500   (---*---)
Location   4    48.750   2.986               (---*---)
                                 ----+------+------+----
Pooled StDev = 3.436              35.0   42.0   49.0
```

12.41 Does it really matter where you grocery shop financially? Are the prices at one grocery store consistently higher or lower than those at another? "Shopping the grocers," an article in the December 19, 2004, *Democrat & Chronicle*, featured a comparison of products bought at four local supermarkets. The data were analyzed using ANOVA techniques, and the results follow.

```
One-way ANOVA: Martins, Tops, Wal-Mart, Wegmans
Source    DF      SS       MS      F       P
Factor    3       0.50     0.17    0.03    0.993
Error     56      330.74   5.91
Total     59      331.24
S = 2.430    R-Sq = 0.15%    R-Sq(adj) = 0.00%
```

```
                        Individual 95% CIs For
                        Mean Based on Pooled
                        StDev
Level     N    Mean    StDev   --+------+------+----+--
Martins  15   2.542    2.241   (-------------*----------)
Tops     15   2.596    2.294   (-------------*---------)
Wal-Mart 15   2.473    2.173   (-------------*---------)
Wegmans  15   2.723    2.935     (-------------*-------)
                                --+------+------+----+--
                                1.40   2.10   2.80 3.50
```

Pooled StDev = 2.430

a. State the null and alternative hypotheses.

b. Based on the information on the computer printout shown, state the decision and conclusion to the hypothesis test.

c. Using the statistics given for each store, does there seem to be a difference among the mean grocery cost for these four supermarkets?

d. Using the statistics given for each store, does there appear to be a difference among the standard deviations for these four supermarkets?

e. How do your answers in parts c and d support your answer in part b? Explain.

12.42 An experiment was designed to compare the lengths of time that four different drugs provided pain relief after surgery. The results (in hours) follow.

Drug				Drug			
A	B	C	D	A	B	C	D
8	6	8	4	2	4	10	
6	6	10	4			12	
4	4	10	2				

Is there enough evidence to reject the null hypothesis that there is no significant difference in the length of pain relief provided by the four drugs at $\alpha = 0.05$?

a. Solve using the *p*-value approach.

b. Solve using the classical approach.

12.43 [EX12-43] The distance required to stop a vehicle on wet pavement was measured to compare the stopping power of four major brands of tires. A tire of each brand was tested on the same vehicle on a controlled wet pavement. The resulting distances follow.

	Brand A	Brand B	Brand C	Brand D
Distance (replicate)	37	33	41	41
	34	40	34	41
	38	37	38	40
	36	42	35	39
	40	38	42	41
	32		34	43

At $\alpha = 0.05$, is there sufficient evidence to conclude that there is a difference in the mean stopping distance?

a. Solve using the *p*-value approach.

b. Solve using the classical approach.

12.44 [EX12-44] Every time you fill your car's gas tank, consider the federal tax that is charged. That money goes for the repair of roads and bridges. Not all things are equal according to "Alaska thanks you," an article in the May 18, 2005, *USA Today*. For example, Alaska receives $6.60 for every dollar paid in federal gas taxes, whereas Texas only gets $0.86 on the dollar. The following table outlines state gas-tax money by region.

Northeast	Southeast	Southwest	North Central	Rocky Mtn	Pacific
2.17	1.69	1.12	2.28	2.22	6.60
1.83	1.02	0.88	2.17	1.46	2.23
1.6	1.02	0.88	1.07	1.40	1.01
1.41	0.98	0.86	1.05	1.08	0.99
1.21	0.95		1.03	1.07	0.91
1.17	0.91		1.01	0.93	
1.08	0.90		0.97		
1.00	0.89		0.96		
0.95	0.88		0.92		
0.95	0.86		0.90		
0.87	0.86		0.89		
			0.89		
			0.88		

a. Do these data show a significant difference in the mean amount of gas-tax money received by U.S. region? Use a 0.05 level of significance.

b. Do these data show a significant difference in the mean amount of gas-tax money received by U.S. region if you use a 0.10 level of significance [$F_{(5, 44, 0.10)} = 1.98$]?

c. What effect did the change in the level of significance have on your decisions in parts a and b? What key step in the hypothesis test procedure does this emphasize?

12.45 [EX12-45] A certain vending company's soft-drink dispensing machines are supposed to serve 6 oz of beverage. Various machines were sampled, and the resulting amounts of dispensed drink (in ounces) were recorded, as shown in the following table.

Machines

A	B	C	D	E
3.8	6.8	4.4	6.5	6.2
4.2	7.1	4.1	6.4	4.5
4.1	6.7	3.9	6.2	5.3
4.4		4.5		5.8

Does this sample evidence provide sufficient reason to reject the null hypothesis that all five machines dispense the same average amount of soft drink? Use $\alpha = 0.01$.

a. Solve using the p-value approach.

b. Solve using the classical approach.

12.46 [EX12-46] Suburbs, each with its own attributes, are located around every metropolitan area. There is always the "rich" one (the most expensive one), the least expensive one, and so on. The dollar amount of county transfer taxes paid on homes from five suburbs follows.

Suburb A	Suburb B	Suburb C	Suburb D	Suburb E
105	101	95	74	79
114	88	107	135	89
85	105	101	165	140
177	100	92	114	114
104	161	91	80	80
135	113	89	115	86
	94			94
				102

a. Do the sample data show sufficient evidence to conclude that the suburbs represented do have a significant effect on the transfer tax of their homes? Use $\alpha = 0.01$.

b. Construct a graph that demonstrates the conclusion reached in part a.

12.47 [EX12-47] Each year when the National Football League playoffs begin, the question arises, "Which division's teams are the toughest, East, North, South, or West?" Two ways to measure the strength of the football teams that play are the number of points they score (Pts F) and the number of points their opponents score (Pts A). The final results for the 16 regular-season games played by each of the eight teams in each division in the 2004 season follow.

East		North		South		West	
Pts F	Pts A	Pts F	Pts A	Pts F	Pts A	Pts F	Pts A
437	260	372	251	522	351	446	313
333	261	317	268	261	280	381	304
395	284	374	372	309	339	483	435
275	354	276	390	344	439	320	442
386	260	424	380	340	337	371	373
303	347	405	395	348	405	319	392
293	405	296	350	355	339	284	322
240	265	231	331	301	304	259	452

Source: http://www.nfl.com/

Complete an ANOVA table for (1) points scored and (2) points scored by opposing teams. In each case, test the null hypothesis that the mean points scored is the same for each of the four divisions. Use the 0.05 level of significance.

a. Solve using the p-value approach.

b. Solve using the classical approach.

12.48 [EX12-48] To compare the effectiveness of three different methods of teaching reading, 26 children of equal reading aptitude were divided into three groups. Each group was instructed for a given period of time using one of the three methods. After completing the instruction period, all students were tested. The test results are shown in the following table.

	Method I	Method II	Method III
Test Scores	45	45	44
(replicates)	51	44	50
	48	46	45
	50	44	55
	46	41	51
	48	43	51
	45	46	45
	48	49	47
	47	44	

Is the evidence sufficient to reject the hypothesis that all three instruction methods are equally effective? Use $\alpha = 0.05$.

a. Solve using the *p*-value approach.

b. Solve using the classical approach.

12.49 [EX12-49] Contact lenses are big business. From 25 to 27 million Americans wear contact lenses according to the University of Michigan Kellogg Eye Center website (http://www.kellogg.umich.edu/patientcare/conditions/contact.lenses.html). Companies must continue research to maintain market share. Suppose five lots of competitive product were obtained and evaluated for a particular lens dimension and compared against its nominal. Twenty-two lenses from each lot were evaluated. Data were coded in two ways:

A	B	C	D	E
−0.020	−0.043	−0.002	0.002	−0.018
−0.016	−0.051	0.024	−0.024	−0.032

•• Remainder of data on Student's Suite CD-ROM

Courtesy: Bausch & Lomb

a. Do the data show a significant difference in the mean nominal comparison for the five competitors? Use a 0.01 level of significance.

b. Draw a graph that demonstrates the results found in part a.

c. Explain the meaning of the results, including an explanation of how the graph portrays the results.

12.50 [EX12-50] The following table gives the number of arrests made last year for violations of the narcotic drug laws in 24 communities. The data given are rates of arrest per 10,000 inhabitants.

Cities (over 250,000)	Cities (under 250,000)	Suburban Communities	Rural Communities
45	23	25	8
34	18	17	16
41	27	19	14
42	21	28	17
37	26	31	10
28	34	37	23

At $\alpha = 0.05$, is there sufficient evidence to reject the hypothesis that the mean rates of arrests are the same in all four sizes of communities?

a. Solve using the *p*-value approach.

b. Solve using the classical approach.

12.51 [EX12-51] Seven golf balls from each of six manufacturers were randomly selected and tested for durability. Each ball was hit 300 times or until failure occurred, whichever came first.

A	B	C	D	E	F
300	190	228	276	162	264
300	164	300	296	175	168
300	238	268	62	157	254
260	200	280	300	262	216
300	221	300	230	200	257
261	132	300	175	256	183
300	156	300	211	92	93

Do these sample data show sufficient reason to reject the null hypothesis that the six different brands tested withstood the durability test equally well? Use $\alpha = 0.05$.

a. Solve using the *p*-value approach.

b. Solve using the classical approach.

12.52 [EX12-52] The U.S. Department of Labor Bureau of Labor Statistics posts a table of average weekly hours worked by production workers on private nonfarm payrolls by major industry. The following data were downloaded from their website on May 18, 2005.

Average Weekly Hours of Production Workers

Year	Jan	Feb	Mar	Apr	to	Dec
1995	34.5	34.4	34.4	34.3		34.2
to						
2005	33.7	33.7	33.7	33.9		

••• Remainder of data on Student's Suite CD-ROM

Source: http://data.bls.gov/cgi-bin/surveymost

a. At the 0.05 level of significance, does the evidence provide sufficient reason to reject the claim that there is no difference in the mean weekly hours worked by month? Show graphical evidence to visually support your conclusion.

b. At the 0.05 level of significance, does the evidence provide sufficient reason to reject the claim that there is no difference in the mean weekly hours worked each year? Show graphical evidence to visually support your conclusion.

12.53 [EX12-53] Ronald Fisher, an English statistician (1890–1962), collected measurements for a sample of 150 irises. Of concern were the following variables: species, petal width (PW), petal length, sepal width (SW), and sepal length (all in mm). (Sepals are the outermost leaves that encase the flower before it has opened.) The goal of Fisher's experiment was to produce a simple function that could be used to classify flowers correctly. A sample of his data follows.

Type	PW	SW	Type	PW	SW
0	2	35	1	24	28
2	18	32	1	19	25
1	19	27	0	1	31
0	3	35	1	23	32
0	3	38	2	13	23
2	12	26	2	15	30
1	20	38	1	25	33
2	15	31	1	21	33
2	15	29	0	2	37
2	12	27	1	18	27
1	22	28	1	17	25
1	13	30	1	24	34
0	2	29	0	2	36
2	16	27	2	10	22
0	5	33	0	2	32

a. Is there a significant difference in the mean petal width for the three species? Use a 0.05 level of significance.

b. Is there a significant difference in the mean sepal width for the three species? Use a 0.05 level of significance.

c. How could Fisher use these outcomes to help him classify irises into the correct species?

12.54 [EX12-54] Cicadas are flying, plant-eating insects. One particular species, the 13-year cicadas *(Magicicada)*, spends five juvenile stages in underground burrows. During the 13 years underground, the cicadas grow from approximately the size of a small ant to nearly the size of an adult cicada. Every 13 years, this species then emerges from their burrows as adults. The adult body weights (BW) in grams and body lengths (BL) in millimeters are given for three different species of these 13-year cicadas in the following table.

BW	BL	Species	BW	BL	Species
0.15	22	*tredecula*	0.18	24	*tredecula*
0.29	26	*tredecim*	0.21	20	*tredecassini*
0.17	24	*tredecim*	0.15	24	*tredecula*
0.18	23	*tredecula*	0.17	23	*tredecula*
0.39	32	*tredecim*	0.13	22	*tredecassini*
0.26	27	*tredecim*	0.17	23	*tredecassini*
0.17	24	*tredecassini*	0.23	25	*tredecassini*
0.16	24	*tredecassini*	0.12	24	*tredecim*
0.14	25	*tredecassini*	0.26	26	*tredecula*
0.14	25	*tredecassini*	0.19	25	*tredecula*
0.28	27	*tredecassini*	0.20	23	*tredecassini*
0.12	29	*tredecim*	0.14	22	*tredecula*

a. Is there any significant difference in the body weights of adult cicadas with respect to species? Construct a one-way ANOVA table and test for the difference at the 0.01 level of significance.

b. Is there any significant difference in the body lengths of adult cicadas with respect to species? Construct a one-way ANOVA table and test for the difference at the 0.01 level of significance.

12.55 [EX12-36] Mr. B, the manager of the retail store in Exercise 12.36, has seen the effect month has on the mean number of customers per day and is now wondering about the effect day of the week has on the mean number of customers per day.

Is the mean number of customers per day affected by the day of the week? Or equivalently, "Is the mean number of customers per day the same for all days of the week?" versus "Is there at least one day of the week when the mean number of customers per day is significantly different from the others?" The following computer output resulted from analysis of the data. (Day Code: 1 = Monday, 2 = Tuesday, ..., 6 = Saturday.)

Dotplot of Customers vs Day

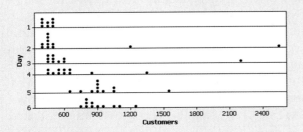

```
One-way ANOVA: Customers versus Day
Source   DF      SS        MS       F       P
Day       5    1604566   320913   2.15   0.072
Error    56    8344274   149005
Total    61    9948840
```

Inspect the preceding dotplot for number of customers per day for the 6 weekdays and the ANOVA output for number of customers versus day of week. Look for evidence that leads to the conclusion, "Not all days have the same mean number of customers per day."

a. Describe the graphical evidence found and discuss how it shows that not all days of the week are the same. Which day or days appear to be different from the others?

b. Describe the meaning and source of the 5 points located to the right and separate from the rest of the data. (*Hint:* Look at the data.)

c. Describe the numerical evidence found and discuss how it shows that not all days are the same.

d. Can you tell which days are different based on the numerical evidence? Explain.

e. Use your calculator or computer to perform the ANOVA shown and verify the results.

12.56 [EX12-36] Mr. B, the manager of the retail store in Exercise 12.55, has seen the effect day has on the mean number of customers per day and is now wondering if day has a similar effect on number of items purchased.

a. Do you think "day of week" has an effect on the mean number of items purchased per day? Equivalently, do you think the mean number of items purchased per day is the same for all days? If not, which days do you believe will be different? Explain.

b. Construct a dotplot for the number of items purchased per day for each different day of the week.

c. Does the dotplot in part b support your conjectures in part a? Explain.

d. Use the ANOVA technique to answer the question, "Does day affect the mean number of items purchased per day?" Use $\alpha = 0.05$.

e. Explain any differences and similarities between your answer in part a and the answer found in d.

12.57 [EX12-36] Mr. B, the manager of the retail store in Exercise 12.55, has seen the effect day has on the mean number of customers per day and is now wondering if day has a similar effect on total cost of items purchased.

a. Do you think "day of week" has an effect on the mean total cost of items purchased per day? Equivalently, do you think the mean total cost of items purchased per day is the same for all days? If not, which days do you believe will be different? Explain.

b. Construct a dotplot for the total cost of items purchased per day for each different day of the week.

c. Does the dotplot in part b support your conjectures in part a? Explain.

d. Use the ANOVA technique to answer the question, "Does day of the week affect the mean total cost of items purchased per day?" Use $\alpha = 0.05$.

e. Explain any differences and similarities between your answer in part a and the answer found in d.

f. Explain any differences and similarities between the answers found for Exercises 12.55, 12.56, and 12.57. Do the similarities seem reasonable? What does this imply about these variables?

12.58 For the following data, find SS(error) and show that

$$SS(error) = (k_1 - 1)s_1^2 + (k_2 - 1)s_2^2 + (k_3 - 1)s_3^2$$

where s_i^2 is the variance for the ith factor level.

Factor Level		
1	2	3
8	6	10
4	6	12
2	4	14

12.59 For the following data, show that

$$SS(factor) = k_1(\bar{x}_1 - \bar{x})^2 + k_2(\bar{x}_2 - \bar{x})^2 + k_3(\bar{x}_3 - \bar{x})^2$$

where $\bar{x}_1, \bar{x}_2, \bar{x}_3$ are the means for the three factor levels and $\bar{x}$ is the overall mean.

Factor Level		
1	2	3
6	13	9
8	12	11
10	14	7

12.60 An article in the *Journal of Pharmaceutical Sciences* discusses the change of plasma protein binding of diazepam at various concentrations of imipramine. Suppose the results were reported as follows.

Diazepam Alone (1.25 mg/mL)	Diazepam with Imipramine		
	1.25	2.50	5.00
97.99	97.68	96.29	93.92

The values given represent mean plasma protein binding, and $n = 8$ for each of the four groups. Find the sum of squares among the four groups.

12.61 A study reported in the *Journal of Research and Development in Education* evaluates the effectiveness of social skills training and cross-age tutoring for improving academic skills and social communication behaviors among boys with learning disabilities. Twenty boys were divided into three groups, and their scores on the Test of Written Spelling (TWS) can be summarized as follows.

Group	n	TWS Mean	St. Dev.
Social skills training and tutoring components	7	21.43	9.48
Social skills training only	7	20.00	8.91
Neither component	6	20.83	9.06

Calculate the entries of the ANOVA table using these results.

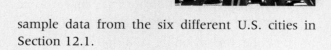

Chapter Project

Time Spent Commuting to Work

How much time did you spend getting to school or work this morning? How much time do your parents spend commuting to work each morning? Do you think that what city you live in has an effect on the average commute time? The graphic "Longest Commute to Work" in Section 12.1, "Time Spent Commuting to Work" (p. 657), seems to suggest that the city has an effect on the commute time. Let's investigate this question using the sample data from the six different U.S. cities in Section 12.1.

Putting Chapter 12 to Work

12.62 [EX12-01] Let $x =$ one-way travel to work in minutes. Independent and random samples were obtained in each of the six different U.S. cities from workers who commute to work during the 8:00 AM rush hour.

Atlanta	Boston	Dallas	Philadelphia	Seattle	St. Louis
29	18	42	29	30	15
21	37	25	20	19	54
20	37	36	33	31	42
15	25	32	37	39	23
37	32	20	42	30	33
26	34	26			35
	48	35			

a. Construct a side-by-side boxplot showing the six cities.

b. Does your graph show visual evidence suggesting that the city has an effect on the average morning commute time? Justify your answer.

c. Using the ANOVA technique learned in this chapter, do these data show sufficient evidence to claim that the city has an effect on the average morning commute time? Use $\alpha = 0.05$.

d. Does the statistical answer found in part c agree with your graphical display in part a and your response in part b? Explain why your answers agree or disagree, citing statistical information learned in this chapter.

e. "Does the sample show that the city has an effect on the *amount of time* spent commuting to work?" "Does the sample show that the city has an effect on the *average amount of time* spent commuting to work?" Are these different questions? Explain.

Your Study

12.63 Design your own "commute time" study.

a. Define a specific population that you will sample based on four age groupings, describe your sampling plan, and collect a random sample of at least 25 observations.

b. Make a descriptive presentation of your sample data, using a chart or graph and a descriptive paragraph.

c. Using the ANOVA technique learned in this chapter, do these data show sufficient evidence to claim that a person's age has an effect on the average amount of time spent commuting to work in the morning? Use $\alpha = 0.05$.

d. Discuss the similarities and differences between results of your study and the results found in Exercise 12.62.

Chapter Practice Test

Part I: Knowing The Definitions

Answer "True" if the statement is always true. If the statement is not always true, replace the words shown in bold with words that make the statement always true.

12.1 To partition the sum of squares for the total is to separate the numerical value of SS(total) into two values such that the **sum** of these two values is equal to SS(total).

12.2 A **sum of squares** is actually a measure of variance.

12.3 **Experimental error** is the name given to the variability that takes place between the levels of the test factor.

12.4 **Experimental error** is the name given to the variability that takes place among the replicates of an experiment as it is repeated under constant conditions.

12.5 **Fail to reject** H_o is the desired decision when the means for the levels of the factor being tested are all different.

12.6 The **mathematical model** for a particular problem is an equational statement showing the anticipated makeup of an individual piece of data.

12.7 The degrees of freedom for the factor are equal to the **number of factors tested.**

12.8 The measure of a specific level of a factor being tested in an ANOVA is the **variance** of that factor level.

12.9 We **need not** assume that the observations are independent to do analysis of variance.

12.10 The rejection of H_o **indicates** that you have identified the level(s) of the factor that is (are) different from the others.

Part II: Applying the Concepts

12.11 Determine the truth (T/F) of each statement with regard to the one-factor analysis of variance technique.

_____ a. The mean squares are measures of variance.

_____ b. "There is no difference between the mean values of the random variable at the various levels of the test factor" is a possible interpretation of the null hypothesis.

_____ c. "The factor being tested has no effect on the random variable x" is a possible interpretation of the alternative hypothesis.

_____ d. "There is no variance among the mean values of x for each of the different factor levels" is a possible interpretation of the null hypothesis.

_____ e. The "partitioning" of the variance occurs when SS(total) is separated into SS(factor) and SS(error).

_____ f. We will reject the null hypothesis and conclude that the factor has an effect on the variable when the amount of variance assigned to the factor is significantly larger than the variance assigned to error.

_____ g. In order to apply the F-test, the sample size from each factor level must be the same.

_____ h. In order to apply the F-test, the sample standard deviation from each factor level must be the same.

_____ i. If 20 is subtracted from every data value, then the calculated value of the $F\star$ statistic is also reduced by 20.

When the calculated value of F, $F\star$, is greater than the table value for F,

_____ j. The decision will be fail to reject H_o.

_____ k. The conclusion will be the factor being tested does have an effect on the variable.

Independent samples were collected to test the effect a factor had on a variable. The data are summarized in this ANOVA table:

	SS	df
Factor	810	2
Error	720	8
Total	1530	10

Is there sufficient evidence to reject the null hypothesis that all levels of the test factor have the same effect on the variable?

_____ l. The null hypothesis could be $\mu_A = \mu_B = \mu_C = \mu_D$.

_____ m. The calculated value of F is 1.125.

_____ n. The critical value of F for $\alpha = 0.05$ is 6.06.

_____ o. The null hypothesis can be rejected at $\alpha = 0.05$.

12.12 Consider this table:

	SS	df	MS	$F\star$
Factor	A	4	18	E
Error	B	18	D	
Total	144	C		

Find the values:

a. A b. B c. C d. D e. E

Part III: Understanding the Concepts

12.13 In 50 words or less, explain what a single-factor ANOVA experiment is.

12.14 A state environmental agency tested three different scrubbers used to reduce the resulting air pollution in the generation of electricity. The primary concern was the emission of particulate matter. Several trials were run with each scrubber. The amount of particulate emission was recorded for each trial.

[PT12-14]

	Amounts of Emission					
Scrubber I	11	10	12	9	13	12
Scrubber II	12	10	12	8	9	
Scrubber III	9	11	10	7	8	

a. State the mathematical model for this experiment.

b. State the null and alternative hypotheses.

c. Calculate and form the ANOVA table.

d. Complete the testing of H_o using a 0.05 level of significance. State the decision and conclusion clearly.

e. Construct a graph representing the data that is helpful in picturing the results of the hypothesis test.

Statistics⊖Now™ Preparing for an exam? Assess your progress by taking the post-test at **http://1pass.thomson.com**.

ⓥMentor Do you need a live tutor for homework problems? Access vMentor on the StatisticsNow website at **http://1pass.thomson.com** for one-on-one tutoring from a statistics expert.

CHAPTER

13

Linear Correlation and Regression Analysis

13.1 Wheat! Beautiful Golden Wheat!

Wheat is a common name for cereal grass of a genus of the grass family. It has been cultivated for food since prehistoric times and is one of our most important grain crops. The common types of wheat grown in the United States are spring wheat, planted in the spring for fall harvest, and winter wheat, planted in the fall for spring harvest. The main use of wheat is in flour for bread and other food products. It's also used to a limited extent in the making of beer, whiskey, industrial alcohol, and other products.

Statistics⟨△⟩Now™

Throughout the chapter, this icon introduces a list of resources on the StatisticsNow website at **http://1pass.thomson.com** that will:

- Help you evaluate your knowledge of the material

- Allow you to take an exam-prep quiz

- Provide a Personalized Learning Plan targeting resources that address areas you should study

THE U.S. WHEAT CROP FOR 2002 IS EXPECTED TO BE THE SMALLEST IN A QUARTER CENTURY

July 12, 2002 (EIRNS)—The area harvested in the United States this year for winter wheat (the predominant wheat variety in U.S. latitudes) is estimated to be only 29.8 million acres (12.06 million hectares)—the same as in 1917! (The United States harvested winter wheat area in recent years has been between 35 and over 40 million acres.) Farmers have abandoned large amounts of sown land because of drought and related pests and disease. Estimates now put the total U.S. wheat harvest (all types) this year at around 1.79 billion bushels (48.9 million metric tons), about the same as in 1974,

and way down from the 64 million ton levels of recent yearly harvests. Western Canada could potentially harvest 19.7 million metric tons of wheat, down from the five-year average of 23.3 million tons, which itself has been declining.

In terms of world trade in basic foodstuffs, the United States and Canada are a major source of world wheat supplies—now severely contracted. Australia's wheat output next season is expected to drop. Argentina is in turmoil. Only Europe (principally France) expects a good harvest. World wheat stocks are way down.

Source: http://committeerepubliccanada.ca/English/News/Slug012.htm

The business decisions a grain farmer makes are not as simple as the statistical relationship between the four variables listed in the table that follows. However, an understanding of the relationship between these variables is an important component of what a grain farmer needs to know in order to make decisions about how many acres to plant, what kind of grain to plant, and so on.

Twenty randomly selected wheat-producing counties in Kansas were identified and data were collected for these variables:

Planted = 1000s of acres planted with winter wheat

Harvested = 1000s of acres harvested (not all planted acres are harvested for a variety of reasons)

Yield = bushels of wheat harvested per acre

Production = 1000s of bushels of wheat harvested

[EX13-01]

County	Planted	Harvested	Yield	Production	County	Planted	Harvested	Yield	Production
Allen	32.0	29.5	49	1,451	McPherson	208.0	192.3	41	7,942
Chautauqua	5.5	4.8	43	206	Miami	10.0	9.0	55	491
Cherokee	64.4	61.1	46	2,791	Nemaha	27.7	24.8	46	1,134
Coffey	26.4	25.1	51	1,281	Neosho	37.2	35.3	43	1,523
Gove	118.0	79.9	34	2,739	Sherman	185.0	171.8	35	5,984
Gray	127.0	103.9	36	3,689	Stafford	141.0	127.4	38	4,781
Greenwood	6.1	5.6	43	242	Sumner	350.0	317.8	40	12,726
Johnson	5.5	5.0	52	261	Thomas	196.0	166.9	44	7,400
Linn	14.6	14.3	54	774	Washington	88.0	80.3	43	3,433
Logan	145.0	84.2	32	2,706	Wilson	41.5	39.1	49	1,932

Source: http://www.usda.gov/nass/graphics/county01/data/ww01.csv

After completing Chapter 13, further investigate the Chapter Project on page 744.

The basic ideas of regression and linear correlation analysis were introduced in Chapter 3. (If these concepts are not fresh in your mind, review Chapter 3 before beginning this chapter.) Chapter 3 was only a first look: a presentation of the basic graphic (the scatter diagram) and descriptive statistical aspects of linear correlation and regression analysis. In this chapter we take a second, more detailed look at linear correlation and regression analysis.

SECTION 13.1 EXERCISES

Statistics ⬡ Now™

Datasets can be found on your Student's Suite CD-ROM or at the StatisticsNow website at **http://1pass.thomson.com**.

13.1 [EX13-01] Refer to the wheat crop data above.

a. Construct a scatter diagram for Planted (*x*) and Harvested (*y*).

b. Construct a scatter diagram of Harvested (*x*) and Production (*y*).

c. Construct a histogram of Yield.

d. Explain what is learned from parts a–c.

13.2 [EX13-01] Refer to the wheat crop data on page 696.

a. Find the linear correlation coefficient and the equation for the line of best fit for Planted (*x*)

and Harvested (*y*). What does the correlation coefficient tell you with respect to the variables?

b. Find the linear correlation coefficient and the equation for the line of best fit for Harvested (*x*) and Production (*y*). What does the correlation coefficient tell you with respect to the variables?

13.2 Linear Correlation Analysis

In Chapter 3 the linear correlation coefficient was presented as a quantity that measures the strength of a linear relationship (dependency). Now let's take a second look at this concept and see how *r*, the coefficient of linear correlation, works. Intuitively, we want to think about how to measure the mathematical linear dependency of one variable on another. As *x* increases, does *y* tend to increase or decrease? How strong (consistent) is this tendency? We are going to use two measures of dependence—covariance and the coefficient of linear correlation—to measure the relationship between two variables. We'll begin our discussion by examining a set of bivariate data and identifying some related facts as we prepare to define covariance.

EXAMPLE 13.1

Understanding and Calculating Covariance

Let's consider the sample of six bivariate data (ordered pairs): (2, 1), (3, 5), (6, 3), (8, 2), (11, 6), (12, 1). See Figure 13.1. The mean of the six *x* values (2, 3, 6, 8, 11, 12) is $\bar{x} = 7$. The mean of the six *y* values (1, 5, 3, 2, 6, 1) is $\bar{y} = 3$.

FIGURE 13.1 Graph of Bivariate Data

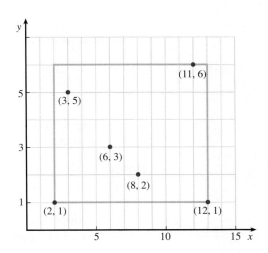

The point $(\bar{x}, \bar{y})$, which is (7, 3), is located as shown on the graph of the sample points in Figure 13.2. The point $(\bar{x}, \bar{y})$ is called the **centroid** of the data. A vertical and a horizontal line drawn through the centroid divide the graph into four sections, as shown in Figure 13.2. Each point (x, y) lies a certain distance from each of these two lines: $(x - \bar{x})$ is the horizontal distance from (x, y) to the vertical line that passes through the centroid, and $(y - \bar{y})$ is the vertical distance from (x, y) to the horizontal line that passes through the centroid. Both the horizontal and vertical distances of each data point from the centroid can be measured, as shown in Figure 13.3. The distances may be positive, negative, or zero, depending on the position of the point (x, y) in relation to $(\bar{x}, \bar{y})$. [Figure 13.3 shows $(x - \bar{x})$ and $(y - \bar{y})$ represented by braces, with positive or negative signs.]

FIGURE 13.2 The Point (7, 3) Is the Centroid

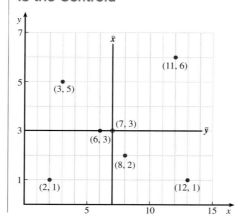

FIGURE 13.3 Measuring the Distance of Each Data Point from the Centroid

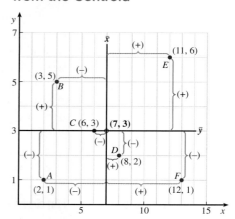

One measure of linear dependency is the covariance. The **covariance of x and y** is defined as the sum of the products of the distances of all values of x and y from the centroid,

$$\sum [(x - \bar{x})(y - \bar{y})], \text{ divided by } n - 1:$$

Covariance of x and y

$$\text{covar}(x, y) = \frac{\sum_{i=1}^{n} (x_i - \bar{x})(y_i - \bar{y})}{n - 1} \tag{13.1}$$

Calculations for the covariance for the data in Example 13.1 are given in Table 13.1. The covariance, written as covar(x, y), of the data is $\frac{3}{5} = \mathbf{0.6}$.

Notes:

1. $\sum (x - \bar{x}) = 0$ and $\sum (y - \bar{y}) = 0$. This will always happen. Why? (See pp. 84–85.)
2. Even though the variance of a single set of data is always positive, the covariance of bivariate data can be negative.

TABLE 13.1

Calculations for Finding covar(x, y) for the Data of Example 13.1

Points	$x - \bar{x}$	$y - \bar{y}$	$(x - \bar{x})(y - \bar{y})$
(2, 1)	−5	−2	10
(3, 5)	−4	2	−8
(6, 3)	−1	0	0
(8, 2)	1	−1	−1
(11, 6)	4	3	12
(12, 1)	5	−2	−10
Total	0 ⓒⓚ	0 ⓒⓚ	3

The covariance is positive if the graph is dominated by points to the upper right and to the lower left of the centroid. The products of $(x - \bar{x})$ and $(y - \bar{y})$ are positive in these two sections. If the majority of the points are to the upper left and the lower right of the centroid, then the sum of the products is negative. Figure 13.4 shows data that represent (a) a positive dependency, (b) a negative dependency, (c) and little or no dependency. The covariances for these three situations would definitely be positive in part a, negative in b, and near zero in c. (The sign of the covariance is always the same as the sign of the slope of the regression line.)

FIGURE 13.4 Data and Covariance

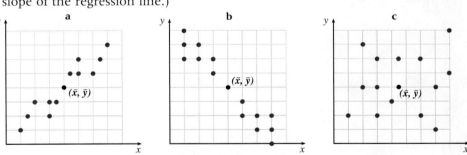

FYI This calculation is assigned in Exercise 13.13 (p. 703).

The biggest disadvantage of covariance as a measure of linear dependency is that it does not have a standardized unit of measure. One reason for this is that the spread of the data is a strong factor in the size of the covariance. For example, if we multiply each data point in Example 13.1 by 10, we have (20, 10), (30, 50), (60, 30), (80, 20), (110, 60), and (120, 10). The relationship of the points to each other is changed only in that they are much more spread out. However, the covariance for this new set of data is 60. Does this mean that the dependency between the *x* and *y* variables is stronger than in the original case? No, it does not; the relationship is the same, even though each data value has been multiplied by 10. This is the trouble with covariance as a measure. We must find some way to eliminate the effect of the spread of the data when we measure dependency.

If we standardize *x* and *y* by dividing the distance of each from the respective mean by the respective standard deviation:

$$x' = \frac{x - \bar{x}}{s_x} \quad \text{and} \quad y' = \frac{y - \bar{y}}{s_y}$$

The Correlation Coefficient

The complete name of the correlation coefficient deceives many into a belief that Karl Pearson developed this statistical measure himself. Although Pearson did develop a rigorous treatment of the mathematics of the Pearson Product Moment Correlation (PPMC), it was the imagination of Sir Francis Galton that originally conceived modern notions of correlation and regression. Galton's fascination with genetics and heredity provided the initial inspiration that led to regression and the PPMC.

FYI Refer to Chapter 3 (pp. 164–165) for an illustration of the use of this formula.

FYI Computer and calculator commands to find the correlation coefficient were presented in Chapter 3 (p. 166).

and then compute the covariance of x' and y', we will have a covariance that is not affected by the spread of the data. This is exactly what is accomplished by the linear correlation coefficient. It divides the covariance of x and y by a measure of the spread of x and by a measure of the spread of y (the standard deviations of x and of y are used as measures of spread). Therefore, by definition, the **coefficient of linear correlation** is:

Coefficient of Linear Correlation

$$r = \text{covar}(x', y') = \frac{\text{covar}(x, y)}{s_x \cdot s_y} \qquad (13.2)$$

The coefficient of linear correlation standardizes the measure of dependency and allows us to compare the relative strengths of dependency of different sets of data. [Formula (13.2) for linear correlation is also commonly referred to as **Pearson's product moment**, r.]

We can find the value of r, the coefficient of linear correlation, for the data in Example 13.1 by calculating the two standard deviations and then dividing:

$$s_x = 4.099 \quad \text{and} \quad s_y = 2.098$$

$$r = \frac{\text{covar}(x, y)}{s_x \cdot s_y} : \quad r = \frac{0.6}{(4.099)(2.098)} = \mathbf{0.07}$$

Finding the correlation coefficient using formula (13.2) can be a very tedious arithmetic process. We can write the formula in a more workable form, however, as it was in Chapter 3:

Shortcut for Coefficient of Linear Correlation

$$r = \frac{\text{covar}(x, y)}{s_x \cdot s_y} = \frac{\dfrac{\sum[(x - \bar{x})(y - \bar{y})]}{n - 1}}{s_x \cdot s_y} = \frac{\text{SS}(xy)}{\sqrt{\text{SS}(x) \cdot \text{SS}(y)}} \qquad (13.3)$$

Formula (13.3) avoids the separate calculations of $\bar{x}$, $\bar{y}$, s_x, and s_y as well as the calculations of the deviations from the means. Therefore, formula (13.3) is much easier to use, and more important, it is more accurate when decimals are involved because it minimizes round-off error.

SECTION 13.2 EXERCISES

Statistics ⬡ Now™

Skillbuilder Applet Exercises must be worked using an accompanying applet found on your Student's Suite CD-ROM or at the StatisticsNow website at **http://1pass.thomson.com**.

Datasets can be found on your Student's Suite CD-ROM or at the StatisticsNow website at **http://1pass.thomson.com**.

13.3 Consider a set of paired bivariate data.

a. Explain why $\sum(x - \bar{x}) = 0$ and $\sum(y - \bar{y}) = 0$.

b. Describe the effect that lines $x = \bar{x}$ and $y = \bar{y}$ have on the graph of these points.

c. Describe the relationship of the ordered pairs that will cause $\sum[(x - \bar{x}) \cdot (y - \bar{y})]$ to be (1) positive, (2) negative, and (3) near zero.

13.4 Skillbuilder Applet Exercise matches correlation coefficients with their scatterplots. After several practice rounds using "New Plots," explain your method of matching.

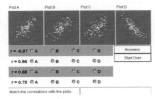

13.5 [EX13-05] The following data values are from a random sample of 40 college students; the data show the students' gender, American College Test (ACT) composite scores, and grade point averages (GPAs) after their first term in college.

Female		Male		Female		Male	
ACT	GPA	ACT	GPA	ACT	GPA	ACT	GPA
23	1.833	33	3.333	15	3.000	13	3.053
28	4.000	17	2.835	22	3.600	16	2.600
22	3.057	26	3.249	20	2.665	27	2.000
20	4.000	25	2.290	17	2.934	19	2.500
23	3.550	20	2.178	21	3.422	22	4.000
19	2.583	23	2.835	18	3.002	33	2.833
20	3.165	19	2.364	17	3.000	17	3.438
29	3.398	21	3.000	25	4.000	26	2.418
27	3.868	22	3.934	25	3.472		
18	2.918	29	3.533	25	3.550		
17	2.360	16	3.313				

Source: http://www.act.org/research/briefs/97-2.html

a. Construct a scatter diagram of the data with ACT scores on the horizontal axis and GPA on the vertical axis, being sure to identify the male and female students.

b. Do the patterns for males and females appear to be the same, or are they different? Identify specific similarities and differences.

c. Assume that a student had an ACT score of 25. What would you predict that student's GPA to be at the end of the first term in college?

d. Does there appear to be any relationship between ACT scores and the first-term GPA?

13.6 [EX13-06] Aerial photographs are one of many techniques used to monitor wildlife populations. Knowing the number of animals and their location relative to areas inhabited by the human population is very useful. It is also important to monitor physical characteristics of the animals. Is it possible to use the length of a bear, as estimated from an aerial photograph, to estimate the bear's age and/or weight? (It would be much safer than asking it to stand on a set of scales! ☺) The data that follow are for age (in months), gender (1 = male, 2 = female), length (in inches), and weight (in pounds).

Age	Gender	Length	Weight
19	1	45.0	65
29	2	62.0	121

••• Remainder of data on Student's Suite CD-ROM.

Source: MINITAB's Bears.mtw

a. Investigate the relationship between the length and age of the bears. Be sure to include the gender variable.

b. Does there seem to be a predictable pattern for the relationship between length and age? How does gender of the bear affect the relationship? Explain. Describe the pattern.

c. Investigate the relationship between the length and weight of the bears. Be sure to include the gender variable.

d. Does there seem to be a predictable pattern for the relationship between length and weight? How does gender of the bear affect the relationship? Explain. Describe the pattern.

e. If the gender of a smaller or younger bear cannot be determined, how will this affect the estimate for age or weight? Explain.

13.7 [EX13-07] a. Construct a scatter diagram of the following bivariate data.

Point	A	B	C	D	E	F	G	H	I	J
x	1	1	3	3	5	5	7	7	9	9
y	1	2	2	3	3	4	4	5	5	6

b. Calculate the covariance.

c. Calculate s_x and s_y.

d. Calculate r using formula (13.2).

e. Calculate r using formula (13.3).

13.8 [EX13-08] a. Draw a scatter diagram of the following bivariate data.

Point	A	B	C	D	E	F	G	H	I	J
x	0	1	1	2	3	4	5	6	6	7
y	6	6	7	4	5	2	3	0	1	1

b. Calculate the covariance.

c. Calculate s_x and s_y.

d. Calculate r using formula (13.2).

e. Calculate r using formula (13.3).

13.9 [EX13-09] A computer was used to complete the preliminary calculations: form the extensions table; calculate the summations $\sum x$, $\sum y$, $\sum x^2$, $\sum xy$, $\sum y^2$; and find the SS(x), SS(y), and SS(xy) for the following set of bivariate data. Verify the results by calculating the values yourself.

x	45	52	49	60	67	61
y	22	26	21	28	33	32

```
MINITAB output:
Row      X        Y        XSQ       XY       YSQ
 1       45       22       2025      990      484
 2       52       26       2704      1352     676
 3       49       21       2401      1029     441
 4       60       28       3600      1680     784
 5       67       33       4489      2211     1089
 6       61       32       3721      1952     1024

Row    sum X    sum Y    sum XSQ   sum XY   sum YSQ
 1      334      162      18940     9214     4498
```

```
SS(X)    347.333
SS(Y)    124.000
SS(XY)   196.000
```

13.10 [EX13-10] Use a computer to form the extensions table; calculate the summations $\sum x$, $\sum y$, $\sum x^2$, $\sum xy$, $\sum y^2$; and find the SS(x), SS(y), and SS(xy) for the following set of bivariate data.

x	11.4	9.4	6.5	7.3	7.9	9.0	9.3	10.6
y	8.1	8.2	5.8	6.4	5.9	6.5	7.1	7.8

13.11 [EX13-11] National Football League (NFL) football enthusiasts often look at a team's total points scored for (Pts F) and total points scored against (Pts A) as a way of comparing the relative strength of teams. The season totals for the 32 teams in the NFL 2004 season are shown here.

Pts F	Pts A	Pts F	Pts A
285	322	339	338

••• Remainder of data on Student's Suite CD-ROM.

Source: http://www.nfl.com/stats

a. Calculate the linear correlation coefficient (Pearson's product moment, r) for the points scored for and against.

b. What conclusion can you draw from the answer in part a?

c. Construct the scatter diagram and comment on how it supports, or disagrees with, your comments in part b.

FYI See page 166 for information about using MINITAB, Excel, or TI-83/84 to find the correlation coefficient.

13.12 [EX13-12] Knowing a horse's weight (measured in pounds) is important information for a horse owner. The amount of feed and medicine dosages all depend on the horse's weight. Most owners do not have the resources to have a scale large enough to weigh a horse, so other measurements are used to estimate the weight. Height (measured in hands) and girth and length (measured in inches) are common measurements for a horse. A sample of Suffolk Punch stallion measurements were taken from the website http://www.suffolkpunch.com/horses/stalnum.html.

Row	Height	Girth	Length	Weight
1	16.0	93	72	1825
2	15.3	78	69	1272
3	16.0	84	70	1515
4	17.0	96	80	2100
5	16.2	86	70	1569
6	16.0	88	72	1690
7	16.0	83	72	1500

a. Calculate the linear correlation coefficient (Pearson's product moment, r) between (1) height and weight, (2) girth and weight, and (3) length and weight.

b. What conclusions might you draw from your answers in part a?

c. Construct a scatter diagram for each pair of variables listed in part a.

d. Do the scatter diagrams support your answer in part b?

e. Based on this evidence, which measurement do you believe has the most potential as a predictor of weight? Explain your choice.

13.13 a. Calculate the covariance of the set of data (20, 10), (30, 50), (60, 30), (80, 20), (110, 60), and (120, 10).

b. Calculate the standard deviation of the six x-values and the standard deviation of the six y-values.

c. Calculate r, the coefficient of linear correlation, for the data in part a.

d. Compare these results to those found in the text for Example 13.1 (pp. 697–699).

13.14 A formula that is sometimes given for computing the correlation coefficient is

$$r = \frac{n(\sum xy) - (\sum x)(\sum y)}{\sqrt{n(\sum x^2) - (\sum x)^2}\sqrt{n(\sum y^2) - (\sum y)^2}}$$

Use this expression as well as the formula

$$r = \frac{SS(xy)}{\sqrt{SS(x) \cdot SS(y)}}$$

to compute r for the data in the following table.

x	2	4	3	4	0
y	6	7	5	6	3

13.3 — Inferences about the Linear Correlation Coefficient

In Section 13.2 we learned that covariance is a measure of linear dependency. Also noted was the fact that its value is affected by the spread of the data; therefore, we standardize the covariance by dividing it by the standard deviations of both x and y. This standardized form is known as r, the coefficient of linear correlation. Standardizing enables us to compare different sets of data, thereby allowing r to play a role much like z or t does for $\bar{x}$. The calculated r value becomes r★, the test statistic for inferences about ρ, the population correlation coefficient. (ρ is the lowercase Greek letter rho.)

Assumptions for inferences about the linear correlation coefficient: The set of (x, y) ordered pairs forms a random sample, and the y values at each x have a normal distribution. Inferences use the t-distribution with n − 2 degrees of freedom.

Caution: Inferences about the linear correlation coefficient are about the pattern of behavior of the two variables involved and the usefulness of one variable in predicting the other. *Significance of the linear correlation coefficient does not mean that you have established a cause-and-effect relationship.* Cause and effect is a separate issue. (See the causation discussion on pages 167–168.)

Confidence Interval Procedure

As with other parameters, a **confidence interval** may be used to estimate the value of ρ, the linear correlation coefficient of the population. Usually this is accomplished by using a table that shows **confidence belts.** Table 10 in Appendix B gives confidence belts for 95% confidence intervals. This table is a bit tricky to read and utilizes n, the sample size, so be extra careful when you use it. The next example demonstrates the procedure for estimating ρ.

EXAMPLE 13.2

Constructing a Confidence Interval for the Population Correlation Coefficient

A random sample of 15 ordered pairs of data has a calculated r value of 0.35. Find the 95% confidence interval for ρ, the population linear correlation coefficient.

SOLUTION

Step 1 **Parameter of interest:** The linear correlation coefficient for the population, ρ.

Step 2 a. **Assumptions:** The ordered pairs form a random sample, and we will assume that the y values at each x have a normal distribution.
    b. **Formula:** The calculated linear correlation coefficient, r.
    c. **Level of confidence:** $1 - \alpha = 0.95$.

Step 3 **Sample information:** $n = 15$ and $r = 0.35$.

Step 4 **Confidence interval:** The confidence interval is read from Table 10 in Appendix B. Find $r = 0.35$ at the bottom of Table 10. (See the arrow on Figure 13.5.) Visualize a vertical line drawn through that point. Find the two points where the belts marked for the correct sample size cross the vertical line. The sample size is 15. These two points are circled in Figure 13.5. Now look horizontally from the two circled points to the vertical scale on the left and read the confidence interval. The values are **−0.20** and **0.72.**

Step 5 **Confidence interval:** The 95% confidence interval for ρ, the population coefficient of linear correlation, is −0.20 to 0.72.

Hypothesis-Testing Procedure

After the linear correlation coefficient, r, has been calculated for the sample data, it seems necessary to ask this question: Does the value of r indicate that there is a linear dependency between the two variables in the population from which the sample was drawn? To answer this question we can perform a **hypothesis test.** The null hypothesis is: The two variables are linearly unrelated ($\rho = 0$), where ρ is the linear correlation coefficient for the population. The alternative hypothesis may be either one-tailed or two-tailed. Most frequently it is two-tailed, $\rho \neq 0$. However, when we suspect that there is only a positive or only a negative correlation, we should use a one-tailed test. The alternative hypothesis of a one-tailed test is $\rho > 0$ or $\rho < 0$.

The area that represents the p-value or the critical region for the test is on the right when a positive correlation is expected and on the left when a negative correlation is expected. The test statistic used to test the null hypothesis is the calculated value of r from the sample data. Probability bounds for the p-value or critical values for r are found in Table 11 of Appendix B (p. 823). The

FIGURE 13.5
Using Table 10 of
Appendix B,
Confidence Belts
for the Correlation
Coefficient

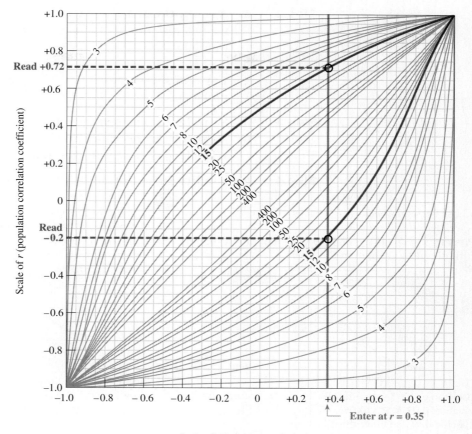

Scale of r (sample correlation)

number of degrees of freedom for the r statistic is 2 less than the sample size, df $= n - 2$. Specific details for using Table 11 follow Example 13.3.

Rejection of the null hypothesis means that there is evidence of a linear relationship between the two variables in the population. Failure to reject the null hypothesis is interpreted as meaning that a linear relationship between the two variables in the population has not been shown.

Now let's look at an example of a hypothesis test.

EXAMPLE 13.3

Two-Tailed Hypothesis Test

In a study of 15 randomly selected ordered pairs, $r = 0.548$. Is this linear correlation coefficient significantly different from zero at the 0.02 level of significance?

SOLUTION

Step 1   a.   **Parameter of interest:** The linear correlation coefficient for the population, ρ.
              b.   **Statement of hypotheses:**
                   H_o: $\rho = 0$
                   H_a: $\rho \neq 0$

Step 2   **a.** **Assumptions:** The ordered pairs form a random sample, and we will assume that the y values at each x have a normal distribution.

b. **Test statistic:** $r\star$, formula (13.3), with df $= n - 2 = 15 - 2 = 13$.

c. **Level of significance:** $\alpha = 0.02$ (given in the statement of the problem).

Step 3   **a.** **Sample information:** $n = 15$ and $r = 0.548$.

b. **Value of the test statistic:** The calculated sample linear correlation coefficient is the test statistic: $r\star = 0.548$.

Step 4   **Probability Distribution:**

p-**Value:**   **OR**   **Classical:**

a. Use both tails because H_a expresses concern for values related to "different from."

$\mathbf{P} = P(r < -0.548) + P(r > 0.548) = 2 \cdot P(r > 0.548)$, with df = 13 as shown in the figure.

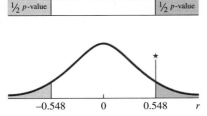

Use Table 11 (Appendix B) to place bounds on the *p*-value: **0.02 < P < 0.05.**
Specific details follow this illustration.
b. The *p*-value is not smaller than the level of significance, α.

a. The critical region is both tails because H_a expresses concern for values related to "different from." The critical value is found at the intersection of the df = 13 row and the two-tailed 0.02 column of Table 11: **0.592.**

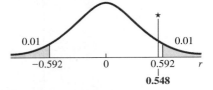

Specific details follow this illustration.
b. $r\star$ is not in the critical region, as shown in **red** in the figure.

Step 5   **a.** **Decision:** Fail to reject H_o.

b. **Conclusion:** At the 0.02 level of significance, we have failed to show that x and y are correlated.

Calculating the *p*-value

Use Table 11 in Appendix B to "place bounds" on the p-value. By inspecting the df = 13 row of Table 11, you can determine an interval within which the *p*-value lies. Locate $r\star$ along the row labeled df = 13. If $r\star$ is not listed, locate the two table values it falls between, and read the bounds for the *p*-value from the top of the table. In this case, $r\star = 0.548$ is between 0.514 and 0.592; therefore, **P** is between 0.02 and 0.05. Table 11 shows only two-tailed values. When the alternative hypothesis is two-tailed, the bounds for the *p*-value are read directly from the table.

Portion of Table 11

df	...	Amount of α in two tails				
		0.05	P	0.02	...——→	0.02 < P < 0.05
⋮	⋮	↑	⋮			
13		0.514	*0.548*	0.592		

Note: When H_a is one-tailed, divide the column headings by 2 to place bounds on the *p*-value.

Use Table 11 in Appendix B to find the critical values. The critical value is at the intersection of the df = 13 and the two-tailed $\alpha = 0.02$ column. Table 11 shows only two-tailed values. Since the alternative hypothesis is two-tailed, the critical values are read directly from the table.

Portion of Table 11

df		Amount of α in two tails 0.02	
	...	0.02	...
⋮	⋮		
13		0.592	

⟶ Critical values: ± 0.592

Note: When H_a is one-tailed, divide the column headings by 2.

Use of Correlation in a Medical Study

CORRELATION OF ACTIVATED CLOTTING TIME AND ACTIVATED PARTIAL THROMBOPLASTIN TIME TO PLASMA HEPARIN CONCENTRATION

Study Objective: Determine the correlation between activated clotting time (ACT) or activated partial thromboplastin time (aPTT) and plasma heparin concentration
Design: Two-phase prospective study
Patients: Thirty patients receiving continuous-infusion intravenous heparin
Interventions: Measurement of ACT, aPTT, and plasma heparin concentrations

Heparin has been administered for over 50 years as an anticoagulant and is known to have a narrow therapeutic range. Underdosing of heparin is associated with recurrent thromboembolism, whereas excessive dosing may increase the risk of hemorrhagic complications. Several clotting time tests are available to monitor heparin, including whole blood clotting time, activated partial thromboplastin time (aPTT), and activated clotting time (ACT).

The study was conducted in two phases. In phase 1 (intraperson phase), sequential blood draws from five patients were evaluated. The goal was to determine if there was a significant relationship between plasma heparin concentrations and clotting time tests within an individual. In phase 2 (interperson phase), single random blood draws from 25 additional patients were evaluated with the same collection technique and analysis as in phase 1. Blood draws were performed within 48 hours after the start of heparin therapy. The goal of phase 2 was to determine the quantitative relationship between ACT or aPTT and plasma heparin concentration between individuals.

For both phases, correlations between ACT or aPTT results and plasma heparin concentrations were performed using the Pearson moment R correlation test. Phase 1: Linear correlation coefficients (r) for the five patients were 0.93 ($p = 0.02$), 0.99 ($p = 0.009$), 0.89 ($p = 0.12$), 0.96 ($p = 0.04$), and 0.90 ($p = 0.10$). Phase 2: Correlation coefficient for these data was 0.58 (linear, $p = 0.008$). The linear regression line formula is $137 + (52.9)$(plasma heparin concentration), which, for a therapeutic heparin range of 0.3–0.7 U/ml (by antifactor Xa), equates to an ACT range of 153–174

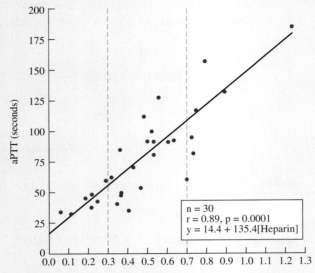

FIGURE 7 Linear aPTT versus plasma heparin concentration for phase 2 (interperson correlation and regression). Vertical dashed lines indicate the therapeutic range for plasma heparin concentration by antifactor Xa.

n = 30
r = 0.89, p = 0.0001
y = 14.4 + 135.4[Heparin]

seconds. Linear regression lines for aPTT versus plasma heparin concentration are shown in Figure 7. Correlation coefficient for these data was 0.89 (linear, $p = 0.0001$). The linear regression line formula was 14.4 + (135.4)(plasma heparin concentration), which, for the same therapeutic heparin range, equates to an aPTT range of 55–109 seconds.

The decision analysis results indicate that a standard clotting time test therapeutic range (not derived from heparin concentration) often results in incorrect patient management decisions. The ACT based on a standard therapeutic range may result in dosage adjustment decisions that may increase the risk of bleeding (in 43% of patients). The aPTT based on a standard therapeutic range may result in dosage adjustment decisions that may increase the risk of thrombosis (in 37% of patients). A larger study in 200 patients is under way to confirm these results using heparin concentration-derived therapeutic ranges for both aPTT and ACT.

Source: John M. Koerber, B.S.; Maureen A. Smythe, Pharm.D.; Robert L. Begle, M.D.; Joan C. Mattson, M.D.; Beverly P. Kershaw, M.S.; and Susan J. Westley, M.T. (ASCP). *Pharmacotherapy,* 19(8):922–931. © Pharmacotherapy Publications, http://www .medscape.com/viewarticle/418017_3. Reprinted with permission.

SECTION 13.3 EXERCISES

Statistics △ Now™

13.15 Using graphs to illustrate, explain the meaning of a correlation coefficient with the following values:

a. −1.0   b. 0.0   c. +1.0   d. +0.5   e. −0.6

13.16 Using Figure 13.5 on page 705, find the 95% interval when a sample of $n = 25$ results in $r = 0.35$.

13.17 a. Using Figure 13.5 on page 705, find the 95% interval when a sample of $n = 100$ results in $r = 0.35$.
b. Compare your answer in part a with the confidence interval formed in Exercise 13.16. Describe what occurred when you increased the sample size.

13.18 Use Table 10 of Appendix B to determine a 95% confidence interval for the true population linear correlation coefficient based on the following sample statistics:

a. $n = 8, r = 0.20$ b. $n = 100, r = -0.40$

c. $n = 25, r = +0.65$ d. $n = 15, r = -0.23$

13.19 Use Table 10 of Appendix B to determine a 95% confidence interval for the true population linear correlation coefficient based on the following sample statistics:

a. $n = 50, r = 0.60$ b. $n = 12, r = -0.45$

c. $n = 6, r = +0.80$ d. $n = 200, r = -0.56$

13.20 [EX13-20] Does a country's gross domestic product (GDP) indicate its level of technology? The 2005 Edition of *The EU-15's New Economy—A Statistical Portrait* provided statistics on its European Union members. The following information was obtained for a random sample of five of the countries.

Country	2003 GDP	2003 % Household Internet Conn.	Country	2003 GDP	2003 % Household Internet Conn.
Denmark	112.9	64	Italy	98.6	31
Spain	87.4	25	Portugal	68.4	22
France	103.8	28			

Find r and set a 95% confidence interval for ρ.

13.21 [EX13-21] The test–retest method is one way of establishing the reliability of a test. The test is administered, and then, at a later date, the same test is readministered to the same individuals. The correlation coefficient is computed between the two sets or scores. The following test scores were obtained in a test–retest situation.

First Score	75	87	60	75	98	80	68	84	47	72
Second Score	72	90	52	75	94	78	72	80	53	70

Find r and set a 95% confidence interval for ρ.

13.22 [EX13-22] Perhaps the size of an animal's brain determines intelligence for that species. Or perhaps the weight of the brain. Or perhaps their body size or weight have a role. The following chart compares several animals' brain and body sizes and weights.

Species	Brain Length (cm)	Brain Weight (g)	Body Length (cm)	Body Weight (g)
Human	15	1400	100	62,000
Baboon	8	140	75	30,000
Monkey	5	100	30	7,000
Camel	15	680	200	529,000
Dolphin	Missing	1700	305	160,000
Kangaroo	5	56	150	35,000
Cat	5	30	60	3,300
Raccoon	5.5	39	80	4,290
Rabbit	5	12	30	2,500
Squirrel	3	6	20	900
Frog	2	0.1	10	18

Source: http://serendip.brynmawr.edu/bb/kinser/Sizechart.html

Calculate the correlation coefficient and use it and Table 10 of Appendix B to determine a 95% confidence interval on ρ for each of the following cases:

a. Brain length and brain weight
b. Brain length and body weight
c. Brain weight and body weight

13.23 [EX13-23] Portugal is known for its sweet port, or "simple rose." To keep with the current trend of dry table wines, Portuguese winemakers are adding new reds to their inventory of wines. Listed in the table are five varieties with their Wine Spectator score and price per bottle. Wine Spectator rates wines on a 100-point scale, and all wines are blind-tasted.

Name	Score	Price
Quinta do Bale Meao Douro 2000	95	$49.00
Quinta do Portal Tinta Roriz Douro 2000	89	$27.00
Casa Santos Lima Alicante Bouschet Estremadura 2000	87	$10.00
Luis Pato Baga Beiras Casta 2001	87	$14.00
Eborae Vitis e Vinus Alentejo Singularis 2002	85	$7.00

a. Calculate r.

b. Set a 95% confidence interval of ρ.

c. Describe the meaning of the answer to part b.

d. Explain the meaning of the width of the interval answer in part b.

13.24 [EX13-24] A new study shows that the death rate on rural roads is higher than that on other roads in the United States. The table that follows is an excerpt taken from "Death rates higher on rural road" in the March 3, 2005, *USA Today,* which gives the death rates for each state on rural, noninterstate roads in 2003 for every 100 million miles of travel and the rates for all other roads in the state.

State	Rural Roads	All Others
AL	2.45	1.21
AK	1.76	2.00
••• Remainder of data on Student's Suite CD-ROM.		

Source: *USA Today*

a. Calculate the correlation coefficient and use it and Table 10 to determine a 95% confidence interval for ρ.

b. What other factors could be having an effect on this relationship? Explain.

13.25 State the null hypothesis, H_o, and the alternative hypothesis, H_a, that would be used to test the following statements:

a. The linear correlation coefficient is positive.

b. There is no linear correlation.

c. There is evidence of negative correlation.

d. There is a positive linear relationship.

13.26 a. State the standard null hypothesis, H_o, for testing the linear correlation coefficient, ρ.

b. What does a decision of "Fail to reject H_o" indicate in a hypothesis test for ρ?

c. What does a decision of "Reject H_o" indicate in a hypothesis test for ρ?

13.27 Place bounds on the p-value resulting from a sample with $n = 18$ and $r = 0.444$ in the following circumstances:

a. H_a is two-tailed.   b. H_a is one-tailed.

13.28 Determine the bounds on the p-value that would be used in testing each of the following null hypotheses using the p-value approach:

a. $H_o: \rho = 0$ vs. $H_a: \rho \neq 0$, with $n = 32$ and $r = 0.41$

b. $H_o: \rho = 0$ vs. $H_a: \rho > 0$, with $n = 9$ and $r = 0.75$

c. $H_o: \rho = 0$ vs. $H_a: \rho < 0$, with $n = 15$ and $r = -0.83$

13.29 Determine the critical values of r for $\alpha = 0.05$ and $n = 20$ in the following circumstances:

a. H_a is two-tailed.   b. H_a is one-tailed.

13.30 Determine the critical values that would be used in testing each of the following null hypotheses using the classical approach.

a. $H_o: \rho = 0$ vs. $H_a: \rho \neq 0$, with $n = 18$ and $\alpha = 0.05$

b. $H_o: \rho = 0$ vs. $H_a: \rho > 0$, with $n = 32$ and $\alpha = 0.01$

c. $H_o: \rho = 0$ vs. $H_a: \rho < 0$, with $n = 16$ and $\alpha = 0.05$

13.31 Referring to Applied Example 13.4:

a. Explain the meaning of "Correlation coefficient for these data was 0.58 (linear, $p = 0.008$)" as reported for Phase 2.

b. Using Table 11, what bounds would be placed on the p-value? How do these bounds compare with the p-value in part a?

c. What is the critical value for a two-tailed test of $\rho = 0.00$ at the $\alpha = 0.01$ level?

d. Is $r = 0.58$ significant?

13.32 a. If a sample of size 10 has a linear correlation coefficient of 0.60, is there significant reason to conclude that the linear correlation coefficient of the population is positive? Use $\alpha = 0.01$.

b. If a sample of size 42 has a linear correlation coefficient of 0.60, is there significant reason to conclude that the linear correlation coefficient of the population is positive? Use $\alpha = 0.01$.

c. Describe the similarities and differences between parts a and b.

13.33 A sample of 20 bivariate data has a linear correlation coefficient of $r = 0.43$. Does this provide sufficient evidence to reject the null hypothesis that $\rho = 0$ in favor of a two-sided alternative? Use $\alpha = 0.10$.

13.34 If a sample of size 18 has a linear correlation coefficient of -0.50, is there significant reason to conclude that the linear correlation coefficient of the population is negative? Use $\alpha = 0.01$.

13.35 When it comes to high-end Japanese eateries featuring sushi, the quality and presentation of the food are no doubt indicators of the cost. What about the décor of the restaurant? Would that also have an effect on the cost? Zagat Survey results, published in the November 11, 2004, *Newsweek,* produced a correlation coefficient of 0.532 between restaurant décor rating and the average cost of dinner. If these results were based on five restaurants, can we conclude the relationship is significant at the 0.05 level of significance?

13.36 Is a value of $r = +0.24$ significant in trying to show that ρ is greater than zero for a sample of size 62 at the 0.05 level of significance?

13.37 [EX13-37] The population (in millions) and the violent crime rate (per 1000) were recorded for 10 metropolitan areas. The data are shown in the following table.

Population	10.0	1.3	2.1	7.0	4.4	0.3	0.3	0.2	0.2	0.4
Crime Rate	12.0	9.5	9.2	8.4	8.2	7.3	7.1	7.0	6.9	6.9

Do these data provide evidence to reject the null hypothesis that $\rho = 0$ in favor of $\rho \neq 0$ at $\alpha = 0.05$?

13.38 [EX13-38] One would think that playing in the National Basketball Association (NBA) regular season, competing at the 2004 Olympics, and then playing in the NBA postseason would tire any player out. A look at six former Olympians still playing in the postseason gives a different impression, as noted in the May 27, 2005, *USA Today* article titled "2004 Olympians giving Herculean effort in playoffs."

Player, team	Regular Season PPG	Postseason PPG
Tim Duncan, Spurs	20.3	24.5
Manu Ginobili, Spurs	16.0	21.8
Dwayne Wade, Heat	24.1	28.5
Amare Stoudemire, Suns	26.0	28.5
Shawn Marion, Suns	19.4	19.9
Carlos Arroyo, Pistons	6.6	1.6

PPG, points per game.

a. Do these data provide evidence to reject the null hypothesis that $\rho = 0$ in favor of $\rho > 0$ at $\alpha = 0.01$?

b. Explain the meaning of the apparent positive correlation.

13.39 [EX13-39] Two indicators of the level of economic activity in a given geographical area are its total personal income and the value of new privately owned housing units. The following table lists the data for seven states for the year 2004.

State	Personal Income ($ millions)	Valuation ($ thousands)
Colorado	165,942	8,050,293
Kansas	84,282	1,925,642
Missouri	176,137	4,286,161
Nebraska	54,756	1,372,588
New Mexico	49,849	1,747,309
Oklahoma	98,974	2,184,108
Wyoming	17,377	524,504

Source: http://www.census.gov/const/C40/Table2/tb2v2004.txt

a. Calculate the correlation coefficient between the two variables.

b. Test for a significant correlation at the 0.05 level of significance and draw your conclusion.

13.40 [EX13-40] Sugar beet growers are interested in realizing higher yields and higher sucrose percentages from their crops. But do they go together? The data that follow are from the Montana 2003 sugar beet crop—values listed are by county, yield is tons per acre, and sucrose is percentage of sucrose.

County	Yield	Sucrose	County	Yield	Sucrose
Northeast			South Central		
Dawson	20.9	18.94	Yellowstone	25.9	16.71
Richland	24.5	19.67	Other	27.4	16.35
Roosevelt	21.0	19.25	Southeast		
South Central			Custer	21.8	18.96
Big Horn	29.7	16.41	Prairie	22.2	19.58
Carbon	22.7	16.56	Rosebud	31.3	17.10
Treasure	29.4	17.07			

Source: http://www.nass.usda.gov/mt/county/crops/sgrcty03.htm

a. What, if any, relationship do you expect to find between the yield per acre and the sucrose percentage for sugar beets?

b. Draw the scatter diagram for yield in tons per acre (x) and sucrose percentage (y) for the Montana data. Describe the relationship as seen on the scatter diagram. Is it what you anticipated?

c. Find the linear correlation coefficient.

d. At the 0.05 level of significance, is the linear correlation coefficient significantly different than zero?

e. One of the ordered pairs appears to be outside of the pattern created by the other 11 ordered pairs. What effect do you think removal of this pair from the data values would have on (1) the appearance of the scatter diagram, (2) the linear correlation coefficient, and (3) the answer in part d?

f. Remove Carbon county from the data and answer parts b–d. Compare the results with your answers to part e.

13.4   Linear Regression Analysis

Recall that the **line of best fit** results from an analysis of two (or more) related quantitative variables. (We will restrict our work to two variables.) When two variables are studied jointly, we often would like to control one variable by controlling the other. Or we might want to predict the value of a variable based on knowledge about another variable. In both cases we want to find the line of best fit, provided one exists, that will best predict the value of the dependent, or output, variable from a value of the independent, or input, variable. Recall that the variable we know or can control is called the *independent,* or input, variable; the variable that results from using the equation of the line of best fit is called the *dependent,* or predicted, variable.

In Chapter 3 we developed the method of least squares. From this concept, formulas (3.7) and (3.6) were obtained and used to calculate b_0 (the **y-intercept**) and b_1 (the **slope of the line of best fit**):

$$b_0 = \frac{1}{n}\left(\sum y - b_1 \cdot \sum x\right) \tag{3.7}$$

$$b_1 = \frac{SS(xy)}{SS(x)} \tag{3.6}$$

Then these two coefficients are used to write the equation of the line of best fit in the form

$$\hat{y} = b_0 + b_1 x$$

When the line of best fit is plotted, it does more than just show us a pictorial representation of the line. It tells us two things: (1) whether or not there really is a linear relationship between the two variables and (2) the quantitative (equation) relationship between the two variables. When there is no relationship between the variables, a horizontal line of best fit will result. A horizontal line has a slope of zero, which implies that the value of the input variable has no effect on the output variable. (This idea will be amplified later in this chapter.)

The result of regression analysis is the mathematical equation of the line of best fit. We will, as mentioned before, restrict our work to the **simple linear** case—that is, one input variable and one output variable where the line of best fit is straight. However, you should be aware that not all relationships are of this nature. If the scatter diagram suggests something other than a straight line, the relationship may be **curvilinear regression.** In cases of this type we must introduce terms to higher powers, x^2, x^3, and so on, or other functions, e^x, log x, and so on; or we must introduce other input variables. Maybe two or three input variables would improve the usefulness of our regression equation. These possibilities are examples of curvilinear regression and **multiple regression.**

The linear model used to explain the behavior of linear bivariate data in the population is:

Linear Model

$$\hat{y} = \beta_0 + \beta_1 x + \epsilon \qquad (13.4)$$

This equation represents the linear relationship between the two variables in a population. β_0 is the y-intercept, and β_1 is the slope. ϵ (lowercase Greek letter epsilon) is the random **experimental error** in the observed value of y at a given value of x.

The **regression line** from the sample data gives us b_0, which is our estimate of β_0, and b_1, our estimate of β_1. The error ϵ is approximated by $e = y - \hat{y}$, the difference between the observed value of y and the predicted value of y, $\hat{y}$, at a given value of x:

Estimate of the Experimental error

$$e = y - \hat{y} \qquad (13.5)$$

FYI Explore Skillbuilder Applet "Residuals & Line of Best Fit" on your CD.

The random variable e (also known as the "residual") is positive when the observed value of y is larger than the predicted value, $\hat{y}$; e is negative when y is less than $\hat{y}$. The sum of the errors (residuals) for all values of y for a given value of x is exactly zero. (This is part of the least squares criteria.) Thus, the mean value of the experimental error is zero; its variance is σ_ϵ^2. Our next goal is to estimate this **variance of the experimental error.**

Before we estimate the variance of ϵ, let's try to understand exactly what the error represents: ϵ is the amount of error in our observed value of y. That is, it is the difference between the observed value of y and the mean value of y at that particular value of x. Since we do not know the mean value of y, we will use the regression equation and estimate it with $\hat{y}$, the predicted value of y at this same value of x. Thus the best estimate that we have for ϵ is $e = y - \hat{y}$, as shown in Figure 13.6.

FIGURE 13.6 The Error, e, Is $y - \hat{y}$

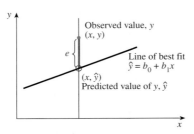

Note: e is the observed error in measuring y at a specified value of x.

If we were to observe several values of y at a given value of x, we could plot a distribution of y values about the line of best fit (about $\hat{y}$, in particular). Figure 13.7 shows a sample of bivariate values that share a common x value. Figure 13.8 shows the theoretical distribution of all possible y values at a given x value. A similar distribution occurs at each different value of x. The mean of the observed y's at a given value of x varies, but it can be estimated by $\hat{y}$.

Before we can make any inferences about a regression line, we must assume that the distribution of y's is approximately normal and that the variances of the distributions of y at all values of x are the same. That is, the standard deviation of the distribution of y about $\hat{y}$ is the same for all values of x, as shown in Figure 13.9.

FIGURE 13.7 Sample of y Values at a Given x

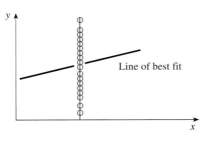

FIGURE 13.8 Theoretical Distribution of y Values for a Given x

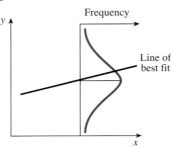

FIGURE 13.9 Standard Deviation of the Distribution of y Values Is the Same for All x

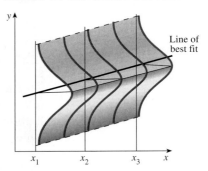

Before we look at the variance of e, let's review the definition of sample variance. The sample variance, s^2, is defined as $\dfrac{\sum(x - \bar{x})^2}{n - 1}$, the sum of the squares of each deviation divided by the number of degrees of freedom, $n - 1$, associated with a sample of size n. The variance of y involves an additional complication: there is a different mean for y at each value of x. (Notice the many distributions in Figure 13.9.) However, each of these "means" is actually the predicted value, $\hat{y}$, that corresponds to the x that fixes the distribution. So the variance of the error e is estimated by the formula:

FYI Variance of y about the line of best fit is the same as the variance of error e.

Variance of the Error, e

$$s_e^2 = \frac{\sum(y - \hat{y})^2}{n - 2}$$

(13.6)

where $n - 2$ is the number of degrees of freedom

Formula (13.6) can be rewritten by substituting $b_0 + b_1 x$ for $\hat{y}$. Since $\hat{y} = b_0 + b_1 x$, we have

$$s_e^2 = \frac{\sum(y - b_0 - b_1 x)^2}{n - 2} \tag{13.7}$$

With some algebra and some patience, this formula can be rewritten once again into a more workable form. The form we will use is

---

Variance of the Error *e*

$$s_e^2 = \frac{(\sum y^2) - (b_0)(\sum y) - (b_1)(\sum xy)}{n - 2} \tag{13.8}$$

For ease of discussion let's agree to call the numerator of formulas (13.6), (13.7), and (13.8) the **sum of squares for error** (SSE).

Now let's see how we can use all of this information.

EXAMPLE 13.5

Determining the Variance of *y* about the Regression Line

Suppose you move to a new city and take a job. You will, of course, be concerned about the problems you will face commuting to and from work. For example, you would like to know how long it will take you to drive to work each morning. Let's use "one-way distance to work" as a measure of where you live. You live *x* miles away from work and want to know how long it will take you to commute each day. Your new employer, foreseeing this question, has already collected a random sample of data to be used in answering your question. Fifteen of your co-workers were asked to give their one-way travel times and distances to work. The resulting data are shown in Table 13.2. (For convenience, the data have been arranged so that the *x* values are in numerical order.) Find the line of best fit and the variance of *y* about the line of best fit, s_e^2.

TABLE 13.2

Data on Commute Distances and Times [TA13-02]

Co-Worker	Miles (x)	Minutes (y)	x^2	xy	y^2	Co-Worker	Miles (x)	Minutes (y)	x^2	xy	y^2
1	3	7	9	21	49	9	13	26	169	338	676
2	5	20	25	100	400	10	15	25	225	375	625
3	7	20	49	140	400	11	15	35	225	525	1225
4	8	15	64	120	225	12	16	32	256	512	1024
5	10	25	100	250	625	13	18	44	324	792	1936
6	11	17	121	187	289	14	19	37	361	703	1369
7	12	20	144	240	400	15	20	45	400	900	2025
8	12	35	144	420	1225	Total	184	403	2,616	5,623	12,493

SOLUTION The extensions and summations needed for this problem are shown in Table 13.2. The line of best fit can now be calculated using formulas (2.9), (3.4), (3.6), and (3.7). From formula (2.9):

$$\text{SS}(x) = \sum x^2 - \frac{(\sum x)^2}{n}: \quad \text{SS}(x) = 2616 - \frac{(184)^2}{15} = 358.9333$$

FYI Use extra decimal place during these calculations.

FIGURE 13.10
The 15 Random Errors as Line Segments

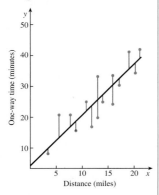

From formula (3.4):

$$SS(xy) = \sum xy - \frac{\sum x \cdot \sum y}{n}: \quad SS(xy) = 5623 - \frac{(184)(403)}{15} = 679.5333$$

We use formula (3.6) for the slope:

$$b_1 = \frac{SS(xy)}{SS(x)}: \quad b_1 = \frac{679.5333}{358.9333} = 1.893202 = \textbf{1.89}$$

We use formula (3.7) for the y-intercept:

$$b_0 = \frac{\sum y - (b_1 \cdot \sum x)}{n}: \quad b_0 = \frac{403 - (1.893202)(184)}{15} = 3.643387 = \textbf{3.64}$$

Therefore, the equation of the line of best fit is

$$\hat{y} = 3.64 + 1.89x$$

The variance of y about the regression line is calculated by using formula (13.8):

$$s_e^2 = \frac{(\sum y^2) - (b_0)(\sum y) - (b_1)(\sum xy)}{n - 2}:$$

$$s_e^2 = \frac{(12,493) - (3.643387)(403) - (1.893202)(5623)}{15 - 2} = \frac{379.2402}{13} = \textbf{29.17}$$

$s_e^2 = 29.17$ is the variance of the 15 e's. In Figure 13.10 the 15 e's are shown as vertical line segments.

FYI Computer and calculator commands to find the regression line for a set of bivariate data can be found in Chapter 3 (pp. 180–181).

Note: Extra decimal places are often needed for this type of calculation. Notice that b_1 (1.893202) was multiplied by 5623. If 1.89 had been used instead, that one product would have changed the numerator by approximately 18. That, in turn, would have changed the final answer by almost 1.4—a sizable round-off error.

In the sections that follow, we will use the variance of e in much the same way as the variance of x (as calculated in Chapter 2) was used in Chapters 8, 9, and 10 to complete the statistical inferences studied there.

SECTION 13.4 EXERCISES

Statistics⬦Now™

Datasets can be found on your Student's Suite CD-ROM or at the StatisticsNow website at **http://1pass.thomson.com**.

13.41 [EX13-41] Ten salespeople were surveyed, and the average number of client contacts per month, x, and the sales volume, y (in thousands), were recorded for each:

x	12	14	16	20	23	46	50	48	50	55
y	15	25	30	30	30	80	90	95	110	130

Refer to the following computer output and verify that the equation of the line of best fit is $\hat{y} = 13.4 + 2.3x$ and that $s_e = 10.17$ by calculating these values yourself.

```
The regression equation is y = - 13.4 + 2.30 x
Predictor     Coef
Constant     -13.414
x             2.3028
s = 10.17
```

13.42 [EX13-42] Nationwide, hailstones cause about $1 billion in property and crop damage each year, according to the April 25, 2005, *USA Today Weather Focus* report "Violent storms create the largest hail." The speed of a thunderstorm's updraft is one of the factors affecting the size (diameter) of a hailstone. The following data were given in the article.

x, wind updraft speed (mph)	22	37	56	100
y, hailstone size (inches)	0.5	0.75	1.75	3.0

Refer to the following computer output and verify that the equation of the line of best fit is $\hat{y} = -0.289 + 0.0333x$ and that $s_e = 0.1904$ by calculating these values yourself.

```
The regression equation is
size = -0.289 + 0.0333 speed
Predictor    Coef      SE Coef    T       P
Constant     -0.2889   0.1989     -1.45   0.284
speed        0.033282  0.003249   10.24   0.009
s = 0.190383   R-Sq = 98.1%   R-Sq(adj) = 97.2%
```

13.43 [EX13-43] The NBA calculates many statistics, just like all other professional sports. Average points per game, average rebounds per game, number of years played, number of titles, number of All-Star appearances, and number of most valuable player (MVP) awards are just some examples that were given in the chart "Comparing the best big men" in the June 3, 2005, *USA Today*. Using the data that follow, investigate the relationship between the average number of points per game and the number of All-Star appearances. Include a scatter diagram, the linear correlation coefficient and line of best fit, and a statement about their meaning.

Player	Points	All Star	Player	Points	All Star
George Mikan	22.6	4	Kareem Abdul-Jabbar	24.6	19
Bill Russell	15.1	12	Hakeem Olajuwon	21.8	12
Wilt Chamberlain	30.1	13	Shaquille O'Neal	27.1	9

13.44 [EX13-44] Thirteen of Minnesota's sweet corn–producing counties were randomly selected, and the following information about their 2004 crop was recorded: acres planted (100s of acres) and total production in 100s of tons of sweet corn.

County	Acres Planted (100 acres)	Production (100 tons)	County	Acres Planted (100 acres)	Production (100 tons)
Waseca	50	353	Kandiyohi	37	237
Freeborn	69	365	Olmsted	86	553
Martin	21	144	Goodhue	45	295
Dakota	34	187	Meeker	13	82
McLeod	20	122	Nicollet	26	178
Redwood	70	483	Sherburne	22	178
Dodge	35	245			

Source: http://www.nass.usda.gov/mn/swtcrn04.pdf

a. Investigate the relationship between the number of acres planted to sweet corn and the total tons of sweet corn produced. Include a scatter diagram, linear correlation coefficient and line of best fit, and a statement about their meaningfulness.

b. If you were advising a Minnesota sweet corn grower, based on the preceding information, how many tons of sweet corn, on average, can the grower expect to produce for each acre planted?

13.45 [EX13-45] Diamonds are often thought of as a cherished item with a personal value well in excess of their monetary value. The monetary value of a diamond is determined by its exact quality as defined by the 4 Cs: cut, color, clarity, and carat weight. The price (dollars) and the carat weight of a diamond are its two most known characteristics. To understand the role carat weight has in determining the price of a diamond, the carat weight and price of 20 loose round diamonds, all of color D and clarity VS1, were obtained October 19, 2002, on the Internet.

Carat Wt	Price
0.58	2791
0.64	2803

••• Remainder of data on Student's Suite CD-ROM.

Source: http://www.overnightdiamonds.com/diamondlist1.htm

a. Draw a scatter diagram of the data: carat weight (x) and price (y).

b. Do the data suggest a linear relationship for the domain 0.50 to 0.66 carat? Discuss your findings in part a.

c. Diamonds smaller than 0.50 carat and diamonds larger than 0.66 carat may not fit the linear pattern demonstrated by these data. Explain.

d. Find the equation for the line of best fit.

e. According to this information, what would be a typical price for a 0.50-carat loose diamond of this quality?

f. On average, by how much does the price increase for each extra 0.01 carat in weight? Within what interval of x values would you expect this to be true?

g. Find the variance of y about the regression line. What characteristics in the scatter diagram support this large value?

13.46 [EX13-46] The computer-science aptitude score, x, and the achievement score, y (measured by a comprehensive final), were measured for 20 students in a beginning computer-science course. The results were as follows. Find the equation of the line of best fit and s_e^2.

x	4	16	20	13	22	21	15	20	19	16	18	17	8	6	5	20	18	11	19	14
y	19	19	24	36	27	26	25	28	17	27	21	24	18	18	14	28	21	22	20	21

13.47 [EX13-47] a. Using the 10 points shown in the following table, find the equation of the line of best fit, $\hat{y} = b_0 + b_1 x$ and graph it on a scatter diagram.

Point	A	B	C	D	E	F	G	H	I	J
x	1	1	3	3	5	5	7	7	9	9
y	1	2	2	3	3	4	4	5	5	6

b. Find the ordinates $\hat{y}$ for the points on the line of best fit whose abscissas are $x = 1, 3, 5, 7,$ and 9.

c. Find the value of e for each of the points in the given data ($e = y - \hat{y}$).

d. Find the variance s_e^2 of those points about the line of best fit by using formula (13.6).

e. Find the variance s_e^2 by using formula (13.8). (Answers to parts d and e should be the same.)

13.48 [EX13-48] The following data show the number of hours studied for an exam, x, and the grade received on the exam, y (y is measured in 10s; that is, $y = 8$ means that the grade, rounded to the nearest 10 points, is 80).

x	2	3	3	4	4	5	5	6	6	6	7	7	7	8	8
y	5	5	7	5	7	7	8	6	9	8	7	9	10	8	9

a. Draw a scatter diagram of the data.

b. Find the equation of the line of best fit and graph it on the scatter diagram.

c. Find the ordinates $\hat{y}$ that correspond to $x = 2, 3, 4, 5, 6, 7,$ and 8.

d. Find the five values of e that are associated with the points where $x = 3$ and $x = 6$.

e. Find the variance s_e^2 of all the points about the line of best fit.

13.5 Inferences Concerning the Slope of the Regression Line

Now that the equation of the line of best fit has been found and the linear model has been verified (by inspection of the scatter diagram), we are ready to determine whether we can use the equation to predict y. We will test the null hypothesis: The equation of the line of best fit is of no value in predicting y given x. That is, the null hypothesis to be tested is: β_1 (the slope of the relationship in the population) is zero. If $\beta_1 = 0$, then the linear equation will be of no real use in predicting y.

Before we look at the confidence interval or the hypothesis test, let's discuss the **sampling distribution** of the slope. If random samples of size n are repeatedly taken from a bivariate population, then the calculated slopes, the

b_1's, will form a sampling distribution that is normally distributed with a mean of β_1, the population value of the slope, and with a variance of $\sigma^2_{b_1}$, where

$$\sigma^2_{b_1} = \frac{\sigma^2_\epsilon}{\sum(x - \bar{x})^2} \tag{13.9}$$

provided there is no lack of fit. An appropriate estimator for $\sigma^2_{b_1}$ is obtained by replacing σ^2_ϵ by s^2_e, the estimate of the variance of the error about the regression line:

$$s^2_{b_1} = \frac{s^2_e}{\sum(x - \bar{x})^2} \tag{13.10}$$

This formula may be rewritten in the following, more manageable form:

Estimate for Variance of Slope $\qquad s^2_{b_1} = \dfrac{s^2_e}{\sum x^2 - \dfrac{(\sum x)^2}{n}}$ $\qquad$ (13.11)

Note: The "**standard error** of ___ " is the standard deviation of the sampling distribution of ___. Therefore, the *standard error of regression* (slope) is σ_{b_1} and is estimated by s_{b_1}.

FYI Recall that we found SS(x) with formula (2.9).

In our example of commute times and distances, the variance among the b_1's is estimated by using formula (13.11):

$$s^2_{b_1} = \frac{s^2_e}{\sum x^2 - \dfrac{(\sum x)^2}{n}}: \quad s^2_{b_1} = \frac{29.1723}{358.9333} = 0.081275 = \mathbf{0.0813}$$

Assumptions for inferences about the linear regression: The set of (x,y) ordered pairs forms a random sample, and the y values at each x have a normal distribution. Since the population standard deviation is unknown and replaced with the sample standard deviation, the t-distribution will be used with $n - 2$ degrees of freedom.

Confidence Interval Procedure

The slope β_1 of the regression line of the population can be estimated by means of a confidence interval.

Confidence Interval for Slope $\qquad b_1 \pm t(n - 2, \alpha/2) \cdot s_{b_1}$ $\qquad$ (13.12)

EXAMPLE 13.6 **Constructing a Confidence Interval for β_1, the Population Slope of the Line of Best Fit**

Find the 95% confidence interval for the population's slope, β_1, for Example 13.5 (p. 715).

Statistics⬡Now™

Watch a video example at
http://1pass.thomson.com
or on your CD.

SOLUTION

Step 1 **Parameter of interest:** The slope, β_1, of the line of best fit for the population.

Step 2 a. **Assumptions:** The ordered pairs form a random sample, and we will assume that the y values (minutes) at each x (miles) have a normal distribution.

b. **Probability distribution and formula:** Student's t-distribution and formula (13.12).

c. Level of confidence: $1 - \alpha = 0.95$.

Step 3 **Sample information:** $n = 15$, $b_1 = 1.89$, and $s_{b_1}^2 = 0.0813$.

Step 4 **a. Confidence coefficients:** From Table 6 in Appendix B, we find $t_{(df, \alpha/2)} = t(13, 0.025) = 2.16$.

b. Maximum error of estimate: We use formula (13.12) to find

$$E = t(n - 2, \alpha/2) \cdot s_{b_1}: \quad E = (2.16) \cdot \sqrt{0.0813} = 0.6159$$

c. Lower and upper confidence limits:

$$b_1 - E \quad \text{to} \quad b_1 + E$$

$$1.89 - 0.62 \quad \text{to} \quad 1.89 + 0.62$$

Thus, 1.27 to 2.51 is the 95% confidence interval for β_1.

Step 5 **Confidence interval:** We can say that the slope of the line of best fit of the population from which the sample was drawn is between 1.27 and 2.51 with 95% confidence. That is, 95% confident that, on average, every extra mile will take between 1.27 minutes (1 min, 16 sec) and 2.51 minutes (2 min, 31 sec) of time to make the commute.

APPLIED EXAMPLE 13.7

Reexamining the Use of Seriousness Weights in an Index of Crime

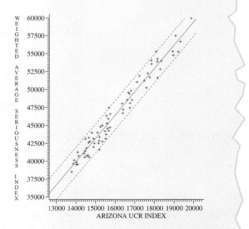

Regression of the Arizona UCR Index on the average seriousness index produces the linear relationship depicted in the figure. Also shown is the 95% confidence interval (3.001, 3.262), which is based upon a standard error of 0.065 on the estimate of the slope. The regression equation for this relationship is
$$S_t = -3953.85 + 3.13A_t$$

Source: Reprinted with permission from the *Journal of Criminal Justice*, Volume 17, Thomas Epperlein and Barbara C. Nienstedt, "Reexamining the Use of Seriousness Weights in an Index of Crime," Pergamon Press, Inc.

Hypothesis-Testing Procedure

We are now ready to test the hypothesis $\beta_1 = 0$. That is, we want to determine whether the equation of the line of best fit is of any real value in predicting y. For this hypothesis test, the null hypothesis is always H_o: $\beta_1 = 0$. It will be tested using Student's t-distribution with df $= n - 2$ and the test statistic $t\star$ found using formula (13.13):

Test Statistic for Slope

$$t\star = \frac{b_1 - \beta_1}{s_{b_1}}$$ (13.13)

EXAMPLE 13.8

One-Tailed Hypothesis Test for the Slope of the Regression Line

Is the slope of the line of best fit significant enough to show that one-way distance is useful in predicting one-way travel time in Example 13.5? Use $\alpha = 0.05$.

SOLUTION

Step 1   **a. Parameter of interest:** β_1, the slope of the line of best fit for the population.

   b. Statement of hypotheses:

   $H_o: \beta_1 = 0$ (This implies that x is of no use in predicting y; that is, $\hat{y} = \bar{y}$ would be as effective.)

   The alternative hypothesis can be either one-tailed or two-tailed. If we suspect that the slope is positive, as in Example 13.5, a one-tailed test is appropriate.

   $H_a: \beta_1 > 0$ (We expect travel time y to increase as the distance x increases.)

Step 2   **a. Assumptions:** The ordered pairs form a random sample, and we will assume that the y values (minutes) at each x (miles) have a normal distribution.

   b. Probability distribution and test statistic: The t-distribution with df $= n - 2 = 13$, and the test statistic $t\star$ from formula (13.13).

   c. Level of significance: $\alpha = 0.05$.

Step 3   **a. Sample information:** $n = 15$, $b_1 = 1.89$, and $s_{b_1}^2 = 0.0813$.

   b. Test statistic: Using formula (13.13), we find the observed value of t:

   $$t\star = \frac{b_1 - \beta_1}{s_{b_1}}: \quad t\star = \frac{1.89 - 0.0}{\sqrt{0.0813}} = 6.629 = \mathbf{6.63}$$

Step 4   **Probability Distribution:**

p-Value:

a. Use the right-hand tail because H_a expresses concern for values related to "positive."

P $= P(t\star > 6.63$, with df $= 13)$ as shown in the figure.

OR

Classical:

a. The critical region is the right-hand tail because H_a expresses concern for values related to "positive." The critical value is found in Table 6:

$t(13, 0.05) = \mathbf{1.77}$

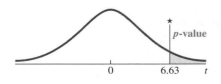

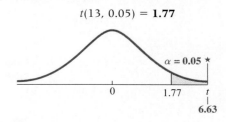

To find the *p*-value, use one of three methods:

1. Use Table 6 (Appendix B) to place bounds on the *p*-value: **P < 0.005.**
2. Use Table 7 (Appendix B) to place bounds on the *p*-value: **P < 0.001.**
3. Use a computer or calculator to find the *p*-value: **P = 0.0000082.**

Specific details are on pages 483–484.

b. The *p*-value is smaller than the level of significance, α.

Specific instructions are on pages 477–479.

b. $t\star$ is in the critical region, as shown in **red** in the figure.

Step 5   **a.** **Decision:** Reject H_o.

b. **Conclusion:** At the 0.05 level of significance, we conclude that the slope of the line of best fit in the population is greater than zero. The evidence indicates that there is a linear relationship and that the one-way distance (*x*) is useful in predicting the travel time to work (*y*).

TECHNOLOGY INSTRUCTIONS: REGRESSION ANALYSIS

Output includes the equation for the regression line, information for a *t*-test concerning the slope of the regression line, the standard deviation of error, *r* and/or r^2, and a scatter diagram showing the regression line.

MINITAB (Release 14)   MINITAB output also includes the predicted *y* values for given *x* values and residuals.

Input the *x*-variable data into C1 and the corresponding *y*-variable data into C2; then continue with:

```
Choose:     Stat > Regression > Regression . . .
Enter:      Response (y): C2
            Predictors (x): C1
Select:     Results
            Regression equation, table of coefficients, s, R-squared, . . .
            > OK
Select:     Storage
            Residuals and Fits > OK > OK
Choose:     Graph > Scatterplot
Select:     With Regression > OK
Enter:      Y variables: C2   X variables: C1
Select:     Labels > Title/Footnotes
Enter:      your title > OK > OK
```

Excel   Excel output also includes predicted *y* values for given *x* values, residuals, and a $1 - \alpha$ confidence interval for the slope.

Input the *x*-variable data into column A and the corresponding *y*-variable data into column B; then continue with:

```
Choose:     Tools > Data Analysis > Regression
Enter:      Input Y Range: (B1:B10 or select cells)
            Input X Range: (A1:A10 or select cells)
Select:     Labels (if necessary)
            Confidence Level:
Enter:      95% (desired level)
```

FYI Additional commands to adjust the window can be found on page 155.

Select:	**Output Range:**
Enter:	**(C1 or select cell)**
Select:	**Line Fits Plots > OK**

To make the output more readable, continue with: Format > Column > Autofit Selection.

TI-83/84 Plus Input the *x*-variable data into L1 and the corresponding *y*-variable data into L2; then continue with the following, entering the apppropriate values and highlighting Calculate:

Choose:	**STAT > TESTS > E:LinRegTTest**

(To enter Y1, use: VARS > YVARS > 1:Function . . . >1:Y1.)

Enter the following to obtain a scatter diagram with regression line:

Choose:	**2nd > STATPLOT > 1:Plot1 . . . On**
Choose:	**ZOOM > 9:ZoomStat > Trace**

Here is the MINITAB printout with explanations for parts of Example 13.5.

Regression Analysis: y, minutes versus x, miles

The regression equation is
y, minutes = 3.64 + 1.89 x, miles

Equation of line of best fit
$\hat{y} = 3.64 + 1.89x$;
see pp. 715–716
Calculated values of b_0 and b_1

Predictor	Coef	SECoeff	T	P
Constant	3.643	3.765	0.97	0.351
x, miles	1.8932	0.2851	6.64	0.000

Calculated value of s_{b_1},
$s_{b_1} = 0.285$; compare to
$s_{b_1}^2 = 0.0813$ see p. 719
($\sqrt{0.0813} = 0.285$)

s = 5.401   R − Sq = 77.2%   R − Sq (adj) = 75.5%

Calculated $t\star$ and *p*-value for
$H_o: \beta_1 = 0$ as found in steps 3 and 4 on pp. 721–722

Calculated value of s_e,
$s_e = 5.4011$: compare to
$s_e^2 = 29.1723$ as found on
p. 716 ($\sqrt{29.1723} = 5.401$)

Given data

Obs	x, miles	y, minute	Fit	Residual
1	3.0	7.00	9.32	−2.32
2	5.0	20.00	13.11	6.89
3	7.0	20.00	16.90	3.10
4	8.0	15.00	18.79	−3.79
5	10.0	25.00	22.58	2.42
6	11.0	17.00	24.47	−7.47
7	12.0	20.00	26.36	−6.36
8	12.0	35.00	26.36	8.64
9	13.0	26.00	28.26	−2.26
10	15.0	25.00	32.04	−7.04
11	15.0	35.00	32.04	2.96
12	16.0	32.00	33.93	−1.93
13	18.0	44.00	37.72	6.28
14	19.0	37.00	39.61	−2.61
15	20.0	45.00	41.51	3.49

Values of $\hat{y}$ for each given
x-value using
$\hat{y} = 3.634 + 1.8932x$

continued

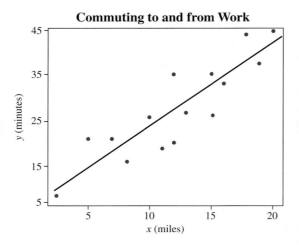

Commuting to and from Work

x (miles)

SECTION 13.5 EXERCISES

13.49 a. The vertical scale in the figure in Applied Example 13.7 on page 720 is drawn at $A_t = 12{,}600$, and the line of best fit appears to intersect the vertical scale at approximately 35,500. Verify the coordinates of this point of intersection.

   b. The article also gives an interval estimate of (3.001, 3.262). Verify this 95% interval using the information given in the article.

13.50 Calculate the estimated standard error of regression, s_{b_1}, for the computer-science aptitude score–achievement score relationship in Exercise 13.46 (p. 718).

13.51 Calculate the estimated standard error of regression, s_{b_1}, for the number of hours studied–exam grade relationship in Exercise 13.48 (p. 718).

13.52 Using the estimated standard error of regression, s_{b_1}, found in Exercise 13.51 for the number of hours studied–exam grade relationship, find the 95% confidence interval for the population slope β_1. The equation for the line of best fit was: $\hat{y} = 3.96 + 0.625x$.

13.53 [EX13-53] An article titled "Statistical Approach for the Estimation of Strontium Distribution Coefficient" *(Environmental Science & Technology)* reports a linear correlation coefficient of 0.55 between the strontium distribution coefficient (mL/g) and the total aluminum (mmol/100 g-soil) for soils collected from the surface throughout Japan. Consider the following data for 10 such samples.

Soil Sample	Strontium Dist. Coeff.	Total Aluminum	Soil Sample	Strontium Dist. Coeff.	Total Aluminum
1	100	200	6	500	400
2	120	225	7	450	375
3	300	325	8	445	385
4	250	310	9	310	350
5	400	350	10	200	290

Let *Y* represent the strontium distribution coefficient and *X* represent the total aluminum.

a. Find the equation of the line of best fit.

b. Find a 95% confidence interval for β_1.

c. Explain the meaning of the interval in part b.

13.54 State the null hypothesis, H_o, and the alternative hypothesis, H_a, that would be used to test the following statements:

a. The slope for the line of best fit is positive.

b. The slope of regression line is not significant.

c. The negative slope for the regression is significant.

13.55 Determine the *p*-value for each of the following situations.

a. $H_a: \beta_1 > 0$, with $n = 18$ and $t\star = 2.4$

b. $H_a: \beta_1 \neq 0$, with $n = 15$, $b_1 = 0.16$, and $s_{b_1} = 0.08$

c. $H_a: \beta_1 < 0$, with $n = 24$, $b_1 = -1.29$, and $s_{b_1} = 0.82$

13.56 Determine the critical value(s) and regions that would be used in testing each of the following null hypotheses using the classical approach:

a. $H_o: \beta_1 = 0$ vs. $H_a: \beta_1 \neq 0$, with $n = 18$ and $\alpha = 0.05$

b. $H_o: \beta_1 = 0$ vs. $H_a: \beta_1 > 0$, with $n = 28$ and $\alpha = 0.01$

c. $H_o: \beta_1 = 0$ vs. $H_a: \beta_1 < 0$, with $n = 16$ and $\alpha = 0.05$

13.57 [EX13-57] An issue of *Popular Mechanics* gives specifications and dimensions for various jet boats. The following table summarizes some of this information.

Model	Base Price	Engine Horsepower
Baja Blast	$8,395	120
Bayliner Jazz	$8,495	90
Boston Whaler Rage 15	$11,495	115
Dynasty Jet Storm	$8,495	90
Four Winds Fling	$9,568	115
Regal Rush	$9,995	90

Model	Base Price	Engine Horsepower
Sea-Doo Speedster	$11,499	160
Sea Ray Sea Rayder	$8,495	90
Seaswirl Squirt	$8,495	115
Suga Sand Mirage	$8,395	120

Using the Excel output at the bottom of the page:

a. Determine the equation for the line of best fit.

b. Verify the calculation of $t\star$ (*t* Stat) for engine horsepower.

c. Determine whether horsepower is an effective predictor of base price.

d. Verify the 95% confidence interval for β_1.

13.58 [EX13-58] The relationship between the diameter of a spot weld, *x*, and the shear strength of the weld, *y*, is very useful. The diameter of the spot weld can be measured after the weld is completed. The shear strength of the weld can be measured only by applying force to the weld until it breaks. Thus, it would be very useful to be able to predict the shear strength based only on the diameter. The following data were obtained from several sample welds.

x, Dia. of Weld (0.001 inch)	190	215	200	230	209	250	215	265	215	250
y, Shear Strength (lb)	680	1025	800	1100	780	1030	885	1175	975	1300

Complete these questions with the aid of a computer.

a. Draw a scatter diagram.

b. Find the equation for the line of best fit.

c. Is the value of b_1 significantly greater than zero at the 0.05 level?

d. Find the 95% confidence interval for β_1.

13.59 [EX13-59] Each student in a sample of 10 was asked for the distance and the time required to commute to college yesterday. The data collected are shown in the table on p. 726.

Table for Exercise 13.57
Excel Summary Output

	Coefficients	Standard Error	t Stat	p-Value	Lower 95%	Upper 95%
Intercept	5936.793025	1929.63032	3.076647876	0.01519394	1487.05465	10386.5314
Engine horsepower	30.73218982	17.15820176	1.791107847	0.111051486	−8.834719985	70.29909963

Distance	1	3	5	5	7	7	8	10	10	12
Time	5	10	15	20	15	25	20	25	35	35

a. Draw a scatter diagram of these data.

b. Find the equation that describes the regression line for these data.

c. Does the value of b_1 show sufficient strength to conclude that β_1 is greater than zero at the $\alpha = 0.05$ level?

d. Find the 98% confidence interval for the estimation of β_1. (Retain these answers for use in Exercise 13.65 [p. 735].)

13.60 [EX13-60] Diopters represent the amount of correction needed to provide 20/20, or normal, vision. The greater the degree of nearsighted or farsighted vision, the higher the corrective prescription in diopters. Measurements in negative diopters refer to nearsighted vision, whereas measurements in positive diopters refer to farsighted vision. A sample of 30 competitive contact lenses was taken from a lot shipped to a company for analyses. Acceptance of the lot depended on the relationship between the lens power, which is measured in diopters, and a certain optical effect labeled C/O. The sample data (coded in two manners) follow.

Group 1

Power	C/O
−0.25	0.105
−0.50	0.106

••• Remainder of data on Student's Suite CD-ROM.

Courtesy of Bausch & Lomb

a. Draw a scatter diagram of these data. The *x*-term is lens power.

b. Calculate the correlation coefficient between the two variables.

c. Test for a significant correlation at the 0.05 level of significance.

d. Find the equation of the line of best fit.

e. Determine whether there is a linear relationship between C/O and lens power for Group 1 by testing the significance of the results (slope of the line of best fit) found in part d. Use $\alpha = 0.05$.

13.61 [EX13-61] The company in Exercise 13.60 must also look at other competitive lenses. It obtains another sample of 30 lenses from a lot shipped for comparison. It labeled these lenses as Group 2. Acceptance of this lot also depended on the relationship between the lens power and the optical effect labeled C/O. The sample data (coded in two manners) follow.

Group 2

Power	C/O
−5.5	0.20
−5.5	0.25

••• Remainder of data on Student's Suite CD-ROM.

Courtesy of Bausch & Lomb

a. Draw a scatter diagram of these data. The *x*-term is lens power.

b. Calculate the correlation coefficient between the two variables.

c. Test for a significant correlation at the 0.05 level of significance.

d. Find the equation of the line of best fit.

e. Determine whether there is a linear relationship between C/O and lens power for Group 2 by testing the significance of the results (slope of the line of best fit) found in part d. Use $\alpha = 0.05$.

13.62 a. Compare and contrast the two samples of lenses in Exercises 13.60 and 13.61. Include comparative descriptions of the data, the correlation analyses, and the regression analyses.

b. Identify one specific noticeable difference between these two samples.

13.6 Confidence Intervals for Regression

Once the equation of the line of best fit has been obtained and determined usable, we are ready to use the equation to make predictions. We can estimate two different quantities: (1) the mean of the population y values at a given value of x, written $\mu_{y|x_0}$, and (2) the individual y value selected at random that will occur at a given value of x, written y_{x_0}. The best point estimate, or **prediction, for both $\mu_{y|x_0}$ and y_{x_0}** is $\hat{y}$. This is the y value obtained when an x value is substituted into the equation of the line of best fit. Like other point estimates, it is seldom correct. The calculated value of $\hat{y}$ will vary above and below the actual values for both $\mu_{y|x_0}$ and y_{x_0}.

Before we develop interval estimates of $\mu_{y|x_0}$ and y_{x_0}, recall the development of confidence intervals for the population mean μ in Chapter 8 when the variance was known and in Chapter 9 when the variance was estimated. The sample mean, $\bar{x}$, was the best point estimate of μ. We used the fact that $\bar{x}$ is normally distributed, or approximately normally distributed, with a standard deviation of $\dfrac{\sigma}{\sqrt{n}}$ to construct formula (8.1) for the confidence interval for μ. When σ has to be estimated, we used formula (9.1) for the confidence interval.

The **confidence interval for $\mu_{y|x_0}$** and the **prediction interval for y_{x_0}** are constructed in a similar fashion, with $\hat{y}$ replacing $\bar{x}$ as our point estimate. If we were to randomly select several samples from the population, construct the line of best fit for each sample, calculate $\hat{y}$ for a given x using each regression line, and plot the various $\hat{y}$ values (they would vary because each sample would yield a slightly different regression line), we would find that the $\hat{y}$ values form a normal distribution. That is, the **sampling distribution of $\hat{y}$** is normal, just as the sampling distribution of $\bar{x}$ is normal. What about the appropriate standard deviation of $\hat{y}$? The standard deviation in both cases ($\mu_{y|x_0}$ and y_{x_0}) is calculated by multiplying the square root of the variance of the error by an appropriate correction factor. Recall that the variance of the error, s_e^2, is calculated by means of formula (13.8).

Before we look at the correction factors for the two cases, let's see why they are necessary. Recall that the line of best fit passes through the point $(\bar{x}, \bar{y})$, the centroid. In Section 13.5 we formed a confidence interval for the slope β_1 (see Example 13.6) by using formula (13.12). If we draw lines with slopes equal to the extremes of that confidence interval, 1.27 to 2.51, through the point $(\bar{x}, \bar{y})$ [which is (12.3, 26.9)] on the scatter diagram, we will see that the value for $\hat{y}$ fluctuates considerably for different values of x (Figure 13.11). Therefore, we should suspect a need for a wider confidence interval as we select values of x that are farther away from $\bar{x}$. Hence we need a correction factor to adjust for the distance between x_0 and $\bar{x}$. This factor must also adjust for the variation of the y values about $\hat{y}$.

First, let's estimate the mean value of y at a given value of x, $\mu_{y|x_0}$. The confidence interval formula is:

$$\hat{y} \pm t(n-2, \alpha/2) \cdot s_e \cdot \sqrt{\frac{1}{n} + \frac{(x_0 - \bar{x})^2}{\sum (x - \bar{x})^2}} \qquad (13.14)$$

Note: The numerator of the second term under the radical sign is the square of the distance of x_0 from $\bar{x}$. The denominator is closely related to the variance of x and has a "standardizing effect" on this term.

FIGURE 13.11
Lines Representing the Confidence Interval for Slope

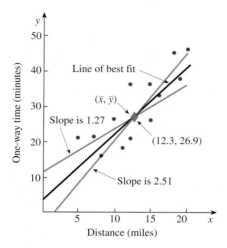

Formula (13.14) can be modified for greater ease of calculation. Here is the new form:

Confidence Interval for $\mu_{y|x_0}$

$$\hat{y} \pm t(n - 2, \alpha/2) \cdot s_e \cdot \sqrt{\frac{1}{n} + \frac{(x_0 - \bar{x})^2}{\text{SS}(x)}} \qquad (13.15)$$

Let's compare formula (13.14) with formula (9.1): $\hat{y}$ replaces $\bar{x}$, and

$$s_e \cdot \sqrt{\frac{1}{n} + \frac{(x_0 - \bar{x})^2}{\sum(x - \bar{x})^2}} \qquad \text{(the \textbf{standard error of} } \hat{y})$$

the estimated standard deviation of $\hat{y}$ in estimating $\mu_{y|x_0}$, replaces $\frac{s}{\sqrt{n}}$, the standard deviation of $\bar{x}$. The degrees of freedom are now $n - 2$ instead of $n - 1$ as before.

These ideas are explored in the next example.

EXAMPLE 13.9

Statistics ◯ Now™

Watch a video example at
http://1pass.thomson.com
or on your CD.

Constructing a Confidence Interval for $\mu_{y|x_0}$

Construct a 95% confidence interval for the mean travel time for the co-workers who travel 7 miles to work (refer to Example 13.5).

SOLUTION

Step 1   **Parameter of interest:** $\mu_{y|x = 7}$, the mean travel time for co-workers who travel 7 miles to work.

Step 2   a. **Assumptions:** The ordered pairs form a random sample, and we will assume that the y values (minutes) at each x (miles) have a normal distribution.

   b. Probability distribution and formula: Student's t-distribution and formula (13.15).

   c. Level of confidence: $1 - \alpha = 0.95$.

Step 3   **Sample information:**

$s_e^2 = 29.17$ (found in Example 13.5)
$s_e = \sqrt{29.17} = 5.40$
$\hat{y} = 3.64 + 1.89x = 3.64 + 1.89(7) = 16.87$

Step 4   **a. Confidence coefficient:** $t(13, 0.025) = 2.16$ (from Table 6 in Appendix B).

   b. Maximum error of estimate: Using formula (13.15), we have

$$E = t(n - 2, \alpha/2) \cdot s_e \cdot \sqrt{\frac{1}{n} + \frac{(x_0 - \bar{x})^2}{SS(x)}}: \quad E = (2.16)(5.40)\sqrt{\frac{1}{15} + \frac{(7 - 12.27)^2}{358.933}}$$

$$= (2.16)(5.40)\sqrt{0.06667 + 0.07738}$$

$$= (2.16)(5.40)(0.38) = 4.43$$

   c. Lower and upper confidence limits:

$$\hat{y} - E \quad \text{to} \quad \hat{y} + E$$

$$16.87 - 4.43 \quad \text{to} \quad 16.87 + 4.43$$

Thus, **12.44 to 21.30** is the 95% confidence interval for $\mu_{y|x=7}$. That is, with 95% confidence, the mean travel time for commuters that travel 7 miles is between 12.44 minutes (12 min, 26 sec) and 21.30 minutes (21 min, 18 sec).

This confidence interval is shown in Figure 13.12 by the dark red vertical line. The confidence belt showing the upper and lower boundaries of all inter-

FIGURE 13.12

Confidence Belts for $\mu_{y|x_0}$

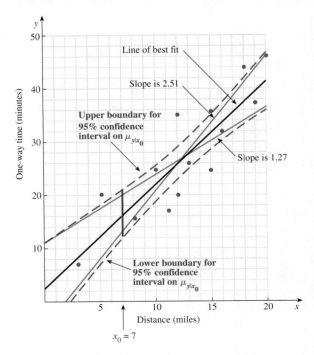

vals at 95% confidence is also shown in red. Notice that the boundary lines for the *x* values far away from $\bar{x}$ become close to the two lines that represent the equations with slopes equal to the extreme values of the 95% confidence interval for the slope (see Figure 13.12).

Often we want to predict the value of an individual *y*. For example, you live 7 miles from your place of business and you are interested in an estimate of how long it will take you to get to work. You are somewhat less interested in the average time for all of those who live 7 miles away. The formula for the prediction interval of the value of a single randomly selected *y* is

Prediction Interval for $y_{x\,=\,x_0}$

$$\hat{y} \pm t(n-2,\,\alpha/2) \cdot s_e \cdot \sqrt{1 + \frac{1}{n} + \frac{(x_0 - \bar{x})^2}{SS(x)}} \qquad (13.16)$$

EXAMPLE 13.10

Constructing a Prediction Interval for $y_{x\,=\,x_0}$

What is the 95% prediction interval for the time it will take you to commute to work if you live 7 miles away?

SOLUTION

Step 1   **Parameter of interest:** $y_{x\,=\,7}$, the travel time for one co-worker who travels 7 miles to work.

Step 2   a. **Assumptions:** The ordered pairs form a random sample, and we will assume that the *y* values (minutes) at each *x* (miles) have a normal distribution.

       b. **Probability distribution and formula:** Student's *t*-distribution and formula (13.16).

       c. **Level of confidence:** $1 - \alpha = 0.95$.

Step 3   **Sample information:** $s_e = 5.40$ and $\hat{y}_{x\,=\,7} = 16.87$ (from Example 13.9)

Step 4   a. **Confidence coefficient:** $t(13, 0.025) = 2.16$ (from Table 6 in Appendix B).

       b. **Maximum error of estimate:** Using formula (13.15), we have

$$E = t(n-2,\,\alpha/2) \cdot s_e \cdot \sqrt{1 + \frac{1}{n} + \frac{(x_0 - \bar{x})^2}{SS(x)}}:$$

$$E = (2.16)(5.40)\sqrt{1 + \frac{1}{15} + \frac{(7 - 12.27)^2}{358.933}}$$

$$= (2.16)(5.40)\sqrt{1 + 0.06667 + 0.07738}$$

$$= (2.16)(5.40)\sqrt{1.14405}$$

$$= (2.16)(5.40)(1.0696) = 12.48$$

c. Lower and upper confidence limits:

$$\hat{y} - E \quad \text{to} \quad \hat{y} + E$$

$$16.87 - 12.48 \quad \text{to} \quad 16.87 + 12.48$$

Thus, **4.39 to 29.35** is the 95% prediction interval for $y_{x=7}$. That is, with 95% confidence, the individual travel times for commuters who travel 7 miles is between 4.39 minutes (4 min, 23 sec) and 29.35 minutes (29 min, 21 sec).

The prediction interval is shown in Figure 13.13 as the blue vertical line segment at $x_0 = 7$. Notice that it is much longer than the confidence interval for $\mu_{y|x = 7}$. The dashed blue lines represent the prediction belts, the upper and lower boundaries of the prediction intervals for individual y values for all given x values.

Can you justify the fact that the prediction interval for individual values of y is wider than the confidence interval for the mean values? Think about "individual values" and "mean values" and study Figure 13.14.

FIGURE 13.14 Confidence Belts for the Mean Value of y and Prediction Belts for Individual y's

FIGURE 13.13 Prediction Belts for y_{x_0}

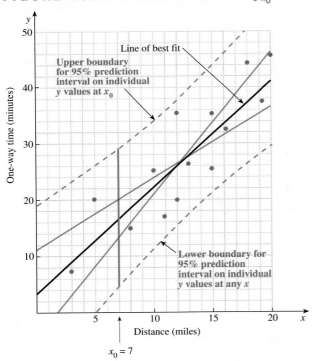

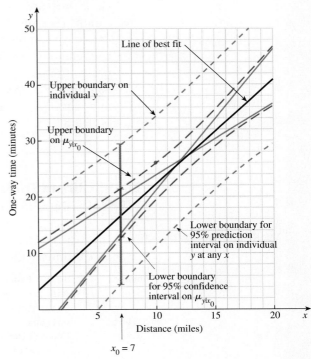

There are three basic precautions that you need to be aware of as you work with regression analysis:

1. Remember that the regression equation is meaningful only in the domain of the x variable studied. Estimation outside this domain is extremely dangerous; it requires that we know or assume that the relationship between x and y remains the same outside the domain of the

sample data. For example, Joe says that he lives 75 miles from work, and he wants to know how long it will take him to commute. We certainly can use $x = 75$ in all the formulas, but we do not expect the answers to have the confidence or validity of the values of x between 3 and 20, which were in the sample. The 75 miles may represent a distance to the heart of a nearby major city. Do you think the estimated times, which were based on local distances of 3 to 20 miles, would be good predictors in this situation? Also, at $x = 0$ the equation has no real meaning. However, although projections outside the interval may be somewhat dangerous, they may be the best predictors available.

2. Don't get caught by the common fallacy of applying the regression results inappropriately. For example, this fallacy would include applying the results of Example 13.5 to another company. But suppose that the second company had a city location, whereas the first company had a rural location, or vice versa. Do you think the results for a rural location would be valid for a city location? Basically, the results of one sample should not be used to make inferences about a population other than the one from which the sample was drawn.

3. Don't jump to the conclusion that the results of the regression prove that *x causes y* to change. (This is perhaps the most common fallacy.) Regressions only measure movement between x and y; they never prove causation. (See pp. 167–168 for a discussion of causation.) A judgment of causation can be made only when it is based on theory or knowledge of the relationship separate from the regression results. The most common difficulty in this regard occurs because of what is called the *missing variable,* or *third-variable, effect.* That is, we observe a relationship between x and y because a third variable, one that is not in the regression, affects both x and y.

TECHNOLOGY INSTRUCTIONS: CONFIDENCE AND PREDICTION INTERVALS CALCULATION AND GRAPH

MINITAB (Release 14)   Input the *x*-variable data into C1 and the corresponding *y*-variable data into C2; then continue with:

```
Choose:     Stat > Regression > Regression . . .
Enter:      Response (y): C2
            Predictors (x): C1
Select:     Options
Enter:      Prediction intervals for new observations:
                x-value or C1 (C1—list of x values)
                Confidence level: 1 − α (ex. 95.0)
Select:     Confidence limits
            Prediction limits         >OK > OK
Choose:     Stat > Regression > Fitted Line Plot
Enter:      Response (y): C2
            Predictor (x): C1
Select:     Type of Regression Model: Linear
```

```
Select:        Options
               Display options:  Confidence limits
                                 Prediction limits
Enter:         Confidence level: 1 − α (ex. 95.0) > OK
Select:        Storage
               Residuals
               Fits > OK > OK
```

Here is the MINITAB printout for parts of Examples 13.5, 13.6, and 13.8.

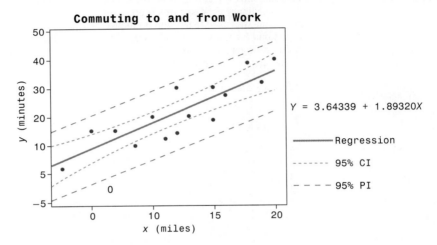

Commuting to and from Work

$Y = 3.64339 + 1.89320X$

———— Regression

------- 95% CI

— — — 95% PI

APPLIED
EXAMPLE 13.11

Using Regression Confidence Intervals in an Environmental Study

Much time, money, and effort are spent studying our environmental problems so that effective and appropriate management practices might be implemented. Here are excerpts from a study in South Florida in which linear regression analysis was an important tool.

METHODOLOGY FOR ESTIMATING NUTRIENT LOADS DISCHARGED FROM THE EAST COAST CANALS TO BISCAYNE BAY, MIAMI–DADE COUNTY, FLORIDA

A major concern in many coastal areas across the nation is the ecological health of bays and estuaries. One common problem in many of these areas is nutrient enrichment as a result of agricultural and urban activities. Nutrients are essential compounds for the growth and maintenance of all organisms and especially for the productivity of aquatic environments. Nitrogen and phosphorus compounds are especially important to seagrass, macroalgae, and phytoplankton. However, heavy nutrient loads transported to bays and estuaries can result in conditions conducive to eutrophication and the attendant problems of algal blooms and high phytoplankton productivity. Additionally, reduced light penetration in the water column because of phytoplankton blooms can adversely affect seagrasses, which many commercial and sport fish rely on for their habitat.

The purpose of this report is to present methodology that can be used to estimate nutrient loads discharged from the east coast canals into Biscayne Bay in southeastern Florida. Water samples were collected from the gated control structures at the east coast canal sites in Miami–Dade County for the purpose of developing models that could be used to estimate nitrogen and phosphorus loads.

An ordinary least-squares regression technique was used to develop predictive equations for the purpose of estimating total nitrogen and total phosphorus loads discharged from the east coast canals to Biscayne Bay. The predictive equations can be used to estimate the value of a dependent variable from observations on a related or independent variable. In this study, load was used as the dependent or response variable and discharge as the independent or explanatory variable. All of the total nitrogen load models had p-values less than 0.05, indicating they were statistically significant at an alpha level of 0.05. Plots showing total nitrogen load as a function of discharge at the east coast canal sites are shown in figure 17. [Sites S25 and S27 from figure 17 are shown here.]

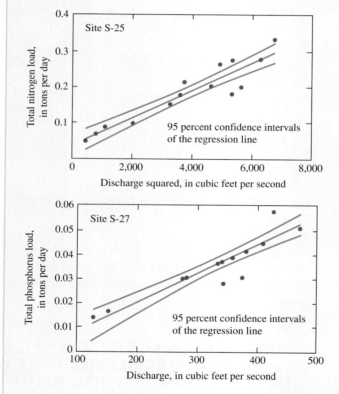

Source: U.S. Geological Survey, Water-Resources Investigations Report 99-4094, by A. C. Lietz

---

SECTION 13.6 EXERCISES

13.63 A study in *Physical Therapy* reports on seven different methods to determine crutch length plus two new techniques using linear regression. One of the regression techniques uses the patient's reported height. The study included 107 individuals.

The mean of the self-reported heights was 68.84 inches. The regression equation determined was $y = 0.68x + 4.8$, where y = crutch length and x = self-reported height. The MSE (s_e^2) was reported to be 0.50. In addition, the standard deviation of the self-reported heights was 7.35 inches. Use this information to determine a 95% confidence interval estimate for the mean crutch length for individuals who say they are 70 inches tall.

13.64 [EX13-64] Cicadas are flying, plant-eating insects. One particular species, the 13-year cicadas *(Magicicada),* spends five juvenile stages in underground burrows. During the 13 years underground, the cicadas grow from approximately the size of a small ant to nearly the size of an adult cicada. The adult body weights (BW) in grams and body lengths (BL) in millimeters are given for three different species of these 13-year cicadas in the following table.

BW	WL	Species	BW	WL	Species
0.15	28	tredecula	0.18	29	tredecula
0.29	32	tredecim	0.21	27	tredecassini
0.17	27	tredecim	0.15	30	tredecula
0.18	30	tredecula	0.17	27	tredecula
0.39	35	tredecim	0.13	27	tredecassini
0.26	31	tredecim	0.17	29	tredecassini
0.17	29	tredecassini	0.23	30	tredecassini
0.16	28	tredecassini	0.12	22	tredecim
0.14	25	tredecassini	0.26	30	tredecula
0.14	28	tredecassini	0.19	30	tredecula
0.28	25	tredecassini	0.20	30	tredecassini
0.12	28	tredecim	0.14	23	tredecula

a. Draw a scatter diagram with body weight as the independent variable and wing length as the dependent variable. Find the equation of the line of best fit.

b. Is body weight an effective predictor of wing length for a 13-year cicada? Use a 0.05 level of significance.

c. Give a 90% confidence interval for the mean wing length for all 0.20-gram cicada body weight.

13.65 Use the data and the answers found in Exercise 13.59 (p. 725) to make the following estimates.

a. Give a point estimate for the mean time required to commute 4 miles.

b. Give a 90% confidence interval for the mean travel time required to commute 4 miles.

c. Give a 90% prediction interval for the travel time required for one person to commute the 4 miles.

d. Answer parts a–c for $x = 9$.

13.66 Refer to Applied Example 13.11 on page 733. The graphs for Site S-25 and Site S-27 display 95% confidence intervals of the regression line. What distinguishing feature would 95% prediction intervals have with respect to these graphs? Explain the difference between confidence intervals and prediction intervals.

13.67 [EX13-67] An experiment was conducted to study the effect of a new drug in lowering the heart rate in adults. The data collected are shown in the following table.

x, Drug Dose in mg	0.50	0.75	1.00	1.25	1.50	1.75	2.00	2.25	2.50	2.75
y, Heart Rate Reduction	10	7	15	12	15	14	20	20	18	21

a. Find the 95% confidence interval for the mean heart rate reduction for a dose of 2.00 mg.

b. Find the 95% prediction interval for the heart rate reduction expected for an individual receiving a dose of 2.00 mg.

13.68 [EX13-68] The relationship between the "strength" and "fineness" of cotton fibers was the subject of a study that produced the following data.

x, Strength	76	69	71	76	83	72	78	74	80	82	90	81	78	80	81	78
y, Fineness	4.4	4.6	4.6	4.1	4.0	4.1	4.9	4.8	4.2	4.4	3.8	4.1	3.8	4.2	3.8	4.2

a. Draw a scatter diagram.

b. Find the 99% confidence interval for the mean measurement of fineness for fibers with a strength of 80.

c. Find the 99% prediction interval for an individual measurement of fineness for fibers with a strength of 75.

13.69 [EX12-36] Mr. B, the manager at a large retail store, is investigating several variables while measuring the level of his business. His store is open every day during the year except for New Year's Day, Christmas, and all Sundays. From his records, which cover several years prior, Mr. B has randomly identified 62 days and collected data for the daily total for three variables: number of paying customers, number of items purchased, and total cost of items purchased.

Day	Month	Customers	Items	Sales
2	1	425	1311	$12,707.00
1	1	412	1123	$11,467.50

••• Remainder of data on Student's Suite CD-ROM.

Data are actual values; store name withheld for privacy reasons.
Day Code: 1 = M, 2 = Tu, 3 = W, 4 = Th, 5 = F, 6 = Sa
Month Code: 1 = Jan, 2 = Feb, 3 = Mar, ..., 12 = Dec

Is there evidence to claim a linear relationship between the two variables number of customers and number of items purchased?
The computer output that follows resulted from analysis of the data.

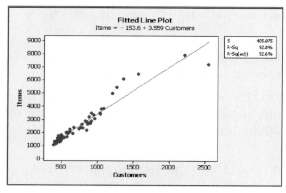

Regression Analysis: Items versus Customers
The regression equation is
Items = − 154 + 3.56 Customers

Predictor	Coef	SE Coef	T	P
Constant	−153.6	108.2	−1.42	0.161
Customers	3.5591	0.1284	27.71	0.000

S = 405.075   R-Sq = 92.8%   R-Sq(adj) = 92.6%

Inspect the preceding scatter diagram and the regression analysis output for number of customers versus number of items purchased. Look for evidence that either supports or contradicts the claim,

"There is a linear relationship between the two variables."

a. Describe the graphical evidence found and discuss how it shows lack of linearity for the entire range of values. Which ordered pairs appear to be different from the others?

b. Describe how the numerical evidence shown indicates the linear model does fit this data. Explain.

c. Some of the evidence seems to indicate the linear model is the correct model, and some evidence indicates the opposite. What months provided the points that are separate from the rest of the pattern? What is going on in those months that might cause this?

13.70 [EX13-70] Mr. B, store manager in Exercise 13.69 (and Exercises 12.36–12.38), found the data from the months of November and December to be different from the data for the other months. Since the data that is separate from the rest in the scatter diagram in Exercise 13.69 is from November and December, let's remove the November and December values and investigate the relationship between the number of customers per day and the number of items purchased per day for the first 10 months of the year.

January to October

Day J-O	Month J-O	Customers J-O	Items J-O	Sales J-O
2	1	425	1311	12707.00

••• Remainder of data on Student's Suite CD-ROM.

Day Code: 1 = M, 2 = Tu, 3 = W, 4 = Th, 5 = F, 6 = Sa
Month Code: 1 = Jan, 2 = Feb, 3 = Mar, ..., 10 = Oct

a. Use your calculator or computer to construct the scatter diagram for the data for January to October.

b. Describe the graphical evidence found and discuss the linearity. Are there any ordered pairs that appear to be different from the others?

c. What is the relationship between the number of customers per day and the number of items purchased per day for the first 10 months of the year?

d. Is the slope of the regression line significant at α = 0.05?

e. Give the 95% prediction interval for the number of items that one would expect to be purchased if the number of customers were 600.

13.71 [EX13-70] Help Mr. B, the store manager in Exercises 13.69 and 13.70 (and Exercises 12.36–12.38), by analyzing the relationship between the numbers of items purchased daily and the total daily sales for the data from the first 10 months of the year.

a. Construct the scatter diagram for the data for January to October.

b. Describe the graphical evidence found and discuss the linearity. Are there any ordered pairs that appear to be different from the others?

c. What is the relationship between the number of items purchased per day and the total daily sales for the first 10 months of the year?

d. Is the slope of the regression line significant at $\alpha = 0.05$?

e. Give the 95% prediction interval for the total daily sales that one would expect if the number of items purchased per day were 3000.

13.72 [EX13-72] Do you think your height and shoe size are related? Probably so. There is a known "quick" relationship that says your height (in inches) can be approximated by doubling your shoe size and adding 50 ($y = 2x + 50$). A random sample of 30 community college students' heights and shoe sizes was taken to test this relationship.

Heights	Shoe Sizes
74	13.0
71	10.0

••• Remainder of data on Student's Suite CD-ROM.

a. Construct a scatter diagram of the data with shoe size as the independent variable (x) and height as the dependent variable (y). Comment on the visual linear relationship.

b. Calculate the correlation coefficient, r. Is it significant at the 0.05 level of significance?

c. Calculate the line of best fit.

d. Compare the slope and intercept from part c to the slope and intercept of $y = 2x + 50$. List similarities and differences.

e. Estimate the height for a student with a size 10 shoe first using the line of best fit found in part c and then using the relationship, $y = 2x + 50$. Compare your results.

f. Construct the 95% confidence interval for the mean height of all community college students with a size 10 shoe using the equation formed in part c. Is your estimate using $y = 2x + 50$ for a size 10 included in this interval?

g. Construct the 95% prediction interval for the individual heights of all community college students with a size 10 shoe using the equation formed in part c.

h. Comment on the widths of the two intervals formed in parts f and g. Explain.

13.73 Explain why a 95% confidence interval for the mean value of y at a particular x is much narrower than a 95% prediction interval for an individual y value at the same value of x.

13.74 When $x_0 = \bar{x}$, is the formula for the standard error of $\hat{y}_{x_0}$ what you might have expected it to be, $s \cdot \dfrac{1}{\sqrt{n}}$? Explain.

13.7 Understanding the Relationship between Correlation and Regression

Now that we have taken a closer look at both correlation and regression analysis, it is necessary to decide when to use them. Do you see any duplication of work?

The primary use of the linear correlation coefficient is in answering the question: Are these two variables linearly related? Other words may be used to ask this basic question—for example: Is there a linear correlation between the annual consumption of alcoholic beverages and the salary paid to fire-fighters?

The linear correlation coefficient can be used to indicate the usefulness of x as a predictor of y in the case where the linear model is appropriate. The test concerning the slope of the regression line (H_o: $\beta_1 = 0$) tests this same basic concept. Either one of the two is sufficient to determine the answer to this query.

The choice of mathematical model can be tested statistically (called a "lack of fit" test); however, these procedures are beyond the scope of this text. We do perform this test informally, or subjectively, when we view the scatter diagram and use the presence of a linear pattern as our reason for using the linear model.

The concepts of linear correlation and regression are quite different because each measures different characteristics. It is possible to have data that yield a strong linear correlation coefficient and have the wrong model. For example, the straight line can be used to approximate almost any curved line if the domain is restricted sufficiently. In such a case the linear correlation coefficient can become quite high, but the curve will still not be a straight line. Figure 13.15 illustrates one interval where r could be significant but the scatter diagram does not suggest a straight line.

FIGURE 13.15
The Value of r Is High but the Relationship Is Not Linear

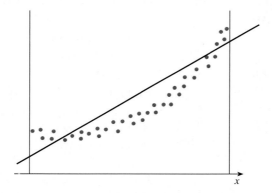

Regression analysis should be used to answer questions about the relationship between two variables. Such questions as "What is the relationship?" and "How are two variables related?" require this regression analysis.

CHAPTER REVIEW

In Retrospect

In this chapter we have made a more thorough inspection of the linear relationship between two variables. Although the curvilinear and multiple regression situations were mentioned only in passing, the basic techniques and concepts have been explored. We would only have to modify our mathematical model and our formulas if we wanted to deal with these other relationships.

Although it was not directly emphasized, we have applied many of the topics of earlier chapters in this chapter. The ideas of confidence interval and hypothesis testing were applied to the regression problem. Reference was made to the sampling distribution of the sample slope b_1. This allowed us to make inferences about β_1, the slope of the population from which the sample was drawn. We estimated the mean value of y at a fixed value of x by pooling the variance for the slope with the variance of the y's. This was allowable because they are independent. Recall that in Chapter 10 we presented formulas for combining the variances of independent samples. The idea here is much the same. Finally, we added a measure of variance for individual values of y and made estimates for these individual values of y at fixed values of x.

Applied Example 13.7 presents the results of regression analysis on data collected to compare two crime-reporting indices. (Take another look at Applied Example 13.7, p. 720.) The scatter diagram very convincingly shows that the two crime indices being compared are related to each other in a very strong and predictable pattern. Thus, as stated in the original article, "the weighted index contributed no further information" because the two indices are basically the same. Thus, the introduction of the weighted index seems unnecessary because the Uniform Crime Reports index is a recognized standard.

As this chapter ends, you should be aware of the basic concepts of regression analysis and correlation analysis. You should now be able to collect the data for, and do a complete analysis on, any two-variable linear relationship.

Vocabulary and Key Concepts

assumptions (pp. 703, 719)
bivariate data (pp. 697, 713)
centroid (p. 698)
coefficient of linear correlation (p. 700)
confidence belts (p. 703)
confidence interval (pp. 704, 719, 727, 730)
covariance (p. 698)
curvilinear regression (p. 713)
experimental error (ϵ or e) (p. 713)

hypothesis tests (pp. 704, 720)
intercept (b_0 or β_0) (p. 712)
line of best fit (p. 712)
linear correlation (p. 697)
linear regression (p. 712)
multiple regression (p. 713)
Pearson's product moment, r (p. 700)
predicted value of μ_y (p. 727)
predicted value of y ($\hat{y}$) (p. 713)
prediction interval (pp. 727, 730)

regression line (p. 713)
rho (ρ) (p. 703)
sampling distribution (pp. 718, 727)
scatter diagram (pp. 713, 718, 738)
slope (b_1 or β_1) (p. 712)
standard error (pp. 719, 728)
sum of squares for error (SSE) (p. 715)
variance (s^2 or σ^2) (p. 713)

Learning Outcomes

✓ Understand what bivariate data, independent variable, and dependent variable are.	pp. 152–153, Ex. 3.13
✓ Understand that the linear correlation coefficient, r, measures the strength of the linear relationship between two variables.	pp. 162–165, Ex. 3.29, 3.41
✓ Understand that the centroid for bivariate data is $(\bar{x}, \bar{y})$.	pp. 177, 698
✓ Understand that the centroid is used in the calculation of the correlation coefficient.	pp. 698–700, Ex. 13.3
✓ Understand that covariance is a measure of linear dependency but that it is affected by the spread of the data.	p. 699, Ex. 13.13

✓ Understand that correlation coefficient, *r*, standardizes covariance so that relative strengths can be compared. — p. 700, Ex. 13.7

✓ Understand that the assumptions for inferences about the linear correlation coefficient are that the ordered pairs form a random sample and the *y* values at each *x* have a normal distribution. Inferences will utilize the *t*-distribution using $(n - 2)$ degrees of freedom. — pp. 703, 714

✓ Compute, describe, and interpret a confidence interval for the population correlation coefficient, ρ, using Table 10 in Appendix B. — EXP 13.2, Ex. 13.19, 13.23

✓ Perform, describe, and interpret a hypothesis test for the population correlation coefficient, ρ, using the *t*-distribution with the *p*-value approach and classical approach. — EXP 13.3, Ex. 13.33, 13.38

✓ Understand that the significance of *r* does not imply a cause-and-effect relationship. — pp. 167–168, 732

✓ Understand that the estimate of the experimental error, *e*, is the difference between the observed *y* and the predicted *y*, $(y - \hat{y})$, at a given value of *x*. — pp. 712–713

✓ Understand that the variance about the line of best fit is the same as the variance of the error, *e*. — pp. 713–715, EXP 13.5, Ex. 13.47

✓ Understand that the line of best fit passes through the centroid. — p. 177

✓ Compute, describe, and interpret a confidence interval for the population slope of the regression line, β_1, using the *t*-distribution. — EXP 13.6, Ex. 13.53

✓ Perform, describe, and interpret a hypothesis test for the population slope of the regression line, β_1, using the *t*-distribution with the *p*-value approach and classical approach. — EXP 13.8, Ex. 13.59

✓ Compute, describe, and interpret a confidence interval for the mean value of *y* for a particular *x*, $(\mu_{y|x_0})$, using the *t*-distribution. — EXP 13.9, Ex. 13.63, 13.65

✓ Compute, describe, and interpret a prediction interval for an individual value of *y* for a particular *x*, (y_{x_0}), using the *t*-distribution. — EXP 13.10, Ex. 13.67

✓ Understand the difference between a confidence interval and a prediction interval for a *y* value at a particular *x* value. — p. 727, Ex. 13.73

Chapter Exercises

Statistics ⊜ Now™

Go to the StatisticsNow website **http://1pass.thomson.com** to

- Assess your understanding of this chapter
- Check your readiness for an exam by taking the Pre-Test quiz and exploring the resources in the Personalized Learning Plan

Datasets can be found on your Student's Suite CD-ROM or at the StatisticsNow website at **http://1pass.thomson.com**.

13.75 Answer the following as "sometimes," "always," or "never." Explain each "never" and "sometimes" response.

a. The correlation coefficient has the same sign as the slope of the least squares line fitted to the same data.

b. A correlation coefficient of 0.99 indicates a strong causal relationship between the variables under consideration.

c. An *r* value greater than zero indicates that ordered pairs with high *x* values will have low *y* values.

d. The *y* intercept and the slope for the line of best fit have the same sign.

e. If x and y are independent, then the population correlation coefficient equals zero.

13.76 [EX13-76] About 2527 athletes competed in the 2002 Olympic Winter Games held at Salt Lake City, Utah, with medals in 78 events. Athletes from 78 nations and territories participated. The following table shows the distribution of gold, silver, and bronze medals awarded to athletes representing the 18 nations who won the most:

Nation	Gold	Silver	Bronze
Germany	12	16	7
USA	10	13	11
Norway	11	7	6
Canada	6	3	8
Austria	2	4	10
Russian Fed.	6	6	4
Italy	4	4	4
France	4	5	2
Switzerland	3	2	6
China	2	2	4
Netherlands	3	5	0
Finland	4	2	1
Sweden	0	2	4
Croatia	3	1	0
Korea	2	2	0
Bulgaria	0	1	2
Estonia	1	1	1
Great Britain	1	0	2

Source: http://www.wikipedia.org/wiki/2002_Winter_Olympic_Games

Calculate the correlation coefficient and use it and Table 10 in Appendix B to determine a 95% confidence interval on ρ for each of the following cases:

a. Gold and silver   b. Gold and bronze

c. Silver and bronze

13.77 A study in the *Journal of Range Management* examines the relationships between elements in Russian wild rye. The correlation coefficient between magnesium and calcium was reported to be 0.69 for a sample of size 45. Is there a significant correlation between magnesium and calcium in Russian wild rye (i.e., is $\rho > 0$)?

13.78 A study concerning the plasma concentration of the drug ranitidine was reported in the

Journal of Pharmaceutical Sciences. The drug was administered (coded I), and the plasma concentration of ranitidine was followed for 12 hours. The time to the first peak in concentration was called T_{max1}. The same experiment was repeated 1 week later (coded II). Twelve subjects participated in the study. The correlation coefficient between T_{max1}, I, and T_{max1}, II, was reported to be 0.818. Use Table 11 in Appendix B to determine bounds on the p-value for the hypothesis test of H_o: $\rho = 0$ versus H_a: $\rho \neq 0$.

13.79 The use of electrical stimulation (ES) to increase muscular strength is discussed in the *Journal of Orthopedic and Sports Physical Therapy.* Seventeen healthy volunteers participated in the experiment. Muscular strength, Y, was measured as a torque in foot-pounds, and ES, X, was measured in mA (microamps). The equation for the line of best fit is given as $Y = 1.8X + 28.7$, and the Pearson correlation coefficient was 0.61.

a. Was the correlation coefficient significantly different from zero? Use $\alpha = 0.05$.

b. Predict the torque for a current equal to 50 mA.

13.80 An article in *Geology* gives the following equation relating pressure, P, and total aluminum content, AL, for 12 Horn blende rims: $P = -3.46(+0.24) + 4.23(+0.13)AL$. The quantities shown in parentheses are standard errors for the y-intercept and slope estimates. Find a 95% confidence interval for the slope, β_1.

13.81 [EX13-81] The following data resulted from an experiment performed for the purpose of regression analysis. The input variable, x, was set at five different levels, and observations were made at each level.

x	0.5	1.0	2.0	3.0	4.0
y	3.8	3.2	2.9	2.4	2.3
	3.5	3.4	2.6	2.5	2.2
	3.8	3.3	2.7	2.7	2.3
		3.6	3.2	2.3	

a. Draw a scatter diagram.

b. Draw the regression line by eye.

c. Place a star, $\star$, at each level approximately where the mean of the observed y values is located. Does your regression line look like the line of best fit for these five mean values?

d. Calculate the equation of the regression line.

e. Find the standard deviation of y about the regression line.

f. Construct a 95% confidence interval for the true value of β_1.

g. Construct a 95% confidence interval for the mean value of y at $x = 3.0$ and at $x = 3.5$.

h. Construct a 95% prediction interval for an individual value of y at $x = 3.0$ and at $x = 3.5$.

13.82 [EX13-82] The tobacco settlement negotiated by a team of eight attorneys general on behalf of 41 states resulted in $206 billion to be paid by the tobacco industry to recoup Medicaid costs the states incurred while treating ill smokers. Payments are to be made in annual increments over a 25-year span from 1998 to 2022. The following table shows an excerpt of the population (in millions of dollars) and the amounts (in billions of dollars) awarded to 46 states, the District of Columbia, and Puerto Rico:

State	Settlement	Population
AL	3.17	4.27
AK	0.67	0.61

••• Remainder of data on Student's Suite CD-ROM.

Sources: Washington State Attorney General Office and Bureau of the Census, U.S. Department of Commerce.

a. Draw a scatter diagram of these data with tobacco settlement as the dependent variable, y, and population as the predictor variable, x.

b. Calculate the regression equation and draw the regression line on the scatter diagram.

c. If your state's population were equal to 11.5 million people, of all 48 observations shown in the table, what would you estimate the tobacco settlement to be? Make your estimate based on the equation and then draw a line on the scatter diagram to illustrate it.

d. Construct a 95% prediction interval for the estimate you obtained in part c.

13.83 [EX13-83] Twenty-one mature flowers of a particular species were dissected, and the number of stamens and carpels present in each flower were counted.

x, Stamens	y, Carpels	x, Stamens	y, Carpels	x, Stamens	y, Carpels
52	20	65	30	45	27
68	31	43	19	72	21
70	28	37	25	59	35
38	20	36	22	60	27
61	19	74	29	73	33
51	29	38	28	76	35
56	30	35	25	68	34

a. Is there sufficient evidence to claim a linear relationship between these two variables at $\alpha = 0.05$?

b. What is the relationship between the number of stamens and the number of carpels in this variety of flower?

c. Is the slope of the regression line significant at $\alpha = 0.05$?

d. Give the 95% prediction interval for the number of carpels that one would expect to find in a mature flower of this variety if the number of stamens were 64.

13.84 [EX13-84] The following set of 25 scores was randomly selected from a teacher's class list. Let x be the prefinal average and y be the final examination score. (The final examination had a maximum of 75 points.)

Student	x	y
1	75	64
2	86	65

••• Remainder of data on Student's Suite CD-ROM.

a. Draw a scatter diagram for these data.

b. Draw the regression line (by eye) and estimate its equation.

c. Estimate the value of the coefficient of linear correlation.

d. Calculate the equation of the line of best fit.

e. Draw the line of best fit on your graph. How does it compare with your estimate?

f. Calculate the linear correlation coefficient. How does it compare with your estimate?

g. Test the significance of r at $\alpha = 0.10$.

h. Find the 95% confidence interval for the true value of ρ.

i. Find the standard deviation of the y values about the regression line.

j. Calculate a 95% confidence interval for the true value of the slope β_1.

k. Test the significance of the slope at $\alpha = 0.05$.

l. Estimate the mean final-exam grade that all students with an 85 prefinal average will obtain (95% confidence interval).

m. Using the 95% prediction interval, predict the grade that John Henry will receive on his final, knowing that his prefinal average is 78.

13.85 [EX13-85] It is believed that the amount of nitrogen fertilizer used per acre has a direct effect on the amount of wheat produced. The following data show the amount of nitrogen fertilizer used per test plot and the amount of wheat harvested per test plot.

x, Pounds of Fertilizer	y, 100 Pounds of Wheat
30	5
30	9

••• Remainder of data on Student's Suite CD-ROM.

a. Is there sufficient reason to conclude that the use of more fertilizer results in a higher yield? Use $\alpha = 0.05$.

b. Estimate, with a 98% confidence interval, the mean yield that could be expected if 50 lb of fertilizer was used per plot.

c. Estimate, with a 98% confidence interval, the mean yield that could be expected if 75 lb of fertilizer was used per plot.

13.86 [EX13-70] Help Mr. B, the store manager in Exercises 13.69–13.71 (and Exercises 12.36–12.38), by analyzing the relationship between the number of customers per day and the total daily sales for the data from the first 10 months of the year.

a. Use your calculator or computer to construct the scatter diagram for the data for January to October.

b. Describe the graphical evidence found and discuss the linearity. Are there any ordered pairs that appear to be different from the others?

c. What is the relationship between the number of customers per day and the total daily sales for the first 10 months of the year?

d. Is the slope of the regression line significant at $\alpha = 0.05$?

e. Give the 95% prediction interval for the total daily sales that one would expect if the number of customers were 600.

13.87 Compare the results obtained in Exercises 13.70, 13.71, and 13.86. Explain the similarities and differences. Why do you think the scatter diagram for number of items purchased and total sales shows less variability about the line of best fit than do the other two scatter diagrams?

13.88 [EX13-70] Investigate the relationship of the variables studied in Exercises 13.69, 13.70, 13.71, 13.86, and 13.87 for the November and December data.

November & December

Day N&D	Month N&D	Customers N&D	Items N&D	Sales N&D
6	11	1049	3799	40,362.70

••• Remainder of data on Student's Suite CD-ROM.

13.89 The correlation coefficient, r, is related to the slope of best fit, b_1, by the equation

$$r = b_1 \sqrt{\frac{SS(x)}{SS(y)}}$$

Verify this equation using the following data.

x	1	2	3	4	6
y	4	6	7	9	12

13.90 The following equation is known to be true for any set of data: $\sum (y - \bar{y})^2 = \sum (y - \hat{y})^2 + \sum (\hat{y} - \bar{y})^2$. Verify this equation with the following data.

x	0	1	2
y	1	3	2

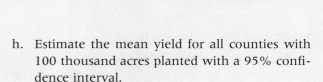

Chapter Project

Wheat! Beautiful Golden Wheat!

The four variables defined in Section 13.1, "Wheat! Beautiful Golden Wheat!" (p. 696), all seem to be interrelated. Upon completion of the Section 13.1 exercises, one finds that a strong linear relationship exists between the amount planted and the amount harvested, as well as the amount harvested and the production. As one increased, so did the other. Intuitively, this makes sense. Let's further our investigation of relationships with the combinations of planted with production and planted with yield to see if they also exist and have meaning.

Putting Chapter 13 to Work

13.91 **[EX13-01]** Refer to the wheat crop data in Section 13.1 on page 696.

a. Construct a scatter diagram and find the linear correlation coefficient and the equation for the line of best fit for Planted (x) and Production (y).

b. Find the 95% confidence interval for the true linear correlation coefficient between Planted and Production.

c. Calculate the 95% confidence interval for the slope, β_1, for the line of best fit relating Planted and Production.

d. Predict the total production for a county with 125 thousand acres planted using a 95% prediction interval.

e. Construct a scatter diagram and find the linear correlation coefficient and the equation for the line of best fit for Planted (x) and Yield (y).

f. Test the significance of the correlation between Planted and Yield. Use $\alpha = 0.05$.

g. Test the significance of the slope, β_1, for the line of best fit relating Planted and Yield. Use $\alpha = 0.05$.

h. Estimate the mean yield for all counties with 100 thousand acres planted with a 95% confidence interval.

Your Study

13.92 Design your own study to investigate another grain. Answer questions similar to those asked in Exercise 13.91. Use the Internet and the U.S. Department of Agriculture's National Agricultural Statistics Service website (http://www.usda.gov/nass/) to find your data. The data sets posted will contain extra information that makes the data in its original format unusable. However, this unwanted information can easily be removed by deleting the unwanted rows and columns. (*Note:* http://www.aragriculture.org/News/ also has many data sets posted.)

Chapter Practice Test

Part I: Knowing the Definitions

Answer "True" if the statement is always true. If the statement is not always true, replace the words shown in bold with words that make the statement always true.

13.1 The error **must be** normally distributed if inferences are to be made.

13.2 Both x and y **must be** normally distributed.

13.3 A high correlation between x and y **proves** that x causes y.

13.4 The value of the input variable **must be** randomly selected to achieve valid results.

13.5 The output variable must be **normally distributed** about the regression line for each value of x.

13.6 **Covariance** measures the strength of the linear relationship and is a standardized measure.

13.7 The **sum of squares for error** is the name given to the numerator of the formula used to calculate the variance of *y* about the line of regression.

13.8 **Correlation** analysis attempts to find the equation of the line of best fit for two variables.

13.9 There are $n - 3$ degrees of freedom involved with the inferences about the regression line.

13.10 $\hat{y}$ serves as the **point estimate** for both $\mu_{y|x_0}$ and y_{x_0}.

Part II: Applying the Concepts

Answer all questions, showing formulas and work.

It is believed that the amount of nitrogen fertilizer used per acre has a direct effect on the amount of wheat produced. The following data show the amount of nitrogen fertilizer used per test plot and the amount of wheat harvested per test plot. All test plots were the same size. [PT13-11]

x, Pounds of Fertilizer	*y*, 100 Pounds of Wheat	*x*, Pounds of Fertilizer	*y*, 100 Pounds of Wheat
30	9	70	19
30	11	70	22
30	14	70	31
50	12	90	29
50	14	90	33
50	23	90	35

13.11 Draw a scatter diagram of the data. Be sure to label completely.

13.12 Complete an extensions table.

13.13 Calculate SS(*x*), SS(*xy*), and SS(*y*).

13.14 Calculate the linear correlation coefficient, *r*.

13.15 Determine the 95% confidence interval estimate for the population linear correlation coefficient.

13.16 Calculate the equation of the line of best fit.

13.17 Draw the line of best fit on the scatter diagram.

13.18 Calculate the standard deviation of the *y* values about the line of best fit.

13.19 Does the value of b_1 show strength significant enough to conclude that the slope is greater than zero at the 0.05 level?

13.20 Determine the 0.95 confidence interval for the mean yield when 85 lb of fertilizer are used per plot.

13.21 Draw a line on the scatter diagram representing the 95% confidence interval found in question 13.20.

Part III: Understanding the Concepts

13.22 "There is a high correlation between how frequently skiers have their bindings tested and the incidence of lower-leg injuries, according to researchers at the Rochester Institute of Technology. To make sure your bindings release properly when you begin to fall, you should have them serviced by a ski mechanic every 15 to 30 ski days or at least at the start of each ski season" (University of California, Berkeley, "Wellness Letter," February 1991). Explain what two variables are discussed in this statement, and interpret the "high correlation" mentioned.

13.23 If a "moment" is defined as the distance from the mean, describe why the method used to define the correlation coefficient is referred to as "a product moment."

13.24 If you know that the value of *r* is very close to zero, what value would you anticipate for b_1? Explain why.

13.25 Describe why the method used to find the line of best fit is referred to as "the method of least squares."

13.26 You wish to study the relationship between the amount of sugar in a child's breakfast and the child's hyperactivity in school during the 4 hours after breakfast. You ask 200 mothers of fifth-grade children to keep a careful record of what the children eat and drink each morning.

Each parent's report is analyzed, and the sugar consumption is determined. During the same time period, data on hyperactivity are collected at school. What statistic will measure the strength and kind of relationship that exists between the amount of sugar and the amount of hyperactivity? Explain why the statistic you selected is appropriate and what value you expect this statistic might have.

13.27   You are interested in studying the relationship between the length of time a person has been supported by welfare and self-esteem. You believe that the longer a person is supported, the lower the self-

esteem. What data would you need to collect and what statistics would you calculate if you wish to predict a person's level of self-esteem after having been on welfare for a certain period of time? Explain in detail.

Statistics⌂Now™   Preparing for an exam? Assess your progress by taking the post-test at **http://1pass.thomson.com**.

Mentor   Do you need a live tutor for homework problems? Access vMentor on the StatisticsNow website at **http://1pass.thomson.com** for one-on-one tutoring from a statistics expert.

CHAPTER
14

Elements of Nonparametric Statistics

14.1

Teenagers' Attitudes

A national survey revealed that teenagers' attitudes toward moral and social values are much more conventional than widely believed.

How Teenagers See Things

One could think of them as Generation "V"—for values. According to The Mood of American Youth study, today's teens are neither as rebellious as adolescents in the 1970s nor as materialistic as those of the 1980s. What they want is not to change the world or to own a chunk of it, but to be happy. Among the teens' greatest concerns: the decline in moral and social values.

The study, conducted by NFO Research, Inc., included 938 young people aged 13 to 17 who are representative of America's adolescent population as a whole. Of those polled, 9 in 10 say they don't drink or smoke, and 7 in 10 say religion is important in their lives. Most of the respondents respect their parents, get along well with them, and consider their rules strict but fair.

Teens' Views on Contemporary Issues

Families and Children	Agree
Teenagers are not prepared to have babies.	91%
A single parent can raise a family.	75%
I am very likely to raise my children differently than I was raised.	55%

In the Schools	Agree
Local school officials should be able to censor the books and materials used in their schools.	47%
School prayer should not be permitted.	32%

Social Concerns	
The "V" chip will unfairly censor what teens can watch on TV.	67%
It is important to control information on the Internet.	60%

. . . And What about the Government?

Government spending on AIDS research should be increased.	83%
Adequate health care for all should be provided through a national health plan.	81%
Being rich is necessary to get elected to high office.	55%

Part of the survey asked the teens to indicate the one thing they wanted most from life. The following table lists the choices and the percentage of teens selecting each.

The One Thing Teens Want Most from Life* [EX14-01]

Choices	All (%)	Boys (%)	Girls (%)	Choices	All (%)	Boys (%)	Girls (%
Happiness	28	23	32	Love	7	6	7
Long, enjoyable life	16	18	14	Personal success	6	6	5
Marriage and family	9	8	11	Personal contribution to society	2	3	2
Financial success	8	11	4	Friends	2	3	1
Career success	8	9	6	Health	2	2	2
Religious satisfaction	8	7	7	Education	2	1	2

*Some teens didn't respond, so figures don't total 100%.

Source: © 1996 Dianne Hales. All rights reserved. From *Parade Magazine*, August 18, 1996, and reprinted with permission.

After completing Chapter 14, further investigate the preceding data and "Teenagers' Attitudes" in the Chapter Project on page 800.

SECTION 14.1 EXERCISES

Statistics⬭Now™

Datasets can be found on your Student's Suite CD-ROM or at the StatisticsNow website at **http://1pass.thomson.com**.

14.1 [EX14-01] Consider the table "The One Thing Teens Want Most from Life" above:

a. Does it appear that teenage boys and girls agree on what they want most from life?

b. Construct a scatter diagram of Boys' Percent Choices versus Girls' Percent Choices.

c. Does the graph constructed in part b support your answer in part a?

14.2 [EX14-01] a. Construct a side-by-side bar chart of the Boys' Percent Choices and the corresponding Girls' Percent Choices.

b. Based on the graph constructed in part a, does it appear that teenage boys and girls agree on what they want most from life?

c. Explain why the chi-square test of homogeneity studied in Chapter 11 cannot be completed using this information.

14.2   Nonparametric Statistics

Most of the statistical procedures we have studied in this book are known as **parametric methods.** For a statistical procedure to be parametric, either we assume that the parent population is at least approximately normally distributed or we rely on the central limit theorem to give us a normal approximation. This is particularly true of the statistical methods studied in Chapters 8, 9, and 10.

The **nonparametric methods,** or **distribution-free methods** as they are also known, do not depend on the distribution of the population being sampled. The nonparametric statistics are usually subject to much less confining restrictions than are their parametric counterparts. Some, for example, require only that the parent population be continuous.

The recent popularity of nonparametric statistics can be attributed to the following characteristics:

1. Nonparametric methods require few assumptions about the parent population.

2. Nonparametric methods are generally easier to apply than their parametric counterparts.

3. Nonparametric methods are relatively easy to understand.

4. Nonparametric methods can be used in situations in which the normality assumptions cannot be made.

5. Nonparametric methods are generally only slightly less efficient than their parametric counterparts.

14.3   Comparing Statistical Tests

This chapter presents only a sampling of the many nonparametric tests. The selections presented demonstrate their ease of application and variety of technique. They represent a very small sampling of the many different nonparametric tests that exist. Many of the nonparametric tests can be used in place of certain parametric tests. The question is, then: Which statistical test do we use, the parametric or the nonparametric? Sometimes there is more than one nonparametric test to choose from.

The decision about which test to use must be based on the answer to the question: Which test will do the job best? First, let's agree that when we compare two or more tests, they must be equally qualified for use. That is, each test has a set of assumptions that must be satisfied before it can be applied. From this starting point we will attempt to define "best" to mean the test that is best able to control the risks of error and at the same time keep the size of the sample to a number that is reasonable to work with. (Sample size means cost—cost to you or your employer.)

Power and Efficiency Criteria

Let's look first at the ability to control the risk of error. The risk associated with a type I error is controlled directly by the level of significance α. Recall that $P(\text{type I error}) = \alpha$ and $P(\text{type II error}) = \beta$. Therefore, it is β that we must control. Statisticians like to talk about *power* (as do others), and the **power of a statistical test** is defined to be $1 - \beta$. Thus, the power of a test, $1 - \beta$, is the probability that we reject the null hypothesis when we should have rejected it. If two tests with the same α are equal candidates for use, then the one with the greater power is the one you would want to choose.

The other factor is the sample size required to do a job. Suppose that you set the levels of risk you can tolerate, α and β, and then you are able to determine the sample size it would take to meet your specified challenge. The test that required the smaller sample size would seem to have the edge. Statisticians usually use the term *efficiency* to talk about this concept. *Efficiency* is the ratio of the sample size of the best parametric test to the sample size of the best nonparametric test when compared under a fixed set of risk values. For example, the efficiency rating for the sign test is approximately 0.63. This means that a sample of size 63 with a parametric test will do the same job as a sample of size 100 will do with the sign test.

The power and the efficiency of a test cannot be used alone to determine the choice of test. Sometimes you will be forced to use a certain test because of the data you are given. When there is a decision to be made, the final decision rests in a trade-off of three factors: (1) the power of the test, (2) the efficiency of the test, and (3) the data (and the sample size) available. Table 14.1 shows how the nonparametric tests discussed in this chapter compare with the parametric tests covered in previous chapters.

TABLE 14.1

Comparison of Parametric and Nonparametric Tests

Test Situation	Parametric Test	Nonparametric Test	Efficiency of Nonparametric Test
One mean	*t*-test (p. 474)	Sign test (p. 752)	0.63
Two independent means	*t*-test (p. 564)	*U* test (p. 765)	0.95
Two dependent means	*t*-test (p. 550)	Sign test (p. 756)	0.63
Correlation	Pearson's (p. 703)	Spearman test (p. 785)	0.91
Randomness		Runs test (p. 776)	Not meaningful; there is no parametric test for comparison

14.4 The Sign Test

The sign test is a versatile and exceptionally easy-to-apply nonparametric method that uses only plus and minus signs. Three sign test applications are presented here: (1) a confidence interval for the median of one population, (2) a hypothesis test concerning the value of the median for one population, and (3) a hypothesis test concerning the median difference (paired difference) for two **dependent samples.** These sign tests are carried out using the same basic confidence interval and hypothesis test procedures as described in earlier chapters. They are the nonparametric alternatives to the *t*-tests used for one mean (see Section 9.2) and the difference between two dependent means (see Section 10.3).

> **Assumptions for inferences about the population single-sample median using the sign test:** The *n* random observations that form the sample are selected independently, and the population is continuous in the vicinity of the median *M*.

Single-Sample Confidence Interval Procedure

The sign test can be applied to obtain a confidence interval for the unknown **population median, M.** To accomplish this we will need to arrange the sample data in ascending order (smallest to largest). The data are identified as x_1 (smallest), x_2, x_3, . . . , x_n (largest). The critical value, *k* (known as the "maximum allowable number of signs"), is obtained from Table 12 in Appendix B, and it tells us the number of positions to be dropped from each end of the ordered data. The remaining extreme values become the bounds of the $1 - \alpha$ confidence interval. That is, the lower boundary for the confidence interval is x_{k+1}, the $(k + 1)$th data value; the upper boundary is x_{n-k}, the $(n - k)$th data value.

In general, the two data values that bound the confidence interval occupy positions $k + 1$ and $n - k$, where *k* is the critical value read from Table 12. Thus,

$$x_{k+1} \text{ to } x_{n-k}, \quad 1 - \alpha \text{ confidence interval for } M$$

The next example will clarify this procedure.

EXAMPLE 14.1 | ## Constructing a Confidence Interval for a Population Median

Suppose that we have a random sample of 12 daily high temperature readings in ascending order, [50, 62, 64, 76, 76, 77, 77, 77, 80, 86, 92, 94], and we wish to form a 95% confidence interval for the population median. Table 12 shows a critical value of 2 ($k = 2$) for $n = 12$ and $\alpha = 0.05$ for a two-tailed hypothesis test. This means that we drop the last two values on each end (50 and 62 on left; 92 and 94 on right). The confidence interval is bounded inclusively by the remaining end values, 64 and 86. That is, the 95% confidence interval is 64 to 86 and is expressed as:

64° to 86°, the 95% confidence interval for the median daily high temperature

Single-Sample Hypothesis-Testing Procedure

The sign test can be used when the null hypothesis to be tested concerns the value of the population median *M*. The test may be either one- or two-tailed. This test procedure is presented in the following example.

EXAMPLE 14.2   **Two-Tailed Hypothesis Test**

A random sample of 75 students was selected, and each student was asked to carefully measure the amount of time it takes to commute from his or her front door to the college parking lot. The data collected were used to test the hypothesis, "The median time required for students to commute is 15 minutes," against the alternative that the median is unequal to 15 minutes. The 75 pieces of data were summarized as follows:

Under 15: 18   15: 12   Over 15: 45

Use the sign test to test the null hypothesis against the alternative hypothesis.

SOLUTION The data are converted to + and − signs according to whether each data value is more or less than 15. A plus sign will be assigned to each larger than 15, a minus sign to each smaller than 15, and a zero to those equal to 15. The sign test uses only the plus and minus signs; therefore, the zeros are discarded and the usable sample size becomes 63. That is, $n(+) = 45$, $n(-) = 18$, and $n = n(+) + n(-) = 45 + 18 = 63$.

Step 1   **a.** **Parameter of interest:** M, the population median time to commute.

        b. **Statement of hypotheses:**

H_o: $M = 15$

H_a: $M \neq 15$

Step 2   **a.** **Assumptions:** The 75 observations were randomly selected, and the variable commute time is continuous.

        b. **Test statistic:** The test statistic that will be used is the number of the less frequent sign: the smaller of $n(+)$ and $n(-)$, which is $n(-)$ for our example. We will want to reject the null hypothesis whenever the number of the less frequent sign is extremely small. Table 12 in Appendix B gives the maximum allowable number of the less frequent sign, k, that will allow us to reject the null hypothesis. That is, if the number of the less frequent sign is less than or equal to the critical value in the table, we will reject H_o. If the observed value of the less frequent sign is larger than the table value, we will fail to reject H_o. In the table, n is the total number of signs, not including zeros. The test statistic = $x\bigstar = n(\textbf{least frequent sign})$.

        c. **Level of significance:** $\alpha = 0.05$ for a two-tailed test.

Step 3   **a.** **Sample information:** $n = 63$; $[n(-) = 18, n(+) = 45]$.

        b. **Test statistic:** The observed value of the test statistic is
$x\bigstar = n(-) = \textbf{18}$.

Step 4   **Probability Distribution:**

p-Value:		Classical:
a. Since the concern is for values "not equal to," the *p*-value is the area of both tails. We will find the left tail and double it: $\mathbf{P} = 2 \times P(x \leq 18,$ for $n = 63)$.		a. The critical region is split into two equal parts because H_a expresses concern for values related to "not equal to." Since the table is for two-tailed tests, the critical

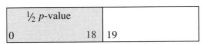

Number of less frequent sign

To find the *p*-value, you have two options:
1. Use Table 12 (Appendix B) to place bounds on the *p*-value. Table 12 lists only two-tailed values (do not double): **P < 0.01.**
2. Use a computer or calculator to find the *p*-value: **P = 0.0011.**
Specific instructions follow this example.
b. The *p*-value is smaller than α.

value is located at the intersection of the $\alpha = 0.05$ column and the $n = 63$ row of Table 12: **23.**

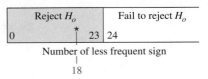

Number of less frequent sign

b. $x\star$ is in the critical region, as shown in the figure.

Step 5 **a.** **Decision:** Reject H_o.

b. **Conclusion:** The sample shows sufficient evidence at the 0.05 level to conclude that the median commute time is not equal to 15 minutes.

Calculating the p-value when using the sign test

Method 1: Use Table 12 in Appendix B to place bounds on the p-value. By inspecting the $n = 63$ row of Table 12, you can determine an interval within which the *p*-value lies. Locate the value of *x* along the $n = 63$ row and read the bounds from the top of the table. Table 12 lists only two-tailed values (therefore, do not double): **P < 0.01.**

Method 2: If you are doing the hypothesis test with the aid of a computer or graphing calculator, most likely it will calculate the *p*-value for you. Specific instructions are described next.

TECHNOLOGY INSTRUCTIONS: SIGN TEST FOR A SINGLE-SAMPLE HYPOTHESIS TEST OF THE MEDIAN

MINITAB (Release 14) Input the set of data into C1; then continue with:

```
Choose:     Stat > Nonparametrics > 1-Sample Sign
Enter:      Variables: C1
Select:     Test median:*
Enter:      M (hypothesized median value)
Select:     Alternative: less than or not equal or greater than > OK
```

*A confidence interval may also be selected.

(If original data are not given, just the number of plus and minus signs, then input data values above and below the median that will compute into the correct number of each sign.)

Excel The following Excel commands will compute the differences between the data values and the hypothesized median. The data will then be sorted so that the number of + and − signs can be easily counted.

Input the data into column A and select cell B1; then continue with:

```
Choose:     Insert function fₓ > All > SIGN > OK
Enter:      Number: A1 − hypothesized median value > OK
Drag:       Bottom right corner of the B1 cell down to give other differences
```

Select the data in columns A and B; then continue with:

Choose: **Data > Sort**
Enter: Sort by: **Column B**
Select: **Ascending > OK**

TI-83/84 Plus   Input the data into L1; then continue with:

Choose: **PRGM > EXEC > SIGNTEST***
Select: **PROCEDURE: 3: HYP TEST**   INPUT? **2:DATA: 1 LIST**
Enter: DATA: **L1**   MEDO: **hypothesized median value**
Select: ALT HYP? **1: >**   or **2: <**   or **3: ≠**

*Program SIGNTEST is one of many programs that are available for downloading from http://www.duxbury.com. See page 42 for specific directions.

Two-Sample Hypothesis-Testing Procedure

The sign test may also be applied to a hypothesis test dealing with the median difference between **paired data** that result from **two dependent samples.** A familiar application is the use of before-and-after testing to determine the effectiveness of some activity. In a test of this nature, the signs of the differences are used to carry out the test. Again, zeros are disregarded.

> **Assumptions for inferences about the median of paired differences using the sign test:** The paired data are selected independently, and the variables are ordinal or numerical.

The following example shows this procedure.

EXAMPLE 14.3   ## One-Tailed Hypothesis Test for the Median of Paired Differences

Statistics ⬡ Now™

Watch a video example at **http://1pass.thomson.com** or on your CD.

A new no-exercise, no-starve weight-reducing plan has been developed and advertised. To test the claim that "you will lose weight within 2 weeks or . . . ," a local statistician obtained the before-and-after weights of 18 people who had used this plan. Table 14.2 lists the people, their weights, and a minus (−) for those who lost weight during the 2 weeks, a 0 for those who remained the same, and a plus (+) for those who actually gained weight.

    The claim being tested is that people lose weight. The null hypothesis that will be tested is, "There is no weight loss (or the median weight loss is zero)," meaning that only a rejection of the null hypothesis will allow us to conclude in favor of the advertised claim. Actually we will be testing to see whether there are significantly more minus signs than plus signs. If the weight-reducing plan is of absolutely no value, we would expect to find an equal number of plus and minus signs. If it works, there should be significantly more minus signs than plus signs. Thus, the test performed here will be a one-tailed test.

(We want to reject the null hypothesis in favor of the advertised claim if there are "many" minus signs.)

TABLE 14.2

Sample Results on Weight-Reducing Plan [TA14-02]

Person	Weight Before	Weight After	Sign of Difference, After — Before	Person	Weight Before	Weight After	Sign of Difference, After — Before
Mrs. Smith	146	142	—	Mr. Carroll	187	187	0
Mr. Brown	175	178	+	Mrs. Black	172	171	—
Mrs. White	150	147	—	Mrs. McDonald	138	135	—
Mr. Collins	190	187	—	Ms. Henry	150	151	+
Mr. Gray	220	212	—	Ms. Greene	124	126	+
Ms. Collins	157	160	+	Mr. Tyler	210	208	—
Mrs. Allen	136	135	—	Mrs. Williams	148	148	0
Mrs. Noss	146	138	—	Mrs. Moore	141	138	—
Ms. Wagner	128	132	+	Mrs. Sweeney	164	159	—

SOLUTION

Step 1   **a. Parameter of interest:** M, the median weight loss.

   b. Statement of hypotheses:

   H_o: $M = 0$ (no weight loss)

   H_a: $M < 0$ (weight loss)

Step 2   **a. Assumptions:** The 18 observations were randomly selected, and the variables, weight before and weight after, are both continuous.

   b. Test statistic: The number of the less frequent sign: the test statistic $= x\star = n$(least frequent sign).

   c. Level of significance: $\alpha = 0.05$ for a one-tailed test.

Step 3   **a. Sample information:** $n = 16$ [$n(+) = 5$, $n(-) = 11$].

   b. Test statistic: The observed value of the test statistic is $x\star = n(+) = 5$.

Step 4   **Probability Distribution:**

OR

p-Value:

a. Since the concern is for values "less than," the *p*-value is the area to the left: $\mathbf{P} = P(x \leq 5$, for $n = 16)$.

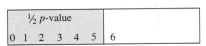

Number of less frequent sign

To find the *p*-value, you have two options:
1. Use Table 12 in Appendix B to estimate the *p*-value. Table 12 lists only two-tailed α (this is one-tailed, so divide α by two): **P ≈ 0.125**.

Classical:

a. The critical region is one-tailed because H_a expresses concern for values related to "less than." Since the table is for two-tailed tests, the critical value is located at the intersection of the $\alpha = 0.10$ column ($\alpha = 0.05$ in each tail) and the $n = 16$ row of Table 12:

$k = 4$

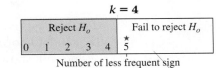

Number of less frequent sign

2. Use a computer or calculator to find the *p*-value: **P = 0.1051.**

For specific instructions, see p. 755.

b. The *p*-value is not smaller than α.

b. $x\star$ is not in the critical region, as shown in the figure.

Step 5   **a.**   **Decision:** Fail to reject H_o.

               b.   **Conclusion:** The evidence observed is not sufficient to allow us to reject the no-weight-loss null hypothesis at the 0.05 level of significance.

TECHNOLOGY INSTRUCTIONS: SIGN TEST FOR THE MEDIAN OF PAIRED DIFFERENCES

MINITAB (Release 14)   Input the paired set of data into C1 and C2; then continue with:

Choose:	**Calc > Calculator**
Enter:	Store result in: **C3**
	Expression: **C1–C2** (whichever order is needed, based on H_a) > **OK**
Choose:	**Stat > Nonparametrics > 1-Sample Sign . . .**
Enter:	Variables: **C3**
Select:	**Test median:***
Enter:	**0** (hypothesized median value)
Select:	Alternative: **less than** or **not equal** or **greater than** > **OK**

*As before, the confidence interval may be selected.

Excel   Input the paired data into columns A and B; then continue with:

Choose:	**Tools > Data Analysis Plus > Sign Test > OK**
Enter:	Variable 1 Range: **(A1:A20 or select cells)**
	Variable 2 Range: **(B1:B20 or select cells)**
Select:	**Labels** (if necessary)
Enter:	Alpha: α (ex. 0.05)

TI-83/84 Plus   Input the paired data into L1 and L2; then continue with:

Highlight:	**L3**
Enter:	**L1–L2** (whichever order is needed, based on H_a)
Choose:	**PRGM > EXEC > SIGNTEST***
Select:	PROCEDURE: **3: HYP TEST**
	INPUT? **2:DATA: 1 LIST**
Enter:	DATA: **L3**
	MEDO: **hypothesized median value**
Select:	ALT HYP? **1:** > or **2:** < or **3:** ≠

*Program SIGNTEST is one of many programs that are available for downloading from http://www.duxbury.com. See page 42 for specific directions.

Normal Approximation

The sign test may be carried out by means of a normal approximation using the standard normal variable *z*. The normal approximation will be used if Table 12 does not show the particular levels of significance desired or if *n* is large.

Notes:
1. x may be the number of the less frequent sign or the more frequent sign. You will have to determine this in such a way that the direction is consistent with the interpretation of the situation.
2. x is really a **binomial random variable,** where $p = 0.5$. The sign test statistic satisfies the properties of a binomial experiment (see p. 287). Each sign is the result of an independent trial. There are n trials, and each trial has two possible outcomes ($+$ or $-$). Since the median is used, the probabilities for each outcome are both 0.5. Therefore, the mean, μ_x, is equal to

$$\mu_x = \frac{n}{2} \quad \left[\mu = np = n \cdot \frac{1}{2} = \frac{n}{2}\right]$$

and the standard deviation, σ_x, is equal to

$$\sigma_x = \frac{1}{2}\sqrt{n} \quad \left[\sigma = \sqrt{npq} = \sqrt{n \cdot \frac{1}{2} \cdot \frac{1}{2}} = \frac{1}{2}\sqrt{n}\right]$$

3. x is a discrete variable. But recall that the normal distribution must be used only with continuous variables. However, although the binomial random variable is discrete, it does become approximately normally distributed for large n. Nevertheless, when using the normal distribution for testing, we should make an adjustment in the variable so that the approximation is more accurate. (See Section 6.6, p. 343, on the normal approximation.) This adjustment is illustrated in Figure 14.1 and is called a **continuity correction.** For this discrete variable the area that represents the probability is a rectangular bar. Its width is 1 unit wide, from $\frac{1}{2}$ unit below to $\frac{1}{2}$ unit above the value of interest. Therefore, when z is to be used, we will need to make a $\frac{1}{2}$-unit adjustment before calculating the observed value of z. So x' will be the adjusted value for x. If x is larger than $\frac{n}{2}$, then $x' = x - \frac{1}{2}$. If x is smaller than $\frac{n}{2}$, then $x' = x + \frac{1}{2}$. The test is then completed by the usual procedure, using x'.

FIGURE 14.1
Continuity Correction

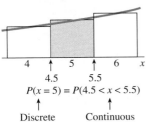

4   5   6 x
4.5   5.5
$P(x = 5) = P(4.5 < x < 5.5)$
Discrete   Continuous

Confidence Interval Procedure
If the normal approximation is to be used (including the continuity correction), the position numbers for a $1 - \alpha$ confidence interval for M are found using the formula:

$$\frac{1}{2}(n) \pm \left(\frac{1}{2} + \frac{1}{2} \cdot z(\alpha/2) \cdot \sqrt{n}\right) \qquad (14.1)$$

The interval is

$$x_L \text{ to } x_U, \quad 1 - \alpha \text{ confidence interval for } M \text{ (median)}$$

where

$$L = \frac{n}{2} - \frac{1}{2} - z(\alpha/2) \cdot \sqrt{n} \qquad \text{and} \qquad U = \frac{n}{2} + \frac{1}{2} + z(\alpha/2) \cdot \sqrt{n}$$

Note: L should be rounded down and U should be rounded up to be sure that the level of confidence is at least $1 - \alpha$.

EXAMPLE 14.4

Constructing a Confidence Interval for a Population Median

Estimate the population median daily high temperature with a 95% confidence interval based on the following random sample of 60 daily high temperature readings. (Note: temperatures have been arranged in ascending order.)

43(x_1)	55(x_2)	59	60	67	73	73	73	73	73
73	75	75	76	78	78	78	79	79	80
80	80	80	80	80	80	82	82	82	82
83	83	83	83	83	84	84	84	85	85
86	86	87	87	88	88	88	88	88	88
88	89	89	89	89	90	92	93	94	98(x_{60})

SOLUTION When we use formula (14.1), the position numbers L and U are

$$\frac{1}{2}(n) \pm \left(\frac{1}{2} + \frac{1}{2} \cdot z(\alpha/2) \cdot \sqrt{n} \right): \quad \frac{1}{2}(60) \pm \left(\frac{1}{2} + \frac{1}{2} \cdot 1.96 \cdot \sqrt{60} \right)$$

$$30 \pm (0.50 + 7.59)$$

$$30 \pm 8.09$$

Thus,

$L = 30 - 8.09 = 21.91$, rounded down becomes 21 (21st data value)

$U = 30 + 8.09 = 38.09$, rounded up becomes 39 (39th data value)

Therefore,

80° to 85°, the 95% confidence interval for the median high daily temperature

Hypothesis Testing Procedure

When a hypothesis test is to be completed using the standard normal distribution, z will be calculated with the formula:

$$z\bigstar = \frac{x' - \dfrac{n}{2}}{\dfrac{1}{2} \cdot \sqrt{n}} \tag{14.2}$$

(See Note 3 on p. 759 with regard to x'.)

EXAMPLE 14.5

One-Tailed Hypothesis Test

Use the sign test to test the hypothesis that the median number of hours, M, worked by students of a certain college is at least 15 hours per week. A survey of 120 students was taken; a plus sign was recorded if the number of hours the student worked last week was equal to or greater than 15, and a minus sign was recorded if the number of hours was less than 15. Totals showed 80 minus signs and 40 plus signs.

SOLUTION

Step 1   **a. Parameter of interest:** M, the median number of hours worked by students.

   b. Statement of hypotheses:

   H_o: $M = 15$ ($\geq$)   (at least as many plus signs as minus signs)

   H_a: $M < 15$ (fewer plus signs than minus signs)

Step 2   **a. Assumptions:** The random sample of 120 adults was independently surveyed, and the variable, hours worked, is continuous.

   b. Probability distribution and test statistic: The standard normal z and formula (14.2).

   c. Level of significance: $\alpha = 0.05$.

Step 3   **a. Sample information:** $n(+) = 40$ and $n(-) = 80$; therefore, $n = 120$ and x is the number of plus signs; $x = 40$.

   b. Test statistic: Using formula (14.2), we have

$$z\star = \frac{x' - \dfrac{n}{2}}{\dfrac{1}{2} \cdot \sqrt{n}}:$$

$$z\star = \frac{40.5 - \dfrac{120}{2}}{\dfrac{1}{2} \cdot \sqrt{120}} = \frac{40.5 - 60}{\dfrac{1}{2} \cdot (10.95)} = \frac{-19.5}{5.475}$$

$$= -3.562 = -3.56$$

Step 4   **Probability Distribution:**

p-Value: **OR**   **Classical:**

a. Use the left-hand tail because H_a expresses concern for values related to "fewer than." **P** $= P(z < -3.56)$ as shown in the figure.

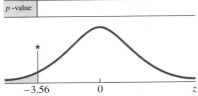

To find the *p*-value, you have three options:
1. Use Table 3 (Appendix B) to calculate the *p*-value: **P** $= 0.5000 - 0.4998 =$ **0.0002.**
2. Use Table 5 (Appendix B) to place bounds on the *p*-value: **P = 0.0002.**
3. Use a computer or calculator to find the *p*-value: **P = 0.0002.**

For specific instructions, see p. 432.

b. The *p*-value is smaller than α.

a. The critical region is the left-hand tail because H_a expresses concern for values related to "fewer than." The critical value is obtained from Table 4A:

$$-z(0.05) = -1.65$$

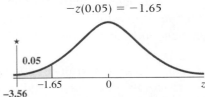

Specific instructions for finding critical values are on page 451.

b. $z\star$ is in the critical region, as shown in **red** in the figure.

FYI See pages 755 and 756 for computer and calculator commands.

Step 5   a. **Decision:** Reject H_o.
b. **Conclusion:** At the 0.05 level, there are significantly more minus signs than plus signs, thereby implying that the median is less than the claimed 15 hours.

SECTION 14.4 EXERCISES

Statistics⊘Now™

Datasets can be found on your Student's Suite CD-ROM or at the StatisticsNow website at **http://1pass.thomson.com**.

14.3 a. Describe why the sign test may be the easiest test procedure of all to use.

b. What population parameter can be tested using the sign test? What property of that parameter allows the sign test to be used? Explain.

14.4 [EX14-04] Ten randomly selected shut-ins were each asked how many hours of television they watched last week. The results are as follows:

| 82 | 66 | 90 | 84 | 75 | 88 | 80 | 94 | 110 | 91 |

Determine the 90% confidence interval estimate for the median number of hours of television watched per week by shut-ins.

14.5 [EX14-05] The following daily high temperatures (°F) were recorded in the city of Rochester, New York, on 20 randomly selected days in December.

| 47 | 46 | 40 | 40 | 46 | 35 | 34 | 59 | 54 | 33 |
| 65 | 39 | 48 | 47 | 46 | 46 | 42 | 36 | 45 | 38 |

Use the sign test to determine the 95% confidence interval for the median daily high temperature in Rochester, New York, during December.

14.6 [EX14-06] Fifteen peanut-producing counties were randomly identified, and the 2004 peanut yield rate, in pounds of peanuts harvested per acre, was recorded.

| 3755 | 2885 | 3080 | 3200 | 3035 | 3365 | 2850 | 3850 |
| 3455 | 3015 | 3310 | 2915 | 3630 | 3765 | 3050 |

Source: http://www.nass.usda.gov:81/ipedbcnty/report2.htm

Use the sign test to determine the 95% confidence interval for the median yield rate for peanuts.

14.7 [EX14-07] Every year sixth-grade students in Ohio schools take proficiency tests. The following list is of sixth-grade reading score changes from the prior year. Negative values indicate a decrease in score, positive values show an increase, and a zero shows no change from the prior year.

−4	−4	−10	−9	−30	6	18	−3	2	−5	−6
−12	−9	1	−1	−2	19	6	−1	−14	−13	5
12	−8	6	−3	−8	−14	−16	−6	2	0	16
−7	6	−11	6	−8	−4	13	9	−12	12	−10

Construct a 95% confidence interval for the median change in reading scores.

14.8 [EX14-08] A sample of the daily rental-car rates for a compact car was collected in order to estimate the average daily cost of renting a compact car.

39.93	41.00	42.99	38.99	42.93	35.00	40.95	29.99	49.93	50.95
34.95	28.99	43.93	43.00	41.99	42.99	36.93	34.95	35.99	31.99
45.93	46.50	34.90	29.80	32.93	29.70	32.99	27.94	53.93	46.00
35.94	34.99	29.93	28.70	34.99	31.48	37.93	37.90	37.92	35.99

Find the 99% confidence interval for the median daily rental cost.

14.9 State the null hypothesis, H_o, and the alternative hypothesis, H_a, that would be used to test the following statements:

a. The median length of vacation time is less than 18 days.

b. The median value is at least 32.

c. The median tax rate is 4.5%.

14.10 State the null hypothesis, H_o, and the alternative hypothesis, H_a, that would be used to test the following statements.

a. Post–self-esteem survey scores were higher than pre–self-esteem survey scores.

b. People prefer the taste of the bread made with the new recipe.

c. There is no change in weight from weigh-in until after 2 weeks of the diet.

14.11 Determine the p-value for the following hypothesis tests involving the sign test:

a. H_o: $P(+) = 0.5$ vs. H_a: $P(+) \neq 0.5$, with $n = 18$ and $x\star = n(-) = 3$

b. H_o: $P(+) = 0.5$ vs. H_a: $P(+) > 0.5$, with $n = 78$ and $x\star = n(-) = 30$

c. H_o: $P(+) = 0.5$ vs. H_a: $P(+) < 0.5$, with $n = 38$ and $x\star = n(+) = 10$

d. H_o: $P(+) = 0.5$ vs. H_a: $P(+) \neq 0.5$, with $n = 148$ and $z\star = -2.56$

14.12 Determine the critical value that would be used to test the null hypothesis for the following situations using the classical approach and the sign test:

a. H_o: $P(+) = 0.5$ vs. H_a: $P(+) \neq 0.5$, with $n = 18$ and $\alpha = 0.05$

b. H_o: $P(+) = 0.5$ vs. H_a: $P(+) > 0.5$, with $n = 78$ and $\alpha = 0.05$

c. H_o: $P(+) = 0.5$ vs. H_a: $P(+) < 0.5$, with $n = 38$ and $\alpha = 0.05$

d. H_o: $P(+) = 0.5$ vs. H_a: $P(+) \neq 0.5$, with $n = 148$ and $\alpha = 0.05$

14.13 An article titled "Graft-versus-Host Disease" (http://www.nature.com/bmt/journal/v35/n10/abs/1704957a.html) gives the median age of 42 years for the 87 patients with acute myeloid leukemia (AML) who received hematopoietic stem cell transplants from unrelated donors after standard conditioning. Clinical outcome after the use of two different antithymocyte globulins (ATG) for the prevention of graft-versus-host disease (GvHD) were then analyzed. Suppose that a sample of 100 patients with AML was recently selected for a study and it found that 40 of the patients were older than 42 and 60 were younger than 42 years of age. Test the null hypothesis that the median age of the population from which the 100 patients were selected equals 42 years versus the alternative that the median does not equal 42 years. Use $\alpha = 0.05$.

14.14 [EX14-07] Every year sixth-grade students in Ohio schools take proficiency tests. The list of sixth-grade reading score changes from the prior year are given in Exercise 14.7. Negative values indicate a decrease in score, positive values show an increase, and a zero shows no change from the prior year. Use the sign test to test the hypothesis that, "On average, reading scores have decreased from the prior year." Use $\alpha = 0.05$.

14.15 According to the June 13, 2005, *USA Today Weather Focus* article "Teen girls more careful in the sun," 48% of teen boys say they wear protective clothing to combat the sun's danger versus 52% of the teen girls. Suppose we wish to test the null hypothesis that one-half of all teen boys wear protective clothing to combat the sun's danger against the alternative hypothesis, "The proportion of teen boys who wear protective clothing to combat the sun's danger differs from one-half."

Furthermore, suppose we asked 75 randomly selected teen boys about their methods of combating the sun's danger. Let $+$ represent wears protective clothing and $-$ represent does not wear protective clothing. Do we have sufficient evidence to show the proportion of teen boys who wear protective clothing is different than one-half in the following cases? Explain.

a. We obtain 20 $(+)$ signs and 55 $(-)$ signs.

b. We obtain 27 $(+)$ signs and 48 $(-)$ signs.

c. We obtain 30 $(+)$ signs and 45 $(-)$ signs.

d. We obtain 33 $(+)$ signs and 42 $(-)$ signs.

14.16 [EX14-16] In "The Annual Report on the Economic Status of the Profession" done by American Association of University Professors, the mean salary of a full professor was reported to be \$83,282. The following table lists the average salary (in dollars) for a random sample of institutions in Colorado.

54,500 63,000 83,600 67,000 49,700 60,800 47,700 82,200 86,800 73,900
57,700 58,200 62,200 82,000 78,500 70,000 96,100 89,700 57,200 55,400

Using the computer output that follows, test the claim that the median salary of full professors in Colorado is lower than the mean for the whole country by writing the hypotheses and verifying the number below and the number above the median. Complete the test by using the *p*-value given and $\alpha = 0.05$. Verify the given *p*-value using Table 12 in Appendix B.

```
Sign Test for Median: C1
Sign test of median = 83282 versus < 83282
          N    Below   Equal   Above    P
C1       20     16       0       4    0.0059
```

14.17 [EX14-17] Part of the results from the Third International Mathematics and Science Study was a comparison of eighth-grade science achievement by nation from 1995, 1999, and 2003. The following table gives the average scores for the nations with all three years.

Nation	1995	1999	2003
Belgium-Flemish	533	535	516
Bulgaria	545	518	479
Cyprus	452	460	441
Hong Kong	510	530	556
Hungary	537	552	543
Iran, Islamic Republic of	463	448	453
Japan	554	550	552
Korea, Republic of	546	549	558
Latvian	476	503	513
Lithuania	464	488	519
Netherlands	541	545	536
New Zealand	511	510	520
Romania	471	472	470
Russian Federation	523	529	514
Singapore	580	568	578
Slovak Republic	532	535	517
United States	513	515	527

Source: http://nces.ed.gov/pubs2005/timss03/tables/table_11.asp

a. Construct a table showing the sign of the difference between the years 1995 and 1999 for each country.

b. Using $\alpha = 0.05$, has there been a significant improvement in science scores?

Repeat parts a and b for the years 1999 and 2003.

14.18 An article titled "Naturally Occurring Anticoagulants and Bone Marrow Transplantation: Plasma Protein C Predicts the Development of Venoocclusive Disease of the Liver" *(Blood)* compared baseline values for antithrombin III with antithrombin II values 7 days after a bone marrow transplant for 45 patients. The differences were found to be nonsignificant. Suppose 17 of the differences were positive and 28 were negative. The null hypothesis is that the median difference is zero, and the alternative hypothesis is that the median difference is not zero. Use the 0.05 level of significance. Complete the test and carefully state your conclusion.

14.19 A blind taste test was used to determine people's preference for the taste of the "classic" cola and "new" cola. The results were as follows:

645 preferred the new

583 preferred the old

272 had no preference

Is the preference for the taste of the new cola significantly greater than one-half? Use $\alpha = 0.01$.

14.20 A taste test was conducted with a regular beef pizza. Each of 133 individuals was given two pieces of pizza, one with a whole-wheat crust and the other with a white crust. Each person was then asked whether he or she preferred whole-wheat or white crust. The results were as follows:

65 preferred whole-wheat to white crust

53 preferred white to whole-wheat crust

15 had no preference

Is there sufficient evidence to verify the hypothesis that whole-wheat crust is preferred to white crust at the $\alpha = 0.05$ level of significance?

14.21 According to a local college's NEWS website (http://www.cincynet.cnyric.org/Computers/Career/news&quotes.htm), 60% of high school seniors have mastered basic math—decimals, fractions, percentages, and simple equations. Suppose we wish to test the null hypothesis, "One-half of all seniors can solve problems involving fractions, decimals, and percentages," against an alternative that the proportion who can solve is greater than one-half. Let + represent passed and − represent failed the test on fractions, decimals, and percentages. If a random sample of 1500 students is tested, what value of x, the number of the least frequent sign, will be the critical value at the 0.05 level of significance?

14.22 Fifty-one percent of grooms say they want to lose weight before their wedding day according to the June 30, 2005, USA Snapshot "Weigh-in before the wedding." Suppose we wish to test the hypothesis that at least one-half of grooms want to lose weight before their wedding. If a random sample of 900 grooms is surveyed, what value of x, the number of the least frequent sign, will be the critical value at the 0.05 level of significance?

14.5	# The Mann–Whitney *U* Test

The Mann–Whitney U test is a nonparametric alternative for the t-test for the difference between two independent means. The usual two-sample situation occurs when the experimenter wants to see whether the difference between the two samples is sufficient to reject the null hypothesis that the two sampled populations are identical.

Hypothesis-Testing Procedure

> **Assumptions for inferences about two populations using the Mann–Whitney *U* Test:** The two **independent random samples** are independent within each sample as well as between samples, and the random variables are ordinal or numerical.

This test is often used in situations in which the two samples are drawn from the same population of subjects but different "treatments" are used on each set. We will demonstrate the procedure in the next example.

EXAMPLE 14.6

Statistics ⌂ Now™

Watch a video example at **http://1pass.thomson.com** or on your CD.

Two-Tailed Hypothesis Test

In a large lecture class, when a 1-hour exam is given, the instructor gives two "equivalent" examinations. It is reasonable to ask: Are these two different exams equivalent? Students in even-numbered seats take exam A, and those in the odd-numbered seats take exam B. To test this "equivalent" hypothesis, two random samples were taken. Table 14.3 lists the exam scores of the two samples.

TABLE 14.3

Data on Exam Scores [TA14-03]

Exam A	52	78	56	90	65	86	64	90	49	78
Exam B	72	62	91	88	90	74	98	80	81	71

If we assume that the odd- or even-numbered seats had no effect, does the sample present sufficient evidence to reject the hypothesis: The exam forms yielded scores that had identical distributions? Test using $\alpha = 0.05$.

SOLUTION

Step 1 a. **Parameter of interest:** The distribution of scores for each version of the exam.

   b. **Statement of hypotheses:**

H_o: Exam A and exam B have test scores with identical distributions.

H_a: The two distributions are not the same.

Step 2 a. **Assumptions:** The two samples are independent, and the random variable, exam score, is numerical.

   b. **Test statistic:** The Mann–Whitney U statistic.

   c. **Level of significance:** $\alpha = 0.05$.

Step 3 a. **Sample information:** The sample data are listed in Table 14.3.

   b. **Test statistic.**

The size of the individual samples will be called n_a and n_b; actually, it makes no difference which way these are assigned. In our example they both have the value 10. The two samples are combined into one sample (all $n_a + n_b$) and ordered from smallest to largest:

49	52	56	62	64	65	71	72	74	78
78	80	81	86	88	90	90	90	91	98

Each is then assigned a **rank** number. The smallest (49) is assigned rank 1, the next smallest (52) is assigned rank 2, and so on, up to the largest, which is assigned rank $n_a + n_b$ (20). Ties are handled by assigning to each of the tied observations the mean rank of those rank positions that they occupy. For example, in our example there are two 78s; they are the 10th and 11th. The mean rank for each is then $\dfrac{10+11}{2} = 10.5$. In the case of the three 90s—the 16th, 17th, and 18th data values—each is assigned 17 because $\dfrac{16 + 17 + 18}{3} = 17$. The rankings are shown in Table 14.4.

TABLE 14.4

Ranked Exam Score Data

Ranked Data	Rank	Source	Ranked Data	Rank	Source	Ranked Data	Rank	Source
49	1	A	72	8	B	98	20	B
52	2	A	74	9	B	91	19	B
56	3	A	78	10.5	A	90	17	B
62	4	B	78	10.5	A	90	17	A
64	5	A	80	12	B	90	17	A
65	6	A	81	13	B	88	15	B
71	7	B	86	14	A			

Figure 14.2 shows the relationship between the two sets of data, first by using the data values and second by comparing the rank numbers for the data.

FIGURE 14.2
Comparing the Data of Two Samples

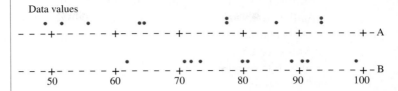

The calculation of the **test statistic U** is a two-step procedure. We first determine the sum of the ranks for each of the two samples. Then, using the two sums of ranks, we calculate a U score for each sample. The smaller U score is the test statistic.

The sum of ranks R_a for sample A is computed as

$$R_a = 1 + 2 + 3 + 5 + 6 + 10.5 + 10.5 + 14 + 17 + 17 = \mathbf{86}$$

The sum of ranks R_b for sample B is

$$R_b = 4 + 7 + 8 + 9 + 12 + 13 + 15 + 17 + 19 + 20 = \mathbf{124}$$

The U score for each sample is obtained by using the following pair of formulas:

Mann–Whitney U Test Statistic

$$U_a = n_a \cdot n_b + \frac{(n_b)(n_b + 1)}{2} - R_b \qquad (14.3)$$

$$U_b = n_a \cdot n_b + \frac{(n_a)(n_a + 1)}{2} - R_a \qquad (14.4)$$

$U\bigstar$, the test statistic, is the smaller of U_a and U_b.

For our example, we obtain

$$U_a = (10)(10) + \frac{(10)(10 + 1)}{2} - 124 = 31$$

$$U_b = (10)(10) + \frac{(10)(10 + 1)}{2} - 86 = 69$$

Therefore, $U\bigstar = \mathbf{31}$.

Before we carry out the test for this example, let's try to understand some of the underlying possibilities. Recall that the null hypothesis is that the distributions are the same and that we will most likely want to conclude from this that the averages are approximately equal. Suppose for a moment that the distributions are indeed quite different; say, all of one sample comes before the smallest data value in the second sample when they are ranked together. This would certainly mean that we want to reject the null hypothesis. What kind of a value can we expect for U in this case? Suppose that the 10 A values had ranks 1 through 10 and the 10 B values had ranks 11 through 20. Then we would obtain

$$R_a = 55 \quad \text{and} \quad R_b = 155$$

$$U_a = (10)(10) + \frac{(10)(10 + 1)}{2} - 155 = 0$$

$$U_b = (10)(10) + \frac{(10)(10 + 1)}{2} - 55 = 100$$

Therefore, $U\bigstar = \mathbf{0}$.

If this were the case, we certainly would want to reach the decision: Reject the null hypothesis.

Suppose, on the other hand, that both samples were perfectly matched; that is, a score in each set is identical to one in the other.

54	54	62	62	71	71	72	72	...
A	B	A	B	A	B	A	B	...
1.5	1.5	3.5	3.5	5.5	5.5	7.5	7.5	...

Now what would happen?

$$R_a = R_b = 105$$

$$U_a = U_b = (10)(10) + \frac{(10)(10 + 1)}{2} - 105 = 50$$

Therefore, $U\bigstar = \mathbf{50}$. If this were the case, we certainly would want to reach the decision: Fail to reject the null hypothesis.

Note: The sum of the two U's ($U_a + U_b$) will always be equal to the product of the two sample sizes ($n_a \cdot n_b$). For this reason we need only to concern ourselves with the smaller U value.

Now, let's return to the solution of Example 14.6.

Step 4   Probability Distribution:

p-**Value:**   OR   **Classical:**

a. Since the concern is for values related to "not the same," the *p*-value is the probability of both tails. It will be found by finding the probability of the left tail and doubling:

$$\mathbf{P} = 2 \times P(U \leq 31 \text{ for } n_1 = 10 \text{ and } n_2 = 10)$$

a. The critical region is two-tailed because H_a expresses concern for values related to "not the same." Use Table 13A for two-tailed $\alpha = 0.05$. The critical value is at the intersection of column $n_1 = 10$ and row $n_2 = 10$: **23.** The critical region is $U \leq 23$.

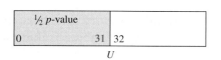

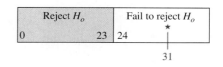

To find the *p*-value, you have two options:
1. Use Table 13 in Appendix B to place bounds on the *p*-value: **P > 0.10.**
2. Use a computer or calculator to find the *p*-value: **P = 0.1612.**

Specific instructions follow this example.

b. The *p*-value is not smaller than α.

b. $U\star$ is not in the critical region, as shown in the figure.

Step 5	**a. Decision:** Fail to reject H_o.
	b. Conclusion: We do not have sufficient evidence to reject the "equivalent" hypothesis.

Calculating the *p*-value when using the Mann–Whitney test

Method 1: Use Table 13 in Appendix B to place bounds on the p-value. By inspecting Table 13A and B at the intersection of column $n_1 = 10$ and row $n_2 = 10$, you can determine that the *p*-value is greater than 0.10; the larger two-tailed value of α is 0.10 in Table 13B.

Method 2. If you are doing the hypothesis test with the aid of a computer or graphing calculator, most likely it will calculate the *p*-value for you. Specific instructions are described on pages 771–772.

Normal Approximation

If the samples are larger than size 20, we may make the test decision with the aid of the standard normal variable, *z*. This is possible because the distribution of *U* is approximately normal with a mean

$$\mu_U = \frac{n_a \cdot n_b}{2} \tag{14.5}$$

and a standard deviation

$$\sigma_U = \sqrt{\frac{n_a \cdot n_b \cdot (n_a + n_b + 1)}{12}} \tag{14.6}$$

The hypothesis test is then completed using the **test statistic $z\star$:**

$$z\star = \frac{U\star - \mu_U}{\sigma_U} \tag{14.7}$$

The standard normal distribution may be used whenever n_a and n_b are both greater than 10.

The normal approximation procedure for the Mann–Whitney *U* test is demonstrated in Example 14.7.

EXAMPLE 14.7   ## One-Tailed Hypothesis Test

A dog-obedience trainer is training 27 dogs to obey a certain command. The trainer is using two different training techniques: (I) the reward-and-encouragement method and (II) the no-reward method. Table 14.5 shows the numbers of obedience sessions that were necessary before the dogs would obey the command. Does the trainer have sufficient evidence to claim that the reward method will, on average, require fewer obedience sessions ($\alpha = 0.05$)?

TABLE 14.5

Data on Dog Training [TA14-05]

Method I	29	27	32	25	27	28	23	31	37	28	22	24	28	31	34
Method II	40	44	33	26	31	29	34	31	38	33	42	35			

SOLUTION

Step 1 a. **Parameter of interest:** The distribution of needed obedience sessions for each technique.

 b. **Statement of hypotheses:**
 H_o: The distributions of the needed obedience sessions are the same for both methods.
 H_a: The reward method, on average, requires fewer sessions.

Step 2 a. **Assumptions:** The two samples are independent, and the random variable, training time, is numerical.

 b. **Test statistic:** The Mann–Whitney U statistic.

 c. **Level of significance:** $\alpha = 0.05$.

Step 3 a. **Sample information:** The sample data are listed in Table 14.5.

 b. **Test statistic:** The two sets of data are ranked jointly and ranks are assigned as shown in Table 14.6.

TABLE 14.6

Rankings for Training Methods

Number of Sessions	Group	Rank		Number of Sessions	Group	Rank	
22	I	1		31	II	15	14.5
23	I	2		31	II	16	14.5
24	I	3		32	I	17	
25	I	4		33	II	18	18.5
26	II	5		33	II	19	18.5
27	I	6	6.5	34	I	20	20.5
27	I	7	6.5	34	II	21	20.5
28	I	8	9	35	II	22	
28	I	9	9	37	I	23	
28	I	10	9	38	II	24	
29	I	11	11.5	40	II	25	
29	II	12	11.5	42	II	26	
31	I	13	14.5	44	II	27	
31	I	14	14.5				

The sums are:

$$R_{\mathrm{I}} = 1 + 2 + 3 + 4 + 6.5 + \cdots + 20.5 + 23 = 151.0$$

$$R_{\mathrm{II}} = 5 + 11.5 + 14.5 + \cdots + 26 + 27 = 227.0$$

The U scores are found using formulas (14.3) and (14.4):

$$U_{\mathrm{I}} = (15)(12) + \frac{(12)(12+1)}{2} - 227 = 180 + 78 - 227 = 31$$

$$U_{\mathrm{II}} = (15)(12) + \frac{(15)(15+1)}{2} - 151 = 180 + 120 - 151 = 149$$

Therefore, $U\bigstar = \mathbf{31}$. Now we use formulas (14.5), (14.6), and (14.7) to determine the z statistic:

$$\mu_U = \frac{n_a \cdot n_b}{2}: \quad \mu_U = \frac{12 \cdot 15}{2} = 90$$

$$\sigma_U = \sqrt{\frac{n_a \cdot n_b \cdot (n_a + n_b + 1)}{12}}: \quad \sigma_U = \sqrt{\frac{12 \cdot 15 \cdot (12 + 15 + 1)}{12}}$$

$$= \sqrt{\frac{(180)(28)}{12}} = \sqrt{420} = 20.49$$

$$z\bigstar = \frac{U\bigstar - \mu_U}{\sigma_U}: \quad z\bigstar = \frac{31 - 90}{20.49} = \frac{-59}{20.49} = -2.879 = \mathbf{-2.88}$$

Step 4   **Probability Distribution:**

p-Value:

a. Use the left-hand tail because H_a expresses concern for values related to "fewer than." $\mathbf{P} = P(z < -2.88)$ as shown in the figure.

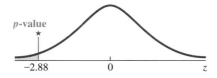

To find the p-value, you have three options:
1. Use Table 3 (Appendix B) to calculate the p-value: $\mathbf{P} = 0.5000 - 0.4980 = \mathbf{0.0020}$.
2. Use Table 5 (Appendix B) to place bounds on the p-value: $\mathbf{0.0019 < P < 0.0022}$.
3. Use a computer or calculator to find the p-value: $\mathbf{P} = \mathbf{0.0020}$.

For specific instructions, see page 432.
b. The p-value is smaller than α.

OR

Classical:

a. The critical region is the left-hand tail because H_a expresses concern for values related to "fewer than." The critical value is obtained from Table 4A:

$$-z(0.05) = \mathbf{-1.65}$$

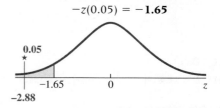

Specific instructions for finding critical values are on page 451.
b. $z\bigstar$ is in the critical region, as shown in red in the figure.

Step 5   **a. Decision:** Reject H_o.

   b. Conclusion: At the 0.05 level of significance, the data show sufficient evidence to conclude that the reward method does, on average, require fewer training sessions.

TECHNOLOGY INSTRUCTIONS: MANN–WHITNEY *U* TEST FOR THE DIFFERENCE BETWEEN TWO INDEPENDENT DISTRIBUTIONS

MINITAB (Release 14)   Input the two independent sets of data into C1 and C2; then continue with:

Choose: **Stat > Nonparametrics > Mann–Whitney**

Enter: First Sample: **C1**   Second Sample: **C2**
              Confidence level: 1−α
Select: Alternative: **less than or not equal or greater than** > **OK**

With respect to the *p*-value approach, the *p*-value is given. With respect to the classical approach, just the sum of the ranks for one of the samples, *W*, is given. Use this to find *U* for that one sample. The *U* for the other sample is found by subtracting *U* from the product of n_1 and n_2.

Excel   Input the two independent sets of data into column A and column B; then continue with:

Choose: **Tools** > **Data Analysis Plus** > **Wilcoxon Rank Sum Test***
Enter: Variable 1 Range: **(A1:A20 or select cells)**
              Variable 2 Range: **(B1:B20 or select cells)**

*The Wilcoxon rank sum test is equivalent to the Mann–Whitney test.

Select: **Labels** (if necessary)
Enter: Alpha: **α** (ex. 0.05)

The sum of the ranks is given for both samples and also the *p*-value.

TI-83/84 Plus   Input the two independent sets of data into L1 and L2; then continue with:

Choose: **PRGM** > **EXEC** > **MANNWHIT**
Enter: XLIST: **L1**
              YLIST: **L2**
              NULL HYPOTHESIS D0 = difference amount (ex. 0)
Select: ALT HYP? **1:U1-U2** > **D0** or **2:U1-U2** < **D0** or **3:U1-U2** ≠ **D0**

*Program MANNWHIT is one of many programs that are available for downloading from http://www.duxbury.com. See page 42 for specific directions.

APPLIED EXAMPLE 14.8

Sea Otters

© Kennan Ward/CORBIS

QUANTITATIVE ASSESSMENT OF SEA OTTER BENTHIC PREY COMMUNITIES WITHIN THE OLYMPIC COAST NATIONAL MARINE SANCTUARY: 1999 RE-SURVEY OF 1995 AND 1985 MONITORING STATIONS

This report summarizes the changes in the distribution and abundance of selected benthic species within sea otter prey communities along the Washington State Olympic coast between 1987 and 1999. During this 12-year period, the Washington otter population has undergone a dramatic increase in both numbers and range, now occupying habitats that were otter free when first sampled in 1987. Invertebrate prey such as com-

mercially harvested sea urchins that were abundant just outside the boundaries of the 1987 sea otter range are now virtually absent along the entire outer rocky coast. Understory foliose red, coralline, and brown algal cover have also undergone changes as otters removed large invertebrate grazers from the newly occupied habitats. In 1995 a test comparison was conducted at Chibahdehl Rocks to compare invertebrate size and abundance data collected using both methods. Results showed no significant difference (*t*-tests, $p = 0.32$ and 0.24 for abundance and size, respectively).

Hypotheses

H_1: As the Washington State sea otter population continues to grow, it will expand north, drawn by and depleting the rich prey resources found there.

H_2: If sea otters move into northern habitats, significant changes in benthic algal cover will occur with reduced abundance of sea urchins and other invertebrate grazers.

H_3: Sea otters will be slower to colonize areas with higher water velocities, resulting in a higher prey biomass in those areas.

Results

For 1999, there was no significant difference in prey abundance between sites. Foliose red, coralline, and brown algal cover were followed at three sites, Neah Bay, Anderson Pt. and Cape Alava for all years. The only significant difference in foliose red cover between 1995 and 1999 was the decline at Anderson Pt. (Mann–Whitney *U* test $p < 0.0001$). Coralline cover continued to drop dramatically and significantly at Neah Bay (100%, 44%, 1%) (Mann–Whitney *U* test $p < 0.0001$) and at Anderson Pt. (18%, 17%, 6%) (Mann–Whitney *U* test $p <$ 0.0001), while fluctuating slightly but significantly at Cape Alava (Mann–Whitney *U* test $p = 0.0006$). Brown algae has increased steadily and significantly from 0% to 33% at Neah Bay since 1987 (Mann–Whitney *U* test $p = 0.009$), fluctuated significantly between 4% and 34% at Anderson Point (Mann–Whitney *U* test $p < 0.0001$), and did not change significantly at Cape Alava (Mann–Whitney *U* test $p = 0.20$).

Conclusions Otter numbers have increased within their range since 1987, and their range has expanded to the north as predicted (H1). Prey abundance and biomass have declined by an order of magnitude to very low levels at newly otter-occupied sites on either side of Cape Flattery by 1995, also as predicted (H2). By 1999, the high prey numbers and biomass found at Cape Flattery and Tatoosh Island in 1995 had also dropped to levels comparable with the other monitoring site, refuting the high current prey refuge hypothesis (H3). The removal of urchin grazers by sea otters was most likely responsible for the rise in cover of more palatable algae at the recently occupied Neah Bay and Anderson Pt. sites. The most dramatic change in algal cover occurred at Neah Bay, the site that experienced the greatest decline in urchin abundance following the movement of sea otters into the area.

Source: Rikk Kvitek, Pat Iampietro, and Kate Thomas, California State University Monterey Bay, Seaside, CA, http://seafloor.csumb.edu/publications/posters/OCNMS.pdf. Reprinted with permission.

SECTION 14.5 EXERCISES

Statistics ⏅ Now™

Datasets can be found on your Student's Suite CD-ROM or at the StatisticsNow website at **http://1pass.thomson.com**.

14.23 a. What parametric test procedure is comparable to the Mann–Whitney *U* test?

b. What characteristic of the data used in a parametric test is not used in the Mann–Whitney *U* test?

14.24 Consider the side-by-side dotplots for the data values and for the ranks in Figure 14.2 on page 767. Do you see a different relationship between the two sets of data? Explain.

14.25 State the null hypothesis, H_o, and the alternative hypothesis, H_a, that would be used to test the following statements:

a. There is a difference in the distributions of the variable between the two groups of subjects.

b. The average value is not the same for both groups.

c. The distribution of blood pressure for group A is higher than that for group B.

14.26 State the null hypothesis, H_o, and the alternative hypothesis, H_a, that would be used to test the following statements:

a. The students in the new reading program scored higher on the comprehension test than the students in the traditional reading programs.

b. Men on the grapefruit diet lose more weight than men not on the grapefruit diet.

c. There is no difference in growth between the use of the two fertilizers.

14.27 Determine the *p*-value that will result when testing the following hypotheses for experiments involving two independent samples:

a. H_o: Average(A) = Average(B)
   H_a: Average(A) > Average(B)
   with $n_A = 18$, $n_B = 15$, and $U = 95$.

b. H_o: Average(I) = Average(II)
   H_a: Average(I) ≠ Average(II)
   with $n_A = 8$, $n_B = 10$, and $U = 13$.

c. H_o: The average height is the same for both groups.
   H_a: Group I average heights are less than those for group II.
   with $n_I = 50$, $n_{II} = 45$, and $z = -2.37$.

14.28 Of the seven Mann–Whitney U test *p*-values given in Applied Example 14.8 on page 772, six are less than 0.001 and the seventh is 0.20. Explain how these *p*-values relate to statements containing phrases such as *significant, drop dramatically, increased steadily,* and *did not change significantly.*

14.29 Determine the critical value that would be used to test the following hypotheses for experiments involving two independent samples, using the classical method:

a. H_o: Average(A) = Average(B)
   H_a: Average(A) > Average(B)
   with $n_A = 18$, $n_B = 15$, and $\alpha = 0.05$.

b. H_o: The average score is the same for both groups.
   H_a: Group I average scores are less than those for group II.
   with $n_I = 78$, $n_{II} = 45$, and $\alpha = 0.05$.

14.30 The Oregon Health & Science University's news website (http://www.ohsu.edu/news/2003/071803smoke.html) gives information on a study that found that some commercial cigarette brands contained 10 to 20 times higher percentages of nicotine in the "free-base" form, that is, the form thought to be the most addictive. Consider another study designed to compare the nicotine content of two different brands of cigarettes. The nicotine content was determined for 25 cigarettes of brand A and 25 cigarettes of brand B. The sum of ranks for brand A equals 688, and the sum of ranks for brand B equals 587. Use the Mann–Whitney U statistic to test the null hypothesis that the average nicotine content is the same for the two brands versus the alternative that the average nicotine content differs. Use $\alpha = 0.01$.

14.31 [EX14-31] Pulse rates were recorded for 16 men and 13 women. The results are shown in the following table.

Males	61	73	58	64	70	64	72	60	65	80	55	72	56	56	74	65
Females	83	58	70	56	76	64	80	68	78	108	76	70	97			

These data were used to test the hypothesis that the distribution of pulse rates differs for men and women. The following MINITAB output printed out the sum of ranks for males ($W = 192.0$) and the *p*-value of 0.0373. Verify these two values by calculating them yourself.

```
Mann-Whitney Confidence Interval and Test
Males          N = 16      Median = 64.50
Females        N = 13      Median = 76.00
W = 192.0
Test of ETA1 = ETA2      vs      ETA1 not = ETA2 is
significant at 0.0373
```

14.32 [EX14-32] The Ohio State Proficiency test results for Toledo, Ohio, fourth-grade students was the highest recorded since the start of the statewide proficiency testing. Although the results were an improvement districtwide, in some subjects there was not as much improvement as for other subjects. The results that follow show the amount of change for reading and writing. Change in scores is indicated by positive for improvement, negative for lower scores, and zero for no change.

Writing	2	0	3	30	10	25	7	17	2
	6	15	−9	−2	6	13	−5	−5	10
	24	6	29	−4	27	16	1	−4	−8
	−6	13	8	5	−23	3	14	−1	7
	16	−12	10	42	−2	4	8	38	24
Reading	23	25	2	6	40	3	3	32	−2
	8	28	−1	8	5	34	−6	7	6
	34	6	19	27	23	6	46	23	35
	−4	10	11	31	−13	10	20	10	−10
	−5	17	22	20	19	11	13	3	21

Using the following Excel output, test the claim that there was equal improvement in fourth-grade writing results as for reading. Use $\alpha = 0.05$. (*Note:* The Wilcoxon rank sum test is equivalent to the Mann–Whitney U test.)

Wilcoxon Rank Sum Test

	Rank Sum	Observations
Writing	1798.5	45
Reading	2296.5	45
z Stat	−2.0094	
P(Z ≤ z) two-tail	0.0444	
z Critical two-tail	1.96	

14.33 A study titled "Factors Leading to Reduced Intraocular Pressure after Combined Trabeculotomy and Cataract Surgery" in the *Journal of Glaucoma* (http://www.glaucomajournal.com/pt/re/jglaucoma/abstract) investigated the influence cataract surgery alone and cataract surgery with trabeculotomy has on eye pressure. Two groups were formed for each type of surgery and were compared for similarities with respect to several factors beforehand. No significant difference was found between the two groups with respect to the number of preoperative glaucoma medications the patient was taking. Suppose a similar study involving six patients receiving the combined surgery

and five patients receiving the cataract surgery alone produced the following values of number of medications.

| Combined Surgery | 3 | 1 | 4 | 0 | 1 | 2 |
| Cataract Surgery Only | 3 | 1 | 0 | 1 | 2 | |

Using the Mann–Whitney U test, determine whether the two groups are the same with respect to number of medications. Use $\alpha = 0.05$.

14.34 [EX14-34] An article in the *International Journal of Sports Medicine* discusses the use of the Mann–Whitney U test to compare the total cholesterol (mg/dL) of 35 adipose (obese) boys with that of 27 adipose girls. No significant difference was found between the two groups with respect to total cholesterol. A similar study involving six adipose boys and eight adipose girls gave the following total cholesterol values.

| Adipose Boys | 175 | 185 | 160 | 200 | 170 | 150 | | |
| Adipose Girl | 160 | 190 | 175 | 190 | 185 | 150 | 140 | 195 |

Use the Mann–Whitney U test to test the research hypothesis that the total cholesterol values differ for the two groups, using the 0.05 level of significance.

14.35 [EX14-35] As part of a study to determine whether cloud seeding increased rainfall, clouds were randomly seeded or not seeded with silver nitrate. The amounts of rainfall that followed are listed here.

Unseeded	4.9	41.1	21.7	372.4	26.3	17.3	36.6	26.1
	47.3	95.0	147.8	321.2	11.5	68.5	29.0	24.4
	1202.6	87.0	28.6	830.1	81.2	4.9	163.0	345.5
	244.3							
Seeded	129.6	334.1	274.7	198.6	430.0	274.7	31.4	115.3
	1656.0	118.3	489.1	302.8	255.0	32.7	119.0	17.5
	242.5	2745.6	7.7	40.6	978.0	200.7	703.4	92.4
	1697.8							

Do these data show that cloud seeding will significantly increase the average amount of rainfall? Use $\alpha = 0.05$.

14.36 [EX14-36] Fifteen Texas and eleven Oklahoma peanut-producing counties were randomly identified, and the 2004 peanut yield rate, in pounds of peanuts harvested per acre, was recorded.

OK County	OK Yield	TX County	TX Yield
Love	2500	Donley	3230
Bryan	2600	Terry	3010
Tillman	2685	Collingsworth	2040
Jackson	2700	Haskell	3100
Caddo	2835	Frio	3480
Washita	2855	Childress	2000
Greer	2865	Bailey	2765

OK County	OK Yield	TX County	TX Yield
Grady	2875	Wheeler	2700
Harmon	2890	Hall	2785
Custer	2900	Hockley	3245
Beckham	3050	Andrews	3635
		Motley	2205
		Gaines	3485
		Dawson	3310
		Mason	3700

Source: http://www.nass.usda.gov:81/ipedbcnty/report2.htm

Use the Mann–Whitney U statistic to test the hypothesis that the average yield is different for the two states. Use $\alpha = 0.05$.

14.6 The Runs Test

The runs test is used most frequently to test the **randomness** of data (or lack of randomness). A **run** is a sequence of data that possesses a common property. One run ends and another starts when an observation does not display the property in question. The **test statistic** in this test is V, the number of runs observed.

The following example illustrates what constitutes a run and how to count the number of runs.

EXAMPLE 14.9 | **Determining the Number of Runs**

To illustrate the idea of runs, let's draw a sample of 10 single-digit numbers from the telephone book, listing the next-to-last digit from each of the selected telephone numbers:

Sample: 2 3 1 1 4 2 6 6 6 7

Let's consider the property of "odd" (o) or "even" (e). The sample, as it was drawn, becomes e, o, o, o, e, e, e, e, e, o, which displays four runs:

e   o o o   e e e e e   o

Thus, $V\bigstar = 4$.

In Example 14.9, if the sample contained no randomness, there would be only two runs—all the evens, then all the odds, or the other way around. We would also not expect to see them alternate—odd, even, odd, even. The maximum number of possible runs would be $n_1 + n_2$ or less (provided n_1 and n_2 are not equal), where n_1 and n_2 are the numbers of data that have each of the two properties being identified.

> **Assumption for inferences about randomness using the runs test:** Each sample data can be classified into one of two categories.

The runs test is generally a two-tailed test. We will reject the hypothesis when there are too few runs because this indicates that the data are "separated" according to the two properties. We will also reject when there are too many runs because that indicates that the data alternate between the two properties too often to be random. For example, if the data alternated all the way down the line, we might suspect that the data had been tampered with. There are many aspects to the concept of randomness. The occurrence of odd and even as discussed in Example 14.9 is one aspect. Another aspect of randomness that we might wish to check is the ordering of fluctuations of the data above or below the mean or median of the sample.

EXAMPLE 14.10

Hypothesis Test for Randomness

Consider the following sample and determine whether the data points form a random sequence with regard to being above or below the median value.

2	5	3	8	4	2	9	3	2	3	7	1	7	3	3
6	3	4	1	9	5	2	5	5	2	4	3	4	0	4

Test the null hypothesis that this sequence is random. Use $\alpha = 0.05$.

SOLUTION

Step 1 **a. Parameter of interest:** Randomness of the values above or below the median.

      b. Statement of hypotheses:

        H_o: The numbers in the sample form a random sequence with respect to the two properties "above" and "below" the median value.

        H_a: The sequence is not random.

Step 2 **a. Assumptions:** Each sample data value can be classified as "above" or "below" the median.

      b. Test statistic: V, the number of runs in the sample data.

      c. Level of significance: $\alpha = 0.05$

Step 3 **a. Sample information:** The sample data are listed at the beginning of the example.

      b. Test statistic: First we must rank the data and find the median. The ranked data are

0	1	1	2	2	2	2	2	3	3	3	3	3	3	3
4	4	4	4	4	5	5	5	5	6	7	7	8	9	9

Since there are 30 data values, the depth of the median is at the $d(\tilde{x}) = 15.5$ position. Thus, $\tilde{x} = \dfrac{3+4}{2} = 3.5$. By comparing each number in the original sample to the value of the median, we obtain the following sequence of **a**'s (above) and *b*'s (below):

$$b\ a\ b\ a\ a\ b\ a\ b\ b\ b\ a\ b\ a\ b\ b\ a\ b\ a\ b\ a\ b\ a\ a\ b\ a\ a\ b\ a\ b\ a\ b\ a$$

We observe $n_a = 15$, $n_b = 15$, and 24 runs. So $V\star = 24$.

If n_1 and n_2 are both less than or equal to 20 and a two-tailed test at $\alpha = 0.05$ is desired, then Table 14 in Appendix B is used to complete the hypothesis test.

Step 4 Probability Distribution:

OR

p-Value:

a. Since the concern is for values related to "not random," the test is two-tailed. The *p*-value is found by finding the probability of the right tail and doubling:

$$\mathbf{P} = 2 \times P(V \geq 24 \text{ for } n_a = 15 \text{ and } n_b = 15)$$

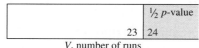

		½ *p*-value
	23	24

V, number of runs

To find the *p*-value, you have two options:
1. Use Table 14 (Appendix B) to place bounds on the *p*-value: **P < 0.05.**
2. Use a computer or calculator to find the *p*-value: **P = 0.003.**

Specific instructions follow this example.

b. The *p*-value is smaller than α.

Classical:

a. Since the concern is for values related to "not random," the test is two-tailed. Use Table 14 for two-tailed $\alpha = 0.05$. The critical values are at the intersection of column $n_1 = 15$ and row $n_2 = 15$: 10 and 22. The critical region is $V \leq 10$ or $V \geq 22$.

Reject H_o	Fail to reject H_0	Reject H_o ★
10	11 21	22

V, number of runs

24

b. $V\star$ is in the critical region, as shown in the figure.

Step 5

a. Decision: Reject H_o.

b. Conclusion: We are able to reject the hypothesis of randomness at the 0.05 level of significance and conclude that the sequence is not random with regard to above and below the median.

Calculating the *p*-value when using the runs test

Method 1: Use Table 14 in Appendix B to place bounds on the p-value. By inspecting Table 14 at the intersection of column $n_1 = 15$ and row $n_2 = 15$, you can determine that the *p*-value is less than 0.05; the observed value of $V\star = 24$ is larger than the larger critical value listed.

Method 2. If you are doing the hypothesis test with the aid of a computer or graphing calculator, most likely it will calculate the *p*-value for you. Specific instructions are given on page 780–781.

Normal Approximation

To complete the hypothesis test about randomness when n_1 and n_2 are larger than 20 or when α is other than 0.05, we will use *z*, the standard normal random variable. *V* is approximately normally distributed with a mean of μ_V and a standard deviation of σ_V. The formulas for the mean and standard deviation of the V statistic and the test statistic $z\star$ follow:

$$\mu_V = \frac{2n_1 \cdot n_2}{n_1 + n_2} + 1 \tag{14.8}$$

$$\sigma_V = \sqrt{\frac{(2n_1 \cdot n_2) \cdot (2n_1 \cdot n_2 - n_1 - n_2)}{(n_1 + n_2)^2(n_1 + n_2 - 1)}} \tag{14.9}$$

$$z\star = \frac{V\star - \mu_V}{\sigma_V} \tag{14.10}$$

EXAMPLE 14.11

Two-Tailed Hypothesis Test for Randomness

Test the null hypothesis that the sequence of sample data in Table 14.7 is a random sequence with regard to each data value being odd or even. Use $\alpha = 0.10$.

TABLE 14.7

Sample Data for Example 14.11 [TA14-07]

1	2	3	0	2	4	3	4	8	1
2	1	2	4	3	9	6	2	4	1
5	6	3	3	2	2	1	2	4	2
3	6	3	5	1	7	3	3	0	1
4	4	1	2	7	2	1	7	5	3

SOLUTION

Step 1   a. **Parameter of interest:** Randomness of odd and even numbers.

        b. **Statement of hypotheses:**

          H_o: The sequence of odd and even numbers is random.
          H_a: The sequence is not random.

Step 2   a. **Assumptions:** Each sample value can be classified as either odd or even.

        b. **Test statistic:** V, the number of runs in the sample data.

        c. **Level of significance:** $\alpha = 0.10$.

Step 3   a. **Sample information:** The data are given at the beginning of the illustration.

        b. **Test statistic:** The sample data when converted to "o" for odd and "e" for even become

o e o e e e o e e o e o e e o o e e e o o e o o e

e o e e e o e o o o o o o o e o e e e o e o e o o o o

and reveals: $n_o = 26$, $n_e = 24$, and 29 runs, so $V\star = 29$. Now use formulas (14.8), (14.9), and (14.10) to determine the z statistic:

$$\mu_V = \frac{2n_1 \cdot n_2}{n_1 + n_2} + 1: \quad \mu_V = \frac{2 \cdot 26 \cdot 24}{26 + 24} + 1 = 24.96 + 1 = 25.96$$

$$\sigma_V = \sqrt{\frac{(2n_1 \cdot n_2) \cdot (2n_1 \cdot n_2 - n_1 - n_2)}{(n_1 + n_2)^2(n_1 + n_2 - 1)}}:$$

$$\sigma_V = \sqrt{\frac{(2 \cdot 26 \cdot 24) \cdot (2 \cdot 26 \cdot 24 - 26 - 24)}{(26 + 24)^2(26 + 24 - 1)}}$$

$$= \sqrt{\frac{(1248)(1198)}{(50)^2 \cdot (49)}} = \sqrt{12.20493} = 3.49$$

$$z\star = \frac{V\star - \mu_V}{\sigma_V}: \quad z\star = \frac{29 - 25.96}{3.49} = \frac{3.04}{3.49} = 0.87$$

Step 4   Probability Distribution:

| *p*-Value: | **OR** | Classical: |

a. A two-tailed test is used:

$$\mathbf{P} = 2 \times P(z > 0.87)$$

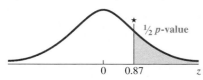

To find the *p*-value, you have three options:
1. Use Table 3 (Appendix B) to calculate the *p*-value:
   P = 2(0.5000 − 0.3078) = 0.3844.
2. Use Table 5 (Appendix B) to place bounds on the *p*-value: **0.3682 < P < 0.3954.**
3. Use a computer or calculator to find the *p*-value:
   P = 0.3844.
For specific instructions, see p. 432.
b. The *p*-value is not smaller than α.

a. A two-tailed test is used. The critical values are obtained from Table 4A:

$$-z(0.05) = -\mathbf{1.65} \quad \text{and} \quad z(0.05) = \mathbf{1.65}$$

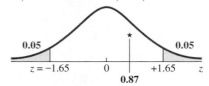

Specific instructions for finding critical values are on page 451.
b. $z\star$ is not in the critical region, as shown in red in the figure.

Step 5   a. Decision: Fail to reject H_o.

b. Conclusion: At the 0.10 level of significance, we are unable to reject the hypothesis of randomness and conclude that these data are a random sequence.

TECHNOLOGY INSTRUCTIONS: RUNS TEST FOR TESTING RANDOMNESS ABOVE AND BELOW THE MEDIAN

MINITAB (Release 14)   Input the set of data into C1; then continue with:

Choose: **Stat > Nonparametrics > Runs Test**
Enter: Variable: **C1**

Select: **Above and below mean > OK**
                 or
                 Above and below:
Enter: **Median value > ok**

Excel   The following commands will compute only the differences between the data values and the median. Then to complete the runs test you will need to count the number of runs created by the sequence of + and − signs.

Input the data into column A; select cell B1; then continue with:

Choose: **Edit Formula (=)**
Enter: **A1 − median((A1:A20 or select cells) > OK**
Drag: **Bottom right corner of the B1 cell down to give other differences**

TI-83/84 Plus   Input the data into L1; then continue with:

Highlight: **L2**
Enter: **L1 − median*(L1)**   **(*2nd LIST > MATH > 4:median( )**
Choose: **PRGM > EXEC > RUNSTEST***
Enter: **n1 = # of observations with particular characteristic**
                 (ex. below median)
                 n2 = # of observations with other characteristic
                 (ex. above median)
                 V = # of runs

*Program RUNSTEST is one of many programs that are available for downloading from http://www.duxbury.com. See page 42 for specific directions.

**APPLIED
EXAMPLE 14.12**

Casino Gaming Rules

Many casino games rely on electronically generated random numbers for "fair" play. Here is a sample of the rules governing these casino games.

BLACKJACK

ROULETTE

CRAPS

SLOTS

REQUIREMENTS RELATING TO ELECTRONIC GAMING DEVICES IN INTERNATIONAL CASINOS

These conditions are drafted pursuant to the Casino Act (Fi1999: 355). The purpose of the conditions is to guarantee the player security in relation to casinos and manufacturers of games, mainly as regards cheating through manipulation of gaming devices. Electronic gaming devices used in a casino must meet the specifications set forth in this rule.

The following conditions apply to randomness events and randomness testing:

(a) A random event has a given set of possible outcomes that has a given probability of occurrence.

(b) Two events are called independent if both of the following conditions exist:
  (i) The outcome of one event does not have an influence on the outcome of the other event.
  (ii) The outcome of one event does not affect the probability of occurrence of the other event.

(c) An electronic gaming device shall be equipped with a random number generator to make the selection process. A selection process is considered random if all of the following specifications are met:

(i) The random number generator satisfies not less than a 99% confidence level using chi-square tests.

(ii) The random number generator does not produce a statistic with regard to producing patterns of occurrences. Each reel position is considered random if it meets not less than 99% confidence level with regard to the runs

test or any similar pattern testing statistic.

(iii) The random number generator produces numbers that are independently chosen without regard to any other symbol produced during that play. This test is the correlation test. Each pair of reels is considered random if the pair of reels meet not less than 99% confidence level using standard correlation analysis.

Source: http://216.239.37.100/search?q=cache:KD66UcpLAzEaC:www.kemaquality .com/en/pages/services/gaming/Downloads/SW/Electronic%2520 Gaming%2520Devices.pdf+testing+randomness,+%22runs+test%22&hl=en&ie= UTF-8

SECTION 14.6 EXERCISES

Statistics △ Now™

Datasets can be found on your Student's Suite CD-ROM or at the StatisticsNow website at **http://1pass.thomson.com**.

14.37 State the null hypothesis, H_o, and the alternative hypothesis, H_a, that would be used to test the following statements:

a. The data did not occur in a random order about the median.

b. The sequence of odd and even is not random.

c. The gender of customers entering a grocery store was recorded; the entry is not random in order.

14.38 Determine the p-value that would be used to complete the following runs tests:

a. H_o: The sequence of gender of customers coming into the gym was random.
H_a: The sequence was not random.
with $n(A) = 10$, $n(B) = 12$, and $V = 5$.

b. H_o: The home prices collected occurred in random order above and below the median.
H_a: The home prices did not occur in random order.
with $z = 1.31$.

14.39 Determine the critical values that would be used to complete the following runs tests using the classical approach:

a. H_o: The results collected occurred in random order above and below the median.
H_a: The results were not random.
with $n(A) = 14$, $n(B) = 15$, and $\alpha = 0.05$.

b. H_o: The two properties alternated randomly.
H_a: The two properties did not occur in random fashion.
with $n(I) = 78$, $n(II) = 45$, and $\alpha = 0.05$.

14.40 Jessica did not believe she was playing a game with a fair die. She thought that if the die was fair, the tossing of the die should result in a random order of even and odd output. She performed her experiment 14 times. After each toss, Jessica recorded the results. The following data were reported (E = 2, 4, 6; O = 1, 3, 5).

O E O O O O E E O O O E E O

Use the runs test at a 5% level of significance to test the claim that the results reported are random.

14.41 A manufacturing firm hires both men and women. The following shows the gender of the last 20 individuals hired (M = male, F = female).

M M F M F F M M M M M M F M M F M M M M

At the $\alpha = 0.05$ level of significance, are we correct in concluding that this sequence is not random?

14.42 In an attempt to answer the question, "Does the husband (h) or wife (w) do the family banking?" the results of a sample of 28 married customers doing the family banking show the following sequence of arrivals at the bank.

w w w w h w h w h h h h w w w w w h h w w w h h h h w h h w

Do these data show lack of randomness with regard to whether the husband or wife does the family banking? Use $\alpha = 0.05$.

14.43 [EX14-43] The following data were collected in an attempt to show that the number of minutes the city bus is late is steadily growing larger. The data are in order of occurrence.

Minutes: 6 1 3 9 10 10 2 5 5 6 12 3 7 8 9 4 5 8 11 14

At $\alpha = 0.05$, do these data show sufficient lack of randomness to support the claim?

14.44 A student was asked to perform an experiment that involved tossing a coin 25 times. After each toss, the student recorded the results. The following data were reported (H = heads, T = tails).

H T H T H T H T H T H H T T H H T T H T H T H T H

Use the runs test at a 5% level of significance to test the student's claim that the results reported are random.

14.45 [EX14-45] According to a new survey—the Information and Communications Technologies in Schools Survey (http://www.statcan.ca/Daily/English/040610/d040610b.htm)—virtually all elementary and secondary schools in Canada had computers and were connected to the Internet during the 2003–2004 school year. The overall median number of students per computer for all of Canada is 5. Each province, starting on the east coast and proceeding clockwise around the country, reported their median number of students per

computer in the sequence that follows: 4.4, 5.4, 4.9, 4.6, 5.9, 5.4, 3.6, 3.7, 4.1, 5.0, 2.9, 3.5, 4.1.

a. Determine the median of these reported values and the number of runs above and below that median.

b. Use the runs test to test these data for randomness about the median. Use $\alpha = 0.05$.

14.46 [EX14-46] According to the March 10, 2005, *USA Today* article titled "Electronic world swallows up kids' time, study finds," U.S. children are spending an average of approximately 6.5 hours a day watching television, using the computer, or enjoying other electronic activities. The average amount of time 8- to 18-year-olds spend playing video games is given as 49 minutes per day. Suppose twenty 8- to 18-year-olds were randomly selected and monitored for a day, with the number of minutes they spent playing video games that day being recorded. The resulting sequence of times is given in ascending order of participant's age:

Minutes: 50 59 16 34 43 47 46 27 43 12 45 50 51 89 63 42 23 39 43 28

a. Determine the median and the number of runs above and below the median.

b. Use the runs test at 5% level of significance to test these data for randomness about the median.

c. State your conclusion.

14.47 [EX14-47] The following are 24 consecutive downtimes (in minutes) of a particular machine.

Downtime: 20 33 33 35 36 36 22 22 25 27 30 30
           30 31 31 32 32 36 40 40 50 45 45 40

The null hypothesis of randomness is to be tested against the alternative that there is a trend. A MINITAB analysis of the number of runs above and below the median follows.

```
Runs Test: Downtime
Runs test for Downtime
Runs above and below K = 32.5
The observed number of runs = 4
The expected number of runs = 13.0000
12 Observations above K  12 below
The test is significant at 0.0002
```

a. Confirm the values reported for the median and the number of runs by calculating them yourself.

b. Compute the value of $z\star$ and the p-value.

c. Would you reject the hypothesis of randomness? Explain.

d. Construct a graph that displays the sample data and visually supports your answer to part c.

14.48 Posted on the Economic Statistics Briefing Room page of the White House's website on June 20, 2005, was the statement, "Median household income in 2002 in the United States was $42,409." A random sample of 250 incomes has a median value different from any of the 250 incomes in the sample. The data contains 105 runs above and below the median. Use the preceding information to test the null hypothesis that the incomes in the sample form a random sequence with respect to the two properties above and below the median value versus the alternative that the sequence reported is not random at $\alpha = 0.05$.

14.49 [EX14-49] The number of absences recorded at a lecture that met at 8 AM on Mondays and Thursdays last semester were (in order of occurrence)

n(absences)	5	16	6	9	18	11	16	21	14	17	12	14	10
	6	8	12	13	4	5	5	6	1	7	18	26	6

Do these data show a randomness about the median value at $\alpha = 0.05$? Complete this test by using (a) critical values from Table 14 in Appendix B and (b) the standard normal distribution.

14.50 [EX14-50] The students in a statistics class were asked if they could be a good random-number generator. Each student was asked to write down a single digit from 0 through 9. The data was collected starting at the front left of the class, moving row by row, to the back right of the class. The sequence of digits were as follows:

7 4 3 6 9 5 4 4 4 3 6 3 3 7 7 7 6 3 6 7 6 9 6 7 3 7 7 3 4 6

Do these data show a randomness about the median value of 4.5 at $\alpha = 0.05$? Complete this test by using (a) critical values from Table 14 in Appendix B and (b) the standard normal distribution.

14.51 Randomness incorporates many different concepts. Referring to Applied Example 14.12:

a. What aspect of randomness will be tested using the chi-square test mentioned in part (i) of rule (c)? Describe how it will be used.

b. What aspect of randomness will be tested using the runs test mentioned in part (ii) of rule (c)? Describe how it will be used.

c. What aspect of randomness will be tested using the correlation analysis mentioned in part (iii) of rule (c)? Describe how it will be used.

d. These gaming rules are written using the phrase "99% confidence level" instead of "0.01 level of significance" as hypothesis tests typically use. Explain why this seems appropriate.

14.52 Research Randomizer is a free service offered to students and researchers interested in conducting random assignment and random sampling. Although every effort has been made to develop a useful means of generating random numbers, Research Randomizer and its staff do not guarantee the quality or randomness of numbers generated by Research Randomizer. Any use to which these numbers are put remains the sole responsibility of the user who generated them.

a. Go to the website http://www.randomizer .org/about.htm and generate one set of 20 random numbers from 1 to 9, where each number can repeat (select "No" for each number to be unique). (Use your computer, calculator, or Table 1 in the back, if you do not have a web connection.)

b. Test your set for randomness above and below the median value of 5. Use $\alpha = 0.05$.

c. Conduct the test again with the same parameters.

d. Test your new set for randomness. Use $\alpha = 0.05$. Did you get the same results?

e. Solve part d using the standard normal distribution. Did you reach the same conclusion?

14.7   Rank Correlation

Charles Spearman developed the rank correlation coefficient in the early 1900s. It is a nonparametric alternative to the linear correlation coefficient (Pearson's product moment, r) that was discussed in Chapters 3 and 13.

The **Spearman rank correlation coefficient, r_s,** is found by using this formula:

Spearman rank correlation coefficient

$$r_s = 1 - \frac{6\sum(d_i)^2}{n(n^2 - 1)} \tag{14.11}$$

FYI The subscript s is used in honor of Spearman, the originator.

where d_i is the difference in the **paired rankings** and n is the number of pairs of data. The value of r_s will range from -1 to $+1$ and will be used in much the same manner as Pearson's linear correlation coefficient, r, was used.

The Spearman rank coefficient is defined by using formula (3.1) with data rankings substituted for quantitative x and y values. The original data may be rankings, or if the data are quantitative, each variable must be ranked separately; then the rankings are used as pairs. If there are no ties in the rankings, formula (14.11) is equivalent to formula (3.1). Formula (14.11) provides us with an easier procedure to use for calculating the r_s statistic.

Assumptions for inferences about rank correlation: The n ordered pairs of data form a random sample, and the variables are ordinal or numerical.

The null hypothesis that we will be testing is: There is no correlation between the two rankings. The alternative hypothesis may be either two-tailed, there is correlation, or one-tailed if we anticipate either positive or negative correlation. The critical region will be on the side(s) corresponding to the specific alternative that is expected. For example, if we suspect negative correlation, then the critical region will be in the left-hand tail.

EXAMPLE 14.13   **Calculating the Spearman Rank Correlation Coefficient**

Let's consider a hypothetical situation in which four judges rank five contestants in a contest. Let's identify the judges as A, B, C, and D and the contestants as a, b, c, d, and e. Table 14.8 lists the awarded rankings.

TABLE 14.8

Rankings for Five Contestants

Contestant	Judge A	B	C	D	Contestant	Judge A	B	C	D
a	1	5	1	5	d	4	2	4	4
b	2	4	2	2	e	5	1	5	3
c	3	3	3	1					

When we compare judges A and B, we see that they ranked the contestants in exactly the opposite order: perfect disagreement (see Table 14.9). From our previous work with correlation, we expect the calculated value for r_s to be exactly -1 for these data. We have:

TABLE 14.9

Rankings of A and B

Contestant	A	B	$d_i = A - B$	$(d_i)^2$	Contestant	A	B	$d_i = A - B$	$(d_i)^2$
a	1	5	-4	16	d	4	2	2	4
b	2	4	-2	4	e	5	1	4	16
c	3	3	0	0					40

$$r_s = 1 - \frac{6\sum(d_i)^2}{n(n^2 - 1)}: \quad r_s = 1 - \frac{(6)(40)}{5(5^2 - 1)} = 1 - \frac{240}{120} = 1 - 2 = \mathbf{-1}$$

When judges A and C are compared, we see that their rankings of the contestants are identical (see Table 14.10). We would expect to find a calculated correlation coefficient of $+1$ for these data:

TABLE 14.10

Rankings of A and C

Contestant	A	C	$d_i = A - C$	$(d_i)^2$	Contestant	A	C	$d_i = A - C$	$(d_i)^2$
a	1	1	0	0	d	4	4	0	0
b	2	2	0	0	e	5	5	0	0
c	3	3	0	0					0

$$r_s = 1 - \frac{6\sum(d_i)^2}{n(n^2 - 1)}: \quad r_s = 1 - \frac{(6)(0)}{5(5^2 - 1)} = 1 - \frac{0}{120} = 1 - 0 = \mathbf{1}$$

By comparing the rankings of judge A with those of judge B and then with those of judge C, we have seen the extremes: total agreement and total disagreement. Now let's compare the rankings of judge A with those of judge D (see Table 14.11). There seems to be no real agreement or disagreement here. Let's compute r_s:

TABLE 14.11

Rankings of A and D

Contestant	A	D	$d_i = A - D$	$(d_i)^2$	Contestant	A	D	$d_i = A - D$	$(d_i)^2$
a	1	5	-4	16	d	4	4	0	0
b	2	2	0	0	e	5	3	2	4
c	3	1	2	4					24

$$r_s = 1 - \frac{6\sum(d_i)^2}{n(n^2 - 1)}: \quad r_s = 1 - \frac{(6)(24)}{5(5^2 - 1)} = 1 - \frac{144}{120} = 1 - 1.2 = \mathbf{-0.2}$$

The result is fairly close to zero, which is what we should have suspected, since there was no real agreement or disagreement.

The test of significance will result in a failure to reject the null hypothesis when r_s is close to zero; the test will result in a rejection of the null hypothesis when r_s is found to be close to $+1$ or -1. The critical values in Table 15 in Appendix B are the positive critical values only. Since the null hypothesis is, "The population correlation coefficient is zero (i.e., $\rho_s = 0$)," we have a symmetrical test statistic. Hence we need only add a plus or minus sign to the value found in the table, as appropriate. The sign is determined by the specific alternative that we have in mind.

When there are only a few ties, it is common practice to use formula (14.11). Even though the resulting value of r_s is not exactly equal to the value that would occur if formula (3.1) were used, it is generally considered to be an acceptable estimate. Example 14.14 shows the procedure for handling ties and uses formula (14.11) for the calculation of r_s.

When ties occur in either set of the ordered pairs of rankings, assign each tied observation the mean of the ranks that would have been assigned had there been no ties, as was done for the Mann–Whitney U test (see p. 766).

EXAMPLE 14.14   **One-Tailed Hypothesis Test**

Students who finish exams more quickly than the rest of the class are often thought to be smarter. Table 14.12 presents the scores and order of finish for 12 students on a recent 1-hour exam. At the 0.01 level, do these data support the alternative hypothesis that the first students to complete an exam have higher grades?

TABLE 14.12

Data on Exam Scores [TA14-12]

Order of Finish	1	2	3	4	5	6	7	8	9	10	11	12
Exam Score	90	78	76	60	92	86	74	60	78	70	68	64

SOLUTION

Step 1   a. **Parameter of interest:** The rank correlation coefficient between score and order of finish, ρ_s.

  b. **Statement of hypotheses:**
  H_o: Order of finish has no relationship to exam score.
  H_a: The first to finish tend to have higher grades.

Step 2   a. **Assumptions:** The 12 ordered pairs of data form a random sample; order of finish is an ordinal variable and test score is numerical.

  b. **Test statistic:** The Spearman rank correlation coefficient, r_s.

  c. **Level of significance:** $\alpha = 0.01$ for a one-tailed test.

Step 3   a. **Sample information:** The data are given in Table 14.12.

  b. **Test statistic:** Rank the scores from highest to lowest, assigning the highest score the rank number 1, as shown. (Order of finish is already ranked.)

92	90	86	78	78	76	74	70	68	64	60	60
1	2	3	4	5	6	7	8	9	10	11	12
			4.5	4.5						11.5	11.5

The rankings and preliminary calculations are shown in Table 14.13.

TABLE 14.13

Rankings of Test Scores and Differences

Order of Finish	Test Score Rank	Difference (d_i)	(d_i)²
1	2	−1	1.00
2	4.5	−2.5	6.25
3	6	−3	9.00
4	11.5	−7.5	56.25
5	1	4	16.00
6	3	3	9.00
7	7	0	0.00
8	11.5	−3.5	12.25
9	4.5	4.5	20.25
10	8	2	4.00
11	9	2	4.00
12	10	2	4.00
			142.00

FYI For comparison, Exercise 14.67 (p. 794) asks you to calculate r_s using formula (3.2).

Using formula (14.11), we obtain

$$r_s = 1 - \frac{6\sum(d_i)^2}{n(n^2 - 1)} : \quad r_s = 1 - \frac{(6)(142.0)}{12(12^2 - 1)} = 1 - \frac{852}{1716} = 1 - 0.4965 = -0.503$$

Thus, $r_s\star = 0.503$.

Step 4   **Probability Distribution:**

p-Value:

a. Since the concern is for values "positive," the *p*-value is the area to the right:

$$\mathbf{P} = P(r_s \geq 0.503 \text{ for } n = 12)$$

To find the *p*-value, you have two options:
1. Use Table 15 (Appendix B) to place bounds on the *p*-value. Table 15 lists only two-tailed α (this test is one-tailed, so divide the column heading by 2): **0.025 < P < 0.05.**
2. Use a computer or calculator to find the *p*-value: **P = 0.048.**

Specific instructions follow this example.
b. The *p*-value is not smaller than α.

OR

Classical:

a. The critical region is one-tailed because H_a expresses concern for values related to "positive." Since the table is for two-tailed, the critical value is located at the intersection of the $\alpha = 0.02$ column ($\alpha = 0.01$ in each tail) and the $n = 12$ row of Table 15: **0.703.**

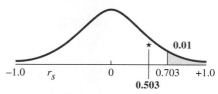

b. $r_s\star$ is not in the critical region, as shown in the figure.

Step 5   **a.   Decision:** Fail to reject H_o.

b.   Conclusion: These sample data do not provide sufficient evidence to enable us to conclude that the first students to finish have higher grades, at the 0.01 level of significance.

Calculating the p-value for the Spearman rank correlation test

Method 1: Use Table 15 in Appendix B to place bounds on the p-value. By inspecting the $n = 12$ row of Table 15, you can determine an interval within which the *p*-value lies. Locate the value of r_s along the $n = 12$ row and read the bounds from the top of the table. Table 15 lists only two-tailed values (therefore, you must divide by 2 for a one-tailed test). We find **$0.025 < P < 0.05$.**

Method 2. If you are doing the hypothesis test with the aid of a computer or graphing calculator, most likely it will calculate the *p*-value for you. Specific instructions are described next. MINITAB and Excel calculate a two-tailed *p*-value; therefore, you must divide by 2 when the test is one-tailed.

TECHNOLOGY INSTRUCTIONS: SPEARMAN'S RANK CORRELATION COEFFICIENT

MINITAB (Release 14)   Input the set of data for the first variable into C1 and the corresponding data values for the second variable into C2; then continue with:

```
Choose:     Data > Rank...
Enter:      Rank data in: C1
            Store ranks in: C3 > OK
```

Repeat the preceding commands for the data in C2 and store in C4.

```
Choose:     Stat > Basic Statistics > Correlation
Enter:      Variables: C3 C4 > OK
```

Excel   Input the set of data for the first variable into column A and the corresponding data values for the second variable into column B; then continue with:

FYI Both Excel and TI-83/84 Plus use the normal approximation to complete the Spearman rank correlation test.

```
Choose:     Tools > Data Analysis Plus > Correlation (Spearman)
Enter:      Variable 1 range: (A1:A10 or select cells)
            Variable 2 range: (B1:B10 or select cells)
Select:     Labels (if necessary)
Enter:      Alpha: α (ex. 0.05)
```

TI-83/84 Plus   Input the set of data for the first variable into L1 and the corresponding data values for the second variable into L2; then continue with:

```
Choose:     PRGM > EXEC > SPEARMAN*
Enter:      XLIST: L1
            YLIST: L2
Select:     DATA?        1:UNRANKED
            ALT HYP? 1:RHO > 0 or 2:RHO < 0 or 3:RHO ≠ 0
```

*Program SPEARMAN is one of many programs that are available for downloading from http://www.duxbury.com. See page 42 for specific directions.

APPLIED
EXAMPLE 14.15   Calculus or Tartar

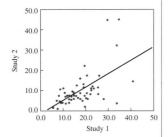

CONSISTENCY OF CALCULUS FORMATION IN CONTROLLED CLINICAL STUDIES

ABSTRACT

Methods for demonstrating clinical efficacy of anti-tartar agents have been well established. These clinical models typically select high tartar-forming populations from short-term pre-test periods. This paper evaluates the consistency of tartar formation in pre-test periods in a common population. Two randomized, controlled 6-month tartar control studies were conducted over a 4-year period at a single center using a common design except for use of different examiners. Both studies used a 2-month pre-test period to evaluate calculus formation of participating subjects with a non-tartar control dentifrice. At the beginning of this period, all subjects received a prophylaxis, and were provided a regular dentifrice and a toothbrush and instructed to brush their teeth twice daily. After 8 weeks, subjects were assessed for the accumu-lation of supragingival calculus on the lingual surfaces of 6 mandibular anterior teeth using the Volpe-Manhold Calculus Index (VMI). The two studies have a total of 58 common subjects which completed the pre-test phase. In the first study, VMI scores at the end of pre-test phase ranged from 5.0–40.5 with a mean of 17.0 $\pm$ 7.8, and in the second study, the scores ranged from 0.5–45.5 with a mean of 9.7 $\pm$ 8.7. While the mean scores are significantly different ($p = 0.0001$), Pearson correlation analysis shows that VMI scores in the two studies were significantly correlated ($r = 0.60$, $p = 0.0001$). Spearman's rank correlation coefficient was also highly significant ($r = 0.57$, $p = 0.0001$). The data demonstrate that VMI scores in pre-test phases can be strongly related within subjects over several years and different examiners.

Source: H. Liu and others, Procter & Gamble Company, Mason, OH. Research presented at the 30th Annual Meeting of the AADR. © The Procter & Gamble Company. Reprinted with permission.

SECTION 14.7 EXERCISES

Statistics⊜Now™

14.53 State the null hypothesis, H_o, and the alternative hypothesis, H_a, that would be used to test the following statements:

a. There is no relationship between the two rankings.

b. The two variables are unrelated.

c. There is a positive correlation between the two variables.

d. Refrigerator age has a decreasing effect on monetary value.

14.54 Determine the p-value that would be used to test the null hypothesis for the following Spearman rank correlation experiments:

a. H_o: No relationship between the two variables.
    H_a: There is a positive relationship.
    with $n = 21$ and $r_s = 0.55$

b. H_o: No correlation.
    H_a: There is a relationship.
    with $n = 27$ and $r_s = 0.71$.

c. H_o: Variable A has no effect on variable B.
    H_a: Variable B decreases as variable A increases.
    with $n = 10$ and $r_s = -0.62$.

14.55 Determine the test criteria that would be used to test the null hypothesis for the following Spearman rank correlation experiments:

a. H_o: No relationship between the two variables.
   H_a: There is a relationship.
   with $n = 14$ and $\alpha = 0.05$.

b. H_o: No correlation
   H_a: Positively correlated
   with $n = 27$ and $\alpha = 0.05$.

c. H_o: Variable A has no effect on variable B.
   H_a: Variable B decreases as variable A increases.
   with $n = 18$ and $\alpha = 0.01$.

14.56 [EX14-56] When it comes to getting workers to produce, money is not everything; feeling appreciated is more important. Do the rankings assigned by workers and the boss show a significant difference in what each person thinks is important? (Ratings: 1 = most important; 10 = least important.) Test using $\alpha = 0.05$.

Component of Job Satisfaction	Worker Ranking	Boss Ranking
Full appreciation of work done	1	8
Feeling of being in on things	2	10
Sympathetic help on personal problems	3	9
Job security	4	2
Good wages	5	1
Interesting work	6	5
Promotion and growth in the organization	7	3
Personal loyalty to employees	8	6
Good working conditions	9	4
Tactful disciplining	10	7

Source: Philadelphia Inquirer

14.57 [EX14-57] Consumer product testing groups commonly supply ratings of all sorts of products to consumers in an effort to assist them in their purchase decisions. Different manufacturers' products are usually tested for their performance and then given an overall rating. *PC World* ranked the top ten 17-inch computer monitors and also supplied the street price (dollars). The ranks of each are shown in the table that follows, with the highest-priced monitor given a rank of 1 and the lowest a rank of 10.

Overall Rating	Street Price Rank	Overall Rating	Street Price Rank
1	3	6	2
2	4	7	8.5
3	6.5	8	6.5
4	8.5	9	10
5	5	10	1

Source: PC World

a. Compute the Spearman rank correlation coefficient for the overall rating and the street price of the 17-inch monitors.

b. Does a higher price yield a higher rating? Test the null hypothesis that there is no relationship between the overall ratings of the monitors and their street prices versus the alternative that there is a positive relationship between them. Use $\alpha = 0.05$.

14.58 [EX14-58] *Reader's Digest* (July 2005) did a special report on "America's Cleanest [and Dirtiest] Cities." The 50 most populous U.S. cities were compared and ranked based on air pollution, water pollution, toxic emissions, hazardous waste, and sanitation force. The top 10 cleanest cities follow. As noted, air pollution is one of the major factors. Does it have a significant influence on the ranking of the cities or is it a combination of factors?

City	City Rank	Air Pollution Score	City	City Rank	Air Pollution Score
Portland, OR	1	49	Denver	6	27
San Jose	2	41	Rochester, NY	7	46
Buffalo	3	34	Austin	8	44
Columbus	4	24	Orlando	9	48
San Francisco	5	47	San Diego	10	13

a. Rank the air pollution scores based on the knowledge that a score of 50 is the highest and cleanest score.

b. Compute the Spearman rank correlation coefficient for the two rankings.

c. At the 0.05 level of significance, determine whether there is a significant relationship between a city's rank and its corresponding air pollution score.

14.59 [EX14-59] The following data represent the ages of 12 subjects and the mineral concentration (in parts per million) in their tissue samples.

Age, x	82	83	64	53	47	50	70	62	34	27	75	28
Mineral Concentration, y	170	40	64	5	15	5	48	34	3	7	50	10

Refer to the following MINITAB output and verify that the Spearman rank correlation coefficient equals 0.753 by calculating it yourself.

Correlations: xRank, yRank
Correlation of xRank and yRank = 0.753,
P-Value = 0.005

14.60 [EX14-60] Many people are concerned about eating foods that have a high sodium content. They are also advised of the benefits of obtaining sufficient fiber in their diets. Do foods high in fiber tend to have more sodium? The following table was obtained by selecting 11 soups from a list published in *Nutrition Action Healthletter*. The soups were measured on the basis of both sodium content and fiber:

Soup	Sodium	Fiber	Soup	Sodium	Fiber
A	480	12	G	420	2
B	830	0	H	290	4
C	510	1	I	450	10
D	460	5	J	430	6
E	490	3	K	390	9
F	580	7			

Source: Nutrition Action Healthletter

a. Rank the soups in ascending order on the basis of their sodium content and on their fiber content. Show your results in a table.

b. Compute the Spearman rank order correlation coefficient for the two sets of rankings.

c. Does higher sodium content accompany foods that are higher in fiber? Test the null hypothesis that there is no relationship between the fiber and sodium content of the soups versus the alternative that there is a relationship between them. Use $\alpha = 0.05$.

14.61 [EX14-61] The *Journal of Professional Nursing* article titled "The Graduate Record Examination as an Admission Requirement for the Graduate Nursing Program" reported a significant correlation between undergraduate grade point average (GPA) and GPA at graduation from a graduate nursing program. The following data were collected on 10 nursing students who graduated from a graduate nursing program.

Undergraduate GPA	3.5	3.1	2.7	3.7	2.5	3.3	3.0	2.9	3.8	3.2
GPA at Graduation	3.4	3.2	3.0	3.6	3.1	3.4	3.0	3.4	3.7	3.8

Compute the Spearman rank coefficient and test the null hypothesis of no relationship versus a positive relationship. Use a level of significance equal to 0.05.

14.62 [EX14-62] The U.S. Department of Transportation tracks and reports a variety of information regarding airlines and airports. It reports the "percent on-time" frequently. Compare the percent on-time for the 33 major airports in the United States for the 4 months of 2004 and 2005.

Ranking of Major Airport On-Time Departure Performance Year-to-date through April 2005 (Percent On-Time)

Rank	January 1– April 30, 2004	%	Rank	January 1– April 30, 2005	%
1	Houston, TX (IAH)	89.49	1	Houston, TX (IAH)	86.77
2	Washington Reagan National, DC (DCA)	88.69	2	Portland, OR (PDX)	84.43

••• Remainder of data on Student's Suite CD-ROM

Source: U.S. Bureau of Transportation Statistics, Airline On-Time Data

At the 0.05 level of significance, test the claim there is no correlation between the 2004 and 2005 percent on-time at these airports.

14.63 [EX14-63] Most people in the United States work in either trade or service jobs. "Trade" is defined as businesses such as retail stores, car dealers, and restaurants—places where goods are sold. The "service" area includes jobs such as those in health care, the hotel industry, and cleaning services—jobs where people perform services rather than selling items. The following table shows nine job classifications comparing Lee County, Florida, with all of Florida and the United States.

Job Category	Lee Co.	Florida	U.S.
Agriculture	2.0%	2.0%	2.6%
Construction	8.0%	5.0%	6.5%
Manufacturing	5.0%	8.0%	16.0%
Trans., comm., and pub. util.	5.0%	5.0%	7.1%
Retail trade	25.0%	21.0%	16.8%

Job Category	Lee Co.	Florida	U.S.
Wholesale trade	3.0%	5.0%	7.1%
Fin., insur., and real estate	6.0%	6.0%	6.5%
Services	30.0%	34.0%	35.9
Government	16.0%	15.0%	NA

Source: Florida Department of Labor; U.S. Department of Labor

a. Construct a new table ranking the percentages for Lee County, Florida, and the United States separately.

b. Using the Spearman rank correlation and a 0.05 level of significance, determine whether there is a relationship between Lee County and all of Florida.

c. Using the Spearman rank correlation and a 0.05 level of significance, determine whether there is a relationship between Lee County and all of the United States.

d. Using the Spearman rank correlation and a 0.05 level of significance, determine whether there is a relationship between all of Florida and all of the United States.

e. Review the results of parts b–d and comment on your combined findings.

14.64 [EX14-64] "Survey of Home Buyer Preferences" was conducted by the National Association of Home Builders to determine the features that home buyers really want. Respondents were to rate each feature desirable as well as if essential. The following table shows the results.

Feature	Desirable	Essential
Laundry room	40	52
Linen closet	56	32
Exhaust fan	44	42
Dining room	43	36
Walk-in pantry	59	19
Island work area	55	16
Separate shower enclosure	49	20
Temperature control faucets	49	18
Whirlpool tub	46	12
White bathroom fixtures	40	16
Ceramic wall tiles	43	12
Solid-surface countertops	48	7
Den/library	43	11
Wood burning fireplace	39	15
Special use storage	47	6

Source: National Association of Home Builders

It is not surprising that the ratings in the "desire" column of the table are considerably higher than the ratings in the "essentials" column. There is no question about there being a difference in the ratings; however, an appropriate question is, "Do the items on the list appear in the same order of preference in both columns?"

a. Use the Mann–Whitney U test to test the hypothesis that the items follow essentially the same distribution using $\alpha = 0.05$.

b. Use the Spearman rank correlation coefficient to test the hypothesis that the rankings of the items are not correlated using $\alpha = 0.05$.

c. State your conclusion.

14.65 [EX14-65] As the following chart shows, what is "good enough" to qualify as "proficient" may vary widely from state to state. *Education Week* (February 20, 2002) compared the percent of students who scored at or above proficient on the National Assessment of Educational Progress (NAEP) and on state assessments in mathematics.

State	Statewide Assessment	NAEP Assessment	State	Statewide Assessment	NAEP Assessment
AR	41	13	NY	65	22
CT	30	32	NC	84	28
GA	62	18	ND	15	25
ID	16	21	RI	28	23
KS	39	30	SC	24	18
LA	12	14	TX	43	27
MA	40	33	VT	38	29
MI	75	29	WY	27	25
MO	37	23			

Source: Education Week, http://www.edweek.com

a. Present the information in the table in the form of a bar graph to visualize any relationship between the two different assessments. Does there appear to be any relationship? Explain.

b. Find the rank numbers for each set of percentages separately.

c. Present the information in the table in the form of a scatter diagram to visualize any relationship between the two different assessments. Does there appear to be any relationship? Explain.

d. Use the Spearman rank correlation coefficient to test the hypothesis that there is no correlation between the two sets of percentages. Use $\alpha = 0.05$.

14.66 Referring to Applied Example 14.15:

a. Explain the meaning of, "the mean scores are significantly different $(p = 0.0001)$." What mean scores are being referred to? What methodology might have been used to establish this significance?

b. Explain the meaning of, "Pearson correlation analysis shows that VMI scores in the two studies were significantly correlated $(r = 0.60, p = 0.0001)$."

c. Explain the meaning of, "Spearman's rank correlation coefficient was also highly significant $(r = 0.57, p = 0.0001)$."

d. What is the relationship between the Pearson correlation and Spearman correlation analysis?

14.67 Using formula (3.2), calculate the Spearman rank correlation coefficient for the data in Example 14.14 (p. 787). Recall that formula (3.2) is equivalent to the definition formula (3.1) and that rank numbers must be used with this formula for the resulting statistic to be the Spearman r_s.

14.68 Refer to the bivariate data shown in the following table.

x	−2	−1	1	2
y	4	1	1	4

a. Construct a scatter diagram.

b. Calculate the Spearman rank correlation coefficient, r_s [formula (14.11)].

c. Calculate Pearson's correlation coefficient, r [formula (3.2)].

d. Compare the two results from parts b and c. Do the two measures of correlation measure the same thing?

CHAPTER REVIEW

In Retrospect

In this chapter you have become acquainted with some of the basic concepts of nonparametric statistics. While learning about the use of nonparametric methods and specific nonparametric tests of significance, you should have also come to realize and understand some of the basic assumptions that are needed when the parametric techniques of the earlier chapters are encountered. You now have seen a variety of tests, many of which somewhat duplicate the job done by others. Keep in mind that you should use the best test for your particular needs. The power of the test and the cost of sampling, as related to the size and availability of the desired response variable, will play important roles in determining the specific test to be used.

Vocabulary and Key Concepts

assumptions (pp. 751, 753, 756, 765, 777, 785)
binomial random variable (p. 759)
continuity correction (p. 759)
correlation (p. 785)

dependent sample (p. 756)
distribution-free test (p. 751)
efficiency (p. 751)
independent sample (p. 765)
Mann–Whitney U test (p. 765)
median, M (p. 753)

normal approximation (pp. 751, 758, 769, 778)
paired data (p. 756)
parametric test (p. 751)
power (p. 757)
randomness (p. 776)
rank (p. 766)

run (p. 776)
runs test (p. 776)
sign test (p. 752)

Spearman rank correlation co-
efficient (p. 785)

test statistic (pp. 754, 760, 767,
769, 776, 779, 787)

Learning Outcomes

✓ Understand that parametric methods are statistical methods that
assume that the parent population is approximately normal or
that the central limit theorem gives (at least approximately) a
normal distribution of a test statistic.

p. 751

✓ Understand that nonparametric methods (distribution-free
methods) do not depend on the distribution of the population
being sampled.

p. 751

✓ Understand that the power of a test $(1 - \beta)$ is its ability to reject
a false null hypothesis.

pp. 751–752

✓ Understand that efficiency of a nonparametric test takes into
account the power of a test and the required sample size.

pp. 751–752

✓ Understand that the sign test is the nonparametric alternative to
the t-test for one mean and the difference between two dependent
means.

pp. 752–753, Ex. 14.3

✓ Compute, describe, and interpret a confidence interval for a population
median using the sign test.

EXP 14.1, Ex. 14.5

✓ Perform, describe, and interpret a hypothesis test for a single median
using the sign test with the p-value approach and classical approach.

EXP 14.2, Ex. 14.13

✓ Perform, describe, and interpret a hypothesis test for median of paired
differences using the sign test with the p-value approach and classical
approach.

EXP 14.3, Ex. 14.17

✓ Understand that the Mann–Whitney U test is the nonparametric
alternative to the t-test for the difference between two independent
means.

p. 765, Ex. 14.23

✓ Perform, describe, and interpret a hypothesis test for the difference
between two means using the Mann–Whitney U test with the p-value
approach and classical approach.

EXP 14.6, Ex. 14.33

✓ Perform, describe, and interpret a hypothesis test for the difference
between two means using the normal approximation to the Mann–
Whitney U test with the p-value approach and classical approach.

EXP 14.7, Ex. 14.35

✓ Perform, describe, and interpret a hypothesis test for the randomness
of data using the runs test with the p-value approach and classical
approach.

EXP 14.9, 14.10,
Ex. 14.41, 14.46

✓ Perform, describe, and interpret a hypothesis test for the randomness
of data using the normal approximation to the runs test with the p-value
approach and classical approach.

EXP 14.11, Ex. 14.49

✓ Understand that the Spearman rank correlation coefficient is the
nonparametric alternative to the Pearson linear correlation coefficient, r.

p. 785, EXP 14.13

✓ Perform, describe, and interpret a hypothesis test for the significance of correlation between two variables using the Spearman rank correlation coefficient with the *p*-value approach and classical approach.

p. 796, EXP 14.14,
Ex. 14.57, 14.61

Chapter Exercises

Statistics⊘Now™

Go to the StatisticsNow website **http://1pass.thomson.com** to

- Assess your understanding of this chapter
- Check your readiness for an exam by taking the Pre-Test quiz and exploring the resources in the Personalized Learning Plan

Datasets can be found on your Student's Suite CD-ROM or at the StatisticsNow website at **http://1pass.thomson.com**.

14.69 [EX14-69] "Because regional-scale atmospheric deposition data in the Rocky Mountains are sparse, a program was designed by the US Geological Survey to more thoroughly determine the quality of precipitation and to identify sources of atmospherically deposited pollution in a network of high-elevation sites. Depth-integrated samples of seasonal snow packs at 52 sampling sites, in a network from New Mexico to Montana, were collected and analyzed each year since 1993." One of a number of chemical characteristics sampled was hydrogen. Following are the 5-year average results from each of the 52 sites.

7.3	3.3	4.3	4.8	4.5	5.1	5.3	6.7	3.5	6.1	5.1	8.3	4.1
9.7	5.4	3.8	5.8	6.1	4.5	8.8	5.2	7.2	4.9	2.0	3.6	6.3
7.8	5.5	11.1	9.7	5.1	15.2	5.0	9.9	3.8	5.4	7.8	9.4	4.5
10.6	3.6	2.7	10.5	12.4	3.1	2.8	5.7	4.3	8.3	5.9	4.6	6.1

Source: U.S. Geological Survey

a. Construct a stem-and-leaf plot of the data.

b. Describe the pattern you see in the stem-and-leaf plot.

c. Using the stem-and-leaf plot and the sign test, find the 95% confidence interval for the population median.

14.70 [EX14-70] Research about the health practices of cardiovascular technologists compared the body mass index (BMI) of the technologists with that of the general population. Weight classification by BMI is as follows: underweight, less than 19; normal, 19 to 24; overweight, 25 to 29; and obesity, 30 and higher. The following list shows the BMI values for a sample of 30 technologists.

BMI	16	50	39	33	33	25	29	30	39	23
	21	24	19	28	26	34	19	20	18	21
	24	24	20	18	26	22	24	18	25	25

a. Construct a stem-and-leaf plot for the BMI of the cardiovascular technologists.

b. Describe the sample of technologists as underweight, normal, overweight, or obese.

c. Find the 95% confidence interval for the median body mass index.

14.71 [EX14-71] A sample of 32 students received the following grades on an exam.

41	42	48	46	50	54	51	42	51	50	45	42	32	45	43	56
55	47	45	51	60	44	57	57	47	28	41	42	54	48	47	32

a. Does this sample show that the median score for the exam differs from 50? Use $\alpha = 0.05$.

b. Does this sample show that the median score for the exam is less than 50? Use $\alpha = 0.05$.

14.72 [EX14-72] Is the absentee rate in the 8 AM statistics class the same as that in the 11 AM statistics class? The following sample of the daily number of absences was taken from the attendance records of the two classes.

	Day											
Class	1	2	3	4	5	6	7	8	9	10	11	12
8 AM	0	1	3	1	0	2	4	1	3	5	3	2
11 AM	1	0	1	0	1	2	3	0	1	3	2	1

Is there sufficient reason to conclude that there are more absences in the 8 AM class? Use $\alpha = 0.05$.

14.73 [EX14-73] Track coaches, runners, and fans talk a lot about the "speed of the track." The surface of the track is believed to have a direct effect on the amount of time that it takes a runner to cover the required distance. To test this effect, 10 runners were asked to run a 220-yard sprint on each of two tracks. Track A is a cinder track, and track B is made of a new synthetic material. The running times (in seconds) are given in the following table. Test the claim that the surface on track B is conducive to faster running times.

					Runner					
Track	1	2	3	4	5	6	7	8	9	10
A	27.7	26.8	27.0	25.5	26.6	27.4	27.2	27.4	25.8	25.1
B	27.0	26.7	25.3	26.0	26.1	25.3	26.7	27.1	24.8	27.1

a. State the null and alternative hypotheses being tested. Complete the test using $\alpha = 0.05$.

b. State your conclusions.

14.74 A candy company has developed two new chocolate-covered candy bars. Six randomly selected people all preferred candy bar I. Is this statistical evidence, at $\alpha = 0.05$, that the general public will prefer candy bar I?

14.75 An article in the journal *Sedimentary Geology* compares a measure called the roughness coefficient for translucent and opaque quartz sand grains. If you measured the roughness coefficient for 20 sand grains of each type (translucent and opaque), for what values of the Mann–Whitney U statistic would you reject the null hypothesis in a two-tailed test with $\alpha = 0.05$?

14.76 [EX14-76] Twenty students were randomly divided into two equal groups. Group 1 was taught an anatomy course using a standard lecture approach. Group 2 was taught using a computer-assisted approach. The test scores on a comprehensive final exam were as follows:

Group 1	75	83	60	89	77	92	88	90	55	70
Group 2	77	92	90	85	72	59	65	92	90	79

Test the claim that a computer-assisted approach produces higher achievement (as measured by final exam scores) in anatomy courses than does a lecture approach. Use $\alpha = 0.05$.

14.77 [EX14-77] The use of nuclear magnetic resonance (NMR) spectroscopy for detection of malignancy is discussed in the journal *Clinical Chemistry*. The line width at the half height of peaks in the NMR spectrum is measured. The spectrum is pro-

duced from assaying plasma from an individual. Suppose the following line widths (measured in Hertz [Hz]) were obtained from a normal group and a group known to have malignancies. Would you reject a two-tailed research hypothesis at the 0.05 level of significance?

Normal Group	35.1	32.9	30.6	30.5	30.9
Malignancy Group	28.5	29.5	30.7	27.5	28.0

14.78 [EX14-78] A firm is currently testing two different procedures for adjusting the cutting machines used in the production of greeting cards. The results of two samples show the following recorded adjustment times (in seconds).

Method 1	17	15	14	18	16	15	17	18	15	14	14	16	15
Method 2	14	14	13	13	15	12	16	14	16	13	14	13	12	15	17	13

Is there sufficient reason to conclude that method 2 requires less time, on average, than method 1 at the 0.05 level of significance?

14.79 [EX14-79] Two statistics that baseball enthusiasts use to compare the overall strengths of one team against another are team batting average (the higher the batting average, the better) and team pitching average (the lower the earned run average, the better). Using the results for the National and American Leagues in 2004, investigate the relationship between team batting averages and earned run averages.

NL Team	Batting Avg.	ERA	AL Team	Batting Avg.	ERA
Arizona Diamondbacks	4.98	0.253	Anaheim Angels	4.28	0.282
••• Remainder of data on Student's Suite CD-ROM					

Source: MLB.com

a. Convert the table to ranks of the (1) batting averages and (2) earned run averages for the National League (N) and the American League (A), showing the league (N or A) represented by a team's rank.

b. Use the Mann–Whitney U test to test the hypothesis that (1) the batting average of the American League is higher and (2) the earned run average of the National League is lower. Use the 0.05 level of significance.

14.80 [EX14-80] Two manufacturers of table-tennis balls have agreed that the quality of their products can be measured by the height to which the balls rebound. A test is arranged, the balls are dropped from a constant height, and the rebound heights are measured. The results (in inches) are shown in the following table. Manufacturer A claims, "The results show my product to be superior." Manufacturer B replies, "I know of no statistical test that supports this claim."

A	14.0	12.5	11.5	12.2	12.4	12.3	11.8	11.9	13.7	13.2
B	12.0	12.5	11.6	13.3	13.0	13.0	12.1	12.8	12.2	12.6

a. What parametric test would be appropriate if we assume rebound heights are normally distributed?

b. Does the parametric test in part a show that A's product is superior? Use a 0.05 level of significance.

c. What nonparametric test would be appropriate if we are unsure about the distribution of rebound heights?

d. Does the nonparametric test in part c show that A's product is superior? Use a 0.05 level of significance.

e. What do both tests tell you about A's claim of a superior product?

14.81 Consider the following sequence of defective parts (d) and nondefective parts (n) produced by a machine.

n n n d n n n n n d n n n n n n n d n d n n n n

Can we reject the hypothesis of randomness at $\alpha = 0.05$?

14.82 [EX14-82] A patient was given two different types of vitamin pills, one that contained iron and one that did not contain iron. The patient was instructed to take the pills on alternate days. To avoid having to remember which pill to take on a particular day, the patient mixed all of the pills together in a large bottle. Each morning the patient took the first pill that came out of the bottle. To see whether this was a random process, for 25 days the patient recorded an "I" each morning that he took a vitamin with iron and an "N" for no iron.

Day	1	2	3	4	5	6	7	8	9	10	11	12	13	14	15	16	17	18	19	20	21	22	23	24	25
Type	I	I	N	I	I	N	N	I	N	N	N	N	N	I	I	I	N	I	I	I	I	N	I	I	N

Is there sufficient reason to reject the null hypothesis that the vitamins were taken in random order at the 0.05 level of significance?

14.83 [EX14-83] What makes one company more attractive to work for than another? One possibility is the growth in new jobs. The editors of *Fortune* developed a list of the top 100 companies to work for in America. Included in the list was the percentage change in full-time positions of each company during the past 2 years. The top 20 are shown in the table that follows.

Company	Job Growth	Company	Job Growth
1	26	11	23
2	54	12	13
3	34	13	17
4	10	14	23
5	31	15	9
6	48	16	3
7	26	17	15
8	22	18	11
9	24	19	1
10	10	20	122

Source: Fortune, "The 100 Best Companies to Work for in America"

a. Determine the median job growth percentage and the number of runs above and below the median.

b. Use the runs test to test whether the growth rates are listed in a random sequence about the median.

c. Do companies ranked higher also have higher job growth rates? State your conclusion.

14.84 [EX14-84] In a study to see whether spouses are consistent in their preferences for television programs, a market research firm asked several married couples to rank a list of 12 programs (1 represents the highest score; 12 represents the lowest). The average ranks for the programs, rounded to the nearest integer, were as follows:

	Program											
Rank	1	2	3	4	5	6	7	8	9	10	11	12
Husbands	12	2	6	10	3	11	7	1	9	5	8	4
Wives	5	4	1	9	3	12	2	8	6	10	7	11

Is there significant evidence of negative correlation at the 0.01 level of significance?

14.85 [EX14-85] Can today's high temperature be effectively predicted using yesterday's high? Pairs of yesterday's and today's high temperatures were randomly selected. The results are shown in the following table. Do the data present sufficient evidence to justify the statement: "Today's high temperature tends to correlate with yesterday's high temperature"? Use $\alpha = 0.05$.

Reading	1	2	3	4	5	6	7	8	9	10	11	12	13	14	15	16	17	18
Yesterday's	40	58	46	33	40	51	55	81	85	83	89	64	73	63	46	58	28	69
Today's	40	56	34	59	46	51	74	77	83	84	85	68	65	60	54	62	34	66

14.86 [EX14-86] U.S. commercial radio stations are classified by the primary format of their broadcasts. As people change their listening preferences, the stations are likely to react to the change by adjusting their formats. The following table shows the percentages of radio stations in 1997, 2002, and 2004, broken down by their primary format:

Primary Format	1997	2002	2004
Country	24.26	20.18	19.22
Adult Contemporary (AC)	14.75	14.34	9.94
News, Talk, Business, Sports	12.73	14.73	20.36
Religion (Teaching and Music)	10.22	9.90	6.95
Rock	9.13	8.51	8.41
Oldies	7.30	7.70	7.66
Spanish and Ethnic	5.32	6.57	7.28
Adult Standards	5.20	5.29	4.80
Urban, Black, Urban AC	3.47	3.69	3.84
Top-40	3.40	4.47	4.67
Easy Listening	0.84	2.61	2.34
Variety	0.49	0.40	0.39
Jazz	0.48	0.78	0.84
Classical, Fine Arts	0.45	0.30	0.32
Preteen	0.35	0.47	0.54
All Other	1.61	0.06	2.43

Source: M Street Corporation, Nashville, TN

a. Construct a table that shows the ranks of the relative frequency of stations within each format for 1997, 2002, and 2004.

b. Use the Spearman rank order correlation coefficient to test at the 0.01 level of significance the hypothesis that there is no correlation between the ranks of the formats offered by radio stations in 1997 and 2002, 1997 and 2004, and 2002 and 2004.

c. Has the distribution of primary formats changed? Describe how the results in part b support your answer.

14.87 [EX14-87] Every year before the college football season starts, *Sports Illustrated* presents its college football preview and ranks the top 25 teams based primarily on scouting reports. As the season progresses, other college football polls provide a weekly ranking of the teams, evaluations that are largely influenced by how well the teams are playing and who they play against. Using the 2002 rankings, let's investigate the similarity, or lack of similarity, of the *Sports Illustrated* preseason, *USA Today*/ESPN, and AP Top 25 polls.

Team	(1) *Sports Illustrated* (Preseason)	(2) *USA Today*/ ESPN (After Regular Season)	(3) AP Top 25 (After Regular Season)
Oklahoma	1	8	8

••• Remainder of data on Student's Suite CD-ROM

Sources: http://sportsillustrated.cnn.com, http://sports.espn.go.com, http://www.sltrib.com

Note: The ranks given to teams that were ranked in the preseason poll but no longer ranked after the season were obtained by $[26 + 27 + 28 + \ldots + (25 + n)] \div n$, where n is the number of teams no longer ranked in the top 25.

a. Compute the Spearman rank correlation coefficient for the *Sports Illustrated* preseason poll and the *USA Today*/ESPN, the *Sports Illustrated* preseason poll and the AP Top 25, and the *USA Today*/ESPN poll and the AP Top 25.

b. Test the null hypothesis that there is no relationship between the polls versus the alternative that there is a relationship between them for each of the three possible paired comparisons. Use $\alpha = 0.05$.

14.88 Nonparametric tests are also called distribution-free tests. However, the normal distributions are used in the inference-making procedures.

a. To what does the *distribution-free* term apply? (The population? The sample? The sampling distribution?) Explain.

b. What is it that has the normal distribution? Explain.

Chapter Project

Teenagers' Attitudes

Does every generation believe that the following generation has a flawed attitude regarding the important factors of life? Perhaps the media plays up these negative factors so that one perceives that a whole generation believes them. The data that follow are from Section 14.1, "Teenagers' Attitudes" (p. 750), which included a study by NFO Research, Inc., surveying American's adolescent population on the things teens want most from life.

The One Thing Teens Want Most from Life [EX14-01]

Choices	All	Boys	Girls
Happiness	28%	23%	32%
Long enjoyable life	16%	18%	14%
Marriage and family	9%	8%	11%
Financial success	8%	11%	4%
Career success	8%	9%	6%
Religious satisfaction	8%	7%	7%
Love	7%	6%	7%
Personal success	6%	6%	5%
Personal contribution to society	2%	3%	2%
Friends	2%	3%	1%
Health	2%	2%	2%
Education	2%	1%	2%

*Some teens didn't respond, so figures don't total 100%.

Putting Chapter 14 to Work

14.89 a. Do responses from the boys and the girls have the same distribution? Use the Mann–Whitney U test and $\alpha = 0.05$.

b. Rank the choices for the boys and girls separately of each other.

c. Do the boys' preferences correlate to the girls' preferences? Use Spearman's rank correlation to test at the 0.05 level of significance.

d. Compare the results obtained in parts b and c.

Your Study

14.90 Define a population of your choice and randomly sample the following variables:

Variable 1: Gender or class level (freshmen, sophomore, upper classes, etc.)

Variable 2: Name "The One Thing Want Most From Life"; choose from happiness, long enjoyable life, marriage and family, financial success, career success, religious satisfaction, love, personal success, personal contribution to society, friends, health, or education

a. Rank the choices for each level of variable 1.

b. Do the responses from the levels of variable 1 have the same distribution? Use the Mann–Whitney U test and $\alpha = 0.05$

c. Do the levels of the responses correlate to each other? Use Spearman's rank correlation to test at the 0.05 level of significance.

d. Write a paragraph comparing and contrasting the results obtained in parts b and c.

Chapter Practice Test

PART I: Knowing the Definitions

Answer "True" if the statement is always true. If the statement is not always true, replace the words shown in bold with words that make the statement always true.

14.1 One of the advantages of the nonparametric tests is the necessity for **less restrictive** assumptions.

14.2 The sign test is a possible replacement for the *F*-test.

14.3 The **sign test** can be used to test the randomness of a set of data.

14.4 If a tie occurs in a set of ranked data, the data that form the tie are **removed from the set.**

14.5 Two dependent **means** can be compared nonparametrically by using the sign test.

14.6 The sign test is a possible alternative to Student's *t*-test for **one mean value.**

14.7 The **runs test** is a nonparametric alternative to the difference between two independent means.

14.8 The **confidence level** of a statistical hypothesis test is measured by $1 - \beta$.

14.9 Spearman's rank correlation coefficient is an alternative to using the **linear correlation coefficient.**

14.10 The **efficiency** of a nonparametric test is the probability that a false null hypothesis is rejected.

PART II: Applying the Concepts

14.11 The weights (in pounds) of nine people before they stopped smoking and five weeks after they stopped smoking are listed here: [PT14-11]

Person	1	2	3	4	5	6	7	8	9
Before	148	176	153	116	128	129	120	132	154
After	155	178	151	120	130	136	126	128	158

Find the 95% confidence interval estimate for the average weight change.

14.12 The following data show the weight gains (in ounces) for 20 laboratory mice, half of which were fed one diet and half a different diet. Test to determine whether the difference in weight gain is significant at $\alpha = 0.05$. [PT14-12]

Diet A	41	40	36	43	36	43	39	36	24	41
Diet B	35	34	27	39	31	41	37	34	42	38

14.13 A large textbook publishing company hired nine new sales representatives 3 years ago. At the time of hire, the nine were ranked according to their potential. Now 3 years later the company president wants to know how well their potential ranks correlate with their sales totals for the 3 years. [PT14-13]

Sales Representative	a	b	c	d	e	f	g	h	l
Potential	2	5	6	1	4	3	9	8	7
Sales Total	450	410	350	345	330	400	250	310	270

Is there significant correlation at the 0.05 level?

14.14 The new school principal thought there might be a pattern to the order in which discipline problems arrived at his office. He had his secretary record the grade levels of the students as they arrived. [PT14-14]

9	10	11	9	12	11	9	10	10	11
10	11	10	10	11	12	12	9	9	11
12	10	9	12	10	11	12	11	10	10

At the 0.05 level, is there significant evidence of randomness?

PART III: Understanding the Concepts

14.15 What advantages do nonparametric statistics have over parametric methods?

14.16 Explain how the sign test is based on the binomial distribution and is often approximated by the normal distribution.

14.17 Why does the sign test use a null hypothesis about the median instead of the mean like a *t*-test uses?

14.18 Explain why a nonparametric test is not as sensitive to extreme data as a parametric test might be.

14.19 A restaurant has collected data on which of two seating arrangements its customers prefer. In a sign test to determine whether one seating arrangement is significantly preferred, which null hypothesis would be used?

a. $M = 0$ b. $M = 0.5$
c. $p = 0$ d. $p = 0.5$

Explain your choice.

Statistics ⬡ Now™ Preparing for an exam? Assess your progress by taking the post-test at **http://1pass.thomson.com.**

vMentor Do you need a live tutor for homework problems? Access vMentor on the StatisticsNow website at **http://1pass.thomson.com** for one-on-one tutoring from a statistics expert.

Working with Your Own Data

The existence of bivariate data is commonplace in everyday life and there are multiple options for analyzing the relationship between the two variables. In Chapter 10, the relationship between the paired data was analyzed as paired differences. In Chapter 12, the analysis of variance methods tested for an effect that one variable might have on a second variable. In Chapter 13, the methods of correlation and regression were used to investigate the relationship between the variables in order to determine whether they have a mathematical relationship that can be approximated by means of a straight line. The following illustrates such a situation.

A The Age and Value of Peggy's Car

Peggy would like to sell her 1994 Corvette, and she wants to determine an asking price for it in order to advertise. Her Corvette features the typical Corvette equipment with no customizing and is in average condition for a well-cared-for 1994 Corvette. She wants to advertise using an average asking price and expects to get an average price for it (average for a Corvette!) when it sells. Presently she must answer the question, "What is an average asking price for a 1994 Corvette?"

Inspection of many classified sections of newspapers turned up only three advertisements for 1994 Corvettes. The prices listed varied a great deal, and Peggy needed more information to determine her asking price. She decided to use the Internet and search for prices. She restricted her search to used Chevrolet Corvettes, 1988 to present, that were in good repair, not customized or show cars, and were being sold by their owner, not a dealer or an auction. Peggy collected the following data on January 8, 2003.

Peggy knows that the price she should ask, and the amount she receives, for her Corvette is affected by its age. The general questions are: "Is the effect of age on price predictable?" and "Can a meaningful relationship between the age and the typical asking price for used Corvettes be established?"

Independent variable, x: The age of the car as measured in years and defined by

$x =$ (present calendar year) −
    (year of manufacture) + 1

Example: During 2003, Peggy's 1994 Corvette is considered to be 10 years old.

$x = (2003 - 1994) + 1 = 9 + 1 = 10$

Dependent variable, y: The advertised asking price

1. Discuss why age should be used instead of manufacture year.

Year	Asking Price	Year	Asking Price	Year	Asking Price	Year	Asking Price	Year	Asking Price
2001	$35,800	1989	$15,700	1989	$18,200	1998	$26,500	1995	$18,900
1995	$19,900	2002	$35,500	1992	$17,900	2001	$36,000	2000	$33,000
1996	$26,700	2001	$35,700	2002	$35,900	1997	$21,000	1994	$23,400
1994	$19,000	1996	$21,900	1997	$28,900	1991	$11,500	1999	$31,500
1991	$16,700	1999	$33,000	1998	$28,600	1994	$16,000	1999	$31,900
1991	$17,500	1992	$15,500	1997	$30,000	1995	$24,000	1998	$30,000
1997	$30,600	1999	$29,000	2000	$30,500	1995	$21,700	1988	$14,400
2000	$32,000	1988	$ 9,500	1993	$17,500	1989	$10,500	1990	$16,000
1998	$33,000	1993	$16,500	1991	$15,500	1990	$17,500	1996	$21,000
1994	$16,000	1992	$16,000	2000	$35,000	1990	$14,800	2001	$34,000
1996	$25,800	1993	$16,300	1997	$25,000	1992	$19,800		
1988	$10,500	2002	$35,000	1993	$14,000	1993	$19,500		

2. Convert manufacture year to the variable age, x, using the preceding formula.

3. Construct side-by-side vertical dotplots of the asking price for each year of age.

4. Does it appear that the asking price for a Corvette is affected by its age? Describe the effect as pictured on the graph shown for question 3.

5. Complete a one-way ANOVA and test the hypothesis that age has no effect on the asking price. Use $\alpha = 0.05$.

6. What effect would using year of manufacture instead of age have on the above analysis?

7. Discuss why using age instead of manufacture year results in a scatter diagram that is more representative of behavior of the price or value of a used car as it ages. Describe how a scatter diagram using year of manufacture as x would differ from one using age as x.

8. Construct and label a scatter diagram of Peggy's data.

9. Discuss the relationship between the side-by-side dotplots drawn in question 3 and the scatter diagram drawn in question 8.

10. Determine the equation for the line of best fit.

11. Draw the line of best fit on the scatter diagram.

12. Test the question of the line of best fit to see whether the linear model is appropriate for the data. Use $\alpha = 0.05$.

13. Construct a 95% confidence interval for the mean advertised price for 1994 Corvettes.

14. Draw a line segment on the scatter diagram that represents the interval estimate found for question 13.

15. What does the value of the slope, b_1, represent? Explain.

16. What does the value of the y-intercept, b_0, represent? Explain.

17. Write a meaningful paragraph answering the general questions of concern:

    a. Is the effect of age on price predictable?

    b. Can a meaningful relationship between the age and the typical asking price for used Corvettes be established?

c. What is an average asking price for a 1994 Corvette that is for sale?

B Your Own Investigation

Identify a situation of interest to you that can be investigated statistically using bivariate data. (Consult your instructor for specific guidance.)

1. Define the population, the independent variable, the dependent variable, and the purpose for studying these two variables as a regression analysis.

2. Collect 15 to 20 ordered pairs of data.

3. Partition the independent variable values into three or more categories that are meaningful or appropriate for your data.

4. Complete a one-way ANOVA and test the hypothesis that the independent variable has no effect on the dependent variable.

5. Construct and label a scatter diagram of your data.

6. Determine the equation for the line of best fit.

7. Draw the line of best fit on the scatter diagram.

8. Test the equation of the line of best fit to see whether the linear model is appropriate for the data. Use $\alpha = 0.05$.

9. Construct a 95% confidence interval for the mean value of the dependent variable at the following value of x: Let x be equal to one-third the sum of the lowest value of x in your sample and twice the largest value; that is,

$$x = \frac{L + 2H}{3}$$

10. Draw a line segment on the scatter diagram that represents the interval estimate found for question 9.

11. What does the value of the slope, b_1, represent? Explain.

12. What does the value of the y-intercept, b_0, represent? Explain.

13. Write a meaningful paragraph comparing and/or contrasting the results from your ANOVA test and the test for the appropriateness of a linear model. What conclusions can you reach based on the results of these tests?

Basic Principles of Counting

Appendix A is available on the Student's Suite CD-ROM.

B Tables

TABLE 1

Random Numbers

10 09 73 25 33	76 52 01 35 86	34 67 35 48 76	80 95 90 91 17	39 29 27 49 45
37 54 20 48 05	64 89 47 42 96	24 80 52 40 37	20 63 61 04 02	00 82 29 16 65
08 42 26 89 53	19 64 50 93 03	23 20 90 25 60	15 95 33 43 64	35 08 03 36 06
99 01 90 25 29	09 37 67 07 15	38 31 13 11 65	88 67 67 43 97	04 43 62 76 59
12 80 79 99 70	80 15 73 61 47	64 03 23 66 53	98 95 11 68 77	12 17 17 68 33
66 06 57 47 17	34 07 27 68 50	36 69 73 61 70	65 81 33 98 85	11 19 92 91 70
31 06 01 08 05	45 57 18 24 06	35 30 34 26 14	86 79 90 74 39	23 40 30 97 32
85 26 97 76 02	02 05 16 56 92	68 66 57 48 18	73 05 38 52 47	18 62 38 85 79
63 57 33 21 35	05 32 54 70 48	90 55 35 75 48	28 46 82 87 09	83 49 12 56 24
73 79 64 57 53	03 52 96 47 78	35 80 83 42 82	60 93 52 03 44	35 27 38 84 35
98 52 01 77 67	14 90 56 86 07	22 10 94 05 58	60 97 09 34 33	50 50 07 39 98
11 80 50 54 31	39 80 82 77 32	50 72 56 82 48	29 40 52 42 01	52 77 56 78 51
83 45 29 96 34	06 28 89 80 83	13 74 67 00 78	18 47 54 06 10	68 71 17 78 17
88 68 54 02 00	86 50 75 84 01	36 76 66 79 51	90 36 47 64 93	29 60 91 10 62
99 59 46 73 48	87 51 76 49 69	91 82 60 89 28	93 78 56 13 68	23 47 83 41 13
65 48 11 76 74	17 46 85 09 50	58 04 77 69 74	73 03 95 71 86	40 21 81 65 44
80 12 43 56 35	17 72 70 80 15	45 31 82 23 74	21 11 57 82 53	14 38 55 37 63
74 35 09 98 17	77 40 27 72 14	43 23 60 02 10	45 52 16 42 37	96 28 60 26 55
69 91 62 68 03	66 25 22 91 48	36 93 68 72 03	76 62 11 39 90	94 40 05 64 18
09 89 32 05 05	14 22 56 85 14	46 42 75 67 88	96 29 77 88 22	54 38 21 45 98
91 49 91 45 23	68 47 92 76 86	46 16 28 35 54	94 75 08 99 23	37 08 92 00 48
80 33 69 45 98	26 94 03 68 58	70 29 73 41 35	54 14 03 33 40	42 05 08 23 41
44 10 48 19 49	85 15 74 79 54	32 97 92 65 75	57 60 04 08 81	22 22 20 64 13
12 55 07 37 42	11 10 00 20 40	12 86 07 46 97	96 64 48 94 39	28 70 72 58 15
63 60 64 93 29	16 50 53 44 84	40 21 95 25 63	43 65 17 70 82	07 20 73 17 90
61 19 69 04 46	26 45 74 77 74	51 92 43 37 29	65 39 45 95 93	42 58 26 05 27
15 47 44 52 66	95 27 07 99 53	59 36 78 38 48	82 39 61 01 18	33 21 15 94 66
94 55 72 85 73	67 89 75 43 87	54 62 24 44 31	91 19 04 25 92	92 92 74 59 73
42 48 11 62 13	97 34 40 87 21	16 86 84 87 67	03 07 11 20 59	25 70 14 66 70
23 52 37 83 17	73 20 88 98 37	68 93 59 14 16	26 25 22 96 63	05 52 28 25 62
04 49 35 24 94	75 24 63 38 24	45 86 25 10 25	61 96 27 93 35	65 33 71 24 72
00 54 99 76 54	64 05 18 81 59	96 11 96 38 96	54 69 28 23 91	23 28 72 95 29
35 96 31 53 07	26 89 80 93 54	33 35 13 54 62	77 97 45 00 24	90 10 33 93 33
59 80 80 83 91	45 42 72 68 42	83 60 94 97 00	13 02 12 48 92	78 56 52 01 06
46 05 88 52 36	01 39 09 22 86	77 28 14 40 77	93 91 08 36 47	70 61 74 29 41
32 17 90 05 97	87 37 92 52 41	05 56 70 70 07	86 74 31 71 57	85 39 41 18 38
69 23 46 14 06	20 11 74 52 04	15 95 66 00 00	18 74 39 24 23	97 11 89 63 38
19 56 54 14 30	01 75 87 53 79	40 41 92 15 85	66 67 43 68 06	84 96 28 52 07
45 15 51 49 38	19 47 60 72 46	43 66 79 45 43	59 04 79 00 33	20 82 66 95 41
94 86 43 19 94	36 16 81 08 51	34 88 88 15 53	01 54 03 54 56	05 01 45 11 76
98 08 62 48 26	45 24 02 84 04	44 99 90 88 96	39 09 47 34 07	35 44 13 18 80
33 18 51 62 32	41 94 15 09 49	89 43 54 85 81	88 69 54 19 94	37 54 87 30 43
80 95 10 04 06	96 38 27 07 74	20 15 12 33 87	25 01 62 52 98	94 62 46 11 71
79 75 24 91 40	71 96 12 82 96	69 86 10 25 91	74 85 22 05 39	00 38 75 95 79
18 63 33 25 37	98 14 50 65 71	31 01 02 46 74	05 45 56 14 27	77 93 89 19 36

For specific details about using this table, see page 22 or the *Student Solutions Manual*.

TABLE 1

Random Numbers (continued)

74 02 94 39 02	77 55 73 22 70	97 79 01 71 19	52 52 75 80 21	80 81 45 17 48
54 17 84 56 11	80 99 33 71 43	05 33 51 29 69	56 12 71 92 55	36 04 09 03 24
11 66 44 98 83	52 07 98 48 27	59 38 17 15 39	09 97 33 34 40	88 46 12 33 56
48 32 47 79 28	31 24 96 47 10	02 29 53 68 70	32 30 75 75 46	15 02 00 99 94
69 07 49 41 38	87 63 79 19 76	35 58 40 44 01	10 51 82 16 15	01 84 87 69 38
09 18 82 00 97	32 82 53 95 27	04 22 08 63 04	83 38 98 73 74	64 27 85 80 44
90 04 58 54 97	51 98 15 06 54	94 93 88 19 97	91 87 07 61 50	68 47 66 46 59
73 18 95 02 07	47 67 72 62 69	62 29 06 44 64	27 12 46 70 18	41 36 18 27 60
75 76 87 64 90	20 97 18 17 49	90 42 91 22 72	95 37 50 58 71	93 82 34 31 78
54 01 64 40 56	66 28 13 10 03	00 68 22 73 98	20 71 45 32 95	07 70 61 78 13
08 35 86 99 10	78 54 24 27 85	13 66 15 88 73	04 61 89 75 53	31 22 30 84 20
28 30 60 32 64	81 33 31 05 91	40 51 00 78 93	32 60 46 04 75	94 11 90 18 40
53 84 08 62 33	81 59 41 36 28	51 21 59 02 90	28 46 66 87 95	77 76 22 07 91
91 75 75 37 41	61 61 36 22 69	50 26 39 02 12	55 78 17 65 14	83 48 34 70 55
89 41 59 26 94	00 39 75 83 91	12 60 71 76 46	48 94 97 23 06	94 54 13 74 08
77 51 30 38 20	86 83 42 99 01	68 41 48 27 74	51 90 81 39 80	72 89 35 55 07
19 50 23 71 74	69 97 92 02 88	55 21 02 97 73	74 28 77 52 51	65 34 46 74 15
21 81 85 93 13	93 27 88 17 57	05 68 67 31 56	07 08 28 50 46	31 85 33 84 52
51 47 46 64 99	68 10 72 36 21	94 04 99 13 45	42 83 60 91 91	08 00 74 54 49
99 55 96 83 31	62 53 52 41 70	69 77 71 28 30	74 81 97 81 42	43 86 07 28 34
33 71 34 80 07	93 58 47 28 69	51 92 66 47 21	58 30 32 98 22	93 17 49 39 72
85 27 48 68 93	11 30 32 92 70	28 83 43 41 37	73 51 59 04 00	71 14 84 36 43
84 13 38 96 40	44 03 55 21 66	73 85 27 00 91	61 22 26 05 61	62 32 71 84 23
56 73 21 62 34	17 39 59 61 31	10 12 39 16 22	85 49 65 75 60	81 60 41 88 80
65 13 85 68 06	87 60 88 52 61	34 31 36 58 61	45 87 52 10 69	85 64 44 72 77
38 00 10 21 76	81 71 91 17 11	71 60 29 29 37	74 21 96 40 49	65 58 44 96 98
37 40 29 63 97	01 30 47 75 86	56 27 11 00 86	47 32 46 26 05	40 03 03 74 38
97 12 54 03 48	87 08 33 14 17	21 81 53 92 50	75 23 76 20 47	15 50 12 95 78
21 82 64 11 34	47 14 33 40 72	64 63 88 59 02	49 13 90 64 41	03 85 65 45 52
73 13 54 27 42	95 71 90 90 35	85 79 47 42 96	08 78 98 81 56	64 69 11 92 02
07 63 87 79 29	03 06 11 80 72	96 20 74 41 56	23 82 19 95 38	04 71 36 69 94
60 52 88 34 41	07 95 41 98 14	59 17 52 06 95	05 53 35 21 39	61 21 20 64 55
83 59 63 56 55	06 95 89 29 83	05 12 80 97 19	77 43 35 37 83	92 30 15 04 98
10 85 06 27 46	99 59 91 05 07	13 49 90 63 19	53 07 57 18 39	06 41 01 93 62
39 82 09 89 52	43 62 26 31 47	64 42 18 08 14	43 80 00 93 51	31 02 47 31 67
59 58 00 64 78	75 56 97 88 00	88 83 55 44 86	23 76 80 61 56	04 11 10 84 08
38 50 80 73 41	23 79 34 87 63	90 82 29 70 22	17 71 90 42 07	95 95 44 99 53
30 69 27 06 68	94 68 81 61 27	56 19 68 00 91	82 06 76 34 00	05 46 26 92 00
65 44 39 56 59	18 28 82 74 37	49 63 22 40 41	08 33 76 56 76	96 29 99 08 36
27 26 75 02 64	13 19 27 22 94	07 47 74 46 06	17 98 54 89 11	97 34 13 03 58
91 30 70 69 91	19 07 22 42 10	36 69 95 37 28	28 82 53 57 93	28 97 66 62 52
68 43 49 46 88	84 47 31 36 22	62 12 69 84 08	12 84 38 25 90	09 81 59 31 46
48 90 81 58 77	54 74 52 45 91	35 70 00 47 54	83 82 45 26 92	54 13 05 51 60
06 91 34 51 97	42 67 27 86 01	11 88 30 95 28	63 01 19 89 01	14 97 44 03 44
10 45 51 60 19	14 21 03 37 12	91 34 23 78 21	88 32 58 08 51	43 66 77 08 83
12 88 39 73 43	65 02 76 11 84	04 28 50 13 92	17 97 41 50 77	90 71 22 67 69
21 77 83 09 76	38 80 73 69 61	31 64 94 20 96	63 28 10 20 23	08 81 64 74 49
19 52 35 95 15	65 12 25 96 59	86 28 36 82 58	69 57 21 37 98	16 43 59 15 29
67 24 55 26 70	35 58 31 65 63	79 24 68 66 86	76 46 33 42 22	26 65 59 08 02
60 58 44 73 77	07 50 03 79 92	45 13 42 65 29	26 76 08 36 37	41 32 64 43 44
53 85 34 13 77	36 06 69 48 50	58 83 87 38 59	49 36 47 33 31	96 24 04 36 42
24 63 73 97 36	74 38 48 93 42	52 62 30 79 92	12 36 91 86 01	03 74 28 38 73
83 08 01 24 51	38 99 22 28 15	07 75 95 17 77	97 37 72 75 85	51 97 23 78 67
16 44 42 43 34	36 15 19 90 73	27 49 37 09 39	85 13 03 25 52	54 84 65 47 59
60 79 01 81 57	57 17 86 57 62	11 16 17 85 76	45 81 95 29 79	65 13 00 48 60

From tables of the RAND Corporation. Reprinted from Wilfred J. Dixon and Frank J. Massey, Jr., *Introduction to Statistical Analysis*. 3rd ed. (New York: McGraw-Hill, 1969), pp. 446–447. Reprinted by permission of the RAND Corporation.

TABLE 2

Binomial Probabilities $[\binom{n}{x} \cdot p^x \cdot q^{n-x}]$

								P							
n	x	0.01	0.05	0.10	0.20	0.30	0.40	0.50	0.60	0.70	0.80	0.90	0.95	0.99	x
2	0	.980	.902	.810	.640	.490	.360	.250	.160	.090	.040	.010	.002	0+	0
	1	.020	.095	.180	.320	.420	.480	.500	.480	.420	.320	.180	.095	.020	1
	2	0+	.002	.010	.040	.090	.160	.250	.360	.490	.640	.810	.902	.980	2
3	0	.970	.857	.729	.512	.343	.216	.125	.064	.027	.008	.001	0+	0+	0
	1	.029	.135	.243	.384	.441	.432	.375	.288	.189	.096	.027	.007	0+	1
	2	0+	.007	.027	.096	.189	.288	.375	.432	.441	.384	.243	.135	.029	2
	3	0+	0+	.001	.008	.027	.064	.125	.216	.343	.512	.729	.857	.970	3
4	0	.961	.815	.656	.410	.240	.130	.062	.026	.008	.002	0+	0+	0+	0
	1	.039	.171	.292	.410	.412	.346	.250	.154	.076	.026	.004	0+	0+	1
	2	.001	.014	.049	.154	.265	.346	.375	.346	.265	.154	.049	.014	.001	2
	3	0+	0+	.004	.026	.076	.154	.250	.346	.412	.410	.292	.171	.039	3
	4	0+	0+	0+	.002	.008	.026	.062	.130	.240	.410	.656	.815	.961	4
5	0	.951	.774	.590	.328	.168	.078	.031	.010	.002	0+	0+	0+	0+	0
	1	.048	.204	.328	.410	.360	.259	.156	.077	.028	.006	0+	0+	0+	1
	2	.001	.021	.073	.205	.309	.346	.312	.230	.132	.051	.008	.001	0+	2
	3	0+	.001	.008	.051	.132	.230	.312	.346	.309	.205	.073	.021	.001	3
	4	0+	0+	0+	.006	.028	.077	.156	.259	.360	.410	.328	.204	.048	4
	5	0+	0+	0+	0+	.002	.010	.031	.078	.168	.328	.590	.774	.951	5
6	0	.941	.735	.531	.262	.118	.047	.016	.004	.001	0+	0+	0+	0+	0
	1	.057	.232	.354	.393	.303	.187	.094	.037	.010	.002	0+	0+	0+	1
	2	.001	.031	.098	.246	.324	.311	.234	.138	.060	.015	.001	0+	0+	2
	3	0+	.002	.015	.082	.185	.276	.312	.276	.185	.082	.015	.002	0+	3
	4	0+	0+	.001	.015	.060	.138	.234	.311	.324	.246	.098	.031	.001	4
	5	0+	0+	0+	.002	.010	.037	.094	.187	.303	.393	.354	.232	.057	5
	6	0+	0+	0+	0+	.001	.004	.016	.047	.118	.262	.531	.735	.941	6
7	0	.932	.698	.478	.210	.082	.028	.008	.002	0+	0+	0+	0+	0+	0
	1	.066	.257	.372	.367	.247	.131	.055	.017	.004	0+	0+	0+	0+	1
	2	.002	.041	.124	.275	.318	.261	.164	.077	.025	.004	0+	0+	0+	2
	3	0+	.004	.023	.115	.227	.290	.273	.194	.097	.029	.003	0+	0+	3
	4	0+	0+	.003	.029	.097	.194	.273	.290	.227	.115	.023	.004	0+	4
	5	0+	0+	0+	.004	.025	.077	.164	.261	.318	.275	.124	.041	.002	5
	6	0+	0+	0+	0+	.004	.017	.055	.131	.247	.367	.372	.257	.066	6
	7	0+	0+	0+	0+	0+	.002	.008	.028	.082	.210	.478	.698	.932	7
8	0	.923	.663	.430	.168	.058	.017	.004	.001	0+	0+	0+	0+	0+	0
	1	.075	.279	.383	.336	.198	.090	.031	.008	.001	0+	0+	0+	0+	1
	2	.003	.051	.149	.294	.296	.209	.109	.041	.010	.001	0+	0+	0+	2
	3	0+	.005	.033	.147	.254	.279	.219	.124	.047	.009	0+	0+	0+	3
	4	0+	0+	.005	.046	.136	.232	.273	.232	.136	.046	.005	0+	0+	4
	5	0+	0+	0+	.009	.047	.124	.219	.279	.254	.147	.033	.005	0+	5
	6	0+	0+	0+	.001	.010	.041	.109	.209	.296	.294	.149	.051	.003	6
	7	0+	0+	0+	0+	.001	.008	.031	.090	.198	.336	.383	.279	.075	7
	8	0+	0+	0+	0+	0+	.001	.004	.017	.058	.168	.430	.663	.923	8

For specific details about using this table, see pages 291–292.

TABLE 2

Binomial Probabilities $[\binom{n}{x} \cdot p^x \cdot q^{n-x}]$ (continued)

n	x	0.01	0.05	0.10	0.20	0.30	0.40	0.50	0.60	0.70	0.80	0.90	0.95	0.99	x
9	0	.914	.630	.387	.134	.040	.010	.002	0+	0+	0+	0+	0+	0+	0
	1	.083	.299	.387	.302	.156	.060	.018	.004	0+	0+	0+	0+	0+	1
	2	.003	.063	.172	.302	.267	.161	.070	.021	.004	0+	0+	0+	0+	2
	3	0+	.008	.045	.176	.267	.251	.164	.074	.021	.003	0+	0+	0+	3
	4	0+	.001	.007	.066	.172	.251	.246	.167	.074	.017	.001	0+	0+	4
	5	0+	0+	.001	.017	.074	.167	.246	.251	.172	.066	.007	.001	0+	5
	6	0+	0+	0+	.003	.021	.074	.164	.251	.267	.176	.045	.008	0+	6
	7	0+	0+	0+	0+	.004	.021	.070	.161	.267	.302	.172	.063	.003	7
	8	0+	0+	0+	0+	0+	.004	.018	.060	.156	.302	.387	.299	.083	8
	9	0+	0+	0+	0+	0+	0+	.002	.010	.040	.134	.387	.630	.914	9
10	0	.904	.599	.349	.107	.028	.006	.001	0+	0+	0+	0+	0+	0+	0
	1	.091	.315	.387	.268	.121	.040	.010	.002	0+	0+	0+	0+	0+	1
	2	.004	.075	.194	.302	.233	.121	.044	.011	.001	0+	0+	0+	0+	2
	3	0+	.010	.057	.201	.267	.215	.117	.042	.009	.001	0+	0+	0+	3
	4	0+	.001	.011	.088	.200	.251	.205	.111	.037	.006	0+	0+	0+	4
	5	0+	0+	.001	.026	.103	.201	.246	.201	.103	.026	.001	0+	0+	5
	6	0+	0+	0+	.006	.037	.111	.205	.251	.200	.088	.011	.001	0+	6
	7	0+	0+	0+	.001	.009	.042	.117	.215	.267	.201	.057	.010	0+	7
	8	0+	0+	0+	0+	.001	.011	.044	.121	.233	.302	.194	.075	.004	8
	9	0+	0+	0+	0+	0+	.002	.010	.040	.121	.268	.387	.315	.091	9
	10	0+	0+	0+	0+	0+	0+	.001	.006	.028	.107	.349	.599	.904	10
11	0	.895	.569	.314	.086	.020	.004	0+	0+	0+	0+	0+	0+	0+	0
	1	.099	.329	.384	.236	.093	.027	.005	.001	0+	0+	0+	0+	0+	1
	2	.005	.087	.213	.295	.200	.089	.027	.005	.001	0+	0+	0+	0+	1
	3	0+	.014	.071	.221	.257	.177	.081	.023	.004	0+	0+	0+	0+	3
	4	0+	.001	.016	.111	.220	.236	.161	.070	.017	.002	0+	0+	0+	4
	5	0+	0+	.002	.039	.132	.221	.226	.147	.057	.010	0+	0+	0+	5
	6	0+	0+	0+	.010	.057	.147	.226	.221	.132	.039	.002	0+	0+	6
	7	0+	0+	0+	.002	.017	.070	.161	.236	.220	.111	.016	.001	0+	7
	8	0+	0+	0+	0+	.004	.023	.081	.177	.257	.221	.071	.014	0+	8
	9	0+	0+	0+	0+	.001	.005	.027	.089	.200	.295	.213	.087	.005	9
	10	0+	0+	0+	0+	0+	.001	.005	.027	.093	.236	.384	.329	.099	10
	11	0+	0+	0+	0+	0+	0+	0+	.004	.020	.086	.314	.569	.895	11
12	0	.886	.540	.282	.069	.014	.002	0+	0+	0+	0+	0+	0+	0+	0
	1	.107	.341	.377	.206	.071	.017	.003	0+	0+	0+	0+	0+	0+	1
	2	.006	.099	.230	.283	.168	.064	.016	.002	0+	0+	0+	0+	0+	2
	3	0+	.017	.085	.236	.240	.142	.054	.012	.001	0+	0+	0+	0+	3
	4	0+	.002	.021	.133	.231	.213	.121	.042	.008	.001	0+	0+	0+	4
	5	0+	0+	.004	.053	.158	.227	.193	.101	.029	.003	0+	0+	0+	5
	6	0+	0+	0+	.016	.079	.177	.226	.177	.079	.016	0+	0+	0+	6
	7	0+	0+	0+	.003	.029	.101	.193	.227	.158	.053	.004	0+	0+	7
	8	0+	0+	0+	.001	.008	.042	.121	.213	.231	.133	.021	.002	0+	8
	9	0+	0+	0+	0+	.001	.012	.054	.142	.240	.236	.085	.017	0+	9
	10	0+	0+	0+	0+	0+	.002	.016	.064	.168	.283	.230	.099	.006	10
	11	0+	0+	0+	0+	0+	0+	.003	.017	.071	.206	.377	.341	.107	11
	12	0+	0+	0+	0+	0+	0+	0+	.002	.014	.069	.282	.540	.886	12

TABLE 2

Binomial Probabilities $[\binom{n}{x} \cdot p^x \cdot q^{n-x}]$ (continued)

n	x	0.01	0.05	0.10	0.20	0.30	0.40	0.50	0.60	0.70	0.80	0.90	0.95	0.99	x
13	0	.878	.513	.254	.055	.010	.001	0+	0+	0+	0+	0+	0+	0+	0
	1	.115	.351	.367	.179	.054	.011	.002	0+	0+	0+	0+	0+	0+	1
	2	.007	.111	.245	.268	.139	.045	.010	.001	0+	0+	0+	0+	0+	2
	3	0+	.021	.100	.246	.218	.111	.035	.006	.001	0+	0+	0+	0+	3
	4	0+	.003	.028	.154	.234	.184	.087	.024	.003	0+	0+	0+	0+	4
	5	0+	0+	.006	.069	.180	.221	.157	.066	.014	.001	0+	0+	0+	5
	6	0+	0+	.001	.023	.103	.197	.209	.131	.044	.006	0+	0+	0+	6
	7	0+	0+	0+	.006	.044	.131	.209	.197	.103	.023	.001	0+	0+	7
	8	0+	0+	0+	.001	.014	.066	.157	.221	.180	.069	.006	0+	0+	8
	9	0+	0+	0+	0+	.003	.024	.087	.184	.234	.154	.028	.003	0+	9
	10	0+	0+	0+	0+	.001	.006	.035	.111	.218	.246	.100	.021	0+	10
	11	0+	0+	0+	0+	0+	.001	.010	.045	.139	.268	.245	.111	.007	11
	12	0+	0+	0+	0+	0+	0+	.002	.011	.054	.179	.367	.351	.115	12
	13	0+	0+	0+	0+	0+	0+	0+	.001	.010	.055	.254	.513	.878	13
14	0	.869	.488	.229	.044	.007	.001	0+	0+	0+	0+	0+	0+	0+	0
	1	.123	.359	.356	.154	.041	.007	.001	0+	0+	0+	0+	0+	0+	1
	2	.008	.123	.257	.250	.113	.032	.006	.001	0+	0+	0+	0+	0+	2
	3	0+	.026	.114	.250	.194	.085	.022	.003	0+	0+	0+	0+	0+	3
	4	0+	.004	.035	.172	.229	.155	.061	.014	.001	0+	0+	0+	0+	4
	5	0+	0+	.008	.086	.196	.207	.122	.041	.007	0+	0+	0+	0+	5
	6	0+	0+	.001	.032	.126	.207	.183	.092	.023	.002	0+	0+	0+	6
	7	0+	0+	0+	.009	.062	.157	.209	.157	.062	.009	0+	0+	0+	7
	8	0+	0+	0+	.002	.023	.092	.183	.207	.126	.032	.001	0+	0+	8
	9	0+	0+	0+	0+	.007	.041	.122	.207	.196	.086	.008	0+	0+	9
	10	0+	0+	0+	0+	.001	.014	.061	.155	.229	.172	.035	.004	0+	10
	11	0+	0+	0+	0+	0+	.003	.022	.085	.194	.250	.114	.026	0+	11
	12	0+	0+	0+	0+	0+	.001	.006	.032	.113	.250	.257	.123	.008	12
	13	0+	0+	0+	0+	0+	0+	.001	.007	.041	.154	.356	.359	.123	13
	14	0+	0+	0+	0+	0+	0+	0+	.001	.007	.044	.229	.488	.869	14
15	0	.860	.463	.206	.035	.005	0+	0+	0+	0+	0+	0+	0+	0+	0
	1	.130	.366	.343	.132	.031	.005	0+	0+	0+	0+	0+	0+	0+	1
	2	.009	.135	.267	.231	.092	.022	.003	0+	0+	0+	0+	0+	0+	2
	3	0+	.031	.129	.250	.170	.063	.014	.002	0+	0+	0+	0+	0+	3
	4	0+	.005	.043	.188	.219	.127	.042	.007	.001	0+	0+	0+	0+	4
	5	0+	.001	.010	.103	.206	.186	.092	.024	.003	0+	0+	0+	0+	5
	6	0+	0+	.002	.043	.147	.207	.153	.061	.012	.001	0+	0+	0+	6
	7	0+	0+	0+	.014	.081	.177	.196	.118	.035	.003	0+	0+	0+	7
	8	0+	0+	0+	.003	.035	.118	.196	.177	.081	.014	0+	0+	0+	8
	9	0+	0+	0+	.001	.012	.061	.153	.207	.147	.043	.002	0+	0+	9
	10	0+	0+	0+	0+	.003	.024	.092	.186	.206	.103	.010	.001	0+	10
	11	0+	0+	0+	0+	.001	.007	.042	.127	.219	.188	.043	.005	0+	11
	12	0+	0+	0+	0+	0+	.002	.014	.063	.170	.250	.129	.031	0+	12
	13	0+	0+	0+	0+	0+	0+	.003	.022	.092	.231	.267	.135	.009	13
	14	0+	0+	0+	0+	0+	0+	0+	.005	.031	.132	.343	.366	.130	14
	15	0+	0+	0+	0+	0+	0+	0+	0+	.005	.035	.206	.463	.860	15

Areas of the Standard Normal Distribution

The entries in this table are the probabilities that a random variable, with a standard normal distribution, assumes a value between 0 and z; the probability is represented by the shaded area under the curve in the accompanying figure. Areas for negative values of z are obtained by symmetry.

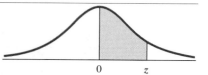

| | | | | | Second Decimal Place in z | | | | | |
z	0.00	0.01	0.02	0.03	0.04	0.05	0.06	0.07	0.08	0.09
0.0	0.0000	0.0040	0.0080	0.0120	0.0160	0.0199	0.0239	0.0279	0.0319	0.0359
0.1	0.0398	0.0438	0.0478	0.0517	0.0557	0.0596	0.0636	0.0675	0.0714	0.0753
0.2	0.0793	0.0832	0.0871	0.0910	0.0948	0.0987	0.1026	0.1064	0.1103	0.1141
0.3	0.1179	0.1217	0.1255	0.1293	0.1331	0.1368	0.1406	0.1443	0.1480	0.1517
0.4	0.1554	0.1591	0.1628	0.1664	0.1700	0.1736	0.1772	0.1808	0.1844	0.1879
0.5	0.1915	0.1950	0.1985	0.2019	0.2054	0.2088	0.2123	0.2157	0.2190	0.2224
0.6	0.2257	0.2291	0.2324	0.2357	0.2389	0.2422	0.2454	0.2486	0.2517	0.2549
0.7	0.2580	0.2611	0.2642	0.2673	0.2704	0.2734	0.2764	0.2794	0.2823	0.2852
0.8	0.2881	0.2910	0.2939	0.2967	0.2995	0.3023	0.3051	0.3078	0.3106	0.3133
0.9	0.3159	0.3186	0.3212	0.3238	0.3264	0.3289	0.3315	0.3340	0.3365	0.3389
1.0	0.3413	0.3438	0.3461	0.3485	0.3508	0.3531	0.3554	0.3577	0.3599	0.3621
1.1	0.3643	0.3665	0.3686	0.3708	0.3729	0.3749	0.3770	0.3790	0.3810	0.3830
1.2	0.3849	0.3869	0.3888	0.3907	0.3925	0.3944	0.3962	0.3980	0.3997	0.4015
1.3	0.4032	0.4049	0.4066	0.4082	0.4099	0.4115	0.4131	0.4147	0.4162	0.4177
1.4	0.4192	0.4207	0.4222	0.4236	0.4251	0.4265	0.4279	0.4292	0.4306	0.4319
1.5	0.4332	0.4345	0.4357	0.4370	0.4382	0.4394	0.4406	0.4418	0.4429	0.4441
1.6	0.4452	0.4463	0.4474	0.4484	0.4495	0.4505	0.4515	0.4525	0.4535	0.4545
1.7	0.4554	0.4564	0.4573	0.4582	0.4591	0.4599	0.4608	0.4616	0.4625	0.4633
1.8	0.4641	0.4649	0.4656	0.4664	0.4671	0.4678	0.4686	0.4693	0.4699	0.4706
1.9	0.4713	0.4719	0.4726	0.4732	0.4738	0.4744	0.4750	0.4756	0.4761	0.4767
2.0	0.4772	0.4778	0.4783	0.4788	0.4793	0.4798	0.4803	0.4808	0.4812	0.4817
2.1	0.4821	0.4826	0.4830	0.4834	0.4838	0.4842	0.4846	0.4850	0.4854	0.4857
2.2	0.4861	0.4864	0.4868	0.4871	0.4875	0.4878	0.4881	0.4884	0.4887	0.4890
2.3	0.4893	0.4896	0.4898	0.4901	0.4904	0.4906	0.4909	0.4911	0.4913	0.4916
2.4	0.4918	0.4920	0.4922	0.4925	0.4927	0.4929	0.4931	0.4932	0.4934	0.4936
2.5	0.4938	0.4940	0.4941	0.4943	0.4945	0.4946	0.4948	0.4949	0.4951	0.4952
2.6	0.4953	0.4955	0.4956	0.4957	0.4959	0.4960	0.4961	0.4962	0.4963	0.4964
2.7	0.4965	0.4966	0.4967	0.4968	0.4969	0.4970	0.4971	0.4972	0.4973	0.4974
2.8	0.4974	0.4975	0.4976	0.4977	0.4977	0.4978	0.4979	0.4979	0.4980	0.4981
2.9	0.4981	0.4982	0.4982	0.4983	0.4984	0.4984	0.4985	0.4985	0.4986	0.4986
3.0	0.4987	0.4987	0.4987	0.4988	0.4988	0.4989	0.4989	0.4989	0.4990	0.4990
3.1	0.4990	0.4991	0.4991	0.4991	0.4992	0.4992	0.4992	0.4992	0.4993	0.4993
3.2	0.4993	0.4993	0.4994	0.4994	0.4994	0.4994	0.4994	0.4995	0.4995	0.4995
3.3	0.4995	0.4995	0.4995	0.4996	0.4996	0.4996	0.4996	0.4996	0.4996	0.4997
3.4	0.4997	0.4997	0.4997	0.4997	0.4997	0.4997	0.4997	0.4997	0.4997	0.4998
3.5	0.4998	0.4998	0.4998	0.4998	0.4998	0.4998	0.4998	0.4998	0.4998	0.4998
3.6	0.4998	0.4998	0.4999	0.4999	0.4999	0.4999	0.4999	0.4999	0.4999	0.4999
3.7	0.4999									
4.0	0.49997									
4.5	0.499997									
5.0	0.4999997									

For specific details about using this table to find: probabilities, see pages 317–320; confidence coefficients, pages 338–339, 340–341; *p*-values, pages 432–433, 435; critical values, pages 317–320, 338–339.

TABLE 4

Critical Values of Standard Normal Distribution

A   ONE-TAILED SITUATIONS

The entries in this table are the critical values for z for which the area under the curve representing α is in the right-hand tail. Critical values for the left-hand tail are found by symmetry.

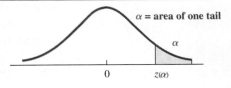

α = area of one tail

Amount of α in one tail

α	0.25	0.10	0.05	0.025	0.02	0.01	0.005
$z(\alpha)$	0.67	1.28	1.65	1.96	2.05	2.33	2.58

One-tailed example:
$\alpha = 0.05$
$z(\alpha) = z(0.05) = 1.65$

B   Two-Tailed Situations

The entries in this table are the critical values for z for which the area under the curve representing α is split equally between the two tails.

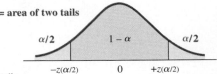

α = area of two tails

Amount of α in two-tails

α	0.25	0.20	0.10	0.05	0.02	0.01
$z(\alpha/2)$	1.15	1.28	1.65	1.96	2.33	2.58
$1 - \alpha$	0.75	0.80	0.90	0.95	0.98	0.99

Area in the "center"

Two-tailed example:
$\alpha = 0.05$ or $1 - \alpha = 0.95$
$\alpha/2 = 0.025$
$z(\alpha/2) = z(0.025) = 1.96$

For specific details about using:

Table A to find: critical values, see page 451.

Table B to find: confidence coefficients, see pages 403, 405; critical values, page 453.

p-Values for Standard Normal Distribution

The entries in this table are the *p*-values related to the right-hand tail for the calculated $z\star$ for the standard normal distribution.

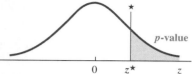

$z\star$	*p*-value	$z\star$	*p*-value	$z\star$	*p*-value	$z\star$	*p*-value
0.00	0.5000	1.00	0.1587	2.00	0.0228	3.00	0.0013
0.05	0.4801	1.05	0.1469	2.05	0.0202	3.05	0.0011
0.10	0.4602	1.10	0.1357	2.10	0.0179	3.10	0.0010
0.15	0.4404	1.15	0.1251	2.15	0.0158	3.15	0.0008
0.20	0.4207	1.20	0.1151	2.20	0.0139	3.20	0.0007
0.25	0.4013	1.25	0.1056	2.25	0.0122	3.25	0.0006
0.30	0.3821	1.30	0.0968	2.30	0.0107	3.30	0.0005
0.35	0.3632	1.35	0.0885	2.35	0.0094	3.35	0.0004
0.40	0.3446	1.40	0.0808	2.40	0.0082	3.40	0.0003
0.45	0.3264	1.45	0.0735	2.45	0.0071	3.45	0.0003
0.50	0.3085	1.50	0.0668	2.50	0.0062	3.50	0.0002
0.55	0.2912	1.55	0.0606	2.55	0.0054	3.55	0.0002
0.60	0.2743	1.60	0.0548	2.60	0.0047	3.60	0.0002
0.65	0.2578	1.65	0.0495	2.65	0.0040	3.65	0.0001
0.70	0.2420	1.70	0.0446	2.70	0.0035	3.70	0.0001
0.75	0.2266	1.75	0.0401	2.75	0.0030	3.75	0.0001
0.80	0.2119	1.80	0.0359	2.80	0.0026	3.80	0.0001
0.85	0.1977	1.85	0.0322	2.85	0.0022	3.85	0.0001
0.90	0.1841	1.90	0.0287	2.90	0.0019	3.90	0+
0.95	0.1711	1.95	0.0256	2.95	0.0016	3.95	0+

For specific details about using this table to find *p*-values, see pages 432, 433, 435.

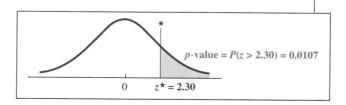

$$p\text{-value} = P(z > 2.30) = 0.0107$$

$z\star = 2.30$

TABLE 6

Critical Values of Student's *t*-Distribution

The entries in this table, $t(df, \alpha)$, are the critical values for Student's *t*-distribution for which the area under the curve in the right-hand tail is α. Critical values for the left-hand tail are found by symmetry.

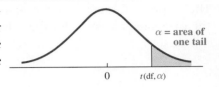

α = area of one tail

$0 \qquad t(df, \alpha)$

df	Amount of α in One Tail					
	0.25	0.10	0.05	0.025	0.01	0.005
	Amount of α in Two Tails					
	0.50	0.20	0.10	0.05	0.02	0.01
3	0.765	1.64	2.35	3.18	4.54	5.84
4	0.741	1.53	2.13	2.78	3.75	4.60
5	0.729	1.48	2.02	2.57	3.37	4.03
6	0.718	1.44	1.94	2.45	3.14	3.71
7	0.711	1.42	1.89	2.36	3.00	3.50
8	0.706	1.40	1.86	2.31	2.90	3.36
9	0.703	1.38	1.83	2.26	2.82	3.25
10	0.700	1.37	1.81	2.23	2.76	3.17
11	0.697	1.36	1.80	2.20	2.72	3.11
12	0.696	1.36	1.78	2.18	2.68	3.05
13	0.694	1.35	1.77	2.16	2.65	3.01
14	0.692	1.35	1.76	2.14	2.62	2.98
15	0.691	1.34	1.75	2.13	2.60	2.95
16	0.690	1.34	1.75	2.12	2.58	2.92
17	0.689	1.33	1.74	2.11	2.57	2.90
18	0.688	1.33	1.73	2.10	2.55	2.88
19	0.688	1.33	1.73	2.09	2.54	2.86
20	0.687	1.33	1.72	2.09	2.53	2.85
21	0.686	1.32	1.72	2.08	2.52	2.83
22	0.686	1.32	1.72	2.07	2.51	2.82
23	0.685	1.32	1.71	2.07	2.50	2.81
24	0.685	1.32	1.71	2.06	2.49	2.80
25	0.684	1.32	1.71	2.06	2.49	2.79
26	0.684	1.32	1.71	2.06	2.48	2.78
27	0.684	1.31	1.70	2.05	2.47	2.77
28	0.683	1.31	1.70	2.05	2.47	2.76
29	0.683	1.31	1.70	2.05	2.46	2.76
30	0.683	1.31	1.70	2.04	2.46	2.75
35	0.682	1.31	1.69	2.03	2.44	2.73
40	0.681	1.30	1.68	2.02	2.42	2.70
50	0.679	1.30	1.68	2.01	2.40	2.68
70	0.678	1.29	1.67	1.99	2.38	2.65
100	0.677	1.29	1.66	1.98	2.36	2.63
df > 100	0.675	1.28	1.65	1.96	2.33	2.58

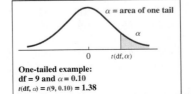

α = area of one tail

α

$0 \qquad t(df, \alpha)$

One-tailed example:
df = 9 and $\alpha = 0.10$
$t(df, \alpha) = t(9, 0.10) = 1.38$

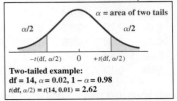

α = area of two tails

$\alpha/2 \qquad\qquad \alpha/2$

$-t(df, \alpha/2) \quad 0 \quad +t(df, \alpha/2)$

Two-tailed example:
df = 14, $\alpha = 0.02$, $1 - \alpha = 0.98$
$t(df, \alpha/2) = t(14, 0.01) = 2.62$

For specific details about using this table to find: confidence coefficients, see pages 477–480; *p*-values, pages 483, 486; critical values, pages 477–478.

TABLE 7

Probability-Values for Student's *t*-distribution

The entries in this table are the *p*-values related to the right-hand tail for the calculated *t★* value for the *t*-distribution of df degrees of freedom.

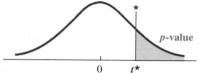

Degrees of Freedom

t★	3	4	5	6	7	8	10	12	15	18	21	25	29	35	df ≥ 45
0.0	0.500	0.500	0.500	0.500	0.500	0.500	0.500	0.500	0.500	0.500	0.500	0.500	0.500	0.500	0.500
0.1	0.463	0.463	0.462	0.462	0.462	0.461	0.461	0.461	0.461	0.461	0.461	0.461	0.461	0.460	0.460
0.2	0.427	0.426	0.425	0.424	0.424	0.423	0.423	0.422	0.422	0.422	0.422	0.422	0.421	0.421	0.421
0.3	0.392	0.390	0.388	0.387	0.386	0.386	0.385	0.385	0.384	0.384	0.384	0.383	0.383	0.383	0.383
0.4	0.358	0.355	0.353	0.352	0.351	0.350	0.349	0.348	0.347	0.347	0.347	0.346	0.346	0.346	0.346
0.5	0.326	0.322	0.319	0.317	0.316	0.315	0.314	0.313	0.312	0.312	0.311	0.311	0.310	0.310	0.310
0.6	0.295	0.290	0.287	0.285	0.284	0.283	0.281	0.280	0.279	0.278	0.277	0.277	0.277	0.276	0.276
0.7	0.267	0.261	0.258	0.255	0.253	0.252	0.250	0.249	0.247	0.246	0.246	0.245	0.245	0.244	0.244
0.8	0.241	0.234	0.230	0.227	0.225	0.223	0.221	0.220	0.218	0.217	0.216	0.216	0.215	0.215	0.214
0.9	0.217	0.210	0.205	0.201	0.199	0.197	0.195	0.193	0.191	0.190	0.189	0.188	0.188	0.187	0.186
1.0	0.196	0.187	0.182	0.178	0.175	0.173	0.170	0.169	0.167	0.165	0.164	0.163	0.163	0.162	0.161
1.1	0.176	0.167	0.161	0.157	0.154	0.152	0.149	0.146	0.144	0.143	0.142	0.141	0.140	0.139	0.139
1.2	0.158	0.148	0.142	0.138	0.135	0.132	0.129	0.127	0.124	0.123	0.122	0.121	0.120	0.119	0.118
1.3	0.142	0.132	0.125	0.121	0.117	0.115	0.111	0.109	0.107	0.105	0.104	0.103	0.102	0.101	0.100
1.4	0.128	0.117	0.110	0.106	0.102	0.100	0.096	0.093	0.091	0.089	0.088	0.087	0.086	0.085	0.084
1.5	0.115	0.104	0.097	0.092	0.089	0.086	0.082	0.080	0.077	0.075	0.074	0.073	0.072	0.071	0.070
1.6	0.104	0.092	0.085	0.080	0.077	0.074	0.070	0.068	0.065	0.064	0.062	0.061	0.060	0.059	0.058
1.7	0.094	0.082	0.075	0.070	0.066	0.064	0.060	0.057	0.055	0.053	0.052	0.051	0.050	0.049	0.048
1.8	0.085	0.073	0.066	0.061	0.057	0.055	0.051	0.049	0.046	0.044	0.043	0.042	0.041	0.040	0.039
1.9	0.077	0.065	0.058	0.053	0.050	0.047	0.043	0.041	0.038	0.037	0.036	0.035	0.034	0.033	0.032
2.0	0.070	0.058	0.051	0.046	0.043	0.040	0.037	0.034	0.032	0.030	0.029	0.028	0.027	0.027	0.026
2.1	0.063	0.052	0.045	0.040	0.037	0.034	0.031	0.029	0.027	0.025	0.024	0.023	0.022	0.022	0.021
2.2	0.058	0.046	0.040	0.035	0.032	0.029	0.026	0.024	0.022	0.021	0.020	0.019	0.018	0.017	0.016
2.3	0.052	0.041	0.035	0.031	0.027	0.025	0.022	0.020	0.018	0.017	0.016	0.015	0.014	0.014	0.013
2.4	0.048	0.037	0.031	0.027	0.024	0.022	0.019	0.017	0.015	0.014	0.013	0.012	0.012	0.011	0.010
2.5	0.044	0.033	0.027	0.023	0.020	0.018	0.016	0.014	0.012	0.011	0.010	0.010	0.009	0.009	0.008
2.6	0.040	0.030	0.024	0.020	0.018	0.016	0.013	0.012	0.010	0.009	0.008	0.008	0.007	0.007	0.006
2.7	0.037	0.027	0.021	0.018	0.015	0.014	0.011	0.010	0.008	0.007	0.007	0.006	0.006	0.005	0.005
2.8	0.034	0.024	0.019	0.016	0.013	0.012	0.009	0.008	0.007	0.006	0.005	0.005	0.005	0.004	0.004
2.9	0.031	0.022	0.017	0.014	0.011	0.010	0.008	0.007	0.005	0.005	0.004	0.004	0.004	0.003	0.003
3.0	0.029	0.020	0.015	0.012	0.010	0.009	0.007	0.006	0.004	0.004	0.003	0.003	0.003	0.002	0.002
3.1	0.027	0.018	0.013	0.011	0.009	0.007	0.006	0.005	0.004	0.003	0.003	0.002	0.002	0.002	0.002
3.2	0.025	0.016	0.012	0.009	0.008	0.006	0.005	0.004	0.003	0.002	0.002	0.002	0.002	0.001	0.001
3.3	0.023	0.015	0.011	0.008	0.007	0.005	0.004	0.003	0.002	0.002	0.002	0.001	0.001	0.001	0.001
3.4	0.021	0.014	0.010	0.007	0.006	0.005	0.003	0.003	0.002	0.002	0.001	0.001	0.001	0.001	0.001
3.5	0.020	0.012	0.009	0.006	0.005	0.004	0.003	0.002	0.002	0.001	0.001	0.001	0.001	0.001	0.001
3.6	0.018	0.011	0.008	0.006	0.004	0.004	0.002	0.002	0.001	0.001	0.001	0.001	0.001	0+	0+
3.7	0.017	0.010	0.007	0.005	0.004	0.003	0.002	0.002	0.001	0.001	0.001	0.001	0+	0+	0+
3.8	0.016	0.010	0.006	0.004	0.003	0.003	0.002	0.001	0.001	0.001	0.001	0.001	0+	0+	0+
3.9	0.015	0.009	0.006	0.004	0.003	0.002	0.001	0.001	0.001	0.001	0+	0+	0+	0+	0+
4.0	0.014	0.008	0.005	0.004	0.003	0.002	0.001	0.001	0.001	0+	0+	0+	0+	0+	0+

For specific details about using this table to find *p*-values, see pages 484, 486.

TABLE 8

Critical Values of χ^2 ("Chi-Square") Distribution

The entries in this table, χ^2 (df, α), are the critical values for the χ^2 distribution for which the area under the curve to the right is α.

area to right

0   χ^2(df, **area to right**)

					Area to the Right							
0.995	0.99	0.975	0.95	0.90	0.75	0.50	0.25	0.10	0.05	0.025	0.01	0.005
		Area in Left-hand Tail				Median			Area in Right-hand Tail			
df 0.005	0.01	0.025	0.05	0.10	0.25	0.50	0.25	0.10	0.05	0.025	0.01	0.005
1 0.0000393	0.000157	0.000982	0.00393	0.0158	0.101	0.455	1.32	2.71	3.84	5.02	6.63	7.88
2 0.0100	0.0201	0.0506	0.103	0.211	0.575	1.39	2.77	4.61	5.99	7.38	9.21	10.6
3 0.0717	0.115	0.216	0.352	0.584	1.21	2.37	4.11	6.25	7.82	9.35	11.3	12.8
4 0.207	0.297	0.484	0.711	1.06	1.92	3.36	5.39	7.78	9.49	11.1	13.3	14.9
5 0.412	0.554	0.831	1.15	1.61	2.67	4.35	6.63	9.24	11.1	12.8	15.1	16.8
6 0.676	0.872	1.24	1.64	2.20	3.45	5.35	7.84	10.6	12.6	14.5	16.8	18.6
7 0.990	1.24	1.69	2.17	2.83	4.25	6.35	9.04	12.0	14.1	16.0	18.5	20.3
8 1.34	1.65	2.18	2.73	3.49	5.07	7.34	10.2	13.4	15.5	17.5	20.1	22.0
9 1.73	2.09	2.70	3.33	4.17	5.90	8.34	11.4	14.7	16.9	19.0	21.7	23.6
10 2.16	2.56	3.25	3.94	4.87	6.74	9.34	12.5	16.0	18.3	20.5	23.2	25.2
11 2.60	3.05	3.82	4.57	5.58	7.58	10.34	13.7	17.3	19.7	21.9	24.7	26.8
12 3.07	3.57	4.40	5.23	6.30	8.44	11.34	14.8	18.5	21.0	23.3	26.2	28.3
13 3.57	4.11	5.01	5.89	7.04	9.30	12.34	16.0	19.8	22.4	24.7	27.7	29.8
14 4.07	4.66	5.63	6.57	7.79	10.2	13.34	17.1	21.1	23.7	26.1	29.1	31.3
15 4.60	5.23	6.26	7.26	8.55	11.0	14.34	18.2	22.3	25.0	27.5	30.6	32.8
16 5.14	5.81	6.91	7.96	9.31	11.9	15.34	19.4	23.5	26.3	28.8	32.0	34.3
17 5.70	6.41	7.56	8.67	10.1	12.8	16.34	20.5	24.8	27.6	30.2	33.4	35.7
18 6.26	7.01	8.23	9.39	10.9	13.7	17.34	21.6	26.0	28.9	31.5	34.8	37.2
19 6.84	7.63	8.91	10.1	11.7	14.6	18.34	22.7	27.2	30.1	32.9	36.2	38.6
20 7.43	8.26	9.59	10.9	12.4	15.5	19.34	23.8	28.4	31.4	34.2	37.6	40.0
21 8.03	8.90	10.3	11.6	13.2	16.3	20.34	24.9	29.6	32.7	35.5	38.9	41.4
22 8.64	9.54	11.0	12.3	14.0	17.2	21.34	26.0	30.8	33.9	36.8	40.3	42.8
23 9.26	10.2	11.7	13.1	14.8	18.1	22.34	27.1	32.0	35.2	38.1	41.6	44.2
24 9.89	10.9	12.4	13.8	15.7	19.0	23.34	28.2	33.2	36.4	39.4	43.0	45.6
25 10.5	11.5	13.1	14.6	16.5	19.9	24.34	29.3	34.4	37.7	40.6	44.3	46.9
26 11.2	12.2	13.8	15.4	17.3	20.8	25.34	30.4	35.6	38.9	41.9	45.6	48.3
27 11.8	12.9	14.6	16.2	18.1	21.7	26.34	31.5	36.7	40.1	43.2	47.0	49.6
28 12.5	13.6	15.3	16.9	18.9	22.7	27.34	32.6	37.9	41.3	44.5	48.3	51.0
29 13.1	14.3	16.0	17.7	19.8	23.6	28.34	33.7	39.1	42.6	45.7	49.6	52.3
30 13.8	15.0	16.8	18.5	20.6	24.5	29.34	34.8	40.3	43.8	47.0	50.9	53.7
40 20.7	22.2	24.4	26.5	29.1	33.7	39.34	45.6	51.8	55.8	59.3	63.7	66.8
50 28.0	29.7	32.4	34.8	37.7	42.9	49.33	56.3	63.2	67.5	71.4	76.2	79.5
60 35.5	37.5	40.5	43.2	46.5	52.3	59.33	67.0	74.4	79.1	83.3	88.4	92.0
70 43.3	45.4	48.8	51.7	55.3	61.7	69.33	77.6	85.5	90.5	95.0	100.0	104.0
80 51.2	53.5	57.2	60.4	64.3	71.1	79.33	88.1	96.6	102.0	107.0	112.0	116.0
90 59.2	61.8	65.6	69.1	73.3	80.6	89.33	98.6	108.0	113.0	118.0	124.0	128.0
100 67.3	70.1	74.2	77.9	82.4	90.1	99.33	109.0	118.0	124.0	130.0	136.0	140.0

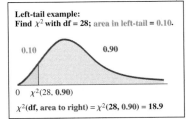

Left-tail example:
Find χ^2 with df = 28; area in left-tail = 0.10.

0.10   0.90

0   χ^2(28, **0.90**)

χ^2(**df**, area to right) = χ^2(28, 0.90) = **18.9**

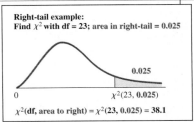

Right-tail example:
Find χ^2 with df = 23; area in right-tail = 0.025

0.025

0   χ^2(23, **0.025**)

χ^2(**df**, area to right) = χ^2(23, 0.025) = **38.1**

For specific details about using this table to find: p-values, see pages 521, 522, 524; critical values, pages 517–518, 521, 523.

TABLE 9A

Critical Values of the *F* Distribution ($\alpha = 0.05$)

The entries in this table are critical values of *F* for which the area under the curve to the right is equal to 0.05.

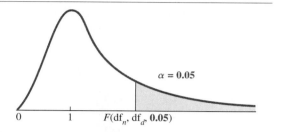

$\alpha = 0.05$

$F(\mathrm{df}_n, \mathrm{df}_d, 0.05)$

Degrees of Freedom for Numerator

		1	2	3	4	5	6	7	8	9	10
Degrees of Freedom for Denominator	1	161.	200.	216.	225.	230.	234.	237.	239.	241.	242.
	2	18.5	19.0	19.2	19.2	19.3	19.3	19.4	19.4	19.4	19.4
	3	10.1	9.55	9.28	9.12	9.01	8.94	8.89	8.85	8.81	8.79
	4	7.71	6.94	6.59	6.39	6.26	6.16	6.09	6.04	6.00	5.96
	5	6.61	5.79	5.41	5.19	5.05	4.95	4.88	4.82	4.77	4.74
	6	5.99	5.14	4.76	4.53	4.39	4.28	4.21	4.15	4.10	4.06
	7	5.59	4.74	4.35	4.12	3.97	3.87	3.79	3.73	3.68	3.64
	8	5.32	4.46	4.07	3.84	3.69	3.58	3.50	3.44	3.39	3.35
	9	5.12	4.26	3.86	3.63	3.48	3.37	3.29	3.23	3.18	3.14
	10	4.96	4.10	3.71	3.48	3.33	3.22	3.14	3.07	3.02	2.98
	11	4.84	3.98	3.59	3.36	3.20	3.09	3.01	2.95	2.90	2.85
	12	4.75	3.89	3.49	3.26	3.11	3.00	2.91	2.85	2.80	2.75
	13	4.67	3.81	3.41	3.18	3.03	2.92	2.83	2.77	2.71	2.67
	14	4.60	3.74	3.34	3.11	2.96	2.85	2.76	2.70	2.65	2.60
	15	4.54	3.68	3.29	3.06	2.90	2.79	2.71	2.64	2.59	2.54
	16	4.49	3.63	3.24	3.01	2.85	2.74	2.66	2.59	2.54	2.49
	17	4.45	3.59	3.20	2.96	2.81	2.70	2.61	2.55	2.49	2.45
	18	4.41	3.55	3.16	2.93	2.77	2.66	2.58	2.51	2.46	2.41
	19	4.38	3.52	3.13	2.90	2.74	2.63	2.54	2.48	2.42	2.38
	20	4.35	3.49	3.10	2.87	2.71	2.60	2.51	2.45	2.39	2.35
	21	4.32	3.47	3.07	2.84	2.68	2.57	2.49	2.42	2.37	2.32
	22	4.30	3.44	3.05	2.82	2.66	2.55	2.46	2.40	2.34	2.30
	23	4.28	3.42	3.03	2.80	2.64	2.53	2.44	2.37	2.32	2.27
	24	4.26	3.40	3.01	2.78	2.62	2.51	2.42	2.36	2.30	2.25
	25	4.24	3.39	2.99	2.76	2.60	2.49	2.40	2.34	2.28	2.24
	30	4.17	3.32	2.92	2.69	2.53	2.42	2.33	2.27	2.21	2.16
	40	4.08	3.23	2.84	2.61	2.45	2.34	2.25	2.18	2.12	2.08
	60	4.00	3.15	2.76	2.53	2.37	2.25	2.17	2.10	2.04	1.99
	120	3.92	3.07	2.68	2.45	2.29	2.18	2.09	2.02	1.96	1.91
	∞	3.84	3.00	2.60	2.37	2.21	2.10	2.01	1.94	1.88	1.83

For specific details about using this table to find: *p*-values, see page 597; critical values, page 594.

TABLE 9A (CONTINUED)

				Degrees of Freedom for Numerator					
	12	15	20	24	30	40	60	120	∞
1	244.	246.	248.	249.	250.	251.	252.	253.	254.
2	19.4	19.4	19.4	19.5	19.5	19.5	19.5	19.5	19.5
3	8.74	8.70	8.66	8.64	8.62	8.59	8.57	8.55	8.53
4	5.91	5.86	5.80	5.77	5.75	5.72	5.69	5.66	5.63
5	4.68	4.62	4.56	4.53	4.50	4.46	4.43	4.40	4.37
6	4.00	3.94	3.87	3.84	3.81	3.77	3.74	3.70	3.67
7	3.57	3.51	3.44	3.41	3.38	3.34	3.30	3.27	3.23
8	3.28	3.22	3.15	3.12	3.08	3.04	3.01	2.97	2.93
9	3.07	3.01	2.94	2.90	2.86	2.83	2.79	2.75	2.71
10	2.91	2.85	2.77	2.74	2.70	2.66	2.62	2.58	2.54
11	2.79	2.72	2.65	2.61	2.57	2.53	2.49	2.45	2.40
12	2.69	2.62	2.54	2.51	2.47	2.43	2.38	2.34	2.30
13	2.60	2.53	2.46	2.42	2.38	2.34	2.30	2.25	2.21
14	2.53	2.46	2.39	2.35	2.31	2.27	2.22	2.18	2.13
15	2.48	2.40	2.33	2.29	2.25	2.20	2.16	2.11	2.07
16	2.42	2.35	2.28	2.24	2.19	2.15	2.11	2.06	2.01
17	2.38	2.31	2.23	2.19	2.15	2.10	2.06	2.01	1.96
18	2.34	2.27	2.19	2.15	2.11	2.06	2.02	1.97	1.92
19	2.31	2.23	2.16	2.11	2.07	2.03	1.98	1.93	1.88
20	2.28	2.20	2.12	2.08	2.04	1.99	1.95	1.90	1.84
21	2.25	2.18	2.10	2.05	2.01	1.96	1.92	1.87	1.81
22	2.23	2.15	2.07	2.03	1.98	1.94	1.89	1.84	1.78
23	2.20	2.13	2.05	2.01	1.96	1.91	1.86	1.81	1.76
24	2.18	2.11	2.03	1.98	1.94	1.89	1.84	1.79	1.73
25	2.16	2.09	2.01	1.96	1.92	1.87	1.82	1.77	1.71
30	2.09	2.01	1.93	1.89	1.84	1.79	1.74	1.68	1.62
40	2.00	1.92	1.84	1.79	1.74	1.69	1.64	1.58	1.51
60	1.92	1.84	1.75	1.70	1.65	1.59	1.53	1.47	1.39
120	1.83	1.75	1.66	1.61	1.55	1.50	1.43	1.35	1.25
∞	1.75	1.67	1.57	1.52	1.46	1.39	1.32	1.22	1.00

Degrees of Freedom for Denominator

From E. S. Pearson and H. O. Hartley, *Biometrika Tables for Statisticians,* vol. 1 (1958), pp. 159–163. Reprinted by permission of the Biometrika Trustees.

Critical Values of the *F* Distribution ($\alpha = 0.025$)

The entries in this table are critical values of *F* for which the area under the curve to the right is equal to 0.025.

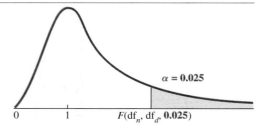

$\alpha = 0.025$

$F(\text{df}_n, \text{df}_d, \mathbf{0.025})$

		Degrees of Freedom for Numerator								
	1	2	3	4	5	6	7	8	9	10
1	648.	800.	864.	900.	922.	937.	948.	957.	963.	969.
2	38.5	39.0	39.2	39.2	39.3	39.3	39.4	39.4	39.4	39.4
3	17.4	16.0	15.4	15.1	14.9	14.7	14.6	14.5	14.5	14.4
4	12.2	10.6	9.98	9.60	9.36	9.20	9.07	8.98	8.90	8.84
5	10.0	8.43	7.76	7.39	7.15	6.98	6.85	6.76	6.68	6.62
6	8.81	7.26	6.60	6.23	5.99	5.82	5.70	5.60	5.52	5.46
7	8.07	6.54	5.89	5.52	5.29	5.12	4.99	4.90	4.82	4.76
8	7.57	6.06	5.42	5.05	4.82	4.65	4.53	4.43	4.36	4.30
9	7.21	5.71	5.08	4.72	4.48	4.32	4.20	4.10	4.03	3.96
10	6.94	5.46	4.83	4.47	4.24	4.07	3.95	3.85	3.78	3.72
11	6.72	5.26	4.63	4.28	4.04	3.88	3.76	3.66	3.59	3.53
12	6.55	5.10	4.47	4.12	3.89	3.73	3.61	3.51	3.44	3.37
13	6.41	4.97	4.35	4.00	3.77	3.60	3.48	3.39	3.31	3.25
14	6.30	4.86	4.24	3.89	3.66	3.50	3.38	3.28	3.21	3.15
15	6.20	4.77	4.15	3.80	3.58	3.41	3.29	3.20	3.12	3.06
16	6.12	4.69	4.08	3.73	3.50	3.34	3.22	3.12	3.05	2.99
17	6.04	4.62	4.01	3.66	3.44	3.28	3.16	3.06	2.98	2.92
18	5.98	4.56	3.95	3.61	3.38	3.22	3.10	3.01	2.93	2.87
19	5.92	4.51	3.90	3.56	3.33	3.17	3.05	2.96	2.88	2.82
20	5.87	4.46	3.86	3.51	3.29	3.13	3.01	2.91	2.84	2.77
21	5.83	4.42	3.82	3.48	3.25	3.09	2.97	2.87	2.80	2.73
22	5.79	4.38	3.78	3.44	3.22	3.05	2.93	2.84	2.76	2.70
23	5.75	4.35	3.75	3.41	3.18	3.02	2.90	2.81	2.73	2.67
24	5.72	4.32	3.72	3.38	3.15	2.99	2.87	2.78	2.70	2.64
25	5.69	4.29	3.69	3.35	3.13	2.97	2.85	2.75	2.68	2.61
30	5.57	4.18	3.59	3.25	3.03	2.87	2.75	2.65	2.57	2.51
40	5.42	4.05	3.46	3.13	2.90	2.74	2.62	2.53	2.45	2.39
60	5.29	3.93	3.34	3.01	2.79	2.63	2.51	2.41	2.33	2.27
120	5.15	3.80	3.23	2.89	2.67	2.52	2.39	2.30	2.22	2.16
∞	5.02	3.69	3.12	2.79	2.57	2.41	2.29	2.19	2.11	2.05

Degrees of Freedom for Denominator

For specific details about using this table to find: *p*-values, see page 597; critical values, page 594.

TABLE 9B (CONTINUED)

Degrees of Freedom for Numerator

	12	15	20	24	30	40	60	120	∞
1	977.	985.	993.	997.	1001.	1006.	1010.	1014.	1018.
2	39.4	39.4	39.4	39.5	39.5	39.5	39.5	39.5	39.5
3	14.3	14.3	14.2	14.1	14.1	14.0	14.0	13.9	13.9
4	8.75	8.66	8.56	8.51	8.46	8.41	8.36	8.31	8.26
5	6.52	6.43	6.33	6.28	6.23	6.18	6.12	6.07	6.02
6	5.37	5.27	5.17	5.12	5.07	5.01	4.96	4.90	4.85
7	4.67	4.57	4.47	4.42	4.36	4.31	4.25	4.20	4.14
8	4.20	4.10	4.00	3.95	3.89	3.84	3.78	3.73	3.67
9	3.87	3.77	3.67	3.61	3.56	3.51	3.45	3.39	3.33
10	3.62	3.52	3.42	3.37	3.31	3.26	3.20	3.14	3.08
11	3.43	3.33	3.23	3.17	3.12	3.06	3.00	2.94	2.88
12	3.28	3.18	3.07	3.02	2.96	2.91	2.85	2.79	2.72
13	3.15	3.05	2.95	2.89	2.84	2.78	2.72	2.66	2.60
14	3.05	2.95	2.84	2.79	2.73	2.67	2.61	2.55	2.49
15	2.96	2.86	2.76	2.70	2.64	2.59	2.52	2.46	2.40
16	2.89	2.79	2.68	2.63	2.57	2.51	2.45	2.38	2.32
17	2.82	2.72	2.62	2.56	2.50	2.44	2.38	2.32	2.25
18	2.77	2.67	2.56	2.50	2.44	2.38	2.32	2.26	2.19
19	2.72	2.62	2.51	2.45	2.39	2.33	2.27	2.20	2.13
20	2.68	2.57	2.46	2.41	2.35	2.29	2.22	2.16	2.09
21	2.64	2.53	2.42	2.37	2.31	2.25	2.18	2.11	2.04
22	2.60	2.50	2.39	2.33	2.27	2.21	2.14	2.08	2.00
23	2.57	2.47	2.36	2.30	2.24	2.18	2.11	2.04	1.97
24	2.54	2.44	2.33	2.27	2.21	2.15	2.08	2.01	1.94
25	2.51	2.41	2.30	2.24	2.18	2.12	2.05	1.98	1.91
30	2.41	2.31	2.20	2.14	2.07	2.01	1.94	1.87	1.79
40	2.29	2.18	2.07	2.01	1.94	1.88	1.80	1.72	1.64
60	2.17	2.06	1.94	1.88	1.82	1.74	1.67	1.58	1.48
120	2.05	1.95	1.82	1.76	1.69	1.61	1.53	1.43	1.31
∞	1.94	1.83	1.71	1.64	1.57	1.48	1.39	1.27	1.00

Degrees of Freedom for Denominator

From E. S. Pearson and H. O. Hartley, *Biometrika Tables for Statisticians,* vol. I (1958), pp. 159–163. Reprinted by persmission of the Biometrika Trustees.

TABLE 9C

TABLE 9C

Critical Values of the *F* Distribution ($\alpha = 0.01$)

The entries in the table are critical values of *F* for which the area under the curve to the right is equal to 0.01

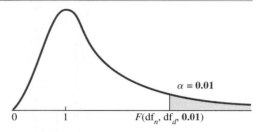

$\alpha = 0.01$

$0 \qquad 1 \qquad F(\text{df}_n, \text{df}_d, 0.01)$

Degrees of Freedom for Numerator

		1	2	3	4	5	6	7	8	9	10
	1	4052.	5000.	5403.	5625.	5764.	5859.	5928.	5982.	6024.	6056.
	2	98.5	99.0	99.2	99.2	99.3	99.3	99.4	99.4	99.4	99.4
	3	34.1	30.8	29.5	28.7	28.2	27.9	27.7	27.5	27.3	27.2
	4	21.2	18.0	16.7	16.0	15.5	15.2	15.0	14.8	14.7	14.5
	5	16.3	13.3	12.1	11.4	11.0	10.7	10.5	10.3	10.2	10.1
	6	13.7	10.9	9.78	9.15	8.75	8.47	8.26	8.10	7.98	7.87
	7	12.2	9.55	8.45	7.85	7.46	7.19	6.99	6.84	6.72	6.62
	8	11.3	8.65	7.59	7.01	6.63	6.37	6.18	6.03	5.91	5.81
	9	10.6	8.02	6.99	6.42	6.06	5.80	5.61	5.47	5.35	5.26
	10	10.0	7.56	6.55	5.99	5.64	5.39	5.20	5.06	4.94	4.85
	11	9.65	7.21	6.22	5.67	5.32	5.07	4.89	4.74	4.63	4.54
	12	9.33	6.93	5.95	5.41	5.06	4.82	4.64	4.50	4.39	4.30
Degrees of Freedom for Denominator	13	9.07	6.70	5.74	5.21	4.86	4.62	4.44	4.30	4.19	4.10
	14	8.86	6.51	5.56	5.04	4.70	4.46	4.28	4.14	4.03	3.94
	15	8.68	6.36	5.42	4.89	4.56	4.32	4.14	4.00	3.89	3.80
	16	8.53	6.23	5.29	4.77	4.44	4.20	4.03	3.89	3.78	3.69
	17	8.40	6.11	5.19	4.67	4.34	4.10	3.93	3.79	3.68	3.59
	18	8.29	6.01	5.09	4.58	4.25	4.01	3.84	3.71	3.60	3.51
	19	8.19	5.93	5.01	4.50	4.17	3.94	3.77	3.63	3.52	3.43
	20	8.10	5.85	4.94	4.43	4.10	3.87	3.70	3.56	3.46	3.37
	21	8.02	5.78	4.87	4.37	4.04	3.81	3.64	3.51	3.40	3.31
	22	7.95	5.72	4.82	4.31	3.99	3.76	3.59	3.45	3.35	3.26
	23	7.88	5.66	4.76	4.26	3.94	3.71	3.54	3.41	3.30	3.21
	24	7.82	5.61	4.72	4.22	3.90	3.67	3.50	3.36	3.26	3.17
	25	7.77	5.57	4.68	4.18	3.86	3.63	3.46	3.32	3.22	3.13
	30	7.56	5.39	4.51	4.02	3.70	3.47	3.30	3.17	3.07	2.98
	40	7.31	5.18	4.31	3.83	3.51	3.29	3.12	2.99	2.89	2.80
	60	7.08	4.98	4.13	3.65	3.34	3.12	2.95	2.82	2.72	2.63
	120	6.85	4.79	3.95	3.48	3.17	2.96	2.79	2.66	2.56	2.47
	∞	6.63	4.61	3.78	3.32	3.02	2.80	2.64	2.51	2.41	2.32

For specific details about using this table to find: *p*-values, see page 597; critical values, page 594.

TABLE 9C (CONTINUED)

Degrees of Freedom for Numerator

		12	15	20	24	30	40	60	120	∞
	1	6106.	6157.	6209.	6235.	6261.	6287.	6313.	6339.	6366.
	2	99.4	99.4	99.4	99.5	99.5	99.5	99.5	99.5	99.5
	3	27.1	26.9	26.7	26.6	26.5	26.4	26.3	26.2	26.1
	4	14.4	14.2	14.0	13.9	13.8	13.7	13.7	13.6	13.5
	5	9.89	9.72	9.55	9.47	9.38	9.29	9.20	9.11	9.02
	6	7.72	7.56	7.40	7.31	7.23	7.14	7.06	6.97	6.88
	7	6.47	6.31	6.16	6.07	5.99	5.91	5.82	5.74	5.65
	8	5.67	5.52	5.36	5.28	5.20	5.12	5.03	4.95	4.86
	9	5.11	4.96	4.81	4.73	4.65	4.57	4.48	4.40	4.31
	10	4.71	4.56	4.41	4.33	4.25	4.17	4.08	4.00	3.91
	11	4.40	4.25	4.10	4.02	3.94	3.86	3.78	3.69	3.60
	12	4.16	4.01	3.86	3.78	3.70	3.62	3.54	3.45	3.36
	13	3.96	3.82	3.66	3.59	3.51	3.43	3.34	3.25	3.17
	14	3.80	3.66	3.51	3.43	3.35	3.27	3.18	3.09	3.00
	15	3.67	3.52	3.37	3.29	3.21	3.13	3.05	2.96	2.87
	16	3.55	3.41	3.26	3.18	3.10	3.02	2.93	2.84	2.75
	17	3.46	3.31	3.16	3.08	3.00	2.92	2.83	2.75	2.65
	18	3.37	3.23	3.08	3.00	2.92	2.84	2.75	2.66	2.57
	19	3.30	3.15	3.00	2.92	2.84	2.76	2.67	2.58	2.49
	20	3.23	3.09	2.94	2.86	2.78	2.69	2.61	2.52	2.42
	21	3.17	3.03	2.88	2.80	2.72	2.64	2.55	2.46	2.36
	22	3.12	2.98	2.83	2.75	2.67	2.58	2.50	2.40	2.31
	23	3.07	2.93	2.78	2.70	2.62	2.54	2.45	2.35	2.26
	24	3.03	2.89	2.74	2.66	2.58	2.49	2.40	2.31	2.21
	25	2.99	2.85	2.70	2.62	2.53	2.45	2.36	2.27	2.17
	30	2.84	2.70	2.55	2.47	2.39	2.30	2.21	2.11	2.01
	40	2.66	2.52	2.37	2.29	2.20	2.11	2.02	1.92	1.80
	60	2.50	2.35	2.20	2.12	2.03	1.94	1.84	1.73	1.60
	120	2.34	2.19	2.03	1.95	1.86	1.76	1.66	1.53	1.38
	∞	2.18	2.04	1.88	1.79	1.70	1.59	1.47	1.32	1.00

Degrees of Freedom for Denominator

From E. S. Pearson and H. O. Hartley, *Biometrika Tables for Statisticians,* vol. I (1958), pp. 159–163. Reprinted by permission of the Biometrika Trustees.

Confidence Belts for the Correlation Coefficient $(1 - \alpha) = 0.95$

The numbers on the curves are sample sizes.

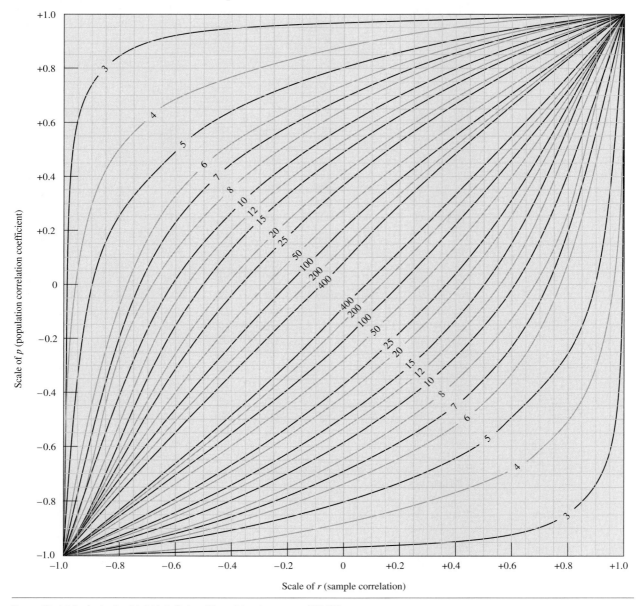

For specific details about using this table to find confidence intervals, see pages 704–705.

TABLE 11

Critical Values of *r* When $\rho = 0$

The entries in this table are the critical values of *r* for a two-tailed test at α. For simple correlation, df = $n - 2$, where *n* is the number of pairs of data in the sample. For a one-tailed test, the value of α shown at the top of the table is double the value of α being used in the hypothesis test.

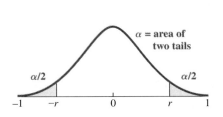

α = area of two tails

df	0.10	0.05	0.02	0.01
1	0.988	0.997	1.000	1.000
2	0.900	0.950	0.980	0.990
3	0.805	0.878	0.934	0.959
4	0.729	0.811	0.882	0.917
5	0.669	0.754	0.833	0.874
6	0.621	0.707	0.789	0.834
7	0.582	0.666	0.750	0.798
8	0.549	0.632	0.716	0.765
9	0.521	0.602	0.685	0.735
10	0.497	0.576	0.658	0.708
11	0.476	0.553	0.634	0.684
12	0.458	0.532	0.612	0.661
13	0.441	0.514	0.592	0.641
14	0.426	0.497	0.574	0.623
15	0.412	0.482	0.558	0.606
16	0.400	0.468	0.542	0.590
17	0.389	0.456	0.528	0.575
18	0.378	0.444	0.516	0.561
19	0.369	0.433	0.503	0.549
20	0.360	0.423	0.492	0.537
25	0.323	0.381	0.445	0.487
30	0.296	0.349	0.409	0.449
35	0.275	0.325	0.381	0.418
40	0.257	0.304	0.358	0.393
45	0.243	0.288	0.338	0.372
50	0.231	0.273	0.322	0.354
60	0.211	0.250	0.295	0.325
70	0.195	0.232	0.274	0.302
80	0.183	0.217	0.256	0.283
90	0.173	0.205	0.242	0.267
100	0.164	0.195	0.230	0.254

For specific details about using this table to find: *p*-values, see pages 706–707; critical values, page 707.

TABLE 12

Critical Values of the Sign Test

The entries in this table are the critical values for the number of the least frequent sign for a two-tailed test at α for the binomial $p = 0.5$. For a one-tailed test, the value of α shown at the top of the table is double the value of α being used in the hypothesis test.

	α					α			
n	0.01	0.05	0.10	0.25	n	0.01	0.05	0.10	0.25
1					51	15	18	19	20
2					52	16	18	19	21
3				0	53	16	18	20	21
4				0	54	17	19	20	22
5			0	0	55	17	19	20	22
6		0	0	1	56	17	20	21	23
7		0	0	1	57	18	20	21	23
8	0	0	1	1	58	18	21	22	24
9	0	1	1	2	59	19	21	22	24
10	0	1	1	2	60	19	21	23	25
11	0	1	2	3	61	20	22	23	25
12	1	2	2	3	62	20	22	24	25
13	1	2	3	3	63	20	23	24	26
14	1	2	3	4	64	21	23	24	26
15	2	3	3	4	65	21	24	25	27
16	2	3	4	5	66	22	24	25	27
17	2	4	4	5	67	22	25	26	28
18	3	4	5	6	68	22	25	26	28
19	3	4	5	6	69	23	25	27	29
20	3	5	5	6	70	23	26	27	29
21	4	5	6	7	71	24	26	28	30
22	4	5	6	7	72	24	27	28	30
23	4	6	7	8	73	25	27	28	31
24	5	6	7	8	74	25	28	29	31
25	5	7	7	9	75	25	28	29	32
26	6	7	8	9	76	26	28	30	32
27	6	7	8	10	77	26	29	30	32
28	6	8	9	10	78	27	29	31	33
29	7	8	9	10	79	27	30	31	33
30	7	9	10	11	80	28	30	32	34
31	7	9	10	11	81	28	31	32	34
32	8	9	10	12	82	28	31	33	35
33	8	10	11	12	83	29	32	33	35
34	9	10	11	13	84	29	32	33	36
35	9	11	12	13	85	30	32	34	36
36	9	11	12	14	86	30	33	34	37
37	10	12	13	14	87	31	33	35	37
38	10	12	13	14	88	31	34	35	38
39	11	12	13	15	89	31	34	36	38
40	11	13	14	15	90	32	35	36	39
41	11	13	14	16	91	32	35	37	39
42	12	14	15	16	92	33	36	37	39
43	12	14	15	17	93	33	36	38	40
44	13	15	16	17	94	34	37	38	40
45	13	15	16	18	95	34	37	38	41
46	13	15	16	18	96	34	37	39	41
47	14	16	17	19	97	35	38	39	42
48	14	16	17	19	98	35	38	40	42
49	15	17	18	19	99	36	39	40	43
50	15	17	18	20	100	36	39	41	43

From Wilfred J. Dixon and Frank J. Massey, Jr., *Introduction to Statistical Analysis,* 3d ed. (New York: McGraw-Hill, 1969), p. 509. Reprinted by permission.

For specific details about using this table to find: confidence intervals, see page 753; *p*-values, pages 754–755; critical values, pages 753–755.

TABLE 13

Critical Values of U in the Mann-Whitney Test

A. The entries are the critical values of U for a one-tailed test at 0.025 or for a two-tailed test at 0.05.

n_2 \ n_1	1	2	3	4	5	6	7	8	9	10	11	12	13	14	15	16	17	18	19	20
1																				
2								0	0	0	0	1	1	1	1	1	2	2	2	2
3					0	1	1	2	2	3	3	4	4	5	5	6	6	7	7	8
4				0	1	2	3	4	4	5	6	7	8	9	10	11	11	12	13	13
5			0	1	2	3	5	6	7	8	9	11	12	13	14	15	17	18	19	20
6			1	2	3	5	6	8	10	11	13	14	16	17	19	21	22	24	25	27
7			1	3	5	6	8	10	12	14	16	18	20	22	24	26	28	30	32	34
8		0	2	4	6	8	10	13	15	17	19	22	24	26	29	31	34	36	38	41
9		0	2	4	7	10	12	15	17	20	23	26	28	31	34	37	39	42	45	48
10		0	3	5	8	11	14	17	20	23	26	29	33	36	39	42	45	48	52	55
11		0	3	6	9	13	16	19	23	26	30	33	37	40	44	47	51	55	58	62
12		1	4	7	11	14	18	22	26	29	33	37	41	45	49	53	57	61	65	69
13		1	4	8	12	16	20	24	28	33	37	41	45	50	54	59	63	67	72	76
14		1	5	9	13	17	22	26	31	36	40	45	50	55	59	64	67	74	78	83
15		1	5	10	14	19	24	29	34	39	44	49	54	59	64	70	75	80	85	90
16		1	6	11	15	21	26	31	37	42	47	53	59	64	70	75	81	86	92	98
17		2	6	11	17	22	28	34	39	45	51	57	63	67	75	81	87	93	99	105
18		2	7	12	18	24	30	36	42	48	55	61	67	74	80	86	93	99	106	112
19		2	7	13	19	25	32	38	45	52	58	65	72	78	85	92	99	106	113	119
20		2	8	13	20	27	34	41	48	55	62	69	76	83	90	98	105	112	119	127

B. The entries are the critical values of U for a one-tailed test at 0.05 or for a two-tailed test at 0.10.

n_2 \ n_1	1	2	3	4	5	6	7	8	9	10	11	12	13	14	15	16	17	18	19	20
1																			0	0
2					0	0	0	1	1	1	1	2	2	2	3	3	3	4	4	4
3			0	0	1	2	2	3	3	4	5	5	6	7	7	8	9	9	10	11
4			0	1	2	3	4	5	6	7	8	9	10	11	12	14	15	16	17	18
5		0	1	2	4	5	6	8	9	11	12	13	15	16	18	19	20	22	23	25
6		0	2	3	5	7	8	10	12	14	16	17	19	21	23	25	26	28	30	32
7		0	2	4	6	8	11	13	15	17	19	21	24	26	28	30	33	35	37	39
8		1	3	5	8	10	13	15	18	20	23	26	28	31	33	36	39	41	44	47
9		1	3	6	9	12	15	18	21	24	27	30	33	36	39	42	45	48	51	54
10		1	4	7	11	14	17	20	24	27	31	34	37	41	44	48	51	55	58	62
11		1	5	8	12	16	19	23	27	31	34	38	42	46	50	54	57	61	65	69
12		2	5	9	13	17	21	26	30	34	38	42	47	51	55	60	64	68	72	77
13		2	6	10	15	19	24	28	33	37	42	47	51	56	61	65	70	75	80	84
14		2	7	11	16	21	26	31	36	41	46	51	56	61	66	71	77	82	87	92
15		3	7	12	18	23	28	33	39	44	50	55	61	66	72	77	83	88	94	100
16		3	8	14	19	25	30	36	42	48	54	60	65	71	77	83	89	95	101	107
17		3	9	15	20	26	33	39	45	51	57	64	70	77	83	89	96	102	109	115
18		4	9	16	22	28	35	41	48	55	61	68	75	82	88	95	102	109	116	123
19	0	4	10	17	23	30	37	44	51	58	65	72	80	87	94	101	109	116	123	130
20	0	4	11	18	25	32	39	47	54	62	69	77	84	92	100	107	115	123	130	138

Reproduced from the *Bulletin of the Institute of Educational Research at Indiana University*, vol. 1, no. 2; with the permission of the author and the publisher.

For specific details about using this table to find: *p*-values, see pages 768–769; critical values, page 768.

TABLE 14

Critical Values for Total Number of Runs (*V*)

The entries in this table are the critical values for a two-tailed test using $\alpha = 0.05$. For a one-tailed test at $\alpha = 0.025$, use only one of the critical values: the smaller critical value for a left-hand critical region, the larger for a right-hand critical region.

The larger of n_1 and n_2

The smaller of n_1 and n_2	5	6	7	8	9	10	11	12	13	14	15	16	17	18	19	20
2								2 / 6	2 / 6	2 / 6	2 / 6	2 / 6	2 / 6	2 / 6	2 / 6	2 / 6
3		2 / 8	2 / 8	2 / 8	2 / 8	2 / 8	2 / 8	2 / 8	2 / 8	2 / 8	3 / 8	3 / 8	3 / 8	3 / 8	3 / 8	3 / 8
4	2 / 9	2 / 9	2 / 10	3 / 10	3 / 10	3 / 10	3 / 10	3 / 10	3 / 10	3 / 10	3 / 10	4 / 10	4 / 10	4 / 10	4 / 10	4 / 10
5	2 / 10	3 / 10	3 / 11	3 / 11	3 / 12	3 / 12	4 / 12	4 / 12	4 / 12	4 / 12	4 / 12	4 / 12	4 / 12	5 / 12	5 / 12	5 / 12
6		3 / 11	3 / 12	3 / 12	4 / 13	4 / 13	4 / 13	4 / 13	5 / 14	5 / 14	5 / 14	5 / 14	5 / 14	5 / 14	6 / 14	6 / 14
7			3 / 13	4 / 13	4 / 14	5 / 14	5 / 14	5 / 14	5 / 15	5 / 15	6 / 15	6 / 16	6 / 16	6 / 16	6 / 16	6 / 16
8				4 / 14	5 / 14	5 / 15	5 / 15	6 / 16	6 / 16	6 / 16	6 / 16	6 / 17	7 / 17	7 / 17	7 / 17	7 / 17
9					5 / 15	5 / 16	6 / 16	6 / 16	6 / 17	7 / 17	7 / 18	7 / 18	7 / 18	8 / 18	8 / 18	8 / 18
10						6 / 16	6 / 16	7 / 17	7 / 17	7 / 18	7 / 18	8 / 19	8 / 19	8 / 19	8 / 20	9 / 20
11							7 / 17	7 / 18	7 / 19	8 / 19	8 / 19	8 / 20	9 / 20	9 / 20	9 / 21	9 / 21
12								7 / 19	8 / 19	8 / 20	8 / 20	9 / 21	9 / 21	9 / 21	10 / 22	10 / 22
13									8 / 20	9 / 20	9 / 21	9 / 21	10 / 22	10 / 22	10 / 23	10 / 23
14										9 / 21	9 / 22	10 / 22	10 / 23	10 / 23	11 / 23	11 / 24
15											10 / 22	10 / 23	11 / 23	11 / 24	11 / 24	12 / 25
16												11 / 23	11 / 24	11 / 25	12 / 25	12 / 25
17													11 / 25	12 / 25	12 / 26	13 / 26
18														12 / 26	13 / 26	13 / 27
19															13 / 27	13 / 27
20																14 / 28

From C. Eisenhart and F. Swed, "Tables for testing randomness of grouping in a sequence of alternatives," *Annals of Statistics*, vol. 14 (1943): 66–87. Reprinted by permission.

For specific details about using this table to find: *p*-values, see page 778; critical values, page 778.

TABLE 15

Critical Values of Spearman's Rank Correlation Coefficient

The entries in this table are the critical values of r_s for a two-tailed test at α. For a one-tailed test, the value of α shown at the top of the table is double the value of α being used in the hypothesis test.

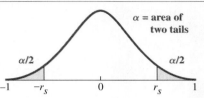

n	$\alpha = 0.10$	$\alpha = 0.05$	$\alpha = 0.02$	$\alpha = 0.01$
5	0.900	–	–	–
6	0.829	0.886	0.943	–
7	0.714	0.786	0.893	–
8	0.643	0.738	0.833	0.881
9	0.600	0.700	0.783	0.833
10	0.564	0.648	0.745	0.794
11	0.536	0.618	0.736	0.818
12	0.497	0.591	0.703	0.780
13	0.475	0.566	0.673	0.745
14	0.457	0.545	0.646	0.716
15	0.441	0.525	0.623	0.689
16	0.425	0.507	0.601	0.666
17	0.412	0.490	0.582	0.645
18	0.399	0.476	0.564	0.625
19	0.388	0.462	0.549	0.608
20	0.377	0.450	0.534	0.591
21	0.368	0.438	0.521	0.576
22	0.359	0.428	0.508	0.562
23	0.351	0.418	0.496	0.549
24	0.343	0.409	0.485	0.537
25	0.336	0.400	0.475	0.526
26	0.329	0.392	0.465	0.515
27	0.323	0.385	0.456	0.505
28	0.317	0.377	0.448	0.496
29	0.311	0.370	0.440	0.487
30	0.305	0.364	0.432	0.478

From E. G. Olds, "Distribution of sums of squares of rank differences for small numbers of individuals," *Annals of Statistics*, vol. 9 (1938), pp. 138–148, and amended, vol. 20 (1949), pp. 117–118. Reprinted by permission.

For specific details about using this table to find: *p*-values, see pages 788–789; critical values, page 788.

Answers to Selected Exercises

Chapter 1

1.1 a. Americans
     b. communication method workers preferred
     c. 63% of those people surveyed
     e. 7.2 fatal crashes per 100 million miles for 19-year-olds

1.3 b. does not appear, Java professionals work a 40-hour week
     c. only if long work hours are desirable

1.7 a. descriptive
     b. inferential

1.9 a. American heads of households
     b. 1000
     c. hardest place to clean
     d. 120
     e. 5% lower or 5% higher
     f. Between 30% and 40% think Venetian blinds.

1.11 a. Americans who file taxes
     b. users of TurboTax
     c. action taken upon receipt of tax refund
     d. majority plan to pay bills; largest portion of the dollar bill

1.13 a. yes, 50%
     b. "50% jump" works at getting people's attention

1.15 a. all U.S. adults
     b. 1200 randomly selected
     c. "allergy status" for each adult
     d. 33.2% based on the sampled adults
     e. percent of all U.S. adults who have an allergy, 36%

1.17 categorical variable

1.19 a. marital status, ZIP code
     b. level of education, rating for first impression

1.21 a. Scores are counted.
     b. Time is measured.

1.23 a. satisfaction level
     b. ordinal

1.25 a. all individuals who have hypertension
     b. the 5000 in the study
     c. proportion of the population for which the drug is effective

d. proportion of the sample for which the drug is effective, 80%
     e. no

1.27 a. all assembled parts from the assembly line
     b. infinite
     c. the parts checked
     d. attribute, attribute, numerical

1.29 a. all people suffering from seasonal allergies
     b. the 679 people
     c. relief status and side effects
     d. qualitative

1.31 a. numerical
     b. attribute
     c. numerical
     d. attribute
     e. numerical
     f. numerical

1.33 a. Population contains all objects of interest; sample contains only those actually studied.
     b. convenience, availability, practicality

1.35 football players, wider range

1.37 Price/standard unit makes price the only variable.

1.39 too easy, too hard, can distinguish among the students' knowledge

1.41 volunteer; yes

1.43 volunteer, bias

1.45 convenience

1.49 probability samples

1.51 Statistical methods assume the use of random samples.

1.53 Randomly select first item between 1 and 25, select every 25th thereafter.

1.55 A proportional sample would work best.

1.57 Only people with telephones and listed phone numbers will be considered

1.59 a. probability
     b. statistics

1.61 a. statistics
     b. probability
     c. statistics
     d. probability

1.63 perform many of the computations and tests quickly and easily

1.65   Calculators do only the calculations they are directed to perform.

1.67   a. color of hair, major, gender, marital status
       b. number of courses taken, height, distance from hometown to college

1.69   a. data value
       b. What is the average of sample?
       c. What is the average for all people?

1.71   a. U.S. adults
       b. flu vaccine status, location where obtained, precaution status, type of precaution
       c. All are attribute.

1.73   a. All U.S. public schools
       b. 1000 principals
       c. probability sample
       d. stratified

1.75   a. Both are increasing at the same rate; the number of nondrivers remains constant, one increasing at a faster rate than the other.
       b. more drivers than vehicles before 1971, more vehicles than drivers after 1973; number of drivers and vehicles was same

1.77   a. observational study
       b. percent of helmet use in children
       c. proportion of sample that wore a helmet, 41%
       d. activity, gender, helmet use—attribute; age—numerical

1.79   a. all Americans
       b. physical activities and nutrition
       c. age—numerical; leisure activity status—attribute; nutrition status—attribute

1.81   a. number of new prescriptions
       b. women, by 1.5 new prescriptions
       c. 75 and older, 13 new prescriptions
       d. yes

Chapter 2

2.3   b. relative proportions as a whole
      c. relative proportions between the individual answers

2.5   a.   The current dress code at my company is . . .

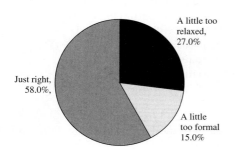

b.   The current dress code at my company is . . .

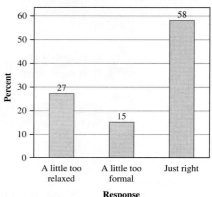

2.7   a.   **Points Scored by Winning Teams**
           Opening Night 2004–2005 NBA Season

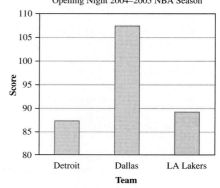

b.   **Points Scored by Winning Teams**
     Opening Night 2004–2005 NBA Season

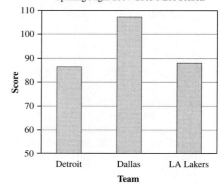

c. bar graph in part a
d. begin vertical scale at zero

2.9

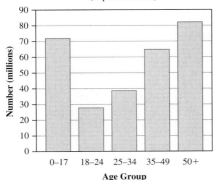

Age Grouping of U.S. Population
(September 2004)

2.11

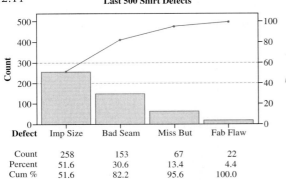

Last 500 Shirt Defects

Defect	Imp Size	Bad Seam	Miss But	Fab Flaw
Count	258	153	67	22
Percent	51.6	30.6	13.4	4.4
Cum %	51.6	82.2	95.6	100.0

2.13 a.

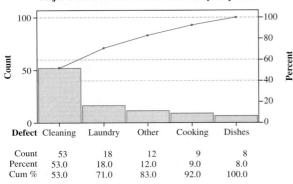

Major Chores Mothers Would Like Family Help With

Defect	Cleaning	Laundry	Other	Cooking	Dishes
Count	53	18	12	9	8
Percent	53.0	18.0	12.0	9.0	8.0
Cum %	53.0	71.0	83.0	92.0	100.0

b. It is a collection of several answers; it needs to be broken down.

2.15 a. 150 defects
  b. 0.30
  c. 37.3 + 30.0 + 15.3 + 8.0
  d. Blem. and Scratch, total 67.3

2.17 points scored per game by basketball team

Points Scored per Game by Basketball Team

2.19 a. **Heights of NBA First Round Picks on 6/24/2004**

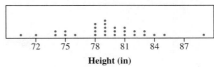

b. 71 inches, 89 inches
  c. 79 inches, 5 players
  d. tallest column

2.21 **Overall Length of Commutators**

2.23 **Points scored per game**

```
3 | 6
4 | 6
5 | 6 4 5 4 2 1
6 | 1 1 8 0 6 1 4
7 | 1
```

2.25 a. **Quik Delivery's delivery charges**

```
2. | 0
2. | 9 8 8 9
3. | 1 1
3. | 5 8 8 5 8 6 6 8 7 7 8
4. | 0 3 1 0 0
4. | 5 5 9 6 8 6
5. | 0 4 0 2 4
5. | 6 7
6. | 0 1
6. | 8
7. |
7. | 8
```

b. skewed right

2.27 a. Place value of the leaves is hundredths.
  b. 16
  c. 5.97, 6.01, 6.04, 6.08
  d. Cumulative frequencies starting at the top and bottom

2.29 a.

x	f
0	2
1	5
2	3
3	0
4	2

b. f is frequency; value of 1 occurred 5 times
  c. 12
  d. number of data, or sample size

2.31 a. bar graph
  c. histogram

2.33   b.

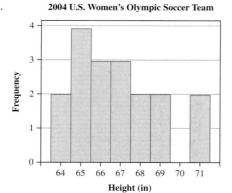

2004 U.S. Women's Olympic Soccer Team

d. 0.667

2.35   a.

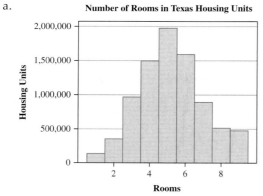

Number of Rooms in Texas Housing Units

b. mounded
c. centered on 5 rooms; 3 to 7 rooms for most

2.37   b.

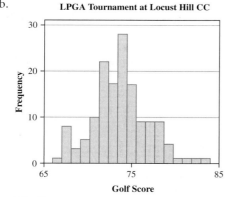

LPGA Tournament at Locust Hill CC

2.39   a. 35–45

d.

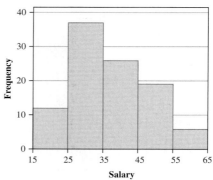

Annual Salary ($1000)

2.41   a. 12 and 16
b. 2, 6, 10, 14, 18, 22, 26
c. 4.0
d. 0.08, 0.16, 0.16, 0.40, 0.12, 0.06, 0.02
e.

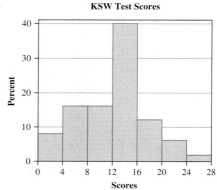

KSW Test Scores

2.43   a. freq.: 1, 14, 22, 8, 5, 3, 2
b. 6
c. 27
d.

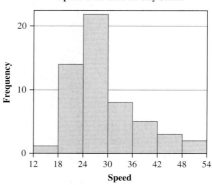

Speed of 55 Cars on City Street

2.45   a. `Third Graders at Roth Elementary School`

```
                 :
                 :
         . . : .     :       .     . : . : : : .  .
      . : : : : :   . : . : . : : : : : : : : : . :
    +---------+---------+---------+---------+---------+-- PhyStren
   0.0       5.0      10.0      15.0      20.0      25.0
```

b. freq.: 6, 10, 7, 6, 8, 11, 10, 6

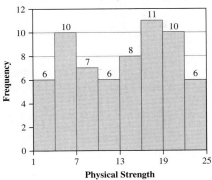

Third Graders, Physical Strength Test

d.

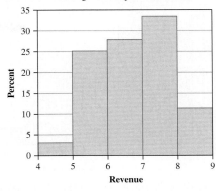

Coal, Nuclear, Electric, and Alternate Fuels Report
Average revenue per kilowatt hour

c. freq.: 3, 10, 4, 9, 7, 11, 11, 7, 2

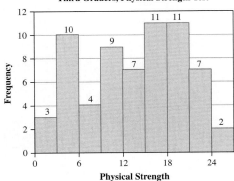

Third Graders, Physical Strength Test

2.51   a. cum. freq.: 12, 49, 75, 94, 100

c.

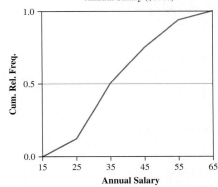

Annual Salary ($1000)

d. freq.: 3, 13, 13, 15, 17, 3

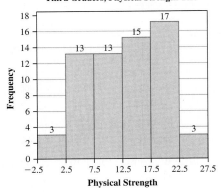

Third Graders, Physical Strength Test

2.53   a. cum. rel. freq.: 0.08, 0.24, 0.40, 0.80, 0.92, 0.98, 1.00

b.

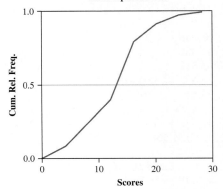

KSW Aptitude Test

f. b and c, bimodal; d, skewed left; dotplot shows mode to be 9; histogram shows two modal classes to be 4–7 and 16–22; mode; not in either modal class

2.47   a. 1, 9, 10, 12, 4
      b. 1
      c. 4.5, 5.5, 6.5, 7.5, 8.5

2.55   a. freq.: 9, 15, 17, 8, 7, 11, 11, 2, 1, 1
      b. rel. freq.: 0.110, 0.183, 0.207, 0.098, 0.085, 0.134, 0.134, 0.024, 0.012, 0.012

c. **Poor Population Living in High-Poverty Neighborhoods**
Percents in U.S. Cities

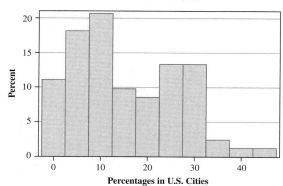

d. cum. rel. freq.: 0.110, 0.293, 0.500, 0.598,
0.683, 0.817, 0.951, 0.975,
0.987, 0.999

e. **Poor Population Living in U.S. High Poverty Neighborhoods**

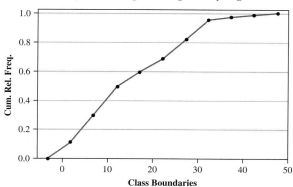

2.57 quantitative variable numbers with which arithmetic can be performed; qualitative variable does not
2.59 a. 9
      b. value = 0
2.61 a. 3.4
      b. 5.2
      c. 5.9
      d. 5.6
      e. 4.95
      f. 5.025
2.63 4.5th; 4.55
2.65 2
2.67 a. 8.2, 8.5, 9.0, 8.0
2.69 a. 6.0
      b. 3.5th; 6.5
      c. 7
      d. 5.5
2.71 a. 40.2
      b. 5.5th; 41.5
      c. 38.5
      d. 48
2.73 a. 71.16%
      b. 16th; 72.66%

c. Stem-and-Leaf of Percentage $N = 31$
Leaf Unit = 1.0

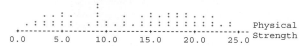

```
  2     5    89
  5     6    244
 10     6    57999
(17)    7    00111223333444444
  4     7    578
  1     8    0
```

d. left tail causes mean to be less than median; 58.60 and 59.25; have a reducing effect on the mean value

2.75 a. **Third Graders at Roth Elementary School**

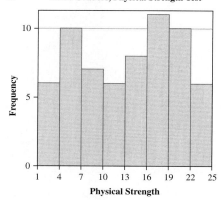

b. 9
c. **Third Graders, Physical Strength Test**

d. appears bimodal; 4–7, 16–19
f. no
g. mode is single data value; modal class, cluster of data values

2.77 a. & b.

	Runs at Home	Runs Away
Mean	4.828	4.797
Median	4.870	4.860
Maximum	6.380	5.570
Minimum	3.630	3.430
Midrange	5.005	4.500

2.79 c. little difference; more male drivers; more female drivers

d.

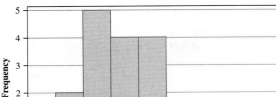

Ratio of Male Drivers to Female Drivers by State

(x-axis: M/F Ratio; y-axis: Frequency)

e. mounded

f. 0.997

2.85 a. $1465

b. 5.2%

2.87 $\Sigma(x - \bar{x}) = \Sigma x - n\bar{x} = \Sigma x - n \cdot (\Sigma x/n) = \Sigma x - \Sigma x = 0$

2.89 a. $\Sigma(x - \bar{x})^2 = 46$; 11.5

b. $\Sigma x^2 = 171$; 11.5

c. same

2.91 a. 5

b. $\Sigma(x - \bar{x})^2 = 16$; 3.2

c. 1.8

2.93 a. $\Sigma(x - \bar{x})^2 = 42.95$; 3.1

b. $\Sigma x^2 = 764$; 3.1

c. 1.8

2.95 a. $n = 6$, $\Sigma x = 37,116$, $\Sigma x^2 = 229,710,344$; 22,153.6

b. $n = 6$, $\Sigma x = 1,116$, $\Sigma x^2 = 318,344$; 22,153.6

2.97 a.

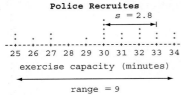

Police Recruites

$s = 2.8$

25 26 27 28 29 30 31 32 33 34

exercise capacity (minutes)

range = 9

b. 30.05

c. 9

d. $\Sigma x^2 = 18,209$; 7.8

e. 2.8

g. Except for the value $x = 30$, the distribution looks rectangular.

2.99 a. 21.7; 5.242

b. Stem-and-Leaf of Percentage $N = 31$
Leaf Unit = 1.0

```
  2    5   89
  5    6   24
 10    6   557999
(17)   7   00111233334444444  ] s  range
  4    7   578
  1    8   0
```

c. skewed left

2.101 incorrect; standard deviation never negative; error in calculations or typographical error

2.105 a. 44th position from low value; 7th position from high value

b. 10.5th; $P_{20} = 64$; 18th; $P_{35} = 70$

c. 10.5th from H; 88.5; 3rd from H; 95

2.107 a.

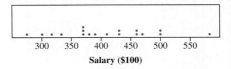

Elementary School Teachers' Salaries

Salary ($100)

b. 2nd from L, 17th from H

c. 5th, $36,700

d. 14th, $45,800

2.109 a. 3.8; 5.6

b. 4.7

c. 3.5th, 3.5; 7th, 4.0; 18.5th, 6.9

2.111

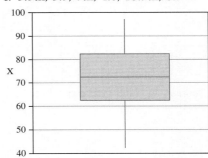

2.113 a.

2005 Men's Teams' Graduation Rates
NCAA Division I Basketball Tournament

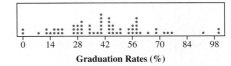

Graduation Rates (%)

b.

Stem-and-leaf of Graduation Rates (%) $N = 64$
Leaf Unit = 1.0

```
  3     0   008
 11     1   11455779
 20     2   055577799
 27     3   0033368
(15)    4   000000334445557
 22     5   003455577888
 10     6   0477
  6     7   135
  3     8
  3     9   2
  2    10   00
```

c. 5-number summary: 0, 27, 40, 55, 100

Graduation Rates for 2005 Men's Teams
NCAA Division I Basketball Tournament

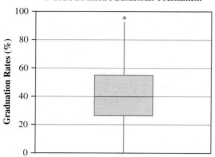

d. 4th, 11; 61st, 75
e. skewed right

2.115 a.

Major Airport On-Time Arrival Performance

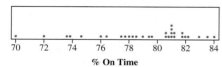

% On Time

b.

```
Stem-and-leaf of % On Time N = 31
Leaf Unit = 0.10
   1    69   9
   1    70
   2    71   9
   2    72
   4    73   58
   5    74   6
   5    75
   7    76   04
  10    77   689
  13    78   239
  15    79   45
  (5)   80   55899
  11    81   0012779
   4    82   29
   2    83   69
```

c. 69.96, 77.62, 80.50, 81.20, 83.93

**Major Airport On-Time
Arrival Performance**

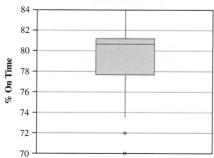

d. 4th, 73.85; 7th, 76.45
e. skewed left
f. best on-time performance rate

2.117 symmetrical
2.119 1.67, −0.75
2.121 a. −1.76
      b. −0.54
      c. 0.42
      d. 1.63
2.123 a. 120
      b. 144.0
      c. 92.0
      d. 161.0
2.125 b. 0.03, 0.14, 0.20, 0.30, 0.55

**State Percentages of Deficient
or Functionally Obsolete Bridges**

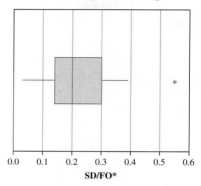

SD/FO*

c. 0.22; 0.16
d. −0.75, 1.66, −1.42, 0.22, 3.20
2.127 1.625, 1.2; A
2.129 175 through 225 words, inclusive
2.131 Nearly all data, 99.7%, lies within 3 standard
      deviations of mean.
2.133 a. 2.5%
      b. 70.4 to 97.6 hours
2.135 a. 50%
      b. 0.16
      c. 0.84
      d. 0.815
2.137 a. at least 75%
      b. at least 89%
2.139 a. at most 11%
      b. at most 6.25%
2.141 a. *f*: 6, 9, 8, 10, 6, 4, 4, 2, 1
      b. 11.2, 6.2
      c. −1.2 to 23.6; 96% (48/50)

2.143 a.

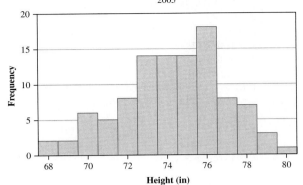

Heights of U.S. Top 100 High School Football Players
2005

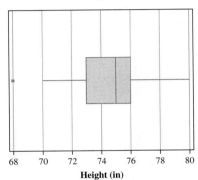

Heights of U.S. Top 100 High
School Football Players
2005

b. 74.4, 2.5
d. 71.9 to 76.9, 68%; 69.4 to 79.4, 97%;
  66.9 to 81.9, 100%
e. 68%, 97% and 100% do agree
f. 97% and 100% satisfy the theorem
2.147 yes
2.153 b. $\Sigma f = 20$; $\Sigma xf = 46$; $\Sigma x^2 f = 128$
2.155 a. 269/80 = 3.4
  b. 3
  c. 1.7277; 1.31
2.157 a. 13.1
  b. 0.0532
  c. 0.23
2.159 11.2; 12.225; 3.5
2.161 16.021; 0.026
2.163 24.2; 14.1

2.165 b.

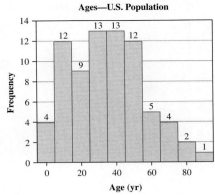

Ages—U.S. Population

c. skewed right
d. 35.2
e. 38th; 30
f. 90
g. 21.1
h.

	Ungrouped	Grouped	Percent Error
Mean	34.37	35.2	2.41%
Median	34	30	−11.76%
Range	90	90	0%
Std. Dev.	20.95	21.08	0.62%

2.167 a. 125
  b.

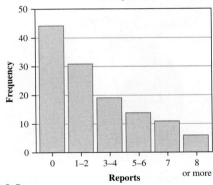

Weekly Number of Sports
Reports Watched on Television
ESPN Sport Poll

c. 2.5
d. 63rd; 1.5
e. 0
2.169 a. first and last classes not same width
  b. last class changed to "85–94 years"

c.

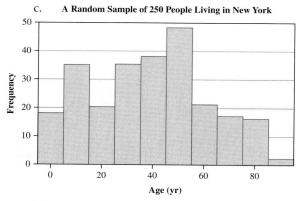

A Random Sample of 250 People Living in New York

d. & e. 38.7 years; 22.6 years

2.171 a.

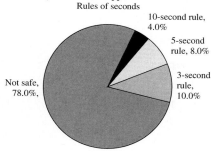

Do you eat food dropped on the floor?
Rules of seconds

10-second rule, 4.0%

5-second rule, 8.0%

3-second rule, 10.0%

Not safe, 78.0%,

b. 30, 24, 12, 234

2.173 a.

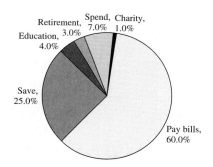

What will you do with your tax refund?

Retirement, 3.0%   Spend, 7.0%   Charity, 1.0%

Education, 4.0%

Save, 25.0%

Pay bills, 60.0%

2.175 a.

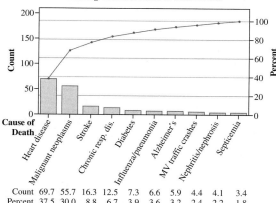

Leading Causes of Death in U.S. for 2002

Cause of Death	Heart disease	Malignant neoplasms	Stroke	Chronic resp. dis.	Diabetes	Influenza/pneumonia	Alzheimer's	MV traffic crashes	Nephritis/nephrosis	Septicemia
Count	69.7	55.7	16.3	12.5	7.3	6.6	5.9	4.4	4.1	3.4
Percent	37.5	30.0	8.8	6.7	3.9	3.6	3.2	2.4	2.2	1.8
Cum %	37.5	67.5	76.2	82.9	86.9	90.4	93.6	96.0	98.2	100.0

b. two leading causes of death, heart disease and malignant neoplasms

2.177 a. numerical
b. attribute
c. numerical
d. attribute
e. numerical

2.179 a, d, e, f, and g increased; b and c unchanged

2.181 $n = 8$, $\Sigma x = 36.5$, $\Sigma x^2 = 179.11$
a. 4.56
b. 1.34
c. very closely to 4%

2.183 $n = 118$, $\Sigma x = 2364$
a. 20.0
b. 59.5th, 17
c. 16
d. 30th, 15; 89th, 21
e. 12th, 14; 113th, 43

2.185 $n = 25$, $\Sigma x = 1997$, $\Sigma x^2 = 163,205$; 79.9; 12.4

2.187 b. data
c. statistic
d. no

2.189 a. $196,861, $62,819
b. $134,042, $259,680
c. 34/50 = 68%
d. $71,223, $322,499
e. 48/50 = 96%
f. $8,404, $385,318
g. 50/50 = 100%
h. agree with Chebyshev's
i. agree with empirical

j.

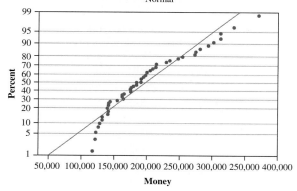

2004 Nationwide Tour Money Leaders
Normal

Mean	196861
StDev	62819
N	50
AD	1.295
p-Value	<0.005

k. Normality test suggests not normal.

2.191 a. 13.15
b. 13.85
c. 15.0
d. 12.95

e. 5.7

f. 25.5th, 10.95; 75.5th, 14.9

g. 12.925

h. 35.5th, 12.05; 64.5th, 14.5

j. **Lengths of 100 Brown Trout—Happy Acres Fish Hatchery**

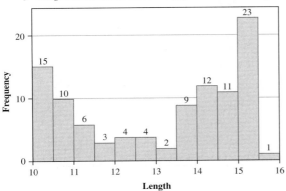

k. shown in part i

l. **Lengths of 100 Brown Trout — Happy Acres Fish Hatchery**

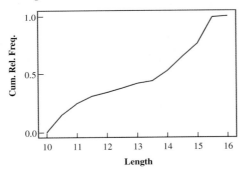

m. $n = 100$; $\Sigma xf = 1314$, $\Sigma x^2f = 17{,}635.26$;
13.14, 1.93

2.193 e. $n = 48$, $\Sigma x = 8503.88$; 177.2; 24.5th, 86.3;
no mode; 539.425

State Population Density per Square Mile—U.S.

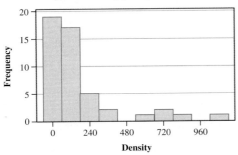

g. NJ, RI, MA, CT, MD; WY, MT, ND, SD, NM

2.195 b. **2003 State Percentages of High School Graduates Who Took the ACT Exam**

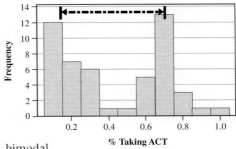

c. bimodal

e. 0.4308

g. 0.2874

h. 53%;28/50 = 56%

i. quite widespread

2.197 a. weight

30 Bags of M&M's

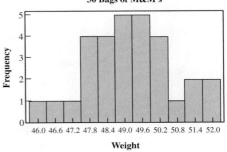

30 Bags of M&M's

b. = 49.215, median = 49.07, s = 1.522,
min = 46.22, max = 52.06

c. no

f. $\bar{x}$ = 57.1, median = 58, s = 2.383, min = 50,
max = 61

g. One bag has "only 50" M&M's in it.

2.199 a. ≈ -0.8 or -0.9

b. $\approx +1.6$ or $+1.7$

2.201 z-scores must be changed to percentiles; P_{97}, P_{84},
P_{16}, P_{50}

2.203 $n = 8$, $\Sigma x = 31{,}825$, $\Sigma x^2 = 126{,}894{,}839$

a. 3978.1

b. 203.9

c. 3570.3 to 4385.9

2.205 a. calories: 40, 111.88, 44.92
sodium (mg): 40, 566.3, 238.4

b. calories: 22.04 and 201.72
sodium: 89.5 and 1043.1

c. sodium: 327.9 and 804.7, 67.5%

2.207 a.

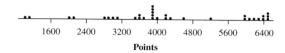

2004 NASCAR Top Standings

b. 4248, 1624
c. 986, 3170, 3902, 6058, 650
d. 1000 and 7496
e. 2624 and 5872; 56.3; no

2.217 $n = 3570$; $\Sigma xf = 55,155$, $\Sigma x^2f = 890,655$

a.

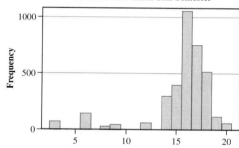

b. 15.4; 16; 16; 11.5; 16
c. 15; 17
d. 14; 14
e. 17; 10.7967; 3.3

2.219 a.

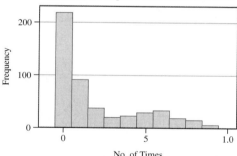

b. $\Sigma f = 500$, $\Sigma xf = 994$, $\Sigma x^2f = 5200$; 1.988, 1, 0, 4.5
c. 6.46; 2.5
d. 0, 4, 6
e. 2
f. 0, 0, 1, 4, 9

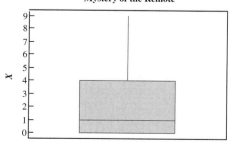

2.221 $n = 220$; $\Sigma xf = 219,100$, $\Sigma x^2f = 224,470,000$

a.

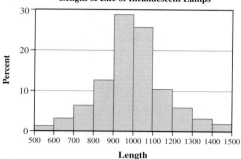

b. 995.9
c. 169.2

2.223 a.

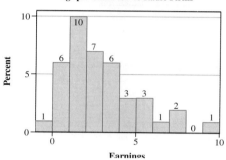

b. \$2.00–\$3.00

Chapter 3

3.1 a. yes
b. somewhat

3.3 a.

	On Airplane	Hotel Room	All Other	Marginal total
Business	35.5%	9.5%	5.0%	50%
Leisure	25.0%	16.5%	8.5%	50%

b.

	On Airplane	Hotel Room	All Other	Marginal total
Business	71.0%	19.0%	10.0%	100%
Leisure	50.0%	33.0%	17.0%	100%

Business and leisure are separate distributions.

c.

	On Airplane	Hotel Room	All Other	Marginal total
Business	58.7%	36.5%	37.0%	50%
Leisure	41.3%	63.5%	63.0%	50%

Each category is a separate distribution.

3.5 a. adults; gender; age would like to remain rest of life

b.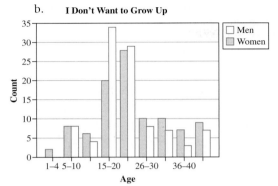

c. no

3.7 a. 3350

b. two variables, political affiliation and television network; both qualitative

c. 880

d. 46.9%

e. 19.2%

f. 5.9%

3.9 East: $\bar{x} = 4.438$, $\tilde{x} = 4.55$; West: $\bar{x} = 4.838$, $\tilde{x} = 4.6$

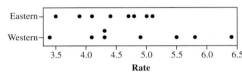

3.11 a.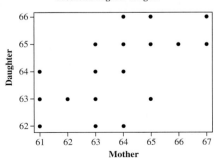

b. mother heights more spread

c. **Mother/Daughter Heights**

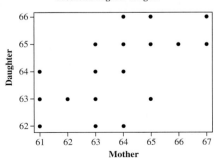

d. As mothers' heights increased, daughters' heights increased.

3.13 height; weight is often predicted

3.15 a.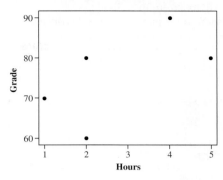

b. As hours studied increased, exam grades increased.

3.17 a. age, height

b. age = 3 years, height = 87 cm

c. Growth is above or below normal.

3.19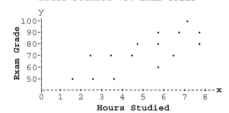

3.21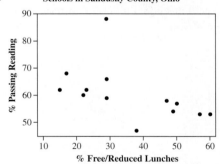

3.23 c.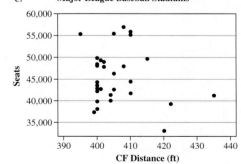

d. no relationship

3.25 b. strong increasing pattern in all three
c. All are strength skills.
e. seem to display a linear relationship

3.27 a.

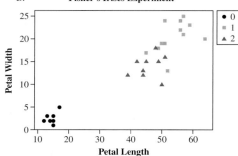

b.

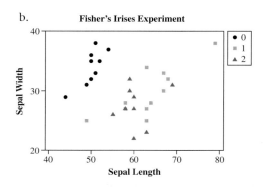

c. Type 0 displays a different pattern than types 1 and 2

d.

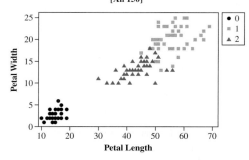

3.29 a. become closer to a straight line with a positive slope
b. become closer to a straight line with a negative slope

3.31 very little or no linear correlation

3.33 a. SS(x) = 10.8; SS(y) = 520; SS(xy) = 46
b. 0.61

3.35 a.

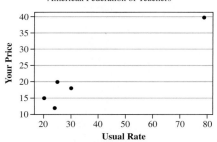

b. SS(x) = 2405.38822
c. SS(y) = 479.44078
d. SS(xy) = 1040.07546
e. 0.97

3.37 a. manatees, powerboats
b. number of registrations, manatee deaths
c. As one increases, the other does also.

3.39 a. near −0.75
b. SS(x) = 82.0; SS(y) = 67.6; SS(xy) = −60.0; −0.81

3.41 positive vs. negative; nearness to straight line, etc.

3.43 a.

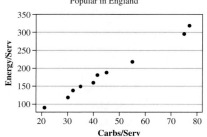

b. yes
c. SS(x) = 3125.511; SS(y) = 48,505.6; SS(xy) = 12,264.84; r = 0.996
d. strong, positive correlation

e.

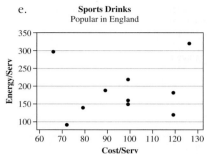

Sports Drinks
Popular in England

no linear relationship
SS(x) = 3794.1; SS(y) = 48,505.6;
SS(xy) = 2044.4; r = 0.15;
little or no correlation

3.45   a.

Interstate 95
per state

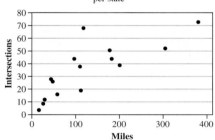

b. somewhat
c. 0.79
d. yes
e. has more intersections than it has miles of interstate
f.

Interstate 95
per state
without Connecticut

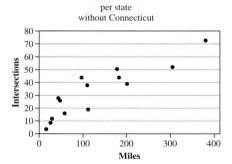

more closely follows a straight line
g. 0.892
h. stronger positive value

3.47   a. **National Highway System—October 2005**
Number of Miles

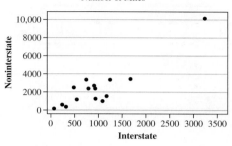

b. linear; quite different; Texas
c. r = 0.91
d.   **National Highway System—October 2005**
Number of Miles
Texas Not Included

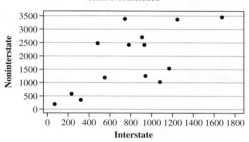

r = 0.67

3.49   a. SS(x) = 700; SS(y) = 3286.7254;
SS(xy) = 1404.4; r = 0.93

b. **Nontobacco Monthly Rates for Life Insurance**

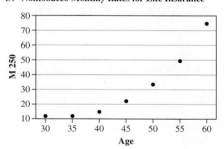

c. no
d. elongated pattern
e. Rate for insurance increases (accelerates) as the insured person's issue age increases, thus the "upward-bending" pattern.

3.51   Bigger fires require more fire trucks.

3.53

Scatter Diagram Ex. 3.53

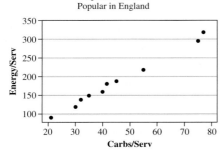

no

3.55 a. $\Sigma x^2 = 13{,}717$, $SS(x) = 1396.9$; $\Sigma y^2 = 15{,}298$,
$SS(y) = 858.0$; $\Sigma xy = 14{,}257$, $SS(xy) = 919.0$
  b. (Σ's) are sums of data, ($SS(\ )$) are parts of formulas
3.57 a. $\hat{y} = 28.1$; $\hat{y} = 47.9$
  b. yes
3.59 a.

Sports Drinks
Popular in England

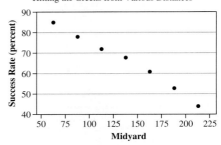

linear
  b. $SS(x) = 3125.511$; $SS(xy) = 12{,}264.84$;
$\hat{y} = 9.55 + 3.924x$
  c. 166.51
  d. 264.61
3.61 a. For each increase in height of 1 inch, weight increased by 4.71 lb.
  b. The scale for the y-axis starts at $y = 95$, and the scale for the x-axis starts at $x = 60$.
3.63 a. 121.14 or \$12,114
  b. 56.58 or \$5658
  c. 21.52(\$100) = \$2152
3.65 a. \$492,411,000
  b. \$990,241,000
  c. \$1,488,041,000
3.67 vertical scale is at $x = 58$ and is not the y-axis
3.69 5.83, -260.61

3.71 a. **Success Rates for 2004 PGA Tour Players**
Hitting the Greens from Various Distances

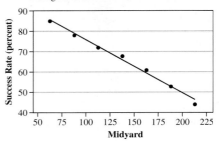

  b. yes
  c. $SS(x) = 17{,}500$; $SS(y) = 1222.8571$;
$SS(xy) = -4600$; $r = -0.994$
  d. very strong negative
  e. yes
  f. $= 102.15 - 0.263x$
  g. **Success Rates for 2004 PGA Tour Players**
Hitting the Greens from Various Distances

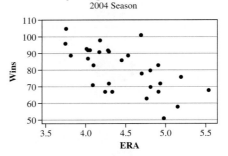

  h. 78.5%
3.73 c. **Major League Baseball Teams**
2004 Season

  d. yes
  e. $\hat{y} = 163 - 18.5x$
  f. Wins decrease by 18.5 for every one increase in ERA.
  g. yes

3.75 a. **Insurance Rates for $250,000 Insurance**

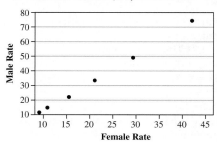

b. SS(x) = 914.117; SS(y) = 3286.725;
SS(xy) = 1732.582; r = 0.9996

c. $\hat{y} = -5.83 + 1.90x$

d. male rate = $22.67

e. males paid a higher rate

3.77 a. **Mishandled Baggage Reports Filed by Airline Passengers—October 2004**
The Office of Aviation Enforcement & Proceedings

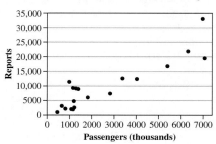

b. yes

c. $\hat{y} = 1427 + 3.47x$

d. **Mishandled Baggage Reports Filed by Airline Passengers—October 2004**
Reports = 1427 + 3466 Passangers

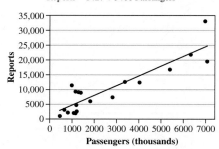

3.79 a. both increasing at the same rate, not parallel, one increasing at a faster rate

b. more drivers than vehicles before 1971, then more vehicles than drivers after 1973; 1971 to 1973, the number of drivers and vehicles was about the same

c. The number of drivers was increasing at a faster rate than the number of cars.

d. The number of motor vehicles is increasing at a faster rate than the number of drivers.

3.81 a. Fear: 138, Do not: 362

b.

	Elem.	Jr. H.	Sr. H.	Coll.	Adult
Fear	7.4%	5.6%	5.0%	5.4%	4.2%
Do not	12.6%	14.4%	15.0%	14.6%	15.8%

d.

	Elem.	Jr. H.	Sr. H.	Coll.	Adult
Fear	26.8%	20.3%	18.1%	19.6%	15.2%
Do not	17.4%	19.9%	20.7%	20.2%	21.8%

e. **Fear of the Dentist**

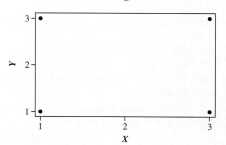

3.87 a. determine whether linearly related; result is r

b. determine the equation of the line of best fit; result is the equation

3.89 a. **Scatter Diagram**

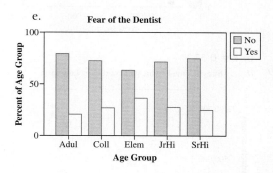

b. SS(x) = 4.0; SS(y) = 4.0; SS(xy) = 0.0;
r = 0.00

c. $\hat{y} = 2.0 + 0.0x$

3.95 a. **2004 Life Expectancies**
The World Factbook

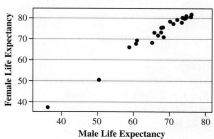

b. yes

c. $\hat{y} = -1.40 + 1.11x$

d. For every one additional male life expectancy year, the female life expectancy increases by 1.11 years.

3.97 a.

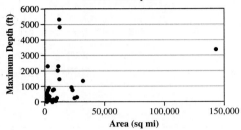

Scatter Diagram of Areas of Lakes and Their Maximum Depths

b. $r = 0.3709$

3.99 b.

Sugar Cane–Producing Counties in Louisiana

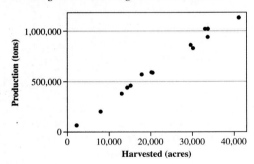

c. very elongated; shows an increasing relationship

d. 167,689.2143; 477,586.4286; $\hat{y} = 12.3 + 2.85x$

e. 2.85, for every 100 acres, 2850 tons of sugar cane produced

3.101 a.

Cicadas (Magicicada)

b. linear

c. $SS(x) = 0.0967333$; $SS(y) = 181.333$; $SS(xy) = 2.71667$; $r = 0.649$

d. $\hat{y} = 23.0 + 28.1x$

e. 28.62 mm

Chapter 4

4.1 a. most: yellow, blue, and orange; least: brown, red, and green

b. not exactly, but similar

4.3 5, 6, 6, 9, 8, 6, respectively

4.7 $P'(5) = 0.225$

4.9 a. 50%

b. 0.50

c. same question but in two different formats, percentage and probability

4.11 a. 0.356

b. 0.389

c. 0.703

d. 0.18

4.13 a. 0.09; 0.95

b. 0.48; 0.27

c. San Diego

4.15 {0, 1, 2, 3, 4, 5, 6, 7, 8, 9}

4.17 $P(5) = 4/36$; $P(6) = 5/36$; $P(7) = 6/36$; $P(8) = 5/36$; $P(9) = 4/36$; $P(10) = 3/36$; $P(11) = 2/36$; $P(12) = 1/36$

4.19 e. theoretical probabilities

4.21 $S = \{$JH, JC, JD, JS, QH, QC, QD, QS, KH, KC, KD, KS$\}$

4.23 a. $S = \{\$1, \$5, \$10, \$20\}$

b.

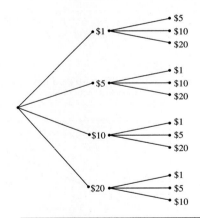

4.25 a.

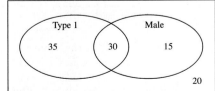

b. 0.55

c. 0.35

4.27 0.04; 4%

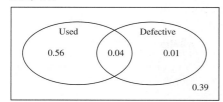

4.29 All are inappropriate.
4.31 a. expect a 1 to occur approximately 1/6th of the time when you roll a single die
b. 50% of the tosses are expected to be heads; the other 50%, tails

4.33

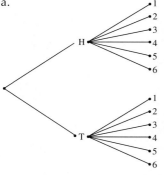

4.35 a.

b. {(H,1),(H,2),(H,3),(H,4),(H,5),(H,6), (T,1),(T,2),(T,3),(T,4),(T,5),(T,6)}
4.39 1/5
4.41 4/5
4.43 a. 1/13
b. 12:1
4.45 a. 0.00000000000629908
b. 0.00000153908
c. 6.29908×10^{-12}, 1.53908×10^{-6}

4.47 a. S = {HH, HT, TH, TT}; equally likely
b. S = {HH, HT, TH, TT}; not equally likely
4.49 a. 0.45
b. 0.40
c. 0.55
4.51 a. 0.35
b. 0.38
c. 0.64
d. 0.03
e. 0.91
f. 0.74
g. 0.98
4.53 a. 0.59
b. 0.41
c. 0.35
d. 0.27
e. 0.30
f. 0.60
g. 0.60
h. different ways of asking same question
4.55 a. Some categories would be counted twice.
b. 0.08
c. 0.78
d. 0.09
e. 0.39
4.57 a. 0.3
b. 0.22
4.59 0.34
4.61 0.8
4.63 0.2
4.65 0.81
4.67 4%
4.69 0.28
4.71 0.5
4.73 0.098
4.75 0.90
4.77 a.

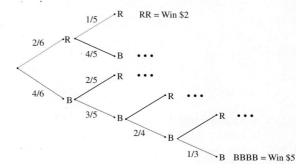

b. 2/5 or 1/5, depending on first pick
c. 0.067
d. 0.067, same probability
4.79 0.62
4.81 0.133
4.83 a. 0.6
b. 0.7
c. 0.5

4.85   a. 0.4
      b. 0.4
4.87   a. not mutually exclusive
      b. not mutually exclusive
      c. not mutually exclusive
      d. mutually exclusive
4.89   There is no intersection.
4.91   a. 0.7
      b. 0.6
      c. 0.7
      d. 0.0
4.93   a. yes
      b. no
      c. no
      d. yes
      e. no
      f. yes
      g. no
4.95   a. A & C and A & E are mutually exclusive.
      b. 12/36, 11/36, 10/36
4.97   a. yes
      b. yes
      c. no
      d. 0.307
      e. 0.587
      f. 0.363
      g. 0.559
      h. 0.145
      i. 0.658
      j. 0.081
4.99   0.54
4.101 a. independent
      b. not independent
      c. independent
      d. independent
      e. not independent
      f. not independent
4.103 0.28
4.105 0.5
4.107 a. 0.12
      b. 0.4
      c. 0.3
4.109 a. 0.5
      b. 0.667
      c. no
4.111 a. independent
      b. independent
      c. dependent
4.113 a. 0.51
      b. 0.15
      c. 0.1326
4.115 a. 0.0289
      b. 0.6889
      c. 0.0008
4.117 0.0741

4.119 a. 0.36
      b. 0.16
      c. 0.48
4.121 a. 3/5
      b. 0.16, 0.48, 0.36
4.123 a. whether or not part-time and graduate are
        independent
      b. no
      c. 0.4074
4.125 a. cannot occur at the same time
      b. Occurrence of one has no effect on the prob-
        ability of the other.
      c. mutually exclusive—whether or not share
        common elements; independence—effect one
        event has on the other event's probability
4.127 a. 0.25
      b. 0.2
      c. 0.6
      d. 0.8
      e. 0.7
      f. no
      g. no
4.129 a. 0.0
      b. 0.7
      c. 0.6
      d. 0.0
      e. 0.5
      f. no
4.131 a. 0.625
      b. 0.25
      c. $P(\text{satisfied} \mid \text{skilled female}) = 0.25$
        $P(\text{satisfied} \mid \text{unskilled female}) = 0.667$
        not independent
4.133 a. 0.17
      b. 0.17
      c. 0.41
      d. without
4.135 0.300
4.137 a. $S = \{$GGG, GGR, GRG, GRR, RGG, RGR,
        RRG, RRR$\}$
      b. 3/8
      c. 7/8
4.139 7/8
4.141 a. 0.40
      b. 0.49
      c. 0.06
      d. 0.82
      e. 0.40
      f. 0.45
4.143 a. Brazil, Spain, India, etc.
      b. "based on countries included"
      c. 44%
      d. 0.44
      e. same question; answer in a different format
4.145 $P(\text{A or B}) = P(\text{A}) + P(\text{B}) - P(\text{A and B}) =$
      $P(\text{A}) + P(\text{B}) - P(\text{A}) = P(\text{B})$

4.147 a. 0.30
      b. 0.60
      c. 0.10
      d. 0.60
      e. 0.333
      f. 0.25

4.149 a. 0.3168
      b. 0.4659
      c. no
      d. no
      e. "Candidate wants job" and "RJB wants candidate" could not both happen.

4.151 a. 0.429
      b. 0.476
      c. 0.905

4.153 a. 0.531
      b. 0.262
      c. 0.047

4.155 a. 0.5087
      b. 0.2076
      c. 0.3336
      d. 0.56018
      e. 0.1993
      f. 0.4989
      g. mutually exclusive

4.157 a. false
      b. true
      c. false
      d. false

4.159 8/30

4.161 a. 1/2, 1/4, 1/8
      b. 9/16, 9/32, 9/64

4.163 0.592

4.165 a. 26/52
      b. 26/52
      c. 32/52
      d. 20/52

4.167 a. 0.60
      b. 0.648
      c. 0.710
      d. (a) 0.70   (b) 0.784   (c) 0.874
      e. (a) 0.90   (b) 0.972   (c) 0.997
      f. "best" team most likely wins more games; greater difference between teams

4.169 a. 1/7
      b. 1/7

Chapter 5

5.1 a. 22%
     b. 4+ cups or cans
     c. number of daily cups or cans of caffeinated beverage
     d. Yes, events (none, one, two, three, four+) are nonoverlapping.

5.3 number of siblings $x = 0, 1, 2, 3, \ldots, n$; length of conversation $x = 0$ to ? minutes

5.5 a. discrete, count; continuous, measurable
     b. discrete, count
     c. continuous, measurable

5.7 a. number of new jobs
     b. discrete, count

5.9 distance, $x = 0$ to n, n = radius of the target, continuous

5.11 a. average amount of time spent on various activities
      b. continuous, measurable

5.13

x	0	1
$P(x)$	1/2	1/2

5.15 a. Events never overlap.
      b. all outcomes accounted for

5.17 a. $P(x)$ is a probability function.

x	$P(x)$
1	0.12
2	0.18
3	0.28
4	0.42

b.

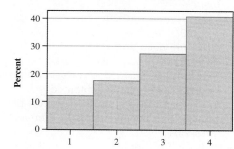

$P(x) = (x^2 + 5)/50$, for $x = 1, 2, 3, 4$

5.19

x	0	1	2	3
$P(x)$	0.20	0.30	0.40	0.10

5.21 a.

Number	Proportion	Number	Proportion
0	0.196	5	0.082
1	0.131	6	0.054
2	0.162	7 to 10	0.072
3	0.168	11 or more	0.014
4	0.121		

     b. Student chooses among multiple acceptances.

5.23 percentages do not sum to 1.00; not a random variable; attribute variable

5.27 $\sigma^2 = \Sigma[(x - \mu)^2 P(x)]$
$= \Sigma[x^2 - 2x\mu + \mu^2)P(x)]$
$= \Sigma[x^2 P(x) - 2x\mu P(x) + \mu^2 P(x)]$
$= \Sigma[x^2 P(x)] - 2\mu\Sigma[xP(x)] + \mu^2[\Sigma P(x)]$
$= \Sigma[x^2 P(x)] - 2\mu[\mu] + \mu^2[1]$
$= \Sigma[x^2 P(x)] - 2\mu^2 + \mu^2$
$= \Sigma[x^2 P(x)] - \mu^2$ or $\Sigma[x^2 P(x)] - \{\Sigma[xP(x)]\}^2$

5.29 nothing of any meaning

5.31 2.0, 1.4

5.33 a. 0.54, 0.79
b.

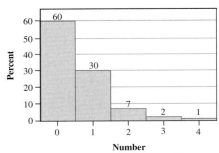

Number of Florida Hurricanes
Annual Probabilities

5.35 11.3, 1.35

5.37 a. 2.0, 1.4
b. −0.8 to 4.8 encompasses the numbers 1, 2, 3, and 4
c. 0.9

5.39 a. vary; close to −$0.20
b. −$0.20
c. close; no; need mean = 0

5.41 a. Each question is a separate trial.
b. four different ways, one correct and three wrong answers can be obtained
c. 1/3 is the probability of success; 4 is the number of independent trials; number of questions

5.43 defective items should be fairly small and easier to count

5.45 a. 24
b. 5040
c. 1
d. 360
e. 10
f. 15
g. 0.0081
h. 35
i. 10
j. 1
k. 0.4096
l. 0.16807

5.47 $n = 100$ trials (shirts), two outcomes (first quality or irregular), $p = P$(irregular), $x = n$(irregular); any integer value from 0 to 100

5.49 a. Trials are not independent.

b. $n = 4$; ace, not ace; $p = P$(ace) = 4/52 and $q = P$(not ace) = 48/52; $x = n$(aces), 0, 1, 2, 3, or 4

5.51 a.

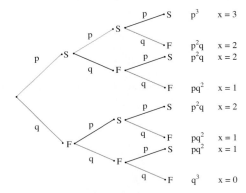

e. $P(x) = \binom{3}{x} p^x q^{3-x}$, for $x = 0, 1, 2, 3$

5.53 $P(x) = \binom{3}{x}(0.5)^x(0.5)^{3-x}$, 0.125, 0.375, 0.125

5.55 a. 0.3585
b. 0.0159
c. 0.9245

5.57 a. 0.4116
b. 0.384
c. 0.5625
d. 0.329218
e. 0.375
f. 0.0046296

5.59 $n = 5$; $p = 1/2$, $q = 1/2$ $(p + q = 1)$; exponents add up to 5; $x =$ any integer from 0 to $n = 5$; binomial

5.61 0.930

5.63 a. 0.444
b. 0.238
c. 0.153

5.65 a. 0.590
b. 0.918

5.67 0.0011

5.69 0.984

5.71 a. 0.4596
b. 0.0457
c. 0.0042

5.73 0.007

5.75 a. 0.03132
b. 0.99962

5.79 a. 0.6858
b. 0.4458
c. 0.8300
d. 0.7414
e. Values of p are almost complements.
f. As p increased, the probability of the interval increased.

5.81 0.9292

5.83 a. P(2 of 2) = 0.308; P(9 of 15) = 0.274
b. P(2 of 2) = 0.687; P(8 of 10) = 0.763

5.85 18, 2.7

5.87   a. $\Sigma[xP(x)] = 0.55$, $\Sigma[x^2P(x)] = 0.819$, 0.549, 0.72

     b. same

5.89   a. 25.0, 3.5

     b. 4.4, 1.98

     c. 24.0, 4.7

     d. 44.0, 2.3

5.91   b. 0.4338

     c. 4.92, 1.7

     d. **In a Random Sample of 12 Airline Travelers**
Probability of number who put on headphone

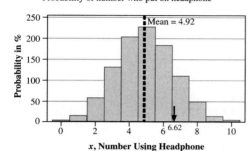

x, **Number Using Headphone**

5.93 $p = 0.5$, $n = 400$

5.95 $\mu = 6$, $\sigma = 1.9$; 0.03383

5.97   b. 0.886385

     c. 0.99383

     d. 0.12, 0.345

     e. −0.225, 0.465; 0.88638

     f. −0.57, 0.81; 0.88638

     g. do not agree with the empirical rule; does agree with Chebyshev's

5.99   1. Each $P(x)$, is a value between 0 and 1 inclusive.

     2. Sum of all the $P(x)$ is exactly 1.

5.101   a. probability function

     b. probability function

     c. NOT a probability function

     d. NOT a probability function

5.103   a. 0.1

     b. 0.4

     c. 0.6

5.105   a. 3.3

     b. 1.187

5.107   a. 0.930

     b. 0.264

5.109   0.103

5.111   a. 0.999

     b. 0.206

     c. 0.279

5.113   no, variable is attribute

5.115   0.063

5.117 P(defective) changes; trials not independent

5.119   a. 0.914

     b. 0.625

     c. Even though the P(defective) changes from trial to trial, if the population is very large, the probabilities are very similar.

5.121   2600

5.123   a. 0.88

     b. $B(50, 0.88)$

5.125   tool shop: mean profit = 7000.0; book store: mean profit = 6000.0

5.127 $\mu = \Sigma[xP(x)]$

$= (1)(1/n) + (2)(1/n) + \ldots + (n)(1/n)$

$= (1/n)[1 + 2 + 3 + \ldots + n]$

$= (1/n)[(n)(n + 1)/2]$

$= (n + 1)/2$

Chapter 6

6.1   a. It is a quotient.

     b. IQ: 100, 16; SAT: 500, 100; standard score: 0, 1

     c. $z = (IQ − 100)/16$; $z = (SAT − 500)/100$

     d. 2, 132, 700

     e. same

6.3   a. proportion

     b. percentage

     c. probability

6.5   a. bell-shaped mean of 0; standard deviation of 1

     b. reference used to determine the probabilities for all other normal distributions

6.7   a. 0.4032

     b. 0.3997

     c. 0.4993

     d. 0.4761

6.9   0.0212

6.11   0.4177

6.13   0.8571

6.15   a. 0.4394

     b. 0.0606

     c. 0.9394

     d. 0.8788

6.17   a. 0.5000

     b. 0.1469

     c. 0.9893

     d. 0.9452

     e. 0.0548

6.19   a. 0.4906

     b. 0.9725

     c. 0.4483

     d. 0.9306

6.21   a. 0.2704

     b. 0.8528

     c. 0.1056

     d. 0.9599

6.23   0.2144

6.25   a. 0.7737

     b. 0.8981

     c. 0.8983

     d. 0.3630

6.27   a. 1.14
      b. 0.47
      c. 1.66
      d. 0.86
      e. 1.74
      f. 2.23

6.29   a. 1.65
      b. 1.96
      c. 2.33

6.31 -1.28 or $+1.28$

6.33 -0.67 and $+0.67$

6.35   a. 0.84
      b. -1.15 and $+1.15$

6.37   a. 1.28
      b. 1.65
      c. 2.33

6.39   a. 0.7606
      b. 0.0386
      c. 0.2689

6.41   2.88

6.43   a. 0.5000
      b. 0.3849
      c. 0.6072
      d. 0.2946
      e. 0.9502
      f. 0.0139

6.45   a. 0.3944
      b. 0.8944

6.47   a. 0.6826 or 68.26%
      b. 0.9544 or 95.44%
      c. 0.9974 or 99.74%
      d. $0.6826 \approx 68\%$; $0.9544 \approx 95\%$;
        $0.9974 \approx 99.7\%$

6.49   a. 0.0397 or 4.0%
      b. 0.1271 or 12.7%
      c. 0.9147 or 91.5%
      d. 0.3729 or 37.3%
      e. 0.0375 or 3.8%

6.51   a. 0.0075
      b. 0.3336

6.53   a. 0.3557
      b. $100(0.3557) \approx 36$ bottles

6.55   a. from 23.6 mm to 24.6 mm
      b. 0.998997 or 99.9%
      c. 0.0001 or 0.01%
      d. 0.9928 or 99.3%

6.57   a. 89.6
      b. 79.2
      c. 57.3

6.59   20.26

6.61   7.664

6.63   a. 20%
      b. 50%
      c. 50.375
      d. 49.55
      e. 51.395

      f. to avoid fines
      g. to avoid putting extra M&M's in bag

6.65   a. 0.056241
      b. 0.505544
      c. 0.438215
      d. 0.0559, 0.5077, 0.4364
      e. Round-off errors in z

6.71   a. $z(0.03)$
      b. $z(0.14)$
      c. $z(0.75)$
      d. $z(0.22)$
      e. $z(0.87)$
      f. $z(0.98)$

6.73   a. $z(0.01)$
      b. $z(0.13)$
      c. $z(0.975)$
      d. $z(0.90)$

6.77   a. 1.96
      b. 1.65
      c. 2.33

6.79   a. 1.41
      b. -1.41

6.81   a. 1.28, 1.65, 1.96, 2.05, 2.33, 2.58
      b. $-2.58, -2.33, -2.05, -1.96, -1.65, -1.28$

6.83   a. area, 0.4602
      b. z-score, 1.28
      c. area, 0.5199
      d. z-score, -1.65

6.85 $np = 2$, $nq = 98$; no

6.87   binomial: 0.829; normal approx.: 0.8133

6.89   0.1812; 0.183

6.91   0.6406; 0.655

6.93 $x = n(\text{survive})$, $\mu = 225$, $\sigma = 4.74$; 0.9999997

6.95 $\mu = 25$, $\sigma = 3.54$
      a. 0.5557
      b. 0.0002

6.97 $\mu = 27.63$, $\sigma = 3.86$; 0.6844

6.99 $\mu = 420$, $\sigma = 16.52$
      a. 0.0367
      b. 0.034687
      c. 0.032952

6.101   at least 3/4; 0.9544

6.103   a. 1.26
       b. 2.16
       c. 1.13

6.105   a. 0.0930
       b. 0.9684

6.107   a. 1.175 or 1.18
       b. 0.58
       c. -1.04
       d. -2.33

6.109   a. 0.1867
       b. 0.4458
       c. 0.1423

6.111   a. 0.0038
       b. 0.5610
       c. 0.0011

6.113 a. 0.3192
    b. 0.1685
    c. 156.0 to 184.2 minutes
    d. 135.5 to 204.7 minutes
6.115 10.033
6.117 a. 0.0143
    b. 619.4
    c. 107.2
    d. 755.4
6.119 a. $np = 7.5$, $nq = 17.5$; both greater than 5
    b. 7.5, 2.29
6.121 b. 0.77023
    c. 0.751779
6.123 a. $P(0) + P(1) + \ldots + P(75)$
    b. 0.9856
    c. 0.9873
6.125 0.0087
6.127 $\mu = 8$, $\sigma = 2.6$
    a. 0.0418
    b. 0.4247
    c. 0.7128
6.129 $\mu = 19.5$, $\sigma = 3.45$
    a. 0.3859
    b. 0.0735
6.131 $\mu = 130$, $\sigma = 11.1$
    a. 0.0322
    b. 0.3085
6.133

Nation	$\mu = np$	$\sigma = \sqrt{npq}$	$P(x \geq 70)*$
China	50	6.98	0.0026
Germany	8	2.82	0.0000+
India	116	10.45	0.9999+
Japan	6	2.45	0.0000+
Mexico	44	6.56	0.0001
Russia	34	5.78	0.0000+
S. Africa	124	10.78	0.9999+
United States	14	3.73	0.0000+

6.135 a. from -0.0392 to $+0.0392$
    b. 0.8727 or 87.3%
    c. 0.8664 or 86.6%

Chapter 7

7.1 a. histogram

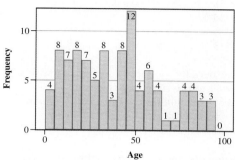

Age of U.S. Citizens
($n = 100$)

    b. mounded from 0 to 60 skewed right
    c. looks like population
7.3 a. no
    b. variability
7.5 a. distribution formed by means from all possible samples of a fixed size taken from a population
    b. It is one element in the sampling distribution.
7.11 a. not all drawn from the same population; each type of vehicle has a different sample size
    b. monitoring transit vehicle populations, which are continually changing
7.17 b. very close to $\mu = 65.15$
    d. approximately normal
    e. took many (1001) samples of size 4 from an approximately normal population; (1) mean of the x-bars $\approx \mu$, (2) $s_{\bar{x}} \approx \sigma/\sqrt{n}$, (3) approximately normal distribution
7.19 a. 1.0
    b. $\sigma_{\bar{x}} = \sigma/\sqrt{n}$; as n increases, the value of this fraction gets smaller
7.21 a. 500
    b. 5
    c. approximately normal
7.23 a. approximately normal
    b. 4.0 hours
    c. 0.133
7.25 a. 17.71 pounds/person
    b. 0.514
    c. approximately normal
7.29 2.69
7.31 0.4772
7.33 a. approximately normal
    b. 50
    c. 1.667
    d. 0.9974
    e. 0.8849
    f. 0.9282
7.35 a. approximately normal, $\mu = 69$, $\sigma = 4$.
    b. 0.4013
    c. approximately normal

d. $\mu_{\bar{x}} = 69$; $\sigma_{\bar{x}} = 1.0$
e. 0.1587
f. 0.0228

7.37 a. 0.3830
b. 0.9938
c. 0.3085
d. 0.0031

7.39 a. 0.2643
b. 0.0294
c. No, especially for part a; wind speed will be skewed to the right, not normal.
d. Actual probabilities are most likely not as high as found.

7.41 a. 0.9821
b. 0.0001
c. Normal distribution should allow for reasonable estimates since $n > 30$.

7.43 38.73 inches

7.45 a. computer: 0.68269; Table 3: 0.6826

7.47 6.067, 3.64, 2.6, 1.82

7.49 a. normally distributed with a mean = $675 and a standard deviation = $85
b. 0.8900
c. approximately normally distributed with a mean = $675 and a standard error = $17
d. 0.8900

7.51 a. $e = 0.49$
b. $E = 0.049$

7.53 a. 0.0060
b. 0.1635
c. skewed distribution

7.55 a. 0.1498
b. 0.0089

7.57 0.0228

7.59 0.0023

7.61 a. Σx total weight; approximately 1.000
b. 0.9772

7.65 a. $\mu = 60$; $\sigma = 6.48$

Chapter 8

8.1 a. female health professionals
b. $\bar{x} = 64.7$, $s = 3.5$; mounded about center, approximately symmetrical

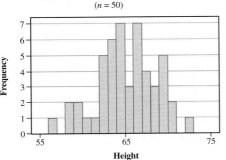

Heights of Females in Health Profession
(n = 50)

8.3 Point estimate is a single number; interval estimate is an interval of some width.

8.5 $n = 15$, $\Sigma x = 271$, $\Sigma x^2 = 5015$
a. 18.1 dollars
b. 8.5
c. 2.9 dollars

8.7 a. II has lower variability.
b. II has a mean value equal to the parameter.
c. Neither is a good choice.

8.9 difficulty, collector fatigue; cost of sampling; destruction of product

8.11 3

8.13 a. 2(0.3997) = 0.7994
b. 2(0.4251) = 0.8502
c. 2(0.4750) = 0.9500
d. 2(0.4901) = 0.9802

8.15 19($17,320) + 6($20,200) = $450,280 for 25 projects; 174($450,280/25) = $3,133,948.80

8.17 a. between 11:17 and 11:37 AM
b. Yes, 11:20:48 AM is within interval 11:17 to 11:37 AM.
c. 90% occur within predicted interval

8.19 Sampling distribution of sample means must be normal.

8.21 a. $z(0.01) = 2.33$
b. $z(0.005) = 2.58$

8.23 a. 25.76 to 31.64
b. Yes, population is normal.

8.25 a. 125.58 to 131.42
b. yes, CLT

8.27 a. 128.5
b. $z(0.05) = 1.65$
c. 1.76845
d. 2.92
e. 125.58
f. 131.42

8.31 a. 15.9; $\approx 68\%$
b. 31.4; $\approx 95\%$
c. 41.2; $\approx 99\%$
d. higher level makes for a wider width

8.33 a. 75.92
b. 0.368
c. 75.552 to 76.288

8.35 a. 14.01 to 14.59
b. 13.89 to 14.71

8.37 487.5 to 520.5

8.39 a. mean length
b. 75.92
c. 75.512 to 76.328

8.41 a. 276.9 to 314.4
b. 284.69 to 344.01
c. Different sample means gave different center points.
d. Decrease in sample size; increase in width
e. 310 is in both intervals, neither interval gives reason to doubt claim

8.43 Numbers are calculated for samples of snow.

8.45 49

8.47 27

8.49 25

8.51 H_o: system is reliable; H_a: system is not reliable

8.53 a. H_a: Special delivery mail takes too much time.
b. H_a: New design is more comfortable.
c. H_a: Cigarette smoke has an effect.
d. H_a: Hair conditioner is effective on "split ends."

8.55 A: Party will be a dud; did not go.
B: Party will be a great time, did go.
I: Party will be a dud, did go.
II: Party will be a great time, did not go.

8.57 a. H_a: The victim is not alive.
b. A: alive, treated as though alive
I: alive, treated as though dead
II: dead, treated as though alive
B: dead, treated as though dead
c. I very serious; victim may die soon without attention. II not serious; victim is receiving attention that is of no value.

8.59 missed a great time

8.65 a. type I b. type II c. type I d. type II

8.67 a. Commercial is not effective.
b. Commercial is effective.

8.69 a. very serious
b. somewhat serious
c. not at all serious

8.71 a. α
b. β

8.73 α is probability of rejecting a TRUE null hypothesis; $1 - \beta$ is probability of rejecting a FALSE null hypothesis.

8.75 a. "See, I told you so."
b. "Okay, not significant. I'll try again tomorrow."

8.77 a. 0.1151
b. 0.2119

8.79 a. 29 corks
b. batch refused; 3 corks do not meet the specification.

8.81 H_o: The mean shearing strength is at least 925 lb.
H_a: The mean shearing strength is less than 925 lb.

8.83 a. H_a: $\mu > 1.25$
b. H_a: $\mu < 335$
c. H_a: $\mu \neq 230{,}000$
d. H_a: $\mu > 210$
e. H_a: $\mu > 9.00$

8.85 II; buy and use weak rivets

8.87 I: reject H_o interpret as "mean hourly charge is less than \$60 when, in fact, it is at least \$60
II: fail to reject H_o interpreted as "mean hourly charge is at least \$60 when, in fact, it is less than \$60.

8.89 a. 1.26
b. 1.35
c. 2.33
d. -0.74

8.91 a. reject H_o, fail to reject H_o
b. p-value is smaller than or equal to α, reject H_o. p-value is larger than α, fail to reject H_o.

8.93 a. reject H_o, $\mathbf{P} < \alpha$
b. fail to reject H_o, $\mathbf{P} > \alpha$
c. reject H_o, $\mathbf{P} < \alpha$
d. reject H_o, $\mathbf{P} < \alpha$

8.95 a. fail to reject H_o
b. reject H_o

8.97 b. ≈ 0.0000
d. reject H_o

8.101 0.2714

8.103 a. 0.0694
b. 0.1977
c. 0.2420
d. 0.0174
e. 0.3524

8.105 a. 1.57
b. -2.13
c. -2.87, $+2.87$

8.107 6.67

8.109 a. H_a: $\mu < 525$
b. fail to reject H_o
c. 9.733

8.111 a. H_a: $\mu \neq 6.25$
b. reject H_o
c. 0.1585
d. 514.488, 3518.3437

8.113 a. mean price for all laptops
b. H_a: $\mu < \$1240$
c. -2.57; $\mathbf{P} = 0.0051$
d. fail to reject H_o

8.115 H_a: $\mu < 12$; $z\star = -2.53$; $\mathbf{P} = 0.0057$; reject H_o

8.117 H_a: $\mu < \$104.63$; $z\star = -2.72$; $\mathbf{P} = 0.0033$; reject H_o

8.119 a. mean accuracy of quartz watches
b. H_a: $\mu > 20$
c. normality assumed, $n = 36$; $\sigma = 9.1$
d. $n = 36$, $\bar{x} = 22.7$
e. $z\star = 1.78$; $\mathbf{P} = 0.0375$
f. $\mathbf{P} < \alpha$; reject H_o

8.123 H_o: The mean shearing strength is at least 925 lb.
H_a: The mean shearing strength is less than 925 lb.

8.125 a. H_a: $\mu < 1.25$
b. H_a: $\mu \neq 335$
c. H_a: $\mu > 230{,}000$

8.127 a. decided average salt content is more than 350 mg when, in fact, it is not
b. decided average salt content is less than or equal to 350 mg when, in fact, it is greater

8.129 I: decided mean minimum is greater than $85 when, in fact, it is not
II: decided mean minimum is at most $85 when, in fact, it is greater

8.131 a. set of all values of test statistic that will cause us to reject H_o
b. Critical value(s) is(are) value(s) of the test statistic that form boundary between critical region and noncritical region; the critical value is in the critical region.

8.133 If one is reduced, the other one becomes larger.

8.135 $z \le -2.33$

8.137 a. $z \le -1.63$, $z \ge 1.65$
b. $z \ge 2.33$
c. $z \le -1.65$
d. $z \le -2.58$, $z \ge 2.58$

8.139 $\bar{x} = 247.1$; $\Sigma x = 21,004.133$

8.141 a. 3.0 standard errors
b. Critical region is $z \ge 2.33$, reject H_o.

8.143 a. reject H_o or fail to reject H_o
b. Calculated test statistic falls in the critical region; reject H_o. Calculated test statistic falls in the noncritical region; fail to reject H_o.

8.145 a. H_o: $\mu = 15.0$ vs. H_a: $\mu \ne 15.0$
b. ± 2.58; reject H_o
c. 0.0913

8.147 a. H_o: $\mu = 72$ ($\le$) vs. H_a: $\mu > 72$
b. fail to reject H_o
c. 2.0

8.149 H_a: $\mu > 79.68$; normality is assumed, $n = 50$; $z\star = 1.07$; $z(0.05) = 1.65$; fail to reject H_o

8.151 H_a: $\mu \ne 55$; normality assumed, $n = 35$; $z\star = -1.70$; $z \le -1.65$, $z \ge 1.65$; reject H_o

8.153 H_a: $\mu > 170.1$; normality indicated; $z\star = 2.73$; $z \ge 1.65$; reject H_o

8.155 H_a: $\mu > 36.8$; normality assumed, $n = 42$; $z\star = 1.45$; $z \ge 1.65$; fail to reject H_o

8.159 a. 32.0
b. 2.4
c. 64
d. 0.90
e. 1.65
f. 0.3
g. 0.495
h. 32.495
i. 31.505

8.161 43.3 to 46.7

8.163 a. 9.75 to 9.99
b. 9.71 to 10.03
c. widened the interval

8.165 a. 69.89 to 75.31
b. yes

8.167 162.04 to 176.16

8.169 b. 2.00; $129.02 \pm (1.96)(2.00)$

8.171 92

8.173 60

8.175 a. "boundary" for decision
b. none

8.177 a. H_o: $\mu = 100$
b. H_a: $\mu \ne 100$
c. 0.01
d. 100
e. 96
f. 12
g. 1.70
h. -2.35
i. 0.0188
j. Fail to reject H_o

8.179 H_a: $\mu > 45$; $z\star = 2.47$; **P** $= 0.0068$; $z \ge 2.05$; reject H_o

8.181 a. H_a: $\mu \ne 0.50$
b. 0.2112
c. $z \le -2.33$, $z \ge 2.33$

8.183 H_a: $\mu > 0.0113$; $z\star = 2.47$; **P** $= 0.0068$; $z \ge 2.33$; reject H_o

8.185 H_a: $\mu > 9$; $z\star = 3.14$; **P** $= 0.0008$; $z \ge 2.05$; reject H_o

8.187 a. H_a: $\mu < 129.2$; $z\star = -2.02$; **P** $= 0.0217$; fail to reject H_o

8.189 a. H_a: $\mu \ne 18$; fail to reject H_o; population mean is not significantly different from 18
b. 0.756; $z\star = -1.04$; p-value $= 0.2984$

8.191 a. 39.6 to 41.6
b. H_a: $\mu \ne 40$; $z\star = 1.20$; **P** $= 0.2302$; fail to reject H_o
c. H_a: $\mu \ne 40$; $z\star = 1.20$; $z \le -1.96$, $z \ge 1.96$; fail to reject H_o

8.193 a. 39.9 to 41.9
b. H_a: $\mu > 40$; $z\star = 1.80$; **P** $= 0.0359$; reject H_o
c. H_a: $\mu > 40$; $z\star = 1.80$; $z \ge 1.65$; reject H_o

8.195 a. H_a: $r > A$, burden on old drug
b. H_a: $r < A$, burden on new drug

Chapter 9

9.1 a. mean amount of physical exercise time per week for women
b. **Cardiovascular Technicians' Weekly Exercise Times**

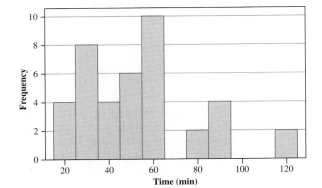

c. skewed right or bimodal
d. yes

9.5 a. 2.68
     b. 2.07
     b. 1.30
     d. 3.36

9.7 a. −1.33
     b. −2.82
     c. −2.03
     d. computer: −2.26; Table 6: −2.14 < t < −2.62;
        interpolation: −2.30

9.9 a. 1.73
     b. ±3.18
     c. −2.55
     d. 1.33

9.11 ±2.18

9.13 a. −2.49
      b. 1.71
      c. −0.685

9.15 df = 7

9.17 0.0241

9.19 a. Symmetric about mean: mean is 0.
      b. Standard deviation of t is greater than 1; t has
         df; t-distribution is family of distributions; one
         z-distribution.

9.21 15.60 to 17.8

9.23 $82.63 to $87.37

9.25 a. $87.37
      b. $15.78
      c. $83.23 to $91.51

9.27 a. 88 oz; 4.68 lb
      b.

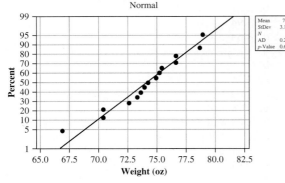

Probability Plot of Weight (oz)
Normal

      c. 74.10; 3.182
      d. 71.95 to 76.25
      e. 71,950 to 76,250 oz

9.31 (6.073, 9.427)

9.33 a.

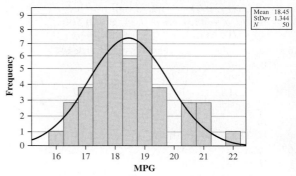

Histogram (with Normal Curve) of MPG

      b. 18.07 to 18.83

9.35 a. H_o: $\mu = 11$ ($\geq$) vs. H_a: $\mu < 11$
      b. H_o: $\mu = 54$ ($\leq$) vs. H_a: $\mu > 54$
      c. H_o: $\mu = 75$ vs. H_a: $\mu \neq 75$

9.37 1.20

9.39 a. 0.025 < **P** < 0.05
      b. 0.025 < **P** < 0.05
      c. 0.05 < **P** < 0.10
      d. 0.05 < **P** < 0.10

9.41 a. 0.10 < **P** < 0.25; fail to reject H_o
      b. 1.75; fail to reject H_o

9.43 0.124

9.45 a. 0.10 < **P** < 0.20; $t = \pm 2.14$; fail to reject H_o
      b. 0.01 < **P** < 0.025; $t \geq 1.71$; reject H_o
      c. 0.025 < **P** < 0.05; $t \leq -1.68$; reject H_o
      d. identical

9.47 H_a: $\mu < 25$; $t\bigstar = -3.25$; **P** < 0.005; $t \leq -2.46$;
      reject H_o

9.49 H_a: $\mu > 130$; $t\bigstar = 1.53$; 0.05 < **P** < 0.10;
      $t \geq 2.50$; fail to reject H_o

9.53 fail to reject H_o

9.55 H_a: $\mu \neq 35$; $t\bigstar = 1.02$; 0.20 < **P** < 0.50;
      $\pm t(5, 0.025) = \pm 2.57$; fail to reject H_o

9.57 a.

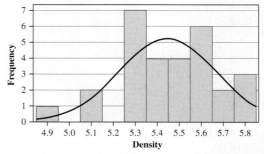

Histogram of Density, with Normal Curve

      b. H_a: $\mu < 5.517$; $t\bigstar = -1.68$; 0.05 < **P** < 0.10;
         $t \leq -1.70$; fail to reject H_o

9.59 a. yes
      b. 593.93 to 598.67
      c. Mean is less than 600 mg.
      d. H_a: $\mu \neq 24.0$; $t\bigstar = 0.69$; **P** = 0.500;
         $\pm t(17, 0.01) = \pm 2.57$; fail to reject H_o
      e. Part d was much smaller.

9.61　a. H_a: $\mu \neq 45.0$; $t\star = -1.95$; $0.05 < \mathbf{P} < 0.10$;
　　　　$\pm t(11, 0.01) = \pm 2.72$; fail to reject H_o
　　　b. H_a: $\mu \neq 45.0$; $t\star = 0.24$; $\mathbf{P} > 0.500$;
　　　　$\pm t(17, 0.01) = \pm 2.57$; fail to reject H_o

9.63　a. number of successes, sample size
　　　b. 0.30
　　　c. 0.096
　　　d. 0.312
　　　e. 0.697

9.65　a. yes
　　　b. The mean of the p' is p.

9.67　a. $z(\alpha/2) = 1.65$
　　　b. $z(\alpha/2) = 1.96$
　　　c. $z(\alpha/2) = 2.58$

9.69　a. 0.02104
　　　b. 0.189 to 0.271

9.71　a. $p = P(\text{did not know})$
　　　b. 0.20, statistic
　　　c. 0.0143
　　　e. 0.186 to 0.214

9.73　0.206 to 0.528

9.75　0.250 to 0.306

9.79　a. 0.028, 0.030, 0.022
　　　b. differing product of pq
　　　c. yes
　　　e. 0.5

9.83　a. 0.5005
　　　b. 0.003227
　　　c. 0.4942 to 0.5068

9.87　2401

9.89　a. 1048
　　　b. 262
　　　c. 2089
　　　d. Increasing maximum error decreases sample size.
　　　e. Increasing level of confidence increases sample size.

9.91　a. H_a: $p > 0.60$
　　　b. H_a: $p > 1/3$
　　　c. H_a: $p > 0.50$
　　　d. H_a: $p < 0.75$
　　　e. H_a: $p \neq 0.50$

9.93　a. 1.78
　　　b. −1.70
　　　c. −0.49
　　　d. 1.88

9.95　a. 0.1388
　　　b. 0.0238
　　　c. 0.1635
　　　d. 0.0559

9.97　a. $z \geq 1.65$
　　　b. $z \leq -1.96$, $z \geq 1.96$
　　　c. $z \leq -1.28$
　　　d. $z \geq 1.65$

9.99　a. 0.017
　　　b. 0.085

c. 0.101
d. 0.004

9.101　a. correctly fail to reject H_o
　　　b. 0.036
　　　c. commit a type II error
　　　d. 0.128

9.103　H_a: $p < 0.90$; $z\star = -4.82$; $\mathbf{P} = 0.000003$;
　　　$z \leq -1.65$; reject H_o

9.105　H_a: $p < 0.60$; $z\star = -2.04$; $\mathbf{P} = 0.0207$;
　　　$z \leq -1.65$; reject H_o

9.107　H_a: $p < 0.72$; $z\star = -1.57$; $\mathbf{P} = 0.0582$;
　　　$z \leq -1.65$; fail to reject H_o

9.109　a. −1.72
　　　b & c. $\mathbf{P} = 0.0427$; $z \leq -2.33$; fail to reject H_o

9.111　a. 0.0438
　　　b. 0.0332
　　　c. 0.04375
　　　d. Coenen's: "struck evenly or balanced" or H_o: $P(\text{H}) = 0.5$; Blight's: "not balanced" or H_a: $P(\text{H}) \neq 0.5$
　　　e. Probability of "getting a result as extreme" is the definition of the p-value.
　　　f. 0.498 to 0.622
　　　g. binomial experiment —no chance of bias due to interviewing process

9.113　b. reject H_o
　　　c. 0.305

9.115　a. A: 1.72; B: 3.58
　　　b. increased
　　　c. quite different than the rest of the data; had a big effect on the standard deviation

9.117　a. 23.2
　　　b. 23.3
　　　c. 3.94
　　　d. 8.64

9.119　a. 30.1
　　　b. 13.3
　　　c. 7.56
　　　d. 43.2
　　　e. 11.6 and 32.7
　　　f. 1.24 and 14.5

9.121　a. $\chi^2(5, 0.05) = 11.1$
　　　b. $\chi^2(5, 0.05) = 11.1$
　　　c. $\chi^2(5, 0.10) = 9.24$

9.123　0.94

9.125　a. 0.8356
　　　b. 0.1644

9.127　a. H_a: $\sigma > 24$
　　　b. H_a: $\sigma > 0.5$
　　　c. H_a: $\sigma \neq 10$
　　　d. H_a: $\sigma^2 < 18$
　　　e. H_a: $\sigma^2 \neq 0.025$

9.129　a. 25.08
　　　b. 60.15

9.131　a. $0.02 < \mathbf{P} < 0.05$
　　　b. 0.01

c. $0.05 < \mathbf{P} < 0.10$

d. $0.025 < \mathbf{P} < 0.05$

9.133 a. $0.05 < \mathbf{P} < 0.10$; fail to reject H_o

b. $\chi^2 \geq 33.4$; fail to reject H_o

9.135 H_a: $\sigma \neq 8$; $\chi^2\star = 29.3$; $0.01 < \mathbf{P} < 0.02$;
$\chi^2 \leq 32.4$, $\chi^2 \geq 71.4$; reject H_o

9.137 b.

Probability Plot of Weights (lbs)
Normal

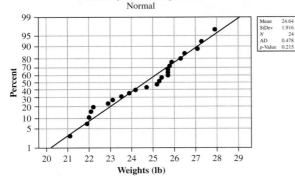

Mean	24.64
StDev	1.916
N	24
AD	0.478
p-Value	0.215

c & d. H_a: $\sigma > 1.0$; $\chi^2\star = (23)(1.916^2)/(1.0^2) =$ 84.43; $\mathbf{P} < 0.005$; $\chi^2 \geq 41.6$; reject H_o

9.139 H_a: $\sigma > 85$; $\chi^2\star = 64.88$; $\mathbf{P} < 0.005$; $\chi^2 \geq 43.8$; reject H_o

9.141 a. H_a: $\sigma \neq 0.3275$; $\chi^2\star = 7.15$; $0.50 < \mathbf{P} < 1.00$; $\chi^2 \leq 2.09$, $\chi^2 \geq 21.7$; fail to reject H_o

b. H_a: $\sigma \neq 0.3275$; $\chi^2\star = 13.97$; $0.20 < \mathbf{P} <$ 0.50; $\chi^2 \leq 2.09$, $\chi^2 \geq 21.7$; fail to reject H_o

c. Larger sample standard deviations increase chi-square value.

9.143 0.0359

9.147 35,524 to 36,476

9.149 a. 8.782, 0.710

b. 8.78

c. 8.64 to 8.92

9.151 72

9.153 a. $\bar{x} = \$908.30$, $s = \$118.50$

Histogram (with Normal Curve) of Textbook Price

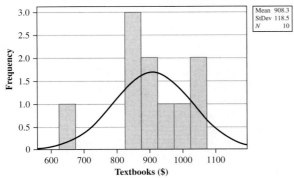

Mean	908.3
StDev	118.5
N	10

c. \$823.61 to \$992.99

9.155 H_a: $\mu > 4.35$; $t\star = 1.63$; $0.05 < \mathbf{P} < 0.10$;
$t \geq 1.76$; fail to reject H_o

9.157 a. 31.45, 8.049

b. H_a: $\mu > 28.0$; $t\star = 1.92$; $0.025 < \mathbf{P} < 0.05$;
$t \geq 1.73$; reject H_o

9.159

Velocity of Light in Air
by Albert Michelson

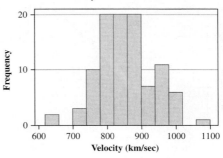

H_a: $\mu \neq 734.5$; $t\star = 14.92$;
$\mathbf{P} = 0.00+$; $t \leq -2.65$, $t \geq 2.65$; reject H_o

9.161 0.122 to 0.278

9.163 a. 0.126 to 0.340

b. overestimated percent of satisfied customers

9.165 a. 343

b. 121

c. 93

9.167 3258

9.169 0.0401

9.175 a. parameter; binomial p, P(success)

b. 0.60 to 0.66

9.177 H_a: $\sigma > 81$; $\chi^2\star = 123.1$; $0.05 < \mathbf{P} < 0.10$;
$\chi^2 \geq 124.0$; fail to reject H_o

9.179 H_a: $\sigma > \$2.45$; $\chi^2\star = 101.5$; $0.005 < \mathbf{P} < 0.01$;
$\chi^2 \geq 90.5$; reject H_o

9.181 b. H_a: $\mu \neq 2.0$; $t\star = 3.08$; $\mathbf{P} < 0.010$; $t \leq -2.04$,
$t \geq 2.04$; reject H_o

c. H_a: $\sigma > 0.040$; $\chi^2\star = 48.96$; $0.01 < \mathbf{P} < 0.25$;
$\chi^2 \geq 43.8$; reject H_o

9.183 a.

Weights of 14-oz Corn Flakes Boxes

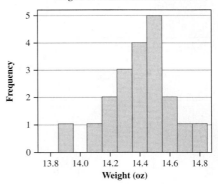

b. $\bar{x} = 14.386$, $s = 0.217$

c. 5%

d.

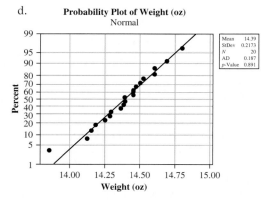

Probability Plot of Weight (oz)
Normal

Mean	14.39
StDev	0.2173
N	20
AD	0.187
p-Value	0.891

e. $\bar{x} = 14.386$, $s = 0.217$; 14.285 to 14.487

f. H_a: $\sigma > 0.2$; $\chi^2 \star = 22.37$; **P** = 0.2662; $\chi^2 \ge 36.2$; fail to reject H_o

9.185 a. 0.8051
  b. 0.1271
  c. 1016.46 boxes

Chapter 10

10.1 a. college students
  b. F: 48.5%; 54%; S: 93.5%; 92%
  d.

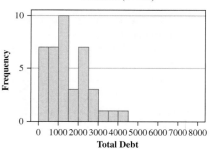

Credit Card Debt
Freshmen ($n = 40$)

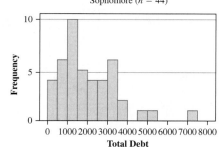

Credit Card Debt
Sophomore ($n = 44$)

Both mounded skewed right; sophomores more dispersed

  e. F: $1519, $1036; S: $2079, $1434

10.5 independent; samples are separate sets of students

10.7 dependent; each person provides one data for each sample

10.9 independent; separate samples

10.13 a. $d = A - B$: 1 1 0 2 −1
  b. 0.6
  c. 1.14

10.15 a. 4.24 to 8.36
  b. narrower confidence interval

10.19 $n = 8$, $\Sigma d = 8$, $\Sigma d^2 = 48$
  a. 1.0
  b. −1.53 to 3.53

10.21 $\bar{d} = I - II$; $n = 10$, $\bar{d} = 0.8$, $s_d = 1.32$; −0.143 to 1.743

10.23 a. H_a: $\mu_d > 0$; $d = $ posttest − pretest
  b. H_a: $\mu_d \ne 0$; $d = $ after − before
  c. H_a: $\mu_d \ne 0$; $d = $ reading1 − reading2
  d. H_a: $\mu_d > 0$; $d = $ postscore − prescore

10.25 a. $P(t > 1.86 \mid df = 19)$; $0.025 < P < 0.05$
  b. $2P(t < -1.86 \mid df = 19)$; $0.05 < P < 0.10$
  c. $P(t < -2.63 \mid df = 28)$; $0.005 < P < 0.01$
  d. $P(t > 3.57 \mid df = 9)$; $P < 0.005$

10.27 H_a: $\mu_d > 0$ (beneficial); $t\star = 3.067$; $P < 0.005$; $t(39, 0.01) = 2.44$; reject H_o

10.31 $t\star = 2.45$; $0.025 < P < 0.05$; $t(4, 0.05) = 2.13$; reject H_o

10.33 $t\star = 1.35$; $0.20 < P < 0.50$; $\pm t(4, 0.005) = \pm 4.60$; fail to reject H_o

10.35 H_a: $\mu_d > 0$ (improvement); $t\star = 0.56$; $P > 0.25$; $t(9, 0.05) = 1.83$; fail to reject H_o

10.37 a. Average difference is zero.
  b. values used to make the decision
  c. Test is two-tailed; t-distribution is symmetrical; absence of negative numbers makes it less confusing.
  d. fail to reject the null hypothesis in 12 of them
  e. Two methods are equivalent.
  f. revised Florida method for sampling accepted and implemented

10.39 4.92

10.41 case I: between 17 and 40; case II: 17

10.43 −6.3 to 16.3

10.45 $4.19 to $19.11

10.49 N. Dakota: $n = 11$, $\bar{x} = 976.2$, $s = 255.7$; S. Dakota: $n = 14$, $\bar{x} = 1370$, $s = 397$; 101.3 to 686.3

10.51 a. H_a: $\mu_1 - \mu_2 \ne 0$
  b. H_a: $\mu_1 - \mu_2 > 0$
  c. H_a: $\mu_S - \mu_N > 0$
  d. H_a: $\mu_M - \mu_F \ne 0$

10.53 a. 1.21
  b. 0.1243
  c. 1.56

10.55 2.64

10.57 a. ±2.13
  b. −2.48

c. 1.42

d. ±2.16

10.61 $2P(t > 1.44 \mid df = 13)$; $0.10 < \mathbf{P} < 0.20$

10.63 H_a: $\mu_1 - \mu_2 > 0$; $t\bigstar = 4.02$; $\mathbf{P} < 0.005$; $t \geq 2.44$; reject H_o

10.65   no

10.67   c. 12 to 26

      d. computer

      e. 12

      f. p-value is less than 0.0005.

10.69   b. $0.554 < \mathbf{P} < 0.624$

      c. $0.560 < \mathbf{P} < 0.626$

10.71 H_a: $\mu_B - \mu_A > 0$; $t\bigstar = 1.98$; $0.025 < \mathbf{P} < 0.05$; $t \geq 1.83$; reject H_o

10.73 H_a: $\mu_{PN} - \mu_{PR} > 0$; $t\bigstar = 1.36$; $0.10 < \mathbf{P} < 0.25$; $t \geq 1.53$; fail to reject H_o

10.79   75, 250, 0.30, 0.70

10.81   a. 0.085

      b. 0.115

10.85   0.196 to 0.384; positive difference indicates that the male teenagers' proportion is significantly greater than the female teenagers' proportion

10.87   0.000 to 0.080

10.89   a. H_a: $p_m - p_w \neq 0$

      b. H_a: $p_b - p_g > 0$

      c. H_a: $p_c - p_{nc} > 0$

10.91   0.076; 0.924

10.93   1.34; 0.0901

10.95   a. $z \geq 1.65$

      b. $z \leq -1.96$, $z \geq 1.96$

      c. $z \leq -1.75$

      d. $z \geq 2.33$

10.97 H_a: $p_m - p_c \neq 0$; $z\bigstar = 1.42$; $\mathbf{P} = 0.1556$; $z \leq -1.96$ and $z \geq 1.96$; fail to reject H_o

10.99   b. H_a: $p_w - p_m \neq 0$; $z\bigstar = 0.64$; $\mathbf{P} = 0.5222$; $z \leq -1.96$ and $z \geq 1.96$; fail to reject H_o

      c. $z\bigstar = 3.18$; $\mathbf{P} = 0.0014$; $z \leq -1.96$ and $z \geq 1.96$; reject H_o

      d. takes a reasonably large sample size to show significance

10.101 H_a: $p_2 - p_1 \neq 0$; $z\bigstar = 1.43$; $\mathbf{P} = 0.1528$; $z \leq -1.96$ and $z \geq 1.96$; fail to reject H_o

10.103   a. H_a: $p_M - p_w \neq 0$; $z\bigstar = 1.82$; $\mathbf{P} = 0.0688$; $z \leq -1.96$ and $z \geq 1.96$; fail to reject H_o

      b. $z\bigstar = 2.57$; $\mathbf{P} = 0.0102$; $z \leq -1.96$ and $z \geq 1.96$; reject H_o

      c. 291

10.105   a. H_a: $\sigma_A^2 \neq \sigma_B^2$

      b. H_a: $\sigma_I / \sigma_{II} > 1$

      c. H_a: $\sigma_A^2 / \sigma_B^2 \neq 1$

      d. H_a: $\sigma_D^2 / \sigma_C^2 > 1$

10.107   Divide inequality by σ_p^2.

10.109   a. $F(9, 11, 0.025)$

      b. $F(24, 19, 0.01)$

      c. $F(8, 15, 0.01)$

      d. $F(15, 9, 0.05)$

10.111   a. 2.51

      b. 2.20

      c. 2.91

      d. 4.10

      e. 2.67

      f. 3.77

      g. 1.79

      h. 2.99

10.113   3.37

10.115   1.52

10.117   0.495; smaller variance in numerator

10.119 H_a: $\sigma_k^2 \neq \sigma_m^2$; $F\bigstar = 1.33$; $\mathbf{P} > 0.10$; $F \geq 3.73$; fail to reject H_o

10.121 H_a: $\sigma_{sc} \neq \sigma_{sb}$; $F\bigstar = 1.12$; $\mathbf{P} > 0.10$; $F \geq 3.10$; fail to reject H_o

10.123   multiply by 2

10.125 $(4.43)^2/(3.50)^2 = 1.60$

10.127   a. $\bar{x}_1 = 0.01525$, $s_1 = 0.00547$; $\bar{x}_2 = 0.02856$, $s_2 = 0.00680$

      b. H_a: $\sigma_1^2 \neq \sigma_2^2$; $F\bigstar = 1.55$; $\mathbf{P} > 0.10$; $F \geq 4.42$; fail to reject H_o

      c. H_a: $\mu_2 - \mu_1 \neq 0$; $t\bigstar = 5.64$; $\mathbf{P} < 0.01$; $t \leq -2.36$, $t \geq 2.36$; reject H_o

10.129   a. men: $\bar{x}_m = \$68.14$, $s_m = \$47.95$; women: $\bar{x}_w = \$85.90$, $s_w = \$63.50$

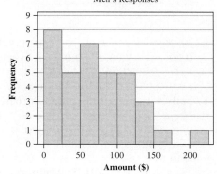

How much should someone spend on you for Valentine's Day?
Men's Responses

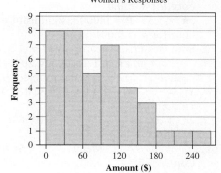

How much should someone spend on you for Valentine's Day?
Women's Responses

      d. H_a: $\mu_w - \mu_m > 0$; $t\bigstar = 1.36$; $0.05 < \mathbf{P} < 0.10$; $t \geq 1.70$; fail to reject H_o

e. H_a: $\sigma_w^2 \neq \sigma_m^2$; $F\star = 1.75$; **P** ≈ 0.10;
$F \geq 2.72$; fail to reject H_o
f. Difference was not significant.

10.133 -8.85 to 16.02

10.135 0.95 to 3.05

10.137 H_a: $\mu_d > 0$; $t\star = 3.70$; **P** < 0.005; $t \geq 2.54$;
reject H_o

10.139 -0.21 to 10.61

10.141 -9.14 to 19.54

10.143 0.012 to 0.072

10.145 H_a: $\mu_2 - \mu_1 > 0$; $t\star = 0.988$; $0.10 < $ **P** < 0.25;
$t \geq 1.72$; fail to reject H_o

10.147 H_a: $\mu_2 - \mu_1 > 0$; $t\star = 1.30$; $0.10 < $ **P** < 0.25;
$t \geq 2.47$; fail to reject H_o

10.149   c. H_a: $\mu_A - \mu_B \neq 0$; $t\star = 5.84$; **P** < 0.01;
$t \leq -2.98$, $t \geq 2.98$; reject H_o

10.151   a. M: $\bar{x} = 74.69$, $s = 10.19$; F: $\bar{x} = 79.83$,
$s = 8.80$

University's Mathematics Placement Exam

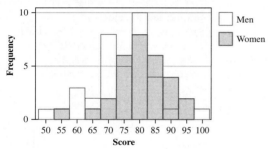

Men
Women

b. M: H_a: $\mu \neq 77$; $t\star = -1.36$; $0.10 < $ **P** $<$
0.20; $t \leq -2.03$, $t \geq 2.03$; fail to reject H_o;
F: H_a: $\mu \neq 77$; $t\star = 1.76$; $0.05 < $ **P** < 0.10;
$t \leq -2.05$, $t \geq 2.05$; fail to reject H_o
c. both not significantly different than 77
d. H_a: $\mu_F - \mu_M \neq 0$; $t\star = 2.19$; $0.02 < $ **P** $<$
0.05; $t \leq -2.05$, $t \geq 2.05$; reject H_o
e & f. asking different questions

10.153 -0.044 to 0.164

10.155   a. $z\star = 2.37$; **P** $= 0.0178$; significant difference for $\alpha \geq 0.02$
b. $z\star = 2.90$; **P** $= 0.0038$; significant difference for $\alpha \geq 0.01$
c. $z\star = 3.35$; **P** $= 0.0008$; significant difference for $\alpha \geq 0.001$
d. Standard error became smaller.

10.157 H_a: $p_a - p_1 \neq 0$; $z\star = 1.26$; **P** $= 0.2076$;
$z \leq -2.58$ and $z \geq 2.58$; fail to reject H_o

10.159 H_a: $\sigma_m^2 > \sigma_f^2$; $F\star = 2.58$; $0.025 < $ **P** < 0.05;
$F \geq 2.53$; reject H_o

10.161 H_a: $\sigma_n^2 \neq \sigma_s^2$; $F\star = 1.28$; **P** > 0.10; $F \geq 1.80$;
fail to reject H_o

10.163   a.

	N	Mean	StDev
Cont	50	0.005459	0.000763
Test	50	0.003507	0.000683

Control or Existing Design

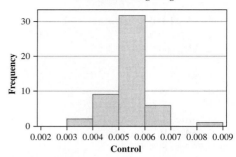

Test or New Designs

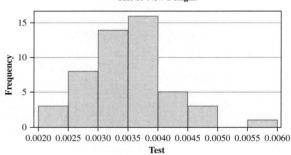

b. both approximately normal
c. one-tailed—looking for a reduction
d. H_a: $\sigma_c^2 > \sigma_t^2$; $F\star = 1.248$; **P** > 0.05;
$F \geq 1.69$; fail to reject H_o
e. H_a: $\mu_c - \mu_t > 0$; $t\star = 13.48$; **P** $= +0.000$;
$t \leq -1.68$, $t \geq 1.68$; reject H_o
f. Mean force has been reduced, but not the
variability.

Chapter 11

11.1   a. preferred way to "cool" their mouth after
hot sauce
b. U.S. adults professing to love eating hot
spicy food, method of cooling
c. 36.5%; 17.5%; 10%; 9.5%; 14.5%; 5.5%;
6.5%
d. fairly similar

11.3   a. 23.2
b. 23.3
c. 3.94
d. 8.64

11.5   a. $\chi^2 (14, 0.01) = 29.1$
b. $\chi^2 (25, 0.05) = 40.1$

11.7   a. asking one person
b. birth day of the week
c. the 7 days of the week

11.9 a. H_o: $P(1) = P(2) = P(3) = P(4) = P(5) = 0.2$
      H_a: not equally likely
   b. H_o: $P(1) = 2/8$, $P(2) = 3/8$, $P(3) = 2/8$, $P(4) = 1/8$
      H_a: at least one is different
   c. H_o: $P(E) = 0.16$, $P(G) = 0.38$, $P(F) = 0.41$, $P(P) = 0.05$
      H_a: percentages different than specified

11.11 a. $\chi^2 \geq 7.82$
     b. $\chi^2 \geq 9.21$

11.15 a. 60
     b. 2
     c. H_a: ratio other than $6:3:1$; $\chi^2\bigstar = 2.67$; **P** = 0.263; $\chi^2 \geq 4.61$; fail to reject H_o

11.17 H_a: proportions different; $\chi^2\bigstar = 7.35$; **P** = 0.062; $\chi^2 \geq 7.82$; fail to reject H_o

11.19 H_a: opinions distributed differently; $\chi^2\bigstar = 213.49$; **P** < 0.005; $\chi^2 \geq 9.49$; reject H_o

11.21 H_a: opinions distributed differently; $\chi^2\bigstar = 10.05$; **P** = 0.123; $\chi^2 \geq 12.6$; fail to reject H_o

11.23 H_a: colors distributed differently; $\chi^2\bigstar = 3.057$; **P** = 0.548; $\chi^2 \geq 9.49$; fail to reject H_o

11.25 a. H_a: proportions different than listed; $\chi^2\bigstar = 44.4928$; **P** < 0.005; $\chi^2 \geq 7.82$; reject H_o
     b. 4th cell

11.27 a. H_a: Voters preference and party affiliation are not independent.
     b. H_a: The distribution is not the same for all three.
     c. H_a: The proportion of yeses is not the same in all categories.

11.29 10

11.33 H_a: size of community of residence is not independent of size of community reared in; $\chi^2\bigstar = 35.749$; **P** < 0.005; $\chi^2 \geq 13.3$; reject H_o

11.35 H_a: response is not independent of years; $\chi^2\bigstar = 3.390$; **P** = 0.335; $\chi^2 \geq 6.25$; fail to reject H_o

11.37 H_a: number of defectives is not independent of day; $\chi^2\bigstar = 8.548$; **P** = 0.074; $\chi^2 \geq 9.49$; fail to reject H_o

11.39 H_a: blog creators and readers are not proportioned the same; $\chi^2\bigstar = 3.954$; **P** = 0.138; $\chi^2 \geq 5.99$; fail to reject H_o

11.41 H_a: Fear and Do not Fear darkness are not proportioned the same; $\chi^2\bigstar = 80.957$; **P** < 0.005; $\chi^2 \geq 13.3$; reject H_o

11.43 Gender: H_a: females and males not proportioned the same for each dosage; $\chi^2\bigstar = 0.978$; **P** = 0.613; $\chi^2 \geq 9.21$; fail to reject H_o
    Dosage: H_a: age groups are not proportioned the same for each dosage; $\chi^2\bigstar = 7.449$; **P** = 0.114; $\chi^2 \geq 13.3$; fail to reject H_o

11.45 a. $\chi^2\bigstar = 4.043$; **P** = 0.257
     b. $\chi^2\bigstar = 8.083$, **P** = 0.044; $\chi^2\bigstar = 12.127$, **P** = 0.007
     c. yes

11.47 H_a: proportions are other than $1:3:4$; $\chi^2\bigstar = 10.33$; **P** = 0.006; $\chi^2 \geq 5.99$; reject H_o

11.49 H_a: percentages different than listed; $\chi^2\bigstar = 6.693$; **P** = 0.153; $\chi^2 \geq 9.49$; fail to reject H_o

11.51 H_a: percentages different than listed; $\chi^2\bigstar = 17.92$; **P** = 0.003; $\chi^2 \geq 15.1$; reject H_o

11.53 $P(x < 130) = 0.0228$, $P(130 < x < 145) = 0.1359$, $P(145 < x < 160) = 0.3413$, $P(160 < x < 175) = 0.3413$, $P(175 < x < 190) = 0.1359$, $P(x > 190) = 0.0228$
    H_a: The weights are not $N(160, 15)$; $\chi^2\bigstar = 5.812$; **P** = 0.325; $\chi^2 \geq 11.1$; fail to reject H_o

11.55 H_a: percentages different than listed; $\chi^2\bigstar = 4.70$; **P** = 0.195; $\chi^2 \geq 7.82$; fail to reject H_o

11.57 a. H_a: distributions are different; $\chi^2\bigstar = 6.1954$; **P** = 0.2877; $\chi^2 \geq 11.1$; fail to reject H_o
     b. $\chi^2\bigstar = 36.761$; **P** < 0.005; $\chi^2 \geq 11.1$; reject H_o
     c. $\chi^2\bigstar = 92.93$; **P** < 0.005; $\chi^2 \geq 11.1$; reject H_o
     d. Chi-square becomes more sensitive to variations as the sample size gets larger.

11.59 H_a: distributions are different; $\chi^2\bigstar = 3.123$; **P** = 0.793; $\chi^2 \geq 12.6$; fail to reject H_o

11.61 H_a: political preference is not independent of age; $\chi^2\bigstar = 23.339$; **P** < 0.005; $\chi^2 \geq 13.3$; reject H_o

11.63 H_a: proportion of popcorn that popped is not the same for all brands; $\chi^2\bigstar = 2.839$; **P** = 0.417; $\chi^2 \geq 7.82$; fail to reject H_o

11.65 a. 2003: 73.2%, 2004: 74.2%
     b. H_a: ratio of organ donors is not the same; $\chi^2\bigstar = 5.955$; **P** = 0.015; $\chi^2 \geq 3.84$; reject H_o
     c. With very large sample sizes, differences must be very small to be considered nonexistent.

Chapter 12

12.1 a. & b.

One-Way Travel Times to Work (in minutes)

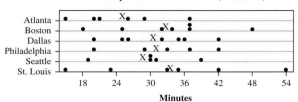

Minutes

c. yes

d. yes

12.3 **Units produced per hour at each temperature level**

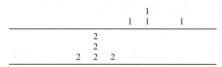

12.5 a. df(error) = 17; SS(Factor) = 123.8;
MS(Factor) = 41.2667;
MS(Error) = 2.3765

b. 41.2667/2.3765 = 17.36

12.7 greater variability between the four levels

12.9 a. Category has a slight effect on the average number.

12.11 a. 0
b. 2
c. 4
d. 31
e. 393

12.13 a. H_o: $\mu_1 = \mu_2 = \mu_3 = \mu_4 = \mu_5$;
H_a: not all equal
b. H_o: $\mu_1 = \mu_2 = \mu_3 = \mu_4$; H_a: not all equal
c. H_o: $\mu_1 = \mu_2 = \mu_3 = \mu_4$; H_a: not all equal
d. H_o: $\mu_1 = \mu_2 = \mu_3$; H_a: not all equal

12.15 a. $F \geq 3.34$
b. $F \geq 5.99$
c. $F \geq 3.44$

12.17 a. depends on whether it is larger or smaller than α
b. reject H_o
c. fail to reject H_o

12.19 a. Test factor has no effect.
b. Test factor does have an effect.
c. $P \leq \alpha$; F in the critical region
d. significant effect
e. $P > \alpha$; F in noncritical region
f. Tested factor does not have a significant effect.

12.21 a. 17
b. df(Group) = 2
f. Most likely fail to reject H_o

12.25 H_a: mean for workers is not all equal

Source	df	SS	MS	F★
Work	2	17.73	8.87	4.22
Error	12	25.20	2.10	
Total	14	42.93		

P = 0.041 or $F \geq 3.89$; reject H_o

12.27 H_a: means ratings are not all equal

Source	df	SS	MS	F★
Factor	2	19.60	9.80	3.68
Error	12	32.00	2.67	
Total	14	51.60		

P = 0.057 or $F \geq 3.89$; fail to reject H_o

12.29 H_a: mean won/loss percentages on the road are not all equal

Source	df	SS	MS	F★
Factor	2	4.3	2.1	0.02
Error	27	2592.8	96.0	
Total	29	2597.0		

P = 0.978; $F \geq 3.39$; fail to reject H_o

12.31 H_a: mean age for groups is not all equal

Source	df	SS	MS	F★
Group	2	254.591	127.30	0.81
Error	55	8621.564	156.76	
Total	57	8876.155		

P = 0.449; $F \geq 3.15$; fail to reject H_o

12.33 a. **"Ideal Age" by Actual Age Groups**

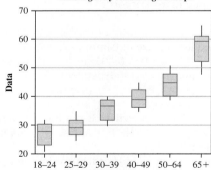

b. H_a: mean "ideal age" is not the same for all age groups

Source	df	SS	MS	F★
Factor	5	3765.3	753.1	42.33
Error	30	533.7	17.8	
Total	35	4299.0		

P < 0.01; $F \geq 2.53$; reject H_o

d. "Ideal ages" were different because the means do not line up horizontally.

12.35 a. H_a: mean hourly wage is not the same for months

Source	df	SS	MS	F★
Factor	11	0.78	0.07	0.03
Error	112	304.27	2.72	
Total	123	305.05		

$\mathbf{P} = 1.000$; $F \geq 1.99$; fail to reject H_o

Boxplots of January through December
(Means are indicated by solid circles)

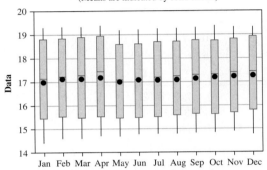

b. H_a: mean hourly wage is not the same for years

Source	df	SS	MS	F★
Factor	10	302.3136	30.2314	1250.59
Error	113	2.7316	0.0242	
Total	123	305.0452		

$\mathbf{P} = 0.000$; $F \geq 1.99$; reject H_o

Boxplots of 1995–2005
(Means are indicated by solid circles)

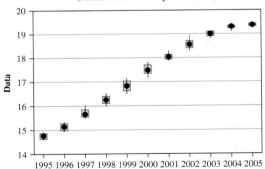

12.37 b. **Mean Number of Items Purchased per Day**

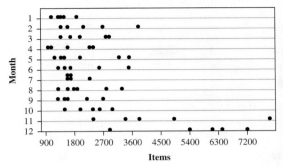

d. H_a: mean number of items purchased is not the same for months

Source	df	SS	MS	F★
Month	11	84,869,019	7,715,365	7.56
Error	50	51,003,447	1,020,069	
Total	61	135,872,465		

$\mathbf{P} < 0.01$; $F \geq 2.08$; reject H_o

12.39 a. H_a: mean amount of salt is not the same
b. random/independent/normal samples, 0.05, F
c. fail to reject H_o; no significant difference

12.41 a. H_a: mean amount spent is not the same
b. fail to reject H_o
c. no
d. no

12.43 H_a: mean stopping distance is affected

Source	df	SS	MS	F★
Brand	3	95.36	31.79	4.78
Error	19	126.47	6.66	
Total	22	221.83		

$\mathbf{P} = 0.012$; $F \geq 3.13$; reject H_o

12.45 H_a: mean amounts dispensed are not all equal

Source	df	SS	MS	F★
Machine	4	20.998	5.2495	31.6
Error	13	2.158	0.166	
Total	17	23.156		

$\mathbf{P} < 0.01$; $F \geq 5.21$; reject H_o

12.47 Points scored: H_a: mean points scored are not all equal

Source	df	SS	MS	F★
Factor	3	3055	1018	0.19
Error	28	147,817	5279	
Total	31	150,872		

$\mathbf{P} = 0.900$; $F \geq 2.99$; fail to reject H_o

Points against: H_a: mean points scored are not all equal

Source	df	SS	MS	F★
Factor	3	22,599	7533	2.40
Error	28	87,827	3137	
Total	31	110,426		

$P = 0.089$; $F \geq 2.99$; fail to reject H_o

12.49　a. H_a: mean nominal comparison is not the same

Source	df	SS	MS	F
Factor	4	0.001830	0.000458	1.05
Error	105	0.045732	0.000436	
Total	109	0.047563		

$P = 0.385$; $F \geq 3.65$; fail to reject H_o

b.

Boxplots of Nominal Comparisons of Contact Lenses for 5 Competitors
(Means are indicated by solid circles)

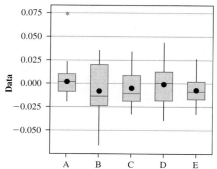

12.51　H_a: brands of golf balls do not withstand durability test equally well

Source	df	SS	MS	F★
Brand	5	75,047	15,009.4	5.30
Error	36	101,899	2,830.5	
Total	41	176,946		

$P = 0.001$; $F \geq 2.48$; reject H_o

12.53　a. H_a: mean petal width is not the same

Source	df	SS	MS	F
Species	2	1671.56	835.78	118.06
Error	27	191.14	7.08	
Total	29	1862.70		

$P < 0.01$; $F \geq 3.37$; reject H_o

b. H_a: mean sepal width is not the same

Source	df	SS	MS	F
Species	2	197.1	98.6	7.78
Error	27	342.2	12.7	
Total	29	539.4		

$P = 0.002$; $F \geq 3.37$; reject H_o

c. Type 0 has the shortest PW and the longest SW. Type 1 has the longest PW and the middle SW. Type 2 has the middle PW and the shortest SW.

12.57　b.

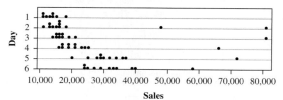

c. yes

d. H_a: mean total cost of items purchased per day is not the same

Source	df	SS	MS	F★
Day	5	2,657,284,622	531,456,924	2.24
Error	56	13,311,874,185	237,712,039	
Total	61	15,969,158,806		

$P = 0.063$; $F \geq 2.45$; fail to reject H_o

12.59　42

Chapter 13

13.1　a.

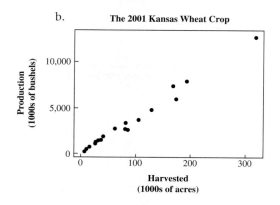

The 2001 Kansas Wheat Crop

b.

c. **The 2001 Kansas Wheat Crop**

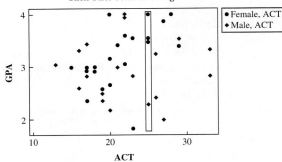

d. As the number of acres planted increases, so does number of bushels harvested; as number of harvested acres increases, so does production; yield rate is approximately normal.

13.3 a. Summation of deviations about the mean was zero.
b. divides data into four quadrants

13.5 a. **Their First Term in College**

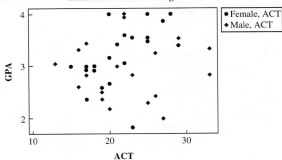

b. somewhat similar

c. **Their First Term in College**

from 1.8 to 4.0
d. no, does not help

13.7 a.6

b. 4.44; $\Sigma x = 50$, $\Sigma y = 35$, $\Sigma x^2 = 330$, $\Sigma xy = 215$, $\Sigma y^2 = 145$
c. 2.981, 1.581
d. 0.943
e. 0.943

13.11 a. 0.008
b. no linear relationship
c. **NFL 2004 Total Points**

no upward or downward trend, no correlation

13.13 a. 60
b. 40.99, 20.98
c. 0.07

13.17 a. 0.17 to 0.52
b. The interval becomes narrower.

13.19 a. 0.40 to 0.74
b. −0.78 to +0.15
c. 0.05 to 0.93
d. −0.65 to −0.45

13.21 $\Sigma x = 746$, $\Sigma y = 736$, $\Sigma x^2 = 57,496$, $\Sigma xy = 56,574$, $\Sigma y^2 = 55,826$; 0.955; 0.78 to 0.98

13.23 a. 0.985
b. $0.55 < \rho < 1.00$

13.25 a. $H_a: \rho > 0$
b. $H_a: \rho \neq 0$
c. $H_a: \rho < 0$
d. $H_a: \rho > 0$

13.27 a. $0.05 < \mathbf{P} < 0.10$
b. $0.025 < \mathbf{P} < 0.05$

13.29   a. ± 0.444
      b. -0.378 if left tail; 0.378 if right tail

13.31   b. $\mathbf{P} < 0.01$
      c. ± 0.537, using table; ± 0.507, using interpolation
      d. significant at $\alpha = 0.01$

13.33 H_a: $\rho \neq 0.0$; $r\star = 0.43$; $0.05 < \mathbf{P} < 0.10$; ± 0.378; reject H_o

13.35 H_a: $\rho \neq 0.0$; $r\star = 0.532$; $\mathbf{P} > 0.10$; ± 0.878; fail to reject H_o

13.37 H_a: $\rho \neq 0.0$; $r\star = 0.798$; $\mathbf{P} = 0.006$; $r \leq -0.632$, $r \geq 0.632$; reject H_o

13.39   a. $r = 0.855$
      b. H_a: $\rho \neq 0.0$; $r\star = 0.855$; $\mathbf{P} = 0.014$; $r \leq -0.754$, $r \geq 0.754$; reject H_o

13.43

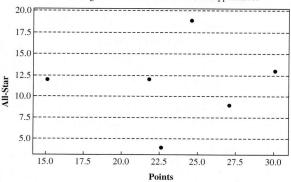

National Basketball Association
Average Points Per Game vs. All-Star Appearances

$r = 0.086$; $\hat{y} = 9.6 + 0.082x$

13.45   a.

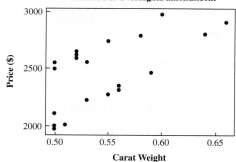

Diamonds at OvernightDamonds.com

      b. linear
      c. cannot predict with confidence outside range
      d. $\Sigma x = 10.92$, $\Sigma y = 49{,}428$, $\Sigma x^2 = 6.005$, $\Sigma xy = 27{,}166.9$, $\Sigma y^2 = 123{,}927{,}308$; $\hat{y} = 179 + 4199x$
      e. \$2278.50
      f. \$41.99, $x = 0.50$ carats to 0.66
      g. 56,596.41

13.47   a. $\Sigma x = 50$, $\Sigma y = 35$, $\Sigma x^2 = 330$, $\Sigma xy = 215$, $\Sigma y^2 = 145$; $\hat{y} = 1.0 + 0.5x$

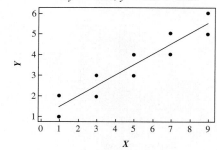

      b. 1.5, 2.5, 3.5, 4.5, 5.5
      c. -0.5, 0.5 alternately
      d. 0.3125
      e. 0.3125

13.51   0.1564

13.53   a. $\hat{y} = -348 + 2.04x$
      b. 1.60 to 2.48

13.55   a. 0.0145
      b. 0.0668
      c. 0.0653

13.57   a. $\hat{y} = 5936.79 + 30.732x$
      c. $\mathbf{P} = 0.111$; not an effective predictor
      d. $30.732 \pm (2.31)(17.158)$

13.59   a.

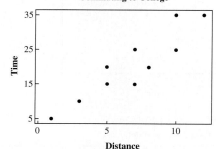

Commuting to College

      b. $\Sigma x = 68$, $\Sigma y = 205$, $\Sigma x^2 = 566$, $\Sigma xy = 1670$, $\Sigma y^2 = 5075$; $\hat{y} = 2.38 + 2.664x$
      c. H_a: $\beta_1 > 0$; $t\star = 6.55$; $\mathbf{P} < 0.005$ or $t \geq 1.86$; reject H_o
      d. 1.48 to 3.84

13.61   a.

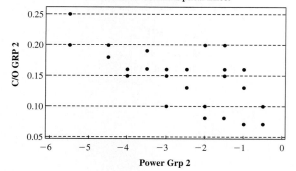

Contact Lenses — Bausch & Lomb
Power vs. Certain Optical Effect

b. -0.674

c. H_a: $\rho \neq 0.0$; **P** < 0.01; $r \leq -0.381$, $r \geq 0.381$; reject H_o

d. $\hat{y} = 0.0881 - 0.0221x$

e. H_a: $\beta_1 < 0$; $t\star = -4.82$; **P** < 0.005 or $t \leq -1.70$; reject H_o

13.63 52.4 ± 0.14; 52.3 to 52.5

13.65 a. 13.04

b. 13.04 ± 3.23; 9.81 to 16.27

c. 13.04 ± 8.35; 4.69 to 21.39

d. 26.36 ± 2.95; 23.41 to 29.31
26.36 ± 8.25; 18.11 to 34.61

13.67 $\Sigma x = 16.25$, $\Sigma y = 152$, $\Sigma x^2 = 31.5625$, $\Sigma xy = 275$, $\Sigma y^2 = 2504$; $\hat{y} = 6.3758 + 5.4303x$

a. 17.24 ± 1.88; 15.4 to 19.1

b. 17.24 ± 5.59; 11.6 to 22.8

13.69 a. Overall pattern is elongated.

b. significant slope

c. November and December—holiday season

13.71 a.

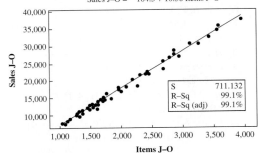

Fitted Line Plot
Sales J–O = −184.5 + 10.36 Items J–O

S	711.132
R–Sq	99.1%
R–Sq (adj)	99.1%

c. $r = 0.995$; $\hat{y} = -185 + 10.4x$

d. H_a: $\beta_1 \neq 0$; $t\star = 73.68$; **P** < 0.01; $t \leq -2.01$, $t \geq 2.01$; reject H_o

e. $30{,}881.5 \pm 1470.94$; 29,410.6 to 32,352.4

13.75 a. always

b. never

c. sometimes

d. sometimes

e. always

13.77 H_a: $\rho > 0.0$; $r\star = 0.69$; **P** < 0.005; $r \geq 0.29$; reject H_o

13.79 a. H_a: $\rho \neq 0.0$; $r\star = 0.61$; **P** < 0.01; $r \leq -0.482$, $r \geq 0.482$; reject H_o

b. 118.7

13.81 a & c.

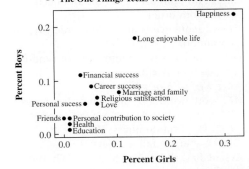

d. $\Sigma x = 37.5$, $\Sigma y = 52.7$, $\Sigma x^2 = 104.75$, $\Sigma xy = 98.75$, $\Sigma y^2 = 159.49$; $\hat{y} = 3.79 - 0.415x$

e. $s_e = 0.21045$

f. -0.502 to -0.328

g. 2.55 ± 0.13; 2.42 to 2.68
2.34 ± 0.16; 2.18 to 2.50

h. 2.55 ± 0.47; 2.08 to 3.02
2.34 ± 0.47; 1.87 to 2.81

13.83 $\Sigma x = 1177$, $\Sigma y = 567$, $\Sigma x^2 = 70{,}033$, $\Sigma xy = 32{,}548$, $\Sigma y^2 = 15{,}861$

a. H_a: $\rho \neq 0.0$; $r\star = 0.513$; $0.01 < $ **P** < 0.02; $r \leq -0.433$, $r \geq 0.433$; reject H_o

b. $\hat{y} = 16.40 + 0.189x$

c. H_a: $\beta_1 > 0$; $t\star = 2.61$; $0.005 < $ **P** < 0.01; $t \geq 1.73$; reject H_o

d. 28.50 ± 9.97; 18.53 to 38.47

13.85 a. H_a: $\rho > 0.0$; $r\star = 0.895$; **P** < 0.005; $r \geq 0.34$; reject H_o

b. $\hat{y} = -2.30 + 0.39x$; 17.23 ± 2.85; 14.38 to 20.08

c. 26.99 ± 2.60; 24.39 to 29.59

13.87 Relationships (customer vs. items, customer vs. sales) are not as strong—can have a customer who does not make a purchase.

13.89 $\Sigma x = 16$, $\Sigma y = 38$, $\Sigma x^2 = 66$, $\Sigma xy = 145$, $\Sigma y^2 = 326$
$b_1 = 1.5811$; $r = 0.9973$

Chapter 14

14.1 a. "general" agreement

b. **The One Things Teens Want Most from Life**

14.3 a. involves only the counts of plus and minus signs

b. median; half of the data are above the median, and half are below

14.5 39 to 47

14.7 -7 to 1

14.9 a. H_a: median < 18

b. H_a: median < 32

c. H_a: median $\neq 4.5$

14.11 a. 0.01
b. $0.025 < P < 0.05$
c. 0.005
d. 0.0104

14.13 H_a: median $\neq$ 42 years; $x = n(+) = 40$;
$0.05 < P < 0.10$ or $x \leq 39$; fail to reject H_o

14.15 H_a: $P(+) \neq 0.5$
If $P \leq \alpha$, reject H_o; if $P > \alpha$, fail to reject H_o.
a. $x = n(+) = 20$, $P < 0.01$
b. $x = n(+) = 27$, $0.01 < P < 0.05$
c. $x = n(+) = 30$, $0.10 < P < 0.25$
d. $x = n(+) = 33$, $P > 0.25$

14.17 b. H_a: $M > 0$, improved; $x = n(-) = 5$;
$P \approx 0.125$ or $x \leq 4$; fail to reject H_o
2003–1999: $x\star = n(-) = 8$; $P > 0.125$ or
$x \leq 4$; fail to reject H_o

14.19 H_a: preference for new; $p > 0.5$; $z\star = 1.74$;
$P = 0.0409$ or $z \geq 2.33$; fail to reject H_o

14.21 H_a: $P(+) > 0.5$; $x' = 718$

14.23 a. difference between two independent means
b. size of data not used, only its rank

14.25 a. H_o: the same; H_a: different
b. H_o: the same; H_a: not the same
c. H_o: the same; H_a: A is higher than B

14.27 a. $P > 0.05$
b. $P < 0.05$
c. $P = 0.0089$

14.29 a. $U \leq 88$
b. $z \leq -1.65$

14.33 H_a: not the same; $U\star = 12.5$; $P > 0.10$ or
$U \leq 3$; fail to reject H_o

14.35 H_a: is higher; $U\star = 178$; $z\star = -2.61$;
$P = 0.0045$ or $z \leq -1.65$; reject H_o

14.37 a. H_o: random order; H_a: did not occur in a random order
b. H_o: sequence is in random order; H_a: not random
c. H_o: order was random; H_a: not random

14.39 a. $V \leq 9$ or $V \geq 22$
b. $z \leq -1.96$ or $z \geq +1.96$

14.41 H_a: sequence not of random order; $V\star = 9$;
$P > 0.05$ or $V \leq 4$, $V \geq 12$; fail to reject H_o

14.43 H_a: Lack of randomness; $V\star = 8$; $P > 0.05$ or
$V \leq 3$, $V \geq 10$; fail to reject H_o

14.45 a. median = 4.4; $V = 5$ (Note: When the median is one of the data values, the two categories are "above the median" and "below or equal to the median.")
b. H_a: did not occur randomly; $V\star = 5$;
$P > 0.05$ or $V \leq 3$, $V \geq 12$; fail to reject H_o

14.47 b. $z\star = -3.76$; $P = 0.0002$
c. yes, reject

d.
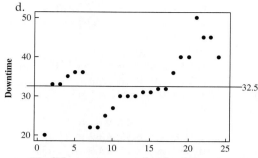

14.49 a. H_a: did not occur randomly; $V\star = 9$;
$P > 0.05$ or $V \leq 8$, $V \geq 20$; fail to reject H_o
b. $z\star = -2.00$; $P = 0.0456$ or $z \leq -1.96$,
$z \geq 1.96$; reject H_o

14.53 a. H_o: no relationship; H_a: a relationship
b. H_o: unrelated; H_a: related
c. H_o: no correlation; H_a: positive correlation
d. H_o: age has no effect; H_a: age has a decreasing effect

14.55 a. $r_s \leq -0.545$ or $r_s \geq 0.545$
b. $r_s \geq 0.323$
c. $r_s \leq -0.564$

14.57 a. 0.133
b. H_a: $\rho_s > 0$; $r_s\star = 0.133$; $P > 0.10$ or
$r_s \geq 0.564$; fail to reject H_o

14.59 $n = 12$, $\Sigma d^2 = 70.5$, 0.753

14.61 H_a: $\rho_s > 0$; $r_s\star = 0.736$; $0.01 < P < 0.025$ or
$r_s \geq 0.564$; reject H_o

14.63 b. H_a: $\rho_s \neq 0$; $r_s\star = 0.8625$; $P < 0.01$ or
$r_s \leq -0.700$, $r_s \geq 0.700$; reject H_o
c. H_a: $\rho_s \neq 0$; $r_s\star = 0.619$; $P > 0.10$ or
$r_s \leq -0.738$, $r_s \geq 0.738$; fail to reject H_o
d. H_a: $\rho_s \neq 0$; $r_s\star = 0.869$; $0.01 < P < 0.02$
or $r_s \leq -0.738$ and $r_s \geq 0.738$; reject H_o

14.65 a.
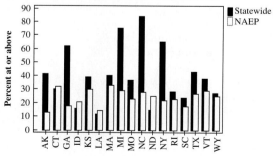
Percent Students at or above Proficient Level

C. Percent Student at or above Proficient Level

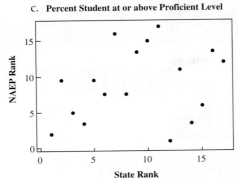

d. H_a: $\rho_s \neq 0$; $r_s\bigstar = 0.272$; **P** > 0.10 or
$r_s \leq -0.490$, $r_s \geq 0.490$; fail to reject H_o

14.67 0.502

14.69 a. Stem-and-leaf of Hydrogen N = 52
Leaf Unit = 0.10

```
    3    2    078
   10    3    1356688
   18    4    13355689
  (12)   5    001112444589
   22    6    11137
   17    7    2388
   13    8    3378
    9    9    4479
    5   10    56
    3   11    1
    2   12    4
    1   13
    1   14
    1   15    2
```

b. skewed right
c. 5.0 to 6.3

14.71 a. H_a: median $\neq 50$; $x = n(+) = 10$;
P $= 0.0987$ or $x \leq 9$; fail to reject H_o
b. H_a: median < 50; $x = n(+) = 10$;
P $= 0.0494$ or $x \leq 10$; reject H_o

14.73 H_a: B is faster; $x = n(-) = 2$; **P** > 0.05 or
$x \leq 1$; fail to reject H_o

14.75 reject for $U \leq 127$

14.77 H_a: there is a difference; $U\bigstar = 2$; **P** ≈ 0.05 or
$U \leq 2$; reject H_o

14.79 b. BA: H_a: AL are higher; $U\bigstar = 71.5$;
P > 0.05 or $U \leq 71$; fail to reject H_o
ERA: H_a: NL is lower; $U\bigstar = 63.5$;
P < 0.025 or $U \leq 71$; reject H_o

14.81 H_a: lack of randomness; $V\bigstar = 9$; **P** > 0.05 or
$V \leq 4$, $V \geq 10$; fail to reject H_o

14.83 a. median $= 22.5$; runs above: 6, below: 6
b. H_a: lack of randomness; $V\bigstar = 12$;
P > 0.05 or $V \leq 6$, $V \geq 16$; fail to reject H_o

14.85 H_a: $\rho_s > 0$; $r_s\bigstar = 0.880$; **P** < 0.01 or
$r_s \geq 0.399$; reject H_o

14.87 a. $r_{12} = 0.321$, $r_{13} = 0.298$, $r_{23} = 0.988$
b. H_o: $\rho_{s} = 0$ vs. H_a: $\rho_s \neq 0$
For (1) vs. (2): $r_s\bigstar = 0.321$, **P** > 0.10 or
$r_s \leq -0.400$, $r_s \geq 0.400$; fail to reject H_o
For (1) vs. (3): $r_s\bigstar = 0.298$, **P** > 0.10 or
$r_s \leq -0.400$, $r_s \geq 0.400$; fail to reject H_o
For (2) vs. (3): $r_s\bigstar = 0.988$, **P** < 0.01 or
$r_s \leq -0.400$, $r_s \geq 0.400$; reject H_o

Answers to Chapter Practice Tests

Part 1: Only the replacement for the word(s) in bold-face type is given. (If the statement is true, no answer is shown. If the statement is false, a replacement is given.)

Chapter 1, Page 36

Part I

1.1 descriptive
1.2 inferential
1.4 sample
1.5 population
1.6 attribute or qualitative
1.7 quantitative
1.9 random

Part II

1.11 a. Nominal **b.** Ordinal **c.** Continuous **d.** Discrete **e.** Nominal
1.12 c, g, h, b, e, a, d, f

Part III

1.13 See definitions; examples will vary. Note: *population* is set of ALL possible, whereas *sample* is the actual set of subjects studied.
1.14 See definitions; examples will vary. Note: *variable* is the idea of interest, whereas *data* are the actual values obtained.
1.15 See definitions; examples will vary. Note: *data value* is the value describing one source, the *statistic* is a value (usually calculated) describing all the data in the sample, the *parameter* is a value describing the entire population (usually unknown).
1.16 Every element of the population has an equal chance of being selected.

Chapter 2, Page 141

Part I

2.1 median
2.2 dispersion
2.3 never
2.5 zero
2.6 higher than

Part II

2.11 a. 30 **b.** 46 **c.** 91 **d.** 15 **e.** 1 **f.** 61
   g. 75 **h.** 76 **i.** 91 **j.** 106 or 114
2.12 a. two items purchased
   b. Nine people purchased 3 items each.
   c. 40 **d.** 120 **e.** 5 **f.** 2 **g.** 3 **h.** 3
   i. 3.0 **j.** 1.795 **k.** 1.34
2.13 a. 6.7 **b.** 7 **c.** 8 **d.** 6.5 **e.** 5 **f.** 6
   g. 3.0 **h.** 1.7 **i.** 5
2.14 a. −1.5 **b.** 153

Part III

2.15 a. 98 **b.** 50 **c.** 121 **d.** 100
2.16 a. $32,000, $26,500, $20,000, $50,000
   b.

```
  :    :  .:   .    .         .                        .
--+---+---+---+---+---+---+---+---+---+---+---+---+---+-
  .   .   .            .
--+---+---+---+---+---+---+---+---+---+---+---+---+---+-
 20      30      40      50      60      70      80
mode  median  mean      midrange
```
Salary ($1000)

   c. Mr. VanCott—midrange; business manager—mean; foreman—median; new worker—mode
   d. The distribution is J-shaped.
2.17 There is more than one possible answer for these.
   a. 12, 12, 12 **b.** 15, 20, 25
   c. 12, 15, 15, 18 **d.** 12, 15, 16, 25 25
   e. 12, 12, 15, 16, 17 **f.** 20, 25, 30, 32, 32, 80
2.18 A is right; B is wrong; standard deviation will not change.

2.19 B is correct. For example, if standard deviation is $5, then the variance, (standard deviation)^2, is "25 dollars squared." Who knows what "dollars squared" are?

Chapter 3, Page 200

Part I

3.1 regression
3.2 strength of the
3.3 +1 or −1
3.5 positive
3.7 positive
3.8 −1 and +1
3.9 output or predicted value

Part II

3.11 a. B, D, A, C **b.** 12 **c.** 10 **d.** 175 **e.** N
f. (125, 13) **g.** N **h.** P
3.12 Someone made a mistake in arithmetic; *r* must be between −1 and +1.
3.13 a. 12 **b.** 10 **c.** 8 **d.** 0.73 **e.** 0.67
f. 4.33 **g.** $\hat{y} = 4.33 + 0.67x$

Part III

3.14 Young children have small feet and probably tend to have less mathematics ability, whereas adults have larger feet and would tend to have more ability.
3.15 Student B is correct. −1.78 can occur only as a result of faulty arithmetic.
3.16 These answers will vary but should somehow include the basic thought:
a. strong negative **b.** strong positive
c. no correlation **d.** no correlation
e. impossible value, bad arithmetic
3.17 There is more than one possible answer for these.
a. (1, 1), (2, 1), (3, 1) **b.** (1, 1), (3, 3), (5, 5)
c. (1, 5), (3, 3), (5, 1) **d.** (1, 1), (5, 1), (1, 5), (5,5)

Chapter 4, Page 266

Part I

4.1 any number value between 0 and 1, inclusive
4.4 simple
4.5 seldom
4.6 sum to 1.0

4.7 dependent
4.8 complementary
4.9 mutually exclusive or dependent
4.10 multiplication rule

Part II

4.11 a. $\frac{4}{8}$ **b.** $\frac{4}{8}$ **c.** $\frac{2}{8}$ **d.** $\frac{6}{8}$ **e.** $\frac{2}{8}$

f. $\frac{6}{8}$ **g.** 0 **h.** $\frac{6}{8}$ **i.** $\frac{1}{8}$ **j.** $\frac{5}{8}$

k. $\frac{2}{4}$ **l.** 0 **m.** $\frac{1}{2}$ **n.** no (e)

o. yes (g) **p.** no (i) **q.** yes (a, k) **r.** no (b, 1)
s. yes (a, m)
4.12 a. 0 **b.** 0.7 **c.** 0 **d.** no (c)
4.13 a. 0.7 **b.** 0.5 **c.** no, $P(E \text{ and } F) = 0.2$
d. yes, $P(E) = P(E \mid F)$
4.14 0.51

Part III

4.15. Student B is right. *Mutually exclusive* means no intersection, whereas *independence* means one event does not affect the probability of the other.
4.16 These answers will vary but should somehow include the basic thoughts:
a. no common occurrence
b. either event has no effect on the probability of the other
c. the relative frequency with which the event occurs.
d. probability that an event will occur even though the conditional event has previously occurred

Chapter 5, Page 311

Part I

5.1 continuous
5.3 one
5.5 exactly two
5.6 binomial
5.7 one success occurring on 1 trial
5.8 population
5.9 population parameters

Part II

5.11 a. Each $P(x)$ is between 0 and 1, and the sum of all $P(x)$ is exactly 1.
    b. 0.2   **c.** 0   **d.** 0.8   **e.** 3.2   **f.** 1.25
5.12 a. 0.230   **b.** 0.085   **c.** 1.2   **d.** 1.04

Part III

5.13 n independent repeated trials of two outcomes; the two outcomes are "success" and "failure"; $p = P(\text{success})$ and $q = P(\text{failure})$ and $p + q = 1$; $x = n(\text{success}) = 0, 1, 2, \ldots , n$.
5.14 Student B is correct. The sample mean and standard deviation are statistics found using formulas studied in Chapter 2. The probability distributions studied in Chapter 5 are theoretical populations and their means and standard deviations are parameters.
5.15 Student B is correct. There are no restrictions on the values of the variable x.

Chapter 6, Page 357

Part I

6.1 its mean
6.4 1 standard deviation
6.6 right
6.7 zero, 1
6.8 some (many)
6.9 mutually exclusive events
6.10 normal

Part II

6.11 a. 0.4922   **b.** 0.9162   **c.** 0.1020   **d.** 0.9082
6.12 a. 0.63   **b.** −0.95   **c.** 1.75
6.13 a. $z(0.8100)$ **b.** $z(0.2830)$
6.14 0.7910
6.15 28.03
6.16 a. 0.0569   **b.** 0.9890   **c.** 537   **d.** 417   **e.** 605

Part III

6.17 This answer will vary but should somehow include the basic properties: bell-shaped, mean of 0, standard deviation of 1.
6.18 This answer will vary but should somehow include the basic ideas: it is a z-score, α represents the area under the curve and to the right of z.
6.19 All normal distributions have the same shape and probabilities relative to the z-score.

Chapter 7, Page 390

Part I

7.1 is not
7.2 some (many)
7.3 population
7.4 divided by $\sqrt{n}$
7.5 decreases
7.6 approximately normal
7.7 sampling
7.8 means
7.9 random

Part II

7.11 a. 0.4364   **b.** 0.2643
7.12 a. 0.0918   **b.** 0.9525
7.13 0.6247

Part III

7.14 In this case each head produced one piece of data, the estimated length of the line. The CLT assures us that the mean value of a sample is far less variable than individual values of the variable x.
7.15 All samples must be of one fixed size.
7.16 Student A is correct. A population distribution is a distribution formed by all x values that make up the entire population.
7.17 Student A is correct. The standard error is found by dividing the standard deviation by the square root of the *sample size*.

Chapter 8, Page 470

Part I

8.1 alpha
8.2 alpha
8.3 sample distribution of the mean
8.7 type II error
8.8 beta
8.9 correct decision
8.10 critical (rejection) region

Part II

8.11 4.72 to 5.88
8.12 a. H_o: $\mu = 245$, H_a: $\mu > 245$
    b. H_o: $\mu = 4.5$, H_a: $\mu < 4.5$
    c. H_o: $\mu = 35$, H_a: $\mu \neq 35$

8.13 a. 0.05, z, $z \leq -1.65$
 b. 0.05, z, $z \geq +1.65$
 c. 0.05, z, $z \leq -1.96$ or $z \geq +1.96$
8.14 a. 1.65 **b.** 2.33 **c.** 1.18 **d.** -1.65
 e. -2.05 **f.** -0.67
8.15 a. $z\star = 2.50$ **b.** 0.0062
8.16 H_o: $\mu = 1520$ vs. H_a: $\mu < 1520$, crit. reg. $z \leq -2.33$, $z\star = -1.61$, fail to reject H_o

Part III

8.17 a. no specific effect
 b. reduces it **c.** narrows it **d.** no effect
 e. increases it **f.** widens it
8.18 a. H_o − (a), H_a − (b) **b.** 3 **c.** 2
 d. P(type I errror) is alpha, decreases: P(type II error) increases
8.19 The alternative hypothesis expresses the concern; the conclusion answers the concern.

Chapter 9, Page 540

Part I

9.2 Student's t
9.3 chi-square
9.4 to be rejected
9.6 t score
9.7 $n - 1$
9.9 $\sqrt{pq/n}$
9.10 z(normal)

Part II

9.11 a. 2.05 **b.** -1.73 **c.** 14.6
9.12 a. 28.6 **b.** 1.44 **c.** 27.16 to 30.04
9.13 0.528 to 0.752
9.14 a. H_o: $\mu = 255$, H_a: $\mu > 225$
 b. H_o: $p = 0.40$, H_a: $p \neq 0.40$
 c. H_o: $\sigma = 3.7$, H_a: $\sigma < 3.7$
9.15 a. 0.05, z, $z \leq -1.65$
 b. 0.05, t, $t \leq -2.08$ or $t \geq +2.08$
 c. 0.05, z, $z \geq +1.65$
 d. 0.05, χ^2, $\chi^2 \leq 14.6$ or $\chi^2 \geq 43.2$
9.16 H_o: $\mu = 26$ vs. H_a: $\mu < 26$, crit. reg, $t \leq -1.71$, $t\star = -1.86$, reject H_o
9.17 H_o: $\sigma = 0.1$ vs. H_a: $\sigma > 0.1$, crit. reg. $\chi^2 \geq 21.1$, $\chi^2\star = 23.66$, reject H_o
9.18 H_o: $p = 0.50$ vs. H_a: $p > 0.50$, crit. reg. $z \geq 2.05$, $z\star = 1.29$, fail to reject H_o

Part III

9.19 If the distribution is normal, 6 standard deviations is approximately equal to the range.
9.20 B
9.21 They are both correct.
9.22 When the sample size, n, is large, the critical value of t is estimated by using the critical value from the standard normal distribution of z.
9.23 Student A
9.24 Student B is right. It is significant at the 0.01 level of significance.
9.25 Student A is correct.
9.26 It depends on what it means to improve the confidence interval. For most purposes, an increased sample size would be the best improvement.

Chapter 10, Page 614

Part I

10.1 two independent means
10.3 F-distribution
10.4 Student's t-distribution
10.5 Student's t
10.7 nonsymmetrical (or skewed)
10.9 decreases

Part II

10.11 a. H_o: $\mu_N - \mu_A = 0$, H_a: $\mu_N - \mu_A \neq 0$
 b. H_o: $\sigma_o/\sigma_m = 1.0$, H_a: $\sigma_o/\sigma_m > 1.0$
 c. H_o: $p_m - p_f = 0$, H_a: $p_m - p_f \neq 0$
 d. H_o: $\mu_d = 0$, H_a: $\mu_d > 0$
10.12 a. z, $z \leq -1.96$ or $z \geq 1.96$
 b. t, $t \leq -2.05$, $t \geq 2.05$
 c. t, df = 7, $t \geq 1.89$
 d. t, df = 37, $t \leq -1.69$
 e. F, $F \geq 2.11$
10.13 a. 2.05 **b.** 2.13 **c.** 2.51 **d.** 2.18
 e. 1.75 **f.** 1.69 **g.** -2.50 **h.** -1.28
10.14 H_o: $\mu_L - \mu_P = 0$ vs. H_a: $\mu_L - \mu_P > 0$, crit. reg. $t \geq +1.83$, $t\star = 0.979$, fail to reject H_o
10.15 H_o: $\mu_d = 0$ vs. H_a: $\mu_d > 0$, crit. reg. $t \geq 1.89$, $t\star = 1.88$, fail to reject H_o
10.16 0.072 to 0.188

Part III

10.17 independent
10.18 One possibility: Test all students before the course starts, then randomly select 20 of those who finish the course and test them afterward. Use the before scores for these 20 as the before sample.

10.19 For starters, if the two independent samples are of different sizes, the techniques for dependent samples could not be completed. They are testing very different concepts, the "mean of the differences of paired data" and the "difference between two mean values."

10.20 It is only significant if the calculated *t*-score is in the critical region. The variation among the data and their relative size will play a role.

10.21 The 80 scores actually are two independent samples of size 40. A test to compare the mean scores of the two groups could be completed.

10.22 A fairly large sample of both Catholic and non-Catholic families would need to be taken, and the number of each whose children attended private schools would need to be obtained. The difference between two proportions could then be estimated.

Chapter 11, Page 653

Part I

11.1 one less than
11.3 expected
11.4 contingency table
11.6 test of homogeneity
11.8 approximated by chi-square

Part II

11.11 a. H_o: Digits generated occur with equal probability.
H_a: Digits do not occur with equal probability.
b. H_o: Votes were cast independently of party affiliation.
H_a: Votes were not cast independently of party affiliation.
c. H_o: The crimes distributions are the same for all four cities.
H_a: The crimes distributions are not all the same.

11.12 a. 4.40   **b.** 35.7

11.13 H_o: $P(1) = P(2) = P(3) = \dfrac{1}{3}$
H_a: preferences not all equal, $\chi^{2\star} = 3.78$;
$0.10 < \mathbf{P} < 0.25$ or crit. reg. $\chi^2 \geq 5.99$;
fail to reject H_o

11.14 a. H_o: The distribution is the same for all types of soil.
H_a: The distributions are not all the same.
b. 25.622   **c.** 13.746
d. $0.005 < \mathbf{P} < 0.01$ **e.** $\chi^2 \geq 9.49$

f. Reject H_o: There is sufficient evidence to show that the growth distribution is different for at least one of the three soil types.

Part III

11.15 Similar in that there are *n* repeated independent trials. Different in that the binomial has two possible outcomes, whereas the multinomial has several. Each possible outcome has a probability and these probabilities sum to 1 for each different experiment, both for binomial and multinomial.

11.16 The test of homogeneity compares several distributions in a side-by-side comparison, whereas the test for independence tests the independence of the two factors that create the rows and columns of the contingency table.

11.17 Student A is right in that the calculations are completed in the same manner. Student B is correct in that the test of independence starts with one large sample and homogeneity has several samples.

11.18 a. If a chi-square test is to be used, the results of the four questions would be pooled to estimate the expected probability.
b. Use a chi-square test for homogeneity.

Chapter 12, Page 691

Part I

12.2 mean square
12.3 SS(factor) or MS(factor)
12.5 reject H_o
12.7 the number of factor levels less one
12.8 mean
12.9 need to
12.10 does not indicate

Part II

12.11 a. T   **b.** T   **c.** F   **d.** T   **e.** T   **f.** T
g. F   **h.** F   **i.** F   **j.** F   **k.** T   **l.** F   **m.** F
n. F   **o.** T

12.12 a. 72   **b.** 72   **c.** 22   **d.** 4   **e.** 4.5

Part III

12.13 This answer will vary but should somehow include the basic ideas: It is the comparison of several mean values that result from testing some statistical population by measuring a variable repeatedly at each of the several levels for which the factor is being tested.

12.14 a. $x_{r,k} = \mu + F$ scrubber $+ \epsilon_{k(r)}$

b. H_o: The mean amount of emissions is the same for all three scrubbers tested.
H_a: The mean amounts are not all equal.

c.

Source	df	SS	MS
Scrubber	2	12.80	6.40
Error	13	33.63	2.59
Total	15	46.44	

d. $F(2, 13, 0.05) = 3.81$, $F\star = 2.47$, fail to reject H_o. The difference in the mean value for the scrubbers is not significant.

e. I
```
           .   .   :  .
  --+--+--+--+--+--+--+--
    7       9      11     13
```

II
```
         .   .       :
  --+--+--+--+--+--+--+--
    7       9      11     13
```

III
```
       .   .   .   .   .
  --+--+--+--+--+--+--+--
    7       9      11     13
```

Chapter 13, Page 744

Part I

13.2 need not be
13.3 does not prove
13.4 need not be
13.6 the linear correlation coefficient
13.8 regression
13.9 $n - 2$

Part II

13.11

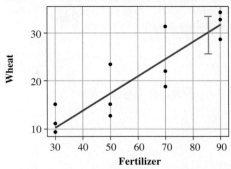

Amount of Wheat Harvest

13.12. $\sum x = 720$, $\sum y = 252$, $\sum x^2 = 49,200$, $\sum xy = 17,240$, $\sum y^2 = 6228$
13.13 $SS(x) = 6000$, $SS(y) = 936$, $SS(xy) = 2120$
13.14 0.895
13.15 0.65 to 0.97
13.16 $\hat{y} = -0.20 + 0.353x$
13.17 See red line in figure in 13.11.
13.18 4.324
13.19 yes; H_o: $\beta_1 = 0$ vs. H_a: $\beta_1 > 0$, $t\star = 6.33$, reject H_o
13.20 25.63 to 33.98
13.21 See blue vertical segment in 13.11.

Part III

13.22 Variable 1: The frequency of skiers having their bindings tested
Variable 2: The incidence of lower-leg injury. The statement implies that as the frequency with which the bindings are tested increases, the frequency of lower-leg injury decreases; thus the strong correlation must be negative for these variables.

13.23 A "moment" is the distance from the mean, and the product of both the horizontal moment and the vertical moment is summed in calculating the correlation coefficient.

13.24 A value close to zero, also. The formulas used to calculate both values have the same numerator, namely $SS(xy)$.

13.25 The vertical distance from a potential line of best fit to the data point is measured by $(y - \hat{y})$. The line of best fit is defined to be the line that results in the smallest possible total when the squared values of $(y - \hat{y})$ are totaled, thus "the method of least squares."

13.26 The strength of the linear relationship could be measured with the correlation coefficient.

13.27 A random sample will be needed from the population of interest. The data collected need to be for the variables length of time on welfare and the measure of current level of self-esteem.

Chapter 14, Page 800

Part I

14.2 t-test
14.3 runs test
14.4 assigned equal ranks
14.7 Mann–Whitney U test
14.8 power
14.10 power

Part II

14.11 -2 to $+7$
14.12 H_o: There is no difference in weight gain.
H_a: There is a difference in weight gain, crit. val.: 23, $U\star = 32.5$, fail to reject H_o.
14.13 H_o: no correlation
H_a: correlated, crit. val.: ± 0.683, $r_s\star = -0.70$, reject H_o. Yes, there is significant correlation.

14.14 $(+)$ = higher grade level than previous problem
$(-)$ = lower grade level than previous problem
H_o: $P(+) = 0.5$
H_a: $P(+) \neq 0.5$, crit. val.: 7, $x = 11$, fail to reject H_o. Ths sample does not show a significant pattern.

Part III

14.15 The nonparametric statistics do not require assumptions about the distribution of the variable.
14.16 The sign test is a binomial experiment of n trials (the n data observations) with two outcomes for each data $[(+)$ or $(-)]$, and $p = P(+) = 0.5$. The variable x is the number of the least frequent sign.
14.17 The median is the middle value such that 50% of the distribution is larger in value and 50% is smaller in value.
14.18 The extreme value in a set of data can have a sizeable effect on the mean and standard deviation in the parametric methods. The nonparametric methods typically use rank numbers. The extreme value with ranks is either 1 or n, and neither changes if the value is more extreme.
14.19 d; $p = P(+) = P$(prefer seating arrangement A) = 0.5, no preference

Index

Formula Card for Johnson & Kuby, ELEMENTARY STATISTICS, Tenth Edition

Sample mean:

$$\bar{x} = \frac{\Sigma x}{n} \quad \textbf{(2.1)} \qquad \text{or} \qquad \frac{\Sigma xf}{\Sigma f} \quad \textbf{(2.13)}$$

Depth of sample median:

$$d(\tilde{x}) = (n + 1)/2 \tag{2.2}$$

Range: $H - L$ **(2.4)**

Sample variance:

$$s^2 = \frac{\Sigma(x - \bar{x})^2}{n - 1} \tag{2.6}$$

or

$$s^2 = \frac{\Sigma x^2 - \frac{(\Sigma x)^2}{n}}{n - 1} \tag{2.10}$$

or

$$s^2 = \frac{\Sigma x^2 f - \frac{(\Sigma xf)^2}{\Sigma f}}{\Sigma f - 1} \tag{2.14}$$

Sample standard deviation:

$$s = \sqrt{s^2} \tag{2.7}$$

Chebyshev's theorem: at least $1 - (1/k^2)$ **(p. 109)**

Sum of squares of x:

$$SS(x) = \Sigma x^2 - ((\Sigma x)^2/n) \tag{2.9}$$

Sum of squares of y:

$$SS(y) = \Sigma y^2 - ((\Sigma y)^2/n) \tag{3.3}$$

Sum of squares of xy:

$$SS(xy) = \Sigma xy - ((\Sigma x \cdot \Sigma y)/n) \tag{3.4}$$

Pearson's correlation coefficient:

$$r = SS(xy)/\sqrt{SS(x) \cdot SS(y)} \tag{3.2}$$

Equation for line of best fit: $\hat{y} = b_0 + b_1 x$ **(p. 174)**

Slope for line of best fit: $b_1 = SS(xy)/SS(x)$ **(3.6)**

y-intercept for line of best fit:

$$b_0 = [\Sigma y - (b_1 \cdot \Sigma x)]/n \tag{3.7}$$

Empirical (observed) probability:

$$P'(A) = n(A)/n \tag{4.1}$$

Theoretical probability for equally likely sample space:

$$P(A) = n(A)/n(S) \tag{4.2}$$

Complement rule:

$$P(\text{not } A) = P(\overline{A}) = 1 - P(A) \tag{4.3}$$

General addition rule:

$$P(A \text{ or } B) = P(A) + P(B) - P(A \text{ and } B) \tag{4.4}$$

General multiplication rule:

$$P(A \text{ and } B) = P(A) \cdot P(B|A) \tag{4.5}$$

Special addition rule for mutually exclusive events:

$$P(A \text{ or } B \text{ or } \ldots \text{ or } E) = P(A) + P(B) + \cdots + P(E) \tag{4.6}$$

Special multiplication rule for independent events:

$$P(A \text{ and } B \text{ and } \ldots \text{ and } E) = P(A) \cdot P(B) \cdot \cdots \cdot P(E) \tag{4.7}$$

Mean of discrete random variable:

$$\mu = \Sigma[xP(x)] \tag{5.1}$$

Variance of discrete random variable:

$$\sigma^2 = \Sigma[x^2 P(x)] - \{\Sigma[xP(x)]\}^2 \tag{5.3a}$$

Standard deviation of discrete random variable:

$$\sigma = \sqrt{\sigma^2} \tag{5.4}$$

Factorial: $n! = (n)(n - 1)(n - 2) \cdot \cdots \cdot 2 \cdot 1$ **(p. 289)**

Binomial coefficient:

$$\binom{n}{x} = \frac{n!}{x! \cdot (n - x)!} \tag{5.6}$$

Binomial probability function:

$$P(x) = \binom{n}{x} \cdot p^x \cdot q^{n-x}, \; x = 0, \ldots, n \tag{5.5}$$

Mean of binomial random variable: $\mu = np$ **(5.7)**

Standard deviation, binomial random variable:

$$\sigma = \sqrt{npq} \tag{5.8}$$

Standard score: $z = (x - \mu)/\sigma$ **(6.3)**

Standard score for $\bar{x}$: $z = \dfrac{\bar{x} - \mu}{\sigma/\sqrt{n}}$ **(7.2)**

Confidence interval for mean, μ (σ known):

$$\bar{x} \pm z(\alpha/2) \cdot (\sigma/\sqrt{n}) \tag{8.1}$$

Sample size for $1 - \alpha$ confidence estimate for μ:

$$n = [z(\alpha/2) \cdot \sigma/E]^2 \tag{8.3}$$

Calculated test statistic for H_o: $\mu = \mu_0$ (σ known):

$$z\star = (\bar{x} - \mu_0)/(\sigma/\sqrt{n}) \tag{8.4}$$

Confidence interval estimate for mean, μ (σ unknown):

$$\bar{x} \pm t(\text{df}, \alpha/2) \cdot (s/\sqrt{n}) \quad \text{with df} = n - 1 \tag{9.1}$$

Calculated test statistic for H_o: $\mu = \mu_0$ (σ unknown):

$$t\star = \frac{\bar{x} - \mu_0}{s/\sqrt{n}} \quad \text{with df} = n - 1 \tag{9.2}$$

Confidence interval estimate for proportion, p:

$$p' \pm z(\alpha/2) \cdot \sqrt{(p'q')/n}, \quad p' = x/n \tag{9.6}$$

Calculated test statistic for H_o: $p = p_0$:

$$z\star = (p' - p_0)/\sqrt{(p_0 q_0/n)}, \quad p' = x/n \tag{9.9}$$

Calculated test statistic for H_o: $\sigma^2 = \sigma_0^2$ or $\sigma = \sigma_0$:

$$\chi^2\star = (n - 1)s^2/\sigma_0^2, \quad \text{df} = n - 1 \tag{9.10}$$

Mean difference between two dependent samples:

Paired difference: $d = x_1 - x_2$ **(10.1)**

Confidence interval for mean, μ_d:

$$\bar{d} \pm t(\text{df}, \alpha/2) \cdot s_d/\sqrt{n} \tag{10.2}$$

Sample mean of paired differences:

$$\bar{d} = \Sigma d/n \tag{10.3}$$

Sample standard deviation of paired differences:

$$s_d = \sqrt{\frac{\sum d^2 - \left[\dfrac{(\sum d)^2}{n}\right]}{n - 1}} \qquad (10.4)$$

Calculated test statistic for H_o: $\mu_d = \mu_o$:

$$t\star = (\bar{d} - \mu_o)/(s_d/\sqrt{n}), \quad \text{df} = n - 1 \qquad (10.5)$$

Difference between means of two independent samples:

Degrees of freedom:

$$\text{df} = \text{smaller of } (n_1 - 1) \text{ or } (n_2 - 1) \qquad \textbf{(p. 565)}$$

Confidence interval estimate for $\mu_1 - \mu_2$:

$$(\bar{x}_1 - \bar{x}_2) \pm t_{(df,\, \alpha/2)} \cdot \sqrt{(s_1^2/n_1) + (s_2^2/n_2)} \qquad (10.8)$$

Calculated test statistic for H_o: $\mu_1 - \mu_2 = (\mu_1 - \mu_2)_o$:

$$t\star = [(\bar{x}_1 - \bar{x}_2) - (\mu_1 - \mu_2)_o]/\sqrt{(s_1^2/n_1) + (s_2^2/n_2)} \quad (10.9)$$

Difference between proportions of two independent samples: $p_1' q_1' \backslash n_1$

Confidence interval for $p_1 - p_2$:

$$(p_1' - p_2') \pm z_{(\alpha/2)} \cdot \sqrt{\frac{p_1' q_1'}{n_1} + \frac{p_2' q_2'}{n_2}} \qquad (10.11)$$

Pooled observed probability:

$$p_p' = (x_1 + x_2)/(n_1 + n_2) \qquad (10.13)$$
$$q_p' = 1 - p_p' \qquad (10.14)$$

Calculated test statistic for H_o: $p_1 - p_2 = 0$:

$$z\star = \frac{p_1' - p_2'}{\sqrt{(p_p')(q_p')\left[\left(\dfrac{1}{n_1}\right) + \left(\dfrac{1}{n_2}\right)\right]}} \qquad (10.15)$$

Ratio of variances between two independent samples:

Calculated test statistic for H_o: $\sigma_1^2/\sigma_2^2 = 1$:

$$F\star = s_1^2/s_2^2 \qquad (10.16)$$

Calculated test statistic for enumerative data:

$$\chi^2\star = \sum [(O - E)^2/E] \qquad (11.1)$$

Multinomial experiment:

Degrees of freedom: $\text{df} = k - 1 \qquad (11.2)$

Expected frequency: $E = n \cdot p \qquad (11.3)$

Test for independence or Test of homogeneity:

Degrees of freedom:

$$\text{df} = (r - 1) \cdot (c - 1) \qquad (11.4)$$

Expected value: $E = (R \cdot C)/n \qquad (11.5)$

Mathematical model:

$$x_{c,\,k} = \mu + F_c + \epsilon_{k(c)} \qquad (12.13)$$

Total sum of squares:

$$\text{SS(total)} = \sum(x^2) - \frac{(\sum x)^2}{n} \qquad (12.2)$$

Sum of squares due to factor:

$$\text{SS(factor)} =$$
$$\left[\left(\frac{C_1^2}{k_1}\right) + \left(\frac{C_2^2}{k_2}\right) + \left(\frac{C_3^2}{k_3}\right) + \cdots\right] - \left[\frac{(\sum x)^2}{n}\right] \qquad (12.3)$$

Sum of squares due to error:

$$\text{SS(error)} =$$
$$\sum(x^2) - [(C_1^2/k_1) + (C_2^2/k_2) + (C_3^2/k_3) + \cdots] \qquad (12.4)$$

Degrees of freedom for total:

$$\text{df(total)} = n - 1 \qquad (12.6)$$

Degrees of freedom for factor:

$$\text{df(factor)} = c - 1 \qquad (12.5)$$

Degrees of freedom for error:

$$\text{df(error)} = n - c \qquad (12.7)$$

Mean square for factor:

$$\text{MS(factor)} = \text{SS(factor)}/\text{df(factor)} \qquad (12.10)$$

Mean square for error:

$$\text{MS(error)} = \text{SS(error)}/\text{df(error)} \qquad (12.11)$$

Calculated test statistic for H_o: Mean value is same at all levels:

$$F\star = \text{MS(factor)}/\text{MS(error)} \qquad (12.12)$$

Covariance of x and y:

$$\text{covar}(x, y) = \sum[(x - \bar{x})(y - \bar{y})]/(n - 1) \qquad (13.1)$$

Pearson's correlation coefficient:

$$r = \text{covar}(x, y)/(s_x \cdot s_y) \qquad (13.2)$$

or

$$r = \text{SS}(xy)/\sqrt{\text{SS}(x) \cdot \text{SS}(y)} \qquad \textbf{(3.2) or (13.3)}$$

Experimental error: $e = y - \hat{y} \qquad (13.5)$

Variance of error ϵ: $s_e^2 = \sum(y - \hat{y})^2/(n - 2) \qquad (13.6)$

or

$$s_e^2 = \frac{(\sum y^2) - (b_0)(\sum y) - (b_1)(\sum xy)}{n - 2} \qquad (13.8)$$

Standard deviation about the line of best fit:

$$s_e = \sqrt{s_e^2} \qquad \textbf{(p. 715)}$$

Square of standard error of regression:

$$s_{b_1}^2 = \frac{s_e^2}{\text{SS}(x)} = \frac{s_e^2}{\sum x^2 - [(\sum x)^2/n]} \qquad (13.11)$$

Confidence interval for β_1:

$$b_1 \pm t_{(df,\, \alpha/2)} \cdot s_{b_1} \qquad (13.12)$$

Calculated test statistic for H_o: $\beta_1 = 0$:

$$t\star = (b_1 - \beta_1)/s_{b_1} \text{ with df} = n - 2 \qquad (13.13)$$

Confidence interval for mean value of y at x_0:

$$\hat{y} \pm t_{(n-2,\, \alpha/2)} \cdot s_e \cdot \sqrt{\frac{1}{n} + \frac{(x_0 - \bar{x})^2}{\text{SS}(x)}} \qquad (13.15)$$

Prediction interval for y at x_0:

$$\hat{y} \pm t_{(n-2,\, \alpha/2)} \cdot s_e \cdot \sqrt{1 + \frac{1}{n} + \frac{(x_0 - \bar{x})^2}{\text{SS}(x)}} \qquad (13.16)$$

Mann–Whitney U test:

$$U_a = n_a \cdot n_b + [(n_b) \cdot (n_b + 1)/2] - R_b \qquad (14.3)$$
$$U_b = n_a \cdot n_b + [(n_a) \cdot (n_a + 1)/2] - R_a \qquad (14.4)$$

Spearman's rank correlation coefficient:

$$r_s = 1 - \left[\frac{6\sum d^2}{n(n^2 - 1)}\right] \qquad (14.11)$$

Index for Computer and Calculator Instructions

Areas of the Standard Normal Distribution

The entries in this table are the probabilities that a random variable, with a standard normal distribution, assumes a value between 0 and z; the probability is represented by the shaded area under the curve in the accompanying figure. Areas for negative values of z are obtained by symmetry.

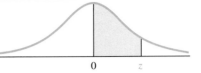

Second Decimal Place in z

z	0.00	0.01	0.02	0.03	0.04	0.05	0.06	0.07	0.08	0.09
0.0	0.0000	0.0040	0.0080	0.0120	0.0160	0.0199	0.0239	0.0279	0.0319	0.0359
0.1	0.0398	0.0438	0.0478	0.0517	0.0557	0.0596	0.0636	0.0675	0.0714	0.0753
0.2	0.0793	0.0832	0.0871	0.0910	0.0948	0.0987	0.1026	0.1064	0.1103	0.1141
0.3	0.1179	0.1217	0.1255	0.1293	0.1331	0.1368	0.1406	0.1443	0.1480	0.1517
0.4	0.1554	0.1591	0.1628	0.1664	0.1700	0.1736	0.1772	0.1808	0.1844	0.1879
0.5	0.1915	0.1950	0.1985	0.2019	0.2054	0.2088	0.2123	0.2157	0.2190	0.2224
0.6	0.2257	0.2291	0.2324	0.2357	0.2389	0.2422	0.2454	0.2486	0.2517	0.2549
0.7	0.2580	0.2611	0.2642	0.2673	0.2704	0.2734	0.2764	0.2794	0.2823	0.2852
0.8	0.2881	0.2910	0.2939	0.2967	0.2995	0.3023	0.3051	0.3078	0.3106	0.3133
0.9	0.3159	0.3186	0.3212	0.3238	0.3264	0.3289	0.3315	0.3340	0.3365	0.3389
1.0	0.3413	0.3438	0.3461	0.3485	0.3508	0.3531	0.3554	0.3577	0.3599	0.3621
1.1	0.3643	0.3665	0.3686	0.3708	0.3729	0.3749	0.3770	0.3790	0.3810	0.3830
1.2	0.3849	0.3869	0.3888	0.3907	0.3925	0.3944	0.3962	0.3980	0.3997	0.4015
1.3	0.4032	0.4049	0.4066	0.4082	0.4099	0.4115	0.4131	0.4147	0.4162	0.4177
1.4	0.4192	0.4207	0.4222	0.4236	0.4251	0.4265	0.4279	0.4292	0.4306	0.4319
1.5	0.4332	0.4345	0.4357	0.4370	0.4382	0.4394	0.4406	0.4418	0.4429	0.4441
1.6	0.4452	0.4463	0.4474	0.4484	0.4495	0.4505	0.4515	0.4525	0.4535	0.4545
1.7	0.4554	0.4564	0.4573	0.4582	0.4591	0.4599	0.4608	0.4616	0.4625	0.4633
1.8	0.4641	0.4649	0.4656	0.4664	0.4671	0.4678	0.4686	0.4693	0.4699	0.4706
1.9	0.4713	0.4719	0.4726	0.4732	0.4738	0.4744	0.4750	0.4756	0.4761	0.4767
2.0	0.4772	0.4778	0.4783	0.4788	0.4793	0.4798	0.4803	0.4808	0.4812	0.4817
2.1	0.4821	0.4826	0.4830	0.4834	0.4838	0.4842	0.4846	0.4850	0.4854	0.4857
2.2	0.4861	0.4864	0.4868	0.4871	0.4875	0.4878	0.4881	0.4884	0.4887	0.4890
2.3	0.4893	0.4896	0.4898	0.4901	0.4904	0.4906	0.4909	0.4911	0.4913	0.4916
2.4	0.4918	0.4920	0.4922	0.4925	0.4927	0.4929	0.4931	0.4932	0.4934	0.4936
2.5	0.4938	0.4940	0.4941	0.4943	0.4945	0.4946	0.4948	0.4949	0.4951	0.4952
2.6	0.4953	0.4955	0.4956	0.4957	0.4959	0.4960	0.4961	0.4962	0.4963	0.4964
2.7	0.4965	0.4966	0.4967	0.4968	0.4969	0.4970	0.4971	0.4972	0.4973	0.4974
2.8	0.4974	0.4975	0.4976	0.4977	0.4977	0.4978	0.4979	0.4979	0.4980	0.4981
2.9	0.4981	0.4982	0.4982	0.4983	0.4984	0.4984	0.4985	0.4985	0.4986	0.4986
3.0	0.4987	0.4987	0.4987	0.4988	0.4988	0.4989	0.4989	0.4989	0.4990	0.4990
3.1	0.4990	0.4991	0.4991	0.4991	0.4992	0.4992	0.4992	0.4992	0.4993	0.4993
3.2	0.4993	0.4993	0.4994	0.4994	0.4994	0.4994	0.4994	0.4995	0.4995	0.4995
3.3	0.4995	0.4995	0.4995	0.4996	0.4996	0.4996	0.4996	0.4996	0.4996	0.4997
3.4	0.4997	0.4997	0.4997	0.4997	0.4997	0.4997	0.4997	0.4997	0.4997	0.4998
3.5	0.4998	0.4998	0.4998	0.4998	0.4998	0.4998	0.4998	0.4998	0.4998	0.4998
3.6	0.4998	0.4998	0.4999	0.4999	0.4999	0.4999	0.4999	0.4999	0.4999	0.4999
3.7	0.4999									
4.0	0.49997									
4.5	0.499997									
5.0	0.4999997									

For specific details about using this table to find: probabilities, see pages 317–320; confidence coefficients, pages 338–339, 340–341; p-values, pages 432–433, 435; critical values, pages 317–320, 338–339.